The Methods
and Materials of Demography

Condensed Edition

This is a volume in

STUDIES IN POPULATION

A complete list of titles in this series appears at the end of this volume.

The Methods and Materials of Demography

Henry S. Shryock
Jacob S. Siegel
and Associates

Condensed Edition by
Edward G. Stockwell

Bowling Green University
Bowling Green, Ohio

ACADEMIC PRESS, INC.

(Harcourt Brace Jovanovich, Publishers)

Orlando San Diego New York London
Toronto Montreal Sydney Tokyo

ACADEMIC PRESS, INC.
Orlando, Florida 32887

United Kingdom Edition published by
ACADEMIC PRESS, INC. (LONDON) LTD.
24/28 Oval Road, London NW1 7DX

Library of Congress Cataloging in Publication Data

Shryock, Henry S
    The methods and materials of demography.

    (Studies in population)
    Includes bibliographies.
    1.   Demography.   I.   Siegel, Jacob S.
II.   Stockwell, Edward G.   III.   Title.   IV. Se-
ries.
HB881.S526   1976   301.32'01'82   76-18312
ISBN 0-12-641150-6

PRINTED IN THE UNITED STATES OF AMERICA

84 85 86 87   9 8 7

Contents

Preface

The present work is a condensed version of the two-volume work *The Methods and Materials of Demography* first published by the U.S. Bureau of the Census in 1971. The original work has enjoyed widespread use both in the United States and abroad.

The idea for a condensed edition of *The Methods and Materials of Demography* was suggested by Professor Halliman H. Winsborough, the editor for Academic Press's Series on Studies in Demography. This condensed edition is intended to serve several purposes. Among these are to reach a wider audience, particularly in academic circles, and to serve the needs of many users and prospective users of the previous work more effectively. It is believed that a one-volume edition of about 555 pages can much more easily serve as a text in a one-semester course in the techniques of demographic analysis and could be covered more thoroughly in a two-semester course, than the original two-volume work of 900 pages.

The authors are pleased that Professor Edward G. Stockwell was willing to undertake the time-consuming and demanding task of abridging the two-volume original. He worked from the latest edition then in print, namely the second printing of 1973; but he had available the record of the numerous but relatively minor corrections that were to be made in the third printing, which was completed in September 1975. In our opinion he has done a consistent and judicious job of selecting parts to be omitted and of providing the necessary editorial modifications.

Like the original two-volume work, this work attempts to present a systematic and comprehensive exposition, with illustrations, of the methods used by technicians and research workers in dealing with demographic data. The book is concerned with how data on population are gathered, classified, and treated to produce tabulations and various summarizing measures that reveal the significant aspects of the composition and dynamics of populations. It sets forth the sources, limitations, underlying definitions, and bases of classification, as well as the techniques and methods that have been developed for summarizing and analyzing the data.

This book is intended to serve both as a classroom text for courses on demographic methods, aimed at instructing students in how to use population data for analytic studies, and as a reference for professional workers who have occasion to use population data; but it is designed primarily as a classroom text. For courses that are focused on the subject matter of population, this condensed edition might well be used for supplementary reading. As a reference work we hope that the book will be helpful to statistical analysts and planners in private industry, city and regional planners, technicians in State and regional development agencies, and also statisticians and analysts in national agencies concerned with population matters.

The book could be employed as a text in a single semester, particularly with the omission of a few topics. On the other hand, with a more complete coverage of the book, and with supplementary readings from other appropriate publications, a full year could be devoted to the book. In addition to copious citations of relevant publications in the footnotes, there are lists of suggested readings at the end of each chapter that will identify for the interested student the more important works on demographic methods.

In the hope of making the book as widely useful as possible, we have presented the methodological material in the simplest mathematical form. For the most part, only high school algebra and elementary statistics are needed for a proper understanding of the book. However, a knowledge of elementary mathematical analysis is required for understanding some parts, particularly sections of Chapters 18 and 24, and Appendix C.

Considerable attention is paid to the data and measurement problems of the less developed countries, the kinds and quality of data that are available for these areas, and the special methods that have been worked out for handling incomplete and defective data. Although their demographic publications tend to be less adequate and less available, numerous illustrations for the less developed countries are given. The materials of the United States are also covered in relatively great detail, especially with respect to sources, definitions, and historical developments. Although national demographic materials differ considerably with respect to availability, definitions, classifications, etc., certain demographic principles and methods are essentially "culture free," and measures worked out for the United States could serve as well as for any other country.

The original work was intended to be comprehensive in its treatment of demographic methods and materials; yet, inevitably, choices had to be made as to what to include and what to exclude. These choices reflected the judgment of the authors as to what is most important and useful to demographers. This condensed version has tried to preserve the broad scope of the original work. As may be seen from the table of contents, the present book includes material on both formal demography and social and economic demography (i.e., the study of many social and economic characteristics of individuals as well as of such social groups as families and households). Several important but peripheral or applied fields are barely touched on, however. These include worker commuting, morbidity, the evaluation of family planning programs, optimum population, population quality, and various types of applied projections (e.g., needs for housing).

The derivation of most demographic measures described is illustrated by step-by-step examples using actual or, occasionally, hypothetical statistics. The official statistics are usually accepted as published for use in these calculations, and the handling of a problem is sometimes simplified, in order to focus attention on the method being illustrated. In actual practice, the official statistics for many countries would need to be evaluated and perhaps adjusted, and related procedures might be required. Accordingly, the results from the illustrative examples given here should not be regarded as necessarily valid for substantive purposes.

The original work was written in the years 1967 to 1970. Thus, it takes account of recommendations made by the United Nations and other international agencies for the 1970 round of population censuses and the plans for the enumeration, processing, tabulation, and publication of the 1970 Census of the United States. In this condensed version, the material on the 1970 Census of the United States has been updated so that the treatment of the census more nearly describes what was actually done.

The condensed edition retains the essential organization of the earlier work. In particular, all the main topics covered in that work are retained, although necessarily many details had to be omitted in the condensation. The condensation did not involve a mere mechanical reduction to achieve the brevity intended, however. Where sections were deleted, the new text was woven together again to achieve logical continuity.

The earlier work consisted of 25 chapters, grouped under five broad headings, and four appendixes. The major structural changes in the abridged edition are (1) the deletion of Chapter 22 on "Selected General Methods" and the consequent renumbering of subsequent chapters; (2) the incorporation of what was Appendix A in the larger two-volume edition into the new Chapter 23 on "Population Projections," and the renumbering of Appendixes B and D as A and B; and (3) an expansion of Appendix C to include some of the material originally covered in the now deleted Chapter 22. As compared with the unabridged edition, the present one contains less emphasis on historical development, the recommendations of international organizations, the practices of national statistics agencies with respect to demographic data, and the more technical and mathematical aspects of demographic analysis.

Either Shryock or Siegel as principal authors wrote, or reviewed and edited, each chapter in the original work. Associate authors contributed drafts of many of the chapters or parts of chapters, either working under the immediate direction of one of the principal authors or independently. These associate authors were Maria Davidson, Paul C. Glick, Elizabeth A. Larmon, Wilson H. Grabill, and Charles R. Kindermann of the Bureau of the Census; Robert D. Grove and Robert A. Israel of the National Center for Health Statistics; Charles B. Nam of Florida State University; Abram J. Jaffe of Columbia University; Francisco Bayo of the Social Security Administration; and Paul Demeny of the University of Hawaii. Elizabeth Larmon was editorial coordinator of the unabridged edition and was generally in charge of the technical production aspects of that work. Adriana Weininger performed these tasks in the final stages. We are grateful to the Bureau of the Census and particularly to the Population Division and the International Statistical Programs Division for its generous support of the work on this book. Acknowledgement is made of the role of the U.S. Agency for International Development in providing major financial support for the original publication.

The bulk of the retyping that was necessary for the condensed edition was done by members of the secretarial staff in the Department of Sociology at Bowling Green State University, with particular acknowledgement being due to Lauretta Lahman and Lynn Schmid. Thanks are also due to Helen S. Curtis, Anne Donnelly, and Mary Hartman of the Center for Population Research, Kennedy Institute, Georgetown University, for helping with the voluminous correspondence in connection with the present edition, and to Georgetown University for other courtesies.

The first-named author would like to acknowledge his appreciation to his wife, Annie Frances King Shryock, for her patience during his work on the various drafts and editions, and for her help with some of the editorial work. The second-named author would also like to express his appreciation to Rose V. Siegel and Lorise V. Siegel for their forbearance and sympathetic support during the long period when the original work was being prepared.

Henry S. Shryock
Jacob S. Siegel

CHAPTER 1

Introduction

THE FIELD OF DEMOGRAPHY

Demography is the science of population. The word was coined by a Belgian, Achille Guillard, who published his *Eléments de statistique humaine, ou démographie comparée* (Elements of human statistics or comparative demography) in 1855.[1] He defined it as the natural and social history of the human species or the mathematical knowledge of populations, of their general changes, and of their physical, civil, intellectual, and moral condition.

Like most other sciences, demography may be defined narrowly or broadly. The narrowest sense is that of "formal demography." Formal demography is concerned with the size, distribution, structure, and change of populations. **Size** is simply the number of units (persons) in the population. **Distribution** refers to the arrangement of the population in space at a given time, that is, geographically or among various types of residential areas. **Structure,** in its narrowest sense, is the distribution of the population among its sex and age groupings. **Change** is the growth or decline of the total population or of one of its structural units. The components of change in total population are births, deaths, and migrations. In analyzing change in structure, however, we have to include the transition from one group to another. In the case of age, this is expressed as a simple function of time.

A broader sense includes additional characteristics of the units. These include ethnic characteristics, social characteristics, and economic characteristics. Ethnic characteristics like race, legal nationality, and mother tongue shade into social characteristics. Other examples of social characteristics are marital and family status, place of birth, literacy, and educational attainment. Economic characteristics include economic activity, employment status, occupation, industry, and income, among others. Other characteristics that might be encompassed are genetic inheritance, intelligence, and health; but the usual sources of demographic data, such as censuses, seldom deal with these directly. Furthermore, demography may look beyond the basic personal units to such customary social groupings as families and married couples.

The widest sense of demography extends to applications of its data and findings in a number of fields including the study of problems that are related to demographic processes. These include the pressure of populations upon resources, depopulation, family limitation, eugenics, the assimilation of immigrants, urban problems, legislative apportionment, manpower, and the maldistribution of income. This book covers very little of these fringe areas of demography.

Hauser and Duncan regard the field of demography as consisting of a narrow scope—demographic analysis—and a wider scope—population studies. "Demographic analysis is confined to the study of components of population variation and change. Population studies are concerned not only with population variables but also with relationships between population changes and other variables—social, economic, political, biological, genetic, geographical, and the like. The field of population studies is at least as broad as interest in the 'determinants and consequences of population trends.' "[2] In the same work, Lorimer also discusses demography as a discipline.[3]

DEMOGRAPHIC DATA AND THEIR USES

The types of demographic data have already been suggested. Counts of persons are obtained from censuses and sample surveys and from the files of continuous population registers. Counts of events are obtained from registered vital events (births, deaths, marriages, divorces, etc.), and also from continuous population registers. Sometimes censuses and surveys inquire about events, e.g., the number of children women have borne in the preceding 12 months. Any of these counts may be shown in the form of multiple classifications, e.g., population by sex and age for urban areas or deaths by age and cause. Various demographic measures, such as percentages, ratios, and averages may be derived from them. Furthermore, numbers of registered vital events can be related to the corresponding population to produce vital rates, for example, the number of deaths per 1,000 of the population or the number of births to married women 20 to 24 years old. Such vital rates can also be derived from continuous population registers without the use of other sources of demographic data.

The resulting demographic statistics can then be used to describe the distribution of the population in space, its density and degree of concentration, the fluctuations in its rate of growth, its movements from one area to another, and the force of natality, nuptiality, and mortality within it. These demographic statistics have many and increasingly varied applications. The fields of application include public health; local planning for land use, school and hospital construction, public

---

[1] Adolphe Landry, *Traité de démographie* (Treatise on demography), Paris, Payot, 1945, p. 7.

[2] Philip M. Hauser and Otis Dudley Duncan, "Overview and Conclusions" in: *The Study of Population,* Hauser and Duncan, eds., Chicago, University of Chicago Press, 1959, pp. 2–3.

[3] Frank Lorimer, "The Development of Demography" in Hauser and Duncan, op. cit., pp. 157–166.

utilities, etc.; marketing; manpower analysis; family planning programs; land settlement; immigration and emigration policy; and many others. An analysis of current demographic levels and past trends is the necessary first step in the construction of population forecasts that in turn form the underpinning of national plans for economic development and other programs, including explicit population policies in some cases.

COVERAGE OF TOPICS

As may be seen from the table of contents, the scope of this book corresponds pretty much to the first and second concentric circles of the field of demography that were described in the opening section, that is, it covers formal demography and many additional social and economic characteristics of the population as well as such social groups as families and households. Several important but peripheral or applied fields are barely touched on. These include worker commuting, morbidity, pregnancy wastage through miscarriages and induced abortions, the evaluation of the effectiveness of family limitation programs, and population quality.

ORGANIZATION OF THIS BOOK

Arrangement of Chapters

The book consists of 24 chapters covering the sources of demographic data, the major topics of population size, distribution, and composition, the basic components of demographic change, and population estimates and projections.

A major problem that had to be faced in the writing was the fact that the subject and methods of a given chapter cannot be adequately described without drawing on those of a later chapter. This problem would arise no matter what order of topics or chapters was followed. In the analysis of age composition in chapter 8, for example, it is essential to make use of survival rates, which are derived by methods explained in chapter 15, "The Life Table." To understand the life table, on the other hand, we have to know how age is defined in censuses and in vital statistics. More particularly, we need to be familiar with the sources of error in the underlying age statistics in order to have some appreciation of the accuracy of a given life table.

A related problem is that a given method may apply to a number of subject fields within demography. Standardization is used with almost all kinds of rates and averages—birth, death, and marriage rates; migration rates; enrollment rates; worker participation rates and unemployment rates; median years of school completed; mean income; etc.

We have tried to cope with these problems in several ways. Frequent use has been made of forward and backward references. Nonetheless, a certain amount of duplication in the exposition was inevitable.

Treatment of Subjects Within Chapters

In addition to the conventional treatment of standard methodological issues and measures, each chapter generally contains a discussion of their uses and limitations, the quality of the statistics, and some of the factors important in their analysis.

Uses and Limitations.—The uses to which the given statistics and measures derived from them are or can be put are sometimes illustrated by actual examples from the literature. These uses often represent contributions to the investigation of basic social and economic problems or of challenging scientific questions.

The limitations of the statistics have to do with their scope and pertinency. Because of the definitions used, the restricted number and form of the questions, the scope of the census, survey, or register, and the extent of cross-classifications in the tabulations, the statistics can at best give only approximate answers to the broad questions that they were designed to answer and may be even less adequate for the purposes of some imaginative analysts. Demographers often find new and previously unforeseen uses to which they put the statistics available to them. When they do so, they need to recognize that the statistics may not apply to just the universe that is of interest to them or may measure obliquely rather than directly the phenomenon they are investigating. In short, they should be aware of these limitations and the implications of them for their analysis and conclusions.

The demographer should also be aware of the assumptions that he has to make when he extends the interpretation of standard, readily available measures or models in certain ways. For example, a life table based on the mortality experience of a given year does not describe the mortality experience of any actual group of persons as they pass through life. Neither does a gross reproduction rate based on the fertility experience of a given year describe the fertility experience of any actual group of females who started life together. With due caution, however, these measures may be used in many important descriptive and analytical applications.

We may cite also the limitations of population projections when they are actually used to predict the future. The fact that, in some situations, this type of use is unavoidable does not relieve the forecaster of considering the realism and appropriateness of the assumptions underlying the projections. Often, too, he may realize that if he had additional data at his disposal, he could express his assumptions in terms of more meaningful metrics.

Quality of the Statistics.—By the "quality" of the statistics is meant the degree to which they measure what was intended to be measured when the questions and procedures were designed. Hence, the term quality as used here is measured **within** the limitations of the statistics and does not include those limitations. "Accuracy" could be used as a synonym for quality.

Factors Important in Analysis.—Only a limited understanding of demographic phenomena is obtained from statistics (percentage distributions, rates, etc.) presented for a topic in isolation. Such so-called "inventory" statistics have mainly a descriptive value. Inventory statistics for a large number of small geographic areas can be very useful in ecological analysis; but here an important additional factor, namely, geographic location has already been introduced. In some countries, racial or ethnic group may be associated with striking variations in the topic (mortality, literacy, family income, etc.) being investigated.

The "factors" treated in these subsections can be regarded in several different ways. In ordinary statistical terms, they are the "independent variables" used to study variations in the "dependent variable" (or the topic under consideration). Hence, they may be introduced into the analysis in various ways as "controls" or as variables to be held "constant." The most important control factors in much demographic analysis are age and sex. Any of these factors may be introduced by means of cross-classification (and then by computation of factor-specific rates), by standardization, or by correlation.

To a lesser extent, the topic in question may itself be treated here as an independent variable. In that case, we study its effects on other demographic variables (or qualitative attributes).

More broadly, these subsections occasionally mention a few comprehensive studies in which the topic is an important element. These studies place the topic in the context of substantive research and give the reader a broader view of how the data and methods he has just examined in detail have been applied in some of the classical demographic works.

DESIDERATA IN DEMOGRAPHIC DATA AND STATISTICS

There are certain properties that demographers would like their data to possess. Taken literally these desiderata are of the nature of ideals in that even in the most advanced countries they are never fully attained. Nonetheless, they are goals that should be kept in view by agencies that produce demographic data. Moreover, whenever compromises are made with these goals because of practical restrictions, the nature of these deviations should be imparted to the public in the resulting statistical publications. Furthermore, when unintended errors of coverage, classification, etc., occur in the data gathering and processing, some resources should be devoted to evaluating them, on a quantitative basis if possible.

We are concerned with all types of demographic data—those from censuses, sample surveys, vital records, continuous population registers, and immigration and emigration controls. In general, the data should completely cover the units of the specified universe or a representative sample of them. The data should be comparable over time and space. They should be published promptly and in adequate detail. Some considerations apply to only censuses and others to vital statistics, and so on. Let us start with more specific consideration of censuses.

Censuses

A recent U.N. publication lists four essential features of a population census:

1. Each individual is enumerated separately and the characteristics of each person are recorded separately.
2. The census covers a precisely defined territory and includes every person present or residing within its scope.
3. The population is enumerated with respect to a well-defined point of time, and data are in terms of a well-defined reference period.
4. The censuses are taken at regular intervals.[4]

The Census as Part of an Integrated Program of Demographic Statistics

In planning a national population census, attention should be paid to its interrelationships with sample surveys, vital statistics, population registers, etc. The census may serve as a frame for sampling surveys to be taken shortly after the field work of the census is completed; such surveys may probe in greater depth topics included in the census, or they may investigate additional topics. In the case of periodic intercensal surveys, the census may serve as a benchmark in various ways. For example, it may serve as an occasion for updating the population "controls" to which the percentage distributions obtained in the survey are applied. As part of the census evalua-

tion program, matching studies can be carried out between a sample of cases in the census and a survey taken at a nearby date. Such matching studies can also be carried out between the census and vital records, universal population registers, etc. Statistical totals can also be compared to measure the net differences. In the case of comparisons between the census, on the one hand, and a survey, population register, etc., on the other, both demographic programs will benefit from these matching studies and statistical comparisons unless one source is vastly superior to the other.

To achieve an integrated program of demographic statistics, it is essential that consistent concepts and definitions be employed throughout. When the respective programs are administered by different agencies, however, consistency may be difficult to achieve in some particulars since different program needs may seem to call for different treatments.

By the same token, international comparability is even more difficult to accomplish. Moreover, there may be such profound cultural, social, and economic differences among countries that, even if the phrasing and definition of questions are identical (and if identity can be achieved through translation), the resulting data from certain questions may not be comparable. Let us cite a few examples. If most of the people in a Far Eastern country are accustomed to figure their ages by the Oriental mode, simply asking for the age in completed years will not necessarily yield the Western-style age. More elaborate artifices may be required, and even these may yield only an approximation to the Western-style age distribution. If plural marriage or consensual unions are widespread, the conventional Western categories of marital status may give an unreal description of the population's distribution by marital status. Finally, if most of the working population are not part of a market economy and if there is considerable underemployment, the conventional labor force questions will yield meaningless statistics on unemployment. As Linder puts it, "A standard international definition for such characteristics can achieve comparable statistics only to the extent that the social and economic features of the country are similar. In two dissimilar countries, it may be that greater comparability of meaning for a census item can be obtained if the census question is asked in distinctly different ways."[5]

As source material, demography requires both regular censuses and the registration of vital events. In theory, a continuous, universal population register might suffice; but, in practice, even the best registers need to be checked periodically by a census. Moreover, a register cannot ordinarily accommodate as many items of social and economic characteristics of the person as a census can nor can new items be introduced so readily. Where census and vital records are not both adequate, as is true in many statistically underdeveloped countries, the demographer will work with what is available, using, for example, the techniques described in chapter 24.

Sample Surveys, Registration Systems, and Other Sources

Current sample surveys like censuses represent periodic stock-takings, whereas registration systems (universal population registers, vital records, etc.) and immigration and emigration control systems involve the continuous recording of data

---

[4] United Nations, *Principles and Recommendations for the 1970 Population Censuses*, Statistical Papers, Series M, No. 44, 1967, pp. 3–4.

[5] Forrest E. Linder, "World Demographic Data," in Hauser and Duncan, eds., op. cit., pp. 323–324.

even though the published statistics are extracted only at periodic intervals. Thus, some of the desiderata in census data are not applicable to registration data.

Sample Surveys. — Unlike censuses, many surveys are taken under private auspices. Although a periodic sample survey constitutes a continuing "micro-census" of a country or other area, a one-time survey can provide useful data for an *ad hoc* purpose. Surveys taken as census pretests or as experiments in survey methodology need not necessarily result in published tables based on the substantive demographic data. In other respects (individual enumeration, universality, and simultaneity), the scope of the survey and the methods of recording the data should be subject to the same principles as are censuses.

Registration Systems. — In the interests of national uniformity, it is best for a register to be operated by an agency of the national government. In some countries with a federal form of government or at least with relatively autonomous state governments, however, good vital statistics have been produced under an arrangement whereby the states handle the registration process and the national government exercises a coordinating function, setting standards and publishing national reports, for example.

The equivalent of the desideratum "individual enumeration" in the census or survey is the individual vital record. As Hauser and Duncan put it, "To provide adequate data for demography, a vital registration system must include procedures to assure the filing of a uniform record for every vital event—for example, live birth, death, stillbirth, and marriage; to provide for complete and usable answers to the inquiries on the record form; and to enable the information in the record form to be processed for statistical purposes— that is, edited, coded, tabulated, and presented, preferably through some central office which provides vital statistics for the nation and its subdivisions on a comparable basis. Principles and procedures for achieving these objectives have been evolved over the years."[6] (See also United Nations *Handbook of Vital Statistics Methods,* Studies in Methods, Series F, No. 7, 1953.)

A registration system should likewise have "universality within a defined territory." The principle of "simultaneity" does not apply, of course; but it suggests another criterion that is pertinent to registration, namely, a specified maximum interval of time between the occurrence of the event and its recording. For most demographic purposes, moreover, the data should be tabulated as of the date of occurrence, not the date of recording. On the other hand, tabulation by the place of occurrence is less useful than tabulation by the place of residence of the person concerned.

Again, "defined periodicity" would not apply to the continuous recording of data, but it does suggest to us that compilations of the records for statistical purposes should be made periodically. (The date—day, month, year—of both occurrence and registration should be on the record form.) The usual interval for publication purposes is the year; but some series should be published quarterly, or even monthly. Finally, the compilers of statistics from registers, like the agencies that collect and tabulate census data, are obligated to publish, evaluate, and, to some extent, analyze them.

SOME BASIC DEMOGRAPHIC METHODS

As was noted above, it is not possible adequately to discuss demographic methods following a simple linear path. At many points, we have to refer to what lies ahead, or to the left or right, as well as to what lies behind. Partly for this reason, an early overview of how demographers organize and analyze their data seems necessary.

A good summary account of the essence of demographic methods is given by Hauser and Duncan.[7] After first pointing out that demography shares many of its methods with science in general and especially with statistics, they list certain groups of techniques that tend to be peculiar to demography. These techniques include: (a) techniques of data collection; (b) techniques of data evaluation and adjustment and of statistical estimation; and (c) techniques of analysis including demographic projections or forecasts. We will not comment here on (a) but rather will focus on (c) with some attention to (b), since the respective techniques are interdependent to some extent.

"For dealing with population 'statics' the demographer depends largely on general statistical descriptive techniques with some special rates and graphic devices which have become widely used. In this latter category are such things as the sex ratio, the dependency ratio, the index of displacement, and the population pyramid . . . To deal with population dynamics, demography has developed a rather comprehensive and elaborate set of 'rates' designed to measure vital events or components of population change, such as natality, mortality, morbidity, marriage, divorce, and migration. . . . On the whole, demographic rates are designed to measure change and are calculated as approximations to a posteriori probability statements."[8]

The Balancing Equation

The most basic method of demography is the decomposition of population change into its components, or, conversely, the synthesis of the components to estimate the total population change. Schematically, we may express this process in terms of the fundamental equation

$$P_t - P_o = B - D + I - O, \qquad (1)$$

where P_t is the population at the end of the period, P_o that at the beginning of the period, B is births, D is deaths, I is in-migration, and O is out-migration.

This simple equation, which is called the "balancing equation" (or the "inflow-outflow relationship" or the "component equation") has many forms and many uses. To be exactly true (i.e., represent a necessary relationship), it must apply to a fixed territory and there must be no measurement errors. In fact, the equation may be used to estimate the net error in this system of demographic statistics. If we find that the right-hand side differs from the left-hand side by an amount e, then we can write,

$$P_t - P_o = B - D + I - O + e \qquad (2)$$

Here e can be called the "residual error" or the "error of closure." On the basis of additional knowledge about the

---

[6] Hauser and Duncan, "The Data and Methods," in Hauser and Duncan, eds., op. cit., p. 62.

[7] Op. cit., pp. 70–73.

[8] Ibid., p. 70.

accuracy of the various terms, one may be able to decide whether e can be attributed as a measurement error almost wholly to a particular term in the equation. For example, if there is evidence that the right-hand terms are all based on very accurate registration data and the population figures come from successive censuses, then e would represent the relative accuracy of coverage of the two censuses. If e is positive, P_t is more nearly complete than P_o; if e is negative, then the reverse would be true.

Let us consider some other illustrations of the uses of (1) or its variations. Suppose that a country has adequate vital statistics and statistics on immigration and emigration. Then t years after the last census but before the next census, it is desired to make a postcensal estimate of the current national population. We have

$$P_t = P_o + B - D + I - O \qquad (3)$$

In this form, the equation may be thought of as the "basic estimating equation," which uses a straightforward book-keeping procedure.

If we are interested in projecting the population to a future date, we can use equation (3) in principle by making assumptions about the future births, deaths, and migration. Especially in the case of births and deaths, however, these assumptions are ordinarily made in the form of fertility and mortality **rates**, not in the form of the absolute numbers of births and deaths. The nature of these rates is another fundamental part of demographic methodology and will be discussed presently.

For another application, suppose that we have two successive population counts for a subnational area (province, county, commune, etc.). We also have vital statistics on births and deaths but no statistics on internal migration (or on the extent to which external migration affects the individual subnational areas). Then we may write

$$M = I - O = (P_t - P_o) - (B - D) \qquad (4)$$

where M is the net migration to or from the area. In other words, to estimate the intercensal net migration for the subnational area, we subtract the natural increase, $B - D$, from the total population change, $P_t - P_o$.

First, however, we should mention another kind of elaboration of the balancing equation, namely, its use for a population subgroup, such as the male population, the female population of childbearing age, the native population, or university graduates. For some subgroups, the males in the native population, for example, we simply have to obtain the corresponding components, i.e., statistics on births, deaths, and migration for that subgroup. This restriction may be expressed by using the superscript i, to denote the subgroup, thus:

$$P_t^i - P_o^i = B^i - D^i + I^i - O^i \qquad (5)$$

In the case of an age group, however, the very specification of the group, i.e., its age, changes over the period. For example, if t is 10 years, then we should compare age x at time 0 with age $x + 10$ at time t. The identification of age is more complicated in the case of the components. For example, we shall need to have death statistics for ages x, $x + 1$, $x + 2$, . . ., depending on the time elapsed since the first census, and then we shall need to redistribute them into other age groupings. For persons aged x at that census, for example, we shall need to

have part of the deaths at age x, part of those at age $x + 1$, $x + 2$, etc., depending on the time elapsed since the census.

In this situation although age changes, there is something about the population subgroup that remains the same. A group having such a common property is what demographers call a "cohort." Here the cohort is all the people who were born in a given year or their survivors. There are other types of demographic cohorts, such as marriage cohorts, or all the marriages that occurred in a given year. A great deal of demographic analysis is carried out in terms of cohorts.

Not all types of population subgroups are entered or left only by means of birth, death, migration, or growing older. The numbers in most social and economic subgroups are also affected by changes of status. For example, the number of citizens of a country is changed by naturalizations and losses of citizenship as well as by demographic factors. The number of married persons is increased by marriages and decreased by the number of "widowings" and divorces. In general, there are at least two gross components, a minimum of one positive and one negative one, that must be added to the right hand side of equation (3) to allow for these changes in status.

We analyze changes in population subgroups not only because we are interested in particular subgroups *per se* but also because this form of decomposition of population change gives us a better understanding of the nature of changes in the total population and of variations over time in these changes. The demographer's interest in the factors to be taken into account in this decomposition process is often limited only by the data that are available to him. The decomposition is more effectively carried out by multiple cross-classification rather than by simply distributing the population by one factor at a time. An example of a multiple cross-classification would be the distribution of the population by age, sex, and marital status, in which one subgroup or cell is the number of single females 15 to 19 years old. "Having accomplished what he regards as a suitable decomposition of population changes (or variations in rates of change from one period or place to another), the demographer may, of course, put the components back together again, e.g., in the form of a 'balance sheet,' a mathematical model, a statement of the relative importance of the several components, or the backward or forward projection of population changes, which may involve assumptions about the several components and the ways in which they are likely to change." [9]

The general topic dealing with deaths is called **mortality** and that with births, **natality** or **fertility**. These terms apply both to absolute numbers and to rates, and both to totals and to births and deaths specific with respect to various characteristics.

Rates and Ratios

As we have mentioned, demographic analysis also makes abundant use of rates. The term "rate" most appropriately applies to the number of demographic events in a given period of time divided by the population at risk during that period. Thus, we may speak of the number of deaths divided by the population as the death rate. The population at risk is usually only approximated. It may be the population at the middle of the period (which is roughly the average population during the period), the population at the beginning of

[9] Hauser and Duncan, "Overview and Conclusions," in Hauser and Duncan, eds., *op. cit.*, p. 4.

the period, or a more complex definition. The period is usually a year and the rate is often expressed per 100 or per 1,000 of the population. For example, the **crude** birth rate, which is based on the midperiod population, is

$$b = \frac{B}{P} \times 1,000 \qquad (6)$$

where B is the number of births in the year and P is the midyear population. Similarly, the **crude** death rate is given by,

$$d = \frac{D}{P} \times 1,000 \qquad (7)$$

The adjective "crude" is especially appropriate in the case of the crude birth rate because obviously men, children, and old people are not at risk of having a baby. (Although rarely used, there are "paternal fertility rates," which relate the number of children sired to the number of adult males.) In the case of the crude marriage rate, persons who are in the married state throughout the year are likewise not at risk. Various refinements are introduced in the population base in order to obtain a more meaningful rate. These may not only omit that part of the population for which the risk is zero, but they may also take account of the fact that the risk is much greater for some population subgroups than for others. For example, all persons are at risk of dying but the risk is much greater at age 90 than at age 10. Accordingly, we have age-specific rates, for example, the number of deaths during the year to persons aged 20 to 24. The general formula is

$$_nr_x = \frac{_nE_x}{_nP_x} \qquad (8)$$

where E is the number of events, x is the initial age, and n is the number of years in the age group. Note that here the numerator is restricted so as to correspond to the age restriction in the population. Rates may be specific for any other characteristic into which the demographic data are subdivided. Common characteristics are sex, race, nativity, and marital status. Moreover, the rates may be specific for two or more characteristics simultaneously.

In addition to the distinction between crude and refined rates, demography follows actuarial science in distinguishing central rates from probabilities. In central rates, the denominator is the population at the midpoint of the period (or the average population during the period). Probabilities, on the other hand, are based on the population at the beginning of the period, which is then viewed as the population at risk of experiencing the event during the period. The distinction is somewhat blurred for short periods when the population is not closed, e.g., when it is subject to immigration or emigration. In an "open" population, the midperiod population may be viewed as the average population at risk during the period, or the number of person-years that the population is at risk.

Suppose we are concerned with deaths at a given age, x, in a given year. The central death rate would be given by the number of deaths occurring at age x in the given year divided by the population of that age at the middle of the year. The probability of death, on the other hand, would be given by the population at exact age x at the beginning of the period divided into the deaths occurring at age x during the year; the deaths are in the same cohort as the population. When the population is not closed, the inclusion of any deaths of immigrants at age x and the exclusion of any deaths of emigrants make the

resulting rate not strictly a probability because the deaths are no longer restricted to those occurring to the initial population.

The term rate is also loosely used to refer to the ratio between a population subgroup and the total population where the definition of the subgroup reflects a prior event. For example, the number of persons reported in the census as migrants to an area during t preceding years divided by the total population of the area is often called the in-migration rate. The underlying justification here is that the number of enumerated migrants roughly approximates the number of persons who migrated into the area during the specified t years—although it always falls short of that number. The population counted in the census is likewise only an approximation to the population at risk of migration during a specified period. A rate like an illiteracy rate is even more similar to a simple proportion or percentage. It is simply the percent of persons in a given population subgroup who are classified as illiterate. The acquirement of literacy may have occurred many years earlier.

Still other types of ratios are called "rates" in demography. For example, the infant (or infantile, in British usage) mortality rate,

$$\text{IMR} = \frac{\text{Deaths to children under 1 year of age during the year}}{\text{Births during the year}} \times 1,000 \quad (9)$$

Refinement of the rate takes account of the facts that some of the deaths in the numerator occurred to births of the preceding year and that some of the births in the denominator are at risk in the following year.

Survival rates are very frequently used in demographic analysis. In a closed population (i.e., a population with no external migration), a survival rate is the ratio of the number of persons in a cohort at one date to the number at an earlier date. Survival is from a given age to a subsequent age. There are 1-year, 5-year, 10-year, etc., survival rates. Survival rates are typically calculated from age distributions in two successive censuses or from a life table. Survival rates, too, may be specific for various characteristics or population subgroups, e.g., we may have age-specific survival rates for females, married men, or the country's aboriginal population.

By multiplying a population subgroup at one census by its appropriate survival rate, we obtain the **expected population** t years later. The expected population can be used in a number of ways—to make estimates or projections of the population, to estimate net migration during the period, to help in estimating the contribution of certain factors to the total change during the period, or to estimate the relative completeness of two successive census counts. By dividing the survival rate into the population, one obtains an estimate of the size of the cohort t years earlier, for example at the preceding census date.

In general, rates are of interest in their own right in describing the dynamics of population. They may also be applied, as just illustrated, by multiplication or division to demographic aggregates to obtain various types of estimates. For example, a death rate for a prior period or one from a life table may be multiplied by an appropriate population to estimate the number of deaths in a more current period. Conversely, on the assumption that the death rate has remained constant, a rate for an earlier period (say the census year) may be divided into the number of registered deaths to obtain a postcensal estimate of the population.

We have shown that the distinction between "rates" and "ratios" in demography, as elsewhere, is somewhat fuzzy.

Restriction of the term "rate" to probabilities or to fractions where the numerator is part of the denominator is not observed in actual practice. There is a tendency for what are called "ratios" (e.g., the sex ratio or the dependency ratio) to be used for descriptive purposes in population studies and for rates to be used in the analysis of change, i.e., in population dynamics: but this is only a tendency not a universal rule. Some writers even use the term "survival ratio."

Life Tables

Several references have already been made to the life table. A whole chapter, chapter 15, is devoted to this subject—the definition and interpretation of life table functions, the construction of life tables, and some of the applications of life tables in demography. Other applications are scattered through the book.

The primary purpose of the life table for the actuary is to measure the expectation of life at each age, $\mathring{e}_x$. The table is built up from age-specific death rates, conventionally those observed in a single year or period of years for all cohorts that were alive and hence subject to the risk of dying in that period. An example of a life table is given in table 15–1. In addition to the expectation of life at age x, it may be observed that the life table displays a number of other functions. These are:

$_nq_x$ proportion of persons alive at beginning of age interval dying during interval

l_x number living at beginning of age interval of 100,000 born alive

$_nd_x$ number dying during age interval

$_nL_x$ number living in the age interval

T_x number living in this and all subsequent age intervals

The subscript n is the number of years of age in the age interval. For a complete life table, it is 1; for an abridged life table, it is usually 5. A more precise definition of $\mathring{e}_x$ is the average number of years of life remaining at the beginning of the age interval. $_nL_x$ may also be defined as the number of man-years lived in the age interval, and $_nT_x$ as the number of man-years lived in this age interval and all subsequent age intervals.

A life table ordinarily starts with 100,000 births, a benchmark which is called the radix. $_nL_x$ may also be viewed as describing the age distribution of a population called the **stationary population** because it is constant over time. This population is generated by the constant 100,000 annual births and the death rates of the life table. The survival rates previously referred to may be derived from the $_nL_x$ values. Thus, the probability of surviving k years for persons in the age group x to $x+n$ is given by

$$\frac{nL_{x+k}}{nL_x} \tag{10}$$

The survival rate is a particularly useful measure in demography.

Unabridged life tables are usually constructed by actuaries, who are concerned with the careful graduation or smoothing of the basic data or the calculated values. There are, of course, many actuarial applications of life tables in the field of insurance. Demographers often construct their own life tables from deficient data, for special population subgroups, or by shortcut methods.

Both actuaries and demographers are interested in what are called **multiple-decrement tables**. In these tables, the popu-

lation is subject to attrition not only from the force of mortality but also from that of some other factor or factors as well, for example, marriage, widowhood, or entry into the labor force. Thus, we might start with 100,000 single persons and apply to their "survivors" at successive ages not only mortality rates ($_nq_x$ values) for single persons but also a proportion that represents the probability of marrying during the interval.

Demographic Models

Increasing use is being made of model-building in demography. The life table is one such model. More complex models link together component models of fertility, mortality, and migration. Most models, however, represent closed populations. In addition to the life table, one of the best-known models is the "stable population."

Stable Population.—Demographers have long been interested in the question of what population structure would result if fixed age-specific schedules of mortality and fertility rates remained in effect indefinitely. More specifically, what would be the age-sex composition of such a population and what rate of growth would it have? The solution was found by Lotka and was presented in a classic paper.[10] He proved that the resulting population would eventually have a fixed age composition and a fixed rate of growth and provided the formulas for these. This population is called the **stable population** because its age composition is stable. Unlike the life table population, which is stationary as well as stable, it may increase or decrease in absolute numbers.

There are other parameters of (measures associated with) the stable population that are widely used in demographic analysis. These include the **true** (or **intrinsic**) **birth and death rates**, the **mean length of generation**, and the **gross and net reproduction rates**. The reproduction rates are particularly important and deserve a brief explanation at this point.

The gross reproduction rate represents the average number of daughters that would be borne by a cohort of females all of whom lived to the end of the childbearing period if the cohort bore children according to a given set of age-specific fertility rates. The net reproduction rate removes the assumption of no mortality before age 50, say, and represents the average number of daughters borne by a cohort of females starting life together, if there were no changes in the age-specific schedules of fertility and mortality.

The concept of stable population has proven itself a powerful tool in demographic analysis. For example, when an underdeveloped country has very inadequate demographic statistics but can be assumed to have had roughly constant fertility for many years, its age structure, life table, gross reproduction rate, etc., can be estimated fairly closely under some circumstances on the basis of a very few simple statistics using stable population theory.

On the other hand, when demographers tried to interpret or even apply such measures as the net reproduction rate in situations where fertility was changing very sharply, certain inconsistencies and limitations in these measures became apparent. Efforts were then made to refine the concepts and measures. Probably the chief modification was the shift from period to generation (cohort) measures, not only for making

[10] Louis I. Dublin and Alfred J. Lotka, "On the True Rate of Natural Increase," *Journal of the American Statistical Association*, 20(150):305–339, September 1925.

projections of reproduction but also for gauging its force more realistically. Another refinement was to take account of the parity of the women and order of birth of the child when computing age-specific fertility rates. **Parity** is the number of children previously borne. Zero-parity (childless) women are at risk of having a first birth, one-parity women of having a second birth, and so on. Measures using the **interval** between marriage and a birth of a given order and between births of successive order have also been developed.

Cohort Analysis

In an earlier section, we defined and illustrated a cohort. Other examples are the persons immigrating in a given year and the persons completing a given year of school in a given year. A great deal of demographic analysis is carried out by arranging the data in cohort form, particularly when changes over time are being studied.

Cohort statistics are distinguished from **period statistics,** the latter applying to a combination of cohorts in a given year or other period. A life table or a conventional reproduction rate describes a period such as a year or group of years. There are corresponding cohort forms, which are called "generation life tables" and "generation reproduction rates," respectively.

The statistical ingredients are the same for cohort as for period analysis, the difference lies in how they are put together. The death rate of a given age group in a given year can be viewed as applying either to a period of time or to a cohort. When we compare the number of persons 40 years old at two different censuses, we are comparing two different cohorts at the same stage of life, whereas when we are comparing the number of persons 40 years old with the number 50 years old 10 years later we are analyzing change within a single cohort. Both comparisons are of value to the demographer.

Cohort analysis may be carried out either forward in time or in the reverse direction. For example, **reverse survival rates** are applied to estimate the size of a cohort at earlier dates.

Controlled and Uncontrolled Factors

Much of demography is concerned with the comparisons of populations and of demographic processes in space and time. In making such comparisons, one quickly encounters a common problem in scientific analysis—that of "uncontrolled" factors. For example, if the force of mortality is being compared for two populations, one must ask whether the difference in the crude death rates can be explained by differences in their age composition—since mortality is so highly correlated with age. The demographer applies a number of methods that are in general use in social statistics in order to cope with this problem. Some of these methods, however, have been adapted in specific ways in demography; indeed, some of them have had their greatest technical elaboration in demographic applications.

One general method, namely that of **cross-classification,** has already been mentioned. Reverting to the previous illustration, we may compare the mortality of two populations using specific rates, age for age. If there are no errors in reporting age in the census or in the death records, the finer the age detail, the better the control on that factor. We may also control or "hold constant" more than one factor at a time by multiple cross-classification. Thus, we can have death rates specific for age, sex, and race.

This method of decomposition enables us to make mortality comparisons without the disturbing influence of the specific factors used in the cross-classifications. We have lost, on the other hand, one advantage of the crude rates, namely an overall summary comparison. By the technique of **standardization,** we can derive from these factor-specific rates a summary measure that, to some extent, holds constant the influence of the factor. Thus, we can compare age-standardized mortality rates for two populations or for the same population at two different dates. What is required here is the age distribution (or distribution by the factor to be controlled) for a **standard population.** The standard population may be one of the two populations concerned, their averaged distributions, or a third population, for example, in the case of comparison of two or more provinces, the total population of their country. The population of England and Wales in 1901 was long used as such a standard, and its distribution by age and sex was referred to as the **standard million.** The stationary population of a life table may also be used as the standard for the corresponding observed population.

What we have just sketched is called the **direct** method of standardization. There is also an indirect method. Suppose we know, for the two areas to be compared, their crude rates and their age distributions and, for a standard population, its factor-specific rates (age-specific, etc.). We can then compute another kind of standardized rate. There are still more types of standardization, and this general technique is closely related to **Westergaard's Method of Expected Cases** used in conjunction with multiple classifications. [11]

The concept of an expected population is a very general one in demography and is very widely used. Suppose we have a population distributed by age and sex and, for each age-sex group, the percent having some other characteristic such as being married, enrolled in school, or in the labor force. Such percentages are sometimes called **participation** rates. Then we assume that the particular participation rate remains constant. By applying such rates to the observed age-sex structure at the later date, we can calculate the expected number of married persons, students, workers, etc., on the assumption of constant participation rates. Comparison of these expected numbers with those actually observed tells us what part of the total change in the number of married persons, etc., was attributable to the change in the participation rates as opposed to change in the age-sex structure of the populations.

Summary demographic factors that are used to control the influence of age include the expectation of life from the life table and gross and net reproduction rates. There are other general statistical methods that hold constant a given factor or that measure the relative effects of two or more factors upon a dependent variable. These include partial correlation, analysis of variance, analysis of covariance, and factor analysis.

In this section we have given a brief overview of the typical approaches in demographic analysis and some of the methods used. This discussion was by no means intended to explain those methods to the extent that the reader would feel equipped to apply them. Detailed definitions and explanations are given in the body of the book. As he reads the following chapters, the reader will sometimes encounter certain terms before he encounters their fullest explanation. We hope then that he will have received a general notion of their meaning and uses and will be able to look ahead, if necessary, by means of the forward references that are given, the table of contents, or the index.

---

[11] Robert M. Woodbury, "Westergaard's Method of Expected Deaths as Applied to the Study of Infant Mortality," *Journal of the American Statistical Association,* 18(139):366–376, September 1922.

SUGGESTED READINGS

Barclay, George W. *Techniques of Population Analysis.* New York. John Wiley & Sons. 1958. Pp. 1–15.

Benjamin, Bernard. *Demographic Analysis.* New York, Frederick A. Praeger, Inc., 1968.

Bogue, Donald J. *Principles of Demography.* New York. John Wiley & Sons. 1969. Pp. 1–31.

Cox, Peter R. *Demography.* Cambridge, England. Cambridge at the University Press, 3rd ed., 1959. Pp. 1–13.

Hauser, Philip M., and Duncan, Otis Dudley, eds. *The Study of Population.* Chicago, University of Chicago Press, 1959. Pp. 1–117; 157–166; 321–360.

Pressat Roland. *Demographic Analysis: Methods, Results, Applications.* Chicago, Aldine-Atherton, 1972. (Translated by Judah Matras.)

Spiegelman, Mortimer. *Introduction to Demography.* Cambridge, Mass., Harvard University Press, 1968, 2nd ed. Pp. 1–7.

United Nations. *Handbook of Population Census Methods.* Studies in Methods, Series F, No. 5, rev. 1, Vol. I. *General Aspects of a Population Census.* 1958. Pp. 3–8.

———. *Handbook of Vital Statistics Methods.* Studies in Methods, Series F, No. 7. April 1955. Pp. 3–18.

———. *Principles and Recommendations for the 1970 Population Censuses.* Statistical Papers, Series M, No. 44. 1967. Pp. 3–9.

CHAPTER 2

Basic Sources of Statistics

TYPES OF SOURCES

The primary sources of demographic statistics may be regarded as the published reports, unpublished worksheets, tally tapes, etc., that are produced by official or private agencies. The reports may contain texts that tell how the report is organized, how the statistics were obtained, and how accurate the statistics are deemed to be. The texts may also contain descriptive or analytical material based on the statistics. Moreover, graphic devices are often incorporated.

The same statistics may be selectively reproduced or rearranged in secondary sources such as compendia, statistical abstracts, and yearbooks. Additional derived figures may be introduced at this stage. Other secondary sources that present some of these statistics are journals, textbooks, and research reports. Occasionally, a textbook or research report may include demographic statistics based on the unpublished tabulations of an official agency.

Many important kinds of demographic statistics are produced by combining census and vital statistics. Examples are vital rates, life tables, and population estimates and projections. Population estimates may also draw on immigration and emigration statistics and even on such local statistics as those on school enrollment, residential building permits, and registered voters.

Primary Demographic Data and Statistics

A country may have a central statistical office, or there may be separate agencies that take the census and compile the vital statistics. Even when both kinds of statistics emanate from the same agency, they are usually published in separate reports, reflecting the fact that censuses are customarily taken decennially or quinquennially and vital statistics are compiled annually or monthly. Both kinds of statistics may be included in the same compendium, of course, along with other kinds of official statistics.

In some countries, the province or state may have important responsibilities in conducting the census or operating a registration system. Even here, however, the national government customarily issues a summary report or series of reports. At one extreme, the central office simply combines the statistics that were tabulated in the provincial offices. On the other hand, it may receive the original records, copies of them, or abstracts from them and make its own tabulations. In either situation, both national and provincial offices may publish their own reports. The statistics from the different governmental sources may differ with respect to their arrangement, detail, and choice of derived figures. Moreover, what purport to be comparable statistics may differ because of variations in

classification or editing rules and because of processing errors.

The *raison d'être* of a census or sample survey is the statistics that it produces. The registration of vital events or a population register, on the other hand, may be at least as much directed toward the legal and administrative uses of its records. In fact, the compilation and publication of statistics from a population register may be rather minimal, partly because these activities tend to disturb the day-to-day operation of the register. Even though the equivalent of census statistics could be compiled from a population register, the countries concerned still find it necessary to conduct censuses through the usual method of delivering forms to all households simultaneously. This partial duplication of data-gathering is justified as a means of making sure that the register is working properly and of including additional items (characteristics) beyond those recorded in the register. There are often restrictions imposed on the public's access to the individual census or registration records in order to protect the interests of the persons concerned.

Statistics Produced from Combinations of Census and Registration Data

Some examples of the measures based on combinations of population figures from a census with vital statistics were given above. Rates or ratios that have the number of some kind of vital event as the numerator and a population as the denominator are the most obvious type. The denominator may be a subpopulation, such as the number of men 65 to 69 years old—divided into the number of deaths occurring at that age —or the number of women 15 to 44 years old—divided into the total number of births. Moreover, the population may come from a sample survey or a population estimate, which in turn was based partly on past births and deaths.

Products of more complex combinations include current population estimates, life tables, net reproduction rates, estimates of net intercensal migration by age, and estimates of relative completeness of enumeration in successive censuses. The computation of population projections by the component method also starts with a population by age and sex, mortality rates by age, and natality rates by age of mother. Here there may be a series of successive computations in which population and vital statistics are introduced at one or more stages.

All of these illustrative measures can be produced by the combination of statistics. A different approach is to relate the individual records. This approach is that of the matching study. By matching birth certificates, infant death certificates, and babies born in the corresponding period of time and counted in the census, one can estimate both the proportion of births that were not registered and the proportion of infants that were not counted in the census. Other statistics of demographic value can be obtained by combining the information from the two sources for matched cases so as to be able to obtain a greater number of characteristics for use in the computation of specific vital rates. For example, if educational attainment is recorded on the census schedule but is not called for on the death certificate, a matching study can yield mortality statistics for persons with various levels of educational attainment. When the same characteristic, such as age, is called for on both documents, the matching studies yield measures of the consistency of reporting.

In a country with a population register, matching studies with the census can also be carried out. Again, the resulting statistics could be either of the evaluative type or could represent cross-classifications of the population in terms of a greater number of characteristics than is possible from either source alone.

Secondary Sources

Secondary sources may be either official or unofficial and include a wide variety of textbooks, yearbooks, periodical journals, research reports, gazetteers, and atlases. In this section we mention only a few of the major sources of international population statistics. The United Nations is today the chief producer of secondary demographic statistics for the countries of the world. Some of its relevant publications include the following:

Demographic Yearbook (since 1948) presenting basic population figures from censuses or estimates, and basic vital statistics yearly, and in every issue, featuring a special topic that is presented in more detail (natality statistics, mortality statistics, population distribution, population censuses, ethnic and economic characteristics of population, marriage and divorce statistics, population trends, etc.). This *Yearbook* now includes a cumulative subject-matter index in English and French.

Statistical Yearbook (since 1948) containing fewer demographic series than the foregoing, but also including four tables of manpower statistics.

Monthly Bulletin of Statistics (since 1947) carrying four tables each on demographic topics and on manpower, for the countries of the world.

Statistical Papers, Series A, *Population and Vital Statistics Reports* (quarterly since 1949).

Statistical Papers, Series K, No. 3, *Compendium of Social Statistics*: 1967, containing tables on population and vital statistics, health, education, labor force, etc.

Epidemiological and Vital Statistics Report, published monthly since June 1947 by the World Health Organization, a specialized agency of the United Nations.

The recently established Social and Economic Statistics Administration publishes several series of reports on international population statistics. In 1969 the U.S. Bureau of the Census inaugurated Series P-96, *Demographic Reports for Foreign Countries*, under an agreement with the U.S. Agency for International Development, in 1973 ISP-30, *Country Demographic Profiles*, and in 1974 ISWP (International Statistical Programs - World Population). The Foreign Demographic Analysis Division (formerly in the Bureau of the Census but now in the Bureau of Economic Analysis, another part of S.E.S.A.) publishes Series P-90, *International Population Statistics Reports* and Series P-91, *International Population Reports*.

Several recent publications by university scholars have pre-

sented derived demographic measures for large numbers of countries, all based on a common computer program. Noteworthy among these is the book by Keyfitz and Flieger, which presents 10 tables covering 63 countries.[1] The first table presents population, births, and deaths, all by age and sex, and from these are derived life tables, population projections, intrinsic rates for the stable population, and a wide variety of other demographic measures, some of them of an evaluative nature.

Population Index, which is published by the Population Association of America and the Office of Population Research at Princeton University, has appeared since 1937. It has always contained a tabular section, which gives more attention to international than domestic demographic statistics.

United States.—The U.S. Bureau of the Census publishes a number of secondary compilations of statistics; although typically demographic statistics usually account for only a minority of the whole, nonetheless they usually have priority of place. These sources include the following:

Statistical Abstract of the United States, published since 1878.

Supplements to the *Statistical Abstract*, such as
   Pocket Data Book. USA, bienially since 1966–67
   Directory of Federal Statistics for Local Areas, 1966, a finding guide, which does not itself present statistics
   County and City Data Book, 1944, 1947, 1949, 1952, 1956, 1962, 1967, 1972 (The 1944 edition was called *Cities Supplement*)
   Congressional District Data Book (Districts of the 87th, 88th, 93rd, and 94th Congresses) and *Supplements, Redistricted States.*
   Historical Statistics of the United States, Colonial Times to 1957
   Historical Statistics of the United States, Colonial Times to 1957; Continuation to 1962 and Revisions

From 1850 through 1930, all but one census had a one-volume *Compendium*, or *Abstract*, or both. Several census studies that were not limited to the statistics of a single census are useful for historical research and contain some interesting material that is not available in the decennial reports. These are:
   A Century of Population Growth, 1909
   Negro Population, 1790–1915, 1918
   Negroes in the United States, 1920–1932, 1935

Health, Education and Welfare Indicators was published monthly April 1958 to February 1967 by the Department of Health, Education, and Welfare. It contained vital statistics, statistics on educational attainment and enrollment, etc. The annual supplement to this publication is called *Health, Education, and Welfare Trends.*

A number of other official publications are less appropriately regarded as secondary sources than as sources of important collections of statistics (population estimates and projections, rates, etc.) that were developed expressly for these publications. In this category, for example, are:

U.S. Bureau of the Census, *Vital Statistics Rates in the United States: 1900–1940*, by Forrest E. Linder and Robert D. Grove, 1943.

United States, National Center for Health Statistics,

[1] Nathan Keyfitz and Wilhelm Flieger, *World Population: An Analysis of Vital Data*, Chicago, University of Chicago Press, 1968.

Vital Statistics Rates in the United States: 1940–1960, by Robert D. Grove and Alice M. Hetzel, 1968.

Various reports of the National Resources Board (later the National Resources Committee and the National Resources Planning Board) during the late 1930's and the 1940's.

A bibliography on "Sources for National Demographic Statistics" is contained in an appendix to one of the 1950 Census Monographs.[2]

CENSUSES AND SURVEYS

The distinction between a census and a survey is far from clear-cut. At one extreme, a complete national canvass of the population would always be recognized as a census. At the other extreme, a canvass of selected households in a village to describe their living conditions would probably be regarded as a social survey. But neither the mere use of sampling nor the size of the geographic area provides a universally recognized criterion. Most national censuses do aim at a complete count or listing of the inhabitants; however, in many modern censuses, sampling is used at one or more stages—in the collection of some of the questions, in the units to be tabulated, etc. When the U.S. Bureau of the Census, at the request and expense of the local government, takes a canvass of the population of a village with 100 inhabitants, it has no hesitation in calling the operation "a special census." The Current Population Survey in the United States and other national sample surveys with equally low rates of sampling are sometimes referred to as "microcensuses." That term might more appropriately apply to very large samplings that provide statistics in considerable geographic detail, such as the 10-percent sample census of Great Britain in 1966. Here, however, we will arbitrarily designate complete official canvasses of an area as censuses but canvasses or mailings made on only a sample basis, as surveys. This convention tends to be the usual practice. Finally, the main objective of a population census is the determination of the number of inhabitants. This is rarely true of a sample survey.

The definition used by the United Nations is as follows: "A census of population may be defined as the total process of collecting, compiling and publishing demographic, economic and social data pertaining, at a specified time or times, to all persons in a country or delimited territory."[3]

The typical scope of a census or demographic survey is the size, distribution, and characteristics of the population. In countries without adequate registration of vital events, however, a population census or survey may include questions about births or deaths of household members in the period (usually the year) preceding the census. Moreover, even when vital statistics of good quality exist, the census or survey may include questions on fertility (children ever born, children still living, date of birth of each child, etc.) because the distribution of women by number of children ever born and by interval between successive births cannot be discovered from birth certificates. Recently, in the United States, the National Center for Health Statistics has mailed questionnaires to mothers identified from a sample of birth statistics for a given year in order to obtain additional particulars concerning the pregnancy and the characteristics of the parents.

International View

History of Census Taking.—Census-taking had its beginning in ancient times—in Egypt, Babylonia, China, Palestine, and Rome. Few of the results have survived, however. Moreover, the counts were undertaken to determine fiscal, labor, and military obligations and were usually limited to heads of households, males of military age, taxpayers, or adult citizens. Women and children were seldom counted.

The first of two enumerations mentioned in the Bible is assigned to the time of the Exodus, 1491 B.C. The second was taken at the order of King David in 1017 B.C. The Roman Census, taken quinquennially, lasted for about 800 years. Citizens and their property were inventoried for fiscal and military purposes. This enumeration was extended to the entire Roman Empire in 5 B.C.

The Domesday inquest ordered by William I of England in 1086 covered landholders and their holdings. The Middle Ages, however, were a period of retrogression in census-taking throughout Europe, North Africa, and the Near East.

As Kingsley Davis points out, it is hard to say when the first census in the modern sense was undertaken since censuses were long deficient in some important respects.[4] Nouvelle France (later Quebec) and Acadia (later Nova Scotia) had 16 enumerations between 1665 and 1754. In Europe, Sweden's census of 1749 is sometimes regarded as the first, but those in some of the Italian principalities (Naples, Sicily, etc.) go back into the 17th century; again, the issue hinges upon one's criteria for a census. The clergy in the established Lutheran Church of Sweden had been compiling lists of parishioners for some years prior to the time when they were required to take annual (or later triennial) inventories. Whereas in Scandinavia this ecclesiastical function evolved into population registers and occasional censuses, the parish registers of baptisms, marriages, and burials in England evolved into a vital statistics system, as will be described later in this chapter. Carr-Saunders, Davis, Lorimer, and Wolfenden mention 18th century censuses in some of the Italian and German States (Sardinia, Parma, Tuscany, Prussia), in Iceland and Denmark, and in British North America.

There were 25 colonial enumerations within what is now the United States, beginning with a census of Virginia in 1624–1625. The first census of the United States (1790) was followed in 1801 by censuses of England and France. The American experience is described in more detail in later sections of this chapter.

Turning again to the Far East, we may cite Irene Taeuber's generalization that the great demographic tradition of that region is that of population registration.[5] This practice began in ancient China with the major function being the control of the population at the local level. Occasionally, the records could be summarized to successively higher levels to yield population totals and vital statistics. The family may be viewed as the basic social unit in this system of record-keeping. In theory, a continuous population register should have resulted, but in practice statistical controls were usually relatively weak and the compilations were either never made or they tended to languish in inaccessible archives.

[2] Conrad and Irene B. Taeuber, *The Changing Population of the United States,* Philadelphia, John Wiley & Sons, 1958, pp. 327–334.

[3] United Nations, *Principles and Recommendations for National Population Censuses,* Statistical Papers, Series M, No. 27, 1958, p. 3.

[4] Kingsley Davis, "Census", *Encyclopaedia Britannica,* 1966, Vol. 5, pp. 167–170.

[5] Irene B. Taeuber, "Demographic Research in the Pacific Area," in *The Study of Population,* Philip M. Hauser and Otis Dudley Duncan, eds., Chicago, University of Chicago Press, 1959, p. 261.

The Chinese registration system diffused gradually to nearby lands. Until the present century, the statistics from this source were intended to cover only part of the total population and contained gross inaccuracies. Japan's adaptation of the Chinese system resulted in the *koseki*, or household registers. These had been in existence for more than a thousand years when, in 1721, an edict was issued that the numbers registered should be reported. Such compilations were made at six-year intervals down to 1852 although certain relatively small classes of the population were omitted. Thus, this use of the population register parallels that in Scandinavia in the same centuries. The first census of Japan by means of a canvass of households was not attempted until 1920; it presumably resulted from the adoption of the western practice that was then more than a century old. Fairly frequent compilations of populations and households were made in Korea during the Yi dynasty; the earliest was in 1395. [6]

In summary, then, the evolution of the modern census was a gradual one. The tradition of household canvasses or population registration often had to continue for a long time before the combination of public confidence, administrative experience, and technology could produce counts that met modern standards of completeness, accuracy, and simultaneity. Beginning with objectives of determining military, tax, and labor obligations, censuses in the 19th century changed their scope to meet other administrative needs as well as the needs of business, labor, education, and academic research. New items included on the census questionnaire reflected new problems confronting state and society.

Of continuing interest are the periodic national sample surveys of households that have been established in a number of countries, somewhat on the model of the Current Population Survey in the United States. These may be conducted monthly, quarterly, or only annually; in some countries, they have been discontinued after one or two rounds because of financial or other problems. Usually the focus is employment and unemployment or consumer expenditures rather than demographic information, but demographic items may be added in particular rounds.

Censuses and Surveys in the United States

Both population censuses and national sample surveys developed relatively early in the United States of America, the former much earlier than the latter, of course. Moreover, the decennial census and the Current Population Survey have never missed a scheduled round since their respective inceptions. Both have also tended to grow in the range of topics covered, in sophistication of procedures, in accuracy of results, and in the volume of statistics made available to the public.

Decennial Censuses.—The census of population has been taken regularly every 10 years since 1790 and was one of the first to be started in modern times. At least as early as the 1940's, there have been demands for a quinquennial census of population, which is the frequency in a fair number of other countries; but, so far, no mid-decade census has been approved by the Congress.

Evolution of the population census schedule.—The area covered by the census included the advancing frontier within continental United States. Each outlying territory and posses-

sion has been included also, but the statistics for these areas are mostly to be found in separate reports.[7] Beginning as a simple list of heads of households with a count of members in five demographic and social categories, the population census has developed into an inventory of many of the personal, social, and economic characteristics of the American people.

A comprehensive account of the content of the population schedule at each census through the 11th in 1890 is included in *The History and Growth of the United States Census* by Wright and Hunt.[8] A list of items included in each census through 1960 is given in table 2-1.

The changing content of the population schedule has reflected the rise and wane of different public problems. Since the Constitution provided that representatives and direct taxes should be apportioned among the States "according to their respective numbers, which shall be determined by adding to the whole number of free persons ... excluding Indians not taxed, three-fifths of all other persons," early attention was directed to free Negroes and slaves. Indians were not shown separately until 1860, and most were omitted until 1890. Increasing detail was obtained on age and color; but it was not until 1850 that single years of age and sex were reported for whites, blacks, and mulattoes. Interest in immigration was first reflected on the census schedule in 1820 in an item on "foreigners not naturalized"; but the peak of attention occurred in 1920 when there were questions on country of birth, country of birth of parents, citizenship, mother tongue, ability to speak English, year of immigration, and year of naturalization of the foreign born.

Internal migration did not become a subject of inquiry until 1850 when State of birth of the native population was asked for, and it was not until 1940 that questions were carried on residence at a fixed date in the past. The first item on economic activity was obtained in 1820 ("number of persons engaged in agriculture," "number of persons engaged in commerce," "number of persons engaged in manufactures"). The items on economic characteristics have increased in number and detail; they have included some on wealth and, more recently, income. Welfare interests in the defective, delinquent, and dependent were first recognized on the 1840 schedule. Such inquiries were expanded over the course of many decades and did not completely disappear from the main schedule until 1920. In 1970, again, an item on disability has been introduced. Education and the veteran both received their first recognition in 1840 also. Attempts were made to collect vital statistics through the census before a national registration system was begun. Interest in public health led to a special schedule on mortality as early as 1850; but questions on marriages and births were carried on the population schedule itself, beginning in 1860 and 1870, respectively. A few questions on real property owned and on housing were included, beginning in 1850; but, with the advent of the concurrent housing census in 1940, such items were dropped from the population schedule. The topic of the journey to work ("commuting") did not receive attention until 1960 when questions on place of work and means of transportation were included. New items added in the 1970 census included major activity

---

[6] Kap Suk Koh, *The Fertility of Korean Women*, Seoul, The Institute of Population Problems, 1966, p. 6. (A volume in the 1960 monograph series.)

[7] Alaska and Hawaii, previously the subjects of separate reports, were included in the national totals in the 1960 census, after they had become States.

[8] Carroll D. Wright and William C. Hunt, *The History and Growth of the United States Census*, Washington, Government Printing Office, 1900.

Table 2–1. —Questions Included in Each Population Census in the United States: 1790 to 1970

CENSUS OF 1970

Information obtained from all persons: Address; name; relationship to head of household; sex; race; age; month and year of birth; marital status; if American Indian, name of tribe.

Information obtained from 20-percent sample: Whether residence is on farm;[1] place of birth; educational attainment; for women, number of children ever born; employment status; hours worked in week preceding enumeration; year last worked; industry, occupation, and class of worker; state or country of residence 5 years ago; activity 5 years ago; weeks worked last year; earnings last year from wages and salary, from self-employment; other income last year.

Information obtained from 15-percent sample: Country of birth of parents; county, and city or town of residence 5 years ago (and whether in city limits or outside); length of residence at present address; language spoken in childhood home; school or college attendance, and whether public, parochial, or other private school; veteran status; place of work--street address, which city or town (and whether in city limits or outside), county, State, ZIP code; means of transportation to work.

Information obtained from 5-percent sample: Whether of Spanish descent; citizenship; year of immigration; whether married more than once and date of first marriage; whether first marriage ended because of death of spouse; vocational training; (for persons of working age) presence and duration of disability; industry, occupation, and class of worker 5 years ago.

Supplemental schedule for Americans overseas.

CENSUS OF 1960

Information obtained from all persons: Address; name; relationship to head of household; sex; race; month and year of birth; marital status.

Information obtained from 25-percent sample: Whether residence is on farm;[1] place of birth; if foreign born, language spoken in home before coming to U.S.; country of birth of parents; length of residence at present address; State, county, and city or town of residence 5 years ago; educational attainment; school or college attendance, and whether public or private school; whether married more than once and date of first marriage; for women ever married, number of children ever born; employment status; hours worked in week preceding enumeration; year last worked; occupation, industry, and class of worker; place of work--which city or town (and whether in city limits or outside), county, State; means of transportation to work; weeks worked last year; earnings last year from wages and salary, from self-employment; other income last year; veteran status.

Supplemental schedule for Americans overseas.

CENSUS OF 1950

Information obtained from all persons: Address; whether house is on farm; name; relationship to head of household; race; sex; age; marital status; place of birth; if foreign born, whether naturalized; employment status; hours worked during week preceding enumeration; occupation, industry, and class of worker.

Information obtained from 20-percent sample: Whether living in same house a year ago; whether living on a farm a year ago; if not in same house, county and State of residence a year ago; country of birth of parents; educational attainment; school attendance; if looking for work, number of weeks; weeks worked last year; for each person and each family, earnings last year from wages and salary, from self-employment; other income last year; veteran status.

CENSUS OF 1950--Continued

Information obtained from 3-1/3 percent sample: For persons who worked last year but not in current labor force: occupation, industry, and class of worker on last job; if ever married, whether married more than once; duration of present marital status; for women ever married, number of children ever born.

Supplemental schedules for persons on Indian reservations; infants born in the first 3 months of 1950; Americans overseas.

CENSUS OF 1940

Information obtained from all persons: Address; home owned or rented; value or monthly rental; whether on a farm; name; relationship to head of household; sex; race; age; marital status; school or college attendance; educational attainment; place of birth; citizenship of foreign born; county, State, and town or village of residence 5 years ago and whether on a farm; employment status; if at work, whether in private or nonemergency government work, or in public emergency work (WPA, NYA, CCC, etc.); if in private or non-emergency government work, number of hours worked during week of March 24-30; if seeking work or on public emergency work, duration of unemployment; occupation, industry, and class of worker; number of weeks worked last year; wage or salary income last year and whether received other income of $50 or more.

Information from 5-percent sample: Place of birth of parents; language spoken in home in earliest childhood; veteran status, which war or period of service; whether wife or widow of veteran; whether a child under 18 of a veteran and, if so, whether father is living; whether has Social Security number, and if so, whether deductions were made from all or part of wages or salary; occupation, industry, and class of worker; of women ever married-- whether married more than once, age at first marriage, and number of children ever born.

Supplemental schedule for infants born during the four months preceding the census.

(All inquiries in censuses from 1790 through 1930 were asked of all applicable persons.)

CENSUS OF 1930

Address; name; relationship to head of family; sex; race; age; marital status; age at first marriage; home owned or rented; value or monthly rental; radio set; whether family lives on a farm; school attendance; literacy; place of birth of person and parents; if foreign born, language spoken in home before coming to U.S.; year of immigration; naturalization; ability to speak English; occupation, industry, and class of worker; whether at work previous day (or last regular working day); veteran status; for Indians, whether of full or mixed blood, and tribal affiliation.

Supplemental schedules for gainful workers not at work on the day preceding the enumeration; blind and deaf-mutes.

CENSUS OF 1920

Address; name; relationship to head of family; sex; race; age; marital status; year of immigration to the U.S.; whether naturalized and year of naturalization; school attendance; literacy; place of birth of person and parents; mother tongue of foreign born; ability to speak English; occupation, industry, and class of worker.

Supplemental schedules for the blind and for the deaf.

---

[1] The question on farm residence was carried on the housing schedule but was available for cross-classification with population characteristics.

Table 2–1 —Questions Included in Each Population Census in the United States: 1790 to 1970—Con.

CENSUS OF 1910

Address; name; relationship to head of family; sex; race; age; marital status; number of years of present marriage; for women, number of children born and number now living; place of birth and mother tongue of person and parents; if foreign born, year of immigration., whether naturalized or alien, and whether able to speak English, or if not, language spoken; occupation, industry, and class of worker; if an employee, whether out of work on census day, and number of weeks out of work during preceding year; literacy; school attendance; home owned or rented; if owned, whether mortgaged; whether farm or house; whether a survivor of Union or Confederate Army or Navy; whether blind or deaf and dumb.

Supplemental schedules for the Indian population; blind; deaf; feeble-minded in institutions; insane in hospitals; paupers in almshouses; prisoners and juvenile delinquents in institutions.

CENSUS OF 1900

Address; name; relationship to head of family; sex; race; age; month and year of birth; marital status; number of years married; for women, number of children born and number now living; place of birth of person and parents; if foreign born, year of immigration to the U.S., number of years in the U.S., and whether naturalized; occupation; months not employed; months attended school during census year; literacy; ability to speak English.

Supplemental schedules for the blind and for the deaf.

CENSUS OF 1890

Address; name; relationship to head of family; race; sex; age; marital status; number of families in house; number of persons in house; number of persons in family; whether a soldier, sailor or marine during Civil War (Union or Confederate) or widow of such person; whether married during census year; for women, number of children born, and number now living; place of birth of person and parents; if foreign born, number of years in the U.S., whether naturalized or whether naturalization papers had been taken out; profession, trade, or occupation; months unemployed during census year; months attended school during census year; literacy; whether able to speak English, and if not, language or dialect spoken; whether suffering from acute or chronic disease, with name of disease and length of time afflicted; whether defective in mind, sight, hearing, or speech, or whether crippled, maimed, or deformed, with name of defect; whether a prisoner, convict, homeless child, or pauper; home rented or owned by head or member of family; if owned by head or member, whether mortgaged; if head of family a farmer, whether farm rented or owned by him or member of his family; if owned, whether mortgaged; if mortgaged, post office address of owner.

Supplemental schedules for the Indian population; for persons who died during the year; insane; feeble-minded and idiots; deaf; blind; diseased and physically defective; inmates of benevolent institutions; prisoners; paupers and indigent persons; surviving soldiers, sailors, and marines, and widows of such; inmates of soldiers' homes.

CENSUS OF 1880

Address; name; relationship to head of family; sex; race; age; marital status; month of birth if born within the census year; married within the year; occupation; number of months unemployed during year; sickness or temporary disability; whether blind, deaf and dumb, idiotic, insane, maimed, crippled, bedridden, or otherwise disabled; school attendance; literacy; place of birth of person and parents.

CENSUS OF 1880--Continued

Supplemental schedules for the Indian population; for persons who died during the year; insane; idiots; deaf-mutes; blind; homeless children; prisoners; paupers and indigent persons.

CENSUS OF 1870

Name; age; sex; race; occupation; value of real estate; value of personal estate; place of birth; whether parents were foreign born; month of birth if born within the year; month of marriage if married within the year; school attendance; literacy; whether deaf and dumb, blind, insane, or idiotic; male citizens 21 and over, and number of such persons denied the right to vote for other than rebellion.

Supplemental schedules for persons who died during the year; paupers; prisoners.

CENSUS OF 1860

Name; age; sex; race; value of real estate; value of personal estate; occupation; place of birth; whether married within the year; school attendance; literacy; whether deaf and dumb, blind, insane, idiotic, pauper, or convict.

Supplemental schedules for slaves; public paupers and criminals; persons who died during the year.

CENSUS OF 1850

Name; age; sex; race; whether deaf and dumb, blind, insane, or idiotic; value of real estate; occupation; place of birth; whether married within the year; school attendance; literacy; whether a pauper or convict.

Supplemental schedules for slaves; public paupers and criminals; persons who died during the year.

CENSUS OF 1840

Name of head of family; age; sex; race; slaves; number of deaf and dumb; number of blind; number of insane and idiotic and whether in public or private charge; number of persons in each family employed in each of six classes of industry and one of occupation; literacy; pensioners for Revolutionary or military service.

CENSUS OF 1830

Name of head of family; age; sex; race; slaves; deaf and dumb; blind; foreigners not naturalized.

CENSUS OF 1820

Name of head of family; age; sex; race; foreigners not naturalized; slaves; industry (agriculture, commerce, and manufactures).

CENSUS OF 1810

Name of head of family; if white, age and sex; race; slaves.

CENSUS OF 1800

Name of head of family; if white, age and sex; race; slaves.

CENSUS OF 1790

Name of head of family; free white males of 16 years and up, free white males under 16; free white females; slaves; other persons.

and occupation five years ago, vocational training, and additional particulars designed to improve the classification of occupation.

Publications.—There is a cumulative list of census publications covering those from the beginning through 1945. [9] Subsequent years are covered by annual catalogues, which are preceded by monthly and quarterly issues.[10]

The results of the 1970 Census of Population and Housing are available in the form of printed reports, microfiche copies of the printed reports, computer summary tapes, computer printouts, and microfilm. The population census statistics are published in two basic volumes:

Volume I, *Characteristics of the Population,* consists of separate reports for the United States, each of the 50 states, the District of Columbia, Puerto Rico, Guam, Virgin Islands of the United States, American Samoa, Canal Zone, and Trust Territory of the Pacific Islands. For each of the 58 areas, the data were first issued in four separate paperbound chapters designated as PC (1)-A, B, C, and D. For the outlying areas other than Puerto Rico, all the data on characteristics of the population are included in chapter B. A brief description of the content of each of these four chapters is as follows:

PC (1)-A, *Number of Inhabitants.* Final official population counts are presented for States, counties (by urban–rural residence), Standard Metropolitan Statistical Areas (SMSA's), urbanized areas, minor civil divisions, census country divisions, all incorporated places, and unincorporated places of 1,000 inhabitants or more.

PC (1)-B, *General Population Characteristics.* These reports contain statistics on age, sex, race, marital status, and relationship to head of household for States, counties (by urban–rural residence), Standard Metropolitan Statistical Areas, urbanized areas, minor civil divisions, census country divisions, and places of 1,000 inhabitants or more.

PC (1)-C, *General Social and Economic Characteristics.* These reports focus on the population subjects collected on a sample basis (see Table 2-1). Each subject is shown for some or all of the following areas: States (by urban, rural–nonfarm, and rural–farm residence), counties, SMSA's, urbanized areas, and places of 2,500 inhabitants or more.

PC(1)-D, *Detailed Characteristics.* These reports cover most of the population subjects collected on a sample basis, presenting the data in considerable detail or cross-classified by age, race, and other characteristics. Each subject is shown for some or all of the following areas: States (by urban, rural–nonfarm, and rural–farm residence), SMSA's, and large cities.

Volume II, *Special Reports.* There are 39 reports scheduled in this volume, each of which concentrates on a particular subject. Detailed information and cross-relationships are generally provided on a national and regional level; in a few reports, data for States or standard metropolitan statistical areas are also shown. Among the characteristics covered are national origin and race, fertility, families, marital status, migration, education, employment, unemployment, occupation, industry, and income. There is also a report on the geographic distribution and characteristics of the institutional population.

A number of reports have also been published that combine population items with those of the housing census. There are three series of these reports based on the 1970 enumeration:

Series PHC(1) *Census Tract Reports.* One report for each SMSA, showing data for most of the population and housing subjects included in the 1970 Census. Some tables are based on the 100-percent tabulation, others on a sample tabulation.

Series PHC (2) *General Dmographic Trends for Metropolitan Areas, 1960 to 1970.* This series consists of one report for each State and the District of Columbia, as well as a national summary report, presenting statistics for the State and for SMSA's and their central cities and constituent counties. Comparative 1960 and 1970 data are shown on population counts by age and race.

Series PHC (3) *Employment Profiles of Selected Low-Income Areas.* This series consists of 75 reports, each presenting statistics on the social and economic characteristics of the residents of a particular low-income area, as well as a United States summary. The data, derived from sample surveys conducted during late 1970 and early 1971, relate to low-income neighborhoods in 51 cities and seven rural areas. Each report provides statistics on employment, education, vocational training, availability for work, job history, and income.

Another type of publication is represented by the occasional census monograph dealing with detailed analysis of specific topics. For example, the following monographs were prepared as part of the 1960 publication program:

Herman P. Miller	*Income Distribution in the United States*
John K. Folger and Charles B. Nam	*Education of the American Population*
Dale E. Hathaway, J. Allan Beegle, and W. Keith Bryant	*People of Rural America*
Daniel O. Price	*Changing Characteristics of the Negro Population*
Irene B. Taeuber and Conrad Taeuber	*People of the United States in the 20th Century*

There are, in addition, a number of methodological publications that are either part of a decennial census program or that are included in one of the series of technical reports. A prominent example of the former is the *Procedural History* of the 1970 Census of Population and Housing.

Available unpublished data.—Unpublished data may be classified according to the stage of processing. Coded schedules are not available directly because of confidentiality considerations. Computer tapes containing individual records are ordinarily subject to the same restrictions, but an important breakthrough occurred in the 1960 census when, without violating confidentiality, 1/1,000 and 1/10,000 samples of the records were marketed at a subsidized price for use in training and special tabulations.[11] Thus, demographers in universities could produce additional cross-classifications of the data as part of their own research, and students were able to learn more about how census data are edited and tabulated. Special tabulations may be contracted for from the record tapes, or more cheaply, from the tally tapes (statistics in the form of tallies that were made in the process of preparing the computer printouts of tables). The provisions for use of unpublished data were greatly expanded in the 1970 Census. See Bureau of the Census, *Catalog 1974,* Part II, "Data Files and Special Tabulations."

Special Censuses Conducted by the Federal Government.—At the request and expense of local governments, many

[9] U.S. Bureau of the Census, *Catalogue of the United States Census Publications: 1790–1945,* by Henry J. Dubester, 1950.

[10] *Bureau of the Census Catalog,* [date] (published as *Census Publications—Catalog and Subject Guide* until 1952). This annual catalogue lists not only the official publications (with a subject index) but also the "outside" publications of staff members. Furthermore, it contains a section on "Data files and unpublished materials."

[11] *U.S. Censuses of Population and Housing: 1960, One-in-a-Thousand Sample, Description and Technical Documentation,* 1964.

complete enumerations have been taken at off-census dates by the Bureau of the Census.

The local government almost invariably chooses to collect only the minimum of items of information—name, relationship to the head of the household, sex, age, and race. (The census is usually taken to obtain a certified count for some fiscal purpose.) Most of the special censuses are requested for cities; but counties, minor civil divisions, and annexations have also been covered and occasionally even an entire State. Results for individual areas are published in *Current Population Reports,* Series P–28 (formerly Series P–SC), *Special Censuses,* for areas with 50,000 inhabitants or more. Statistics are shown by age, sex, and color for all political areas of this size; if the area is tracted, these statistics are also published by tracts. Quarterly and annual summaries present abridged data for all the special censuses. Furthermore, all the special censuses taken in the preceding decade have been listed in the U.S. Summaries of the 1950, 1960, and 1970 Censuses.[12]

State and Local Censuses.—The trend in the number of censuses taken by States and localities has been quite unlike that in the number of special censuses taken by the federal government. In or around 1905, 15 States took their own census; in 1915, 15; in 1925, 9; in or around 1935, 6; in 1945, 4; and in 1955 and 1965, only 2. The last survivors are Kansas and Massachusetts.

Censuses conducted by cities and other local governments are not currently and never have been very plentiful. State agencies in three Pacific Coast States that have had official responsibility for preparing population estimates have implemented that function partly by conducting censuses, particularly of small incorporated places. These agencies are the Washington State Census Board, the Oregon Center for Population Research and Census, and the Financial and Population Research Section of the California State Department of Finance. They began conducting such special censuses, on the request of local governments, in 1943, 1955, and 1960, respectively.

Sample Surveys in the United States.—Beginning in the second third of the 20th century, sample surveys of many kinds have had a tremendous vogue in the United States. Demography has shared in this movement.

Surveys conducted by the Bureau of the Census.—The sample check on the 1937 unemployment census was a methodological forerunner of the many sample surveys that have been conducted by the Federal government, but demographic data in the narrow sense were not collected in it. The procedures inaugurated in the 1940 census of asking selected questions of a representative sample of persons and of tabulating for only a sample of persons certain data that could have been tabulated for all persons should be distinguished from surveys where only a sample of households is interviewed. The ancestry of the sample surveys of the Bureau of the Census traces rather to the Monthly Report on the Labor Force of the Works Progress Administration.

This survey was designed to provide an official national series on the labor force, employment, and unemployment. The total population and its distributions by age, color, and

sex are not obtained from this survey but are independent estimates based on other demographic data, and the survey results are inflated to these totals. After the liquidation of the Works Progress Administration in August 1942, this survey was transferred to the Census Bureau. In June 1947, it was renamed the Current Population Survey. This sample survey provides many demographic data, some from special tabulations of the contents of the regular monthly forms and others by means of the supplementary questions that are carried from time to time. The first of the special tabulations was for marital status in February 1944, and the first supplementary demographic question dealt with migration status as of March 1945.[13]

There are now regular annual series on marital status and family status, households and families, metropolitan and non-metropolitan residence, population mobility, farm population, school enrollment, educational attainment, Negro population, and consumer income.[14] In addition, there are supplements about every two years on fertility and voting behavior. Furthermore, there are occasional supplements on other demographic topics. Many of these supplements are added at the request of another Federal agency. Here the contracting agency usually publishes the data, but sometimes data from the same supplement are also published by the Bureau of the Census. These special topics have included reasons for moving, religion, educational change in a generation, literacy, occupational mobility, financial resources and major expenditures of the elderly, outdoor recreational activity, and poverty.[15]

Prior to July 1, 1959, the Census Bureau's *Current Population Reports* also included Series P–50, *Labor Force,* and Series P–57, *Labor Force,* "Monthly Report on the Labor Force." P–50 comprised an annual report on the labor force and reports on special supplements relating to employment and unemployment as well as projections of the labor force. Although such data are still collected through the Current Population Survey, they are now published by the Bureau of Labor Statistics—the equivalent of the P–50 reports as articles in its *Monthly Labor Review* and the erstwhile P–57 statistics in *Employment and Earnings.*

The Bureau of the Census also operates other national sample surveys of households. For example, statistics based on its National Health Survey are published by the U.S. Public Health Service.

Surveys conducted by other federal agencies.—Since most other federal agencies do not have their own national field

---

[12] *U.S. Census of Population: 1950,* Vol. I, *Number of Inhabitants,* Washington, 1952, table 32, pp. 1–80 to 1–84; and *U.S. Census of Population: 1960,* Vol. I, *Characteristics of the Population,* Part A, *Number of Inhabitants,* 1961, table 40, pp. 1–127 to 1–138; U.S. Census of Population: 1970, Vol. 1, *Characteristics of the Population,* Part A: *Number of Inhabitants,* 1972, table 45, pp. 1–223 to 1–245.

---

[13] U.S. Bureau of the Census, *Population—Special Reports,* Series P–S, No. 1, "Marital Status of the Civilian Population: February, 1944," June 20, 1944; idem, *Population—Special Reports,* Series P–S, No. 5, "Civilian Migration in the United States: December, 1941, to March, 1945," September 2, 1945.

[14] For convenience, all the current publication series of the Population Division, U.S. Bureau of the Census, will be listed at this point, even though some of them are not based, or are not entirely based, on current surveys.

CURRENT POPULATION REPORTS
P-20 *Population Characteristics*
P-23 *Special Studies*
P-25 *Population Estimates and Projections*
P-26 *Federal–State Cooperative Program for Population Estimates*
P-27 (also Series Census-ERS), *Farm Population* (Jointly with Department of Agriculture)
P-28 *Special Censuses*
P-60 *Consumer Income*

[15] Gertrude Bancroft, "Special Uses of the Current Population Survey Mechanism," *Estadística,* 12(43):198–205, June 1954; Daniel B. Levine and Charles B. Nam, "The Current Population Survey: Methods, Content, and Sociological Uses," *American Sociological Review* 27(4):585–590, August 1962.

organizations for conducting household surveys, they tend to turn to the Bureau of the Census as the collecting agency when social or economic data are needed for their research or administrative programs. An exception that is of interest here is the Department of Agriculture. From a mail-questionnaire survey conducted for the Economic Research Service by the Department's Statistical Reporting Service, data are collected annually about the number of persons moving to and from a sample of farms, from which the Economic Research Service constructs estimates of the farm population by geographic divisions.[16] These estimates are adjusted to the national estimates of farm population, which are based on the Census Bureau's Current Population Survey and are published jointly by the two agencies in Series P–27, *Farm Population,* as mentioned above. Furthermore, the Current Population Survey includes every December, for the Department of Agriculture, supplementary questions on hired farm workers.[17] A feature of particular demographic interest from this source is the data on migratory agricultural workers, which are sometimes also the subject of special reports. Other topics concerning farm population or agricultural manpower are covered by occasional supplements to the Current Population Survey.

The Office of Education and the American Association of Collegiate Registrars and Admissions Officers have conducted several surveys on the migration of college students.[18] The information was collected from the colleges and universities themselves on the basis of their registration records, and not from the individual students.

A rather different kind of survey using a probability sample has been conducted by the National Center for Health Statistics. It is called the National Natality Survey. Beginning with the births of 1963, a sample of about 3,700 families in which a birth had occurred was sent a questionnaire that contained a number of items concerning the mother's and father's social and economic characteristics. (This is information that was not called for on the standard birth certificate.)[19]

Surveys by State and local governments and by private agencies.—Sample surveys of population conducted by State or local governments were very rare until after 1960.

In recent years such surveys have proliferated, partly in connection with programs in the fields of manpower, unemployment, health, education, and welfare. Federal grants have been made in large numbers to State and city agencies, and especially to universities, for surveys and research. Few of the surveys are concerned directly with population but may include "background" questions on the demographic characteristics of the persons in the sample.

There are a great many private survey organizations in the United States, and some of them conduct national sample surveys in which demographic data are collected. In fact, much of what

we know about the determinants of differential fertility behavior in this country today has been derived from one or another sample surveys. The Survey Research Center at the University of Michigan includes among its many programs the Detroit Area Survey, which was established in 1951. These surveys frequently include questions on migration, family income, and other population topics.[20] In the past it has also conducted two surveys (in 1955 and 1960) underlying the study on the Growth of American Families.[21] Likewise, the National Opinion Research Center at the University of Chicago and National Analysts have been data-collectors for important demographic studies. The latter, for example, carried out the field work for Princeton's fertility study, popularly identified as the Third Child Study, and the fieldwork for the 1973 round of the Family Growth Survey for the National Center for Health Statistics; and it has been the major data collecting agent for the National Fertility Studies of 1965, 1970, and 1975.[22] The classic Indianapolis Study, which has engaged the research efforts of so many analysts of human fertility at so many different institutions over the course of a generation, was, however, carried out in the field by an *ad hoc* survey organization.[23] Demographic surveys conducted by universities in particular communities are legion, and their number grows at an accelerated pace. Moreover, some teams of demographers and other social scientists have conducted population surveys abroad or have given the essential supervision of technical assistance to such surveys; and, of course, many foreign surveys have been financed in whole or part by American foundations. Demographic survey data from a number of countries are being included in a network of archives, the Social Science Data Archives.[24]

Public opinion surveys sometimes include attitudinal questions regarding population problems. For example, the Gallup Poll (American Institute of Public Opinion) has inquired about what was considered the ideal family size in 10 polls from 1936 to 1966.

REGISTRATION SYSTEMS

This major section of the chapter treats not only the statistics that are produced by registration of vital events and the recording of arrivals and departures at international boundaries but also universal population registers and registers of parts of the population (workers employed in jobs covered by social insurance plans, aliens, members of the armed forces, voters, etc.). In most cases one's name is inscribed in a register as the result of the occurrence of a certain event (birth, entering the country, attaining military age, entering gainful

---

[16] For example: U.S. Department of Agriculture, Economic Research Service, *Farm Population—Estimates for 1968,* ERS 427, Sept. 1969.

[17] The latest regular report from this source is: U.S. Department of Agriculture, Economic Research Service, *The Hired Farm Working Force of 1967,* by Robert C. McElroy, Agricultural Economic Report No. 148, September 1968.

[18] The latest report is: U.S. Office of Education, *Residence and Migration of College Students, Fall 1963: State and Regional Data,* by Mabel C. Rice and Paul L. Mason, OE–54033, Circular No. 783, 1965; There has been a long series of such surveys under various auspices. See: H. Theodore Groat, "Internal Migration Patterns of a Population Subgroup: College Students, 1887–1958," *American Journal of Sociology,* 69(4):383–394, January 1964.

[19] National Center for Health Statistics, *Vital and Health Statistics,* Series 22, No. 3, *Methods and Response Characteristics: National Natality Survey, United States, 1963,* by Gooloo S. Wunderlich, September 1966.

[20] University of Michigan, Survey Research Center, *The Detroit Area Study—A Bibliography of Material Based on Detroit Area Study Research: September 1, 1951—December 1, 1960,* Ann Arbor, Michigan, Detroit Area Survey, 1960; University of Michigan, Institute for Social Research, *List of Publications: 1961–1965,* compiled by William Goodrich Jones, Ann Arbor, Michigan, Survey Research Center, 1965.

[21] Ronald Freedman, Pascal K. Whelpton, Arthur A. Campbell, *Family Planning, Sterility, and Population Growth,* New York, McGraw-Hill Book Co., 1959; and Pascal K. Whelpton, Arthur A. Campbell, and John E. Patterson, *Fertility and Family Planning in the United States,* Princeton, N.J., Princeton University Press, 1966.

[22] Charles F. Westoff, et al., *Family Growth in Metropolitan America,* Princeton, N.J., Princeton University Press, 1961; idem, *The Third Child,* Princeton University Press, 1963. As of this writing, only the results of the 1965 National Fertility Survey had been published. See Norman B. Ryder and Charles F. Westoff, *Reproduction in the United States: 1965,* Princeton, N.J., Princeton University Press, 1971.

[23] P. K. Whelpton and Clyde V. Kiser, eds., *Social and Psychological Factors Affecting Fertility,* 5 vols., New York, Milbank Memorial Fund, 1946 to 1958.

[24] Ralph L. Bisco, "Social Science Data Archives: A Review of Developments," *American Political Science Review,* 60(1):93–109, March 1966.

employment, etc.). Some registers are completed at a single point in time, some are repeated periodically, others are cumulative. The cumulative registers may be brought up to date by recording the occurrence of other events (death, migration, naturalization, retirement from the labor force, etc.).

Vital Statistics

International View.—The tangled origins of civil registration, population censuses, and population registers have been noted above. The chronology of important events in the development of civil registration and of the vital statistics derived from it are outlined in a very useful United Nations publication.[25] Skipping over the few known registration systems in antiquity, we may note that the recording of marriages, christenings, and burials in parish registers developed as an ecclesiastical function in Christendom but gradually evolved into a secular system for the compulsory registration of births, marriages, deaths, etc., that extended to the population outside the country's established church. The 1532 English ordinance that required weekly "Bills of Mortality" to be compiled by the parish priests in London is a famous landmark.[26] In 1538 every Anglican priest was required by civil law to make weekly entries in a register for weddings and baptisms as well as for burials, but these were not compiled into statistical totals for all of England. In fact, it was not until the Births, Marriages and Deaths Registration Act became effective in 1837 that these events were registered under civil auspices and a central records office was established. Meanwhile, the Council of Trent in 1563 made keeping of registers of marriages and baptisms a law of the Catholic Church, and registers were instituted not only in many European countries but also in their colonies in the New World. Registration of vital events began relatively early in Protestant Scandinavia; the oldest parish register in Sweden goes back to 1608. Compulsory civil registration of births, deaths, stillbirths, and marriages was enacted in Finland (1628), Denmark (1646), Norway (1685), and Sweden (1686).[27] The first regular publication of vital statistics by a government office is credited to Dr. William Farr, who was appointed Compiler of Abstracts in the General Register Office in 1839, shortly after England's Registration Act of 1837 went into effect.

Outside the western cultural sphere, registration of vital events had independent origins in China. From China, the registration system diffused to Japan.[28] The civil register there had its roots in the previously mentioned *koseki*, or community-family archives, of the seventh century A.D. Civil Registration also diffused to other Asian and African countries, either having been introduced by European colonial governments or having been adopted by independent countries.

After this very brief sketch of the development of vital statistics around the world, we may turn to vital statistics as they exist today. According to the United Nations *Handbook,* " . . . a vital statistics system can be defined as including the

legal registration, statistical recording and reporting of the occurrence of, and the collection, compilation, analysis, presentation, and distribution of statistics pertaining to 'vital events', which in turn include live births, deaths, foetal deaths, marriages, divorces, adoptions, legitimations, recognitions, annulments, and legal separations."[29] The end products of the system that are used by demographers are, of course, the vital statistics and not the legal issues of the documents.[30]

Events registered.—The events registered may include live births, deaths, fetal deaths (stillbirths), marriages, divorces, annulments, adoptions, legitimations, recognitions, and legal separations. Not all countries with a civil registration system register all these types or publish statistics on their numbers. Moreover, some types are of marginal interest to demographers.

As is pointed out in the United Nations *Handbook,* other demographic events, such as migration and naturalization, are not generally considered as part of the vital statistics system because they are not usually recorded by civil registration.[31] Moreover, these events might not be considered "vital" events.

Items on the certificates.—In discussing the items of information on the certificate or other statistical report of the vital event, we may distinguish between those that are of demographic value and those that are of legal or medical value only. The former include the date of occurrence, the usual place of residence of the decedent or of the child's mother, age and sex of the decedent, sex of the child (birth), age and marital status of the mother, occupation of the father, order of the marriage (first, second, etc.), date of marriage for the divorce, etc. The latter include such items as hour of birth, name of physician in attendance, name of person certifying the report, and date of registration. Some items such as weight at birth, period of gestation, and place of occurrence (instead of usual place of residence) are of marginal demographic utility but may be used in specialized studies.

Publications.—Recommended annual tabulations of live births, deaths, fetal deaths, marriages, and divorces are outlined in the U.N. *Handbook.*[32] Rates and indexes, which are essential to even the most superficial demographic analysis, are also treated in the *Handbook.*[33]

Unfortunately, there is no readily available summary, from the United Nations or any other source, of the tables currently published by the various countries. The "Bibliography" in the U.N. *Demographic Yearbook* lists the national publications from which vital statistics were obtained. In most years, only the new sources are listed. In 1953, however, there was a cumulative summary that was separately reprinted.[34] Fortunately, there are few changes in the national agencies that publish the official annual reports on vital statistics except in the newly independent states.

Federal-State Registration System in the United States.—In the United States, the collection of facts about individual births, deaths, marriages, and divorces has been the responsibility primarily of States and a few cities and not of the Fed-

[25] United Nations, *Handbook of Vital Statistics Methods,* Studies in Methods, Series F, No. 7, Annex 1, pp. 213–216, April 1955. A revised edition of this handbook is now being prepared.

[26] John Graunt's, *Natural and Political Observations Mentioned in a Following Index, and Made Upon the Bills of Mortality* (1662), based on these London bills, was the first real study of vital statistics.

[27] For a discussion of the various shades of meaning of "civil registration," the roles of local registration offices, ecclesiastical authorities, public health services, etc., see: United Nations, *Handbook of Vital Statistics Methods,* pp. 33–34.

[28] Irene B. Taeuber, op. cit., pp. 261–263.

[29] United Nations, *Handbook of Vital Statistics Methods,* p. 218.

[30] We need to digress to warn the unwary English-speaking reader that what is called "vital statistics" in English is roughly equivalent to the French *mouvement de la population* and the Italian *movimento naturale della popolazione. Mouvement* is used in the sense of change, not migration.

[31] P. 218.   [32] Pp. 164–183.   [33] Pp. 184–199.

[34] United Nations Statistical Office, *Bibliography of Recent Official Demographic Statistics,* Statistical Papers, Series M, No. 18, March 1954.

eral government. At a relatively late date, the Federal government began to set standards, compile statistics from these individual records, and publish reports. In fact, the United States is far from having shown the early leadership in vital statistics that it achieved in census-taking.

It has been mentioned that keeping records of baptisms, weddings, and burials was the function of the clergy in 17th century England. This practice was carried over to the English colonies in North America but was mostly under secular auspices. As early as 1639, the judicial courts of the Massachusetts Bay Colony issued orders and decrees for the reporting of births, deaths, and marriages as part of an administrative-legal system, so that this colony may have been the first state in the Christian world in which such records were a function of officers of the civil government.[35] Massachusetts also had the first State registration law (1842); but even under this, registration was voluntary and incomplete. By 1865 deaths were fairly completely reported, however.

The other States gradually fell into line, and since 1919 all of the States have had birth and death records on file for their entire area even though registration was not complete. Several of the present States provided for compulsory registration while they were still territories. Most of the States and the District of Columbia now publish an annual or biennial report on vital statistics, but there is considerable range in the scope and quality of these publications. The registration of marriages and divorces has lagged even more than that of births and deaths, as will be described below.

Births and deaths.—As previously mentioned, statistics of births and deaths (in the preceding year) were collected in some of the national censuses of the latter half of the 19th century. Earlier in that century, the Surgeon General of the Army had begun a series of reports on mortality in the army.[36] From the standpoint of civil registration systems, the role of the Federal government begins with its setting up of the Death Registration Area in 1900.[37]

It has been pointed out that the American system is fairly unusual in that States (and a few cities with independent registration systems) collect certificates of births and deaths from their local registrars and are paid to transmit copies to the Federal government. In the beginning, the Federal government recommended a model State law, obtained the adoption of standard certificates, and admitted States to the Registration Areas as they qualified. Only 10 States and the District of Columbia were in the Original Death Registration Area of 1900. The Bureau of the Census set up its Birth Registration Area in 1915, with 10 States and the District of Columbia initially qualifying. In theory, 90 percent of deaths, or births, occurring in the State had to be registered; but ways of measuring performance were very crude. By 1933 all the present States except Alaska had been admitted to both registration areas. The Territory of Alaska was admitted in 1950, Hawaii in 1917 for deaths and 1929 for births, Puerto Rico in 1932 for deaths and 1943 for births, and the Virgin Islands in 1924.

The chief Federal publications on vital statistics, and for many years practically the only ones, are in the form of annual reports. Those through 1944 were published by the Bureau of the Census, but those beginning in 1945 by the United States Public Health Service, since the Division of Vital Statistics of the Census Bureau was transferred in 1946 to the Public Health Service as the National Office of Vital Statistics. The National Office of Vital Statistics was reorganized in 1960 and became part of the newly established National Center for Health Statistics.

The volumes of mortality statistics began with 1900 and those of natality statistics with 1915. In 1937 the two series were fused into *Vital Statistics of the United States.*

The present organization of these annual reports is as follows:
    Volume I, *Natality*
    Volume II, *Mortality*
        Part A. *General Mortality*
        Part B. *Geographic Detail for Mortality*
    Volume III, *Marriage and Divorce*

Volume I, *Natality,* is divided into three sections. Section 1 devotes 59 tables to statistics for the United States and the larger geographic areas (States, standard metropolitan statistical areas, and urban places of 50,000 or more). There are fairly extensive historical and analytical data for the Nation as a whole. Topics covered include, among others: age-specific birth rates (both maternal and paternal), stable-population measures, probability of birth by age and parity of women, cumulative birth rates by order and age of mother, sex ratios of births, completeness of birth registration, month of occurrence (with and without adjustment for seasonality), illegitimate births by age and color of mother, and plural births. Section 2 contains the statistics for small areas—counties and urban places of 10,000 or more. The detail is limited; the number of births occurring in the area and to mothers resident in the area is given by color and whether or not the birth occurred in a hospital. Section 3, "Technical Appendix," is brief but contains valuable information on the registration system and on the quality and interpretation of the statistics. National and State population estimates, provided by the Bureau of the Census, are included to permit the computation of additional rates. There is a very detailed finding-chart, a feature common to all volumes of the annual report.

The two parts of Volume II, *Mortality,* are really continuous and are bound separately mainly because of the size of this volume. Part A, Sections 1 to 6, contains the statistics for the larger areas, a section on life tables, including an annual life table, and the technical appendix.

Part B consists of Section 7 and contains more detailed cross-tabulations for the Nation as a whole, as well as statistics for geographic areas below the national level.

In addition to the annual reports, there have also been a few special bound volumes. Noteworthy among these are the report giving mortality rates for the preceding decade that was published after the census of 1920[38] and the report containing vital rates back to 1900 that appeared after the 1940 census.[39] A report published in 1968 brings these rates forward to 1960.[40]

The life tables published by the Bureau of the Census and

---

[35] Hugh H. Wolfenden, *Population Statistics and Their Compilation,* rev. ed., published for the Society of Actuaries by the University of Chicago Press, 1954, pp. 22–23.

[36] U.S. Bureau of the Census, *Introduction to the Vital Statistics of the United States: 1900 to 1930,* by Walter F. Willcox, 1933, p. 1.

[37] This is the year when the annual series began. Several States and cities had made transcripts of death certificates for the Census Bureau in 1880 and 1890.

---

[38] U.S. Bureau of the Census, *Mortality Rates: 1910–20,* 1923.

[39] U.S. Bureau of the Census, *Vital Statistics Rates in the United States: 1900–1940,* 1943. A second printing was published by the U.S. Public Health Service in 1947.

[40] U.S. National Center for Health Statistics, *Vital Rates in the United States: 1940–1960,* 1968.

later by the Public Health Service are extremely important demographic data not only for what they tell us about the chances of survival but also because of their many uses in demographic analysis. The National Center for Health Statistics of the Public Health Service has recently issued a *Guide to United States Life Tables: 1900–1959,* which lists and discusses all the Federal life tables.[41]

From the early twenties until the Census Bureau's Division of Vital Statistics was reorganized in the mid-thirties, federal vital statistics were in a period of relative stagnation. Very few new features were introduced into the annual volumes. The volumes were sometimes published as much as four years after the year to which their statistics applied.

The use of offset reproduction instead of letterpress printing, beginning with the 1934 volumes, narrowed the gap to about two years. Even more important has been the publication of advance data in several series of current reports. The earlier publication of selected data was made possible not only through the use of these new media but also through the use of such reporting techniques as microfilm of certificates, monthly telegraphic statistical summaries from the States, and sample counts in State offices. By the mid-1970's, the gap had widened to about four or five years again; but concerted efforts are being made to improve the schedule.

In the more than 30 years since the introduction of these unbound, relatively brief photo-offset reports, a number of series have flourished for some time only to be replaced by new series. The current reports series now being published that are of most demographic interest are the following:

Monthly Vital Statistics Report, January 1952 to present. Current provisional statistics (two or three months' lag), on births, deaths, marriages, and divorces, by place of occurrence, and month of receipt in registration office, for States. Based on counts transmitted by State offices to the National Center for Health Statistics. National figures adjusted on basis of past differences between provisional and final figures. Deaths by cause from 10-percent sample. Also annual summaries.

Vital and Health Statistics, 1963 to present
    Series 1–4, General Series
        Series 1, Programs and collection procedures
        Series 2, Data evaluation and methods research
        Series 3, Analytical studies
        Series 4, Documents and committee reports
    Series 10–12, Data from the National Health Survey
    Series 20–22, Data from the National Vital Statistics System
        Series 20, Data on mortality
        Series 21, Data on natality, marriage, and divorce
        Series 22, Data from the National Natality and Mortality Surveys

Marriages and divorces.—The registration of marriages and divorces in the United States has lagged even more than that of births and deaths. Indeed, national registration areas for marriages and divorces were not established until 1957 and 1958, respectively.[42]

Data on the characteristics of these vital events and of the persons concerned are drawn from samples of the records for States in the registration areas. Simple counts of total events, however, are available for all States and for years before the establishment of the registration areas. In fact, an annual national series, partly estimated, is available back to 1867 for marriages and back to 1887 for divorces. Some of the underlying data represent marriage licenses issued rather than marriages performed.

The inclusion of statistics on marriages and divorces in a number of the annual volumes and current report series published by the National Center for Health Statistics and its predecessor agencies has already been indicated. Inclusion in the bound annual volumes began in 1946. Here it may suffice to outline briefly the current contents of Volume III of *Vital Statistics of the United States.* The organization into sections is similar to that in Volume I:

    Section 1, Marriages
    Section 2, Divorces
    Section 3, Technical Appendix

In both Sections 1 and 2, one set of tables is for the entire United States and the other for the registration area. In the first set, there are national and State rates, totals for the nation, States, and counties, and some historical statistics. The number of marriages by months and the number of annulments are also shown for States. Marriage statistics for the Registration States only include those on type of ceremony (civil or religious); and State of residence, age, color, and previous marital status of groom and bride. Many of the statistics are shown separately for first marriages and remarriages. Divorce statistics shown for the States in the Registration Area include those on grounds for divorce, duration of marriage, number of marriages, number of children involved, State where marriage was performed, and age and color of husband and wife.

State and local sources of vital statistics.—Since some State and local governments were active in the field of vital statistics long before the Federal government, it is not surprising that they also published the first reports. The State of Massachusetts inaugurated an annual report in 1843.[43] Until 1949 the only tables giving the characteristics of brides and grooms were those published by a number of the States. Iowa is still the only State in America to have published marriage statistics by religion of the bride and groom. (Indiana does include religion of the bride and groom on its marriage record, however; and Christensen and others have compiled and analyzed statistics from these records.)[44] State health departments and State universities have also prepared and published a number of life tables. On the whole, however, the annual reports on vital statistics published by State and city health departments do not represent a major additional source of demographic information. They are usually much less detailed than the Federal reports. The corresponding figures in State and Federal reports may differ somewhat because of such factors as the inclusion of more delayed certificates in the tabulations made in the State offices, different definitions and procedures, sampling errors when tabulations are restricted to a sample, and processing errors in either or both offices.

Immigration, Emigration, and Naturalization

National governments often publish statistics on the basis of the records of immigrants arriving at and emigrants departing from the official ports of entry and stations on land

[41] Public Health Service Publication No. 1086, Public Health Bibliography Series No. 42, 1961.

[42] For a discussion of the development of federal statistics on marriages and divorces in this country, see: *Vital Statistics of the United States: 1960,* Volume III, *Marriage and Divorce,* Section 7, "Technical Appendix," Washington, 1964.

[43] Robert Gutman, *Birth and Death Registration in Massachusetts: 1639–1900,* New York, Milbank Memorial Fund, 1959, "Bibliography."

[44] Harold T. Christiansen and Kenneth E. Barber, "Interfaith versus Intrafaith Marriages in Indiana," *Journal of Marriage and the Family,* 29(3): 461–469, August 1967.

borders. Statistics may also be based on numbers of passports issued, local registers, and miscellaneous sources. All such records tend to be most complete and detailed for aliens arriving for purposes of settlement and least so for the arrivals and departures of the country's own citizens. The United Nations has published information on the scope of international migration statistics, categories of international travelers, and types of organizational arrangements for collecting and processing data in this field.[45]

Some of the sources of statistics are discussed in the first section of a U.N. publication, *International Research on Migration*.[46] The *U.N. Demographic Yearbooks* of 1948 to 1952, 1954, 1957, 1959, 1962, 1966, 1968, 1970, and 1972 have carried from three to seven tables on international migration. The present policy is to carry four tables biennially. Usually, statistics have been given by countries, on major categories of arrivals and departures, long-term immigrants by country of last permanent residence, long-term emigrants by country of intended permanent residence, and long-term immigrants and emigrants by age and sex. From these *Yearbooks* and other sources, the United Nations prepared several summary reports.[47]

Prior to the entry of the United Nations into this field, the International Labour Office had compiled a number of statistical, or partly statistical, volumes on international migration.[48]

United States.—There are only a few fragmentary statistics on immigration from abroad during the colonial period. The continuous series of federal statistics begins with 1820. The statistics were compiled by the Department of State from 1820 to 1874, by the Bureau of Statistics of the Treasury Department from 1867 to 1895, and by the Office or Bureau of Immigration, now the Immigration and Naturalization Service, from 1892 to the present, although publication was in abridged form or omitted from 1933 to 1942. Over this period the coverage of the statistics has tended to become more complete, especially for immigrant aliens (those admitted for permanent settlement).[49]

The series for emigrants began more recently—aliens deported (1892), aliens voluntarily departing (1927), and emigrant and nonemigrant aliens (1908). Moreover, statistics on emigrant and nonemigrant aliens were discontinued in 1957 and 1956, respectively.[50]

At the present time, the data from visa applications are compiled and published for immigrant aliens only. Another source of statistics, which applies to both aliens and citizens, is the passenger lists supplied by vessels and aircrafts, except those arriving from or departing to Canada. These lists are the only current source of direct information on the arrivals and departures of civilian citizens and, of course, they do not cover overland travel. Throughout history, arrivals and departures of civilian citizens have been relatively incompletely recorded. Data on migration between the outlying possessions (such as Puerto Rico) and mainland United States are no longer available from the passenger records that were obtained by the Immigration and Naturalization Service, but since the early 1950's the Puerto Rico Planning Board has attempted to compile such data from the airlines. The only statistics by age and sex now available for international migrants to and from the United States are for the immigrant aliens. All modern statistics on immigration and emigration are on a fiscal-year basis (July 1 to June 30).

The bulk of the *Annual Report of the Immigration and Naturalization Service* is devoted to statistical tables. The 1973 report contained historical statistics on immigration and current statistics on arrivals and departures by month; immigrants by port of entry, by classes under the immigration law, by quota to which charged, by country of last permanent residence, by country of birth, by State of intended residence, by occupation, by sex and age, and by marital status; aliens previously admitted for a temporary stay who changed their status to that of permanent residents; refugees; temporary visitors; alien and citizen border crossers over land boundaries; aliens excluded and deported by cause; aliens who reported under the alien address program (see "Alien registration" in next major section); and naturalization by country of former allegiance, sex, age, marital status, occupation, and year of entry. The Immigration and Naturalization Service also publishes the quarterly *I and N Reporter*, which includes a table on selected subjects and an occasional article of a demographic nature.

Population Registers

The universal population register should be distinguished from official registers of parts of the population. It is true that the modern universal registers may have evolved from registers that excluded certain classes of the population (members of the nobility, etc.), but the intent of the modern registers is usually to cover all age and sex groups, all ethnic groups, all social classes, etc.

The partial registers, on the other hand, are established for specific administrative purposes and cover only those persons directly affected by the particular program. Examples are registers of workers or other persons covered by national social insurance schemes, of males eligible for compulsory military service, of persons registered as eligible to vote, of aliens, and of licensed automobile drivers. Most such registers are continuous, but some are periodic or exist only during a particular emergency. For example, there have been wartime registrations for the rationing of consumer goods. These may indeed include all or nearly all of the people; but, unlike the universal registers, they are temporary rather than permanent.

Universal Registers.—Of universal population registers, population censuses, and vital statistics, the first is now the least common. In fact, until the twentieth century, it flourished in only two widely separated regions—northwestern Europe (mainly Scandinavia) and the Far East. More recently, some of the eastern European countries have established population registers.

An agenda document prepared for the Twelfth Session of

[45] United Nations Statistical Office, Statistical Papers, Series M, No. 20, *International Migration Statistics*, 1953.

[46] United Nations, Population Division, ST/SOA/*Miscellaneous*, No. 18, 1953, "Periodicals and Yearbooks", p. 1.

[47] U.N. Dept. of Social Affairs, Population Division, Population Studies, No. 11, *Sex and Age of International Migrants: Statistics for 1918–1947*, (ST/SOA/Ser.A/11), 1953; idem, No. 12, *Economic Characteristics for Selected Countries, 1918–1954*, (ST/SOA/Ser.A/12), 1958; idem, No. 24, *Analytical Bibliography of International Migration Statistics*, (ST/SOA/Ser.A/24), 1955.

[48] The most noteworthy of these was probably: *International Migrations*, Vol. I, *Statistics*, compiled on behalf of the International Labor Office by Imre Ferenczi, New York, National Bureau of Economic Research, 1929. The statistical series presented here extend forward mainly to 1924 but go back, in some cases, to 1821.

[49] For selected historical series and a good discussion of the development of the data, see: U.S. Bureau of Census, *Historical Statistics of the United States: Colonial Times to 1957*, 1960, pp. 48–66; idem, *Historical Statistics of the United States: Continuation to 1962 and Revisions*, 1965, pp. 10–11; Gertrude D. Krichefsky, "International Migration Statistics as Related to the United States," Part 1, *I and N Reporter*, 13(1):8–15, July 1964.

[50] Gertrude D. Krichefsky, "International Migration Statistics as Related to the United States," Part 2, *I and N Reporter*, 13(2):21, October 1964.

the United Nations Statistical Commission by the Secretariat gave some general information about population registers.[51] First we will quote their definition. "For purposes of this study, a true population register system has been defined as a mechanism which will provide for the continuous recording of information about the population in such a manner that data on particular events that occur to each individual, as well as selected characteristics describing him, are maintained on a current basis."[52] This is regarded as the "ideal type," to which some of the national registers described are only approximations.

The U.N. document sketches the history of population registers, their uses, general features (coverage, documents, information recorded, and administrative control), and their accuracy. Its table 1 lists, by countries, both the date of establishment of the original register and the date of establishment of the register as then organized. This list, however, also contains a number of what we have called "partial registers" including some that exclude half or more of the population. The countries listed whose registers have virtually complete coverage and are used for demographic purposes comprised: Taiwan, Israel, Japan, Belgium, Bulgaria, Czechoslovakia, Denmark, Finland,[53] Federal Republic of Germany, Gibraltar, Iceland, Italy, Netherlands, Norway, Sweden[53] and the U.S.S.R.

Population registers have been established primarily for identification, control, and police purposes; and often little use has been made of them for the compilation of population statistics. In a number of countries, data from the registers are used to produce one or more of the following: (1) current estimates of population for provinces and local areas; (2) statistics of internal migration; (3) vital statistics.

Partial Registers.—It is best to consider each type of partial register separately since the various types do not have many features in common. Although the United States has never had a universal population register, it has had several types of partial registers.

Social insurance and welfare.—Modern social insurance and social welfare systems (unemployment, retirement, sickness, public assistance, family allowances, etc.) had their origins in Europe and the British Dominions in the latter half of the 19th century. From the millions of records accumulated, statistics are compiled for administrative purposes. Some few of these tables are of demographic interest, especially those relating to employment, unemployment, the aged, widows and orphans, mortality (including life tables for the population covered by certain programs), and births. From these records, moreover, special tabulations with a demographic orientation can be made; frequently such tabulations are based on a sample of the records. Finally, the statistics may be used in the preparation of population estimates or estimates of the total labor force. Likewise life tables for a "covered" population may be used to estimate corresponding life tables for the total population.

The *Statistical Abstract of the United States* indicates the official sources of such statistics in this country. A few publi-

cations only will be mentioned here. A comprehensive monthly report by the Bureau of Labor Statistics brings together the statistics on employment and unemployment from payroll reports of establishments covered by the Old-Age, Survivors, Disability, and Health Insurance (OASDHI) system and on insured unemployment.[54] These statistics originate in State agencies that transmit them to the Bureau of Employment Security and the Bureau of Labor Statistics.

The new Programs of Health Benefits for the Aged (MEDI-CARE and MEDICAID) in the United States are providing statistics on registered persons 65 years old and over by county of residence beginning with 1966.[55] It is thought that virtually all persons in this age group have registered.

A potentially very useful feature of the OASDHI record system is the 1-percent Continuous Work History Sample. This is already being used to a limited extent for studies of various types of mobility, for other longitudinal studies, and for population estimates. An early illustrative study based on special tabulations for two States was carried out by Bogue.[56]

Military service.—Countries that have compulsory military service ordinarily provide for the registration of persons attaining military age, and the person's record is maintained in the register until he passes beyond the prescribed maximum age. The records of the Selective Service in the United States have been the source of some studies on the physical and mental health and the intelligence test scores of draftees, but they have not been otherwise an important source of demographic information, partly because the decentralized custody of the files makes special studies difficult. When the initial registrations began in World War II, however, the numbers of males registering in various age groups were used to measure the completeness of enumeration of the corresponding cohorts in the 1940 Census of Population.[57]

The records maintained by the military services themselves may also yield statistics that are useful to the demographer.[58] The demographic characteristics of members of the armed forces are perhaps of limited interest; but, in the United States, they have been used in the construction of population estimates for the total and civilian populations. (See following major section on "Estimates and Projections.") The useful characteristics have included age, sex, and race; geographic area in which stationed; and geographic area from which inducted. In estimating current migration, whether international or internal, it has been found desirable to distinguish military from civilian migration.

Consumer rationing.—Rationing of food, articles of clothing, gasoline, and other consumer goods ordinarily represents an emergency national program in time of war, famine, etc.

[51] United Nations Economic and Social Council, Statistical Commission, Twelfth Session, *Methodology and Evaluation of Continuous Population Registers*, E/CN.3/293, February 7, 1962.

[52] Ibid., p. 3.

[53] This table should have indicated that the data of the Finnish and Swedish registers *are* used for demographic and other statistical purposes.

[54] U.S. Bureau of Labor Statistics, *Employment and Earnings and Monthly Report on the Labor Force* (title varies and earlier periodicals have been consolidated).

[55] Social Security Administration, Office of Research and Statistics, *Health Insurance Enrollment under Social Security: July 1, 1966*, Washington, Government Printing Office, (no date).

[56] Donald J. Bogue, *A Methodological Study of Migration and Labor Mobility in Michigan and Ohio in 1947*, Oxford, Ohio, Scripps Foundation for Research in Population Problems, Studies in Population Distribution, No. 4, June 1952.

[57] Daniel O. Price, "A Check on Underenumeration in the 1940 Census," *American Sociological Review*, 12(1):44–49, February 1947; R. J. Myers, "Underenumeration in the Census as Indicated by Selective Service Data," *American Sociological Review*, 13(3):320–325, June 1948.

[58] Jacob S. Siegel and Meyer Zitter, "Demographic Aspects of Military Statistics," *Proceedings of the Social Statistics Section, 1958*, Washington, American Statistical Association, pp. 216–220.

Hence, registration of the population for rationing purposes is not to be considered as a permanent source of demographic statistics. Nonetheless, some rationing programs have continued for a fairly long term of years, and important demographic uses have been made of the records. There are sometimes problems in the form of exempt classes and illegal behavior (duplicate registration, failure to notify the authorities of a death or removal, etc.); but these are often small, or appropriate adjustments can be made to the statistics.

Voters.—In countries where voting is compulsory for adults or where a very high proportion of all adults are registered as eligible voters, statistics of demographic value may be compiled.

Aliens.—Registers of aliens may be of some value in studying immigration and emigration, assimilation through naturalization, and the characteristics of those foreign-born persons who have not become citizens. For the most part, a register mainly supplements other sources of information on these subjects (from the census and migration records). In the United States since 1950, aliens have been required to register every January. Two tables on the registrants are published in the *Annual Report of the Immigration and Naturalization Service.* In the 1973 report, one table gives aliens by nationality by selected States of present residence; the other gives aliens by State of residence by selected nationalities. Although the registration card calls for age, sex, and other characteristics, this information is not compiled for statistical purposes. Aliens are also required by law to give notice of change of address within 10 days; but no statistical use is made of these notifications, and, indeed, there is considerable doubt about the quality of the reporting.

School enrollment and school censuses.—Statistics compiled from lists of children enrolled in school are widely used in the United States, because of their universality and pertinency, to make estimates of current population. Some schools keep these lists continuously up to date. The enrollment statistics are usually given by grade, less often by age. The geographic areas shown are those of attendance not of residence. The statistics are published by departments of education of States, cities, etc. and often cover only the public schools. Publications of the U.S. Office of Education bring together the statistics reported by the State departments.[59] For statistics of enrollment in the Catholic schools, see: the *Official Catholic Directory,* New York, P.J. Kenedy and Sons.

Since school census statistics are sometimes substituted for school enrollment statistics in the same general method of making population estimates, this source is mentioned here. The school census is really a partial census rather than a register, however. There is a canvass of households either by direct interview or by means of forms sent home through the school children. Often the preschool children as well as the children of compulsory school age are covered.

MISCELLANEOUS SOURCES OF DATA

Here are listed some of the partial official registers that are less widely used for demographic studies, registers or other records maintained by private agencies, records that apply directly to things but indirectly to people, and the like. Again,

statistics from these sources are sometimes used for population estimates. These include the following:

Tax office records of taxpayers and their dependents
City directories (addresses of householders published by private companies)
Church membership records
Postal delivery stops
Permits for new residential construction and for demolition
Telephone subscribers
Water, electric, and gas meters at residential addresses

ESTIMATES AND PROJECTIONS

Even though population estimates have been alluded to a number of times, their importance as demographic source material seems to call for separate, systematic discussion. They are discussed in the last section of this chapter because they are not primary data but are derived from the other source materials already treated. The methodology of making population estimates is treated fully in chapter 22. Here it may be useful to indicate simply that estimates may be produced in a variety of ways, including: (1) from continuous population registers, (2) from sample surveys, (3) by carrying forward the population from the last census either (*a*) by mathematical extrapolation of past trends, or (*b*) by a component method making use of direct measurements of change (e.g., birth and death registration) and symptomatic data reflecting migration, and (4) by computations using ratios of a postcensally observed datum to the population. These postcensal ratios may be those observed at the last census, or, in the absence of a prior census, they may be purely speculative.

By population projections we mean estimates made for *future* dates. (Methodologically, there is no difference between projections and current estimates made by method (3)(*a*).) Both estimates and projections may be made by age, sex, marital status, etc., as well as for the total population. Moreover, estimates and projections are made for other demographic categories—marriages, households, the labor force, school enrollment, etc.

International View

Estimates.—Many of the international compilations of demographic statistics that were mentioned early in this chapter (the U.N. *Demographic Yearbook, Population Index,* etc.) contain annual estimates of total population, mainly for countries. The tables of the *Demographic Yearbook* have copious notes indicating the source of the estimate, the type of method used, and a qualitative characterization of accuracy.

More detailed estimates (especially in greater geographic detail) are usually published in national reports. These may range from national statistical yearbooks, in which only a small part of the content is devoted to these estimates, to unbound periodicals that are restricted to population estimates. Some of these estimates are postcensal. Others are intercensal, that is they are made for (past) intercensal years after a new census is taken.

Projections.—International compilations of population projections are considerably rarer than those of estimates. The United Nations has at various times compiled projections made by national governments, modified them to conform to a global set of assumptions of its own devising, or made projections for regions or countries entirely on its own.

---

[59] U.S. Office of Education, *Education Directory,* Part 2, "Public School Systems" (annually).

Many countries publish their own population projections and projections for other demographic units. Included are population by age, sex, and race; households, families, married couples; marriages, births, and deaths; urban and rural population; population for geographic areas; school and university enrollment; educational attainment of the population; and economically active population, total and by occupational distribution.[60]

Estimates and Projections in the United States

Estimates.—One of the first problems that confronted the United States Census when it was organized as a permanent bureau in 1902 was the making of official postcensal estimates of population. Previously, the Treasury Department had been issuing estimates. The first annual report of the Census Bureau in 1903 described plans for estimates and gave their projected frequency and scope.[61]

Figures were to be issued as of the first of June for each year after 1900. These were to be for continental United States as a whole, the several States, cities of 10,000 or more population, the urban balance in each State, and the rural part of each State. County estimates were also published for some years. This relatively ambitious program was predicated on the method of arithmetic progression, and the program gradually broke down as its inadequacies became apparent.[62] The last city and county estimates under this program were published for 1926. After that year, efforts were concentrated on better estimates of national and State population by component methods that used postcensal data. A good deal of experimentation went on during the 1930's. In the 1970's, with more experience and more resources, the program is being extended with the objective of covering all the standard metropolitan statistical areas and their constituent counties. Furthermore, contracts with other Federal agencies have made it possible to make occasional estimates in much more detail, such as the estimates for all counties as of 1966.[63]

Over the years, the population estimates have been published in a number of different reports series. From July 1947, until June 1969, however, they all appeared in *Current Population Reports,* Series P-25, *Population Estimates.* The new Series P-26, *Federal-State Cooperative Program for Population Estimates,* contains county estimates made by qualifying State agencies according to prescribed standards. At the present time, the regular publication program encompasses the following:

> Total population of the United States—total including armed forces overseas, total resident, and civilian resident (monthly)
> Estimates of the population of the United States and components of change (annual)—estimates of components are for semi-annual periods

> Estimates of the population of the United States by age, sex, and race (published annually as of July 1 but prepared by single years of age)
> Estimates of the total population of Puerto Rico and other outlying areas (annual)
> Estimates of the total population of States—resident and civilian resident (provisional and revised figures are published annually)
> Estimates of the population of States by broad age groups —resident population (provisional and revised figures annually)
> Estimates of total population of metropolitan areas and their constituent counties (annual provisional and revised figures; all SMSA's included in 1965 but present, 1970, coverage is 100 largest)
> Estimates of the number of households by State (annual beginning with 1965)

Intercensal estimates are also computed and published for the preceding decade when the new census figures become available. There are several publications containing such estimates that are of value for historical studies.[64]

Many current population estimates are prepared by State, county, and municipal statistical agencies; but the detail and the methodology are not uniform from one agency to another. The Bureau of the Census has conducted four mail surveys of the work performed by State and large-city agencies and has reported information on the name of the agency, areas and detail available, method or types of data used, and publication title and frequency.[65] A recent development is the Federal-State Cooperative Program for annual estimates of population by counties.[66] The ultimate objective is to have a State agency, designated by the Governor, to produce annual estimates of county population by methods approved by the Bureau of the Census. The Bureau would then accept these estimates as official and publish them.

Projections.—Official projections or forecasts of the population were essentially a much later development in the United States, although there were a few modest beginnings in the nineteenth century that did not develop into a continuing program. The Compendium of the 1850 Census contained a table on the "Future Progress of the United States" that gave projections of the total population decennially to 1950 on eight different assumptions, as well as the regional distribution in 1950 alone.[67] This is attributed to J. D. B. DeBow. The Cen-

---

[60] A number of these national reports are discussed or cited in two background papers prepared for the World Population Conference in Belgrade: United Nations, *World Population Conference* (Belgrade, August 30 to September 10, 1965), Background paper WPC/WP/454, "Projections of Total Population and of Age-Sex Structure," by Henry S. Shryock, Jr., and Background paper WPC/WP/494, "Projections of Urban and Rural Population and Other Socio-Economic Characteristics," by Jacob S. Siegel.

[61] U.S. Bureau of the Census, *Report of the Director to the Secretary of Commerce and Labor,* 1903, pp. 12–14.

[62] Henry S. Shryock, Jr., "Population Estimates in Postcensal Years," *Annals of the American Academy of Political and Social Science,* 188:167–177, November 1936.

[63] U.S. Bureau of the Census, *Current Population Reports,* Series

P-25, No. 427, "Estimates of the Population of Counties and Metropolitan Areas, July 1, 1966: A Summary Report," July 31, 1969.

[64] U.S. Bureau of the Census, *Current Population Reports,* Series P-25, No. 368, "Estimates of the Population of the United States and Components of Change: 1940 to 1967," June 27, 1967; and U.S. Bureau of the Census, *Current Population Reports,* Series P-25, No. 311, "Estimates of the Population of the United States, by Single Years of Age, Color, and Sex: 1900 to 1959," July 2, 1965. (Report No. 368 gives monthly estimates back to Jan. 1, 1950, and midyear estimates back to 1900. Other reports in this series giving intercensal estimates are Nos. 139 and 304, which give annual State estimates for 1900–1949 and 1950–1960, respectively, and Nos. 98 and 310, which give annual estimates by age, sex, and color, with components of change for 1940 to 1950 and 1950 to 1960, respectively.)

[65] U.S. Bureau of the Census, *Current Population Reports,* Series P-25, No. 328, "Inventory of State and Local Agencies Preparing Population Estimates: Survey of 1965," March 9, 1966.

[66] Meyer Zitter, "Federal-State Cooperative Program for Local Population Estimates," *The Registrar and Statistician* (U.S. National Center for Health Statistics) 33(1):4–8, January 1968.

[67] For an account of the early published forecasts, including those made by such notable figures as Benjamin Franklin, Thomas Jefferson, and Abraham Lincoln, see: Harold F. Dorn, "Pitfalls in Population Forecasts and Projections," *Journal of the American Statistical Association* 45(251):315–317, September 1950.

sus of 1860 also contained forecasts to 1900 of the population of the United States by color.[68] For the most part, these projections were based on the assumption of the continuation of a past rate of growth or used a relatively simple mathematical function that provided for a declining rate of growth.

The modern era of population projections might be considered to have begun in the 1920's with two widely used sets of figures prepared by two pairs of eminent demographers with private organizations. They were Pearl and Reed at The Johns Hopkins University and Thompson and Whelpton of the Scripps Foundation for Population Research.

The rediscovery of the logistic curve by Pearl and Reed and their application of it to projecting many different kinds of populations have had a worldwide influence; and the use of this method, although waning, has by no means disappeared.[69] Their original projections for the United States, which extended to the year 2000, were obtained by fitting Logistic I to the census counts from 1790 to 1910, and were published in the *Proceedings of the National Academy of Science*. After the 1940 census count became available, Logistic II was fitted to these additional figures.[70]

The methodology of projections at the U.S. Bureau of the Census, however, has as its ancestry the projections made by the Scripps Foundation using the component method (using separate assumptions concerning fertility, mortality, and net immigration). By their method, the future distribution of the population by age and sex was obtained as an integral product of the computations. The first published projections from this source were presented in a 1928 article by Whelpton.[71] Three of the subsequent sets of projections (1934, 1937, and 1943) were published by the National Resources Board and its successor agencies. After the war, Whelpton joined forces with the Bureau of the Census staff to produce another set of projections.[72] Thereafter, the Bureau of the Census assumed an active role in the field of national population projections. All the national projections published by Pearl and Reed, the Scripps Foundation for Population Research, and the Bureau of the Census down through 1950 are listed in an article by Shryock.[73]

All of the population projections independently prepared and published by the Bureau of the Census have appeared in the previously mentioned Series P-25, *Population Estimates*. Current practice is to publish new national projections every two or three years, while monitoring demographic developments for indications of unexpected changes.

The reports on demographic projections that are dependent on the basic population projections have been produced even more on an *ad hoc* basis, reflecting the availability of these national "controls," the expressed needs of users, and the extent to which earlier projections were out-of-line with subsequent demographic changes. All the reports on State projections have also been carried in Series P-25. They began later (January 1952) than the national reports, have been less frequent although their frequency has lately been increasing, and cover a shorter range of years, no more than 20 years into the future. The first State projections for broad age groups were presented in August 1957 and the first for age groups and sex, in October 1967.

The first population projections for metropolitan areas appeared in 1969.[74] These projections are for the standard metropolitan statistical areas as defined in 1960.

The other types of demographic projections published by the Bureau of the Census (except those for foreign countries) are now all published in Series P-25, but some types were published in Series P-20 or P-50 in earlier years. Here again, the projections tend to be for relatively short periods. The topics covered include school enrollment, the population by educational attainment and marital status, and households and related units.

Projections of the labor force have been published by both the Bureau of the Census and the Bureau of Labor Statistics. All the recent reports have been published by the latter agency.

Many State, county, and city planning departments, university departments, and private firms prepare demographic projections of these sorts, often for very small geographic areas.[75]

The advent of the electronic computer has notably facilitated the kinds of computations that are employed in making population projections. This technological change is leading to a great expansion in the frequency, detail, and complexity of projections in those agencies that have such equipment.

[68] Census Office, 1860, *Population of the United States in 1860*, Washington, Government Printing Office, 1864, p. ix.

[69] Raymond Pearl and Lowell J. Reed, "On the Rate of Growth of the Population of the United States since 1790 and Its Mathematical Representation," *Proceedings of the National Academy of Sciences*, 6:275–288, 1920.

[70] Raymond Pearl, Lowell J. Reed, and Joseph F. Kish, "The Logistic Curve and the Census Count of 1940" *Science*, 92:486–488, 1940.

[71] P. K. Whelpton, "Population of the United States, 1925 to 1975," *American Journal of Sociology* 34(2):253–270, September 1928.

[72] U.S. Bureau of the Census, *Forecasts of the Population of the United States: 1945–1975*, by P. K. Whelpton assisted by Hope Tisdale Eldridge and Jacob S. Siegel, 1947.

[73] Henry S. Shryock, "Accuracy of Population Projections for the United States," *Estadística* (Washington) 12(45):590–591, December 1954.

[74] U.S. Bureau of the Census, *Current Population Reports*, Series P–25, No. 415, "Projections of the Population of Metropolitan Areas: 1975," January 31, 1969.

[75] Just two of the more outstanding sources will be singled out by way of illustration: Ira S. Lowry, *Metropolitan Populations to 1985: Trial Projections*, Santa Monica (California), RAND Corporation, Memorandum RM–4125-RC, September 1964; and National Planning Association, *Economic and Demographic Projections for Two Hundred and Twenty-four Metropolitan Areas*, by Joe Won Lee, Regional Economic Projections Series, Report No. 67-R-1, May 1967.

SUGGESTED READINGS

Chasteland, J.-C. *Démographie: Bibliographie et analyse d'ouvrages et d'articles en français* (Demography: Bibliography and analysis of works and articles in French). Paris, L'Institut national d'études démographiques et l'Union international pour l'étude scientifique de la population, 1961.

Cox, Peter R. *Demography.* Cambridge, England, published for the Institute of Actuaries and the Faculty of Actuaries, Cambridge University Press, 1955. Pp. 11–67.

Eldridge, Hope T. *The Materials of Demography: A Selected and Annotated Bibliography.* New York, International Union for the Scientific Study of Population and Population Association of America, 1959.

Hauser, Philip M., and Duncan, Otis Dudley. "The Data and Methods," in *The Study of Population.* Hauser and Duncan, eds. Chicago, University of Chicago Press, 1959. Pp. 45–75.

Istituto di demografia, Università di Roma. *Bibliografia delle opere demografiche in lingua italiana.* (Bibliography of demographic works in the Italian language). Antonio Golini, ed. Rome, 1966.

Linder, Forrest E. "World Demographic Data," in *The Study of Population.* Philip M. Hauser and Otis Dudley Duncan, eds. Chicago, University of Chicago Press, 1959. Pp. 321–360.

Lorimer, Frank. *Demographic Information on Tropical Africa.* Boston, Boston University Press, 1961.

————. "The Development of Demography," in *The Study of Population,* Philip M. Hauser and Otis Dudley Duncan, eds. Chicago, University of Chicago Press, 1959. Pp. 124–179.

Pearl, Raymond. *Introduction to Medical Biometry and Statistics.* 2nd ed. Philadelphia, W. B. Saunders Co., 1930. Pp. 42–62.

Population Research Center, University of Texas. *International Population Census, Bibliography, Latin America and the Caribbean,* No. 1; *Africa,* No. 2; *Oceania,* No. 3; *North America,* No. 4; *Asia,* No. 5; and *Europe,* No. 6. Austin, Bureau of Business Research, University of Texas, 1965.

United Nations. *Analytical Bibliography of International Migration Statistics.* Series A, Population Studies No. 24, 1955.

————. *Bibliography of Recent Official Demographic Statistics.* Statistical Papers, Series M. No. 18, March 1954.

————. *Demographic Yearbook,* 1949–50 to date. Text and index.

————. *Handbook of Vital Statistics Methods.* Studies in Methods, Series F, No. 7, April 1955. Pp. 3–11 and 213–216.

————. *International Research on Migration.* Miscellaneous, No. 18, 1953.

————. *World Population Prospects as Assessed in 1968,* Series A, Population Studies, No. 53, 1973.

U.S. Bureau of the Census. *Bureau of the Census Catalog.* (Annually, quarterly, and monthly.)

————. *Historical Statistics of the United States: Colonial Times to 1957.* 1960. Pp. 1–30 and 39–78.

————. *Historical Statistics of the United States: Continuation to 1962 and Revisions.* 1965. Pp. 1–6 and 8–14.

————. *Introduction to the Vital Statistics of the United States: 1900 to 1930,* by Walter F. Willcox, 1933.

Wolfenden, Hugh H. *Population Statistics and Their Compilation.* rev. ed published for the Society of Actuaries by the University of Chicago Press. 1954. pp. 4–31.

Wright, Carroll D. and Hunt, William C. *The History and Growth of the United States Census,* Washington, Government Printing Office, 1900.

CHAPTER 3

Collection and Processing of Demographic Data

This chapter deals with the collection and processing of demographic data. It is closely related to the preceding chapter, which treated the important kinds of demographic statistics and their availability. The discussion covers censuses and surveys and also registration systems for the collection of vital statistics. Practices differ considerably from country to country and it would not be practicable to cover in this chapter all the important differences in data collection methods. Instead, we have chosen to cite some illustrative examples and to discuss this subject mainly in terms of the norms as they are recognized by countries with a long history of censuses or registration systems and as they are presented in publications of the United Nations and other international organizations.

POPULATION CENSUSES AND SURVEYS

Introduction

Since many of the procedures and problems of data collection in censuses and surveys are common to both, these two sources of data are treated together. Some distinctions between censuses and surveys were mentioned in chapter 2.

Essential Features of a Population Census. — The essential features of a population census, as stated in a recent United Nations publication are "individual enumeration, universality within a defined territory, simultaneity, and defined periodicity." [1]

Individual enumeration. — The principle to be observed here is to list persons individually along with their specified characteristics. In some primitive types of censuses, the "group enumeration" method is employed, however; that is, the number of adult males, adult females, and children is tallied within each tribe or family. This procedure was widely followed in most of the enumerations of the native population in tropical Africa. The first few censuses of the United States represented only a slight refinement of this group enumeration. The main disadvantage of this method is that no greater detail on characteristics can be provided in the tabulations than that contained in the tally cells themselves. Tabulation becomes a process of mere summation. It is impossible to cross-classify characteristics unless they were tallied in cross-classification during the enumeration.

Universality within a defined territory. — Ideally, a national census should cover the country's entire territory and all people resident or present (depending upon whether the basis of enumeration is *de jure* or *de facto*). When these ideals cannot be achieved for some reason (enemy occupation of part of the country in wartime, hostility of isolated tribes, etc.), then the type of coverage attempted and achieved should be fully described in the census publications.

Simultaneity. — Ideally, again, a census is taken as of a given day. The canvass itself need not be completed on that day, particularly in the case of a *de jure* census. Often the official time is midnight of the census day. The more protracted the period of the canvass, however, the more difficult it becomes to avoid omissions and duplications. Some of the topics in a census may refer not to status on the census day but to that at a specified date or period in the past, e.g., residence 5 years ago, labor force status in the week preceding the census day, and income in the preceding calendar year.

Defined periodicity. — The United Nations recommends, "Censuses should be taken at regular intervals so that comparable information is made available in a fixed sequence. A series of censuses makes it possible to appraise the past, accurately describe the present and estimate the future." [2] If the censuses are spaced exactly 5 or 10 years apart, cohort analysis can be carried out more readily and the results can be presented in more conventional terms. In the interests of international comparability, the United Nations suggests that population censuses be taken as closely as feasible to the years ending in "0".

The United Nations publications also emphasize the importance of sponsorship of the census by the national government. [3] A national census is conducted by the national government, perhaps with the active cooperation of State or provincial governments. It is feasible to have a national sample survey conducted by a private survey organization or to have a small-scale census (for a limited area) conducted by a city government, university department, training center, etc.; but only national governments have the resources to support the vast organization and large expenditures of a full-scale census.

Advantages and Uses of Sample Surveys. — As vehicles for the collection of demographic statistics, the sample surveys have certain advantages and disadvantages, and their purposes and applications differ somewhat from those of censuses. One advantage of sample surveys is the possibility of experimenting

---

[1] United Nations, *Principles and Recommendations for the 1970 Population Censuses,* Statistical Papers, Series M, No. 44, 1967, p. 3.

[2] Ibid. p. 4.
[3] United Nations, *Handbook of Population Census Methods,* Vol. I, *General Aspects of a Population Census,* Studies in Methods, Series F, No. 5, rev. 1, p. 4.

with new questions. The fact that a new question is not altogether successful is less critical in the case of a sample survey than in that of a census, where the investment is much larger and where failure cannot be remedied until after the lapse of 5 or 10 years. In a continuing survey new features can be introduced not only in the questions proper but also in the instructions to the canvassers, the coding, the editing, and the tabulations. Since a national population census is a multi-purpose statistical project, a fairly large number of different topics must be investigated and no one of them can be explored in any great depth. In a survey, even when there is a nucleus of items that have to be included on the form every time, it is feasible in supplements, or occasional rounds, to probe a particular topic with a "battery" of related questions at relatively moderate additional cost.

Among disadvantages of surveys, sampling error is probably the major one. This disadvantage is offset to some extent, however, by the ability to compute the sampling error for estimates of various sizes and thus describe the limits of reliability, whereas the magnitude of some other types of errors in both censuses and surveys may remain undetermined. The size of the sample is usually such that reliable statistics can be shown only in very limited geographic detail and for relatively broad cross-tabulations. For this reason, the census returns are the principal source of data for small areas and detailed cross-classifications of population characteristics. There is also usually some sampling bias arising from the design of the survey or from failure to carry out the design precisely. As to the design, it may not be practical to sample the entire population even when that is desirable, so that coverage is not extended to certain population subgroups (nomadic or tribal populations, persons living in group quarters, etc.). The public may not cooperate as well in a sample survey as in a national census, which receives a great deal of publicity with attendant patriotic appeal.

On the other hand, the data from a regular survey program may be superior in some respects to those from a census. The field staff is retained from month to month or year to year. The smaller size of the survey operation makes it possible to do the work with a smaller and, therefore, more select staff and to maintain closer surveillance and control of procedures.

The shorter time interval between surveys makes them more suitable for studying population growth and household formation and those population characteristics which change frequently in some countries, such as fertility and employment status. With observations taken more frequently, it is much more feasible to analyze time trends in the statistics. The analyst can delineate seasonal movements if the survey is conducted monthly or quarterly. Even when the survey data are available only annually, cyclical movements can be delineated more precisely than from censuses, and turning points in trends are more accurately located. The response of demographic phenomena to economic changes and to political events can also be studied more satisfactorily.

The uses of censuses and surveys are sometimes interrelated. The use of the sample survey for testing new questions has already been mentioned. New procedures may also be tested. Census statistics may serve as benchmarks for analyzing and evaluating survey data. The census can be used as a sampling frame for selecting the population to be included in a survey or may be a means of selecting a specific population group, such as persons in specified occupations, for a later special survey.

International Recommendations

Methods of data collection vary among countries according to their cultural and technical advancement, the amount of data-collecting experience, and the resources available. Both the methods used and the practices recommended by international agencies are covered in a number of sources. The Statistical Office of the United Nations has produced a considerable body of literature on the various aspects of the collection and processing of demographic statistics from censuses and surveys. Most of this material is contained in two series of publications titled *Statistical Papers* and *Studies in Methods* and in the annual issues of the *Demographic Yearbook*. Two basic documents in this collection are *Handbook of Population Census Methods*, Studies in Methods, Series F, No. 5, rev. 1, and *Handbook of Household Surveys*, Studies in Methods, Series F, No. 10, which contains detailed descriptions of data collection methods and international recommendations for the conduct of censuses and surveys. Of particular pertinence to the planning of the 1970 censuses is the more recent publication referred to above, *Principles and Recommendations for the 1970 Censuses*, published in 1967. (A reprinting of this manual was issued in 1969 with some changes in page numbers. The references to the manual in the present work, however, are to the 1967 printing.)

The U.S. Bureau of the Census has prepared a series of case studies for use in training programs in countries where statistical activities are being developed. The first of these, *Florencia: A Case Study in Data Processing of Population and Housing Censuses*, is applicable specifically to census taking in Latin American countries but may serve as a guide for census methods in other areas of the world.[4] The material covers all steps of processing from receipt of the questionnaires to preparation of the final publication tables. A revised case study on census taking, *New Florencia: A Case Study for the 1970 Censuses of Population and Housing*, was prepared in 1969, updating processing procedures and widening the scope to include all aspects of a census program—from determination of objectives to analysis and publication of results. The latter series was used by the U.S. Bureau of the Census in workshops on the 1970 censuses, conducted under the sponsorship of the Agency for International Development. A similar case study, titled *Atlantida: A Case Study in Household Sample Surveys*, covers all aspects of a household sample survey program, from determination of the topics to be investigated and selection of the sample households to the tabulation and analysis of the results.[5] Some recent textbooks on demography such as those by Bogue and Spiegelman also include chapters on data collection and processing.[6]

Organization of National Statistical Offices. — The statistical programs of a country may be largely centered in one national statistical office, which conducts the census and the major sample surveys, or they may be scattered among a number of government agencies each with specific interests and responsibilities.

National census offices. — Considerable differences exist among countries in the organization and permanence of the national census office, which may be an autonomous agency or part of the central statistical office. The U.N. *Handbook of*

---

[4] U.S. Bureau of the Census, 1962.
[5] U.S. Bureau of the Census, 1966.
[6] Donald J. Bogue, *Principles of Demography*, New York, John Wiley and Sons, 1969, pp. 100–113; Mortimer Spiegelman, *Introduction to Demography*, Cambridge, Mass., Harvard University Press, 1968, pp. 8–42.

Population Census Methods groups countries into three categories according to types of central organizations:[7] ,1) those with a permanent census office and subsidiary offices in the provinces, (2) those with a permanent central office but no continuing organization of regional offices so that they depend on provincial services or officials or field organizations of other national agencies, and (3) those that have no permanent census office, but create an organization for the taking of each census and dissolve it when the census operations are complete.

There are many advantages to maintaining a permanent census office. Much of the work, including analysis of the data from the past census and plans and preparation for the next census, can best be accomplished by being spread throughout the intercensal period. The basic staff retained for this purpose forms a nucleus of experienced personnel to assume administrative and supervisory responsibilities when the organization is expanded for taking the census. Many records such as maps and technical documents relating to census operations need to be maintained and brought up to date.

Planning and Preparatory Work. — Any national census or major survey involves a vast amount of preparatory work, some aspects of which may be in process a number of years before the enumeration date. Preliminary activities include geographic work, such as preparing maps and lists of places; determining the data needs of the national and local governments, business, labor, and the public; choosing the questions to be asked and the tabulations to be made; deciding on the method of enumeration; designing the questionnaire; testing the forms and procedures; planning the data processing procedures; and acquiring the equipment to be used. Proper publicity for the census is important to the success of the enumeration, especially in countries where a census is being taken for the first time and the citizens may not understand its purpose. The public should also be assured of the confidentiality of the census returns, that is, that personal information will not be used for other than statistical purposes, and will not be revealed in identifiable form by census officials. In the words of the U.N. principles for the 1970 censuses, ". . . the confidentiality of the individual information should be strongly and clearly established in the census legislation and guaranteed by adequate sanctions so as to form a basis for the confident cooperation of the public." [8]

Development of procedures for evaluation of the census should be part of the early planning to assure that they are included at the appropriate stages of the field work and data processing and to assure that funds will be set aside for them. The funding of the census itself is but one of a host of administrative responsibilities involved in the taking of a national census. Legislation must be passed to provide a legal basis; funds must be appropriated and a budget prepared; a time schedule of census operations must be set up; and a huge staff of census workers must be recruited and trained. Since these administrative problems are not of direct concern to the demographer, they will be omitted from the detailed treatment of census and survey procedures which follows.

Geographic work. — In a national census the geographic work has a two-fold purpose: (a) to assure a complete and unduplicated count of the population of the country as a whole and of the many subdivisions for which data are to be published; (b) to delineate the enumeration areas to be assigned to individual enumerators.

The boundaries which must be observed include the national boundaries and the boundaries of the administrative and political subdivisions such as states or provinces and the smaller political units. Often data are to be shown also for statistical areas which have little relation to political boundaries but are established for purposes of publication of the census returns. In countries which have a well-established census program, the geographic work is continuous and involves updating maps for changes in boundaries (e.g., annexations), redefining of statistical areas, and so forth. When maps are not available from a previous census, they may be developed from existing maps obtained from various sources such as military organizations, school systems, ministries of health or interior, or highway departments, or they may be prepared from aerial photographs. The materials from these various sources are put together to produce working maps for the enumeration.

Once the maps have been prepared, the enumeration areas are delineated. There are two requirements for the establishment of enumeration areas. First, the enumeration area must not cross the boundaries of any tabulation area; and, second, the population of the enumeration area as well as its physical dimensions must be such that one canvasser can complete the enumeration of the area in the time allotted for the census.

In some countries the preparation of adequate maps is not feasible because of the lack of qualified personnel or because of the cost of producing the maps, and a complete listing of all inhabited places, villages, and so forth made by field workers is used as a substitute for maps. Such lists are prepared by many countries as either a substitute for or a supplement to the census maps.

The geographic work is sometimes supplemented with a precanvassing of the enumeration areas shortly before enumeration. A precanvass serves to prepare the way for the enumeration by filling in any missing information on the map, providing publicity for the census, arranging with village chiefs for the enumerator's visit, determining the time necessary for covering the area, and planning the itinerary.

Geographic work is equally important as a preparatory phase of sample surveys. The selection of the sample usually depends upon the delineation of certain geographical areas to serve as primary sampling units, then subdivisions of those areas, and finally delineation of small area segments of suitable size for the interviewer to cover in the alloted time period. In South Korea sample surveys often use the enumeration districts of the previous census as primary sampling units.

Content and tabulations. — The subjects to be included in a census are a balance between needs for the data and resources for carrying out the census program. National and local needs are of primary importance, but some consideration may also be given to achieving international comparability in the topics chosen. As a rule, the list of topics included in the previous census or censuses provides the starting point from which further planning of content proceeds. In general, it is desirable that most questions be retained from census to census in essentially the same form to provide a time series which can serve for analysis of the country's progress and needs. Some changes in content are necessary, however, to meet the changing needs of the country. Advice is usually sought from various national and local government agencies. Advisory groups including experts covering a wide range of interest may be

[7] Volume 1, p. 22.

[8] United Nations, *Principles and Recommendations for the 1970 Censuses,* pp. 11–12.

organized and invited to participate in the formulation of the questionnaire content.

The United Nation's list of recommended items for censuses is valuable as an indicator of the basic items which have proved useful in many countries and as a guide to international comparability in subjects covered.[9] Their list of "recommended" (*) and "useful" topics to be included on the census questionnaire is as follows:

Geographic characteristics

*Place where found at time of census
              and/or
*Place of usual residence
*Place of birth
 Duration of residence
 Place of previous residence
 Place of work

Personal and household characteristics

*Sex
*Age
*Relationship to head of household
 Relationship to head of family
*Marital status
 Age at marriage
 Duration of marriage
 Marriage order
*Children born alive
*Children living
 Citizenship
*Literacy
*School attendance
*Educational attainment
 Educational qualifications
 National and/or ethnic group
 Language
 Religion

Economic characteristics

*Type of activity
*Occupation
*Industry
*Status (as employer, employee, etc.)
 Main source of livelihood

The following derived topics are also listed as recommended or useful:

Geographic characteristics

*Total population
*Locality
*Urban and rural

Personal and household characteristics

*Household composition
 Family composition

Economic characteristics

Socio-economic status
Dependency

Regional interests are another consideration in the planning of census content. Four regional groups (Economic Commission for Europe, Economic Commission for Asia and the Far East, Economic Commission for Africa, and the Inter-American Statistical Institute), by arrangement with the United Nations, hold conferences at which census content and methods are considered and make recommendations for the forthcoming census period. Neighboring countries sometimes cooperate in census planning through regional conferences or advisory groups for census subject matter and practices. Jamaica, for example, in the 1960 census "agreed in principle to participate with other territories in the British Caribbean in a programme of tabulations which may be considered to be more extensive than its normal administrative requirements, but which would facilitate research workers who are interested in investigating various social conditions as they exist in the West Indies."[10]

Public reaction to a subject also may influence the choice of census topics, since some questions may be too difficult or complicated for the respondent, or the public may object to the substance of the question. Limitations as to the number of questions respondents are willing to answer and the quantity of material that can be processed with the resources available also force some otherwise desirable topics to be dropped.

Although some sample surveys, like the National Sample Survey of India, for example, are multisubject surveys, it is more common for the survey to be restricted to one field, such as demographic characteristics or events, health, family income and expenditures, or labor force characteristics. One way in which sample surveys achieve multisubject scope is to vary the content from time to time. The U.N. *Handbook of Household Surveys* presents a list of recommended items for demographic surveys.[11]

Closely related to the choice of subjects to be included in a census or survey is the planning of the tabulation program. Potential cross-tabulations in a census are boundless. Therefore, the selection of material is dictated partly by the uses that are going to be made of the results. The capacity of the manpower and equipment for processing the data, and the available facilities for publishing the results, e.g., page space available, place some restrictions on the material to be tabulated. The tabulation plans as well as the choice of subjects on the questionnaire should undergo review by the major potential users of the statistics, both within and outside the government. Recommended tabulations for each of the subjects covered in national censuses and in various types of surveys are listed in the U.N. manuals previously referred to.

Part of the planning of the tabulation program involves a determination of the number of different levels of geographic detail to be presented. Data are usually presented for administrative divisions and subdivisions of the country and for cities in various size categories as well as for the country as a whole. For the smallest geographic areas, such as small villages, the results as a rule are limited to a report of the number of inhabitants or perhaps the population by sex only. At the next higher level, which might be secondary administrative divisions, the tabulations may provide only "inventory statistics"; that is, simply a count of persons in the various categories of age, marital status, economic activity, etc., with but little cross-

---

[9] United Nations, *Principles and Recommendations for the 1970 Population Censuses*, pp. 40–41.

[10] Jamaica, Department of Statistics, *Census of Jamaica, 7th April, 1960*, Vol. I, Part A, Kingston, 1963, p. 2.

[11] United Nations, *Handbook of Household Surveys*, Studies in Methods, Series F, No. 10, p. 34.

classification with other characteristics. For the primary administrative divisions and major cities, most subjects are cross-tabulated by age and sex, and often there are also cross-classifications with other social and economic characteristics, such as educational attainment by economic activity, or employment status by occupation. Also, more detailed categories may be shown on such subjects as country of birth, mother tongue, or occupation. The greatest degree of detail, sometimes termed "analytical" tabulations as opposed to "inventory" statistics, is that in which cross-tabulations involve detailed categories of each of the three of four characteristics involved.

Definitions of concepts. — One requirement of a well-planned and executed census or survey is the development of a set of concepts and classes to be covered, and adherence to these definitions throughout all stages of the collection and processing operations. These concepts provide the basis for development of question wording, instructions for the enumerators, and specifications for editing, coding, and tabulation of the data. Only when concepts are carefully defined in operational terms and consistently applied can there be a firm basis for later analysis of the results. Definitions of all of the recommended topics for national censuses and household surveys are presented in the manuals of the United Nations and are recognized by many countries as international standard definitions for the various population characteristics.[12]

Treatment of nonresponses. — Planning the tabulations includes making some basic decisions about the treatment of nonresponses. Nonresponses may be published in a separate category as "not reported," or they may be distributed among the specific categories according to some plan, ideally on the basis of other available characteristics of the person. Practices vary on the extent to which responses are allocated, but the elimination of "unknowns" before publication is a growing practice, partly because the greater capabilities of modern tabulating equipment have improved the possibilities of assigning a reasonable entry without prohibitive cost and partly because convenience to the user of the data favors the elimination of nonresponses.

Pretesting. — Pretesting of census content and methods has been found to be very useful in providing a basis for decisions that must be made during the advance planning of the census. Such tests vary in scope. They may be limited to testing a few new items of data, alternate wording of a question, different types of questionnaires, or different enumeration procedures. Most census testing includes at least one full-scale test, containing all questions to be asked on the census itself and sometimes covering part or all of the processing phases as well. The suitability of topics which have not been tried before may be determined from a small-scale survey in two or three localities with enough other questions on the questionnaire to achieve something close to a normal census situation and to provide other personal characteristics which might be pertinent to an evaluation of the results obtained from the question being tested. A test involving only the employees of the census office and their families may sometimes suffice for this purpose. Countries having an annual sample survey sometimes use this survey as a vehicle for testing prospective census questions.

Collecting the Data. — The crucial phase of a census or survey comes when the questionnaires are taken into the field and the task of obtaining the required information begins. The kinds of problems encountered and the procedures used for collecting the data are similar for censuses and surveys.

De jure and de facto counts. — In a census the procedures for enumeration are affected by the type of population count to be obtained. The census may be designed to count persons where they are found on census day (a *de facto* count) or according to their usual residence (a *de jure* count).

In a *de facto* census, the method is to list all persons present in the household or other living quarters at midnight of the census day or all who passed the night there. In this type of enumeration, there is a problem of counting persons who happen to be traveling on census day or who work at night and consequently would not be found in any of the places where people usually live. It may be necessary to count persons on trains and boats or to ask households to include such members on the census form as well as those persons actually present. In some countries all persons are requested to stay in their homes on the census day or until a signal announces the completion of the enumeration.

In a *de jure* census, all persons who usually live in the household are listed on the form whether they are present or not. Visitors who have a usual residence elsewhere are excluded from the listing but are counted at their usual residence. Provision must be made in a *de jure* census for counting persons away from home if they think it is likely that no one at their usual residence will report them. The usual practice is to enumerate such persons on a special form which is forwarded to the census office of their home address. The form is checked against the returns for that area and is added to the count there if the person is not already listed. This is a complicated and expensive procedure, and there still remains some chance that some persons will be missed and some counted twice.

Methods of enumeration. — There are two major types of enumeration, the direct interview or canvasser method and the self-enumeration or householder method. In the direct interview method, a census agent visits the household, lists the members living there, and asks the required questions in order for each person, usually by interviewing one member of the household. This method has the advantage that the enumerator is a trained person who is familiar with the questions and their interpretation and may assume a high degree of responsibility for the content of the census. Also, the difficulty of obtaining information in an area where there is a low level of literacy is reduced by this method. For these reasons it is considered possible to include more complex forms of questions in the direct interview type of enumeration.

In self-enumeration, the census forms are distributed, usually one to each household, and the information is entered on the form by one or more of the household members for all persons in the household. With this method of enumeration there is less need for highly trained enumerators. The census enumerator may distribute the forms and later collect them, or the mails may be used for either the distribution or collection of the forms or for both. If the enumerator collects the forms, he reviews them for completeness and correctness and requests additional information when necessary. In a mail census, the telephone may be used to collect information found to be lacking on the forms mailed in, or the enumerator may visit the household to obtain the missing information. In some cases he may complete an entire questionnaire if the household is unable to do so.

Self-enumeration has the advantage of giving the respondents

---

[12] *Principles and Recommendations for the 1970 Population Censuses*, pp. 41–63, and *Handbook of Household Surveys*, pp. 18–33.

more time to obtain the information and to consult records if necessary. Each person can supply the information about himself; e.g., the wage earner himself can report the information about his occupation rather than having it reported by his wife, who may not have a good understanding of what his job is. The possibility of enumerator bias resulting from an erroneous interpretation of the questions by a single enumerator is minimized in this method of enumeration. It is also more feasible to achieve simultaneity with self-enumeration because all respondents can be asked to complete their questionnaires as of the census day. Thus, in this respect, self-enumeration is the more suitable method if a *de facto* count is desired.

Self-enumeration has played an important role in the decennial census of the United States since 1960. Part of the 1970 questionnaire, which was used both in self-enumeration and in direct interviewing, is presented as Figure 3-1.

Self-enumeration is the more frequently used method in the European countries and in Australia and New Zealand, whereas direct interview is the usual method in other countries. A combination of these two main types of enumeration is often used. The self-enumeration method may be considered appropriate for certain areas of the country and the interviewer method for others, or some of the information may be obtained by interview and the remainder by self-enumeration. In a census which uses the interviewer method as its basic procedure, self-enumeration may be used for some individuals, such as roomers, when the head of the household cannot supply the information or when confidentiality is desired.

Enumeration of special groups.—Some special procedures for enumeration are required for certain groups of the population such as nomads, tribal people living in inaccessable jungle areas, and other indigenous groups who have little relation to the normal social structure of the country. The level of literacy may be low among these groups, and they may have little understanding of the purpose of a census or interest in its objectives. A procedure sometimes followed is to request that all the members of such groups assemble in one place on a given day, since enumerating them at their usual place of residence might require from 4 to 5 months. For some of these, a method of group enumeration has been used. Rather than obtaining information for each individual, the enumerator obtains from the head of the group a count of the number of persons in various categories, such as marital status, sex, and age groups. In the 1966 census of Iran, the "unsettled population" of each Shahrestan was enumerated on a special form which recorded only the number of persons in broad age-sex groups.

Enumeration of persons in hotels, pensions, missions, hospitals, and similar group living quarters usually requires special procedures. Since some are transients, inquiry must be made to determine whether they have already been counted elsewhere. If a *de jure* count is being made, steps must be taken to assure that they are counted at their usual residence. Special individual census forms are usually used in group quarters, since the proprietor or other residents of the place could not provide the required information about each person. Another segment of the population which presents a problem in census taking is the population with no fixed address who sleep in parks, in automobiles, or on streets, or who drift from one abode to another. Such persons comprise a significant element of the population in slum areas of many large cities.

Noninterview households.—In some households the enumerator is unable to interview anyone even after repeated visits, because no one is at home or, more rarely, because the occupants refuse to be enumerated. In a census, since the primary purpose is to obtain a count of the population, an effort is made to obtain information from neighbors about the number and sex of the household members. Neighbors may also be able to supply information about family relationship and marital status which may, in turn, provide a basis for estimating age. Reliable information on other subjects usually cannot be obtained except from the members themselves and these questions are left blank perhaps to be supplied during processing according to procedures which are discussed in a later section, "Processing the Data."

In a sample survey it is less practical to get information from neighbors because the emphasis is on characteristics more than on a count of the population. The usual procedure is to base the results on the cases interviewed and adjust the basic weighting factors to allow for noninterview cases when the final estimates are derived from the sample returns. The effect of this procedure is to impute to the population not interviewed the same characteristics reported by the interviewed population. Since this assumption may not be very accurate, the presence of a great many noninterview households may bias the sample.

Types of questionnaires.—Questionnaires may be classified into three general types: first, the single individual questionnaire containing information for only one person; second, the single household questionnaire, which contains information for all the members of the household or housing unit; and third, the multihousehold questionnaire, which contains information for as many persons as can be entered on the form, including members of several households. Each of these has certain advantages. The single individual questionnaire is more flexible for compiling information if the processing is to be done without the help of mechanical equipment. This type of questionnaire has been used in India, where it proved to be convenient for hand sorting. In New Zealand the individual questionnaire, originally introduced for use in group living quarters such as hotels, was adopted generally because of its convenience and privacy. The individual had the privilege of sealing his census form in an envelope if he chose. The single household questionnaire has the advantage of being easy to manage in an enumeration and is especially convenient for obtaining a count of the number of households and for determining the relationship of each member of the household. If part of the questions in a census are to be confined to a sample of households, a single household schedule is required. The multihousehold questionnaire is more economical from the standpoint of printing costs and is convenient for processing on conventional or electronic tabulating equipment, but it may be awkward to handle because of its size.

Another type of questionnaire is that described above for group enumeration of nomadic people, when only the number of persons by broad age-sex groups is recorded. Although these summarized data do not provide census data in the strictest sense of the term, the group enumeration procedure has been used to enumerate classes of the population for whom conventional enumeration methods are not practical.

Field checking.—Various procedures are followed for quality control of the enumeration. Supervisors review samples of each enumerator's work for completeness and acceptability of entries and accompany him on some of his visits. Special forms for recording the results of review of enumerators' work and procedures for dismissing those whose work was unacceptable were introduced in the 1960 census in the United States. Progress reporting of the enumeration enables census officials to know when an individual enumerator or the enumerators in a given area are falling seriously behind schedule and thus jeopardizing the completion of the census within the allotted time. Hand tallies of the population counted in each small area are compared with advance estimates, and review of the enumeration is made if the results vary too widely from the expected number.

Reinterviewing is a common technique used for quality control of the data collection process in sample surveys. A sample of households visited by the original interviewer is reinterviewed by his supervisor, and the results of the check-interview are compared with the original responses. Such checking determines whether the recorded interview actually took place and reveals any shortcomings of the interviewer.

Receipt and Control.—When the questionnaires have been completed in the field, they are assembled into bundles, usually corresponding to the area covered by one enumerator. The number of documents, the geographic identification of the area, and other appropriate information are recorded on a control form which accompanies the set of documents throughout the various stages of processing. The tremendous volume of records involved in a census or large survey makes the receipt and control of material a very important function. The identification of the geographic area provides a basis for filing the documents and a means of locating a particular set of documents at any stage of the processing.

Processing the Data.—The processing of the data includes all the steps, whether carried out by hand or by machine, which are required to produce from the information on the original document the final published reports on the number and characteristics of the population. The extent to which these operations are accomplished by mechanical or electronic equipment or by hand varies among countries and among surveys and censuses within countries.

Methods of processing.—A purely manual operation may be suitable for a survey of limited size when the tabulations to be made are fairly simple; but in a large-scale census, or in a survey from which many cross-classifications are desired, a hand operation is cumbersome and time-consuming. However, few censuses are processed entirely by machine. Usually, some of the data, such as preliminary counts of the population for geographic areas, are obtained primarily from a hand-count; and even data which are produced primarily by machine must undergo some manual processing to correct omissions or inconsistencies on the questionnaire and to convert certain types of entries into appropriate input for the automatic equipment. The output of the tabulating machines also may undergo a certain amount of hand processing before it is ready for reproduction as a published report. Such factors as cost of equipment and manpower, availability of the various human and mechanical resources, and the goals in terms of tabulations to be made, reports to be published, and time schedules to be met determine the choice between hand or machine operations. The data processing operations to be performed in a census or survey usually consist of the following basic steps: editing, coding, card punching or an equivalent operation, sorting, and tabulating.

Editing.—Questionnaires are edited to correct inconsistencies and to eliminate omissions. Census or survey procedures often include some editing of the questionnaires in the field offices, since errors in the information can then more easily be corrected by checking with the respondent himself, and systematic errors made by the enumerator can more easily be rectified. Whether the editing is done in the field office or is part the central office processing, elimination of omissions and inconsistencies is a necessary step preliminary to tabulation.

A "not reported" category is permitted in some classifications of the population, but it is desirable to minimize the number of such cases. It was noted earlier in this chapter that in many classifications the tabulation plans provide for eliminating unknowns entirely. Where information is lacking, a reasonable entry can often be supplied by examining other information on the questionnaire. For example, a reasonable assumption of the relationship of a person to the head of the household can be made by checking names, ages, and marital status; or an entry of "married" may be assigned for marital status of a person whose relationship entry is "wife." Similar comparisons of various items of information on the questionnaire reveal inconsistencies which may be corrected in the editing operation. A person's entry for years of school completed, for example, may be inconsistent with the age reported.

The editing may be a manual operation performed prior to coding, a mechanical operation following the coding process, or a combination of these two. If the determination of the correct entry depends on reading other written entries, such as names, the determination must be made by a clerk. The degree of complexity of the editing is also a consideration in deciding whether to instruct clerks to do the work or to write a computer program for the purpose.

In manual editing, the clerks are given detailed specifications for assigning characteristics. Nonresponse cases may be assigned to a modal category; e.g., persons with place of birth not reported may be classified as native; or they may be distributed according to a known distribution of the population based on an earlier census.

Since much of the editing for blanks and inconsistencies is accomplished by applying fairly mechanical rules, the use of automatic equipment for performing this operation is becoming increasingly common. The machine processing is designed to reject or to correct a card or tape record with missing or inconsistent data and assign a reasonable response on the basis of other information. Depending on the capacity of the equipment used and the technical skill of the staff designing the specifications, mechanical editing may be either a fairly simple procedure of examining one or two other items or a very sophisticated procedure involving the examination of several other variables, including entries for other persons in the household. In the former case, the advantage of the machine operation over the manual process is greater speed and accuracy in the application of simple rules; but in the latter, the machine is also able to apply more complicated instructions and thus to arrive at an estimate which is much more likely to be consistent with other information for the person.

When mechanical editing is used extensively, manual editing may be limited to a few items which are known to be more complicated and, consequently, more prone to generate errors on the part of the enumerator or respondent. When the processing method involves heavy reliance on mechanical editing, it may include a screening operation to detect work units where the quality of enumeration is especially poor, in order that they may be subjected to some manual repair before going through the mechanical operation. The machine edit also may be so planned that enumeration areas with a high error level are rejected and routed to a unit where special processing can be done. In any case, the number of erroneous entries corrected and the number of allocations for nonresponse should be tabulated so that this information will be available for evaluation of the final results. This record of the editing process is one of the advantages of mechanical editing over manual editing, which does not automatically provide a count of edited entries.

Coding. — Coding is the conversion of entries on the questionnaire into symbols which can be used as input to the tabulating equipment. Many of the questions on a census or survey require no coding. For questions which have a small number of possible answers, such as sex or marital status, and questions which are answered in terms of a numerical entry, such as years of school completed, the appropriate code can be punched directly from the questionnaire onto the punch card. Entries may also be "precoded" by having the code for each written entry printed on the schedule. Then a mark made on the schedule to indicate the correct entry indicates the correct code also. The more complicated coding, such as that for occupation, which involves large numbers of categories, requires specially trained clerks.

Transference of data to tabulator. — In most data processing systems there must be some means of transferring the data from the original document to the tabulating equipment. The data on the questionnaire, after going through editing and coding may be transferred to punch cards by means of a hand-operated key punch machine. The cards are then fed into an electrical accounting machine, or the data may be transferred from cards to magnetic tape for processing by an electronic computer. In some modern equipment the data are punched directly onto computer tape, or a character reading device transfers the data automatically from the document to the tape. Special equipment developed for the 1960 census in the United States "reads" microfilmed copies of the questionnaires and transfers the data directly to computer tape. This equipment, known as FOSDIC (Film Optical Sensing Device for Input to Computer), reads the schedule by means of a moving beam of light, decides which codes have been marked, and records them on magnetic tape.

Tabulation. — It was noted earlier that during the planning stage of a census or survey, decisions are made about the tabulations to be produced, and outlines are prepared showing how the data are to be classified and the cross-tabulations to be made. The outlines may be quite specific, showing in detail the content of each proposed table.

On the basis of these outlines, specifications for computer programs are written for the various operations of sorting, adding, subtracting, counting, comparing, and other arithmetic procedures to be performed by the tabulating equipment. The input is usually punched cards or computer tape, and the output is the printed results in tabular arrangement. In the most advanced systems of tabulation, the final results include not only the absolute numbers in each of the prescribed categories but derived numbers such as percentage distributions, medians, means, and ratios as well.

Verification. — Verification of the operation is an important element of each stage in the processing. Verification is not done for the purpose of removing all errors, for this is virtually impossible and does not justify the expense of time and resources. The purpose rather is (1) to detect systematic errors throughout the operation which can be remedied by changes in the instructions or by additional training of

personnel; (2) to detect unsatisfactory performance on the part of an individual worker; and (3) to determine whether the general error rate of the operation is within tolerance. Therefore, it is seldom necessary to have 100-percent verification. A procedure often followed is to verify an individual's work until he is found to be qualified in terms of a maximum allowable error rate, and thereafter to verify only a sample of his work. If during the operation he is found to have dropped below the acceptable level of accuracy, his work units may be subjected to a complete review and correction process.

Verification may be "dependent," in which the verifier reviews the work of the original clerk and determines whether it is correct; or "independent," in which two persons do the same work independently and then a comparison is made of the results. Tests have shown that in dependent verification, a large proportion of the errors are missed. Independent verification, in which the verifier is not influenced by what was done by the original worker, has been found to be more successful in discovering errors.

Review of data. — The statistical tables produced by the tabulating equipment are usually subjected to editorial and statistical review before being prepared for publication. On the basis of advance estimates and data from previous surveys or other independent sources, judgments are made regarding the reasonableness of both absolute and relative numbers. Figures which are radically different from the expected magnitude may indicate an error in the specifications for tabulation. Review at this stage may show the need for expansion of the editing procedure. For example, early tabulations of educational statistics showing occasional impossible combinations of age and educational attainment might lead to an addition to the editing specifications to eliminate spurious cases of this nature. Tables are reviewed for internal consistency. It is not necessary that corresponding figures in different tables agree perfectly to the last digit, since minor differences are common in tables produced by different passes through the tabulating equipment. Arbitrary corrections for all small differences are not feasible, and such changes would add little to the accuracy of the data. If the tables printed out by the tabulating equipment are to be used for publication, the spelling, punctuation, spacing, indentation, etc., are also carefully reviewed so that corrections can be made before reproduction of the table.

Table preparation. — The output of the tabulation equipment may be used as the final statistical tables suitable for reproduction in the published reports, or it may be an interim tabular arrangement of the data from which the final tables will be produced. In the latter situation, typing of the final tables is either done directly from the machine print-outs or requires preliminary hand posting of the data on worksheets to arrange them as required for the publication tables. These additional steps, of course, require verification, proofreading, and machine-checking.

The advent of the electronic computers with their high-speed printer components has made the automatic production of final tables both feasible and economical. The elimination of one or more manual operations in the production process reduces the burden of quality control, improves the timeliness of publication, and reduces manpower requirements.

The use of high-speed printer output demands very precise advance planning of the content of each table, the wording of captions and stubs, and the spacing of lines and columns.

The technical skill involved and the lead time required for such planning have led some countries to use a compromise procedure in which the machine printout is used for the body of the table but the stubs and captions are provided by means of preprinted overlays. The programing of the computer printout in these instances is designed to produce a display of the data in the desired arrangement and to include rudimentary captions which serve to identify the numbers.

Evaluation of Results. – The evaluation of census results is cited by the Statistical Office of the United Nations as a requirement of good census practice. A definite program for measuring the extent of error in the census, followed by publication of the results and a description of the procedure of evaluation, is recommended.[13]

Census errors may occur at any of the various stages of enumeration and processing and may be either *coverage* errors, resulting from persons being missed or counted more than once, or *content* errors, that is, errors in the characteristics of the persons counted, resulting from incorrect reporting or recording or from failure to report. Methods for measuring the extent of error include reenumeration of a sample of the population covered in the census; overall comparison of census results with data from independent sources, usually administrative records; matching of census documents with other documents for the same person; and demographic analysis, which includes the comparison of statistics from successive censuses, analysis of the consistency of census statistics with birth, death, and immigration statistics, and the analysis of census data for internal consistency and reasonableness. These methods are described in some detail in the U.N. *Demographic Yearbook, 1962,* which also cites examples of their use in various countries of the world.

Nonresponse rates. – One of the most obvious sources of information about the quality of the data from a census or survey is the nonresponse rates. Even when a nonresponse category is not published and characteristics are allocated for those persons for whom information is lacking, a count of the nonresponse cases should be obtained during processing. One advantage of performing the edit in the computer is that not only the number of nonresponses on a given subject but also the known characteristics of the nonrespondents may be recorded to provide a basis for analyzing nonresponses and judging the effects of the allocation procedures.

The nonresponse rate for a given item has more meaning if it is based on the population to which the question applies or to which analysis of that subject is limited. The base for nonresponse rates on date of first marriage, for example, would exclude the single population; and nonresponse rates for country of birth would be limited to the foreign born. A problem arises in the establishment of a population base if the qualifying characteristic also contains a substantial number of nonresponses.

Use of Sampling in Censuses. – Although censuses as a rule involve a complete count of the number of inhabitants by certain basic demographic characteristics, sampling is often used as an integral part of the enumeration to obtain additional information. As noted by the United Nations:

"The rapidly growing needs in a number of countries for extensive and reliable demographic data have made sampling methods a very desirable adjunct of any complete census.

Sampling is increasingly being used for broadening the scope of the census by asking a number of questions of only a sample of the population. Modern experience in the use of sampling techniques has confirmed that it is not necessary to gather all demographic information on a complete basis; the sampling approach makes it feasible to obtain required data of acceptable accuracy when factors of time and cost might make it impracticable, or other considerations make it unnecessary, to obtain the data on a complete count basis."[14]

Waksberg emphasizes the use of sampling as a means of improving the timeliness of the data and recommends that only a bare minimum of information be collected on a 100-percent basis.[15] Some items of data may have to be collected on a complete-count basis because of legal requirements or because of the need for a high degree of precision in the data on basic topics so as to establish benchmarks for subsequent studies. When these needs have been met, the collection of other items on a sample basis saves time and money not only at the enumeration stage but throughout the processing as well.

Even when data collection is on a 100-percent basis, a representative sample of the schedules may be selected for advance processing to permit early publication of basic information for the country as a whole and for large areas. In the 1966 census of Iran, for example, a one-percent sample of the schedules provided the basis for a preliminary report presenting summary statistics for the total country only 6 months after the enumeration. Some or even most of the final tabulations in a census may be limited to a sample of the population, thus reducing the cost of tabulation considerably, especially when detailed cross-classifications are involved. In addition to its use in the enumeration and processing, sampling is important in the testing of census questionnaires and methods prior to enumeration, in the application of quality control procedures during enumeration and processing, and in the evaluation of the census by means of a postenumeration survey.

Sample Survey Methods. – The role of sample survey methods in the collection of demographic data is well-established, and some of the uses and advantages of sample surveys were discussed earlier in this chapter. The quality of the statistics from this source depends heavily upon the design of the sample and its faithful execution; and the usefulness of the statistics obtained is enhanced by a knowledge of their degree of reliability, as expressed in the standard deviation of the sample estimates. It is not practical to include in this chapter a treatment of the subject of sampling methods and practices nor to go into a discussion of sampling theory. Some basic references on these subjects are included in the list of readings at the end of this chapter.

Sample Estimation. – The derivation of a final estimate from the sample returns requires an additional processing step for sample surveys and for those portions of a census that are based on a sample. The sampling ratio itself determines the basic weights to be applied to the record for each person in the sample (e.g., a sample of one in five leads to a weight of five). The figures produced by the application of these weights, however, are often subjected to other adjustments to obtain the final estimates. The adjustments may be made

---

[13] United Nations, *Principles and Recommendations for the 1970 Population Censuses,* p. 19.

[14] United Nations, *Principles and Recommendations for the 1970 Population Censuses,* p. 25.
[15] Joseph Waksberg, "The Role of Sampling in Population Censuses: Its Effect on Timeliness and Accuracy," *Demography,* 5(1):362–373, 1968.

to account for population not covered because of failure to obtain an interview. Also, independent population controls often are available to which the sample results are adjusted. In a census, the data obtained on a sample basis may be adjusted to the 100-percent population counts by means of a ratio estimation procedure. The ratios of complete-count figures for specified age-sex categories to the sample figures for the same groups are computed and used for adjusting the tabulations based on the sample. Similarly, the results of sample surveys may be adjusted to independent controls which are postcensal estimates made by applying the balancing equation to popula ion figures from the last census.

The United States Census of Population

Nation-wide population censuses have been taken in the United States every 10 years since 1790, as a requirement of the Constitution for apportionment of the House of Representatives. The changing scope of the decennial census is discussed in chapter 2. During this time, methods of processing have advanced from a mere hand-count to the use of highly technical computer facilities for tabulation and publication of information in very detailed cross-classifications. The evolution of census methods is treated in detail in a number of other sources.[16] Here we will merely cite the years in which a few important innovations occurred:

Census	Innovation
1850	Listing of each person by name with one line of information for each person
1890	Large-scale use of mechanical equipment
1940	Some items collected for a representative national sample
1950	Tabulation by electronic computer (The entire census was not tabulated by computer until 1960.)
1960	Use of self-enumeration in densely settled areas
1970	Use of mail-out–mail-back procedure in large metropolitan areas

Current Population Survey

History and Purpose. — Some information on the history and purpose of the Current Population Survey (CPS) of the U.S. Bureau of the Census was presented in chapter 2. This survey, since its organization in 1940, has been the primary source of current information on the labor force in the United States and has become the principal means of obtaining postcensal statistics on many other population characteristics as well. During this time, the sample has undergone repeated expansion. Since 1943, when it was established in essentially its present form, it has grown from about 25,000 units spread over 68 sample areas comprising 125 counties and independent cities to its present size of approximately 55,000 housing units (47,000 occupied units) in a 461-area sample in 923 counties and independent cities. Part of a recent Current Population Survey form is presented as Figure 3-2.

VITAL STATISTICS

Dual Functions of a Vital Statistics System

In general, vital statistics are major products of relatively complex procedures designed primarily to accomplish the registration of vital events. They have sometimes been called "by-products" of registration systems, but this term gives a misleading impression. It is true that registration of births, deaths, marriages, and divorces originated to meet public and private needs for permanent legal records of these events, and these needs continue to be very important. However, equally important are the demands for useful statistics that have come from the fields of public health, life insurance, medical research, and population analysis.

Viewed as one of several general methods of collecting demographic statistics, registration has certain advantages and disadvantages. If events and their associated characteristics are registered near the time of occurrence, completeness of reporting and accuracy of information are potentially greater than if reporting depends upon a later visit or letter from an enumerator or other interviewer and recall of the facts by the respondent. Also, continuous availability of the data file tends to be assured by the dual uses of the information—for legal and for statistical, individual, and public purposes.

There are also certain limitations of the registration method. The fact that the vital record is a legal document limits the amount and kind of nonlegal information that can be included in it. The method is also affected by the number and variety of persons involved in registering the events. For example, birth registration in some countries requires actions by thousands or millions of individual citizens and hundreds of local officials. In other countries thousands of physicians, nurses, or hospital employees may be involved, and all of these people have other duties which they consider more urgent. It seems inevitable that, for the most part, these many and diverse persons will have less training and expertness in data collection than the enumerators who interview in censuses or other population surveys. The latter are usually given intensive training in which the importance, purposes and exact specifications of the information sought are thoroughly explained.

Satisfactory performance of registration, in terms of both the legal and the statistical requirements, is closely related to the completeness and promptness with which events are registered and the accuracy of the information in the registration records. Certain functions, e.g., indexing and filing of certificates, issuance of copies, and amendment of records are important for their legal uses but do not significantly affect the statistics. However, if the legal functions are poorly performed, the statistical program will suffer because public pressures will demand that first priority be given to serving the people's needs for copies of their personal records.

International Standards and National Practices

The *Handbook of Vital Statistics Methods*, published by the United Nations Statistical Office, is the principal source of the material presented in this section on international recommendations for the collection and processing of vital statistics.[17] The *Handbook* is scheduled to be rewritten by that office.

Definitions of Vital Events. — As in all systems of data collection, clear, precise definitions of the phenomena measured are prerequisites for accurate vital statistics. Use of standard

---

[16] Carroll D. Wright and William C. Hunt, *The History and Growth of the United States Census*, Washington, Government Printing Office, 1900; and Leon E. Truesdell, *The Development of Punch Card Tabulation in the Bureau of the Census, 1890–1940*, Washington, Government Printing Office, 1965.
  A detailed description of the methods and procedures of the most recent censuses is contained in the following two sources: (1) U.S. Bureau of the Census, *United States Censuses of Population and Housing: 1960, Procedural History*, Washington, 1966; and (2) U.S. Bureau of the Census, *1970 Census Users' Guide, 1970, Part I*, Washington, 1970. See also, U.S. Bureau of the Census, *Census Use Study*, various reports.

[17] United Nations, *Handbook of Vital Statistics Methods*, Studies in Methods, Series F, No. 7, 1955.

Figure 3-1.—100-Percent Questionnaire Used in the U.S. Census of Population: 1970

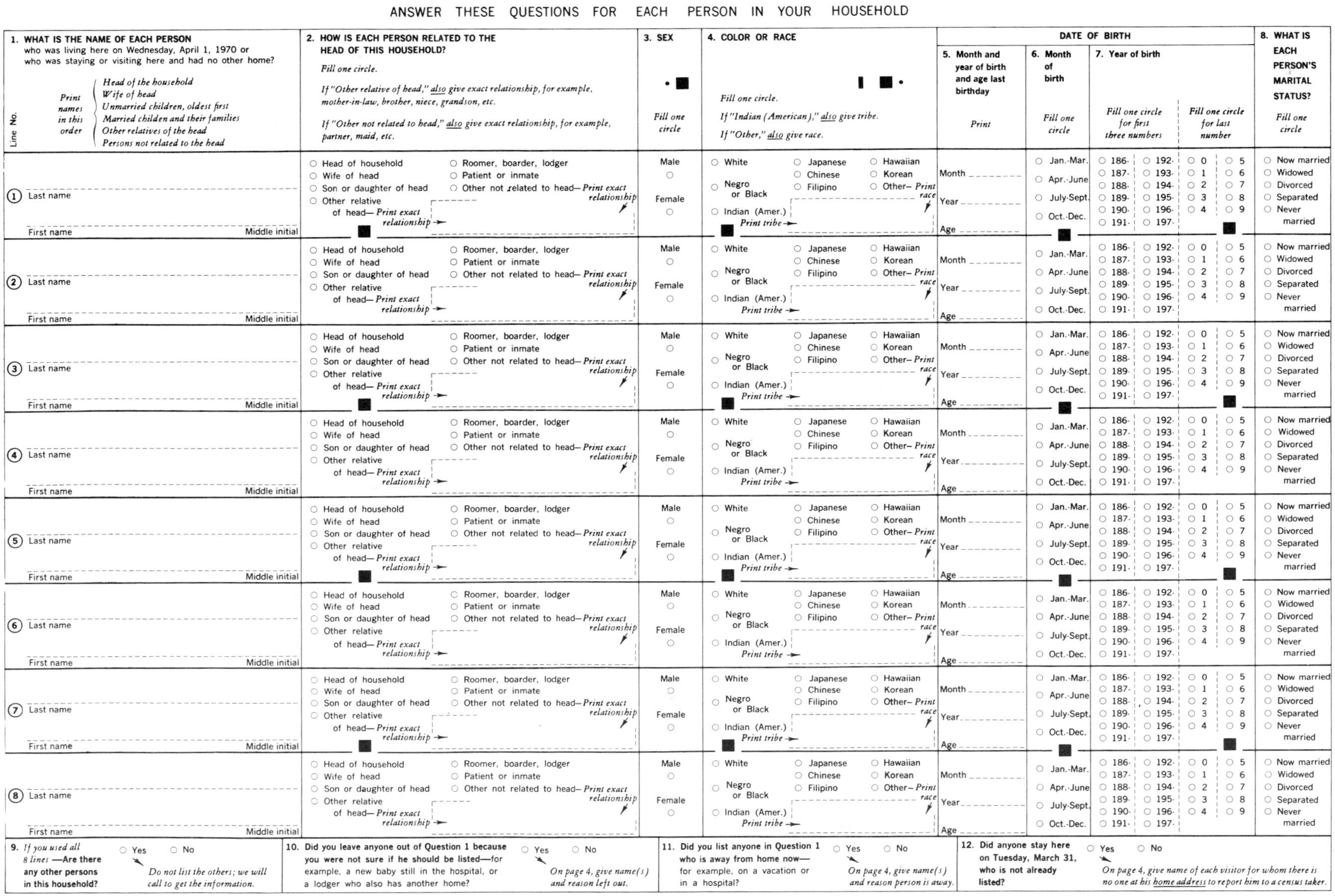

definitions of vital events is essential for comparability of statistics for different countries.

Live birth. — Most countries now follow the definition of live birth recommended by the World Health Assembly in May 1950 and by the United Nations Statistical Commission in 1953. which is as follows:

"Live birth is the complete expulsion or extraction from its mother of a product of conception, irrespective of the duration of pregnancy, which after such separation, breathes or shows any other evidence of life, such as beating of the heart, pulsation of the umbilical cord, or definite movement of voluntary muscles, whether or not the umbilical cord has been cut or the placenta is attached; each product of such birth is considered live-born." [18]

Under this definition a birth should be registered as a live birth regardless of its "viability" or death soon after birth or death before the required registration date. Although variations in the statistical treatment of "nonviable" live births (defined by low birthweight or short period of gestation) do not significantly affect the statistics of live births, they can have a substantial effect on fetal death and infant death statistics.

Death. — Until very recently, there has been less difficulty with respect to the definition of death than with definitions of live birth and fetal death. For statistical purposes, the United Nations has recommended the following definition of death:

"Death is the permanent disappearance of all evidence of life at any time after live birth has taken place (postnatal cessation of vital functions without capability of resuscitation). This definition therefore excludes foetal deaths." [19]

Fetal death. — The definition of fetal death recommended by the World Health Organization (WHO) and the United Nations Statistical Commission is:

"Foetal death is death prior to the complete expulsion or extraction from its mother of a product of conception, irrespective of the duration of pregnancy; the death is indicated by the fact that after such separation the foetus does not breathe or show any other evidence of life, such as beating of the heart, pulsation of the umbilical cord, or definite movement of voluntary muscles." [20]

Prior to promulgation of the above definition most countries had required registration only of those "stillbirths" which had completed 28 or more weeks of gestation. Many countries have not yet adopted the WHO definition for legal reporting purposes. Fetal death or stillbirth statistics are affected very greatly by differences in registration requirements. Fetal deaths of 28 or more weeks of gestation constitute a small percentage of all fetal deaths.

Marriage. — The Statistical Commission of the United Nations has recommended the following definition of marriage for statistical purposes:

"Marriage is the legal union of persons of opposite sex. The legality of the union may be established by civil, religious, or other means as recognized by the laws of each country; and irrespective of the type of marriage, each should be reported for vital statistics purposes." [21]

This definition includes so-called "consensual unions," "common-law" marriages, and other stable *de facto* unions. However, there is no adequate method of counting such marriages. They are not registered because the legal requirements have not been observed. In some countries they constitute a majority of all marriages.

Divorce. — The United Nations Statistical Commission's recommended definition of divorce is:

"Divorce is the final legal dissolution of a marriage, that is, the separation of husband and wife by a judicial decree which confers on the parties the right to civil and/or religious remarriage, according to the laws of each country." [22]

This definition excludes petitions, provisional divorces, and legal separations since they do not imply final dissolution of marriage and the right to remarry. In some countries, legal annulment is a statistically significant method of marriage termination. It is desirable in such countries to include annulments with divorces in determining the statistics of marriage dissolution. The *Handbook* defines annulment as "the invalidation or voiding of a marriage by a competent authority, according to the laws of each country, which confers on the parties the status of never having been married to each other."

Collection of Vital Statistics. — Vital statistics systems differ in the amount of authority given to the collecting agency, the degree of national centralization of its organization, and the type of agency carrying out the program. The basic features of a vital statistics collection system are discussed in the following sections.

Compulsory registration. — The position of the United Nations on the matter of compulsory registration is strongly stated. "The compulsion or legal obligation to register a vital event may be accepted as the basic premise of the entire vital-statistics system. When registration is voluntary rather than compulsory, there can be no complete or accurate vital records or statistics." Legal requirement for registration of vital events has, therefore, been made one of the United Nations' "principles" for the development of a vital statistics system. [23]

Governmental organization. — The registration systems may be classified as organized under centralized or decentralized control. In 65 countries examined by the United Nations as of 1950, 28 had their local registration units under direct central control; whereas, in the remaining 37 countries, a national agency exercised only general control, which was sometimes merely nominal. [24] Within these 37, the federated countries, such as the United States of America, Canada, Australia, Argentina, and Brazil, represent a special type. Here the States or Provinces enact registration laws, appoint and supervise local registrars, file certificates of registration, etc. The pertinent national office sets standards in cooperation with the State offices and compiles national statistics. Even the compilation and publication of vital statistics may be decentralized to some extent; but, in every modern system, a national agency does at least part of the compilation and issues statistical reports.

It is desirable that there be one national agency with the legal authority and responsibility for registration and statistical reporting in the country. The Statistical Commission of the United Nations makes the following recommendation:

"Responsibility for the establishment or development of a national vital statistics system should be the function of a national governmental agency or agencies." [25]

Most, but not all, nations have established a national authority over registration. In some countries, it is the civil registration office, in others, the department of public health, and in others, the central department of statistics. Again, in some countries the same national agency is responsible for both registration and statistics, but in others these two functions

---

[18] Ibid., p. 46.   [19] Ibid., p. 53.   [20] Ibid., p. 60.

[21] Ibid, p. 60.

[22] Ibid., p. 62.   [23] Ibid., p. 19.   [24] Ibid., p. 33.

[25] Ibid., p. 40.

are controlled by two or occasionally three separate departments.

Local registration areas are the basic units of a vital registration system. They must have clearly defined geographic boundaries and be small enough for the registrar to give the area the attention required to produce good registration and for persons reporting vital events to come to or communicate with the registration office without excessive difficulty.

One of the most important responsibilities of the local registrar is to encourage the general population, physicians, midwives, and others to report occurrences of vital events promptly and to supply complete and accurate information about them.

Informants and reporters. — The person responsible by law for reporting the occurrence of a vital event may or may not also be the source of the information concerning the facts associated with the event. In most countries, for example, the father or mother is responsible for reporting the occurrence of a live birth or fetal death, together with certain personal information, but the attendant physician or midwife is also responsible for reporting the event along with certain medical information. The nearest relative is usually legally responsible for reporting the occurrence of a death, but the attending physician is required also to report the event together with a certification of the cause. The officiant, civil or religious, at the marriage is required to report it in about one-half of the countries. In the other countries, the participants, bride and groom, are responsible. Reporting of divorces is the responsibility of the court in some countries, one or both of the parties to the divorce in others.

Place of registration. — The United Nations recommends and, with few exceptions, the countries of the world require registration of vital events in the local registration area where the event occurred. Statistics tabulated by the United Nations for 65 countries as of 1950 showed that nearly all vital events are registered where they occurred.[26] In some countries arrangements are made to send a copy of the certificate to or otherwise notify the registration office in the place of residence when it is other than the place of occurrence. Tabulations are frequently made by area of usual residence of the mother, decedent, etc.; these are generally regarded as more useful for demographic purposes than tabulations by place of occurrence.

Time allowed for current registration. — The registration record usually calls for both the date of the event and the date of registration. National laws usually specify the maximum interval permitted between these two dates for each type of vital event. The United Nations surveyed practices in 65 countries in 1950.[27] In the case of births, the grace period varied from one day to 90 days. The period for deaths tends to be shorter; only eight countries reported a period exceeding five days. The law specifies the fine or other penalty for failure to register the event within the prescribed period.

The United Nations recommends that final tabulations for any calendar period should be based on events which occurred during that period and not on those registered. However, the latest information available indicates that among countries reporting to the U.N. Statistical Office in 1954, 44 tabulated the rrecords by date of occurence and 27 by date of registration.

Types of registration records and statistical reports. — Sometimes the registration document, or an exact copy of it, is also used as the statistical report of an individual vital event.

In other cases, the statistical report consists of an abstract or summary of the registration record, plus additional medical information supplied by the physician or hospital.

Registration records are generally maintained in one of two forms — (1) separate individual documents for each birth, death, or marriage which are kept in binders that can be opened to release individual records if needed, or (2) books in which each page is devoted to recording one or several events and the pages are not detachable. The United Nations has not made an official recommendation on type of registration records. However, it has recommended that the statistical report be an individual document for each event. If the original registration record is also an individual form for each event, compliance with the United Nations recommendation is made easier. However, a large majority of countries use the bound-book method of recording registrations of vital events. For statistical purposes, a majority of countries use the individual record, but a sizable minority use listings in which information on a number of events is given on the same page.

Contents of statistical reports. — In some countries the statistical report is identical in content to the registration record. This practice permits making copies for statistical uses and avoids having to match two or more separate reports of a particular vital event. However, it has two disadvantages. It limits the amount of information for statistical purposes to what can be accommodated on a record designed primarily for legal uses. Also, it makes difficult inclusion of any information that would be regarded as confidential or private. In some countries medical information related to a live birth or fetal death (stillbirth) or to causes of death, and information on fertility history, attitudes toward and practice of contraception, more detailed socioeconomic characteristics, etc. are obtained in separate reports submitted by the attending physician or the hospital in which the birth, death, or fetal death occurred.

The United Nations has recommended lists of statistical items that should be included in the records of live births, fetal deaths, deaths, marriages, and divorces. (The World Health Organization recommended the form of the medical certificate of cause of death.) The recommended items are subdivided into two groups, first-priority items, which all countries should include, and second-priority items, which may not be feasible for all countries at this time.[28] Parallel listings of first-priority items for the various vital statistics records are shown in table 3–1.

The kinds of information ordinarily contained on a birth certificate is illustrated by the Philippine certificate in Figure 3-3.

It will be noted that some of these items are not of first priority from the demographic standpoint. Some of the second priority items are definitely of interest to demographers; for example, in the case of births, date of birth of father, date of parents' marriage, and education and occupation of parents; in the case of deaths, age of surviving spouse (for married), marital status, and number of children ever born (for women); and, in the case of marriages, the number of previous marriages.

The "1970 World Programme for the Improvement of Vital Statistics," recommended by the United Nations, has as its goal "the establishment, in every country of the world by 1979, of a vital statistics system capable of producing reliable measures of population growth and the statistics needed for demographic research."[29]

---

[26] Ibid., p. 78.

[27] Ibid., pp. 84–86.

[28] Ibid., pp. 114–115 and 174.

[29] United Nations, Economic and Social Council, Statistical Commission, 15th Session, *Progress Report on Demographic Statistics*, E/CN.3/377, January 5, 1968, p. 10.

Figure 3–2. —Portion of Demographic Supplement to the Current Population Survey Questionnaire in the United States: 1969

FILL FOR ALL HOUSEHOLD MEMBERS 14 YEARS OLD AND OVER—INCLUDING THOSE IN ARMED FORCES

34. Was ... living in this house on March 1, a year ago?
No ○ *(Ask 35)*
Yes ○ *(Skip to 37)*

If "No" in 34
35. Was ... living in this same county on March 1, a year ago?
No ○ *(Ask 36)*
Yes ○ *(Skip to 37)*

If "No" in 35
36. What State *(or foreign country)* **was ... living in on March 1, a year ago?**
(Enter state or foreign country and mark circle below.)
This State.... ○
Different State ○
Abroad...... ○

OFFICE USE ONLY
NE ○ NC ○ S ○ W ○
C ○ N ○

FOR PERSONS EVER MARRIED
37. When did ... first get married?
(mark month, decade, and year)
Jan.-Mar. ○
Apr.-June ○
July-Sept. ○
Oct.-Dec. ○
Decade 0 1 2 3 4 5 6 7 8 9
Year 0 1 2 3 4 5 6 7 8 9

INCOME IN THE YEAR 1968 *(Ask for all persons 14 years of age and over. Write actual amount in space provided in items 38 - 41e and mark appropriate dotted numbers. Mark $10,000+ circle if write-in entry is $10,000 or more.)*

LAST YEAR *(1968)* How much did ... receive:

38. In wages or salary before any deductions?
$
$10,000+ ○
None ○

39. In net income from his own business or professional practice or partnership?
GROSS INCOME $
MINUS
BUSINESS EXPENSE $
EQUALS
NET INCOME $
(Mark net income)
$10,000+ ○
None ○
Lost money ○

40. In net income from his own farm?
GROSS INCOME $
MINUS
BUSINESS EXPENSE $
EQUALS
NET INCOME $
(Mark net income)
$10,000+ ○
None ○
Lost money ○

41. During 1968, did ... receive any money from:

41a. —Social Security or Railroad Retirement checks from the U.S. Government?
Yes ○ No ○
HOW MUCH ALTOGETHER? $

41b. —Dividends?
Yes ○ No ○
—Interest on savings accounts or bonds?
Yes ○ No ○
—Net rental income or income from estates or trusts?
Yes ○ No ○
HOW MUCH ALTOGETHER? $
$10,000+ ○
Lost money ○

41c. —Welfare payments or other public assistance *(aid to families with dependent children, old age assistance, or aid to the blind or totally disabled)*?
Yes ○ No ○
HOW MUCH ALTOGETHER? $
$10,000+ ○

41d. —Unemployment compensation?
Yes ○ No ○
—Workmen's compensation?
Yes ○ No ○
—Government employee pensions?
Yes ○ No ○
—Veteran's payments?
Yes ○ No ○
HOW MUCH ALTOGETHER? $
$10,000+ ○

41e. —Private pensions or annuities?
Yes ○ No ○
—Alimony?
Yes ○ No ○
—Regular contributions from persons not living in this household?
Yes ○ No ○
—Anything else?
Yes ○ No ○
HOW MUCH ALTOGETHER? $
$10,000+ ○

In rotations 1, 2, 3, 5, 7, 8, be sure you have filled items 18-41e for civilian household members 14 years old and over, and items 18 and 25-41e for all Armed Forces members.

42. INTERVIEWER CHECK ITEM
Who reported on income for this person?
Self ○ Proxy ○

Figure 3–3. — Birth Registration Form of the Philippines: 1956

CODES
(Do not fill out)

1(a). ——————
1(b). ——————
1(c). ——————
2. ——————
2(f). ——————
4. ——————
5(a). ——————
5(b). ——————
6. ——————
8. ——————
9. ——————
11(a). ——————
11(b). ——————
13. ——————
14. ——————
16(a). ——————
16(b). ——————
16(c). ——————
18(b). ——————
22(a). ——————
22(b). ——————
23. ——————

RESERVE FOR BINDING CODES

MUNICIPAL FORM No. 102 — (Revised, July, 1956)

CERTIFICATE OF LIVE BIRTH
(To be accomplished in Duplicate)

Register Number:
(a) Civil Registrar General No. ——————
(b) Local Civil Registrar No. ——————

Province: ——————————————
City or Municipality: ——————————————

1. PLACE OF BIRTH a. PROVINCE	**2. USUAL RESIDENCE OF MOTHER** (Where does mother live?) a. PROVINCE
b. CITY, TOWN, OR LOCATION	b. CITY, TOWN, OR LOCATION c. STREET
c. NAME OF HOSPITAL OR INSTITUTION (If not in hospital, give street address)	d. DATE AND PLACE OF MARRIAGE (for legitimate birth)
d. IS PLACE OF BIRTH INSIDE CITY LIMITS? YES ☐ NO ☐	e. IS RESIDENCE INSIDE CITY LIMITS? YES ☐ NO ☐ — f. IS RESIDENCE ON A FARM? YES ☐ NO ☐

CHILD

3. NAME (Type or print) First	Middle	Last	
4. SEX	**5a. THIS BIRTH** Single ☐ Twin ☐ Triplet ☐	**5b. IF TWIN OR TRIPLET, WAS CHILD** 1st ☐ 2d ☐ 3rd ☐	**6. DATE OF BIRTH** Month Day Year

FATHER

7. NAME First Middle Last	**8. NATIONALITY** **8a. RACE**
9. AGE (At time of this birth) Years	**10. BIRTHPLACE** — **11a. USUAL OCCUPATION** — **11b. KIND OF BUSINESS OR INDUSTRY**

MOTHER

12. MAIDEN NAME First Middle Last	**13. NATIONALITY** **13.a RACE**
14. AGE (At time of this birth) Years	**15. BIRTHPLACE** — **16. PREVIOUS DELIVERIES TO MOTHER** (Do Not Include this Birth)

	a. How many other children are now living?	b. How many other children were born alive but are now dead?	c. How many fetal deaths (fetuses born dead at any time after conception)?
17a. INFORMANT'S SIGNATURE: b. NAME IN PRINT: c. ADDRESS:			

18. MOTHER'S MAILING ADDRESS

I hereby certify that this child was born alive on the date stated above.	**18a. SIGNATURE.** b. NAME IN PRINT	**18c. ATTENDANT AT BIRTH** M.D. ☐ MIDWIFE ☐ OTHER (Specify)
	18d. ADDRESS	**18e. DATE SIGNED**
19. DATE RECEIVED BY LOCAL CIVIL REGISTRAR	**20a. REGISTRAR'S SIGNATURE** b. NAME IN PRINT c. POSITION	**21. DATE ON WHICH GIVEN NAME ADDED** (Registered) By (Registrar)

FOR MEDICAL AND HEALTH USE ONLY
(This section must be filled out)

22a. LENGTH OF PREGNANCY (COMPLETED WEEKS)	**22b. WEIGHT AT BIRTH** lbs. —— oz. ——	**23. LEGITIMATE** YES ☐ NO ☐

(Space for addition of Medical and Health items for special Purposes)

The notification of birth must be given or the certificate be sent to the Local Civil Registrar of the city, municipality or municipal district where the child was born within 30 days from the date of birth. Failure to do so is punishable by a fine of not less than ₱10 nor more than ₱200. (Act 3753, sections 5 and 17.)

59612

TO BE FILLED ACCURATELY

Table 3–1. — **First-Priority Items Recommended for Inclusion in Statistical Reports of Live Birth, Death, Fetal Death, Marriage, and Divorce**

Live birth	Death	Fetal death	Marriage	Divorce
Date of occurrence	Date of occurrence	Date of occurrence	Date of occurrence	Date of occurrence
Date of registration	Date of registration	Date of registration	—	—
Place of occurrence	Place of occurrence	Place of occurrence	Place of occurrence	Place of occurrence
Place of residence of mother	Place of residence	Place of residence of mother	Place of residence of bride and groom	Place of residence of divorcees
Sex	Sex	Sex	—	—
Legitimacy	Cause	Legitimacy	Date of birth	Date of birth
Type of birth	Certifier	Type of birth	Previous marital status	Date of marriage
Date of birth of mother	Date of birth	Date of birth of mother		Number of dependent children
Number of children born to mother		Number of children born to mother		
Attendant at birth		Period of gestation		

Source: United Nations, *Handbook of Vital Statistics Methods*, Studies in Methods, Series F, No. 7, 1955, table F, p. 174.

Compilation and Tabulation of Vital Statistics. — The underlying purpose of a vital statistics system is to make available useful statistics for the planning, administration, and evaluation of public health programs and to provide basic statistics for demographic research. The documents undergo much the same processing that is required for census and survey data, and similar planning is required to produce the desired tabulations.

In a majority of countries the central statistical office has been given responsibility for compilation of national vital statistics. In some countries, including the United States, this function has been located in the national public health agency. In other countries, responsibility has been divided between the health agencies and the statistical and registration agencies.

Of equal or greater importance is the degree of national authority over the processes of record collection and statistics compilation. The United Nations has recommended that (1) "reports on vital events for national statistical purposes should be collected centrally by the agency which is responsible for the statistical compilation"[30] and (2) "national vital statistics should be compiled on a centralized basis by a national agency specifically charged with this statistical function."[31] The principal advantages of centralized processing of national vital statistics are increased accuracy and uniformity of classification, greater flexibility in making changes in coding rules and content of tabulations, and more timely availability of national statistics.

Because the basic statistics needed are very similar from country to country, and because international comparability of statistics is desirable, the United Nations has prepared lists of first-priority and second-priority subjects to be tabulated for each type of vital event.[32] Separate tabulations for each population group and for geographic areas are recommended in countries where large segments of the population have very different social and economic characteristics and where the level of completeness and accuracy may vary considerably.

Since a certain amount of time is allowed for registering events and additional time is required for transmittal of the individual reports, there is a considerable lag between the end of the reporting period and the receipt of the last reports. It is recommended that a "cut-off" date be specified consistent with time required for registration and reporting and that tabu-

lations be based on reports received before that date. For many uses of vital statistics, the data need to be converted to rates or ratios expressing the relation between the number of vital events and the corresponding population. Such rates are described in chapters 14, 16, and 19.

Other Methods of Obtaining Vital Statistics. — Every nation has as a goal the coverage of all its states or other areas in its vital statistics system. This objective is often not achieved without a long period during which the registration system is being developed and its coverage gradually extended. In the United States, 33 years intervened between admission of the first 10 States to the national death registration area and the last two in 1933. For births, the interval between the establishment of the registration area and complete coverage of all States was 18 years. The marriage and divorce registration areas were initiated in 1957 and 1958, respectively, and are still incomplete. Prior to admission of all states, statistics for the incomplete registration areas have been used as the only available substitute for complete national data. This method tends to encourage the national office to work hardest at development of parts of the country that seem most likely to be ready for a complete registration system. However, it has the disadvantage that the sections that can qualify first are not likely to be demographically representative of the entire country. For example, in the early years, the United States death and birth registration areas did not include States in the southern and western sections of the country.

Sample registration area. — A promising method of obtaining national statistics before the registration system is operating effectively in all parts of the country is the use of a "representative sample registration area."

In addition to earlier production of national statistics, a sample registration area can be used to develop and perfect the organizational structure and the various registration procedures which will eventually be adopted in all parts of the country. There can be experimentation in different local areas with various reporting sources, various types of local officials as registrars, different time allowances, and a variety of enforcement procedures, educational methods, and registration incentives. Methods of communication between registration officials, hospitals, physicians, midwives, statisticians, and health officials can be developed and tested.

Surveys, censuses, and population registers. — Other data collection methods may supplement or be a substitute for the registration system for vital statistics. Vital statistics may be

---

[30] United Nations, *Handbook of Vital Statistics Methods*, p. 102.
[31] Ibid., p. 37.
[32] Ibid., pp. 174–179 and 180–182.

obtained from a household sample survey by interrogating members of the household regarding vital events which occurred in that household in some specific past period. This method can be implemented in a relatively short time, if the necessary technical skills can be mobilized to plan and conduct it, and can be expected to provide some statistics rather speedily. It need not be very expensive if most of the manpower can be recruited within the country. Its success depends heavily on the willingness of persons in the sample to supply the information and on their ability to recall the vital events occurring during some past period of time, and the date, place of occurrence, and other facts about the events. Also, the considerable skills required for sample design, survey organization and operation, and questionnaire construction need to be available on a continuing basis.

Information on vital events is sometimes obtained in the population census. Statistics on births in the previous year, marriages, and more rarely, deaths are available from this source in some countries.

In countries which maintain a continuous population register, birth, death, marriage, and divorce registration may be an integral part of the register. The information obtained in the registration of the vital events must not only serve the needs for statistics on these subjects but also be consistent in definitions and classifications with the information to be kept in the population register on the entire population.

It is also essential that statistics of births, deaths, and marriages be based on definitions and classifications that are identical to or consistent with those used in the population census. Computation of valid vital rates and use of these rates in population estimation depend upon consistent treatment of vital statistics and population data. This objective is sometimes difficult of attainment, however, especially when different agencies are responsible for the two programs of statistics.

The United States Vital Statistics System

National-State Relationships. — The United States system for collecting vital records is decentralized in that the legal authority over registration is located in each of the 50 States and the District of Columbia. Each State is divided into local registration districts, for each of which a registrar is appointed. There are about 10,000 such registrars, appointed by the State governments or locally elected. Each State separately processes the statistics which it wishes for its own area and population. New York City is an independent registration area and has its own laws and regulations and publishes its own reports, as do Puerto Rico and the Virgin Islands of the United States. In contrast, the processing of national vital statistics is centralized in the National Center for Health Statistics, a Federal agency located in the U.S. Public Health Service.

Uniformity of reporting. — Although registration of vital events is governed by State laws, a considerable degree of uniformity has been achieved in definitions, organization, procedures, and forms. Uniformity has been promoted, primarily, by the development of model laws and certificate forms which have been recommended for State use. The Model State Vital Statistics Law has been followed with variations in the laws enacted in the various States. It was first promulgated in 1907 and has been revised and reissued several times. The most recent version was promulgated in 1959.[33] Standard certificates of the several vital events, issued by the responsible national agency, have been the principal means of achieving uniformity in the certificates of the individual States, which provide the information upon which national vital statistics are based.

The responsible national vital statistics agency (the Bureau of the Census from 1903 to 1946, and the Public Health Service, thereafter) has actively assisted the State agencies to achieve complete, prompt, and accurate registration of vital events. Tests of registration completeness and intensive educational campaigns to promote registration have been joint Federal-State efforts. The national office has developed and recommended to the States model handbooks designed to instruct physicians, hospitals, coroners and medical examiners, funeral directors, and marriage license clerks on current registration procedures and the meaning of the information requested in the certificates.[34]

Functions Performed by State Offices. — In the decentralized registration system of the United States, the primary responsibility for the collection of vital records rests with each State. This responsibility encompasses a number of functions which are carried out in each State vital statistics office.

Planning content of forms. — It is the responsibility of the State vital statistics office to recommend the format and content of the vital records used in their jurisdiction. These recommendations are usually based to a large extent on the United States Standard Certificates but also often reflect special interests or needs not encompassed in the Federal model forms. In spite of the efforts of the Federal government to promote national uniformity, State and local uses of vital records, especially in the health field, produce differences in record content and format which have an effect on the statistics. Some of the States have not included all of the standard demographic items on their vital records. For example, Massachusetts did not include birth order until 1961. Many of the birth, death, and fetal death certificates in use in New Jersey in 1962 and 1963 did not call for the item on race. Sixteen States do not report the legitimacy of the birth.

Confidentiality of records. — It is the responsibility of each State or other registration area to determine the need for confidentiality and to maintain confidentiality of the vital records. In some areas, vital records are considered to be public documents; in other areas the vital statistics laws and administrative regulations permit the release of information or certified copies of the record only to certain authorized persons.

Receipt and processing of records. — One of the major functions of a State office is to serve as the repository for vital records of events occurring within the State, thus providing a central source within each State for both the legal and statistical uses of the records. This function entails a number of related responsibilities, such as the handling of corrections, missing data, name changes, and adoptions and legitimations, as well as issuing certified copies of records on file. The State office is also responsible for various aspects of records management including indexing, binding, filing, and preservation of the records.

Among these responsibilities, the correction of errors and the obtaining of missing data have the greatest impact on the

[33]U.S. Public Health Service, *Model State Vital Statistics Act,* Publication No. 794, 1960. More recent changes in the standard certificates are discussed in: U.S. Public Health Service, *Model State Vital Statistics Regulations,* Document No. 616.6, August 7, 1973.

[34] For example, National Center for Health Statistics. *Hospital Handbook on Birth and Fetal Death Registration,* Public Health Service Publication No. 593A, 1967.

quality of the statistical data derived from the records. A major objective of any statistical system is to minimize the number of items where the response is missing, unknown, or apparently erroneous. State vital statistics offices work toward this objective through a query program in which letters of inquiry, when required, are directed to those responsible for originally supplying the information on the vital record or to those responsible for filling out the certificate.

Tabulation and publication of the data.—The majority of State vital statistics offices use some form of mechanical data processing equipment to reduce the masses of data into useful summaries, but an increasing number of offices have converted their operation to electronic processing. The use of electronic computers allows the tabulations to proceed on a much faster schedule but requires much more careful planning to take advantage of the versatility and speed of the equipment. Furthermore, no State office has found the volume and complexity of its operations sufficiently large to justify the acquisition of a computer for vital statistics work alone.

Just as each State prepares and processes its own vital statistics data, so does each State prepare an annual summary of its vital statistics. These summaries vary in complexity and comprehensiveness, but almost all States publish some kind of annual vital statistics report. Some of these reports merely present selected vital statistics data, whereas others contain, in addition to tabular material, an analysis and interpretation of the statistics.

Another activity of the State vital statistics offices is the transmittal of data to the National Center for Health Statistics (NCHS) for the purpose of assembling national statistics there. This transmittal of data is now in the form of microfilm copies of the original records on file in the State offices. The microfilm is usually sent to the National Center monthly; the records in each transmittal primarily represent the events of the previous month, although some States follow a different schedule. The NCHS purchases the data from each registration area through a contractual arrangement, which includes a guarantee of confidentiality prohibiting the Center from releasing any data other than statistical summaries without the written consent of the State vital statistics office.

In order to issue provisional statistics in its *Monthly Vital Statistics Report*, the National Center for Health Statistics receives reports from the States on the total number of records (birth, death, infant death, marriage, and divorce) received during the month regardless of date of occurrence. There is a lag of about 2 months after the data month before publication. In addition, a 10-percent sample of deaths is drawn in State offices to provide estimated deaths and death rates by cause and by age for publication in the same report series. These figures appear about two months after the date of receipt of the certificates in the State offices and three months after the month in which the events occurred.

Functions Performed by the National Center for Health Statistics.—The NCHS performs a variety of functions designed to improve the national vital statistics system. It exercises leadership in the revision of the standard certificates and in evaluating the completeness of birth registration; represents the United States in international conferences on the standard classification of causes of death; conducts a training program on vital and health statistics; helps the States in developing forms, procedures, draft legislation, definitions, and tabulations; and acts as a clearinghouse for certificates of nonresident births and deaths. In the clearinghouse operation,

copies of nonresident certificates are sent to the States of usual residence for inclusion in their own tabulations.

The National Center for Health Statistics serves as the focal point for the collection, analysis, and dissemination of national vital statistics for the United States. Because of the diversity of practices and procedures existing in the decentralized U.S. system, the production of national statistics involves more than the combining of statistics from each registration area to produce nation-wide vital statistics.

As indicated previously, raw data are obtained through the purchase of microfilm copies of original records. The microfilm is reviewed for clarity, legibility, and omission of records. The omission of records is checked by comparing the number of records on the film with a separate report submitted by the vital statistics offices indicating the number of records they received during the period covered by the microfilm. Marriage and divorce records are sampled, with a different sample size for each registration area; birth records are also sampled, 50 percent of all records on the microfilm being taken for each area. Because of the detail involved in tabulating mortality data by cause of death and by a variety of demographic characteristics, a sample of death records is not taken. Shipments received by NCHS from the States after a cut-off date are not processed by NCHS. This cut-off date varies from 6 to 9 months after the end of the calendar year. Different cut-off dates used by the national and State offices explain part of the difference between published figures in their respective reports.

After the appropriate records are selected, they are coded and transferred to magnetic tape. The most complex coding is that of the cause of death. The classification of causes is revised from time to time in the light of international biometric experience and of new medical knowledge.[35]

The coding of vital records and the transfer of the data to magnetic tape are checked for errors through verification procedures similar to those performed in State offices. In some instances, codes which were previously entered on the records by a State office are used by NCHS following sample verification of their accuracy.

The magnetic tapes which are produced are subjected to a series of computer editing operations which eliminate certain inconsistencies in the data. The computer also imputes missing data for certain items. This is generally done only when the number of items with missing data comprises a very small proportion of the total. Sex, race, and geographic classification are assigned if not reported on the birth or death certificates, and age of mother is assigned if not reported on the birth certificate.

The final computer tabulations of national vital statistics appear in various publications prepared by NCHS and mentioned in chapter 2, "Basic Sources of Statistics." In addition, unpublished material and resource data for special investigations are maintained by NCHS.

OTHER DEMOGRAPHIC RECORD SYSTEMS

We will not describe the collection and processing for other demographic record systems (immigration and emigra-

---

[35] National Center for Health Statistics. *Eighth Revision, International Classification of Diseases, Adapted for Use in the United States*, Vol. I, *Tabular List*, and Vol. II, *Alphabetical Index*, Public Health Service Publication No. 1693, December 1968.

tion records, universal population registers, partial registers, etc.) in any detail here. Collection systems for immigration and emigration are treated in chapter 20.

The administration of universal, continuous population registers differs somewhat from country to country, but basically it calls for registration at birth and the entering of specified subsequent events (marriage, change of residence, death, etc.) upon the individual or household record. A copy of this record, or an extract thereof, may be required to follow the person when he moves from one local jurisdiction to another. There are always local registers, and there may also be a central national register. The discussion of population registers in chapter 2 gave an indication of their general nature and cited a number of publications concerning them. Broadest in scope of these publications was the U.N. document, *Methodology and Evaluation of Continuous Population Registers*.[36] This document treats, *inter alia*, uses of such registers, general features, structure and content, accuracy

of information, and costs of the system. Two copiously annotated tables give, for 46 countries, the year of legislation and year of establishment of the country's original register system and the register system as now organized, the national authority in charge of the system, the population excluded from coverage, and types of statistical uses; and, for 20 countries, the degree of centralization of the register, the level and approximate number of administrative units maintaining registers, the types of persons covered by the system, whether basic documents are for persons, families, or households, the physical type of the basic document, and the sequence of the documents in the registers.

Descriptions of the central population register systems of a number of individual countries are included in papers presented at a symposium on automation of population registers held in 1967.[37] Other papers from those meetings treat some of the problems of maintaining population registers and the statistical applications of data from registers.

---

[36] United Nations Economic and Social Council, Statistical Commission, 12th Session. E/CN.3/293, February 7, 1962.

[37] *Proceedings of the International Symposium on Automation of Population Register Systems,* Jerusalem, Information Processing Association of Israel, September 25–28, 1967.

SUGGESTED READINGS

Bogue, Donald J. *Principles of Demography.* New York, John Wiley and Sons, 1969. Pp. 100–113.

Canada, Dominion Bureau of Statistics. *Canadian Labour Force Survey (Methodology).* Ottawa, 1966. Pp. 28–30.

Hansen, Morris H., Hurwitz, William N., and Bershad, Max A. "Measurement Errors in Censuses and Surveys." *Bulletin of the International Statistical Institute,* 38(2):359–374, 1961. (Proceedings of the 32d Session, Toyko, 1960.)

Muniruzzaman, A. N. M. *Demographic Survey in East Pakistan, 1961–62,* Part 1, "Survey Methodology and Operational Procedures." University of Dacca, 1965, Pp. 40–42.

Proceedings of the International Symposium on Automation of Population Register Systems. Jerusalem, Information Processing Association of Israel, September 25–28, 1967.

Spiegelman, Mortimer. *Introduction to Demography.* Cambridge, Mass., Harvard University Press, 1968. Pp. 8–42.

Truesdell, Leon E. *The Development of Punch Card Tabulation in the Bureau of the Census, 1890–1940.* Washington, Government Printing Office, 1965.

Turkey, Ministry of Health and Social Welfare. *Vital Statistics from the Turkish Demographic Survey.* Ankara, 1967. Pp. 3–4.

United Nations. *Handbook of Household Surveys.* Studies in Methods, Series F, No. 10, 1964.

————. *Handbook of Population Census Methods:* Vol. 1, *General Aspects of a Population Census;* Vol. II, *Economic Characteristics of the Population;* Vol. III, *Demographic and Social Characteristics of the Population.* Studies in Methods, Series F, No. 5, rev. 1, 1958–59.

————. *Handbook of Vital Statistics Methods.* Studies in Methods, Series F, No. 7, 1955.

————. *Methodology and Evaluation of Population Registers and Similar Systems.* Studies in Methods, Series F. No. 15. 1969.

————. *Principles and Recommendations for the 1970 Population Censuses.* Statistical Papers, Series M, No. 44. 1967.

United Nations, Economic and Social Council. *Methodology and Evaluation of Continuous Population Registers.* Statistical Commission, 12th Session, E/CN. 3/293, February 7, 1962.

U.S. Bureau of the Budget. *Household Survey Manual: 1969.*

U.S. Bureau of the Census. *Atlantida: A Case Study in Household Sample Surveys.* 1966.

————. *Current Population Reports,* Series P-23, No. 22, "Concepts and Methods Used in Manpower Statistics from the Current Population Survey." June 1967.

————. *Florencia: A Case Study in Data Processing of Population and Housing Censuses.* 1962.

————. *New Florencia: A Case Study for the 1970 Censuses of Population and Housing.* 1970.

————. *1960 Censuses of Population and Housing: Procedural History.* 1966.

CHAPTER 4

Population Size

The size of a population is usually the first demographic fact that a government tries to obtain.[1] The earliest censuses of a people are often a mere headcount. Particularly in premodern times, the emphasis in census taking was upon fiscal and military potentials. Hence, women, children, aliens, slaves, or aborigines were usually relatively undercounted or omitted altogether. For many decades, the United States census omitted "Indians not taxed." At the present time, the only resident group it does not attempt to cover is foreign diplomatic personnel living on the premises of an embassy or consulate. Similarly, the standards laid down by the Population Commission of the United Nations formerly called for the exclusion of foreign military forces stationed in the country under consideration.[2]

CONCEPTS OF TOTAL POPULATION

In general, however, modern censuses are designed to include the "total population" of an area. This concept is not so simple as may at first appear. There are two ideal types of total population counts, the *de facto* and the *de jure*. The former comprises all the people actually present in a given area at a given time. The latter is more ambiguous. It comprises all the people who "belong" to a given area at a given time by virtue of legal residence, usual residence, or some similar criterion. In practice, modern censuses call for one of these ideal types with specified modifications, and it is difficult to avoid some mixture of the two approaches.

International Standards and National Practices

The position of the United Nations has undergone some modification in recent years. It formerly favored the so-called "international conventional total" (see below). In its *Principles and Recommendations for the 1970 Population Censuses*, however, the "place where found at the time of census" and the "place of usual residence" are given equal billing on the list of recommended topics.[3]

The United Nations recommendation for the national total in the 1960 round of population censuses was: "For international purposes, it is desirable that, where consonant with the type of census taken, and in addition to any other total computed for national purposes, each country compile the total number of persons present in the country at the time of the census, **excluding** foreign military, naval, and diplomatic personnel and their families located in the country but **including** military, naval, and diplomatic personnel of the country and their families located abroad and merchant seamen resident in the country but at sea at the time of the census. This is neither a *de facto* nor a *de jure* population, but provides the information needed for international comparisons and for regional and world totals. For convenience, this figure, which may also be the desired total for national purposes, may be referred to as the 'international conventional total'.[4]

"The composition of the total population figure compiled by each country and used as the total for its census tabulations should be described in detail in the published report of the census. It is not sufficient to state that the figure is a *de facto* or *de jure* population count. The description should clearly show which of the following groups have been included and which have been excluded; where feasible, counts or estimates of the size of these groups should be given:

    A. Autochthonous inhabitants and nomadic tribes.
    B. Military, naval and diplomatic personnel and their families located abroad.
    C. Merchant seamen resident in the country but at sea at the time of the census.
    D. Other civilian national residents temporarily abroad at the time of the census.
    E. Foreign military, naval and diplomatic personnel and their families located in the country.
    F. Other civilian aliens temporarily in the country at the time of the census."[5]

National Practices. – The practice followed in 52 national censuses in the decade 1945–1954 is summarized in the U.N. *Demographic Yearbook, 1960*, table A, page 16. It should be emphasized that the classification merely shows which of the two ideal types, *de jure* or *de facto*, the actual practice approximated more closely.

Nevertheless, it can be said fairly confidently that the *de facto* type of census is considerably more common than the *de jure*, on a worldwide basis. In Africa, the latest census was

---

[1] The significance of national population size is discussed in the following: Katherine and A.F.K. Organski, *Population and World Power*, New York, Alfred A. Knopf, 1961; E.A.G. Robinson, ed., *Economic Consequences of the Size of Nations*, New York, St. Martin's Press, 1960; Jack Sawyer, "Dimensions of Nations: Size, Wealth, and Politics," *American Journal of Sociology* 73(2): 145–172, September 1967.

[2] United Nations, *Studies in Methods*, Series F, No. 5, Rev. 1, *Handbook of Population Census Methods*, Vol. III, *Demographic and Social Characteristics of the Population, 1959, p. 57.

[3] United Nations, *Statistical Papers*, Series M, No. 44, *Principles and Recommendations for the 1970 Population Census*, 1967, p. 40.

[4] On other occasions, the United Nations has referred to this as the "modified *de facto*" population.

[5] Ibid.; This recommendation appeared first in: *Statistical Papers*, Series M, No. 27, *Principles and Recommendations for National Population Censuses*, 1958, p. 10.

conducted on a *de jure* basis in only Morocco. Bechuanaland. and several Portuguese colonies. South American. Asian. and Oceanic censuses are almost all *de facto*. The situation is mixed in North and Central America: those using the *de jure* basis only comprising Canada. Cuba. Haiti. Nicaragua. and the United States. A mixed situation also exists in Europe. The *de jure* practice is followed in Austria. Belgium. Czechoslovakia. Denmark. Finland. France. both Germanys. Iceland. Luxemburg. the Netherlands. Norway. Sweden. Switzerland. and Yugoslavia.

For many countries. the distinction between *de jure* and *de facto* would not be very important for the national total. Usually. however. the choice would appreciably affect the count for many geographic subdivisions. The effect would also vary according to the census date. (See section "Time Reference" below.)

The United Nations regards the method used to allocate persons to a geographic subdivision of the country as being best determined by national needs. At first it seemed to favor the *de facto* principle here. but later it recognized the complications of that approach for family statistics. migration statistics. the computation of resident vital rates. etc.

The concept *de jure* seems to be rather ambiguous. Legal residence. usual residence. and still other criteria could be used to define the people who "belong" to a given area at a given time. In the United States. moreover. there is no unique definition of "legal residence." A person may have certain rights or duties (voting. public assistance. admission to a public institution. jury duty. certain taxes. etc.) in one State or community and other rights or duties in another State or community. A citizen who has recently moved may not have some of these rights in any State.[6] In certain Far Eastern societies. the people have sometimes been enumerated according to their familial or even ancestral home – at which they actually may have lived only in childhood or never at all. The relative difficulties of the *de facto* and *de jure* methods and their relative accuracy depend to some extent upon the particular country. The most appropriate method may represent a mixture of the *de jure* and *de facto* principles. People may be listed in the field in a particular manner; but. when the tabulations are made. some of them may be reassigned to other areas on the basis of recorded facts about where they spent the previous night. their usual residence. etc. There was some discussion of this problem in chapter 3.

Whatever coverage method is used, it must be clearly spelled out for the benefit of those who report in the census, those who process the data, and those who use the statistics. There should be rules for the treatment of any doubtful classes of the population and for ambiguous situations. These results should be understandable by enumerators and respondents; they should provide for a complete and unduplicated count; and they should produce useful statistics.

Evidence of a Person. – In addition to questions of whether certain classes of people are to be included in the national census count, and where a particular person should be counted, problems arise in actual practice as to whether there is sufficient evidence of a person. For example, even after repeated attempts to obtain the information by mail, telephone, or personal visit, there may remain a number of marginal cases

where the only evidence consists of (1) names copied by the enumerator from mailboxes or (2) information from a neighbor that one or more people live at a given address. Decisions must then be made as to whether there is enough information to warrant listing these persons on the schedule. In the United States, a criterion such as a minimum of two known characteristics (name, race, sex, etc.) has been used. In the 1960 census, there were a number of housing units reported as occupied for which no persons were listed. It was decided to impute households to these in the computer processing by replicating reported households in neighboring units. A total of 776,665 persons were thus imputed, or 0.4 percent of the official national count.

Total, Civilian, and Military Population. – In some ways, the civilian and military populations constitute separate economies. There are constraints on free movement from one to another. Moreover, they have different components of change and their geographic distributions are very different. The most feasible methods of enumerating them may also differ. All these considerations have led a few countries to publish separate statistics for their civilian and military populations.[7]

Concept Used in the United States: Usual Residence

The Constitution of the United States requires merely that "Representatives . . . shall be apportioned among the several States which may be included within this Union, according to their respective members . . ." The constitution does not provide a unique prescription for the type of enumeration to be made. Ordinarily, there would not be a great deal of difference at the national level, but at the peak of activity in World War II, a *de facto* count would have yielded about 9 million fewer persons than a count taken on a strict *de jure* basis. In the eighteenth century, there was considerably less difference between the *de jure* and *de facto* populations of an area than there is in the twentieth, because the limited transportation facilities and the way of life tended to keep people at home. Hence, the Founding Fathers probably were unaware of the ambiguity of their directive.

"The Census has never been taken on a *de facto* basis, however; and it has come to be considered that such a basis would be inconsistent with the spirit, if not the letter, of the Constitution. The basic principle followed in American censuses is that of 'usual residence.' This type of census more nearly approximates the *de jure* than the *de facto*."[8]

Definition. – The definition of "usual residence" itself is not a simple matter and has to be spelled out in some detail for the benefit of enumerators and respondents. The following quotation indicates why there is concern about the general method and the specific rules. By analogy. it has applicability to census counts in other countries as well.

"Where people are counted in the United States Census of Population. that is. the area in whose statistics they are included. is important from many standpoints. Of greatest consequence is the effect on the apportionment of representatives to the Congress and to State legislatures. There are also other legal uses of population counts that are affected by the way in which persons enumerated are assigned to a geographic area. These uses include. for example. the distribution of certain State funds to municipalities on the basis of population and the requirements in State laws that a munici-

---

[6] Henry S. Shryock. Jr.. "The Concepts of *De facto* and *De jure* Population: the Experience in Censuses of the United States." *Proceedings of the World Population Conference,* 1954. Vol. IV. United Nations, 1955., E/CONF. 13/416, pp. 877-878.

[7] For example: New Zealand, Census and Statistics Department. *Population Census, 1945,* Vol. I, *Increase and Location of Population.* Wellington, 1947, p xiii. (Number by race and area in which situated.)

[8] Shryock, "The Concepts of De facto and De jure Population." p. 877.

pality attain a specified population before it is granted certain powers. Moreover, the description of an area's population and the levels of per caput rates that are computed for it will be influenced by the enumeration procedure followed; and these statistics, in turn, affect the use of the data in administration and in scientific research. . . . the size of some types of areas determines whether certain kinds of statistics will be published, not only in census reports but also in other statistical reports." [9]

Determination of Usual Residence in the Decennial Census. — The 1960 *Enumerator's Reference Manual* had 20 numbered sections on "How to Determine 'Usual Residence'." [10] The general principle is, "usual place of residence is, ordinarily, the place a person regards as his home. As a rule, it will be the place where he usually sleeps." The first two paragraphs defined the census date (12:01 a.m. on April 1), specified that babies born after that time should be omitted, whereas persons dying after that time should be included, and stated that persons moving in after April 1 for permanent residence should be enumerated unless they had already been enumerated in the enumeration district (ED) whence they came. Members of the household temporarily absent on vacation, visiting, on business, or in a general hospital were to have been counted, as were boarders or lodgers who usually slept in the housing unit. The instructions also covered a number of special cases, some of which are discussed in the sections below.

Members of the armed forces within the United States. — Persons in the Army, Navy, Air Force, Marine Corps, or Coast Guard of the United States were supposed to have been counted as residents of the place where they were stationed, not at the place from which they were inducted or at their parental home. Those members who lived off post were to be counted at their nearby homes (with families, if any), whereas those who lived in barracks or similar quarters were considered as residents of those quarters.

College students. — Beginning with the census of 1950, a student attending college has been considered a resident of the ED in which he lives while attending college. [11] That was also apparently the rule down to 1850; but, in most of the intervening censuses, he was counted at his parental home. Students away from home attending schools below the college level have been counted at their parental homes, however.

The present method is deemed to be more consistent with the usual residence principle. Students who go away to college are away from home for a longer period, on the average, than military draftees of similar age; one-fifth of them are wholly self-supporting and another two-fifths partially self-supporting; and one-quarter of them are married. Moreover, most of those who go away to college will never again live regularly with their parents. Leaving the parental home is a gradual process for many young people, and going away to college is often the most significant event in this process. According to the Current Population Survey of October 1958, the address at which college students would be enumerated would be unchanged for at least 55 percent of them if there were a reversion to the earlier rule. [12] The rule chosen has an important effect upon the

population size of college towns but relatively little effect upon other areas.

Institutional inmates. — Inmates in types of institutions where inmates usually remain for long periods of time (regardless of the length of stay of the person considered) were enumerated as residents of the institution. These include prisons, reformatories, jails, and detention homes; homes for retired members of the armed forces, orphans, or the aged; asylums or hospitals for the insane, incurables, and tubercular; nursing homes and convalescent homes and the like — but not general hospitals or other institutions for medical care where patients usually stay for only a short period. Since, however, in vital statistics, deaths of even long-resident inmates are allocated back to the area of residence prior to institutionalization, a bias is introduced into the death rates for affected areas.

Persons with more than one residence. — "Perhaps the most difficult decisions to be made concern persons with more than one residence. These include those who change their homes with the seasons. We think first of the idle rich who move from Newport to Palm Beach and of the hotel employees and other service workers who move because of them. But, in our increasingly affluent and mobile society with its long vacations and early retirement, the occupancy of more than one home during the year is increasingly a mass phenomenon. Of course, there have also long been classes of workers who changed their residences seasonally with the jobs — lumbermen, fishermen, agricultural laborers, canners, etc. In the cases that involve an annual cycle, the ordinary rule is to choose the residence where the person lives the greatest part of the year — although there is always the temptation to play safe and count him where you find him. Another class of dual residence consists of persons who work and live away from their homes and families, perhaps returning on weekends. In their case the need for meaningful family statistics clashes with the need to include persons in the area where they are living most of the time. Obviously, we must take into account the length of stay, not only past but prospective.

Persons with no usual residence. — Persons with no usual residence anywhere (migratory agricultural workers, vagrants, some traveling salesmen, etc.) have been counted where they were found according to a provision that goes back to the Act of 1790. To obtain a complete and unduplicated count of such persons, canvassing procedures like "T-night" ("T" for transient) and "M-night" ("M" for mission) were introduced some decades ago.

Americans abroad. — This and the following category are of especial interest from the standpoint of the United Nations' "international conventional total." It may be recalled that that recommendation called for the *inclusion* of the country's own military and diplomatic personnel stationed abroad or at sea. Until the 1970 census the United States, in fact, excluded them from the count that is used for apportionment purposes and from the population for which the basic statistics are presented. The United States does, however, make an effort to count such persons and report them separately. Moreover, in recent times, it has increasingly attempted to enumerate other types of Americans resident abroad at the time of the census and to report their number and characteristics in special tables of the census reports. (See subsection below on "Population Abroad".) Americans abroad only temporarily (as tourists, visitors, persons on short business trips, etc.) were supposed to be counted at their usual place of residence in the United States, whereas those away for longer periods (employed

[9] Henry S. Shryock, Jr., "The Concept of 'Usual Residence' in the Census of Population," *Proceedings of the Social Statistics Section, 1960*, Washington, American Statistical Association, Aug. 23–26, 1960, p. 98.

[10] U.S. Bureau of the Census, *Enumerator's Reference Manual: 1960 Census of Population and Housing*, F–210, 1959, pp. 5–7. The 1970 rules are similar.

[11] In American usage, a "college" is ordinarily an institution of higher learning, i.e., of the university level.

[12] Shryock, "The Concept of 'Usual Residence' in the Census of Population," p. 100.

abroad, enrolled in a foreign university, living in retirement, etc.) were excluded from the basic count.

In the enumeration of the population overseas in 1970, however, it has been decided that members of the armed forces, civilian Federal employees, and their dependents will be allocated back to their State of recorded residence. It is expected that most of the tabulations will relate to the conventional population concept, however.

Foreign citizens temporarily in the United States. — The United States census adheres partially to the principle of the "international conventional total" by excluding foreign military and diplomatic personnel and their families who are stationed here if they are living in embassies or similar quarters. In fact, such persons are not even listed. On the other hand, it fails to list all other citizens of foreign countries temporarily in this country. The American rules for inclusion or exclusion parallel those given in the preceding subsection, e.g., foreigners working or studying in the United States are counted. Failure to provide statistics on foreigners temporarily present and not on official assignments represents one of the points at which it is impossible to construct the "international conventional total" for the United States by the combination of published statistics.

Doubtful cases. — It may be apparent by now that these rules require a certain amount of judgment in some cases because such words as "temporary" and "usual" are not precisely defined. Attempts to use a time criterion, such as at least 60 days, have not been very satisfactory. For one thing, both past and prospective length of stay must be considered. It may also be apparent that the nature and purpose of the stay are just as important considerations as the duration.

It should be noted that most of the specific decisions concerning where a person should be enumerated are not made in the central office, as is the case in some other countries, such as the U.S.S.R., but are made in the field. In the United States, the Bureau of the Census formulates the general principles but leaves their application to the enumerator or the respondent. Except in the case of groups canvassed in certain special operations, the central office does not have the facts that would be needed to change the area to which the person is allocated.

Effect of Self-Enumeration. — Since 1960 the procedures for conducting the census have depended more on self-enumeration and less on the canvasser method. In these circumstances, the set of detailed rules provided to the enumerators has less impact on the decisions concerning whether and where a person should be counted. It is patently not feasible to expect the householder to read through such long instructions.

Special Censuses and the Current Population Survey. — For the most part, "usual residence" is defined the same way in special censuses and current household surveys as in the decennial census. A noteworthy exception in the residence rules is the enumeration of college students in the CPS. They are still counted at their parental homes, partly because counting them in the college communities during the academic year and at home (or where they are employed) during the vacation period would lead to seasonal variations in enumeration procedures and in the resulting statistics.

Population Abroad. — At least as early as 1910, the United States compiled a simple count of its citizens in "military and naval services, etc., abroad." [13] These counts were included in

a table that gave the total population of the United States and its outlying areas. The counts were based on reports sent in by local commanders, embassies, consulates, etc., and were not based on an actual enumeration. That this procedure led to an underreporting is illustrated by the fact that the published figure for the overseas military population in the 1940 census was 118,933 persons, whereas, on the basis of independent records of the armed forces, it was estimated later that this group numbered 151,000.

In 1950 and again in 1960, the reporting procedures were improved somewhat and efforts were made to include additional classes of Americans overseas. In 1950 some basic characteristics of this population were also recorded and published; in 1960, social and economic characteristics were added. The classes added in 1950 were: (1) civilian Federal employees stationed abroad; (2) dependents of Federal employees, military or civilian; (3) crews of merchant vessels. In 1960, all other Americans usually resident abroad (businessmen, students, retired persons, etc.) were given an opportunity to report voluntarily through United States consulates. It is felt that this group is the least completely reported, however, especially those Americans living in Canada and Mexico, which have land borders with the United States. [14] The available statistics on the American population living abroad are summarized in table 4-1. The coverage of Americans overseas planned for the 1970 census is similar to that for 1960 except that, as previously described, members of the armed forces, civilian Federal employees, and their dependents will be allocated back to the several States.

Table 4-1. — **United States Population Enumerated Abroad: 1940 to 1960**

Population enumerated abroad	1960	1950[1]	1940
Total[2]............................	1,374,421	481,545	118,933
Federal employees.....................	645,499	328,505	(NA)
Armed forces......................	610,174	301,595	118,933
Civilians........................	35,325	26,910	(NA)
Dependents of Federal employees.....	505,752	107,350	(NA)
Crews of merchant vessels...........	32,464	45,690	(NA)
Other citizens.....................	[3]190,706	(NA)	(NA)

NA Not available.
[1] Based on 20-percent sample of reports received.
[2] Excludes U.S. citizens temporarily abroad on private business, travel, etc., for all years. Such persons were enumerated at their usual place of residence in the United States as absent members of their own households. Additional groups omitted in 1950 and 1940 as noted.
[3] Represents U.S. citizens abroad for extended periods. Since this population was enumerated on a voluntary basis, its coverage is probably less complete than that of other categories of Americans abroad.

Source: *U.S. Census of Population, 1960* Vol. I, *Characteristics of the Population,* Part 1, *United States Summary,* table 1 as corrected.

The statistical coverage of American citizens abroad in the current population estimates has a somewhat different history. During World War II, it became important to know the population of the United States both including and excluding the armed forces overseas. There was also an interest in the civilian population resident in the United States. These needs led to the development of a basic typology in the current estimates that has continued down to the present time. This typology is discussed in the next section.

[13] U.S. Bureau of the Census, *Fifteenth Census of the United States: 1930,* Vol. I, *Number and Distribution of Inhabitants,* 1931, table I, p. 5.

[14] *U.S. Census of Population: 1960, Selected Area Reports, Americans Overseas,* PC(3)-1C, 1964.

Civilian and Military Population. – The national sample surveys conducted by the U.S. Bureau of the Census have mostly been limited to the civilian noninstitutional population, but the annual March supplement to the Current Population Survey covers the institutional population and members of the armed forces living off post or with their families on post. That survey's focus on the labor force leads it to exclude people who are outside the market economy. The wartime population estimates for counties based on registration for rationing were also restricted to civilians.

Monthly independent estimates are published currently and have been extended back to January 1944, with semiannual estimates to 1940, for: (1) the total population including armed forces overseas; (2) the total resident population (resident within the 50–State, formerly 48–State, area); and (3) the resident civilian population.[15]

Household Population. – There have been demands for the size and characteristics of the "normal" population of an area which would exclude not only members of the armed forces stationed there and living in barracks or aboard ship but also persons living in institutions, college dormitories, and other group quarters. Although the concept of the "normal" population of an area may be a bit chimerical, it is true that the presence of a relatively large "nonnormal" population will distort the demographic composition and vital rates so as to obscure comparisons with other areas not containing such a population. In table 31 of the 1960 State population reports, therefore, the household population was presented by age, race, and sex for urban places and counties with a population of 1,000 or more living in group quarters. Consider, for example, the case of Leavenworth County, Kansas, which contains a Federal prison and a military post.

	Total population	Household population
Population size	48,524	40,663
Sex ratio	126.2	96.4
Median age	32.0	30.0

TIME REFERENCE

As noted by the United Nations, one of the essential features of a population census is simultaneity. A national census should not be taken by a crew of enumerators that moves from one district to another as it completes its work; nor, in general, should the enumeration begin on different dates in different parts of the country. Occasionally, however, exceptional starting dates may be justified by such considerations as gross variations in climate or the annual dispersal of nomads to isolated grazing grounds. The census of Alaska was conducted as of October 1, 1929 and 1939, when that of conterminous United States was taken as of the following April 1. The object was to avoid canvassing in Alaska during the spring thaw.

The *de facto* population of an area is obviously subject to diurnal and seasonal fluctuations. These are relatively in-significant for most national totals, but particular areas could be greatly affected. There is some worker commuting over land borders, for example, between Windsor, Canada, and Detroit, U.S.A. If a census has a specific official hour, it is usually midnight, a time when most persons are at home. There are seasonal variations in tourism and in migratory labor.

In summary, the actual population present is particularly affected by the diurnal factor in downtown areas and in other concentrations of shopping and employment. Resort areas and certain types of agricultural areas are particularly affected by the seasonal factor.

The enumeration in the earliest historic censuses and in contemporary censuses of relatively primitive countries typically extended over many months. If a day or month was cited, it meant nothing more than the time when the field work began. At a more advanced stage of census taking, there are specifications like "as of 2400 hours on 31 December." In many cases a serious attempt is made to complete the enumeration in a day's time. These censuses are characteristically on a *de facto* basis. Such rapid censuses are by no means limited to the more industralized societies or the householder (self-enumeration) method of enumeration. The most dramatic numerations are those in which normal business activities cease and the populace must stay at home until the end of the census day or until cannon are fired to announce the completion of the canvass. The "one-day census" usually turns out to be an ideal or a figure of speech, however. Thus, in the 1964 census of Guatemala, it was attempted for the urban areas only. In the 1956 census of Taiwan, there had been a listing of household members by the enumerator 5 to 7 days earlier; and, on the census day, he simply brought the facts up to date. Even in the 1965 census of Turkey, some clean-up work had to be conducted after the "blitz" canvass.

In other countries, with a lower ratio of enumerators to population, the canvass may extend over several weeks or months. The disadvantages of such protracted enumerations are that omissions and duplications are more difficult to avoid and that it becomes increasingly difficult to relate the facts to the official census date. The rather incomplete reports on national practices have been summarized by the United Nations; even these tend to give the planned rather than the actual duration, which was frequently considerably longer.[16]

COMPLETENESS OF COVERAGE

As has been brought out already, countries exclude from their censuses certain relatively small classes of population on the basis of the type of census being taken – *de jure, de facto*, or some modification of one of these. Other exclusions are on ground of feasibility – cost, danger to census personnel, or considerations of national security. Finally, some persons will inadvertently be omitted from the population as defined and others will be counted twice. Omissions by design then will be discussed separately from the net underenumeration (or overenumeration) that always occurs to some extent in counting a sizeable population.

[15] U.S. Bureau of the Census, "Estimated Population of the United States, for Selected Dates: July 1, 1940, to July 1, 1945," *Population – Special Reports*, Series P–46, No. 1, January 16, 1946; and "Estimates of the Population of the United States and Components of Change: 1940 to 1969," *Current Population Reports*, Series P–25, No. 418, March 14, 1969.

[16] United Nations, *Studies in Methods*, Series F, No. 5, Rev. 1, *Handbook of Population Census Methods*, Vol. 1, "General Aspects of a Population Census," 1958, pp. 92–93.

Deliberate Exclusion of Territory or Group

It is not unusual for various population subgroups to be excluded from a census for one reason or another. In some countries, for example, either the indigenous or nonindigenous population may be omitted from the census count, or the two may be enumerated at different times. Some examples are as follows:

Country	Census date	Exclusion
Madagascar	1956	Africans
Malawi	1961	Africans not employed in European money economy
Mauritania	1956	Africans
Zambia	1963	Non-Africans (enumerated in 1961)
Southern Rhodesia	1962	Non-Africans (enumerated in 1961)
New Guinea (Australian Trust Territory)	1961	Indigenous population
Papua	1961	Indigenous population
New Hebrides	1957	Indigenous population

In addition to tribal jungle areas, censuses may omit parts of the country that are under the control of alien enemies or of insurgents. These are essentially temporary arrangements and rarely occur in practice because the usual action is to postpone the entire census.

Attempts have also been made to estimate the population of the excluded territory or groups, and the more credible estimates are cited in the U.N. *Demographic Yearbooks*. The sources vary from sample surveys, projections from past counts, reports of tribal or village chiefs, and aerial photographs, down to guesses by officials, missionaries, or explorers.

Unintentional Exclusions — Underenumeration

In many countries the public probably has an exaggerated notion of the accuracy of the census count. As Deming has pointed out, it also has an exaggerated notion of the accuracy needed.[17] This is not to say that, for most countries, there is not now a need to improve the accuracy of the census count and of the postcensal estimates of total population. Even in the countries of Northwestern Europe with a long history of census taking and registration, occasional measures of accuracy are desirable even though there may be considerable confidence that coverage errors are minimal.

The more sophisticated users of census data have long been aware that even the size of the population of a given area is not exact to the last person. The reader who has followed this discussion of the definition of population size will appreciate some of the uncertainties and the opportunities for omission or duplication. Some familiarity with field surveys will confirm the fact that it is not possible to make the count for a fair-sized area with absolute accuracy.

According to Hauser and Duncan, "The degree of accuracy required in data is a relative thing and a function of the use to which the data are put."[18] Truesdell, Mauldin, and Depoid have also discussed this issue.[19]

Although there is some danger of duplicate counting and

there have been historical instances of deliberate padding, the major danger is that of underenumeration. Most measures of the accuracy of census totals have been confined to net rather than to gross error. When we consider that a person may be counted but listed at, or allocated to, the incorrect geographic subdivision, the magnitude of gross errors assumes more importance.

Attempts to measure the accuracy of census counts are a relatively modern development. For a long time, indeed, most census officials were smug and defensive about the official figures on population size. As statisticians acquired experience through scientific sample surveys of households, they realized that many of the types of errors encountered there would also occur in a "complete" census.

Sources of Error. — The sources or types of errors in figures on population size can be classified from several different standpoints. First, we may distinguish population size as given by universal population registers, censuses, sample surveys, and "independent" estimates. Next, within each of these major sources, we can examine the nature and causes of error.

In population registers. — Here there are, to begin with, errors in the register from failures to report births, deaths, and changes of residence or failure to record these events properly even when they are reported. Then, when counts are made for a specific date, there may be processing errors arising from the personnel or the machines. The accuracy of the register may be checked against a census count, which is itself an imperfect standard. Errors are also reflected in "books of the missing" that are maintained in Sweden and other countries — lists of registered persons who can no longer be located. Having many persons remain in the register past the age of 100 would indicate failures to remove many decedents and emigrants. Errors of omission in the civil register in South Korea are apparent when many children try to enter school for the first time, and there is no record of them in the register.

In population censuses. — Here the errors may be attributable to the design of the census (lack of simultaneity, inadequate instructions, missing or erroneous maps), the field operations (enumerators, field office editors, etc.) the respondents, loss of schedules in shipment, central office operations, tabulation equipment, right down to the typing and printing of the final tables.[20] Errors are sometimes deliberate, but more often they arise from carelessness, lack of understanding of the procedures, or lack of knowledge of the facts. It has also been found helpful in measuring faulty omissions or faulty inclusions to distinguish omission (or duplication, etc.) of entire households from omission (or erroneous inclusion) of persons within properly included households. The section on "Methods of Evaluating Census Coverage" and the references cited in that section, go into some of these types of error in more detail. See also the more general discussion of sources of error on pages 185 to 197 of Depoid's article and in Zarkovich's book.[21]

In sample surveys. — Population size as measured in national

[17] W. Edwards Deming, "On Errors in Surveys," *American Sociological Review*, 9(4):369, August 1944.

[18] Philip M. Hauser and Otis Dudley Duncan, "The Data and Methods," in *The Study of Population*, Hauser and Duncan, eds., Chicago, The University of Chicago Press, 1959, p. 65.

[19] Leon E. Truesdell, "The Problem of Quality in Census Data," *Estadistica*, 9(31):163–171, June 1951; W. Parker Mauldin, "Comments on Possible Na-

tional and International Standards for the Determination and Indication of Degree of Accuracy of Demographic Statistics" in *Proceedings of the World Population Conference, 1954*, Vol. IV, United Nations, 1955 (E/CONF. 13/416):174–175; Pierre Depoid, "Rapport sur le degré de précision des statistiques démographiques" (Report on the degree of accuracy of demographic statistics), *Bulletin de l'Institut international de statistique*, 35(3):203–206, Rio de Janeiro, 1957 (29th session of the I.S.I., Petropolis, Brazil, 1955).

[20] Giorgio Mortara is cited as reporting that lost schedules contained an estimated 16,629 and 31,597 names in the 1940 and 1950 censuses, respectively, of Brazil. Depoid, op. cit., p. 140.

[21] S.S. Zarkovich, *Quality of Statistical Data*, Rome, Food and Agriculture Organization of the United Nations, 1966.

sample surveys or samples of the population of particular areas is affected by all the kinds of errors to which censuses are subject and, in addition, to sampling error and sampling bias. The nature and measurement of sampling errors are the subject of a very extensive literature. A brief list is given below.[22]

Where there are relatively accurate postcensal independent estimates of population, the sample survey results may be inflated to these population controls, rather than being simply inflated by the sampling ratio or ratios. In this case, the total population will not be subject to sampling error, although it will be affected by errors in the last census total and in the postcensal components of change.[23] This step leads us to the next subsection.

In independent population estimates. — The types and methods of construction of population estimates are described in detail in later chapters of this book. What needs to be noted here, however, is that errors in all the base data from which population estimates are derived contribute to errors in the resulting estimates.

Methods of Evaluating Census Coverage. — Let us concentrate now on the accuracy of published figures on population size in complete censuses, since these are the most important and have been the object of most of the research. Most measures are of national totals; relatively few are available for geographic subdivisions. Although the present discussion concentrates mainly on the accuracy of the total count and not on that of population subgroups or characteristics, some of the most effective methods of measurement require the use of statistics by age and sex. Since the actual calculations often require knowledge of techniques that have not yet been presented in this book, most details and illustrative computations are given in later chapters.

A number of publications classify the methods that have been used.[24] The present classification focuses more narrowly on the accuracy of census figures on population size. We will discuss:

(1) Reenumeration
(2) Comparison of successive censuses
(3) Internal consistency within a single census
(4) Check against independent aggregates
    (a) Universal population register

(b) Partial register
(5) Matching against individual records
    (a) Successive censuses
    (b) Periodic sample survey
    (c) Independent records
(6) Postenumeration sample survey
(7) Combinations of these

Reenumeration. — A complete repetition of the canvass does yield two counts that can be compared, but a national reenumeration would never be undertaken unless the government were convinced that the first count was unacceptable. Hence, it would not ordinarily be made public. The 1937 census of the U.S.S.R. was rejected, and a new census was taken in 1939. Also, the 1962 census of Nigeria was declared invalid and repeated in 1963. Political as well as statistical factors were involved in both cases, however.

Some partial reenumerations have been conducted in American censuses, usually before final certification but occasionally after the final census figures had been published. In such comparisons, allowance should be made for the lapse of time between the two counts.

Comparison of successive censuses. — The relative completeness of two censuses is frequently estimated by starting with the first count, computing the expected population at the second date, and comparing it with the second count. The expected population is computed by some variety of the inflow-outflow method (or the "balancing equation," as it is often called). Here it is necessary to measure or estimate the intercensal population change; and the accuracy of the measure of relative completeness obviously depends upon the accuracy of the measure of change.

Some ways of computing the expected population are the following:

(1) Adding births, subtracting deaths, adding immigrants, and subtracting emigrants.

(2) Estimating the survivors by applying life-table survival rates to the population at the first census, distributed by age and sex, and taking account of registered or estimated births and net immigration.

(3) An iterative method that requires age-sex distributions for two or more censuses and previous estimates of net enumeration error for some age groups (say of children from birth statistics).[25] In practice, survivors of recent births have also been employed in this method. There is a basic hypothesis that the pattern of net undercounts or overcounts by age and sex is similar from one census to the next.

(4) A method of reconstructing births from age statistics in a series of censuses, establishing a "true" series of births, and then reconstructing the census populations.[26]

(5) An evaluation of the intercensal growth rate for reasonableness.

Methods (1) and (2) overlap somewhat. Method (1) requires reasonably complete vital statistics although adjustments may be made for under-registration. Method (2) can be applied in the absence of adequate vital statistics by using model life tables or suitable age-specific fertility rates. For either method, there must be statistics on immigration and emigration unless it is known that one or the other or both are negligible. In

[22] United Nations, *A Short Manual on Sampling*, Vol. 1, "Elements of Sample Survey Theory," Studies in Methods, Series F, No. 9, 1960 (especially pp. 74–83); United Nations, *Handbook of Household Surveys*, Studies in Methods, Series F, No. 10, 1964, pp. 32–34, and 129–132; U.S. Bureau of the Census, Sampling Staff, *A Chapter in Population Sampling*, March 1947, pp. 42–53; Morris H. Hansen, William N. Hurwitz, and William G. Madow, *Sample Survey Methods and Theory*, Vol. 1, Methods and Applications, New York, John Wiley & Sons, 1953; S. S. Zarkovich, op cit. (also contains detailed discussion of response errors and other sources of error in survey statistics).

[23] Estimates of the underenumeration of the inflated population from the U.S. Census Bureau's Current Population Survey prior to adjustment to the control totals are presented in an article by Siegel. The underenumeration is partitioned into the components: (1) CPS relative to current estimate based on the last census and (2) current estimate relative to "true" population. See Jacob S. Siegel, "Completeness of Coverage of the Nonwhite Population in the 1960 Census and Current Estimates, and Some Implications," pp. 13–54, in *Social Statistics and the City*, David M. Heer, ed., Cambridge, Mass., Joint Center for Urban Studies of the Massachusetts Institute of Technology and Harvard University, 1968.

[24] For example, Mauldin, op. cit., pp. 175–178; Mortimer Spiegelman, *Introduction to Demography*, Cambridge, Mass., Harvard University Press, rev. ed., 1968, pp. 46–49; United Nations, *Handbook of Population Census Methods*, Vol. 1, *General Aspects of a Population Census*, Series F, No. 5, rev. 1, 1958, pp. 104–111; United Nations, *Manuals on Methods of Estimating Population*, Manual II, *Methods of Appraisal of Quality of Basic Data for Population Estimates*, Series A, Population Studies, No. 23, 1955, pp. 4–17; United Nations, Economic and Social Council, *Methods of Evaluating the Reliability of Population and Housing Census Data*, Seminar on the Organization and Conduct of Population and Housing Censuses for Latin America, Santiago, Chile, May 1968, ST/ECLA/CONF.32/L.11, March 26, 1968, pp. 10–24.

[25] Ansley J. Coale, "The Population of the United States in 1950 Classified by Age, Sex, and Color — a Revision of Census Figures," *Journal of the American Statistical Association*, 50(269):16–54, March 1955.

[26] Ansley J. Coale and Melvin Zelnik, *New Estimates of Fertility and Population in the United States*, Princeton, N.J., Princeton University Press, 1963.

method (2) the deaths of the immigrants must also be estimated.

The difference between the enumerated and the expected population at the second date may be called the "error of closure."[27]

An illustration of (1) may be given for Australia:

Census population, June 30, 1954..............	8,986,530
Births, June 30, 1954, to June 30, 1961......	1,544,240
Deaths, June 30, 1954, to June 30, 1961.....	600,551
Arrivals, June 30, 1954, to June 30, 1961.....	1,766,858
Departures, June 30, 1954, to June 30, 1961.	1,182,104
Expected population, June 30, 1961...........	10,514,973
Census population, June 30, 1961..............	10,508,186
Error of closure, amount........................	−6,787
Percent (of 1961 census population)...	−0.06

Since the error of closure relates to the comparison of two censuses, use of the later census population as the base is purely arbitrary. The first census or the average of the two might have been used.

Methods (2) to (4) all involve statistics by age. Hence, they are considered further in chapter 8.

Regarding (5), the United Nations points out that, in the absence of appreciable net immigration (or emigration), average annual rates of growth have seldom been recorded outside the range of 0 to 3 percent. Hence, a rate outside this range suggests that either one census was an underenumeration or the other was an overenumeration. The average annual rate for Honduras between 1926 and 1930 appeared to be 5.8, suggesting much less complete coverage in 1926 than in 1930.[28] A later survey showed that both census counts were probably in error. This method by itself does not yield a measure of relative coverage, but it may lead to a more intensive investigation, perhaps by the use of method (1) or (2). Furthermore, even in the absence of statistics on the components of change, an expected population can be computed on the basis of various hypotheses, such as the continuation of the last intercensal growth rate or some specified modification thereof. (The trend of the growth rate can be gauged better if there has been a series of preceding censuses.) This procedure, in turn, requires careful study of the available evidence on recent growth tendencies, and the resulting expected population will constitute only a very rough check on the reasonableness of the latest census count.

When a comparison of a recent census with the population expected on the basis of the previous census indicates an apparent overenumeration in the former, the alternative conclusion of an underenumeration in the previous census is usually more plausible. Thus, "a preliminary partial estimate of the net relative coverage error (overenumeration) ranging between 0.6 percent and 2.5 percent of the total population" for the 1960 census of Ghana was officially interpreted as confirming gross underenumeration in the 1948 census.[29]

Internal consistency within a single census. — Various tests are used to examine the reported age-sex structure. In some of these tests, it is necessary to assume that fertility and mortality

rates have long been constant. Relative net undercounts in particular age-sex groups can then be inferred. Characterizations of the accuracy of the size of the total population is mostly qualitative, although estimated gross deficits in particular groups (young children, males of military age, etc.) may be also imputed to some extent to the total population. This approach may involve applications of stable, or quasi-stable, population theory as described in chapter 24. Further suggestions are given in United Nations Manual II.[30]

Check against independent aggregates. — Although, as has been stated, censuses are often used for periodic updating of universal population registers, there have been occasions when population size as given in the register has been used to measure the accuracy of the census count. For example, the 1960 census of Aruba in the Netherlands Antilles was stated to be about 4 percent too low in comparison with the population register.[31]

The United Nations makes the following suggestions regarding sources and procedures of this type: "Such sources as police records, church records, rationing records, tax lists, or counts of persons eligible for military service can provide some estimate of the population. Other censuses, such as censuses of housing and agriculture, also supply relevant information on households and farm population. Sometimes the information mentioned refers to the total population of the area considered; in other instances the data relate only to parts or segments of the population and can be converted into estimates of the total population. The estimates so obtained are compared with the census counts for the whole country and each region. If there is considerable disagreement, it may prove necessary to check further in an effort to appraise the two figures. It should be added that, in certain instances, data on a particular characteristic, such as school attendance, are collected by governmental agencies on a current basis, which makes it easier to check the accuracy of the particular census figure (in this case the census total of persons attending school is checked against the number provided by school statistics)."[32]

Comparisons with the aggregates from **partial** registers by their very nature yield estimates of net underenumeration for only the pertinent population subgroups. Some such registers cover a substantial majority of the population, however; or a net underenumeration estimated for the subgroup concerned may be assumed to apply to the total population under some circumstances. Moreover, estimates for two or more subgroups may be used to build up an estimate of net underenumeration for the population of all classes combined. Examples of such comparisons are checks of the number of children against the estimated number of survivors of recent births, of males of military age against the number registering for compulsory military service, of the very old (generations that will be extinct or nearly extinct shortly after the census) as enumerated against the numbers of subsequent decedents of the same cohorts,[33] and of persons of working age against registered workers.

Matching against individual records. — One advantage of this and the following method is that they yield measures of gross differences (erroneous omissions and inclusions) rather than of net differences only. Moreover, the matching of names from the census

[27] U.S. Bureau of the Census, "Estimates of the Population of the United States and Components of Change: 1940 to 1968," *Current Population Reports,* Series P–25, No. 398, July 31, 1968, table 1.

[28] United Nations, *Handbook of Population Census Methods,* Vol. 1, p. 110.

[29] E. N. Omaboe and K. T. de Graft-Johnson, "Possibilities for Evaluation of Census or Survey Data in Developing Countries," *Bulletin of the International Statistical Institute,* 42(1):91–101. (Proceedings of the 36th Session, Sydney, 1967), Sydney, 1969.

[30] Pp. 5–9.

[31] United Nations, *Demographic Yearbook, 1965,* p. 110.

[32] United Nations, *Handbook of Population Census Methods,* Vol. 1, p. 110.

[33] For this last, see: Paul Vincent, "La mortalité des vieillards," (Mortality of the aged), *Population* (Paris), 6(2):189–190, April–June 1951.

against those in another file can be used to generate a list of names that is more complete than either source taken alone.

A sample of names from the preceding census may be checked against the names of the census being investigated. This procedure was part of the coverage check for the 1960 Census of Population in the United States.[34] One of the great difficulties in this procedure is the tracing of movers over a period of years. Furthermore, since names may be changed or inconsistently reported and since more than one person may have the same name, there are often uncertainties as to whether a pair of names has been matched. Rules have to be set up to define a match and a probable match. An unmatched person is not necessarily a missed person. The 1960 census report just referred to contains several different estimates of net underenumeration based on alternative assumptions.

Sometimes when the census listings are checked against the universal population register, complete files are used from both sources rather than a sample from one. In this case, the procedure may be used to correct both files for omissions and duplications. Something of this sort has been done in Belgium, the Netherlands, and Norway, and possibly several other countries as well. This procedure gives as a by-product the number of persons not enumerated.

Other types of records that were matched as part of the 1960 census of the United States are listed below in the subsection dealing specifically with that country.

Postenumeration sample surveys. — In general, these checks involve (1) reenumerating designated sample segments or enumeration districts and (2) reenumerating the persons in a sample of previously enumerated households. Table 4–2 summarizes some of the available results from national post-

Table 4–2.— **Net Coverage Error as Measured by Post–Enumeration Surveys: Selected Countries**

Country and census year	Net error (percent)	Area detail	Remarks
Ceylon, 1953.........	-0.7	Urban and rural	Within households only
India, 1951..........	-1.1	Regions; urban and rural	Within households only; 0.1 percent of households missed
Israel, 1961.........	-0.1		
Japan, 1955..........	-0.4	Urban and rural	
Japan, 1950..........	-0.1		
South Korea, 1960....	-1.2	Urban and rural	
United States, 1960..	-1.7		
United States, 1950..	-1.4	Regions; size of place	
West Germany, 1956...	+0.4 to +0.5		Population reported in housing census
Yugoslavia, 1953.....	+0.01 ·	Urban and rural	

Source: From official national statistics, except India and Yugoslavia which are from Pierre Depoid, "Rapport sur le degré de précision des statistiques démographiques", *Bulletin de l' Institut international de statistique*, (29th session, Petropolis, Brazil, 1955), Rio de Janeiro, 1957, Vol. 35, Part 3, pp. 123–126.

[34] U.S. Bureau of the Census, *Evaluation and Research Program of the U.S. Censuses of Population and Housing, 1960: Record Check Studies of Population Coverage*, Series ER 60, No. 2, 1964, p. 2.

enumeration surveys, conducted around 1950 or 1960. Since the estimates of net underenumeration themselves vary in accuracy, conclusions about the relative accuracy of the given national censuses must be drawn with great caution. In view of the sources of error in this procedure as mentioned above, these estimates are probably understatements of the true underenumeration. The Annex, "National Reports on the Methodology of Ad Hoc Postenumeration Sample Field Checks," to the United Nations 1968 seminar document, *Methods of Evaluating the Reliability of Population and Housing Census Data* lists 18 reports that give the results as well as the methodology of postenumeration surveys in censuses taken from 1955 to 1964.

Miscellaneous available measures. — Table 2 of the U.N. *Demographic Yearbook, 1965* gives an estimate of net underenumeration, or a qualitative statement, for a number of additional countries, mainly in Latin America. The method of estimation is not indicated there, however. These data are as follows:

	Net under-enumeration (percent)
North America:	
Honduras, 1961...................................	5.3
Jamaica, 1960 [a]...................................	0.9
South America:	
Argentina, 1960...................................	3.6
Bolivia, 1950.....................................	8.4
Chile, 1960.......................................	5.4
Ecuador, 1962....................................	3.9
Peru, 1961..	4.2
Uruguay, 1963....................................	2.05
Venezuela, 1961..................................	5.8

[a] Results from adjustment for ages 0 to 4 only.

The Depoid article contains, as Annexe III, a statement by José Carneiro Felippe and Octávio Alexander de Moraes on the efficacy of the enumeration of the 1940 census of Brazil.[35] After the preliminary count was available, omissions were ferreted out by several different methods, namely, (1) a special questionnaire regarding enumeration status sent to employees of government and private agencies, (2) a check carried out by the post office in 191 localities, (3) revisits to households in areas where the preliminary counts had been challenged, (4) offers of payment for reporting missed persons or households, and (5) assistance from local judiciary officials. Persons discovered to have been missed were included in the final count. The total omissions constituted 1.7 percent of the preliminary count, with a range among the States from 0.1 to 2.4 percent.

Evaluation of Coverage in the United States. — Although President Washington expressed his conviction that the first census, that of 1790, represented an undercount, no estimate of its accuracy was attempted.[36] Although, for more than one hundred years thereafter, most census officials never admitted publicly that the census could represent an underenumeration,

[35] Depoid, op. cit., pp. 214–215.
[36] Leon Pritzker and N. D. Rothwell, "Procedural Difficulties in Taking Past Censuses in Predominantly Negro, Puerto Rican, and Mexican Areas," in *Social Statistics and the City*, David M. Heer, ed., p. 56.

there were a few wise exceptions such as General Francis A. Walker, in his introduction to the 1870 census.[37] He complained of the "essential viciousness of a protracted enumeration" because it led to omissions and duplications.

Walter F. Wilcox wrote as follows in 1906: "A careful census is like a decision by a court of last resort—there is no higher or equal authority to which to appeal. Hence there is no trustworthy means of determining the degree of error to which a census count of population is exposed, or the accuracy with which any particular census is taken. But no well-informed person believes that the figures of a census, however carefully taken, may be relied upon as accurate to the last figures. There being no test available, the opinions of competent experts may be put in evidence in support of this conclusion. Thus Francis A. Walker, Superintendent of the Ninth and Tenth Censuses, testified in 1892 to a select committee of the House: 'I should consider that a man who did not come within half of 1 percent of the population had made a great mistake and a culpable mistake.' Hon. Carroll D. Wright, Commissioner of Labor, who completed the work of the Eleventh Census, wrote in July, 1897: 'I think that the Eleventh Census came within less than 1 percent of the true enumeration of the inhabitants,' and authorized the publication of this opinion."[38] Later evidence, however, indicates that these contemporary guesses as to the level of accuracy were too optimistic.

Later comparisons of successive censuses.—Estimates of relative completeness of enumeration based on the comparison of successive censuses have been made more recently for other censuses down to 1960. In addition to methods that compare pairs of censuses, there are those that use the more elaborate iterative methods, (3) and (4), described in the section "Comparison of successive censuses" above. These methods use some extrinsic data and hence yield estimates of absolute rather than relative error. There are still other variations, but these relate to the handling of the age-sex detail.

The results of the first of these methods are available for 1930 versus 1940, 1940 versus 1950, and 1950 versus 1960. They are as follows:[39]

Coverage of later relative to earlier

Census	Number	Percent of later census
1930 vs. 1940	− 1,303,000	− 1.0
1940 vs. 1950	+ 202,000	+ 0.2
1950 vs. 1960	− 3,000	−

The computation may be illustrated for the 1950–1960 decade:

Census population, April 1, 1950	151,326,000
Births, April 1, 1950, to March 31, 1960	40,963,000
Deaths, April 1, 1950, to March 31, 1960	15,608,000
Net migration, April 1, 1950, to March 31, 1960	2,645,000
Expected population, April 1, 1960	179,326,000
Census population, April 1, 1960	179,323,000
Error of closure	− 3,000

The method of measuring net underenumeration by the type of demographic analysis that is sometimes called the iterative method does more than measure relative underenumeration among censuses. This is because it starts out by comparing the youngest age groups in the most recent censuses with the corresponding numbers of registered births. Ultimately, an estimate of deviation from true population size can be computed for the population of all ages.

The first comprehensive published example of the application of an iterative method to American censuses for the purpose of measuring underenumeration appeared in the previously cited article by Coale.[40] He made several different assumptions about relative age-sex-color specific errors among the censuses of 1930, 1940, and 1950. Coale's final estimates for the 1950 census, however, drew on the Post-Enumeration Survey measures for persons 65 years old and over. Thus, for an enumerated total of 150,697,000, he estimated that about 5.4 million persons were missed, or 3.5 percent of the true total. The results tend to indicate again that the 1950 census was more complete than the 1940 census.

Estimates of the completeness of enumeration of the white population in the censuses from 1880 to 1960 by another type of iterative method (the method of birth reconstruction) are presented in the previously cited book by Coale and Zelnik. The estimates for the total white population are summarized in table 4–3. The estimates for 1960 were preliminary, and later Zelnik published revised 1960 estimates but these were restricted to the native white population.[41]

Table 4–3. — Estimated Net Underenumeration of the White Population of Conterminous United States: 1880 to 1960

[Numbers in thousands]

Year	Census count	Estimated true population	Estimated error	
			Number	As percent of true population
1960	158,455	[1]161,800	[1]3,345	[1]2.1
1950	134,942	138,500	3,558	2.6
1940	118,215	123,200	4,985	4.0
1930	110,287	115,400	5,113	4.4
1920	94,821	100,500	5,679	5.7
1910	81,732	86,700	4,968	5.7
1900	66,809	71,200	4,391	6.2
1890	55,101	58,700	3,599	6.1
1880	43,403	46,200	2,797	6.1

[1] Preliminary.

Source: Ansley J. Coale and Melvin Zelnik, *New Estimates of Fertility and Population in the United States,* copyright(c) 1963 by Princeton University Press, Princeton, N.J., tables 16 to 18 (reprinted by permission of the publisher); and *U.S. Census of Population: 1960,* Part 1, *United States Summary,* table 44.

Considerable research has been undertaken at the U.S. Bureau of the Census on the estimation of net underenumeration in the 1960 census, using both demographic methods and combinations of these with the results of record checks, etc.[42]

[37] Ibid., pp. 56–57.

[78] U.S. Bureau of the Census, *Special Reports of the Twelfth Census, Supplementary Analysis and Derivative Tables,* 1906, p. 16.

[39] U.S. Bureau of the Census, *Current Population Reports,* Series P–25, No. 418, "Estimates of the Population of the United States and Components of Change: 1940 to 1969," table C, March 14, 1969; unpublished estimates of the Bureau of the Census for 1930–40.

[40] Coale, op. cit., p. 44.

[41] Melvin Zelnik, "Errors in the 1960 Census Enumeration of Native Whites," *Journal of the American Statistical Association* 59(306): 437–459, June 1964.

[42] Conrad Taeuber and Morris H. Hansen, "A Preliminary Evaluation of the 1960 Census of Population and Housing," *Demography,* 1(1):2, 1964; Jacob S. Siegel and Melvin Zelnik, "An Evaluation of Coverage in the 1960 Census of Population by Techniques of Demographic Analysis and by Composite Methods," *Proceedings of the Social Statistics Section: 1966,* American Statistical Association, Washington, pp. 71–85; Siegel, "Completeness of Coverage of the Nonwhite Population," 1968.

The most definitive estimates are from Siegel's 1968 study. They are 3.1 percent net underenumeration for the 1960 census and 3.6 for 1950. The absolute errors are both 5.7 million. These may be regarded as the Census Bureau's "best estimates."

The foregoing estimates were all obtained by examining the internal consistency of successive censuses and by using available statistics of births, deaths, immigration, and emigration. These latter are themselves deficient; and, in some cases, rather rough allowances had to be made for these deficiencies. Let us next consider some comparisons that represent checks against administrative records or the results of reinterviews or other sample surveys.

Checks against registrations for military service. — During World Wars I and II, registrations of men of military age were conducted. These strongly indicate some underenumeration in the censuses of 1920 and 1940, respectively.

Checks against registrations for rationing. — The total number of registrants for ration books in World War II were also compared with the number expected from the 1940 census. Allowances had to be made for postcensal births and deaths and for the size of the armed forces and of the institutional population. Ration books could be obtained only for civilians not in institutions and for members of the armed forces who did not take their meals at military establishments. For two of the registrations the number of registrants is available not only as of the initial date but also as of one month later so as to include delayed registrations. The comparisons are as follows:

Deviation of expected number from registered number

War ration book	Registration date	Initial		Final	
		Number (thous.)	Percent	Number (thous.)	Percent
One.........	May 1, 1942	−700	−0.5	(NA)	(NA)
Two........	Mar. 1, 1943	−1,100	−0.9	−3,510	−2.7
Three.......	July 1, 1943	−2,000	−1.5	(NA)	(NA)
Four........	Nov. 1, 1943	+400	+0.3	−2,850	−2.2

The direction of the differences is consistent with underenumeration in the 1940 census. These amounts, however, are merely suggestive since there are reasons why the registration figures themselves may not have measured the eligible population exactly. These reasons include, for example, mistakes in compiling the count in local boards of the Office of Price Administration (which conducted the registrations), the inclusion or exclusion of late registrants, duplicate or fraudulent registrations, and failure of some persons eating only in restaurants to register since ration coupons were not required for restaurant meals. After a thorough study of the causes of differences, Eldridge concluded that about 700,000 of the differences for War Ration Books Two and Four could be explained by the registration of military personnel, who were not included in the estimates of expected registrants.[43] (The number of members of the armed forces who did not take their meals at military messes and were entitled to ration books was known only very approximately and hence, this number had not been included in the expected number of registrants.) Incidentally, when the Census Bureau prepared estimates of the population of States and counties on the basis of the

registration statistics, the results were adjusted to be consistent with the level of enumeration in the 1940 census.

Public opinion polls. — Shortly after the conclusion of the 1940 census field work, the Gallup Poll asked a sample of respondents whether they thought they had been missed in the census. About 4 percent replied affirmatively. Their names and addresses were supplied to the Census Bureau, which was able to find all but about one quarter of the cases in its schedules, reducing the percent of missed persons to 1 percent. (This "find" rate is fairly typical for persons who claim they have been missed. Only a minority of the population is actually interviewed by the enumerators, and some of these do not understand the auspices of the interview.) The 1-percent underenumeration is probably minimal since the quota sample then used by the American Insititute of Public Opinion was likely to underrepresent the types of persons missed by census enumerators.

The first census of the United States, then, to be followed by a postenumeration survey was that of 1950.[44] The results on coverage are summarized in table 4–4. The results were also published by size of place and by region. As is usual in such checks, the gross error was much larger than the net error. Part of the gross error here, however, is attributable to persons who were counted in the wrong enumeration district. The census coverage was found to be "more seriously deficient than had generally been believed"; but particularly when Coale's demographic analysis was published, it was realized that the PES estimate probably understated the true underenumeration.

Table 4–4. — Post–Enumeration Survey Estimates of Error in Population Coverage: United States, 1950

Item	Number (thousands)	Percent of enumerated population	Sampling error of PES estimate [1]	
			Number (thousands)	Percent of enumerated population
Population enumerated in census..................	150,697	100.0	(X)	(X)
Omitted persons..........	3,400	2.3	197	0.1
Erroneous inclusions.....	1,309	0.9	140	0.1
Net undercoverage........	2,091	1.4	242	0.2
Estimated population.....	152,788	101.4	242	0.2

X Not applicable.
i Range of 2 chances out of 3.

Source: U.S. Bureau of the Census, *The Post-Enumeration Survey: 1950,* page 5.

On the basis of the available evidence, the Bureau of the Census set its "minimum reasonable estimate" at 3,715,000, as compared with 2,091,000 from the PES, and 5,429,000 as then proposed by Coale.[45] These represent 2.4, 1.4, and 3.5 percent, respectively, of the estimated true population. The chief shortcoming of the 1950 PES, and of the corresponding survey in 1960 as well, was that it grossly underestimated the number of persons missed **within** enumerated living quarters. This net component was estimated at 126,000. On the other

[43] Hope Tisdale Eldridge, "Problems and Methods of Estimating Postcensal Population," *Social Forces,* 24(1):41–46, Oct. 1945.

[44] The most definitive account of the procedure and findings is contained in: U.S. Bureau of the Census, *The Post-Enumeration Survey: 1950,* Technical Paper No. 4, 1960. (See especially pages 1 to 12.)

[45] Ibid., p. 6.

hand, the net number of persons estimated by PES as having lived in quarters that were missed altogether by the census was 1,975,000, which seems fairly reasonable.

1960 evaluation program.—Checks on population coverage as part of the Evaluation and Research Program of the 1960 Census were more varied and complex than those in the 1950 census; and, so far, the results are scattered over several official reports. There is, however, a summary account in a paper by Marks and Waksberg.[46] The 1960 coverage checks included (1) "reverse" record checks that started with samples drawn from an independent frame of categories of persons who should have been enumerated in the 1960 census and (2) two reinterview studies.

The frame for (1) consisted of:

  (a) persons enumerated in the 1950 census
  (b) persons missed in the 1950 census but detected in the 1950 PES
  (c) children born during the intercensal period (as given in birth certificates)
  (d) aliens who registered with the Immigration and Naturalization Service in January 1960.

This frame was logically incomplete at several points, such as persons missed in both the 1950 census and its PES, unregistered births, and 1950–60 immigrants who were naturalized before January 1960 or else failed to register. It is thought that the bias in the estimated net underenumeration rate attributable to these deficiencies was not very great. Other errors occurred, however, which in the process of tracing and matching, also affected the results. The objective, of course, was to establish whether the person being checked had died or emigrated during the intercensal decade, was enumerated in the 1960 census, or remained within the United States but was missed in the census.

The reinterview studies consisted of (*a*) a reenumeration of housing units in an area sample of 2,500 segments and (*b*)

a reenumeration of persons and housing units in a list sample of 15,000 living quarters enumerated in the census. The purpose of the first reinterview study was to estimate the number of missed households and the population in them. The primary purpose of the second study was to check on the accuracy of census coverage of persons in enumerated units.

Depending on the assumption regarding the classification of uncertain matches and of noninterviews, six alternative estimates were published from the record check study.[47] The estimates of missed persons ranged from 2.7 to 4.8 percent of the enumerated population. The standard errors on these were 0.4 and 0.5, respectively. Note that these are estimates of gross, not net, underenumeration since this study did not identify erroneously enumerated persons. Hence the true population was not estimated. A combination of the two reinterview studies yielded an estimate of gross omissions of 5,653,000, or 3.2 percent, of the enumerated population. This is within the range of the six alternative estimates from the record check.

The net underenumeration for 1960 from reinterview studies was 1.8 percent of the estimated "true" population (1.9 percent of the enumerated population). The corresponding figure for the 1950 census was 1.4 percent, but it is possible that all of the difference is attributable to the better design of the 1960 reinterview study. On the basis of all the evidence, in fact, the Bureau of the Census has concluded that the net underenumeration rate was probably lower in 1960 than in 1950. This conclusion is consistent with the estimates made by demographic analysis shown in table 4–3. Finally, what are now regarded as the best estimates of net underenumeration for 1960 and 1950, all things considered, are those made by demographic analysis as presented in the Siegel paper. Nonetheless, the case-by-case studies have added a great deal to our knowledge of such things as gross errors, area differentials, and causes of error.

[46] Eli S. Marks and Joseph Waksberg, "Evaluation of Coverage in the 1960 Census of Population Through Case-by-Case Checking," *Proceedings of the Social Statistics Section: 1966*, Washington, American Statistical Association, pp. 62–70.

[47] U.S. Bureau of the Census, *Evaluation and Research Program of the U.S. Censuses of Population and Housing, 1960: Record Check Studies of Population Coverage*, Series ER 60, No. 2, 1964, pp. 6–9.

SUGGESTED READINGS

Barclay, George W. *Techniques of Population Analysis.* New York, John Wiley & Sons, 1958. Chapter 3, "Accuracy and Error," pp. 56–92.

Depoid, Pierre. "Rapport sur le degré de précision des statistiques démographiques," (Report on the degree of accuracy of demographic statistics). *Bulletin of the International Statistical Institute,* 35(3):119–230 (29th session of the International Statistical Institute, Petropolis, Brazil, 1955), Rio de Janeiro, 1957.

Hauser, Philip M., and Duncan, Otis Dudley. "The Data and Methods," in: *The Study of Population,* Hauser and Duncan, eds. Chicago, University of Chicago Press, 1959. Pp. 65–70.

Mauldin, W. Parker. "Comments on Possible National and International Standards for the Determination and Indications of Degree of Accuracy of Demographic Statistics," in *Proceedings of the World Population Conference, 1954* (Rome), Vol. IV. New York, United Nations, 1955. Pp. 173–180.

Pritzker, Leon, and Rothwell, N. D. "Procedural Difficulties in Taking Past Censuses in Predominantly Negro, Puerto Rican, and Mexican Areas," in *Social Statistics and the City.* David M. Heer, ed. Cambridge, Mass., Joint Center for Urban Studies of the Massachusetts Institute of Technology and Harvard University, 1968. Pp. 55–79.

Shryock, Henry S., Jr. "The Concepts of *De facto* and *De jure* Populations: The Experience in Censuses of the United States." *Proceedings of the World Population Conference, 1954,* Vol. IV. United Nations, 1955. Pp. 877-878.

————. "The Concept of 'Usual Residence' in the Census of Population," *Proceedings of the Social Statistics Section, 1960.* Washington, American Statistical Association. Pp. 98–102.

Siegel, Jacob S. "Completeness of Coverage of the Nonwhite Population," in *Social Statistics and the City.* David M. Heer, ed. Cambridge, Mass., Joint Center for Urban Studies

of the Massachusetts Institute of Technology and Harvard University, 1968. Pp. 13–54.

Taeuber, Conrad, and Hansen, Morris H. "A Preliminary Evaluation of the 1960 Census of Population and Housing," *Demography,* 1(1):2, 1964.

United Nations. *Demographic Yearbook, 1962.* Pp. 1–9; 17–20.

————. *Handbook of Population Census Methods,* Vol. I, *General Aspects of a Population Census.* Studies in Methods, Series F, No. 5, rev. 1, 1958. Pp. 92–93; 104–111; 131–136. Vol. III, *Demographic and Social Characteristics of the Population.* 1959. Pp. 55–59.

————. *Manuals on Methods of Estimating Population,* Manual II, *Methods of Appraisal of Quality of Basic Data for Population Estimates.* Series A, Population Studies, No. 23, 1955. Pp. 4–17.

————. *Principles and Recommendations for the 1970 Population Censuses.* Statistical Papers, Series M, No. 44, 1967. Pp. 60–61.

U.S. Bureau of the Census. *Enumerator's Reference Manual: 1960 Census of Population and Housing.* F–210, 1959. Pp. 5–7.

————. *1960 Censuses of Population and Housing: Procedural History,* 1966. Pp. 59–60; 359–360.

————. *Evaluation and Research Program of the U.S. Censuses of Population and Housing, 1960: Record Check Studies of Population Coverage.* Series ER 60, No. 2, 1964.

————. *The Post-Enumeration Survey: 1950.* Technical Paper No. 4, 1960. Pp. 1–12.

Zarkovich, S. S. *Quality of Statistical Data.* Rome, Food and Agriculture Organization of the United Nations, 1966. Pp. 41–44; 487–513.

CHAPTER 5

Population Distribution Geographic Areas

For many purposes information on the size and characteristics of the total population of a nation is not sufficient. Population data are often needed for geographic subdivisions of a country and for other classifications of areas in which people live. In most countries, the geographic distribution of the population is not even but is dense in some places and sparse in others. There are urban centers where millions of people live within a few square kilometers, and there are also vast stretches of mountains and deserts where the population averages only one or two persons per square kilometer. This chapter treats the geographic distribution of the population by political areas and by several other types of geographic areas.

ADMINISTRATIVE OR POLITICAL AREAS

Political areas are not ordinarily created or delineated by a country's central statistical agency or its census office but instead are established by national constitutions, laws, decrees, regulations, or charters. In some countries, the primary political subdivisions are empowered to create secondary and tertiary subdivisions.

The present discussion is confined to the "major and minor civil divisions" and to cities proper. ("Urban agglomerations" and "urban and rural" areas are discussed in the next major section of this chapter and in chapter 6, respectively.)

Primary Divisions. — The generic names of the major civil divisions are given in several of the U.N. *Demographic Yearbooks* — most recently the 1971 *Yearbook*. The generic names appear in English and French, and often in the national language as well. The most common names in English for the primary areas are district, province, island, department, parish, state, and region.

It is fairly common for the capital city to constitute a primary division in its own right and often with a distinctive name. Examples are Nairobi District in Kenya, the Distrito Federal in Mexico, Brazil, and Venezuela, the District of Columbia in the U.S.A., the Capital Federal in Argentina, Asunción in Paraguay, Seoul Special City in the Republic of Korea, Ulan Bator (Mongolia), Manila (P.I.), Damascus (Syria), Tiranë (Albania), Copenhagen, Greater Athens, Budapest, Stockholm, Warsaw, Bucharest, and the Australian Capital Territory. In a few countries, some of the larger cities are also primary political divisions, (e.g., in Morocco, United Arab

Republic, Trinidad and Tobago, Cambodia, Taiwan, Bulgaria, Hungary, Poland, and Romania).

Countries that have been settled relatively recently or countries that contain large areas of virtually uninhabited land or land inhabited mainly by aborigines may have a different kind of primary subdivision that has a distinctive generic name and a rudimentary political character. Examples are the Frontier Districts of the United Arab Republic, some of the Territories of Canada and Australia, the Federal Territories of Venezuela, and some of the Union Territories of India.

Secondary and Tertiary Divisions. — The intermediate or secondary political divisions also have a wide variety of names. These include county, district, and commune. Some small countries have only primary divisions. Some large countries have three or more levels. Examples of tertiary divisions are the townships in the United States, the *myun* and *eup* in Korea, and the *hsiang* and *chen* in Taiwan. For different administrative functions, a province, state, etc., may be divided into more than one set of political areas. The census of Canada shows, for the Province of Ontario, figures for both counties and electoral districts. In other census publications, such alternative administrative divisions may be ignored.

Municipalities. — It is difficult to find a universal, precise term for the type of political area discussed in this subsection. The ideal type is the city; but smaller types of municipalities such as towns and villages are also included. (Incidentally, in Puerto Rico, a *municipio* is the equivalent of a county in mainland United States.) In some countries, these areas could be described as incorporated places or localities. In some countries, again, these municipalities are located within secondary or tertiary divisions; but, in other countries, they are simply those territorial divisions that are administratively recognized as having an urban character.

The larger municipalities are frequently subdivided for administrative purposes into such areas as boroughs or wards (Britain and some of its former colonies), *arrondissements* (France), *ku* (Japan and Korea). and *chu* (China, Taiwan). These subdivisions of cities, in turn, may be divided into precincts (United States), *chun* (China, Taiwan), or *dong* (Korea). In China and Korea, even a fifth level exists — the *lin* and *ban*, respectively — for which "urban neighborhood" is as close as one could come in English. These smaller types of administrative areas are ordinarily not used for the presentation of official demographic statistics, but they are sometimes used as units in sample surveys.

Sources

Population totals for the major (primary) civil divisions are published in several of the *Demographic Yearbooks* of the United Nations; and fairly frequently there is a table showing the total population of capital cities and cities of 100,000 or more inhabitants.

The U.N. *Demographic Yearbooks* present statistics for neither smaller cities or other municipalities nor for the secondary, tertiary, etc., divisions. For these one must usually refer to the national publications.

Uses and Limitations

Statistics on the distribution of the population among political areas are useful for many purposes. For example, they may be used to meet legal requirements for determining the apportionment of representation in legislative bodies; they are needed for studies of internal migration and population distribution in relation to social, economic, and other administrative planning; and they provide base data for the computation of subnational vital statistics rates and for preparing local population estimates and projections.

A limitation of these political areas from the standpoint of the analysis of population distribution, and even from that of planning, is the fact that the boundaries may be rather arbitrary and may not consider physiographic, economic, or social factors. Moreover, the areas officially designated as cities may not correspond very well to the actual physical city in terms of population settlement or to the functional economic unit. Furthermore, in some countries the smallest type of political areas does not provide adequate geographic detail for ecological studies or city planning. Therefore, various types of statistical and functional areas have been defined, in census offices and elsewhere, to meet these needs. These may represent groups or subdivisions of the political areas, or they may disregard them altogether. Such areas are the subject of the second major section ("Statistical Areas") of this chapter and of chapter 6.

Quality of the Statistics

Most of what was said in chapter 4 about the accuracy of total national population applies also to the country's geographic divisions. Furthermore, given a set of rules on where people should be enumerated within a country, there will be errors in applying these. Hence, some people will be counted in the wrong area, and the accuracy of the counts for the areas affected will be impaired. For example, the 1950 Post-Enumeration Survey of the United States found that an estimated 1,024,000 persons had been counted in the wrong enumeration district.[1] The corresponding figure for the 1960 census was 800,000, according to an unpublished estimate.

Political Areas of the United States

The primary purpose of the census of the United States is the determination of the number of residents in each State for the purpose of apportioning the representatives to the Congress of the United States among the States. Within States, population must be obtained by smaller areas for determining Congressional Districts and for setting up districts (by various

methods) for electing representatives to the individual States' legislative body or bodies and for other purposes required by State or local laws.

States. — There are now 50 States and the District of Columbia within the United States proper. The number of States and some of their boundaries have changed in the course of American history; but from 1912 to 1959, there were 48 States. That area is called the "Conterminous United States." The primary divisions of States are usually called counties. These in turn are subdivided into political units collectively known as minor civil divisions ("MCD's"). In most States, the places incorporated as municipalities are subordinate to minor civil divisions; but in some States the incorporated places are themselves minor civil divisions of the counties. As will be shown below, there are fairly numerous differences among the States in the nature and nomenclature of their political areas.

Counties. — The primary divisions of the States are termed counties in all but two States although three States also contain one or more independent cities. The county equivalents in Louisiana are the parishes. At the time of the 1960 census, the primary divisions in the new State of Alaska were the 24 election districts; but, as of April 1970, they are the 29 census divisions. The latter are defined cooperatively by the Bureau of the Census and the State. The independent cities were Baltimore (Maryland), St. Louis (Missouri), and 38 in Virginia.

All in all, there were 3,141 counties or county equivalents in the United States in 1970.

Minor Civil Divisions. — The minor civil divisions of counties have many kinds of names. "Township" is the most frequent. In New England States, New York, and Wisconsin, most MCD's are called "towns"; these are unlike the incorporated towns in other States in that they are not necessarily densely settled population centers. Some tertiary divisions have no local governmental organizations at all and may be uninhabited. Furthermore, in many States, some or all of the incorporated municipalities are also minor civil divisions. A further complication, in some of the New England States, is that all of the MCD's, be they cities or towns, are viewed locally as "incorporated." In the usage of Census Bureau publications, however, the term "incorporated place" has been reserved for localities or nucleated settlements and is not applied to other areal subdivisions.

In addition to the minor civil divisions shown in the census volumes, there are thousands of school and other taxation units for which separate figures are not published. Where more than one kind of primary subdivision exists in a county, the Census Bureau tries to select the more stable kind. In some States, however, no type of minor civil division has much stability. In some of the Western States, for example, the election precincts may be changed after each election on the basis of the number of votes cast. Obviously, such units have practically no other statistical value. Even in States where the minor civil divisions do not change very often, they may have so little governmental significance that data published for them are also of limited usefulness. Here too the minor civil divisions may be so unfamiliar locally that it is very difficult for enumerators in the field to observe their boundaries. This is the situation in some Southern States. At the other extreme are the stable towns of New England, which are of more political importance than the counties. A statistical solution to the problem of the evanescent or little known minor civil divisions is the "census county division," which was first

[1] U.S. Bureau of the Census, *The Post-Enumeration Survey: 1950*, Technical Paper No. 4, 1960, p. 7.

introduced in one State, Washington, in 1950 and then in many more States in the 1960 reports. The census county divisions, then, are the geographic-statistical equivalents of minor civil divisions; but, since they are not political areas, they are discussed in the next major section.

In 1970 there were 28,130 minor civil divisions recognized in the census reports and 7,068 census county divisions, making a total of 35,198 county subdivisions.

Incorporated Places. — The generic definition of a "place" is a concentration of population regardless of the existence of legally prescribed limits, powers, or functions. By definition, the incorporated places are the only ones that are political areas. All States contain incorporated places known as "cities," and some States also have incorporated "towns," "boroughs," and "villages." [2] Where a State has more than one kind of municipality, cities tend to be larger places than the other types. Here practices vary widely.

From the political standpoint, an incorporated place may thus be: (1) itself a minor civil division, (2) coextensive with a minor civil division, on which it is dependent, or (3) dependent on one or more minor civil divisions. In the case of (3), the incorporated place is usually physically contained within one minor civil division; but it may extend into two or more MCD's and a few large places physically contain several entire MCD's. Even in this last situation, however, the MCD is politically part of the county not of the incorporated place. An incorporated place may also be located in two or more counties but never in more than one State. Denver, New Orleans, Philadelphia, and San Francisco are coextensive with a county or parish. These relationships are shown in table 10 of the State population reports for 1970.

Wards of Cities. — Finally, the larger incorporated places are usually divided into political areas known as wards. Here again, in some cities wards have fairly permanent boundaries and are statistically useful. In others, ward boundaries change frequently. (See the discussion of "Census Tracts" below.) There is a tendency for wards to disappear from the American municipal scene. New York City no longer has wards. Its five vast boroughs are coextensive with counties. The table for wards that was carried in previous census reports was dropped from the reports of 1960. The demand for such population figures was so great, however, that a special compilation had to be published. [3]

Annexations. — The data shown for any area in a census report refer to the area at the census date. There are a great many annexations to incorporated places during a decade, and many new places are incorporated, for which the population at previous dates is not available. Until the census of 1960, no attempt was made to measure the population in these annexations. [4] Table 8 of the State reports gave the 1970 population in the 1960 area and in the annexed area for incorporated places of 2,000 or more in 1960. The chief purpose of these statistics is to permit the measurement of population changes in a **fixed** area.

Congressional and Legislative Districts. — Congressional Districts are the districts represented by a congressman in the U.S. House of Representatives whereas Legislative Districts are those represented by lawmakers serving in the State legislatures. Except for Nebraska, every State legislature consists of two houses, each with its own districts. These are all political areas, but they are not administrative areas.

General Considerations. — The political uses of census data are so important that they go far to determine the basis of census tabulations of population for geographic areas. Fortunately, political units serve very well as statistical units of analysis in many demographic problems. In the realm where they are less satisfactory, other types of area or residence classifications of population data have been provided by the Census Bureau with increasing satisfactoriness over recent decades. A new tool in 1970, the address register, in conjunction with the electronic computer, should make it possible to provide census statistics, at least in large metropolitan areas, as needed for such additional types of areas as school districts, traffic zones, ZIP-code areas, and, indeed, for any other areas, political or otherwise, that can be defined or satisfactorily approximated in terms of combinations of city blocks. Most of these data will be supplied in the form of machine print-outs or tapes rather than published tables.

The usefulness of demographic data by political areas of the United States for analysis of trends is greatest for the largest political subdivisions, namely the States. Counties and cities are probably next in order. Least satisfactory are the minor civil divisions or wards of cities, which have been stated to change their boundaries frequently in some States. Another reason that the amount of analytical work done on population data for minor civil divisions has been quite limited is that many other types of data that one might wish to relate to census data are not available for geographic areas smaller than cities or counties. Moreover, the amount of detail and cross-classification of population data published by the census for minor civil divisions are quite limited.

For cross-sectional analyses that do not involve changes over time, counties, cities, and minor civil divisions, as well as States, may be very useful as units of analysis. In general, the smaller the geographic area with which one deals, the more homogeneous will be the population living in the area. Rates, averages, and other statistical summarizing measures are usually more meaningful if they relate to a relatively homogeneous population. However, if the geographic area and the population residing in it are very small, rates such as a migration rate or a death rate may be so unstable as to be meaningless, as the total population exposed to the risk of migration or death may be too small for the statistical regularity of demographic events that is manifested when large populations are observed.

In publications of population statistics, data are often shown for a combination of political and nonpolitical areas, such as for the States and major geographic divisions, for counties and their urban and rural populations, or for incorporated and unincorporated places. The Census Bureau and other statistics-producing agencies present data for the States sometimes listed in alphabetical order and sometimes in a geographic order. The usual geographic order conforms to the regions and divisions that are defined below. It is shown in table 5-1, for example. This order facilitates the preparation of regional totals, shows the composition of the regions and geographic divisions, and facilitates comparisons of neighboring States. It will be noted that States in the same geographic division often have similar characteristics.

---

[2] Strictly speaking, there are no incorporated places in Hawaii, but Honolulu and Hilo have long been treated in the census reports as "cities." Beginning with the 1970 census, places defined by the Hawaii Department of Planning and Economic Development will be recognized as "incorporated" for purposes of presentation in census reports.

[3] U.S. Bureau of the Census, *1960 Census of Population, Supplementary Reports*, PC(S1)–6. "Population of Cities of 10,000 or More by Wards: 1960," June 16, 1961.

[4] Beginning in 1947, however, fairly accurate estimates based on the special censuses have been given in U.S. Bureau of the Census, *Current Population Reports*, Series P–28, *Special Censuses*.

Table 5–1.—Selected Characteristics of the Population, by Regions, Divisions, and States: 1960

Region, division, and State	Percent foreign born	Persons 5 years old and over — Percent migrant [1]	Percent who completed 4 years of high school or more [2]	Cumulative fertility rate [3]	Region, division, and State	Percent foreign born	Persons 5 years old and over — Percent migrant [1]	Percent who completed 4 years of high school or more [2]	Cumulative fertility rate [3]
United States....	5.4	17.4	41.1	1,746	WEST NORTH CENTRAL-- Continued				
					North Dakota......	4.7	17.4	38.8	2,069
REGIONS:					South Dakota......	2.7	19.6	42.2	2,042
Northeast.........	10.2	13.0	41.0	1,551	Nebraska..........	2.9	18.6	47.7	1,865
North Central......	4.4	15.2	41.7	1,792	Kansas............	1.5	21.5	48.2	1,830
South.............	1.8	19.2	35.3	1,837	SOUTH ATLANTIC:				
West.............	6.9	25.3	50.8	1,797	Delaware..........	3.3	16.8	43.3	1,703
					Maryland..........	3.0	20.7	40.0	1,697
NORTHEAST:					Dist. of Columbia..	5.1	16.2	47.8	1,312
New England.......	10.1	12.9	44.5	1,613	Virginia..........	1.2	24.4	37.9	1,700
Middle Atlantic....	10.3	13.1	40.0	1,532	West Virginia......	1.3	12.3	30.6	1,835
NORTH CENTRAL:					North Carolina.....	0.5	15.0	32.3	1,767
East North Central.	5.2	13.8	41.2	1,773	South Carolina.....	0.5	14.6	30.4	1,910
West North Central.	2.6	18.4	42.9	1,842	Georgia...........	0.6	19.0	32.0	1,882
SOUTH:					Florida...........	5.5	32.9	42.6	1,776
South Atlantic.....	2.1	21.0	36.6	1,770	EAST SOUTH CENTRAL:				
East South Central.	0.5	14.6	29.5	1,892	Kentucky..........	0.6	14.1	27.6	1,888
West South Central.	2.1	19.6	37.2	1,903	Tennessee.........	0.4	14.2	30.4	1,738
WEST:					Alabama...........	0.5	14.6	30.3	1,914
Mountain..........	3.7	28.4	49.7	1,979	Mississippi........	0.4	16.1	29.8	2,138
Pacific...........	7.9	24.4	51.1	1,739	WEST SOUTH CENTRAL:				
					Arkansas..........	0.4	18.2	28.9	2,028
NEW ENGLAND:					Louisiana..........	0.9	15.8	32.3	2,032
Maine............	6.2	13.7	43.2	1,878	Oklahoma..........	0.9	21.2	40.5	1,807
New Hampshire......	7.4	16.3	42.9	1,730	Texas............	3.1	20.7	39.5	1,861
Vermont..........	6.0	15.4	42.9	1,821					
Massachusetts......	11.2	12.2	47.0	1,565	MOUNTAIN:				
Rhode Island.......	10.0	12.8	35.0	1,561	Montana...........	4.5	23.1	47.8	2,029
Connecticut.......	10.9	12.7	43.8	1,573	Idaho............	2.3	25.2	48.5	2,124
MIDDLE ATLANTIC:					Wyoming..........	2.9	28.1	52.0	2,007
New York..........	13.6	13.7	40.9	1,478	Colorado..........	3.4	30.9	52.0	1,827
New Jersey.........	10.1	17.1	40.7	1,544	New Mexico........	2.3	29.7	45.5	2,107
Pennsylvania.......	5.3	10.0	38.1	1,606	Arizona...........	5.4	32.8	45.7	1,964
EAST NORTH CENTRAL:					Utah.............	3.6	19.8	55.8	2,083
Ohio.............	4.1	14.0	41.9	1,761	Nevada...........	4.6	36.2	53.3	1,760
Indiana...........	2.0	15.5	41.8	1,814					
Illinois...........	6.8	12.8	40.4	1,670	PACIFIC:				
Michigan..........	6.8	13.8	40.9	1,854	Washington........	6.3	22.3	51.5	1,838
Wisconsin.........	4.3	14.0	41.5	1,855	Oregon...........	4.0	24.8	48.4	1,860
WEST NORTH CENTRAL:					California.........	8.5	24.5	51.5	1,700
Minnesota.........	4.2	17.0	43.9	1,887	Alaska............	3.6	49.0	54.7	2,050
Iowa.............	2.0	15.9	46.3	1,864	Hawaii...........	10.9	19.9	46.1	1,854
Missouri..........	1.8	19.3	36.6	1,728					

[1] Persons who lived in different counties in the United States in 1955 and 1960.
[2] Persons 25 years old and over.
[3] Children ever born per 1,000 women 15 to 44 years old of all marital classes.

Source: *U.S. Census of Population: 1960*, Part 1, *United States Summary*, 1962, table 105, p. 1–248.

STATISTICAL AREAS

For many purposes, data are needed for areas other than those recognized as political entities by law. Nonpolitical areas in common use for statistical purposes include both combinations and subdivisions of political areas. The most general objective in delineating such statistical areas is to attain relative homogeneity within the area; and, depending on the particular purpose of the delineation, the homogeneity sought may be with respect to geographic, demographic, economic, social, historical, or cultural characteristics. Also, groups of noncontiguous areas meeting specified criteria, such as all the urban areas within a State, are frequently used in presentation and analysis of population data.

International Recommendations and National Practices

There are several types of such statistical areas; for example, regions or functional economic areas; metropolitan areas, urban agglomerations, or conurbations; localities; and urban census tracts (which are small subdivisions of cities and their adjacent area).

Regions or Functional Economic Areas.—The terminology for this kind of geographic area is not too well standardized, but as used here, a "region" means a large area. It ordinarily means something more, however, namely some kind of functional economic area or a cultural area.[5] A region may represent a grouping of a country's primary divisions (states or provinces) or a grouping of secondary or tertiary divisions

---

[5] As used in geography, a "region" may be an area of any size so long as it possesses homogeneity or cohesion. See, Derwent Whittlesey, "The Regional Concept and the Regional Method," in *American Geography: Inventory and Prospect*, Preston E. James and Clarence F. Jones, eds., published for the Association of American Geographers by Syracuse University Press, 1954, pp. 21–22; and Conrad Taeuber, "Regional and Other Area Statistics in the United States," *Bulletin of the International Statistical Institute* (Proceedings of the 35th Session, Belgrade, 1965), 41(1): 161–162. For a more sociological view of the concept, see Howard W. Odum and Harry Estill Moore, *American Regionalism: A Cultural-Historical Approach to National Integration*, New York, Henry Holt and Company, 1938.

that cuts across the boundaries of the primary divisions. (There are, of course, also international regions, which are either combinations of whole countries or which cut across national boundaries.) Among the factors on which regions are delineated are physiography, climate, type of soil, type of farming, culture, and economic levels and organizations. The cultural and economic factors include ethnic or linguistic differences, type of economy, and level of living. The objective may be homogeneous regions—with minimal differences within regions and maximal differences among regions—or relatively self-contained regions where a large city or urban complex is economically dominant over a hinterland. Some delineations may be based on statistical manipulations of a large number of indexes, for example, by factor analysis. The regions defined and used by geographers, anthropologists, etc., are somewhat more likely than those defined by demographers and statisticians to ignore political areas altogether. The latter users have to be more concerned with the units for which their data are readily available and to use such units as building-blocks in constructing regions. There may be also a hierarchy of regions; the simplest type consists of the region and the subregion.

Large Urban Agglomerations.—The concept of an urban agglomeration is defined by the United Nations as follows: "A large locality of a country (i.e., a city or a town) is often part of an urban agglomeration, which comprises the city or town proper and also the suburban fringe or thickly settled territory lying outside of, but adjacent to, its boundaries. The urban agglomeration is, therefore, not identical with the locality but is an additional geographic unit which includes more than one locality." [6] (It will be seen below that this concept is broad enough to encompass both the standard metropolitan statistical areas and the urbanized areas used in the United States.)

A more detailed discussion of the concept is provided by Kingsley Davis and his associates. [7] According to them, the city as officially defined and the urban aggregate as ecologically conceived may differ because the city is either under-bounded or overbounded. Cities in Pakistan, for example, usually are "truebounded," that is, they approximate the actual urban aggregate fairly closely. The underbounded city is the most common type. Most of the cities in the Philippines are stated to be overbounded, in that they include huge areas of rural land within their boundaries. The *shi* in Japan are also of this type. To define an urban aggregate or agglomeration, one may move in the direction of either an urbanized area or a metropolitan area. The former represents the territory "settled continuously in an urban fashion"; the latter includes some rural territory as well. Urbanized areas have been delineated in only a few countries. Their boundaries ignore political lines for the most part. In addition to the urbanized area in the United States, the conurbation in England and Wales is of this type. Metropolitan areas, which use political areas as building blocks, are regarded by Davis and associates as theoretically less desirable but more feasible for international comparisons. Even if they have not been officially defined, they can be constructed in most countries from the available statistics, following standard principles; and, in fact, this work has been carried out by International Urban Research. It can be seen that this discussion merges into that of the urban-rural classification which is treated in chapter 6.

Localities.—A "locality" is a distinct population cluster (inhabited place, settlement, population nucleus, etc.) of which the inhabitants live in closely adjacent structures. The locality usually has a commonly recognized name, but it may be named or delineated for purposes of the census. Localities are not necessarily the same as the smallest civil divisions of a country.

Localities, places, or settlements may be incorporated or unincorporated; thus, it is only the latter, or the sum of the two types, that is not provided for by the conventional statistics on political areas. The problem of delineating an unincorporated locality is similar to that of delineating a large urban agglomeration; but, with the shift to the lower end of the scale, the areas required often cannot be approximated by combining several political areas since small localities are often part of the smallest type of political area. Just how small the smallest delineated locality should be for purposes of studying population distribution is rather arbitrary in countries where there is a size-continuum from the largest agglomeration down to the isolated dwelling unit. In view of the considerable work required for such delineations, 200 inhabitants seems about as low a minimum as is reasonable. [8] In countries where there is essentially no scattered rural population but all rural families live in a village or hamlet, the answer is automatically provided by the settlement pattern.

Urban Census Tracts.—The urban census tract is a statistical subdivision of a relatively large city, especially delineated for purposes of showing the internal distribution of population within the city and the characteristics of the inhabitants of the tract as compared with those of other tracts. Once their boundaries are established, not only census data but also other kinds of data, such as vital and health records, can be assembled for these areas. In Far Eastern countries and others where there are well-established small administrative units within cities, such special statistical subdivisions are unnecessary. So far, the census tract is not a very widespread type of statistical area. Originating in the United States (*q.v.*), it has so far been taken over only in Canada.

Statistical Areas of the United States

Regions.—Two types of regional delineations of the United States are common—those which are groupings of whole States and those which cut across State lines. An illustration of the former is the set of six regions developed by Howard W. Odum, [9] and an illustration of the latter (which happens not to have the label "regions" in its identification) is the set of major type-of-farming areas developed in the Bureau of Agricultural Economics. [10] The greater convenience of the group-of-State regions for statistical compilations has led to their rather general adoption for presenting census data, although the greater homogeneity of regions that cut across State lines is well recognized.

Combinations of States.—The U.S. Bureau of the Census for its population publications uses two levels of State groupings. The 50 States and the District of Columbia are combined into nine groups, identified as "**geographic divisions**," and these in turn are further combined into three or four groups,

[6] United Nations, *Principles and Recommendations for the 1970 Population Censuses*, Statistical Papers, Series M, No. 44, 1967, p. 51.

[7] International Urban Research, *The World's Metropolitan Areas*, Berkeley, University of California Press, 1959, pp. 1–17.

[8] This is the class-mark between the lowest and the next to the lowest intervals in the table recommended by the United Nations, the lowest interval having no minimum.

[9] Howard W. Odum, *Southern Regions of the United States*, Chapel Hill, The University of North Carolina Press, 1936.

[10] U.S. Bureau of Agricultural Economics, *Generalized Types of Farming in the United States*, Agricultural Information Bulletin No. 3, Department of Agriculture, February 1950.

identified as "regions." Statistics may be presented for regions when the size of the sample does not permit publication for areas as small as States or when it is too expensive to tabulate detailed cross-classifications for States. Table 5-1 indicates the States included in each division and each region.

The present geographic divisions have been used since 1910, and the four regions since 1942. The objective in establishing these State groupings is described as follows: "The states within each of these divisions are for the most part fairly homogeneous in physical characteristics, as well as in the characteristics of their population and their economic and social conditions, while on the other hand each division differs more or less sharply from most others in these respects. In forming these groups of states the lines have been based partly on physical and partly on historical conditions," [11]

Economic subregions and State economic areas. — The term "subregion" or "subarea" has been used in two senses in the United States: (1) to denote the subparts of a region (larger than a State) which may cut across State lines, [12] and (2) to denote subparts of a State. [13] In either case, the delineation of the subregions may be based on any one or any combination of a number of types of criteria — agricultural, demographic, economic, social, cultural, etc. Moreover, subregional boundaries may be coincident with county lines or they may cut across county lines.

The idea of the decennial census as a national inventory can be adequately implemented only by having material for examination and analysis for areas more appropriate for certain types of data than the conventional political units. This is especially important in the United States because of its large area, the great mobility of its population, and the fact that political boundaries in the United States offer little impediment to the flow of commerce and population across them. Because the political boundaries have so little effect in shaping the spatial patterns of economic and population phenomena, they are inadequate for delineating the most meaningful areas for portraying and analyzing these phenomena. Ideally, the delineation of economic areas should not have to follow county or even township lines, but areas that did not do so would not be practicable or feasible for census purposes.

The Bureau of the Census and the Bureau of Agricultural Economics commissioned Donald J. Bogue to develop a set of county groupings for the presentation of certain statistics from the 1950 Censuses of Population and Agriculture. [14] The resulting State economic areas (SEA's) were submitted to many State agencies and to individual scholars for their comments, and it was originally hoped that they would be accepted as general purpose statistical areas. They have met with a mixed reception, however.

Shortly after the delineation of the State economic areas, Bogue and Beale combined them into a smaller number of economic subregions, which disregarded State lines. [15] Still

later these were further combined into 13 economic regions and five economic provinces, and Bogue and Beale described the whole system in a monumental volume of more than 1,100 pages. [16] These sources should also be consulted for a full account of the methodology and the definition of each area in terms of constituent areas down to the individual counties.

In brief, the State economic areas are relatively homogeneous subdivisions of the States. They consist of single counties or groups of counties that have similar economic and social characteristics. There are two principal types of SEA's — the metropolitan and the nonmetropolitan. The former consist of the larger standard metropolitan statistical areas except that when an SMSA was located in two or more States or economic subregions, each part became a separate metropolitan SEA. The New England SMSA's, which are defined in terms of towns, were approximated on a whole-county basis for this purpose. In nonmetropolitan areas, demographic, climatic, physiographic, and cultural factors, as well as factors pertaining more directly to the production and exchange of agricultural and nonagricultural goods, were considered.

For 1960 there was a limited revision of 1950 areas, mainly to take account of the two new States and of changes in the standard metropolitan statistical areas but without reexamining the original principles or applying them to more recent data. As a result of the revisions, the number of SEA's was increased to 509 and that of economic subregions to 121. To provide comparability in the migration statistics for 1965–1970 with those for 1955-1960, the SEA's in 1970 were kept just about as they were in 1960. The recognition of one more metropolitan SEA brought the grand total to 510.

Standard Metropolitan Statistical Areas. — The responsibility for the definition of standard metropolitan statistical areas (SMSA's) is lodged in the Office of Statistical Standards of the U.S. Bureau of the Budget. [17]

1970 criteria. — The criteria used in the 1970 Census publications are sufficiently important to be quoted in full with the omission of footnotes, which clarify a few technical points.

"The definition of an individual standard metropolitan statistical area involves two considerations; first, a city or cities of specified population to constitute the central city and to identity the county in which it is located as the central county; and, second, economic and social relationships with contiguous counties which are metropolitan in character, so that the periphery of the specific metropolitan area may be determined. Standard metropolitan statistical areas may cross State lines, if this is necessary in order to include qualified contiguous counties.

Population Criteria

"1. Each standard metropolitan statistical area must include at least:
    "(a) One city with 50,000 or more inhabitants, or
    "(b) Two cities having contiguous boundaries and constituting, for general economic and social purposes, a single community with a combined population of at least 50,000, the smaller of which must have a population of at least 15,000.
"2. If two or more adjacent counties each have a city of 50,000 inhabitants or more (or twin cities under 1(b)) and the cities are

[11] U.S. Bureau of the Census, *Thirteenth Census of the United States, Abstract of the Census*, 1913, p. 13.

[12] For example, T. J. Woofter, Jr., "Subregions of the Southeast," *Social Forces*, 13(1):43–50, October 1934.

[13] A set of intrastate functional areas that could appropriately be called "subregions" is described in: Illinois, Board of Economic Development, *Suggested Economic Regions in Illinois by Counties*, by Eleanor Gilpatrick, Springfield (Illinois), March 12, 1965.

[14] U.S. Bureau of Census, *State Economic Areas*, by Donald J. Bogue, 1951. See also Calvin L. Beale, "State Economic Areas—A Review After 17 Years," *Proceedings of the Social Statistics Section of the American Statistical Association*, 1967, pp. 82-85.

[15] U.S. Bureau of the Census and U.S. Bureau of Agricultural Economics, "Economic Subregions of the United States," by Donald J. Bogue and Calvin L. Beale, Series Census-BAE, No. 19, *Farm Population*, July 27, 1953.

[16] Donald J. Bogue and Calvin L. Beale, *Economic Areas of the United States*, New York, Free Press of Glencoe, 1961.

[17] Office of Statistical Standards, *Standard Metropolitan Statistical Areas*, Washington, Government Printing Office, 1967.

within 20 miles of each other (city limits to city limits), they will be included in the same area unless there is definite evidence that the two cities are not economically and socially integrated.

Criteria of Metropolitan Character

"The criteria of metropolitan character relate primarily to the attributes of the county as a place of work or as a home for a concentration of nonagricultural workers. Specifically, these criteria are:

"3. At least 75 percent of the labor force of the county must be in the nonagricultural labor force.

"4. In addition to criterion 3, the county must meet at least one of the following conditions:

"(a) It must have 50 percent or more of its population living in contiguous minor civil divisions with a density of at least 150 persons per square mile, in an unbroken chain of minor civil divisions with such density radiating from a central city in the area.

"(b) The number of nonagricultural workers employed in the county must equal at least 10 percent of the number of nonagricultural workers employed in the county containing the largest city in the area, or be the place of employment of 10,000 nonagricultural workers.

"(c) The nonagricultural labor force living in the county must equal at least 10 percent of the number of the nonagricultural labor force living in the county containing the largest city in the area, or be the place of residence of a nonagricultural labor force of 10,000.

"5. In New England, the city and town are administratively more important than the county, and data are compiled locally for such minor civil divisions. Here, towns and cities are the units used in defining standard metropolitan statistical areas. In New England, because smaller units are used, and more restricted areas result, a population density criterion of at least 100 persons per square mile is used as the measure of metropolitan character.

Criteria of Integration

"The criteria of integration relate primarily to the extent of economic and social communication between the outlying counties and central county.

"6. A county is regarded as integrated with the county or counties containing the central cities of the area if either of the following criteria is met:

"(a) If 15 percent of the workers living in the county work in the county or counties containing central cities of the area, or

"(b) If 25 percent of those working in the county live in the county or counties containing central cities of the area.

Area Titles

"7. The following general guidelines are used for determining titles for standard metropolitan statistical areas:

"(a) The name of the standard metropolitan statistical area is that of the largest city.

"(b) The addition of up to two city names may be made in the area title, on the basis and in the order of the following criteria:

"(1) The additional city or cities have at least 250,000 inhabitants.

"(2) The additional city or cities have a population of one-third or more of that of the largest city and a minimum population of 25,000, except that both city names are used in those instances where cities qualify under criterion 1(b).

"(c) In addition to city names, the area titles will contain the name of the State or States included in the area.

Data Sources

"The definitions and titles of standard metropolitan statistical areas are established by the Bureau of the Budget with the advice of the Federal Committee on Standard Metropolitan Statistical Areas. This Committee is composed of representatives of the major statistical agencies of the Federal Government. In applying the foregoing criteria, data from the following sources are used by the Committee:

"Population, labor force, density, and occupational data: Bureau of the Census and Bureau of Employment Security

"Employment by place of work: Bureau of Old-Age and Survivors Insurance, Department of Labor, Department of Defense, and Civil Service Commission.

"Volume of commuting: Bureau of the Census and Bureau of Employment Security." [18]

In the actual application of these criteria, however, a fair number of exceptions have been made. The SMSA's have been so popular with the public that there is a great deal of essentially nonstatistical interest in them on the part of chambers of commerce, national advertisers, newspaper publishers, radio and television station owners, etc.[19] The local pride and financial interest result in pressures for creating new areas that do not meet the given criteria, adding additional counties, or splitting off independent areas. In these Brobdingnagian clashes, the demographer has a relatively feeble voice. On the whole, the tendency has been to create more SMSA's—note the "twin-city" provision in criterion 1(b)—and to enlarge the existing SMSA's by the addition of more outlying counties.

History.—The standard metropolitan area (as the standard metropolitan statistical area was first called) replaced three other types of statistical areas in time to be used in the 1950 Censuses of Population and Housing—the metropolitan district of the Censuses of Population and Housing, the industrial area of the Census of Manufactures, and the labor market area of the Bureau of Employment Security. The objective was to define standard areas, by the uniform application of a set of criteria, that would be used by the Federal statistical agencies. The SMSA's were supposed to be general-purpose areas rather than areas delineated solely for the presentation of demographic data.

In response to the very great public interest in the SMSA's, these played an ever more prominent role in the 1950 and 1960 publications. This greater public acceptance undoubtedly reflected the sanction of the SMSA's by the Bureau of the Budget, their use by a number of Federal agencies in presenting a wide variety of statistics, and the fact that the use of whole counties as building blocks made possible the assembly of many statistical series for them by State, local, and private agencies. At the time of the 1960 census, there were 212 SMSA's in the United States and three in Puerto

[18] Office of Statistical Standards, *Standard Metropolitan Statistical Areas,* pp. 1–3.
[19] Henry S. Shryock, Jr., "The Natural History of Standard Metropolitan Areas," *American Journal of Sociology,* 63(2):163–170, September 1957.

Rico. By 1970 the Bureau of the Budget had increased the number in the United States proper to 243 and in Puerto Rico to 4.

Presentation. — The population composition of individual SMSA's is shown in two different ways, for example:

	1960 population
Chicago, Ill., SMSA	6,220,913
Cook County	5,129,725
Du Page County	313,459
Kane County	208,246
Lake County	293,656
McHenry County	84,210
Will County	191,617
Chicago, Ill., SMSA	6,220,913
Chicago	3,550,404
Outside central city	2,670,509

The part of an SMSA outside the central city is also referred to as the "outlying part" or the "metropolitan ring."

Standard Consolidated Areas .—In view of the special importance of the metropolitan complexes around New York and Chicago, the Nation's two largest cities, several contiguous SMSA's and additional counties that do not appear to meet the formal integration criteria but do have strong interrelationships of other kinds have been combined into the New York-Northeastern New Jersey and the Chicago-Northwestern Indiana Standard Consolidated Areas, respectively. The former consists of Middlesex and Somerset Counties in New Jersey and the following SMSA's: New York, Newark, Jersey City, and Paterson-Clifton-Passaic. The latter consists of the following SMSA's: Chicago and Gary-Hammond-East Chicago.

Urbanized Areas. — The urban agglomeration known as the metropolitan district was replaced in 1950 not only by the standard metropolitan area but also by the urbanized area. The distinction between these two concepts was explained above in the section on "Urban Agglomerations." In brief, the latter may be viewed as the physical city, the built-up area that would be identified from an aerial view, whereas the former also includes the more thinly settled area of the day-to-day economic and social influence of the metropolis in the form of worker commutation, shopping, newspaper circulation, etc. Probably the greatest justification for setting up still another type of urban agglomeration, however, was the resulting improvement of the urban-rural classification.

Each urbanized area consists of a central city or cities and a densely settled residential belt outside the city limits that is called the "urban fringe." There is generally one urbanized area in each standard metropolitan statistical area. Sometimes, however, there are two because there exists another qualifying city with 50,000 inhabitants or more whose surrounding urban fringe is separated from the urban fringe of the larger central city or cities. (The Chicago metropolitan area had three urbanized areas in 1970). In other cases, a single urbanized area covers portions of two or more standard metropolitan statistical areas. One metropolitan area (New London-Groton-Norwich, Conn.) had no urbanized area in 1970.

Unincorporated Places. — Persons living in areas outside incorporated places fall into three main groups (with marginal zones between): those who live in unincorporated population centers, those who live just outside the limits of cities in closely built-up suburban areas, and those who live in the open country with dwellings relatively dispersed. The unincorporated population centers or places are towns, villages,

or hamlets. They are not recognized as entities by law, and they have no legally established boundaries or limits, as incorporated places do. These unincorporated places were settled just like incorporated places, and they physically resemble incorporated places of the same size.

"Each unincorporated place possesses a definite nucleus of residences and has its boundaries drawn so as to include, if feasible, all the surrounding closely settled area." These unincorporated places are thus conceptually similar to urbanized areas but are located at the lower end of the size scale. By the nature of the delineation process, they should qualify as "true-bounded." Also like the urbanized areas, they have permitted a more consistent and meaningful urban-rural classification in that unincorporated places of 2,500 or more are defined as urban.

Census County Divisions. — The origin and nature of census county divisions ("CCD's") were described briefly in the subsection on the complementary "minor civil divisions." The relationship between these two types of county subdivisions is analogous to that between unincorporated and incorporated places, the former being statistical artifacts and the latter, political, or administratively defined, areas. From another standpoint census county divisions could be viewed loosely as the nonmetropolitan counterpart of the census tracts in metropolitan areas. (For a more specific definition of CCD's see: *U.S. Census of Population: 1960,* Vol. I, *Characteristics of the Population,* Part 1, *U. S. Summary,* 1964, p. XXIX).

Census Tracts. — Wards, election precincts, and other types of political areas within cities are not usually the most appropriate areas for statistical analysis of variation by locality within cities. Many of these areas are frequently changed so that variations over time cannot be studied. To meet this need, most large cities and their metropolitan areas have been divided into small areas called "census tracts." Most tracts have a population between 3,000 and 6,000. They are designed to include an area fairly homogeneous with respect to ethnic origin, economic status, and living conditions. Although tract boundaries are intended to be permanent, changes are sometimes made, particularly in the case of tracts that have had such a large population growth that they need to be subdivided. On the whole, the preservation of fixed boundaries is regarded as more basic than the preservation of homogeneity within a tract.

The census tract idea began with Walter Laidlaw, who divided New York City into tracts for the census of 1910. Tracts are laid out by local committees representative of statistical consumers, but the plans must meet certain standards set by the Bureau of the Census before their acceptance. Tract statistics from the 1970 census are published for 241 areas, 238 in the United States and 3 in Puerto Rico. For a further discussion of census tract data and their uses, see U.S. Bureau of the Census, *Census Tract Manual,.* 5th ed., 1966.

Enumeration Districts. — Enumeration districts, the administrative field areas used in taking a decennial census, are described in chapter 3. No data have been published by the Census Bureau for ED's, but these areas were used in the first regular tabulation of recent censuses, initially as a machine control but later to meet some of the needs for small-area statistics.

Blocks. — Beginning with the 1940 Census of Housing, blocks in cities of 50,000 inhabitants or more at the preceding census were numbered, and statistics and analytical maps were published using the block as a unit. In 1960, under special arrangements, the block statistics program was extended to 172 smaller cities as well. There was a total of about 737,000 blocks in the block-numbered areas. For the first time, the population total was also tabulated for blocks.[20] From the 1970 census, four demographic measures (percent of the total population that is Negro, percent living in group quarters, percent under 18 years old, and percent 62 years old and over) are also included. In the 1970 census program, 278 Block Statistics Reports were published for urbanized areas or for "selected areas" specified by local authorities.

Conclusions. — We think of geographic factors as being relatively stable and unchanging. And yet this section has reported a picture of continuous change over the last few decades in the ways developed for presenting data on the geographic distribution of population in the United States. The next chapter describes still further changes. With a highly developed, expanding economy and a highly mobile population, the significant classifications for examining population distribution cannot remain static if they are to be functionally adequate. And yet some degree of comparability of classification used in successive decades must be maintained to afford a basis of revealing trends and permitting historical analyses. This is an ever present dilemma in the planning of population censuses, which faces statistical agencies in other countries as well. If there were no changes made, the concepts and definitions would increasingly fail to describe the current situation. If each census were planned afresh, with no regard to what had been done in the past, there would be no basis for studying trends. An intermediate alternative is to introduce improvements, but in the year they are introduced to make at least some data available on both the old and the new basis. This is the policy the Census Bureau adopted with respect to several changes introduced in 1950. In the case of other changes where such a procedure is not feasible, the Current Population Survey has been used not only to pretest the feasibility of new proposals, but also to provide a basis for linking the old and the new classifications.

POPULATION DENSITY

The density of population is a simple concept much used in studies relating population size to resources and in ecological studies. This simple concept has a number of pitfalls, however; some of which are discussed below.[21] Density is usually computed as population per square kilometer, or per square mile, of land area rather than of gross area (land and water).[22] The 1972 *Demographic Yearbook* of the United Nations gives population per square kilometer for continents and regions (table 1) and for countries (table 2), as estimated for 1971. Table 5–2 is abstracted from table 1 in the *Yearbook*. A few populous countries (Taiwan, Japan, South Korea, Belgium, and England and Wales) now have density in excess of 250 persons per square kilometer. From 500 to 2,000, the country is likely to be a relatively small island (Barbados, Bermuda, the Channel Islands, and Malta); beyond 2,000, the country is essentially a city (Singapore, Hong Kong, Macao, East and West Berlin, Monaco, and Gibraltar).

At the other extreme, countries with considerable parts of their land area in deserts, mountains, tropical rain forests, icecaps, etc., have very low densities. The most thinly settled

Table 5–2. – **Estimated Population, Area, and Density, for Major Areas of the World: 1971**

Major area	Estimated midyear population (millions)	Area[1] (square kilometers in thousands)	Density
World Total.	3,706	135,783	27
Africa.	354	30,320	12
America.	522	42,083	12
Asia.	2,104	27,532	76
Europe.	466	4,936	94
Oceania.	20	8,510	2
U.S.S.R.	245	22,402	11

[1] Comprising land area and inland waters but excluding uninhabited polar regions and some uninhabited islands.

Source: United Nations, *Demographic Yearbook, 1972*, table 1.

countries of all tend to be close to the Arctic or Antarctic circles. Even if we use the area of the ice-free portion of Greenland, its density is only 0.1 per square kilometer. Even Canada has a density of only 2.

These illustrations suggest that, for some purposes, more meaningful densities are obtained for a country or region by relating the size of its population to the amount of agricultural land. On this basis, the densities are often much greater, of course.

Another measure of population density has been suggested by George.[23] His measure relates to the "ratio between the requirements of a population and the resources made available to it by production in the area it occupies." [24] The ratio is $\Delta e = \dfrac{Nk}{Sk'}$, where N is the number of inhabitants, k the quantity of requirements per caput, S the area in square kilometers, and k' the quantity of resources produced per square kilometer. George concludes, however, that, "It is impossible to make a valid calculation of economic density in an industrial economy . . .". Duncan, Cuzzort, and Duncan have discussed the conceptual difficulties in comparing the population density of different areas.[25]

If there have been no changes in boundaries, the change in population density for a different date is, of course, simply proportionate to the change in population size. Thus, if the population has increased 10 percent, the density has also increased 10 percent.

United States

The land area and population of the United States in 1970 were as follows:

United States

Land area (square miles).	3,536,855
Population per square mile.	57.5
Population per square kilometer.	22.2

---

[20] *U.S. Census of Housing: 1960*, Vol. III, *City Blocks*, Series HC(3), Nos. 1 to 421, 1961 to 1962, table 2.

[21] See also: John I. Clarke, *Population Geography*, Oxford (England), Pergamon Press, 1965, pp. 28–32.

[22] 1 square kilometer (km²) = 0.386103 square miles; 1 square mile = 2.58998 km².

[23] Pierre O. L. George. "Sur un project de calcul de la densité économique de la population." (On a project for calculating the economic density of the population.), in United Nations, *Proceedings of the World Population Conference, 1954* (Rome), Vol. IV, New York, 1955, pp. 303–313.

[24] Ibid. p. 313.

[25] Otis Dudley Duncan, Ray P. Cuzzort, and Beverly Duncan, *Statistical Geography*, Free Press of Glencoe, Illinois, 1961, pp. 35–38.

METHODS OF ANALYSIS

There are a number of measures for describing the spatial distribution of a population and many graphic devices for portraying population distribution and population density. The interdisciplinary nature of demography is particularly displayed in this field. Geographers, statisticians, sociologists, and even physicists have contributed to it. Duncan gives the following classification, which is not claimed to be exhaustive or free of overlapping:

A. Spatial measures
    (1) Number and density of inhabitants by geographic subdivisions
    (2) Measures of concentration
    (3) Measures of spacing
    (4) Centrographic measures
    (5) Population potential
B. Categorical measures
    (1) Rural-urban and metropolitan-nonmetropolitan classification
    (2) Community size distribution
    (3) Concentration by proximity to centers or to designated sites [26]

In this book, topics B (1) and (2) are treated more fully in chapter 6 than in chapter 5.

Percentage Distribution

A simple way of ordering the statistics that is appropriate for any demographic aggregate is to compute the percentage distribution living in the geographic areas of a given class. Table 5-3 is an illustration. Note that the change given in the last column is in terms of percentage points, i.e., the numerical difference between the two percentages. The percentages as rounded may not add exactly to 100.0. In such cases, however, it is conventional not to force the distribution to add exactly or to show the total line as 99.9, 100.1, etc. Where there is a very large number of geographic areas and many would contain less than 0.1 percent of the population, the percentages should be carried out to two decimal places. Alternatively, the figures may be shown per 1,000 rather than per 100.

Table 5-3. — **Percent Distribution by Provinces of the Population of Cuba: 1953 and 1943**

Province (Provincia)	1953		1943		Change in percent, 1943 to 1953
	Number	Percent of total	Number	Percent of total	
Cuba, total.......	5,829,029	100.0	4,778,583	100.0	(X)
Pinar del Rico........	448,422	7.7	398,794	8.3	-0.6
La Habana............	1,538,803	26.4	1,235,939	25.9	+0.5
Matanzas............	395,780	6.8	361,079	7.6	-0.8
Las Villas...........	1,030,162	17.7	938,581	19.6	-1.9
Camagüey............	618,256	10.6	487,701	10.2	+0.4
Oriente.............	1,797,606	30.8	1,356,489	28.4	+2.4

X Not applicable.

Source: Adapted from, Cuba, Oficina nacional de los censos demografico y electoral, *Censos de población viviendas y electoral, 1953, Informe general,* 1955, table 2, p. 1.

Rank

Another common practice is to include a supplementary table listing the geographic areas of a given class in rank order. Again, the rankings can be compared from one census to another and the changes in rank indicated. Table 5-4 gives an illustration for the "urban areas" of New Zealand. In cases of an exact tie, it is conventional to assign all tying areas the average of the ranks involved, for example, if two areas tied for seventh place, they would both be given a rank of 7½. The choice of sign for the change in rank requires a little reflection. It was decided to use the sign resulting from the literal arithmetic operation even though a minus sign denotes a move toward the "top" rankings.

Table 5-4. — **Population and Rank of Urban Areas in New Zealand (Excluding Maoris): 1945 and 1936**

Urban area	1945		1936		Change in rank, 1936 to 1945
	Population	Rank	Population	Rank	
Auckland................	258,467	1	210,393	1	-
Wellington..............	172,320	2	149,382	2	-
Christchurch............	149,570	3	132,282	3	-
Dunedin.................	83,191	4	81,848	4	-
Invercargill............	27,464	5	25,682	5	-
Palmerston North........	27,091	6	23,953	7	-1
Hamilton................	25,945	7	19,373	8	-1
Wanganui................	25,767	8	25,312	6	+2
New Plymouth............	20,229	9	18,194	11	-2
Napier..................	19,821	10	18,443	10	-
Hastings................	19,741	11	17,715	12	-1
Timaru..................	19,569	12	18,805	9	+3
Nelson..................	16,483	13	13,545	14	-1
Gisborne................	16,111	14	15,521	13	+1

— Represents zero.

Source: Adapted from, New Zealand, Census and Statistics Department, *Population Census, 1945,* Vol. 1, *Increase and Location of Population,* 1947, p. ix.

Measures of Average Location and of Concentration

There has long been an interest in calculating some sort of average point for the distribution of population within a country or other area. Both European and American statisticians have contributed to this concept.[27] The most popular measures are the median point or location; the mean point, often called the "center of population"; and the point of minimum aggregate travel. A somewhat different concept is that of the point of maximum "population potential." There has been somewhat less scientific interest in measuring the concentration, or dispersion, of the population. Here we will describe Bachi's "standard distance." Average positions and dispersion, density surfaces, etc., are treated systematically by Warntz and Neft.[28]

Median Lines and Median Point. — The "median lines" are two orthogonal lines (at right angles to each other), each of which divides the area into two parts having equal numbers of inhabitants. The "median point" is the intersection of these two lines. The median lines are conventionally the north-

---

[26] Otis Dudley Duncan, "The Measurement of Population Distribution," *Population Studies* (London) 11(1):27, July 1957.

[27] Roberto Bachi, "Graphical Representation and Analysis of Geographical-Statistical Data", *Bulletin of the Internation Statistical Institute* (Proceedings of the 35th Session, Belgrade, 1965), 41(1):225, Belgrade, 1966.

[28] William Warntz and David Neft, "Contributions to a Statistical Methodology for Areal Distributions," *Journal of Regional Science,* 2(1): 47–66, Spring 1960.

south and east-west lines, but the location of the median point depends slightly upon how these axes are rotated.[29]

Hart and others also mention that, in addition to median lines that divide a territory into halves in terms of population, other common fractions may be used, such as quarters and tenths. For the population and area of the United States in 1960, equal tenths ("decililes") have been computed in the north-south and the east-west directions.[30] These devices describe population distribution rather than central tendency, of course.

Center of Population. — The center of population, or mean point of the population distributed over an area, may be defined as the center of population gravity for the area, ". . . in other words, the point upon which the [area] would balance, if it were a rigid plane without weight and the population distributed thereon, each individual being assumed to have equal weight and to exert an influence on the central point proportional to his distance from the point. The pivotal point, therefore, would be its center of gravity . . ."[31] The formula for the coordinates of the center of population may be written

$$\bar{x} = \frac{\Sigma p_i x_i}{\Sigma p_i} \text{ and } \bar{y} = \frac{\Sigma p_i y_i}{\Sigma p_i}, \tag{1}$$

where p_i is the population at point i and x_i and y_i are its horizontal and vertical coordinates, respectively.

Thus, the mean point, unlike the median point, is influenced by the distance of a person from it. It is greatly affected by extreme items and is influenced by *any* change of the distribution over the total area. In the United States, for example, a change in Alaska or Hawaii, which is far removed from the center of population, exerts a much greater leverage than a change in Illinois, the State where the center is now located.

Hart outlines a simple method of calculating the center of population from a map, which is parallel to his method for locating the median point.[32] This graphic method is suitable for only a relatively small area where a map projection like a Mercator projection does not distort too much the relative distances along different parallels of latitude; i.e., where it may be assumed that equal distances in terms of degrees represent equal linear distances.

A more exact method for computing the center of population and one that is required when dealing with a very large area is described by the set of equations shown below.

$$\bar{x} = \frac{\Sigma p_a (x_a - \bar{x}') - \Sigma p_b (\bar{x}' - x_b)}{\Sigma p_i} + \bar{x}' \tag{2}$$

$$\bar{y} = \frac{\Sigma p_c (y_c - \bar{y}') - \Sigma p_d (\bar{y}' - y_d)}{\Sigma p_i} + \bar{y}' \tag{3}$$

where $\bar{x}'$ and $\bar{y}'$ are the coordinates of the assumed mean, x_a is any point east of that mean, x_b, any point west of it, y_c, any point north of it, y_d, any point south of it; and p_a, p_b, p_c, p_d are the populations in areas east, west, north, and south of the assumed mean, respectively.

The procedure is described in several publications of the U.S. Bureau of the Census as follows: "Through this point [the assumed center] a parallel and a meridian are drawn, crossing the entire country.

"The product of the population of a given area by its distance from the assumed meridian is called an east or west moment. In calculating north and south moments the distances are measured in minutes of arc; in calculating east and west moments it is necessary to use miles on account of the unequal length of the degrees and minutes in different latitudes. The population of the country is grouped by square degrees—that is, by areas included between consecutive parallels and meridians—as they are convenient units with which to work. The population of the principal cities is then deducted from that of the respective square degrees in which they lie and treated separately. The center of population of each square degree is assumed to be at its geographical center except where such an assumption is manifestly incorrect; in these cases the position of the center of population of the square degree is estimated as nearly as possible. The population of each square degree north and south of the assumed parallel is multiplied by the distance of its center from that parallel; a similar calculation is made for the principal cities; and the sum of the north moments and the sum of the south moments are ascertained. The difference between these two sums, divided by the total population of the country, gives a correction to the latitude. In a similar manner the sums of the east and of the west moments are ascertained and from them the correction in longitude is made."[33]

For a large area, adjustments should be made for the sphericity of the earth.

The location of the center of population, unlike that of the median point, is independent of the particular axes chosen. The calculation of the center of population for a large country is well suited to programming for an electronic computer. There it is feasible to introduce an additional refinement for the sphericity of the earth.

For illustrative computations of the center of population (and the median point), see the unabridged edition of *The Methods and and Materials of Demography*, pp. 136–141.

Table 5–5, shows the movement of the center of population of the United States from 1790 to 1960. Note the difference between the locations for the "United States" (50 States) and "conterminous United States" (48 States). Table 5–6 gives the movement of the center of the Jewish population of Israel from 1922 to 1962. The definition of the **center of area** is analogous to that of the center of population, but the computation is somewhat simpler. In some countries those two centers may be a great distance apart. Thus, in 1960, the center of population of the conterminous United States was in Illinois, whereas the center of area was 822 kilometers to the northwest in Kansas.[34]

In 1970 the center of population of the United States was located in St. Clair County, Illinois at a point about five miles east-southeast of the city of Mascoutah.

In the last decades of the nineteenth and the early decades of the twentieth century, there was great interest in the concept of center of population and in the mean location of many other units that are reported in censuses. For example, the *Statistical Atlas* published as part of the 1920 census of the United

[29] John Fraser Hart, "Central Tendency in Areal Distributions," *Economic Geography*, 30(1):54, January 1954.

[30] U.S. Bureau of the Census, "Zones of Equal Population in the United States: 1960," *Geographic Reports*, GE–10, No. 3, October 1963.

[31] U.S. Bureau of the Census, *Statistical Atlas of the United States, 1924*, p. 7.

[32] Hart, op. cit., pp. 50–54.

[33] U.S. Bureau of the Census, *Statistical Atlas of the United States, 1924*, pp. 7–8.

[34] U.S. Geological Survey, *Geographic Centers of the United States*, M10–8, October 1967 (pamphlet).

Table 5–5. — **Center of Population of the United States: 1790 to 1960**

Census year	North latitude	West longitude	Approximate location
	° ′ ″	° ′ ″	
United States: 1960.........	38 35 58	89 12 35	In Clinton County, Ill., 6 1/2 miles northwest of Centralia.
1950.........	38 48 15	88 22 8	3 miles northeast of Louisville, Clay County, Ill.
Conterminous United States: 1960.........	38 37 57	88 52 23	4 miles east of Salem, Marion County, Ill.
1950.........	38 50 21	88 9 33	8 miles north-northwest of Olney, Richland County, Ill.
1940.........	38 56 54	87 22 35	2 miles southeast by east of Carlisle, Haddon township, Sullivan County, Ind.
1930.........	39 3 45	87 8 6	3 miles northeast of Linton, Greene County, Ind.
1920.........	39 10 21	86 43 15	8 miles south-southeast of Spencer, Owen County, Ind.
1910.........	39 10 12	86 32 20	In the city of Bloomington, Ind.
1900.........	39 9 36	85 48 54	6 miles southeast of Columbus, Ind.
1890.........	39 11 56	85 32 53	20 miles east of Columbus, Ind.
1880.........	39 4 8	84 39 40	8 miles west by south of Cincinnati, Ohio (in Kentucky).
1870.........	39 12 0	83 35 42	48 miles east by north of Cincinnati, Ohio.
1860.........	39 0 24	82 48 48	20 miles south by east of Chillicothe, Ohio.
1850.........	38 59 0	81 19 0	23 miles southeast of Parkersburg, W. Va.[1]
1840.........	39 2 0	80 18 0	16 miles south of Clarksburg, W. Va.[1]
1830.........	38 57 54	79 16 54	19 miles west-southwest of Moorefield, W. Va.[1]
1820.........	39 5 42	78 33 0	16 miles east of Moorefield, W. Va.[1]
1810.........	39 11 30	77 37 12	40 miles northwest by west of Washington, D.C. (in Virginia).
1800.........	39 16 6	76 56 30	18 miles west of Baltimore, Md.
1790.........	39 16 30	76 11 12	23 miles east of Baltimore, Md.

[1] West Virginia was set off from Virginia Dec. 31, 1862, and admitted as a State June 19, 1863.

Source: *U.S. Census of Population: 1960*, Vol. I, *Characteristics of the Population*, Part A, *Number of Inhabitants*, 1961, p. XI, table B.

States, gave the center of population for individual States, of the Negro population, and of the urban and rural population, and the mean point of the number of farms. This tradition has been revived to some extent by the Israeli demographer, Roberto Bachi, who has computed or compiled centers of population for a variety of countries and population subgroups.[35]

The center of population, being merely the arithmetic mean of the population distribution, need not fall in a densely settled part of the country. In fact, the center of population of an archipelago may be in the sea. This is one of the circumstances that has led the astronomer John Q. Stewart and the geographer William Warntz to regard the concept of center of population as being more misleading than useful.[36] Stewart's concept of "population potential" is discussed below. Nevertheless,

[35] Roberto Bachi, "Standard Distance Measures and Related Methods for Spatial Analysis," *Regional Science Association: Papers X, Zurich Congress,* 1962, pp. 83–132.

[36] John Q. Stewart and William Warntz, "Some Parameters of the Geograph-

there seems to be real merit in Hart's view that the center of population is a useful summary measure for studying the shifts of population over time.[37]

Point of Minimum Aggregate Travel. — This centrographic measure, sometimes called the "median center," is defined as "that point which can be reached by all items of a distribution with the least total straight line travel for all items," or "the point from which the total *radial* deviations of an areal distribution are at a minimum."[38] Hart gives a graphic method for locating this point. This concept has fairly obvious applications to location theory, e.g., to estimating the optimum central location for a public or private service of some sort.

Table 5–6. — **Center of Population and Standard Distance for Jews in Israel: 1922 to 1967**

Date	Center of population[1]		Standard distance (in km.)
	$\bar{x}$	$\bar{y}$	
1922............................	158.7	169.0	50.9
1931............................	151.8	170.6	46.4
1948 (Nov. 8)...................	143.8	182.0	45.0
1949 (Dec. 31)..................	143.7	181.9	43.1
1950 (Dec. 31)..................	144.2	181.7	46.9
1951 (Dec. 31)..................	144.3	180.9	47.6
1952 (Dec. 31)..................	144.1	180.1	47.7
1953 (Dec. 31)..................	144.0	179.8	47.8
1954 (Dec. 31)..................	143.8	179.4	48.1
1955 (Dec. 31)..................	143.8	178.9	49.1
1956 (Dec. 31)..................	143.8	178.1	49.9
1957 (Dec. 31)..................	143.6	177.7	50.6
1958 (Dec. 31)..................	143.6	177.3	51.0
1959 (Dec. 31)..................	143.5	177.2	51.3
1961 (Census-May 22)............	143.2	175.6	51.8
1962 (Dec. 31)..................	144.9	173.8	52.4
1963 (Dec. 31)..................	144.9	173.1	53.4
1964 (Dec. 31)..................	144.8	172.6	53.9
1965 (Dec. 31)..................	144.7	172.2	54.1
1966 (Dec. 31)..................	144.6	171.8	54.1
1967 (Dec. 31)..................	144.6	171.6	54.2

[1] According to "conventional coordinates" in kilometers, i.e., as measured from an arbitrary point within the country.

Source: Israel, Central Bureau of Statistics, *Statistical Abstract, 1968*, No. 19, 1968.

Standard Distance. — Measures of the dispersion of population have been proposed from time to time, but the most recent exposition and the one that has been most thoroughly developed is Bachi's "standard distance."[39] The standard distance bears the same kind of relationship to the center of population that the standard deviation of any frequency distribution bears to the arithmetic mean. In other words, it is a measure of the dispersion of the distances of all inhabitants from the center of population.

If $\bar{x}$ and $\bar{y}$ are the coordinates of the center of population, say its longitude and latitude, then the distance from any item i, with coordinates x_i and y_i, to the center is given by

$$D_{ic} = \sqrt{(x_i - \bar{x})^2 + (y_i - \bar{y})^2} \qquad (4)$$

and the standard distance,

ical Distribution of Population," *The Geographical Review,* 49(2):270–272, April 1959; William Warntz, "Macrogeography and the Census," *The Professional Geographer,* 10(6):6–10, November 1958.

[37] Hart, op. cit., p. 59.

[38] Hart, op. cit., pp. 56 and 58, respectively.

[39] Bachi, op. cit.; idem, "Statistical Analysis of Geographic Series," *Bulletin de l'Institut international de statistique* (Proceedings of the 30th Session, Stockholm, 1957), 36(2):229–240, Stockholm, 1958.

$$D = \sqrt{\frac{\sum_{i=1}^{n} D_{ic}^2}{n}} \qquad (5)$$

In practice, the distance would not be measured individually for each person but rather we should use data grouped by political areas (or square degrees), and it would then be assumed that the population of a unit area is concentrated at its geographic center. Here, then,

$$D = \sqrt{\frac{\sum_i f_i(x_i - \bar{x})^2}{n} + \frac{\sum_i f_i(y_i - \bar{y})^2}{n}} \qquad (6)$$

where f_i is the number of persons in a particular unit of area.

It is pointed out by Duncan, Cuzzort, and Duncan that the standard distance is much less influenced by the set of areal subdivisions used than are other measures of population dispersion, such as the Lorenz curve.[40] In general, however, the smaller the type of area used as a unit, the more closely will the computed standard distance approach the value computed from the locations of individual persons.

Some standard distances in kilometers were given for Israel in table 5-6. Standard distances can also be drawn on a map. Representing the standard distance by a line segment, we know the length of the line and its origin at the center of the population, but the direction in which it is drawn is purely arbitrary. One could appropriately draw a circle with the standard distance as its radius about the center of population. Since the standard distance is equivalent to one standard deviation (1σ), the circle would indicate the area in which about two-thirds of the population is concentrated. The exact proportion would vary with the specific distribution.

Population Potential. —The concept of population potential as developed by Stewart applies to the accessibility to the population, or "level of influence" on the population, of a point on a map or of a small unit of area.[41] "If the 'influence' of each individual at a point is considered to be inversely proportional to his distance from it, the total potential of population at a point, Lo, is the sum of the reciprocals of the distances of all individuals in the population from the point. In practice, of course, the computation is made by assuming that all the individuals within a suitably small area are equidistant from Lo, whence,

$$\text{Potential at } Lo = \sum_{i=1}^{n} \frac{P_i}{D_i} \qquad (7)$$

where the P_i are the populations of the n areas into which a territory is divided and the D_i are the respective distances of these areas from Lo (usually measured from the geographic centre or from the approximate centre of gravity of the population, in each area)."[42] Like the center of population, the population potential at any point in the territory is thus affected by the distribution of population over the entire territory. When the potential has been computed for a sufficient number of points, those of equal potential may be joined on the map to show contours or isopleths. It can be well appreciated that each computation involves a good deal of labor so that, with a hand calculator, it is feasible to compute only a limited num-

ber of values and to use only relatively large areal units. This process produces a coarse-grained map. To produce a fine-grained map, the computations would need to be performed on an electronic computer. On such a fine-grained map, there would be peaks of potential around every city that are not brought out on most of the available maps showing this measure.

To illustrate, we will show only the first few computations needed to compute the population potential at one particular point. This is a hypothetical case. Let the "point" in question be a capital city A with a population of 100,000. Let this population be P_1. Assume that the population is evenly distributed over the city. Since this "point" is a relatively populous area, it is necessary to take into account the average distance of its own population from its geographic center. Let us say that this has been estimated from the city's map at 3 kilometers.

Then measure the distance from the geographic center of every other political unit in the set being used to the center of the capital city. This set of units should account for all the national territory unless population potential is being studied for some other kind of area, such as a region. These geographic centers can be plotted by inspection; but, where a primary unit has a very large and unevenly distributed population, the secondary divisions within it can be used for increased accuracy. Suppose we then have

Area	P	D	P/D
1	100,000	3	33,333
2	25,000	8	3,125
3	10,000	10	1,000
n	15,000	500	30

The population potential for the city is the sum of the last column. One does not have to work outwards from the area in question while listing the areas; any systematic listing is acceptable. If the latitudes and longitudes of all the centers of geographic area (or, ideally, the centers of population of all the areas) are known, these can be programmed for a computer so that the distances to any point can be computed by triangulation.

Figures 5-1 and 5-2 are maps of population potential for the United States and Europe. Stewart notes that the contours for Europe were based on 93 control points whereas those for the United States were based on only 24. Warntz and Neft point out that, "The peak of population potential coincides with the modal center on the smoothed density surface for the United States."[43] The statement applied to 1950 but presumably it would still hold good.

The large-scale map produced by Warntz for the United States in 1960 was based on 3,105 control points (counties, etc.).[44] The handling of such a large input and the resultant detail in the isopleths shown on the map were made possible by the use of a digital computer.

The concept of population potential is more useful than that of aggregate travel distance, and has sometimes proved valuable as an indicator of geographical variations in social and economic phenomena, e.g., rural population density, farmland values, miles of railway track per square mile, road density, density of wage earners in manufacturing, and death

[40] *Statistical Geography*, p. 93.
[41] John Q. Stewart and William Warntz, "Macrogeography and Social Science," *Geographical Review*, 48(2):170, April 1958.
[42] Duncan, op. cit., pp. 35 and 36.

[43] Warntz and Neft, op. cit., p. 65.
[44] William Warntz, "A New Map of the Surface of Population Potentials for the United States, 1960," *Geographical Review*, 54(2):170-184, April 1964.

Figure 5–1. — Contours of Population Potential for the United States: 1940

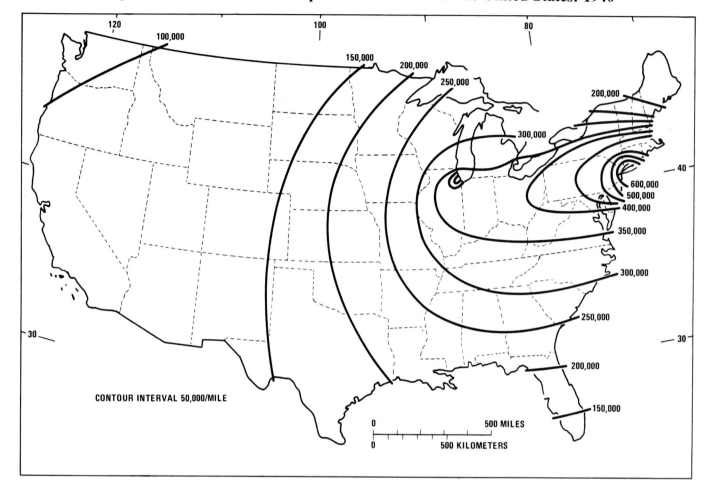

Source: John Q. Stewart, "Empirical Mathematical Rules Concerning the Distribution and Equilibrium of Population," *Geographic Review*, 37(3): 476, July 1947, copyright, 1947, by the American Geographical Society of New York.

rates. Rural density. for example, tends to be proportional to the square of the potential.[45]

Population in Distance Zones from Map Features. — Still another way of summarizing the population distribution of a nation, province, etc., is to compute the absolute or relative number of people within a specified distance from its borders, from its seacoast, or from selected points. The measure can be extended to successive distance zones, e.g., within 10 km., 20 km., etc., inland from the coast. For the population of the United States in 1960, the population living within 50 miles (80 kilometers) of 686 points (mostly cities of 25,000 or more) has been computed.[46] The point with the maximum population within 50 miles is near the point of maximum population potential; but, because a large fraction of the circle centered on New York City lies in the Atlantic Ocean, the most populous circle is farther inland, around White Plains, N.Y., and encompasses almost 16 million persons.

Mapping Devices

There is a voluminous literature on the mapping of demographic data to which demographers, geographers, and mem-

bers of other disciplines have contributed.[47] Here we are concerned with mapping just the distribution of population and of population density.

Population Distribution. — The commonest method of representing the distribution of the absolute number of inhabitants is a dot map, in which a small dot or spot of constant size represents a round number such as 100 or 1,000. (Special pen points are available for this purpose.) The dot is one of the devices used in figure 5–3 . If a general impression is all that is wanted, the dots may be plotted more or less uniformly within the units of area given on the map. For a more exact portrayal, regard should be paid to any actual concentrations of population within the unit areas. This procedure calls for referral to figures for geographic subdivisions below the level of those outlined on the map. For example, with a county outline map of the United States, one could refer to the published figures for minor civil divisions or for incorporated places.

In maps of population distribution for a country or other area containing both thinly settled rural territory and large

[45] Clarke, op. cit., p. 39.
[46] U.S. Bureau of the Census, "Population Within 50 Miles of Selected Points: 1960," *Geographic Reports*, GE–10, No. 1, April 1963 (revised).

[47] Roberto Bachi, "Geographical Representation and Analysis of Geographical-Statistical Data," *Bulletin of the International Statistical Institute*, Belgrade, 1966 (Proceedings of the 35th Session, Belgrade, 1965), 41(1):219–238; Calvin F. Schmid, *Handbook of Graphic Presentation*, Ronald Press Company, New York, 1954, pp. 184–222.

Figure 5–2. — **Contours of Population Potential per Kilometer in Europe: About 1939**

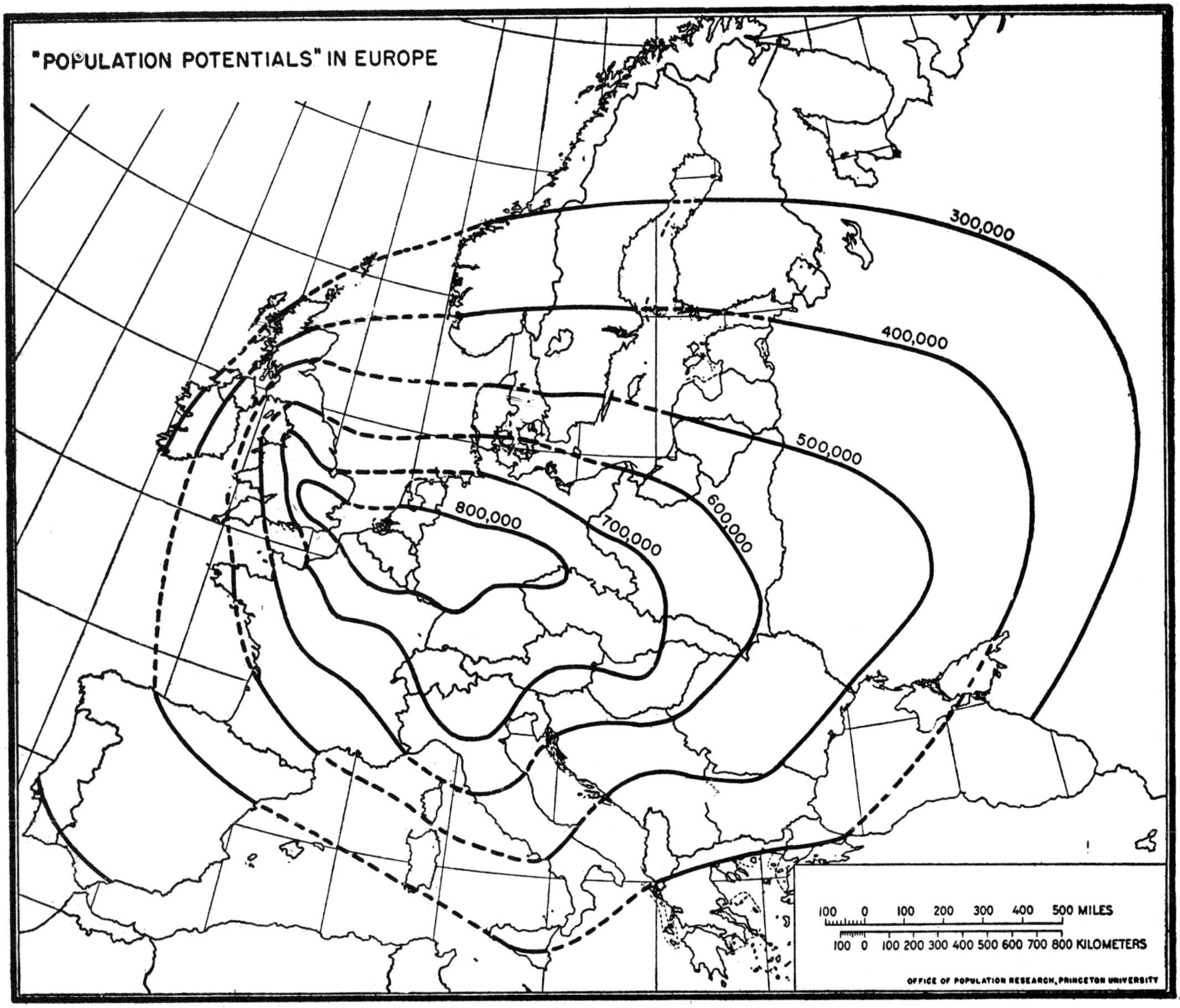

Source: Dudley Kirk, *Europe's Population in the Interwar Years*, Princeton, N.J., Princeton University Press for the League of Nations, 1946, p. 9.

urban agglomerations, there is a real problem in the application of the conventional dot method. A dot that represents few enough people to show the distribution of the rural population requires so many plottings within the limits of large cities that one sees only a solid black area and even that will grossly underrepresent the actual number of dots required. To portray the population of large cities, one could use a dot of the same size but of a different color to which a higher value is assigned, for example, the black dot could represent 100 people and the red dot, 10,000. Another variation is to use circles of varying size for specific urban places. (See figure 5–3.) Such circles (or other graphic symbols) may be chosen in a limited number of sizes, or forms, e.g.,

- •   2,500 to 10,000
- ●   10,000 to 25,000
- ☉   25,000 to 50,000

or, especially for larger cities, the circle may be drawn with the area proportional to the size of the population.

In the latter case, it is best to start with the largest place and determine the size of circle that can reasonably be accommodated on the map. (Since a number of the circles will overlap and will extend beyond the areas to which they apply, they should be either "open," i.e., unshaded, or shaded in a light tint so that boundary lines can show through.) Suppose a circle with a diameter of 5 cm. is chosen to represent a city of 500,000. Then, since the population is drawn proportionate to the **area** of the circle, and the area is πr^2, the radius required for a smaller population is solved by the following equation:

$$\frac{\pi r^2}{\pi \times 6.25} = \frac{P}{500,000} \qquad (8)$$

or,

Figure 5–3. – **Population Distribution of the United States: 1960**

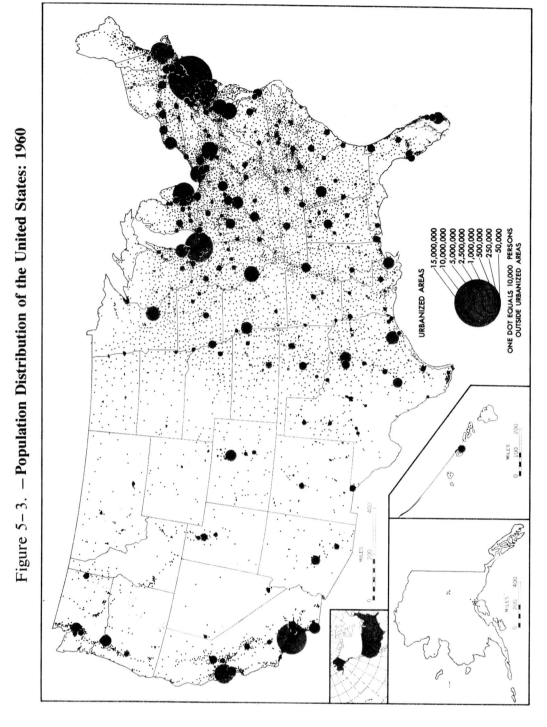

Source: *U.S. Census of Population: 1960*, Vol. 1, *Characteristics of the Population*, Part 1, *United States Summary*, p. S 17, figure 18.

Figure 5–4. – Population of El Salvador, by Departments: May 1961

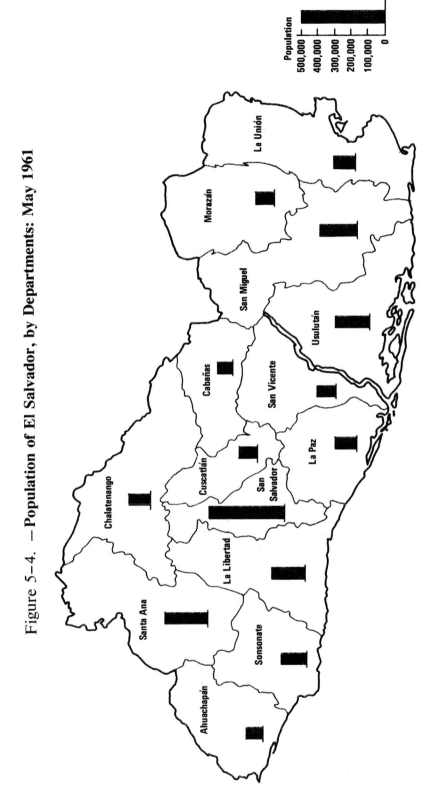

Source of data: El Salvador, Dirección general de estadística y censos, *Tercer censo nacional de poblacion, 1961*, June 1965, table 1 (map traced from *El Salvador en graficas*, 1966. p. 3).

$$r = \sqrt{\frac{P}{80,000}}\, cm. \qquad (9)$$

so that, for a population of 100,000, a circle with a radius of 1.12 cm. is needed. (Note that the radius varies with the square root of the population.)

To represent very wide ranges of population size, spherical symbols can be used instead of circles for the largest localities. The population of the large localities would then be proportional to the volume of the sphere implied.

Other graphic devices are sometimes used to denote the population in a geographical area, for example, the heights of a rectangle (two-dimensional bar) or of a three-dimensional column shown in perspective. (See figure 5–4.) Such devices are convenient for only a relatively small number of areal units, such as the primary divisions of a country.

Population Density.—A conventional way of indicating population density is that of shading or hatching, with the darker shadings representing the greater densities.[48] Such shadings may gloss over considerable internal variation within an area since they represent simply the area's average density. Figure 5–5 gives population density in 1960 for South Korea using the country's secondary administrative divisions (gun). The contour or isopleth map also lends itself to the presentation of geographic regularities in population density. Some of the problems, considerations, and techniques in the construction of such maps are discussed by Duncan[49] and by Schmid.[50]

FACTORS AFFECTING POPULATION DISTRIBUTION

Much has been written about the factors determining the geographic distribution of population—among international regions, among countries, among national divisions and subdivisions, and within urban agglomerations. Some of this literature was summarized in a United Nations study about 15 years ago, and the numerous references contained therein are also useful.[51] Further consideration has been given by Clarke, who as a geographer, gives particular attention to altitude and describes the "vertical distribution" of population.[52]

The following list of factors is adapted from *The Determinants and Consequences of Population Trends*, with a few additions and other changes:

Climate
    Temperature
    Precipitation
Landforms
    Topography (altitude and slope)
    Swamps, marshes, deserts
Soils
Energy resources and mineral raw materials
Space relationships (accessibility)
    Distance from sea coast, natural harbors, navigable
        rivers
    "Fall lines" (heads of river navigation)
Cultural factors
    Historical
        Recency of discovery and settlement

---

[48] For types of shadings available, see Schmid, op. cit., pp. 187–198.
[49] Duncan, op. cit., p. 28.
[50] Schmid, op. cit., pp. 212–219.
[51] United Nations, *The Determinants and Consequences of Population Trends,* Volume I, Series A, Population Studies No. 50, 1973, pp. 159–173.
[52] Clarke, op. cit., pp. 17–18.

Figure 5–5.—Population Density of South Korea, by Gun: December 1, 1960

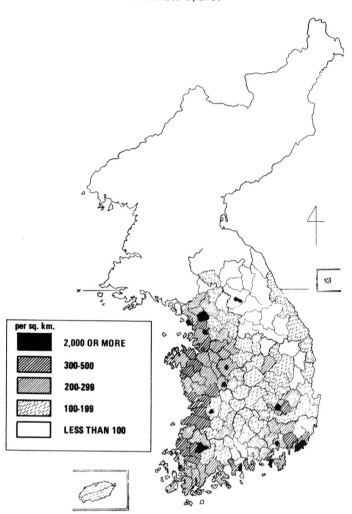

per sq. km.	
■	2,000 OR MORE
▨	300–500
▨	200–299
▨	100–199
□	LESS THAN 100

Source: Republic of Korea, Economic Planning Board, *1960 Population and Housing Census of Korea,* Vol. I, *Complete Tabulation Report,* 11–1, *Whole Country,* Seoul, 1963, p. 4.

Political
    Political boundaries
        Buffer zones
        Controls on migration and trade
    Government policies
Types of economic activities
Technology
    State of the arts
    Type of farming
    Highway, rail, water, and air transportation
        facilities
Social organization
Demographic factors
    Variations in natural increase
    Variations in net migration

These factors are obviously not all of coordinate importance. Thus, the cultural factors modify the effects of the natural and physical factors, and all of them operate through the demographic factors. A discussion of how to measure the relationships among all these factors, on the one hand, and population distribution, on the other, as well as the causal sequences involved, would require more space than can be spared in this book.

Study of the relationship between population density and some of these factors, notably the climatic and physiographic ones, is facilitated by the fact that, in some census reports, there are tables cross-classifying population or density with these factors. See, for example, the classification of the Italian population by topographic zones (mountains, hills, plains) in the census of 1961.[53] Whole communes were used as the units in this classification. A set of tables in a report of the U.S. Census of 1890 presented population and population density by: drainage basins, topographical features, altitude, mean annual temperature, mean annual rainfall, mean relative humidity of atmosphere, and latitude and longitude.[54]

A rather specialized literature deals with the distribution of population within cities or metropolitan areas. In addition to some of the publications already mentioned in this chapter, we may mention a couple of illustrative analyses given elsewhere. Using small-area statistics of the census-tract type for various large cities in Europe, America, and Australia, Clark found that the falling off of population density (except in the central business zone) from the center of the city followed a simple exponential equation. If A and b are constants, e is the base of the system of natural logarithms, x is the distance from the center of the city, and y is the density of resident population, then,

$$y = Ae^{-bx} \tag{10}$$

Expressing distances in miles and density in thousands of people per square mile, Clark has computed A and b for a number of cities at dates back into the nineteenth century.[55] Duncan and Duncan have related residential distribution to "workplace potential."[56] Workplace potential is analogous to Stewart's population potential.

[53] Italy, Istituto centrale di statistica, *10° Censimento generale della popolazione, 15 ottobre 1961*, Vol. I, *Dati riassuntivi comunali e provinciali sulla popolazione e sulle abitazioni* (Summary facts for communes and Provinces on population and dwellings), Rome, 1963; table 3.
[54] U.S. Census Office, *Report on Population of the United States at the Eleventh Census: 1890*, Part I, 1895, pp. xxxvii–lxiv.

[55] Colin Clark, "Urban Population Densities," *Journal of the Royal Statistical Society*, Series A, 114(4):490–496, 1951.
[56] Beverly Duncan and Otis Dudley Duncan, "The Measurement of Intracity Locational and Residential Patterns," *Journal of Regional Science*, 2(2):37–54, 1960.

SUGGESTED READINGS

Bachi, Roberto, "Graphical Representation and Analysis of Geographical-Statistical Data." *Bulletin of the International Statistical Institute* (Proceedings of the 35th session, Belgrade, 1965), 41(1):225, Belgrade, 1966.

———. "Standard Distance Measures and Related Methods for Spatial Analysis." *Regional Science Association, Papers X, Zurich Congress*, 1962. Pp. 83–132.

———. "Statistical Analysis of Geographic Series." *Bulletin of the International Statistical Institute* (Proceedings of the 30th session, Stockholm, 1957), 36(2):229–240, Stockholm, 1958.

Beale, Calvin L. "State Economic Areas—A Review After 17 Years." Washington, *American Statistical Association, Proceedings of the Social Statistics Section*, 1967. Pp. 82–85.

Bénard, Edouard. "Contribution à l'étude des agglomérations françaises." *Population* (Paris), 7(1):95–108, Jan.–March 1952.

Bogue, Donald J., and Beale, Calvin L. *Economic Areas of the United States*. New York, Free Press of Glencoe, 1961.

Clark, Colin. "Urban Population Densities." *Journal of the Royal Statistical Society*, Series A, 114(4):490–496, 1951.

Clarke, John I. *Population Geography*. Oxford, England, Pergamon Press, 1965.

Duncan, Otis Dudley. "The Measurement of Population Distribution." *Population Studies* (London), 11(1):27–45, July 1957.

Duncan, Otis Dudley, Cuzzort, Ray P., and Duncan, Beverly. *Statistical Geography*. Glencoe, Illinois, Free Press of Glencoe, 1961.

Hart, John Fraser. "Central Tendency in Areal Distributions." *Economic Geography*, 30(1):54, January 1954.

India, Registrar-General. Census of India [1951], Paper No. 2. *Population Zones, Natural Regions, Sub-Regions and Districts*. 1952.

International Urban Research. *The World's Metropolitan Areas*. Berkeley, University of California Press, 1959.

Odum, Howard W. *Southern Regions of the United States*. Chapel Hill, University of North Carolina Press, 1936.

Odum, Howard W., and Moore, Harry Estill. *American Regionalism: A Cultural-Historical Approach to National Integration*. New York, Henry Holt and Co., 1938.

Schmid, Calvin F. *Handbook of Graphic Presentation*. New York, Ronald Press Co., 1954. Pp. 184–222.

Shryock, Henry S., Jr. "The Natural History of Standard Metropolitan Areas." *American Journal of Sociology*, 63(2):163–170, September 1957.

Stewart, John Q., and Warntz, William. "Macrogeography and Social Science." *Geographical Review*, 48(2):167–184, April 1958.

Stewart, John Q., and Warntz, William. "Some Parameters of the Geographical Distribution of Population." *Geographical Review*, 49(2):270–272, April 1959.

Taeuber, Conrad. "Regional and Other Area Statistics in the United States," *Bulletin of the International Statistical Institute* (Proceedings of the 35th session, Belgrade, 1965), 41(1):161–162.

United Nations. *Demographic Yearbook, 1971*. Pp. 15–16, 22–23.

SUGGESTED READINGS – Continued

————. *The Determinants and Consequences of Population Trends, Vol. I, Series A, Population Studies No. 50*, 1973. Pp. 159-173.

————. *Proceedings of the World Population Conference, 1954* (Rome), Vol. IV:

"Utilité d'un système international de répartition en région et subdivisions de région, pour l'élaboration et l'analyse des statistique démographiques," by Donald J. Bogue. Pp. 481–490.

"Urban Theory and Concepts in Relation to the Definition of Urban Agglomerations," by Hope T. Eldridge. Pp. 581–588.

"Conurbations in England and Wales," by L. M. Feery. Pp. 615–626.

"Notes on the Concept of 'City' and of 'Agglomeration,'" by G. Goudswaard and J. Schmitz. Pp. 685–695.

"The Definition of the Urban Agglomeration and Its Influence Upon Urban Population Volume," by Miloš Macura. Pp. 741–758

"Introduction of a System for Defining Agglomerations in French Demographic Statistics," by Jean Porte. Pp. 835–843.

————. *Principles and Recommendations for the 1970 Population Censuses*, Statistical Papers, Series M, No. 44, 1967 (Second Printing, 1969).

U.S. Bureau of the Budget, Office of Statistical Standards. *Standard Metropolitan Statistical Areas*, 1967.

U.S. Bureau of the Census. *Census Tract Manual*. 5th ed., 1966, and subsequent appendices

————. *Metropolitan Area Definition: A Re-Evaluation of Concept and Statistical Practice*, by Brian J. L. Berry. Working Paper No. 28. June 1969.

————. *Statistical Atlas of the United States*, 1924. Pp. 7–24.

————. *U.S. Census of Population: 1970.* Vol. 1, *Characteristics of the Population*, Part 1, United States Summary - Section 2, 1973. Pp. App.-1 to 12.

Warntz, William. "A New Map of the Surface of Population Potentials for the United States, 1960." *Geographical Review*, 52(2): 170–184, April 1964.

Warntz, William, and Neft, David. "Contributions to Statistical Methodology for Areal Distributions." *Journal of Regional Science*, 2(1): 47–66, Spring 1960.

Whittlesey, Derwent. "The Regional Concept and the Regional Method," in *American Geography: Inventory and Prospect*. Preston E. James and Clarence F. Jones, eds. Published for the Association of American Geographers by Syracuse University Press, 1954.

CHAPTER 6

Population Distribution—Classification of Residence

This chapter is more or less a continuation of the preceding one, and the allocation of some topics between them is somewhat arbitrary. The emphasis here is on classes of geographic areas of residence that are formed primarily for statistical purposes rather than on individual areas as entities. The resulting groupings are not contiguous pieces of territory. The main classifications discussed are the urban-rural; the suburban; size of locality, of metropolitan area, etc.; farm-nonfarm; and combinations of these.

THE URBAN-RURAL CLASSIFICATION

As in the case of certain other phenomena, there is no problem in classifying extreme cases; the population of Calcutta is undoubtedly urban; the population of Buffalo County, South Dakota, with only 1,700 inhabitants is undoubtedly rural. It is in the case of intermediate situations that somewhat arbitrary rules have to be set up for making the urban-rural classification.

International Recommendations

The latest recommendations of the United Nations on this subject are those made for the 1970 censuses, under "Topics to be investigated:" Because of national differences in the characteristics which distinguish urban from rural areas, the distinction between urban and rural population is not yet amenable to a single definition which would be applicable to all countries. For this reason, each country should decide for itself which areas are urban and which are rural.

National Practices

What then are these different definitions of urban that are used by the several national statistical organizations? (It can be said at the outset that the rural population is almost always not defined directly but is simply the residual population after the urban population is distinguished.) The Statistical Office of the United Nations has classified the definitions in use into five principal types.[1] The underlying concept is that of (1) administrative area, (2) population size, (3) local government area, (4) urban characteristics, or (5) predominant economic activity.

Administrative Area.—This concept treats as urban the administrative divisions (municipios, shi, districts, communes, Gemeinden, etc.) that have been so classified by the national government or such parts of them as their administrative centers, capitals, or principal localities. This classification is based primarily on historical, political, or administrative considerations, rather than on statistical considerations. It tends to be relatively static and is not automatically changed after each census to recognize the decreasing size of formerly important places or the increasing size of places that have recently become important.

Population Size.—This concept treats as urban those places (cities, towns, agglomerations, localities, etc.) having either a specified minimum number of inhabitants or a specified minimum population density. ("Population density" could well be regarded as an independent concept in this set.) The discussion of this concept in the U.N. *Handbook* cited recognizes that suburbs of large places, densely populated fringes around incorporated municipalities, and the like are sometimes classified as urban in this approach. This type of classification can readily be brought up-to-date at the time of each census, but it is almost impossible to do so in each postcensal year for classifications used in tabulations of data from sample surveys or registrations of vital events.

Within a fairly broad range, the choice of the minimum qualifying value is arbitrary, although efforts may have been made to base the cutting-score originally on statistics regarding the presence or absence of various facilities or functions in places of given sizes. The variety of minimum sizes that are or have been in use in national statistical programs reflects the lack of consensus on these matters, however. It also reflects the fact that a place of a given size in a small country with a subsistence-agricultural economy will be relatively more important in that country than in a highly industrialized populous country like Japan. Similarly, a size that might have represented a fairly important place in feudal Japan might not be very important in modern Japan. At one time the minimum required size used by New Zealand to record the growth of its urban population increased from census to census.[2] Over a

[1] United Nations, *Handbook of Population Census Methods*, Vol. III, *Demographic and Social Characteristics of the Population*, Studies in Methods, Series F, No. 5, rev. 1, 1959, pp. 60–62.

[2] New Zealand, Census and Statistics Office. *Population Census, 1926*, Vol. I, *Increase and Geographical Distribution*, Wellington, 1927, p. 5.

time period of at least medium duration, however, it seems better to retain the same minimum size in a country's urban definition so as to facilitate the study of urbanization in the longitudinal sense.[3] Other supplementary measures discussed below may be used to bring out the historical shifts in the meaning of the specific definition of urban.

Local Government Area.—This concept defines urban in terms of those places, agglomerations, or localities "possessing some form of local government." For some countries, at least, this phrase might be interpreted as referring to a particular form of local government, especially one having relatively great autonomy. The American equivalent would be the incorporated municipality (city, town, borough, or village). Other terms cited by the *Handbook* include "chartered towns," "local government areas," "municipal communities," and "burghs." Note that no minimum population size is used in this definition. The remarks about the relatively static character of classifications under concept (1) apply here also.

Urban Characteristics.—This concept requires an urban place to possess specific types of urban characteristics, such as established street patterns, contiguously aligned buildings, and public services (sewer system, piped water supply, electric lighting, police station, hospital, school, court of law, and local transportation system). A classification of this sort would need to be developed during or shortly before the census canvass. This concept has the possibility of regular up-dating, therefore. Some of the characteristics, like electric lighting, could only have been used in modern times.

Predominant Economic Activity.—Here places or other areas qualify as urban if they have at least a specified proportion of their economically active population engaged in nonagricultural activities. It would be difficult to base the urban-rural classification on the statistics regarding economic activity from the current census unless that classification were presented only in a report or reports appearing relatively late in the publication program.

Use of These Concepts.—For 51 censuses taken in the period

1945–1954, the United Nations has classified, as far as possible, the urban-rural definition used in terms of these five concepts. It will be noted from table 6–1 that most of the countries—32 out of 51—used definitions based on combinations of two or more concepts.

To what extent these national differences in definitions represent true differences in culture, settlement patterns, or system of administrative areas and to what extent they are simply a matter of historical tradition or of a somewhat understandable bureaucratic inertia, would be difficult to determine.

A better idea of the range of variation in the national definitions of "urban" is provided in Table 6-2, which has been extracted from table 9 of the 1962 *Demographic Yearbook* of the United Nations.

Several writers have contended that a threefold classification would describe the situation better than a dichotomy,[4] and a few countries have actually used such a classification in their official statistics.[5] This use of a third, "semiurban," category should be distinguished from classifications that subdivide the urban or rural total. If the intermediate category is defined purely in terms of population size, however, the resulting threefold classification may be viewed as a simple case of the distribution of localities by size. (See below.)

Sources

International statistics on the urban population of the countries of the world, total and by selected characteristics, are published in various issues of the United Nations *Demographic Yearbook*.

Weber's *The Growth of Cities in the Nineteenth Century* is a convenient source of early historical figures for various countries.[6] There are also historical data in a French book of about the same date.[7] Of course, the national census volumes themselves and other official national statistical reports often give more detailed statistics and more explanatory material than is available from these secondary sources.

Concepts of Urban and Rural in Demographic Theory

The writings of demographers, sociologists, etc., about the concept of urban population or of urbanization, from the theoretical standpoint may help us in our consideration of operational definitions. Eldridge (formerly Tisdale), for example, defines urbanization as ". . . a process of population concentration. It proceeds in two ways: the multiplication of points of concentration and the increase in size of individual concentrations . . . Consistent with the definition of urbanization, cities may be defined as points of concentration . . . It is convenient from time to time arbitrarily to name certain levels beyond which concentrations are designated as cities."

Table 6–1.—**Number of Countries Using Each Urban-Rural Concept or Combination of Concepts, by Region: Censuses in Period 1945 to 1954**

[See text for definitions of concepts]

Urban-rural concepts	Total[1]	Africa	America	Asia	Europe	Oceania
Total............	51	3	20	8	18	2
(1)..................	8	1	7			
(2)..................	9		5		4	
(3)..................	2			1	1	
(1)+(2)..............	8			1	7	
(1)+(3)..............	6			2	4	
(1)+(4)..............	3		3			
(2)+(3)..............	2		1	1		
(2)+(4)..............	4	1	3			
(1)+(2)+(3)..........	5	1		2		2
(1)+(2)+(5)..........	1				1	
(2)+(3)+(4)..........	2			1	1	
(1)+(2)+(4)+(5)......	1		1			

[1] For 51 censuses only. No definition was available for one European country.

Source: United Nations, *Handbook of Population Census Methods*, Vol. III, *Demographic and Social Characteristics of the Population*, 1959, p. 62.

---

[3] For a discussion of the cross-sectional and longitudinal aspects of urbanization, see: Otis Dudley Duncan and Albert J. Reiss, Jr., *Social Characteristics of Urban and Rural Communities, 1950*, New York, John Wiley & Sons, 1956, pp. 19–20.

---

[4] For example, Willcox suggested the trichotomy—city (urban), village (semiurban), and country, with the definition being based on density and with the class-marks being 1,000 per square mile and 100 per square mile. In a small-scale study in New York State, he thought he had found some evidence of these two natural "breaks" in the density continuum. Walter F. Willcox, *Studies in American Demography*, Ithaca (New York), Cornell University Press, 1940, p. 117.

[5] The 1962 *Demographic Yearbook* mentions an intermediate category consisting of populated centers of 1,000 to 2,499 inhabitants used by Venezuela. Intermediate categories were used, at least at one time, in the classifications for the Netherlands (communes of 5,000 to 20,000) and Spain (municipios of 2,000 to 10,000). See: United Nations, *Data on Urban and Rural Population in Recent Censuses*, Series A, Population Studies, No. 8, July 1950, p. 8.

[6] Adna Ferrin Weber, *The Growth of Cities in the Nineteenth Century*, Ithaca (New York), Cornell University Press, 1963. Originally published in 1899 for Columbia University by the Macmillan Company, New York, as Vol. XI of *Studies in History, Economics and Public Law*.

[7] Paul Meuriot, *Des agglomérations urbaines dans l'Europe contemporaine* (Urban agglomerations in contemporary Europe), Paris, Belin Frères, 1898.

Table 6-2. — Urban Population of Selected Countries at Latest Census in Period 1955 to 1962

[Numbers in thousands]

Country and year	Population			Definition
	Total	Urban		
		Number	Percent	
AFRICA				
Algeria, 1960.........................	10,205	3,314	32.5	Fifty-five most important communes having local self-government.
Ghana, 1960..........................	6,727	1,551	23.1	Localities of 5,000 or more inhabitants.
Morocco, 1960........................	11,626	3,411	29.3	Prefectures, municipalities and autonomous, delimited, and other urban centers.
Tanganyika, 1957.....................	8,788	364	4.1	Thirty-three gazetted (declared) townships (towns).
Uganda, 1959.........................	6,537	159	2.4	The fifteen largest towns.
United Arab Republic, 1960...........	25,771	9,719	37.7	Cities, including the five largest cities, which are also governorates, and the capitals of provinces and districts.
NORTH AMERICA				
El Salvador, 1961....................	2,511	967	38.5	Administrative centers of departments, districts, and municipios.
Honduras, 1961.......................	1,883	575	30.5	Localities of 1,000 or more inhabitants having essentially urban characteristics.
Jamaica, 1960........................	1,614	381	23.6	Kingston metropolitan area.
United States, 1960..................	179,323	125,269	69.9	Incorporated and unincorporated places of 2,500 or more inhabitants, including the urbanized zones around cities of 50,000 or more inhabitants.
SOUTH AMERICA				
Brazil, 1960.........................	70,967	31,991	45.1	Urban and suburban zones of administrative centers of municipios and distritos.
Venezuela, 1961......................	7,524	5,079	67.5	Populated centers (centros poblados) of 1,000 or more inhabitants, of which those having 1,000 to 2,499 inhabitants are classified as intermediate between urban and rural.
ASIA				
Federation of Malaya, 1957...........	6,279	2,680	42.7	Towns and villages of 1,000 or more inhabitants.
Indonesia, 1960......................	97,019	14,358	14.8	NA.
Israel, 1961.........................	2,179	1,698	77.9	All settlements of more than 2,000 inhabitants, except those where at least one-third of the heads of households, participating in the labor force, earn their living from agriculture.
Pakistan, 1961.......................	93,721	12,295	13.1	Municipalities, civil lines, cantonments not included within municipal limits, any other continuous collection of houses inhabited by not fewer than 5,000 persons and having urban characteristics, and also a few areas having urban characteristics but fewer than 5,000 population.
Viet-Nam, North, 1960................	15,917	1,519	9.5	Cities.
EUROPE				
Albania, 1960........................	1,391	383	27.5	Towns, and other industrial centers with more than 400 inhabitants.
Austria, 1961........................	7,073	3,536	50.0	Communes (Gemeinden) of more than 5,000 inhabitants.
Finland, 1960........................	4,446	2,487	55.9	Nonadministrative agglomerations, i.e., almost all groups of buildings occupied by at least 200 people and with usually not more than 200 meters between houses.
Greece, 1961.........................	8,389	3,628	43.3	Municipalities and communes having 10,000 or more inhabitants in the largest population center.
Ireland, 1956........................	2,898	1,285	44.3	Cities and towns, including suburbs, of 1,500 or more inhabitants.
Poland, 1960.........................	29,361	14,112	48.1	Towns and settlements of urban type, e.g., workers' settlements, fishermen's settlements, health resorts.
United Kingdom, 1961 England and Wales.................	46,072	36,838	80.0	Area classified as urban for local government purposes, i.e., county boroughs, municipal boroughs and urban districts.
OCEANIA				
New Zealand, 1961....................	2,415	1,556	64.4	Cities, boroughs, and town districts.
U.S.S.R.				
U.S.S.R., 1959.......................	208,827	99,978	47.9	Cities, towns, and urban-type localities.

Source: Adapted from United Nations, *Demographic Yearbook, 1962*, table 9. Footnotes omitted.

[Authors' note, i.e., as "urban".] [8] She rejects definitions of cities ". . . as ways of life, states of mind, collections of traits, types of occupation and the like. Such definitions are bound to get us in trouble sooner or later because none of the attributes named are constants of the city and all of them spill over into other areas. Traits change, occupations change, political organization changes, the economic system changes. The only trait that is constant is that the city is different from what is not the city." [9]

Dewey takes a very similar position. He shows that there is very little consensus in the literature regarding matters that distinguish urbanism from ruralism. "However, it is not logical to hold that the rural-urban continuum has no universal or general referents, merely because of the mistaken assignment to urbanism of a welter of cultural items which can be, and are, independent of city environments. In the first place, it is probably futile to argue for the abandonment of the terms 'rural' and 'urban' as indicators of size and density of population . . . The inclusion of both population and cultural bases in the term 'urbanism' renders it useless except for labeling timebound phenomena." [10]

Charles T. Stewart, however, rejected size of place as the proper criterion of "urban" versus "rural" and preferred an economic or a sociological concept such as the population of workers in the area who are engaged in nonagricultural pursuits or the sociocultural folk-urban distinction. "Demographic concepts of urban and rural fail because they apply the same rules of numbers to advanced and backward countries, to farmers and peasants, to villagers and townsmen." [11] Some European social scientists share Stewart's views.[12]

The present authors find that the urban-rural concept advanced by Eldridge and Dewey is useful in demographic analysis. In many, if not most countries, it is associated with fairly sharp differentials in demographic, social, and economic characteristics for persons or families. The precise population size chosen to separate urban from rural is obviously fairly arbitrary within a wide range. Nevertheless, there would be some advantages in a uniform definition among all countries.

A fixed classification in terms of size of place (or population density) can serve as a framework against which differences in various characteristics viewed as representing contemporary urban character (industrial composition, presence of specified municipal services, smaller families, etc.) and the changes over time in these characteristics can be measured. The argument that the urban-rural distinction is outmoded in some industrialized countries, including the United States, because the entire population is already "urbanized" with respect to attitudes or way of life becomes somewhat irrelevant to this approach. As a matter of fact, however, there seem to be no valid national data that show that such a homogenization of mental outlook has actually taken place, and the statistics that relate to way of life and demographic characteristics show that appreciable differences still exist between the inhabitants of urban and rural territory even though there may have been

measurable convergence over time.[13] Urban-rural differences with respect to certain demographic characteristics in the opposite direction from those observed in industralized countries, or in some cases the absence of such differences, have been reported for contemporary Egypt.[14]

Urban-Rural Classification in the United States

In discussing the definitions of urban and rural territory used in official American statistics, we will first describe briefly the historical development of the classification in census statistics, then give the 1970 definition, and finally give the situation in the Current Population Survey and in vital statistics. The definition used in the United States has always been based on the principle of size of locality, but refinements have been introduced in an attempt to obtain a better and more inclusive definition of a locality.

Development of the classification. — The most comprehensive account of the early development of the urban-rural classification in the United States is given by Truesdell.[15] Although such a classification had been considered as early as the time of the census of 1850, the first such figures actually published were contained in the 1874 publication, *Statistical Atlas of the United States,* and even these were simply the by-product of the preparation of maps showing population density by counties. The urban population was defined as that living in incorporated places of 8,000 or more, and a summary table was prepared showing that population back to 1790. This work was done under the direction of Francis A. Walker, who made many contributions to demographic statistics. The fact that only incorporated places (cities, towns, boroughs, and villages) were considered for classification as urban should be emphasized since this was a limitation that obtained, with a few exceptions, until 1950.

The problem of how to classify the New England town was explicitly examined in the census of 1880. As was explained in chapter 5, towns in that section of the country are the equivalent not of incorporated places in other sections of the country but of minor civil divisions (e.g., townships). Some towns, it is true, contained within their borders one or more population agglomerations that would have qualified as urban had they been incorporated. According to Truesdell, all towns of 8,000 or more in their entirety seem to have been classified as urban in 1880.

In 1880 the qualifying size for an urban place was lowered from 8,000 to 4,000; but the former value was given some attention in the presentation, particularly in historical tables, through 1930. It was not until the publication of the *Supplementary Analysis* volume in 1906, which reorganized some of the data previously published from the census of 1900, that the present minimum size of 2,500 for individual places classed as urban was introduced. It was also in 1880 that the urban population was first made available by States. Urban and rural totals by counties were published in the 1910 reports, and

[8] Hope Tisdale, "The Process of Urbanization," *Social Forces,* 20(3):311, March 1942.

[9] Ibid., p. 312.

[10] Richard Dewey, "The Rural-Urban Continuum: Real but Relatively Unimportant," *American Journal of Sociology,* 66(1):63-64, July 1960.

[11] Charles T. Stewart, Jr., "The Urban-Rural Dichotomy: Concepts and Uses," *American Journal of Sociology,* 64(2):158, September 1958.

[12] For example: Francisco Benet, "Sociology Uncertain: The Ideology of the Rural-Urban Continuum," *Comparative Studies in Society and History* (The Hague), 6(1):5-6, October 1963.

[13] For evidence regarding the persistence of attitudinal differences between urban and rural residents in modern societies, see: Norval D. Glenn, "Massification Versus Differentiation: Some Trend Data from National Surveys," *Social Forces,* 46(2):172-179, December 1967.

[14] Janet Abu-Lughod, "Urban-Rural Differences as a Function of the Demographic Transition: Egyptian Data and an Analytical Model," *American Journal of Sociology,* 69(5):476-490, March 1964.

[15] U.S. Bureau of the Census, *Current Population Reports, Population Characteristics,* "The Development of the Urban-Rural Classification in the United States: 1874 to 1949," by Leon E. Truesdell, Series P-23, No. 1, August 5, 1949

comparative figures for 1900 were also included. Characteristics of the urban and rural populations of the larger types of areas were also first published in the 1910 reports.

Little or no change in the urban-rural classification occurred in the U.S. censuses of 1920 and 1940; and the relatively slight change in 1930 concerned the classification as urban under special rules of large and densely settled minor civil divisions, mostly in the Northeast, that did not contain any incorporated places. Similar special rules were introduced for the 1960 census but were abandoned again in 1970 when a more discriminating definition in terms of unincorporated localities (agglomerations) was substituted.

Fundamental improvements in the definition were made for the 1950 census. It had become increasingly obvious that confining the territory officially classified as urban almost entirely to incorporated places of 2,500 or more left in the rural classification millions of people who lived in urban communities according to any realistic view of the matter. The special rules used in 1930 and 1940 merely scratched the surface of this problem and employed rather crude units (the MCD's) for that purpose. The special rules had added only 1,667,000 persons, or 2.4 percent, to the urban total in 1930, whereas the new definition in 1950 added 9,917,000 persons, or 11.5 percent, above the number living in incorporated places of 2,500 or more.

There were two main developments in the pattern of population settlement that cried out for recognition in the urban-rural classification. First, the built-up areas of large cities were bursting beyond their legal boundaries in the process known as "urban sprawl." Unincorporated suburban communities with tens of thousands of inhabitants (e.g., Silver Spring, Md., adjacent to Washington) were still classified as rural. Second, there were outside these suburban areas, about 400 agglomerations of 2,500 inhabitants or more that had never been incorporated because of the laws or customs of the States in which they were located. The solution was to delineate the urban fringes around cities of 50,000 or more and to delineate the outlying unincorporated villages and hamlets down to the size level of about 800 inhabitants— although only those localities with a population of at least 2,500 enumerated in the census were classified as urban. These delineations required very careful geographic fieldwork and mapping. Determining the boundaries was not entrusted to the enumerators; that had been tried before with unsatisfactory results.

The urban fringes and the unincorporated urban places represented to some extent, net rather than gross gains to the urban territory and population. Contained within the former were some satellite incorporated places of 2,500 or more. Both types of areas included all or parts of the minor civil divisions previously classified as urban under the special rules. The relationships for 1960 are shown in table 6-3.

The 1970 census definition. - The following definition is quoted in full from the text of U.S. Census of Population: 1970, *Number of Inhabitants,* Final Report PC (1) - Al, *United States Summary* (p.1X).

"According to the definition adopted for use in the 1970 census, the urban population comprises all persons living in urbanized areas and in places of 2,500 inhabitants or more outside urbanized areas. More specifically, the urban population consists of all persons living in (a) places of 2,500 inhabitants or more incorporated as cities, villages, boroughs (except Alaska), and towns (except in the New England States, New York, and Wisconsin), but excluding those persons living in the rural portions of extended cities; (b) unincorporated places of 2,500 inhabitants or more; and (c) other territory, incorporated or unincorporated, included in urbanized areas.

"In censuses prior to 1950, the urban population comprised all persons living in incorporated places of 2,500 or more and areas (usually minor civil divisions) classified as urban under special rules relating to population size and density. The most important component of the urban territory in any definition is the group of incorporated places having 2,500 inhabitants or more. A definition of urban territory restricted to such places, however, would exclude a number of large and densely settled areas merely because they are not considered "incorporated places." Prior to 1950, an effort was made to avoid some of the more obvious omissions by inclusion of selected areas which were classified as urban under special rules. Even with these rules, however, many large and closely built-up areas were excluded from the urban territory.

"To improve its measure of the urban population, the Bureau of the Census adopted, in 1950, the concept of the urbanized area and delineated, in advance of enumeration, boundaries for unincorporated places. With the adoption of the urbanized area and unincorporated place concepts for the 1950 census, the urban population was defined as all persons residing in urbanized areas and, outside these areas, in all places incorporated or unincorporated, which had 2,500 inhabitants or more. With the following two exceptions, the 1950 definition of urban was continued substantially unchanged to 1960 and 1970. In 1960 (but not 1970), certain towns in the New England States, townships in New Jersey and Pennsylvania, and counties elsewhere were designated as urban. However, most of the population of these "special rule" areas would have been classified as urban in any event because they were residents of an urbanized area or an unincorporated place of 2,500 or more. Second, the introduction of the concept of "extended cities" in 1970 has very little impact on the urban and rural figures generally.

"In all urban and rural definitions, the population not classified as urban constitutes the rural population.

"*Extended cities.* - Over the 1960-1970 decade there has been an increasing trend toward the extension of city boundaries to include territory essentially rural in character. Examples are city-county consolidations such as the creation of the city of Chesapeake, Virginia, from South Norfolk city and Norfolk County and the extension of Oklahoma City, Oklahoma, into five counties. The classification of all the inhabitants of such cities as urban would include in the urban population persons whose environment is primarily rural in character. In order to separate these people from those residing in the closely settled portions of such cities, the Bureau of the Census examined patterns of population density and classified a portion or portions of each such city as rural. These cities—designated as extended cities—thus consist of an urban part and a rural part. An extended city contains one or more areas, each of at least 5 square miles in extent and with a population density of less than 100 persons per square mile according to the 1970 census. The area or areas constitute at least 25 percent of the land area of the legal city or total 25 square miles or more. The delineation of extended cities was limited to cities in urbanized areas."

Urban-rural classification in the CPS. — For many years the U.S. Census Bureau's reports on demographic subjects that were based on the Current Population Survey included numerous cross-classifications with urban-rural residence. The urban-rural classification was perforce that of the last census, which meant not only that the same definition was applied but that almost exactly the same territory was classified as urban (and as rural) since there were no firm figures on sub-

Table 6–3.—**Population in Groups of Incorporated and Unincorporated Places and Urban Towns and Townships Classified According To Size, for the United States: 1960**

Size of place and urban-rural classification	All places		Incorporated places		Unincorporated places		Urban towns and townships[1]	
	Number	Population	Number	Population	Number	Population	Number	Population
Total......................	19,790	125,808,073	18,088	115,910,865	1,576	6,583,649	126	3,313,559
500,000 or more....................	21	28,595,050	21	28,595,050	...	...	...	...
250,000 to 500,000.................	30	10,765,881	30	10,765,881	...	...	...	...
100,000 to 250,000.................	81	11,652,426	79	11,384,755	1	104,270	1	163,401
50,000 to 100,000..................	201	13,835,902	180	12,511,961	9	585,104	12	738,837
25,000 to 50,000...................	432	14,950,612	366	12,720,406	26	858,459	40	1,371,747
20,000 to 25,000...................	200	4,450,070	165	3,670,604	17	379,067	18	400,399
10,000 to 20,000...................	934	13,118,216	813	11,391,075	84	1,186,783	37	540,358
5,000 to 10,000....................	1,394	9,779,714	1,282	9,030,786	104	688,135	8	60,793
2,500 to 5,000.....................	2,152	7,580,028	1,763	6,237,739	379	1,304,265	10	38,024
2,000 to 2,500.....................	896	1,997,875	734	1,635,978	162	361,897	...	...
1,500 to 2,000.....................	1,334	2,307,124	1,048	1,812,513	286	494,611	...	...
1,000 to 1,500.....................	2,241	2,742,745	1,733	2,121,687	508	621,058	...	...
500 to 1,000.......................	3,267	2,341,061	3,267	2,341,061	...	...	...	...
200 to 500.........................	4,153	1,389,190	4,153	1,389,190	...	...	...	...
Under 200..........................	2,454	302,179	2,454	302,179	...	...	...	...
In urbanized areas..................	2,430	85,997,382	2,204	80,177,353	109	2,679,492	117	3,140,537
500,000 or more....................	21	28,595,050	21	28,595,050	...	...	...	...
250,000 to 500,000.................	30	10,765,881	30	10,765,881	...	...	...	...
100,000 to 250,000.................	81	11,652,426	79	11,384,755	1	104,270	1	163,401
50,000 to 100,000..................	201	13,835,902	180	12,511,961	9	585,104	12	738,837
25,000 to 50,000...................	232	8,015,421	172	5,969,619	23	772,178	37	1,273,624
20,000 to 25,000...................	101	2,240,476	69	1,529,779	15	331,785	17	378,912
10,000 to 20,000...................	423	6,090,162	328	4,705,102	61	886,155	34	498,905
5,000 to 10,000....................	399	2,862,099	392	2,809,001	...	...	7	53,098
2,500 to 5,000.....................	346	1,250,219	337	1,216,459	...	...	9	33,760
2,000 to 2,500.....................	112	249,559	112	249,559	...	...	...	...
1,500 to 2,000.....................	86	149,220	86	149,220	...	...	...	...
1,000 to 1,500.....................	122	152,177	122	152,177	...	...	...	...
500 to 1,000.......................	140	101,825	140	101,825	...	...	...	...
200 to 500.........................	91	32,620	91	32,620	...	...	...	...
Under 200..........................	45	4,345	45	4,345	...	...	...	...
Other urban territory..............	3,611	29,420,263	3,091	26,820,650	511	2,426,591	9	173,022
25,000 to 50,000...................	200	6,935,191	194	6,750,787	3	86,281	3	98,123
20,000 to 25,000...................	99	2,209,594	96	2,140,825	2	47,282	1	21,487
10,000 to 20,000...................	511	7,028,054	485	6,685,973	23	300,628	3	41,453
5,000 to 10,000....................	995	6,917,615	890	6,221,785	104	688,135	1	7,695
2,500 to 5,000.....................	1,806	6,329,809	1,426	5,021,280	379	1,304,265	1	4,264
Rural territory....................	13,749	10,390,428	12,793	8,912,862	956	1,477,566	...	...
2,000 to 2,500.....................	784	1,748,316	622	1,386,419	162	361,897	...	...
1,500 to 2,000.....................	1,248	2,157,904	962	1,663,293	286	494,611	...	...
1,000 to 1,500.....................	2,119	2,590,568	1,611	1,969,510	508	621,058	...	...
500 to 1,000.......................	3,127	2,239,236	3,127	2,239,236	...	...	...	...
200 to 500.........................	4,062	1,356,570	4,062	1,356,570	...	...	...	...
Under 200..........................	2,409	297,834	2,409	297,834	...	...	...	...

[1] Includes one urban county (Arlington, Va.) with a population of 163,401.

Source: *U.S. Census of Population: 1960,* Vol. 1, *Characteristics of the Population,* Part 1, *United States Summary,* 1964, table 6.

sequent population growth in specific localities or on incorporations, annexations, etc.

The nature of the current statistics on urban-rural residence was clearly described in the Current Population Reports, (Series P–20, etc.), but some users nonetheless interpreted the statistics as showing that the rural population had begun to grow faster than the urban population. In fact, the staff of the Bureau of the Census was itself surprised by the wide margin by which the urban population as measured in the 1960 census exceeded the level that had been estimated from the Current Population Survey. Of course, it might be argued that the cross-classifications of other characteristics with urban-rural residence were still useful in studying trends in differentials and relationships; but, even here, the omission

from the urban population of several millions of suburbanites living in the as yet unrecognized outer zone of the urbanized areas and the classification of those suburbanites as rural would undoubtedly bias the comparison somewhat. On the other hand, it was useful to be able to compare the 1960 figures representing population in the 1950 urban territory and in the up-dated territory. These were 108.6 million and 125.3 million, respectively.[16] The lower figure indicated that the rural population was growing more than twice as fast as the urban population, whereas the higher one indicated

---

[16] Henry S. Shryock, Jr. "Some Results of the 1960 Census of the United States," *Rural Sociology,* 27(4):463, December 1962; and unpublished estimates.

that the urban population had increased by 29 percent whereas the rural population had decreased very slightly. Comparisons of these figures, then, threw some light on the very great extent to which the actual growth of the urban population was attributable to reclassification of territory. Reclassification accounted for about three-fifths of this growth and natural increase and net migration, for only two-fifths.

In view of these considerations, the regular publication in the current reports of even percentage distributions of various characteristics within urban and rural totals was discontinued in 1963. For 1970, unpublished CPS éstimates give 135,947,000 for the civilian noninstitutional population in the 1960 urban territory as compared with 146,177,000 derived from 1970 census data. Thus comparison indicates that, of the total 24 million increase in the urban population from 1960 to 1970, somewhat more than 10 million was due to net reclassification of territory from rural to urban.

Urban-rural classification in vital statistics.—The annual volumes and other reports presenting vital statistics have rarely used an urban-rural definition that agreed with the contemporary definition in the census of population. From 1909 through 1929, the former was defining the population in incorporated places of 10,000 or more as urban when the latter was using 2,500 or more. The complex current definition that was introduced in the 1950 census was not feasible for use in vital statistics and other sources of statistical data since the necessary geographic coding from addresses would have been prohibitively expensive. In the period from 1950 to 1963, the 1940 census definition, which did not include urbanized areas or unincorporated places, was used in classifications of vital statistics. Because of the cost of coding the individual incorporated places of 2,500 to 10,000 and some concern about the ability to code accurately from information available on the certificates, the National Center for Health Statistics reverted to the 10,000-and-over criterion in the 1964 reports. As was the case in the Current Population Survey, the size of place is always that at the last census and hence lags up to nine years behind the current size. Of course, only those places whose size would have risen above, or dropped below, the critical size during the decade are affected by this circumstance.

SUBURBS

Although many people think of suburbs as a modern phenomenon, they had their prototypes in mediaeval times.

On the other hand, suburban areas have certainly had their efflorescence in modern times, and the end of "urban sprawl" is not yet in sight. It was the development of the steam-railway commuter line in the late nineteenth century, followed by the electric line and all-weather automobile road, that brought about the growth of the suburb as we know it today.

The ideal type of a suburb, then, is a residential area that extends out from the central city following transportation arteries in a stellate shape, is dependent on the central city for many of its services, has an intermediate population density, is growing rapidly by horizontal expansion, and is peopled mainly by upper and middle class families of commuters. But there are exceptions. There are, for example, industrial as well as residential suburbs. The shantytowns outside Seoul and the favelas of Rio de Janeiro creep up the rugged hills, without the help of good transportation, to provide homes for squatters. Relatively inexpensive public housing may also be developed on the outskirts of Asian and Latin American cities to provide homes for working class and lower

middle-class families. A very different type of exception from the stereotype of suburban communities of single-family dwellings each in its own yard is afforded by the high-rise apartments for middle-class families that spring out of the fields 10 kilometers or more beyond Washington. In many old cities, the upper-class families still prefer the convenience and prestige of long-established neighborhoods in the central city so that it is the poorer workers who have the inconvenience of long daily rides on crowded buses. Both types of suburbanization, that of the affluent and that of the poor, are now occuring around some Latin American cities.[17] Mehta also finds the upper middle classes in Indian cities holding on to the desirable residential locations close to the city center and the recent poor in-migrants from villages forced ". . . to settle on outlying tracts of vacant land . . ."[18]

Standards and Definitions

International Standards.—One of the United Nations handbooks recognizes that, "The urban agglomeration has been defined as including the suburban fringe or thickly settled territory lying outside of, but adjacent to, the city boundaries."[19] The United Nations' recommendations for the 1970 population censuses include a table that would give the population of both the city proper and its urban agglomeration.[20] Although it is not suggested, subtraction of the former from the latter would yield a figure that could be regarded as the suburban population of the given city. Summing these remainders would give the total suburban population for these "principal localities."

Attempts at Operational Definition.—A bridge between sociological and demographic theory and an operational definition of a suburb is provided by Dobriner: "A rough 'content analysis' of the relevant literature reveals that suburbs are (1) communities (2) outside the central city and politically independent of it (3) culturally and economically dependent upon the central city, and (4) in general, highly specialized communities particularly along familistic and residential lines. Accordingly, a working definition of the suburb might be: *those urbanized, residential communities which are outside the corporate limits of a large central city, but which are culturally and economically dependent upon the central city.*"[21] Obviously, terms such as "urbanized," "residential," "communities," "large central city," and "culturally and economically dependent" would have to be translated into specific metrics and quantified. Duncan and Reiss give a similar definition and contend that, although peripheral parts of the central city may otherwise have a suburban character, "the only workable criterion yet discovered for identifying suburbs is their political separation from the central municipality."[22] We might add that some contiguous

[17] Marshall Wolfe, "Some Implications of Recent Changes in Urban and Rural Settlement Patterns in Latin America" in: United Nations, *Proceedings of the World Population Conference,* (Belgrade, 1965), Vol. IV, New York, 1967, pp. 457–460.

[18] Surinder K. Mehta, "Patterns of Residence in Poona (India) by Income, Education, and Occupation (1937–65)," *American Journal of Sociology,* 73 (4):508, January 1968.

[19] United Nations, *Handbook of Population Census Methods,* Volume III, p. 63.

[20] United Nations *Principles and Recommendations for the 1970 Population Censuses,* p. 75.

[21] William M. Dobriner, ed., *The Suburban Community,* New York, G. P. Putnam's Sons, 1958, p. XVII.

[22] Duncan and Reiss, op. cit., p. 7.

industralized areas resemble the central city more than they do the suburb in all respects except that they are not within the corporate limits of the central city. A contrasting view to that of Duncan and Reiss is that of Douglass, who writes, "Sociologically speaking, however, political status is no final criterion of a suburb, which should include all areas, whether located within or without city boundaries, that reflect the basic suburban characteristics."[23] The difference is partly conceptual and partly a matter of feasibility—to apply Douglass's approach to American census statistics would require painstaking manipulations in terms of census tracts or enumeration districts.

National Practices

Denmark.—Denmark is among the few countries to show an official classification of areas as suburban in its national statistics. According to Goldstein, *"The suburbs of the Capital* consist of nine surrounding municipalities, in which a large proportion of the resident population earn their living in industrial and commercial activities located in the capital. . . . *The suburbs of the provincial towns* represent the built-up areas outside these urban centers and are functionally tied to the centers in the same way in which the Capital's suburbs relate to the Capital."[24]

United States.—It has already been remarked that the United States Bureau of the Census has not as yet provided an official definition of the term "suburbs." It has published statistics, however, for two types of areas that have been used informally to measure the extent of suburbanization. These are: (1) the urban fringe of the urbanized area (i.e., that part of the urban area outside the central city or cities), (2) the metropolitan ring or outlying area (i.e., that part of the metropolitan area outside the central city or cities). The basic terms were defined in chapter 5. If the suburbs are viewed as a peripheral part of the physical city, and therefore entirely urban, rather than as a transitional zone between urban and rural territory, then the former, more restrictive definition would be the preferable one. The zone beyond the urban fringe but within the metropolitan area could be viewed as the "rural-urban" or "rurban" fringe. (See below.) The relationship between the two kinds of peripheral zones is illustrated in table 6–4.

RURAL-URBAN FRINGE

Closely related to the concept of suburbs, and no less vaguely defined, is that of the "rural-urban fringe." It is essentially an American sociological concept that has some similarity to the category "semiurban" that has been described above as used in a few other countries. A critical review of the various definitions that have been propounded is contained in an article by Kurtz and Eicher.[25]

Although akin to the concept of "rurban" developed by the rural sociologist C. J. Galpin, the term "fringe" is credited by Blizzard and Anderson with having first been used in 1937

by T. Lynn Smith.[26] Duncan and Reiss, who devote an entire chapter (ch. 12) to the subject, define it as ". . . that area in which the countryside is in process of transition from a rural to an urban mode of settlement."[27] The definition arrived at by Kurtz and Eicher is more detailed: "location beyond the limits of the legal city, in the 'agricultural hinterland', exhibiting characteristics of mixed land use, with no consistent pattern of farm and nonfarm dwellings. The residents are involved in rural and urban occupations. The area is unincorporated, relatively lax zoning regulations exist, and few, if any, municipal services are provided. The area shows potentialities for population growth and increasing density ratios. Present density ratios are intermediate between urban and rural."[28] Duncan and Reiss include the unincorporated part of the Census Bureau's "urban fringe" in their rural-urban fringe, whereas Kurtz and Eicher tend to regard the rural-urban fringe as being located beyond the urbanized area.[29] These and other writers also exclude even the smallest incorporated places. The present authors find most satisfactory a concept that is implied in Dobriner's idealized diagram, which is reproduced here as figure 6–1. This shows concentric circles beginning with the city proper and proceeding successively to suburban and rural-urban zones. In these outer zones, both incorporated and unincorporated territory is included. In the case of metropolitan cities, the rural-urban fringe would be confined to the metropolitan area, for the most part.

Table 6–4.—**Population of the United States by Residence in Standard Metropolitan Statistical Areas, Urbanized Areas, and Their Component Parts: 1960**

[Numbers in thousands]

Metropolitan-nonmetropolitan residence	Total	In urbanized areas			Other urban	Rural
		Total	Central cities	Urban fringe		
United States......	179,323	95,848	57,975	37,873	29,420	54,054
In metropolitan areas..	112,885	95,076	57,975	37,101	4,486	13,323
Central cities.......	58,004	57,777	57,777	–	227	–
Outside central cities.............	54,881	37,299	198	37,101	4,259	13,323
Outside metropolitan areas...............	66,438	772	–	772	24,934	40,732
PERCENT						
United States......	100.0	53.5	32.3	21.1	16.4	30.1
In metropolitan areas..	63.0	53.0	32.3	20.7	2.5	7.4
Central cities.......	32.3	32.2	32.2	–	0.1	–
Outside central cities.............	30.6	20.8	0.1	20.7	2.4	7.4
Outside metropolitan areas...............	37.0	0.4	–	0.4	13.9	22.7

– Represents zero.

Source: *U.S. Census of Population: 1960*, Vol. 1, Part 1, tables O, 5, 18, and 33.

[23]H. Paul Douglas, "Suburbs," in *Encyclopedia of the Social Sciences*, New York, Macmillan, 1937, vol. 14, p. 433.

[24] Sidney Goldstein, "Rural-Suburban-Urban Population Redistribution in Denmark," *Rural Sociology*, 30(3):268–269, September 1965.

[25] Richard A. Kurtz and Joanne B. Eicher, "Fringe and Suburb: A Confusion of Concepts," *Social Forces*, 37(1):32–37, October 1958.

[26] Samuel W. Blizzard and William F. Anderson II, "Problems in Rural-Urban Fringe Research: Conceptualization and Delineation," Pennsylvania Agricultural Experiment Station Progress Report No. 89, State College, Pa., November 1952, p. 7. They refer to: T. Lynn Smith, *The Population of Louisiana: Its Composition and Changes*, Louisiana Agricultural Experiment Station Bulletin 293, Baton Rouge, 1937, p. 26. See also, Glenn V. Fuguitt, "The Rural-Urban Fringe" in *Proceedings of the American Country Life Association*, Chicago, 1962, p. 1.

[27] Duncan and Reiss, op. cit., p. 6.

[28] Kurtz and Eicher, op. cit., pp. 36–37.

[29] Ibid., p. 35.

Figure 6–1. – A Model Metropolitan Area Showing the Ecological Relationship of a Central City and the Suburban Zone, Suburban Community (SC), and Rural-Urban Fringe

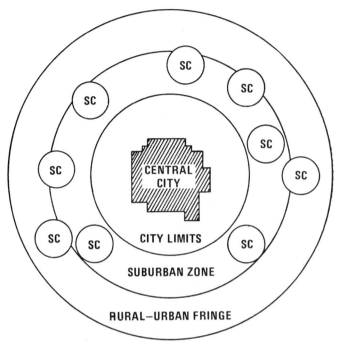

Source: William M. Dobriner, *The Suburban Community*, G. P. Putman's Sons, New York, 1958, p. xviii; copyright © 1958 by G. P. Putman's Sons, all rights reserved.

SIZE OF LOCALITY

If the urban-rural classification is based on size of locality, then a classification by multiple size categories can be regarded as an extension of the dichotomous classification that recognizes the fact that there is an urban-rural continuum; size differences and their associated demographic characteristics may be just as important within the urban, or within the rural, part of the range as are the differences between the smaller urban and the larger rural localities. Statistics on the number of people living in each size-of-place category, then, tell us a great deal about the settlement pattern in a country or region that cannot be appreciated directly from statistics for individual administrative areas. Moreover, as Duncan and Reiss point out, "The sheer physical contrast between the urbanized area of 2,500,000 inhabitants and the town of 2,500 is so striking that one would be amazed if there were not also important social contrasts. The aggregate population of a thousand separate towns, therefore, is expected to have characteristics quite different from those of an equally large population residing in a single urban agglomeration." [30] They suggest that these social contrasts rest mainly on three complexes of factors: (1) persistent currents of selective migration, (2) family organization and functions, and (3) economic structure and functions.

International Standards

For the 1970 census, the United Nations has recommended the equivalent of the tabulation by size of locality illustrated in table

---

[30] Duncan and Reiss, op. cit., p. 3.

6-5. Such a tabulation is more useful for purposes of international comparability than a table showing the distribution of the population by urban-rural residence, using national definitions of urban and rural. However, it cannot be said that full comparability will be achieved because the definition of the basic unit (i.e., the locality) is only very loosely standardized.

The current stress on the locality or place rather than on the administrative unit (commune, minor civil division, etc.) is noteworthy, however. The U.N. *Handbook of Population Census Methods* makes this point very plain: [63] "For census purposes, it is recommended that a 'locality' be defined as a distinct and indivisible population cluster (also designated as agglomeration, inhabited place, populated centre, settlement, etc.) of any size, having a name or a locally recognized status and functioning as an integrated social entity. This definition embraces population clusters of all sizes, with or without legal status, including fishing hamlets, mining camps, ranches, farms, market towns, communes, villages, towns, cities and many others" (Vol. III, p. 63).

National Practices

Table 11 of the 1962 *Demographic Yearbook* of the United Nations, supplemented by table 32 of the 1963 *Yearbook*, gives the available distributions by size of locality from censuses taken in the period 1955 to 1963. A letter code indicates which of three types of definitions was used for "locality," as follows:

Type a – agglomerations without regard to fixed boundaries

Type b – localities with fixed boundaries, commonly under the jurisdiction of local or "urban" forms of government

Type c – relatively small or smallest administrative subdivisions having fixed boundaries and, in sum, comprising the entire country.

Table 6–5, drawn from this source, presents illustrative statistics using the "Type a" definition.

Table 6–5. – Population of Venezuela in Localities By Size-Class: 1961

Size of locality (Number of inhabitants)	Number of localities (Type a)	Population[1]
Total........................	(X)	7,523,999
In localities......................	24,177	7,426,743
500,000 and over..................	1	1,101,147
100,000 to 499,999................	4	919,811
50,000 to 99,999..................	11	762,879
20,000 to 49,999..................	24	765,386
10,000 to 19,999..................	33	437,749
5,000 to 9,999....................	51	352,600
2,000 to 4,999....................	142	451,150
1,000 to 1,999....................	217	287,902
500 to 999.......................	567	389,878
200 to 499.......................	2,372	696,880
Under 200........................	20,755	1,261,361
Outside localities................	(X)	97,256

X Not applicable.
[1] Excluding Indian jungle population, estimated at 31,800.

Source: Adapted from, United Nations, *Demographic Yearbook, 1962*, p. 385.

The United States

The classification of the total urban population into size classes began with the census of 1880 when 10 classes were shown by States. The categories, which began with 4,000 to 8,000 are convertible at only one point to the U.N. recommendation for the 1970 census. The classification used in one of the tables of the 1960 census, however, is exactly convertible to the standard. (See table 6–3).The incorporated place has always been the modal unit used in the American classification so that this classification corresponds most closely to the U.N.'s "Type b." Supplementation of this unit with specially delineated unincorporated localities, beginning with the 1950 census, has introduced some units of "Type a," however, particularly in the range from 1,000 to 2,500. Finally, the use of a relatively few "urban towns and townships" under special rules introduced some units of "Type c". (These last were eliminated from the 1970 classification.)

FARM POPULATION

The farm, or the agricultural, population is sometimes regarded as being equivalent to the rural population; but it is obvious that most of the definitions given above for the latter do not confine it to the population economically dependent on agriculture or to the population living on farms or in agricultural villages. Conversely, as will be seen, the definitions that have been used to define the farm population usually treat it as only part of the rural population or as overlapping the urban-rural classification. The more highly developed a country is economically the more its rural population will tend to exceed its farm population.

A fundamental difference between the classification of the population as farm or nonfarm and the residence classifications previously considered is that the farm-nonfarm classification is usually applied to individual households or families and not to geographic areas. In a sense then, this topic is intermediate between the territorial distribution of population and the composition of population in terms of personal, social, and economic characteristics. In some countries, the farm population has been treated as a final category in the distribution by size of place, as if it approximated the population living in the smallest places or in the open country.

The importance and uniqueness of the farm population are both well recognized. "Statistics of the population dependent on agriculture are especially valuable, not only because of the importance of agriculture as the primary source of livelihood for most of the world's peoples, but also for other reasons. The agricultural population tends to be set off more distinctly from other population groups, in its cultural traits, living standards and social institutions, than the population dependent on any other major branch of economic activity. This is so because, in most countries, the people who make their living from agriculture are, to a comparatively great extent, geographically isolated; and because the home life of a farming household is intimately connected with the operation of the farm." [31]

International Recommendations

In view of this quotation, it may seem surprising that the recommendations of the United Nations and of its regional economic commissions do not list farm population even as an optional topic for 1970 and, in fact, say almost nothing about

it. This omission was not the situation with respect to the 1950, or even the 1960, censuses of population, however. Moreover, when one looks at the recommendations of the U.N. Food and Agriculture Organization (FAO) for the 1970 censuses of agriculture, one finds that the subject is still alive although not necessarily very robust. Part of the explanation is that attempts to develop and apply definitions have encountered serious conceptual and operational problems. Furthermore, national practices vary quite widely, to some extent reflecting differences in the settlement patterns of farm people, with many countries ignoring the topic altogether in their national censuses. For example, the attention given to the population living on farms in the censuses of the United States and Canada reflects the settlement of the agricultural population of those countries in open-country farmsteads. In most of the other parts of the world, farmers live in agricultural villages.

In many highly developed countries, one or more members of a farmer's family often works at nonagricultural activities and contributes significantly, sometimes predominantly, to the family's income, thereby making it difficult to delimit the population economically dependent upon agriculture. This problem is much less serious in those areas of the world (e.g., many sections of Latin America) where agricultural activities still constitute the principal basis of the economy.

From the foregoing description of the difficulties with the topic and the relatively few countries that have published statistics thereon, it is not surprising that not much is available on farm population in international compilations. The *Demographic Yearbook* of the United Nations, for example, has never covered the topic. On the other hand, national totals of the agricultural population are given in the *Production Yearbook* of FAO.[32] These are intended to represent all persons depending for their livelihood on agriculture; that is, all persons actively engaged in agriculture and their nonworking dependents. Agriculture is understood to include forestry, hunting, and fishing. The 1965 estimates in the 1967 *Yearbook* were prepared by FAO itself, mostly on the basis of the fact that the ratio of the agricultural population to the total population had been found to approximate closely the ratio of the number of workers engaged in agriculture to the total economically active population. Knowing three terms of this proportionality, the first term could be computed. Table 6–6 is derived from estimates in the 1967 *Production Yearbook* and gives the percentage of the total population that is agricultural population for regions of the world.

The United States

Statistics of the farm population have been a prominent part of the reports of American censuses and current surveys for four or five decades and especially in the period from 1930 to 1950. In the latter part of this period, however, the correspondence between the population living on farms and the population dependent on agriculture has become weaker, the relative importance of the farm population has decreased sharply, and the conceptual and operational difficulties in classifying a household as farm or nonfarm have become more apparent. The United States has made very little effort to measure directly the population dependent on agriculture, partly because the responsible experts have had serious doubts about the value of the concept.

[31] United Nations, *Population Census Methods*, Population Studies, No. 4, November 1949, p. 145.

[32] For example, Vol. 21 (1967), table 5, gives figures for 1965.

Table 6-6. — Agricultural Population — Continental and World Estimates: 1950 and 1965

[Population in millions. Totals and derived figures were obtained before rounding]

Continent	1950			1965		
	Total population	Agricultural population[1]		Total population	Agricultural population[1]	
		Number	Percent		Number	Percent
World total............................	2,501	1,403	56	3,359	1,752	52
World excluding China, Mainland...........	1,954	1,026	53	2,595	1,271	49
Europe.....................:.....	392	129	33	445	103	23
U.S.S.R................................	181	90	50	231	73	32
North and Central America.....................	218	52	24	294	55	19
North America..........................	166	23	14	214	13	6
Central America..........................	52	29	56	80	42	53
South America..........................	110	65	59	166	76	46
Asia (excluding China, Mainland)..............	824	522	64	1,131	731	65
Africa..	217	165	76	311	230	74
Oceania..	12	3	22	17	3	19
China, Mainland............................	547	377	69	764	481	63

[1] Data must be regarded as only rough estimates.

Source: United Nations, Food and Agriculture Organization, *Production Yearbook, 1967*, Vol. 21, table 6.

Historical Development of the Classification. — The most definitive account of the development of the farm-nonfarm classification in the United States and of statistics on the farm population was written by Leon E. Truesdell, who in his long service in the Bureau of the Census, where he was chief first of the Agriculture Division and then of the Population Division, played a key role in the history he describes.[33]

In 1890, the census classified families as farm and nonfarm, primarily as part of an inquiry on tenure of homes, but no use was made of this distinction as a basis for the classification of population. Beginning with the 1920 census, provisions were made at the urgent request of C. J. Galpin, then Chief of the Division of Farm Population and Rural Life, Bureau of Agricultural Economics, for a classification of households by whether or not the household lived on a farm.

Using data from the 1920 census as a benchmark, the Bureau of Agricultural Economics undertook a program of annual mailed questionnaire surveys for measuring the changes that took place in the farm population each year, and in about 1923 began issuing annual estimates of the farm population for the United States and major geographic divisions. These series of annual farm population estimates have been issued by the Bureau of Agricultural Economics (or its successor agency) each year since. Beginning in April 1944, annual (or quarterly) estimates of the U.S. farm population, based on the Current Population Survey, have been issued cooperatively by the Bureau of the Census and the Bureau of Agricultural Economics.

In 1926, Truesdell published a monograph, *Farm Population of the United States*, presenting the 1920 data on farm population by States; these had not been released in the regular volumes of the 1920 Census of Population but they had been included in the *General Report on Agriculture*. By 1930, the urban-rural and the farm-nonfarm classifications were combined into a three-way classification:
1. Urban (including the small urban-farm population)
2. Rural-farm
3. Rural-nonfarm

Rather extensive tabulations were made by this classification, and the differences that were revealed among these three groups in various characteristics proved that a farm residence subdivision within the rural population was highly useful. This three-way classification has been continued in subsequent censuses.

Since 1960 the "farm" segment of the rural population has been determined as those persons living on places of 10 or more acres from which sales of farm products amounted to $50 or more in the preceding calendar year or on places of less than 10 acres from which sales of farm products amounted to $250 or more in the preceding year. Other persons in rural territory are classified as "nonfarm."

Farm and nonfarm residence has been obtained in each census since 1930. For 1970, as in 1960, the figures are for the farm population residing in rural territory. In all other censuses, farm residence was obtained in cities and other territories classified as urban; but since 1960 it has arbitrarily been assumed that the urban population is completely nonfarm.

The changing definition. — Many of the practical and research uses of farm population data had assumed that the farm population could be treated as if it were the agricultural population. This assumption has become gradually less correct in the decades following 1920 when the farm population was originally defined. In fact, the very concept of an agricultural population or a population dependent upon agriculture has become impossible to apply precisely in the United States, because of part-time farmers, nonagricultural employment of members of farm operators' families, nonagricultural employment for a part of the year by farm wage workers and members of their families, etc. Even so, the groups mentioned and their families do have some connection with agriculture and could be covered in a concept of population wholly or partly dependent upon agriculture. But there is an additional group of families who live on farms but have no other connection with agriculture. Such families frequently rent a second or third house on a farm that originally was built to house a farm laborer or tenant. Or in case of consolidation of adjoining or nearby farms, the house of one of the original operators may be rented by an essentially "nonfarm" family.

Since the farm population has always been defined essentially as the population living on a farm, it may seem surprising

[33] U.S. Bureau of the Census, *Farm Population: 1880 to 1950*, by Leon E. Truesdell (Bureau of the Census Technical Paper No. 3), 1960.

that no definition of a farm was provided in the census of population *per se*. From 1920 through 1950, however, the decennial censuses of population and agriculture were conducted simultaneously by the same enumerators. There were instructions as to when an agriculture schedule had to be filled, and presumably this procedure had some influence on the enumerator's entry on the population schedule.

The typical question during this period was, "Is this house on a farm (or ranch)?", and often the enumerator made an entry on the basis of his observations without asking the question. In 1950, there was an instruction that a household living in a house for which cash rent was paid for the house and yard only, and persons living in an institution, summer camp, motel, or tourist cabin were always to receive an entry of "No" in the question regarding farm residence.

Current Population Survey.—The first estimates of the farm population to be based on the Current Population Survey were those for April 1944. Through April 1949, quarterly estimates were published; but since that date the estimates have been annual, as of April 1. The estimates from 1951 onwards, however, represent an average of five quarterly observations centered on April, thus the preceding and following October estimates are each given a weight of one-eighth and January, April, and July are each given a weight of one-fourth. From the beginning, the estimates from the Current Population Survey have been published jointly by the Bureau of the Census and the Department of Agriculture. The title of the series has changed somewhat, but the reports are numbered consecutively. The title now reads, *Current Population Reports*, P–27 (Census-ERS), *Farm Population*. In addition to totals, the reports now contain classifications of the farm population by age, sex, color, employment status, and region.

In the Current Population Surveys prior to 1950, the farm population consisted of all persons living on tracts of land of 3 acres or more (or less than 3 acres if agricultural products valued at $250 or more were produced from it during the preceding 12 months) upon which agricultural operations were being currently performed or were contemplated for the next crop season. From 1950 to 1960, the 1950 census definition of farm residence was followed in the CPS. Revisions back to 1910 were prepared to bring the earlier figures in line with the 1950 definition.[34] The rather radical change in definition adopted for the 1960 census was used after 1960, but no attempt was made to revise the earlier figures. A study of the effect of the change in the definition showed that the CPS estimate of the farm population in April 1960 was 15.7 million using the new definition and 19.8 million using the old definition.[35] Although the current estimates of the farm population are not comparable with those for years prior to 1960, a continuing, though not consistent, series has been prepared by the Economic Research Service of the U.S. Department of Agriculture.[36] The previous way of classifying farm people is thought to represent adequately the conditions prevailing in 1940 and earlier years, when farming for subsistence purposes was much more common than it is today. Thus, no revisions have been made of estimates for these years.

Historically, the CPS estimates of the farm population have exceeded the census figures even when the same definition has been used. The differences are as follows:

	Census	CPS
1950	23,300,000	25,100,000
1960	13,500,000	15,700,000
1970	8,287,000	9,425,000

It is thought that the CPS estimate is more nearly correct inasmuch as investigations based on other subject matter have demonstrated that the more experienced and better trained CPS enumerators generally treat marginal cases more correctly. Some of the factors contributing to the difference are discussed in the 1960 population reports.[37]

Census of Agriculture.—Farm population figures were obtained from mid-decade censuses of agriculture taken in 1925, 1935, and 1945; but they do not agree very well with figures derived from population census data.

"There is a basic reason why data on the farm-resident population obtained in a census of agriculture are likely to differ from those obtained in a census of population. In a population census, the emphasis is on counting people; each household is visited or receives a schedule to be filled, and an effort is made to secure first-hand information about every individual in each household. The identification of farms, if it is undertaken at all in a population census, is only incidental to one part of the census schedule. In an agricultural census, the identification of farms is paramount; the population data, if any, are only incidental and they are often obtained second-hand from a person only slightly acquainted with the people for whom he reports."[38]

No information on farm population was collected in the 1954 Census of Agriculture, but in 1964 persons in the households of farm operators were counted and some of their characteristics recorded.[39]

Department of Agriculture—The cooperation between what is now the Economic Development Division, Economic Research Service, U.S. Department of Agriculture and the Bureau of the Census in the publication of farm population estimates has already been discussed. "The Economic Research Service and its predecessor agencies have conducted an annual survey of the farm population and its components of change since 1923. Utilizing the Crop Reporting System of the Department of Agriculture, reports are collected through a mailed questionnaire. . .The respondents report on the number of persons who are living on their own and neighboring farms at the beginning and end of a specified 12-month period. They also report on births, deaths, and changes through migration which occurred during this period.

"From these data, ratios of each component of change during the specified period to the sample farm population at the beginning of the period are computed for each State. By applying ratios from the sample to previously obtained expanded State estimates from the beginning of the period, estimates are obtained of each component of change during a year. These expanded estimates of births, deaths, and movement of persons to and from farms are summed to the populations at the beginning of the period to obtain a set of preliminary estimates of the farm population at the end of the year. Final estimates are developed by adjusting the preliminary State estimates so that

[34] U.S. Bureau of the Census and U.S. Bureau of Agricultural Economics, *Farm Population*, Series Census–BAE, No. 16, "Revised Estimates of the Farm Population of the United States: 1910 to 1950," March 9, 1953, pp. 5–6.

[35] U.S. Bureau of the Census and Agricultural Marketing Service, *Farm Population*, Series Census–AMS (P–27), No. 28, "Effect of Definition Changes on Size and Classification of the Rural-Farm Population: April 1960 and 1959," April 17, 1961.

[36] U.S. Department of Agriculture, Economic Research Service, *Farm Population: Estimates for 1910–62*, by Vera J. Banks, Calvin L. Beale, and Gladys K. Bowles, ERS–130, October 1963.

[37] *U.S. Census of Population: 1960*, Vol. I, *Characteristics of the Population*, Part 1, *United States Summary*, 1964, p. XXXVIII.

[38] United Nations, *Population Census Methods*, p. 153.

[39] U.S. Bureau of the Census, *Census of Agriculture, 1964, Statistics by Subjects*, chapter 5, *Characteristics of Farm Operators and Persons Living on Farms*, 1967.

they will sum to the independent estimate for the United States provided by Current Population Surveys of the Bureau of the Census, and likewise adjusting components of change to account for differences shown." [40]

From the data gathered in the survey under the Crop Reporting Survey, estimates of the farm population are published for geographic divisions and estimates of the components of change are also given in the report series on farm population of the Department of Agriculture. This additional information for the years from 1910 to 1962 appears in the report previously cited. For subsequent years, the same information is given in an annual series (for example: U.S. Department of Agriculture, Economic Research Service, *Farm Population—Estimates for 1967*, ERS-410, May 1969). Annual State estimates are also published for each intercensal decade.

Consistency and Quality of the Statistics.—Perhaps the most important question with regard to the quality of the official statistics on farm population, or perhaps one should say with regard to their interpretation, relates to the meaningfulness of the concept as operationally defined. There was considerable discussion of this issue above. How meaningful is farm residence *per se*, particularly as more and more farm residents are becoming engaged in nonagricultural pursuits and more farm operators and their families live off the farm?

A question like "Does this household live on a farm?" when used without a definition of a farm was obviously somewhat attitudinal in nature, and it is not surprising that it had a fairly high response error. Particularly in marginal cases, the answer would depend on the choice of respondent for the household and even the same respondent could give different answers in different interviews. This phenomenon may be illustrated from the results of a supplement to the March 1957 Current Population Survey.

	Number (thousands)	Percent
Persons originally reported as living on farms but place did not satisfy census of agriculture criteria	4,631	100.0
With no farm income	3,149	68.0
Reported as living on farm in second interview	1,550	33.5
Farm residence status uncertain	572	12.4
Reported as not living on farm in second interview	1,027	22.2
With some farm income	1,483	32.0

This example relates to an unrepresentative part of the farm population, of course; but it illustrates the gross differences that can occur in a classification.

Another slightly disturbing factor is seasonality. The number of persons engaged in agriculture tends to reach peaks in the planting and harvesting seasons. Thus, when the census was taken in January 1920, the farm population was probably near its seasonal low. To control for seasonality, an effort has been made to standardize on an April reference point, especially when an adjusted series is produced.

Although the farm population figures from the censuses of 1960 and 1970 are based on a sample, the samples are so large as to contribute a very minor part of the total error at the national level. For certain counties and socioeconomic subgroups, the sampling error is fairly appreciable, however. Sampling error definitely has to be taken into account in assessing annual changes in the farm population as measured by the Current Population Survey. Given the sample design used in the Current Population Survey, farm residence tends to have higher sampling errors than many other characteristics (holding constant the size of the observed figure) because all persons in a household receive the same classification as farm or nonfarm and farm households are highly clustered rather than being distributed at random throughout a primary sampling unit.

The annual estimates of the farm population by geographic divisions, although tied to national controls from the Current Population Survey, are based on replies to mail questionnaires that had been sent to "crop reporters." These reporters provide information, *inter alia*, on birth and deaths on their own and neighboring farms and on movers to and from these farms. This is not a probability sample, less than half of the questionnaires are returned, and consequently, various adjustments have to be made on the basis of judgment and other pertinent and available data.[41] Errors from these sources would far outweigh the contribution of sampling error.

OTHER CLASSIFICATIONS

A few other ways of grouping statistical areas, disregarding geographical contiguity, are also to be found in official publications, mostly in the United States. Table 6–7 shows the population in standard metropolitan statistical areas by size of SMSA. Instead of classifying metropolitan areas by the size of the areas themselves, they may be classified by the size of their largest city (primary central city). Schnore and Varley have organized statistics from the 1950 census of the United States in this manner.[42]

The stub of table 6–8 indicates another way of organizing areas by type of residence, namely, in terms of cross-classification of typologies. This table gives for 1960 an example of the elaborate cross-classifications in recent specialized census reports.

Tables 6-3 and 6-4 above are still other examples. The three-way classification of urban, rural nonfarm, and rural farm is so familiar in American census reports that it is not always recognized as a 2x2 classification of urban–rural by farm-nonfarm from which the very small urban-farm category has been eliminated.

A relatively simple cross-classification of urban-rural residence by size of locality is given in table 6–9 for six American countries from the 1950 Census of the Americas. It will be recalled that in some countries urban localities are not defined strictly on the basis of population size. Thus, in Brazil, all defined localities, even those of less than 1,000 inhabitants, were classified as urban.

There have been a number of studies of the urban or metropolitan influence upon rural areas. One approach is to classify counties by the size of the largest place, to cumulate the population of the counties into these categories, and to compute various demographic, social, and economic measures for the

[40] U.S. Department of Agriculture, Economic Research Service, *Farm Population Estimates for 1910–62*, ERS–130, p. 11.

[41] U.S. Department of Agriculture, *Major Statistical Series of the U.S. Department of Agriculture*, Vol. 7, *Farm Population, Employment, and Level of Living*, Agriculture Handbook No. 118, p. 3, September 1957.

[42] Leo F. Schnore and David W. Varley, "Some Concomitants of Metropolitan Size," *American Sociological Review*, 20(4):408–414, August 1955.

Table 6–7. —**Population of Standard Metropolitan Statistical Areas, by Size of Area, for the United States: 1960**

Size and component parts of SMSA	Population
ALL SMSA's...........................	112,885,178
Central cities...........................	58,004,334
Outside central cities...................	54,880,844
SMSA's of 3,000,000 or more...........	31,763,499
Central cities...........................	17,828,227
Outside central cities...................	13,935,272
SMSA's of 1,000,000 to 3,000,000.....	29,818,571
Central cities...........................	12,707,503
Outside central cities...................	17,111,068
SMSA's of 500,000 to 1,000,000.......	19,214,817
Central cities...........................	10,126,684
Outside central cities...................	9,088,133
SMSA's of 250,000 to 500,000.........	15,829,067
Central cities...........................	7,750,597
Outside central cities...................	8,078,470
SMSA's of 100,000 to 250,000.........	14,497,817
Central cities...........................	8,235,553
Outside central cities...................	6,262,264
SMSA's under 100,000.................	1,761,407
Central cities...........................	1,355,770
Outside central cities...................	405,637

Source: Adapted from, *U.S. Census of Population: 1960*, Vol. I, *Characteristics of the Population*, Part A, *Number of Inhabitants*, 1961, table Q, p. XXVI.

Table 6–8. —**Population of the United States by Size of Place, Distinguishing Central Cities From Urban Fringe and Farm From Nonfarm: 1960**

[Based on 25-percent sample]

Size of place	Population	
	Number	Percent
Total.............................	179,325,671	100.0
In urbanized areas.....................	95,821,832	53.4
Areas of 3,000,000 or more.............	33,742,618	18.8
Central cities........................	19,140,690	10.7
Urban fringe.........................	14,601,928	8.1
In incorporated places of 2,500 or more..	9,715,916	5.4
Other..............................	4,886,012	2.7
Areas of 1,000,000 to 3,000,000........	18,014,524	10.0
Central cities........................	8,747,775	4.9
Urban fringe.........................	9,266,749	5.2
In incorporated places of 2,500 or more..	5,908,458	3.3
Other..............................	3,358,291	1.9
Areas of 250,000 to 1,000,000..........	25,991,480	14.5
Central cities........................	16,500,359	9.2
Urban fringe.........................	9,491,121	5.3
In incorporated places of 2,500 or more..	4,220,377	2.4
Other..............................	5,270,744	2.9
Areas of less than 250,000.............	18,073,210	10.1
Central cities........................	13,592,456	7.6
Urban fringe.........................	4,480,754	2.5
In incorporated places of 2,500 or more..	1,636,300	0.9
Other..............................	2,844,454	1.6
Urban outside urbanized areas..............	29,461,951	16.4
Places of 25,000 or more...............	6,933,862	3.9
In SMSA's..........................	836,636	0.5
Outside SMSA's.....................	6,097,226	3.4
Places of 10,000 to 25,000.............	9,254,452	5.2
In SMSA's..........................	1,262,528	0.7
Outside SMSA's.....................	7,991,924	4.5
Places of 2,500 to 10,000.............	13,273,637	7.4
In SMSA's..........................	2,422,705	1.4
Outside SMSA's.....................	10,850,932	6.0
Rural..................................	54,041,888	30.1
Nonfarm..............................	40,596,990	22.6
In SMSA's..........................	11,675,448	6.5
Outside SMSA's.....................	28,921,542	16.1
Farm.................................	13,444,898	7.5
In SMSA's..........................	1,636,829	0.9
Outside SMSA's.....................	11,808,069	6.6

Source: Adapted from, *U.S. Census of Population: 1960, Selected Area Reports, Size of Place*, PC(3)–1B, 1964, table 1.

Table 6–9. —**Population by Size of Locality and Urban-Rural Residence, for Selected Latin American Countries: Around 1950**

[In thousands]

Urban-rural residence and size of locality	Brazil, 1950	Cuba, 1953	Ecuador, 1950	Nicaragua, 1950	Panama, 1950	Venezuela, 1950
Urban, total....	[1]18,783	3,125	914	369	290	2,709
Less than 1,000......	1,435	–	4	35	1	–
1,000 to 4,999.......	3,415	504	140	105	51	596
5,000 to 9,999.......	1,782	236	88	28	33	290
10,000 to 99,999.....	5,270	1,106	214	92	77	988
100,000 and over.....	6,873	1,278	469	109	128	836
Rural, total.....	33,162	2,704	2,289	688	516	2,325
Less than 1,000......	–	–	1,867	459	467	2,325
1,000 to 4,999.......	–	–	396	210	48	–
5,000 to 9,999.......	–	–	26	6	–	–
10,000 to 24,999....	–	–	–	13	–	–

— Represents zero.
[1] Includes 7,000 persons in localities not classified by size.

Source: Adapted from, Inter-American Statistical Institute, *La éstructura demografica de las naciònes americanas*, Vol. I, *Caracteristicas generales de la poplación*, Part I, *Población censada y estimada; agrupaciónes básicas de la población censada*, Washington, Pan American Union, 1960, table 2–05.

resulting populations.[43] Another approach is to group the counties according to their distance from a standard metropolitan statistical area and according to the size of the SMSA.[44]

There have been quite a number of classifications of localities or other communities by type of functional specialization or their patterns of economic production. For example, Duncan and Reiss identified the following types of specialization: manufacturing, trade, higher education, public administration, transportation, military, and entertainment and recreation. Many cities are of a diversified type, of course. Using data from a variety of sources (not limited to the census of population), they presented classifications, in appendix tables, for the metropolitan areas, their central cities, and other cities of 10,000 inhabitants or more. Duncan and his associates updated this classification a few years later for 50 major cities.[45] A functional classification of America's SMSA's has appeared in the *Municipal Yearbook*.[46] Similarly, Boustedt published a table showing the communities of Bavaria classified by economic structure, degree of dominance, and central functions, all cross-classified by size of community.[47]

MEASURES

Some of the measures described in the previous chapters are also applicable to classifications by type of residence. There are a few additional ones that are particularly appropriate to the present classifications. This book barely touches on the complex topic of the structure of urban communities that is analyzed in terms of very small units of area. Such

[43] Otis Dudley Duncan, "Gradients of Urban Influence on the Rural Population," in *Urban Research Methods*, Jack Gibbs, ed., Princeton (New Jersey), D. Van Nostrand Company, 1961, pp. 550–556.

[44] Dale E. Hathaway, J. Allan Beegle, and W. Keith Bryant, *People of Rural America*, a 1960 Census Monograph, U.S. Bureau of the Census, 1968, pp. 7–15.

[45] O. Dudley Duncan et al., *Metropolis and Region*, (Published for Resources for the Future), Baltimore, Johns Hopkins Press, 1960, pp. 279–550.

[46] Victor Jones and Richard L. Forstall, "Economic and Social Classification of Metropolitan Areas," *Municipal Yearbook, 1963*, Chicago, International City Managers' Association, 1963, pp. 31–44.

[47] Olaf Boustedt, "Urban Population, Urban Areas, and the Problem of Dominance in West-German Statistics," in *Proceedings of the World Population Conference, 1954* (Rome) Vol. IV, United Nations, New York, 1955, p. 503.

studies involve a number of highly specialized measures. Again, this book merely recognizes that there is an important and growing field that can be loosely termed "microdemography."

Percentage Distributions

As with any demographic variate or attribute, more meaning can be comprehended from the statistics by type of residence when the absolute numbers are reduced to percentage distributions. Comparisons can then be made more effectively between two areas of widely different sizes or between observations on the same area at two points in time.

In a complex classification with one or more subtotals, a choice has to be made of the most appropriate base, or perhaps percentages based on more than one denominator are of interest and should be computed. For example, from table 6–9, the total population of 1,057,000 can be obtained by addition for Nicaragua. Then the percent of the total population that is urban can be computed at 34.9 and the percent in places of 100,000 and over, at 10.3. If we want to know the percent of the urban population that is in this largest size class, we use 369,000 as our denominator and compute this percentage at 29.5. Table 6–8 permits more alternatives. The percentage of the total population of the United States living in incorporated places of 2,500 or more in the urban fringes of urbanized areas of 3 million or more is shown as 5.4 percent. This population of 9,716,000 represents 10.1 percent of the population in urbanized areas, 28.8 percent of the population of urbanized areas of 3 million or more, and 66.5 percent of the fringe population of these largest urbanized areas. Sometimes the cross-classified types of residence are displayed in two dimensions, as in table 6–4 instead of in one dimension only. Here all the percentages were computed using the grand total as the base, but it would be useful to have the horizontal or vertical percentage distributions as well. For example, that part of the total rural population that lived outside metropolitan areas in 1960 may be computed at 75.4 percent.

Extent of Urbanization

The most obvious measure of the extent of urbanization is the percent of the total population living in urban territory. But is this the best measure? We have already discussed the great variety of urban definitions in use among the countries of the world. For purposes of international comparisons, the percentage living in agglomerations of at least a given size is certainly an improvement. Some writers go further and regard the use of measures based on the distribution of the population by size of place as being preferable even where there is a uniform definition of urban. Such measures do not depend upon an arbitrary decision as to what minimum size of place is to be chosen for the definition of urban.

As Gibbs puts it, "Even if a rationale could be formulated for a universally applicable minimum urban size limit, the conventional measure of the degree of urbanization would still reveal only one dimension of urbanization. Assume for the sake of argument that the minimum size limit is 2,000 or more inhabitants. Two countries, X and Y, have a degree of urbanization of 50.0, meaning that in both instances exactly one-half of the total population resides in points of population concentration of that size range. Yet, it is **logically** possible for all the urban population of country X to reside in urban units of 1,000,000 or more inhabitants, while in country Y there is no urban unit exceeding 5,000 inhabitants. In other words, countries having the same degree of urbanization

may be quite different with regard to urban size structure or, as designated here, 'scale.' The suggestion is not that the conventional measure of the degree of urbanization be abandoned, but it does need to be supplemented with a measure of scale."[48]

In the article just cited, Gibbs proposed two supplementary measures. These have been the object of considerable criticism, followed by rebuttals, and further criticisms, in *Social Forces*.[49] This discussion became fairly heated, and we will not attempt to summarize all the claims and counterclaims. We introduce Gibbs's measures here mainly to illustrate recent attempts to find a summary measure of urbanization that goes beyond the simple percent urban in the total population.

Gibbs's measures, then, are the "scale of urbanization" (S_u) and the "scale of population concentration" (S_p). If X_i is the proportion of the urban population in urban size class (i) and all size classes above it, and Y_i is the proportion of the total population in urban size class (i) and all size classes above it, then,

$$S_u = \Sigma X_i Y_i \qquad (1)$$

If Z_i is the proportion of the total population in size class (i) and all sizes above it, then,

$$S_p = \Sigma Z_i \qquad (2)$$

Both of these indexes measure the extent to which population is concentrated at the upper end of the scale of size of locality. These two measures, plus what Gibbs called the "degree of urbanization," which is merely the percentage of the total population classed as urban using a uniform definition in terms of population in localities of 2,000 inhabitants or more, are shown for selected countries in table 6–10. The basic statistics on size of locality were taken from the *Demographic Yearbook* of the United Nations.

As Gibbs himself recognizes, however, the scale of urbanization, although it is affected by the full distribution, is not independent of the arbitrary population size used to define "urban." Moreover, it is also affected by the number of categories in the distribution so that international comparisons require the use of class intervals that are nominally identical. For the countries examined, the two measures have very high product-moment correlation coefficients with the ordinary "degree of urbanization." Hence, despite their use of the full distribution by size of locality, very little additional information about the relative extent of urbanization seems to be conveyed by the proposed measures. Of these two proposed measures, however, the scale of population concentration seems to be the superior one.

Jones, of the Australian National University, has written a fuller critique.[50] He also suggests the use of the Lorenz curve and the associated Gini Concentration Ratio as more appropriate measures for the description and summarization, respectively, of urbanization. These are quite general measures; and, before turning to them in the next subsection, we will describe another measure that is restricted to the phenomenon of the distribution of population by size of locality or, more loosely, to the extent of urbanization. This is the size of place of the median inhabitant.

---

[48] Jack P. Gibbs, "Measures of Urbanization," *Social Forces*, 45(2):170–171, December 1966.

[49] A succession of articles by Gibbs, F. Lancaster Jones, and S. Mitra appeared in the issues of December 1967 (Vol. 46, No. 2), March 1968 (Vol. 46, No. 3), and December 1968 (Vol. 47, No. 2).

[50] F. Lancaster Jones, "A Note on 'Measures of Urbanization,' with a Further Proposal," *Social Forces*, 46(2):275–279, December 1967.

Table 6–10. — **Measures of the Degree of Urbanization, Scale of Urbanization, and Scale of Population Concentration, 18 Countries: Around 1960**

Country and year	Degree of urbanization	Scale of urbanization	Scale of population concentration
Brazil, 1960.....................	40.5	1.440	2.919
Canada, 1961....................	55.8	1.932	3.535
Denmark, 1960...................	62.5	2.582	4.249
Ghana, 1960.....................	36.1	0.641	2.031
Honduras, 1961..................	22.5	0.573	1.827
Ireland, 1961...................	45.3	1.740	3.285
Israel, 1961....................	84.3	2.908	4.672
Morocco, 1960...................	29.1	1.312	2.566
Netherlands, 1960...............	75.9	2.655	4.548
New Zealand, 1961...............	70.7	2.843	4.405
Norway, 1960....................	48.8	1.756	3.314
Panama, 1960....................	42.6	1.691	3.052
Sweden, 1960....................	60.9	2.013	3.698
Turkey, 1961....................	34.7	0.945	2.399
United Kingdom, 1961............	78.2	3.342	4.974
United States, 1960............	65.0	2.352	4.061
Venezuela, 1961.................	63.6	2.308	3.994
Yugoslavia, 1961................	41.2	0.887	2.361

Source: Adapted from, Jack P. Gibbs, "Measures of Urbanization," *Social Forces*, 45(2):173, December 1966, table 2; copyright © 1966, University of North Carolina Press.

Size of Place of Median Inhabitant. — The median is the value that divides the distribution into two equal parts, one-half the cases falling below this value and one-half the cases exceeding this value. In the present application, the size of place of the median inhabitant is the size of locality that divides the population in half. It is quite distinct from the median size of locality. The first median applies to the distribution of people, the second to the distribution of places.

This measure is systematically affected neither by the definition of urban nor by the number of class intervals, although, like all measures of urbanization, it is affected by the way localities are defined. Since there are ordinarily more places at the lower than at the upper end of a size interval, the assumption that is usually made, namely, that there is a rectangular distribution within the class interval containing the median, is not very appropriate. If the median falls in a class containing very few localities (say near the top of the size distribution), the populations of the individual cities contained therein may be arrayed and the median computed from these discrete data. If the number of localities in the interval containing the median is large, then, as Duncan recommends, one could interpolate on the logarithms of locality size. In other words, if we take the logarithms of the class marks, interpolate between them, and find the antilog of the interpolated value as our median, this value will be lower, and more nearly accurate, than the median interpolated between the absolute values of locality size.[51]

If statistics are available on both agglomerations and cities proper (incorporated places), then a choice must be made concerning which to use for the distribution of localities by size. If the agglomerations represent the built-up area in and around the city, as is the case with the urbanized areas of the United States, then these would seem to be the preferable units.

Instead of the size of place of the median inhabitant, Arriaga proposes the corresponding mean value.[52] A weighted average of this mean with the percent urban produces his "new index of urbanization."

Lorenz Curve and Gini Concentration Ratio

The Lorenz Curve, first expounded in 1905, has long been used to measure inequalities in the distribution of wealth or income.[53] It has also been used to depict the state (as opposed to the process) of concentration of population and of other demographic aggregates. To plot the curve, the units are first either arrayed individually or grouped in class intervals according to the appropriate independent variate. Then the cumulative percentage of the number of areas (Y_i) is plotted against the cumulative percentage of population (X_i). For comparison a diagonal line is drawn at 45° to show the condition of equal distribution.

To illustrate from data on urbanization, consider the cumulative percentage of population up through each size class. (We have to omit that part of the population, if any, reported as not living in localities.)

The Gini Concentration Ratio measures the proportion of the total area under the diagonal that lies in the area between the diagonal and the Lorenz Curve. This proportion may be computed thus:

$$Gi = \left(\sum_{i=1}^{n} X_i Y_{i+1}\right) - \left(\sum_{i=1}^{n} X_{i+1} Y_i\right) \qquad (3)$$

where X_i and Y_i are respective cumulative percentage distributions and n is the number of class intervals (or units). The computations for Venezuela, 1961, are given in table 6–11, and the corresponding Lorenz Curve is shown in figure 6–2.

Steps 1 and 2. Post the number of localities and the population for each size-class in columns (1) and (2), respectively.

Step 3. Compute the proportionate distribution of localities from column (1) and post in column (3), e.g., $1 \div 24,177$ is less than .00005 so we enter a dash, meaning that the number rounds to zero.

Step 4. Compute the proportionate distribution of the population from column (2) and post in column (4), e.g., $1,101,147 \div 7,426,743 = 0.1483$.

Step 5. Cumulate the proportions of column (3) downward and post the results in column (5), e.g., $0 + .0002 = .0002$.

Step 6. Cumulate the proportions of column (4) downward and post the results in column (6), e.g., $0.1483 + 0.1239 = 0.2722$.

Step 7. Multiply the first line in column (6) by the second line in column (5), the second line by the third line, etc., to obtain column (7), e.g., $0.1483 \times .0002$ is less than .00005, so we enter a dash.

Step 8. Multiply the first line in column (5) by the second line in column (6), etc., to obtain column (8), e.g., $0 \times 0.2722 = 0$.

Step 9. Sum col. (7) and post the total, 0.9889, in the column. Sum col. (8) and post the total, 0.2068.

Step 10. Subtract the total of col. (8) from the total of col. (7), $(0.9889 - 0.2068 = 0.7821)$.

---

[51] Otis Dudley Duncan, "The Measurement of Population Distribution," *Population Studies*, 11(1):40, July 1957.

---

[52] Eduardo E. Arriaga, "A Methodological Note on Urbanization Indices with Applications," in International Union for the Scientific Study of Population, *Contributed Papers*, Sydney Conference (Australia), August 21 to 25, 1967 pp. 675–683.

[53] M. O. Lorenz, "Methods of Measuring the Concentration of Wealth." *Quarterly Publications of the American Statistical Association*, 9(70):209–219, June 1905.

Table 6–11.—Computation of Gini Concentration Ratio for Persons Living in Localities in Venezuela in 1961

[See text for further explanation]

Size of locality	Number of localities	Population	Proportion		Cumulative proportion		X_iY_{i+1}	$X_{i+1}Y_i$
			Localities	Population	Localities (Y_i)	Population (X_i)		
	(1)	(2)	(3)	(4)	(5)	(6)	(7)	(8)
All localities.........	24,177	7,426,743	1.0000	1.0000	–	–	–	–
500,000 and over..........	1	1,101,147	–	0.1483	–	0.1483	–	–
100,000 to 499,999.........	4	919,811	0.0002	0.1239	0.0002	0.2722	0.0002	0.0001
50,000 to 99,999..........	11	762,879	0.0005	0.1027	0.0007	0.3749	0.0006	0.0003
20,000 to 49,999..........	24	765,386	0.0010	0.1031	0.0017	0.4780	0.0015	0.0009
10,000 to 19,999..........	33	437,749	0.0014	0.0589	0.0031	0.5369	0.0028	0.0018
5,000 to 9,999............	51	352,600	0.0021	0.0475	0.0052	0.5844	0.0065	0.0034
2,000 to 4,999............	142	451,150	0.0059	0.0607	0.0111	0.6451	0.0130	0.0076
1,000 to 1,999............	217	287,902	0.0090	0.0388	0.0201	0.6839	0.0298	0.0148
500 to 999................	567	389,878	0.0235	0.0525	0.0436	0.7364	0.1043	0.0362
200 to 499................	2,372	696,880	0.0981	0.0938	0.1417	0.8302	0.8302	0.1417
Under 200.................	20,755	1,261,361	0.8585	0.1698	1.0000	1.0000	–	–
							0.9889	0.2068
Gini Ratio (difference of sums).................							0.7821	

– Represents zero or rounds to zero.   Source of basic data: See table 6–5.

Figure 6–2.—Lorenz Curve for Measuring Population Concentration in Venezuela, 1961, in Relation to the Number of Localities

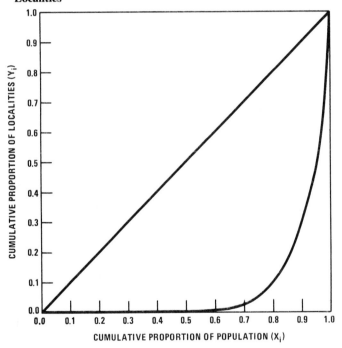

Source: See table 6–11.

As was mentioned above, the Lorenz Curve and the Gini Ratio were recommended by Jones as measures of urbanization. By including the population living outside localities of 2,000 or more and assigning an arbitrary average size to the lowest level, he computes a Gini Concentration Ratio which he names the "coefficient of population concentration." Omitting this class, the index becomes the "coefficient of urban concentration." Multiplying the degree of urbaniza-

tion (percent urban) by the coefficient of urban concentration, Jones obtains a measure that he calls the "scale of urbanization."

Another type of application of the Lorenz Curve and the Gini Ratio is more common. This application deals with the concentration of population in terms of area. A country's geographic units are arrayed in order of population density, and the cumulative percentage of area is plotted against the cumulative percentage of population. A good discussion of this application is available in the previously cited article by Duncan.

Duncan has also introduced in the same article another measure based on the Lorenz Curve, the "index of concentration," Δ. This measure ". . . algebraically is simply the maximum of the set of k values of $(X_i - Y_i)$. Geometrically, it is the maximum vertical distance from the diagonal to the curve." [54] (Authors' note: The "k values" refers simply to all the values of this difference). Δ is also algebraically equivalent to the Index of Dissimilarity, which is the sum of the positive differences between the two percentage distributions.

Algebraically, therefore,

$$\Delta = \frac{1}{2} \sum_{i=1}^{k} |x_i - y_i|, \tag{4}$$

where x_i and y_i are uncumulated percentages of the two distributions. Table 6–12 gives, for Peru, illustrative computations for the Gini Index (cumulative population versus cumulative area) and the index of concentration (Δ), and figure 6–3 shows the corresponding Lorenz Curve and geometric representation of Δ.

The maximum value of $|X_i - Y_i|$ is read from table 6–12 as .4681. This is Δ. The corresponding value of X_i is .8712, so Δ is drawn with that abscissa. (On the graph the proportions X_i and Y_i are converted into percents by multiplying them by 100.)

[54] Duncan, "The Measurement of Population Distribution," p. 30.

Table 6–12.—Computation of Gini Concentration Ratio and Index of Concentration (Δ) for Peruvian Departamentos Arrayed by Population Density: 1961

| Departamento | Population | Area in km.^{2}[1] | Density | Proportion | | $\lvert x_i-y_i \rvert$ | Cumulative proportion | | $X_i \cdot Y_i$ | $X_{i+1}Y_i$[2] | X_iY_{i+1}[2] |
				Population (x_i)	Area (y_i)		Population (X_i)	Area (Y_i)			
Total....................	9,906,746	1,285,216	7.7	1.0000	1.0000	0.9362				5.7572	6.3840
Madre de Dios..............	14,890	78,403	0.2	0.0015	0.0610	0.0595	1.0000	1.0000	–	0.9392	0.9987
Loreto.....................	337,094	478,336	0.7	0.0340	0.3722	0.3382	0.9987	0.9392	0.0595	0.5663	0.9060
Amazonas...................	118,439	41,297	2.9	0.0120	0.0321	0.0201	0.9647	0.5670	0.3977	0.5160	0.5402
San Martin.................	161,763	53,064	3.0	0.0163	0.0413	0.0250	0.9527	0.5349	0.4178	0.4703	0.5009
Moquegua...................	51,614	16,175	3.2	0.0052	0.0126	0.0074	0.9364	0.4936	0.4428	0.4504	0.4596
Tacna......................	66,024	14,767	4.5	0.0067	0.0115	0.0048	0.9312	0.4810	0.4502	0.4372	0.4447
Arequipa...................	388,881	63,528	6.1	0.0393	0.0494	0.0101	0.9245	0.4695	0.4550	0.3884	0.4156
Pasco......................	138,369	21,854	6.3	0.0140	0.0170	0.0030	0.8852	0.4201	0.4651	0.3568	0.3660
Cuzco......................	611,972	76,225	8.0	0.0618	0.0593	0.0025	0.8712	0.4031	0.4681	0.2995	0.3263
Huanuco....................	328,919	35,314	9.3	0.0332	0.0275	0.0057	0.8094	0.3438	0.4656	0.2560	0.2669
Ayacucho...................	410,772	44,181	9.3	0.0415	0.0344	0.0071	0.7762	0.3163	0.4599	0.2188	0.2324
Puno.......................	686,260	72,382	9.5	0.0693	0.0563	0.0130	0.7347	0.2819	0.4528	0.1657	0.1876
Tumbes.....................	55,812	4,732	11.8	0.0056	0.0037	0.0019	0.6654	0.2256	0.4398	0.1477	0.1489
Junin......................	521,210	43,384	12.0	0.0526	0.0338	0.0188	0.6598	0.2219	0.4379	0.1241	0.1347
Ica........................	255,930	21,251	12.0	0.0258	0.0165	0.0093	0.6072	0.1881	0.4191	0.1042	0.1094
Apurimac...................	288,223	20,655	14.0	0.0291	0.0161	0.0130	0.5814	0.1716	0.4098	0.0904	0.0948
Huancavelica...............	302,817	21,079	14.4	0.0306	0.0164	0.0142	0.5523	0.1555	0.3968	0.0768	0.0811
Ancash.....................	582,598	36,308	16.0	0.0588	0.0283	0.0305	0.5217	0.1391	0.3826	0.0578	0.0644
Piura......................	668,941	33,067	20.2	0.0675	0.0257	0.0418	0.4629	0.1108	0.3521	0.0394	0.0438
Lambayeque.................	342,446	16,586	20.6	0.0346	0.0129	0.0217	0.3954	0.0851	0.3103	0.0285	0.0307
Cajamarca..................	746,938	35,418	21.1	0.0754	0.0276	0.0478	0.3608	0.0722	0.2886	0.0161	0.0206
La Libertad................	582,243	23,241	25.1	0.0588	0.0181	0.0407	0.2854	0.0446	0.2408	0.0076	0.0101
Lima.......................	2,031,051	33,895	59.9	0.2050	0.0264	0.1786	0.2266	0.0265	0.2001	(Z)	0.0006
Prov. Const. Del Callao....	213,540	74	2,892.7	0.0216	0.0001	0.0215	0.0216	0.0001	0.0215	–	–

$$\Sigma \lvert x_i-y_i \rvert = .9362$$
$$1/2 \Sigma \lvert x_i-y_i \rvert = .4681$$
$$\Delta = 46.81 \text{ percent}$$

$$\Sigma X_iY_{i+1} = 6.3840$$
$$\Sigma X_{i+1}Y_i = -5.7572$$
$$\text{Gini Concentration Ratio} = 0.6268$$

Z Less than 0.00005.

[1] From. Peru, Dirección nacional de estadística y censos, *Resultados de los censos nacionales*, Vol. I, Part 1, *Cuadros comparativos, distribución geográfica, edad y sexo, lugar de nacimiento,* Lima. March 1965.   [2] Multiplications follow order of cumulations, i.e. from bottom to top of arrays.

Figure 6–3.—Lorenz Curve for Measuring Population Concentration in Peru, 1961, in Relation to Density (Departamento Basis) and Index of Concentration, Δ

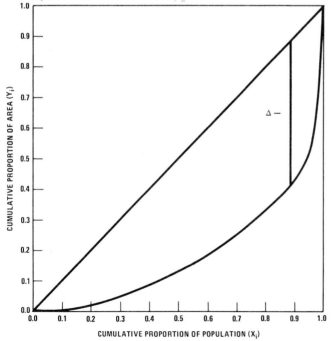

Source: See table 6–12.

Duncan also points out that the value of his concentration index depends upon the set of geographic units used. For a fixed territory, the larger the number of geographic units, the higher the value. Thus, for the United States in 1950, he obtained the following:

Type of geographic unit	Number	Concentration Index (Δ)
County..............................	3,103	58.9
Economic subregion.............	119	50.1
Geographic division.............	9	39.2

Rank-Size Rule

A number of students have noted that, in any fairly large country, the number of localities in a size interval tends to be inversely related to the size level. Apparently Auerbach was the first to observe that, when the largest cities of a country are ranked by size, the product of a city's rank by its size tends to be a constant. [55]

The most thorough exposition of this relationship is in the writings of Zipf, however. [56] Several variants of the simple relationship $P_i = \dfrac{K}{r_i}$ (where P_i is population, r_i is rank, and K is the size of the largest city) have been developed to secure better fits for various sets of statistics. These formulations are discussed by Duncan, who critically examined the general

[55] Felix Auerbach, "Das Gesetz der Bevölkerungskonzentration" (The law of population concentration), *Petermanns Mitteilungen*, 59:74–76, 1913.
[56] George Kingsley Zipf, *National Unity and Disunity*, Bloomington (Indiana), Principia Press, 1941, especially Chapters I and II.

concept. He concludes "A careful appraisal of the theoretical significance of the Pareto or the 'rank-size' rule would probably assign it a position midway between two extremes: on the one hand, a merely empirical curiosity, and, on the other, a 'law' rigorously deduced from an accepted theoretical scheme and verified under fully specified conditions."[57] Many of the tests of the rank-size rule have been applied to merely the k (100, 50, etc.) largest cities; if all identified localities are included, obviously it becomes necessary to use grouped data. It has also been contended that true agglomerations provide better fits than cities proper (incorporated places).

Stewart writes the rank-size rule more generally as

$$P_i = \frac{K}{r_i^n}, \tag{5}$$

where n is a positive constant.[58] He evaluated n at between 0 and 1 for India and Japan. Tables 6–13 and 6–14, respectively, show the actual and expected sizes of ranked cities of 100,000 or more in Mexico and of the 30 largest urbanized areas of the United States. In both cases, n is assumed to be 1.

Table 6–13.—Rank and Size of Cities of Mexico: 1960

[Expected size from Rank-Size Rule]

City	Rank	Actual population[1]	Expected population 2,832,133÷(1)	Actual minus expected (2) - (3)
	(1)	(2)	(3)	(4)
Mexico............	1	2,832,133	2,832,133	-
Guadalajara.........	2	736,800	1,416,066	−679,266
Monterrey...........	3	596,939	944,044	−347,105
Puebla de Zaragoza...	4	289,049	708,033	−418,984
Ciudad Juárez........	5	261,683	566,426	−304,743
Leon...............	6	209,469	472,022	−262,553
Torreon.............	7	179,955	404,590	−224,635
Merida.............	8	170,834	354,017	−183,183
San Luis Potosi......	9	159,980	314,681	−154,701
Tijuana.............	10	151,939	283,213	−131,274
Chihuahua..........	11	150,430	257,467	−107,037
Veracruz Llave.......	12	144,232	236,011	−91,779
Aguascalientes.......	13	126,617	217,856	−91,239
Tampico............	14	122,197	202,295	−80,098
Villa de Guadalupe Hidalgo............	15	102,602	188,809	−86,207

[1] From, United Nations, *Demographic Yearbook, 1962*, table 10.

It is obvious that in neither case does the rank-size rule describe the distribution very well. A Chi-Square test for Goodness of Fit is hardly necessary; but, when Chi-Square was computed, the corresponding probability of much less than .001 confirmed the poor fits seen by visual inspection. Figure 6–10, however, seems to show that the urbanized areas give a better fit to the rank-size rule (here called Zipf's theoretical curve) than either the cities (urban places) themselves or the standard metropolitan statistical areas when the 30 largest areas of each type are plotted.

The usual practice of confining the test to the top of the range has often resulted in a preoccupation with the relative size of the largest city in the country. A considerable literature has thus developed about "primate" cities. As originally defined by Mark Jefferson, the primate city was simply the largest city in the country.[59] Although he regarded any such city as supereminent in both size and national influence, attention has tended to focus more recently on largest cities that, when compared with other cities in the same country, are far larger than one would expect from the rank-size rule, and the concept of "primate city" has tended to shift toward this definition. Some demographers believe that many underdeveloped countries in Asia and Latin America are "over-urbanized" in the sense that they contain one extremely large city that greatly overshadows the other cities and is "parasitic" on the nation's economic life. According to Berry, ". . . the rank-size regularity applies throughout the world for countries which are highly developed with high degrees of urbanization, for large countries, and for countries such as India and China which in addition to being large also have long urban traditions; conversely, 'primate cities' or some stated degree of primacy obtains if a country is very small, or has a 'dual economy.' Moreover, additional studies have recently shown that many distributions with some degree of primacy take on more of a rank-size form as level of development and degree of urbanization increase."[60]

Table 6–14.—Rank and Size of the 30 Largest Urbanized Areas of the United States: 1960

Urbanized area	Rank	Actual population[1]	Expected population	Actual minus expected
New York............	1	14,114,927	14,114,927	-
Los Angeles.........	2	6,488,791	7,057,464	−568,673
Chicago............	3	5,959,213	4,704,976	+1,254,237
Philadelphia........	4	3,635,228	3,528,732	+106,496
Detroit............	5	3,537,709	2,822,985	+714,724
San Francisco.......	6	2,430,663	2,352,488	+78,175
Boston.............	7	2,413,236	2,016,418	+396,818
Washington.........	8	1,808,423	1,764,366	+44,057
Pittsburgh..........	9	1,804,400	1,568,325	+236,075
Cleveland..........	10	1,784,991	1,411,493	+373,498
St. Louis...........	11	1,667,693	1,283,175	+384,518
Baltimore..........	12	1,418,948	1,176,244	+242,704
Minneapolis........	13	1,377,143	1,085,764	+291,379
Milwaukee..........	14	1,149,997	1,008,209	+141,788
Houston............	15	1,139,678	940,995	+198,683
Buffalo............	16	1,054,370	882,183	+172,187
Cincinnati..........	17	993,568	830,290	+163,278
Dallas.............	18	932,349	784,163	+148,186
Kansas City........	19	921,121	742,890	+178,231
Seattle............	20	864,109	705,746	+158,363
Miami..............	21	852,705	672,139	+180,566
New Orleans........	22	845,237	641,588	+203,649
San Diego..........	23	836,175	613,692	+222,483
Denver.............	24	803,624	588,122	+215,502
Atlanta............	25	768,125	564,597	+203,528
Providence.........	26	659,542	542,882	+116,660
Portland, Oreg......	27	651,685	522,775	+128,910
San Antonio........	28	641,965	504,104	+137,861
Indianapolis........	29	639,340	486,722	+152,618
Columbus..........	30	616,743	470,496	+146,247

[1] From, *U.S. Census of Population: 1960*, Vol. I, *Characteristics of the Population*, Part 1, *United States Summary*, p. 50.

Upon investigation of the situation in 87 countries, S. Mehta concluded, however: "In summary, the 'primate city' urban structure does not appear to be a function of the level of economic development, industrialization or urbanization.

[57] Duncan, *Measurement of Population Distribution*, p. 43.

[58] John Q. Stewart, "Empirical Mathematical Rules Concerning the Distribution and Equilibrium of Population," *Geographical Review*, 37(3):462, July 1947. His notation has been transformed into that used in formula (5).

[59] Mark Jefferson, "The Law of the Primate City," *Geographical Review*, 29(2):226–232, April 1939.

[60] Brian J. L. Berry, "Cities as Systems within Systems of Cities," in: John Friedmann and William Alonso, editors, *Regional Development and Planning*, Cambridge (Mass.), the M.I.T. Press, 1964, p. 119.

It is a phenomenon by no means limited to or characteristic of the underdeveloped countries of the world. Primacy appears to be to some significant extent a function of small areal and population size. To the extent that we were able to explore the hypothesis of the 'parasitic' effect of 'primate cities,' the results do not warrant a clear negative judgement of primacy." [61]

This concern with "overdeveloped" primate cities had led International Urban Research (of the University of California at Berkeley) to suggest a "primacy rate," i.e., the percentage of the population of a country's four largest cities that is contained in the largest city. Ginsburg's *Atlas of Economic Development* contains a table and map giving this primacy rate for 104 countries.[62] Extreme values are 32.1 percent for Italy and 94.2 for Thailand; but "primacy is not a simple correlate of national income" although "the primate city . . . does appear to be more common" in poorer, newer countries.

Population Cumulated by Size of Area

A simple way of measuring population concentration is to cumulate the population in a type of area starting with the largest unit and working down. This process can be carried out for primary or secondary divisions of the country, for localities, etc. Table 6–15 illustrates this procedure for the counties of the United States. Where the number of units has changed appreciably over the period covered by the

Table 6–15. — **Minimum Number of Counties Required to Include One-Quarter, One-Half, and Three-Quarters of the Total Population of the United States: 1910 to 1960**

Census year	One-quarter	One-half	Three-quarters
1960...............	29	138	695
1950...............	28	162	798
1940...............	33	213	919
1930...............	27	189	862
1920...............	33	250	992
1910...............	39	312	1,068

Sources: 1960 – Unpublished data supplied by letter from Amos H. Hawley, January 8, 1968; 1950 – Conrad and Irene Taeuber, *The Changing Population of the United States*, Philadelphia, John Wiley & Sons, 1958, p. 23; 1910 to 1940 – Amos H. Hawley and Don J. Bogue, "Recent Shifts in Population: The Drift Toward the Metropolitan District, 1930–40," *Review of Economics and Statistics*, 24(3):143, August 1942, copyright, 1942, by the President and Fellows of Harvard College.

[61] Surinder K. Mehta, "Some Demographic and Economic Correlates of Primate Cities: A Case for Reevaluation", *Demography*, 1(1):147, 1964.

[62] Norton Ginsburg, *Atlas of Economic Development*, Chicago, University of Chicago Press, 1961, p. 36.

table, it is better to express the statistics as percentages of the total number of units at the given date.

FACTORS IMPORTANT IN ANALYSIS

A study by the Population Division, Bureau of Social Affairs, United Nations treats the relationship between urban growth and the recency of urbanization, population density, and economic development.[63] Urban population is defined as that living in localities of 20,000 or more, and recency of urbanization is classified into three categories—25 percent urban by 1920, by 1960, and not yet. Another study for the same conference examined the relationship between the level of urbanization on the one hand, and various social and economic indicators, on the other hand.[64] In a narrower demographic focus, internal migration is, of course, usually the most important factor in the process of urbanization. The report of a recent conference sponsored by ECAFE explores that relationship.[65] Earlier, there had been chapters on demographic aspects of urbanization and on economic causes and implications of urbanization, both in the ECAFE region, in the report of a joint UN/UNESCO seminar held at Bangkok in 1956.[66]

There is a vast literature on urbanization, suburbanization, metropolitanization, etc. Much of it is a part of economic history and relates the growth of cities to industrialization and economic development. A more theoretical part of the literature deals with the function of cities, with why cities develop at particular locations, with the concept of a "central place," with the structure of metropolitan communities, and with the urban hierarchy. The *Economics of Location* by August Losch is a classic in this field.[67] A real understanding of these phenomena requires an analysis of the dynamic processes involved and not just an attempted explanation of a static situation in terms of correlated variables.

[63] United Nations, *World Urbanization Trends, 1920–1960*, Inter-Regional Seminar on Development Policies and Planning in Relation to Urbanization, Working Paper No. 6, University of Pittsburgh (Pennsylvania, U.S.A.), October 24 to November 7, 1966.

[64] United Nations, *Urbanization and Economic and Social Change*, Inter-Regional Seminar on Development Policities and Planning in Relation to Urbanization, Working Paper No. 10.

[65] United Nations, Economic Commission for Asia and the Far East, *Report of the Expert Working Group on Problems of Internal Migration and Urbanization and Selected Papers*, held at Bangkok, May 24 to June 5, 1967, SA/DEM/EGIM/L.22.

[66] Philip M. Hauser, ed., *Urbanization in Asia and the Far East*, Proceedings of the Joint UN/UNESCO Seminar on Urbanization in the ECAFE Region (Bangkok, August 8–18, 1956), UNESCO, Calcutta Research Centre on the Social Implications of Industrialization in Southeast Asia, 1957.

[67] New Haven, Yale University Press, 1954.

SUGGESTED READINGS

Abu-Lughod, Janet. "Urban-Rural Differences as a Function of the Demographic Transition: Egyptian Data and an Analytical Model." *American Journal of Sociology*, 69(5) : 476–490. March 1964.

Anderson, Isabel B. *Internal Migration in Canada, 1921–1961*. Ottawa, Economic Council of Canada, Staff Study No. 13, March 1966. Pp. 71–82.

Arriaga, Eduardo E. "A Methodological Note on Urbanization Indices with Applications" in International Union for the Scientific Study of Population. *Contributed Papers*. Sydney Conference (Australia). August 21 to 25, 1967. Pp. 675–683.

Blizzard, Samuel W., and Anderson, William F. II. "Problems in Rural-Urban Fringe Research: Conceptualization and

SUGGESTED READINGS – Continued

Delineation." Pennsylvania Agricultural Experiment Station Progress Report No. 89. State College, Pa., November 1952.

Boustedt, Olaf. "Urban Population, Urban Areas, and the Problem of Dominance in West-German Statistics" in *Proceedings of the World Population Conference, 1954* (Rome), Vol. IV. United Nations, 1955. Pp. 491–505.

Bunle, Henri. "Rapport de la commission pour la définition de la 'population rurale'". *Bulletin of the International Statistical Institute* (Proceedings of the 24th session, Prague, 1938), 30(2) : 158–173. The Hague, 1938.

Dewey, Richard. "The Rural-Urban Continuum: Real but Relatively Unimportant." *American Journal of Sociology,* 66(1) : 60–66. July 1960.

Dobriner, William M., ed. *The Suburban Community.* New York, G. P. Putnam's Sons, 1958.

Douglass, H. Paul. "Suburbs," in *Encyclopedia of the Social Sciences.* New York, Macmillan, 1937. Vol. 14. P. 433.

Duncan, Otis Dudley. "Gradients of Urban Influence on the Rural Population," in *Urban Research Methods.* Jack Gibbs, ed. Princeton, New Jersey, D. Van Nostrand Co., 1961. Pp. 550–556.

Duncan, Otis Dudley, et al. *Metropolis and Region.* (Published for Resources for the Future) Baltimore, Johns Hopkins University Press, 1960. Pp. 279–550.

Duncan, Otis Dudley, and Reiss, Albert J., Jr. *Social Characteristics of Urban and Rural Communities, 1950.* New York, John Wiley & Sons, 1956.

Encyclopaedia Britannica, Article, "Metropolitan Area." Chicago, 1966. Vol. 15. P. 315.

Febvay, Maurice. "Population agricole et population active agricole, population active et population inactive par catégorie socio-professionnelle." *Proceedings of the World Population Conference, 1954* (Rome). United Nations, 1955. Vol. IV. Pp. 606–613.

Fuguitt, Glenn V. "The Rural-Urban Fringe," in *Proceedings of the American Country Life Association.* Chicago, 1962. Pp. 88–98.

Gibbs, Jack P. "Measures of Urbanization." *Social Forces,* 45(2):170–177. December 1966.

Goldstein, Sidney. "Rural-Suburban-Urban Population Redistribution in Denmark." *Rural Sociology,* 30(3):268–269. September 1965.

Hathaway, Dale E., Beegle, J. Allan, and Bryant, W. Keith. *People of Rural America,* a 1960 Census Monograph. U.S. Bureau of the Census, 1968. Pp. 1–18.

Hauser, Philip M., and Schnore, Leo F., eds. *The Study of Urbanization.* New York, John Wiley & Sons, 1965.

Idenburg, Ph. J., and Schmitz, J. "A New Approach to the Urban-Rural Classification." *Bulletin of the International Statistical Institute.* (Proceedings of the 32nd session. Tokyo, 1960), 38(2):529–538. Tokyo, 1961.

Inter-American Statistical Institute. *Report of the First Session of the Subcommittee on Demographic and Housing Statistics to the Committee on Improvement of National Statistics.* Washington, Pan American Union, 1967. Pp. 49, 51, 57–59.

————. *Revised Recommendations for the Census of Population, Housing and Agriculture Under COTA–1960.* Washington, Pan American Union (Supplement to the Report of the V session of the Census Subcommittee of COINS, San José, Costa Rica, July 14–26, 1958). Pp. 39 and 41–42.

Jefferson, Mark. "The Law of the Primate City." *Geographical Review,* 29(2):226–232. April 1939.

Jones, Victor, and Forstall, Richard L. "Economic and Social Classification of Metropolitan Areas," *Municipal Yearbook, 1963.* Chicago, International City Managers' Association, 1963. Pp. 31–44.

Kaufman, Harold F., and Singh. Avtar. "The Rural-Urban Dialogue and Rural Sociology." *Rural Sociology,* 34(4):546–551, December 1969.

Kurtz, Richard A., and Eicher, Joanne B. "Fringe and Suburb: A Confusion of Concepts." *Social Forces,* 37(1):32–37. October 1958.

Linge, G. J. R. *The Delimitation of Urban Boundaries.* Canberra, Australian National University, Research School of Pacific Studies, Department of Geography Publication G/2(1965), July 1965.

Lorenz, M. O. "Methods of Measuring the Concentration of Wealth." *Quarterly Publications of the American Statistical Association,* 9(70):209–219, June 1905.

Macura, Miloš. "The Influence of the Definition of the Urban Place on the Size of the Urban Population," in *Urban Research Methods.* Jack P. Gibbs, ed., Princeton, New Jersey, D. Van Nostrand, 1961. Pp. 21–31.

Mehta, Surinder K. "Patterns of Residence in Poona (India) by Income, Education, and Occupation (1937–65)." *American Journal of Sociology,* 73(4):496–508, January 1968.

————. "Some Demographic and Economic Correlates of Primate Cities: A Case for Reevaluation." *Demography,* 1(1):136–147, 1964.

Meuriot, Paul. *Des agglomérations urbaines dans l'Europe contemporaine.* Paris, Belin Frères, 1898.

Morita, Yuzo. "A New Method of Delimiting Urbanized Areas in Population Census Statistics." *Bulletin of the International Statistical Institute* (Proceedings of the 35th session, Belgrade, 1965), 41(1):178–192. Belgrade, 1966.

Shryock, Henry S., Jr. "Population Redistribution Within Metropolitan Areas: Evaluation of Research." *Social Forces,* 35(2):154–159, December 1956.

Stewart, Charles T., Jr. "The Urban-Rural Dichotomy: Concepts and Uses." *American Journal of Sociology,* 64(2):158, September 1958.

SUGGESTED READINGS — Continued

Stone, Leroy O. *Urban Development in Canada, 1961*. Census Monograph. Ottawa, Dominion Bureau of Statistics. 1967.

Tisdale, Hope. "The Process of Urbanization." *Social Forces,* 20(3):311–316, March 1942.

United Nations. *Data on Urban and Rural Population in Recent Censuses.* Series A, Population Studies No. 8, July 1950.

———. *Demographic Yearbook, 1962.* Pp. 33–36.

———. *Demographic Yearbook, 1972.* Pp. 5–12.

———. *Handbook of Population Census Methods,* Vol. III, *Demographic and Social Characteristics of the Population.* Studies in Methods, Series F, No. 5, rev. 1, 1959. Pp. 60–66.

———. *Population Census Methods,* Series A, Population Studies, No. 4. November 1949. Pp. 145–158; 168–171.

———. *Principles and Recommendations for the 1970 Population Censuses* (first printing). Statistical Papers, Series M, No. 44. 1967. Pp. 51, 63, and 72–75.

———. *Tabulations of the Agricultural Population and the Population Dependent on Other Economic Activities in Recent Censuses.* Studies of Census Methods, No. 13, August 1949.

———. *World Population Conference, 1965* (Belgrade), Vol. I. *Summary Report,* Meeting A.8, "Demographic Aspects of Urban Development and Housing." Statement by the Moderator (M. V. Sovani) and by the Rapporteur (Hope Eldridge). New York, 1966. Pp. 174–178.

U.S. Bureau of the Census. *Census of Agriculture, 1964, Statistics by Subjects.* Chapter 5, *Characteristics of Farm Operators and Persons Living on Farms.* 1967.

———. *Current Population Reports,* Series P–23, No. 1. "The Development of the Urban-Rural Classification in the United States," by Leon E. Truesdell. August 5, 1949.

———. *Farm Population: 1880 to 1950,* by Leon E. Truesdell, Bureau of the Census. Technical Paper No. 3, 1960.

———. *U.S. Census of Population: 1960. Selected Area Reports, Size of Place,* PC(3)–1B, 1964.

U.S. Department of Agriculture. Economic Research Service. *Farm Population: Estimates for 1910–62,* by Vera J. Banks, Calvin L. Beale, and Gladys K. Bowles. ERS–130. October 1963.

U.S. Bureau of the Census and Agricultural Marketing Service. *Farm Population,* Series Census–AMS (P–27), No. 28. "Effect of Definition Changes on Size and Classification of the Rural-Farm Population: April 1960 and 1959." April 17, 1961.

Weber, Adna Ferrin. *The Growth of Cities in the Nineteenth Century.* Ithaca, New York, Cornell University Press, 1963. Originally published in 1899 for Columbia University by the Macmillan Co., N.Y., as Vol. XI of *Studies in History, Economics and Public Law.*

Willcox, Walter F. *Studies in American Demography.* Ithaca, New York, Cornell University Press, 1940. Pp. 105–119.

Wolfe, Marshall. "Some Implications of Recent Changes in Urban and Rural Settlement Patterns in Latin America," in *Proceedings of the World Population Conference* (Belgrade, 1965). Vol. IV. United Nations, 1967. Pp. 457–460.

Zipf, George Kingsley. *National Unity and Disunity.* Bloomington, Indiana, Principia Press, 1941. Especially chapters I and II.

CHAPTER 7

Sex Composition

USES OF DATA

The personal characteristic of sex holds a position of prime importance in demographic studies. Separate data for males and females are important in themselves, for the analysis of other types of data, and for the evaluation of the completeness and accuracy of the census counts of population.

Many types of planning, both public and private, such as military planning, planning of community institutions and services, particularly health services, and planning of sales programs require separate population data for males and females. The sociologist and the economist have a vital interest in such data. The balance of the sexes affects social and economic relationships within a community. Social roles and cultural patterns may be affected. For example, under the condition of the male "shortage" brought about by World War II, the extent of labor force participation and the occupational distribution of women in the United States, Canada, and several European countries underwent considerable change and a new pattern of social relationships between the sexes emerged, including a greatly increased independence of women from the home.[1]

For such subjects as natality, mortality, migration, marital status, and economic characteristics, statistics are sometimes shown only for both sexes combined; but the ordinary and more useful practice is to present and analyze the statistics separately for males and females. In fact, a very large part of the usefulness of the sex classification in demographic statistics lies in its cross-classification with other classifications in which one may be interested. For example, the effect of variations in the proportion of the sexes on measures of natality is considerable. This effect may make itself felt indirectly through the marriage rate, of course. Generally, there are substantial differences between the death rates of the sexes; hence, the effect of variations in sex composition from one population group to another should be taken into account in comparative studies of general mortality. The analysis of labor supply and military manpower requires separate information on males and females cross-classified with economic activity and age. In fact, a cross-classification with sex is useful for the effective analysis of nearly all types of data obtained in censuses and surveys, including data on racial and ethnic composition, educational status, and citizenship status, as well as the types of data mentioned above.

Data on sex composition serve another important analytic purpose. Because the expected proportion of the sexes can often be independently determined within a narrow range, the tabulations by sex are useful in the evaluation of census and survey data, particularly with respect to the coverage of the population by sex and age.

CLASSIFICATION

The definition and classification of sex present no statistical problems. It is a readily ascertainable characteristic and the data are easy to obtain. The situation with respect to sex is in contrast to that of most other population characteristics, the definition and classification of which are much more complex because they involve numerous categories and are subject to alternative formulation as a result of cultural differences, differences in the uses to which the data will be put, differences in the interpretations of respondents and enumerators, etc.

SOURCES OF DATA

The importance of this classification in censuses, surveys, and registrations has been widely recognized.[2] Wherever national population censuses have been taken, sex has nearly always been included among those subjects for which information was secured. Census or survey data for males and females are presented for nearly all countries of the world in a table annually included in the U.N. *Demographic Yearbook*. Recent census data or estimates of the age-sex distribution are also presented for most countries in another table of the *Yearbook*.

A classification by sex has been part of the U.S. census from its very beginning.[3] At first, data were collected and tabulated on the number of males and females in the white population only; but, from 1820 on, the total population and each identified racial group were classified by sex. Regional detail is available from 1820, and data by size of community from 1890. The first classification of sex by single years of age was published in 1880. Estimates of the sex distribution of the population (cross-classified with age and color), for the United States as a whole, are available for each year since 1900, and

---

[1] For some notes on the sociological aspects of sex composition, see: Hans Von Hentig, "The Sex Ratio: A Brief Discussion Based on United States Census Figures," *Social Forces*, 30(4):443–449, May 1952.

[2] United Nations, *Principles and Recommendations for National Population Censuses*, Statistical Papers, Series M, No. 27, 1958, p. 9; idem, *Principles and Recommendations for the 1970 Population Censuses*, Statistical Papers, Series M, No. 44, 1967, pp. 40 and 67–69.

[3] See U.S. Bureau of the Census, *Historical Statistics of the United States: Colonial Times to 1957*, 1960, Series A 23, 24, 34, and 35; idem, *Historical Statistics of the United States: Continuation to 1962 and Revisions*, 1965, Series A 23 and 24.

projections of the population by sex (also by age and color) are available to 2015. Such estimates and projections appear currently in the P–25 Series of *Current Population Reports* of the U.S. Bureau of the Census. Almost every characteristic for which data are shown in the 1960 U.S. census reports was cross-classified with sex. This is true also of the U.S. Current Population Survey. Cross-classification with sex is also a common practice in the U.S. vital statistics tabulations.

QUALITY OF DATA

The principal problem relating to the quality of the data on sex collected in censuses concerns the differential completeness of coverage of the two sexes. At least in the statistically developed countries, misreporting of sex is negligible; there appears to be little or no reason for a tendency for one sex to be reported at the expense of the other. The reports on sex in the 1960 census of the United States and in the accompanying reinterview study differed by about one percent of the matched population; the reports on sex in the 1960 census and the Current Population Survey differed by a similar percentage, as shown by the Census-CPS Match Study.[4] Because of misreporting of sex in both directions, the net reporting error in the 1960 census indicated by the match studies is less than 0.5 percent. In some countries deliberate misreporting of sex may be more serious. Parents may report young boys as girls to avoid the attention of evil spirits or so that they may be overlooked when their cohort is called up for military service. The same factors may contribute to differential underenumeration of the two sexes.

How complete are the census counts of males and females? Although there are no ideal standards against which the accuracy of census data can be measured, it is possible to derive some indication of both the relative and absolute completeness of enumeration of males and females. For the most part, these techniques are essentially the same as those used to evaluate total population coverage, and would include reinterview studies, the use of external checks (e.g., Selective Service registration data, Social Security account holders), and various techniques of demographic analysis, such as the application of the population balancing equation separately for each sex. Illustrative results for the United States in 1960 and 1950 are given in Table 7-1.

TECHNIQUES OF ANALYSIS

Numerical Measures

The numerical measures of sex composition are few and simple to compute. They are: (a) the percentage of males in the population, or the masculinity proportion; (b) the sex ratio or the masculinity ratio; and (c) the ratio of the excess or deficit of males to the total population. The mere excess or deficit of males is affected by the size of the population and is not, therefore, a very useful measure for making comparisons of one population group with another. The three measures listed are all useful for interarea or intergroup comparisons, or comparisons over time, since in one way or another they remove the effect of variations in population size.

The **masculinity proportion** is the measure of sex composi-

Table 7-1.— Estimates of Net Underenumeration, by Sex, in the 1950 and 1960 Censuses of the United States

[Numbers in thousands]

Year and sex	Post-enumeration survey[1]		Analytic studies[2]	
	Number	Percent[3]	Number	Percent[3]
1960				
Total..............	3,328	1.8	5,702	3.1
Male...............	1,636	1.8	3,489	3.8
Female.............	1,693	1.8	2,213	2.4
1950				
Total..............	2,093	1.4	5,675	3.6
Male...............	1,026	1.4	3,342	4.3
Female.............	1,067	1.4	2,333	3.0

[1] Eli S. Marks and Joseph Waksberg, "Evaluation of Coverage in the 1960 Census of Population Through Case-By-Case Checking," *Proceedings of the Social Statistics Section: 1966*, American Statistical Association, p. 69; U.S. Bureau of the Census, *The Post-Enumeration Survey: 1950*, Technical Paper No. 4, 1960, p. 12.
[2] Jacob S. Siegel, "Completeness of Coverage of the Nonwhite Population in the 1960 Census and Current Estimates, and Some Implications," *Social Statistics and the City*, David M. Heer, ed., Joint Center for Urban Studies, Massachusetts Institute of Technology and Harvard University, Cambridge, Mass., 1968, p. 41.
[3] Base is corrected population.

tion most commonly used in nontechnical discussions. The formula for the masculinity proportion is:

$$\frac{P_m}{P_t} \times 100 \qquad (1)$$

where P_m represents the number of males and P_t the total population.[5] Let us apply the formula to Morocco in 1960. The 1960 census showed 5,809,172 males and a total population of 11,626,232. Therefore, the masculinity proportion is $\frac{5,809,172}{11,626,232} \times 100 = 50.0$ percent. Fifty is the point of balance of the sexes, or the standard, according to this measure. A higher figure denotes an excess of males and a lower figure denotes an excess of females. The masculinity proportion of national populations varies over a rather narrow range, usually falling just below 50.0, unless exceptional historical circumstances have prevailed.

The **sex ratio** is the principal measure of sex composition used in technical studies. The sex ratio is usually defined as the number of males per 100 females, or

$$\frac{P_m}{P_f} \times 100 \qquad (2)$$

where P_m, as before, represents the number of males and P_f the number of females. The formula may be computed for Morocco in 1960 as follows, given the male population as 5,809,172 and the female population as 5,817,060:

$$\frac{5,809,172}{5,817,060} \times 100 = 99.9$$

[4] U.S. Bureau of the Census, *Evaluation and Research Program of the U.S. Censuses of Population and Housing, 1960: Accuracy of Data on Population Characteristics as Measured by Reinterviews*, Series ER 60, No. 4, 1964, p. 10; idem, *Evaluation and Research Program of the U.S. Censuses of Population and Housing, 1960: Accuracy of Data on Population Characteristics as Measured by CPS-Census Match*, Series ER 60, No. 5, 1964, p. 9.

[5] The multiple of 10, or the k factor, employed to shift the decimal in this and other formulas, is often arbitrary and conventional. The particular k factor employed in a given formula may sometimes vary from one reference to another in this volume where there is no conventional k factor. Where there is a conventional k factor for a given formula, this factor has ordinarily been accepted for use here.

One hundred is the point of balance of the sexes according to this measure. A sex ratio above 100 denotes an excess of males; a sex ratio below 100 denotes an excess of females. Accordingly, the greater the excess of males the higher the sex ratio; the greater the excess of females, the lower the sex ratio.

This form of the sex ratio is sometimes called the masculinity ratio. The sex ratio is also sometimes defined as the number of females per 100 males. This is or has been the official practice in many countries, particularly in Eastern Europe (e.g., Poland, Bulgaria, Hungary, India, Pakistan, Iran, and Yugoslavia), but the United Nations as well as most countries follow the former definition.

The sex ratio of the Moroccan population might be described as "typical" or a little above the typical level. In general, national sex ratios tend to fall in the narrow range from about 95 to 102, barring special circumstances, such as a history of heavy war losses or heavy immigration; national sex ratios outside the range of 90 to 105 are to be viewed as extreme.

Variations in the sex ratio are similar to those in the masculinity proportion, but the sex ratio is a more sensitive indicator of differences in sex composition because it has a relatively smaller base. It has the disadvantage, however, of being difficult to manipulate mathematically.

The third measure of sex composition, the **excess (or deficit) of males as a percent of the total population**, is given by the following formula:

$$\frac{P_m - P_f}{P_t} \times 100 \qquad (3)$$

Again, employing the data for Morocco in this formula, we obtain:

$$\frac{5,809,172 - 5,817,060}{11,626,232} \times 100 = -0.1 \text{ percent}$$

This figure indicates that the deficit of males amounts to 0.1 percent of the total population. The point of balance of the sexes according to this measure, or the standard, is zero; a positive value denotes an excess of males and a negative value denotes an excess of females.

It may be evident that the various measures of sex composition convey essentially the same information. Sometimes it is desired to convert the masculinity proportion into the sex ratio or the percent excess (or deficit) of males, or the reverse, in the absence of the basic data on the numbers of males and females. These conversions may be effected by use of the following formulas, the application of which is illustrated with figures for Morocco in 1960. [6]

$$\text{Masculinity proportion} = \frac{\text{sex ratio}}{1 + \text{sex ratio}} \times 100 = \qquad (4)$$

$$\frac{.9986}{1.9986} \times 100 = 50.0 \text{ percent}$$

$$\text{Sex ratio} = \frac{\text{masculinity proportion}}{1 - \text{masculinity proportion}} \times 100 = \qquad (5)$$

$$\frac{.4997}{1 - .4997} \times 100 = \frac{.4997}{.5003} \times 100 = 99.9$$

Percent excess or deficit of males = [masculinity proportion − (1 − masculinity proportion)] × 100 = [.4997 − (1 − .4997)] × 100 = (.4997 − .5003) × 100 = −.0006 × 100 = −0.1 percent

$$(6)$$

Thus, if we divide the masculinity proportion (omitting the k factor) for Morocco in 1960, .4997, by its complement, .5003, and multiply by 100, we obtain 99.9 as the sex ratio, the same value obtained earlier by direct computation. Or, if we divide the sex ratio, .9986 by 1 plus the sex ratio, 1.9986, and multiply by 100, we obtain 50.0 as the masculinity ratio.

There are few graphic devices which are designed specifically for description and analysis of sex composition. Principal among these is the so-called population "pyramid." Inasmuch as age is ordinarily combined with sex in the "content" of these devices, particularly in the case of the population pyramid, discussion of their construction and interpretation is postponed until the next chapter. The standard graphic devices, including bar charts, line graphs, pie charts, etc, are available, however, for depicting differences in sex composition from group to group or over time for a particular group.

Since the sex ratio is the most widely used measure of sex composition, we will give primary attention to it in the remainder of this chapter.

Analysis of Sex Ratios in Terms of Population Subgroups

Because the sex ratio may vary widely from one population subgroup to another, it is frequently desirable to consider separately the sex ratios of the important component subgroups in any detailed analysis of the sex composition of a population group. Account might be taken of these variations in the analysis of the overall level of the sex ratio at any date and of the differences in the sex ratio from area to area or from one population group to another.

Table 7-2. — **Sex Ratios by Region and Residence, for the United States: 1960**

[Sex ratios are shown as males per 100 females]

Residence	United States			North-east	North Central	South	West
	Population (in thousands)		Sex ratio				
	Male	Female	[(1)÷(2)] x100				
	(1)	(2)	(3)	(4)	(5)	(6)	(7)
Total.....	88,331	90,992	97.1	94.7	97.4	97.0	100.6
Urban.........	60,733	64,536	94.1	93.0	94.4	93.0	97.1
Rural.........	27,598	26,456	104.3	101.7	104.4	102.9	113.5

Source: Adapted from *U.S. Census of Population: 1960*, Vol. I, *Characteristics of the Population*, Part 1, *United States Summary*, 1964, table 58.

For the United States in 1960, notably different sex ratios were recorded for the separate color, race, nativity, residence, regional, age, and marital status groups in the population (see tables 7-2, 7-3, and 7-4 for illustrative figures). The marked deficit of males in the urban population may be compared with the marked excess of males in the rural population. The urban

---

[6] In general, correct intermediate algebraic manipulation of the formulas presented requires that this manipulation be done on the basis of formulas omitting the k factor. For example, the sex ratio should be represented merely by $P_m \div P_f$, and the masculinity proportion by $P_m \div P_t$. The appropriate k factor may then be applied at the end.

In general, in numerically applying a formula, one should carry in the intermediate calculations at least one additional significant figure beyond the number of significant figures to be shown in the result. Then, the "result" figure may be rounded as desired.

population has lower sex ratios principally because of the greater migration of females to cities. The sex ratio also varies widely among regions. Thus, the sex ratio is quite low in the Northeast and in approximate balance in the West. The marked excess of females among Negroes may be compared with the moderate excess of females among whites.

Sex ratios for age groups vary widely around the sex ratio for the total population. For many analytic purposes, this variation may be considered the most important. The sex ratio tends to be high at the very young ages and then tends to decrease with increasing age. "Young" populations and populations with high birth rates tend to have higher overall sex ratios than "old" populations and populations with low birth rates because of the excess of boys among births and children and the deficit of men among older persons.

Table 7–3.—**Sex Ratios by Nativity and Race and by Age, for the United States: 1960**

[Sex ratios are shown as males per 100 females]

Age	Sex ratio	Age	Sex ratio
Total..............	97.1	Total..............	97.1
White..................	97.4	Under 5 years...........	103.4
Native...............	97.5	5 to 9 years............	103.4
Foreign born..........	94.2	10 to 14 years..........	103.3
Nonwhite...............	94.3	15 to 24 years..........	98.3
Negro.................	93.4	25 to 34 years..........	96.1
Indian................	101.2	35 to 44 years..........	95.4
Japanese..............	93.9	45 to 54 years..........	97.1
Chinese...............	133.2	55 to 64 years..........	93.8
Filipino..............	175.4	65 to 74 years..........	87.0
All other.............	111.4	75 years and over.......	75.2

Source: Adapted from *U.S. Census of Population: 1960*, Vol. I, *Characteristics of the Population*, Part 1, *United States Summary*, 1964, tables 44, 45, and 156.

Table 7–4.—**Sex Ratios by Nativity and Residence, for the United States: 1960**

[Sex ratios are shown as males per 100 females]

Nativity	Total	Urban	Rural-nonfarm	Rural-farm
Total....................	97.1	94.0	103.2	107.7
Native...................	97.1	94.1	103.1	107.1
Foreign born.............	95.6	93.6	104.3	144.0

Source: Adapted from *U.S. Census of Population: 1960*, Vol. I, *Characteristics of the Population*, Part 1, *United States Summary*, 1964, table 156.

Analysis of Changes

It is frequently desired to explain in demographic terms the change in the sex composition of the population from one census to another. What is called for is a quantitative indication of how the various components of population change—births, deaths, immigrants, and emigrants—contributed to the change.

Unfortunately, such an analysis is complicated by the lack of perfect consistency between the data on the components of change and census data with respect to the intercensal change implied. It was pointed out earlier that coverage of males and females is likely to be different in a particular census and between censuses. Errors in the census data as reported and in the data on components of change affect the apparent change to be explained. It is desirable, therefore, in any analysis of changes shown by census figures, to take into account the errors in the census data and in the data on components. The errors in the census data cannot usually be determined very

closely, however. If it can be assumed that the estimates of the components are satisfactory, the "error of closure" for each sex may be used as an estimate of change in the net coverage of each sex in the two censuses.

For simplicity, and in view of the lack of adequate information, we will generally assume in our treatment of the problem below that the data on components are substantially correct and reasonably consistent with the census figures as observed.

Change in Excess or Deficit of Males.—The formula for analyzing the change between two censuses in the excess or deficit of males in terms of components may be developed from the separate equations representing the male and female populations at a given census (P_m^1 and P_f^1) in terms of the male and female population at the preceding census (P_m^0 and P_f^0) and the male and female components of change (B_m and B_f for births, D_m and D_f for deaths, I_m and I_f for immigrants or in-migrants and E_m and E_f for emigrants or out-migrants):

$$P_m^1 = P_m^0 + B_m - D_m + I_m - E_m \qquad (7)$$

$$P_f^1 = P_f^0 + B_f - D_f + I_f - E_f \qquad (8)$$

These are merely the usual intercensal or component equations expressed separately for males and females. Solving these equations for $P_m^1 - P_m^0$ and $P_f^1 - P_f^0$ (that is, the increase in the male and female population, respectively) and taking the difference between them, we have, for the intercensal change in the difference between the numbers of males and females:

$$(P_m^1 - P_f^1) - (P_m^0 - P_f^0) = (B_m - B_f) - (D_m - D_f) + (I_m - I_f)$$
$$- (E_m - E_f) \qquad (9)$$

Table 7–5 illustrates the application of this equation to the data for the United States in the period 1950–60. Each item in formula (9) is represented in table 7–5, except that immigration and emigration are combined as net immigration. The table shows first that the excess of females increased from 953,000 in 1950 to 2,660,000 in 1960, or by 1,707,000. The factor of deaths was the greatest contributor to this shift. While 1,019,000 more males than females were being

Table 7–5.—**Component Analysis of the Change in the Difference Between the Number of Males and Females in the United States: 1950–1960**

[Numbers in thousands]

Population or component of change	Male	Female	Difference[1]
Population (census):			
April 1, 1950....................	75,187	76,139	-953
April 1, 1960....................	88,331	90,992	-2,660
Change during decade:			
Net change......................	13,145	14,852	-1,707
Births..........................	20,991	19,972	+1,019
Deaths..........................	-8,904	-6,704	-2,200
Net immigration.................	1,150	1,495	-345
Civilian......................	1,478	1,497	-19
Military......................	-328	-2	-326
Residual[2].....................	-92	89	-181

[1] Excess of males. A minus sign denotes an excess of females.
[2] Difference between the intercensal change based on the two census counts and the intercensal change based on the "component" data, i.e., the error of closure.

Source: U.S. Bureau of the Census, *Current Population Reports*, Series P–25, No. 310, "Estimates of the Population of the United States and Components of of Change, by Age, Color, and Sex: 1950 to 1960," June 30, 1965, table 12.

added through birth, 2,200,000 more males than females were being removed through death. Migration contributed only a small part (345,000) of the total change and the remainder (181,000) represents the difference between males and females in the error of closure.

Change in Sex Ratios in Terms of Components. — It is of interest to analyze the difference, in terms of components, between the current sex ratio and a sex ratio of 100 representing a balance of the sexes (such as might result from the action of births and deaths in the absence of heavy migration).

Sex ratio of births. — From an examination of the sex ratios of registered births for a wide array of countries, it is apparent that the component of births tends to bring about or to maintain an excess of males in the general population. The sex ratio of births is above 100 for nearly all countries for which relatively complete data are available and between 104 and 107 in most such countries (see table 7-6). [7]

Careful analysis relating to the sex ratio of births should take into account significant variations in the sex ratio of births according to the demographic characteristics of the child and the parents. Among the important demographic characteristics which appear to distinguish births with respect to their sex ratio are age of mother, order of birth of child, and race. It has been shown, on the basis of United States data, that there is a pronounced inverse relationship between the sex ratio by age of mother and order of birth of child, dominated by the order of birth, and that there is a notable excess of the sex ratio of white births over that for Negroes. [8] Visaria's analysis of the sex ratios of births covering a large number of countries identifies a high proportion of Negroes in the population as a principal characteristic associated with a low sex ratio. [9]

Another factor that may affect the sex ratio of births is the socioeconomic status of the parents. A predominance of male births has been observed among higher socioeconomic groups in Western countries. [10] It may be explained by the predominance of lower order births when fertility is low and the lower rate of prenatal deaths. Similar information on the relationship between socioeconomic status and the sex ratio of births is not available for the less developed countries.

For areas with incomplete reporting of births, the observed sex ratio of births may be suspect. Direct observations or estimates, from vital statistics, censuses, and surveys, on the sex ratio at birth in India showed wide regional variation. [11] For example, the National Sample Survey had given 125 for the Northwest zone and 100 for the South zone. More accurate statistics on births occurring in hospitals and health centers gave an All-India average of 106 with very little regional variation. The values shown by the other sources are evidently subject to considerable reporting errors of various sorts and to sampling error.

Sex ratio of deaths. — The sex ratio of deaths is much more

Table 7-6. — **Sex Ratios at Birth in Various Countries With Relatively Complete Registration**

Country	Period	Sex ratio	Country	Period	Sex ratio
South Africa, Asiatics	1949-57	101.2	South Africa, Whites	1949-57	104.9
South Africa, Colored	1949-57	101.7	Costa Rica	1949-58	105.2
			Finland	1949-58	105.2
British Honduras	1949-58	102.2	Malaya	1949-57	105.2
United States, Nonwhites	1949-58	102.2	Ireland	1949-58	105.5
			Japan	1949-58	105.5
Jamaica	1949-56	102.8	Taiwan	1949-58	105.5
Chile	1949-56	103.0	United States, Whites	1949-58	105.7
Panama	1950-58	103.9			
Venezuela	1949-56	104.2	Romania	1954-58	105.9
El Salvador	1949-58	104.5	Norway	1949-57	106.1
Uruguay	1949-54	104.5	Bulgaria	1949-58	106.4
Guatemala	1949-58	104.8	Mexico	1949-57	106.5
France	1949-58	104.9			

Source: Pravin M. Visaria, "Sex Ratio at Birth in Territories with a Relatively Complete Registration," *Eugenics Quarterly*, 14(2):134-135, June 1967. © 1967 by the American Eugenics Society, Inc. All rights reserved.

variable from country to country than the sex ratio of births. Data for a wide range of countries indicate sex ratios well above 100.0 in most cases. Since this factor operates in a negative fashion, the component of deaths has tended to depress the sex ratio of most populations. High sex ratios of deaths (over 125) occurred in recent years in Canada, Cuba, Australia, and the United States. Low ratios (100-105) occurred in the United Arab Republic, Austria, Portugal, Spain, Yugoslavia, and the United Kingdom. Intermediate ratios (105-125) occurred in Mexico, Venezuela, India, Israel, Japan, Czechoslovakia, Hungary, and Sweden. National differences in the sex ratio of deaths may be accounted for partly by differences from country to country in the age-sex structure of the population and partly by differences in death rates for each age-sex group.

Demographic characteristics important in the further analysis of the sex ratio of deaths include age, race or other ethnic grouping, residence status, and marital status. Sex ratios of deaths in the United States for two broad classes defined by each of these four characteristics for 1964 are as follows:

Under 65 years	171.6	White	131.1
65 and over	109.4	Nonwhite	125.9
Metropolitan	126.4	Married[a]	232.0
Nonmetropolitan	137.4	All other[a]	77.1

[a] 15 years old and over for 1959-61.

There are also pronounced regional variations in the sex ratio of deaths in the United States. Figures for the several States ranged from 112.8 for Massachusetts to 201.5 for Alaska. As for countries, these variations are associated with differences in the composition of the population with respect to age, sex, and other characteristics, as well as with differences in death rates for these categories.

An important analytic question relates to the basis for the difference between male and female death rates. The separate contribution of biological and cultural factors needs to be determined. Differences in the occupational distribution of the sexes illustrate the role of cultural factors; generally men work at the more hazardous, strenuous, and stressful occupations. On the other hand, women are generally exposed to the special risks of childbearing, and some countries (e.g., India) still show higher death rates for females than for males. The weight of biological forces is reflected in the higher mortality of male infants and fetuses. Studies of the mortality of males

[7] Pravin M. Visaria, "Sex Ratio at Birth in Territories With a Relatively Complete Registration," *Eugenics Quarterly*, 14(2): 132-142, June 1967. See also Pravin M. Visaria, "The Sex Ratio of the Indian Population," unpublished Ph. D. dissertation submitted to Princeton University, Princeton, N.J.

[8] C. A. McMahan, "An Empirical Test of Three Hypotheses Concerning the Human Sex Ratio at Birth in the United States, 1915-48," *Milbank Memorial Fund Quarterly*, 29(3):273-293, July 1951, esp. p. 288; Robert J. Myers, "The Effect of Age of Mother and Birth Order on Sex Ratio at Birth," *Milbank Memorial Fund Quarterly*, 32(3):275-281, July 1954; Brian Macmahon and Thomas F. Pugh, "Influence of Birth Order and Maternal Age on the Human Sex Ratio at Birth," *British Journal of Preventive and Social Medicine*, 7(2):83-86, London, April 1953.

[9] Visaria, op. cit., pp. 136-139.

[10] Sanford Winston, "Birth Control and Sex Ratio at Birth," *American Journal of Sociology*, 38:225-231, 1932; idem, "The Influence of Social Factors upon the Sex Ratio at Birth," *American Journal of Sociology*, 37:1-21, 1931.

[11] K. V. Ramachandran and Vinayak A. Deshpande, "The Sex Ratio at Birth in India by Regions," *Milbank Memorial Fund Quarterly*, 42(2), Part 1:84-95, April 1964.

engaged in protected activities (e.g., monks) help to isolate the role of the biological factor among adults.[12]

A special aspect of the relation of mortality to the sex ratio of the population is the effect of war. The sex ratios of the population of the U.S.S.R. in 1959—81.9—and of the Soviet Zone of Germany in 1950—80.2—illustrate the lingering effect of war-caused mortality on the balance of the sexes in an area. For the most part, males have suffered the heaviest casualties because they alone directly participated in battle. Changes in the technology and conduct of wars, including particularly the bombing of industrial and administrative centers, may tend to equalize somewhat the extent of military casualties between the sexes. Further analysis of the relation of war to the sex ratio of deaths, designed to show the effect of the shifting number of males in the population at risk, would compare the sex ratio of deaths in the war years and in the immediate postwar period of various belligerent countries.

Special practices may affect the sex ratio of deaths. Female infanticide (mainland China), the slaughter of male captives of another race or tribe, the provision of better care to the children of one sex than another as dictated by the culture, and the suttee (India) illustrate types of practices which have prevailed historically in various areas of the world.

Sex ratio of migrants.—The sex ratio of migrants has been less uniform from area to area and has often shown more extreme values (above or below 100.0) than the sex ratio of either births or deaths. Immigrants to Brazil, the Federal Republic of Germany, and Italy in 1961 had sex ratios of 135, 198, and 135, respectively.[13] The corresponding figures for Canada and the United States are 81 for each country. Most countries reporting immigration by sex now receive more males than females.

One or the other sex may be attracted in greater numbers to certain areas within countries, depending largely on the types of occupational opportunities and on various cultural factors, particularly customs regarding the separation of family members. Patterns of sex-selectivity of migrants to cities appear to differ in the countries of the West and the less industrialized countries of the East. Men predominate in the migration streams to large cities in Southeast Asia, for example. The Indian peasant who goes from the village to the city to work is normally not accompanied by other members of his family. In Latin America, on the other hand, the rates of cityward migration for women appear to exceed or approximately equal those for men.[14]

In the United States the many office jobs and light factory jobs available in cities attract women particularly. The factor of migration has been an important element in the widely different sex ratios of the rural-farm population, the rural-nonfarm population, and the urban population of the United States. As shown in table 7–4, these values were 107.7, 103.2, and 94.0 in 1960, respectively. Hence, in the migration from farm to towns and cities and from rural to urban areas females must have substantially outnumbered males.

Specific cities show considerable variation in sex composition, largely as a result of differences in type of major economic activity. In 1960, the population sex ratio was 87.3 for Albany, N.Y., a State capital; 93.0 for Hartford, Conn., a State capital and insurance center; and 107.5 for Anchorage, Alaska, the largest city of a "frontier" State.

The sex ratio of an area may be affected by certain special features of the area which select certain classes of "migrants." A large military installation, a college for men or women, or an institution confining mainly or entirely persons of a particular sex, may be located in the area. The sex ratios of Harrison County, Miss. (115.4), and Grafton County, N.H. (107.6), in 1960 illustrate, in part, the effect of the presence of a large military installation (Keesler Air Force Base) and a college for men (Dartmouth College), respectively.

It should be clear that the narrow bounds for acceptability of a national sex ratio do not apply to regional or local population or residence categories. Wide deviations from 100.0 should, however, be explainable in terms of the sex-selective character of migration to and from the specific area, and the particular industrial and institutional "makeup" of the area.

Use of Sex Ratios in Evaluation of Census Data

Because of the relatively limited variability of the national sex ratio and its independence of the absolute numbers of males and females, it is employed in various ways in measuring the quality of census data on sex, particularly in cross-classification with age.

The simplest approach to evaluation of the quality of the data on sex for an area consists of observing the deviation of the sex ratio for the area as a whole from 100.0, the point of equality of the sexes. With, say, a fairly constant sex ratio at birth of about 105 and a sex ratio of deaths in the range 105–125, the sex ratio of a population will fall near 100 in the absence of migration. A sex ratio deviating appreciably from 100—say, below 90 or above 105—must be accounted for in terms of migration (both the volume and sex composition of the migrants being relevant) or a very high death rate, including war mortality. A sex ratio deviating even further from 100—say 'above 110 or below 85—must be accounted for in terms of some unusual feature of the area, such as the location of a military installation in the area.

A theoretically more careful evaluation of the data on sex composition of an area at a census date would involve a check of the consistency of the sex ratio shown by the given census with the sex ratio shown by the previous census. For a country as a whole, a direct check can be made by use of the reported data on the components of population change during a decade.

Comparison can also be made between the observed sex ratio in the census and that shown by a postenumeration survey. In 1960 for the United States the two values were the same, 97.1.

---

[12] Some studies which treat this general question are: L. I. Dublin, A. J. Lotka, and M. Spiegelman, *Length of Life,* New York, Ronald Press Company, 1949, esp. pp. 129–134; Francis C. Madigan, "Are Sex Mortality Differentials Biologically Caused?," *Milbank Memorial Fund Quarterly,* 35(2):202–223, April 1957. For a study that reaches a conclusion opposite Madigan's see Philip Enterline, "Causes of death responsible for recent increases in sex mortality differentials in the United States," *Milbank Memorial Fund Quarterly,* 39(2): 312-328, April 1961.

[13] United Nations, *Demographic Yearbook, 1962,* table 27.

[14] UNESCO, *Urbanization in Asia and the Far East,* Philip M. Hauser, ed. (Proceedings of the joint UN/UNESCO seminar on urbanization in the ECAFE region, Bangkok, August 8-18, 1956), Calcutta, 1957, pp. 107–110; Kingsley Davis, *The Population of India and Pakistan,* Princeton, N.J., Princeton University Press, 1955, pp. 135–136 and 139–141; Juan C. Elizaga, *Migración Diferencial en Algunas Regiones y Ciudades de la América Latina, 1940–1950* (Differential migration in some regions and cities of Latin America, 1940-1950), Centro Latinoamericano de Demografía, Series A, No. 8, Santiago, Chile, 1963.

SUGGESTED READINGS

Eldridge, Hope T., and Siegel, Jacob S. "The Changing Sex Ratio in the United States." *American Journal of Sociology*, 52(3):224–234, November 1946.

Enterline, Philip. "Causes of death responsible for recent increases in sex mortality differentials in the United States." *Milbank Memorial Fund Quarterly* 39 (2): 312-328, April 1961.

Goodman, Leo A. "Population Growth of the Sexes." *Biometrics*, 9(2):212–225, June 1953.

Hawley, A. H. "Population Composition." In *The Study of Population*, P. M. Hauser and O. Dudley Duncan eds. Chicago, University of Chicago Press, 1959, chapter 16.

Hocking, W. S. "The Balance of the Sexes in Great Britain." *Journal of the Institute of Actuaries*, 74:340–344, 1948.

Myers, Robert J. "The Effect of Age of Mother and Birth Order on Sex Ratio at Birth." *Milbank Memorial Fund Quarterly*, 32(3):275–281, July 1954.

Nag, A. C. "Variation of the Sex-Ratio Under Different Conditions." *International Statistical Conference*, 33 (Part IV):75–82, December 1951, India.

Ramachandran, K. V., and Deshpande, Vinayak A. "The Sex Ratio at Birth in India by Regions." *Milbank Memorial Fund Quarterly*, 42(2): Part I, 84–95, 1964.

Rubin, Ernest. "The Sex Ratio at Birth." *American Statistician*, 21(4):45–48, October 1967.

Stern, Curt. *Principles of Human Genetics*. 2nd ed., San Francisco, W. H. Freeman and Co., 1960. Pp. 400–423.

Tarver, James D., and Lee, Che-Fu. "The Sex Ratio of Registered Live Births in the United States, 1942–63." *Demography*, 5(1): 374–381. 1968.

Thompson, Warren, and Lewis, David T. *Population Problems*, 5th ed. New York, McGraw-Hill Book Company, 1965. Pp. 73–84.

Visaria, Pravin M. "The Sex Ratio at Birth in Territories With a Relatively Complete Registration." *Eugenics Quarterly*, 14(2): 132–142, June 1967.

Von Hentig, Hans. "The Sex Ratio: A Brief Discussion Based on United States Census Figures." *Social Forces*, 30(4): 443–449, May 1952.

Winston, Sanford. "The Influence of Social Factors Upon the Sex Ratio at Birth." *American Journal of Sociology*, 32(1):1–21, July 1931.

Yerushalmy, J. "The Age-Sex Composition of the Population Resulting from Natality and Mortality Conditions." *Milbank Memorial Fund Quarterly*, 21(1):37–62, January 1943.

CHAPTER 8

Age Composition

INTRODUCTION

Uses of Data

Much of the discussion of the uses of data on sex composition given at the beginning of chapter 7 pertains also to a consideration of the uses of data on age composition. As with data on sex composition, data on age are important in themselves, for the description and analysis of other types of demographic data, and for the evaluation of the quality of the census counts of population.

Social scientists of many types have a special interest in the age structure of a population since social relationships within a community are considerably affected by the relative numbers at each age. Many types of planning, particularly planning of community institutions and services, require data on age composition. Age is an important variable in measuring potential school population, the potential voting population, potential manpower, etc. Age data are required for preparing current population estimates and projections; projections of households, school enrollment, labor force, etc.; and projections of requirements for schools, teachers, health services, food, and housing.

Age is the most important variable in the study of mortality, fertility, nuptiality, and certain other areas of demographic analysis. Tabulations on age are essential in the computation of basic measures relating to the factors of population change, in the analysis of the factors of labor supply, and in the study of the problem of economic dependency. The importance of census data on age in studies of population growth is even greater when adequate vital statistics from a registration system are not available.[1] As with data on sex, a large part of the usefulness of the age classification lies in its cross-classifications with other demographic characteristics in which one may be primarily interested. For example, the cross-classifications of age with marital status, labor force, migration, etc., make possible a much more effective use of census data on these subjects. Since these social and economic characteristics vary so much with age and since age composition also varies in time and place, populations cannot be meaningfully compared with respect to these other characteristics unless age has been "controlled."

Data on age serve still another important analytic purpose. Because the expected number of children, the expected number in certain older age groups, and the relative number of

males and females at given ages can be determined closely or at least approximately, either on the basis of data external to the census or from census data themselves, the tabulations by age and sex are very useful in the evaluation of the quality of the returns from the census.

Definition

The age of an individual in censuses is commonly defined in terms of the age of the person at his last birthday. Other definitions are possible and have been used. In some cases age has been defined in terms of the age at the nearest birthday or even the next birthday, but these definitions are no longer employed in national censuses.

In much of the Orient, e.g., China, Korea, Hong Kong, and Singapore, age is reckoned in a still different way.[2] At birth an individual is assigned an age of 1, and he then becomes a year older on each Chinese New Year's day. Furthermore, the lunar year is used, which is a few days shorter than the solar year. Accordingly, he may be as much as three years older under this definition, and is always at least one year older, than under the Western definition. In some reports age may be shown in both ways. Early attempts in censuses of Singapore, South Korea, etc., to determine the Western age, either by asking for it directly or by trying to convert the Chinese age, encountered many difficulties. In the 1960 census of South Korea, for example, some respondents gave one type of age and the rest, the other type.[3] The solution is to ask for an additional item of information about the birth date (or age) according to the Chinese system, namely, the "animal year," which occurs in a 12-year cycle. Given the oriental age, the animal year of birth, and whether or not the birthday is located between New Year's day and the census date, the Western age can be determined.[4]

The U.N. recommendation favors the Western approach, defining age as "the estimated or calculated interval of time between the date of birth and the date of the census, expressed in completed solar years."[5] Nevertheless, the elderly and the less literate residents of Oriental countries would have difficulty in supplying this information.

---

[1] United Nations, *National Programmes of Analysis of Population Census Data as an Aid to Planning and Policy-Making*, Population Studies, Series A, No. 36, 1964.

[2] Swee Hock Saw. "Errors in Chinese Age Statistics," *Demography*, 4(2): 859–875, 1967.

[3] Republic of Korea, Bureau of Statistics, *Population Distribution and Internal Migration in Korea*, by Ehn Hyun Choe, 1966, pp. 122–125.

[4] The Chinese New Year always falls in either January or February; hence there are always two animals in a Western solar calendar year. The first lasts for about 20 to 50 days and the second for the rest of the year. The "era," "year of era," and "cycle" shown in table 8–1 can be used to determine Western age when the oriental age is not reported.

[5] United Nations, *Principles and Recommendations for the 1970 Population Censuses*, p. 41.

Whatever the definition, the age actually recorded in a census may vary depending on whether the definition is applied as of the reference date of the census or as of the date of the actual enumeration, which may spread out over several days, weeks, or even months. If, as in the U.S. census of 1950, age is secured by a question on age and is recorded as of the date of the enumeration, the age distribution as tabulated, in effect, more nearly reflects the situation as of the median date of the enumeration than of the official census date. In the 1950 census of the United States, the median date of enumeration was about 1½ months after the official reference date. In those censuses in which the enumeration is confined to a single day, week, or even month (as, for example, by self-enumeration) or where age is ascertained on the basis of census reports on date of birth (e.g., United States, 1960), the age distribution given in the census reports reflects the situation on the census date quite closely.

Basis of Securing Data

Data on age may be secured by asking a direct question on age, by asking a question on date of birth, or month and year of birth (satisfactory if census day is on the first day of the month), or by asking both questions in combination. Inquiry regarding date of birth occurs mostly in European countries, but elsewhere a direct question on age is used.

In general, the information on age in the censuses of the United States has been secured by asking a direct question on age. In the 1900, 1960, and 1970 censuses, however, the information was obtained by a question on age and date (or month and year) of birth (1900 and 1970) or by a question on date of birth only (1960). The Current Population Survey secures information on age through a question on date of birth.

The U.N. recommendations allow for securing information on age either by inquiring about date of birth or by asking directly for age at last birthday. The United Nations recommends asking date of birth for children reported as "1 year of age," even if a direct question on age is used for the remainder of the population, to obviate the tendency to report "1 year of age" for persons "0 years of age."

Direct reports on age are simpler to process but appear to give less accurate information on age than reports on date of birth, possibly because a question on age more easily permits approximate replies. On the other hand, the proportion of the population for which date of birth is not reported is ordinarily higher than for age, and the date-of-birth approach is hardly applicable to relatively illiterate populations. In some "indigenous" or "native" populations of Africa, in fact, the inquiry regarding age attempts to distinguish merely children, working-age population, and elderly persons, or even only children and adults (e.g., Federation of Rhodesia and Nyasaland, 1961). In such situations, where concepts of age have little meaning, individuals may be assigned to broad age groups on the basis of birth before or after certain major historical events affecting the population.

Classification

Age data collected in censuses or national sample surveys may be tabulated in single years of age, 5-year age groups, or broader groups. The U.N. recommendations for the 1970 population censuses call for tabulations of the national total, urban, and rural populations in single years of age to 100, and tabulations of each major and minor civil division (separately for their urban and rural parts), and each principal locality, in 5-year age

groups (under 1, 1–4, 5–9, . . . 80–84, 85 and over).[6] In order to fill the many demands for age data, both for specific ages and special combinations of ages, it is necessary to have tabulations by single years of age. Moreover, detailed age is required for cross-classification with several characteristics which change sharply from age to age over parts of the age range (e.g., school enrollment, labor force status, marital status). However, 5-year data in the conventional age groups are satisfactory for most cross-classifications (e.g., nativity, country of birth, ethnic groups, socioeconomic status). Broader age groups may be employed in cross-tabulations for smaller areas or in cross-tabulations containing a large number of variables. For some countries, as noted, it has been impossible to secure more detailed information during the enumeration.

When date-of-birth information is collected, the recommended method for converting it to age at last birthday is to subtract the exact date of birth from the date of the census. The resulting ages, in whole years, could then be tabulated by single years or classified into age groups, as desired. Some countries do not use this method of "completed years," however. They first classify the data by year of birth and then determine age by taking the difference between the year of birth and the year of the census. Such countries include France (1962), Federal Republic of Germany (1961), Netherlands (1960), and Sweden (1960). In such a classification, many infants would tend to be classified as 1 year of age if the census was taken early in the calendar year, and hence the number of infants would tend to be greatly understated.

It is also useful for some purposes to tabulate and publish the data in terms of calendar year of birth.[7] Such tabulations are of particular value for use in combination with vital statistics (deaths, marriages) tabulated by year of birth.

Sources of Data

Census counts or estimates of population age distributions, both for single years of age and for broader age groups, are published in various issues of the U.N. *Demographic Yearbooks*.

The U.S. Bureau of the Census has published data on the age-sex distribution of the population of the United States from almost the very beginning of the country's existence. Data for five broad age groups by sex are available for 1800. The amount of age detail increased with subsequent censuses until 1880, when, for the first time, data for 5-year age groups and for single years of age were published. Data by race and sex in broad age groups first became available in 1820, and subsequently the age detail shown was tabulated by sex and race. Tabulations for States accompanied the national tabulations in each census year.

EVALUATION AND ADJUSTMENT OF AGE DATA

Concepts and Types of Age Errors

The errors in the reporting of age have probably been examined more intensively than the reporting errors for any

---

[6] United Nations, *Principles and Recommendations for the 1970 Population Censuses*, pp. 67 and 80–83.

[7] Finland, *General Census of Population, 1960*, Vol. II, *Population by Age, Marital Status, Main Language, etc.*, 1963, tables 2 and 3; and France, Institut National de la Statistique et des Études Economiques, *Recensement Général de la Population de 1962, Résultats du Sondage au 1/20ᵉ, Récapitulation pour la France Entière*, 1964, table R2.

other question in the census. Three factors may account for this intensive study: Many of these errors are readily apparent, measurement techniques can be more easily developed for age data, and the actuaries have had a special practical need to identify errors and to refine the reported data for the construction of life tables. Errors in the tabulated data on age may arise from the following types of errors of enumeration: Coverage errors, failure to record age, and misreporting of age. There is some tendency for the types of errors in age data to offset one another; the extent to which this occurs depends not only on the nature and magnitude of the errors but also on the grouping of the data, as will be described more fully below.

Before discussing the specific methodology of measuring errors in data on age, it is useful to consider the general features of errors in age data in somewhat more detail. The defects in census figures for a given age or age group due to coverage errors and misreporting of age may each be considered further in terms of the component errors. Coverage errors are of two types. Individuals of a given age may have been missed by the census or erroneously included in it (e.g., counted twice). The first type of coverage error represents **gross underenumeration** at this age and the balance of the two types of coverage errors represents **net underenumeration** at this age. (Since underenumeration commonly exceeds overenumeration, we shall typically designate the balance in this way.)

In addition, the ages of some individuals included in the census may not have been reported, or may have been erroneously reported by the respondent, erroneously estimated by the enumerator, or erroneously allocated by the census office. A complete array of census reports of age in comparison with the true ages of the persons enumerated would show the number of persons at each age for whom age was correctly reported in the census, the number of persons incorrectly reporting "into" each age from lower or higher ages, and the number of persons incorrectly reporting "out of" each age into higher or lower ages. Such tabulations permit calculation of measures of **gross misreporting** of age, referred to, in general, as **response variability** of age. If, however, we disregard the identity of individuals and allow for the offsetting effect of reporting "into" and reporting "out of" given ages, much smaller errors are found than are shown by the gross errors based on comparison of reports for individuals. Such **net misreporting** of a characteristic is, in general, referred to as **response bias**. The combination of net underenumeration and net misreporting for a given age is termed **net census undercount** (net census overcount, if the number in the age is overstated) or **net census error.**

For example, the group of persons reporting age 42 in the census consists of (1) persons whose correct age is 42 and (2) those whose correct age is over or under 42 but who erroneously report age 42. The latter group is offset partly or wholly by (3) the number erroneously reporting "out of" age 42 into older or younger ages. The difference between groups (2) and (3) represents the net misreporting error for age 42. In addition, the census count at age 42 is affected by net underenumeration at this age, i.e., by the balance of the number of persons aged 42 completely omitted from the census and the number of persons aged 42 who are erroneously included in the census.

Where the data are grouped into 5-year groups or broader groups, both the gross and net misreporting errors are smaller than the corresponding errors for single ages since misreporting of age within the broader intervals has no effect. On the other hand, the amount of net underenumeration will tend to accumulate and grow as the age interval widens, since

omissions will tend to exceed erroneous inclusions at each age. For the total population, the amount of net underenumeration and the amount of net census undercount are the same since net age misreporting balances out to zero over all ages.

Many of the measures of error do not serve directly as a basis for adjusting the errors in the data. One may distinguish between the degree of precision required to evaluate a set of age data and the degree of precision required to correct it. Yet, a sharp distinction cannot be made between the measurement of errors in census data and procedures for adjusting the census data to eliminate or reduce these errors; accordingly, these two subjects are best treated in combination. Some of the measures of error in age data are simply indexes describing the relative level of error for an entire distribution or most of it. The indexes may refer to only a small segment of the age distribution, to various ages, or to particular classes of ages (e.g., ages with certain terminal digits). Other procedures provide only estimates of relative error for age groups (i.e., the extent of error in a given census relative to the error in an earlier census in the same category, or relative to another category in the same census). Still other measures of error involve the preparation of alternative estimates of the population for an age or age group which presumably are free of the types of errors under consideration. A carefully developed index for a particular age or age group, or an alternative estimate of the actual population or of its relative size may then serve as the basis for adjusting the erroneous census count.

We shall consider the types of deficiencies in census tabulations of age under four general headings: (1) Errors in single years of age, (2) errors in grouped data, (3) reporting of extreme old age, and (4) failure to report age.

Single Years of Age

Measurement of Age and Digit Preference.—A glance at the single-year-of-age data for the population of the Philippines in 1960 (table 8-1) reveals some obvious irregularities. For example, almost without exception, there is a clustering at ages ending in "0" and corresponding deficiencies at ages ending in "9" and "1". Less marked concentrations are found on ages ending in "5" and ages ending in even numbers such as "2" and "8", especially ages 12 and 18.

The figures for adjacent ages should presumably be rather similar. Even though past shifts in the annual number of births, deaths, and migrants can produce fluctuations from one single age to another, the fluctuations observed immediately suggest faulty reporting. The tendency of enumerators or respondents to report certain ages at the expense of others is called **age heaping**, age preference, or **digit preference**. The latter term refers to preference for the various ages having the same terminal digit. Age heaping is most pronounced among populations or population subgroups having a low educational status. The causes and patterns of age or digit preference vary from one culture to another, but preference for ages ending in "0" and "5" is quite widespread.[8] In some cultures certain numbers are specifically avoided, e.g., 13 in the West and 4 in the Orient. Heaping is the principal type of error in single-year-of-age data, although single ages are also affected by other types of age misreporting, net underenumeration, and nonreporting or misassignment of age. Age 0 is underreported often, for example, because "0" is not regarded as a number by many

[8] See India, Indian Statistical Institute, *National Sample Survey*, No. 12, "A Technical Note on Age Grouping," by Ajit Das Gupta, 1958.

Table 8-1.—**Population of the Philippines, by Single Years of Age: 1960**

Age	Number	Age	Number
Total..............	27,087,685	45 years.............	319,118
Under 1 year.........	786,464	46 years.............	160,329
1 year...............	888,180	47 years.............	160,855
2 years..............	963,230	48 years.............	237,287
3 years..............	969,309	49 years.............	155,094
4 years..............	965,232	50 years.............	313,636
5 years..............	957,698	51 years.............	78,534
6 years..............	928,673	52 years.............	128,935
7 years..............	938,899	53 years.............	93,279
8 years..............	841,636	54 years.............	95,715
9 years..............	702,492	55 years.............	163,093
10 years.............	841,356	56 years.............	87,754
11 years.............	581,400	57 years.............	71,828
12 years.............	796,786	58 years.............	93,049
13 years.............	619,293	59 years.............	72,206
14 years.............	596,592	60 years.............	275,436
15 years.............	565,714	61 years.............	31,299
16 years.............	566,942	62 years.............	49,634
17 years.............	538,891	63 years.............	40,154
18 years.............	651,318	64 years.............	34,381
19 years.............	491,441	65 years.............	102,440
20 years.............	565,801	66 years.............	26,445
21 years.............	494,895	67 years.............	35,311
22 years.............	515,823	68 years.............	40,711
23 years.............	456,892	69 years.............	20,921
24 years.............	425,212	70 years.............	136,771
25 years.............	522,203	71 years.............	13,000
26 years.............	358,549	72 years.............	28,017
27 years.............	376,221	73 years.............	16,662
28 years.............	395,766	74 years.............	14,490
29 years.............	300,610	75 years.............	50,558
30 years.............	535,924	76 years.............	15,010
31 years.............	222,086	77 years.............	11,878
32 years.............	318,481	78 years.............	23,353
33 years.............	246,260	79 years.............	9,212
34 years.............	233,700	80 years.............	73,741
35 years.............	401,936	81 years.............	5,532
36 years.............	242,659	82 years.............	9,331
37 years.............	242,462	83 years.............	5,653
38 years.............	316,210	84 years.............	5,089
39 years.............	225,207	85 years.............	18,604
40 years.............	434,156	86 years.............	4,803
41 years.............	126,632	87 years.............	5,617
42 years.............	217,881	88 years.............	4,388
43 years.............	169,167	89 years.............	4,000
44 years.............	151,142	90 years and over....	57,111

Source: United Nations, *Demographic Yearbook, 1962*, table 6.

people and because parents may tend not to think of newborn infants as regular members of the household. In this section we shall confine ourselves to the topic of age heaping, that is, age preference or digit preference.

In principle, a post-enumeration survey or a sample reinterview study should provide considerable information on the nature and causes of errors of reporting in single ages. A tabulation of the results of the check re-enumeration by single years of age. cross-classified by the original census returns for single years of age. could not only provide an indication of the net errors in reporting both of specific terminal digits and of individual ages but could also provide the basis for an analysis of the errors in terms of the component directional biases characteristic of reporting at specific terminal digits and ages. In practice, however, the size of sample of the reinterview survey ordinarily precludes any evaluation in terms of single ages.

Indexes of age preference.—In place of sample reinterview studies, various arithmetic devices have been developed for measuring heaping on individual ages or terminal digits. These devices depend on some assumption regarding the form of the true distribution of population by age over a part or all of the

age range. On this basis an estimate of the true number or numbers is developed and compared with the reported number or numbers.

The simplest devices assume, in effect, that the true figures are **rectangularly** distributed (i.e., that there are equal numbers in each age) over some age range (such as a 3-year, 5-year, or 11-year age range) which includes and, preferably, is centered on the age being examined. For example, an index of heaping on age 30 in the 1960 census of the Philippines may be calculated as the ratio of the enumerated population aged 30 to one-third of the population aged 29, 30, and 31 (per 100):

$$\frac{P_{30}}{1/3(P_{29}+P_{30}+P_{31})} \times 100 = \frac{535,924}{1/3(300,610+535,924+222.086)}$$
$$\times 100 = 151.9 \qquad (1)$$

or, alternatively, as the ratio of the enumerated population aged 30 to one-fifth of the population aged 28, 29, 30, 31, and 32 (per 100):

$$\frac{P_{30}}{1/5(P_{28}+P_{29}+P_{30}+P_{31}+P_{32})} \times 100$$
$$= \frac{535,924}{1/5(395,766+300,610+535,924+222,086+318,481)}$$
$$\times 100 = 151.1 \qquad (2)$$

In this case the two indexes are approximately the same whether a 3-year group or a 5-year group is used; both indicate considerable heaping on age 30. The higher the index the greater the concentration on the age examined; an index of 100 indicates no concentration on this age. Particularly where the age range is large (e.g., 11 years), the assumption regarding the true form of the distribution may alternatively be regarded as an assumption of **linearity** (that is, that the true figures form an arithmetic progression, or that they decrease by equal amounts from age to age over the range) if the age under consideration is centered in the age range selected.

Whipple's index.—Indexes have been developed to reflect preference for or avoidance of a particular terminal digit or of each terminal digit. For example, employing again the assumption of rectangularity in a 10-year range, heaping on terminal digit "0" in the range 23 to 62 may be measured by comparing the sum of the populations at the ages ending in "0" in this range with one-tenth of the total population in the range:

$$\frac{\sum (P_{30}+P_{40}+P_{50}+P_{60})}{1/10 \sum (P_{23}+P_{24}+P_{25} \ldots +P_{60}+P_{61}+P_{62})} \times 100 \qquad (3)$$

Similarly, employing either the assumption of rectangularity or of linearity in a 5-year range, heaping on multiples of five (terminal digits "0" and "5" combined) in the range 23 to 62 may be measured by comparing the sum of the populations at the ages in this range ending in "0" or "5" and one-fifth of the total population in the range:

$$\frac{\sum (P_{25}+P_{30}+ \ldots +P_{55}+P_{60})}{1/5 \sum (P_{23}+P_{24}+P_{25}+ \ldots +P_{60}+P_{61}+P_{62})} \times 100$$
$$= \frac{\sum_{23}^{62} P_a \text{ ending in 0 or 5}}{1/5 \sum_{23}^{62} P_a} \times 100 \qquad (4)$$

Table 8–2.—Calculation of Preference Indexes for Terminal Digits by Myers' Blended Method, for the Philippines: 1960

[Age range covered here is 10 to 89 years. Commonly the same number of ages is included in the two sets of populations being weighted (cols. 1 and 2), and the second set of populations (col. 2) could be extended to age 99 if the figures for single ages were available. Above age 99 the population may be disregarded]

Terminal digit, a	Population with terminal digit a		Weights for--		Blended population		Deviation of percent from 10.00[1]
	Starting at age 10+a	Starting at age 20+a	Column 1	Column 2	Number (1)x(3)+(2)x(4)=	Percent distribution	(6)-10.00=
	(1)	(2)	(3)	(4)	(5)	(6)	(7)
0...........................	3,176,821	2,335,465	1	9	24,196,006	16.06	6.06
1...........................	1,553,378	971,978	2	8	10,882,580	7.22	2.78
2...........................	2,064,888	1,268,102	3	7	15,071,378	10.00	-
3...........................	1,647,360	1,028,067	4	6	12,757,842	8.47	1.53
4...........................	1,556,321	959,729	5	5	12,580,250	8.35	1.65
5...........................	2,143,666	1,577,952	6	4	19,173,804	12.72	2.72
6...........................	1,462,491	895,549	7	3	12,924,084	8.58	1.42
7...........................	1,443,063	904,172	8	2	13,352,848	8.86	1.14
8...........................	1,762,082	1,110,764	9	1	16,969,502	11.26	1.26
9...........................	1,278,691	787,250	10	0	12,786,910	8.49	1.51
Total...................	(X)	(X)	(X)	(X)	150,695,204	100.00	20.07
Summary index of age preference= Total÷2=...................	(X)	(X)	(X)	(X)	(X)	(X)	10.04

— Represents zero. X Not applicable. [1] Signs disregarded.

Source: Adapted from Robert J. Myers, "Errors and Bias in the Reporting of Ages in Census Data," *Transactions of the Actuarial Society of America*, 41. Part 2(104):413. October–November 1940.

For the Philippines in 1960, we have,

$$\frac{2,965,502}{1/5(9,506,437)} \times 100 = \frac{2,965,502}{1,901,287} \times 100 = 156.0$$

The corresponding figure for the United States in 1960 is 100.9. This measure is known as Whipple's index. It varies between 100, representing no preference for "0" or "5", and 500, indicating that only digits "0" and "5" were reported. Accordingly, the Philippine figures show considerable heaping on multiples of "5" compared with the U.S. figure. In fact, the population tabulated at these ages for the Philippines may be said to overstate the corresponding unbiased population by more than 50 percent.

The choice of the range 23 to 62 is largely arbitrary. In computing indexes of heaping, the ages of early childhood and extreme old age are often excluded since they are more strongly affected by other types of errors of reporting than by preference for specific terminal digits and the assumption of equal decrements from age to age is less applicable.

The procedure described can be extended to provide an index for each terminal digit (0, 1, 2, etc.). The population ending in each digit over a given range, say 23 to 82, or 10 to 89, may be compared with one-tenth of the total population in the range, as was done for digit "0" above, or it may be expressed as a percentage of the total population in the range. In the latter case, an index of 10 percent is supposed to indicate an unbiased distribution of terminal digits, and hence, presumably, accurate reporting of age. Indexes in excess of 10 percent indicate a tendency toward preference for a particular digit, and indexes below 10 percent indicate a tendency toward avoidance of a particular digit.

Myers' blended method.—Myers has developed a "blended" method to avoid the bias in indexes computed in the way just described that is due to the fact that numbers ending in "0" would normally be larger than the following numbers ending

in "1" to "9" because of the effect of mortality.[9] The principle employed is to begin the count at each of the 10 digits in turn and then to average the results. Specifically, the method involves determining the proportion which the population ending in a given digit is of the total population 10 times, by varying the particular starting age for any 10-year age group. Table 8–2 shows the calculation of the indexes of preference for terminal digits in the age range 10 to 89 for the Philippine population in 1960 based on Myers' blended method. In this particular case, the first starting age was 10, then 11, and so on, to 19. The abbreviated procedure of calculation calls for the following steps:

Step (1) Sum the populations ending in each digit over the whole range, starting with the lower limit of the range (e.g., 10, 20, 30, . . . 80; 11, 21, 31, . . . 81).

Step (2) Ascertain the sum excluding the first population combined in step (1) (e.g., 20, 30, 40, . . . 80; 21, 31, 41, . . . 81).

Step (3) Weight the sums in steps (1) and (2) and add the results to obtain a blended population (e.g., weights 1 and 9 for the 0 digit; weights 2 and 8 for the 1 digit).

Step (4) Convert the distribution in step (3) into percents.

Step (5) Take the deviation of each percent in step (4) from 10.0, the expected value for each percent.

The results in step (5) indicate the extent of concentration on or avoidance of a particular digit.[10] The weights in step (3)

---

[9] Robert J. Myers, "Errors and Bias in the Reporting of Ages in Census Data," *Transactions of the Actuarial Society of America*, 41, Pt. 2 (104):411–415, October–November 1940.

[10] The effectiveness of the blending procedure is demonstrated by the results obtained by applying it to a life table stationary population (L_x), which is not directly affected by misreporting of age. If blending is not employed, the results are very sensitive to the choice of the particular starting age and the frequency of the digits shows a substantial decline from 0 to 9. With blending, the frequency of the digits is about equal. See Mortimer Spiegelman, *Introduction to Demography*, rev. ed., Cambridge, Mass., Harvard University Press, 1968, p. 72.

represent the number of times the combination of ages in step (1) or (2) is included when the starting age is varied from 10 to 19. Note that the weights for each terminal digit would differ if the lower limit of the age range covered were different. For example, if the lower limit of the age range covered were 23, the weights for terminal digit 3 would be 1 (col. 1) and 9 (col. 2) and for terminal digit 0 would be 8 (col. 1) and 2 (col. 2).

The method thus yields an index of preference for each terminal digit, representing the deviation, from 10.0 percent, of the proportion of the total population reporting on the given digit. A summary index of preference for all terminal digits is derived as one-half the sum of the deviations from 10.0 percent, each taken without regard to sign. If age heaping is nonexistent, the index would approximate zero. This index is an estimate of the minimum proportion of persons in the population for whom an age with an incorrect final digit is reported. The theoretical range of Myers' index is 0, representing no heaping, and 90 which would result if all ages were reported at a single digit, say zero. A summary preference index of 10.0 for the Philippines in 1960 is obtained.

Very small deviations from 100.0, 10.0, or 0 shown by various measures of heaping are not necessarily indicative of heaping and should be disregarded. The "true" population in any single year of age is by no means equal to exactly one-fifth of the 5-year age group centering around that age (nor one-tenth of the 10-year age group centering around the age), nor is there necessarily a gradual decline in the number of persons from the youngest to the oldest age in a broad group, as is assumed in the common formulas. The age distribution may have small irregular fluctuations, depending largely on the past trend of births, deaths, and migration. Extremely abnormal bunching should be most readily ascertainable in the data for the older ages (but before extreme old age) where mortality takes a heavy toll from age to age but the massive errors in the data for extreme old age do not yet show up; past fluctuations in the number of births and migrants may still affect the figures, however. In short, it is not possible to measure digit preference precisely, because a precise distinction between the error due to digit preference, other errors, and real fluctuations cannot be made.

Other summary indexes of digit preference. — A number of other general indexes of digit preference have been proposed, e.g., the Bachi index, the Carrier index, and the Ramachandran index.[11] These have some theoretical advantages over the Whipple and Myers indexes, but they are more laborious to compute than either and, as indicators of the general extent of heaping, differ little from them. The Bachi method, for example, involves applying the Whipple method repeatedly to determine the extent of preference for each final digit. Like the Myers index the Bachi index equals the sum of the positive deviations from 10 percent. It has a theoretical range from 0 to 90, and 10 percent is the expected value for each digit. The results obtained by the Bachi method resemble those obtained by the Myers method. The Carrier and Ramachandran indexes take account of age structure and of the varying incidence of digit preference over the age range.

[11] Roberto Bachi, "The Tendency to Round Off Age Returns: Measurement and Corrections," *Bulletin of the International Statistical Institute* (Proceedings of the 27th Session, Calcutta, 1951), 33(4): 195–222, Calcutta; idem; "Measurement of the Tendency to Round Off Age Returns," *Bulletin of the International Statistical Institute* (Proceedings of the 28th Session. Rome. 1953), 34(3):129–137, Rome. 1960; N. H. Carrier, "A Note on the Measurement of Digital Preference in Age Recordings," *Journal of the Institute of Actuaries* (Cambridge, England), 85:71–85, 1959; and K. V. Ramachandran, "An Index to Measure Digit Preference Error in Age Data," Summary in United Nations, *Proceedings of the World Population Conference,* 1965 (Belgrade), Vol. III, New York, 1967, pp. 202–203.

Reduction of and Adjustment for Age and Digit Preference. — In the preceding section relating to the measurement of heaping, we were concerned primarily with those measures which described an entire distribution or an important segment of it. We treat here those measures of heaping and procedures for reducing or eliminating heaping that are primarily applicable to individual ages. These measures and procedures include modifying the census schedule, such as by varying the form of the question or questions used to secure the data on age, and preparing alternative estimates or carefully derived corrections for individual ages, such as by use of annual birth statistics or mathematical interpolation to subdivide the 5-year totals established by the census and by calculation of refined age ratios for single ages. In some situations, it is also desirable to consider handling the problem by presenting only grouped data over part or all of the age distribution. In this case, the question of the optimum grouping of ages for tabulation and publication arises.

Question on date of birth. — At the enumeration stage, a question on date of birth may be employed instead of a question on age, or both may be used in combination. As suggested by the U.S. census data for 1960, the resulting pattern of age heaping is likely to be different, with preference for ages which correspond to years of birth ending in 0 or 5. Although the heaping on a few ages may continue to be considerable, the evidence suggests that the use of a question on date of birth, especially in combination with a question on age, contributes to the accuracy of the age data obtained. In many cases the enumerator may not ask both questions, but derives the answer to one by calculation from the answer to the other; yet, it is believed that having both questions on the schedule seems to make the enumerator and the respondent more conscientious in the handling of the questions on age. (The age question is also a useful source of an approximate answer when the respondent is unable or unwilling to estimate the date of birth.)

Calculation of corrected census figures. — Single-year-of-age data as reported may be "adjusted" following tabulation by developing alternative single-year-of-age figures directly. These alternative figures may replace the census counts entirely or, as is more common, provide a pattern by which the census totals for 5-year age groups may be redistributed by single years of age. There are several ways of developing the alternative estimates. These may involve the relatively direct use of annual birth statistics, "surviving" annual births to the census date, use of life table populations, combining birth, death, and migration statistics to derive actual population estimates, and use of various forms of mathematical interpolation.

The first method alluded to involves use of an annual series of past births, in the cohorts corresponding to the census ages, for distributing the 5-year census totals. For this purpose annual birth statistics that have a fairly similar degree of completeness of registration over several years are required. The second procedure is quite similar, but the births employed are first reduced by deaths prior to the census date. A third procedure for replacing the tabulated single-year-of-age figures involves use of the life table "stationary" population (L_x column) from an unabridged life table (i.e., one showing single ages). The specific steps for distributing the 5-year totals according to three special sets of single-year-of-age estimates are illustrated for Costa Rica in table 8–3.

The use of birth statistics or of the life table stationary population to distribute 5-year census totals can easily result in discontinuity in the single ages at the junctions of the 5-year age groups, as may be seen by examining the age-to-age differences of the estimates in table 8–3. A number of mathematical interpolation or graduation devices can be used to subdivide the 5-year census totals into single years of age in such a way as to effect a smooth transition from one age to another, while maintaining the 5-year totals and removing erratic fluctuations in the numbers (see last column in table 8–3). In effect, these devices typically fit various mathematical curves to the totals for several adjacent 5-year age groups in order to arrive at the constituent single ages for the central 5-year age group in the set. The principal types of mathematical curves employed for this purpose are of the osculatory and polynomial form. In this method, various multipliers are ordinarily applied to the enumerated 5-year totals to obtain the required figures directly. It is important to note that each of the methods described also removes some true fluctuations implicit in the original single-year-of-age figures, i.e., fluctuations not due to errors in age misreporting but to actual changes in past years in the number of births, deaths, and migration.

Residual digit preference in grouped data. — In view of the magnitude of the errors which may occur in single ages, it may be preferable to combine the figures into 5-year age groups for publication purposes. This approach eliminates the irregularities, but the question is raised as to the optimum grouping of ages for tabulations from the point of view of minimizing heaping. (The optimum grouping so defined may still not be very practical for demographic analysis.) The concentration on multiples of five and other ages may have but slight effect on grouped data or the effect may be quite substantial. The effect of heaping is certain to remain to some extent in the conventional age grouping if the heaping particularly distorts the marginal ages like 0, 4, 5, and 9.

Serious obstacles exist to the introduction of the "optimum" grouping of data as a general practice. Different population groups (e.g., sex groups, urban-rural residence groups), different population censuses, and different types of demographic data (e.g., population data, death statistics) may require different optimum groupings, so that difficulties arise in the cross-classification of data, in the computation of rates, and in the analysis of data over time; and the data may not be regularly tabulated in the necessary detail (as in the case of population data for counties in the United States and death statistics for the United States and its geographic subdivisions). In view of the fact particularly that the "decimal" grouping of data is the conventional grouping over much of the world, it may be expected that use of this grouping in the principal census tabulations of each country will continue. Illustrative calculations show, moreover, that there may be little difference between the 0–4 (5–9) grouping and other groupings in the extent of residual heaping and that the conventional grouping may show a relatively high level of accuracy even where preference for digit "0" is large.

Grouped Data

Types of Errors and Methods of Measurement. — As indicated earlier, several important types of errors remain in age data even when the data are grouped. In addition to some residual error due to digit preference, 5-year or 10-year data are affected by other types of age misreporting and by net underenumeration. Absolute net underenumeration would tend to cumulate as the age band widens. On the other hand, the percent of net underenumeration would be expected to vary fairly regularly over the age distribution, fluctuating only moderately up and down. Absolute net age misreporting error and the per-

Table 8–3. — Calculation of the Distribution of the Population 25 to 29 and 30 to 34 Years Old by Single Years of Age, by Various Methods, for Costa Rica: 1963

[In each case the census totals for age groups 25–29 and 30–34 are maintained. These are taken, then, as the numerators of the distribution factors, F_1, F_2, and F_3; the denominators are registered births, survivors of births, and life table stationary population in these groups, respectively. See footnotes]

Age	Census counts	Estimates based directly on births		Estimates based on survivors of births			Estimates based on life table population		Estimates derived by mathematical interpolation [5]
		Registered births	Estimated population $F_1{}^1 x(2)=$	Survival rate [2]	Survivors $(2)x(4)=$	Estimated population $F_2{}^3 x(5)=$	Life table stationary population [2]	Estimated population $F_3{}^4 x(7)=$	
	(1)	(2)	(3)	(4)	(5)	(6)	(7)	(8)	(9)
25 to 29 years...............	84,833	124,234	84,833	(X)	107,383	84,833	432,141	84,833	84,833
25 years....................	19,664	25,928	17,705	.86770	22,498	17,773	86,770	17,034	16,973
26 years....................	17,650	25,494	17,409	.86606	22,079	17,442	86,606	17,002	17,230
27 years....................	16,608	25,063	17,114	.86435	21,663	17,114	86,435	16,968	17,356
28 years....................	17,065	24,127	16,475	.86258	20,811	16,441	86,258	16,933	16,974
29 years....................	13,846	23,622	16,130	.86072	20,332	16,062	86,072	16,897	16,301
30 to 34 years...............	77,093	117,029	77,093	(X)	100,008	77,093	427,282	77,093	77,093
30 years....................	19,633	23,632	15,568	.85878	20,295	15,645	85,878	15,495	15,929
31 years....................	10,445	23,794	15,674	.85675	20,386	15,715	85,675	15,458	15,734
32 years....................	16,283	23,697	15,610	.85465	20,253	15,612	85,465	15,420	15,496
33 years....................	16,339	22,909	15,091	.85246	19,529	15,054	85,246	15,381	15,166
34 years....................	14,393	22,997	15,149	.85018	19,552	15,072	85,018	15,340	14,768

[1] F_1 for 25–29 is $\frac{84,833}{124,234} = .68285$; F_1 for 30–34 is $\frac{77,093}{117,029} = .65875$.

[2] Life table for Costa Rica, 1962–1964.

[3] F_2 for 25–29 is $\frac{84,833}{107,383} = .79000$; F_2 for 30–34 is $\frac{77,093}{100,008} = .77087$.

[4] F_3 for 25–29 is $\frac{84,833}{432,141} = .19631$; F_3 for 30–34 is $\frac{77,093}{427,282} = .18043$.

[5] The specific method involved use of Sprague "osculatory" multipliers applied to 5 consecutive 5-year age groups.

Source: Basic data from official national sources.

cent of net age misreporting error should tend to take on positive and negative values alternately over the age scale, dropping to zero for the total population of all ages combined. For the total population, therefore, net census error and net underenumeration are identical. In general, as the age band widens, net age misreporting tends to become less important and net underenumeration tends to dominate as the type of error in age data.

The particular form that these types of errors take varies from country to country and from census to census. We may cite some of the specific types of errors that have been identified or described. Young children, particularly infants, and young adult males are omitted disproportionately in many censuses. The liability for military service may be an important factor in connection with the understatement of young adult males. It is quite possible that the laws and practices relating to age for school attendance, child labor, voting, marriage, purchase of alcoholic beverages, etc., may induce overstatement of age for young people, so that they may share in the privileges accorded under the law to persons who have attained the higher age. Responses regarding age may also be affected by the social prestige accorded certain members of a population, for example, the aged in some tribal societies.

Wolfenden, referring to data for the United States and Canada, mentions the tendency to overstate age until attainment of age 21, to understate age at adult ages, especially on the part of women, and to overstate age at the advanced ages.[12] Young adult women have frequently been suspected of understating their age. Net omissions among women over 50 as a group may be excessive. Persons just under 65 may report themselves as 65 or older, particularly under the influence of social security legislation. Elderly persons tend to exaggerate their ages, and the number of centenarians may be particularly exaggerated.

In some censuses and surveys, as in those of the countries of tropical Africa, the number of females in their teens tends to be understated and the number of females in the adult age groups to be overstated. This bias has been attributed to a tendency among interviewers systematically to "age" those women who are already married or mothers from the teen ages into the adult age groups on the assumption of a higher "typical" age of marriage than actually prevails.[13]

It is quite difficult to measure the errors in grouped data on age with any precision. It may be extremely difficult or impossible, in fact, to determine the separate contribution of each of the types of errors affecting a given figure and to separate the errors from real fluctuations (e.g., fluctuations due to migration); and, further, to identify the errors in relation to their causes. Some of the measures of error for age groups measure net age misreporting and net underenumeration separately, whereas others measure these types of errors only in combination or measure only one of them. Some of the procedures provide only indexes of error for entire age distributions or only estimates of relative error for age groups (i.e., relative to the error in the same category in an earlier census or relative to another category in the same census), whereas other procedures provide estimates of the actual extent of error for age groups.

As in the case of measuring coverage of the total population, the methods for determining the existence of such errors and

their approximate magnitude may be classified into two broad types: first, case-by-case checking techniques employing data from reinterviews and independent lists or administrative records and, second, techniques of demographic analysis. The former techniques relate to studies in which data collected in the census are matched on a case-by-case basis with data for a sample of persons obtained by reinterview or from independent records. The latter techniques involve (1) the development of estimates of expected values for the population by age or in other categories, or for various population ratios, by use and manipulation of (a) data from the census itself or an earlier census or censuses and (b) such data as birth, death, and migration statistics; and (2) the comparison of these expected values with the corresponding figures from the census. This method may also be extended to encompass comparison of aggregated administrative data with census counts.

Measurement by Reinterviews and Record Matching Studies. — We consider first case-by-case checking techniques based on reinterviews and matching against independent lists and administrative records for the light they may throw on errors in grouped data. Case-by-case matching studies permit the separate measurement of the two components of net census error or net census undercounts in age data — net coverage error (or net underenumeration) and net age misreporting. Furthermore, this type of study theoretically permits separating each of these components into its principal components — net coverage error into omissions and erroneous inclusions at each age, and net misreporting error into the various directional biases which affect each age group. Thus, the results of a reinterview study or a record matching study may be cross-classified with the results of the original enumeration by 5-year, 10-year, or broader age groups, to determine the number of persons who were omitted from, or erroneously included in, the census, for the same age groups, or who reported in the same, higher, or lower age group.

When the matching study is employed to measure misreporting of age, the comparison is restricted to persons included both in the census and in the sample survey or the record sample used in the evaluation (that is, "matched persons"), and the age of each person interviewed in the census is compared with his age obtained by more experienced interviewers in the "check" sample. (It may be desirable, also, to exclude from the analysis persons whose age was not reported in either interview.) Differences arise primarily in reporting, but also may occur in the recording and processing of the data. Because of problems relating to the design of the matching study, sample size and sample variability, and matching the census record and the "check" record, it is difficult to establish reliably the patterns of coverage error or age misreporting, or their combination, net census error, for 5-year age groups, or to separate net coverage error reliably into omissions and erroneous inclusions by age.

Reinterview studies designed to measure the extent of net coverage error and net misreporting error by age were conducted following both the 1950 and 1960 censuses of the United States. To evaluate the accuracy of age reporting in these two censuses, the March 1950 and the March 1960 Current Population Survey (CPS) data on age, and the data on age from the 1950 Post-Enumeration Survey (PES) and the Content Evaluation Study of the 1960 census reinterview program (CES) were compared with the corresponding census data. To measure net coverage error by age, only 1950 PES

---

[12] Hugh H. Wolfenden, *Population Statistics and Their Compilation*, rev. ed., Chicago, University of Chicago Press, 1954, pp. 53–58.
[13] William Brass et al., *The Demography of Tropical Africa*, Princeton, N.J., Princeton University Press, 1968, pp. 48–49.

Table 8-4. — Indexes of Response Bias and Variability for the Reporting of Age of the Population of the United States: 1960 and 1950

[CPS represents the Current Population Survey in April 1960, CES represents the Content Evaluation Study of the 1960 census reinterview program and PES represents the 1950 census Post-Enumeration Survey]

Age	1960 census-Current Population Survey match		1960 census-Content Evaluation Study match		1950 census-Post Enumeration Survey match		Difference between 1960 census-CES match and 1950 census-PES match	
	Index of net shift relative to CPS class[1]	Percent in CPS class differently reported	Index of net shift relative to CES class[1]	Percent in CES class differently reported	Index of net shift relative to PES class[1]	Percent in PES class differently reported	Index of net shift[2] $/(3)/ - /(5)/ =$	Percent in class differently reported[3] $(4) - (6) =$
	(1)	(2)	(3)	(4)	(5)	(6)	(7)	(8)
Under 5 years...................	+0.66	1.45	-0.04	1.82	-1.64	2.98	-1.60	-1.16
5 to 14 years...................	-0.38	1.80	+0.36	1.36	+0.54	1.58	-0.18	-0.22
15 to 24 years..................	-1.43	3.55	-0.85	2.57	+0.93	2.59	-0.08	-0.02
25 to 34 years..................	+1.28	2.95	+0.44	2.39	+0.22	3.67	+0.22	-1.28
35 to 44 years..................	-0.42	4.47	+1.00	3.85	+1.07	4.48	-0.07	-0.63
45 to 54 years..................	-0.03	5.38	-0.63	5.31	+0.11	6.42	+0.52	-1.11
55 to 64 years..................	+0.15	6.39	-0.11	5.83	-2.18	6.91	-2.07	-1.08
65 years and over..............	+0.74	2.32	-0.79	3.21	-0.51	2.99	+0.28	+0.22

[1] A minus sign indicates that the census count is lower than the CPS, CES, or PES figure.
[2] Represents the excess of the absolute figure (without regard to sign) in col. (3) over the absolute figure (without regard to sign) in col. (5). Minus sign indicates lower level of error in 1960 census than in 1950 census; plus sign indicates higher level of error in 1960 census.
[3] Minus sign indicates lower level of error in 1960 census than in 1950 census; plus sign indicates a higher level of error in 1960 census.

Source: U.S. Bureau of the Census, *Evaluation and Research Program of the U.S. Censuses of Population and Housing, 1960: Accuracy of Data on Population Characteristics as Measured by CPS-Census Match*, Series ER 60, No. 5, 1964, table 12; idem, *Evaluation and Research Program of the U.S. Censuses of Population and Housing, 1960: Accuracy of Data on Population Characteristics as Measured by Reinterviews*, Series ER 60, No. 4, 1964, table 1; idem, *The Post-Enumeration Survey: 1950*, Technical Paper No. 4, 1960, tables 1A-1E.

data and 1960 reinterview studies data were compared with the corresponding census data.[14]

Table 8-4 illustrates how this type of data may be employed in the analysis of response errors in age data. The 1950 and 1960 census counts of the population by age are compared with the 1950 PES data and the 1960 CPS and CES data, respectively.

Measurement by Demographic Analysis. — As mentioned earlier, numerous techniques of demographic analysis can be employed in the evaluation of census data for age groups. These techniques include such procedures as intercensal cohort analysis based on age data from an earlier census, derivation of estimates based on birth and death statistics, use of expected age ratios and sex ratios, mathematical graduation of census age data, comparison with various types of population models, and other more elaborate techniques involving data from several censuses. In addition, we may include here direct comparison of census counts with estimates of population in particular sex-age categories based on counts from administrative records.

Ordinarily these techniques do not permit the separate measurement of net underenumeration and net age misreporting for any age group; net census error is measured in combination. Some of the techniques measure net age misreporting primarily and net underenumeration only secondarily or partly. Most of the techniques of evaluating grouped age data do not provide

absolute estimates of net census error by which census data can be corrected. Only a few rather elaborate techniques are designed to determine the amount of net census error at each age and to produce substitute census figures corrected for net census error.

The methods of measuring net census error as such can give some suggestive information regarding the nature and extent of net age misreporting, since, as we have previously noted, net coverage error should tend to be in the same direction from age to age and to vary rather regularly over the age distribution. A division of net census error into these two parts may also be possible by employing two or more methods of evaluation in combination. An estimate of net census error is itself subject to error since the corresponding estimate of the corrected population contains errors resulting from, for example, net undercount of the census figure for an age cohort in a previous census; error in the reported or estimated number of births; underreporting and age misclassification in the death statistics; and omission, understatement, or overstatement of the allowance for net migration.

The present discussion of the errors in grouped data on age by the methods of demographic analysis does not attempt to treat the measurement and reduction or elimination of errors separately because, as we have noted, they are often two facets of the same operation. We will, however, particularly note those methods which directly provide corrections of census figures for net undercounts. The latter methods will be illustrated principally by a review of recent U.S. studies of net undercounts using demographic analysis. First, however, we consider the basic methods under the headings of (1) intercensal cohort analysis, (2) comparisons with estimates based on birth statistics, (3) age ratio analysis, (4) sex ratio analysis, (5) mathematical graduation of census data, and (6) comparison with population models. We also consider briefly the method of (7) comparison with aggregated administrative data.

---

[14] The 1950 Census-PES statistics are given in U.S. Bureau of the Census, *The Post-Enumeration Survey: 1950*, Technical Paper No. 4, 1960. The 1960 Census-CES statistics appear idem, *Evaluation and Research Program of the U.S. Censuses of Population and Housing, 1960: Accuracy of Data on Population Characteristics as Measured by Reinterviews*, ER 60, No. 4, 1964; and in Eli S. Marks and Joseph Waksberg, "Evaluation of Coverage in the 1960 Census of Population through Case-by-Case Checking," *Proceedings of the Social Statistics Section, 1966*, American Statistical Association, p. 69. The 1960 Census-CPS statistics appear in No. 5 of the ER 60 series, *Evaluation and Research Program of the U.S. Censuses of Population and Housing, 1960: Accuracy of Data on Population Characteristics as Measured by CPS-Census Match*, 1964.

Table 8-5. —Calculation of the "Error of Closure" for the Population of the United States, by Age: April 1, 1950 to 1960

[Small discrepancies in the consistency of the numbers result from the use of unrounded numbers in the original calculations]

Age in 1950	Census population, April 1, 1950 (+) (1)	Components of change, 1950 to 1960				Population, April 1, 1960		Error of closure		Age in 1960
		Births[1] (+) (2)	Deaths[2] (-) (3)	Net civilian immigration[3] (+) (4)	Net movement of Armed Forces[4] (+) (5)	Expected (1)+(2)-(3)+(4)+(5)= (6)	Enumerated (census) (7)	Amount[5] (7)-(6)= (8)	Percent of expected population, 1960 [(8)÷(6)] x100= (9)	
Total............	151,325,798	40,963,000	15,608,000	2,975,000	-330,000	179,325,000	179,323,175	-3,000	(Z)	Total
Births, 1955 to 1960..	(X)	21,345,000	598,000	73,000	(X)	20,820,000	20,320,901	-498,000	-2.4	Under 5 years
Births, 1950 to 1955..	(X)	19,618,000	658,000	211,000	(X)	19,171,000	18,691,780	-480,000	-2.5	5 to 9 years
Under 5 years.........	16,243,141	(X)	121,000	250,000	(X)	16,372,000	16,773,492	+402,000	+2.5	10 to 14 years
5 to 9 years..........	13,262,123	(X)	80,000	231,000	-113,000	13,300,000	13,219,243	-82,000	-0.6	15 to 19 years
10 to 14 years........	11,167,478	(X)	106,000	333,000	-257,000	11,137,000	10,800,761	-338,000	-3.0	20 to 24 years
15 to 19 years........	10,671,321	(X)	136,000	418,000	-59,000	10,894,000	10,869,124	-25,000	-0.2	25 to 29 years
20 to 24 years........	11,549,355	(X)	170,000	398,000	59,000	11,836,000	11,949,186	+112,000	+0.9	30 to 34 years
25 to 29 years........	12,305,951	(X)	221,000	315,000	5,000	12,405,000	12,481,109	+76,000	+0.6	35 to 39 years
30 to 34 years........	11,572,337	(X)	302,000	209,000	11,000	11,490,000	11,600,243	+110,000	+1.0	40 to 44 years
35 to 39 years........	11,294,478	(X)	446,000	178,000	11,000	11,037,000	10,879,485	-158,000	-1.4	45 to 49 years
40 to 44 years........	10,240,671	(X)	636,000	136,000	7,000	9,748,000	9,605,954	-142,000	-1.5	50 to 54 years
45 to 49 years........	9,101,778	(X)	863,000	95,000	3,000	8,337,000	8,429,865	+93,000	+1.1	55 to 59 years
50 to 54 years........	8,295,580	(X)	1,152,000	63,000	2,000	7,209,000	7,142,452	-66,000	-0.9	60 to 64 years
55 to 59 years........	7,252,524	(X)	1,500,000	35,000	1,000	5,789,000	6,257,910	+470,000	+8.1	65 to 69 years
60 to 64 years........	6,074,363	(X)	1,811,000	18,000	–	4,281,000	4,738,932	+458,000	+10.7	70 to 74 years
65 years and over.....	12,294,698	(X)	6,809,000	12,000	–	5,498,000	5,562,738	+66,000	+1.2	75 years and over
55 years and over.....	25,621,585	(X)	10,120,000	ᵇ 65,000	1,000	15,568,000	16,559,580	+994,000	+6.4	65 years and over

— Entry represents zero or rounds to zero.
X Not applicable.
Z Less than 0.05 percent.
[1] Adjusted for underregistration.
[2] Deaths under 1 year have been adjusted for underregistration.
[3] Minus sign denotes net emigration.
[4] Minus sign denotes net movement of armed forces from the United States.
[5] Minus sign denotes that census count is less than expected figure and plus sign denotes that census count is greater than expected figure.

Source: Adapted from, U.S. Bureau of the Census, *Current Population Reports, Series P-25, No. 310*, "Estimates of the Population of the United States and Components of Change, by Age. Color. and Sex: 1950 to 1960," June 30, 1965, *table 12.*

Intercensal cohort analysis.—In this procedure the counts of one census are, in effect, employed to evaluate the counts at a later census. Ordinarily, the principal demographic factor at the national level accounting for the difference between the figures for the same cohort at the two census dates is mortality. Migration will usually play a secondary, if not a minor, role, although even in this case the number of migrants may exceed the number of deaths at some of the younger ages. The figures from both the earlier and later censuses are affected by net census undercounts. In addition, it is possible that the level of migration and mortality may have been affected by such special factors as movement of military forces into and out of a country, epidemic or famine, war deaths, etc. The method of intercensal cohort analysis is illustrated with data for the United States, 1950 to 1960.

Table 8-5 sets forth the data and steps by which estimates of the expected population by age in April 1960 for the United States were derived. In this case statistics on deaths, net civilian immigration, and net movement of armed forces by age are available and have been compiled in terms of age cohorts for the period April 1950 through March 1960. The expected population in 1960, derived by combining the 1950 census figures with the estimates of change by age cohorts during 1950 to 1960, is compared with the 1960 census counts by age. The details of this compilation are explained in chapter

22.[15] The results reflect the combined effect of underenumeration and age misreporting (i.e., net errors) in the 1960 census, as well as the net errors in the 1950 census data and errors in the data on intercensal change, particularly age misreporting errors in death statistics and coverage errors in the migration statistics. In more general terms, the method measures relative net census error for a cohort defined by age at two successive censuses.

The 3.0 percent deficit at ages 20 to 24 in 1960 (col. 9) suggests an underenumeration of persons of this age in this census on the principal assumption that children 10 to 14 were rather well enumerated in 1950 (col. 1). The error of closure (col. 9) for the population 65 and over in 1960—6.4 percent of the population expected in 1960—suggests a large net overstatement of this group in this census resulting from age misreporting favoring this age group. (The term "error of closure" was defined in chapters 4 and 7.) A similar error of closure was noted in 1950 and 1940. However, the excess of aged persons in the 1960 census is not offset by a correspondingly large deficit in the 60–64 group or 55–64 group which would reflect reporting out of this group into the 65-and-over group, as would be expected. The discrepancy in 1960 could

[15] See also, U.S. Bureau of the Census, *Current Population Reports*, Series P–25, No. 310, "Estimates of the Population of the United States and Components of Change, by Age, Color, and Sex, 1950 to 1960," June 30, 1965.

Table 8-6. — Evaluation of Consistency of Age Data From the 1956 and 1966 Censuses of Iran, by Sex

[Numbers in thousands]

Age in 1956	Age in 1966	Population (census) 1956 Male (3)÷(1)= (1)	1956 Female (2)	1966 Male (3)	1966 Female (4)	Proportion surviving Census Male (3)÷(1)= (5)	Census Female (4)÷(2)= (6)	Model life table[1] Male (7)	Model life table[1] Female (8)	Percent difference Male (5)-(7)/(7) x100= (9)	Percent difference Female (6)-(8)/(8) x100= (10)	Male/female proportion surviving Iran data (5)÷(6)= (11)	Model life table data (7)÷(8)= (12)
All ages.......	All ages......	9,645	9,310	12,982	12,097	(X)	(X)	(X)	(X)	(X)	(X)	(X)	(X)
(X)................	Under 5 years.....	(X)	(X)	2,308	2,129	(X)	(X)	(X)	(X)	(X)	(X)	(X)	(X)
(X)................	5 to 9 years.....	(X)	(X)	2,128	1,978	(X)	(X)	(X)	(X)	(X)	(X)	(X)	(X)
Under 5 years......	10 to 14 years....	1,684	1,664	1,594	1,423	.94656	.85517	.95401	.95677	-0.78	-10.62	1.11	1.00
5 to 9 years.......	15 to 19 years....	1,419	1,464	1,060	1,069	.74700	.76140	.97520	.97580	-23.40	-21.97	0.98	1.00
10 to 14 years.....	20 to 24 years....	975	848	793	889	.81333	1.04835	.96838	.97050	-16.01	+8.02	0.78	1.00
15 to 19 years.....	25 to 29 years....	710	710	802	848	1.12958	1.19437	.95843	.96229	+17.86	+24.12	0.95	1.00
20 to 24 years.....	30 to 34 years....	699	798	863	805	1.23462	1.00877	.95189	.95564	+29.70	+5.56	1.22	1.00
25 to 34 years.....	35 to 44 years....	1,441	1,463	1,502	1,238	1.04233	.84621	.93932	.94556	+10.97	-10.51	1.23	0.99
35 to 44 years.....	45 to 54 years....	1,057	891	850	735	.80416	.82492	.90266	.92326	-10.91	-10.65	0.97	0.98
45 to 54 years.....	55 to 64 years....	742	704	568	530	.76550	.75284	.82275	.86499	-6.96	-12.97	1.02	0.95
55 to 64 years.....	65 to 74 years....	523	464	367	320	.70172	.68966	.66561	.72706	+5.43	-5.14	1.02	0.92
65 to 74 years.....	75 to 84 years....	263	237	111	98	.42205	.41350	.40740	.45710	+3.60	-9.54	1.02	0.89
75 years and over..	85 years and over.	129	123	37	35	.28682	.28455	.10775	.14316	+166.19	+98.76	1.01	0.75

X Not applicable.

[1] The model life table employed here is "Model West, mortality level 16," in Ansley J. Coale and Paul Demeny, *Regional Model Life Tables and Stable Populations,* Princeton, N.J., Princeton University Press, p. 17.

Source: Basic data from census reports of Iran.

arise also from understatement of the population 55 and over in the 1950 census, net overstatement of deaths for this age cohort due to misreporting of age, and underreporting of immigration for the cohort.

The method of intercensal cohort analysis may be applied in another way to evaluate the consistency of the data on age in two successive censuses when net immigration or emigration is negligible and death statistics are lacking or defective. Table 8-6 illustrates this method for Iran, for the 1956 and 1966 censuses. For Iran, adequate death statistics or a life table to measure mortality between the censuses is not available; net migration is assumed to be negligible. First, the proportion surviving at each age between 1956 and 1966 (cols. 5 and 6) is calculated by dividing the 1966 population at a given age (terminal age) by the 1956 population 10 years younger (initial age). For example:

$$\frac{P^{1966}_{m20-24}}{P^{1956}_{m10-14}} = \frac{793,000}{975,000} = .8133$$

Second, the reasonableness of these proportions in themselves or in comparison with an actual set or a model set of life table survival rates is examined as a basis for judging the adequacy of the census data.

In the absence of net migration, proportions surviving in excess of 1.00 are unacceptable and suggest either net understatement in the 1956 census or net overstatement (presumably due to age misreporting) in the 1966 census. This irregularity applies to the proportions for males with terminal ages 25 to 29, 30 to 34, and 35 to 44, and to the proportions for females with terminal ages 20 to 24, 25 to 29, and 30 to 34. Very different proportions surviving for males and females or higher proportions surviving for males than females except at the childbearing ages, as is shown for terminal ages 10 to 14, 20 to 24, 30 to 34, and 35 to 44 in table 8-6, are suspect.

Comparison with estimates based on birth statistics. — Estimates of net undercounts of children may be derived by comparison of the census counts and estimates of children based on birth statistics, death statistics or life table survival rates, and migration statistics. If possible, the birth and death statistics, particularly the former, should be adjusted to include an allowance for underregistration. The method was illustrated with U.S. data in table 8-5. Birth statistics for April 1, 1950, to April 1, 1955, and April 1, 1955, to April 1, 1960, are combined with death and immigration statistics for the same cohorts to derive estimates of the expected population under 5 and 5 to 9 years old in 1960. The difference between the expected population and the census count is then taken as the estimate of net undercount.

For the age group 0-4 in 1960:

$$B^{1955-60} - D^{1955-60} + M^{1955-60} = P^{e1960}_{0-4} \qquad (5)$$

$$P^{c1960}_{0-4} - P^{e1960}_{0-4} = E^{1960}_{0-4} \qquad (6)$$

where P^c represents the census count, P^e the expected population, and E the estimated net undercount. The corresponding figures are:

$$21,345,000 - 598,000 + 73,000 = 20,820,000$$

$$20,321,000 - 20,820,000 = -498,000$$

The census count of children under 5 years old, 20,321,000, falls below the expected population 20,820,000 by about 498,000, or 2.4 percent of the expected figure; this becomes the

Table 8-7. — Comparison of Survivors of Births With Census Counts Under 10 Years of Age, for Jamaica, by Sex: 1960

Sex and year of birth	Births		Survival rate from birth to census age[2]	Expected population	Census count	Deficit or excess of census		Age in 1960
	Registered	Adjusted to "census year"[1]				Amount	Percent	
		(1) redistributed=		(2)x(3)=		(5)-(4)=	(6)÷(4)=	
	(1)	(2)	(3)	(4)	(5)	(6)	(7)	
MALE								
1960......................	34,529	(X)	(X)	(X)	(X)	(X)	(X)	
1959......................	32,254	32,823	.95408	31,316	29,330	-1,986	-6.3	Under 1 year
1955-1958.................	120,940	121,897	(X)	[3]111,836	105,624	-6,212	-5.6	1 to 4 years
1958.....................	32,350	32,326	.93059	30,082	24,862	-5,220	-17.4	1 year
1957......................	30,561	31,008	.91736	28,445	27,197	-1,248	-4.4	2 years
1956......................	29,602	29,842	.91190	27,213	27,403	+190	+0.7	3 years
1955......................	28,427	28,721	.90859	26,096	26,162	+66	+0.3	4 years
1950-1954.................	125,379	[4]126,605	.90580	114,679	110,922	-3,757	-3.3	5 to 9 years
FEMALE								
1960......................	33,884	(X)	(X)	(X)	(X)	(X)	(X)	
1959......................	31,620	32,186	.96278	30,988	29,086	-1,902	-6.1	Under 1 year
1955-1958.................	116,966	118,036	(X)	[3]109,600	103,851	-5,749	-5.2	1 to 4 years
1958.....................	31,167	31,280	.93992	29,401	24,306	-5,095	-17.3	1 year
1957......................	29,884	30,205	.92837	28,041	26,397	-1,644	-5.9	2 years
1956......................	28,575	28,902	.92365	26,695	27,978	+1,283	+4.8	3 years
1955......................	27,340	27,649	.92094	25,463	25,170	-293	-1.2	4 years
1950-1954.................	122,664	[4]123,790	.92666	114,711	109,774	-4,937	-4.3	5 to 9 years

X Not applicable.
[1] Figures apply to period from April of year indicated to April of following year. Census was taken as of April 7, 1960.
[2] 1950-52 life table for Jamaica, adjusted to a 1960 basis.
[3] Obtained by summation.
[4] Equals sum of one-fourth births in 1955, births in 1951-54, and three-fourths births in 1950.

Source: Basic data from, United Nations, *Demographic Yearbook, 1959*, table 10; *1961*, tables 24 and 25; *1962*, table 6.

estimate of the net undercount of children under 5 in the census.

A special problem of calculation and interpretation of the difference between the expected population and the census count of children exists when the birth statistics or the death statistics are probably or definitely incomplete. In the absence of immigration, the comparison then gives a minimum estimate of the net undercount of children when the expected population exceeds the census count (and a minimum estimate of the underregistration of births when the census figure exceeds the estimate based on births). It may be desirable or even preferable in this case to employ life table survival rates in lieu of death statistics because of the inadequacies of the reported death statistics or the convenience of using a life table.

The procedure is illustrated in table 8-7, comparing the expected population under 10 years of age (single years under 5 and the age group 5 to 9), by sex, with the corresponding counts from the census of Jamaica taken on April 7, 1960. Registered births (col. 1), tabulated by calendar year of registration for most years (rather than by year of occurrence), were first redistributed to conform to "census" years, i.e., April to April, on the basis of the assumption that the distribution is rectangular (i.e., even) within each calendar year. Survival rates, representing the probability of survival from birth to the age at the census date, were then calculated from an abridged life table for Jamaica for 1950-52 adjusted to a 1960 basis (col. 3; see ch. 15). The expected population excluding the effect of immigration (col. 4) was then derived as the product of the births in column 2 and the survival rates in column 3. The 6-percent deficit of the census count for children under 1 and the 5-percent deficit for children 1 to 4 years old in comparison with the corresponding expected populations may be taken as minimum

estimates of the net undercounts of these groups. The method also suggests a substantial net census undercount of children 5 to 9 years old (about 4 percent). However, the survival rates from the 1960 life table may be too high, and, hence, the estimate of survivors may be too high. Even allowing for this possibility and the possibility of net emigration, the actual net undercounts are probably greater than those shown because of some underregistration of births.

Age ratio analysis.—The quality of the census returns by age groups may also be evaluated by comparing so-called **age ratios,** calculated from the census data, with expected or standard values. An age ratio may be defined as the ratio of the population in the given age group to one-third of the sum of the populations in the age group itself and the preceding and following groups, times 100.[16] The age ratio for a 5-year age group, $_5P_a$, is defined then as follows:

$$\frac{_5P_a}{1/3\ (_5P_{a-5}+_5P_a+_5P_{a+5})} \times 100 \qquad (7)$$

Barring extreme fluctuations in past births, deaths, or migration, the three age groups should form a nearly linear series. Age ratios should then approximate 100.0, even though actual historical variations in these factors would produce deviations from 100.0 in the age ratio for most ages. Inasmuch as, over a period of nearly a century, most countries have experienced

---

[16] Alternatively, age ratios have been defined as the ratio of the population in an age group to one-half the sum of the population in the preceding and subsequent groups, times 100. The definition given above is preferred.

not only minor fluctuations in population changes but also major upheavals, age ratios for some ages may deviate substantially from 100.0 even where reporting of age is good. The assumption of an expected value of 100.0 also implies that coverage errors are about the same from age group to age group and that age reporting errors for a particular group are offset by complementary errors in adjacent age groups. In sum, age ratios serve primarily as measures of net age misreporting, not net census error, and they are not to be taken as valid indicators of error for particular age groups.

An overall measure of the accuracy of an age distribution, an age-accuracy index, may be derived by taking the average deviation (without regard to sign) from 100.0 of the age ratios over all ages. This is illustrated on the basis of data for Taiwan in 1964 in table 8–8.

Table 8–8. — Calculation of Age-Accuracy Index, for Taiwan: 1964

Age	Population[1] Male (1)	Population[1] Female (2)	Male Ratio[2] (3)	Male Deviation from 100 (3)-100= (4)	Female Ratio[2] (5)	Female Deviation from 100 (5)-100= (6)
Under 5 years.........	1,022,001	965,554	(X)	(X)	(X)	(X)
5 to 9 years..........	964,956	916,269	102.4	+2.4	102.5	+2.5
10 to 14 years........	841,146	799,199	108.0	+8.0	108.1	+8.1
15 to 19 years........	531,428	503,091	94.3	−5.7	85.2	−14.8
20 to 24 years........	318,826	470,153	73.8	−26.2	100.6	+0.6
25 to 29 years........	445,106	429,185	112.5	+12.5	100.8	+0.8
30 to 34 years........	423,287	377,645	100.1	+0.1	100.0	−
35 to 39 years........	400,660	325,684	103.6	+3.6	100.8	+0.8
40 to 44 years........	336,099	265,734	100.1	+0.1	99.6	−0.4
45 to 49 years........	270,331	209,086	96.2	−3.8	94.9	−5.1
50 to 54 years........	236,988	186,234	106.0	+6.0	103.7	+3.7
55 to 59 years........	163,501	143,373	96.3	−3.7	98.9	−1.1
60 to 64 years........	108,910	105,412	96.2	−3.8	97.8	−2.2
65 to 69 years........	67,351	74,716	94.9	−5.1	98.1	−1.9
70 to 74 years........	36,587	48,300	(X)	(X)	(X)	(X)
Total (irrespective of sign)..........	(X)	(X)	(X)	81.0	(X)	42.0
Mean...............	(X)	(X)	(X)	6.2	(X)	3.2

− Represents zero.
X Not applicable.
[1] Source: United Nations, *Demographic Yearbook, 1965*, table 6.
[2] The age ratio is defined as $\frac{{}_sP_a}{\frac{1}{2}({}_sP_{a-5}+{}_sP_a+{}_sP_{a+5})} \times 100$.

The sum of the deviations from 100.0 of the age ratios for males is 81.0, and the mean deviation for the 13 age groups is, therefore, 6.2. The average (4.7) of the mean deviation for males (6.2) and the mean deviation for females (3.2) is a measure of the overall accuracy of the age data of Taiwan in 1964, which can be compared with the same kind of measure for other years or other areas. The lower the age-accuracy index, the more adequate the census data on age would appear to be. The results suggest that reporting of age is less satisfactory for males in Taiwan than for females. The results of similar calculations carried out for Greece, the Philippine Islands, Sweden, Turkey, and the United States suggest that the quality of age reporting in Taiwan occupies an intermediate position:

Country (census year)	Age-accuracy index
United States (1960).....................	2.3
Sweden (1963)............................	3.3
Taiwan (1964)............................	4.7
Greece (1961)............................	6.5
Philippines (1960)........................	7.0
Turkey (1960)............................	9.7

Sex ratio analysis.—Several methods of evaluating census age data employ age-specific sex ratios from the census. One compares expected sex ratios for each age group, developed principally from vital statistics, with the census sex ratios. The expected figures may be carefully developed estimates of the actual sex ratios at each age or theoretical figures based on a population model. Another judges the census age-specific sex ratios in terms of their age-to-age differences.

The first method involves developing estimates of the actual sex ratios at each age at a census date on the basis of the sex ratios of each of the components of change, particularly the sex ratio of births and the sex ratios of survival rates (i.e., the ratio of the male survival rate at a given age to the corresponding female rate, derived from life tables).[17] The basic calculations may be illustrated by the procedure for deriving the expected sex ratios at ages 0 to 4 and 5 to 9 at the census date. If the contribution of net migration is disregarded, the expected sex ratio at ages 0 to 4 equals the product of the sex ratio of births in the 5 years preceding the census date and the ratio of (a) the male survival rate from birth to ages 0 to 4 $\left(\overset{0-4}{\underset{b}{R^m}}\right)$ to (b) the corresponding female survival rate $\left(\overset{0-4}{\underset{b}{R^f}}\right)$:[18]

$$\left(\frac{B^m}{B^f}\right)_{y-5\text{ to }y} \times \frac{\left(\overset{0-4}{\underset{b}{R^m}}\right)_{y-5\text{ to }y}}{\left(\overset{0-4}{\underset{b}{R^f}}\right)_{y-5\text{ to }y}} = \left(\frac{P_{0-4}^m}{P_{0-4}^f}\right)'_y \quad (8)$$

where "y" designates a given year and "$y-5$ to y" the preceding 5 years. The survival rates should be calculated from life tables pertaining to the actual calendar years when the cohort had the ages indicated. The expected sex ratio for the age group 5 to 9 would be derived theoretically as the joint product of the sex ratio of births 5 to 10 years earlier, the sex ratio of survival rates from birth to ages 0 to 4, 5 to 10 years earlier, and the sex ratio of survival rates from ages 0 to 4 to ages 5 to 9 in the previous 5 years:

$$\left(\frac{B^m}{B^f}\right)_{y-10\text{ to }y-5} \times \frac{\left(\overset{0-4}{\underset{b}{R^m}}\right)_{y-10\text{ to }y-5}}{\left(\overset{0-4}{\underset{b}{R^f}}\right)_{y-10\text{ to }y-5}} \times \frac{\left(\overset{5-9}{\underset{0-4}{R^m}}\right)_{y-5\text{ to }y}}{\left(\overset{5-9}{\underset{0-4}{R^f}}\right)_{y-5\text{ to }y}} = \left(\frac{P_{5-9}^m}{P_{5-9}^f}\right)'_y \quad (9)$$

---

[17] Full development of the estimates of expected sex ratios of this type requires a knowledge of the use of life tables and of techniques of population estimation. Both of these topics are treated in later chapters.

[18] The expressions in parentheses are calculated as units or are treated as single numbers in the calculations. For example, the sex ratio of births for a given period, whether calculated on the basis of reported births or assumed on the basis of the sex ratio of births for a later period, is treated as a single number in the "survival" calculations; and the sex ratio of the population is derived as a direct result, without intermediate figures for the absolute numbers of males and females.

"Expected" sex ratios calculated in this way can then be compared to those calculated directly from the census data. An illustration of this procedure is presented in the unabridged edition of this book.

The results of the method are directly applicable for judging the relative magnitude of the net census error of the counts of males and females; they do not indicate the absolute level of net census error for either sex. If the results of this method are to be used to derive absolute estimates of corrected population for either sex or both sexes combined, an acceptable, independently determined set of estimates of net undercounts or corrected census figures by age for one sex is required. For example, if corrected census figures for females are available, the expected sex ratios are applied to them to derive corrected figures for males. Because of the greater likelihood of deficiencies in the basic data and the greater dependence on the various assumptions made as one goes back in time, the estimates of expected sex ratios are subject to greater and greater error as one goes up the age scale.

When the detailed data required to develop a set of estimated actual sex ratios (e.g., historical series of life tables, historical data on births by sex, net immigration or nativity of the population by age and sex, war deaths) are not available or it is not practical to develop them, the expected pattern of sex ratios by age may be approximated by employing a single current life table to measure survival from birth to each age, in conjunction with the current reported or estimated sex ratio at birth. This method, in effect, assumes that there has been no net migration, either civilian or military, or excess mortality due to war or widespread epidemic. In addition, it assumes that the sex ratio of births and the differences in mortality between the sexes at each age have remained unchanged. To the extent that these conditions prevail, the approximation to the actual sex ratios will be closer.

The patterns of the expected sex ratios are quite similar from country to country, and especially so, if one disregards the varying level of the sex ratio at birth (ch. 7).[19] Expected sex ratios at the early childhood ages are not far below the sex ratio at birth. Then, commonly, they fall gradually throughout life, not dipping below 100 until age 40 or later. The decline is gentle at first but becomes steeper at the older ages. The general pattern described results from the usual small excess of boys among births and the usual excess of male over female mortality.

The regularity of the change in the expected sex ratio from age to age which we have just seen provides a basis for elaborating the age-accuracy index based solely on age ratios described earlier to incorporate some measure of the accuracy of sex ratios. The United Nations has proposed such an age-sex accuracy index.[20] In this index the mean of the differences from age to age in reported sex ratios, without regard to sign, is taken as a measure of the accuracy of the observed sex ratios, on the assumption that these age-to-age changes should approximate zero. The U.N. age-sex accuracy index combines the sum of (a) the mean deviation of the age ratios for males from 100.0, (b) the mean deviation of the age ratios for

females from 100.0 and (c) three times the mean of the age-to-age differences in reported sex ratios. In the U.N. procedure, an age ratio is defined as the ratio of the population in a given age group to one-half the sum of the populations in the preceding and following groups. The calculation of the U.N. age-sex accuracy index is illustrated in table 8–9 for Greece in 1961. The mean deviations of the age ratios for males and females are 10.4 and 9.1, respectively, and the mean age-to-age difference in the sex ratios is 5.4. Applying the U.N. formula, we have: $10.4 + 9.1 + 3(5.4) = 35.5$. Comparable indexes for a few other countries are:

Country (census year)	U.N. age-sex accuracy index
United States (1960)	12.2
Sweden (1963)	15.1
Philippines (1960)	32.8
Greece (1961)	35.5
Taiwan (1964)	49.3
Turkey (1960)	70.6

Census age-sex data were described by the United Nations as "accurate," "inaccurate," or "highly inaccurate" depending on whether the U.N. index was under 20, 20 to 40, or over 40.

The U.N. index has a number of questionable features as a summary measure for comparing the accuracy of the age-sex data of various countries. Among these are the failure to take account of the expected decline in the sex ratio with increasing age and of real irregularities in age distribution due to migration, war, and epidemic as well as normal fluctuations in births and deaths; the use of a definition of an age ratio which omits the central age group and which, therefore, has an upward bias; and the considerable weight given to the sex-ratio component in the formula. In addition, the index is primarily a measure of net age misreporting and, for the most part, does not measure net underenumeration by age. An allowance for the typical decline in the sex ratio from childhood to old age can be made by adjusting the mean difference of the census sex ratios downward by the mean difference between the expected sex ratio for ages under 5 and, say, 70 to 74, derived from life tables. In spite of its limitations, however, the U.N. index appears to be a useful measure for making approximate distinctions between countries with respect to the accuracy of reporting age by sex in censuses.

Mathematical graduation of census data. — **Mathematical graduation** of census data can be employed to derive figures for 5-year age groups that are corrected primarily for net reporting error. What these graduation procedures do, essentially, is disregard the original 5-year totals and "fit" different curves to the original 10-year totals. Among the major graduation methods that the mathematically inclined student may wish to look into are the *Carrier-Farrag ratio method,*[21] quadratic interpolation (also called Newton's method of halving an age group), [22]and *osculatory interpolation.*

---

[19]The variations in the theoretical pattern of expected sex ratios by age resulting solely from variations in the level of mortality, holding the sex ratio at birth constant and excluding the effect of civilian migration and military movements, may be shown by employing model life tables which have very different levels of mortality, such as those given in Coale and Demeny, op. cit.

[20]United Nations, *Methods of Appraisal of Quality of Basic Data for Population Estimates,* Manual II, Population Studies, Series A, No. 23, pp. 42–43; idem, "Accuracy Tests for Census Age Distributions Tabulated in Five-Year and Ten-Year Groups," *Population Bulletin,* No. 2, October 1952, pp. 59–79.

[21]N. H. Carrier and A. M. Farrag, "The Reduction of Errors in Census Populations for Statistically Underdeveloped Countries," *Population Studies* (London), 12(3):240–285, March 1959, esp. pp. 251–252 and 265–266.

[22]Margaret J. Hagood, *Statistics for Sociologists,* New York, Reynal and Hitchcock, Inc., pp. 760–763; Carrier and Farrag, op. cit., pp. 250–251.

Table 8-9. — Calculation of the United Nations Age-Sex Accuracy Index, for Greece: 1961

Age	Population[1]		Analysis of sex ratios		Analysis of age ratios			
					Male		Female	
	Male	Female	Ratio	Successive differences	Ratio[2]	Deviation from 100	Ratio[2]	Deviation from 100
			$[(1) \div (2)] \times 100 =$	$\triangle(3) =$		$(5) - 100 =$		$(7) - 100 =$
	(1)	(2)	(3)	(4)	(5)	(6)	(7)	(8)
Under 5 years	493,600	385,100	128.17	(X)	(X)	(X)	(X)	(X)
5 to 9 years	364,900	346,900	105.19	+22.98	83.69	-16.31	93.43	-6.57
10 to 14 years	378,400	357,500	105.85	-0.66	111.94	+11.94	109.65	+9.65
15 to 19 years	311,200	305,200	101.97	+3.88	84.06	-15.94	84.72	-15.28
20 to 24 years	362,000	363,200	99.67	+2.30	110.27	+10.27	107.01	+7.01
25 to 29 years	345,400	373,600	92.45	+7.22	97.57	-2.43	101.32	+1.32
30 to 34 years	346,000	374,300	92.44	+0.01	117.17	+17.17	116.80	16.80
35 to 39 years	245,200	267,300	91.73	+0.71	87.89	-12.11	87.01	-12.99
40 to 44 years	212,000	240,100	88.30	+3.43	85.90	-14.10	91.15	-8.85
45 to 49 years	248,400	259,500	95.72	-7.42	111.99	+11.99	109.01	+9.01
50 to 54 years	231,600	236,000	98.14	-2.42	102.82	+2.82	101.55	+1.55
55 to 59 years	202,100	205,300	98.44	-0.30	105.73	+5.73	97.79	-2.21
60 to 64 years	150,700	183,900	81.95	+16.49	99.11	-0.89	111.32	+11.32
65 to 69 years	102,000	125,100	81.53	+0.42	87.14	-12.86	84.76	+15.24
70 to 74 years	83,400	111,300	74.93	+6.60	(X)	(X)	(X)	(X)
Total (irrespective of sign)	(X)	(X)	(X)	74.84	(X)	134.56	(X)	117.80
Mean	(X)	(X)	(X)	5.35	(X)	10.35	(X)	9.06

Index = 3 times mean difference in sex ratios plus mean deviations of male and female age ratios.
= 3 x 5.35 + 10.35 + 9.06 = 35.46.

X Not applicable.
[1] From United Nations, *Demographic Yearbook, 1962*, table 5.
[2] An age ratio is defined here as $\dfrac{{}_5P_a}{\frac{1}{2}({}_5P_{a-5} + {}_5P_{a+5})} \times 100$.

Comparison with population models. — Still another basis of evaluating the census data on age is to compare the actual percent distribution of the population by age with an expected age distribution corresponding to various population models. One such model is the **stable population model.** In the absence of migration. if fertility and mortality remain constant over several decades. the age distribution of a population would assume a definite unchanging form called stable (ch. 18). Such model age distributions are pertinent in the consideration of actual age distributions since nearly constant fertility and nearly constant or moderately declining mortality are characteristic of many less industralized countries. The declines in mortality which have been occurring in many populations affect the age distribution to only a small extent. Accordingly. many countries of the world have a relatively stable distribution (with constant mortality) or a quasi-stable age distribution (with moderately declining mortality); these are represented rather well by the stable age distributions which would result from the persistence of their current fertility and mortality rates. The stable age distribution may then be used as a standard for judging the adequacy of reported age distributions.[23]

With the limitations implied. the inadequacies of the age distribution in particular countries may be measured by comparing the percentage age distributions in these countries with the age distributions of the corresponding stable populations. Specifically. for each age group an index may be calculated by dividing the percent in the age group in a given country by the corresponding percent in the stable population. The illustrative calculations in table 8-10 relate to various countries of Africa and Asia which would be expected to have about the same age distribution and which. therefore. may be represented by a single stable age distribution. (The choice of stable age distribution is discussed in ch. 24.) The deviations of the indexes from 1.00 reflect the extent to which a particular age group is relatively overstated or understated as a result of net coverage error or age misreporting. For example. the indexes 0.66. 0.76. and 1.23 for ages 10 to 14. 15 to 19. and 25 to 34 in Indonesia (like those in Senegal and Ghana) indicate a large relative understatement of women 10 to 19 years old and a large relative overstatement of women 25 to 34. In contrast. the indexes for the Philippines imply an age distribution more consistent with the expected pattern.

Comparison with aggregated administrative data. — Finally, we note the possible use of various types of aggregated data compiled primarily for administrative purposes in evaluating census data in particular age groups. This procedure assumes that the administrative records are free of the types of errors of coverage and age reporting that characterize household inquiries. It is assumed. for example. that a registration from which the aggregated data are derived is complete and accurate (without omissions. duplications. or inactive records. i.e.. records for persons who died or are no longer eligible or obligated to remain in the file) and is based on accurate age information. possibly involving formal proof of age. In these comparisons. no attempt is made at matching records for individuals; only aggregates are employed. The aggregates may require a substantial amount of adjustment. however. to insure

[23]Ansley J. Coale, "Estimates of Various Demographic Measures Through the Quasi-Stable Age Distribution," *Emerging Techniques in Population Research*, Milbank Memorial Fund, 1963, pp. 175–193; and Etienne van de Walle, "Some Characteristic Features of Census Age Distributions in Illiterate Populations," *American Journal of Sociology*, 71(5):549–557, March 1966, esp. pp. 549–551.

Table 8-10. — Indexes Relating the Reported Age Distribution to the Stable Age Distribution for Females in Various Countries of Asia and Africa: Around 1960

Age	Percent distribution					Ratio of reported to stable distribution			
	Indonesia, 1961 (1)	Ghana, 1960 (2)	Senegal, 1960-61 (3)	Philippines, 1960 (4)	Stable (5)	Indonesia, 1961 (1)÷(5)= (6)	Ghana, 1960 (2)÷(5)= (7)	Senegal, 1960-61 (3)÷(5)= (8)	Philippines, 1960 (4)÷(5)= (9)
Total.....................	100.0	100.0	100.0	100.0	100.0	1.00	1.00	1.00	1.00
Under 5 years..............	17.6	19.7	18.6	16.5	17.1	1.03	1.15	1.09	0.96
5 to 9 years...............	15.7	15.2	14.9	15.8	13.7	1.15	1.11	1.09	1.15
10 to 14 years.............	7.9	9.8	7.5	12.4	12.0	0.66	0.82	0.63	1.03
15 to 19 years.............	7.9	7.9	8.3	10.6	10.4	0.76	0.76	0.80	1.02
20 to 24 years.............	8.9	9.6	9.0	9.4	9.0	0.99	1.07	1.00	1.04
25 to 34 years.............	17.5	16.7	17.5	13.4	14.2	1.23	1.18	1.23	0.94
35 to 44 years.............	11.0	9.7	10.3	9.5	10.0	1.10	0.97	1.03	0.95
45 to 54 years.............	7.1	5.3	6.6	6.4	6.9	1.03	0.77	0.96	·0.93
55 years and over..........	6.3	6.1	7.2	6.0	6.9	0.91	0.88	1.04	0.87

Source: Adapted from, Etienne van de Walle, "Some Characteristic Features of Census Age Distributions in Illiterate Populations," *American Journal of Sociology,* 71(5):549-557, March 1966. For reported distributions, see United Nations, *Demographic Yearbook, 1963,* table 5. The stable population was selected from Ansley J. Coale and Paul Demeny, *Regional Life Tables and Stable Populations,* Princeton, J. J., Princeton University Press, 1965, p. 38. The stable population selected corresponds to Model West, mortality level 7, with an expectation of life at birth of 35 years for females, a growth rate of 2 percent, a birth rate of 48.4 per 1,000 population, and a death rate of 28.4 per 1,000 population.

agreement with the intended census coverage. These data may be a product of the social security system, the military registration system, the educational system, etc.[24]

Taking advantage of earlier studies Siegel has compiled a preferred set of estimates of corrected population and of net undercounts for the United States in 1960, by age, sex, and color, representing a composite of the analytic estimates.

The preferred composite set of estimates of net census undercounts in 1960 by age and sex is shown in table 8-11. These figures indicate net census undercounts in all age-sex groups, with considerable variation over the age distribution.

Extreme Old Age and Centenarians

Census age distributions at advanced ages, say for those 85 years old and over by age, are not too meaningful. The extent of misreporting of age of household members due to ignorance of the true age on the part of the respondent in the household is probably considerable in this age range. There is a notable tendency, in particular, to report an age over 100 for persons of very advanced age, stemming from a desire to share in the esteem generally accorded extreme old age or from a gross ignorance of the true age. The exaggeration of the number of centenarians in census statistics is suggested by a number of considerations. First, if death rates at the later ages are projected

to the end of life, the chance of death at age 100 would be extremely high and few persons would remain alive past 100. For example, at age 100 the probability of death in one year is in the vicinity of 0.4 according to several life tables (e.g., United States, 1959-61; Sweden, 1962). Second, the number of survivors 100 to 109 years old at a given census date, of the population 90 to 99 at the earlier census, tends to be far smaller than the current census count of the population 100 and over. For example, the 1960 U.S. census count of the population 100 and over (10,326) exceeded the number expected on this basis by 328 percent.[25] The same problem appears in the 1959 U.S.S.R. census.[26] Third, the number of centenarians is greatest among the groups and in the areas where health conditions are poorest. Over one-quarter of the 10,326 persons reported as 100 or over in the 1960 U.S. census were nonwhite, whereas nonwhites made up only 11 percent of the total population and only 8 percent of the population 85 and over.

The census count of persons of extreme old age may also be evaluated by Vincent's "method of extinct generations."[27] The population 85 and over in a census taken in 1950 would have almost completely died off by 1970, so that it should be possible, by cumulating the appropriate statistics of deaths in the period 1950-70, to reconstruct the "true" population 85 and over in 1950.

The thinness of the figures in the range 85 and over results in considerable random fluctuation in rates based on them. However, it is necessary to compute rates by age until the end of the life span for many purposes, such as, to develop certain measures for the whole population or some particular age (e.g., computation of the value for life expectancy at birth or at age 40). In this case the rates should not be taken as correct in themselves, but as necessary to develop the other measures.

Age Not Reported

Age is not always reported in a census, even though the enumerator may be instructed to secure an estimate from the

[24] Ansley J. Coale, "The Population of the United States in 1950 Classified by Age, Sex, and Color—A Revision of Census Figures," *Journal of the American Statistical Association,* 50(269):16-54, March 1955; Ansley J. Coale and Melvin Zelnik, *New Estimates of Fertility and Population in the United States,* Princeton, N. J., Princeton University Press, 1963; Conrad Taeuber and Morris Hansen, "A Preliminary Evaluation of the 1960 Censuses of Population and Housing," *Demography,* 1(1):1-14, 1964; Donald J. Bogue, Bhaskar D. Misra, and D. P. Dandekar, "A New Estimate of the Negro Population and Negro Vital Rates in the United States, 1930-60," *Demography,* 1(1):339-358, 1964; *U.S. Census of Population: 1960,* Vol. 1, *Characteristics of the Population,* Pt. 1, *United States Summary,* 1964, p. XL; Melvin Zelnik, "Errors in the 1960 Census Enumeration of Native Whites," *Journal of the American Statistical Association,* 59(2):437-459, June 1964; U.S. Bureau of the Census, *Current Population Reports,* Series P-25, No. 310, app. C and table 12; Jacob S. Siegel and Melvin Zelnik, "An Evaluation of Coverage in the 1960 Census of Population by Techniques of Demographic Analysis and by Composite Methods," *Proceedings of the Social Statistics Section, 1966,* American Statistical Association, pp. 71-85; Jacob S. Siegel, "Completeness of Coverage of the Nonwhite Population in the 1960 Census and Current Estimates, and Some Implications," *Social Statistics and the City,* David M. Heer, ed., Joint Center for Urban Studies of the Massachusetts Institute of Technology and Harvard University, Cambridge, Mass., 1968, pp. 13-54.

[25] A similar calculation is described in Robert J. Myers, "Validity of Centenarian Data in the 1960 Census," *Demography,* 3(2):470-476, 1966.
[26] Robert J. Myers, "Analysis of Mortality in the Soviet Union according to 1958-59 Life Tables," *Transactions of the Society of Actuaries,* 16:309-317, November 1964.
[27] Paul Vincent, "La Mortalité des Vieillards" (The mortality of the aged), *Population* (Paris), 6(2):181-204, April-June 1951.

Table 8–11.—**Percent Net Undercount of the Population of the United States in the 1950 and 1960 Censuses, by Age and Sex**

[Percents relate to the total resident population. Base of percents is the corrected population. Minus sign (−) denotes a net overcount in the census]

Age	1960			1950		
	Both sexes	Male	Female	Both sexes	Male	Female
Total.............	3.1	3.8	2.4	3.6	4.2	3.0
Under 5 years.........	2.1	2.3	2.0	4.7	5.0	4.4
5 to 9 years.........	2.5	2.9	2.0	3.7	4.0	3.3
10 to 14 years........	2.4	2.9	1.8	1.8	1.7	1.8
15 to 19 years........	4.2	5.0	3.4	4.3	5.5	3.0
20 to 24 years........	4.7	6.1	3.4	4.6	6.8	2.3
25 to 29 years........	4.3	6.3	2.4	3.8	6.3	1.2
30 to 34 years........	3.1	5.0	1.3	3.4	5.7	1.0
35 to 39 years........	2.2	3.9	0.6	1.1	3.0	−0.9
40 to 44 years........	1.8	3.1	0.5	4.2	4.8	3.5
45 to 49 years........	2.1	2.7	1.6	2.8	3.2	2.5
50 to 54 years........	5.4	5.1	5.7	3.3	2.9	3.7
55 to 59 years........	1.7	1.0	2.4	7.2	5.8	8.4
60 to 64 years.......	4.4	3.6	5.1	6.7	4.8	8.4
65 years and over.....	3.2	3.7	2.9	0.9	−0.9	2.4

Source: Jacob S. Siegel, "Completeness of Coverage of the Nonwhite Population in the 1960 Census and Current Estimates, and Some Implications," *Social Statistics and the City*, David M. Heer, ed., Joint Center for Urban Studies of the Massachusetts Institute of Technology and Harvard University, Cambridge, Mass., 1968, table 2.

respondent or to estimate it as best he can while enumerating. In most national censuses, persons whose age was not reported by the respondent have been assigned an age on the basis of an estimate made by the enumerator or on the basis of an estimate made in the processing of the census; or the category of unknown ages was distributed arithmetically prior to publication. As a result, census age distributions presented in recent U.N. *Demographic Yearbooks* usually do not show a category of unknown age. The method used in national censuses to eliminate frequencies in this category is not always known and hence it is not usually indicated in the U.N. tables. About one-third of the age distributions (for 197 countries) shown in the 1966 *Demographic Yearbook* have frequencies in a category of unknown age.

In the 1940, 1950, and 1960 censuses of the United States, ages were assigned to persons whose age was not reported, on the basis of related information on the schedule for the person and other members of the household, such as the age of other members of the family (particularly the spouse), school attendance, marital status, and employment status.[51] As noted in chapter 3, in the 1960 procedure the allocation of age was made by electronic computer on the basis of the record of an individual just previously enumerated in the census who had characteristics similar to those of the person whose age was not reported, whereas in 1950 and 1940 the allocation was made on the basis of distributions derived from the previous census. The vast majority of persons whose age was not reported in 1960 were assigned ages over 20. Since the age allocations are based on actual age distributions of similar population groups or the actual characteristics of the same individuals, the resulting assignments of age should be reasonable and show relatively little error. The procedure of allocating age in the 1970 census will be similar to that in the 1960 census.

The proportion of the total population whose age was not reported in the field enumeration in the decennial censuses of the United States has been quite low until recently. The reported proportions for each census since 1900 are as follows:

1900....................	0.3	1940....................	0.2
1910....................	0.2	1950....................	0.2
1920....................	0.1	1960....................	2.2
1930....................	0.1	1970....................	2.7

The change to a question on date of birth in 1960 resulted in an increase in the extent of nonreporting of age. Birth date was estimated or allocated for 2.2 percent of the population in the complete-count statistics.[28]

The recent procedures used in the U.S. censuses are superior to those used generally in the censuses before 1940, when the number of persons whose age was not reported was shown in the published tables as a separate category, or in the 1880 census, when the "unknown ages" were distributed before printing in proportion to the ages reported. The former procedure creates inconveniences in the use of the data, results in less accurate age data, and contributes to the cost of publication. Although simple prorating, like that in 1880, has its limitations (e.g., the results are subject to error and the procedure can be applied to only a few principal age distributions), it is about the only method feasible for eliminating the unknown ages from the age distributions of the censuses before 1940. This elimination is desirable not only for the reasons implied above but also for making comparisons of the age statistics of two censuses.

To accomplish the arithmetic distribution of the unknown ages, it may be assumed that those of unknown age have the same percent distribution by age as those of known age. The application of this assumption simply involves multiplying the number reported at each age by a factor equal to the ratio of the total population to the number whose age was reported; that is,

$$\frac{\sum_0^\infty P_a + P_u}{\sum_0^\infty P_a} \times P_a \qquad (10)$$

where P_a represents the number reported at each age and P_u the number whose age was not reported.[29] Table 8–12 illustrates this procedure for distributing unknown ages in the case of the population of Morocco in 1960. On the basis of the results of the assignment of unknown ages in recent censuses, it may be more appropriate to distribute the unknowns among adults only. Table 8–12 also illustrates the procedure for distributing the unknowns among the population 20 and over for the population of Morocco in 1960.

The relative magnitude of this category reflects in a rough way the quality of the data on age. The existence of a very large proportion of persons of unknown age may raise a question as to the validity of the reported age distribution, although, as stated, this situation is quite uncommon.

---

[28] Birth date was estimated for 1.7 percent of the enumerated population in the complete-count statistics (and for 1.0 percent in the sample statistics) on the basis of other information regarding the same person reported on the census questionnaire. Birth date was allocated for an additional 0.5 percent of the population as a part of the process of substituting persons with reported characteristics for persons not tallied because of the enumerator's failure to interview households or because of mechanical failure in processing.

[29] The numbers so obtained are the same as the numbers obtained by the longer procedure of computing the percent distribution of persons of known age, distributing the number of unknown age according to this percent distribution, and adding the two absolute distributions together.

Table 8-12.—**Procedure for Prorating Unknown Ages, for Morocco: 1960**

Age	Population as enumerated (1)	Population with unknowns distributed over all ages (1) x f$_1$ = (2)	Population with unknowns distributed over ages 20 years and over (1) x f$_2$ (ages 20 and over) = (3)
Total..............	11,626,232	11,626,232	11,626,232
Under 5 years...........	2,194,810	2,199,539	2,194,810
5 to 9 years............	1,872,850	1,876,884	1,872,850
10 to 14 years..........	1,082,720	1,085,052	1,082,720
15 to 19 years..........	722,320	723,876	722,320
20 to 24 years..........	907,526	909,481	911,485
25 to 34 years..........	1,797,133	1,801,004	1,804,973
35 to 44 years..........	1,240,097	1,242,768	1,245,507
45 to 54 years..........	769,870	771,528	773,228
55 to 64 years..........	544,984	546,158	547,361
65 years and over.......	468,932	469,942	470,978
Unknown age............	24,990	(X)	(X)

Factors f$_1$ and f$_2$ are based on data in col. (1):

$$f_1 = \frac{\text{total population}}{\text{total population of known age}} = \frac{\sum_0^\infty P_a + P_u}{\sum_0^\infty P_a} = \frac{11,626,232}{11,601,242} = 1.00215408.$$

$$f_2 = \frac{\text{population 20 years and over + unknown ages}}{\text{population 20 years and over}} = \frac{\sum_{20}^\infty P_a + P_u}{\sum_{20}^\infty P_a} = \frac{5,753,532}{5,728,542} = 1.00436237.$$

X Not applicable.

Source: Basic data from, United Nations, *Demographic Yearbook, 1964*, table 5.

ANALYSIS OF AGE COMPOSITION

General Techniques of Numerical and Graphic Analysis

Nature of Age Distributions.—Data on age are most commonly tabulated and published in 5-year groups (0-4, 5-9, etc.). This detail is sufficient to provide an indication of the form of the age distribution and to serve most analytic uses. For some types of analysis, however, data for single years may be needed. In some parts of the age range (i.e., the late teens, early twenties, late middle age) changes in some of the characteristics of the population (i.e., labor force status, marital status, school enrollment status) are so rapid that single-year-of-age data are required to present them adequately. For other analytic purposes age data may be combined to obtain figures for various broader groups than 5-year groups. Age distributions consisting of combinations of 5-year age groups and 10-year age groups, or 10-year age groups only, may sometimes be published so as to achieve consolidation of masses of data and the reduction of sampling error, yet to provide sufficient detail to indicate variations by age and permit alternative combinations of age groups, etc.

Further consolidation or special combinations are desirable to represent special age groups. For fertility analysis the total number of women 15 to 44 years of age (the childbearing ages) is significant; the population 5 to 17 (school ages) is important in educational research and planning; and the group 18 to 24 as a whole roughly defines the college-age group, the group of prime military age, the principal ages of labor force entry and marriage, etc. For many purposes the number of persons 18 and over or 21 and over (age of legal majority) is useful. A

classification of the total population into several mutually exclusive broad age groups having general functional significance may be found useful for a wide variety of analytic purposes. One classification, employed in some of the publications of the U.S. Census Bureau, is as follows: Under 5 years, the preschool ages; 5 to 17 years, the school ages; 18 to 44 years, the earlier working years; 45 to 64 years, the later working years; 65 years and over, the period of retirement. Any grouping of the ages into working ages, school ages, retirement ages, etc., is admittedly arbitrary and requires some adaptation to the customs and institutional practices of different areas or some modifications as these practices change. For example, in the early nineteenth century in the United States, the period of labor force participation was considerably longer than today, extending back into the current ages of compulsory school attendance and forward into the current ages of retirement.

Special interest also attaches to the numbers reaching certain "threshold" ages in each year. These usually correspond to the initial ages of the functional groupings described in the previous paragraph. On reaching these ages, new social roles are assumed or new stages in the life cycle are begun, e.g., birth and reaching age 5 or 6, 18, 21, and 65.

Percent Distributions.—In the simplest kind of analysis of age data, the magnitude of the numbers relative to one another is examined. If the absolute numbers distributed by 5-year age groups are converted to percents, a clearer indication of the relative magnitudes of the numbers in the distribution is obtained. Conversion to percents is necessary if the age distributions of different countries of quite different population size are to be conveniently compared, either numerically or graphically. The percent distribution by age of the population of Ghana in 1960, for example, was quite different from that of the United States:

	Total	Under 5	5 to 14	15 to 24	25 to 34	35 to 44	45 to 64	65+
Ghana ...	100.0	19.3	25.2	16.8	16.0	10.2	9.4	3.1
United States	100.0	11.3	19.8	13.4	12.8	13.5	20.2	9.2

Table 8-13.—**Population of the United States, 1950 and 1960, and Percent Change, by Age, 1950 to 1960**

Age	Population 1960 (1)	Population 1950 (2)	Increase Amount (1)-(2)= (3)	Increase Percent[1] [(3)÷(2)]x100= (4)
Total............	179,323,175	151,325,798	27,997,377	18.5
Under 5 years........	20,320,901	16,243,141	4,077,760	25.1
5 to 9 years..........	18,691,780	13,262,123	5,429,657	40.9
10 to 14 years.......	16,773,492	11,167,478	5,606,014	50.2
15 to 19 years.......	13,219,243	10,671,321	2,547,922	23.9
20 to 24 years.......	10,800,761	11,549,355	-748,594	-6.5
25 to 29 years.......	10,869,124	12,305,951	-1,436,827	-11.7
30 to 34 years.......	11,949,186	11,572,337	376,849	3.3
35 to 44 years.......	24,081,352	21,535,149	2,546,203	11.8
45 to 54 years.......	20,485,439	17,397,358	3,088,081	17.8
55 to 64 years.......	15,572,317	13,326,887	2,245,430	16.8
65 years and over....	16,559,580	12,294,698	4,264,882	34.7

---

[1] A minus (—) sign denotes a decrease.

Source: *U.S. Census of Population: 1960*, Vol. 1, *Characteristics of the Population*, Part 1, *United States Summary*, 1964, table 47.

Table 8–14.—Calculation of Index of Relative Difference and Index of Dissimilarity of Age Distributions for Sweden and Ghana Compared with the United States

Age	United States (1960) Percent of total (r_{1a}) (1)	Sweden (1960) Percent of total (r_{2a}) (2)	Sweden (1960) Index $[(2) \div (1)] \times 100 =$ (3)	Sweden (1960) Difference from United States, 1960 $(2) - (1) =$ (4)	Ghana (1960) Percent of total (r_{2a}) (5)	Ghana (1960) Index $[(5) \div (1)] \times 100 =$ (6)	Ghana (1960) Difference from United States, 1960 $(5) - (1) =$ (7)
Total........................	100.00	100.00	100.00	–	100.00	100.00	–
Under 5 years.................	11.33	6.74	59.49	−4.59	19.28	170.17	+7.95
5 to 14 years.................	19.78	15.26	77.15	−4.52	25.27	127.76	+5.49
15 to 24 years................	13.39	14.15	105.68	+0.76	16.82	125.62	+3.43
25 to 34 years................	12.72	12.08	94.97	−0.64	15.96	125.47	+3.24
35 to 44 years................	13.43	14.27	106.25	+0.84	10.24	76.25	−3.19
45 to 54 years................	11.42	14.08	123.29	+2.66	5.90	51.66	−5.52
55 to 64 years................	8.68	11.45	131.91	+2.77	3.35	38.59	−5.33
65 to 74 years................	6.13	7.73	126.10	+1.60	1.75	28.55	−4.38
75 years and over.............	3.10	4.24	136.77	+1.14	1.43	46.13	−1.67
(1) Sum of percent differences without regard to sign = $\Sigma /$Index − 100$/$.......			198.39			407.84	
(2) Mean percent difference = $\left(\dfrac{\Sigma /\text{Index} - 100/}{9}\right)$			22.0			45.3	
(3) Index of relative difference = Half of mean percent difference = (2) ÷ 2...			11.0			22.7	
(4) Sum of absolute differences without regard to sign = $\Sigma /r_{2a} - r_{1a}/$........				19.52			40.20
(5) Index of dissimilarity = Half of summed absolute differences = $1/2 \; \Sigma /r_{2a} - r_{1a}/$....................				9.8			20.1

— Represents zero.

Source: Basic data from, United Nations, *Demographic Yearbook*, and *U.S. Census of Population, 1960*.

Percent Changes by Age. – An important phase of the analysis of age data relates to the measurement of changes over time. Most of the methods of description and analysis of age data to be considered below are applicable not only to the comparison of different populations but also to the comparison of the same population at different dates.

The simplest measure of change by age is given by the amount and percent of change at each age. Table 8–13 shows the amounts and percents of change for the U.S. population for 5-year age groups between 1950 and 1960.

Use of Indexes. – Comparison between two percentage age distributions is facilitated by calculating indexes for each age group or overall indexes for the distributions. Age distributions for different areas, for population subgroups in a single area, and for the same area at different dates may be compared in this way.

Index of relative difference. – The magnitude of the differences between any two age distributions, whether for different areas, dates, or population subgroups, may be summarized in single indexes from the individual age-specific indexes as defined above, i.e., the index of relative difference and the index of dissimilarity. In the former procedure, (1) the deviations of the age-specific indexes from 100 are summed without regard to sign, (2) one-*n*th (*n* representing the number of age groups) of the sum is taken to derive the mean of the percent differences at each age, and (3) the result in (2) is divided by 2 to obtain the index of relative difference. The formula is:

$$IRD = \frac{1}{2} \times \frac{\sum \left| \left(\dfrac{r_{2a}}{r_{1a}} \times 100 \right) - 100 \right|}{n} \qquad (11)$$

To reduce the likelihood of very large percent differences at the oldest ages, which are given equal weight in the average, a broad terminal age group should be used. The procedure is illustrated in table 8–14 with the calculation of the index of relative difference between the age distribution of the United States and those of Sweden and Ghana around 1960.

Index of dissimilarity. – Another summary measure of the difference between two age distributions – the index of dissimilarity – is based on the absolute differences between the percents at each age. In this procedure, the differences between the percents for corresponding age groups are determined, they are summed without regard to sign, and one-half of the sum is taken.[30] (Taking one-half the sum of the absolute differences is equivalent to taking the sum of the positive differences or the sum of the negative differences.) The general formula is then:

$$ID = \frac{1}{2} \sum \left| r_{2a} - r_{1a} \right| \qquad (12)$$

It should be apparent that the magnitude of these indexes is affected by the number of age classes in the distribution as well as by the size of the differences and, hence, that the results are of greatest value in comparison with similarly computed indexes for other populations.

---

[30] See O. Dudley Duncan, "Residential Segregation and Social Differentiation," *International Population Conference, Vienna, 1959*, International Union for the Scientific Study of Population, Vienna, 1959, pp. 571–572; or O. Dudley Duncan and Beverly Duncan, "Residential Distribution and Occupational Stratification," *American Journal of Sociology*, 60(5):494, March 1955.

Median Age. — The analysis of age distributions may be carried further by computing some measure of central tendency. The choice of measure of central tendency of a distribution depends, in general, on the logic of employing one or another measure, the form of the distribution, the arithmetic problems of applying one or another measure, and the extent to which the measure is sensitive to variations in the distribution. The most appropriate measure of central tendency for an age distribution is the median. The median age may be defined as the age which divides the population into two equal-size groups, one of which is younger and the other of which is older than the median. It corresponds to the 50-percentile mark in the distribution. The median age must not be thought of as a point of concentration in age distributions of the general population, however.

The arithmetic mean may also be considered. The use of an arithmetic mean as a measure of central tendency for age distributions is less appropriate than the median both because of the marked skewness of the distribution and because the arithmetic mean may be viewed as a less meaningful way of describing the age distribution of the general population. In addition, the calculation of the arithmetic mean is often complicated by the fact that many age distributions end with open-end intervals, such as 65 and over or 75 and over. Because calculation of the mean takes account of the entire distribution, however, it is more sensitive to variations in it.

Because the general form of the age distribution of the general population (i.e., reverse logistic and right skewness) appears also in many other important types of demographic distributions (e.g., families by size, births and birth rates by birth order, birth rates by age for married women, age of the population enrolled in school, age of the single population), the median is commonly used as a summarizing measure of central tendency in demographic analysis.

Because of the importance of the median in demographic analysis, it is desirable to review here the method of computing it. The formula for computing the median age from grouped data, as well as for computing the median of any continuous quantitative variable from grouped data,[31] may be given as:

$$Md = l_{Md} + \left(\frac{\frac{N}{2} - \sum f_x}{f_{Md}} \right) i \qquad (13)$$

where l_{Md} = the lower limit of the class containing the middle, or $N/2^{th}$, item; N = the sum of all the frequencies; $\sum f_x$ = the sum of the frequencies in all the classes preceding the class containing the $N/2^{th}$ item; f_{Md} = frequency of the class containing the $N/2^{th}$ item; and i = size of the class interval containing the $N/2^{th}$ item. If there is a category of unknown age, N would exclude the frequencies of this class. We may illustrate the application of the formula by computing the median age of the population of India in 1961, using the following data:

Age (years)	Population[a] (in thousands)	Age (years)	Population[a] (in thousands)
Total........	438,775	50 to 54.........	17,105
0 to 4.........	66,082	55 to 59.........	9,829
5 to 9.........	64,652	60 to 64.........	11,236
10 to 14......	49,285	65 to 69.........	4,850
15 to 19......	35,866	70 to 74.........	4,414
20 to 24......	37,316	75 and over.....	4,204
25 to 29......	36,568	Unknown age..	176
30 to 34......	30,830		
35 to 39......	25,455	0 to 14..........	180,019
40 to 44......	22,852	15 to 64.........	245,112
45 to 49......	18,053	65 and over.....	13,468

[a] Source: United Nations, *Demographic Yearbook, 1965*, table 6.

The $\frac{N}{2}$th, or "middle," person falls in the class interval 20 to 24 years. The formula may be evaluated as follows:

$$Md = 20.0 + \left(\frac{\frac{438,599}{2} - 215,885}{37,316} \right) 5$$

$$= 20.0 + \left(\frac{3,415}{37,316} \right) 5$$

$$= 20.0 + (.0915)5 = 20.0 + 0.5$$

$$= 20.5$$

Medians are regularly shown for the principal age distributions published in the decennial census reports of the U.S. Bureau of the Census, but this is not common practice in national census volumes elsewhere.

Measures of Old and of Aging Populations. — The **median age** is often used as a basis for describing a population as "young" or "old" or as "aging" or "younging" (i.e., "growing younger"). An examination of the medians for a wide variety of countries around 1960 suggests a current range from 16 years to 36 years (table 8–15). Populations with medians under 20 may be described as "young," those with medians 30 or over as "old," and those with medians 20 to 29 as of "intermediate" age. Syria (17.2 years) and Honduras (16.1) are in the first category. Sweden (36.2) and France (33.3) are in the second, and India (20.5), the U.S.S.R. (26.6), and Chile (23.3) are in the third. The U.S. population, with a median age of 29.5 years in 1960, is among the populations of intermediate age. When the median age rises, the population may be said to be "aging," and when it falls, the population may be said to be "younging."

The **proportion of aged persons** has also been regarded as an indicator of a young or old population and of a population that is aging or younging (table 8–15). On this basis, populations with 10.0 percent or more 65 years old and over may be said to be old (e.g., France, 12.1 percent, and Sweden, 12.0 percent) and those with under 5.0 percent may be said to be young (e.g., Honduras, 2.4 percent, and Syria, 4.7 percent). The U.S.S.R. had 6.2 percent. India had only 3.1 percent and Chile only 4.3 percent. The examples of India and Chile reflect the fact that the degree of "youth" or "age" depends to some extent on the measure employed and the classification categories of that measure. A still different indication of the degree to which a population is old or young and is aging or younging is given by the **proportion of children under 15.** Again, let us suggest some "round" limits for the proportion under 15 for characterizing a

---

[31] A continuous quantitative variable is a quantitative variable which may assume values at any point on the numerical scale within the whole range of the variable (e.g., age, income, birth weight). This type of variable should be distinguished from discontinuous, or discrete, quantitative variables, which may assume only integral values within the range of the variable (e.g., size of family, order of birth, children ever born).

Table 8-15.—**Summary Measures of Age Composition, for Various Countries: Around 1960**

Country and year	Median age	Percent of total population		Ratio of aged persons to children (per 100)
		Under 15 years	65 years and over	
	(1)	(2)	(3)	(4)
Chile (1960).....................	23.3	39.6	4.3	10.8
France (1961)...................	33.3	25.4	12.1	47.8
Ghana (1960)...................	18.4	44.6	3.2	7.1
Honduras (1961)...............	16.1	47.8	2.4	5.1
India (1961).....................	20.5	41.0	3.1	7.5
Iran (1966)......................	17.3	46.1	3.9	8.4
Italy (1961).....................	31.6	24.6	9.5	38.8
Japan (1960)....................	25.6	30.0	5.7	19.1
Sweden (1960).................	36.2	22.0	12.0	54.4
Syria (1960)....................	17.2	46.3	4.7	10.1
Taiwan (1956)...................	17.9	44.2	2.5	5.6
U.S.S.R. (1959)................	26.6	30.4	6.2	20.5
United Arab Republic (1960)...	19.4	42.7	3.5	8.1
United States (1960)..........	29.5	31.1	9.2	29.7
Venezuela (1961)..............	17.8	44.8	2.8	6.2
Yugoslavia (1961).............	26.4	31.5	6.1	19.3

Source: Basic data from, United Nations, *Demographic Yearbook, 1962*, table 5; and *1964*, table 5.

population as young or old: under 30 percent as old (e.g., France, 25.4 percent, and Sweden, 22.0 percent) and 40 percent and over as young (e.g., India, 41.0 percent, Syria, 46.3 percent, and Honduras, 47.8 percent). The U.S.S.R. (30.4 percent) and Chile (39.6 percent) fall, respectively, just at the lower and upper limits of the intermediate category.

A fourth measure, the **ratio of the number of elderly persons to the number of children,** or the **aged-child ratio,** takes into account the numbers and changes at both ends of the age distribution simultaneously. It may be represented by the following formula:

$$\frac{P^{65+}}{P_{0-14}} \times 100 \qquad (14)$$

For India in 1961, the value of this measure is:

$$\frac{13,468,000}{180,019,000} \times 100 = 7.5$$

Populations with aged-child ratios under the value 15, like India's, may be described as young (e.g., Honduras, 5.1, Syria 10.1) and populations with aged-child ratios over the value 30 may be described as old (e.g., France, 47.8, and Sweden, 54.4). Chile (10.8) now appears as definitely young and the U.S.S.R. (20.5) as definitely of intermediate age. Many less developed countries have so small a proportion of persons 65 and over and so large a proportion of children under 15 that it seems desirable to broaden the range of the numerator and narrow that of the denominator. If the age groups under 10 and 50 and over are used for India in 1961, the value of this ratio is: (51,638,000 ÷ 130,734,000) × 100, or 39.5. When the aging of a population has progressed rather far, as in the more developed countries today, the aged-child ratio following this definition tends to approximate or even exceed 100. The U.S. figure

for 1960 is 107.0. Of the summary indicators of aging we have mentioned—increase in median age, increase in proportion of aged persons, decrease in porportion of children, and increase in ratio of aged persons to children—the last measure, in one or another variant, is most sensitive to differences or changes in age composition affecting the number or proportion of children and aged persons, and it may be considered the best index of "aging."

The four criteria of aging described may not give a consistent indication as to whether the population is aging or not. Since changes in the median age over some period depend merely on the relative magnitude of the growth rates of the total age segments above and below the initial median age during the period, the median age may hardly change while the proportions of aged persons and of children may both increase or both decrease. Accordingly, a population may in some cases appear to be aging and younging at the same time. A combination of a rise in the proportion 65 and over and a rise in the proportion under 15 would, of course, be accompanied by a decline in the proportion in the intermediate ages.

Aging of a population should be distinguished from the aging of individuals, an increase in the longevity of individuals, or an increase in the average length of life pertaining to a population. The latter two types of changes result from improvements in the quality of the environment and from medical advances among other factors. The aging of a population is a characteristic of an age distribution and is importantly affected by the trend of the birth rate.

Age Dependency Ratios.—The variations in the proportions of children, aged persons, and persons of "working age" are taken account of jointly in the so-called age dependency ratio. The age dependency ratio represents the ratio of the combined child population and aged population to the population of intermediate age. One formula for the age dependency ratio useful for international comparisons relates the number of persons under 15 and 65 and over to the number 15 to 64: [32]

$$\frac{P_{0-14} + P_{65+}}{P_{15-64}} \times 100 \qquad (15)$$

Applying the formula to the data for India in 1961, we have:

$$\frac{180,019,000 + 13,468,000}{245,112,000} \times 100 = 78.9$$

Separate calculation of the child-dependency ratio, or the component of the age dependency ratio representing children under 15 (i.e., the ratio of children under 15 to persons 15 to 64), and the old-age dependency ratio, or the component representing persons 65 and over (i.e., the ratio of persons 65 and over to persons 15 to 64), gives values of 73.4 and 5.5 (table 8-16). The corresponding figures for the total-, child-, and aged- dependency ratios for France in 1961 are 60.1, 40.6, and 19.4. As suggested by the figures for India and France, differences (and changes) in age dependency ratios reflect largely differences (and changes) in the proportion of the population under 15 rather than in the proportion of the population 65 and over.

---

[32] An alternative formula employs the population under 18 for child dependents and the population 18 to 64 for adults of working age. This formula is more applicable to the more developed countries where entry into the work force typically comes late. Still another formula employs the population 60 and over for the adult dependents and the population 15 to 59 for adults of working age.

Table 8-16.—**Age Dependency Ratios for Various Countries: Around 1960**

[Ratios per 100]

Country and year	Total dependency ratio	Child dependency ratio	Aged dependency ratio
Chile (1960).....................	78.3	70.7	7.7
France (1961).....................	60.1	40.6	19.4
Ghana (1960).....................	91.3	85.2	6.1
Honduras (1961).....................	101.1	96.1	4.9
India (1961).....................	78.9	73.4	5.5
Iran (1966).....................	99.8	92.1	7.7
Italy (1961).....................	51.7	37.3	14.5
Japan (1960).....................	55.7	46.8	8.9
Sweden (1960).....................	51.4	33.3	18.1
Syria (1960).....................	104.1	94.5	9.6
Taiwan (1956).....................	87.4	82.8	4.6
U.S.S.R. (1959).....................	57.8	48.0	9.8
United Arab Republic (1960)......	86.0	79.5	6.5
United States (1960).............	67.6	52.1	15.5
Venezuela (1961).................	90.7	85.4	5.3
Yugoslavia (1961)................	60.2	50.4	9.7

Source: Basic data from, United Nations, *Demographic Yearbook, 1962*, table 5; and *1964*, table 5.

Age dependency ratios for a number of countries around 1960 are shown in table 8-16. Some ratios exceed 100 (e.g., Honduras, 101; Syria, 104); others are only about 50 (e.g., Italy, 52; Sweden, 51). These figures reflect the great differences from country to country in the burden of dependency which the productive population must bear—differences which are principally related to differences in the proportion of children and hence to differences in fertility rates.

Variations in the age dependency ratio reflect in a general way the contribution of variations in age composition to variations in economic dependency. The age dependency ratio is a measure of age composition, not of economic dependency, however. The economic dependency ratio may be defined as the ratio of the economically inactive population to the active population over all ages, or of nonworkers to workers (ch. 12).

Special Graphic Measures

In this section are described a few graphic measures which are particularly applicable to the analysis of age composition, supplementing those previously illustrated in this and earlier chapters.

Time Series Charts.—The first, called the **one-hundred percent surface chart,** is employed in depicting temporal changes in percentage age composition.[33] Figure 8-1 shows the change in the percent distribution of the population in broad age groups for the United States from 1900 to 1960.

Population Pyramid.—A very effective and quite widely used method of graphically depicting the age-sex composition of a population is called a population pyramid. A population pyramid is designed to give a detailed picture of the age-sex structure of a population, indicating either single ages, 5-year groups, or other age combinations. The basic pyramid form consists of bars, representing age groups in ascending order from the lowest to the highest, pyramided horizontally on one another (see fig. 8-2). The bars for males are given on the left of a central vertical axis and the bars for females on the right of the axis.

---

[33] See Calvin F. Schmid, *Handbook of Graphic Presentation*, New York, The Ronald Press, 1954, pp. 71–72.

Figure 8-1.—**100-Percent Surface Chart Showing Percent Distribution of the Population by Broad Age Groups for the United States: 1900 to 1960**

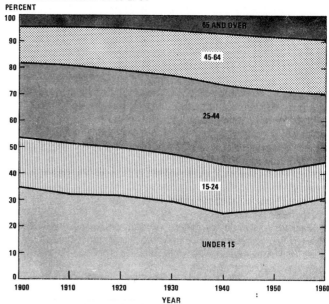

Source: *U.S. Census of Population, 1960*, Vol. 1, Part 1, table 47.

The number of males or females in the particular age group is indicated by the length of the bars from the central axis. The age scale is usually shown straddling the central axis although it may be shown at the right or left of the pyramid only, or both on the right and left, perhaps in terms of both age and year of birth. In general, the age groups in a given pyramid must have the same class interval and must be represented by bars of equal thickness. Most commonly pyramids show 5-year age groups.

Figure 8-2.—**Population Pyramid for Japan: 1960**

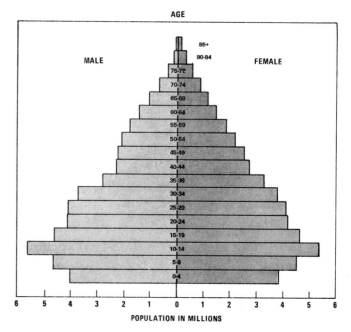

Source: United Nations, *Demographic Yearbook, 1964*, table 4.

A special problem is presented in the handling of the oldest age groups. If data are available for the oldest age groups in the standard class interval (e.g., 5-year age groups) until the end of the life span, the upper section of the pyramid would have an elongated needlelike form and convey little information for the space required. On the other hand, the bar for a broad terminal group is not normally used because it would not ordinarily be visually comparable with the bars for the other age groups. For this reason, pyramids are usually truncated at an age group where the data begin to run thin, e.g., 70 to 74 years, 75 to 79 years, or 80 to 84 years.

Pyramids may be constructed on the basis of either absolute numbers or percents. A special caution to be observed in constructing a "percent" pyramid is to be sure to calculate the percents on the basis of the grand total for the population, including both sexes and all ages (i.e., including the population in a terminal age group not shown in the pyramid). A "percent" pyramid is similar, in the geometric sense of the word, to the corresponding "absolute" pyramid. With an appropriate selection of scales, the two pyramids are identical. The choice of one or the other type of pyramid is more important when pyramids for different dates, areas, or subpopulations are to be compared. Only absolute pyramids can show the differences or changes in the overall size of the total population and in the numbers at each age. Percent pyramids show the differences or changes in the proportional size of each age-sex group. In general, absolute or percent pyramids to be compared should be drawn with the same horizontal scale and with bars of the same thickness. In any case, to minimize visual distortion, for pyramids being compared with respect to their general configuration, the ratio of the thickness of a bar to the distance for a unit amount or one percent on the horizontal scale should be the same.

Comparisons between pyramids for the same area at different dates and between pyramids for different areas or subpopulations may be facilitated by superimposing one pyramid on another either entirely or partly. The pyramids may be distinguished by use of different colors or cross-hatching schemes. Occasionally in absolute pyramids and invariably in percent pyramids, the relative length of the bars in the two superimposed pyramids reverses at some ages. The drafting then becomes more complicated. For example, if one pyramid is to be drawn exactly over another and if the first pyramid is shown entirely in one color or cross-hatching scheme, then the parts of the bars in the second pyramid extending beyond the bars for the first pyramid would be shown in a second color or cross-hatching scheme, and the parts of the bars in the first pyramid extending beyond the bars for the second pyramid would be shown in a third color or cross-hatching scheme (fig. 8–3). An alternative design is to show the second pyramid wholly or partly off-set from the first one. In this design the first pyramid is presented in the conventional way except that the bars are separated from age to age. The second pyramid is drawn partially superimposed on the first, using the space between the bars wholly or in part.

Any characteristic which varies by age and sex (e.g., marital status or urban-rural residence) may be added to a general population pyramid, to develop a pyramid which reflects the age-sex distribution of both the general population and the population of the categories of the characteristic (fig. 8–4). Where additional characteristics beyond age and sex are included in the pyramid, the principles of construction are essentially the same. The bar for each age is subdivided into parts representing each category of the characteristic (e.g., single, married, widowed, divorced; urban, rural). It is important that each category shown separately occupy the same position in

Figure 8–3. — Section of the Pyramid for the Population of the United States: 1950 and 1960

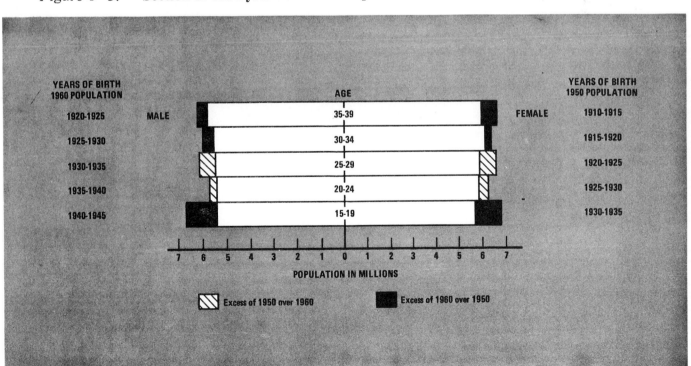

Source: Table 8–13.

every bar relative to the central axis and to the other category or categories shown. Again, if percents are used, all percents are calculated on a single base, the total population. Various cross-hatching schemes or coloring schemes may be used to distinguish the various categories of the characteristic represented in the pyramid. When characteristics are added to a population pyramid, the age-sex distribution is shown most clearly for the innermost category in the pyramid and for the total population covered; the distribution of the other categories is harder to interpret. Population pyramids may also be employed to depict the age-sex distribution of demographic events, such as deaths, marriages, divorces, and migration, during some period.

Pyramids may be analyzed and compared in terms of such characteristics as the relative magnitude of the area on each side of the central axis of the pyramid (the symmetry of the pyramid) or a part of it, the length of a bar or group of bars in relation to adjacent bars, and the steepness and regularity of the slope. (A pyramid may be described as having a steep slope when the sides of the pyramid recede very gradually and rise fairly vertically, and a gentle slope when the sides recede rapidly.) These characteristics of pyramids reflect, respectively, the proportion of the sexes, the proportion of the population in any particular age class or classes, and the general age structure of the population.

Populations with rather different age-sex structures are illustrated by the several pyramids shown in figure 8–5. The pyramid for Costa Rica (1963) has a very broad base and narrows very rapidly. This pyramid illustrates the case of an age-sex structure with a very large proportion of children, a very small proportion of elderly persons, and a low median age, i.e., a relatively "young" population. The pyramid for Sweden (1960) has a relatively narrow base and a middle section of nearly the same dimensions; it does not begin to converge to the vertex until after age 55. This pyramid illustrates the case of an age-sex structure with a very small proportion of children, a very large proportion of elderly persons, and a high median age, i.e., a relatively "old" population. The pyramids for India (1961) and the United States (1960) illustrate configurations inter-mediate between those for Costa Rica and Sweden. The pyramid for the population of France given in figure 8-6 reflects various irregularities associated with that country's special history.

The pyramids of geographically very small countries and of subgroups of national populations — geographic subdivisions or socioeconomic classes — may have quite different configurations, i.e., they may vary considerably from the relatively smooth triangular and semi-elliptical shapes we have identified. For example, the pyramid for the foreign-born population of Hong Kong in 1961 has the general shape of a top (fig. 8–7). It has an extremely narrow base (i.e., a trivial percentage of children), a considerable bulge in the upper middle section (i.e., a high percentage of older adults), and a substantial asymmetry (in this case, a large excess of males). The age-sex pyramids of the married population, the labor force, heads of households, etc., have their characteristic configurations.

Analysis of Age Composition in Terms of Demographic Factors

Amount and Percent of Change by Age. — In this section we extend the analysis of age composition to consider in a preliminary manner the role of the factors of birth, mortality, and net immigration. These factors all operate on the population in an age-selective fashion. Births in a given year directly determine the size of the population under one year old at the end of that year, and because of the nature of the birth component and its magnitude relative to the other components, it is also often the principal determinant of the size of older age groups in the appropriate later years. The deaths and migrants of a given year affect the entire distribution in that year directly, although deaths are concentrated among young children and aged persons and there is usually a disproportionately large number of young adults among migrants.

Number in age group. — The number of persons in a given age group at a census date, and changes in the numbers between census dates by age, may be analyzed in terms of the past numbers of births, deaths, and "net immigrants." The number of persons in a given age group, x to $x+4$ years of age, at a given date represents the balance of the number of births occurring x to $x+4$ years earlier in the area, the number of deaths occurring to this cohort between the years of birth and the census date, and the number of migrants entering or leaving in this period with ages corresponding to this cohort. Any analysis of the factors underlying the census figures must also take into consideration the net undercount of the census figures. We may represent this relationship as follows:

$$P_a = B - D + M - E_a \qquad (16)$$

where P_a is the population in the age group; B, births in the cohort; D, deaths occurring to the cohort; M, net migrants joining the cohort; and E_a, net undercount in the census figure. For purposes of the present analysis, we shall assume that there are no errors in the data on the components or that they have been satisfactorily corrected.

Generally, the younger the age group the smaller the impact of death and migration in determining the size of the group and the greater the relative importance of births. Thus, the size of the population under five years of age on April 1, 1960, in the United States is very largely determined by the number of births occurring between April 1, 1955, and April 1, 1960; this number is modified only by the relatively small number of deaths occurring to these newborn children during the period and by the relatively small number of immigrant children in this cohort entering the country in the period. The net undercount of the census figure increases the potential gap between

Figure 8–4. — **Percent Distribution of the Population of Ghana by Urban-Rural Residence, Age, and Sex: 1960**

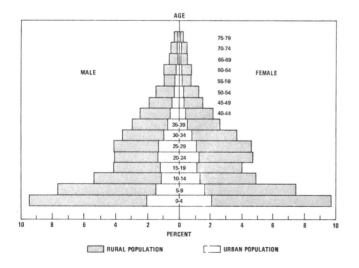

Source: United Nations, *Demographic Yearbook, 1963,* table 5.

Figure 8–5. — Percent Distribution by Age and Sex of the Populations of Sweden, United States, India, and Costa Rica: Around 1960

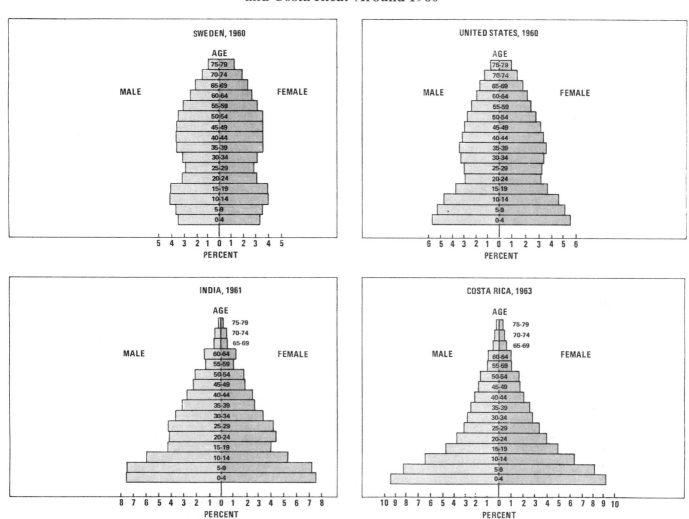

Source: Table 8–13 and United Nations, *Demographic Yearbook, 1964 and 1961*, table 5.

the births and the census figure somewhat further. Similarly, the population 5 to 9 years old on April 1, 1960, represents the births of April 1, 1950, to April 1, 1955, modified by deaths and migration up to April 1, 1960; the population 10 to 14 years old stems from the births of April 1, 1945, to April 1, 1950, etc.

The paramount importance of the past trend of births also applies generally in interpreting short-term changes in the number of aged persons. For example, the increase of 4.3 million, or 35 percent, in the population 65 and over in the United States between 1950 and 1960 may be explained largely in terms of the rise in the number of births between 1865–75 and 1885–95.[34] Since the population 65 and over

is largely comprised of persons 65 to 84, we have used these ages to determine the birth years of the two groups of cohorts and then eliminated the years common to them. Once again, in spite of the considerable importance of deaths in determining the number of aged persons at each census date, the factor of mortality tends to be relatively secondary in explaining the change in the number of aged persons over a period as short as 10 years. As stated earlier, both groups of cohorts being compared have been exposed to the risk of death for the same number of years, albeit the considerable period of 65 to 84 years, and although some improvement in mortality may have occurred as the two groups of cohorts moved through life from infancy to old age 10 calendar years apart, this improvement was much less important than the fact that the number of years of exposure was the same. Because mortality rates in the United States have, in fact, declined less rapidly at the older ages than at the younger ages in the last century, declines in mortality rates have contributed relatively less to increases in the aged population than to increases at the younger ages.

[34] The reference here is specifically to the number of births rather than to the birth rate. The absolute number of births may be rising while the birth rate may be falling because the total population is growing. For example, the U.S. birth rate decreased generally in the period 1865 to 1915, but the number of births increased rapidly during this period. These are the cohorts which have been entering, or will soon enter, the 65-and-over group.

Figure 8-6. — **Population of France, by Age and Sex: January 1, 1967**

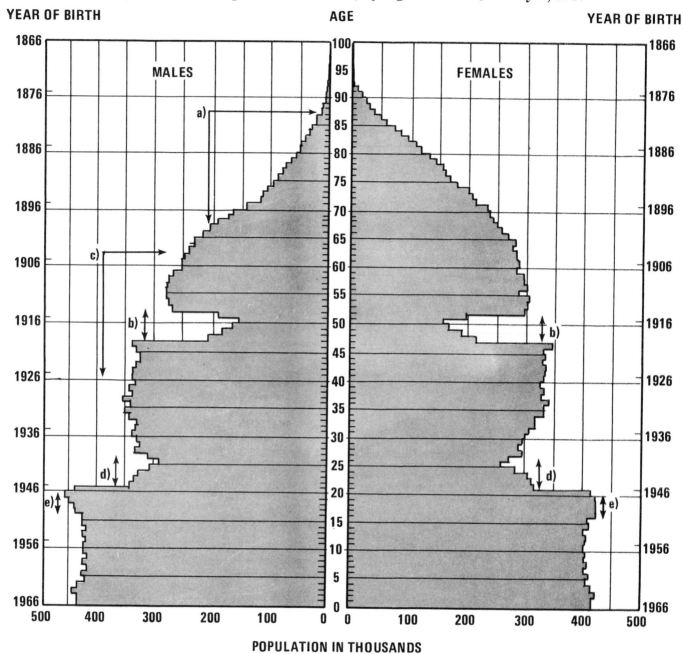

(a) Military losses in World War I
(b) Deficit of births during World War I
(c) Military losses in World War II
(d) Deficit of births during World War II
(e) Rise of births due to demobilization after World War II

Source: Based on France, Institut national de la statistique et des études économiques *Annuaire statistique de la France, 1967,* 1968, p. 33.

Figure 8-7. — **Total and Foreign-Born Populations of Hong Kong, by Age and Sex: 1961**

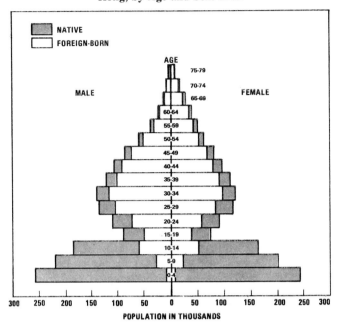

Source: *Hong Kong Report on the 1961 Census*, Vol. II, tables 104 and 150.

The contribution of migration to the change in the aged population may be analyzed to some extent in the same way as the contribution of mortality. The two groups of cohorts have been "exposed to the risk" of migration for the same period of years (65 or more), but there may have been sudden shifts from year to year in the number of migrants affecting the two groups of cohorts.

The age profile may be strongly affected by the events of war and postwar demobilization, as well as by peacetime mobilization.[35] The war may be directly responsible for a temporary or semi-permanent deficit of males of military age in the resident population, an increase in the death rate, a rise or fall in the birth rate, and displacements of civilian populations. To the extent that there are battle losses of military personnel, some of the temporary deficit of males of military age will become permanent and may remain visible not only through the life span of the particular cohorts concerned but may also be reflected in later cohorts through the effect on the birth rate. To the extent that the civilian population is directly involved in military activities and there are war losses of civilians (e.g., from bombing), the impact of the war will be more evenly distributed over the age-sex classes.

The nature and extent of the effect of the war on the birth rate will depend in part on the duration and magnitude of the war, the degree to which families are separated by military service or destruction of civilian communities, and the impact of the war on the economy.[36] Although the birth rate may rise

[35] See Philip M. Hauser, "Population and Vital Phenomena," *American Journal of Sociology*, 48(3):309–322, November 1942; Metropolitan Life Insurance Co., "The Cumulative Effect of Successive Wars on Age Composition of Population," *Statistical Bulletin*, 21(4):2–5, April 1940; and idem, "Post-War Depletion of Ranks of Men at Marrying Ages," *Statistical Bulletin*, 21(2):1–4, February 1940; T. Lynn Smith, *Fundamentals of Population Study*, New York, J. B. Lippincott, 1960, pp. 157–159.

[36] Metropolitan Life Insurance Company, "War and the Birth Rate," *Statistical Bulletin*, 21(3):3–6, March 1940.

or fall during wartime, it is especially likely to rise during the immediate postwar demobilization period because of the reunion of married couples and a rise in the number of marriages. The sharp fluctuations in the birth rate usually associated with war tend to produce marked variations in the size of successive age groups in a population. These "hollow" and "swollen" age groups will grow older with the passage of time and reappear as hollow and swollen groups at successively older ages at various subsequent dates, possibly for decades to come (fig. 8–6).

As a direct result of war the volume of ordinary civilian immigration or emigration is likely to fall off to a mere trickle while movements of military personnel into and out of the country and of refugees and expellees may be voluminous and erratic. With the end of the war, massive demobilization of military personnel or displacements of population may account for a tremendous rise in the volume of movement into or out of the belligerent countries. The changes described may have a considerable impact on the age profile of the population as well as its sex composition.

Percent of change. — We turn next to a formulation of the contribution of each component of change to the total percentage change in an age group between two dates. We may represent the population at a given age in an earlier (P^0) and later (P^1) census in terms of its underlying components as follows:

$$P_a^0 = B^0 - D^0 + M^0 - E_a^0 \qquad (17)$$

$$P_a^1 = B^1 - D^1 + M^1 - E_a^1 \qquad (18)$$

Note that all the elements in each equation relate to the same birth cohorts. (These equations are of the same form as equation (16).) The relative contribution of each component to the total change in population is obtained by taking the difference between these equations, grouping components, and dividing each side by: P_a^0:

$$\frac{P_a^1 - P_a^0}{P_a^0} = \frac{B^1 - B^0}{P_a^0} - \frac{D^1 - D^0}{P_a^0} + \frac{M^1 - M^0}{P_a^0} - \frac{E_a^1 - E_a^0}{P_a^0} \qquad (19)$$

Each term here, multiplied by 100, represents the contribution of a component in percentage points to the total percentage change.

Table 8–17 illustrates the application of this formula in analyzing the increase in the U.S. population for selected age groups between 1950 and 1960. First, determine the numbers of births, deaths, and net immigrants, and the net undercount pertaining to each cohort (cols. 1 and 2); second, take the difference between the amounts for the two cohorts (col. 3); and, third, divide the difference by the population in the age group at the earlier census and multiply by 100 (col. 6). According to this calculation, of the 41 percent increase in the 5 to 9 age group, the increase in births contributed 39 percentage points, the decrease in deaths 0.6 percentage point, the increase in migrants 1.0, and the reduction in net undercount 0.2.

If only the contribution of births to total percentage change is to be measured, or can be measured because data on births alone can be developed, it is necessary simply to (1) determine the numbers of births corresponding to each cohort from the registration data or by estimation, (2) subtract the births of each cohort from the corresponding census counts, to obtain estimates for all other components combined, including the

Table 8–17.—Calculation of the Contribution of the Components of Change to the Percent Change for Selected Age Groups in the Population of the United States Between 1950 and 1960

[Numbers in thousands]

Age and component	Number[1]		Change		Weight	Contribution of each component
	1950[2]	1960	Amount	Percent		$[(3) \div P_a^{50}] \times 100$ or $(4) \times (5) =$
			$(2)-(1)=$	$[(3) \div (1)] \times 100 =$	$(1) \div P_a^{50} =$	
	(1)	(2)	(3)	(4)	(5)	(6)
Under 5 years:						
Births..........................	17,622	21,345	3,723	21.13	1.085	22.9
Deaths[3].......................	-615	-598	17	-2.76	-0.038	0.1
Net migration..................	42	73	31	73.81	0.003	0.2
Net undercount.................	-806	-498	308	-38.21	-0.050	1.9
Population or total change.....	16,243	20,321	4,078	25.11	(1.000)	25.1
5 to 9 years:						
Births..........................	14,445	19,618	5,173	35.81	1.089	39.0
Deaths[3].......................	-742	-658	84	-11.32	-0.056	0.6
Net migration..................	59	211	152	257.63	0.004	1.1
Net undercount.................	-500	-480	20	-4.00	-0.038	0.2
Population or total change.....	13,262	18,692	5,430	40.94	(1.000)	40.9
15 to 19 years:						
Births..........................	12,264	14,445	2,181	17.78	1.149	20.4
Other components[3].............	-1,593	-1,226	367	-23.04	-0.149	3.4
Population or total change.....	10,671	13,219	2,548	23.88	(1.000)	23.9
25 to 29 years:						
Births..........................	14,828	12,264	-2,564	-17.29	1.205	-20.8
Other components[3].............	-2,522	-1,395	1,127	-44.69	-0.025	9.2
Population or total change.....	12,306	10,869	-1,437	-11.68	(1.000)	-11.7
35 to 39 years:						
Births..........................	14,353	14,828	475	3.31	1.271	4.2
Other components[3].............	-3,059	-2,347	712	-23.28	-0.271	6.3
Population or total change.....	11,294	12,481	1,187	10.51	(1.000)	10.5
65 to 84 years:						
Births[4].......................	30,726	35,833	5,107	16.62	2.622	43.6
Other components[3].............	-19,009	-20,203	-1,194	6.28	-1.622	-10.2
Population or total change.....	11,717	15,630	3,913	33.40	(1.000)	33.4

[1] Each figure includes data for Alaska and Hawaii.
[2] For 1950 each component was increased by the ratio of the population in the age group indicated for 50 States to the population for 48 States in 1950.
[3] In the present table, as a negative component, "deaths" and "other components" are assigned a negative sign. A positive amount of change in the number of deaths indicates a decrease in the number of deaths, or in effect, an addition to the population.
[4] For the cohorts 65–84 years of age in 1950 and 1960, births 1865–95 were derived on the basis of census data on young children.

Source: Basic data from U.S. Bureau of the Census and U.S. Public Health Service official reports.

net census undercount, and (3) proceed as before (see table 8–17).

An alternative procedure for computing the contribution of the components, which produces the same results, views the overall percent change in the population as the weighted average of the percent changes in the components. The percent change in each component is shown in column 4 of table 8–17. The weight attached to each component is derived by taking the ratio of a particular component for the earlier cohort to the census population at the earlier date and is shown in column 5. The contribution of each component in percentage points (col. 6) is obtained by taking the product of the percents in column 4 and the weights in column 5. The weights for births and net immigration are positive, and the weights for deaths and net undercount are negative. In the young ages, the component of births receives a relatively high weight in comparison with the other components. As one goes up the age scale, the **relative** weight of the birth component tends to fall and the **relative** weight of the other components taken together, particularly the death component, tends to rise, even though all the **absolute** weights tend to rise. For example, at ages 0 to 4 the birth component has an absolute weight of 1.085 and the other components combined a weight of (−) .085; but at ages

65 to 84 the birth component has an absolute weight of 2.622 compared with a weight of (−) 1.622 for the other components combined. These figures indicate relative weights for births of 93 percent and 62 percent, respectively.[37]

Changes in Proportions by Age and in Overall Age-Sex Structure. — We turn next to a brief consideration of the demographic factors affecting the changes in the proportion of the population in a given age group between two dates or in the age-sex structure as a whole.

Changes in the proportion of the population in a given age group between two dates depend on the relative rates of growth of the age group in question and of the total population. In evaluating the role of natality in particular, it is useful to compare the percent difference between the numbers of births corresponding to the two cohorts under study with the percent difference in the numbers of births during the century or so prior to each of the two dates. For example, the proportion of the population under 5 years old in the United States increased from 10.7 to 11.3 percent between 1950 and 1960 largely

---

[37] The relative weight of births may be given by $\frac{W_b}{2W_b - 1}$, where W_b represents the absolute weight for births. The value of this function declines with the rise in W_b.

because (1) the relative excess of births occurring in 1955–60 over births occurring in 1945–50 (21 percent) was greater than (2) the relative excess of births in 1860–1960 over births in 1850–1950 (about 11 percent). The ratio of births ratios (1.21 ÷ 1.11) equals 1.09, suggesting a contribution of 9 percent due to births in the increase in the proportion of children under 5 between 1950 and 1960; this proportion actually rose by 6 percent. The increase in the proportion of aged persons in the United States between 1950 and 1960 resulted in large part from (1) the relative excess of births between 1875 and 1895 over births between 1865 and 1885 (when most persons 65 and over in 1960 and 1950, respectively, were born) in comparison with (2) the slower average increase of births during the last century, which is responsible for the moderate rate of growth of the total population in the 1950–60 decade. Thus, the long-time downward trend in the birth rate in the United States contributed to the increase in the proportion of persons in the older ages in recent decades.

A reduction in death rates contributes, as was stated earlier, to an increase in the **number** at an age group over a particular time interval (e.g., 10 years), but whether or not reductions in death rates contribute to an increase in the **proportion** of the population in any age group during such a period depends on the age pattern of improvements in death rates. More specifically, the effect of mortality on changes in proportions by age depends on the difference between (1) relative improvement in survival rates from birth up to the specified age for the periods ending on each census date (e.g., 10 years apart) and (2) relative improvement in survival rates from birth to census age over all age cohorts included in the total population at the census dates. In general, if the savings in lives are confined to childhood, the proportion of children will tend to rise and the proportion of older persons to fall; if the savings in lives are confined to older persons, the proportion of children will tend to fall and the proportion of older persons to rise. Since, in fact, in the experience of the United States as well as of many other countries, much more improvement in mortality has taken place at the younger ages than at the older ages, this factor has contributed little or nothing to the increase in the proportion of the aged.[38]

The role of immigration can be analyzed in a similar way, but the specific impact on the age distribution may be quite different from that of deaths because the migrants have their own characteristic age pattern as well as a natural increase. Migrants change the prevailing age composition of the population to the extent that their age composition plus their natural increase differs from that of the general population. Because immigrants tend to be relatively young on arrival and to have a relatively high natural increase, the usual short-term effect of immigration is to reduce the proportion of older adults and aged persons in the population. This "younging" effect will continue if the volume of immigration is maintained and the new arrivals continue to have a young age distribution. The younging effect will tend to decline, however, if the volume of immigration falls off or if the new arrivals begin having an older age distribution, as occurred generally in the United States in the first half of this century.

Since the median age is the 50-percentile mark in the age distribution, changes in the median age may be analyzed in much the same manner as changes in the proportion at some age. The median age will rise during a period if the population above the median age at the beginning of the period grows at a faster rate than the population below the median (or the total population), and the median age will fall under the opposite conditions. Here again, the factors of mortality and migration tend, typically, to be of considerably less importance than the factor of fertility in explaining a change in the median age.

Analysis of population pyramids.—It is possible to infer a good deal about the demographic history of an area simply by examining the population pyramid depicting its age-sex structure. Because a pyramid does not identify the specific demographic factors which shaped it in a particular case, however, and because more than one factor contributes to the determination of the number of persons at each age, it is not possible to state exactly, especially without a knowledge of the history of the area, how demographic factors operated to give a particular pyramid its special form.

A pyramid with a very broad base suggests a population with a relatively high birth rate, and if, in addition, the pyramid narrows rapidly from the base up—i.e., has a generally triangular shape with concave sides—a combination of high birth rates and high death rates is suggested. The pyramid for Costa Rica in 1963 shown in figure 8–5 reflects such a demographic history. The beehive- or barrel-shaped pyramid for Sweden in 1960 shown in the same chart suggests a history of low fertility and low mortality. Irregular variations in the lengths of the bars point to past sharp fluctuations in the number of births or in the volume of migration, but they may also reflect a temporary rise in the number of deaths resulting from war or epidemic. The pyramid for France in 1967 shown in figure 8–6 has a number of irregularities which may be explained on the basis of the fluctuations in the number of births and deaths caused by World Wars I and II. The pyramid has been annotated to indicate the age bands that have been affected by military losses in World Wars I and II, deficits of births during these wars, and the post-World War II rise in births due to demobilization. If the bars for the youngest ages are shorter than those for the next higher ages, a recent decline in the number of births is suggested (fig. 8–2 for Japan in 1960) although some of the shortage may be due to relatively greater underenumeration of the youngest age groups.

Where the bars in the middle or the upper part of the pyramid are excessively long or short, heavy immigration or emigration of persons in these age groups in recent years, or of younger persons in the same age cohorts at an earlier date is indicated. A pyramid with a mushroom shape (i.e., an extremely narrow base and a wide top) represents a population which is not replenished by births or by recent immigrants, e.g., the foreign-born population of Hong Kong in 1961 shown in figure 8-7.

Analysis of age-sex structure by use of population models.—Earlier we referred to the use of various models of population structure in the evaluation and correction of census data on age-sex composition. These models are also useful in the

---

[38] A number of studies have been concerned with the theoretical and empirical analysis of the effect of mortality and natality trends on age composition. Some are noted here; additional references are given in later chapters, particularly chs. 18 and 25. Some general studies are: Ansley J. Coale, "The Effects of Changes in Mortality and Fertility on Age Composition," *Milbank Memorial Fund Quarterly*, 34(1):79–114, January 1956; United Nations, "The Cause of the Aging of Populations: Declining Mortality or Declining Fertility?," *Population Bulletin of the United Nations*, No. 4, December 1954, pp. 30–38; George J. Stolnitz, "Mortality Declines and Age Distribution," *Milbank Memorial Fund Quarterly*, 34(2):178–215, April 1956. The relative contribution of fertility, mortality, and immigration to changes in the age composition of the U.S. population between 1900 and 1960 is analyzed in: Vasilios G. Valaoras, "Patterns of Aging of Human Populations," *The Social and Biological Challenge of Our Aging Population*, Proceedings of the Eastern States Health Education Conference (1949), New York, Columbia University Press, 1950; and Albert I. Hermalin, "The Effect of Changes in Mortality Rates on Population Growth and Age Distribution in the United States," *Milbank Memorial Fund Quarterly*, 44(4–1):451–469, October 1966. The countries of Latin America are considered in: John V. Grauman, "Effects of Population Trends upon Age Structure, with Application to the Americas," *Estadistica*, 14(51):271–287, June 1956.

analysis of the age-sex structure of populations. Comparisons of the structure of actual populations with the theoretical structure generated by various constant or variable conditions of fertility and mortality represented by these population models permits a fuller understanding of the role of these demographic conditions in age-sex structure. The pyramids of these population models may also be used as graphic standards for interpreting the pyramids of actual populations. It may be recalled that the stable population model indicates the fixed age-sex structure of a population resulting from and consistent with constant birth and death rates (i.e.. a constant growth rate. positive or negative). and no migration (a "closed population"). The quasi-stable population model corresponds to the stable population model except that mortality is assumed to be declining moderately. The stationary population model.

identified previously as a special case of the stable population model. assumes a zero growth rate (i.e.. constant and equal numbers of births and deaths each year and. hence. constant and equal birth and death rates each year). The pyramids of these models are smooth and regular.

Populations with high and relatively unchanging birth rates and with moderately high death rates, a type of population which is not uncommon today, resemble the stable population model with a high growth rate (e.g., fig. 8-5, Costa Rica). Populations with low birth and death rates are likely to resemble the stable model with a low growth rate or a stationary population with low birth and death rates (e.g., fig. 8-5, Sweden).The deviations of the age-sex structure of actual populations from the closely related models will give some indications of past fluctuations in fertility, the role of migration and war mortality, etc.

SUGGESTED READINGS

Part A. Methods of Measuring and Adjusting for Errors in Grouped Age Data, and General Findings

Barclay. George W. *Techniques of Population Analysis.* New York, John Wiley and Sons, 1958. Pp. 65–79.

Carrier, N. H., and Farrag, A. M. "The Reduction of Errors in Census Populations for Statistically Underdeveloped Countries." *Population Studies* (London), 12(3):240–285, esp. pp. 251–252 and 265–266. March 1959.

Das-Gupta, Ajit. "Accuracy Index of Census Age Distributions." In United Nations, *World Population Conference, 1954* (Rome). Vol. IV, pp. 63–74. 1955.

Durand. John D. "Adequacy of Existing Census Statistics for Basic Demographic Research." *Population Studies* (London), 4(2):179–199, esp. pp. 187–194. Sept. 1950.

Japanese National Commission for UNESCO. *An Outlook of Studies on Population Problems in Japan,* III, *Errors in Reporting Ages for Censuses,* by Yuzo Morita. 1959.

Miró. Carmen A. *Algunos Problemas Relativos a la Evaluación de los Resultados de los Censos de Población* (Some Problems Relating to the Evaluation of the Census Data on Population). Series A, No. 6, United Nations Demographic Center for Latin America. Santiago, Chile. 1963.

Myers, Robert J. "Errors and Bias in the Reporting of Ages in Census Data." *Transactions of the Actuarial Society of America,* 41:Part II (104):395–415, esp. pp. 395–401. Oct.–Nov. 1940. (Reproduced in U.S. Bureau of the Census, *Handbook of Statistical Methods for Demographers,* pp. 115–125, 1951.)

Spiegelman, Mortimer. *Introduction to Demography* (rev. ed.). Cambridge, Mass., Harvard University Press, 1968. Pp. 58–68.

United Nations. "Accuracy Tests for Census Age Distributions Tabulated in Five-Year and Ten-Year Groups." *Population Bulletin,* (2):59–79. Oct. 1952.

United Nations. Chapter 3 in *Methods of Appraisal of Quality of Basic Data for Population Estimates.* Manuals on Methods of Estimating Population, Manual II. Series A, Population Studies, No. 23, 1955. Pp. 31–54.

United Nations. *Methods of Evaluating the Reliability of Population and Housing Census Data.* Economic Commission for Africa, Seminar on Organization and Conduct of Censuses of Population and Housing, Addis Ababa, June 17–29, 1968. E/CN.14/CPH/4 or ST/STAT/19, Feb. 28, 1968.

U.S. Bureau of the Census. *Handbook of Statistical Methods for Demographers.* By A. J. Jaffe. 1951. Pp. 86–92 and 94–96.

————. *Infant Enumeration Study: 1950.* Washington, 1953.

Wolfenden, H. H. *Population Statistics and Their Compilation* (rev. ed.). Chicago, Ill., University of Chicago Press, 1954. Pp. 32–39 and 68–76.

Part B. Measurement of Age Preference in Census Data, and Illustrative Studies

Arias B., Jorge. "Algunos errores en la declaración de edad en censos de población de 1950 de Centro América y México" (Some Errors in the Reporting of Age in the 1950 Population Censuses of Central America and Mexico). *Estadística,* 14(52):403–425. Sept. 1956.

Bachi, Roberto. "Measurement of the Tendency to Round off Age Returns." *Bulletin de l'Institut International de Statistique* (Proceedings of the 28th Session, Rome, Sept. 6–12, 1953), 34(3):129–138. 1954.

————. "The Tendency to Round off Age Returns: Measurement and Correction." *Bulletin of the International Statistical Institute,* 33(4):195–221. International Statistical Conferences, India, Dec. 1951.

Barclay, George W. *Techniques of Population Analysis.* New York, John Wiley & Sons, 1958. Pp. 65–70.

SUGGESTED READINGS—Continued

Belarmino, Isagani C. "Some Notes on Population Age Grouping in the Philippines." *Statistical Reporter*, 9(4):1–14. Oct.-Dec. 1965. Office of Statistical Coordination and Standards, National Economic Council, Manila, Philippines.

Carrier, N. H. "A Note on the Measurement of Digital Preference in Age Recordings." *Journal of the Institute of Actuaries* (Cambridge, Eng.), 85:71–85. 1959.

Myers, Robert J. "Accuracy of Age Reporting in the 1950 United States Census." *Journal of the American Statistical Association*, 49(268):826–831. Dec. 1954.

———. "Errors and Bias in the Reporting of Ages in Census Data." *Transactions of the Actuarial Society of America*, 41(104):395–415, esp. pp. 402–415. Part II, Oct.-Nov. 1940.

Ramachandran, K. V. "An Index to Measure Digit Preference Error in Age Data." Summary in United Nations, *World Population Conference, 1965* (Belgrade). Vol. III, pp. 202–203. 1967.

Spiegelman, Mortimer. *Introduction to Demography* (rev. ed.), Cambridge, Mass., Harvard University Press, 1968. Pp. 68–77.

United Nations. *Methods of Appraisal of Quality of Basic Data for Population Estimates*, pp. 33–36 and 40–42. Manuals on Methods of Estimating Population, Manual II. Series A, Population Studies, No. 23. 1955.

U.S. Bureau of the Census. *Handbook of Statistical Methods for Demographers*. By A. J. Jaffe. 1951. Pp. 89–90, 94–100.

Wolfenden, H. H. *Population Statistics and Their Compilation* (rev. ed.). Chicago, Ill., University of Chicago Press, 1954. Pp. 42–53.

Zelnik, Melvin. "Age Heaping in the United States Census: 1880–1950." *Milbank Memorial Fund Quarterly*, 39(3): 540–573. July 1961.

Part C. Illustrative Studies of the Measurement of Errors in Grouped Age Data for Particular Areas, and Findings

Brass, William, *et al. The Demography of Tropical Africa*. Princeton, N.J., Princeton University Press, 1968. Pp. 13–52.

Brazil, IBGE—Conselho Nacional de Estatística. *Análises Críticas de Resultados dos Censos Demográficos* (Critical Analysis of the Results of Population Censuses). Estudos de Estatística Teórica e Aplicada, Estatística Demográfica, No. 21. 1956. Esp. pp. 7–47.

Caldwell, John C. "Study of Age Misstatement Among Young Children in Ghana." *Demography*, 3(2):477-490. 1966.

Coale, Ansley J. "The Population of the United States in 1950 Classified by Age, Sex, and Color—A Revision of Census Figures." *Journal of the American Statistical Association*, 50(269):16–54. Mar. 1955.

Coale, Ansley J., and Zelnik, Melvin. *New Estimates of Fertility and Population in the United States*. Princeton, N.J., Princeton University Press, 1963.

Costa Rica, Dirección General de Estadística y Censos. "Proyección de la población de Costa Rica por sexo y grupos de edad, 1965–1990" (Projection of the Population of Costa Rica by Sex and Age Groups, 1965–1990), by Ricardo Jiménez J. *et al. Revista de Estudios y Estadísticas*, No. 8, Oct. 1967, *Serie Demográfica*, No. 5. Pp. 3–76, esp. pp. 34–35 and 45–57.

El Badry, M. A. "Some Demographic Measurements for Egypt Based on the Stability of Census Age Distributions." *Milbank Memorial Fund Quarterly*, 33(3):268–305, esp. pp. 7–19. July 1955.

Eldridge, Hope T., and Siegel, Jacob S. "The Changing Sex Ratio in the United States." *American Journal of Sociology*, 52(3):224–234. Nov. 1946.

Ghana, Census Office. 1960 Census of Ghana, Vol. V, *Administrative Report, Post Enumeration Survey*, Part II, and *Census Evaluation*, Part III. Pp. 313–401. Accra, 1964.

India, Indian Statistical Institute. "A Technical Note on Age Grouping." *National Sample Survey*, No. 12. Calcutta, India, 1958.

Kappas Barrientos, Hector, and Marks, Eli S. "Evaluation of the Accuracy of the 1960 Censuses of Chile." *Estadística*, 39(72):492–505. Sept. 1961.

Marks, Eli S., and Waksberg, Joseph. "Evaluation of Coverage in the 1960 Census of Population through Case-by-Case Checking." *Proceedings of the Social Statistics Section, 1966*. Washington, D.C., American Statistical Association, 1966. Pp. 62–70.

Morita, Yuzo. "Errors in Ages for Population Statistics." *Archives of the Population Association of Japan* (English Edition). The Population Association of Japan, 1963. Pp. 47–48.

———. "The Accuracy of Age-Reporting in the Population Census." *Bulletin de l'Institut International de Statistique*, 36:(2):183–189. 1957.

Rosenwaike, Ira. "On Measuring the Extreme Aged in the Population." *Journal of the American Statistical Association*, 63(321):29–40. Mar. 1968.

Saw, Swee Hock. "Errors in Chinese Age Statistics." *Demography*, 4(2):859–875. 1967.

Siegel, Jacob S., and Zelnik, Melvin. "An Evaluation of Coverage in the 1960 Census of Population by Techniques of Demographic Analysis and by Composite Methods." *Proceedings of the Social Statistics Section, 1966*. Washington, D.C., American Statistical Association, 1966. Pp. 71–85.

Siegel, Jacob S. "Completeness of Coverage of the Nonwhite Population in the 1960 Census and Current Estimates, and Some Implications." In David M. Heer (ed.), *Social Statistics and the City*, Joint Center for Urban Studies of the Massachusetts Institute of Technology and Harvard University, Cambridge, Mass., 1968. Pp. 13–54.

SUGGESTED READINGS—Continued

Somoza, Jorge, and Llano, Luis. *Proyección de la Población de Bolivia*. Series C, No. 9, United Nations Demographic Center for Latin America. Santiago, Chile, 1963. Pp. 46–59.

Taeuber, Conrad, and Hansen, Morris. "A Preliminary Evaluation of the 1960 Censuses of Population and Housing." *Demography*, 1(1):1–14. 1964.

U.S. Bureau of the Census. *Evaluation and Research Program of the U.S. Censuses of Population and Housing, 1960: Accuracy of Data on Population Characteristics as Measured by Reinterviews*. ER 60, No. 4, 1964.

———. *Evaluation and Research Program of the U.S. Censuses of Population and Housing, 1960: Accuracy of Data on Population Characteristics as Measured by CPS-Census Match*. ER 60, No. 5, 1964.

van de Walle, Etienne. "Some Characteristic Features of Census Age Distributions in Illiterate Populations." *American Journal of Sociology*, 71(5):549–557. Mar. 1966.

Yerushalmy, J. "The Age-Sex Composition of the Population Resulting from Natality and Mortality Conditions." *Milbank Memorial Fund Quarterly*, 21(1):37–63. Jan. 1943.

You, Poh Seng. "Errors in Age Reporting in Statistically Underdeveloped Countries." *Population Studies* (London), 13(2):164–182. Nov. 1959.

Zelnik, Melvin. "Errors in the 1960 Census Enumeration of Native Whites." *Journal of the American Statistical Association*, 59(2):437–459. June 1964.

Part D. Analysis of Age Data

Chaddock, Robert E. "Age and Sex in Population Analysis." *Annals of the American Academy of Political and Social Science*, 188:185–193. *The American People*, Nov. 1936. (Reprinted in Joseph J. Spengler and Otis Dudley Duncan (eds.), *Demographic Analysis*, Glencoe, Ill., Free Press, 1956, pp. 443–451.)

Coale, Ansley J. "How a Population Ages or Grows Younger." In Ronald Freedman (ed.), *Population: The Vital Revolution*. Garden City, N.Y., Doubleday-Anchor, 1964. Pp. 47–58.

———. "The Effects of Changes in Mortality and Fertility on Age Composition." *Milbank Memorial Fund Quarterly*, 34(1):79–114. Jan. 1956.

———. "How the Age Distribution of a Human Population is Determined." *Cold Springs Harbor Symposium on Quantitative Biology*, 22:83–89. 1957.

Coulson, Michael R. C. "The Distribution of Population Age Structures in Kansas City." In George J. Demko,

Harold M. Rose, and George A. Schnell (ed.), *Population Geography: A Reader*, New York, McGraw-Hill Book Co., 1970. Pp. 408–430.

Grauman, John V. "Effects of Population Trends upon Age Structure with Application to the Americas." *Estadística*, 14(51):271–287. June 1956.

Hermalin, Albert I. "The Effect of Changes in Mortality Rates on Population Growth and Age Distribution in the United States." *Milbank Memorial Fund Quarterly*, 44(4):451–469, Part 1. Oct. 1966.

Lorimer, Frank. "Dynamics of Age Structure in a Population with Initially High Fertility and Mortality." *Population Bulletin of the United Nations*, 1:31–41. Dec. 1951.

Notestein, Frank W. "Mortality, Fertility, and Size-Age Distribution and the Growth Rate." In National Bureau of Economic Research, *Demographic and Economic Changes in Developed Countries*, Princeton, N.J., Princeton University Press, 1960. Pp. 261–289.

Sauvy, Alfred. "Le vieillissement des populations et l'allongement de lá vie" (The Aging of Populations and the Prolongation of Life). *Population* (Paris), 9(4):675–682. 1954.

Smith, T. Lynn. *Fundamentals of Population Study*. New York, J. B. Lippincott, 1960. Pp. 155–159.

Stolnitz, George J. "Mortality Declines and Age Distribution." *Milbank Memorial Fund Quarterly*, 34(2):178–215. Apr. 1956.

United Nations. *The Aging of Populations and its Economic and Social Implications*. Series A, Population Studies, No. 26, 1956. See esp. "Introduction."

———. "The Cause of the Aging of Populations: Declining Mortality or Declining Fertility." *Population Bulletin of the United Nations*, (4):30–38. Dec. 1954.

———. "Population Structure as a Factor in Manpower and Dependency Problems of Underdeveloped Countries," by John D. Durand. *Population Bulletin of the United Nations*, (3):1–8. Oct. 1953.

Valaoras, Vasilios G. "Patterns of Aging of Human Populations." Eastern States Health Education Conference (1949). *The Social and Biological Challenge of Our Aging Population*. New York, Columbia University Press, 1950. Pp. 67–85.

———. "Population Profiles as a Means for Reconstructing Demographic Histories." In International Union for the Scientific Study of Population, *International Population Conference, 1959* (Vienna). Pp. 62–72.

CHAPTER 9

Racial and Ethnic Composition

There are only a few races—from three to five according to the classification used—but many subraces or ethnic groups. Much nonsense has been written on the subject of race and racial differences. Nonetheless, there are physical, cultural, linguistic, and other differences among various large groups of people that are relatively persistent over time. These groups are often of interest to national policymakers, and they have different geographic distributions and different demographic characteristics. Consequently, such groups are often identified in demographic statistics.

There is no doubt also that certain ethnic groups are often separately identified in censuses or registration systems because of discrimination that is practiced against them. Indeed, groups that are officially regarded as "backward" or "alien" may be omitted from the records and statistics altogether. In other cases, however, such as some of the Latin American countries vis-à-vis their Indian tribal population, there has been recognition of the need to measure the conditions of the disadvantaged groups in order to raise their level of living and to integrate them into the national society and economy. In the United States, in recent years, many of the demands for special census tabulations of ethnic data have come from antidiscrimination boards at various governmental levels and from the minority groups' own organizations. These agencies are interested in measuring the current differentials (in housing, education, occupation, income, etc.) and in determining whether and to what extent handicaps are being overcome. They are also interested in pinpointing the areas of residential concentration so that their funds can be more effectively allocated.

RACIAL, ETHNIC, AND NATIONALITY CHARACTERISTICS AND THEIR INTERRELATIONS

The definition of race and of specific races and subraces is properly the domain of the physical anthropologist. He recognizes that in the modern world most ethnic groups are of mixed racial origin and that fresh racial mixture is constantly occurring. In compiling demographic records, we have to be content with much grosser classifications than those of the anthropologist. These tend to be based on how members of groups identify themselves and how they are regarded by their compatriots. Racial or ethnic groups so defined, however, are usually more meaningful from the standpoints of social policy or demographic analysis than those defined by purely anthropometric criteria. The resulting classification is neither scientific nor objective, but it is reasonably con-

sistent and reproducible in that most persons who know the person would place him in the same category or categories. All things considered, then, our definitions of some of the terms used in this chapter will be less precise than those of other demographic terms or concepts.

Perhaps the primary meaning of "ethnic group" is a subrace, so that there are many times more ethnic groups recognized than races. The term is used loosely, however, to cover racial, nationality, cultural, or linguistic groups. To varying extents such a group will have a common descent, history, and habitat. Often the characteristics are highly correlated. Thus, the Croats tend to be Roman Catholic and use the Roman alphabet to write their South Slav dialect, whereas the Serbs are of Orthodox religion and use the Cyrillic alphabet. These two ethnic groups are concentrated in different regions of Yugoslavia.

"Nationality" has two meanings. One definition is in terms of country of present citizenship. The other refers to the country or other area of origin of one's ethnic group, sometimes to an extinct country or a country that once had very different boundaries. It is usually a matter of affiliation with a group. Flemings characteristically live in Belgium and Slovaks, in Czechoslovakia; but not all the residents of those countries belong to those respective nationalities. Since the unqualified word has this ambiguity, we will seek to avoid using it here.

RACE AND ETHNIC GROUP

International Recommendations

This topic is described in very general terms by the United Nations in its proposal for the 1970 population censuses and is listed among "other useful topics," rather than "recommended topics." "The national and/or ethnic groups of the population about which information is needed in different countries are dependent upon national circumstances. Some of the bases on which ethnic groups are identified are: ethnic nationality (i.e., area of origin or group affiliation as distinct from citizenship or country of legal nationality), race, colour, language, religion, customs of dress or eating, tribe or various combinations of these characteristics. In addition, some of the terms used, such as 'race', 'origin', or 'tribe', have a number of different connotations. The definitions and criteria applied by each country investigating ethnic characteristics of the population must, therefore, be determined by the type of groups which it desires to identify. By the nature of the subject, these groups will vary widely from country to country, so that no internationally accepted criteria can be recommended.

"Because of the interpretative difficulties which may occur, it is important that, where such an investigation is undertaken, the basic criteria used should be clearly explained in the census report so that the meaning of the classification will be readily apparent."[1]

As noted by Hawley, a problem in the analysis of population composition by ethnic group is the lack of a simple, standard classification system.[2] Neither the United Nations nor any of its affiliated agencies has developed a standard classification of ethnic groups, paralleling the International Standard Classification of Occupations. The United States does not have such a standard classification in its statistical programs, and the authors have not been able to discover one that was sponsored by any official coordinating body. When the topic is included in a national census, however, there is often a coding list so that entries not recognized in the tabulation scheme can be assigned to a recognized category.

National Practices

The international recommendations, however, do not tell. us what items the respective countries will actually include on their census schedules or what they have included in past rounds.

The 1963 *Demographic Yearbook* of the United Nations throws more light on these practices and also covers countries that were not members of the regional economic commissions. Of 77 counties represented in the period 1955 to 1963, 70 had included an ethnic item in a census and the remainder, in a national survey only. Only a few sample surveys are listed. Table 9–1 summarizes the information for this period and for the 1945–1955 period, the latter on the basis of data from the 1956 *Demographic Yearbook*. The classification is by the present authors and is judgmental to a considerable degree because of the large number of marginal cases. A system that one country would officially designate as by "race" another would designate as by "ethnic nationality," and so on.

United States.—The racial classifications and definitions used in the census of population, special censuses, and sample surveys conducted by the Bureau of the Census are essentially alike in any period, although there have been changes over time in the census. The classifications by race in vital statistics and in statistics on immigration and emigration have some unique features, however.

Censuses and surveys.—Historically, the determination of race in the U.S. censuses and current surveys has mainly been made by the enumerator on the basis of observation although he was encouraged to inquire when in doubt. With the increasing use of self-enumeration, beginning with the 1960 census, the classification has been based largely on self-identification in censuses—but not in current surveys.

Prior to the 1970 census it was common practice in census reports to distinguish two basic racial groups based on color, white and non-white. The color group designated as "nonwhite" included Negroes, American Indians, Japanese, Chinese, Filipinos, Koreans, Hawaiians, Asian Indians, Malayans, Eskimos, Aleuts, etc. Of the nonwhite ethnic groups just mentioned, all but Asian

[1] United Nations, *Principles and Recommendations for the 1970 Population Censuses,* Statistical Papers, Series M, No. 44, 1967, p. 53.

[2] Amos H. Hawley, "Population Composition," in *The Study of Population,* Philip M. Hauser and Otis Dudley Duncan, eds., Chicago, The University of Chicago Press, 1959, p. 369.

Indians and Malayans were used as categories of race in the 1970 Census. Persons of Mexican birth or ancestry who were not definitely Indian or of other nonwhite race were classified as white. Beginning in 1969, however, the term "nonwhite" has been dropped from Federal statistical reports and has been replaced by "Negro and other races."

Negroes, Indians, Orientals, etc., living in the United States are quite different with respect to some demographic and economic characteristics; but, beginning in 1950, since Negroes contributed more than nine-tenths of all nonwhites, most of the tables in census and current survey reports were shown for the nonwhite races combined in order to effect savings in tabulations and publications. In some parts of the country, especially in the West, members of other nonwhite races have outnumbered Negroes. Since there was also increasing public concern with the status of certain specific ethnic minorities, it was decided about 1967 to put more stress on Negroes in publications of the Bureau of the Census when statistics are shown by color or race and to limit the nonwhite total to a few historical comparisons.

Most of the Negroes in the United States are descendants of persons brought from Africa as slaves although there has been some voluntary immigration in modern times, particularly from the West Indies. From 1850 through 1920, an attempt was made to identify mulattoes–persons with mixed Negro and white ancestry—as a group separate from full-blooded Negroes. This practice was abandoned because of the difficulty of making the distinction and because mulattoes were still customarily regarded as belonging to the Negro race by both whites and Negroes. In certain quarters in the United States, the term "black" is coming to be favored over "Negro."

American Indians, who are descendants of the aboriginal inhabitants of the United States, constitute the second largest racial minority group in that country but represented only 0.4 percent of the total population in 1970. It is estimated that over half of the Indians lived on reservations, but this proportion is declining as they become more widely dispersed within the general population and become a part of the urban, industrial society. They still remain the most rural of the sizable American ethnic groups.

Almost all of the remainder of the American population consists of persons of Mongolian, Polynesian, or Malay race. Altogether these ethnic groups constituted only 0.9 percent of the population in 1970.

There are other important ethnic groups that are not part of the classification by race. Two that are partly of Hispanic stock are the **Mexican-Americans** ("Mexicans") and the **Puerto Ricans.** As compared with other ethnic groups of European, or part European, origin, they have maintained their mother tongue and other cultural traits tenaciously and have tended to be highly endogamous. The Mexicans are concentrated in the Southwest although they have diffused somewhat, particularly to some of the large cities in the Middle West. They have immigrated fairly continuously from the adjacent States of Mexico, but a sizable minority descends from colonial settlement in what is now the American State of New Mexico. Racially, they range from white to American Indian with the majority probably being a fairly stable mestizo group. Since many persons who regard themselves as Mexican-Americans from a cultural standpoint would not be included in a count restricted to those of Mexican birth or parentage, attempts have been made to classify them by other means. The attempt to identify them by a direct question in 1930 and by Spanish

Table 9–1.—Statistics in National Census Reports on Racial and Ethnic Groups, by Continents: 1955 to 1963 and 1945 to 1955

Period and continent	All countries having reports	Subject covered					
		Race	Ethnic group	Race and tribe, ethnic group and tribe, or tribe only	Mixed categories (race, ethnic group, etc.)	Nonindigenous population by race or ethnic group	Indigenous population only
1955 TO 1963							
Total..........................	70	27	13	3	23	3	1
Africa........................	27	15	3	[1]2	3	3	1
North America.................	19	10	1	–	8	–	–
South America.................	1	–	–	–	1	–	–
Asia..........................	8	–	3	1	4	–	–
Europe........................	3	–	3	–	–	–	–
Oceania.......................	11	2	2	–	7	–	–
U.S.S.R.......................	1	–	1	–	–	–	–
1945 TO 1955							
Total..........................	75	24	11	–	37	3	–
Africa........................	23	7	3	–	12	1	–
North America.................	18	10	1	–	7	–	–
South America.................	3	1	1	–	1	–	–
Asia..........................	13	1	3	–	9	–	–
Europe........................	2	1	1	–	–	–	–
Oceania.......................	16	4	2	–	8	2	–

—Represents zero.
[1] Tribe only for Ghana.

Source: United Nations, *Demographic Yearbook, 1963*, table 9; *1964*, table 30; and *1956*, table 7.

mother tongue in 1940, both produced an apparent undercount. Beginning in 1950, the surnames of the population in five Southwestern States (Arizona, California, Colorado, New Mexico, and Texas) were examined, and persons of Spanish surname were counted. This procedure has resulted in more credible approximations.

The Puerto Rican immigration became numerically important only after World War II, and in 1960 relatively few Puerto Ricans on the mainland yet belonged to the third or older generation of immigrants. The first and still the chief focus of this immigration is New York City; very few Puerto Ricans have reached the Southwest so that the two major ethnic groups with Spanish surnames can be distinguished fairly well by their respective geographic locations. The aboriginal Indian population left few survivors in Puerto Rico, but there was a large importation of African slaves in the early colonial period. Consequently, Puerto Ricans may be white (mostly of Spanish descent), Negro, or mulatto.

Recently, there has been a demand for statistics on all persons of Hispanic background. Accordingly, the question on birthplace (see below) was supplemented in the 5-percent sample of the 1970 census by the following item:

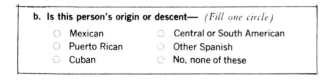

b. **Is this person's origin or descent—** *(Fill one circle)*
- ○ Mexican
- ○ Puerto Rican
- ○ Cuban
- ○ Central or South American
- ○ Other Spanish
- ○ No, none of these

There is an increasing number of Americans of **mixed racial ancestry.** The statistical treatment of the Spanish-Indian

mestizos and of mulattoes has already been indicated. The classification of the children of mixed marriages who do not live as members of a distinct ethnic community or that of descendants of three or more racial groups is particularly difficult. Race mixture may, of course, progress in several more generations to a point where a general classification of the population is both conjectural and meaningless. In the past, the difficult cases have been handled by the enumerator's observation and a few mechanical rules. With the increasing use of self-enumeration, more scope is given to the respondent's self-identification.

Vital statistics.—In current annual volumes of *Vital Statistics of the United States,* many of the statistics are shown by color—white and nonwhite. Here these terms are defined essentially in the same way as they are by the Bureau of the Census.[3] Some statistics are shown by race, using, at a maximum, the categories white; Negro; American Indian; Chinese; Japanese; Hawaiian and Part-Hawaiian; and other. Beginning with 1964, Aleuts and Eskimos are included with American Indians. Cases of race not stated are allocated between "white" and "Negro." The term "colored" used in vital statistics reports in the period 1915 to 1934 is equivalent to "nonwhite."

Immigration and emigration statistics.—The last Annual Report of the Immigration and Naturalization Service that classed immigrant aliens by race or ethnic group was that for the fiscal year 1960 to 1961. The classification in those statistics was slightly different from that used in census and vital statistics, the categories being white, Negro, Chinese, East Indian, Filipino, Japanese, Korean, and Pacific Islander.

---

[3] U.S. National Center for Health Statistics, *Vital Statistics of the United States, 1966*, Vol. I, *Natality,* 1968, pp. 4–6–4–7.

Sources

International statistics on the population by ethnic group are available in the U.N. *Demographic Yearbooks* of 1956, 1963, and 1971. As will be discussed below, however, these statistics are not very useful for making international comparisons. More detailed statistics are of course available in national publications—those devoted to censuses, surveys, vital statistics, and population registers.

Uses and Limitations

In countries that are not homogeneous from the racial or ethnic standpoint, statistics by race or ethnic group are useful from many standpoints—in analyzing past demographic trends, in making population projections, in evaluating the quality of demographic statistics, and in planning social, economic, and health programs. For many of these purposes, of course, one needs not just the distribution of population by this characteristic but also census tables giving other characteristics cross-classified by race, vital rates specific for race, immigration statistics by race, and so on. For example, race-specific fertility rates may have such different levels and trends that more realistic assumptions can be made about future fertility by considering the major races separately. The welfare of aboriginal groups is often of special concern to national governments, and information on the size and characteristics of such groups is a prerequisite of intelligent policy-making and planning.

As has been mentioned, statistical agencies in different countries use the same term ("race," etc.) to denote not only different sets of categories but also what appear to be different concepts. "Some countries, such as the United States, have explained the sense in which the word is used but in most cases there is no such explanation and, in general, the term appears to cover a variety of concepts." [4] Hence, meaningful international comparisons are often risky.

Quality of the Statistics

"In addition to their heterogeneity, there are probably wide differences in the reliability of most of the basic data. Where the investigations have been concerned with more or less endogamous groups which have existed for many generations within a country, each person is usually well aware of the group to which he belongs, and there is little difficulty in obtaining the information. This would apply to responses concerning tribal affiliation and indigenous Indian populations and other aboriginal peoples. In other cases, however, the adequacy of individual responses may be seriously affected by the clarity of the question used and by the explanatory material provided. Aside from the possibilities of misinterpretation of the intent of the question, there is always a considerable chance of deliberate falsification in connection with questions having to do with matters affecting social prestige. In responding to a question on colour, for instance, a person of mixed colour may—if he can—affiliate himself with the group which has the highest social standing in the community. Similarly, he may intentionally misstate the national origin of his ancestors." [5]

These generalizations by the U.N. Statistical Office relate to errors in classification. In addition, there may be differential errors in coverage in the population census and in the completeness of registration of vital events. In extreme cases, an ethnic

group, or the less assimilated part of it, may be omitted from a population census altogether or may not be covered by the registration system. Where a racial or ethnic group is discriminated against or regarded as socially inferior, it may receive less adequate attention in statistical programs and hence the statistics for it will probably be of poorer quality. Even when legal disabilities are removed, their effects may linger in the educational and economic status of the group, and these disabilities in turn may result in poorer demographic statistics.

The quality of other demographic characteristics reported in censuses often differs among the ethnic groups living in the same country. The extent of the heaping of reported ages on favored terminal digits is readily measured. One of the better measures is Myers' Blended Index (see chap. 8). Some illustrative results using this measure of age heaping are as follows:

New Zealand, 1945	
European	1.4
Maori	3.0
New Zealand, 1961	
European	0.9
Maori	1.4
South Africa, 1960	
White	0.7
Bantu	11.3
Ceylon, 1953	
Low-country Sinhalese	19.1
Kandyan Sinhalese	21.3
Indian Tamils	12.5
Ceylon Tamils	18.5

(In all cases the age range used is 10 to 89 years. The higher the index, the more the heaping on favored ages.)

Coverage. - There was some discusssion in chapters 4 and 8 of the coverage of recent censuses and surveys. Table 9-2 gives estimates of net undercount in 1960 by age, sex, and color. The net undercount is obviously much greater for nonwhites and reaches rather high levels in some nonwhite age-sex groups.

Measures

There are not very many statistical measures that are peculiar to the description or analysis of the distribution of the population by race or by ethnic group. Simple percentage distributions are most frequently used.

Various conventional devices are also used for showing the geographic distribution of an ethnic group or for comparing the growth and characteristics of ethnic groups.

Index of Dissimilarity. — A particularly pertinent measure is the Index of Dissimilarity, which was defined in chapter 8. This may be used to compare the distribution by race (or ethnic group) of two areas or two groups of another type, or, conversely, to compare the distribution of two racial groups by some other characteristic, such as age.

Segregation Indexes. — In the United States, there has been a great interest in the extent to which the residences of various racial and ethnic groups (Negroes, European nationalities, etc.) are segregated from each other in the great cities and metropolitan areas of that country. For this purpose, the geographic unit used has been mainly the census tract (see chap. 5). A considerable literature has appeared, both on the extent of such segregation and on the appropriate ways of measuring it, that is, of the "overall unevenness of . . . residential

---

[4] United Nations, *Demographic Yearbook, 1963*, p. 39.

[5] Ibid., p. 39.

Table 9–2. —**Estimated Net Understatement and Corrected Population, by Color and Sex, for the United States: 1967, 1960, and 1950**

[Numbers in thousands. Figures relate to the total resident population. Base of percent is corrected population]

Year, color, and sex	Current estimates or census counts[1]	Corrected population	Net understatement Amount	Net understatement Percent
1967				
All classes..........	197,863	203,565	5,702	2.8
White.................	173,920	177,480	3,560	2.0
Male...............	85,127	87,400	2,273	2.6
Female.............	88,793	90,080	1,287	1.4
Nonwhite............	23,942	26,084	2,142	8.2
Male...............	11,567	12,783	1,216	9.5
Female.............	12,375	13,301	926	7.0
1960				
All classes..........	179,323	185,025	5,702	3.1
White.................	158,832	162,392	3,560	2.2
Male...............	78,367	80,640	2,273	2.8
Female.............	80,465	81,752	1,287	1.6
Nonwhite............	20,491	22,633	2,142	9.5
Male...............	9,964	11,180	1,216	10.9
Female.............	10,527	11,453	926	8.1
1950[2]				
All classes..........	151,327	157,001	5,675	3.6
White.................	135,150	138,728	3,578	2.6
Male...............	67,255	69,407	2,152	3.1
Female.............	67,895	69,321	1,426	2.1
Nonwhite............	16,177	18,274	2,097	11.5
Male...............	7,932	9,122	1,190	13.0
Female.............	8,245	9,152	907	9.9

[1] Estimates as of July 1, 1967, and census counts as of April 1, 1960 and 1950.
[2] Figures relate to 50 States.

Source: Jacob S. Siegel, "Completeness of Coverage of the Nonwhite Population in the 1960 Census and Current Estimates, and Some Implications," *Social Statistics and the City* table 1. Figures for 1967 have been revised to include Armed Forces abroad.

distributions." This literature is summarized by Taeuber and Taeuber.[6] According to the Taeubers, Jahn, Schmid, and Schrag inaugurated the series of sociological articles explicitly concerned with the construction of "segregation indexes."[7] There are a number of indexes for measuring the extent to which areal distributions for two racial or other demographic groups differ from each other, and the Taeubers consider the strongpoints and weaknesses of each of these indexes. (Incidentally, the Index of Dissimilarity discussed above was the index chosen for most of the analysis of the Taeubers' book.) Duncan and Duncan have related the segregation indexes to the more general measures describing spatial distribution which we discussed in chapter 5 of this book.[8]

Factors Important in Analysis

The facts that different racial and ethnic groups are often concentrated in different parts of country and that some groups are characteristically urban whereas others are characteristically rural, have already been mentioned. Ethnic groups also may often have very different rates of growth reflecting differences in their fertility, mortality, and migration rates. Past differences in the components of population change will also produce different age-sex structures. These, in turn, result in different crude vital rates for the ethnic groups.

Large differences in the vital rates of certain racial or ethnic groups may also reflect in considerable part group differences in socio-economic status. Thus, differences in infant mortality may reflect major differences in level of living, education of parents, etc. Multiple cross-classifications shed further light on the nature of ethnic differences, but the analyst must be careful not to oversimplify the temporal or causal relationships. In a multiple-factor analysis involving infant mortality, race, and region of residence, careful study that brings in knowledge beyond the particular cross-classified statistics is required in order to be able to interpret the relative roles of race and region of residence in determining infant mortality. There may be interactions among the factors in that one may both influence and be influenced by the other.

COUNTRY OF BIRTH

An item on place of birth is often included in population censuses. It is usually asked of the entire population. For the foreign born, the country of birth is obtained; and, for the native population, the state, province, etc., of birth. The latter type of detail yields data on internal migration, which are discussed in chapter 21. The use of data on country of birth to measure immigration is discussed in chapter 20. In this chapter, we are concerned more with the status of the population, i.e., with country of birth as a characteristic of persons, rather than with international flows of population. Country of birth of the decedent may also be recorded on the death certificate and country of birth of the parents on the birth certificate. Country of birth is also ordinarily a matter of record in a continuous population register.

International Recommendations

Country of birth has been included on the United Nation's recommended list of items for the world census programs of 1950, 1960, and 1970.

Although country of birth was not one of the items in the list adopted by the U.N. Statistical Commission in 1953 for inclusion on the vital records, it is mentioned as an optional item in the recommendations that have been circulated by the Statistical Office for the consideration of national governments.[9]

For purposes of compiling immigration and emigration statistics, the United Nations has also recommended that the statistical slip contain a question on "country of birth (according to current national boundaries)."[10] More specifically, this information is to be obtained for permanent immigrants and emigrants and, if required, for temporary immigrants as well.

International statistics on country of birth are presented in various issues of the U.N. *Demographic Yearbooks*.

[6] Karl E. Taeuber and Alma F. Taeuber, *Negroes in Cities*, Chicago, Aldine Publishing Company, 1965, pp. 195–245.
[7] Julius A. Jahn, Calvin F. Schmid, and Clarence Schrag, "The Measurement of Ecological Segregation," *American Sociological Review*, 12(3):293–303, June 1947.
[8] Otis Dudley Duncan and Beverly Duncan, "A Methodological Analysis of Segregation Indexes," *American Sociological Review*, 20(2):210–217, April 1955.

[9] United Nations, *Handbook of Vital Statistics Methods*, Studies in Methods, Series F, No. 7, April 1955, pp. 222–223; United Nations, Statistical Commission, *Demographic and Housing Statistics, Recommendations for the Improvement and Standardization of Vital Statistics: Draft Proposals*, Report of the Secretary General, 15th Session, E/CN.3/388/Add.1, pp. 31–32, January 31, 1968.
[10] United Nations, *International Migration Statistics*, Statistical Papers, Series M, No. 20, 1953, p. 19.

National Practices

Country of birth is one of the older and most frequently included items in national population censuses. The usual question is of the form, "Where was this person born?" with an indication on the schedule or in the instructions to enumerators of the type of specificity required in the answer. The following summary for the period 1955 to 1964 is based on national census schedules on file in the Statistical Office of the United Nations and on the 1963 and 1964 volumes of the *Demographic Yearbook*. National sample surveys and partial censuses (nonindigenous population only, etc.) are omitted from our tally. The indicated number of countries having published statistics from censuses conducted in the decade is presumably incomplete since it is based on information available in 1964. A few countries, mostly small ones, had published statistics by nativity only, i.e., the classification of the population as native or foreign born.

Continent	Countries collecting	Countries publishing
Total.....................	131	95
Africa.................	25	16
North America......	34	26
South America......	13	7
Asia...................	21	16
Europe................	24	16
Oceania	14	14

United States Practices. — Because the United States was settled primarily by waves of immigrants from many different European countries, there was long a great interest in information on the composition of the population by nativity and ethnic nationality. This interest related to the size, location, and rate of assimilation of the various groups. After laws were passed in the early 1920's that severely limited foreign immigration, there was some diminution in popular and official interest in the subject and in the attention paid to it in the population census. The size of the foreign-born population decreased in the last four intercensal decades and stood at less than 10 million in 1970, or about 5 percent of the total population. The foreign-born population of the United States is largely an "old" population since it represents mainly the survivors of persons who immigrated years ago.

Fresh waves of immigration, however, were generated in the Nazi era, in the period just after World War II, and at the time of various civil wars abroad and other more recent political events. Temporary relaxations of the immigration laws made it possible for many Hungarian and Cuban refugees to be admitted outside the quota limits. The 1965 Act liberalized the 1924 quota system and led to both a larger actual annual number of immigrant aliens and to a reinforcement of the new tendency for a greater share of the immigrants to come from regions other than northern and western Europe.

Items on the schedule. — Country of birth has been ascertained in every census of the United States since 1850. In addition, beginning in 1870, the United States has inquired about the country of birth of each parent of a native person in order to identify and characterize the second immigrant generation. (In some cases, of course, the person's parents are members of the same household so that their country of birth will already have been ascertained.)

Definitions. — The definitions of the principal terms relating to country of origin used by the U.S. Bureau of the Census are as follows: [11]

Native. — This category comprises persons born in the United States, the Commonwealth of Puerto Rico, or a possession of the United States. Also included in this category is the small number of persons who, although they were born in a foreign country or at sea, have at least one native American parent.

Foreign born. — This category comprises all persons not classified as native. Therefore, this group includes persons who reported a foreign country as their place of birth (with the exception stated above) and those persons with place of birth not reported who answered the question on language spoken prior to coming to the United States.

Native of native parentage. — This category comprises native persons both of whose parents are also natives of the United States.

Native of foreign or mixed parentage. — This group consists of native persons, one or both of whose parents are foreign born. The rules for determining the nativity of parents are substantially the same as those for determining the nativity of the persons enumerated. A limited amount of data is presented separately for the three groups comprising this category; namely, native of foreign parentage (that is, native with both father and mother foreign born), native of mixed parentage — father (but not mother) foreign born, and native of mixed parentage — mother (but not father) foreign born.

Foreign stock. — The foreign-born population is combined with the native population of foreign or mixed parentage in a single category termed the "foreign stock." This category comprises all first and second generation Americans. Third and subsequent generations are described as "native of native parentage."

Country of origin. — The question on place of birth and parents' birthplace specified that country of birth be reported according to international boundaries as recognized by the United States at the time of the census.

Tabulations. — From 1910 through 1950, most tables on country of birth were restricted to the foreign-born white population of the United States because very few nonwhites were foreign born. For other censuses, however, the statistics on country of birth are for the total foreign-born population, with both universes being used in 1960 as a "link" or "bridge." Furthermore, in 1960, distributions by country of origin were published for the native of foreign parentage, by color, so that this distribution can be obtained for the total foreign stock. Abridged distributions for the foreign stock were published for areas as small as counties, cities of 10,000 inhabitants, and census tracts. Table 9–3 presents summary historical figures by nativity and parentage. (Note that it was feasible to reconstruct historical statistics for only the white population and only for conterminous United States.)

The vital statistics of the United States were formerly published by nativity. Two national tables showing births for native white and foreign-born white mothers by age of mother, sex of child, and live-birth order were discontinued in 1966. Deaths have not been published by nativity of the decedent since 1950 except in some of the 1960 Vital and Health

---

[11] *U.S. Census of Population: 1960, Subject Reports, Nativity and Parentage,* PC(2)–1A, 1965, pp. viii–ix.

Table 9–3. — Nativity and Parentage of the Population, by Color for the United States, 1960, and of the White Population for Conterminous United States, 1900 to 1960

[Minus sign (−) denotes decrease]

Census year and color	Total	Native									Foreign born	
		Total native		Native parentage		Foreign or mixed parentage						
		Number	Percent	Number	Percent	Total	Foreign parentage (both parents foreign)	Mixed parentage			Number	Percent
								Total	Father foreign	Mother foreign		
UNITED STATES												
1960................	179,325,675	169,587,520	94.6	145,275,233	81.0	24,312,287	14,105,139	10,207,148	6,451,715	3,755,433	9,738,155	5.4
White..............	158,837,679	149,543,634	94.1	125,759,263	79.2	23,784,371	13,790,446	9,993,925	6,306,239	3,687,686	9,294,045	5.9
Nonwhite...........	20,487,996	20,043,886	97.8	19,515,970	95.3	527,916	314,693	213,223	145,476	67,747	444,110	2.2
CONTERMINOUS UNITED STATES--WHITE												
1960................	158,460,699	149,181,380	94.1	125,444,612	79.2	23,736,768	13,769,054	9,967,714	6,290,540	3,677,174	9,279,319	5.9
1950................	134,478,365	124,382,950	92.5	100,804,575	75.0	23,578,375	14,815,760	8,762,615	5,746,410	3,016,205	10,095,415	7.5
1940................	118,701,558	107,282,420	90.4	84,124,840	70.9	23,157,580	15,183,740	7,973,840	5,267,140	2,706,700	11,419,138	9.6
1930................	110,286,740	96,303,335	87.3	70,400,952	63.8	25,902,383	17,407,527	8,494,856	5,547,325	2,947,531	13,983,405	12.7
1920................	94,820,915	81,108,161	85.5	58,421,957	61.6	22,686,204	15,694,539	6,991,665	4,539,776	2,451,889	13,712,754	14.5
1910................	81,731,957	68,386,412	83.7	49,488,575	60.5	18,897,837	12,916,311	5,981,526	3,923,845	2,057,681	13,345,545	16.3
1900................	66,809,196	56,595,379	84.7	40,949,362	61.3	15,646,017	10,632,280	5,013,737	3,346,652	1,667,085	10,213,817	15.3
Increase over preceding census:												
1960................	23,982,334	24,798,430	(X)	24,640,037	(X)	158,393	−1,046,706	1,205,099	544,130	660,969	−816,096	(X)
1950................	15,776,807	17,100,530	(X)	16,679,735	(X)	420,795	−367,980	788,775	479,270	309,505	−1,323,723	(X)
1940................	8,414,818	10,979,085	(X)	13,723,888	(X)	−2,744,803	−2,223,787	−521,016	−280,185	−240,831	−2,564,267	(X)
1930................	15,465,825	15,195,174	(X)	11,978,995	(X)	3,216,179	1,712,988	1,503,191	1,007,549	495,642	270,651	(X)
1920................	13,088,958	12,721,749	(X)	8,933,382	(X)	3,788,367	2,778,228	1,010,139	615,931	394,208	367,209	(X)
1910................	14,922,761	11,791,033	(X)	8,539,213	(X)	3,251,820	2,284,031	967,789	577,193	390,596	3,131,728	(X)
Percent of increase:												
1960................	17.8	19.9	(X)	24.4	(X)	0.7	−7.1	13.8	9.5	21.9	−8.1	(X)
1950................	13.3	15.9	(X)	19.8	(X)	1.8	−2.4	9.9	9.1	11.4	−11.6	(X)
1940................	7.6	11.4	(X)	19.5	(X)	−10.6	−12.8	−6.1	−5.1	−8.2	−18.3	(X)
1930................	16.3	18.7	(X)	20.5	(X)	14.2	10.9	21.5	22.2	20.2	2.0	(X)
1920................	16.0	18.6	(X)	18.1	(X)	20.0	21.5	16.9	15.7	19.2	2.8	(X)
1910................	22.3	20.8	(X)	20.9	(X)	20.8	21.5	19.3	17.2	23.4	30.7	(X)

X Not applicable.

Source: Adapted from. *U.S. Census of Population: 1960, Subject Reports, Nativity and Parentage,* PC(2)-1A. 1965. table 1.

Statistics Monographs of the American Public Health Association, for which special tabulations were made. The basic separation in immigration or emigration statistics is not by nativity but by citizenship (citizen or alien). Aliens are classified by country of birth but naturalized citizens are not.

Uses and Limitations

Statistics on country of birth can be used to identify and describe immigrant minorities and throw light on the ethnic composition of the areas shown in the census reports. Cross-classified with other characteristics, such as legal nationality or citizenship, and languages spoken, these statistics can be used to study assimilation of the various immigrant groups. In making such comparisons, however, allowances should be made for the length of time that the group has been resident in the host country. Here a distribution by period of immigration is needed. A cross-classification of country of birth of father by that of mother contributes to the study of the intermarriage of ethnic groups. If nativity or country of birth is tabulated by age of woman from the census schedules and births are tabulated by nativity or country of birth (and perhaps also age of the mother) for years centering on the census date, the relative

fertility of the groups can be measured. From census data alone, one can tabulate ratios of children to women of childbearing age (see chap. 17), relating the number of children with a foreign mother to the number of foreign-born women.

The estimation from census statistics on nativity and country of birth of the amount of immigration, total or by country of origin, for past time periods, is a moderately complicated procedure and is subject to fairly large errors. For best results, the data should be tabulated by age and have been published for two or more successive censuses.

A drawback of the data on country of origin in studying ethnic composition is that a number of countries are inhabited by several different ethnic groups. The homogeneity or heterogeneity, in this respect, of the country of emigration is generally known, however, from its own census statistics or from the historical or anthropological literature. An ancillary census question on mother tongue can often provide the necessary resolution, moreover.

Another problem is created by changes in national boundaries during the lifetimes of present immigrants or of their parents. However, international comparability is probably less affected by different national practices in handling problems of changing boundaries than it is by different practices in defining and grouping countries. For example, the United States still recognizes the Baltic States—Lithuania, Latvia, and Estonia—which are *de facto* parts of the U.S.S.R. Other countries do not distinguish between the two Germanys.

The Statistical Office of the United Nations points out that not all countries employ its recommended definition of nativity.[12] The practice of the United States, for example, of classifying as native, persons who were born at sea or abroad of parents who were American citizens, confuses the concepts of legal nationality and of birthplace. It can perhaps be rationalized, however, by regarding it as the extension of the *de jure* principle to birthplace classification; but no attempt is made to assign the child to his parents' state of *de jure* residence. Incidentally, countries conducting a *de facto* enumeration of population will count, *ceteris paribus*, more foreign-born persons than those using the *de jure* principle because visiting foreigners will be included in the official count.

The definition of "native" as used by the U.S. Bureau of the Census is in terms of birthplace anywhere within the jurisdiction of the United States. In effect, outlying areas are treated like States. For example, of the 57,345 "natives" living in Guam in 1960, only 36,193 were born in Guam. Of the remainder, 21,030, were born in the United States and 122 in other outlying areas. It is when this detail is dropped and a simple dichotomy by nativity is used that the statistics may be confusing. (It should be clear by now that, in demographic statistics, "natives" does **not** mean aboriginal inhabitants.)

Quality of the Statistics

In addition to bias in the data arising from the respondent's ignorance as to the country in which his birthplace is now located, there may be bias due to deliberate misreporting. Some of the foreign born who have long lived in the country may claim that they are natives thereof.

Comparability from one census to another for the same country may be affected by the wording of the question on country of birth, the accompanying instructions, and the editing rules employed. We can illustrate how changes in the first two produced fluctuations in the census statistics on American residents with Ireland or Northern Ireland given as the country of birth. The published figures were as follows:

	Ireland	*Northern Ireland*
1940	572,000	106,000
1950	505,000	15,000
1960	338,000	68,000

(The 1940 and 1960 figures are for whites and the 1950, for the total population.) The 1940 schedule itself had contained an instruction "Distinguish . . . Irish Free State (Eire) from Northern Ireland." When this instruction was omitted from the schedule in 1950, even though it was retained in the enumerator's manual, the number of those born in Northern Ireland fell off precipitously while that of the Irish-born decreased less than expected. When an equivalent instruction was restored to the 1960 schedule, a more credible distribution between the two countries was achieved.

In all U.S. censuses subsequent to the Treaty of Versailles, some entries of Austria-Hungary have persisted despite the partitioning of that empire. Various devices have been

used to allocate these entries to the successor countries such as Austria, Hungary, Czechoslovakia, and Yugoslavia. In 1950 the allocation was based on the surname of the person. In 1960 a question on mother tongue of the foreign born had been added to the schedule, and that item seemed to be a better basis of allocation.[13] For example, persons reporting Czech as their mother tongue were allocated to Czechoslovakia. The change was probably an improvement but produced some lack of comparability between the statistics for the two censuses.

Nativity was not one of the items included in the Content Evaluation Study of the 1960 census of the United States, but it was included in the Post-Enumeration Survey of the 1950 census. According to that reinterview survey, only 0.6 percent of the population were misclassified by nativity in the census. More meaningfully, the foreign born in the census had a gross classification error (as native) of 8.9 percent but a net error of only 0.6 percent. The native population was affected relatively less. The complexity of the analysis underlying these simple 2×2 comparisons is suggested by table 9-4.

A match of death certificates with U.S. 1960 Census reports compared entries on race, nativity, and country of birth.[14] As with race, there was very close agreement on nativity—98 percent. The agreement was less close on some of the individual countries of birth. As an extreme example, 20 percent of the entries of "Yugoslavia" on the census record had an entry of another country on the death certificate (14 percent had "Austria"). The net difference rate for Yugoslavia was –6 percent relative to the census.[15]

Another factor affecting the quality of nativity statistics is the *nonresponse rate*. Some countries indicate the level of nonresponse on country of birth, whereas in other countries nonresponses may be prorated or otherwise allocated prior to publication. In the United States, for example, persons with blanks on the schedule in 1960 were assumed to be native "unless their census report contains contradictory information, such as the entry of a language spoken prior to coming to the United States." The number of these allocations was tallied in the computer and recorded in appendix tables.

Country of origin of the foreign born or the foreign stock has not been allocated in the U.S. censuses, however. The nonresponse rate on these items appears to be moderate. For 1970 the percent not reporting was:

Foreign born	3.3
Native of foreign parentage	3.0

Measures and Graphic Devices

The percent distribution by nativity is among the more elementary measures to be derived from a distribution by country of birth. Table 9-5 gives the percent foreign born in the population of selected countries.

Most of what was said above concerning measures of ethnic composition also applies here. In addition, statistics on nativity and country of birth may be used to measure "lifetime" immigration or immigration in an intercensal period. Methods of doing so are described in chapter 20.

The composition of an area's population, or its foreign-born population, by country of birth is frequently depicted graphically in census reports. A favorite graphic device for this

---

[12] United Nations, *Demographic Yearbook, 1963*, p. 36.

[13] *U.S. Census of Population: 1960, Subject Reports, Nativity and Parentage*, p. x.

[14] U.S. National Center for Health Statistics, *Vital and Health Statistics*, Series 2, No. 34, pp. 23–29.

[15] Ibid., table P, p. 26.

purpose is the so-called "pie chart," which is illustrated in figure 9–1.

Factors Important in Analysis

Here again, the discussion under "Race and Ethnic Group" is applicable. Differences among countries with respect to the proportion of foreign born (first generation immigrants) in their population, as well as in the specific foreign countries contributing, reflect differences in national laws, regulations, and policies as much as they do differences in more measurable factors such as climate, population density, level of living, and labor scarcity. For studies of the impact of immigration on a country's social and economic structure, census data on the foreign born, especially by age and sex, are particularly valuable. Thus, Eckler and Zlotnick use the proportion foreign born among gainful workers to measure the minimal impact of

immigration on the American labor force over several decades.[16] They compute similar proportions for particular industries to indicate the tendency of immigrants to concentrate in certain lines of work. In studies of the social and economic status of immigrants and of the assimilation of immigrants, involving comparisons of the characteristics of the immigrants and the general population, the use of census data is necessary or preferable to that of immigration statistics. Census data must

Table 9–4. — **Estimates of Error in Reporting Nativity: Post-Enumeration Survey of the United States, 1950**

[Numbers in Thousands]

PES classification	PES total	Omitted from census	Census classification		
			Total	Native[1]	Foreign born
	(1)	(2)	(3)	(4)	(5)
1. Census total.............	...	...	150,697	140,298	10,399
2. Erroneously included in census...................	...	...	1,309	1,136	173
3. Properly included in census..................	152,788	3,400	149,388	139,162	10,226
4. Native....................	142,388	3,169	139,219	138,733	486
5. Foreign born.............	10,400	231	10,169	429	9,740

Analysis of errors	Classification	Census or PES classification		
	(2)	(3)	(4)	(5)
a. Census total.................	Census	150,697	140,298	10,399
b. Erroneously included in census.....	Census	1,309	1,136	173
e. Classified by nativity in both census and PES..........	Census	149,388	139,162	10,226
f. Erroneously incl. in nativity class in census..............	Census	915	429	486
g. Correctly classified in census.....	Both	148,473	138,733	9,740
h. Erroneously excl. from nativity class in census.................	PES	915	486	429
i. Classified by nativity in both census and PES..........	PES	149,388	139,219	10,169
m. Omitted from census..............	PES	3,400	3,169	231
n. PES total......................	PES	152,788	142,388	10,400
q. Net deficiency in census........(line n-a)		2,091	2,090	1
1. Coverage.....................(m-b)		2,091	2,033	58
4. Misclassification by nativity...(h-f)		...	57	-57
r. Gross overstatement in census........(b+f)		2,224	1,565	659
s. Gross understatement in census........(h+m)		4,315	3,655	660
t. Total gross errors in census........(r+s)		6,539	5,220	1,319
1. Coverage.....................(b+m)		4,709	4,305	404
4. Misclassification by nativity...(h+f)		1,830	915	915
u. Net deficiency as percent of census count.....................(q/a)		1.39	1.49	.01
v. Census percentage distribution..........(a)		100.00	93.10	6.90
w. PES percentage distribution.............(n)		100.00	93.19	6.81
x. Difference in percentage points.......(w-v)		...	.09	-.09

[1] Includes nativity not reported.

Source: U.S. Bureau of the Census, *The Post-Enumeration Survey: 1950*, Technical Paper No. 4, 1960, table 2.

Table 9–5. — **Percent Foreign Born in the Population of Selected Countries: Around 1960**

Country and census date	Population	Foreign born	
		Number	Percent
Argentina, 1960.................	20,005,691	2,565,267	12.8
Australia, 1961.................	10,508,186	1,778,780	16.9
Canada, 1961...................	18,238,247	2,844,263	15.6
Chile, 1960.....................	7,374,115	104,853	1.4
Cyprus, 1960....................	573,566	30,762	5.4
Ghana, 1960....................	6,726,815	559,711	8.3
Jamaica, 1960...................	1,609,814	20,334	1.3
Japan, 1965.....................	98,274,961	594,038	0.6
Korea, 1960.....................	24,431,000	179,000	0.7
Mauritius, 1962.................	681,619	12,742	1.9
Mexico, 1960....................	34,923,129	222,513	0.6
New Zealand, 1961..............	2,414,984	340,475	14.1
Norway, 1960...................	3,591,234	62,196	1.7
Panama, 1960...................	1,075,541	44,978	4.2
Puerto Rico, 1960...............	2,349,544	64,279	2.7
South Africa, 1960..............	16,002,797	930,910	5.8
Sweden, 1960...................	7,495,316	299,879	4.0
Switzerland, 1960...............	5,429,061	733,439	13.5
Turkey, 1960....................	27,754,820	952,514	3.4
United Kingdom, 1961...........	46,104,548	3,147,839	6.8
United States, 1960.............	179,325,657	9,738,091	5.4
Venezuela, 1961.................	7,523,999	556,875	7.4

Sources: Official national reports and United Nations, *Demographic Yearbook*.

be employed where the immigrants are to be classified according to their status following arrival with respect to a characteristic which necessarily changes or which may change as a result of migration (e.g., citizenship, employment status, occupation), or where the immigration data are not tabulated for the characteristic (e.g., educational attainment).[17] A collection of papers on the cultural assimilation of immigrants given at the meetings of the International Union for the Scientific Study of Population in Geneva in 1949 was published as a supplement to *Population Studies*.[18] In particular the paper by Max Lacroix and Edith Adams on "Statistics for the Cultural Assimilation of Migrants" suggests how the types of statistics discussed in the present chapter can be applied.[19]

---

[16] A. Ross Eckler and Jack Zlotnick, "Immigration and the Labor Force," *The Annals of the American Academy of Political and Social Science*, 262:94 and 100, March 1949.

[17] For other illustrations of the use of census data in the analysis of migration and the characteristics of migrants, see: Walter W. Willcox, *Studies in American Demography*, Ithaca, New York, Cornell University Press, 1940, pp. 408–418; and Ewan Clague, "America's Future Manpower Needs," *The Annals of the American Academy of Political and Social Science*, 262:109, March 1949.
An example of the use of census tabulations on the foreign-born and on mother tongue in the study of cultural assimilation of immigrants is given in: C. V. Kiser, "Cultural Pluralism," *The Annals of the American Academy of Political and Social Science*, 262:117–130, March 1949.

[18] Population Investigation Committee, *Cultural Assimilation of Immigrants*, Supplement to *Population Studies*, March 1950.

[19] Ibid., pp. 69–108.

Figure 9–1. – **Distribution by Country of Birth of Foreign-Born Population of Australia: 1961 and 1947**

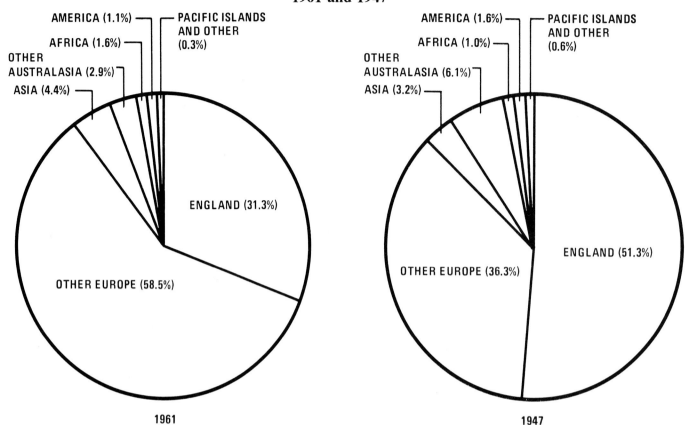

Source: Adapted from, Australia, Commonwealth Bureau of Census and Statistics, *Census of the Commonwealth of Australia, 30th June, 1961,* Vol. VIII, *Australia,* Part I, *Cross-Classification of the Characteristics of the Population,* 1965, table 15.

LEGAL NATIONALITY OR CITIZENSHIP

A citizen, as defined here, is a legal national of the country producing the statistics under consideration. An alien is a person included in the census, register, etc., of the given country but a noncitizen thereof. An alien, however, is usually the citizen of another country. That country represents his **legal nationality**.

Citizenship may be acquired by birth or naturalization. Almost all persons born in some countries, such as the United States, are automatically citizens, or at least have that option. In other countries, however, this is not so. For example, in the 1963 census of Ceylon, 996,440 persons were reported as citizens of India.[20] The number of persons born in India was not given, but the total number of foreign born was only 249,070.[21] In the wake of World Wars I and II, millions of persons were uprooted and left their native countries. Some of them have remained "stateless" persons, i.e., they are not recognized as citizens by any country.

A distribution of aliens by country of legal nationality may differ from that of the foreign born by country of birth for several reasons. (1) The person may have been naturalized and become a citizen of the country in which he now lives. (2) He may have come to his country of present residence via an intermediate country where he acquired citizenship. (3) His country of birth may have been incorporated into another country of which he is now a citizen, but his country of residence has not recognized that incorporation so that in the birthplace statistics he is attributed to a country that has lost its *de facto* existence (for example, a person born in Lithuania who is enumerated in a census of the United States since 1940).

The countries paying considerable attention to this topic are not necessarily the same as those giving considerable detail on country of birth. In fact, to some extent, the two topics seem to be treated as alternatives in censuses. Panama, Sweden, and Australia are unusual then in showing more than the average number of categories on both items. Southern Rhodesia (non-African population only), Costa Rica, Israel, Jordan, France, and Monaco are among the other countries that have particularly exploited country of citizenship in their recent censuses. Of course, a country that has little or no immigration would not have much interest in either item.

United States Practices. – With minor exceptions (e.g., children of diplomatic personnel), anyone born in the United States, Puerto Rico, or the Virgin Islands (or born abroad or

---

[20] Ceylon, Department of Census and Statistics, *Census of Population, Ceylon: 1963,* Vol. I, *Tables Based on a 10% Sample,* Part 1, *General Characteristics,* 1967, p. 34.

[21] Ibid., p. 42.

Table 9–6. — Population of Conterminous United States by Citizenship: 1920 to 1950

[1950 figures based on 20-percent sample]

Year	Total population	Native population	Foreign-born population				
			Total	Naturalized	Having first papers	No papers	Unknown citizenship
BOTH SEXES							
1950	150,216,110	139,868,715	10,347,395	7,562,970	2,052,640		731,785
1940	131,669,275	120,074,379	11,594,896	7,280,265	924,524	2,555,128	834,979
1930	122,775,046	108,570,897	14,204,149	7,919,536	1,266,419	4,518,341	499,853
1920	105,710,620	91,789,928	13,920,692	6,489,883	1,222,553	5,406,780	801,476
MALE							
1950	74,200,085	68,941,830	5,258,255	4,033,070	875,720		349,465
1940	66,061,592	59,939,945	6,121,647	4,137,027	581,713	1,008,071	394,836
1930	62,137,080	54,489,990	7,647,090	4,365,403	955,942	2,081,710	244,035
1920	53,900,431	46,224,996	7,675,435	3,449,547	1,137,021	2,695,042	393,825
FEMALE							
1950	76,016,025	70,926,885	5,089,140	3,529,900	1,176,920		382,320
1940	65,607,683	60,134,434	5,473,249	3,143,238	342,811	1,547,057	440,143
1930	60,637,966	54,080,907	6,557,059	3,554,133	310,477	2,436,631	255,818
1920	51,810,189	45,564,932	6,245,257	3,040,336	85,532	2,711,738	407,651

Source: Adapted from, U.S. Bureau of the Census, *Historical Statistics of the United States, Colonial Times to 1957*, 1960, p. 65; *U.S. Census of Population: 1950*, Vol. II, *Characteristics of the Population*, Part 1, *United States Summary*, 1953, p. 178.

at sea of American citizen parents), has the privilege of citizenship in the United States. Consequently, the number of non-citizens among the "native" population of the United States, as defined above, is negligible (there being voluntary expatriations; and persons born in American Samoa, the Canal Zone, or Guam being "nationals," for example); and information on citizenship has been obtained and tabulated only for the "foreign born." The process of obtaining citizenship for an alien takes five years except in the case of groups affected by special legislation, such as spouses of American citizens. An item on citizenship was included in the population census in 1820, 1830, and 1870 and from 1890 through 1950. Until the 1920 census, this inquiry was limited to males 21 years old and over because the main interest was in determining the maximum number of potential voters. (The Nineteenth Amendment to the U.S. Constitution, which granted the suffrage to women, was adopted in 1920.) Table 9–6 summarizes the statistics obtained since 1920. Note that, through 1940, a distinction was made between aliens who had taken out their first papers, in the process of being naturalized, and those who had not. In only one census, that of 1920, was the year of naturalization asked for. The country of citizenship of aliens has never been asked for, however.

In view of the decline in the alien population and the availability of annual statistics on aliens when the Alien Registration Act went into effect in 1952, the topic was dropped entirely in the 1960 census. It was restored in the 1970 census, however, as one of the items of the 5-percent sample. Two related questions were also asked of the foreign born, namely, "Is this person naturalized?" and "When did he come to the United States to stay?"

Uses and Limitations

"Population statistics by country of citizenship and sex are useful primarily in connection with national problems of legal status and civil rights of immigrants, and for studies of naturalization and assimilation of alien populations." [22] For the latter purpose, a cross-classification by country of birth, or country of previous citizenship of naturalized citizens, enhances the analysis. When cross-classified by age, the data can be used to indicate the number of potential voters in each geographic subdivision; but, to the extent that there are other qualifications for voting, the resulting numbers will be maximal estimates. Whether all of these uses are purely statistical or whether they extend to the identification of individual residents will depend upon whether the confidentiality of the census or registration data is ensured by national law.

The limitations of the statistics on this subject are particularly noticeable when international comparisons are attempted. The ambiguity of the unmodified term "nationality" has led to some confusion. As we stated in an earlier section of this chapter, "nationality" has been used in demographic literature and in official publications of some countries to mean ethnic origin [23] rather than to indicate country of citizenship. To avoid this ambiguity, either the term "ethnic nationality" or the term "legal nationality" will be used here, or their respective synonyms, "ethnic origin" and "country of citizenship." Actually, country of citizenship is not a very satisfactory indicator of ethnic origin.

The item of citizenship is not only sometimes confused with country of birth, but it is also evidently a delicate subject in countries where members of some minority groups are insecure in their status.

International comparisons are impaired by the wide variations in the categories used in tabulations and by differences in the treatment of stateless persons and persons with dual

---

[22] United Nations, *Demographic Yearbook, 1963*, p. 37.
[23] For an example of this use. see: Zdeněk Jureček. "Narodnostni složeni obyvatelstva" (Structure of the population by nationality) in *Demografie* (Prague). 10(2):109. 1968.

citizenship. Sometimes the categories are not specific countries but are adjectival forms such as Austrian, German, or even "British" to cover such Commonwealth countries as Canada, Pakistan, and Kenya. This last use is unsatisfactory, particularly if the statistics are used to infer the ethnic origin of the alien population.

Some national census organizations allocate stateless persons to the country of their last allegiance (or even that of their parents), whereas others show them as a separate class in published tables. The treatment of persons of dual nationality is also variable; but, if one's citizenship relates to the country taking the census, one is usually simply shown as a citizen thereof. In view of the complexity of national laws, a full description of the citizenship status of the population is obviously not the function of census tables, however.

As with the statistics on country of birth, the coverage of those on country of citizenship may differ appreciably depending on whether the census is taken on a *de jure* or on a *de facto* basis. The latter would almost invariably count more aliens.

Quality of the Statistics

The possible ambiguities that have been described as well as the temptation for aliens in some countries to report themselves as citizens undoubtedly affect the quality of the statistics. The alien population is likely to be underreported, and the country of citizenship may be reported erroneously by the reported aliens. There is little quantitative evidence, however, on the extent of the biases.

Measures

In computing the percentage distribution by citizenship, one must consider the appropriate base. If there is interest in the relative importance of the alien or naturalized element in the population, then the total population should be used. For example, we have for Sweden in 1960:

Total population	7,495,129
Aliens	190,621
Percent of total	2.5

If, however, we are interested in the extent to which the foreign born become naturalized, the foreign-born population is the appropriate base. For Sweden again:

Foreign born	299,879
Naturalized	109,258 (by subtraction)
Percent naturalized	36.4

For the latter calculation, the assumption is that all of the native population are citizens. This would not be an appropriate assumption for a country like Ceylon.

The proportion naturalized among persons born in a given foreign country is a useful index of the assimilation of immigrants from that country. Alternatively, such an index could be computed for the foreign born by previous country of citizenship. Table 9-7 illustrates the derivation of the percent naturalized for the United States in 1950 using the foreign born by country of birth. This is defined as:

$$\text{Percent naturalized} = \frac{N_i}{F_i} \times 100, \qquad (1)$$

where N_i is the number of naturalized citizens born in country i and F_i is the total number of persons born in the same country.

Table 9-7. — Computation of the Percent Naturalized for the Foreign-Born White Population of the United States, by Country of Birth: 1950

[Based on 20-percent sample]

Country of birth	Foreign-born population reporting on citizenship (1)	Naturalized Number (2)	Naturalized Percent (3)	Ratio to total for all countries (4)
Total	9,395,885	7,466,335	79.5	100
Austria	383,585	325,805	84.9	107
Canada-French[1]	219,795	171,925	78.2	98
Canada-Other[1]	672,450	533,820	79.4	100
Czechoslovakia	254,955	215,520	84.5	106
England and Wales	544,360	437,485	80.4	101
Germany	922,505	787,445	85.4	107
Ireland	462,160	402,950	87.2	110
Italy	1,347,445	1,127,600	83.7	105
Mexico	411,800	118,950	28.9	36
Norway	188,870	166,310	88.1	111
Poland	811,520	619,170	76.3	96
Sweden	304,270	274,965	90.4	114
U.S.S.R.[2]	810,815	706,005	87.1	110
Other	2,061,355	1,578,385	76.6	96

[1] Based on language spoken before coming to the United States.
[2] Excludes persons reported as of Ukrainian origin.

Source: Basic data from *U.S. Census of Population: 1950*, Vol. IV, *Special Reports*, Part 3, Chapter A, *Nativity and Parentage*, 1954, table 17.

Factors Important in Analysis

The percent naturalized is likely to vary considerably by sex, with male immigrants usually being more likely to be naturalized, or to be naturalized earlier, than females from the same foreign country. The length of time that an immigrant has been in the country, as measured by his year of immigration, is also important. Furthermore, the percent naturalized may differ depending upon whether the immigrant settled in an urban or in a rural area or upon his region of residence. Related to these residential factors is the question of whether the immigrant lives in an area where there are many of his former compatriots or whether he is relatively isolated from them.

LANGUAGE

Several kinds of questions about language spoken have been included in censuses to meet various objectives. First, language has been used as an indicator, or a supplementary indicator, of ethnic origin. For this purpose the language spoken by the person at home in his early childhood or the language of his parents is better than the language spoken in his mature life. The second interest is in current ability to speak various languages. This form of the question yields data on the linguistic skills of the population, both native and foreign born. A third interest is in the ability of the foreign stock or of specific ethnic minorities to speak the official language of the country. The underlying interest here is in assimilation and integration. Still another form of the question may be in terms of the language usually spoken by the person at the time of the census. Unfortunately, statistics based on a question phrased to meet one of those objectives is sometimes used to meet another objective, often with misleading results.

International Recommendations

For the 1970 censuses, language is on the United Nations list of "other useful topics."

The detailed U.N. proposal for 1970 is as follows:

"There are three types of language data which can be collected in censuses. These are:

(a) **Mother tongue,** defined as the language usually spoken in the individual's home in his early childhood.

(b) **Usual language,** defined as the language currently spoken, or most often spoken by the individual in his present home.

(c) Ability to speak one or more **designated languages.**

Each of these types of information serves a distinctly different analytical purpose. Each country should decide which, if any, of these types of information is applicable to its own needs. International comparability of tabulations is not a major factor in determining the form of the data to be collected on this topic.

In compilation of data on usual language or on mother tongue, it is desirable to show each language that is numerically important in the country, and not merely the dominant language.

Information on language should be collected for all persons. In the tabulated results for children under five years of age, the criterion for determining language for children not yet able to speak should be clearly indicated." [24] It seems to us that, with regard to the last of these proposals, where the question refers to usual language or ability to speak one or more designated languages (i.e., not an interest in ethnic origin), there is no reason to ask the question of children who would not yet have learned to talk.

National Practices

The 27 countries reporting a national census item on language in the 1955–64 period published statistics on the following questions:

Mother tongue............11
Usual language............13
Designated language......6
Other........................3

Six countries asked two different questions about language in their censuses.

United States.—The first question on language to be carried in the U.S. census was on "Ability to speak English." This was asked from 1890 through 1930. At the latter date, only 8.7 percent of the foreign-born population (and, of course, a much smaller percentage of the total population) 10 years old and over were unable to speak English. This question has survived, however, in the schedules of the outlying areas of the United States (Puerto Rico, etc.).

An ability to speak English does not necessarily imply that English is one's mother tongue or even that English is the language one usually speaks. Interest in mother tongue as a census question developed about the turn of the century with the increased volume of immigration and the shift toward different countries of origin. More specifically, the interest was in the ethnic origins of the foreign stock in relation to various other characteristics. Since 1910 mother tongue has been included in every census save that of 1950, but it has had a rather checkered career. The form and coverage of the question have

changed to reflect the nature and intensity of public and official interest.

This history may be summarized as follows:

Date	Question	Tabulated for—	Sampling percentage
1910	(Mother tongue written in with place of birth)	Foreign white stock	100
1920	Mother tongue	Foreign white stock	100
1930	Language spoken in home before coming to the United States	Foreign-born white	100
1940	Language spoken in home in earliest childhood	White population	5
1960	What language was spoken in his home before he came to the United States?	Foreign born	25
1970	What language, other than English, was spoken in this person's home when he was a child?	Total population	15

Additional interpretations of the question were contained in the instructions to the enumerators, and these also differed in a few particulars.

The extension of the item to the native population of native parentage reflects the existence of sizable ethnic groups that have held on to their original language for generations. Early in this chapter, we described the attempt made in the 1940 census to identify the Mexican-American population of the Southwest by means of Spanish mother tongue. There are other important foreign language groups, such as the Louisiana French and the so-called "Pennsylvania Dutch" (who speak a German dialect) who cannot be identified when the question is confined to the foreign stock (first and second immigrant generation).

The American data on mother tongue have been published in considerable detail of categories and cross-classifications, and separate subject reports have been devoted to the topic. The cross-classification for the foreign born with country of birth is especially useful.

Uses and Limitations

The principal uses of the various questions on language have already been mentioned. Regarding the determination of questions on ethnic origin, the United Nations points out, "Language, and particularly mother tongue, is probably a more sensitive index for this purpose then either country of birth or country of citizenship because linguistic differences tend to persist until complete cultural assimilation has taken place. Common ancestral customs may be reflected in the mother tongue of individuals long after these persons have changed their citizenship. Thus, important ethnic groups, not only among foreign born alone but also among native born or second generation population groups, may be distinguished by language differentials." [25]

Some immigrant groups in some countries, however, are **rapidly assimilated and lose their ancestral language even in the second generation.** Thus, in Panama in 1950, almost

---

[24] United Nations, *Principles and Recommendations for the 1970 Population Censuses*, pp. 49–50.

[25] *Demographic Yearbook, 1963*, p. 39.

half of the foreign-born population currently spoke a foreign language, whereas all but 5 percent of the native population spoke Spanish.[26] For the foreign born, however, mother tongue is very useful in making ethnic distinctions that cannot be made from country of birth, for example, Flemings from Walloons within Belgium, Czechs from Slovaks within Czechoslovakia, and Tamils from Sinhalese within Ceylon.

Listings of mother tongue in census reports frequently group the languages by linguistic affinity. From the standpoint of ethnic origin, however, such affinities may be misleading. Thus, Yiddish derives mainly from a German dialect, but its Jewish speakers are of Near Eastern origin to a considerable extent and without strong cultural ties to the Germans. Welsh and Gaelic are Celtic languages and English is basically Teutonic, but for hundreds of years the English, Scotch, and Welsh have shared a common national life and have intermarried extensively. Considerations of international comparability in lists of languages, such as grouping and ordering languages in tables, the treatment of residual categories, and the distinction between languages and dialects are dealt with in an early study by the United Nations.[27] No standard nomenclature was proposed, however.

Another use of the question on mother tongue is that it is a rough indicator of religion in certain cases. Most Italian-speaking persons are Roman Catholic and most Serbian-speaking persons are Greek Orthodox. Persons of Yiddish mother tongue are almost certainly Jews, but only those Jews originating in Eastern Europe spoke Yiddish. Thus, mother tongue may be used as a proxy variable in studying the effects of religion upon fertility, but there are a number of cautions that must be observed in such an analysis.

As an indicator of assimilation, usual language, ability to speak the country's official language (or languages), and mother tongue can all be used. The relative value depends somewhat upon the immigrant generation. For example, mother tongue is not a particularly valuable indicator for the immigrants; but it is for subsequent generations. In comparing the immigrants from various countries, it would be interesting to correlate the proportion who are able to speak the official language with the naturalization rate, holding constant year of immigration. The first two measures, along with others, could also be used to construct an index of assimilation.

In countries like Belgium, Switzerland, and India, where linguistic pluralism is a major factor in national life and in political institutions, extensive use may be made of classifications by language. There are official linguistic regions in Belgium, for example, for which statistics are published.

Usual language, or the language currently spoken, is probably the most meaningful question on ability *per se* to speak various languages. There is not very much demographic interest in this topic, however. Reported ability to speak designated languages may cover a very wide range of fluency, paralleling in this respect the simple question on literacy discussed in chapter 11. Still less validity attaches to the reporting of volunteered languages, which the person has studied at one time in secondary school or university. A mass operation like a census is probably not a good means for studying the distribution of relatively rare abilities in a population.

Quality of the Statistics

Relatively few measures of the quality (accuracy) of the statistics on language are available. In the U.S. census of 1960, the nonresponse rate from the initial tabulation on mother tongue was 7.3 percent of the foreign born. This was without the benefit of any allocations, however. In later tabulations, some allocations were made on the basis of country of birth, selecting those countries from which 90 percent or more of the immigrants spoke a single mother tongue, as found in an earlier census. (For example, almost all of the French-born immigrants were also of French mother tongue.) This procedure reduced the nonresponse rate to 3.9 percent of the foreign born.[28]

RELIGION

Religion is another item mentioned by the United Nations in its consideration of ethnic origin. This item is of considerable sociological interest in its own right and is, indirectly, also of demographic interest.

International Recommendations

For the 1970 censuses, religion is listed as one of the "other useful topics." The recommendation contains brief definitions of two alternative concepts.

"For census purposes, **religion** may be defined as either (a) religious or spiritual belief or preference, regardless of whether or not this belief is represented by an organized group, or (b) affiliation with an organized group having specific religious or spiritual tenets. Each country which investigates religion in its census should use the definition most appropriate to its needs and should set forth, in the census publication, the definition which has been used.

"The amount of detail collected on this topic is also dependent upon the requirements of the country. It may, for example, be sufficient to inquire only about the religion of each person; on the other hand, respondents may be asked to specify, if relevant, the particular sect to which they adhere within a religion.

"For the benefit of users of the data who may not be familiar with all of the religions or sects within the country, as well as for purposes of international comparability, the classifications of the data should show each sect as a subcategory of the religion of which it forms a part. A brief statement of the tenets of religion or sects which are not likely to be known beyond the country or region is also helpful."[29]

National Practices

Despite the lukewarm endorsement by international organizations, religion has been and continues to be fairly widely represented in population censuses. South America appears to be the continent where it is given least attention. According to the 1963 and 1964 *Demographic Yearbooks* of the United Nations, the number of countries in each continent that had published census statistics on the subject for the 1955–1964 decade was as follows:

[26] Panama, Direccion de estadística y censo, *Censos nacionales de 1950, Quinto censo de poblacion*, Vol. I, *Caracteristicas generales*, 1954, table 117.

[27] United Nations, *Tabulations of Data on Languages in Population Censuses*, Studies of Census Methods, No. 14, August 1949, pp. 6–9.

[28] *U.S. Census of Population: 1960, Subject Reports*, PC(2)–1E, *Mother Tongue of the Foreign Born*, 1966, p. VIII.

[29] United Nations, *Principles and Recommendations for the 1970 Population Censuses*, pp. 57–58.

World total	79
Africa	12
North America	19
South America	3
Asia	18
Europe	13
Oceania	14

The statistics in the *Yearbook* are for major religious faiths, but separate statistics are available in many of the national sources on specific denominations or sects.

More recent statistics on religion are published in the 1971 *Demographic Yearbook* of the United Nations. This volume also contains international statistics pertaining to the topics discussed in preceding sections—country of birth, citizenship, nationality, and language.

United States.—The United States has never included a question on religion in its decennial census although such a question was seriously considered for the 1960 census. The way in which the negative decision was arrived at is described in great detail by two writers.[30]

In its Current Population Survey for March 1957, however, the U.S. Bureau of the Census did include the question, "What is your religion?" Some of the statistics from this survey were published in Series P-20, No. 79, and a cross-classification with fertility (children ever born) was presented in the *Statistical Abstract of the United States, 1958.*

Uses and Limitations

As an indicator of ethnic origin, religion is essentially of supplementary value. If race or country of origin is also known, religion can be used to make further distinctions among ethnic groups. Thus, among those originating in Northern Ireland, "Roman Catholic" would tend to represent descendants of the long-established Celtic population whereas "Protestant" would tend to represent descendants of immigrants who came from Great Britain, notably in the seventeenth century, many of whom were of Anglo-Saxon or Norman stock. Of course, there are some Protestants of old Irish descent and some Catholics of British descent and processes of conversion and intermarriage are continually going on. In a few instances, however, religion alone is an adequate ethnic indicator in demographic statistics. For example, French censuses of Algeria used the category "Muslim" to indicate the indigenous North African population as distinguished from immigrants from France and other countries bordering the Mediterranean on the European side.

From the standpoint of analysis in social science, religion is a useful personal characteristic because it is associated with a variety of differences in attitudes, statuses, and behavior. Religion as an essentially ascribed characteristic functions usually as the independent variable in the analysis. In most cases, however, religion can be changed at the will of the person, even though such changes are relatively infrequent in most cultures. In the specific field of demography, there has been great interest in the relationship between religion and fertility. The relatively high fertility level of Roman Catholics in such economically developed countries as the Netherlands, Canada, and the United States is perhaps the most outstanding object of interest; but there are other churches that encourage or discourage large numbers of offspring on the part of their members. The Hutterites and the Amish, relatively small Christian sects in North America, have very high fertility levels.

Differences in religion are also associated with differences in mortality, nuptiality, and migration. Differences in mortality and morbidity arise not so much through genetic differences among religious groups as through the mediation of differences in diet, living habits, and occupation.[31] Likewise, religious tenets and practices affect celibacy (through monasticism, etc.), age at marriage, the permissibility of divorce or annulment of marriage, the remarriage of widowed or divorced persons, toleration of consensual marriage, and season of marriage. Finally, the accessibility of a congregation of one's own faith may affect migration and hence residential distribution. An extreme case of the effect of religious beliefs on migration is given by the prohibition of travel abroad from India to certain caste Hindus down to recent times.[32]

From the standpoint of international comparability, one defect in the census statistics on religion is the fact that some countries ask for membership, others for religious belief or preference, and still others do not specify the nature of the question.

---

[30] Charles R. Foster, *A Question on Religion*, University of Alabama Press for the Inter-University Case Program, No. 66, 1961; and Dorothy Good, "Questions on Religion in the United States Census," *Population Index*, 25(1): 3–16, January 1959.

[31] Saxon Graham, "Social Factors in Relation to Chronic Illnesses," in *Handbook of Medical Sociology*, Howard E. Freeman, Sol Levine, and Leo G. Reeder, eds., Englewood Cliffs (New Jersey), Prentice-Hall, 1963, pp. 80–89.

[32] M. K. Gandhi, *The Story of My Experiment With the Truth*, Vol. 1, Ahmedabad, Navajvan Publishing House, 1927, p. 100 et passim.

SUGGESTED READINGS

Canada. Dominion Bureau of Statistics. *1961 Census of Canada.* Series 1.3, *Population.* Bulletin 1.3–8, *Religion by Ethnic Groups.* 1964.

Clarke, John I. *Population Geography.* Oxford, Pergamon Press, 1966. Pp. 93–102.

Damon, Albert. "Race, Ethnic Group, and Disease." *Social Biology,* 16(2):69–80. June 1969.

Davis, Kingsley. *The Population of India and Pakistan.* Princeton, N.J., Princeton University Press, 1951. Pp. 156–159; 162–194.

Duncan, Otis Dudley, and Duncan, Beverly. "A Methodological Analysis of Segregation Indexes." *American Sociological Review,* 20(2):210–217. April 1955.

Eckler, A. Ross, and Zlotnick, Jack. "Immigration and the Labor Force." *Annals of the American Academy of Political and Social Science,* 262:92–101. March 1949.

Foster, Charles R. *A Question on Religion.* University of Alabama Press for the Inter-University Case Program. No. 66, 1961.

Good, Dorothy. "Questions on Religion in the United States Census." *Population Index,* 25(1):3–16. January 1959.

International Labour Office. *World Statistics of Aliens: A Comparative Study of Census Returns, 1910–1920–1930.* Studies and Reports. Series O (Migration) No. 6. Geneva, 1936.

Lacroix, Max, and Adams, Edith. "Statistics for the Cultural Assimilation of Migrants." In Population Investigation Committee (London), *Cultural Assimilation of Immigrants,* supplement to *Population Studies.* March 1950.

Lorimer, Frank. *The Population of the Soviet Union: History and Prospect.* Geneva, League of Nations, 1946. Pp. 50–65.

New Zealand. Department of Statistics. *Population Census, 1961.* Vol. 6. *Birthplaces and Duration of Residence of Persons Born Overseas.* 1964.

Ryder, N. B. "The Interpretation of Origin Statistics." *Canadian Journal of Economics and Political Science,* 21(4):466–479. November 1955.

Schmelz, U. O., and Glikson, P. *Jewish Population Studies.* Jerusalem, Hebrew University, 1970.

Srinivas, M. N. *Social Change in Modern India.* Berkeley, California, University of California Press, 1966. Pp. 94–100.

Taeuber, Irene B. "Hawaii." *Population Index,* 28(2):97–125. April 1962.

Taeuber, Karl E., and Taeuber, Alma F. *Negroes in Cities.* Chicago, Aldine Publishing Co., 1965. Pp. 195–245.

United Nations. *Birthplace, Nationality or Citizenship, and Language: Problems of Census Enumeration and Definition.* Studies of Census Methods, No. 6. May 10, 1948.

———. *Birthplace Tabulations in Recent Censuses.* Studies of Census Methods, No. 7. March 1949.

———. *Demographic Yearbook,* 1971. Pp. 26–32.

———. *Handbook of Population Census Methods.* Vol. III, *Demographic and Social Characteristics of the Population.* Studies in Methods, Series F, No. 5, rev. 1. 1959. Pp. 12–24; 45–50.

———. *Population Census Methods.* Population Studies, No. 4, 1949. Pp. 45–57; 60–78.

———. *Tabulations of Data on Languages in Population Censuses.* Studies of Census Methods, No. 14. August 1949.

U.S. Bureau of the Census. *U.S. Census of Population: 1960.* Vol. I, *Characteristics of the Population.* Part 1, *United States Summary.* 1964. Pp. XLI–XLVI.

———. *U.S. Census of Population: 1960. Subject Reports,* PC(2)–1E. *Mother Tongue of the Foreign Born.* 1966.

———. *U.S. Census of Population: 1960. Subject Reports,* PC(2)–1A. *Nativity and Parentage.* 1965.

———. *U.S. Census of Population: 1960. Subject Reports,* PC(2)–1B. *Persons of Spanish Surname.* 1963.

van den Boomen, J. H. "La variable etno-demográfica." *Revista interamericana de ciencias sociales.* Vol. 2, special number, *Tipología socioeconómica de los países latinoamericanos.* Washington, Pan American Union, 1963. Pp. 85–135.

Willcox, Walter F. *Studies in American Demography.* Ithaca, New York, Cornell University Press, 1940. Pp. 408–418.

CHAPTER 10

Marital Characteristics and Family Groups

As we go from the more biological characteristics of the individual, such as sex, age, and race, to those characteristics involving social or economic relationships with other individuals, organizations, and institutions, the problems of definition become more complex. Because these relationships are affected by and often subject to control by customs and laws that vary among nations, the objective of attaining international comparability among the census statistics on these aspects of population composition becomes more difficult to achieve. Marital status is a demographic characteristic involving biological, social, economic, legal, and, in many cases, religious aspects. Marital status is a most important factor in population dynamics as it affects fertility tremendously and mortality and migration to a lesser extent. Also, its effect on other social and economic characteristics, such as school attendance and labor force participation, is very important in the late adolescent and young adult age groups.

The type of information on marital status that can be obtained from a census or survey is the subject of the first section of this chapter. In some countries, continuous population registers are also a source of data on marital status. Data on marriage and divorce obtained through a registration system for vital events, likewise provide numerators for rates. They are treated in chapter 19.

The principal sources of statistical information on family groups are the same as those on marital status, namely, censuses, surveys, and population registers. Family groups are the subject of the second major section of this chapter. A change in marital status is the usual prelude to the formation or dissolution of a family or subfamily.

MARITAL STATUS

Concepts and Classifications

Systems of classifying the population by marital status vary from country to country in accordance with prevailing marriage laws and customs. Nonetheless, efforts have been made to bring consistency into the use of terms and classifications by the several countries of the world. The results of these efforts have been set forth in the recommendations of the Population Commission of the United Nations for the 1970 censuses of population [1] and are incorporated in the following paragraphs.

Basic Categories of Marital Status. — The United Nations includes the following categories in its minimum list: (a) Single (never married), (b) married and not legally separated, (c) widowed and not remarried, (d) divorced and not remarried, and (e) married but legally separated. In published reports, an additional category for persons with marital status not reported is appropriate if the edited and tabulated data contain frequencies for such persons; in many countries with advanced census procedures, nonresponses on marital status are replaced with entries in the editing operation.

In countries where extra-legal (*de facto*, consensual) or common-law unions are numerous and are readily acknowledged by respondents, the Statistical Office of the United Nations recommends that a category be added to distinguish them from legal unions. Otherwise it would classify them as "married," as indicated in its *Demographic Yearbook*. Where the number of persons with annulled marriages is numerous, the United Nations recommends that still another category may be established for them; otherwise, these persons should be classified according to their marital status before the annulment (usually single). Separated persons who are not legally separated are ordinarily included as a subdivision of married persons. If this procedure is followed, it should be noted in the country's statistical publication.

Information on marital status is generally presented for persons above a minimum age in accordance with the usual lower limit of age at marriage in the country. The United Nations recommends that tabulations of marital status should distinguish at least between those under 15 years of age and those 15 years of age and over; but in some countries other minimum ages are used, such at 14, 16, or 18.

The categories set forth in this section up to this point are generally sufficent except in countries where national needs require information on such special practices as concubinage, polygamy, and inherited widows. Modifications to satisfy these needs should be made within the framework of the basic classification which was presented. Moreover, in countries where the number of *de facto* unions is large and the degree of stability of these unions varies over a wide range, separate data on duration of the union would be very useful.

The frequencies observed in the marital status categories depend not only upon such demographic factors as age-sex structure and mortality but also upon legal and cultural factors. For example, the remarriage of widows (but not of widowers) is quite rare in India. Some countries do not permit divorce, and elsewhere variations in the proportion divorced reflect the strictness or laxity of the divorce laws. Even within the United States, differences in State laws regarding divorce and annulment affect the census statistics on marital status for States, and erroneous

---

[1] United Nations, *Principles and Recommendations for the 1970 Population Censuses*, Statistical Papers, Series M, No. 44, 1967, pp. 52–53.

conclusions can be drawn on the frequency of marital dissolution from gross comparisons of the proportion divorced among States.

Additional Marital Status Concepts. – The analytical usefulness of data on marital status can be increased by making subdivisions or combinations of the standard categories. For example, the categories married (including separated), widowed, and divorced are often combined into the category "ever married," which includes all persons who have ever been married and is the complement of the single population.

The terms "bachelor" and "spinster" are defined in the U.N. *Multilingual Demographic Dictionary* as "those who have never been married," with no mention of age. However, in some countries these terms are used primarily in referring to single persons well above the usual age at first marriage, say persons approximately 35 years old and over for men and 30 years old and over for women.

Marital history is a term which refers to dates of changes in the marital status of a person in the course of his entire life or up to a particular point in life, such as the present. Information required to trace the marital history of a person includes answers to questions on number of times married, when each marriage occurred, and when and how each previous marriage ended.

Age at first marriage is probably the most useful fact about women's marital history for the study of their fertility. Information on this subject can be obtained by asking a single question, but the reporting is probably improved by asking a preliminary question on whether the woman has been married more than once. Data on the spacing of children in relation to the date of the first marriage, or the number of years since the first marriage, are also highly useful for the analysis of fertility.

Information for men regarding their marital history is very valuable for studying factors in the permanence of marriage when it is related to their social and economic characteristics. Information for children on the marital status of their parents provides data on orphans (with both parents dead), half orphans (with one parent dead), children of divorce (with parents divorced), and illegitimate children (with parents not married to each other).

In most Latin American countries, consensually married persons are classified separately. Around 1950 the percent of the population 15 years old and over living in stable consensual unions in Latin America varied from 10 percent or less in Bolivia, Colombia, Costa Rica, and Chile to about 40 percent in Haiti and Guatemala.[2]

United States. – Information on marital status has been published in the census reports of the United States for persons 15 years old and over since 1890, and 14 years and over since 1940, with relatively minor changes in the published categories. Those of the 1960 census were: single (never married); married, spouse present; married, spouse absent – separated; married, spouse absent – other; widowed; and divorced. (The category "married, spouse absent – other" was identified from inspection of the household roster.)

The question on marital status as it appeared on the census schedule in 1970 was generally the same as in 1960. It was phrased as follows: What is each person's marital status? (Now married, widowed, divorced, separated, never married.)

No census statistics on the separated were shown prior to the census of 1950. These persons were subsumed within "married." In present practice persons reported as separated include those legally separated and those otherwise absent from the spouse because of marital discord.

Uses and Limitations

Information on marital status classified by age and sex is used to show the extent to which people of given ages are married, have failed to marry, or have become widowed or divorced and have not remarried. This information has value for the light it throws on the participation by adults in marital behavior, with special reference to the degree to which adults spend their mature years as married persons. The analysis of data on marital status becomes especially fruitful when these data are classified further by measures of ethnic origin, education, occupation, and income. This type of detailed analysis shows the difference in population growth and composition from one marital status group to another and from one point in time to another. Such differences have meaning in the study of sociological and medical problems related to bachelorhood, spinsterhood, widowhood, and divorce. They have meaning also in the study of such economic problems as the factors which affect the supply of labor and the relation of dependency to marital adjustment of the person or members of his family.

In addition to these uses of the census and survey data in their own right, an extremely important use in countries without an adequate vital statistics system is to furnish a basis for estimating the number of marriages and divorces in past years.

Absolute numbers on marital status for the total population or for the "adult" population (e.g., 15 years old and over) are of some value for inventory purposes, e.g., knowing the total number of widows in the population in an area. However, such statistics have limited usefulness, and if meaningful comparisons are to be made among areas with respect to the prevalence of widowhood, marriage, etc., a fairly detailed cross-classification by age is required.

International comparability is impaired by different national practices, as described above, with respect to the categories used in the classification. (We should not lose sight of the fact that some of the differences reflect unusual complexities of the area's marital institutions.) Distributions by marital status are affected by the concept employed in taking the census, i.e., *de facto* or *de jure*. The *de facto* principle may yield statistics on marital status (as well as on household relationship, size and composition of household, etc.) that do not reflect the usual situation of the persons concerned. In the case of marital status, the categories "married, spouse present" and "married, spouse absent" would be particularly affected, with the former tending to be understated and the latter overstated.

Quality of the Statistics

Response Bias and Response Variance. – In reporting marital status, respondents frequently introduce several types of biases which tend to affect the quality of the statistics adversely. These biases usually result from conscious efforts on the part of the respondent to conceal unpleasant facts about difficulties in the marital experience of themselves or others. The direction of the bias is generally toward reporting more persons as married and fewer as unmarried than should be so reported.

Two ways to detect and to estimate the biases in data on

---

[2] Jacob S. Siegel, "Informaciones demográficas para la formulación de programas de vivienda con especial referencia a America Latina," (Demographic information required for housing programs with special references to Latin America), *Estadística*, 21 (79): 241, table 7, June 1963.

marital status are (1) analyzing internal consistency of marital status statistics, (2) comparing distributions of vital statistics and census statistics for consistency. Matching responses on marital status from two or more sources for identical persons yields information on both response variance and response bias since it produces measures of both gross and net differences.

The number of persons reported as "separated" should be identical for men and women in a monogamous society, but actually the number of women reported as separated (in nearly all countries that use this category) exceeds that of men. As an extreme example, half again as many women as men are so reported in the United States.

Matching studies.—Evidence from the matching of records from the 1960 census of the United States and the Current Population Survey shows that the categories "married" and "single" are very consistently reported but that other categories are not.

Caution is appropriate in the interpretation of marriage, birth, or death rates by marital status, because these rates may be seriously affected if the vital data in the numerators have different biases in the reporting of marital status from those of the census data in the denominators. Death rates by marital status in 1959–1961 were corrected by Kitagawa and Hauser,[3] so as to bring the numerators (deaths by marital status) into agreement with the denominators (population by marital status) through the device of matching vital and census records and accepting the census entry for marital status in both numerator and denominator. For persons between the ages of 35 and 64 years by marital status, color, and sex, the magnitude of the corrections of the death rates varied from 2 to 16 percent, with all the corrections raising the death rates of the single and married and lowering them for widowed and divorced.

Another example of biases in reporting that are not the same in vital statistics and census statistics should be cited. This is the tendency for persons below the legal limit for marriage to report erroneously high ages on applications for marriage license forms in order to marry without parental consent, whereas persons with premarital conceptions tend to report erroneously early dates of marriage on census forms so as to conceal the fact that the first child was conceived before marriage.

Nonresponse and Inconsistent Responses.—The quality of data on marital status is lowered if a false entry is made on the subject, if no entry is made, or if the entry is obviously inconsistent with entries on other characteristics of the person. Little harm is done, however, if the missing or inconsistent information on marital status can be deduced from closely related information on other items. Thus, the entry "wife" of household head is accepted by the U.S. Bureau of the Census as evidence that the person is a married woman and she is so coded (except in the rare event that two or more persons in the household have the entry "wife").

Measures and Analysis of Changes

Age and Sex as Variables.—Meaningful measurement and analysis of changes in marital behavior require that the basic data be available by age and sex. The reason is that shifts from one marital status to another are so very highly associated with the period of life through which the person is passing. Moreover, because of the usual role of the man as chief breadwinner

(despite increasing proportions of women in paid employment) and because of the usual role of the woman as the one in charge of childrearing and homemaking, the social and economic attributes which tend to maximize duration of marriage for men are generally not the same as those for women.

Marital status may be used as either a dependent or independent variable in demographic analysis in cross-classification with any other variable, such as age. The usual situation is to present a percent distribution by marital status for each year of age or each age group of the population; in this situation, it may be said that marital status is being regarded as the dependent variable and age as the independent variable. But if the age distribution is analyzed for persons within each marital status category, age is the dependent variable and marital status is the independent variable. Thus, when the interest in the variable is substantive it is considered dependent. But when the interest in the variable is as a control, the variable is considered independent.

The number of married men tends to be approximately equal to the number of married women in a monogamous country, except in atypical cases where a large proportion of one sex may be living away from their usual place of residence (such as members of the armed forces living abroad). However, at the young and old extremes of age, the numbers of married men and married women are usually quite different, largely because women tend to marry at younger ages and to survive to older ages than men. (We are comparing here the numbers of married men and women in the same age group, not the ages of husbands and wives.) These differences in age at marriage cause the sex ratio of married persons at young adult ages to be relatively low. Both the differences in age at marriage and in death rates cause the sex ratio at the upper middle and old ages to be relatively high, on the other hand.

We will conclude this discussion of age and sex as factors in the analysis of marital status by presenting some contrasting situations in table 10–1. This table shows, for a young age group and an old age group, percent distributions by marital status for each sex, and sex ratios by marital status. These data are given for India, where age at marriage is very low, the death rate is high, the remarriage of widows is very rare, and the divorce rate is low; for Ireland, where age at marriage is very late, bachelorhood and spinsterhood are frequent, and no divorces are reported; for Puerto Rico, where migration (especially of young men) to the United States mainland is heavy, consensual marriage is frequent, and formal marriage is relatively late; and for the United States, where age at marriage is fairly young, the death rate is low, the divorce rate is high, and the remarriage rates for widowed and divorced persons are high.

Choice of a Marital Status Group as the Base for a Measure.—Trends or variations in marital behavior can be measured by methods involving various degrees of refinement. Each successive step in the process of refining the measure may involve limiting the denominator (base) to those who are "exposed to risk" of entering the marital status. Thus, only single persons enter first marriage; only married persons can become separated, divorced, or widowed; and only widowed or divorced persons can remarry. A crude marriage rate (or divorce rate), however, disregards these facts by lumping all marriages (or divorces) together and relating them to the entire population of all ages, including infants who cannot marry for many years in most cultures and old unmarried persons who are very unlikely to marry. Despite these shortcomings, crude rates can still be used effectively for gross

[3]Evelyn M. Kitagawa and Philip M. Hauser, *Differential Mortality in the United States: A Study in Socioeconomic Epidemiology,* Vital and Health Statistics Monographs, Cambridge, Mass. Harvard University Press, 1973, pp. 219–222.

analyses where the statistics needed for refined bases are either not available or where only a simple measure seems preferable for other reasons, such as ease of computation. Typically, the denominators of the rates in question come from census statistics and the numerators from vital statistics.

Table 10-1.—Population 20 to 24 and 65 to 69 Years Old by Marital Status and Sex, for India and Ireland, 1961, and Puerto Rico and the United States, 1960

Country and marital status	20 to 24 years old			65 to 69 years old		
	Percent by marital status		Sex ratio	Percent by marital status		Sex ratio
	Male	Female		Male	Female	
India..............	100.0	100.0	95.1	100.0	100.0	103.9
Single...............	43.8	6.0	697.8	2.8	0.4	732.0
Married[1].............	54.6	91.8	56.6	70.7	27.2	270.1
Widowed.............	0.9	1.3	66.0	25.9	71.8	37.4
Divorced and separated.	0.6	0.9	64.2	0.6	0.5	121.1
Not reported..........	0.1	0.1	74.9	0.1	0.1	90.9
Ireland.............	100.0	100.0	103.7	100.0	100.0	97.7
Single..............	92.5	78.2	122.6	27.4	25.0	107.2
Married..............	7.5	21.8	35.6	60.3	40.0	147.3
Widowed..............	(Z)	(Z)	54.5	12.3	35.0	34.3
Divorced.............	-	-	-	-	-	-
Puerto Rico........	100.0	100.0	85.9	100.0	100.0	106.5
Single...............	63.1	39.6	136.8	6.7	5.4	133.4
Married[2].............	28.3	45.2	53.7	68.8	36.2	202.6
Consensually married...	6.5	9.3	60.4	6.4	3.6	190.5
Widowed.............	0.1	0.4	15.7	12.9	48.0	28.6
Divorced.............	0.5	1.6	259.5	2.1	2.6	86.9
Separated............	1.6	3.9	34.7	3.1	4.3	75.7
United States......	100.0	100.0	95.7	100.0	100.0	87.3
Single...............	53.1	28.4	179.1	7.7	7.9	85.3
Married[3].............	44.7	67.2	63.6	77.8	50.2	135.3
Widowed.............	0.1	0.3	27.7	10.2	37.9	23.6
Divorced.............	1.0	1.8	52.5	2.7	2.7	87.1
Separated............	1.1	2.3	46.9	1.6	1.4	102.7

-Represents zero.   Z Less than 0.05 percent.

[1] Excluding "Separated," shown with "Divorced."
[2] Excluding "Consensually married" and "Separated," shown separately.
[3] Excluding "Separated," shown separately.

Source: United Nations. *Demographic Yearbook, 1962,* table 13; and 1963, table 34.

Standardization and Method of Expected Cases.— As is explained in chapter 14, several statistical devices are available for use in standardizing sets of data that are being compared. Standardizing serves the purpose of eliminating extraneous sources of variation in the data (such as differences in age composition) that may seriously affect the analysis of the subject under investigation.

An illustration of the procedure for standardizing marital status for age is given in table 10-2. This uses the method of direct standardization.

The method for obtaining the standardized figures in the table is outlined in the following steps. These steps show how to prepare the percent distribution by marital status in 1890 for males standardized for age, where the number of males in 1960 in each age group is used as the standard. Analogous steps are required to prepare a corresponding standardized marital status distribution for females.

A. Record the number of males in each age group 14 years and over in 1960 in column (1) of table 10-2.
B. Record the proportionate distribution by marital status for males in each age group in *1890* (r_a) in columns (2) to (6).

Table 10-2.—Calculation of Percent Distribution by Marital Status for Males in 1890 Standardized by Age With 1960 Age Distribution as the Standard, for the United States

Age	Male population, 1960[1] (P_a)	Distribution by marital status, 1890[2] (r_a)				
		Total	Single	Married	Widowed	Divorced
	(1)	(2)	(3)	(4)	(5)	(6)
14 to 19 years..........	7,754	1.0000	.9957	.0042	(Z)	(Z)
20 to 24 years..........	4,860	1.0000	.8081	.1889	.0025	.0005
25 to 29 years..........	5,279	1.0000	.4607	.5278	.0099	.0016
30 to 34 years..........	5,834	1.0000	.2655	.7140	.0181	.0024
35 to 44 years..........	11,749	1.0000	.1537	.8102	.0327	.0035
45 to 54 years..........	10,081	1.0000	.0915	.8440	.0602	.0043
55 to 64 years..........	7,532	1.0000	.0683	.8245	.1024	.0048
65 years and over.......	7,492	1.0000	.0561	.7063	.2335	.0040
Males, 14 years and over (ΣP_a)...........	60,581	(X)	(X)	(X)	(X)	(X)
Expected number in marital status ($\Sigma r_a P_a$)	(X)	60,580	19,292	37,431	3,682	175
Percent in marital status, standardized[3] $\left(\frac{\Sigma r_a P_a}{\Sigma P_a} \times 100\right)$........	(X)	100.00	31.84	61.79	6.08	0.29
Percent in marital status (actual)[3].......	(X)	100.00	43.67	52.30	3.80	0.23

X Not applicable.   Z Less than .00005.

[1] From U.S. Bureau of the Census, *Current Population Reports,* Series P-20, No. 159, "Marital Status and Family Status: March 1966," January 25, 1967, table 1.
[2] Adapted from *U.S. Census of Population: 1960,* Vol. I, *Characteristics of the Population,* Part 1, *United States Summary,* 1964, table 177.
[3] For ages 14 and over, 1890.

C. For each marital status, cumulatively multiply each number in step A by the corresponding proportion for that marital status in step B. The results represent the total for each marital status ($\Sigma r_a P_a$). Record these totals on line "Expected number in marital status."
D. Compute the percent distribution by marital status from the numbers obtained in step C. ($\Sigma r_a P_a \div \Sigma P_a$). Record on the line "Percent in marital status standardized."

The results in step C are interpreted as the number of males who would have been in each marital status in 1890 if the number of males in each age group in 1890 had been the same as in 1960 but the distribution of these males by marital status at each age was that of 1890. The results in step D are the percentages of males in each marital status in 1890 standardized for age according to the age distribution in 1960.

The method of expected cases is another way to increase comparability between data for two or more groups or dates. As in the other method of standardization described above, it involves holding constant one or more sources of potential disturbance in the analysis of a variable. For example, Carter and Glick calculated the excess or deficit of U.S. population 14 years old and over in each marital status category in 1960, as compared with numbers expected if the percent distribution by marital status in each age group (by race and sex) had been the same in 1960 as in 1930.[4]

The first step in applying the method of expected cases in

[4] Hugh Carter and Paul C. Glick, "Trends and Current Patterns of Marital Status Among Nonwhite Persons," *Demography,* 3(1):276-288, 1966. See especially tables 7, 8, and 9.

this study was to calculate the expected number of persons in each marital category, by age, race, and sex, by multiplying the number of persons (by race, age, and sex) for 1960 by the proportion in each marital status (by corresponding detail on race, age, and sex) for 1930.

The second step was to obtain totals by marital status, by adding the resulting products for each age, race, and sex group. The total in each marital status category was the number "expected" in 1960 if the 1930 marital status distribution within each age-sex-color group had remained unchanged in 1960.

The final step was to subtract the expected total number in each marital status category for 1960 of the second step from the corresponding observed number as shown by the 1960 census. The amount by which the observed number for 1960 exceeded or fell short of the expected number for 1960 may be summarized as follows:

Marital status of persons 14 years old and over	Excess (+) or deficit (−) in 1960 as compared with numbers expected in 1960 using 1930 standard		
	Both sexes	Men	Women
Single and widowed.....	− 10,740,000	− 5,784,000	− 4,956,000
Single.................	− 7,234,000	− 4,203,000	− 3,031,000
Widowed.............	− 3,506,000	− 1,581,000	− 1,925,000
Married and divorced...	+ 10,740,000	+ 5,784,000	+ 4,956,000
Married...............	+ 9,129,000	+ 5,189,000	+ 3,940,000
Divorced............	+ 1,611,000	+ 595,000	+ 1,016,000

This exhibit brings out the fact that the number of married persons in the United States was 9.1 million larger in 1960 than it would have been if the percent married in each age-sex-race group had not risen during the period of approximately one generation between 1930 and 1960. In fact, the number married would actually have been 10.7 million larger in 1960 if it had not been for the simultaneous excess of 1.6 million divorced persons in 1960 over the number expected at that date according to 1930 standards of divorce and remarriage. The "gross" excess of 10.7 million in the number of married and divorced persons was independent of the growth of population during the 30-year interval. Thus, the method of expected cases, as illustrated here, amounts to standardizing for age and for population growth in analyzing changes in marital status over time.

A simple and economical method of controlling on age is to limit the analysis of trends or variations in marital status to a critical age range which is most relevant to the purpose at hand as, for example, ages 35 to 44 in fertility analyses, or 15 to 19 in studies of school enrollment patterns.

Average Age at First Marriage. — The average age at marriage may be measured in a number of different ways involving different types of data and different methods and assumptions, with resulting differences in the interpretation of the figures. The median and mean age at first marriage may be measured on the basis of vital (marriage) records or on the basis of census data on marital status and age or date of (first) marriage. The figures may be derived from data relating to events occurring in a single year or period (period data) or from data relating to many years and describing the experience of birth cohorts. Finally, the method may be direct or indirect depending on whether marriage data or marital status data are employed. A discussion of some of the ways of measuring the average age at marriage will serve to illustrate the use of the alternative data and methods.

Period and cohort data. — The median age at marriage for all persons **who marry in a given year** is an example of a measure based on period data. By contrast, the median age at marriage for all persons **who were born in a given year** is an example of the use of cohort data. (The distinction between "period" and "cohort" data is explained more fully in chapter 16.) If the marriage experience of a birth cohort is cumulated throughout the life-span of the cohort, the median age at marriage can be computed for this "real" birth cohort. However, if data on age at marriage are not available for a sufficiently long period of time, the median age at marriage can be estimated for real cohorts by asking a question on age at marriage in a census or survey and reconstructing the marriage experience of persons born in each previous year or group of years from this source. (Ideally, we should include the ages at marriage of cohort members who have died or emigrated and exclude those of immigrants.)

A table based on answers to such a direct question is given in one of the subject reports of the 1960 U.S. census.[5] The distribution by age at first marriage is given for ever married males and females, 14 to 79 years old, by single years of age. The median age at first marriage is also given (table 10-3).

Table 10-3. — **Median Age at First Marriage, for Ever Married Males and Females, 14 to 79 Years Old by Single Years of Age, for the United States: 1960**

Age	Male	Female	Age	Male	Female
14 years.........	14.5	14.5	47 years.........	25.1	22.0
15 years.........	15.1	15.1	48 years.........	25.1	22.2
16 years.........	16.0	15.8	49 years.........	25.3	22.1
17 years.........	17.0	16.5	50 years.........	25.4	21.7
18 years.........	17.8	17.3	51 years.........	25.4	21.7
19 years.........	18.6	18.0	52 years.........	25.4	21.6
20 years.........	19.2	18.4	53 years.........	25.2	21.6
21 years.........	19.9	18.8	54 years.........	25.1	21.7
22 years.........	20.5	19.1	55 years.........	25.0	21.6
23 years.........	21.0	19.4	56 years.........	25.1	21.6
24 years.........	21.3	19.6	57 years.........	25.1	21.6
25 years.........	21.6	19.7	58 years.........	25.3	21.7
26 years.........	21.9	19.9	59 years.........	25.5	22.0
27 years.........	22.3	19.9	60 years.........	25.3	21.8
28 years.........	22.5	20.0	61 years.........	25.1	21.7
29 years.........	22.7	20.1	62 years.........	25.1	21.9
30 years.........	22.7	20.2	63 years.........	25.2	22.0
31 years.........	22.7	20.3	64 years.........	25.4	22.0
32 years.........	22.8	20.4	65 years.........	25.6	22.1
33 years.........	22.9	20.6	66 years.........	25.8	22.2
34 years.........	22.9	20.9	67 years.........	25.6	22.1
35 years.........	23.2	21.1	68 years.........	25.8	22.2
36 years.........	23.5	21.2	69 years.........	26.0	22.4
37 years.........	23.8	21.1	70 years.........	25.9	22.3
38 years.........	24.1	21.1	71 years.........	25.7	22.2
39 years.........	24.2	21.4	72 years.........	25.8	22.2
40 years.........	24.1	21.4	73 years.........	25.8	22.3
41 years.........	24.2	21.7	74 years.........	26.0	22.4
42 years.........	24.3	21.8	75 years.........	25.9	22.1
43 years.........	24.5	21.8	76 years.........	26.0	22.3
44 years.........	24.8	21.9	77 years.........	26.1	22.5
45 years.........	25.0	22.0	78 years.........	26.0	22.5
46 years.........	25.0	22.0	79 years.........	26.3	22.6

Since each current age defines a birth cohort (there are 66 cohorts), we should expect the corresponding median ages to rise rapidly at first and then to level off asymptotically. Very few more first marriages will occur to the older cohorts, so their medians are essentially at the ultimate level. We can also

─────────

[5] *U.S. Census of Population: 1960, Subject Reports, Age at First Marriage,* PC(2)-4D, 1966, table 2.

regard the medians over the entire age range as if they described the experience of a single cohort over its lifetime. The values for this synthetic cohort behave just about the way we expect the values for a real cohort to behave.

The chief disadvantage of using cohort data to measure the median age at marriage is that the marital experience of a group is not completed until the members are in old age. Most of the marriages of this group of old persons occurred many years earlier. Moreover, age at marriage is certainly not independent of the probability of surviving to a given advanced age. Those who use cohort data can attempt to overcome these difficulties by making estimates of future marriages by age during the remainder of life for a relatively young cohort (such as persons in their thirties or forties) and by the use of life tables to adjust for those who have died. [6]

Estimates of median age at first marriage by indirect method. — Census or survey data on marital status by age and sex can be used to approximate the median age at first marriage for the census or survey year, as outlined below:

1. Estimate the proportion of young people who will ever marry during their lifetime. (A typical value is about 90 percent in many countries. Thus, about 10 percent never marry.) The proportion who will ever marry is estimated by selecting the percent ever married for persons currently in age beyond which very few persons subsequently marry for the first time. For convenience, the age group 45 to 54 years is often used for this purpose; but for a particular country or part of a country an older age group should be selected if the percent single continues to decline substantially among persons at older ages. (See discussion below about situations where the percent single may rise at the older ages.) Experience has shown that an error of 3 or 4 percentage points in the value chosen for this purpose generally does not affect the final estimate of the median age at first marriage by more than one-tenth or two-tenths of a year.

2. Calculate one-half of the proportion who will ever marry. Using the illustrative value of 90 percent who ever marry, we arrive at 45 percent. In this situation, the median age at first marriage is reached when 45 percent of the young people have ever married (i.e., 55 percent are still single).

3. Determine the current age of young people who are at this halfway mark. This age in most countries is between 20 and 29 years. From the assumptions made and the procedures outlined, it follows that the date of the census or survey is the reference data for this median. Half of the young people of the given age who will ever marry have done so prior to the census or survey date and half are expected to marry in years to come. The estimated median age may, therefore, be assigned to the year of the census or survey. The method assumes that the mortality of single and ever-married persons is the same between the ages of 14 and 50.

A similar procedure can be followed in estimating first and third quartiles of age at first marriage.

An example of the actual calculation of the median age at first marriage by the indirect method is presented below, on the basis of data for the United States from the March 1968

Current Population Survey. [7] The calculations in step 3 are presented two ways: one using data by single years of age and the other using data by 5-year age groups.

Step 1. For males, 93.0 percent of those 45 to 54 years old had ever married. For females, the corresponding value was 95.1 percent.

Step 2. One-half the value in step 1 is 46.50 percent for males and 47.55 percent for females. Subtracting these values from 100.00 yields 53.50 percent single for males at the halfway mark and 52.45 percent single for females.

Step 3. The following figures representing the percent single by age obtained from the survey are used in determining the median age of males and females at first marriage:

Data by single years of age

| | Percent single | |
Age	Male	Female
18 years old	95.2	83.8
19 years old	91.0	69.9
20 years old	82.2	56.5
21 years old	70.2	42.0
22 years old	54.4	32.1
23 years old	45.5	24.5
24 years old	36.5	19.0

Data for age groups

| | Percent single | |
Age	Male	Female
5 to 9 years	100.0	100.0
10 to 14 years old	100.0	100.0
15 to 19 years old	97.2	89.2
20 to 24 years old	59.1	35.9
25 to 29 years old	21.4	10.3
30 to 34 years old	11.2	5.9
35 to 39 years old	7.4	5.1

Two estimates for the median age at first marriage are derived from these distributions below using the method of linear interpolation:

a. Linear interpolation of survey data on marital status by single years of age. This solution assumes a linear progression of values within a single year of age. In our illustration using 1968 data for the United States, the target is the age at which 53.5 percent of males are single. This target age is found among males at least 22 years old and not yet 23. Hence, the "median interval" is males 22 years old. In the interpolation process, it is assumed that the percent remaining single for an age applies to the midpoint of the age interval. Thus, the 54.4 percent is taken to apply to males at exact age 22.5. Interpolating linearly between the midpoint values, 22.5 and 23.5 years of age, yields an estimated median age at first marriage of 22.6 years for males. For females the corresponding value is 20.8 years.

b. Linear interpolation of survey data on marital status by 5-year age groups. This solution also assumes a linear progression of the percent single within age intervals, but over 5 years rather than a single year; and it assumes again that the percent single given for an age group applies to its midpoint. Although

[6] Such estimates are included in: U.S. Bureau of the Census, *Current Population Reports*, Series P–20, No. 108, "Marriage, Fertility, and Childspacing: August 1959." July 12, 1961, p. 9. This report also includes supplementary estimates of marriages for women who died between the date of marriage and the date of the survey.

[7] See U.S. Bureau of the Census, *Current Population Reports*, Series P–20, No. 187, "Marital Status and Family Status: March 1968." August 11, 1969. The distributions were adjusted to include estimates for members of the armed forces not covered by the survey, namely, those stationed in barracks in the United States or stationed overseas.

this procedure has serious limitations when used to study trends in age at marriage during a period of rapid change or to study differences in age at marriage among groups with quite different age distributions, the results show 23.2 years as the median age at first marriage for men and 20.9 years for women. In this illustration, the two solutions for females were similar, because the median age at first marriage is so close to age 20 (the juncture of age groups) that little interpolation is required in calculating it. By contrast the solution for males that made use of data by single years of age yielded a median about one-half year lower than that making use of data by 5-year age groups since the more exact estimate was near the center of the 5-year age group.

The median age at remarriage cannot be estimated by the indirect method by any known procedure that is analogous to that described above for estimating median age at first marriage. The way to measure median age at second, third, etc., marriage, as well as median age at marriage for all marriages regardless of order, is to ask relevant questions on marriage certificates or census forms and to tabulate the answers.

Even having marital status statistics by single years of age, however, does not ensure that the method will yield adequate results. Hajnal presents the following figures on the percent single for Swedish women in 1945: [8]

15 to 19 years	97.0
20 to 24 years	63.6
25 to 29 years	30.4
30 to 34 years	20.4
35 to 39 years	19.0
40 to 44 years	20.4
45 to 49 years	21.0
50 to 54 years	21.0

The changes shown after 35 to 39 years would not be likely to occur in an actual cohort; apparently a transition in marital behavior had been taking place. An estimate of the median age at first marriage from these figures could apply neither to a single cohort nor to a single calendar year.

For a country with a long series of quinquennial, or even decennial censuses, the median age at first marriage can be estimated for a number of true cohorts. Assume that the percent ever married is available from the census tables in 5-year age groups for each sex at 5-year time intervals. Array the percent ever married by censuses and age as follows:

1930	1935	1940	1945	1950	1955	1960	1965
15-19	15-19	15-19	15-19	15-19	15-19	15-19	15-19
20-24	20-24	20-24	20-24	20-24	20-24	20-24	20-24
25-29	25-29	25-29	25-29	25-29	25-29	25-29	25-29
30-34	30-34	30-34	30-34	30-34	30-34	30-34	30-34
35-39	35-39	35-39	35-39	35-39	35-39	35-39	35-39
40-44	40-44	40-44	40-44	40-44	40-44	40-44	40-44
45-49	45-49	45-49	45-49	45-49	45-49	45-49	45-49
50-54	50-54	50-54	50-54	50-54	50-54	50-54	50-54

Start with girls (or boys) 15 to 19 years old in 1930, that is, the cohorts of 1915-1919. Follow their history by means of the underlined cells. Suppose that when they reach 50 to 54 years of age 94 percent are or have been married. Half of this percent is 47 percent. Locate the two percentages for this 5-year age cohort that bracket 47, and compute the exact age corresponding to 47 percent, by interpolation as described above.

Estimates of mean age at first marriage by indirect

method. — The mean age at first marriage can also be approximated by the indirect method from data on marital status by age. The result has been called the "singulate mean age at marriage." [9] The singulate mean age at marriage is an estimate of the mean number of years lived by a cohort of women before their first marriage. The statistic is calculated from the proportion of males or females single in successive age groups as reported on a census schedule. The basic assumption involved in the calculation is that the change in the proportion single from age x to age x + 1 is a measure of the proportion of a birth cohort who married at that age. The percentage single among women in Israel in 1966 serves as an example of the requisite input.

Age	Percentage single
15–19 years	88.9
20–24 years	43.1
25–29 years	13.5
30–34 years	5.1
35–39 years	2.7
40–44 years	1.9
45–49 years	2.0
50–54 years	2.1
Sum	157.2

Assuming that no woman dies between her fifteenth and fifty-fifth birthdays the problem is to compute the mean age at marriage of the women marrying before they reach 50. The steps involved in the calculation may be summarized as follows:

(1) Sum the percentages single through age group 45–49 and multiply the sum by 5 $157.2 \times 5 = 786.0$

(2) Add 1,500 $+ 1,500.0$ $2,286.0$

(3) Average the percentages for 45–49 and 50–54 $1/2(2.0 + 2.1) = 2.05$

(4) Multiply the result of (3) by 50 $2.05 \times 50 = 102.5$

(5) Subtract the result of (4) from (2) $2,286.0$ -102.5 $2,183.5$

(6) Subtract result of (3) from 100 $100.0 - 2.05 = 97.95$

(7) Divide the result of (5) by result of (6) $2,183.5/97.95 = 22.3$

The procedure results in an estimate of the average number of years lived in the single state by those who marry before age 50. The estimate of the number who have married by age 50 was obtained by averaging the percentages single in the age groups 45 to 49 and 50 to 54 and subtracting the average from 100 to arrive at an estimate of the percentage who have ever married by age 50.

The considerations underlying the computations are the following. The number of years lived in the single state between the ages 15 and 50 by the entire cohort is the sum of the proportions single multiplied by 5 (the use of 5 is required by the grouping into 5-year age groups). To this figure are added the 1,500 (15 × 100) years lived by the cohort before the fifteenth birthday. The number of years lived by those

---

[8] John Hajnal, "Age at Marriage and Proportions Marrying," *Population Studies* (London), 7(2):115, November 1953.

[9] Hajnal, op. cit. pp. 111–136.

who did not marry before age 50 must be subtracted from the total. The total is then divided by the number of women who have married. The result of the division is the singulate mean age at marriage.

Proportion Who Never Marry. — In presenting table 10–1 above, attention was drawn to the very wide differences from country to country in the proportion of persons who remain single until old age. These differences were attributed mainly to contrasting patterns of lifetime marriage in various parts of the world. A more systematic analysis of data on the proportions of persons in the United States who are bachelors and spinsters is given here in order to show different approaches to the question as to what proportion of the population never marries and also to point out and explain some peculiarities in the data on percent single at the older ages. The term bachelors is assumed here to refer to single men 35 years old and over and the term spinster to single women 30 years old and over.

The values in table 10–4 can be used to trace changes in the percent single on both a period and a cohort basis for persons 30 to 74 years old during the 1930's, 1940's, and 1950's, an overall span of a little more than a generation. On a **period** basis, the proportion of men who had reached bachelorhood during the preceding five years amounted to 15.4 percent in 1930 but only 8.8 percent in 1960. The corresponding figures for women who had newly become spinsters were a drop from 13.2 percent in 1930 to only 6.9 percent in 1960. Thus, these changes approached a decline of one-half in about a generation. On a **cohort** basis, the proportion who continued to be bachelors or spinsters from one census period to another is shown by the underlined values. Thus, 15.4 percent of the men who were age 35 to 39 in 1930 were bachelors; of this same cohort, 11.2 percent were still bachelors in 1940, 8.3 percent in 1950, and 7.7 percent in 1960. Similarly, 13.2 percent of the women who were age 30 to 34 in 1930 were spinsters, and of the same cohort 7.7 percent were still spinsters in 1960. In other words, close to one-half of those who became bachelors or spinsters in 1930 and who survived until 1960 had not married in the intervening 30 years and had become so old that they would probably never marry.

If the assumption is made that the observed cohort changes in the percent single over the 30-year period will continue at the same rate during the next 30 years, a set of projected percentages single in 1990 can be obtained. Thus, if the 8.8

percent of men 35 to 39 years old who were single in 1960 is reduced by one-half during the remainder of the marriageable years of this 1960 cohort (like the corresponding 1930 cohort), only 4.4 percent of the 1960 cohort would still be bachelors when they reach age 65 to 69 in 1990. Similarly, if the 6.9 percent of women 30 to 34 who were single in 1960 is reduced at the same rate that the corresponding 1930 cohort was reduced, only 4.0 percent of the 1960 cohort would still be spinsters in 1990.

Several values for women in the columns for 1940 and 1960 are marked with a plus (+) sign to call attention to the anomaly that these percentages single were actually higher than the corresponding percentages for the same cohorts in the preceding census when the women were 10 years younger. To illustrate, 8.9 percent of the women 60 to 64 years old in 1930 were reported as single, but 9.5 percent of the same cohort were reported as single 10 years later.

Graphic Devices. — An instructive device for showing the vast differences in the distribution of the single, married, widowed, and divorced population from one age group to another is to plot the number of persons by marital status and age in a population pyramid. (See discussion of the age pyramid in chapter 8.) As can be seen in figure 10–1, the marital status-age-sex pyramid for the United States in 1960 is dominated at mature adult ages by the lung-shaped portion representing married persons, because being married is the

Table 10–4. — **Percent Single, for Persons 30 to 74 Years Old, by Age and Sex, for Conterminous United States: 1930 to 1960**

Age	Male				Female			
	1930	1940	1950	1960	1930	1940	1950	1960
30 to 34 years........	21.2	20.7	13.2	11.9	<u>13.2</u>	14.7	9.3	6.9
35 to 39 years........	<u>15.4</u>	15.3	10.1	8.8	10.4	11.2	8.4	*6.1
40 to 44 years........	13.1	12.6	9.0	7.3	9.5	<u>9.5</u>	8.3	*6.1
45 to 49 years........	11.9	<u>11.2</u>	8.7	*7.1	9.0	*8.6	7.9	6.5
50 to 54 years........	10.9	11.0	*<u>8.3</u>	7.6	9.2	8.7	*<u>7.7</u>	7.6
55 to 59 years........	10.3	10.8	*<u>8.3</u>	8.2	9.0	8.7	*7.7	8.2+
60 to 64 years........	9.9	10.5	8.6	7.6	8.9	9.3+	8.2	<u>7.7</u>
65 to 69 years........	9.3	10.3	8.7	<u>7.7</u>	*8.4	9.4+	8.4	7.9+
70 to 74 years........	*8.6	*9.9	*8.3	7.8	*8.4	9.5+	9.0	8.4+

*Lowest percent single for date.

NOTE. — See text for explanation of underline and plus sign (+).

Source: Based on *U.S. Census of Population: 1960*, Vol. I, *Characteristics of the Population*, Part 1, *United States Summary*, table 177.

Figure 10–1. — **Marital Status of Persons 14 to 74 Years Old, by Single Years of Age and Sex, for the United States: 1950**

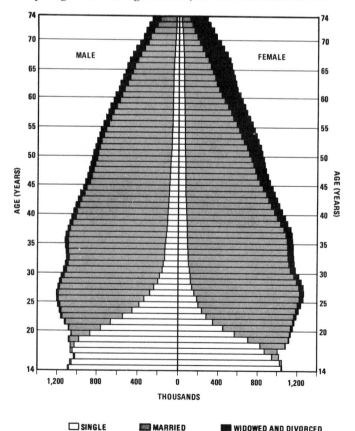

Source: *U.S. Census of Population: 1950*, Vol. IV, *Special Reports*, Part 2, Chapter D, *Marital Status*, p. 2.

norm of adulthood, and by the spindle-shaped center section with its wide base at youthful ages, representing single persons, because being single is characteristic of youth. The numbers of divorced persons were so small that they were combined with widowed persons. The considerably wider bands of "widowed and divorced" persons for women than for men reflect the larger proportions of women than men of a given age who have dissolved marriages but who do not remarry; the difference is also a reflection of higher mortality rates for men than women at each age. The entire structure of the pyramid is affected, moreover, by the somewhat older average age of men than women at first marriage and the considerably older average age of men at remarriage.

Group Variations

Marital status is recognized not only as a useful indicator of the marital experience of the adult population but also as a demographic factor which interacts with various other important aspects of demographic behavior. In addition to the basic tabulation of marital status by age and sex, then, it is also desirable to show statistics on the marital status characteristics of various social and economic subgroups within the population.

Other Factors Important in Analysis

A study in depth of the demographic aspects of marital status calls for data on several closely related variables listed below. These variables provide information on the circumstances surrounding entry into, or exit from, a given marital status; define those at risk of entering or leaving a given marital status; or otherwise reveal particulars about the person's marital history. Among the variables are several that were mentioned in earlier sections of this chapter.

Order of marriage (number of times married, or whether married more than once)
Age at, duration of, and date of marriage, for first marriage and most recent marriage
Age at, duration of, and date of widowhood, separation, divorce, and remarriage
Marital status prior to current marriage
Marital history of spouse—age at marriage, order of marriage, etc.
Type of marriage—monogamous or polygamous (by number of wives); formal or consensual union.

The first two items in the above list are probably the most often used in demographic analyses of marriage and of the marital factor in fertility. The item on date of first marriage is used to identify first marriage cohorts; that is, persons who entered marriage for the first time in a specified period such as a calendar year.

Persons who have ever married may be asked in a census or survey: (1) Date of birth (or age), (2) date of first marriage (or age at first marriage), and (3) whether the person has been married more than once (or number of times married). The answers to these questions can be used to determine several related marital characteristics:

For all persons ever married:
Date of first marriage (from a question on the subject, or by subtracting from census date the difference between current age and age at first marriage)
Age at first marriage (from a question on the subject, or by subtracting date of birth from date of first marriage)

For persons married once:
Duration of first marriage (from a question on the subject, or by subtracting date of first marriage from census date)
For remarried persons:
Interval since first marriage (subtract date of first marriage from census date).

The information on date of first marriage as well as date of birth may be recorded by year only, by year and quarter of year, by year and month, etc. The greater the precision of the date recorded, the greater the precision of the measures that are derived, but the greater the cost of processing the data. In recent censuses of the United States, year and quarter of year have been used as a compromise in tabulations on date of birth and date of first marriage. Similarly, if the data are reported in terms of completed years of age at census or at marriage or birth of successive children, the derived statistics on duration of marriage and child spacing will be imprecise.

FAMILY GROUPS

For many decades, the main use that was made in decennial censuses of the grouping of population by households was to serve as the unit of visitation in the process of enumeration. Inasmuch as the focus was on enumerating persons rather than family groups, little concern was expressed about the demarcation of individual households under circumstances of complex living arrangements in one or more structures. Gradually, however, users of census data became increasingly aware of the practical value of studying groupings of the population in terms of living arrangements, and, accordingly, these groupings came to be more sharply defined and more closely studied.

The household came to be recognized as the logical population unit for analyzing nome ownership and other aspects of housing, and the family was acknowledged as a basic economic and social unit. A host of other subjects related to the welfare of the nation, the community, and the people became better understood in the light of recurring statistics on the characteristics of family groups: living arrangements of people throughout their life span; migration of families as a unit; education of children in the context of the educational achievement of the parents; crowding in housing space; poverty and dependency; family allotments; consumer expenditures for items which are generally purchased in units of one per household; and so on.

A balanced program of demographic statistics provides data on many of the same subjects in terms of both persons and family groups. Thus, data are needed on both persons in poverty and families in poverty. Families in different ethnic or socioeconomic groups may vary widely in size (i.e., number of members).

Demographic data on family groups could be obtained from population registers or from administrative records, such as those used in carrying out social insurance or welfare programs, but by far the most frequent source of such data where the entire population is covered is the household survey or census.

United Nations Concepts and Classifications

The central concepts in the present context are "household" and "family." For most of the people, these two concepts are identical. However, some persons are not members of (private)households—persons living in group quarters such as military barracks or college dormitories, for example; and some who are members of households are not members of families (e.g., boarders). A more detailed discussion of these concepts and of the systems of classifying people in terms of their living arrangements is presented in the U.N. volume, *Principles and Recommendations for the 1970 Population Censuses.* Some idea of the issues involved in such a classification system is provided by the following discussion of United States practices.

Concepts Used in the United States

Household. — According to concepts used in recent censuses of the United States, "A household consists of all the persons who occupy a housing unit. A house, an apartment or other group of rooms or a single room, is regarded as a housing unit when it is occupied or intended for occupancy as separate living quarters. Separate living quarters are those in which the occupants do not live and eat with any other persons in the structure and in which there is either (1) direct access from the outside or through a common hall, or (2) a kitchen or cooking equipment for the exclusive use of the occupants.

"All persons who are not members of households are regarded as living in group quarters. **Group quarters** are living arrangements for institutional inmates or for other groups containing five or more persons unrelated to the person in charge. Group quarters are located most frequently in institutions, lodging and boarding houses, military and other types of barracks, college dormitories, fraternity and sorority houses, hospitals, homes for nurses, convents, monasteries, and ships. Group quarters are also located in a house or an apartment in which the living quarters are shared by the person in charge and five or more persons unrelated to him." [10]

Family and Related Concepts. — The family terminology currently used by the U.S. Bureau of the Census was developed in 1947, and most of its features have remained unchanged since that time. (Prior to that time, the term "family" denoted what is now called "household.")

Family. — By U.S. definition, a family is the entire group of (two or more) persons in a household who are related by blood, marriage, or adoption. According to this terminology, two related married couples, a couple and a related parent-child group, or any other group of two or more persons related to each other is counted as one family if the members occupy the same living quarters and eat together as one household.

Subfamily. — The U.S. Bureau of the Census defines a subfamily as a married couple with or without children, or one parent with one or more single children under 18 years old, living as members of a household and related to, but not including, the head of the household or his wife. The most common example is a young married couple sharing the home of the husband's or wife's parents. Members of a subfamily are counted as members of the family of the household head. Therefore, in statistics in the United States, the number of subfamilies is not included in the number of families.

Married couple and parent-child group. — A "married couple" is defined, for U.S. census purposes, as a husband and his wife enumerated as members of the same household. A "parent-child group" is a parent with no spouse present but with one or more single sons or daughters under 18 years old living as members of the same household. An example of a parent-child family is a widow and her single son who constitute a two-person household. Parent-child groups are included in the count of families or subfamilies on the basis of the relationship of the parent to the head of the household.

Married couple with or without own household. — In the tabulation process, married couples who are members of subfamilies or secondary families are classified as "married couples without own household" and their living arrangements are sometimes described in family analyses as "doubling." Heads of households and their wives are classified as "with own household." The proportion of couples who are without own household is often used as a measure of the need for additional housing.

Primary and secondary families. — Since 1947 the U.S. Bureau of the Census has defined a primary family as a family which includes among its members the head of a household and a secondary family as one which does not. Members of secondary families include persons such as lodgers and resident employees and their relatives who share the living quarters of a household head.

The number of secondary families in the United States diminished sharply during the 1950's and 1960's until only about one in every 500 families was a secondary family. Therefore, the U.S. census of 1970 did not recognize any families except those of household heads, and the unmodified term "families" was used for these population groupings. Statistics from the Current Population Survey, however, will continue to show separate data for primary and secondary families.

Numbers of families by U.N. and U.S. definitions. — Although available statistics do not permit exact comparisons, very close approximations can be made to the difference in count of families in the United States by U.N. and U.S. definitions. Table 10-5 shows that, using 1960 census figures, 3 percent of the families by the U.N. definition of a nuclear family would not be counted as families under the U.S. definition, and, at the same time, 5 percent of the groups counted as families in U.S. statistics would not be recognized as family nuclei under the U.N. definition.

Table 10-5. — Estimated Number of Families by Type, by United Nations and United States Definitions of a Family, for the United States: 1960

[Numbers in thousands]

Type of family	Number		Percent	
	U.N. definition	U.S. definition	U.N. definition	U.S. definition
All families..............	44,170	45,149	100.0	100.0
Nuclear families:				
U.S. families.................	42,743	42,743	96.8	94.7
Husband-wife................	39,657	39,657	89.8	87.8
Parent-child................	[1]3,085	[1]3,085	7.0	6.8
U.S. subfamilies.............	1,428	–	3.2	–
Husband-wife................	834	–	1.9	–
Parent-child................	[2]594	–	1.3	–
Other groups of related persons.	–	[3]2,406	–	5.3

– Represents zero.
[1] Includes an estimate of 893,000 parents who were family heads with single sons or daughters 18 years old and over only and 2,193,000 with single sons or daughters under 18.
[2] Limited to parents with at least one single child under 18 years old.
[3] Includes about 140,000 U.S. census families comprising only one person other than members of subfamilies; such residual persons would not by themselves constitute families.

Source: Based on *U.S. Census of Population: 1960, Subject Reports, Families,* PC(2)-4A, tables 4, 5, 6, 9, 21, 23, and 58; and *Persons by Family Characteristics,* PC(2)-4B, tables 2 and 19.

---

[10] *U.S. Census of Population: 1960,* Vol. I, *Characteristics of the Population,* Part 1, *United States Summary,* p. LV.

Primary and secondary individuals. — Persons not living in family groups are classified in the U.S. census reports as either "unrelated individuals" or "inmates of institutions," according to their living arrangements. Unrelated individuals are, in turn, classified as "primary individuals" if they are heads of households and as "secondary individuals" if they are not. An example of a primary individual is a widow who maintains a house or apartment alone or with nonrelatives only. Examples of secondary individuals are lodgers and boarders who pay the household head for their shelter and maids and other resident employees who generally receive their shelter in partial return for work they perform for the head of the household. Secondary individuals also include all persons in group quarters except inmates of institutions. The concept "unrelated individual" was designed to relate to persons in the housing and labor markets and therefore excludes institutional inmates.

Inmate of an institution. — In reports of the U.S. Bureau of the Census, inmates of institutions are persons who live in such group quarters (not housing units) as prisons and jails; hospitals for the treatment of mental illness, tuberculosis, and other chronic diseases; homes for the aged and dependent; homes for the mentally and physically handicapped; homes for dependent, neglected, and delinquent children; and homes for unwed mothers.

Other resident of group quarters. — Residents of group quarters other than institutions include residents in such living quarters as rooming and boarding houses, college dormitories, military barracks, convents or monasteries, workers' dormitories, missions, ships, general hospitals, and institutional dormitories for staff members. As noted above, all of these persons are classified by the U.S. Bureau of the Census as secondary individuals.

Relationship to Head of Household or Family. — In the United States, the designation of the head of the household and the determination of each person's relationship to the head is made at the time of enumeration. Illustrative of the categories of relationship to head of household are the following: Head, wife, child (son or daughter), grandchild, parent of head or wife, son- or daughter-in-law, brother or sister of head or wife, other relative, lodger, partner, and resident employee.

Head of household, family, or subfamily. — One person in each household, family, or subfamily is designated as the "head." The household head is the person reported as the head by members of the group, except that married women are not classified as heads if their husbands are living with them at the time of the enumeration.

Both the U.N. and U.S. principles of classification are based on the assumption that the household respondent will decide which member is the head. In complex households, however, the classification of the members by relationship depends on which person is reported as the head. Consider, for example, a three-person household consisting of an elderly widower who owns his house and shares his home with his son and his son's wife. If the widower still actively manages his business affairs, or if he is a member of an ethnic group with a partriarchal family structure, he will probably be reported as the head of the household. If so, he will be an "other male head" and his family will include a subfamily. However, if the son is reported as the head, the son will be the head of a "husband-wife" household, and the household will include no subfamily. In either case, both the U.N. procedure and the U.S. procedure would classify the household as containing one "family."

Although leaving to the household members the decision as

to who should be designated as the head may introduce a subjective element into the concept of household head, this procedure usually has the merit of describing social reality as the household members see it. In this sense, it may be superior to the arbitrary designation as the head, the household member who has the largest income or who is the oldest member.

Family (or Household) Coding and Transcription

The coding operation known as family (or household) transcription is not required if the family data are processed on an electronic computer or if punchcard equipment is used but only a few family items are to be tabulated. The operation is generally used if many family items are to be tabulated on punchcard equipment.

If punchcard equipment is used and few items on family groups are to be tabulated, the required items can be manually coded on the portion of the questionnaire used for the household (or family) head and punched on the card for this person. Thus, a highly useful minimum program of family tabulations could be built around special codes for the primary family head showing size of head's family (or household), the number of his children or dependents of the most relevant age (such as under 18 years old), and (if the head is a man) whether his wife is present. All of the head's characteristics would be available for cross-classification.

If punchcard equipment is used and many items on family groups are to be tabulated, the recommended procedure is to have a **family transcription** operation after the manual coding of occupation, place of birth, etc., has been completed. This operation would consist of "transcribing" (generally to a special transcription form) a line of coded information about each family unit (household, family, subfamily, or unrelated individual) and punching a card containing the information on the line.

In the United States, census statistics on households go back to 1890, whereas those on units now defined as families begin with the census of 1950. Figures on the basis of essentially the present definition have been reconstructed for 1940. The Current Population Survey series on families began in 1944. Differences in the contemporaneous census and CPS definitions have been very minor.

Limitations and Quality

International comparability is limited by some fairly wide differences in national definitions of households or families. Given an official definition, moreover, the statistics are also affected by how faithfully enumerators and respondents observe it. Even within the United States slight changes in definition from one census to another limit comparability. Finally, as with statistics on marital status, household and family data may be limited by the extent of response bias and variance and by nonresponses.

Types of Household and Family Statistics

Three types of statistics on family groups are most frequently provided: (1) family composition, (2) characteristics of the head of the group, and (3) characteristics of the wife of the head (if any) and of other members of the group. Statistics on household composition also frequently include size of household and the number of families (or of nuclear family groups) in the household.

Size of Private Household or Collective Household — The term size of household ordinarily relates to the entire number of persons—related and unrelated—who comprise one private household. As pointed out in the section above on "Household," countries differ with regard to the classification of lodgers, partners, boarders, and resident employees as members of the household of the person who owns or rents the housing unit used as their living quarters. Moreover, in some countries statistics on size of household are published only for "family households" comprising two or more related persons living together. For these and other reasons, the illustrative data in table 10–6 on size of household for 14 countries are not entirely comparable, as footnotes on this table and in the original sources will show. The countries selected for presentation here illustrate the extremely wide range in size of households and in average (mean) size of household.

In computing the **mean** size of household, the preferred figure for the numerator is the population in households, but the only available figure for some of the areas or dates that are being compared may be the total population including persons in collective households (group quarters). Thus, mean size of household

$$= \frac{\text{Population in households}}{\text{Number of households}}, \text{ or } \frac{\text{Total population}}{\text{Number of households}}$$

Generally the difference between the two averages is small.

In computing the **median** size of household (or family), the appropriate procedure is to use 1.0, 2.0, 3.0, etc., as the midpoint, rather than the lower limit, of the median interval. Otherwise, the median size of household will be unrealistically low. This procedure is proper because the units of measure are discrete rather than continuous variables.

Type of Household, Family, or Subfamily. – A classification by "type of household" is used in U.S. census reports and is based on the sex and marital status of the head. A "husband-wife household" is a household with a head who is married and living with his wife. The other two basic types of households are "other households with male head" and "households with female head." By analogy primary families, secondary families, and subfamilies are classified into three types (husband-wife, other male head, and female head).

Number of Families in the Household.—Some countries tabulate statistics on number and structure of families in the household. One such classification, recommended by the Conference of European Statisticians[11], is shown in condensed form here:

Nonfamily households
  One-person households
  Multi-person households
    With related persons only, with unrelated persons only, with both related and unrelated persons
One-family households
  Without any other persons present
  With relatives only—with or without ascendants (parents or grandparents)
  With nonrelatives only
  With relatives and nonrelatives—with or without ascendants
Two-family households
  (Similar to the above, but specifying in addition whether the families are of direct ascent or descent or other relationship)
Three-or-more family households
  (These rare groups are classified only as to whether any other persons besides family members are present.

Number of Generations in the Household. – Some countries compile statistics on the number of generations of kinsmen who live together as members of one household. Usually these members constitute representatives of two or more of the following four relationship categories: Parent, head, child, and grandchild. Occasionally they include one or more older or younger generations. In statistics for the Republic of Korea on the distribution of private households by number of generations, generations are defined so as to include representatives

---

[11] United Nations, Statistical Commission and Economic Commission for Europe, Conference of European Statisticians, *Report of the Sixth Session of the Working Group on Censuses of Population and Housing,* Conf. Eur. Stats/WG.6/112, March 30, 1966, pp. 9–10.

Table 10–6. —Number of Private Households and Percent Distribution by Size of Household, for Selected Countries: Selected Years, 1955 to 1962

Country	Year	All private households (thousands)	Total	1	2	3	4	5	6	7 or more	Mean size of household
Canada	1961	4,555	100.0	9.3	22.2	17.8	18.4	13.3	8.2	10.9	3.9
Mexico[1]	1960	6,429	100.0	(X)	12.7	14.0	14.4	16.9	12.3	29.7	5.4
United States	1960	53,021	100.0	13.3	28.0	18.9	17.2	11.1	5.9	5.5	3.3
Peru[2]	1961	1,974	100.0	7.2	10.1	12.2	13.7	14.0	11.5	31.4	4.9
Venezuela	1961	1,339	100.0	8.8	10.0	12.4	12.9	12.0	11.4	32.5	5.3
United Arab Republic	1960	5,177	100.0	7.8	11.6	13.6	14.6	14.4	12.5	25.6	5.0
China (Taiwan)	1956	1,642	100.0	7.7	7.2	9.9	12.5	14.0	13.9	34.8	5.7
Japan	1955	17,383	100.0	3.5	10.8	14.5	16.6	16.7	14.1	23.8	5.0
Republic of Korea	1960	4,345	100.0	2.5	7.2	11.4	14.5	16.3	15.3	32.8	5.5
Czechoslovakia	1961	4,398	100.0	14.2	26.8	22.1	20.2	9.9	4.0	2.8	3.1
France	1962	14,562	100.0	19.6	26.8	18.7	14.7	9.4	5.2	5.5	3.1
Sweden	1960	2,645	100.0	21.9	26.6	21.5	17.2	7.9	3.1	1.9	2.8
West Berlin	1961	1,029	100.0	37.7	32.5	17.8	8.2	2.5	0.8	0.4	2.1
U.S.S.R.[1]	1959	50,333	100.0	(X)	26.0	26.0	21.7	13.4	7.2	5.7	3.7

X Not applicable.
[1] Data refer to family households only, that is, two or more related persons living together and having a common budget.
[2] Households with size not reported were distributed like those reporting.

Source: United Nations, *Demographic Yearbook, 1962,* table 12, and *1963,* table 33.

of collateral generations as well as representatives of generations in direct line of ascent or descent (in relation to the household head).[12] Recent publications by Burch and Petersen,[13] as well as by others, present evidence that only a relatively small proportion of families anywhere in the world are classifiable as "extended."

Size of Family. — The most elementary measure of family composition is size, expressed as the entire number of members living together. This number is limited to the head, wife, and children, according to the U.N. definition, but where the family is defined, as in the United States, to include also all other persons in the household related to the head by blood, marriage, or adoption, these additional members are also included in the size of family. The term "size of family" is often used especially by European demographers as synonymous with the number of children of the family head and wife; but that usage is discouraged by demographers who are interested in the analysis of family composition.

Factors affecting the size of family include not only the birth rate but also the death rate, the age at which children leave the parental home, the doubling rate, and so on. These factors, in turn, are influenced by the level of the economy, the housing supply, prevailing health conditions, degree of urbanization, and other aspects of the environment.

Number and Ages of Children. — Some of the most useful family data show the number of children of dependent ages. These ages may differ from country to country, as may be seen from the distributions of families by number of children in table 10–7. Moreover, the population to be classified as "children" may involve different criteria, such as all young

members of the household, as in the data for Japan; all young relatives, as in the data for the United States; all young children of the head, as in the data for Sweden; all young children of the head (or grandchildren or great-grandchildren without their parents present) plus older children who are enrolled in school full-time, as in the data for England and Wales.

Functional ages of children are generally adopted in the more detailed classifications of families by number of children present. In recent reports of the U.S. Bureau of the Census, for example, children under 6 years of age are regarded as of preschool age; those 6 to 11, of elementary (or primary) school age; those 12 to 17, of high (or secondary) school age; and those 18 to 24, of college age. The employment of mothers outside the home is more closely correlated with the presence of preschool children in the home than almost any other factor. Especially important, in this respect, are very young children, say, under 3 years of age.

Characteristics of Families as Social and Economic Units . — The status of all members of the family may be examined and recorded with respect to a given social or economic characteristic. Illustrations are: Family income (the sum of the income of all family members), family mobility status (such as whether all family members had the same mobility status), family educational level (such as whether any family member has had a college education), as well as characteristics of the housing quarters occupied by the family. Family statistics on these subjects are of obvious relevance to government, business, and academic research workers who are seeking a deeper understanding of the national or local economy or social structure.

Characteristics of Persons by Characteristics of Their Families. —The statistics discussed so far relate to either the "household" or the "family" as the basic unit. By contrast, subjects discussed in the present section feature the *person* as the unit of measurement. More specifically, attention is concentrated on tables that show various characteristics of persons by characteristics of the family in which they live. This type of table is useful because the person is not treated as an isolated entity but as a person living in a given type of family setting. Analyses based on these tables are of particular value when they show data on categories of persons who are usually dependent on the family head, such as wives, children, young mothers with marriages not intact, and elderly relatives.

One of the most frequent applications of the principle of relating the person's and his family's characteristics is the presentation of a cross-classification of husbands and their wives with respect to the same characteristic. Examples of such characteristics are age, marital history, education, ethnic origin, religion, and work experience. Demographic characteristics such as these, plus the socioeconomic status of the person's parents, tend to have considerable importance in mate selection and may be prognostic of duration of marriage.[14]

Several subjects lend themselves to analysis on the basis of a comparison of other family members and the head. The number of family members in each age group may be shown by age of the family head, with a further cross-classification by age of the wife for husband-wife families. This information provides an overview of the age composition of family members and is useful in gauging needs for various types of housing.

Table 10–7. — Number of Families and Percent Distribution by Number of Children of Specified Age Living at Home, for Selected Countries: 1960 and 1961

Number of children living at home	Japan: Related household members under 15	United States: Related children under 18	Sweden: Children under 16 (married head)	England and Wales: Children under 15 or in school
All families.....thous..	18,579	45,062	1,718	12,590
Percent............	100.0	100.0	100.0	100.0
No children............	27.3	40.1	46.2	51.3
1 child................	25.2	19.7	25.8	23.0
2 children.............	26.1	18.7	18.7	16.2
3 children.............	14.9	11.5	6.6	6.1
4 children.............	4.9	5.3	(NA)	2.1
5 children.............	1.3	2.4	(NA)	(NA)
6 or more.............	0.4	2.3	(NA)	(NA)
4 or more.............	6.6	10.0	2.8	3.3
5 or more.............	1.7	4.7	(NA)	1.2
All children.....thous..	[1]27,911	64,015	[1]1,664	11,219
Children per family.....	1.50	1.42	0.97	0.89
Children per family with children.........	2.07	2.37	1.80	1.83

NA Not available.

[1] For Japan, 7 or more children estimated to be 8 children, and for Sweden, 4 or more children estimated to be 5 children.

Source: Official national reports.

---

[12] South Korea, Economic Planning Board, *1960 Population and Housing Census of Korea*, Vol. II, *20% Sample Tabulation Report*, Part 1, *Whole Country*, 1963, table 7.

[13] Thomas K. Burch, "The Size and Structure of Families: A Comparative Analysis of Census Data," *American Sociological Review*, 32(3):347–363, June 1967; Karen Kay Petersen, "Demographic Conditions and Extended Family Households: Egyptian Data," *Social Forces*, 46(4):531–537, June 1968.

[14] Some of the most recent discussion and demographic documentation on this subject is presented in Hugh Carter and Paul C. Glick, *Marriage and Divorce: A Social and Economic Study*, Cambridge, Mass., Harvard University Press, 1970. See especially chapter 5, "Intermarriage Among Educational, Ethnic, and Religious Groups."

Other tables in which the same principle may be applied include marital status of son or daughter (by age) by marital status of parent(s) with whom he (she) lives; residential mobility of family member by residential mobility of the head; occupation and industry of the family member and of the family head; and income of family member by total income for the family. The table on marital status throws light on such topics as the extent to which persons who experience marital disruption (or who delay first marriage) live with parents whose marriage has been disrupted. The table on residential mobility throws light on families who move as a unit; the table on industry reveals the degree of dependence of family members on agriculture or on some other industry; and the table on income shows such things as how many persons with small incomes live in homes with a substantial family income.

Dynamics of Households and Families

This section deals with growth in numbers of households and related family units, changes in the composition of these units, and the typical life cycle of a family. Documentation of these dynamic aspects of family groups provides historical perspective for use in making meaningful analyses of the current status of family life and in making reasonable projections of family trends.

Changes in Numbers of Households and Families. — Change in the number of households over time is obviously significant for the national economy because of its relation to demand for items which are consumed on the basis of one per household. Thus, the demand for housing means a demand for shelter for more households. On the other hand, the increase in households does not fully indicate the need for the construction of new housing, since demolished units must be replaced and the quality of available housing needs to be upgraded in many areas. Moreover, improvement in the housing supply coupled with a steady increase in real income may encourage the "undoubling" of family groups so that more of the "potential household heads" become actual household heads. Migration may also be important, for migrants may create a demand for housing in the new area of residence, even though the former residence remains vacant.

The growth in number of households is not necessarily a simple reflection of growth in population. The rate of increase in the number of households is likely to exceed the rate of population increase in a period when the birth rate is dropping steadily and families are becoming smaller, whereas it will fall below the rate of population increase in a period when the birth rate is rising.

Since the number of household heads is the same as the number of households, the change in number of households may be viewed as a matter of measuring change in the number of household heads. Two components of change in number of households can be identified: growth or decline in population of each adult age group and increase or decrease in the "household headship rate" for each of these age groups.

Net change in number of households—or net household formation—is generally calculated by the simple process of obtaining the difference between the numbers of household heads at the beginning and end of the period in question. However, the analysis of such net change may be enriched by computing it by marital status, age, ethnic group, and sex of the head and by type of residence or tenure of housing unit.

Good use could be made of data on **gross** increases and decreases in numbers of household heads by various demographic and housing characteristics, for such purposes as estimating demand for the purchase, financing, and furnishing of new homes. Such statistics have not been readily available for the United States because some of the elements needed to construct estimates of gross changes do not exist. Even when relevant vital statistics are available, they may not be reconcilable with similar census data. Thus, annual data are available on marriages, but not all newly-wed couples establish a new household. Likewise, annual data are available on deaths by marital status, but the death of the husband does not necessarily mean that a household will be disestablished. Again, when a couple obtains a divorce, both former spouses may become heads of households, or one or both may move in with someone else. Moreover, no comprehensive data are available on the annual additions and subtractions of "bachelor apartments." [15]

Some beginnings on the problem of measuring gross changes in number of households by survey techniques have been made by the U.S. Bureau of the Census. Experimental tabulations of Current Population Survey data for March 1964 and March 1965 were made of persons living at given addresses on the two survey dates, to investigate such questions as how many entire households had moved away, how many of those with some remaining members still had the same head, and how many changed from having a family head to a primary individual as the household head. This study by Turner brought to light many difficult problems in the matching of households and persons for successive annual dates and revealed many gaps in the data, especially with respect to the reasons why persons had left the household during the year.[16]

In connection with the 1950, 1960, and 1970 Censuses of Housing, the U.S. Bureau of the Census made studies of gross changes in the housing inventory during the preceding decade. Since, in the practice of this Bureau, the number of occupied housing units is identical with the number of households, the results reflect gross changes in the number of households.

There are parallel components for families, unrelated individuals, and other family units. Measures of some of these components may be available from a population register. If not, one may start with a list of family groups enumerated in a given survey, and, in a follow-up survey or surveys, attempt to trace the changing status of these groups over the intervening period and to identify new groups that have moved into the sample segment or the listed housing units.[17]

Changes in Household and Family Composition. — The demographer's problems are magnified still further when he attempts to measure such phenomena as past changes in household or family composition in a manner which would be useful in devising methods for projecting these changes into the future. An example discussed here is the changing size of household.

Changes in size of household. — Household or family composition can be very readily studied in terms of net changes if the appropriate distributions (by size, number of children, number of working-age adults, number of elderly persons, etc.) have been obtained from censuses or surveys on a comparable basis for successive dates. But if changes in such items of household composition are to be projected, information on the items should, ideally, be available for a substantial number of points in time.

---

[15] For a fuller discussion of the problems of measuring gross changes in number of households, see Paul C. Glick, *American Families,* New York, John Wiley and Sons, 1957, pp. 178–179.

[16] Marshall L. Turner, Jr., "A New Technique for Measuring Household Changes," *Demography,* 4(1):341–350, 1967.

[17] For further discussion of the problems raised here, see treatment of nuptiality tables in chapter 19, "Marriage and Divorce."

Characteristics of persons who enter or leave the household.—The study by Turner, cited above, found that the size of family changed during one year for 15 percent of the nonmobile families in the United States. Longitudinal studies of this type could be made by comparing the roster of household members at given addresses at standard intervals, such as one year or even one month. Needed information includes the answer to the question as to why a person or family entered or left the household since the interviewer's previous visit. Reasons for such moves include vital events (birth, death, marriage, or divorce), service in the armed forces, college attendance, working elsewhere, and residence in an institution; reasons for movement of the entire household include change in employment or health of the head and the need for more appropriate housing space.

Consideration could be given to asking questions in a survey about whether the person who moved to this address had been the head of the household where he previously lived and how long the head had been the head. The relevance of information on changes in household composition to a thorough study of migration and local mobility should be obvious.

The Life Cycle of the Family.—The social and economic characteristics of families vary widely from family formation through family building to the launching of the younger generation into their marriages and on through the "post-children" stage to the ultimate dissolution of the family. Because of this highly dynamic nature of families, the conceptual framework provided by the idea of a "life cycle of the family" has been used extensively by many demographers, sociologists, and economists to add meaning and order to their analyses of family behavior.

Demographers recognize as stages in the life cycle of the family the usual sequence of vital events—marriage, birth of the first and last children, marriage of last child, and death of one spouse (usually the husband). In many populations, the last stage may typically precede the next to last in this sequence. These stages are generally characterized by the median age of the wife when the events occur, and the corresponding ages of the husband at successive stages are assigned on the basis of the average difference in the ages of husband and wife at marriage (table 10–8).

Table 10–8.—**Median Age of Women at Selected Stages of the Family Life Cycle, for Women Born From 1880 to 1939, by Year of Birth, for the United States**

Event	Year of birth (birth cohort) of women					
	1880 to 1889	1890 to 1899	1900 to 1909	1910 to 1919	1920 to 1929	1930 to 1939
First marriage...........	21.6	21.4	21.1	21.7	20.8	19.9
Birth of first child.....	22.9	22.9	22.6	23.7	23.0	21.5
Birth of last child......	32.9	31.1	30.4	31.5	30.0–31.0	(NA)
First marriage of last child..................	56.2	53.5	51.9	53.0	51.5–52.5	(NA)
Death of one spouse:						
For all couples........	57.0	59.4	62.3	63.7	64.4	64.4
For couples surviving to first marriage of last child...........	69.2	68.1	67.8	68.2	67.2–68.2	(NA)

NA Not available.

Source: Paul C. Glick and Robert Parke, Jr. "New Approaches in Studying the Life Cycle of the Family," *Demography*, 2:187–202, table 1, 1965.

The family life cycle approach is most readily applied in the cross-sectional analysis of couples with unbroken marriages. Largely remaining to future research is the longitudinal study of large numbers of families, so that the present status of the family can be interpreted in the light of marital, reproductive, residential, and work histories of the adults who are being studied.

SUGGESTED READINGS

Agarwala, S.N. *Some Problems of India's Population.* Bombay, Vora and Co., 1966, pp. 50-73.

Burch, Thomas K. "The Size and Structure of Families: A Comparative Analysis of Census Data." *American Sociological Review* 32(3):347–363. June 1967.

Carter, Hugh, and Glick, Paul C. *Marriage and Divorce: A Social and Economic Study.* Cambridge, Mass., Harvard University Press, 1970.

Glick, Paul C. *American Families.* (A volume in the census monograph series.) New York, John Wiley & Sons, 1957.

Hajnal, John. "Age at Marriage and Proportions Marrying." *Population Studies (London)* 7(2):111–136, November 1953.

Jamison, Ellen Linna, and Akers, Donald S. "An Analysis of Differences between Marriage Statistics from Registration and Those from Censuses and Surveys." *Demography* 5(1):460–474, 1968.

Lansing, John B., and Kish, Leslie. "Family Life Cycle as an Independent Variable." *American Sociological Review* 22(5):512–519, October 1957.

Roberts, G. W., and Braithwaite, L. "A Gross Mating Table for a West Indian Population." *Population Studies* 14(3): 198–217, March 1961.

Saveland, Walt, and Glick, Paul C. "First-Marriage Decrement Tables by Color and Sex for the United States in 1958-60." *Demography* 6(3):243–260, August 1969.

Turner, Marshall, Jr. "A New Technique for Measuring Household Changes." *Demography* 4(1):341–350, 1967.

United Nations. *Demographic Yearbook, 1962.* Pp. 10, 36–38, and 398–467.

———. *Principles and Recommendations for the 1970 Population Censuses.* Statistical Papers, Series M, No. 44, 1967. Pp. 41–42, 48–49, 52–53, 76–79, 84–85, 109–112, and 148–149.

U.S. Bureau of the Census. *Current Population Reports.* Series P-20, No. 198. "Marital Status and Family Status: March 1969." March 1970.

———. *U.S. Census of Population: 1960.* Vol. I, *Characteristics of the Population*, Part 1, *United States Summary.* 1964. Pp. LIV–LIX.

———. *U.S. Census of Population: 1960, Subject Reports, Persons by Family Characteristics*, PC(2)–4B. 1964.

———. *U.S. Census of Population: 1960, Subject Reports, Age at First Marriage*, PC(2)–4D. 1966.

———. *U.S. Census of Population: 1960, Subject Reports, Marital Status*, PC(2)–4E. 1966.

CHAPTER 11

Educational Characteristics

INTRODUCTION

The analysis of educational characteristics is included in this volume both because education is an important variable in accounting for demographic behavior and because education is one of the social characteristics of persons covered frequently in population censuses and demographic surveys and occasionally in registration systems. The kinds of education statistics discussed in this chapter will, therefore, be useful in analyses relating education to changes in fertility, mortality, migration, and the labor force, as well as in the study of educational development.

Classification of Measures

A number of measures of education have been used in the demographic analysis of the subject. In order to provide a framework for discussing these measures, we will view them in relation to the educational process. Three broad categories can be distinguished:

(1) *Measures of educational input.*—These are concerned with actual enrollment in school or in any grade or level of school, with types of school and fields of study, and with enrollment expectations.

(2) *Measures of educational progression.*—These have to do with retention in school from one grade or age to another, with graduation, and with grade standing relative to age.

(3) *Measures of educational output.*—These relate to eventual educational status, such as literacy, educational attainment, and educational qualifications, as well as to attainment expectations.

Sources of Data

Data on educational characteristics of the population can be found in a variety of sources. A distinction can be made between reports of school systems, on the one hand, and censuses and surveys, on the other, as original sources of the statistics. The resulting data may be published in local or national reports and may be summarized in international publications.

In most countries, education statistics from school systems are collected by the national government through a reporting process in which individual schools report to local governmental units which, in turn, report school information through a hierarchy of governmental units to the national government. The information so gathered and published usually refers to enrollments, graduates, teachers and other staff, educational finances, and schools and classrooms.[1] Even the data on enrollments and graduates are often of limited value for demographic analysis, however, since they are rarely reported for demographic categories of the population other than sex and geographic division.

Census reports for many nations include statistics on the educational characteristics of the population. Most typically, these cover enrollment, illiteracy, and level of school completed.[2] Since the time span between censuses is usually quite long, reports based on sample demographic surveys, which ask questions about educational status as well as other population subjects, may provide current data on educational characteristics. Moreover, because the sample survey is a more flexible data-collection instrument than the census, it is possible to include questions on education in the survey that would be deemed marginal to, or inappropriate for, a census inquiry.[3]

Recognition must also be made of *ad hoc* surveys dealing with educational topics. These are usually carried out by university or other independent survey organizations but are sometimes conducted by local governments.

Data from these different sources often lack comparability because of variations in the population covered, in definitions employed, and in the completeness and accuracy of reporting. Comparisons of institutional and census or survey data in the United States reveal some discrepancies in the reporting of enrollment data which can probably be accounted for by the factors mentioned.[4]

The U.N. *Demographic Yearbook* includes summaries of education data from time to time. These have dealt with illiteracy (in the 1948 and 1955 editions); level of education and school enrollment (in 1956); historical trends in illiteracy (in 1960); and illiteracy, level of education, and school enrollment (in 1963 and 1964).

---

[1] For a useful summary of education statistics from school systems in the United States, see the annual report by the U.S. Office of Education, *Digest of Educational Statistics.*

[2] A guide to United States census data on education can be found as Appendix D of the 1960 census monograph by John K. Folger and Charles B. Nam, *Education of the American Population,* Washington, U.S. Government Printing Office, 1967.

[3] For illustrations of how the Current Population Survey of the U.S. Bureau of the Census functions in this way in the field of education, see Daniel B. Levine and Charles B. Nam, "The Current Population Survey: Methods, Content, and Sociological Uses," *American Sociological Review,* 27(4):585–590, August 1962.

[4] Charles B. Nam, "Some Comparisons of Office of Education and Census Bureau Statistics on Education," *Proceedings of the Social Statistics Section, 1962,* Washington, American Statistical Association.

EDUCATIONAL INPUT

School Enrollment

Definition and Coverage.—School enrollment, according to the United Nations, refers to enrollment in any regular educational institution, public or private, for systematic instruction at any level of education during a well-defined and recent time period. Instruction in particular skills, which is not part of the recognized educational structure of the country (e.g., in-service training courses in factories), is not considered "school enrollment" for this purpose.[5]

Very often school enrollment statistics will distinguish between enrollment in public or private educational institutions, between full-time and part-time enrollment, and between different levels of schools (primary, secondary and college or university). It is also common to find statistics being shown by various types of educational institutions (e.g., college preparatory, vocational, teacher training), and by fields of study within a given level (e.g., law, engineering, medicine, social sciences).

Consideration must be given to the time reference for enrollment questions. An important factor in this regard is the opening and closing dates of the school year. If the question refers to enrollment at any time during the previous 12 months or calendar year, such a broad time reference may result in two different school years being covered. On this basis, counts of enrollment will be higher than would be expected at any point in time or during any single school year. It is preferable, therefore, to have the question refer to the current school year or school term.

An inquiry on school enrollment is usually directed toward persons within certain age limits. If these age limits are narrow, it is likely that many enrolled persons will be excluded; if, on the other hand, the age limits are wide, the question on enrollment will be asked of many persons to whom it does not apply. It is necessary, consequently, to weigh the advantages and disadvantages of questioning some age segments of the population among whom there are few enrollees in order to count all enrollees, as opposed to limiting the enrollment question to age groups having a substantial number enrolled and thereby missing some enrollees.

Where possible, the United Nations recommends that tabulations of school enrollment data be made according to age, sex, geographic division, and level of schooling.

United States.—A question on school enrollment has been included in the census of the United States since 1840. There were no age limits for the enrollment questions in many of the censuses, but increasing emphasis in the tabulations was placed on the customary ages of school and university enrollment. In the censuses of 1950 and 1960, the question was confined to the population under 30 and 5 to 34 years old, respectively.

In 1970 there were no age limits for this item, but most of the tabulations were limited to the age range from 3 through 34 years. As earlier, enrollment data are shown for fairly detailed age groups; and they are also cross-classified by level of school attended (nursery school, kindergarten, elementary, high school, college or university) and by type of control (public or private).

Data on school enrollment have been collected in the Current Population Survey every October since 1945. The resulting statistics are published in Series P–20 of Current Population Reports.

Uses and Limitations of Enrollment Statistics.—Data on school enrollment are used to measure the extent of participation in the school systems of an area by persons of school age, the potential future participation of such persons, and the relative participation of different segments of the population. Those involved in educational planning utilize current enrollment statistics to indicate the trend in school participation in both absolute and relative terms.

Educational statistics in most countries include more complete coverage of enrollment in regular, graded general public educational institutions than of enrollment in specialized, private technical and vocational educational programs. As a consequence, they often understate the total involvement of the population in the educational system.

Differences in reporting of enrollment status limit international comparability of the statistics. As is brought out in the U.N. *Demographic Yearbook, 1963,* the time reference of the census enrollment question not only varies from country to country, but also is not stated at all in many national publications.[6] Enrollment is not always limited to that in regular schools; for example, enrollment in commercial schools, dancing schools, or language schools is sometimes included. Variations in the age range to which the enrollment item applies have already been mentioned.

Quality of Data.—Enrollment data from school systems vary in quality depending upon the attention given to statistical collection and reporting systems in the country and the adequacy of the number and skills of personnel assigned to amass the data. The quality of census and survey data on enrollment depends greatly on the completeness and accuracy of the census or survey as a whole and upon the attention devoted to defining and clarifying terms used in the inquiry on enrollment. Before using enrollment data, it is advisable to examine the questions used, the population covered, the information provided to assist interviewers in asking the questions and in answering respondents' questions, the response rates to the questions, and other aspects of the data. As with other types of data collected in censuses and surveys, much depends on the knowledge and cooperation of the respondents.

Errors in population coverage and in age reporting are especially important. They affect not only the count of the total persons enrolled for the age range covered but also the age-specific enrollment rates discussed below.

Measures.—Measures of school enrollment usually relate to a point in time or a very short period of time. They may take many forms, however, and may depend on either census or survey data alone or on a combination of such data with statistics from educational systems.

Crude and general enrollment rates.—The first measure, the **crude enrollment rate,** may be expressed symbolically as

$$\frac{E}{P} \times 100, \qquad \text{where} \qquad \begin{aligned} E &= \text{Total enrollment at all levels and ages covered} \\ P &= \text{Total population} \end{aligned} \qquad (1)$$

The constant multiplier employed with the various kinds of enrollment rates is usually 100; in most cases, therefore, the numerical results are labeled as percents. Here the total population is employed in the denominator because no age limitation

---

[5] United Nations, *Principles and Recommendations for the 1970 Population Censuses,* Statistical Papers, Series M, No. 44, 1967, p. 58.

[6] Pp. 44–45.

has been placed on the enrollment question or because the rate for the total population is specifically wanted.

Preferably, the denominator of this rate should be the population eligible to be included in the numerator, or the same age group as was covered by the enrollment question. Whether or not an age limitation has been placed on the enrollment question, the population in ages at which persons are customarily enrolled may be employed in the denominator. In this case, the measure calculated is called the **general enrollment rate**. Using ages 5 to 34 as the arbitrary age range, it may be expressed symbolically as

$$\frac{E}{\sum_{a=5}^{34} P_a} \times 100, \quad \text{where} \quad E = \text{Total enrollment}$$

$$\sum_{a=5}^{34} P_a = \text{Population in ages 5 to 34} \qquad (2)$$

Age-specific and level-specific enrollment rates. — Comparisons based on crude or even general enrollment rates may be misleading because age distributions differ from one population to another. Caution must be exercised in interpreting enrollment trends on the basis of crude and general enrollment rates since they may fail to reveal changes already underway among specific groups; that is, overall trends may change very little while some age segments, for example, are undergoing significant change. Crude measures may also be responsive to modifications in population composition which falsely create the impression of changing patterns of enrollment. For example, a shift toward a more youthful population can raise the crude enrollment rate by placing more persons in the typical enrollment ages.

Age-specific enrollment rates are better measures of effective enrollment than crude or general enrollment rates since they focus on particular ages or age groups. The **age-specific enrollment rate** may be expressed as

$$\frac{E_a}{P_a} \times 100, \quad \text{where} \quad \begin{array}{l} E_a = \text{Enrollment at age } a \\ P_a = \text{Population at age } a \end{array} \qquad (3)$$

One type of age-specific enrollment rate that may be of special interest would represent the percent of the population in the typical enrollment ages, but beyond the ages of compulsory school attendance, that is still enrolled in school. For example, where school attendance is required up through age 13 and very few persons are enrolled after age 24. one could relate enrollment at ages 14 to 24 to the population at these ages.

The **level-specific enrollment rate** may be expressed as:

$$\frac{E_l}{P_a} \times 100, \quad \text{where} \quad \begin{array}{l} E_l = \text{Enrollment at school level } l \\ P_a = \text{Population in age group } a \text{ corresponding to school level in the numerator} \end{array} \qquad (4)$$

In this measure the numerator is not necessarily wholly included in the denominator. Although most persons enrolled in secondary school, for example, may be in the age range 14 to 17 years, some will be below and some above that age. Furthermore, some persons aged 14 to 17 in the population may be enrolled in school but not at the secondary level. (See table 11–1 for the U.S. figures in 1964.) An appropriate age range for the denominator can be selected by examining cross-classifications of age and school grade, and identifying the ages

which are typical for the school grade or level in the numerator. [7] The level-specific enrollment rate can be calculated for other levels of school in addition to the principal level at which an age group is attending. **Grade-specific enrollment rates** can also be calculated; these would relate individual grades to the population in an age group. In much the same way that a series of age-specific enrollment rates in successively older ages can be analyzed to get at lifetime changes in enrollment participation by age, level-specific and grade-specific enrollment rates analyzed from lower to higher grades will provide a view of lifetime changes in enrollment participation through stages of a school system.

The next measure, the **age-level-specific enrollment rate,** in effect, combines the specificity of the level-specific enrollment rate and the age-specific rate. It can be computed when both enrollment by age and enrollment by level are available. It may be expressed as:

$$\frac{E_{al}}{P_a} \times 100, \quad \text{where} \quad \begin{array}{l} E_{al} = \text{Enrollment at age } a \text{ and school level } l \\ P_a = \text{Population at age } a \end{array} \qquad (5)$$

This rate tells us the relative frequency for persons aged a to be enrolled at level l. For the most part, this rate would be computed for a particular age range in combination with a particular school level, e.g., elementary school level and ages 5 to 13. (The selection of the age range follows the same principle as for the level-specific enrollment rate.) However, it would also be appropriate to compute it for a number of different grades, each in combination with a single age. We have then an **age-grade-specific enrollment rate.**

The advantage of the age-level-specific enrollment rate over the level-specific enrollment rate is that it eliminates some of the inconsistency between school levels and ages. To this extent, it achieves a further degree of refinement.

Age-standardized or age-adjusted rate. — It is often desirable to have a single overall refined measure of enrollment rather than a number of specific measures. To derive such a measure, the general enrollment rate can be "standardized" to take account of the age distribution of the population within the age range for enrollment. This adjustment is advisable if the rate is to be used to compare groups at one point in time or to compare the same group at different points in time. What may appear to be differences or changes in enrollment participation when the general rate is used may be partly or wholly a function of differences or changes in the distribution of the population by age within the age range for enrollment.

The age-standardized enrollment rate may be expressed as:

$$\frac{\Sigma (E_a/P_a) \times P_{sa}}{P_s} \times 100, \quad \text{where} \quad \begin{array}{l} E_a = \text{Enrollment in age group } a \\ P_a = \text{Population in age group } a \\ P_{sa} = \text{Standard population in age group } a \\ P_s = \text{Total standard population} \end{array} \qquad (6)$$

---

[7] The level-specific enrollment rate is analogous to various measures which have different names, suggested by the United Nations: (a) *total school enrollment ratio,* which is the total enrollment in all schools below the third level as a percentage ratio to the population aged 5–19 inclusive, (b) *primary school enrollment ratio,* which is the total enrollment in schools at the first level as a percentage ratio to the population aged 5–14 inclusive, and (c) *secondary school enrollment ratio,* which is the total enrollment in all schools at the second level as a percentage ratio to the population aged 15–19 years inclusive.

Table 11–1. — Fall School Enrollment of the Civilian Noninstitutional Population 5 to 34 Years Old, by Age, Sex, and Level of School, for the United States: October 1964

[Numbers in thousands]

Age	Male					Female				
	Population	Enrolled in school				Population	Enrolled in school			
		Total enrolled	Elementary school	High school	College		Total enrolled	Elementary school	High school	College
Total, 5 to 34 years......	[1]43,114	26,852	17,505	6,459	2,888	[1]44,888	24,806	16,698	6,353	1,755
5 and 6 years................	4,170	3,478	3,478	–	–	4,042	3,364	3,364	–	–
7 to 13 years................	13,710	13,549	13,351	198	–	13,283	13,177	12,912	265	–
14 to 17 years...............	7,055	6,659	669	5,825	165	6,923	6,355	416	5,813	126
18 and 19 years.............	2,434	1,239	4	369	866	2,842	958	2	206	750
20 to 34 years...............	15,745	1,927	3	67	1,857	17,798	952	4	69	879

– Represents zero.

[1] The population of all ages: male, 94,739,000, and female, 97,381,000.

Source: U.S. Bureau of the Census, *Current Population Reports*, Series P-20, No. 148, "School Enrollment: October 1964," February 8, 1966, tables 2 and 5.

The standard population may be one of the groups being compared or an average of them, or a specially selected population (e.g., national population when geographic subdivisions are being compared). (For a more detailed description of the standardization procedure, including variations such as indirect standardization, see chapter 14.) Enrollment rates can be standardized for additional factors, of course, such as sex, ethnic group, or urban-rural residence, depending upon the purpose of the comparison.

Measurement of enrollment differentials. — The investigation of the problem of equality of educational opportunity has led to studies of disparities in school and college enrollments among population groups. Here it becomes necessary not only to obtain comparable measures of enrollment for various geographic, ethnic, and socioeconomic groups, and between the sexes, but also to define what constitutes widening, maintenance, or narrowing of disparities. For instance, is narrowing of disparities better indicated by a closing of the gap in **absolute** percentage points of enrollment or by a reduction in the **ratio** of percentages enrolled? If the former, then an appropriate measure of group disparity or dispersion would be the standard deviation; if the latter, it might be the coefficient of variation.[8]

EDUCATIONAL PROGRESSION

Data on educational progression provide a basis for seeing to what extent population groups persist in school and to what extent continuation in school is related to normal grade progression. We are concerned here with the concepts of school retention and dropout and of scholastic retardation and acceleration.

School Retention

Nature of Data.—School retention refers to the continuation of persons enrolled in school from one school grade or level to another or from one age to an older age. Dropping out of

school can be viewed as the obverse of school retention.

Since measures of school retention are based on enrollment data, the international recommendations about school enrollment affect what can be derived on school retention. It is much less customary, however, for nations to publish data that permit the calculation of school retention measures than for them to publish enrollment data in the form that permits calculation of the rates previously discussed. The reason for this is that the measures of school enrollment already cited are period measures that relate to a point in time, whereas measures of school retention are cohort measures that depend upon data at two or more closely spaced points in time.

The statistics used in analyzing school retention are subject to the same limitations as those used in analyzing school enrollment. Additionally, caution needs to be exercised in measuring school retention to assure that the data for different points in time are comparable and relate to the same cohort of persons. In analyzing retention in, or dropping out of, school, it is necessary to specify clearly the "population at risk." Is the interest in the number or in the proportion of an age group which stays in or leaves school by an older age? Is it in the number or proportion of enrollees in a school grade who continue on to a higher grade or drop out? Or is it some combination of these, such as the number or proportion of those in an age group who leave school before attaining a certain grade level?

Measures.—Measures of school retention can be based on data from either school systems, or censuses or surveys. Censuses are not likely to be a useful source for retention statistics unless they are taken at frequent intervals of time. Since censuses ordinarily are taken at 5- or 10-year intervals, their utility for this purpose is limited. Annual demographic surveys which obtain data on school enrollment by grade or age provide the necessary statistics for the computation of retention measures. Finally, the most frequently found source of retention data are the reports of school systems that give annual distributions of enrollment by grade and annual statistics on the number graduating from high school.

---

[8] The properties of these statistics are discussed in standard statistics textbooks. For example, see Hubert M. Blalock, *Social Statistics*, New York, McGraw-Hill, 1960, chapter 6.

Table 11−2.—Age and Grade in Which Enrolled for Persons 5 to 34 Years Old Enrolled in Elementary School, for the United States: 1960

[The heavy lines under the enrollment figures identify the grades which may be assumed to be, normal for enrolled persons 8, 9, and 10 years of age. The heavy lines to the right of the enrollment figures identify the ages which may be assumed to be normal for persons enrolled in grades 2, 3, and 4]

Age	Total	Grade in which enrolled							
		1	2	3	4	5	6	7	8
Total..................	28,987,577	4,076,178	3,926,237	3,790,449	3,555,656	3,519,093	3,486,642	3,532,344	3,100,978
Under 7 years..............	2,821,347	2,559,424	224,537	29,735	7,651	−	−	−	−
7 years...................	3,656,112	1,303,063	2,133,464	196,253	19,107	4,225	−	−	−
8 years...................	3,542,439	142,791	1,270,505	1,973,513	140,465	11,830	3,335	−	−
9 years...................	3,404,916	27,130	194,492	1,234,372	1,823,095	114,302	9,017	2,508	−
10 years..................	3,412,440	8,997	40,761	227,777	1,182,452	1,803,683	130,176	13,899	4,695
Over 10 years..............	12,150,323	34,773	62,478	128,799	382,886	1,585,053	3,344,114	3,515,937	3,096,283

− Represents zero.

Source: *U.S. Census of Population: 1960*, Vol. 1, *Characteristics of the Population*, Part 1, *United States Summary*, table 168.

Scholastic Retardation and Acceleration

Nature of Data.— Scholastic retardation and scholastic acceleration are defined here in terms of the relationship between an enrollee's age and the grade in which he is enrolled. A student is scholastically retarded if the grade in which he is enrolled is below the grade which is normally expected for his age. A student is scholastically accelerated if the grade in which he is enrolled is above the grade which is normally expected for his age.

These measures assume that a person enrolled in school advances one grade each year. In defining the grade range which is considered normal, one must allow for the fact that the age at which children first enter school varies because of the relationship of their birthdays to the entrance ages permitted by law or otherwise commonly observed. Defining a two-grade span as normal serves to cover the variation in ages of school entrance. It should be noted, however, that the broader the normal span of grades, the less will be the relative amount of scholastic retardation or acceleration indicated by the measures.

Scholastic retardation and acceleration, defined this way, measure the relative amount of progress in school grade advancement. In studying scholastic acceleration and retardation, one must be careful to distinguish between the type of age-grade inconsistency identified by the acceleration and retardation measures discussed below and mental ability or scholastic achievement, as indexed by test scores in general areas or specific fields. A reduction in scholastic retardation may be a function of changing school policies and practices through which some children are administratively promoted to a higher grade even though their scholastic performance would not justify promotion. These measures are probably associated, but at best imperfectly, therefore, with measures of scholastic achievement such as performance ratings and achievement test scores.

Since these measures are calculated for the population still enrolled, it should be noted that they are affected by the retention pattern. An increasing scholastic retardation rate may be due, in part, to increased school retention of low-ability youths. On the other hand, the scholastic retardation rate may decline partly because of an increasing proportion of school dropouts.

Measures.— Although the phenomena of scholastic retardation and acceleration are usually measured by examining the grade distribution of enrollees in a given age, they are sometimes measured in the reverse way, namely, by examining the age distribution of enrollees in a given grade. Both approaches are identified below.

The **scholastic retardation rate** may be expressed as

$$\frac{E_a^r}{E_a} \times 100, \quad \text{where} \quad \begin{aligned} E_a^r &= \text{Enrollees at age } a \text{ who are in} \\ &\quad \text{grades below those regarded as} \\ &\quad \text{normal for the age} \qquad (7) \\ E_a &= \text{All enrollees at age } a \end{aligned}$$

$$\frac{E_g^r}{E_g} \times 100, \quad \text{where} \quad \begin{aligned} E_g^r &= \text{Enrollees in grade } g \text{ who are in} \\ &\quad \text{ages above those regarded as} \\ &\quad \text{normal for the grade} \qquad (8) \\ E_g &= \text{All enrollees in grade } g \end{aligned}$$

The **scholastic acceleration rate** is parallel to the scholastic retardation rate. The numerator in this case, however, is the number of enrollees of a given age who are in grades above those regarded as normal for the age (or, alternatively, the number of enrollees in a given grade who are in ages below those regarded as normal for the grade). The intensity of scholastic retardation and scholastic acceleration can be measured by including in the numerator those persons retarded or accelerated by more than 1 year, or by calculating the average number of years retarded or accelerated.

Given the data in table 11−2 showing the age and grade of enrolled persons for the United States in 1960, and assuming grades 3 and 4 to be normal for enrollees at age 9, the scholastic retardation rate at age 9 would be

$$\frac{27,130 + 194,492}{3,404,916} \times 100 = \frac{221,622}{3,404,916} \times 100 = 6.5$$

and the corresponding scholastic acceleration rate would be

$$\frac{114,302 + 9,017 + 2,508}{3,404,916} \times 100 = \frac{125,827}{3,404,916} \times 100 = 3.7$$

Assuming that ages 8 and 9 are normal for grade 3, the scholastic retardation rate for grade 3 would be

$$\frac{227,777 + 128,799}{3,790,449} \times 100 = \frac{356,576}{3,790,449} \times 100 = 9.4$$

and the corresponding scholastic acceleration rate would be

$$\frac{29,735 + 196,253}{3,790,449} \times 100 = \frac{225,988}{3,790,449} \times 100 = 6.0$$

EDUCATIONAL OUTPUT

Literacy

Definition. — Literacy, according to the United Nations, is defined as the ability of a person to both read and write, with understanding, a short simple statement on his everyday life.[9] A person who cannot meet this criterion is regarded as illiterate. An illiterate person, therefore, may not read and write at all, or he may read and write only figures and his own name, or he may only read and write a ritual phrase which has been memorized. The language or languages in which a person can read and write are not a factor in determining literacy. A resident of England who can read and write in French but not English would still be considered literate.

The term "illiteracy," as defined here, must be clearly distinguished from "functional illiteracy." The latter term has been used to refer to persons who have completed no more than a few years of primary schooling. In the case of industrialized countries, it is assumed that such persons have probably not acquired the reading and writing skills needed to function effectively. Cross-tabulations of literacy and years of schooling completed indicate that not all persons reported as illiterate lack formal schooling, and not all persons without schooling are illiterate. In the United States in 1959, 5 percent of persons with a fourth-grade education and 2 percent with a fifth-grade education were reported as illiterate. At the same time, 26 percent of those with no schooling were reported as literate.[10]

Literacy is viewed by some as being differentiated along a continuum, from the barest minimum level to a quite fluent level. It is ordinarily not the practice in demographic censuses and surveys to attempt to make such distinctions, however. Rather, literacy is treated as a dichotomous variable which provides one index of the minimum level of educational output.

International Recommendations and National Practices. — The United Nations lists an inquiry on literacy as a recommended question to be included in national censuses in 1970. It further recommends that data on illiteracy be collected for the population 10 years of age and over. Since reading and writing ability ordinarily is not achieved until one has had some schooling or has at least had time to develop these skills, it is not useful to ask the question for young children. The United Nations recommends that data on illiteracy be tabulated by age, sex, and major civil divisions, distinguishing urban and rural areas within divisions; tabulations of literacy not cross-classified by detailed age should distinguish between persons under 15 years of age and those 15 and over.

The standard practice in obtaining literacy data is to ask the respondent if he can read and write. His answer to this question is usually accepted at face value. Some countries ask separate questions about reading and writing ability, and classify a person as semiliterate if he can read but not write. In some instances, a reading and writing test has been administered to individuals, as in Yugoslavia in 1950, but this practice is infrequent and does not lend itself readily to census and survey field procedures.[11]

United States. — A question on literacy was included in the decennial census of the United States from 1840 through 1930. This question was dropped in the 1940 census in favor of the more informative items on educational attainment. In 1947, 1952, and 1959, a question on illiteracy was carried in the Current Population Survey for persons who had not completed 6 years of primary school (5 years in 1947). In 1959, the question, which applied to persons 14 years old and over, read, "Is . . . able to read and write?" The instructions to the enumerator stated that persons who could not both read and write a simple message either in English or any other language were to be classified as illiterate. The implied definition was thus substantially the same as that recommended by the United Nations.

Quality of Data. — To the extent that there is misreporting on this topic, one might expect understatement of illiteracy. However, in a country with high illiteracy there is presumably no real hesitation in classifying oneself as illiterate. On the other hand, in a country with a high level of literacy, most people may be very hesitant to identify themselves as in the inferior group.

Where tests of reading and writing ability have been administered in addition to a simple inquiry on illiteracy, the general accuracy of the simple inquiry has been upheld.[12] Analysis of the reported illiteracy of age cohorts in a sequence of censuses of the United States showed a high degree of consistency from census to census.[13] It would be helpful, however, to have more evaluation of reported data on illiteracy for a variety of countries.

Estimates by Reference to a Standard Distribution. — Estimates of the number of illiterates for a country on a periodic basis or for geographic subdivisions of a country may be made by the method of reference to a standard national distribution. The method in this case assumes the availability of a cross-tabulation of illiteracy and other characteristics with which illiteracy is closely related, e.g., age, sex, and educational attainment. We would expect changes in the overall number of illiterates to be closely related to the changes in distribution of the population by educational attainment, age, and sex, although other factors might be involved. The method applies the percentages illiterate by age, sex, and educational attainment from the standard distribution to the corresponding population figures for the specific population under study. The calculations are confined to persons who have completed less than a certain number of years of school, the illiteracy rate being assumed to be zero at this grade and higher.

---

[9] United Nations, *Principles and Recommendations for the 1970 Population Censuses,* p. 50.

[10] U.S. Bureau of the Census, *Current Population Reports,* Series P-20, No. 99, "Literacy and Educational Attainment: March 1959," February 4, 1960, table 7.

---

[11] Discussion of the accuracy of reports on literacy as well as gradations of literacy can be found in S. S. Zarkhovic, "Sampling Control of Literacy Data," *Journal of the American Statistical Association,* 49(267):510–519, September 1954; and Charles Windle, "The Accuracy of Census Literacy Statistics in Iran," *Journal of the American Statistical Association,* 54(287):578–581, September 1959.

[12] See footnote 11.

[13] Folger and Nam, op. cit., p. 121.

Table 11−3. − Calculation of Illiteracy Rates for the United States and Louisiana for April 1960, by Method of Reference to a Standard National Distribution

[Relating illiteracy and educational attainment as reported in the March 1959 Current Population Survey]

Age and area	Years of school completed--males						Years of school completed--females					
	None		1 and 2		3 and 4		None		1 and 2		3 and 4	
	Illiteracy rate (R_a)	Population (P_a)	Illiteracy rate (R_a)	Population (P_a)	Illiteracy rate (R_a)	Population (P_a)	Illiteracy rate (R_a)	Population (P_a)	Illiteracy rate (R_a)	Population (P_a)	Illiteracy rate (R_a)	Population (P_a)
UNITED STATES												
Total, 14 years and over......	(X)	1,256,602	(X)	959,506	(X)	2,633,996	(X)	1,182,620	(X)	709,899	(X)	2,146,984
14 to 24.........................	1.000	91,523	.500	84,434	.207	167,968	.762	72,886	.289	64,339	.175	105,875
25 to 34.........................	.863	89,963	.549	92,467	.179	235,294	.818	71,468	.563	58,879	.104	159,971
35 to 44.........................	.767	114,613	.559	134,125	.129	354,449	.824	89,590	.368	86,800	.108	263,176
45 to 54.........................	.754	153,779	.643	168,712	.134	468,876	.728	126,592	.396	111,388	.073	354,863
55 to 64.........................	.757	234,208	.435	190,044	.099	552,963	.693	236,368	.292	149,159	.083	480,639
65 years and over...............	.639	572,516	.366	289,724	.065	854,446	.763	585,716	.303	239,334	.063	782,460
Number illiterate = $\Sigma R_a P_a$ =	916,152		465,147		295,723		890,685		243,868		178,681	
LOUISIANA												
Total, 14 years and over......	(X)	56,975	(X)	44,602	(X)	95,693	(X)	54,550	(X)	34,052	(X)	82,079
14 to 24.........................	1.000	2,034	.500	2,397	.207	7,503	.762	1,914	.289	1,532	.175	3,776
25 to 34.........................	.863	2,945	.549	4,414	.179	10,906	.818	2,375	.563	2,502	.104	7,242
35 to 44.........................	.767	6,023	.559	7,392	.129	16,373	.824	4,695	.368	5,046	.108	13,950
45 to 54.........................	.754	10,611	.643	10,386	.134	21,486	.728	8,820	.396	7,434	.073	18,092
55 to 64.........................	.757	13,657	.435	9,753	.099	19,401	.693	13,012	.292	7,649	.083	17,378
65 years and over...............	.639	21,705	.366	10,260	.065	20,024	.763	23,734	.303	9,889	.063	21,641
Number illiterate = $\Sigma R_a P_a$ =	41,404		22,430		11,719		40,817		11,882		7,047	

ESTIMATED ILLITERACY RATES--UNITED STATES:

$$\frac{\text{Total number illiterate}}{\text{Population 14 and over with less than 5 years of school completed}} \times 100 = \frac{\Sigma R_a P_a}{\Sigma P_a} = \frac{2,990,256}{8,889,607} \times 100 = 33.6 \text{ percent.} \qquad \frac{\text{Total number illiterate}}{\text{Total population 14 years and over}} \times 100 = \frac{2,990,256}{126,276,551} \times 100 = 2.4 \text{ percent.}$$

ESTIMATED ILLITERACY RATES--LOUISIANA:

$$\frac{\text{Total number illiterate}}{\text{Population 14 and over with less than 5 years of school completed}} \times 100 = \frac{\Sigma R_a P_a}{\Sigma P_a} = \frac{135,299}{367,951} \times 100 = 36.8 \text{ percent.} \qquad \frac{\text{Total number illiterate}}{\text{Total population 14 years and over}} \times 100 = \frac{135,299}{2,164,411} \times 100 = 6.3 \text{ percent.}$$

X Not applicable.

Source: U.S. population from *U.S. Census of Population: 1960*, Vol. 1, *Characteristics of the Population*, Part 1, *U.S. Summary*, table 173, and Part 20, *Louisiana*, table 103; illiteracy rates from U.S. Bureau of Census, *Current Population Reports, Population Characteristics*, Series P-20, No. 99, "Literacy and Education Attainment: March 1959," Feb. 1960, table 7.

One basic procedure is as follows:

(1) Assume that all persons who had completed 5 years of school or more were literate.

(2) For persons below this level of attainment, compute the proportion illiterate from the standard population by age, sex, and year of school completed.

(3) Multiply the age-sex-attainment specific proportions from the standard distribution by the corresponding "current" or "local" population figures. The result is the estimated number of illiterates in each specific age-sex-attainment group in the current or local population.

(4) Sum the results over all groups to obtain the total number of illiterates in the current or local population.

This procedure may be repeated at each subsequent date or for each geographic subdivision for which tabulations by age, sex, and educational attainment are available. Estimates of illiteracy for a nation and all its principal geographic sub-divisions (e.g., States) may be obtained by (1) assuming that the standard national proportions apply to each State, (2) multiplying the standard proportions by the corresponding population distribution for each State, (3) summing the results in (2) over all groups for each State to obtain the estimated total number of illiterates for each State, and (4) summing the results in (3) over all States to obtain the estimated total number of illiterates in the nation. The latter procedure was employed in preparing the estimates of illiteracy in 1960 for the United States and States.[14] The March 1959 Current Population Survey provided the standard illiteracy proportions for

[14] U.S. Bureau of the Census, *Current Population Reports*, Series P-23, No. 8, "Estimates of Illiteracy by States: 1960," Feb. 12, 1963; and Series P-23, No. 6, "Estimates of Illiteracy by States: 1950," November 1959.

the United States, and the 1960 census provided the State distributions to which the U.S. proportions could be applied. Table 11–3 illustrates the application of the method for estimating the number and percentage of illiterates in Louisiana in 1960. In this example comparable figures for the United States were derived by applying the CPS proportions for 1959 directly to 1960 census data for the United States rather than by calculating separate estimates for the individual States and combining them.

The method of reference to a standard distribution may be applied to the estimation of other social and economic characteristics. One might, for example, prepare current or regional estimates of the total number of households on the basis of a standard distribution of the population by age, sex, marital status, and number of household heads, using current or regional data on the age, sex, and marital distribution of the population. Here the age-sex-marital status specific proportions of household heads are applied to the corresponding current or regional population figures.

Measures.—General measures of illiteracy provide some indication of the educational status of the population, as well as an indication of the country's socioeconomic level, with which illiteracy is highly correlated. Illiteracy measures for subcategories of the population provide a basis for analyzing changes in literacy, particularly its spread from one segment of the population to another. We may define a crude illiteracy rate and an age-specific illiteracy rate.

The **crude illiteracy rate** may be expressed as

$$\frac{I}{P} \times 100, \quad \text{where} \quad \begin{array}{l} I = \text{Number of illiterates in population} \\ \quad \text{covered} \\ P = \text{Population covered} \end{array} \quad (9)$$

Age coverage needs to be specified but is usually about ages 10 and over or 15 and over. The **age-specific illiteracy rate** may be expressed as

$$\frac{I_a}{P_a} \times 100, \quad \text{where} \quad \begin{array}{l} I_a = \text{Number of illiterates in age group } a \\ P_a = \text{Population in age group } a \end{array} \quad (10)$$

In countries where great advances in schooling have been made in recent years, the crude illiteracy rate may still be high because of the inclusion of the less literate cohorts of earlier years. Specification of illiteracy rates by age permits identification of the magnitude of the illiteracy problem among different age segments of the population and gives some indication of historical change in illiteracy.

Using the data on the number of illiterates in Honduras by age and sex in 1961 shown in table 11–4, we may illustrate the computation of illiteracy rates as shown below. As with enrollment rates, the numerical expression of these rates may be labelled percents.

Crude illiteracy rate (10 and over, both sexes) =
$$\frac{641,550}{1,218,262} \times 100 = 52.7$$

Age-specific illiteracy rates (males)

$$10 \text{ to } 14 \text{ years} = \frac{53,791}{121,024} \times 100 = 44.4$$

$$25 \text{ to } 34 \text{ years} = \frac{59,383}{116,896} \times 100 = 50.8$$

$$45 \text{ to } 54 \text{ years} = \frac{30,799}{55,799} \times 100 = 55.2$$

$$65 \text{ years and over} = \frac{11,697}{22,166} \times 100 = 52.8$$

The crude illiteracy rate for both sexes in Honduras in 1961 is 52.7 percent, that is, more than half of the population 10 years old and over were illiterate. Although the youngest age group has the lowest percentage illiterate, its rate is still relatively high (44.4 percent for boys 10 to 14 years).

The fairly steady rise in age-specific illiteracy rates for Honduras from the youngest to the oldest age groups shown in table 11–4 indicates a general historical increase in literacy in the country and suggests the pattern and pace of this development. On the basic assumption that few persons become literate after age 10, we may arbitrarily describe the most recent literacy achievement of Honduras in terms of the illiteracy rate at ages 10 to 14 (43 percent). Then, the achievement characteristic of a period of about 50 years ago was the rate for the group currently 55 to 64 years (62 percent). The pace of improvement for females was much greater than for males. This type of analysis depends on the assumption that there has been little or no difference in the mortality level of literate and illiterate persons over this period and little or no selective migration, as well as on the assumption that literacy is not achieved after a certain age.

Table 11–4.—Illiteracy Rates by Age and Sex, for Honduras: 1961

Age	Both sexes			Male			Female		
	Total	Illiterate		Total	Illiterate		Total	Illiterate	
		Number	Percent		Number	Percent		Number	Percent
10 years and over....................	1,218,262	641,550	52.7	600,857	299,933	49.9	617,405	341,617	55.3
10 to 14 years......................	235,262	100,915	42.9	121,024	53,791	44.4	114,238	47,124	41.3
15 to 19 years......................	184,173	84,126	45.7	89,485	14,993	46.9	94,688	42,133	44.5
20 to 24 years......................	157,767	78,149	49.5	76,066	36,071	47.4	81,701	42,708	51.5
25 to 34 years......................	241,123	134,585	55.8	116,896	59,383	50.8	124,232	75,202	60.5
35 to 44 years......................	169,841	130,961	60.6	83,762	46,164	55.1	86,079	56,797	66.0
45 to 54 years......................	112,315	68,874	61.3	55,799	30,799	55.2	56,516	38,075	67.4
55 to 64 years......................	71,707	44,684	62.3	35,659	20,035	56.2	36,048	24,649	68.4
65 years and over...................	46,069	27,256	59.2	22,166	11,697	52.8	23,903	15,559	65.1
Unknown............................	1,026	472	46.0	668	286	42.8	358	186	52.0

Source: United Nations, *Demographic Yearbook, 1964*, table 33.

Educational Attainment

Definition.—Educational attainment, according to the United Nations, is the highest grade completed within the most advanced level attended in the educational system of the country where the education was received.[15] For this purpose, a grade is defined as a stage of instruction usually covered in the course of the school year.

Educational attainment is measured not by the number of calendar years which a person has spent in school but by the highest grade which he was able to complete. If the person received his education in the school system of another country than his country of present residence, it is necessary to convert that schooling into the equivalent highest grade completed in the country of present residence.

International Recommendations and National Practices.—The United Nations recommends an inquiry on educational attainment as basic to a census. At least 70 countries investigated the question of the educational attainment of their populations in a census or national sample survey in the 1955–63 period. The inclusion of a question on educational attainment in the U.S. census dates back to 1940. In this country it has been found that a supplementary question "Did he finish the highest grade (or year) he attended?" reduces the tendency to report an unfinished grade, which produced an upward bias in the statistics.

For a long time, there were inquiries about educational attainment in the Current Population Survey only at irregular intervals, but beginning in March 1964, there has been a regular annual report on the educational attainment of the population in the Census Bureau's *Current Population Reports*, Series P–20, *Population Characteristics*. In comparing decennial census and CPS statistics on enrollment and educational attainment for subdivisions of the country, it should be borne in mind that in the census, college students are counted where they actually live while attending college, whereas CPS counts unmarried students at their parental homes (ch. 4).

Quality of data.—Evaluation studies made in the United States following the 1950 and 1960 censuses indicate that misreporting of highest grade or year of school completed is considerable. Yet, three facts stand out: (1) Most misreporting involves a difference of only one or two grades, (2) only a small proportion of the errors affect data on the level completed, and (3) overreporting is somewhat greater than underreporting and, hence, there is a small degree of net overreporting.

Misreporting of educational attainment can be intentional, as when a higher level than actually attained is reported for reasons of prestige, or it may be unintentional, as when recall of older persons is faulty, information is supplied regarding others by someone who did not have reliable information, or highest grade attended is mistaken for highest grade completed.

About 5 percent of the population 25 years old and over in the 1960 census of the United States did not report completely on highest grade of school completed. The figure has been somewhat less in the Current Population Survey.

Measures.—The measures of educational attainment considered here include both measures of distribution and measures of central tendency. Three measures of distribution and

two measures of central tendency are described below. In computing each of these measures, some account must be taken of nonresponses to the question on educational attainment where these are not allocated before publication. Unless there is a valid basis for distributing the nonresponses over the reported categories in a special way, it is customary to distribute them *pro rata*, or, in effect, base the derived measures on the distribution for persons for whom reports on educational attainment have been received. In interpreting the measures of educational attainment, it should be kept in mind that persons under 25 years of age may still be attending school and that the measures for these persons would tend to understate their eventual educational attainment to some degree.

Grade attainment rates.—The **specific grade attainment rate** may be expressed as

$$\frac{C_a^g}{P_a} \times 100, \quad \text{where} \quad \begin{aligned} C_a^g &= \text{Persons at age } a \text{ who completed} \\ &\quad \text{exact grade } g \\ P_a &= \text{Population at age } a \end{aligned} \quad (11)$$

This measure provides information about the percentage of the population which completed an exact grade (or level) of school, no more or less. For example, it may be of interest to know the proportion of persons who finished the 10th grade but did not continue their education to finish another grade of school. Such rates may be shown in the form of a full percentage distribution over all values of g.

The **cumulative grade attainment rate** may be expressed as

$$\frac{C_a^{g+}}{P_a} \times 100 \quad \text{where} \quad \begin{aligned} C_a^{g+} &= \text{Persons at age } a \text{ who completed} \\ &\quad \text{grade } g \text{ or beyond} \\ P_a &= \text{Population at age } a \end{aligned} \quad (12)$$

This measure is the more common statistic on the educational stock of the population. It indicates the proportion of a population at age a that completed a given grade (or level) of school or beyond, or the proportion who ever completed that grade (or level). For example, the rate might be computed for the population 25 to 29 years of age who had ever completed high school or college.

Educational attainment ratio.—The educational attainment ratio may be expressed as

$$\frac{C_a^{g+}}{C_a^{b-}} \times 100 \quad \text{where} \quad \begin{aligned} C_a^{g+} &= \text{Persons at age } a \text{ who completed} \\ &\quad \text{grade } g \text{ or beyond} \\ C_a^{b-} &= \text{Persons at age } a \text{ who completed} \\ &\quad \text{grade } b \text{ or less} \end{aligned} \quad (13)$$

This ratio relates the number of persons completing a high level of education in a population to those completing a low level of education. The levels selected will reflect what levels are significant in a given society. For example, the ratio might relate the number of persons ever completing the third level of education to the number not completing more than the first level. In a population having an increasing number of highly educated persons but a continuing large number of poorly educated persons, the ratio will increase slowly, whereas in a population having an increasing number of highly educated persons and a declining number of poorly educated persons, the ratio will increase rapidly.

Illustrative computations of attainment rates.—The several measures of educational attainment described above are illustrated below for males and females in selected age groups

---

[15] United Nations, *Principles and Recommendations for the 1970 Population Censuses*, p. 45.

using the data on single years of school completed for Honduras shown in table 11-5.

The specific grade attainment rate for fifth year of secondary school is obtained by dividing the frequencies in "secondary level—5" by the number in "total" less the number in "unknown":

$$\text{Males 10 to 24} = \frac{2,539}{286,575 - 4,764} \times 100 = 0.9$$

$$\text{Females 10 to 24} = \frac{2,944}{290,627 - 4,608} \times 100 = 1.0$$

$$\text{Males 25 and over} = \frac{4,368}{314,950 - 10,492} \times 100 = 1.4$$

$$\text{Females 25 and over} = \frac{3,704}{327,136 - 7,136} \times 100 = 1.2$$

The cumulative grade attainment rate for the fifth grade of secondary school or higher level is obtained by cumulating the frequencies in grades "secondary level—5" and "higher level—1 through 5+" and dividing by the "total" minus the "unknown" category:

$$\text{Males 10 to 24} = \frac{3,213}{286,575 - 4,764} \times 100 = 1.1$$

$$\text{Females 10 to 24} = \frac{3,160}{290,627 - 4,608} \times 100 = 1.1$$

$$\text{Males 25 and over} = \frac{7,750}{314,950 - 10,492} \times 100 = 2.5$$

$$\text{Females 25 and over} = \frac{4,198}{327,136 - 7,136} \times 100 = 1.3$$

The educational attainment ratio relating "higher level" to "none" is obtained by cumulating the frequencies in grades "higher level—1 through 5+" and dividing by the "none" category.

$$\text{Males 10 to 24} = \frac{674}{131,855} \times 100 = 0.5$$

$$\text{Females 10 to 24} = \frac{216}{131,335} \times 100 = 0.2$$

$$\text{Males 25 and over} = \frac{3,382}{168,364} \times 100 = 2.0$$

$$\text{Females 25 and over} = \frac{494}{210,468} \times 100 = 0.2$$

Median and mean years of school completed.—The distribution of the population by years of school completed can be summarized in terms of two "averages"—the median years of school completed and the mean years of school completed. The **median years of school completed** may be defined as the value which divides the distribution of the population by educational attainment into two equal parts. one half of the cases falling below this value and one half of the cases exceeding this value. It is preferable to have single years or grades of school in the distribution in order to be able to calculate the median with as high a degree of precision as the quality of the reported data permits.

In calculating the median years of school completed, it is necessary to make assumptions about the boundaries of classes and about the distribution of persons within them. In reports of the U.S. Bureau of the Census, it is assumed, for example, that persons who reported completing the ninth grade are distributed rectangularly between 9.0 and 9.9; that is, educational attainment is treated as a continuous quantitative variable. (For a description of the procedure for computing the median, see ch. 8.)

This assumption is not entirely realistic and, hence, leads to a statistic which is sometimes subject to misinterpretation. There is the basic question whether enrollment in a grade not completed is worth crediting for purposes of calculating educational attainment. That is, should years of schooling be treated as a discrete variable rather than a continuous variable? In the United States, there is a tendency for persons who do not complete the highest grade they attend to drop out early in the school year and, therefore, to complete only a small fraction of the grade. Thus, although the stated class boundaries may describe accurately the limits of attainment (in the previous example, 9 years up to, but not including, 10 years), they do not describe accurately the distribution within the grade.

The form of the actual distribution for a reported year of school differs for those who are still attending at that grade level, on the one hand, and those who have graduated or dropped out of school, on the other hand. The concentration of the former within the grade depends pretty much upon the interval between the beginning of the school year and the date of the census enumeration. In this case the number of years of school completed can be measured to a decimal. For those who are no longer attending school, in view of the tendency to leave school on finishing a grade or to drop out early in the school year, it is reasonable to assume the "exact" grade (e.g., 9.0 years) as the midpoint of the grade interval. Hence, for a reported 9 grades the class limits for computing the median would be 8.5 to 9.4 for this group. Ideally, then, different assumptions should be made for the different groups.

In any event, care must be taken in interpreting the median number of school years completed as conventionally calculated. It is intended primarily for comparative analysis—between population groups and over time; its absolute value is to be interpreted only approximately. Given the assumption about the rectangular distribution of persons within the class limits, a median of, say, 9.3 years should not be interpreted to mean that the average person in the population group for which the median is computed has completed three-tenths of the 10th grade. Instead, it should be interpreted to mean that the average person in the group has completed the ninth grade and that some persons completing the ninth grade have attended the 10th grade. A median of about 12, as in the United States, reflects primarily a very high concentration at high school graduation. In some other countries, a median of 1 or even 0, as in Honduras, likewise reflects a very high concentration at the initial grades of elementary school. If more precision in a summary measurement of attainment is desired, then consideration should be given to using a cumulative grade attainment rate.

The **mean years of school completed** may be defined as the arithmetic average of the years of school completed by all persons in a population. In contrast, the median is a positional measure showing at what educational level the middle person in a distribution is located.

The procedure used in computing the mean years of school completed is to (1) multiply the number of persons in each educational class by the midpoint of the number of years of school covered by the class, (2) sum the products for all

Table 11 – 5. – Male and Female Population of Selected Ages by Years of School Completed, and Corresponding Grade Attainment Rates, for Honduras: 1961

Sex and age	Total	None	Years of school completed																	Un-known	Median years of school com-pleted	Mean years of school com-pleted
			Primary level					Secondary level					Higher level								
			1	2	3	4	5	1	2	3	4	5	1	2	3	4	5 or more			
Male, 10 and over......	601,525	300,219	38,737	83,100	65,477	29,620	43,852	3,469	4,088	3,563	3,181	6,907	360	423	379	420	2,474	15,256	1.0	2.2
10 to 24 years[1]......	286,575	131,855	22,788	45,211	33,461	15,451	21,851	2,546	2,262	1,834	1,339	2,539	213	181	108	76	96	4,764	1.4	2.2
25 years and over[2]...	314,950	168,364	15,949	37,889	32,016	14,169	22,001	923	1,826	1,729	1,842	4,368	147	242	271	344	2,378	10,492	0.9	2.1
Female, 10 and over....	617,763	341,803	30,978	74,821	61,947	28,298	44,468	3,263	4,394	6,018	2,671	6,648	91	146	124	134	215	11,744	0.9	2.0
10 to 24 years[1]......	290,627	131,335	20,295	45,966	35,352	16,280	23,792	2,465	2,707	3,596	1,071	2,944	61	66	42	17	30	4,608	1.6	2.3
25 years and over[2]...	327,136	210,468	10,683	28,855	26,595	12,018	20,676	798	1,687	2,422	1,600	3,704	30	80	82	117	185	7,136	0.8	1.7
PERCENT																				
Male, 10 and over......	100.0	51.2	6.6	14.2	11.2	5.1	7.5	0.6	0.7	0.6	0.5	1.2	0.1	0.1	0.1	0.1	0.4	(X)	(X)	(X)
10 to 24 years[1]......	100.0	46.8	8.1	16.0	11.9	5.5	7.8	0.9	0.8	0.7	0.5	0.9	0.1	0.1	(Z)	(Z)	(Z)	(X)	(X)	(X)
25 years and over[2]...	100.0	55.3	5.2	12.4	10.5	4.7	7.2	0.3	0.6	0.6	0.6	1.4	(Z)	0.1	0.1	0.1	0.8	(X)	(X)	(X)
Female, 10 and over...	100.0	56.4	5.1	12.3	10.2	4.7	7.3	0.5	0.7	1.0	0.4	1.1	(Z)	(Z)	(Z)	(Z)	(Z)	(X)	(X)	(X)
10 to 24 years[1]......	100.0	45.9	7.1	16.1	12.4	5.7	8.3	0.9	0.9	1.3	0.4	1.0	(Z)	(Z)	(Z)	(Z)	(Z)	(X)	(X)	(X)
25 years and over[2]...	100.0	65.8	3.3	9.0	8.3	3.8	6.5	0.2	0.5	0.8	0.5	1.2	(Z)	(Z)	(Z)	(Z)	0.1	(X)	(X)	(X)

X Not applicable.
Z Less than 0.05 percent.
[1] Includes persons attending school whose educational attainment could still increase.
[2] Includes unknown ages.

Source: United Nations, *Demographic Yearbook, 1964*, table 35.

classes, and (3) divide by the total population. The same considerations with respect to the determination of the boundaries of classes apply here as for the calculation of the median. If educational attainment is treated as a continuous quantitative variable, the midpoint of each class would then be at the center of each grade (e.g., 9.5). In calculating the mean years of school completed, it is necessary to assign a value to the highest educational attainment class if it is an open-ended class. For instance, if the highest level is 5 or more years of higher education (or 17 or more years of school), a value must be assigned which represents the midvalue for persons in that category (perhaps 18.0).

Although the mean is generally more sensitive to variations or changes in the educational distribution than is the median, the median is simpler to compute and is nearer the point of greatest concentration in the distribution. The median is the more commonly used summary measure of educational distribution.

The median and mean years of school completed are illustrated below for males 25 years and over using the data for Honduras in table 11–5. The median years of school completed is calculated by (1) dividing the "total" minus "unknown" in half and (2) dividing the result in (1) by the "none" category, which is larger than the result in (1):

$$\text{Males 25 and over} = \frac{314,950 - 10,492}{2} = 152,229$$

$$\frac{152,229}{168,364} = 0.90$$

$$\text{Median} = 0 + 0.90 = 0.90 = 0.9$$

For purposes of computing the median here, the distribution of years of school completed is regarded as continuous. The median for the group 25 years and over falls in the "none" category and, on the basis of the assumption of continuity, the median value is 0.9 year. Commonly, as for the group 10 to 24 years, the median falls in a higher category and the frequencies have to be cumulated up to the median class.

The mean years of school completed by males 25 years old and over in Honduras is calculated by (1) multiplying the frequencies in each category in table 11–5 cumulatively by the midpoint of each category, and (2) dividing by the frequencies in "total" minus "unknown."

$$\text{Males 25 and over} = \frac{648,814.0}{314,950 - 10,492} = 2.1$$

Because the distribution of the population by years of school completed is sharply skewed to the right when educational attainment is very low, as for Honduras, the mean has a substantially higher value than the median. In general, as educational attainment rises, the gap between the two types of averages tends to fall until, for populations like the urban population of the United States in 1960, the median exceeds the mean.

As suggested by the discussion of illiteracy rates, calculation of measures of educational attainment by age, for the ages beyond those at which formal education is normally obtained, from a single census, may provide an indication of the historical changes in the level of schooling of a population. The measures used may be a cumulative attainment rate (e.g., completion of elementary school or higher), the educational attainment ratio, or the median or mean years of school completed. For example, the median (years of school completed) for males in Puerto Rico in 1960 shows a steady upward progression from the oldest to the youngest ages among those 25 and over. The assumption of stability of these figures over time is corroborated by comparison with the medians for the same age cohorts in 1950.

Age	Median years 1960	1950
25 to 34	8.3	5.7
35 to 44	6.1	4.5
45 to 54	4.5	3.5
55 to 64	3.5	0.9
65 to 74	0.9	0.7
75 and over	0.7	0.7

This type of analysis assumes that there is little or no difference in mortality levels or migration rates according to educational attainment and that the educational attainment of individuals does not change after about age 25.

Field of Specialization

Field of specialization refers to the field of study in which a student specialized while attending school. It represents, therefore, a student's major area of training and presumably also of his competence. Field of specialization is a topic that applies to analysis of enrollment data but it is of even wider interest in connection with data on educational attainment and educational qualifications.[16] UNESCO recommends that fields of study be classified into nine categories, as follows: humanities, education, fine arts, law, social sciences, natural sciences, engineering, medical science, and agriculture.[17]

Data on field of specialization for persons completing four or more years of secondary school were collected in the 1966 census of Iran. Such data are a rich source of information on the training which the population has received in different fields at the secondary level and can be used in manpower planning as well as in assessing the output of the educational system.

Educational Qualifications

Educational qualifications, according to the United Nations, are the qualifications (i.e., degrees, diplomas, certificates, etc.) which an individual has acquired, whether by full-time study, part-time study, or private study; whether conferred in his home country or abroad; and whether conferred by educational authorities, special examining bodies, or professional bodies. The acquisition of an educational qualification, therefore, implies the successful completion of a course of study.[18]

Information on educational qualifications is considered "useful" by the United Nations, but it is not a "recommended" item. A number of countries have collected such data in their censuses in recent decades, but these constitute only a small minority of countries taking censuses.

Cross-national comparisons of educational qualifications are difficult to make because of different types of degrees and diplomas awarded in different countries. Within a country, problems of comparability are usually not too severe, but the quality of the data collected needs to be evaluated carefully. Where the data on educational qualifications come from censuses, it

is possible to check them against data on educational attainment. Where the data on educational qualifications come from school systems or individual colleges or universities, aggregate statistics can likewise be compared with those on educational attainment in the census for comparable age groups. For example, in the United States, the bachelor's degree usually corresponds to the completion of the fourth year of college whereas the high school diploma corresponds to the completion of the fourth year of high school.

Numerous types of educational qualifications are recognized in various countries. Major group categories include those relating to completion of high school, bachelor of higher education, master of higher education, and doctorate in higher education. No conventional measures of educational qualifications have been developed, but it is advisable to relate the number of such qualifications in a group to the population eligible to receive the qualification (e.g., the population 22 and over in calculating the rate of qualification for a college degree) when calculating an educational qualifications rate, particularly in comparing population subgroups. It is also possible to relate the number of diplomas, degrees, etc., for a group to the number of persons in the original cohort who successfully completed a lower level of educational attainment.

Uses and Limitations of Data on Educational Output

Data on educational output may be used to study the productivity of the school and college systems in a country, the disparities in literacy and educational attainment among segments of the population, the association of education with employment and occupational placement, the characteristics of the educated manpower supply, and the economic returns to education, as well as the effects of education on fertility, mortality, migration, urbanization, and other demographic processes.

The United Nations has stressed the importance of data on educational status in government planning: "Information on the educational status of populations is important both at the national and the international level in connexion with planning for the education of the population and in setting up or developing programmes for economic and social development, for which indicators of the general level of education and the size of the educated population segment are essential."[19]

Golden has developed the thesis that the literacy rate is a useful index of a country's level of socioeconomic development.[20] Kamerschen later refined her thesis, concluding that the statistics support a threshold theory rather than a continuous one.[21] Such a relationship may be reflected in the fact that a number of the more industrialized countries no longer collect statistics on illiteracy since the problem has already reached a nearly minimal level.

In using statistics on educational attainment and educational qualifications, it should be realized that they are simple quantitative indicators of educational output and do not necessarily indicate the quality of education received or the resulting competencies of the persons involved. For example, historical analysis of educational attainment is complicated because we do not know how much better prepared scholastically a person may be after completing a given school level today than was

[16] United Nations, *Principles and Recommendations for the 1970 Population Censuses*, p. 47.

[17] UNESCO, *Manual of Educational Statistics*, Paris, 1961, p. 158.

[18] United Nations, *Principles and Recommendations for the 1970 Population Censuses*, p. 46.

[19] United Nations, *Demographic Yearbook, 1963*, p. 41.

[20] Hilda Hertz Golden, "Literacy and Social Change in Underdeveloped Countries," *Rural Sociology*, 20(1):1–7, March 1955.

[21] David R. Kamerschen, "Literacy and Socioeconomic Development," *Rural Sociology*, 33(2):175–188, June 1968.

his counterpart at an earlier period of time. Likewise, at any point in time, there may be variations among areas of a country in the types of school attended, the kinds of courses taken, and the quality of teaching, all of which make comparisons among groups difficult. Moreover, there are variations in the abilities and learning of persons completing the same school level or receiving the same educational qualification.

In the case of literacy, a question on ability to read and write is obviously subject to a variety of interpretations and the collection of the data in the census may be handled with varying degrees of conscientiousness relative to any official standard.

The definition of illiteracy in the 1961 census of Senegal as inability to read and write French was not in line with either the international standard or usual national practices. In the case of educational levels, an 8-year primary course in one country obviously cannot fairly be compared with a 4-year course in another country. In sum, demographic statistics on education, apart from inadequacies of the data, reflect the effects of a variety of cultural, social, and psychological factors which must be considered in analyzing the data. All these considerations necessitate interpretation of published statistics on education in only general terms.

SUGGESTED READINGS

Bogue, Donald J. *Principles of Demography*. New York, John Wiley and Sons, 1969. Pp. 181–202.

Duncan, Otis Dudley, and Hodge, Robert W. "Education and Occupational Mobility: A Regression Analysis." *American Journal of Sociology*, 68(6):629–644. May 1963.

Folger, John K., and Nam, Charles B. *Education of the American Population*. Washington, U.S. Government Printing Office, 1967. Especially Appendix.

International Union for Scientific Study of Population. *Proceedings of the Conference, 1969, London*. "Meeting 6.2, Education and Demography." Statement by the Organizer, H. V. Muhsam.

Liu, B. A., "Measuring Progress of Literacy in the General Population." *Proceedings of the World Population Conference, Rome, 1954*. New York, 1955. Vol. IV. Pp. 1015–1025.

Liu, B. A., and Pineda-Espinosa, A. "Measuring the Educational Level of the Population: A Methodological Study." *Bulletin de l'Institut International de Statistique* (Brussels), 37(2):161–174. 1960.

Nam, Charles B. "Some Comparisons of Office of Education and Census Bureau Statistics on Education." *Proceedings of the Social Statistics Section, 1962*. American Statistical Association, 1962. Pp. 258–269.

Sadie, Johannes L. A. "Análisis demográfico del estado de la educación en la América Latina" (Demographic Analysis of the Educational Situation in Latin America). United Nations Demographic Center for Latin America (CELADE). Series A. Santiago, 1962.

Spiegelman, Mortimer. *Introduction to Demography*. Cambridge, Mass., Harvard University Press, 1968. Chapter 11.

Stockwell, Edward G., and Nam, Charles B. "Illustrative Tables of School Life." *Journal of the American Statistical Association*, 58:(304):1113–1124. December 1963.

United Nations. Handbook of Population Census Methods. Studies in Methods, Series F, No. 5, rev. 1. Vol. III. *Demographic and Social Characteristics of the Population*. Pp. 25–32, 1959.

———. *World Population Conference, 1965*. Vol. I: Summary Report. "Meeting A.6, Demographic Aspects of Educational Development." Statement by the Moderator, P. J. Idenburg, and Statement by the Rapporteur, E. Solomon. Pp. 206–218, 1966.

UNESCO. *Manual of Educational Statistics*. Paris, 1961.

U.S. Bureau of the Census. *U.S. Census of Population: 1960. Subject Reports. Educational Attainment*, PC(2)–5B, 1963. Pp. VII–XVI.

———. *U.S. Census of Population: 1960. Subject Reports. School Enrollment*, PC(2)–5A. 1964. Pp. VII–XVII.

U.S. Office of Education. *Report of the Advisory Panel of Educational Statistics*. OE 20061, Circular 737. Department of Health, Education, and Welfare. December 1963.

Windle, Charles. "The Accuracy of Census Literacy Statistics in Iran." *Journal of the American Statistical Association*, 54(287):578–581. September 1959.

Zarkovic, S. S. "Sampling Control of Literacy Data." *Journal of the American Statistical Association*, 49(267):510–519. September 1954.

CHAPTER 12.

Economic Characteristics

This chapter treats the economic activity of persons and the income received by families and unrelated individuals. One reason for including these topics in this book is that they (and economic activity especially) represent items collected in population censuses and household surveys. Indeed, primary attention is given here to data on economic characteristics from those censuses and surveys. Another reason is that many cross-tabulations with more strictly demographic characteristics are available for analysis and the demographer should know how to deal with these intelligently. On the other hand, the fields of economic activity, manpower, and consumer income represent major areas of specialization within the broader field of economics. Many excellent texts have been devoted entirely to one or more of these specializations.[1] The present chapter represents a relatively brief treatment, emphasizing the demographic aspects and citing references to more detailed treatments.

ECONOMIC ACTIVITY AND THE WORKING FORCE

Basic Concepts and Definitions

Although all persons consume goods and services, only a part of the entire population of a country is engaged in producing such goods and services. Most obviously, the youngest and oldest and the physically or mentally incapacitated do not engage in such economic activity because of inability to do so. The **manpower** of a nation, then, is the totality of persons who could produce the goods and services if there were a demand for their labors and they desired to participate in such activity. The **economically active** (sometimes also called the **working force**) is that part of the manpower which actually engages, or attempts to engage, in the production of economic goods and services.[2]

At any given time, an economically active person may be either "employed" or "unemployed." (As we shall see, the distinction is not always clear-cut; and some of the employed may be classified as "underemployed.") Those not economically active may be subdivided according to their major type of activity. The economically active, again, may be classified according to the nature of their current, last, or usual job

(occupation, industry, status, name of employer, place of work, etc.). Other related items are weeks worked in the past year, hours worked in the last week or other reference period, and duration of unemployment for the unemployed.

The United Nations recommends that census information be collected that would allow each person in the population to be classified according to **type of activity,** that is, his relationship to current economic activity. The following classification is recommended:[3]

Economically active population
    Employed
    Unemployed
Not economically active population
    Homemakers
    Students
    Income recipients
    Others

The definitions of the above categories recommended for use in the 1970 censuses of population follow:

"**Economically active population** comprises all persons of either sex who furnish the supply of labour for the production of economic goods and services during the time-reference period chosen for the investigation. It includes both persons in the civilian labour force and those serving in the armed forces. In compilations of the data, a separate category of 'members of the armed forces' may be maintained, so that the category can be deducted from the total labour force whenever desirable. The civilian labour force comprises both persons employed and those unemployed during the time-reference period. These two groups should be distinguished in accordance with the following criteria:

(a) **Employed**

The **employed** comprise all persons, including family workers, who worked during the time-reference period established for data on economic characteristics . . . or who had a job in which they had already worked but from which they were temporarily absent because of illness or injury, industrial dispute, vacation or other leave of absence, absence without leave, or temporary disorganization of work due to such reasons as bad weather or mechanical breakdown.

(b) **Unemployed**

The **unemployed** consist of all persons who, during the reference period, were not working but who were seeking work for pay or profit, including those who never worked

[1] A recent example is a manual prepared by John D. Durand and Ann R. Miller of the University of Pennsylvania: United Nations, *Methods of Analysing Census Data on Economic Activities of the Population,* Population Studies No. 43, 1968.
[2] See also A. J. Jaffe and Charles D. Stewart, *Manpower Resources and Utilization,* chapter 2, "Definitions and Concepts," and chapter 3, "Socio-Economic Development and the Working Force," New York, John Wiley & Sons, 1951.

[3] United Nations, *Principles and Recommendations for the 1970 Population Censuses,* Statistical Papers, Series M. No. 44, 1967, pp. 62–63.

before. Also included are persons who, during the reference period, were not seeking work because of temporary illness, because they made arrangements to start a new job subsequent to the reference period, or because they were on temporary or indefinite lay-off without pay. Where employment opportunities are very limited, the unemployed should also include persons who were not working and were available for work, but were not actively seeking it because they believed that no jobs were open. The recorded data on the unemployed should distinguish persons who never worked before.

"In classifying the population by type of economic activity, participation in an economic activity should always take precedence over a non-economic activity; hence, employed and unemployed persons should not be included in the not economically active population, even though they may also be, for example, student or home-makers.

"Not economically active population comprises the following functional categories:

(a) **Home-makers:** persons of either sex, not economically active, who are engaged in household duties in their own home; for example, housewives and other relatives responsible for the care of their home and children. (Domestic servants working for pay, however, are classified as economically active.)

(b) **Students:** persons of either sex, not economically active, who attend any regular educational institution, public or private, for systematic instruction at any level of education.

(c) **Income recipients:** persons of either sex, not economically active, who receive income from property or other investment, royalties, or pensions from former activities.

(d) **Others:** persons of either sex, not economically active, who are receiving public aid or private support, and all other persons not falling in any of the above categories, such as children not attending school…

"The minimum age-limit adopted for the census questions on economic activity should be set in accordance with the conditions in each country, but never higher than fifteen years." [4]

Labor Force and Gainful Worker Concepts. – The term **economically active** can be considered as a more general term which includes **labor force** and **gainful worker.** The heart of all these terms involves the carrying on of an activity from which the person derives, or attempts to derive, pay or profit. Thus, all persons engaged in nonremunerative work, such as housewives, are excluded unless they qualify as "unpaid family workers" (q.v.). There are a variety of procedures which can be used for determining who is and who is not economically active. The term **labor force** refers specifically to those procedures evolved in the United States in the late 1930's. These procedures have been changed from time to time within the United States,[5] and differ in some degree from the procedures used in other countries.

The term **gainful worker** has been used in a number of countries;[6] in the decennial census of the United States it was last used in 1930. It too refers to economic activity as just defined; the main difference between this term and the

definition of the economically active population recommended by the United Nations and International Labour Office is that the gainful worker term had no explicit time reference. It referred to the person's "usual" activity, which could have encompassed any period from a day to a lifetime. Retired persons were often included among the gainful workers but persons seeking work for the first time were not included. Because of this time vagueness, meaningful statistics on employment and unemployment could not be derived.

The general idea of the labor force can be illustrated by noting the main features of the U.S. collection procedures as now used in the Current Population Survey. All persons who are aged 16 and over and are neither inmates of institutions nor members of the armed forces living in a military installation are asked whether they worked for pay or profit during a specified week. Those who did so are classified as **employed.** Those who had a job or business during that week but were absent from it because of vacation, illness, etc., are also classified as employed. Those who did not do so are then queried as to whether they sought work for pay or profit during the past four weeks; those who did so are classified as the **unemployed.** The remaining persons who were neither employed nor unemployed are classified as **not in the labor force.**

It is clear, even from this brief description, that it is easiest to determine economic activity in a country or other area where practically everyone receives money for his labors. In situations where money does not change hands, as in a subsistence economy, the main criterion for differentiating economic from noneconomic activity is not available and it is very difficult to decide who is and who is not economically active.

Job Characteristics. – Three items of information which describe the economically active population are almost invariably obtained when a census or sample survey is conducted. These are occupation, industry, and status (as employer, employee, etc.). Other items of information about the employment or unemployment of the individual are sometimes collected simultaneously, but these three are the only ones specifically recommended by the United Nations for the census of population.

Occupation. – According to the U.N. recommendations for the 1970 censuses of population, "occupation refers to the kind of work done during the time-reference period established for data on economic characteristics by the person employed (or performed previously by the unemployed) irrespective of the industry or the status (as employer, employee, etc.) in which the person should be classified." [7] Examples of occupations are: teacher, sales clerk, farmer, car washer, bus driver, and typist.

For purposes of international comparisons, the United Nations recommends that countries compile their data in accordance with the latest edition of the *International Standard Classification of Occupations* (ISCO) prepared by the International Labour Office. As the United Nations recognizes that many countries will want to use occupational classification systems which they believe are more useful for national purposes, the U.N. recommendation continues with the statement that "If this is not possible, provision should be made for the categories of the classification employed to be convertible to the ISCO or at least to the minor (two digit)

[4] Ibid., p. 61.
[5] For the evolution of the concepts and procedures, see U.S. Bureau of the Census, *Current Population Reports*, Series P–23, No. 22, "Concepts and Methods Used in Manpower Statistics from the Current Population Survey," June 1967, pp. 15 ff.
[6] Apparently many countries are still using the gainful worker approach, or some variant of it.

[7] United Nations, *Principles and Recommendations for the 1970 Population Censuses*, p. 53.

groups of this classification. If national data are not classified in conformity with the ISCO, an explanation of the differences should be given."

Industry (branch of economic activity). — In addition to information on the kind of work or occupation in which the individual is engaged, it is also important to know the kind of activity carried out by the establishment in which the person works. The United Nations defines "industry" as follows:

"Industry refers to the activity of the establishment in which an economically active person worked during the time reference period established for data on economic characteristics or last worked, if unemployed."[8] Activity of the establishment means the kinds of goods or services produced. Goods-producing establishments include, for example, petroleum refinery, pulp and paper factory, and fruit or vegetable canning plant. Examples of service-producing establishments are hospital, railroad, barber shop, elementary school, and retail electric-appliance store.

For purposes of international comparability, the United Nations recommends that "countries compile their data in accordance with the *International Standard Industrial Classification of All Economic Activities* (ISIC) most recently approved by the United Nations. If this is not possible, provision should be made for the categories of the classification employed to be convertible to the ISIC or at least to the major (two digit) groups of this classification. If the national data are not classified in accordance with the ISIC, an explanation of the differences should be given." [9]

Status (as employer, employee, etc.). — "Status[10] (as employer, employee, etc.) refers to the status of an economically active person with respect to his employment, that is, whether he is (or was, if unemployed) an employer, own-account worker, employee, unpaid family worker, or a member of a producers' cooperative." [11]

The international recommendations for classification of this economic characteristic and the definition of the categories follow:

"(a) **Employer:** a person who operates his or her own economic enterprise or engages independently in a profession or trade, and hires one or more employees. Some countries may wish to distinguish among employers according to the number of persons they employ.

(b) **Own-account worker:** a person who operates his or her own economic enterprise or engages independently in a profession or trade, and hires no employees.

(c) **Employee:** a person who works for a public or private employer and receives remuneration in wages, salary, commission, tips, piece-rates, or pay in kind.

(d) **Unpaid family worker:** a person who works a specified minimum amount of time (at least one third of normal working hours), without pay, in an economic enterprise operated by a related person living in the same household. If there are a significant number of unpaid family workers in enterprises of which the operators are members of a producers' co-operative who are classified in category (e), these unpaid family workers should be classified in a separate sub-group.

(e) **Member of producers' co-operative:** a person who is an active member of a producers' co-operative, regardless of the industry in which it is established. Where this group is not numerically important, it may be excluded from the classification and members of producers' co-operatives should be classified to other headings, as appropriate.

(f) **Persons not classifiable by status:** experienced workers with status unknown or inadequately described and unemployed persons not previously employed." [12]

The United States and other countries often combine "employer" and "own account worker" and do not show separately "member of producers' cooperative." In the United States, unlike such countries as the socialist countries of Eastern Europe, so few workers would fall under this last category, that it is excluded.

Subtotals from distributions by occupation and industry. — The detailed categories of occupation or industry are frequently consolidated into conventionally defined subtotals in the following situations: (1) in the presentation of statistics for small areas; (2) in the presentation of simultaneous cross-classifications of occupation (or industry) with several other variables including social variables like educational attainment; (3) in the presentation of statistics from sample surveys where the reliability of the detailed categories is inadequate. For example, in the United States in 1960, the occupational classification system consisted of 297 items, which were grouped in certain tables into an intermediate classification of 161 categories for males and 70 for females, and, further, into 12 major groups.[13] The major groups are as follows:

Professional, technical, and kindred workers
Farmers and farm managers
Managers, officials, and proprietors, except farm
Clerical and kindred workers
Sales workers
Craftsmen, foremen, and kindred workers
Operatives and kindred workers
Private household workers
Service workers, except private household
Farm laborers and farm foremen
Laborers, except farm and mine
Occupation not reported

This list, which was developed in essentially its present form about 1930 by Alba M. Edwards of the U.S. Bureau of the Census, is often treated by sociologists as if it represented a hierarchy in terms of socioeconomic status, at least for the nonfarm sector. Although this is roughly the case, the listing is probably multidimensional. Nonetheless, these major groups are an important tool in demographic and sociological analysis, in the form of multivariate analysis (where each category is treated as a "dummy variable"), of indexes of dissimilarity between population groups, etc.

In view of some of the criticisms of the 1960 occupational classification system, a revised system was developed for the 1970 census. There are 441 detailed categories instead of 298. There is one more major group—"Transport equipment operatives" is split off from "Operatives and kindred workers." The new major groups will be arranged in terms of four broad occupational areas, as follows:

[8] United Nations, *Principles and Recommendations for the 1970 Population Censuses,* p. 49.

[9] Ibid. p. 49.

[10] The term used in the United States is "class of worker."

[11] United Nations, *Principles and Recommendations for the 1970 Population Censuses,* p. 59.

[12] Ibid., pp. 59–60.

[13] *U.S. Census of Population: 1960,* Vol. 1, *Characteristics of the Population,* Part 1, *United States Summary,* 1964, pp. LXVII–LXXI.

White-collar workers
    Professional, technical, and kindred workers
    Managers and administrators, except farm
    Sales workers
    Clerical and kindred workers
Blue-collar workers
    Craftsmen and kindred workers
    Operatives, except transport
    Transport equipment operatives
    Laborers, except farm
Farm workers
    Farmers and farm managers
    Farm laborers and farm foremen
Service workers
    Service workers, except private household
    Private household workers

Similarly, the 150 categories of industries used in the 1960 census of the United States were expanded to 226 categories for the 1970 census. There are still 14 major industrial groups, however.[14]

Other Economic Characteristics of Workers. — In addition to the basic questions on economic activity, occupation, industry, and status almost universally included in population censuses and sample surveys that deal with economic characteristics of individuals, there are a large number of other economic characteristics that are sometimes included in household surveys. Generally, population censuses can only accommodate questions on a few, if any, additional economic characteristics. On the other hand, special labor force sample surveys or sample economic surveys can and often do obtain information on additional economic characteristics of individuals. Examples of the additional items which may be included are:

— time actually worked in reference week;
— normal or scheduled hours (or days) of work per week;
— whether engaged in a household enterprise or a non-household enterprise;
— type of employing establishment;
— number of employees (employers only);
— secondary occupation;
— seasonal variations in time worked (persons employed whole of last year);
— reason for short-time work during the reference week (only persons who worked under 35 hours or less than 5 days);
— whether looked for more hours of work (ditto);
— whether wanted more hours of work (ditto);
— kind of job sought (unemployed persons);
— duration of unemployment;
— average monthly earnings from enterprise (own-account workers only);
— average weekly wages or salary last month (employees with jobs only);
— migration for employment.

Underemployment. — Some of the items listed above are designed to elicit information on the problem of underemployment. Though neither conceptually clearcut nor easy to measure, the problem of underemployment is often much more serious a problem in underdeveloped countries than is unemployment.

Between "full" employment and complete lack of employment (i.e., unemployment) lies a continuum of working behavior. Any amount of work at any point on this continuum can be called "underemployment," i.e., less than "full" employment. This was expressed by the I.L.O. as:

> Underemployment is the difference between the amount of work performed by persons in employment and the amount of work they would normally be able and willing to perform.[15]

Attempts were made to make this general definition more concrete (at I.L.O. meetings in 1957 and 1963) by subdividing underemployment into two major categories: **visible,** when persons involuntarily work part time or for shorter periods than usual; and **invisible,** when persons work full time but the work is inadequate because earnings are too low or the job does not permit exercise of fullest skills.[16]

Despite the large amount of research that has been carried on,[17] it has not been possible to develop procedures as precise as those used in measuring economic activity, i.e., employment and unemployment. In large measure this difficulty stems from the fact that there is no uniquely correct measure or definition of "full" employment. For example, in the United States, 35 hours of work is considered a "full" week and large numbers work little if any longer. In Taiwan, on the other hand, a "full" week is defined as 42 hours or more of work.[18]

It is also impossible to discover a uniquely correct procedure for ascertaining invisible underemployment. The degree to which work is adequate or inadequate does not lend itself to easy definition and measurement.

Labor Mobility. — Labor mobility is any change in a person's status that involves his economic activity, or more specifically his job. The most common forms of labor mobility are:

    a. Entering or leaving the labor force
    b. Shifting employment status
    c. Changing occupation
    d. Changing industry
    e. Changing class of worker (status)
    f. Changing employer
    g. Moving from one geographic area to another (this last is also treated under geographic mobility, of course; see chapter 21).

Any change in economic activity involving related persons of two or more generations is **intergenerational labor mobility.** The most common comparison is that of father and son. Any of the items listed under labor mobility above, may be compared.

Sources of Data and Collection Procedures

Information on economic activity is collected from households in population censuses or sample surveys, or from

---

[14]U.S. Bureau of the Census, *Census of Population: 1970, Detailed Characteristics,* PC (1) - D1, *United States Summary,* 1973, pp. App. 17-18.

[15]International Labour Organisation, Ninth International Conference of Labour Statisticians, Report No. 4. *Measurement of Underemployment,* Geneva, 1957, p. 17.

[16]See also, Kailas C. Doctor, "Recent Progress in Underemployment Statistics and Analysis," in United Nations, *World Population Conference, 1965,* (Belgrade), 1967, p. 348 ff.

[17]For example, Puerto Rico has measured underemployment continuously since 1952 and was probably the first country to begin such a continuous series. See Fernando Sierra Berdecia and A. J. Jaffe, "The Concept and Measurement of Underemployment," *Monthly Labor Review,* 78(3):283–287, March 1955.

[18]Taiwan, Department of Social Affairs, *Quarterly Report on the Labor Force Survey in Taiwan, October 1968,* prepared by the Labor Force Survey Research Institute; see for example, p. 13 of the January 1967 report for minimum length of work week used for measuring underemployment.

establishments through regular reporting in certain kinds of administrative programs or through special sample surveys. The data from sources other than households pertain to employment and unemployment but not to the economically inactive population.

Censuses. — Available national census statistics on the characteristics of the economically active population have been summarized in the U.N. *Demographic Yearbook,* most recently in 1972 (tables 8 to 15). Availability of these economic characteristics is roughly comparable with that of educational statistics or statistics of households but is markedly less than that of the basic demographic statistics. Statistics on industry of workers are more generally available than those on their occupation, whereas, for most kinds of demographic analysis, occupation is the more meaningful characteristic. Less than half of the population of Africa and one-half of that of Asia lived in countries with even minimally acceptable labor force statistics from censuses.

Efforts have been made by the United Nations to select, arrange, and edit the statistics so as to increase international comparability. The extent and nature of the footnotes on the tables in the *Demographic Yearbook* give one indication of the degree to which this objective could not readily be accomplished. There are variations in definitions, population covered, categories tabulated, and the like.

Household Surveys. — Those few countries that do conduct sample household surveys on a recurrent basis, tend to use concepts and procedures rather similar, but not identical, to the U.N. recommendations and the procedures developed in the United States. Comparison of the main outlines of the procedures used in Taiwan and Panama will illustrate some of the similarities and differences.[19]

Coverage. — Both countries cover the civilian population, excluding those who are inmates of institutions. Taiwan excludes the members of the armed forces (presumably those not living in private dwellings) and reports on the civilian labor force only. Panama has no armed forces as such; the National Guard serves as the combined police and military force. Any member of the Guard living at home is included; members living in barracks are part of the institutional population and are excluded.

Age. — Taiwan includes those aged 12 and over, Panama those aged 15 and over.

Time reference. — Both use one week as the time period, but differently. Taiwan conducts four surveys a year, in January, April, July, and October; interviewing is done in one week of each these months covering the week containing the 15th of the month. Panama conducts interviews throughout the year and cumulates the data to obtain annual figures.

Sample size. — Panama interviews about 10,000 households during the course of a year; Taiwan about 10,000 households each quarter.

Basic sample design. — The design is similar in these two countries. However, Panama obtains half of its cases from its major urban center, whereas Taiwan obtains only about one-fifth of its cases from its five main cities. Hence, Panama can produce more detailed cross-tabulations for its major urban area whereas Taiwan could do this reliably only by summing the four quarters into an annual average; apparently Taiwan does not do this.

Employed. — In Taiwan persons who worked for pay or profit or as unpaid family workers, for at least 18 hours during the reference week, are defined as employed. In Panama a person is employed if he worked any amount of time during the reference week.

Unemployed. — In Taiwan the unemployed are the persons who are able and willing to work but worked 18 hours or less during the survey week, and those without any work who are looking for a job. Panama defines as unemployed any one who did not work for pay or profit and sought work during the reference week.

Underemployed. — In Taiwan in the standard week of the survey, those who worked less than 42 hours in remunerative work, those who worked more than 18 and less than 42 hours in unpaid family work, and those who think their work is not suitable to their taste and ability are classified as underemployed. (Whether other persons who worked less than 42 hours a week were also considered underemployed was not specifically stated in the source available to the authors.) In Panama wage and salary workers are treated separately from the self-employed; the unpaid family workers are shown separately and are treated statistically as neither fully employed nor underemployed. The following items are taken into consideration in determining underemployment: hours worked per week, whether or not the person sought work in addition to working, earnings, and whether or not he wanted to change jobs and the reasons therefor. The resulting statistics on underemployment for the two countries therefore cannot be meaningfully compared.

United States Practices

Statistics From Households. — Although attempts were made in the U.S. census as early as 1820 to collect statistics on the economically active population, the 1870 population census provided the first body of data adequate for more than merely gross descriptions of the working force. In the 1890 and 1910 censuses, substantial revisions were made in the collection and classification procedures so that it is difficult to construct a comparable series from 1870 to 1930. Miller and Brainerd have painstakingly prepared revised series of labor force estimates for decennial dates, on a reasonably consistent basis, back to 1870.[20] A similar reconstruction of the occupational statistics was carried out by Kaplan and Casey of the U.S. Bureau of the Census.[21]

In the late 1930's, the Works Progress Administration developed the labor force concepts and procedures. These were used in the 1940 and subsequent population censuses and also in the monthly sample surveys, which have been taken continuously since 1940.

---

[19] Information about Taiwan from *Quarterly Report on the Labor Force Survey in Taiwan.* For Panama, see Dirección de estadística y censo, *Estadística Panameña,* Año XXVI, Suplemento, "Encuestas de mano de obra en Panamá: Años 1962 a 1967, Informe metodológico." (Some changes made subsequent to that publication are incorporated here.)

[20] Everett S. Lee, et al., *Population Redistribution and Economic Growth: United States, 1870–1950,* Vol. I, *Methodological Considerations and Reference Tables,* Philadelphia, American Philosophical Society, 1957, pp. 363–633.
[21] U.S. Bureau of the Census, *Working Paper* No. 5, "Occupational Trends in the United States: 1900 to 1950," by David L. Kaplan and M. Claire Casey, 1958.

In the population censuses since 1940, every one aged 14 and over (except for inmates of institutions) has been asked about his activities in the week preceding the census or the week preceding enumeration. The 1970 census of population included the following questions on economic activity:

QUESTIONS 29 THROUGH 41 ARE FOR ALL PERSONS BORN BEFORE APRIL 1956 INCLUDING HOUSEWIVES, STUDENTS, OR DISABLED PERSONS AS WELL AS PART-TIME OR FULL-TIME WORKERS ■

29a. Did this person work at any time last week?

○ **Yes**– *Fill this circle if this person did full- or part-time work. (Count part-time work such as a Saturday job, delivering papers, or helping without pay in a family business or farm; and active duty in the Armed Forces)*

○ **No**– *Fill this circle if this person did not work, or did only own housework, school work, or volunteer work.*

Skip to 30

b. How many hours did he work last week (at all jobs)?
Subtract any time off and add overtime or extra hours worked. ■

○ 1 to 14 hours	○ 40 hours
○ 15 to 29 hours	○ 41 to 48 hours
○ 30 to 34 hours	○ 49 to 59 hours
○ 35 to 39 hours	○ 60 hours or more

---

30. Does this person have a job or business from which he was temporarily absent or on layoff last week?

○ Yes, on layoff

○ Yes, on vacation, temporary illness, labor dispute, etc.

○ No

31a. Has he been looking for work during the past 4 weeks?

○ Yes   ○ No— *Skip to 32*

b. Was there any reason why he could not take a job last week?

○ Yes, already has a job

○ Yes, because of this person's temporary illness

○ Yes, for other reasons (in school, etc.)

○ No, could have taken a job

---

32. When did he last work at all, even for a few days?

○ In 1970	○ 1964 to 1967	○ 1959 or earlier	*Skip*
○ In 1969	○ 1960 to 1963	○ Never worked	*to 36*
○ In 1968 ■		■	

33–35. Current or most recent job activity

Describe clearly this person's chief job activity or business last week, if any. If he had more than one job, describe the one at which he worked the most hours.

If this person had no job or business last week, give information for last job or business since 1960.

---

33. Industry

a. For whom did he work? *If now on active duty in the Armed Forces, print "AF" and skip to question 36.*

_____
(Name of company, business, organization, or other employer)

b. What kind of business or industry was this?
Describe activity at location where employed.

_____
(For example: Junior high school, retail supermarket, dairy farm, TV and radio service, auto assembly plant, road construction)

c. Is this mainly— *(Fill one circle)*

○ Manufacturing	○ Retail trade
○ Wholesale trade	○ Other *(agriculture, construction, service, government, etc.)*

---

34. Occupation

a. What kind of work was he doing?

_____
(For example: TV repairman, sewing machine operator, spray painter, civil engineer, farm operator, farm hand, junior high English teacher)

b. What were his most important activities or duties?

_____
(For example: Types, keeps account books, files, sells cars, operates printing press, cleans buildings, finishes concrete)

c. What was his job title? _____

---

35. Was this person— *(Fill one circle)*

Employee of private company, business, or individual, for wages, salary, or commissions... ○

Federal government employee .... ○
State government employee.................. ○
Local government employee *(city, county, etc.)*... ○

Self-employed in own business, ■
professional practice, or farm—
    Own business not incorporated .... ○
    Own business incorporated .... ○

Working without pay in family business or farm ○

---

37. In April 1965, was this person— *(Fill three circles)*

a. Working at a job or business *(full or part-time)*?
    ○ Yes   ○ No

b. In the Armed Forces?
    ○ Yes   ○ No

c. Attending college? ■
    ○ Yes   ○ No

Table 12-1. -Major Occupation Group of Husband — White Women 20 to 24 Years Old, Married and Husband Present, by Number of Children Ever Born and Age, for the United States: 1960

[Based on 5-percent sample]

Employment status and occupation of husband	White women								Children ever born		
	Total	By number of children ever born							Number	Per 1,000 women	Per 1,000 mothers
		Child-less	1	2	3	4	5 and 6	7 or more			
Total.....................	3,173,211	780,664	1,064,584	849,160	340,889	103,271	31,749	2,894	4,385,845	1,382	1,833
Husband employed........................	2,789,708	659,313	941,530	764,174	303,509	90,909	27,823	2,450	3,907,057	1,401	1,834
Professional, technical, & kindred workers.	351,097	121,400	128,979	75,201	20,245	4,045	1,067	160	362,991	1,034	1,580
Farmers and farm managers..................	85,088	15,268	26,077	27,121	11,838	3,514	1,093	177	136,910	1,609	1,961
Mgrs., off'ls., & propr's., exc. farm......	180,172	45,469	62,361	48,493	17,736	4,810	1,203	100	238,931	1,326	1,774
Clerical and kindred workers..............	227,758	65,777	84,892	54,776	17,364	3,724	1,185	40	267,877	1,176	1,654
Sales workers...........................	189,098	54,657	66,953	47,818	15,151	3,735	704	80	227,244	1,202	1,690
Craftsmen, foremen, and kindred workers....	578,664	116,255	191,766	172,664	70,338	21,653	5,469	519	866,707	1,498	1,874
Operatives and kindred workers.............	740,843	140,476	240,969	221,181	96,816	30,976	9,802	623	1,153,184	1,557	1,921
Service workers, incl. priv. household.....	98,589	22,859	33,495	27,092	11,080	2,934	1,089	40	138,681	1,407	1,831
Farm laborers and foremen..................	54,721	9,987	16,367	14,563	8,444	3,462	1,676	222	95,246	1,741	2,129
Laborers, except farm and mine............	177,652	36,300	55,396	50,732	23,465	8,226	3,166	367	279,366	1,573	1,976
Occupation not reported...................	106,026	30,865	34,275	24,533	11,032	3,830	1,369	122	139,920	1,320	1,862
Husband unemployed........................	115,650	26,255	35,890	29,114	15,506	6,528	2,156	201	179,621	1,553	2,009
Husband in armed forces....................	174,436	53,350	58,365	42,100	16,002	3,765	793	61	210,167	1,205	1,736
Husband not in labor force................	93,417	41,746	28,799	13,772	5,872	2,069	977	182	89,000	953	1,722

Source: *U.S. Census of Population: 1960, Subject Reports, Women by Number of Children Ever Born*, PC-2(3A), 1964, table 31, p. 147.

The U.S. census reports contain a great many labor force statistics—detailed categories, historical comparisons, economic characteristics, etc. We have chosen only one illustration for presentation in tabular form. Table 12-1 is an example of how employment status and major occupation group are used to study class differentials in a demographic variable in this case, fertility.

The labor force concepts in the Current Population Survey (CPS) are discussed in a joint Census Bureau-BLS report.[22] The CPS enumerator asks certain questions about each person aged 14 years and over.

With these questions, it is possible to separate the employed into: at work full time (35 hours per week or more); at work part time for noneconomic reasons (vacation, illness, usually work part time, etc.); worked part time for economic reasons (35 hours or more of work was not available); had job but did not work, by reason for not working (bad weather, industrial dispute, vacation, illness, other). The unemployed can be classified by duration of unemployment, occupation and industry of last job if they had worked before, and whether they are seeking full-time or part-time jobs. Persons who are not in the labor force can be classified as housewives, students, and other (including retired, unable to work, etc.). Analysis of these subdivisions is useful for study of the labor reserve and for making labor force projections.

Table 12-2 shows some of the regular statistics on labor force and employment status from the CPS. The annual figures represent averages of the 12 monthly figures in the given

year. The unemployment rate (columns 9 and 10) is shown both as computed directly and as adjusted for seasonality. Actually, many other figures are also seasonally adjusted since labor force statistics show wide seasonal fluctuations.

Sometimes additional questions are asked in the Current Population Survey as: nature of job held at some past date, why women work or do not work, where the person received his training for the job, or whether persons seeking work want full- or part-time jobs.

Labor mobility.—Information on labor mobility is occasionally obtained by asking the respondent about his economic characteristics at some previous date. He can be asked whether he was employed or unemployed, his occupation, industry, etc. By comparing the previous and current activity, changes in economic activity can be noted. Information on gross changes, i.e., all movements, is obtained by such direct questioning. For a variety of reasons, including problems of memory recall, comparatively little use has been made of such questions.

Another method used to measure gross changes in labor force status from CPS data consists of comparing the status of matched persons in consecutive months. Thus, for example, the number of persons who were unemployed in a given month and employed in the following month could be estimated. Such figures were published for several years by the U.S. Bureau of the Census in Series P-59; but the series was discontinued in 1951 after 29 issues because of concern about the accuracy of figures derived by this method. It was suspected that the number of changes was exaggerated.[23] Pearl and others concluded on the basis of later analysis, however,

[22] U.S. Bureau of the Census and U.S. Bureau of Labor Statistics. *Current Population Reports*, Series P-23, No. 22, *BLS Report* No. 313, "Concepts and Methods Used in Manpower Statistics from the Current Population Survey," June 1967.

[23] U.S. Bureau of the Census, *Current Population Reports*, Series P-59, "Gross Changes in the Labor Force, April 8, 1949–July 20, 1951."

Table 12-2.—Employment Status of the Noninstitutional Population 16 Years Old and Over, United States: 1964 to June 1969

[Numbers in thousands]

Year and month	Total noninstitutional population	Total labor force		Civilian labor force							Not in labor force
				Total	Employed			Unemployed			
		Number	Percent of population		Total	Agriculture	Nonagricultural industries	Number	Percent of labor force		
									Not seasonally adjusted	Seasonally adjusted	
1964	127,224	75,830	59.6	73,091	69,305	4,523	64,782	3,786	5.2	–	51,394
1965	129,236	77,178	59.7	74,455	71,088	4,361	66,726	3,366	4.5	–	52,058
1966	131,180	78,893	60.1	75,770	72,895	3,979	68,915	2,875	3.8	–	52,288
1967	133,319	80,793	60.6	77,347	74,372	3,844	70,527	2,975	3.8	–	52,527
1968	135,562	82,272	60.7	78,737	75,920	3,817	72,103	2,817	3.6	–	53,291
1968: June	135,440	84,454	62.4	80,887	77,273	4,516	72,757	3,614	4.5	3.7	50,986
July	135,639	84,550	62.3	80,964	77,746	4,476	73,270	3,217	4.0	3.7	51,088
August	135,839	83,792	61.7	80,203	77,432	4,107	73,325	2,772	3.5	3.5	52,047
September	136,036	82,137	60.4	78,546	75,939	3,838	72,103	2,606	3.3	3.6	53,900
October	136,221	82,477	60.5	78,874	76,364	3,767	72,596	2,511	3.2	3.6	53,744
November	136,420	82,702	60.6	79,185	76,609	3,607	73,001	2,577	3.3	3.4	53,718
December	136,619	82,618	60.5	79,118	76,700	3,279	73,421	2,419	3.1	3.3	54,001
1969: January	136,802	81,711	59.7	78,234	75,358	3,165	72,192	2,876	3.7	3.3	55,091
February	136,940	82,579	60.3	79,104	76,181	3,285	72,896	2,923	3.7	3.3	54,361
March	137,143	82,770	60.4	79,266	76,520	3,327	73,193	2,746	3.5	3.4	54,373
April	137,337	83,137	60.5	79,621	77,079	3,607	73,471	2,542	3.2	3.5	54,200
May	137,549	83,085	60.4	79,563	77,264	3,894	73,370	2,299	2.9	3.5	54,464
June	137,737	85,880	62.4	82,356	78,956	4,367	74,589	3,400	4.1	3.4	51,857

Source: U.S. Bureau of Labor Statistics, *Employment and Earnings,* 16(1):25, July 1969, table A-1.

that the gross change figures were fairly accurate; and, in any case, the consensus is that the gross change for any category of labor force status is indeed much larger than the net change.[24] In countries with a well-developed market economy, the labor force is not a static aggregate changing only as the result of population growth, but it is subject to a great amount of turnover in the short run—about 5 percent each month in the case of the United States.

Statistics From Establishments.—There are two major U.S. series in which information is collected from establishments or other institutional sources rather than households to provide data on employment. The first, available since 1919, comes from establishment reports compiled each month by the Bureau of Labor Statistics via mail questionnaires. A sample of establishments employing about 25 million nonfarm wage and salary workers reports each month on the number of workers, full- or part-time, who received pay during the payroll period nearest the fifteenth of the month. The resulting data are inflated to yield State and national monthly totals on employment of nonfarm wage and salary workers. Each year the inflated estimates of total employment by industry, as obtained from the sample of establishments, are adjusted to the totals obtained from the OASI (Old Age and Survivors Insurance) records.

The second major series on employment comes from the OASI records. Each employer covered by Social Security provides quarterly information about each of his employees for the purpose of credit toward the employee's pension fund—the OASI. Summation of the numbers of workers provides county, State, and national statistics on employment, quarterly, as well as annually. In addition, since each employee's records must be kept separately, a Continuous Work History Sample for one percent of covered workers has been made available. Because the record is continuous, it has been possible to get data on changes for each individual worker. This Continuous Work History Sample contains information for each quarter for a number of years; hence, it is possible to study changes between covered employment and other activities, mobility among employers, industries, and geographic localities of employment; estimated earnings; etc.

There are a number of other series on employment that have specialized uses. One is that of the unemployment insurance system; when the employer pays his insurance premium, he reports the number of workers on his payroll. These figures then serve as the denominator for calculating the "percentage of insured unemployment." Another series dealing with farm labor is collected by the U.S. Department of Agriculture from a sample of farm operators.[25] Censuses of agriculture, manufacturing, mining, etc., also ask the respective establishments

[24] Robert B. Pearl, "Gross Changes in the Labor Force: A Problem in Statistical Measurement," in U.S. Bureau of Labor Statistics, *Employment and Earnings,* 9(10):IV–X, April 1963.

[25] U.S. Department of Agriculture, Statistical Reporting Service, *Farm Labor,* (monthly).

for the numbers of their employees at the time covered by the census.

The main supplementary series on unemployment is that obtained from the unemployment insurance records. All States have unemployment insurance laws although the coverage differs considerably from one to another. Each week each State reports the numbers of persons receiving unemployment insurance, to the U.S. Bureau of Employment Security.[26] It is thus possible to have frequent information on the volume of statutory unemployment for counties, States, and the nation.

Comparison of Household and Establishment Data. — The major difference between household and establishment data is that the former cover all persons 14 years old and over in the United States, whereas the latter cover only part of this population. For example, the BLS payroll series on employment excludes the self-employed, unpaid family workers, and farmers, and does not succeed in obtaining good coverage of persons employed in very small enterprises. Even the OASI records do not yet embrace 100-percent of the employed population.

As regards unemployment, the household data are intended to cover anyone who reported that he sought work and was available for work. The unemployment insurance series, on the other hand, counts only those who qualify for benefits under the various State laws. In practice, about half or more of the unemployed at any one time, do not qualify for benefits and hence do not appear in the unemployment insurance series, although they do appear in the household series.

The household surveys provide more demographic information about each person in the labor force — occupation, age, sex, education, marital status, etc. Such information generally is not obtained from establishment sources.

Information for intercensal years can be obtained for small geographic areas only from the establishments reports; the household surveys can furnish data for only a limited number of the larger areas. Certain other types of information such as industry of employment, hours worked, and earnings may be obtained somewhat more accurately from the establishment reports.

There are various other differences between the two sources of statistics. By and large, however, they supplement each other so that the best analysis of employment trends can be made by judicious use of data from several sources. For most types of demographic analysis, however, the data obtained from the population censuses and the Current Population Survey are generally sufficient. Data from such sources are used, for example, as benchmarks for labor force projections.

Uses and Limitations

Census and survey statistics on the economically active population are especially useful in those countries that do not have a highly developed system of economic statistics from establishment sources. Planning for economic development is a very important use of such statistics, but it is only one of many uses. Labor force statistics provide much of the description of a nation's or a region's human resources. Statis-

tics on occupational characteristics, particularly when cross-tabulated by educational attainment, provide an inventory of skills available among the people of a country. Other uses, in various types of economies, are described, for example, in the U.N. *Handbook of Population Census Methods.*[27]

The analysis of statistics on the economically active population is subject to important limitations, both with respect to the definitions employed and to the methods of compiling the data. Data on economic characteristics of the population derived from population censuses may lack comparability with the results of sample household surveys because of different concepts, definitions, field procedures, and processing procedures such as editing rules. For similar reasons, successive population censuses in the same country may also not be strictly comparable.

Differences among countries in definition of the economically active population may seriously affect the comparability of statistics on this subject. The practice varies among countries with regard to treatment of marginal categories such as inmates of institutions, members of armed forces, persons living in special areas such as reservations, persons seeking work for the first time, persons engaged in part-time economic activities, and seasonal workers. In some countries, all or parts of these groups are included among the inactive population whereas in other countries the same categories are counted among the economically active population. Even in the same country, the treatment of these categories may have changes from one census to another or may differ between a census and a labor force sample survey.

Three major sources of noncomparability of census or survey data which the analyst should be aware of are: (1) variation in length of the time period to which questions on economic activity refer, (2) variation in definition and inclusion of unpaid family workers, and (3) variation in lower age limit of persons covered by economic activity questions.

Variations in Length of Time Period. — Some questions on economic activity refer to the employment situation of the person on the day of the census or survey or during a brief period such as a specific week. In other cases the question may refer to a longer period such as a month or year, whereas in still other cases (i.e., under the "gainful worker" concept) the questions refer to the "usual" position of the person without reference to any specific time period. This variation affects measures of the size of the economically active population since the longer the reference period the larger will be the number of persons who report having been engaged in economic activity at some time during the period. Characteristics of the active population will also be affected by variation in the time reference period, since the persons who are included by virtue of a longer reference period are more likely to be marginal members of the working force, i.e., seasonal workers, the old, the young, or married women. Such marginal members of the working force are likely to have different occupation, industry, and status characteristics from those of the core of the working force who would be included under even a very short time-reference period.

Variation in Definition and Inclusion of Unpaid Family Workers. — Variations in the classification of persons who are engaged in both economic activities and noneconomic activities such as housekeeping or attending school is another major source of noncomparability in international statistics

---

[26] U.S. Bureau of Labor Statistics, *Employment and Earnings* (monthly).

[27] Vol. II, *Economic Characteristics of the Population*, pp. 6–9; 66–69.

on the economically active population. Perhaps the most important variation of this sort is the treatment of unpaid family workers, that is, persons who assist without pay in economic enterprises operated by other members of their households.

The question of receipt of money wages is a particularly important factor affecting the classification of persons as economically active. In industrialized countries and in the large cities of most other countries, the large majority of the workers receive money wages and salaries. However, in many parts of the world, the rural and small town people live a subsistence, or semisubsistence, life, and very large numbers are unpaid family workers. Without the receipt of money, it is very difficult to determine who is and who is not economically active. Furthermore, since so many of the unpaid family workers are farmers' wives, attitudes toward the acknowledgment of women working become involved. Often, it is readily admitted that women "do things" but not that they are "employed."

Variation in Age Group Covered by Statistics on Economic Activity. — Countries differ in the age group of the population for which the questions on economic activity are to be asked. In almost all, if not all, instances, it is the minimum age limit that varies. The U.N. Statistical Office reported that "according to the latest census, 24 countries have chosen 15 years as the minimum; 38 have selected 14 years; 3 countries use 12 years; 32 countries have placed the limit at 10 years." [28]

In order to permit international comparisons of data on the economically active population, the United Nations recommends that the minimum age limit adopted be "never higher than fifteen years," and that "any tabulations of economic characteristics not cross-classified by detailed age should at least distinguish between persons under fifteen years of age and those fifteen years of age and over." [29] (After using 14 as the minimum age for many years, the United States recently switched to 16 years. Supplementary labor force statistics are published for 14 and 15 year olds, however.)

Subsistence and Semisubsistence Economies. — In addition to the foregoing problems of international comparability, a fundamental question can be raised regarding the applicability of the labor force concepts as developed in the United States and Canada to people living in subsistence economies. [30] At one extreme, the Statistical Office of the United Nations characterizes the situation as follows, "For population groups living almost entirely in a subsistence economy, no information specifically centered on the economically active population is required." [31] Note, however, that the entire population is living in a subsistence economy in relatively few underdeveloped countries. Moreover, many countries are in a transitional stage between a subsistence economy and a market, or money, economy.

The role of a "sense of time". — It was noted previously that the United Nations recommended that unpaid family workers be included in the employed only if they worked at least one-third of the normal working time. Panama, it was noted, imposed no minimum number of hours, whereas Taiwan used 18 hours per week. (Actually, it is not clear from the source whether this is 18 or 19.) It is Jaffe's conviction (based on

extensive interviewing and observation) that in the agricultural sector where most unpaid family workers are found, and in particular under subsistence or semisubsistence conditions, there is but a little developed sense of time, and no one knows how many hours he really works per week. Further, he is not sure exactly what the interviewer (a literate outsider) means by the word "work." Should time spent on gathering firewood or wild fruits or hunting game to be eaten by the family be counted as part of "work"?

Even if this semisubsistence farmer should work occasionally for a neighbor, he customarily works a "day" or "helps with a job" for which he receives cash, irrespective of the precise number of hours actually worked. Determining the length of his work week, or that of his son who is an unpaid family worker, is very difficult.

Visible underemployment involves involuntarily working something less than a normal work period. (See previous section on "Underemployment.") Accurate measurement of visible underemployment, then, also requires a reasonably accurate sense of time on the part of respondents.

Quality of the Statistics

Like other demographic statistics, statistics on the working force are subject to errors from various causes. (These errors are in addition to the inadequacies of scope, definition, etc., that have just been discussed.) Moreover, the types of methods for measuring these errors that were described earlier in the book—comparisons of aggregate statistics, matching studies, reinterview studies, etc.—can be and have been applied to economic characteristics of the population. Despite their limitations and their errors of measurement, labor force statistics have proven themselves highly useful in the industrialized countries, at least. In the underdeveloped countries, the conceptual difficulties seem to be a more basic problem than are the response errors in classifying the population.

The quality of the labor force statistics of the United States has been thoroughly investigated by Durand, Bancroft, and others; [32] and considerable information is available in various reports of the U.S. Bureau of the Census. It should be remembered that comparisons of aggregates from different sources reflect any differences in definitions, coverage, time reference, and reporting errors.

Household data as obtained from the U.S. population census and the monthly sample surveys differ slightly for a number of reasons. Among these are the more extensive training, control, and experience of the CPS enumerators than of the census enumerators; differences in the time period to which the labor force data apply (the CPS covering the week containing the 12th of the month; the census, the week prior to the date the respondent filled his questionnaire or was interviewed); differences in question wording and format of the schedules; differences in coverage (the CPS including in the total labor force members of the Armed Forces overseas, excluding inmates of institutions from the population covered, and excluding members of the Armed Forces from the annual work experience data); enumeration of unmarried college students in the CPS at their parental home but in the census at their residence while attending college; differences in the methods used to process the original data into statistical tables;

[28] United Nations, *Demographic Yearbook, 1964*, p. 27.

[29] United Nations, *Principles and Recommendations for the 1970 Population Censuses*, p. 61.

[30] Wilbert E. Moore, "The Exportability of the Labor Force Concept," *American Sociological Review*, 18(1):68–72, February 1953.

[31] United Nations, *Handbook of Population Census Methods*, Vol. II, p. 8.

[32] John D. Durand, *The Labor Force in the United States: 1890–1960*, New York, Social Science Research Council, 1948, Appendix A; Gertrude Bancroft, *The American Labor Force*, New York, John Wiley & Sons, 1958, Appendix A.

differences in weighting procedure and in the population controls; differences in the extent and allocation of noninterview cases; and in the sampling variability in the CPS and in the 5-percent sample used in this report.

Measures

Measures of Economic Activity.—These measures relate to the economically active population, the labor force, or gainful workers depending on the type of data available; but we will refer to them generally as activity rates. As with other demographic characteristics, crude, general, age-specific, and age-standardized rates of economic activity may be computed. These rates may be calculated for both sexes combined, but it is customary to show most rates separately for males and females.

Crude activity rate.—The crude activity rate represents the number of economically active persons as a percent of the total population. It is also referred to as the crude labor force participation rate, in countries where the labor force concept is applicable. For example, in the 1960 census of Ghana, the total population was enumerated at 6,726,815, of whom 2,723,026 were classified as economically active. Thus the crude activity rate for Ghana in 1960 was 2,723,026 divided by 6,726,815 times 100, or 40.5 percent. The crude activity rate is greatly influenced by the age composition of the population. It is primarily useful in comparisons where the analyst wishes to indicate simply the relative number of persons in a population who are working irrespective of the factors involved. It is also useful in comparisons where the analyst wishes to highlight the effect of different levels of natural increase and migration upon economic activity.

General activity rate.—This is simply the activity rate for persons of working age. Except for the simplest kinds of comparison, the researcher will probably wish to use more refined rates of economic activity in his analyses.

In the following example, we calculate an activity rate for all persons 14 years old and over in the United States in 1960, since persons below age 14 were not asked the questions on economic activity. The total population 14 years old and over was 126,277,000, of whom 69,877,000 were classified as in the labor force. Thus the activity rate for persons 14 years and over was 55.3 percent.

Inasmuch as the minimum age for inclusion varies from country to country, the analyst should take account of the possibility that a considerable number of economically active children may have been arbitrarily excluded from the enumerated labor force because the minimal age may have been placed too high.

Sex-specific activity rate.—Both of the above two measures of economic activity, the crude activity rate and the activity rate for persons above the minimum labor force age, are usually calculated separately for males and females. When calculated in this fashion, they are termed sex-specific activity rates. Sex-specific activity rates are calculated for two different reasons. First, the level of such rates is usually much higher for men than for women. A second reason often given is that variations in definitions and operational procedures that most often affect measures of economic activity have their greatest impact on figures for women because women's attachment to the labor force is often marginal and intermittent. Thus, rates for men are less subject to temporary or spurious variations than rates for women, and international and historical comparison of rates for men are more valid. The objection to re-

stricting such comparisons to men, however, is that the female labor force is quite often the most dynamic part of the labor force, and the analyst may be ignoring really important trends by restricting his analysis to the male labor force.

The same warning should be given to analysts who attempt international and historical comparisons by restricting themselves to figures on activities for men in the prime working ages, say 20 to 59 years for example. The justification for such a restriction is that, just as for women, differences in definition and operating procedures (e.g., in regard to unpaid family workers or part-time workers), often produce irregular or spurious variations in activity rates for the young and the old. Although the analyst can make more valid comparison by restricting himself to the prime working ages, once again as in the case for women, he may be ignoring just those groups for which the important real changes and differences in economic activity do occur.

Age-sex-specific activity rate.—The activity rate calculated for a specific age-sex group is a far more widely used measure of economic activity in demographic analyses than any other measure of economic activity. Age-sex-specific activity rates, or labor force participation rates as they are called when the labor force concept is used, are frequently the basic rates that are studied and projected in analyses of the economically active population.

An age-sex specific activity rate is calculated by the formula:

$$\frac{\text{active population in age-sex group}}{\text{total population in age-sex group}} \times 100$$

Thus, for males 25 to 29 years of age in Ghana in 1960, the rate is calculated as follows:

$$\text{rate} = \frac{\text{active men 25–29 years old}}{\text{total men 25–29 years old}} \times 100$$

$$= \frac{268,851}{278,601} \times 100$$

$$= .965 \times 100$$

$$= 96.5$$

Activity rates may be calculated for population groups defined in terms of various characteristics in addition to age and sex, as for example, educational attainment, marital status, ethnic grouping, and economic level of household. The degree to which the activity rates are made specific for the above-mentioned characteristics depends on the problems being studied as well as on the availability of data for computing them.

Manpower analysts have found some fairly typical curves of age-sex specific activity rates in the statistics of different populations to which the national statistician should compare the curves for his own country.[33] For males between 25 and 54 years old, there is universally almost complete participation in the labor force. Variations among countries occur at ages above and below this range, however. Durand and Miller write that "It is useful to define these features as clearly as possible by computing specific activity rates in the finest age groupings which the census provides at the two extremes of the range of working ages—preferably by single years of age below 20 and five-year groups from 55 up to the age at which

---

[33] See United Nations, *Demographic Aspects of Manpower*, Report No. 1.

the number of men in the labor force drops to a negligible figure." [34] The U.N. report *Demographic Aspects of Manpower* documents the more varied patterns which the age-specific activity rates for women can take. [35]

Durand and Miller also point out that analysis of age-specific activity rates may sometimes be as useful for the insights it provides regarding census definitions and procedures as it is for the information it gives. For example, the relatively high activity rates for women in Thailand and the low rates in Iran lead us to try to find the reasons for such abnormal rates. One possible reason is that the rates may reflect unusual census definitions and procedures as much as they do true differences in the economic activity of women.

Age-sex adjusted, or standardized, activity rate.— An age-sex adjusted, or standardized, activity rate is the rate that would result for the total population if the rates for a given date or geographic area, were standardized for age and sex (see chapter 14). Often the standard used is the national age-sex distribution for some specified date.

Dependency ratio.— A common practice in demographic analysis of manpower problems is the calculation of a dependency ratio from the population age statistics without regard to actual participation in economic activities. For example, the commonly used dependency ratio (defined in chapter 8),

$$\frac{\text{population under 15 years old} + \text{population 65 and over}}{\text{population 15 to 64 years old}} \times 100,$$

considers all persons in the age group 15 to 64 as producers and all persons above and below these ages as dependents. This ratio is equivalent, of course, to the simple measure of age structure, the proportion of the total population that is 15 to 64 years old, but the term "dependency ratio" adds an aura of economic significance that is considered by leading analysts as unjustified. The ratio of the economically inactive population to the economically active population is much more meaningful as a measure of economic dependency since it reflects not only the area's age-sex structure but also the economic activity participation rates. For comparison, here are the values of the two kinds of ratios for the United States in 1968:

$$\frac{\text{population under 15} + \text{population 65 and over}}{\text{population 15 to 64}} \times 100 = 64.4$$

$$\frac{\text{population not in the labor force}}{\text{population in the labor force}} \times 100 = 144.4$$

Note that the values of the two measures may be quite different.

Replacement ratios and rates for working ages.— Replacement ratios and rates for working ages are indicative of potential population replacement during a specified future period if assumptions relating to survival rates obtain and if no net migration into or out of the given area takes place.

The "replacement ratio" is the number of expected entrants into the specified working age group during the period per 100 expected departures resulting from death or from reaching retirement age during the period.

Table 12-3.— **Males, 15 to 64— Computational Procedure for Replacement Ratios and Rates, for Panama: 1950-60**

Age in 1950	Males, 1950 (1)	Survival or mortality rates (2)	Survivors, deaths, and retirements, 1960 (1)x(2)= (3)
(a). 5 to 9 years	53,788	.97973	52,698
(b). 10 to 14 years	43,464	.97853	42,531
(c). 15 to 19 years	35,633	.02706	964
(d). 20 to 24 years	33,189	.03241	1,076
(e). 25 to 29 years	30,599	.03677	1,125
(f). 30 to 34 years	26,604	.04401	1,171
(g). 35 to 39 years	24,632	.05726	1,410
(h). 40 to 44 years	17,973	.07794	1,401
(i). 45 to 49 years	14,276	.11056	1,578
(j). 50 to 54 years	12,401	.15980	1,982
(k). 55 to 59 years	8,821	1.00000	8,821
(l). 60 to 64 years	9,258	1.00000	9,258

Working Age 15 to 64 Years

(1). Entrants (a + b, 1960)	5,229
(2). Deaths and retirements ($\sum_c^1$ 1960)	28,786
(3). Net available ((1) - (2))	66,443
(4). Males 15 to 64, 1950 ($\sum_c^1$ 1950)	213,386
(5). Ratio (1) ÷ (2) x 100	330.8
(6). Rate (3) ÷ (4) x 100	31.1

Source: Basic data from United Nations, *Demographic Yearbook, 1960*, table 5, and *1966*, table 23.

The "replacement rate" is the number of entrants minus the number of departures as a percentage of the number in the specified working ages at the beginning of the period.

Many such ratios have been computed by the U.S. Department of Agriculture for rural (or rural-farm) males entering and leaving working ages during intercensal decades. [36] In these ratios the expected entrants are the males 10 to 19 years old at the census who would be expected to survive to ages 20 to 29 at the end of the decade; the departures are the persons leaving the working ages by death or by reaching age 65 by the end of the decade. In the present work we have also computed refined measures at the national level based on actual ages of labor force entrance and departure and labor force participation by age and sex as shown at the preceding census.

We may illustrate the basic method by applying it to the 1950 census data for males in Panama. The work is summarized in table 12-3. Because of the relatively early average age of entry into the labor force in that country, we begin the

---

[34] United Nations, *Methods of Analyzing Census Data on Economic Activities of the Population*, p. 18.

[35] United Nations, *Demographic Aspects of Manpower*, pp. 21-35.

[36] See, for example, U.S. Department of Agriculture, Economic Research Service, *Statistical Bulletin* No. 378, "Potential Supply and Replacement of Rural Males of Labor Force Age: 1960-70," by Gladys K. Bowles, Calvin L. Beale, and Benjamin S. Bradshaw, October 1966; United Nations, Economic Commission for Latin America, *Human Resources of Central America, Panama and Mexico, 1950-1980, in Relation to Some Aspects of Economic Development*, by Louis J. Ducoff, 1960, pp. 75-80.

working ages with 15 rather than 20. The survival and mortality rates are derived by averaging Panamanian life tables for the 1952–1954 and 1960–1961 periods, which are given in the U.N. *Demographic Yearbook* for 1966. The mortality rates are the complements of the survival rates.

Measures of Employment.—Just as there are crude, age-sex specific, standardized, etc., rates of economic activity, there are parallel rates for various aspects of employment. We will not define these specifically in each case since the parallels should be obvious.

Employment rate.—The number of employed persons is expressed as a percent of the number economically active, or in the civilian labor force.

Unemployment rate.—The number of unemployed persons is expressed as a percent of the number economically active, or in the civilian labor force. Note that, for any specified population, the employment rate plus the unemployment rate must equal 100 percent.

Proportion fully employed.—Although this term is not widely used, the recommended definition is simply the number fully employed—i.e., the total employed minus the number underemployed—expressed as a percent of those in the total labor force (or more commonly in the civilian labor force). The number fully employed can be expressed also as a percent of the total number employed. These rates also parallel the activity rates; they can be calculated by age and sex and standardized.

Proportion underemployed.—This term is probably used somewhat more widely than the previous one. The number of underemployed is expressed as a percent of the labor force.

Tables of Working Life.—Tables of working life (also referred to as "labor force life tables" or "tables of economically active life") are extensions of conventional life tables. They incorporate mortality rates and labor force participation rates and describe the variation by age in the probability of entering or leaving the labor force, the average number of years of working life remaining, and other phenomena. The method of constructing tables of working life is described in chapter 15.[37] The tables facilitate study of the labor force, including probable changes in its size and age-sex composition. They help in estimating expected life-time earnings, in evaluating the probable effects of investment upon education and training programs, and in measuring economic effects of changes in the age structure of the population or changes in activity rates.

As mentioned, a measure that can be derived directly from the table of working life is the expectation of active working life at birth or at any age.[38] Working life tables may be used to measure aggregate entrances to, and exits from, the labor force, and to measure the ages at which these events are likely to occur.[39] The expected active life at birth can be subtracted from life expectancy at birth to give a measure of the expectation of inactive life.

Factors Important in Analysis

The nature of the problem being studied determines which demographic factors are the important ones to be included in an analysis of economic activity. For almost any analysis, employment and unemployment data should be tabulated and studied by age and sex. For specific problems, other population characteristics may be added.

For example, if the central problem is that of migration of the labor force and the reasons therefor, it would be necessary to consider labor force and employment status separately for migrants and nonmigrants. If the emphasis is on women in the labor force, then the analysis could consider their marital status, probably also by the number and age of their children, the employment status and earnings of husband, and such other population characteristics as may seem relevant, e.g., urban-rural residence or educational attainment. If the interest is in the relationship of education to employment status and type of job, many of the characteristics already noted, as well as school attendance and educational attainment, would be relevant.

In countries containing both people living in a subsistence economy and those living in a market economy, it is important to distinguish the two sectors in the tabulations. This distinction may be approximated in terms of different ethnic groups, regions, urban-rural residence, or the primate city compared with the balance of the country.

In summary, many characteristics about which information is obtained in a census or sample household survey are useful in one or another analysis of economic activity. The primary problem toward which the study is directed will dictate the specific demographic and economic variables to be considered.

INCOME

Income statistics are of direct value to economists and others interested in wealth, the distribution and sources of consumer income, wage and salary rates, and the effective employment of manpower. Income is one of the best measures of economic well-being, and it vies with educational attainment and occupation as a measure of socioeconomic status. Although the person's or family's income has long been recognized as having important relationships to many of the characteristics such as mortality, fertility, migration, educational level, and occupation, that the demographer studies, income is an item introduced onto population census schedules only recently. It is still not included in any of the lists recommended by international agencies for inclusion in population censuses. The United States and Canada are among the few countries to collect information on consumer income in the population census. Income data are much more widely collected in household surveys, however, especially sample surveys on household expenditures.

Concepts and Definitions

According to the United Nations, there is as yet no international standard concept and definition of income for purposes of household income and expenditure surveys. Despite the lack of internationally accepted standards in regard to income data, there are some common principles that the demographer should be cognizant of in dealing with this subject. Most of these principles are elucidated in chapter VIII, "Economic Level of the Household", prepared by the International Labour Office and published by the Statistical Office of the United Nations

---

[37] Tables of working life are also discussed in United Nations, *Demographic Aspects of Manpower*, Report 1, pp. 17–20; idem, *Methods of Analyzing Census Data on Economic Activities of the Population*, by Durand and Miller, p. 19 and following. An illustration for Malaya is given in: Swee Hock Saw, "Malaya: Tables of Male Working Life, 1957," *Journal of the Royal Statistical Society*, Series A, 128(3):421–438, 1965.

[38] The expectation of working life is used in a historical analysis of the length of working life in the United States by Stuart H. Garfinkle in "The Lengthening of Working Life and Its Implications," *World Population Conference, 1965* (Belgrade), Vol. IV, United Nations, 1967, pp. 277–282. Garfinkle also used the table to measure the expected number of job changes per person.

[39] See U.S. Department of Labor, Manpower Administration, *Manpower Report* No. 8, "The Length of Working Life for Males, 1900–60," by Stuart H. Garfinkle.

in *Handbook of Household Surveys.* Much of the present section is based on this source.

In principle, income from all sources should be counted, both cash income and income in kind. So far as international comparisons of income are concerned, it is preferable to collect information on gross income, i.e., before deductions of taxes and social security contributions.

Specifications for reporting income are given by the United Nations as follows: "The main source of household cash income, comprising the personal incomes of all the members, are: salaries, wages, net earnings from self-employment, business profits, investment income (rent, interest, dividends), royalties and commissions. Periodic payments received regularly from an inheritance or trust fund are also regarded as income, as are the following: alimony, pensions, annuities, social security cash benefits such as children's allowances and unemployment benefits, sick pay, scholarships and various other periodic receipts. The following items should not be regarded as income: receipts from sale of possessions, withdrawals from savings, loans obtained, loan repayments received, windfall gains, inheritances (lump sum), tax refunds, maturity payments on insurance policies, lump-sum compensation for injury, legal damages received, etc., even though the proceeds may sometimes be spent on consumption. To cash income must be added income in kind." [40]

The reference period for income data should be a year. This practice is especially important in countries where a large part of the population are self-employed or unpaid family workers and only a small proportion are wage and salary workers.

The noncash component of total income, or "income in kind," is of greater importance in less developed countries than in the more developed countries and consequently makes even more difficult the task of obtaining precise figures on total income in less developed countries.

In subsistence economies, of course, practically all income is income in kind. Procedures for collecting information about income in these circumstances are necessarily very different from those used in cash economies. In actual practice, censuses and surveys of some industrialized countries ignore income in kind, because it is a small part of total consumer income and very difficult to measure.

Personal and Family Income. — Income data may be collected for each person in the household and then added to produce totals and subtotals for households, families, married couples, etc. Less satisfactorily, family income may be obtained by asking directly for the total income received by all family members combined. The income items are rarely asked of children below working age; and adults in certain types of group quarters (collective households), especially institutional inmates, are often excluded too. Statistics on personal income are most useful for males of working age and, to a lesser extent, for females of those ages. The socioeconomic status of a dependent child or wife is indicated better by the family income than by that of the person himself. Where the household contains one or more persons unrelated to the head, income statistics for the family are generally considered to be more useful than those for the entire household.

---

[40] United Nations, *Handbook of Household Surveys,* Series F, No. 10, 1964, p. 90.

Sources and National Practices

Censuses and household surveys have already been mentioned as sources of income data. The main purpose of the household expenditure survey is to provide information on the amount of money that families spend for specific goods and services in order that the government may prepare retail (consumer) price indexes. Since family expenditures are closely related to family income, data on the latter must also be collected. A consumer expenditure survey differs from other household surveys only in that many more questions are asked about income and expenditures, and household accounts are examined by the enumerator insofar as available. In the United States, such surveys are conducted by the Bureau of Labor Statistics. One shortcoming of income data from expenditure surveys is that the samples are usually too small to permit much cross-classification. These surveys are also often restricted to urban areas, especially in less developed countries.

Data on personal and family income are also available from establishment records (e.g., payrolls), income tax returns, and social security records. Statistics from these sources are of limited value to demographers, however, because there are few demographic and social characteristics recorded in these sources, because the coverage is usually confined to segments of the population that are not defined in demographic terms, and because not all income is covered. For example, there is generally a cut-off point for income below which a tax return does not have to be filed. Social security records in the United States omit income above a legally specified amount and do not cover workers in certain industries or in enterprises with fewer than a specified number of employees. Establishment payrolls are of value, for our purposes, mainly as a source of information about hourly, weekly, or monthly earnings. This information is generally available only in the form of averages, and distributions are not shown. These other sources of income data are useful, however, for evaluating the quality of the data collected in household interviews, as will be discussed later in this chapter.

The United States. — The 1940 census of the United States contained a question on wage or salary income together with a question on whether the person had received $50 or more from other sources in the preceding calendar year. Income from all sources was first investigated in the census in 1950.

The 1970 income questions provide for two types of self-employment income (farm and nonfarm) and three types of income other than earnings (Social Security or Railroad Retirement, public assistance or welfare, and "other" sources such as interest, dividends, veterans' payments and pensions).

Questions on income were first asked in the Current Population Survey in 1945 for the calendar year 1944. The income concepts used in the CPS agree very closely with those used in the contemporary decennial census.

The amounts reported represent income before deductions for personal taxes, Social Security, savings bonds, etc. It should be noted that although the income statistics refer to receipts during the preceding calendar year the characteristics of the person, such as age, marital status, and labor force status, and the composition of the family refer to the survey date. The income of the family does not include amounts received by persons who were members of the family during all or part of the preceding calendar year if these persons no longer resided with the family at the time of enumeration. On the other hand, family income includes amounts reported by related persons who did not reside with the family last year but who were members of the family at the time of enumeration.

Data on consumer income collected by the Bureau of the Census cover money income (exclusive of certain money receipts such as capital gains) and exclude income in kind. The fact that some farm families receive part of their income in the form of rent-free housing and goods produced and consumed on the farm, rather than in money, should be taken into consideration in comparing the income of farm and nonfarm residents. Agriculture in the United States, however, has become so commercialized that income in kind is now a minor factor in the income of most farm families. It should be noted that nonmoney incomes are also received by some nonfarm residents. They often take the form of business expense accounts, use of business transportation and facilities, full or partial compensation by business for medical and educational expenses, rent-free housing, etc. In analyzing size distributions of income, it should be recognized that capital gains tend to be concentrated more among higher income units than among lower ones.

Total money income is defined in the United States as the algebraic sum of money wages and salaries, net income from self-employment, and income other than earnings. The total income of a family is the algebraic sum of the amounts received by all income recipients in the family.

The income tables for families and unrelated individuals include in the lowest income group those that were classified as having no income in the preceding year and those reporting a loss in net income from farm or nonfarm self-employment or in rental income. Many of these were living on income "in kind," savings, or gifts; or were newly constituted families, unrelated individuals who had recently left families, or families in which the sole breadwinner had recently died or had left the household. However, many of the families and unrelated individuals who reported no income probably had some money income which was not recorded in the survey.

Family income and personal income are cross-classified with a wide variety of demographic, social, and economic characteristics, not only in the decennial census reports but also in *Current Population Reports* (Series P–20, P–60, etc.). Table 12-4 is excerpted from a table in a 1960 census report and shows the relationship between personal income and age for males. (The full table shows a cross-classification by urban-rural residence, sex, and color also.) Figure 12-1 illustrates the cross-classification of personal income with a social characteristic, in this case, educational attainment. The cross-classification is intended to show how additional education enhances annual income. Elsewhere the analysis has been carried further to show the relationship between education and lifetime earnings.

Problems in Collection of Income Data

Although the concept of total income is not too difficult to relate to the experience of persons and households in most countries, the problem of obtaining the data necessitates adapting this concept to the practicalities of statistical life. In some countries it is likely to be difficult to obtain satisfactory information on household income. It cannot be too strongly emphasized that a small pilot study to ascertain the possibilities of obtaining utilizable information on income by source be conducted before any large-scale survey to obtain income data is undertaken. Income data obtained from household surveys should be verified against other available sources of income data. In the case of wage and salary income, such data can be checked against employers' records or establishment statistics based on such records. Unfortunately, for many other sources of income, there may be no means of verifying the responses.

Of the various segments of the population, the two groups that pose the major practical problems of income measurement are the farmers and the self-employed nonagricultural workers. The farmer's home, the food that he raises and consumes, and sometimes other items, are part of the total income that he receives for being a farmer, yet they are not cash. Sometimes the imputed value of such noncash items can be estimated and added to his cash profits to get his total income from his farm.

As previously mentioned, the extreme case is that of the subsistence, or semi-subsistence, farmer. To begin with, it is very difficult to fit such a farmer even into the scheme of economic activity. Practically speaking, about all that can be done is to ascertain the approximate value of any produce that he has sold during a specified period. In some countries many such farmers will have sold no produce whatsoever; whatever cash income they have during the year may be obtained from an occasional period of work off the farm, or from some other member of the family who makes a money contribution. About all that can be done statistically with such subsistence farmers is to tabulate and analyze them separately from the nonagricultural workers, most of whom receive cash for their efforts.[41]

The self-employed persons, whether in agriculture or non-agricultural pursuits, present a difficult problem, inasmuch as the true income from their work is the total amount of cash received minus the cost of doing business. Generally, the more developed the economy and the greater the amount of business that the self-employed person has, the more likely he is to have records that indicate the approximate cost of doing business so that his true income is more nearly measureable. In many, if not most, countries it is probably almost impossible to obtain reliable information on income for the self-employed worker (including those who use their own unpaid family labor and a very small number of employees who receive money wages).

Quality of the Statistics

Attempts to measure the quality of the census statistics on income are just about limited to the United States and Canada. The nonresponse rate is relatively high for income in the United States—4.9 percent of all families in the 1950 census and perhaps 10.6 percent in 1960.[42] The nonresponse rate is now considerably higher in the Current Population Survey. Income is one of the characteristics that was allocated in the 1960 census and in the CPS so that the "unknowns" do not appear separately in the regular tables.

In spite of recognized inadequacies, it has been amply demonstrated, at least in the United States and Canada, that very useful

---

[41] See Panama, *Estadística Panameña*, "Mano de obra: Años 1963 y 1964," esp. tables 24 and 27.
[42] For the uncertainty about the 1960 nonresponse rate, see: *U.S. Census of Population: 1960*, Vol. 1, *Characteristics of the Population*, Part 1, *United States Summary*, 1964, p. LXXXVI.

Table 12-4. — Income in 1959 of Males by Age, for the United States: 1960

Age	Males 14 years old and over	Without income	With income											Median income (dollars)
			Total	$1 to $999 or loss	$1,000 to $1,999	$2,000 to $2,999	$3,000 to $3,999	$4,000 to $4,999	$5,000 to $5,999	$6,000 to $6,999	$7,000 to $9,999	$10,000 and over	
Total..........	61,315,362	6,151,125	55,164,237	8,151,664	6,599,128	5,807,135	6,252,137	6,960,363	6,951,384	4,790,443	6,000,312	3,651,671	4,111
14 to 19 years......	8,101,561	4,001,525	4,100,036	2,847,801	690,511	252,180	124,868	109,540	40,595	15,297	16,374	2,870	720
20 to 24 years......	5,283,232	363,037	4,920,195	855,274	1,091,222	895,918	824,304	625,415	365,137	147,213	93,962	21,750	2,573
25 to 34 years......	11,173,569	326,336	10,847,233	563,953	729,757	1,107,508	1,502,904	1,847,191	1,917,592	1,293,095	1,397,017	488,216	4,823
35 to 44 years......	11,739,191	294,714	11,444,477	537,238	594,187	846,365	1,225,737	1,629,662	1,913,645	1,479,739	2,025,202	1,192,702	5,465
45 to 54 years......	10,139,671	346,468	9,793,203	667,614	699,098	867,838	1,130,481	1,387,713	1,485,811	1,067,949	1,439,279	1,047,420	5,097
55 to 64 years......	7,569,153	385,892	7,183,261	779,610	785,643	776,057	880,388	973,182	938,038	609,983	793,053	647,307	4,380
65 years and over...	7,308,985	433,153	6,875,832	1,900,174	2,008,710	1,061,269	563,455	387,660	290,566	177,167	235,425	251,406	1,766
65 to 74 years....	5,022,410	213,245	4,809,165	1,012,902	1,371,785	838,180	459,540	326,853	248,680	149,854	197,799	203,572	2,024
75 years and over.	2,286,575	219,908	2,066,667	887,272	636,925	223,089	103,915	60,807	41,886	27,313	37,626	47,834	1,229

Source: *U.S. Census of Population: 1960*, Vol. I, *Characteristics of the Population*, Part 1, *United States Summary*, 1964, table 219, p. 580.

Figure 12-1. — Percent Distribution by Educational Attainment for Males 25 Years Old and Over, by Income in 1959, for the United States: 1960

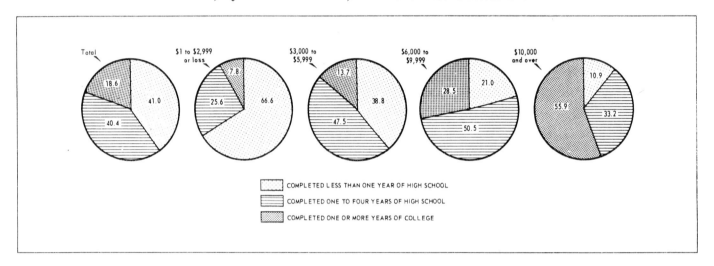

NOTE: Total excludes persons without income.

Source: *U.S. Census of Population: 1960*, Vol. I, *Characteristics of the Population*, Part 1, *United States Summary*, 1964, figure 116, p. S63.

income data can be collected in censuses and household surveys.[43] The demographer will find that cross-tabulations by income reveal very interesting differentials in vital rates, social indexes, etc.

Analytic Measures

Income is almost always shown in the form of distributions. This permits measuring the extent of inequalities, the extent of poverty (whatever may be taken as the limit below which the people are considered to be impoverished), etc. When data are presented in a frequency distribution, however, two analytic problems are presented: how to summarize a frequency distribution and how to measure inequality.

Aggregate Income. — Aggregate income is the sum of the income of all income recipients. From household statistics, the aggregate is computed from the frequency distribution, multiplying the number of persons in the class by the midvalue of the class. The aggregation for any given area or group can be performed in the electronic computer (instead of by a clerical operation). Allocations of the nonresponse cases would ordinarily have been done in an earlier computer operation.

Wherever the aggregation is done, an assumption must be made concerning the mean value for the upper open-end interval provided by the tabulated distribution, or by the census schedule. A value may be assigned arbitrarily (for the United States in 1960, the Bureau of the Census assigned $50,000 for the interval $25,000 and over), or it may be

[43]For a discussion of the quality of U.S. income data, as well as a detailed analysis of trends and differentials, see the following two volumes by Herman P. Miller: *Income Distribution in the United States*, a 1960 census monograph, Washington, U.S. Bureau of the Census, 1966; and *Income of the American People*, New York, John Wiley and Sons, 1955.

estimated by fitting a Pareto curve to the cumulative distribution.[44] The curve may also be used to introduce greater refinement in estimating the average for other extreme intervals, e.g., $19,000, instead of the midvalue $20,000, for the interval $15,000 to $24,999. Persons with a negative income ("Loss") constitute an open-end interval at the lower end of the distribution; but they are conventionally combined with the lowest positive class, say "under $500" and the mid-interval value of $250 is used for the combined classes. This procedure introduces an upward bias in the mean value for this lowest class.

Mean Income.—The mean income is the aggregate income divided by the number of income recipients in the case of mean personal income and by the number of families in the case of mean family income. Since the base for the former is income recipients, persons with no money income in the given year are omitted; in fact, they are shown separately in the income distribution rather than being subsumed in the lowest class. Table 12–5 gives the distribution of family income for Canada in 1961, the aggregate income, the mean income, and the quintiles.

Table 12–5.—**Distribution by Total Income and Selected Income Measures, for Nonfarm Families, Canada: 1961**

Family income	Midpoint of interval m_i	Number of families f_i
Total..........................	–	3,656,968
Under $1,000..........................	500	163,590
$1,000 to $1,499..........................	1,250	146,346
$1,500 to $1,999..........................	1,750	155,769
$2,000 to $2,499..........................	2,250	179,238
$2,500 to $2,999..........................	2,750	202,997
$3,000 to $3,499..........................	3,250	271,790
$3,500 to $3,999..........................	3,750	285,576
$4,000 to $4,499..........................	4,250	321,301
$4,500 to $4,999..........................	4,750	281,891
$5,000 to $5,499..........................	5,250	291,584
$5,500 to $5,999..........................	5,750	210,880
$6,000 to $6,999..........................	6,500	353,405
$7,000 to $7,999..........................	7,500	242,729
$8,000 to $9,999..........................	9,000	262,787
$10,000 to $14,999..........................	12,500	194,303
$15,000 and over..........................	[1]23,399	92,782

Aggregate income...................... $\Sigma m_i f_i$ = $19,983,442,018

Mean = $\frac{\text{Aggregate income}}{\text{No. of families}}$ = $\frac{\$19,983,442,018}{3,656,968}$ = $5,464

Median $4,681

Quintiles:
Lowest fifth — Under $2,713
Second fifth — $2,713 to $4,089
Third fifth — $4,089 to $5,318
Fourth fifth — $5,318 to $7,252
Highest fifth — Over $7,252

[1] Estimated by Pareto curve.

Source: Canada, Dominion Bureau of Statistics, *1961 Census of Canada, Population Sample*, Series 4.1, Bulletin 3, *Family Incomes by Size, Type, and Composition of Family*, May 21, 1964, table C2.

[44] U.S. Bureau of the Census, *Trends in the Income of Families and Persons in the United States: 1947 to 1960*, by Mary F. Henson, Technical Paper No. 17, 1967, pp. 33–35.

Payroll data are always aggregates representing the total payroll. In this case, the mean income is the only parameter that can be calculated; the total payroll is divided by the total number of employees reported. In the United States, the establishment payroll data and those obtained from economic censuses are used in this manner. Another area in which mean income is calculated is in connection with national accounts statistics, where the total national income (or that of a State) may be divided by the total population to obtain per capita income, or by the total employed in order to obtain average income per worker.

Median Income.—The median income is defined like other medians, i.e., it is that income value that divides income recipients (or families) into two parts, one higher and one lower than the median. It is highly unlikely that the median would fall in an open-end interval so that the problem posed by the mean of assigning an average value for that interval does not occur here. The median is usually preferable to the mean for comparing population groups.

Measures of Dispersion.—An income distribution is usefully described by a measure of dispersion as well as by a measure of central tendency. A number of measures of dispersion are in use—standard deviation, quartile points, etc. From the standpoint of welfare economics, it is customary to study the concentration of income, which is just another way of looking at the dispersion.

Quintiles, or fifths of families.—Although quartiles are frequently used to describe frequency distributions because of their close relationship to the median (the median is the second quartile point), quintiles have come to be used extensively in the description of family income distributions. Attention focuses on the lowest fifth and on the share of the nation's aggregate income that it receives. The computation of the four values defining the quintiles is similar to that for computing the median. One-fifth, two-fifths, etc., of the total number of families are first computed. In the case of table 12–5, the number of Canadian families in 1961 is 3,656,968. We then proceed as follows:

$$\frac{3,656,968}{5} = 731,394$$

Adding cumulatively, we obtain 644,943 families with income below $2,500. (The next frequency 202,997 would put us over 731,394.)

$$731,394 - 644,943 = 86,451$$

$$\frac{86,451}{202,997} = .4259$$

The width of interval containing the first quintile value is $500.

$$500 \times .4259 = 213$$

$Q_1 = 2,500 + 213 = \$2,713$, which is the first quintile value.

Lorenz curve and Gini concentration ratio.—The Lorenz curve and the related concentration ratio were described in chapter 6 in connection with measuring the concentration of population among States, city-size classes, etc. These measures, however, were originally developed to measure the concentration of income, and they have received by far their major use in this field. The reader is referred to chapter 6 for definitions and methods of computation and plotting.

Factors Important in Analysis

Sex is almost always used as a "control" in the presentation and study of personal income statistics. Income also varies markedly with age, urban-rural residence, size of place, and, in some countries, with ethnic group. Many studies concentrate on the occupation of workers in specific occupations and industries. There has recently been great interest in the relationship between education and income, particularly lifetime income.[45] Studies of family income often include consideration of characteristics of the head (sex, age, marital status, etc.) and of the family itself (size, composition, number of earners, number of income recipients, etc.). For example, the number of children of various ages should be taken into account when considering both the potential sources and the adequacy of the family income.

Finally, when changes in personal or family income over time are studied, allowances should be made for changes in the cost of living. In the United States, for example, the

annual P–60 reports (Consumer Income) of the Bureau of the Census include time series in constant dollars using the most recent reference year as the base. It can be seen from table 12–6 that the series in current dollars gives an exaggerated impression of the postwar rise in consumer income.

Table 12-6. — **Median Income of Families in Constant (1967) and Current Dollars for the United States: 1947, 1950, and 1956 to 1967**

Income year	Median income		Ratio of (1) to (2)
	Current dollars (1)	Constant dollars (2)	
1967.............................	7,974	7,974	1.000
1966.............................	7,500	7,651	.980
1965.............................	6,957	7,357	.946
1964.............................	6,569	7,073	.929
1963.............................	6,249	6,822	.916
1962.............................	5,956	6,587	.904
1961.............................	5,737	6,418	.894
1960.............................	5,620	6,350	.885
1959.............................	5,417	6,210	.872
1958.............................	5,087	5,871	.866
1957.............................	4,971	5,889	.844
1956.............................	4,783	5,882	.813
1950.............................	3,319	4,611	.720
1947.............................	3,031	4,531	.669

Source: U.S. Bureau of the Census, *Current Population Reports, Consumer Income,* Series P–60, No. 59, April 18, 1969, tables 1 and 2.

---

[45] Miller, *Income Distribution in the United States,* pp. 123–167.

SUGGESTED READINGS

Statistical Compilations

International Labour Office. *Year Book of Labour Statistics.* Geneva,

———. *Quarterly Bulletin of Labour Statistics.* Geneva. United Nations. *Demographic Yearbook, 1964.*

———. *Statistical Yearbook.*

U.S. Bureau of Labor Statistics. *Employment and Earnings.* (Monthly).

International Recommendations and Standards

International Labour Office. *The International Standard Classification of Occupations.* 1968 revision. Geneva, 1969.

———. *International Standardization of Labour Statistics.* Studies and Reports, New Series No. 53. Geneva, 1959.

———. *Technical Guide.* Section III, "Employment" and Section II, "Unemployment." Geneva, 1954.

International Labour Organisation. Ninth International Conference of Labour Statisticians. Report No. 4, *Measurement of Underemployment.* Geneva, 1957.

Organization for European Economic Cooperation. *Labor Force Statistics: Sample Survey Methods.* Paris, 1954.

United Nations. *Application of International Standards to Census Data on the Economically Active Population.* Population Studies No. 9. ST/SOA/Series A. New York, 1951.

———. *Handbook of Household Surveys.* Series F, No. 10, 1964. Pp. 73–111.

———. *Handbook of Population Census Methods.* Volume II, *Economic Characteristics of the Population.* Studies in Methods, Series F, No. 5, rev. 1. New York, 1958.

———. *The International Standard Industrial Classification of All Economic Activities.* Statistical Papers, Series M, No. 4, rev. 1. New York, 1958.

———. *Methods of Analysing Census Data on Economic Activities of the Population.* Series A, Population Studies, No. 43. 1968.

General Studies

Jaffe, A. J. "Suggestions for a Supplemental Grouping of the Occupational Classification System." *Estadística,* 15(54):13–23. March 1957.

Jaffe, A. J., and Stewart, Charles D. *Manpower Resources and*

Utilization. Principles of Working Force Analysis. New York, John Wiley & Sons, 1951.

Kuznets, Simon. "Quantitative Aspects of the Economic Growth of Nations, II, Industrial Distribution of National Product and Labor Force." *Economic Development and Cultural Change.* Supplement to Vol. V, No. 4. July 1957.

Layard, P. R. G., and Sengal, J. C. "Educational and Occupational Characteristics of Manpower: An International Comparison," *British Journal of Industrial Relations,* 4(2):222–266. July 1961.

Pressat, Roland. *Principes d'analyse. Cours d'analyse démographique de l'I.D.U.P.* Paris, Institut national d'études démographiques, 1961.

————. *L'analyse démographique. Méthodes — resultats — applications.* Paris, Institut national d'études démographiques, 1961.

Seklani, Mahmoud. "Variations de la structure par age et charges de la population active dans les pays sous-développés." *International Population Conference, New York 1961.* Vol. II. London, 1963. Pp. 527–532.

United Nations. *Demographic Aspects of Manpower.* Report 1, *Sex and Age Patterns of Participation in Economic Activities.* Population Studies, No. 33. New York, 1962.

————. *World Population Conference, 1965* (Belgrade). Vol. IV, *Migration, Urbanization, Economic Development.* Meeting B.11, "Definition and Measurement of Economically Active Population, Employment and Underemployment." 1967. Pp. 339–395.

Wolfbein, Seymour L. "The Length of Working Life." *Population Studies* (London), 3:(3):286–294. December 1949.

Studies of Specific Countries or Regions

Bancroft, Gertrude. *The American Labor Force: Its Growth and Changing Composition.* New York, John Wiley & Sons, 1958. Especially pp. 151–200.

Canada. Department of Labour, Economics & Research Branch. Occasional Paper No. 2, *Changes in the Occupational Composition of the Canadian Labour Force, 1931–1961.* By Noah H. Mettz.

————. Dominion Bureau of Statistics. *Technical Memorandum, General Series.* No. 24, "History of Census Economic Characteristics Questions, 1931–1967." By M. B. Ismaily. Ottawa, November 1968.

Cumper, G. E. "A Comparison of Statistical Data on the Jamaican Labour Force 1953–61." *Social and Economic Studies* (Jamaica). December 1964.

Durand, John D. *The Labor Force in the United States, 1890–1960.* New York, Social Science Research Council, 1948. Especially pp. 1–16; 191–279.

Jaffe, A. J. *People, Jobs, and Economic Development. A Case History of Puerto Rico Supplemented by Recent Mexican Experiences.* Glencoe, Illinois, Free Press. 1959.

Jaffe, A. J., and Carleton, R. O. *Occupational Mobility in the United States 1930–1960.* New York, King's Crown Press, 1954. Pp. 60–103.

Kalra, B. R. "A Note on Working Force Estimates, 1901–61." In *Census of India (1961) Paper No. 1 of 1962. Final Population Totals.* Appendix I. New Delhi, 1962.

Kpedekpo, G. M. K. "Working Life Tables for Males in Ghana 1960." *Journal of the American Statistical Association,* 64(325):102–110. March 1969.

Lee, Everett S., et al. *Population Redistribution and Economic Growth: United States, 1870–1950.* Vol. I, *Methodological Considerations and Reference Tables.* Philadelphia, American Philosophical Society, 1957. Pp. 363–633.

Long, Clarence D. *The Labor Force Under Changing Income and Employment.* Princeton, N.J., Princeton University Press, 1958.

Miller, Herman P. *Income Distribution in the United States.* Washington, U.S. Bureau of the Census, 1966. Pp. 169–227.

————. *Income of the American People.* New York, John Wiley & Sons, 1955. Pp. 129–164.

Moore, Wilbert E. "The Exportability of the Labor Force Concept." *American Sociological Review,* 18(1):68–72. February 1953.

Pallós, Emil, and Valkovics, Emil. "A Gazdaságilag aktív és inaktív élettartam." (Length of economically active and inactive life.) *Demográfia* (Budapest), 8(1):30–59. 1965.

Panama. Dirección de estadística y censo. *Estadística Panameña,* Año XXVI, Supplemento. "Encuestas de mano de obra en Panamá: Años 1962 a 1967. Informe metodológico."

Pearl, Robert B. "Gross Changes in the Labor Force: A Problem in Statistical Measurement." In U.S. Bureau of Labor Statistics. *Employment and Earnings,* 9(10):IV–X. April 1963.

Saw, Swee Hock. "Malaya: Tables of Male Working Life, 1957." *Journal of the Royal Statistical Society.* Series A, 128(3):421–438. 1965.

Sierra Berdecia, Fernando, and Jaffe, A. J. "The Concept and Measurement of Underemployment." *Monthly Labor Review,* 78(3):283–287. March 1955. (concerning Puerto Rico)

Taiwan. Department of Social Affairs. *Quarterly Report on the Labor Force Survey in Taiwan. October 1968.* Prepared by the Labor Force Survey Research Institute.

United Nations. Economic Commission for Latin America. *Human Resources of Central America, Panama and Mexico, 1950–1980, in Relation to Some Aspects of Economic Development.* By Louis J. Ducoff. New York, 1960.

————. Centro Latino-Americano de Demografía (CELADE) A/5. *Población y mano de obra en Chile, 1930–1975.* By J. L. Sadie. Santiago de Chile, 1964.

————. *Population Growth and Manpower in the Philippines: A Joint Study by the United Nations and the Government of the Philippines.* Series A, Population Studies, No. 32. New York, 1960.

U.S. Bureau of the Census. *Current Population Reports.* Series P-23, No. 22. "Concepts and Methods Used in Manpower Statistics From the Current Population Survey." June 1967.

————. *Evaluation and Research Program of the U.S. Censuses of Population and Housing, 1960: Accuracy of Data on Population Characteristics as Measured by Census–CPS Match.* Series ER 60, No. 5. 1964.

————. *Sixteenth Census of the United States: 1940. Population. Comparative Occupation Statistics for the United States, 1870 to 1940.* By Alba M. Edwards. 1943.

————. *Technical Paper* No. 17. "Trends in the Income of Families and Persons in the United States: 1947 to 1960," by Mary F. Henson, 1967.

————. *Technical Paper* No. 19. "The Current Population Survey Reinterview Program, January 1961 Through December 1966." 1968.

————. *United States Census of Population: 1960,* Vol. I, *Characteristics of the Population.* Part 1, *United States Summary.* 1964. Pp. LXI to LXXXII.

————. *Working Paper* No. 5. "Occupational Trends in the United States 1900 to 1950." By D. L. Kaplan and M. C. Casey. Washington, 1958.

U.S. Bureau of Labor Statistics. *Special Labor Force Report.* No. 105, "Effect of the Census Undercount on Labor Force Estimates." By Denis F. Johnston and James W. Wetzel. March 1969.

U.S. Department of Agriculture. Economic Research Service. *Statistical Bulletin.* No. 378, "Potential Supply and Replacement of Rural Males of Labor Force Age: 1960–70." By Gladys K. Bowles, Calvin L. Beale, and Benjamin S. Bradshaw. October 1966.

U.S. Department of Labor. Manpower Administration. *Manpower Report.* No. 8. "The Length of Working Life for Males, 1900–60." By Stuart Garfinkle. 1963.

United States. President's Committee to Appraise Employment and Unemployment Statistics. *Measuring Employment and Unemployment.* Washington, 1962.

Wolfbein, Seymour, and Jaffe, A. J. "Demographic Factors in Labor Force Growth." *American Sociological Review,* 2(4):392–396. August 1946.

You, Poh Sen. "Growth and Structure of the Labour Force in the Countries of Asia and the Far East." In United Nations, Economic Commission for Asia and the Far East. *Report of the Asian Population Conference and Selected Papers.* New York, 1964.

CHAPTER 13

Population Change

DEFINITION AND TYPES

Direction

Change is measured by the difference between population sizes at different dates. Partly because of the prevailing modern trend of population in the countries of the world, people often speak loosely of population "growth" when actually either increase or decrease is possible. In census reports, increases are usually shown without a plus sign, but decreases are indicated by a minus sign.

Absolute and Percentage Change

The absolute amount of change is obtained by subtracting the population at the earlier date from that at the later date. The percent of change is obtained by dividing the absolute change by the population at the earlier date. Many census tables, particularly those on total population, include columns with the following type of arrangement of captions:

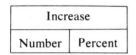

Using the population at the earlier census as the base of the period of change is dictated partly by logic and partly by convenience. Analogy with the calculation of vital rates would suggest the use of the midperiod population, but that would have to be estimated. There are many other types of demographic rates that are based on the population at the beginning of the period, e.g., survival rates and migration propensity rates.

PITFALLS IN MEASUREMENT

In measuring population change, one should be sure that the population of the area or group is comparable over the period in question. The area should be constant, the group should be defined on a consistent basis, and the accuracy of coverage or classification should not vary appreciably. And yet as a practical matter, adjustments are rarely practical; and, unless the modification is fairly extreme, the change is usually computed on the basis of the published figures. Knowledge of these differences is important for analysis, however; and it is often at that stage, rather than in the official report, that adjustments are attempted for the more serious cases. On the other hand, Davis takes the position that, in studying the growth of urban

or metropolitan territory, the expanding area provides the more realistic basis. He writes, "It follows that any model of growth which is to be used for prediction and as a basis for planning should be a two-variable model—that is, should predict the growth in both population and territory." [1]

We should broaden our perspective then to encompass the following: (a) the facts about changing territory should be published and should be kept in mind by users of the statistics, (b) whether to compute population change for the constant or expanding territory will depend upon the purpose of the report or study and upon feasibility, (c) computations of change for a set of areas should be made on a consistent basis.

Change in Territory

The national territory in which enumeration takes place may change as the result of conquest, treaties, advancement of the frontiers of settlement, and similar events or processes. Within national boundaries, water areas may be transformed into land areas, as in the case of part of the Zuyder Zee in the Netherlands, or vice versa. Often, it is possible to show the population within constant boundaries and then to compute the change for this constant area. Ordinarily, it is the present territory that is used.

There are some cases when the use of an expanding area seems clearly preferable to that of a constant area. The total population of the United States is a case in point. Relatively little interest would attach to the growth of population of the United States within the territory of 1790. Fortunately, the new territories purchased or annexed contained but few inhabitants prior to their inclusion and first enumeration. The territorial shifts in Europe and Asia have created much more complex problems for the examination of population change because the areas concerned had been settled for centuries.

In many countries, there are changes in the boundaries of states or provinces, secondary geographic subdivisions, cities, and communes or townships, and in those of areas defined mainly for statistical purposes, such as conurbations or metropolitan areas. In order to estimate intercensal net migration for the states of India, Zachariah first had to estimate intercensal population change within fixed state boundaries, there having been many changes in the period covered by his study. [2] Likewise, Kirk allowed for Euro-

[1] Kingsley Davis, "Conceptual Aspects of Urban Projections in Developing Countries," in: United Nations, *World Population Conference, 1965*, Vol. III, *Projections, Measurement of Population Trends*, 1967, p. 62.

[2] K. C. Zachariah, *A Historical Study of Internal Migration in the Indian Sub-Continent, 1901–1931*, New York, Asia Publishing Company, 1964, pp. 10–20 and 138–142.

Table 13-1. — Cities That Had 50,000 Inhabitants or More Living in Territory Annexed Between 1950 and 1960, for the United States: 1960

Rank	City	Population living in annexed territory in 1960	
		Number	Percent of 1960 population of city
1	Phoenix, Ariz....................	332,398	75.7
2	Houston, Tex.....................	251,193	26.8
3	Dallas, Tex......................	192,707	28.4
4	Atlanta, Ga......................	171,467	35.2
5	Tucson, Ariz.....................	167,112	78.5
6	Tampa, Fla.......................	140,331	51.0
7	San Antonio, Tex.................	139,492	23.7
8	El Paso, Tex.....................	124,155	44.9
9	Milwaukee, Wis...................	123,870	16.7
10	Tulsa, Okla......................	101,325	38.7
11	San Jose, Calif..................	99,380	48.7
12	Norfolk, Va......................	88,321	28.9
13	Seattle, Wash....................	86,079	15.5
14	Wichita, Kans....................	83,490	32.8
15	Columbus, Ohio...................	75,635	16.0
16	Newport News, Va.................	71,429	62.8
17	Oklahoma City, Okla..............	70,176	21.6
18	Memphis, Tenn....................	69,095	13.9
19	San Diego, Calif.................	65,843	11.5
20	Louisville, Ky...................	62,229	15.9
21	Long Beach, Calif................	59,159	17.2
22	Fort Worth, Tex..................	56,799	15.9
23	Charlotte, N.C...................	56,706	28.1
24	Mobile, Ala......................	54,452	27.9
25	Sacramento, Calif................	52,672	27.5
26	Odessa, Tex......................	51,171	63.7

Source: Adapted from, *U.S. Census of Population: 1960*, Vol. 1, *Characteristics of the Population*, Part A, *Number of Inhabitants*, 1961, table M.

pean boundary changes in studying population change between World Wars I and II.[3] The change in the population of an individual city reflects any annexations or retrocessions of territory. Since cities grow by extending their corporate limits as well as by becoming more densely settled, the increase or decrease, in the expanding area is meaningful. Indeed, this gross change is ordinarily all that is measured.

In its 1960 census reports, however, the United States for the first time showed some explicit statistics on the population of areas annexed to incorporated places during the preceding decade. (Approximate figures of this type are also now published in Series P-28, *Special Censuses*.) Table 13-1 shows the magnitude of the largest annexations, and table 13-2 shows how the intercensal growth of central cities and of the outlying parts of standard metropolitan statistical areas was affected by annexations. There were 8.7 million persons in 1960 living in territory annexed between 1950 and 1960 by incorporated places of 2,500 or more in 1950.

In summary, then, there may often be uncertainty or disagreement as to whether fixed territory or changing territory is conceptually more appropriate when measuring change in a particular situation. A measurement of the part of the population change that is attributable to the change in territory is always of analytical interest, however.

Table 13-2. — Population of Standard Metropolitan Statistical Areas, for the United States: 1960 and 1950

[Minus sign (−) denotes decrease]

Year	In SMSA's	In central cities	Outside central cities
1960...........................	112,885,178	58,004,334	54,880,844
1950...........................	89,316,903	52,371,379	36,945,524
Change, 1950 to 1960:			
Total:			
Number.....................	23,568,275	5,632,955	17,935,320
Percent....................	26.4	10.8	48.5
Based on 1950 limits of central cities:			
Number.....................	23,568,275	767,208	22,801,066
Percent....................	26.4	1.5	61.7
From annexations:			
Number.....................	−	4,851,483	−4,851,483
Percent....................	−	9.3	−13.1

— Represents zero.

Source: Adapted from, *U.S. Census of Population: 1960*, Vol. 1, *Characteristics of the Population*, Part A, *Number of Inhabitants*, 1961, table P.

Change in Definition

There are a number of ways in which a national population or the population of a subnational area may be redefined. A change from a *de facto* to a *de jure* basis in taking a census will not ordinarily have a great deal of effect on the intercensal change at the national level, but it will often affect the change in the population of certain geographic subdivisions. In the 1960 census of South Korea, members of the armed forces were allocated to their preservice residence, whereas in the 1955 census they were counted where they were stationed. The two provinces just south of the Demilitarized Zone, where troops are concentrated, contained a much larger percentage of the total armed forces in 1955 than in 1960.[4]

Province	1955	1960
Gyeonggi...........................	22.9	12.3
Kangweon...........................	35.2	6.6

Choe adjusted the 1960 population for comparability with the 1955 population preparatory to measuring population change and then estimating net migration.

As was described in chapter 4, college students living away from home were counted at their college residences in the 1950 census of the United States. The change in definition of "usual residence" is reflected in the following census figures for the town of Chapel Hill, the seat of the University of North Carolina:

Year	Population	Percent change over preceding census
1930.................	2,699	82.0
1940.................	3,654	35.4
1950.................	9,177	151.1
1960.................	12,573	37.0

[3] Dudley Kirk, *Europe's Population in the Interwar Years*, Princeton, N.J., League of Nations, 1946, p. 21 et passim.

[4] Ehn Hyun Choe, *Population Distribution and Internal Migration in Korea*, Seoul, Bureau of Statistics, 1966, p. 126.

According to the U.S. Office of Education, there were 7,419 students enrolled in the university in the fall of 1949. If we assume that all of these students would not have been counted under the old rule, we obtain an adjusted count of 1,758 for 1950. This would indicate a population loss, 1940 to 1950, using a fixed definition. This loss looks unreasonable, however, in the light of our knowledge of the place and the times. There is always the danger of overadjustment when we have recourse to a noncensus aggregate. Here it is certainly possible that (1) a fairly large number of students also had their parental home in the town and hence should not be subtracted, (2) some students had no usual place of residence elsewhere and hence should not be subtracted, or (3) a fairly large number of students lived in the outskirts of the town and hence were properly not included in the 1950 total for the town. Finally, it is also possible that some of the university students living in the town were missed in the 1950 census.

If we are examining the change in a particular class of the population, we wish to be sure that the class is defined consistently over the period considered. To illustrate, in the Netherlands, 93.4 percent of the population was classified as "urban" in 1930 and only 54.4 percent in 1946. The difference is due to a definition change; in the earlier year the census classified as "urban" all places with 2,000 or more inhabitants, but in 1946 only those with a population of 20,000 or more were so classified.[5] Change should not be computed when there has been such an appreciable alteration in the definition.

Change in Accuracy

Our last consideration in the appropriateness of measuring population change is whether the completeness of coverage and the accuracy of classification have remained consistent. Less information is available and published on this than on the other two considerations; but, with the increasing frequency of the post-enumeration surveys and other methods of evaluation (see chapter 4 and elsewhere), this deficiency is being remedied somewhat.

Adjusted national population figures for the United States are available for both 1960 and 1950.[6] The population change based on the adjusted and enumerated figures may be compared as follows:

	Population (thousands)		Change, 1950 to 1960	
	1960	1950[a]	Amount	Percent
Enumerated.......	179,323	151,327	+27,996	+18.5
Adjusted for net undercount.....	185,025	157,001	+28,024	+17.8

[a]50 States and hence comparable with 1960.

DESCRIPTION OF POPULATION CHANGE

It is frequently desired to describe the change that has taken place over more than one period (involving three or more dates on which the population has been measured) or to describe the course of the change within a given period. In the first case, we want such a description because it may simplify and generalize a series of successive changes and thus enable us to see more clearly what has happened.[7] In the second case, we may want to compute the average change within a period or to estimate how the change has been distributed. Some years ago, many demographers were interested in the "laws of growth" that governed various populations. Less naively, we may think of various logical, mathematical ways in which a population may change and then, using these logical formulations as models, see how actual change has corresponded to them. Thus, the description of change is a particular example of curve-fitting to time series or of interpolation. The techniques used in such description are similar to those used in making intercensal estimates and projections by purely mathematical means.

As in all time series, the observations will reflect secular, cyclical, and random movements. If the observations are made at different times of the year, seasonal movements may also be apparent. When we are trying to describe the growth of a population over a relatively longer period of time (for example, India from 1872 to 1961), we are generally interested in the secular trend only. We may also wish our description to include the cyclical movements. In the correlation of population growth with various social and economic series, it is usually desirable to find the secular trend, compute deviations of the actual population from it, and correlate these deviations with deviations from the trend in the other series. In other words, it is usually the association between cyclical changes in growth for two series that is of interest.

Linear Change.—Computing the average **amount** of change a year, a month, etc., is quite straightforward. One simply divides the total change by the number of years, months, etc. Suppose we want to compare the average annual amounts of population growth of a country where the intercensal periods are of unequal length. Let us take the last three censuses of the United Arab Republic (U.A.R.) as an example:

Census date	Population (1)	Increase since preceding date (2)	Inter-censal years (3)	Average annual increase (4)
March 26, 1937.....	15,920,694	(X)	(X)	(X)
March 26, 1947.....	18,966,767	3,046,073	10	304,607
September 20, 1960	26,085,326	7,118,559	13.49	527,692

X Not applicable.

In computing an average amount of increase within a period, we implicitly make an assumption about how the growth is distributed over that period. So far we have implicitly assumed that the growth is linear, or follows an arithmetic progression, i.e., that there is a constant amount of increase per unit of time. It is difficult, however, to hypothesize a set of demographic conditions (see the section on "Components of Population Change") under which population would increase or decrease by arithmetic progression. Nonetheless a straight line has been frequently used not only to describe population

[5] W. S. and E. S. Woytinsky, *World Population and Production*, New York, Twentieth Century Fund, 1953, p. 115.

[6] Jacob S. Siegel, "Completeness of Coverage of the Nonwhite Population in the 1960 Census and Current Estimates, and Some Implications" in: *Social Statistics and the City*, David M. Heer, ed., Report of a Conference Held in Washington, D.C., June 22–23, 1967, Cambridge (Mass.), Joint Center for Urban Studies of the Massachusetts Institute of Technology and Harvard University, 1968, p. 41.

[7] As Pearl put it, "The worker in practically any branch of science is more or less frequently confronted with this sort of problem: he has a series of observations in which there is clear evidence of a certain orderliness, on the one hand, and evident fluctuations from this order, on the other hand. What he obviously wishes to do, on the basis of a quite sound instinct, is to emphasize the orderliness and minimize the fluctuations about it." Raymond Pearl, *Introduction to Medical Biometry and Statistics*, 2nd ed., W.B. Saunders Co., Philadelphia, 1930, p. 407.

growth but also to project it into the future. The general equation for a straight line is

$$y = a + bx \qquad (1)$$

In the case of a single period, we may express this as

$$P_n = P_o + bn, \qquad (2)$$

where P_o is the initial population, P_n is the population at the end of the period (n years), n is the time in years, and b is the annual amount of change.

There are a few circumstances under which it is not particularly culpable to use a straight line in describing population change; for example, when the period of time between population figures is very short and the change rather small, little error can be introduced by assuming that the population was changing arithmetically. When monthly estimates are available, as in the U.S. Census Bureau's Series P-25, *Population Estimates and Projections,* it may usually be assumed that population change within a month followed a straight line.

Rates of Change. — A more meaningful comparison, however, would be of the average annual **rates** of growth. Let us continue with the Egyptian (UAR) figures. First, there is the question of how to express the rate — per unit, per 100, per 1,000, etc. For consistency, we may continue with the convention of expressing the rate as a percent.

Arithmetic approximation to rate of change. — Using the average amount of change that we have already computed (col. (4) in the table above), we may approximate an average rate of change that is logically defensible but requires a little more time for its computation. First, we must choose an appropriate base. Column (5) has been computed using the population at the beginning of the intercensal period (col. (1)).

Intercensal period	$\dfrac{(4)}{(1)} \times 100$	*Average population*	$\dfrac{(4)}{(6)} \times 100$
	(5)	(6)	(7)
1937 to 1947..................	1.91	17,443,731	1.75
1947 to 1960..................	2.78	22,526,047	2.34

These rates are comparable, however, only when the periods are of equal length — in which case we could compare the rates for the total period directly without computing the average rates. With unequal periods and increasing population, the longer period tends to have the highest approximate rate. An obvious choice for the base, to eliminate this bias, is the average of the populations at the beginning and end of the period. This average population is shown in column (6), and the new rate, derived from this population and the average annual amount of change, is shown in column (7). (When we compare (5) and (7), we see that the excess of the later over the earlier average rate is reduced from 0.87 to 0.59.) More important, neither of these "average rates" when applied annually to the successive populations would produce the population at the end of the intercensal period. As we shall see, they represent fairly close approximations to genuine average rates, however. Since from formula (2)

$$b = \frac{P_n - P_0}{n}, \qquad (3)$$

the arithmetic approximation to an average rate is given by

$$r' = \frac{b}{1/2(P_0 + P_n)} \times 100 \qquad (4)$$

Geometric change. — A geometric series is one in which the population increases, or decreases, at the same rate over each unit of time, for example, each year. If this constant rate of change is represented by r and the initial population is represented by P_0, then after n years the final population is given by

$$P_n = P_0(1 + r)^n \qquad (5)$$

Suppose we have the observed population at two dates and wish to find the annual rate of change on the assumption that it is constant throughout the period. Let us return to our census data for the United Arab Republic. For the latest intercensal period, $n = 13.49$ years, $P_0 = 18,966,767$ and $P_n = 26,085,326$. Then,

$$1 + r = \sqrt[n]{\frac{P_n}{P_0}} \qquad (6)$$

$$\log(1 + r) = \frac{\log P_n - \log P_0}{n}, \text{ or} \qquad (7)$$

$$= \frac{\log\left(\dfrac{P_n}{P_0}\right)}{n} \qquad (8)$$

$$= \frac{\log(26,085,326 \div 18,966,767)}{13.49}$$

$$= \frac{\log 1.375317}{13.49}$$

$$= \frac{0.1384028}{13.49} = 0.0102597$$

$$1 + r = \text{antilog } 0.0102597 = 1.0239$$

$$r = .0239, \text{ or about 2.39 percent per annum.}$$

For the 1937–1947 period, the assumption of geometric change yields a rate of about 1.77 percent per annum.

If we had annual observations of the size of the population, and hence annual rates of population change, we could find the geometric mean,

$$\sqrt[n]{\frac{P_1}{P_0} \times \frac{P_2}{P_1} \times \ \cdots \ \frac{P_n}{P_{n-1}}},$$

of these annual rates.

This geometric mean would be identical with the geometric rate of change figured from the population at the beginning and end of the intercensal period, which we have just described.

Exponential change. — Geometric change is a compound-interest type of change in which the compounding takes place at certain constant intervals, such as a year. The annual compounding is arbitrary, however. Why not compound semi-annually or monthly? In the limiting case, the compounding takes place continuously, i.e., a constant rate of change is applied at every infinitesimal of time. For a fixed period, this exponential growth curve can be expressed as

$$P_n = P_0 e^{rn} \qquad (9)$$

Solving for r, we get

$$r = \frac{\log \left(\dfrac{P_n}{P_0} \right)}{n \log e} \qquad (10)$$

(e is a mathematical constant, like π, namely, 2.71828 . . . Its logarithm to the base 10 is 0.4342945 . . .)
We will use this equation to solve for the average rate of growth, r, for the U.A.R. in its two last intercensal periods. For the 1937–1947 period, the computations are as follows:

$$r = \frac{\log \left(\dfrac{18,966,767}{15,920,694} \right)}{10 \times .4342945}$$

$$= \frac{\log 1.1913279}{4.342945} = \frac{.076313}{4.342945}$$

$$= .0175, \text{ or } 1.75 \text{ percent, per annum}$$

For the 1947–1960 intercensal period, the average annual growth rate, compounded continuously, is 2.36 percent. Note that the average rate of growth when the compounding is instantaneous is always less than when the compounding is for a finite time interval.

To the mathematically trained demographer, another short-cut in solving this equation is obvious. One may make use of the fact that the exponential function e^x has been tabulated.[8] $\ln \frac{P_n}{P_0} = e^{rn}$, e^{rn} is the form of e^x. We calculate $\frac{P_n}{P_0}$, and look up the corresponding value of x in the tables, using the "values of the ascending exponential" since the population is increasing. Finding that $x = rn$ is 0.1751 for the 1937–1947 period, $r = .01751$. Likewise for the 1947–1960 period, $rn = 0.3187$ and $r = .02361$.

We will close this subsection by recapitulating the various estimates of the average rate of growth that we have computed by various formulas for the last two intercensal periods of the United Arab Republic. These are as follows:

| | Growth Rate | |
Formula	1937–1947	1947–1960
Arithmetic approximation (mid-period base).........................	1.75	2.34
Geometric — Annual compounding..	1.77	2.39
Exponential — Continuous compounding.........................	1.75	2.36

Note that the arithmetic approximation gives a little lower average annual rate than the other two for the second period. It would be lower than the exponential rate for 1937–1947 too if the rates were carried to more decimal places.

Multiple Periods. — The preceding section dealt with the computation of the average annual rate of change within a single period, especially an intercensal period, and with the interpolation of the population at a date or dates within the period. Now we turn to the description of change over a number of periods. Here most of the curves describing growth that we have already considered can also be applied, as can additional ones.

Description of Trend. — Describing the trend of population from a number of observations is a particular case of curve fitting, a topic that is dealt with more generally in **Appendix C**. In choosing the type of growth curve to fit to the population data, it is desirable both to examine the plotted points (an empirical guide) and to develop a demographic theory as to why population should be expected to change in the hypothesized way given a particular set of historical conditions and in the light of demographic knowledge about the factors affecting the components of population change. The trend curve is of interest in its own right. It may also be used for purposes of interpolation (to find the trend value at a given date within the period of observation) or for purposes of extrapolation (to find the trend value at a date beyond the observed period).

Linear trend. — A straight line (polynomial of first degree) is represented by the following equation and can be fitted to three or more observations by the method of least squares:

$$P_t = a + bt \qquad (11)$$

Other polynomial trends. — Population change is occasionally fairly well described by a polynomial of second or higher degree.

The equation of a second-degree polynomial (parabola) is

$$P_t = a + bt + ct^2 \qquad (12)$$

and that of an nth degree polynomial by

$$P_t = a + bt + ct^2 \ldots . kt^n \qquad (13)$$

Exponential curves. — The general equation for an exponential curve is

$$y = ab^x \qquad (14)$$

For population growth, this could be represented as

$$P_t = a(1 + r)^t \qquad (15)$$

where P_t is the population at time t and r is the rate of change. (The geometric growth equation, $P_n = P_0(1 + r)^n$ can be seen to be a special case of this.)

When the rate of change is positive, the population described by such a curve would increase without any limit. Such a trend manifestly cannot continue indefinitely. An exponential curve may, however, describe population growth over a particular period of time. French demographers sometimes speak of such growth as "Malthusian growth" because of the well-known generalization by Malthus that population tends to increase in a geometric ratio.

Logistic curve. — By far the best known growth curve in demography is the logistic curve. It was first derived by Verhulst about 1838 but was rediscovered and popularized by Pearl and Reed about 1920. Some species of animals and some bacterial cultures have been observed to grow rapidly at first when placed in a limited environment with ideal conditions of food supply and space for their initially few numbers, and then to grow slowly as the population approaches a point where there is pressure on available resources. There may be an upper

---

[8] U.S. National Bureau of Standards, *Tables of the Exponential Function e^x*, Applied Mathematics Series 14, 1951.

limit to the numbers that can be maintained, whereupon the population ceases to grow. Considerations such as these have given rise to attempts to predict what the upper limit might be, by fitting asymptotic growth curves to observed data. The equation of this curve is

$$Y_c = \frac{k}{1 + e^{a+bX}} \qquad (16)$$

Y_c represents the computed value of Y. This curve is very similar to the Gompertz curve. One dissimilarity is that the growth increments of the logistic are symmetrical when plotted, closely resembling a normal curve, whereas those of the Gompertz curve are skewed. For a logistic curve, the differences between the reciprocals of the population values decline by a constant percentage.[9] (See pages 382–384 of the unabridged book for an illustration of the method of fitting a logistics curve).

A very considerable literature has appeared on the use of the logistic curve in the description and projection of organic growth in general and of population growth in particular. Like the other growth curves, the logistic is used for projecting population growth into the future as well as for describing its past course. (See chapter 23.) Some of the more important books and articles on this subject are listed at the end of this chapter. Pearl and Reed have written at length on the rationale behind the logistic curve, and they and their disciples have applied it to many kinds of situations. Other writers have criticized the logistic as being too mechanistic in its assumptions (e.g., for not allowing for voluntary control over fertility and for migration) and for other reasons. The logistic cannot describe a population that is decreasing, of course.

Time Required for Population to Double.—There is often interest in the time required for a population to double in size if a given annual rate of increase were to continue. This interest reflects the problems confronting many rapidly growing countries, especially the economically underdeveloped ones.

For this purpose let us go back to the formula for a constant growth rate, continuously compounded,

$$P_t = P_0 e^{rt}$$

We want to find when P_t will equal $2P_0$

$$2P_0 = P_0 e^{rt}$$

$$\log 2 = rt \log e$$

$$t = \frac{\log 2}{r \log e} \qquad (17)$$

Take, for example, 4.1 percent, the annual rate of growth for Costa Rica in the period 1958 to 1966, as estimated by the United Nations. The ratio of log 2 to log e is a constant 0.693, so we have

$$t = \frac{0.693}{0.041} = 16.9 \text{ years}$$

Figure 13–1 shows the general relationship between annual rate of growth and number of years for the population to double, in both graphic and tabular form.

Cyclical Change.—Most of the foregoing discussion has been concerned with the description of secular trends in population size. In computing the average population change within an intercensal period, we were, however, dealing with the combined effects of secular and cyclical movements. Not much attention has been paid to the separate cyclical analysis of population change. There are several reasons for this in the United States. In the first place, cyclical analysis requires observations of at least annual, and preferably monthly, frequency; and accurate monthly estimates can be made only for the past 35 years or so. This is a fairly short period for cyclical analysis, particularly when population growth has been disturbed by two world wars. Another important reason is that the business cycle affects the components of population change in different ways. A more meaningful analysis can therefore be made by examining the cyclical variations in each of the components. Thomas, Easterlin, and others have done important work in this field, particularly with Swedish, British, and American data. [10]

Seasonal Change.—Because each of the components of population growth is subject to seasonal change, the growth of total population may exhibit some degree of seasonal fluctuation. To observe this, we need fairly accurate estimates of population on a monthly, or at least a quarterly, basis.

For this purpose, we have chosen the monthly population estimates for Japan over the period 1952 to 1961.[11] From these, the following seasonal indexes were computed:

January	130.5
February	109.6
March	106.5
April	104.5
May	88.9
June	82.6
July	95.0
August	100.4
September	105.9
October	91.8
November	87.9
December	98.3

These indicate a tendency toward above-average population growth in the first quarter of the year with below-average growth in May, June, and November. This pattern may not be typical of other countries. In fact, the indexes for individual years suggest a changing pattern within Japan itself during this period. The *de facto* population of particular areas and, to a lesser extent, the "usual population" as defined in some censuses, has a more marked seasonal change. Resort areas are a good case in point. The seasonal factor affects the size of the farm population of the United States as seen in table 13–3. In the North Temperate Zone, farming activities are slack during the winter months.

[9] Frederick E. Croxton, Dudley J. Cowden, and Sidney Klein, *Applied General Statistics*, 3rd ed., Englewood Cliffs, N.J., Prentice-Hall, 1967, pp. 280–282.

[10] Richard A. Easterlin, "Economic-Demographic Interactions and Long Swings in Economic Growth," *American Economic Review*, 56(5):1063–1104, Dec. 1966; Virginia L. Galbraith and Dorothy S. Thomas, "Birth Rates and the Interwar Business Cycles," *Journal of the American Statistical Association*, 36(216):465–476, Dec. 1941; Harry Jerome, *Migration and Business Cycles*, New York, National Bureau of Economic Research, Inc., 1926; Dudley Kirk, "The Influence of Business Cycles on Marriage and Birth Rates," *Demographic and Economic Change in Developed Countries*, National Bureau of Economic Research, Princeton, New Jersey, Princeton University Press, 1960, pp. 241–260; Dudley Kirk and Dorothy L. Nortman, "Business and Babies: The Influence of Business Cycles on Birth Rates," *Proceedings of the Social Statistics Section* (1958) Washington, American Statistical Association, pp. 151–160; Morris Silver, "Births, Marriages, and Business Cycles in the United States," *Journal of Political Economy*; 73(3):237–255, June 1965; Dorothy Swaine Thomas, *Social and Economic Aspects of Swedish Population Movements: 1750–1933*, New York, Macmillan Co., 1941; idem, *Social Aspects of the Business Cycle*, London, George Routledge & Sons, Ltd., 1927.

[11] Japan, Bureau of Statistics, "Monthly Estimates of All Japan Population, [various dates]," *Monthly Report on Current Population Estimates*.

Figure 13–1.—Number of Years Necessary for the Population to Double at Given Annual Rates of Growth

[Computed by use of the formula for continuous compounding $P_t = P_0 e^{rt}$]

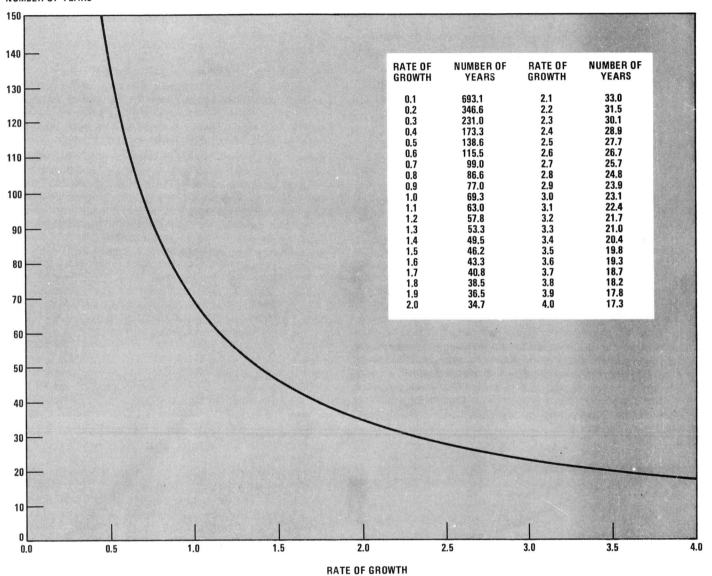

NUMBER OF YEARS

RATE OF GROWTH

RATE OF GROWTH	NUMBER OF YEARS	RATE OF GROWTH	NUMBER OF YEARS
0.1	693.1	2.1	33.0
0.2	346.6	2.2	31.5
0.3	231.0	2.3	30.1
0.4	173.3	2.4	28.9
0.5	138.6	2.5	27.7
0.6	115.5	2.6	26.7
0.7	99.0	2.7	25.7
0.8	86.6	2.8	24.8
0.9	77.0	2.9	23.9
1.0	69.3	3.0	23.1
1.1	63.0	3.1	22.4
1.2	57.8	3.2	21.7
1.3	53.3	3.3	21.0
1.4	49.5	3.4	20.4
1.5	46.2	3.5	19.8
1.6	43.3	3.6	19.3
1.7	40.8	3.7	18.7
1.8	38.5	3.8	18.2
1.9	36.5	3.9	17.8
2.0	34.7	4.0	17.3

ACCURACY OF MEASUREMENT

Since population change may be derived by subtraction from population counts or estimates, we must first consider the accuracy of these counts or estimates (see chapter 4). If the estimates came from sample surveys (or from a sample of a complete census or a population register) and if sampling error were the only source of error, then the error of the change would be given by the ordinary formula for the standard error of a difference:

$$\sigma_{P_1 - P_2} = \sqrt{\sigma_{P_1}^2 + \sigma_{P_2}^2} \qquad (18)$$

(The variance of a difference is the **sum** of the variances of the quantities themselves.) In most situations, however, the bias of the population figure is much more important than its sampling error. Very often there is serial correlation among the biases, that is, since the successive population censuses and surveys are taken by essentially the same procedures under essentially the same environmental conditions, the bias is likely to be of roughly the same sign and percentage at successive censuses. The figures given for the United States in the section above on "Change in Accuracy" imply that the enumerated population was 3.1 and 3.6 percent below the adjusted population in 1960 and 1950, respectively, whereas

Table 13-3. — Seasonal Indexes for the Farm Population 14 Years Old and Over, in the United States: 1963 to 1967

Quarter and month	Index	Quarter and month	Index
QUARTER		MONTH--Continued	
First.....................	99.88	April....................	100.42
Second..................	100.03	May.....................	99.47
Third....................	101.01	June....................	100.20
Fourth..................	99.07	July....................	100.69
		August..................	101.22
MONTH		September..............	101.12
January.................	99.79	October.................	100.17
February...............	99.58	November...............	99.18
March..................	100.26	December...............	97.86

Source: Unpublished figures from the Current Population Survey, April 1962 to December 1968.

the original amount of intercensal change was only 0.1 percent below the change in the adjusted population counts.

COMPONENTS OF POPULATION CHANGE

There are only four components of change in the total population, viz., births, deaths, immigration, and emigration. The algebraic excess of births over deaths is called **natural increase.** (Natural decrease can occur, of course.) The algebraic excess of immigration over emigration is called **net migration,** either net immigration or net emigration. Usually, as is the case in this book, the terms immigration and emigration are reserved for movements across international boundaries, with movements from and to other parts of the same country being called in-migration and out-migration, respectively. In-migration and out-migration are collectively called **internal migration.** Sometimes we are interested in the total migration to or from an area, regardless of whether it is to (or from) another area in the same country or to (or from) abroad. The following chapters are concerned with the components of population change.

These components, of course, may be measured and studied not only for the total population but also for subgroups of the population such as an age-sex group, the employed population, or the single population. In many such cases, we have to deal also with changes in classification. The exceptions are fixed attributes like sex, place of birth, and race. Some of the changes (as in age) are entirely predictable if we are dealing with true rather than reported characteristics. Others, such as those in marital status, have restricted paths, e.g., a change from single status is irreversible and the only possible change from widowed or divorced is to married. With some characteristics, however (occupation, income, or religion), it is theoretically possible to change from any category to any other category.

FACTORS IMPORTANT IN ANALYSIS

A great deal of demographic analysis is carried out in terms of comparisons of population change among different political areas, residence areas, and groups having various personal,

Figure 13-2. — Population Growth of Canada, 1851 to 1961, and of Conterminous United States, 1850 to 1960

[Semilogarithmic scale]

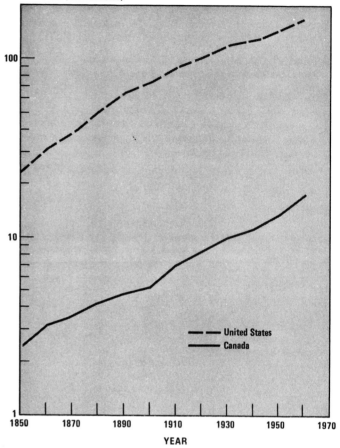

social, and economic characteristics. Comparisons of percentage change are usually more meaningful than those of absolute change. If two countries with intercensal periods of different lengths are being compared, annual average rates should be computed, by the use of the assumption of geometric increase, for instance.

If two areas, or groups, of radically different population size (e.g., the United States and Canada) are being compared graphically, a better indication of relative rates of change is given by a semilogarithmic graph than by an ordinary arithmetic graph (fig. 13-2). Here population is plotted to a logarithmic scale on the vertical axis, and time, to an arithmetic scale on the horizontal axis. Then equal vertical distances will represent equal rates of change no matter what the size of the population. Furthermore, on such a "semilogarithmic" grid, a population increasing at a constant rate is represented by a straight line.

In a broad sense, much of the U.N. report, *The Determinants and Consequences of Population Trends,* first published in 1953 and revised in 1973, is concerned with the factors associated with the rate and nature of population change.

SUGGESTED READINGS

Croxton, Frederick E., Cowden, Dudley J., and Klein, Sidney. *Applied General Statistics*, 3rd ed. Englewood Cliffs, N.J., Prentice-Hall, 1967. Pp. 263–274.

Davis, Kingsley. "Conceptual Aspects of Urban Projections in Developing Countries." In United Nations. *World Population Conference, 1965*. Vol. III, *Projections, Measurement of Population Trends*. 1967. Pp.62–63.

Easterlin, Richard A. "Economic-Demographic Inter-actions among Long Swings in Economic Growth." *American Economic Review*, 56(5):1063–1104. December 1966.

Galbraith, Virginia L., and Thomas, Dorothy S. "Birth Rates and the Interwar Business Cycles." *Journal of the American Statistical Association,* 36(216): 465–476. December 1941.

Hagood, Margaret Jarman. *Statistics for Sociologists*. New York, Reynal and Hitchcock, 1941. Pp. 271–288.

Hogben, Lancelot. *Genetic Principles in Medicine and Social Science*. London, Williams, and Norgate, 1931. Pp. 177–184.

Jerome, Harry. *Migration and Business Cycles*. New York, National Bureau of Economic Research, 1926.

Kirk, Dudley. "The Influence of Business Cycles on Marriage and Birth Rates." *Demographic and Economic Change in Developed Countries*. National Bureau of Economic Research, Princeton, N.J., Princeton University Press, 1960. Pp. 241–260.

Kirk, Dudley, and Nortman, Dorothy L. "Business and Babies: The Influence of Business Cycles on Birth Rates." *Proceedings of the Social Statistics Section (1958)*. Washington, American Statistical Association. Pp. 151–160.

Matras, Judah. "Demographic Trends in Urban Areas." In *Handbook for Social Research in Urban Areas*. Philip M. Hauser, ed., Paris, United Nations Educational, Scientific and Cultural Organization, 1965. Pp. 16–139.

Pearl, Raymond. *The Biology of Population Growth*. New York, Alfred A. Knopf, Rev. ed. 1930. Pp. 1–25.

Pearl, Raymond, Reed, Lowell J., Kish, Joseph F. "The Logistic Curve and the Census Count of 1940." *Science*, 92(2395):486–488. November 22, 1940.

Rasor, Eugene A. "The Fitting of Logistic Curves by Means of a Nomograph." *Journal of the American Statistical Association*, 44(248):548–553. December 1949.

Siegel, Jacob S. "Approximación de la tasa annual promedio de cambio," *Estadística*, 14(52): 426–431. September 1956.

Silver, Morris. "Births, Marriages, and Business Cycles in the United States." *Journal of Political Economy,* 73(3): 237–255. June 1965.

Spurr, William A., and Arnold, David R. "A Short-Cut Method of Fitting a Logistic Curve." *Journal of the American Statistical Association*, 43(241):127–134. March 1948.

Thomas, Dorothy Swaine. *Social and Economic Aspects of Swedish Population Movements: 1750–1933*, New York, Macmillan Co., 1941. Pp. 383–413.

Thomas, Dorothy Swaine. *Social Aspects of the Business Cycle*, London, George Routledge Sons, Ltd., 1927. Pp. 165–181.

United Nations. *The Determinants and Consequences of Population Trends*. Series A, Population Studies No. 50. 1973.

White, Helen R., Siegel, Jacob S., Rosen, Beatrice M. "Short Cuts in Computing Ratio Projections of Population." *Agricultural Economics Research* (U.S. Department of Agriculture) 5(1):6–7. January 1953.

Wilson, Edwin B., and Puffer, Ruth R. "Least Squares and Laws of Population Growth." *Proceedings of the American Academy of Arts and Sciences*, 68(9):285–382. August 1933.

Woofter, T. J., Jr. "Interpolation for Populations Whose Rate of Increase is Declining." *Journal of the American Statistical Association*, 27(178):180–182. June 1932.

CHAPTER 14

Mortality

INTRODUCTION

Nature and Uses of Mortality Statistics

Death is a principal "vital event" for which vital statistics are collected and compiled by the vital statistics registration system. The other principal vital events for which vital statistics are collected and compiled are live births, fetal deaths, marriages, and divorces. Secondarily, adoptions, legitimations, annulments, and legal separations may be included. The vital statistics system includes the legal registration, statistical recording and reporting of the occurrence of vital events, and the collection, compilation, analysis, presentation, and distribution of vital statistics.[1] The vital statistics system employs the registration method of collecting the data on vital events, which typically involves the reporting to government officials of events as they occur and the recording of the occurrence and the characteristics of these events.

Broadly speaking, death statistics are needed for purposes of demographic studies and for public health administration. The most important uses of death statistics include: (1) analysis of the present demographic status of the population as well as its potential growth; (2) filling the administrative and research needs of public health agencies in connection with the development, operation, and evaluation of public health programs; (3) determination of administrative policy and action in connection with the programs of government agencies other than those concerned with public health; and (4) filling the need for information on population changes in relation to numerous professional and commercial activities. Death statistics are needed to make the analyses of past population changes which are required for making projections of population and other demographic characteristics. The latter are employed in developing plans for housing and educational facilities, managing the social security program, and producing and providing services and commodities for various groups in the population. Analysis of mortality statistics is essential to programs of disease control. Local health authorities use mortality statistics reports to determine administrative action needed to improve public health in local areas. All these uses are in addition to the demands of individuals for documentary proof of death.

Definition of Concepts: Statistical Life and Death

Adequate compilation and measurement of vital events require that the concepts of life and death be given formal definition, even though the meaning of these concepts may appear obvious, at least in a physical sense, to most persons. Recent developments in the medical field have gone a long way to blur even the clinical distinction between life and death. When does life begin and end for statistical purposes, that is, for the counting of births and deaths? In fact, vital statistics systems identify three basic, complementary, and mutually exclusive categories—live births, deaths, and fetal deaths.

The United Nations and the World Health Organization have proposed the following definition of death: "Death is the permanent disappearance of all evidence of life at any time after birth has taken place (post-natal cessation of vital functions without capability of resucitation)."[2] A death can occur only after a live birth has occurred. The definition of a death can be understood, therefore, only in relation to the definition of a live birth (see Chapter 16). For one thing it excludes the entire category of fetal mortality, or deaths that occur prior to the completion of the birth process.

The definition of a "death" excludes deaths prior to (live) birth. These are so-called fetal deaths. A fetal death is formally defined as "death (disappearance of life) prior to the complete expulsion or extraction from its mother of a product of conception irrespective of the duration of pregnancy; the death is indicated by the fact that after such separation the fetus does not breathe or show any other evidence of life, such as beating of the heart, pulsation of the umbilical cord, or definite movement of voluntary muscles."

The term fetal death is employed in present demographic practice to embrace the events variously called stillbirths, miscarriages, and abortions in popular, medical, or legal usage. The term stillbirth is often employed as synonymous with late fetal deaths, say a fetal death of 20 or 28 completed weeks of gestation or more, although the term is sometimes employed to refer to all fetal deaths.

The term miscarriage is popularly employed to refer to spontaneous or accidental terminations of fetal life occurring early in pregnancy. The term abortion is popularly used to refer to induced early fetal deaths, including both those which are legal and those which are illegal. In medical usage, an abortion is the expulsion of the fetus prematurely, particularly at any time before it is viable or capable of sustaining life. From a technical viewpoint, abortion and miscarriage can hardly be distinguished. The recommendation of the United Nations and

[1] United Nations, *Principles for a Vital Statistics System*, Statistical Papers, Series M, No. 19, August 1953, p. 4.

[2] Ibid., p. 6; World Health Organization, *Official Records of the World Health Organization*, No. 28, *Third World Health Assembly, Geneva, 8 to 27 May 1950*, December 1950, p. 17.

the World Health Organization is to group all of these events, miscarriages and abortions as well as stillbirths, under the heading "fetal death" and to classify them as early, intermediate, and late according to the months of gestation.

Sources of Data for Mortality Studies

The basic data on deaths for mortality studies, for the statistically developed areas, come from vital statistics registration systems and, less commonly, from national population register systems, as we have noted. The analysis of the death statistics from the vital statistics registration system depends on the availability of appropriate population data from a census or survey, or population estimates, to be used as bases for computing rates of various kinds. This dependence on a second data collection system is avoided where an adequate national population register system is in effect.

The vital statistics registration system is likely to be inadequate in the underdeveloped countries; for these areas, other sources of data for measuring mortality have to be considered. The principal alternative sources are (1) national censuses and (2) national sample surveys. National censuses and sample surveys may provide (a) data on age composition from which the level of recent mortality can be inferred and (b) direct data on mortality. The national sample surveys may also provide additional detailed data permitting a more complete analysis of mortality. A sample survey may be employed to "follow-up" a sample of decedents reported in the registration system. The inference regarding the level of mortality is easier and firmer where two censuses are available, but special techniques permit this inference with only one census in certain cases. Direct data on deaths are very infrequently obtained in censuses, but in recent decades sample surveys have been used in a number of areas to measure mortality levels. The sample survey is not an acceptable substitute for the registration system with respect to the legal aspects of registrations, however, since every vital event should be registered for legal purposes.

Vital statistics from civil registers are very infrequently available in Africa. The availability of vital statistics is only slightly greater for much of Asia and parts of Latin America. In these continents, coverage is uneven and the statistics generally incomplete and unreliable. For these areas, however, estimated death rates are becoming increasingly available from sample surveys or from analysis of census returns.

Mortality statistics are ordinarily presented in the statistical yearbooks of countries or in special volumes presenting vital statistics. In the United States a volume on mortality statistics has been published each year since 1900 (the year when the annual collection of mortality statistics for the Death Registration Area began) by the U.S. Census Bureau or the U.S. Public Health Service. In addition, provisional statistics as well as certain final data are published in brief reports in advance of the publication of the bound volumes. The United Nations' *Demographic Yearbooks* for 1957, 1961, 1966, and 1967 emphasized tabulations of mortality and fetal mortality. Other special compilations of mortality data include *Annual Epidemiological and Vital Statistics,* published by the World Health Organization since 1962 and *World Health Statistics Annual* published by the World Health Organization from 1948 to 1961.

QUALITY OF DEATH STATISTICS

We may consider the deficiencies of death statistics based on vital statistics registration systems under three headings: (1) accuracy of the definition of death applied, (2) completeness of registration, and (3) accuracy of allocation of deaths by place and time.

Application of Definition

Not all countries follow the definition of death recommended by the United Nations in obtaining a count of deaths. In some countries (e.g., Equatorial Guinea, Northern Zone of Morocco) infants who die within 24 hours after birth are classified not as deaths but as "stillbirths," or failing provisions for this, are disregarded altogether, i.e., they are classified neither as a live birth, death, or fetal death. In other countries (e.g., Algeria, French Guiana, and Syria) infants who are born alive but who die before the end of the registration period, which may last a few months, are considered "stillbirths" or are excluded from all tabulations.

Completeness of Registration

Basis and Extent of Underregistration. — The registration of deaths for a country may be incomplete because of either (1) failure to cover the entire geographic area of the country or all groups in the population or (2) failure to register all or nearly all of the vital events in the established registration area. Typically, both of these deficiencies apply if the first is true. Many less developed countries currently collect death statistics only in a designated registration area (e.g., Indonesia, Ghana, Turkey, Brazil, Burma) or try to secure complete registration only in designated sample registration areas (e.g., India) for a variety of reasons. The registration area may exclude parts of the country where good registration is a practical impossibility in view of the country's economic and social development and financial condition. Remote rural, mountainous, or desert areas in a country may not be serviced by registrars. The national government may not have complete administrative control over certain parts of the territory it claims, as a result of civil disorder (e.g., Burma) or a territorial dispute with another country (e.g., India, Pakistan). Nomadic and indigenous groups (e.g., nomadic Indians in Ecuador, jungle Indians in Venezuela, full-blooded aborigines in Australia) or particular ethnic or racial groups (Palestinian refugees in Syria, African population in former Bechuanaland, Bantus in South Africa) may be excluded or hardly covered. Commonly, in the underdeveloped countries babies dying before the end of the legal registration period are registered neither as births nor deaths. In Taiwan, for example, deaths must be registered within 5 days and births within 15 days. A sample test conducted about 1965 showed that a large proportion of babies dying before 15 days were not registered in any way at all.

The United Nations finds death registration to be substantially incomplete for most countries of the world, including nearly all of those considered underdeveloped. According to an analysis of data which the United Nations carried out for 1951–55, only about 33 percent of the world's deaths were being registered, the percentage varying regionally from 7 percent for East Asia to 100 percent in North America and

Europe. Carefully derived estimates of the completeness of registration are rarely, if ever, available. Several countries have provided estimates to the United Nations that are given in the *Demographic Yearbook*. These estimates are typically below 90 percent; even so, they probably overstate the actual extent of registration.

In the United States, annual collection of mortality statistics was begun in 1900, when the registration area consisted of 10 States, the District of Columbia, and a number of cities, accounting for only 41 percent of the population of the United States. Complete geographic coverage of the United States was not achieved until 1933, when the last State joined the registration area.

Even so, registration of deaths could hardly have been complete. Only 90 percent completeness was a condition for a State's joining the Death Registration Area, and the testing was apparently quite crude. The extent of the completeness of death registration in the United States is not known and can only be surmised; no systematic national test has ever been conducted.

Measurement of Underregistration. — There are a number of approaches to the measurement of the completeness of death registration, none of which is highly satisfactory. The results of studies of underregistration of births, including birth registration tests and estimates made by demographic analysis, have general implications for the completeness of death reporting. An appreciable underregistration of births suggests a substantial underregistration of deaths, even though the types of inducements to comply with the laws regarding registration differ for these events.

Chandra Sekar (now Chandrasekaran) and Deming, have presented a mathematical theory which, when applied to a comparison (i.e., matching) of individual entries on a registrar's lists of deaths with the individual entries on the lists obtained in a house-to-house canvass, gives national unbiased estimates of the completeness of registration.[3] The basic procedure is to divide the area covered (either geographically or by a combination of characteristics) into subgroups, each of which is highly homogeneous. (A highly homogeneous population is defined as one in which each individual has an equal probability of being enumerated.) Within such subgroups, the correlation between unregistered and unenumerated events would be very low. The percent of "enumerated" deaths which are registered in each subgroup is assumed, therefore, to apply to the unenumerated deaths which are registered in the subgroup to derive the total number of deaths which occurred in the subgroup. An estimate of the total number of deaths (registered and unregistered) in the area could then be derived by cumulating the total number of deaths corrected for underregistration for subgroups. Relating the figures for registered deaths to this total nationally would then give an unbiased estimate of the completeness of registration.

This procedure represents one of the more promising ways of evaluating the reported number of deaths or death rates, or of estimating them, particularly for the statistically underdeveloped areas. The various procedures for accomplishing this include, in addition to the combined use of registration data and survey data represented by "population-growth estimation studies," use of stable population models, repetitive and overlapping sample household surveys, use of sample registration areas, and analysis of statistics by age from successive censuses.[4]

Accuracy of Allocation by Place and Time

Accuracy of Allocation by Place. — Part of the difficulty in the geographic allocation of deaths is due to the utilization of hospitals in large cities by residents of the surrounding suburban and rural areas whose death in these hospitals results in an excessive allocation of deaths to the large cities. This bias is more likely to occur in the more industrialized countries where there is greater access to hospital facilities and a higher proportion of deaths occur in hospitals. However, there are many causes contributing to the movement of a patient to some area other than his usual place of residence; it cannot be assumed that the movement is always from rural or suburban areas to large cities. Deficiencies in the geographic allocation of deaths may also result from fatal accidents or unexpected deaths which occur away from the usual place of residence. This is more likely to occur in the more industrialized areas where modern means of transportation have contributed to high population mobility. A special problem relates to the assignment of residence of decedents who have lived in an institution outside their area of origin for a considerable period of time. In the United States, all deaths which occur in institutions of all types are allocated to the place of residence before entry into the institution, regardless of the length of time spent by the decedent in the institution. Because of difficulties in the allocation of deaths to the place of usual residence of the decedent, even death statistics tabulated on a usual-residence basis often have a pronounced "occurrence" bias, particularly for central cities of large metropolitan areas.

Deaths tabulated by place of occurrence and place of residence are useful for different purposes although the United Nations recommends that tabulations for geographic areas within countries should be made according to place of usual residence. The occurrence data represent the 'service load;' the residence data reflect the incidence of death in the population living in an area.

Accuracy of Allocation by Time. — Most countries tabulate their death statistics in terms of the year of occurrence of the death, reassigning the deaths from the year of registration. Some countries, including a number of underdeveloped ones, however, fail to make this reassignment. The death statistics published for these countries for any year relate to events which happened to be registered in that year. The United Nations has recommended that the principal tabulations of deaths should be on a "year-of-occurrence" basis rather than on a "year-of-registration" basis. Few countries have tabulated the data in both ways so as to establish the difference in the results. The death statistics for any year based on events registered in that year may be lower or higher than the number which actually occurred in the year because some events occurring in the year may not be registered until many years have passed and the events registered during the year may include events of many earlier years.

FACTORS IMPORTANT IN ANALYSIS

Mortality shows significant variations in relation to certain characteristics of the decedent and certain characteristics of

[3] C. Chandra Sekar and W. E. Deming, "On a Method of Estimating Birth and Death Rates and the Extent of Registration," *Journal of the American Statistical Association*, 44(245):101–115, March 1949. See also Ansley J. Coale, "The Design of an Experimental Procedure for Obtaining Accurate Vital Statistics," *International Population Conference, New York, 1961*, London, International Union for the Scientific Study of Population, 1963, pp. 372–375.

[4] W. Parker Mauldin, "Estimating Rates of Population Growth." *Family Planning and Population Programs: A Review of World Developments* (Proceedings of the International Conference of Family Planning Programs, Geneva, 1965), Chicago, University of Chicago Press, 1966, pp. 642–647.

the event. These characteristics of the event and the decedent define the principal characteristics which are important in the demographic analysis of mortality.

In view of the very close relation between age and the risk of death, age may be considered the most important demographic variable in the analysis of mortality. No other general characteristic of the decedent or of the event offers so definite a clue as to the risk of mortality. (This is true only for the general population; for special population groups, other factors may be more important than age, e.g., duration of disability for disabled persons, cause of illness for the hospitalized population.) The other characteristics of the decedent of primary importance are his or her sex and his or her usual place of residence. Elements of primary importance characterizing the event are the cause of death, place of occurrence of the death, and date of occurrence and of registration of the death.

Other characteristics of the decedent important in the analysis of mortality are marital status, socioeconomic status (e.g., occupation, literacy, educational attainment), hospitalization, and urban-rural (or size-of-locality) residence. The United Nations also includes in its second priority list the age of the surviving spouse (for married persons), industry and class of worker (as employer, employee, etc.) of the decedent, legitimacy of decedent (for infants), and number of children born (for females of childbearing age or older). The list can be extended to cover other factors which would permit analysis of the significant social and economic factors in mortality; for example, the race or other ethnic characteristic of the decedent, such as nativity, country of birth, religion, language, or citizenship.

Mortality also varies with the characteristics of the community and the physical environment. These characteristics include the climate, the altitude, the quality of health facilities, environmental conditions, such as the type of water supply, degree of air pollution, and the quantity and quality of food available. We could include here also those characteristics previously cited relating to the geographic location of the death—place of occurrence of death or usual place of residence of the decedent—in terms of specific geographic subdivision, urban-rural residence, or size of locality.

MEASURES BASED ON DEATH STATISTICS

Observed Rates

There are a great number of measures of mortality based on death statistics. They vary in the aspect of mortality they describe, their degree of refinement or elaboration, whether they are summary measures or specific measures, and whether they are measures of mortality *per se* or merely mortality-related measures. We can distinguish among the measures of mortality so-called "observed rates" from so-called "adjusted rates." The distinction is only approximate. The observed rates are typically the simpler rates and are computed directly from actual data in a single brief calculation. The adjusted rates are more complex both with respect to method of calculation and to interpretation. They are often hypothetical representations of the level of mortality for a given population group, involving the use of various assumptions to derive summary measures based on sets of specific death rates.

Crude Death Rate.—The simplest and commonest measure of mortality is the crude death rate. The crude death rate is defined as the number of deaths in a year per 1,000 of the midyear population; that is,

$$\frac{D}{P} \times 1,000 \qquad (1)$$

The crude death rate for Costa Rica in 1960, for example, is calculated by dividing the count of deaths in Costa Rica in 1960 by the estimated population of Costa Rica on July 1, 1960:

$$\frac{10,063}{1,171,000} \times 1,000 = 8.6$$

The midyear population is employed as an approximation to the average population "exposed to risk" of death during the year. The midyear population may be approximated by combining data on births, deaths, and immigration for the period between the census date and the estimate date with the count from the last census; as an arithmetic or geometric mean of the population estimated directly on the basis of these components for two successive January 1 dates; and in other ways.

An array of crude death rates for a wide range of countries around 1960 described as having rather complete death registration indicates a low of 5.7 for Israel: [5]

Country and year	Rate
Chile (1961)	12.9
Costa Rica (1960)	8.6
El Salvador (1961)	11.3
England and Wales (1961)	11.9
Federal Republic of Germany (1961)	11.0
Israel (1960)	5.7
Japan (1960)	7.6
Puerto Rico (1960)	6.7
Sweden (1960)	10.0
Taiwan (1960)	6.9
United States (1960)	9.5
Yugoslavia (1961)	9.0

In various countries of Africa, reported death rates may be quite high, even though death registration may be incomplete:

Algeria (1960)	16.0
Dahomey (1960–61)	26.0
Gabon (1960–61)	30.0
Senegal (1960–61)	16.7
United Arab Republic (1960)	16.9

The population base may also be low, however. Other countries may have had slightly higher or lower rates than those shown above. The range of historical variation has been much greater, however. In the United States, the crude death rate was around 25 in 1800 and about 9.4 in 1967.

Crude death rates may be computed for any period, but typically they are computed for the calendar year or the "fiscal" year, i.e., the 12-month period from July 1 to June 30. In the latter case, the population figure should relate to January 1 of the fiscal year. Crude death rates are calculated for 12-month periods such as calendar years or fiscal years so as to eliminate the effect of seasonal or monthly variations on the comparability of the rates. In the calculation of the crude death rate, as well as of other measures of mortality, for the census year, the census counts of population are commonly employed, even though the census may have been taken as of some date in the year other than July 1.

---

[5] United Nations, *Demographic Yearbook, 1966*, table 17. See source for special notes.

Sometimes an annual average crude death rate covering data for two or three years is computed in order to represent the longer period with a single figure, or to add stability to rates based on small numbers or intended for use in extensive comparative analysis. These annual average rates may be computed in several different ways. Formulas are given for 3-year averages. In one procedure, crude rates are computed for each year and averaged, i.e.,

$$\frac{1}{3}\left[\frac{D_1}{P_1}\times 1{,}000+\frac{D_2}{P_2}\times 1{,}000+\frac{D_3}{P_3}\times 1{,}000\right] \qquad (2)$$

The average annual death rate for 1959–61 for Costa Rica according to this formula is:

$$\frac{1}{3}\left[\frac{10{,}176}{1{,}126{,}000}\times 1{,}000+\frac{10{,}063}{1{,}171{,}000}\times 1{,}000+\frac{9{,}726}{1{,}225{,}000}\times 1{,}000\right]$$
$$=\frac{1}{3}(9.0+8.6+7.9)=8.5$$

This procedure gives equal weight to the rates for the three years. In the second procedure the average number of deaths is divided by the average population:

$$\frac{1/3\ (D_1+D_2+D_3)}{1/3\ (P_1+P_2+P_3)}\times 1{,}000 \qquad (3)$$

The 1959–61 rate for Costa Rica according to this formula is also 8.5:

$$\frac{1/3(10{,}176+10{,}063+9{,}726)}{1/3(1{,}126{,}000+1{,}171{,}000+1{,}225{,}000)}\times 1{,}000$$
$$=\frac{1/3(29{,}965)}{1/3(3{,}522{,}000)}\times 1{,}000=\frac{9{,}988}{1{,}171{,}000}\times 1{,}000=8.5$$

This procedure in effect weights the three annual crude death rates in relation to the population in each year and, hence, may be viewed as more exact than the first procedure for measuring mortality in the period. In a third procedure the average number of deaths for the three years is divided by the midperiod population:

$$\frac{1/3\ (D_1+D_2+D_3)}{P_2}\times 1{,}000 \qquad (4)$$

This is the most commonly used and most convenient procedure and, in effect, assumes that P_2 represents the average population exposed to risk of death in this period, or the average annual number of "person-years of life" lived in the period (ch. 15). The assumption of an arithmetic progression in the population over the 3-year period is consistent with this general assumption. Normally, annual population figures are not available for small areas; there are only the census counts. According to this formula the average annual death rate for 1959–61 for Costa Rica is:

$$\frac{1/3(10{,}176+10{,}063+9{,}726)}{1{,}171{,}000}\times 1{,}000$$
$$=\frac{1/3(29{,}965)}{1{,}171{,}000}\times 1{,}000=\frac{9{,}988}{1{,}171{,}000}\times 1{,}000=8.5$$

Note that all three formulas give the same result to a single decimal in the present illustration. Similar results may be ex-pected from the three procedures unless there are sharp fluctuations in the size of the population.

The "crude death rate" has a specific and a general meaning. In its specific meaning, it refers to the general death rate for the total population of an area, corresponding to the definition given earlier. More generally, however, it may be used to refer to the general death rate for any population group in an area, such as the male population, the native population, or the urban population of a country. For example, the crude death rate for the rural population of El Salvador in 1962 is represented by:

$$\frac{D_r}{P_r}\times 1{,}000=\frac{16{,}383}{1{,}784{,}371}\times 1{,}000=9.2 \qquad (5)$$

The formula calls for dividing the deaths in rural areas during a year by the midyear population of rural areas. In this more general sense, the principal characteristic of a "crude" rate is that **all ages** are represented in the rate.

The crude death rate should be adjusted for underregistration of the deaths and for underenumeration of the population if satisfactory estimates of these errors are available, according to the following formula:

$$\frac{D\div C_d}{P\div C_p}\times 1{,}000 \qquad (6)$$

where C_d represents the percent completeness of registration of deaths and C_p represents the percent completeness of the census counts or population estimates (as decimals). Generally, in the underdeveloped countries or in the earlier stages of the development of a death registration system in a country, the extent of underreporting of deaths is much greater than the extent of underenumeration of the population in the census. In the statistically more developed situation, the registration of deaths has usually improved to the point where the completeness of death registration exceeds the completeness of census enumeration. It may be prudent to correct neither the deaths nor the population when the percent completeness is believed to be about the same for each. If only death corrections are available, it is advisable to apply them only if they are much larger than the presumed understatement of the population figures; and the same is true in the opposite situation.

Monthly Death Rate. — There is an interest in examining variations in mortality over periods shorter than a year. Monthly death rates computed directly would not be comparable from month to month, however, because of differences in the number of days. Ordinarily, therefore, monthly figures on deaths are converted to an annual basis before rates are computed. Monthly deaths may be "annualized" by inflating the number of deaths for a given month (D_m) by the ratio of the total number of days in the year to the number in the particular month (n_m); the corresponding rate is then computed by dividing the annualized number of deaths by the population for the month.

$$\frac{\dfrac{365}{n_m}\times D_m}{P_m}\times 1{,}000 \qquad (7)$$

where n_m represents the number of days in the month. The

annualized death rate in the United States for June 1960 is obtained as follows:

$$\frac{\frac{365}{30} \times 132,461}{179,788,000} \times 1,000 = \frac{1,611,609}{179,788,000} \times 1,000 = 9.0$$

Monthly death rates are also affected by seasonal variation and excess mortality due to epidemics. It is desirable to remove these influences also from monthly data if the underlying trend of the death rate is to be ascertained. Relatively severe periods of epidemic influenza may distort the variations due to seasonality and the trend in death rates, so it is considered inadvisable to use the observed death rates for these epidemic months. Instead, substitute rates are calculated serving as estimates of what might have been expected had not epidemic conditions intervened. A monthly death rate adjusted for epidemic excess may be further adjusted for seasonal variation by dividing the adjusted rate for the month by an appropriate seasonal adjustment factor for the month. The variations in the monthly crude death rate due to seasonality result largely from some very large differences in seasonality by causes of death. In the United States today the monthly curve of the crude death rate is shaped to an important extent by the fluctuations characteristic of the respiratory causes of death and the monthly pattern of the major cardiovascular-renal diseases. Several causes, such as certain diseases of early infancy and motor vehicle accidents, exhibit seasonal patterns unlike that of the overall crude rate. In many other countries (and in the United States until well into the twentieth century), such diseases of childhood as diarrhea and enteritis made a great contribution to seasonality.

Various procedures for calculating seasonal indexes are described in the standard statistical texts, and particular procedures that may be used for calculating seasonal indexes for deaths are described in various journal articles.[6]

Another procedure which permits analysis of changes on a monthly basis involves direct calculation of annual death rates. Rates are computed for successive 12-month periods ending in each successive month; that is, for example, from January 1966 to December 1966, from February 1966 to January 1967, from March 1966 to February 1967, etc. The deaths for each 12-month period are divided by the population at the middle of the period, to derive the death rates. In this way the monthly "trend" of the death rate can be observed and analyzed on a current basis without distortion by seasonal variations and without need to apply seasonal adjustment factors. Rates of this kind based on provisional data are published monthly with a lag of less than two months following the close of a particular 12-month period by the U.S. National Center for Health Statistics for the United States.[7]

Year-to-year percentage changes in the death rate on a current basis can also be measured by comparing the rates for the same month in successive years. Such rates do not require adjustment for the number of days in the month (except for February) or for seasonal variations. Annual relative changes can also be measured on a current basis by comparing cumulative rates covering all the months in a year to date with the corresponding period in the preceding year; that is, for example, January to May 1967, and January to May 1966; January to June 1968, and January to June 1967, etc.

Specific Death Ratios and Rates. — The crude death rate gives only a very general indication of the level of mortality and its changes. There is also need for measures that describe the specific components of the overall number of deaths and the crude rate. Various types of specific death ratios and rates are of interest both in themselves and for their value in the analysis of the total number of deaths and the crude rate.

Specific death ratios and rates refer to specific categories of deaths and population. These categories may be subdivisions of the population or deaths according to sex, age, occupation, educational level, ethnic group, cause of death, etc. We have already described the general (all ages) death rates for certain very broad classes of the population as crude rates (e.g., crude rate for male population, urban population, native population). These same general rates, however, may also be viewed as specific rates (e.g., sex-specific, urban-rural specific, nativity-specific, etc.). Except in the case of age, which always defines a rate as specific, there is, then, no clear dividing line between a crude rate and a specific rate; a rate is understood to be "crude" unless specifically indicated otherwise, or unless it is specific with respect to age.

Age-specific death rates. — As we have stated, age is the most important variable in the analysis of mortality. Most tabulations of deaths require cross-classification with age if they are to be useful. The United Nations recommends tabulation of deaths by age, sex, and cause in the following age detail: Under 1, 1, 2, 3, 4, 5-year age groups to 80–84, 85 and over, and not stated.[8] The United Nations also recommends tabulations of infant deaths by age (under 28 days, under 1 year, not stated) and month of occurrence; of infant deaths by sex and detailed age (under 1 day, 1, 2, 3, 4, 5, 6 days; 7 to 13, 14 to 20, 21 to 27, 28 days to under 2 months; 2, 3, 4, ... 11 months; not stated); and of total infant deaths by cause of death.

The tabulations of deaths by age are subject to a number of deficiencies. The principal ones are substantial and variable underregistration of deaths by age, extensive misreporting of the age of the decedent, and an excessive proportion of "age not stated."

Reporting of age of decedents among the extreme aged (say, 5-year age groups 85 and over) is believed to be quite inaccurate. The exact or even approximate age of most decedents at these ages is not known to surviving relatives, friends, or neighbors, and their report tends to be a guess, with a tendency toward exaggeration of age. Because of serious errors in the population at these ages also, observed death rates among the extreme aged are highly unreliable.

A distribution of deaths by age typically shows a bimodal pattern, as illustrated in figure 14–1 and table 14–1 with recent data for several countries. The number and proportion of deaths is at a peak at age under 1 year; there are a trough at about ages 5 to 14 and a second peak at about ages 65 to 74,

[6] U.S. Public Health Service, *Public Health Reports*, Vol. 80, No. 3, "Seasonal Adjustment of Vital Statistics by Electronic Computer," by Harry M. Rosenberg, March 1965, pp. 201–210; Ira Rosenwaike, "Seasonal Variation of Deaths in the United States, 1951–60," *Journal of the American Statistical Association*, 61(315):706–719, September 1966; and Harry M. Rosenberg, "Recent Developments in Seasonally Adjusting Vital Statistics," *Demography*, 3(2):305–318, 1966.

[7] U.S. National Center for Health Statistics, *Monthly Vital Statistics Report (Provisional Statistics)*, "Births, Marriages, Divorces, and Deaths for [month] and year]."

[8] United Nations, *Principles for a Vital Statistics System*, p. 21.

after which the numbers fall off rapidly to the terminal ages of the life span. A combination of high rates and a large proportion of children in the population, as occurs in many underdeveloped countries, results in a tremendous proportion of deaths among children under 5 (e.g., 53 percent in Costa Rica in 1963); on the other hand, under the opposite conditions, as in certain developed countries, the proportion of deaths of young children. is very small (e.g., 3 percent in Sweden in 1960). At the older ages, the opposite relations prevail; the proportion of deaths 65 and over may vary from less than one-fifth to over two-thirds of the total.

The figures in table 14-1 describe the relative importance of deaths at each age, allowing jointly for the effect of the distribution of the population by age and the variation in death rates from age to age. In the absence of counts or estimates of population by age, calculation of the percent distribution of deaths by age serves as a simple way of analyzing the role of age as a factor in mortality. To the extent that deaths at each age have the same percent net reporting error, proportions of deaths by age are relatively free of error and are valid for international comparisons.

The principal way of measuring the variation in mortality by age, however, is in terms of age-specific death rates. An age-specific death rate is defined conventionally as the number of deaths of persons of a given age during a year (D_a) per 1,000

of the midyear population at that age (P_a), that is,

$$\frac{D_a}{P_a} \times 1{,}000 \qquad (8)$$

Table 14-1. — **Percent Distribution of Deaths by Age, for Various Countries: Around 1960**

Age	Ceylon, 1961[1]	Costa Rica, 1963	Mexico, 1960[2]	Sweden, 1960	United States, 1960
Total[3].........	100.0	100.0	100.0	100.0	100.0
Under 1 year........	23.2	39.6	29.7	2.3	6.5
1 to 4 years........	13.6	13.1	16.7	0.5	1.0
5 to 14 years.......	5.1	3.9	5.1	0.6	1.0
15 to 24 years......	3.7	2.7	4.4	1.0	1.5
25 to 34 years......	4.2	3.0	5.1	1.2	2.0
35 to 44 years.....	4.6	3.7	5.5	2.6	4.2
45 to 54 years.....	6.0	4.8	6.0	6.2	9.1
55 to 64 years.....	8.5	6.7	7.6	13.4	15.8
65 to 74 years.....	11.1	8.9	7.9	25.6	24.6
75 to 84 years.....	10.1	8.8	6.8	32.8	23.7
85 years and over...	9.8	4.9	5.2	13.9	10.8

[1] Data are from civil registers which are incomplete or of unknown reliability.
[2] Data tabulated by year of registration rather than occurrence.
[3] Deaths of unknown age excluded from total.

Sources: United Nations, *Demographic Yearbook, 1966*, table 18, and *1962*, table 19; and U.S. National Center for Health Statistics, *Vital Statistics of the United States, 1960*, Vol. II, *Mortality*, Part A, 1963, tables 5–9.

Figure 14-1. — **Percent Distribution of Deaths by Age, for Selected Countries: Around 1960**

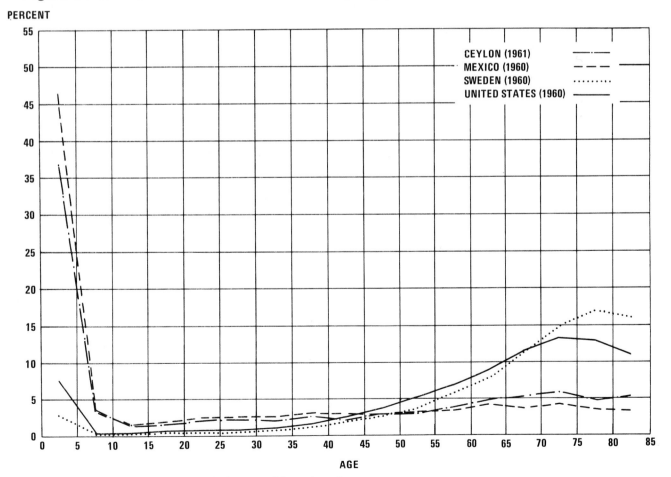

NOTE. — Points are plotted for 5-year age groups in the center of the age interval.
Source: See table 14-1.

Age-specific death rates are usually computed for 5- or 10-year age groups; but, because of the relatively great magnitude of the death rates among infants, separate rates are usually shown for the age groups under 1 and 1 to 4.

The calculation of a set or schedule of age-specific death rates is illustrated for Costa Rica in 1963 in table 14–2. The rate for the age group 75 and over is derived by dividing deaths at ages 75 and over by the population aged 75 and over:

$$\frac{1,544}{15,205} \times 1,000 = 101.5$$

This is a type of central rate, i.e., a rate relating events in a given category during a year to the average population of the year in the category. The average population is ordinarily taken as the midyear population but, often, the census counts are employed in the calculation of rates for the census year. (Two- or three-year average age-specific death rates may be computed in the same way as average crude death rates, using the mid-period population.)

Table 14–2 also illustrates the calculation of changes in age-specific death rates from one date to another. Absolute changes and percent changes are shown. An increase is indicated by a plus (+) sign and a decrease by a minus (−) sign. In calculating the percent change (col. 6), the absolute change (col. 5) is divided by the rate at the earlier date (col. 4). Because of the often very low level of death rates at the younger ages above earliest childhood, and the very high level of death rates at the older ages, absolute figures tend to "minimize" the indication of change at the younger ages and to "maximize"

Table 14–2. — Calculation of Age-Specific Death Rates for Costa Rica, 1963, and of Change in Rates, 1950 to 1963

Age	1963			1950 Death rates	Change in death rates	
	Population[1]	Deaths			Amount	Percent
		Number	Rates [(2)÷(1)] x1,000=		(3)-(4)=	[(5)÷(4)] x100=
	(1)	(2)	(3)	(4)	(5)	(6)
Total..........	[2]1,336,274	[2]11,376	8.5	12.2	-3.7	-30
Under 1 year.......	51,175	4,456	87.1	111.0	-23.9	-22
1 to 4 years.......	197,772	1,476	7.5	15.5	-8.0	-52
5 to 14 years......	387,718	438	1.1	2.1	-1.0	-48
15 to 24 years.....	233,350	303	1.3	2.3	-1.0	-43
25 to 34 years.....	161,926	336	2.1	3.9	-1.8	-46
35 to 44 years.....	121,119	414	3.4	6.6	-3.2	-48
45 to 54 years.....	86,679	540	6.2	10.7	-4.5	-42
55 to 64 years.....	52,185	758	14.5	23.5	-9.0	-38
65 to 74 years.....	27,045	1,001	37.0	52.2	-15.2	-29
75 years and over..	15,205	1,544	101.5	126.9	-25.4	-20

[1] Census population as of April 1, 1963.
[2] Includes unknowns.

Source: United Nations, *Demographic Yearbook, 1966*, tables 5 and 18; *1957*, table 10; and *1955*, table 10.

it at the older ages, whereas percents tend to have the opposite effect.[9]

It is important to recognize the relation of a crude death rate to the underlying age-specific death rates. A crude rate may be viewed as the weighted average of a set of age-specific death rates, the weights being the proportions of the total

population at each age, i.e., as the cumulative product of a set of age-specific death rates and an age distribution for a given population.

$$\sum_{a=0}^{\infty} \left(\frac{D_a}{P_a}\right)\left(\frac{P_a}{P}\right) = \sum_{a=0}^{\infty} \frac{\left(\frac{D_a}{P_a} \times P_a\right)}{P} = \frac{\sum_{a=0}^{\infty} D_a}{P} = \frac{D}{P} \qquad (9)$$

where D and P represent total deaths and population, respectively, D_a and P_a, deaths and population at each age, respectively, D_a/P_a an age-specific death rate, and P_a/P the proportion of the population in an age group. Similarly, the change in the crude death rate between two dates may be viewed as a joint result of changes in age-specific rates and changes in age composition. A crude death rate may have declined because age-specific death rates tended to fall or because the proportion of the population in childhood and the young adult ages, where age-specific death rates are low, increased.

Similarly, age-specific death rates may vary from country to country in a quite different degree from the corresponding crude death rates. The various age-specific death rates may generally be much higher in one country than in another while the crude rates in the two countries may be nearly the same or have an opposite relation.

Death rates may be made specific for sex and for other characteristics in combination with age. An age-sex specific death rate is defined as the number of deaths of males or females in a particular age group per 1,000 male or female population in the age group:

$$\frac{D_a^m}{P_a^m} \times 1,000, \text{ or } \frac{D_a^f}{P_a^f} \times 1,000 \qquad (10)$$

Next to age, sex may be considered the most important variable in the analysis of mortality. In fact, age-specific death rates are almost always sex-specific also. A comparison of male and female death rates by age is given in the form of the percent excess of the male rate over the female rate for a few countries in table 14–3. Death rates of males show a marked excess over death rates for females in all or most age groups, particularly in areas of low mortality. The lower the general level of mortality the greater the excess of male over female mortality tends to be. Thus, the gap between the rates for the sexes tends to be smaller in the less developed countries. It may even be completely erased or the female rates may be higher in older childhood or the reproductive ages, as is true in a number of countries such as Ceylon, India, and Pakistan. Generally, too, the time trends in the differences between male and female age-specific death rates have followed the above patterns as mortality declined. A number of studies have considered the causes of the sex differentials in mortality.[10]

Death rates may be limited on the basis of other characteristics in addition to age and sex, for example, nativity, race,

---

[9] In computing percent differences between death rates, particularly by age, because the base is often small, special attention should be given to the number of significant digits to be shown in the result, taking account of the number of significant digits in the basic rates.

[10] See, for example, Francis C. Madigan, "Are Sex Mortality Differentials Biologically Caused?", *Milbank Memorial Fund Quarterly*, 35(2):202–223, April 1957; G. Herdan, "Causes of Excess Male Mortality in Man," *Acta Genetica et Statistica Medica* (Basel), 3(4):351–376, 1952; W. J. Martin, "A Comparison of the Trends of Male and Female Mortality," *Journal of the Royal Statistical Society*, Series A (General), 114(3):287–306, 1951; S. Koller, "The Development of the Excess of Male Mortality," *International Population Conference, New York, 1961*, Vol. 1, International Union for the Scientific Study of Population, London, 1963, pp. 675–684; and U.S. Public Health Service, *Public Health Reports*, Vol. 69, No. 9, "Why is the Sex Difference in Mortality Increasing?", by Wilson T. Sowder, September 1954, pp. 860–864. See also Philip Enterline, "Causes of Death Responsible for Recent Increases in Sex Mortality Differentials in the United States," *Milbank Memorial Fund Quarterly*, 39(2):312–320, April 1961; and Robert D. Retherford, "Tobacco Smoking and the Sex Mortality Differential," *Demography*, 9(2):203–216, May 1972.

Table 14–3.—Comparison of Male and Female Age-Specific Death Rates, for Various Countries: Around 1960

Age	United States, 1960			El Salvador, 1961			Taiwan, 1960 [1]		
	Male	Female	Excess of male rate [2] (2)-(1)=	Male	Female	Excess of male rate [2] (4)-(5)=	Male	Female	Excess of male rate [2] (7)-(8)=
	(1)	(2)	(3)	(4)	(5)	(6)	(7)	(8)	(9)
All ages...............	11.0	8.1	2.9	12.3	10.4	1.9	7.4	6.5	0.9
Under 5 years...............	7.1	5.5	1.6	33.9	30.7	3.2	13.0	12.8	0.2
Under 1 year...............	30.6	23.2	7.4	98.9	80.7	18.2	34.1	30.7	3.4
1 to 4 years...............	1.2	1.0	0.2	16.7	16.3	0.4	7.6	8.2	-0.6
5 to 14 years...............	0.6	0.4	0.2	2.5	2.5	–	1.1	0.9	0.2
15 to 24 years...............	1.5	0.6	0.9	3.6	2.3	1.3	2.1	1.5	0.6
25 to 34 years...............	1.9	1.1	0.8	5.7	3.3	2.4	2.9	2.2	0.7
35 to 44 years...............	3.7	2.3	1.4	7.0	5.2	1.8	4.9	3.5	1.4
45 to 54 years...............	9.9	5.3	4.6	10.8	8.9	1.9	10.1	6.4	3.7
55 to 64 years...............	23.1	12.0	11.1	19.2	18.8	0.4	25.4	15.5	9.9
65 to 74 years...............	49.1	28.7	20.4	41.1	33.7	7.4	61.8	41.1	20.7
75 years and over...........	118.5	96.6	21.9	86.7	77.5	9.2	157.9	121.6	36.3

– Represents zero.
[1] Data tabulated by year of registration rather than year of occurrence.
[2] Absolute difference in the rates. A minus sign denotes an excess of the female rate.

Sources: U.S. National Center for Health Statistics, *Vital Statistics of the United States, 1960*, Vol. II, *Mortality*, Part A, 1963, table I–E; and United Nations, *Demographic Yearbook, 1966*, tables 5 and 18, and *1961*, tables 5 and 15.

various socioeconomic characteristics, and cause of death. Some other types of specific death rates are considered in the next sections.

Cause-specific death ratios and rates.—An important aspect of the analysis of mortality relates to the classification by cause of death. The classification of deaths according to cause employed by national governments generally follows the International Classification of Diseases as promulgated by the World Health Organization. The classification has regularly gone through a decennial revision under the auspices of the WHO or its predecessor organizations in order that it may be consistent with the latest diagnostic practice and with medical advances. The Eighth Revision of the International Classification was adopted by the World Health Organization in 1965 and was first employed in modified form in the United States in 1968.[11] The Seventh Revision of the International Classification was adopted by the World Health Organization in 1955 and was used for compiling mortality data in the United States from 1958 to 1967.[12] The Sixth Revision was adopted by the World Health Organization in 1948 and was used for compiling mortality data in the United States from 1949 through 1957.

Comparability of cause-of-death statistics has been affected by each decennial revision of the International Classification.

The Manual for the Eighth Revision, like that for the Seventh Revision, includes several special lists recommended for mortality tabulations: the Detailed List, consisting of all three-digit categories; List A, the Intermediate List of 150 Causes; and List B, the Abbreviated List of 50 Causes. The U.S. Public Health Service has adapted these lists in the form of three alternative lists for purposes of tabulation and publication in the volumes of *Vital Statistics in the United States* for 1968 and later years.

According to the Eighth Revision of the International Classification, the causes of death are classified under 17 major headings, several of which represent the major physiological systems in the body. The major classes included, with their code numbers from the Detailed List, are as follows:

I Infective and parasitic diseases (000–136)
II Neoplasms (140–239)
III Endocrine, nutritional, and metabolic diseases (240–279)
IV Diseases of the blood and blood-forming organs (280–289)
V Mental disorders (290–315)
VI Diseases of the nervous system and sense organs (320–389)
VII Diseases of the circulatory system (390–458)
VIII Diseases of the respiratory system (460–519)
IX Diseases of the digestive system (520–577)
X Diseases of the genitourinary system (580–629)
XI Complications of pregnancy, childbirth, and the puerperium (630–678)
XII Diseases of the skin and cellular tissue (680–709)
XIII Diseases of the musculoskeletal system and connective tissue (710–738)
XIV Congenital anomalies (740–759)
XV Certain causes of mortality in early infancy (760–779)
XVI Symptoms and ill-defined conditions (780–796)
E XVII Accidents, poisonings, and violence (external cause) (E800–E999)
or
N XVII Accidents, poisonings, and violence (nature of injury) (N800–N999)

List B, the Abbreviated List of 50 Causes, with code numbers from the Detailed List, identification of the major class to which each listed cause belongs, and comparability ratios between the Eighth and Seventh Revisions for selected causes (based on U.S. data), is presented in table 14–4.

[11] World Health Organization, *Manual of the International Statistical Classification of Diseases, Injuries, and Causes of Death*, 1965 Revision, Vols. I and II, 1967; and National Center for Health Statistics, *Eighth Revision, International Classification of Diseases, Adapted for Use in the United States*, Vols. I and II, 1967.
[12] World Health Organization, *Manual of the International Statistical Classification of Diseases, Injuries, and Causes of Death*, 1955 Revision, Vol. 1, 1957.

Table 14–4.—List B, Abbreviated List of 50 Causes of Death According to the Eighth (1965) Revision of the International Classification of Diseases, and Comparability Ratios between the Eighth and Seventh Revisions for Selected Causes

Major class and abbreviated list numbers	Cause of death	Detailed list numbers	Comparability ratios[1]	Major class and abbreviated list numbers	Cause of death	Detailed list numbers	Comparability ratios[1]
I				**VIII**			
B 1.	Cholera	000		B31.	Influenza	470–474	0.960
B 2.	Typhoid fever	001		B32.	Pneumonia	480–486	0.994
B 3.	Bacillary dysentery and amoebiasis	004, 006		B33.	Bronchitis, emphysema, and asthma	490–493	0.994
B 4.	Enteritis & other diarrhoeal diseases	008, 009	1.185	**IX**			
B 5.	Tuberculosis of respiratory system	010–012	} 0.950	B34.	Peptic ulcer	531–533	0.979
B 6.	Other tuberculosis, including late effects	013–019		B35.	Appendicitis	540–543	
B 7.	Plague	020		B36.	Intestinal obstruction and hernia	{ 550–553, 560	} 0.757
B 8.	Diptheria	032		B37.	Cirrhosis of liver	571	1.003
B 9.	Whooping cough	033		**X**			
B10.	Streptococcal sore throat and scarlet fever	034		B38.	Nephritis and nephrosis	580–584	0.880
B11.	Meningococcal infection	036		B39.	Hyperplasia of prostate	600	0.904
B12.	Acute poliomyelitis	040–043		**XI**			
B13.	Smallpox	050		B40.	Abortion	640–645	
B14.	Measles	055		B41.	Other complication of pregnancy, childbirth and the puerperium. Delivery without mention of complication	{ 630–639, 650–678	
B15.	Typhus and other rickettsioses	080–083		**XIV**			
B16.	Malaria	084		B42.	Congenital anomalies	740–759	1.008
B17.	Syphilis and its sequelae	090–097		**XV**			
B18.	All other infective and parasitic diseases	Remainder of 000–136		B43.	Birth injury, difficult labour and other anoxic and hypoxic conditions	} 764–768, 772, 776	
II				B44.	Other causes of perinatal mortality	{ 760–763, 769–771, 773–775, 777–779	
B19.	Malignant neoplasms, including neoplasms of lymphatic and haemotopoietic tissues	140–209	0.988	**XVI**			
B20.	Benign neoplasms and neoplasms of unspecified nature	210–239	0.968	B45.	Symptoms and ill-defined conditions	780–796	0.994
III				B46.	All other diseases	{ Remainder of 240–738	
B21.	Diabetes mellitus	250	0.994	**EXVII**			
B22.	Avitaminoses and other nutritional deficiency	260–269		BE47.	Motor vehicle accidents	E810–E823	0.974
IV				BE48.	All other accidents	{ E800–E807, E825–E949	} 0.884
B23.	Anemias	280–285	0.944	BE49.	Suicide and self-inflicted injuries	E950–E959	0.939
VI				BE50.	All other external causes	E960–E999	0.993
B24.	Meningitis	320	0.959	**NXVII**			
VII				BN47.	Fractures, intracranial and internal injuries	} N800–N829, N850–N869	
B25.	Active rheumatic fever	390–392	} 1.138	BN48.	Burn	N940–N949	
B26.	Chronic rheumatic heart disease	393–398		BN49.	Adverse effects of chemical substances	N960–N989	
B27.	Hypertensive disease	400–404	1.063	BN50.	All other injuries	{ Remainder of N800–N999	
B28.	Ischemic heart disease	410–414					
B29.	Other forms of heart disease	420–429					
B30.	Cerebrovascular disease	430–438	0.981				

[1] Based on U.S. data. From U.S. National Center for Health Statistics, *Monthly Vital Statistics Report*, Vol. 17, No. 8, Supplement, "Provisional Estimates of Selected Comparability Ratios Based on Dual Coding of 1966 Death Certificates by the Seventh and Eighth Revisions of the International Classification of Diseases," October 25, 1968, table 1.

Source: World Health Organization, *Manual of the International Statistical Classification of Diseases, Injuries, and Causes of Deaths*, Vol. I, 1967, pp. 445–446.

Three special aspects of the quality of cause-of-death tabulations are worth special consideration. These relate to the determination of the cause to be tabulated when more than one cause of death is reported, the problems of historical and international comparability, and the use of the category, "senility, ill-defined, and unknown causes" or "symptoms and ill-defined conditions." The World Health Organization's specification of the form of medical certification and of the coding procedures to be used in processing death certificates serves to deal with these problems. Since the Sixth Revision in 1948, the rule has been that the certification must make clear the train of morbid events leading to death and, when more than one cause of death is reported, the cause designated by the certifying medical attendant as the underlying cause of death is the cause tabulated.[13] This is the cause that the medical examiner judges to be the one that started the train of events leading directly to death. The rule is important in affecting the tabulations of deaths because a large proportion of death certificates filed may report more than one condition as the cause of death, as is true in the United States.[14]

The comparability of the tabulations of deaths by cause, from area to area, depends on the skill of the certifying medical attendant in diagnosing the cause of death and in describing

[13] J. Střiteský, M. Šantruček, and M. Vacek, Summary of "The Train of Morbid Events Leading Directly to Death—A Practical and Methodological Problem," in United Nations, *Proceedings of the World Population Conference, 1965* (Belgrade), Vol. II, New York, 1967, pp. 453–454.

[14] Guralnick, analyzing multiple causes of death in the United States in 1955, reports that the proportion of two-cause deaths was 34 percent and the proportion of deaths of three causes or more, 24 percent. See Lillian Guralnick, Summary of "Multiple Causes of Death, United States, 1955," in United Nations, *Proceedings of the World Population Conference, 1965* (Belgrade), Vol. II, New York, 1967, p. 452.

the cause on the death certificate, and, to a more limited extent, on the skill of the coder in classifying the cause on the basis of the information on the certificate. A study by Puffer and Griffith, which reexamined the certificates of deaths due to cardiovascular diseases in 12 cities (10 in Latin America; San Francisco, Calif.; and Bristol, England) in 1962–63, suggests considerable lack of comparability among tabulations for this important cause of death owing both to the medical certification and to the coding.[15] Other studies in the United States confirm this finding.[16] The comparability of tabulations would vary with the cause of death. International comparability is also affected by the fact that the year a revision of the International Classification is put into actual use by the member nations varies from country to country. The comparability of cause-of-death tabulations over time is affected not only by revisions in the classification scheme but also by improved diagnosis and changing "fashions" in diagnosis. These last factors may account, in part, for the recent rise in mortality attributable to the degenerative diseases of later life in the developed countries.

Although some deaths are extremely difficult to classify, care in the certification of cause of death keeps the number of deaths of ill-defined or unknown cause to a minimum. The proportion of deaths assigned to "symptoms, senility, and ill-defined conditions" (Seventh Revision category numbers 780–795 and Eighth Revision category numbers 780–796) may be employed as one measure of the quality of reporting cause of death. For example, this category accounts for only 1 percent of the deaths in Australia and the United States (1960), but for 18 percent in Colombia (1960), 28 percent in El Salvador (1961), and 56 percent in Thailand (1960). Quite clearly the cause-of-death tabulations for Thailand are almost useless, those for El Salvador are of limited value, and even the figures for Colombia must be used with caution.

Mortality by cause of death may be analyzed in terms of two observed measures, death ratios specific for cause and death rates specific for cause. The former measure requires simply a distribution of deaths by cause and hence can be computed for intercensal years even when population figures are lacking. A **cause-specific death ratio** represents the percent of all deaths due to a particular cause or group of causes. The death ratio for cause C is:

$$\frac{D_c}{D} \times 100 \qquad (11)$$

where D_c represents deaths from a particular cause or group of causes, and D represents all deaths. Death ratios by cause are shown for several countries in table 14–5. A set of death ratios readily permits comparisons from country to country, or from one year to another for the same country, of the relative importance of a particular cause or group of causes of death. For example, the death ratio for heart diseases is considerably lower in some countries than in others (e.g., 6 percent in Mexico, 1960, and 39 percent in the United States, 1960).

The death ratios for congenital malformations and diseases of early infancy tend to have the opposite relationship (e.g., 13 percent in Mexico and 5 percent in the United States).

The population exposed to risk of death from a particular cause or group of causes may be directly taken account of, in rough terms, by calculating "crude" cause-of-death rates (table 14–6). Separate death rates may be calculated for each cause or group of causes of death. A **cause-specific death rate** is conventionally defined as the number of deaths from a given cause or group of causes during a year per 100,000 of the midyear population.

$$\frac{D_c}{P} \times 100,000 \qquad (12)$$

A cause-specific death rate employs a larger constant, or k, factor than a crude death rate or an age-specific death rate because there are relatively few deaths from many of the causes.

It should be noted that a cause-specific death rate is unlike other types of death rates in one important respect; that is, the base population cannot logically be limited to the class defined in the numerator (a particular cause). In an age-specific death rate, for example, the deaths at a given age are related to the population **at that age.** In a cause-specific rate, however, the general population cannot be classified by cause of death even though our interest is in defining the population at risk of death from the given cause. (The conventional "case-fatality rate," for example, indicates the risk of dying from a particular cause during a period among persons who have incurred the illness or injury during the period. Another type of case-fatality rate could be defined in relation to the entire population suffering from a particular illness, injury, or disability.) This limitation can be partly overcome, and the usefulness of cause-specific death rates greatly increased, by making them age-specific or age-sex specific also, especially because many causes of death are largely or wholly confined to a particular age or age-sex group in the population. For example, deaths due to scarlet fever, measles, whooping cough, and diphtheria are largely confined to childhood; infantile diarrhea and enteritis, congenital malformations, immaturity, etc., are virtually confined to infancy; deaths due to diabetes, cerebrovascular lesions, nephritis, cirrhosis, etc., affect largely persons over 50. It may be argued that because certain causes have relevance only to a very restricted part of the age distribution or to only one sex, general rates for these causes have no significance and only age-sex-cause-specific rates should be computed.[17]

Cause-specific death rates have no particular advantage over death ratios in indicating the relative importance of the various causes of death for a particular area in a given year. Their advantage lies in comparisons from area to area of the relative frequency of a particular cause of death "adjusted for" the size of the population in each area. We can say, for example, that the death rate from heart disease in the United States (366.2 per 100,000 population) is over four times the rate in Chile (87.5). (See table 14–6). The cause-specific death rates in a set covering all causes sum to the crude death rate (per 100,000).

[15] Ruth R. Puffer and G. Wynne Griffith, "The Inter-American Investigation of Mortality," in the United Nations, *Proceedings of the World Population Conference, 1965* (Belgrade), pp. 426–432.
[16] Iwao M. Moriyama, Thomas R. Dawber, and William B. Kannel, "Evaluation of Diagnostic Information Supporting Medical Certification of Deaths from Cardiovascular Disease," in National Cancer Institute, Monograph No. 19, *Epidemiological Study of Cancer and Other Chronic Diseases,* January 1966, pp. 405–419. See also Iwao M. Moriyama and others, "Inquiry into Diagnostic Evidence Supporting Medical Certifications of Death," *American Journal of Public Health,* 48(10):1376–1387, October 1958.

[17] Even in its general tables, the U.N. *Demographic Yearbook* (1967) calculates the death rate from hyperplasia of prostate (B39:610) on the basis of the male population 50 and over, the death rate from deliveries and complications of pregnancy, childbirth, and the puerperium (B40:640–689) on the basis of the female population only, and the death rate for diseases peculiar to early infancy (B42–44:760–776) on the basis of births.

Table 14–5. —Deaths and Death Ratios Specific by Cause, for Various Countries: 1960

[Deaths are classified according to the Seventh (1955) Revision of the International Statistical Classification of Diseases, Injuries, and Causes of Death. Ratios are the number of deaths from each cause per 100 deaths from all causes]

Cause of death [1]	United States		Australia		Taiwan		Mexico		United Arab Republic [2]	
	Deaths	Ratios	Deaths	Ratios	Deaths	Ratios	Deaths	Ratios	Deaths	Ratios
All causes...................	1,711,982	100.0	88,464	100.0	72,485	100.0	402,545	100.0	223,107	100.0
Tuberculosis (001-019).............	10,866	0.6	489	0.6	4,927	6.8	9,525	2.4	2,504	1.1
Other infective and parasitic diseases (020-188).....................	10,521	0.6	472	0.5	3,663	5.1	35,200	8.7	5,942	2.7
Malignant neoplasms (140-205).............	267,627	15.6	13,299	15.0	4,250	5.9	12,516	3.1	3,355	1.5
Diabetes mellitus (260)..................	29,971	1.8	1,188	1.3	198	0.3	2,787	0.7	1,021	0.5
Heart diseases (400-443)...............	661,712	38.7	31,864	36.0	5,010	6.9	24,252	6.0	9,257	4.1
Influenza, pneumonia, bronchitis (480-502).	71,146	4.2	4,637	5.2	8,982	12.4	47,313	11.8	31,278	14.0
Deliveries and complications of pregnancy, childbearing, and the puerperium (640-689)	1,579	0.1	121	0.1	443	0.6	3,102	0.8	601	0.3
Congenital malformations and diseases of early infancy (750-776).............	88,954	5.2	3,923	4.4	5,504	7.6	50,692	12.6	23,446	10.5
Symptoms, senility, and ill-defined conditions (780-795)..................	20,385	1.2	894	1.0	6,916	9.5	48,943	12.2	30,242	13.6
All other diseases..................	427,872	25.0	24,931	28.2	26,636	36.7	142,908	35.5	107,761	48.3
Accidents, suicide, and homicide (E800-E999).....................	121,349	7.1	6,646	7.5	5,956	8.2	25,307	6.3	7,700	3.5

[1] Numbers in parentheses represent the code numbers in the Seventh (1955) Revision of the International Classification.
[2] Data are from civil registers which are incomplete or unreliable.

Source: United Nations, *Demographic Yearbook, 1966*, table 20.

In analyzing mortality in terms of cause of death, it is useful to distinguish two broad classes of deaths, designated as **endogenous** and **exogenous**, and to compute separate rates for these classes of deaths.[18] Generally, the former class of deaths is presumed to arise from the genetic makeup of the individual and from the circumstances of prenatal life and the birth process; and the latter class is presumed to arise from purely environmental or external causes. Endogenous mortality would include mortality from such causes as the degenerative diseases of later life (e.g., heart disease, cancer, diabetes) and certain diseases peculiar to early infancy (e.g., immaturity, birth injuries, postnatal asphyxia). Exogenous mortality would include mortality mainly from infections and accidents. Endogenous mortality has a typically biological character and is resistant to scientific progress, whereas exogenous mortality is viewed as relatively preventable and treatable.

The endogenous death rate is derived as follows:

$$\frac{D_{en}}{P} \times 1,000 \qquad (13)$$

where D_{en} represents deaths due to endogeneous causes and P represents midyear population. The exogenous death rate is derived as follows:

$$\frac{D_{ex}}{P} \times 1,000 \qquad (14)$$

where D_{ex} represents deaths due to exogenous causes and P represents midyear population. Table 14–7 presents statistics on deaths and death rates for endogenous and exogenous causes ("ill-defined conditions" and "all other diseases" of the B List being assigned in part to endogenous causes and in part to exogenous causes) for Norway, the United States, and Costa Rica. Data are shown separately for the total population, infants, aged persons, and persons of intermediate age. In Norway and the United States, where both crude and age-specific death rates are relatively low, exogenous causes contribute only a small part of the total number of deaths and the exogenous death rate is quite low. In Costa Rica, where age-specific death rates are relatively high, exogenous causes contribute a large part of the total number of deaths. Generally, endogenous causes dominate in infancy and exogenous causes dominate in adolescence and early adulthood. The exogeneous causes fall off with increasing age until there is a relative dominance again of endogenous causes in old age. The lower the general level of mortality, the more pronounced is the relative role of endogenous causes.

Death rates specific by social or economic characteristics. — Important variations in mortality are or may be associated with a number of social and economic characteristics of the decedent, among them ethnic group, marital status, educational attainment, occupation, income, and socioeconomic class. Such variations have implications for an understanding of the physical and sociological factors in health and for the planning of public health and other public welfare programs. Because the age and sex distribution of the specific subgroups of the population classified by a socioeconomic characteristic (e.g., single vs. widowed persons) may be quite different and some age groups hardly figure in the distributions (e.g., population under 15 in an occupational classification), it is desirable to calculate the death rates for socioeconomic groups separately by age and sex. We will give direct consideration here to death rates for only a few of the possible characteristics, particularly marital status, occupation, and socioeconomic status.

[18] Jean Bourgeois-Pichat, "Essai sur la Mortalité 'Biologique' de l'Homme," *Population* (Paris), 7(3):381–394, July-Sept. 1952.

Table 14–6. — Death Rates Specific by Cause, for Various Countries: 1960

[Causes of death are classified according to the Seventh (1955) Revision of the International Statistical Classification of Diseases, Injuries, and Causes of Death. Rates are the number of deaths from each cause per 100,000 of the population]

Cause of death[1]	United States	Australia	Taiwan	Mexico	Netherlands	Chile[2]	Colombia[2,3]
All causes........................	947.5	861.0	683.0	1,150.5	762.1	1,241.9	1,187.7
Tuberculosis (001-019).................	6.0	4.8	46.4	27.2	2.8	52.5	26.4
Other infective and parasitic diseases (020-138)........................	5.8	4.6	34.5	100.6	4.4	54.8	111.1
Malignant neoplasms (140-205)...........	148.1	129.4	40.0	35.8	168.4	99.5	44.1
Diabetes mellitus (260)................	16.6	11.6	1.9	8.0	15.1	7.2	4.6
Heart diseases (400-443)................	366.2	310.1	47.2	69.3	208.6	87.5	67.3
Influenza, pneumonia, and bronchitis (480-502)........................	39.4	45.1	84.6	135.2	42.4	247.1	142.0
Deliveries and complications of pregnancy, childbearing, and the puerperium (640-689)................	0.9	1.2	4.2	8.9	0.8	10.3	9.5
Congenital malformations and diseases of early infancy (750-776).............	49.2	38.2	51.9	144.9	31.5	194.4	116.0
Symptoms, senility, and ill-defined conditions (780-795).................	11.3	8.7	65.2	139.9	33.1	109.0	207.5
All other diseases...................	236.8	242.6	251.0	408.4	211.2	292.0	383.0
Accidents, suicide, and homicide (E800-E999)........................	67.2	64.7	56.1	72.3	43.6	87.5	76.1
Population........................	180,684,000	10,275,000	10,612,000	34,988,000	11,487,000	7,689,000	15,416,000

[1] Numbers in parentheses represent the code numbers in the Seventh Revision of the International Classification.
[2] Data tabulated by year of registration rather than occurrence.
[3] Data are from civil registers which are incomplete or unreliable.

Source: Table 14–5 and United Nations, *Demographic Yearbook, 1966*, table 20.

Death rates specific for marital status, age, and sex are useful not only for the analysis of mortality patterns but also for the study of the dissolution of marriage. To compute such rates, both deaths distributed by age, sex, and marital status and the population distributed by these characteristics are needed.

Such rates are affected by differences in the errors in, and by the comparability of, the information on marital status on death certificates and in the census. These are general problems affecting vital rates for socioeconomic groupings of the population since the numerator of the rate comes from the vital statistics system and the denominator comes from the census.

Few data of a satisfactory quality on deaths by occupation and socioeconomic status are available. The principal tabulations of deaths for occupational categories have been made for the United States and the United Kingdom, although several other countries have compiled such data. National mortality data by occupation have appeared decennially from 1851 to date for England and Wales, and such data have been tabulated for scattered years beginning in 1890 for the United States.

Death statistics and death rates by occupation are subject to a number of important types of errors. They are subject to the errors of death statistics in general and to errors in the reporting, coding, and classification of occupation. Whenever a high proportion of deaths is unclassified or not clearly classified by occupation, the quality of the statistics on deaths is in question. Like death rates by marital status, death rates by occupation are affected by differences in the nature and accuracy of the information returned on the death certificate as compared with that returned in the population census. The problem of disagreement between reporting of occupations in the two sources can be sharply reduced by combining occupations into major groups. The tabulations should be limited to males; the corresponding data for females are almost useless because of the relatively small proportion of women who work.

In the United States the lack of correspondence between numerator and denominator of the death rate is a result chiefly of the large proportion of decedents reported as "not in the labor force" or "retired" in the census (current activity) for whom a report of usual occupation is given on the death certificate. This problem can largely be eliminated by limiting the tabulations to the age range 20 to 59 or 20 to 64.

A major study of the variation of mortality by occupation and industry among men 20 to 64 years of age in the United States in 1950 resulted in the publication of a number of reports on this subject. Death rates (per 100,000 population) are presented for specific occupations, by age and color of decedent, and for specific industry groups, by age of decedent. Standardized mortality ratios are presented for specific occupations by color for major occupation groups, by selected causes of death and color; and for specific occupations and industries, by selected causes of death.[19]

A **standardized mortality ratio** for a particular occupation or industry among males compares the tabulated numbers of deaths in that occupation or industry with the number to be expected had the age-specific death rates for the total male population with work experience prevailed in that occupation or industry. The method of calculation is illustrated with

[19] U.S. Public Health Service, National Vital Statistics Division, *Vital Statistics—Special Reports*, Vol. 53, No. 2, "Mortality by Occupation and Industry Among Men 20 to 64 Years of Age: United States, 1950," September 1962; idem, *Vital Statistics—Special Reports*, Vol. 53, No. 3, "Mortality by Occupation and Cause of Death Among Men 20 to 64 Years of Age: United States, 1950," September 1963; idem, *Vital Statistics—Special Reports*, Vol. 53, No. 4, "Mortality by Industry and Cause of Death Among Men 20 to 64 Years of Age: United States, 1950," September 1963. All three reports were prepared by Lillian Guralnick.

Table 14–7. — Endogenous and Exogenous Death Rates for Norway, 1959, the United States, 1966, and Costa Rica, 1963, by Age

[Causes of death for Norway, 1959, are classified according to the Sixth (1948) Revision of the International Statistical Classification of Diseases, Injuries, and Causes of Death; causes of death for the United States, 1966, and Costa Rica, 1963, are classified according to the Seventh (1955) Revision]

Country, year, and cause of death	All ages			Under 1 year			1 to 64 years			65 years and over		
	Deaths	Population	Death rate [(1)÷(2)] x1,000=	Deaths	Births	Mortality rate[1] [(4)÷(5)] x1,000=	Deaths	Population	Death rate [(7)÷(8)] x1,000=	Deaths	Population	Death rate [(10)÷(11)] x1,000=
	(1)	(2)	(3)	(4)	(5)	(6)	(7)	(8)	(9)	(10)	(11)	(12)
Norway (1959):												
All causes.........	31,761		8.9	1,177		18.7	8,508		2.7	22,073		57.8
Endogenous.........	24,757	3,555,543	7.0	905	63,005	14.4	6,205	3,111,809	2.0	17,646	381,768	46.2
Exogenous..........	7,004		2.0	272		4.3	2,303		0.7	4,427		11.6
United States (1966):												
All causes.........	1,863,149		9.5	85,516		23.7	636,755		3.6	1,140,245		61.8
Endogenous.........	1,520,802	196,842,000	7.7	65,888	3,606,274	18.3	465,542	174,720,000	2.7	988,951	18,457,000	53.6
Exogenous..........	342,347		1.7	19,628		5.4	171,213		1.0	151,294		8.2
Costa Rica (1963):												
All causes.........	11,376		8.5	4,456		74.1	4,265		3.4	2,545		60.2
Endogenous.........	5,332	1,336,274	4.0	1,543	60,171	25.6	1,935	1,240,749	1.6	1,800	42,250	42.6
Exogenous..........	6,044		4.5	2,913		48.4	2,330		1.9	745		17.6

[1] See discussion of the infant mortality rate in a later section. Births are used instead of population under 1 year of age in the calculation of the infant mortality rate.

Sources: United Nations, *Demographic Yearbook, 1961*, table 18; U.S. National Center for Health Statistics, *Monthly Vital Statistics Report*, Vol. 16, No. 12, Supplement, "Advance Report, Final Mortality Statistics, 1966," March 12, 1963, table 5; Costa Rica, Dirección general de estadística y censos, *Anuario estadístico de Costa Rica, 1963*, 1964, table 24.

data on deaths from tuberculosis of white miners in the United States in 1950: [20]

Age	Number of white miners p_a	Death rate from tuberculosis, all working males M_a
	(1)	(2)
20–24 years.............................	74,598	.0001226
25–29 years.............................	85,077	.0001612
30–34 years.............................	80,845	.0002154
35–44 years.............................	148,870	.0003396
45–54 years.............................	102,649	.0005682
55–59 years.............................	42,494	.0007523
60–64 years.............................	30,037	.0008237
Total, 20–64 years................	564,570	.0009565

(a) Expected deaths from tuberculosis for white miners = $\Sigma M_a p_a = \Sigma$ col. (1) × col. (2) = 206

(b) Tabulated deaths from tuberculosis for white miners = $d = 540$

(c) Standardized mortality ratio =

$$\frac{d}{\Sigma M_a p_a} \times 100 \text{ or } \frac{\Sigma m_a p_a}{\Sigma M_a p_a} \times 100 = \frac{540}{206} \times 100 = 263 \qquad (15)$$

In this calculation the age-specific death rates from tuberculosis for all working males (col. 2) are assumed to apply to white miners (col. 1) in order to derive the expected number of deaths from tuberculosis for white miners. The standardized mortality ratio represents the ratio of the number of actual deaths from tuberculosis for white miners, 540, to the expected number, 206, expressed as an index (i.e., per 100). As the above formulas show, the standardized mortality ratio is equivalent to the relative mortality of an occupation group computed by the method of indirect standardization. Like all summary measures of this kind, the standardized ratios may mask differences found in the underlying age-specific rates.[21]

Occupational differentials in mortality may largely represent the effect of socioeconomic status rather than occupational differentials per se. Daric's review of occupational mortality data up to 1949 suggests that large mortality differentials by social class are attributable more to socioeconomic factors than to direct occupational risks.[22] The effect of socioeconomic levels apart from occupation may be examined by grouping the occupation data for males into very broad classes or by calculating mortality rates for wives and families by occupation of husband, if possible. The former procedure was, in effect, employed in the United States in 1950.[23]

[20] U.S. Public Health Service, National Vital Statistics Division, *Vital Statistics—Special Reports*, Vol. 53, No. 2.

[21] H. Silcock, "The Comparison of Occupational Mortality Rates," *Population Studies* (London), 13(2):183–192, November 1959.
[22] U.S. National Office of Vital Statistics, *Vital Statistics—Special Reports*, Vol. 33, No. 10, p. 186.
[23] U.S. Public Health Service, National Vital Statistics Division, *Vital Statistics—Special Reports*, Vol. 53, No. 5.

A wealth of new information on the social and economic factors in mortality in the United States has come out of the 1960 census-death certificate study carried out at the University of Chicago.[24] Consideration is here given to differentials in education, income, work status, occupational class, and other social and economic factors. This research is part of the general monograph program of the American Public Health Association (APHA). Other volumes include tabulations of death rates by age and sex, for marital classes, the native and foreign-born population, and the foreign-born population by country of birth, as well as special analyses of infant mortality and of particular causes of death.

A very valuable indirect method of studying the effect of socioeconomic status on mortality (or on fertility, etc.) does not depend on the availability of data on the socioeconomic status of decedents. It depends merely on the availability of statistics on the total number of deaths and total population for small geographic units which can be grouped into several economic strata. For example, comparisons of mortality by socioeconomic class have been carried out for a number of large cities in the United States by grouping deaths by census tracts, and grouping census tract data into economic strata on the basis of median monthly rental or median income. The classification into economic strata may also employ various measures of housing quality or density of occupancy. Such ecological studies may be refined by use of age-standardized death rates or separate rates for broad age groups. Simple and multiple correlation of the area death rates and measures of socioeconomic status may be employed to derive a formal expression of the relationship between these variables, but the limitations of ecological correlation must be kept in mind.[25] The APHA monograph cited above will include a regression and correlation analysis of total mortality in the 202 standard metropolitan statistical areas in relation to a large number of social and economic factors.[26]

Adjusted Rates

Rates Adjusted to a Probability Basis. — In general, the rates we have considered so far may be described as "central rates." Although they give a satisfactory indication of the relative frequency of the event of death, they do not describe precisely the risk of dying for any actual cohort, i.e., they are not true probabilities. Such probabilities express the chance that death will occur during a particular period to a person in a particular population group alive at the beginning of the period. When the probability is expressed as a ratio, the denominator represents the initial population "exposed to risk" and the numerator the frequency with which the event of death occurs in this particular population over the period. The most commonly computed probabilities of death are specific by age and relate to a one-year period. An age-specific probability of dying might answer the question, for example, what is the chance that a new-born child will die before it reaches its first birthday? Such probabilities are referred to as "mortality rates," to distinguish them from the "central death rates" or, simply, "death rates" which we have previously described.

Conventional infant mortality rate. — Analysis of infant mortality has commonly been carried out in terms of the "infant mortality rate" rather than the infant death rate, in order to approximate the probability of death among infants in a given year. The accuracy of the approximation varies from one situation to another but depends in general on the annual fluctuations in the number of births. We shall refer to this rate as the "conventional infant mortality rate" so as to distinguish it from certain other types of infant mortality rates, to be described, which are more akin to true probabilities. The conventional infant mortality rate is defined as the number of infant deaths per year per 1,000 live births during the year.

$$\frac{D_0}{B} \times 1,000 \qquad (16)$$

where D_0 represents deaths of infants during a year and B represents live births during the same year.
Infant mortality rates varied in 1965 from under 15 (13 for Sweden) to possibly over 150 (150 for Burundi):[27]

Country	Rate	Country	Rate
Australia	18.5	Philippines	72.9
Burundi	150.0	Portugal	41.8
Chile	107.1	Sweden	13.3
Costa Rica	75.1	Taiwan	22.2
El Salvador	70.6	Thailand [b]	37.8
England and Wales	19.0	United Arab	
Japan	18.5	Republic [a]	118.6
Pakistan [a]	145.6	United States	24.7
Peru	90.7	Venezuela	47.7
		Yugoslavia	71.5

[a] 1963.   [b] 1964.

Rates far in excess of 150 have been observed in various countries in earlier years. Ordinarily, the conventional infant mortality rate gives a sufficiently close approximation to the chance of dying between birth and attainment of the first birthday for the year to which the basic data on deaths relate. It has been widely used as an indicator of the health conditions of a community and, hence, of its level of living, although it may not be particularly appropriate for this purpose in "developed" areas.

Because of the very high level of mortality in the first hours, days, and weeks of life and the difference in the causes accounting for infant deaths at the earlier and later ages of infancy, the conventional infant mortality rate may usefully be "broken up" into a rate covering the first month or so and a rate for the remainder of the year. The rate for the first period is called the neonatal mortality rate and the rate for the second period is called the post-neonatal mortality rate. The neonatal mortality rate is defined as the number of deaths of infants under 4 weeks of age (28 days) or under 1 month of age during a year per 1,000 live births during the year:

$$\text{Neonatal mortality rate} = \frac{D_{0-3 \text{ weeks}}}{B} \times 1,000 \qquad \text{or}$$

$$\frac{D_{< 1 \text{ month}}}{B} \times 1,000 \qquad (17)$$

[24]Evelyn M. Kitagawa and Philip M. Hauser, *Differential Mortality in the United States: A Study in Socioeconomic Epidemiology*, Vital and Health Statistics Monographs, Cambridge, Mass., Harvard Univeristy Press, 1973.

[25]See, for example, C. Horace Hamilton, "Ecological and Social Factors in Mortality Variation," *Eugenics Quarterly*, 2(4): 212–223, December 1955.

[26] A comprehensive summary of studies concerned with the relation between mortality and social class is given in Aaron Antonovsky. "Social Class. Life Expectancy, and Overall Mortality," *Milbank Memorial Fund Quarterly*, 45(2):31–73, Part 1, April 1967.

[27] United Nations, *Demographic Yearbook, 1965*, table 14. See source for special notes.

Because nearly all (over 95 percent) deaths of infants under 1 month of age occur to babies born in the same year, the resulting neonatal mortality rate is close to a probability of neonatal death. The post-neonatal mortality is defined as the number of infant deaths at 4 through 51 weeks of age or 1 through 11 months of age during a year per 1,000 live births during the year:

$$\text{Post-neonatal mortality rate} = \frac{D_{\text{4-51 weeks}}}{B} \times 1,000$$

or

$$\frac{D_{\text{1-11 months}}}{B} \times 1,000 \qquad (18)$$

The formula for the post-neonatal mortality rate is quite far from expressing a true probability because more than a third of the infant deaths over 1 month of age in a year may occur to births of the previous year.

Adjusted infant mortality rates.—The conventional infant mortality rate is not a true probability, as has been noted. We can recognize this immediately by noting that not all of the infant deaths in a given year occurred to births in the same year; some occurred to births of the previous year. Consider, for example, the year of birth of the older infants (e.g.. 10 months old) dying toward the beginning of the calendar year (e.g., February) as compared with the younger infants (e.g., 2 months old) dying toward the end of the year (e.g., October). If the number of births does not fluctuate much from year to year, the conventional infant mortality rate for a year will represent rather well the probability of an infant dying during the year. If there are sharp fluctuations in the number of births between years and within years, however, the conventional infant mortality rate will give a distorted indication of the level and trend of infant mortality.[28] It is desirable then to adjust the conventional rate to allow for the true population exposed to risk. Three alternative ways of calculating adjusted infant mortality rates are described below.[29]

First, it is useful to consider the diagram below which schematically shows the relationship among births in three successive year ($y-1$, y, and $y+1$), deaths under 1 in these years, and the number of survivors reaching their first birthday in these years (l_1).

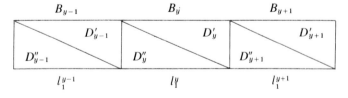

D'_y and D''_y together make up infant deaths in year y; D'_y is the portion occurring to births in year y (B_y) and D''_y is the portion occurring to births of the previous year (B_{y-1}). The cohort of births in year y (B_y) loses D'_y members through death in year y and D''_{y+1} members through death in year $y+1$ while moving toward its first birthday (l_1^{y+1}).

[28] U.S. Bureau of the Census, *Vital Statistics—Special Reports*, Vol. 19, No. 21, "Effect of Changing Birth Rates upon Infant Mortality Rates," Nov. 10, 1944.
[29] See the parallel discussion in George W. Barclay, *Techniques of Population Analysis*, New York, John Wiley and Sons, 1958, pp. 138–143.

Table 14-8 illustrates the calculation of the adjusted infant mortality rate with data for Venezuela from 1961 to 1963 by several procedures. All the procedures allow for the fact that some of the infant deaths in a year occur to births of the same year and some occur to births of the previous year or, conversely, that some of the babies born in a given year who die before their first birthday die in the same year as the year of birth and some die in the following year. The separation of the deaths by birth cohort is readily possible in the case of Venezuela because the deaths are tabulated by year of birth. Normally, this is not the case, and the **separation factors** for infant deaths, that is, the proportions for dividing the deaths according to birth cohort, must be estimated or, more commonly, assumed on the basis of statistics for other countries.

Table 14-8. — **Calculation of Conventional and Adjusted Infant Mortality Rates, for Venezuela: 1961–1963**

Item	Symbols and formulas	Year (y)			
		1960	1961	1962	1963
Births.......	B	338,199	344,989	341,324	353,546
Infant deaths $\quad D'_y$	D'_y	(X)	13,812	12,183	12,997
D''_y	D''_y	(X)	4,434	3,847	3,953
Conventional infant mortality rate.	$\dfrac{D'_y + D''_y}{B_y} \times 1{,}000$	(X)	52.9	47.0	47.9
Adjusted infant mortality rates:					
A.........	$\dfrac{D'_y + D''_{y+1}}{B_y} \times 1{,}000$	(X)	51.2	47.3	(2)
B.........	$\left[\dfrac{D'_y}{B_y} + \dfrac{D''_y}{B_{y-1}}\right] \times 1{,}000$	(X)	53.1	46.8	48.3
C.........	$\dfrac{D_y}{f'B_y + (1-f')B_{y-1}} \times 1{,}000^1$	(X)	53.1	46.8	48.3

X Not applicable.
[1] Separation factor (f') = $\dfrac{D'_y}{D'_y + D''_y}$. Factors: 1961, .7570; 1962, .7600; 1963, .7668.
[2] Outside the scope of this table.

Source: Basic data from Venezuela, Dirección general de estadística y censos nacionales, *Anuario estadístico de Venezuela, 1957–1963*, Vol. 1, 1964, tables 394–396; United Nations, *Demographic Yearbook, 1965*, table 11.

In the first procedure for calculating an "adjusted infant mortality rate" (formula A in the table), the portion of deaths under 1 in year y and the portion of deaths under 1 in year $y+1$ occurring to births in year y are combined and divided by the births in year y. Deaths in 1961 and 1962 are related to births in 1961 for Venezuela:

Formula A:

$$\frac{D'_y + D''_{y+1}}{B_y} \times 1,000 \quad \text{or} \quad \frac{13,182 + 3,847}{344,989} \times 1,000 = 51.2 \qquad (19)$$

In the next procedure (formula B), each portion of the infant deaths occurring in a given year is related to the births in the appropriate year and cohort. The formula is illustrated with deaths in 1961 and births in 1961 and 1960 for Venezuela:

Formula B:

$$\left(\frac{D_y'}{B_y}+\frac{D_y''}{B_{y-1}}\right)\times 1,000 \quad \text{or} \quad \left(\frac{13,812}{344,989}+\frac{4,434}{338,199}\right)$$

$$\times 1,000 = (.04004 + .01311)\times 1,000 = 53.1 \qquad (20)$$

In the third procedure for calculating an adjusted infant mortality rate for a particular year (formula C), infant deaths in year y are divided by a weighted average of births in years y and $y-1$. The weights may be taken as the same (separation) factors by which infant deaths in any year are divided into two parts according to birth cohort. Given statistics on deaths in a calendar year distributed by age and year of birth of the decedent, the separation factors for infant deaths are computed as follows:

$$f'=\frac{D_y'}{D_y'+D_y''} \quad \text{and} \quad f''=1-f' \qquad (21)$$

Evaluating the formulas with data for Venezuela in 1961, we have

$$f'=\frac{13,812}{18,246}=.7570 \quad \text{and} \quad f''=1-.7570=.2430$$

The formula for the infant mortality rate, illustrated with figures on deaths for Venezuela in 1961, is:

Formula C:

$$\frac{D_y}{f'B_y+f''B_{y-1}}\times 1,000 \qquad (22)$$

where f' and f'' are the weights or separation factors

$$\frac{13,812+4,434}{(.7570)\,344,989+(.2430)\,338,199}\times 1,000$$

$$=\frac{18,246}{261,157+82,182}\times 1,000 = 53.1$$

The values of D_y' and D_y'' in the present case were obtained directly by tabulation. In most cases, however, it is necessary to estimate or assume the factors. Accurate estimates of the factors can be made on the basis of tabulations of deaths by detailed age at death under 1 year (for example, days under 1 week, weeks under 1 month, months under 1 year). The procedure consists of (1) assuming that within each tabulation cell the deaths are rectangularly (evenly) distributed over time and age, (2) determining the proportions for separating deaths in each cell according to year of birth, (3) estimating the number in each cell which occurred to births of the previous year, (4) cumulating the numbers in (3) over all tabulation cells, and (5) dividing the sum in (4) by the total number of infant deaths in the year. The result is f'', and f' is obtained as the complement of f''.

One procedure for deriving separation factors from a tabulation by detailed age is illustrated in Table 14-9. This is the kind of tabulation recommended by the United Nations. Assuming a rectangular distribution over the calendar year and within each age, we can set down the proportion of deaths at each age which occurred to births of the previous year. Specifically, it was assumed that one-twelfth of the deaths in the year at any particular age occurred in a month and 1/365 of the deaths in the year at

any age occurred in a single day. Thus, of the deaths 11 months old in 1965, all from January through November and one-half of those in December were assumed to occur to the births of 1964; since only the annual total is known for this age, we take 11.5/12 of the total. Proportions for separating deaths of females at each age under 1 into two birth cohorts are derived in the same general way. These proportions are then applied cumulatively to the total number of deaths at each age in the calendar year to derive the estimates of all infant deaths in the year belonging to the two birth cohorts. The resulting separation factors for deaths of female infants in the United States in 1965 are .8868 and .1132.

For most countries such tabulations of infant deaths by detailed age are lacking and it is necessary to estimate the separation factors for infant deaths from more limited data for the countries or even to assume them on the basis of information for other countries. Fortunately, the separation factors for infants vary only over a moderate range and small variations in the factors make little difference in the infant mortality rate. Separation factors may be observed to vary in relation to the general level of infant mortality. Hence, given a rough estimate of the infant mortality rate, appropriate separation factors may be determined. Separation factors corresponding approximately to various levels of the infant mortality rate are as follows: [30]

	Separation factor	
Infant mortality rate	f'	f''
200	.60	.40
150	.67	.33
100	.75	.25
50	.80	.20
25	.85	.15
15	.95	.05

In view of the limitations of the procedures for adjusting infant mortality rates, when annual figures are not required, annual average rates based directly on infant deaths and births as reported for the years covered may serve adequately as adjusted measures for this period. The preferred formula for a 3-year annual average infant mortality rate is:

$$\frac{D_0^{y-1}+D_0^y+D_0^{y+1}}{B_{y-1}+B_y+B_{y+1}}\times 1,000 \qquad (23)$$

where the symbols identify infant deaths and births for three successive years. The influence of births in adjacent years on deaths of a given year is allowed for by this averaging formula except at the fringes of the period; the relative error from this formula is small unless there are extremely sharp changes in the annual numbers of births.

Mortality rates at ages above infancy. — Death rates at the ages above infancy can also be adjusted to a probability basis, i.e., expressed as mortality rates. Let us consider the problem first in terms of single ages. For those few countries (e.g., Japan, Venezuela) for which tabulations of deaths by single ages and years of birth are available, the task is simple. One can then immediately adapt any of the formulas given above for deriving infant mortality rates. Tables 14-10 and 14-11 indicate two ways of calculating mortality rates in single ages, allowing for the fact that deaths at a given age during a year belong to two different birth cohorts or, conversely, that a

---

[30] Based on a regression analysis relating infant mortality rates and separation factors implicit in the series of model life tables shown in A. J. Coale and Paul Demeny, *Regional Model Life Tables and Stable Populations*, Princeton, N.J., Princeton University Press, 1966.

Table 14–9. —Calculation of Separation Factors for Deaths of Female Infants on the Basis of Annual Data on Deaths by Detailed Age, for the United States: 1965

[One-twelfth of the annual total of deaths is assumed to occur in each month]

Age at death	Assumed proportion born in previous year (1)	Infant deaths		
		Total (2)	Born in 1964 (1)x(2)= (3)	Born in 1965 (2)-(3)= (4)
Total......	(X)	39,447	4,465	34,982
11 months......	$\frac{11.5}{12}$ = .9583	354	339	15
10 months......	$\frac{10.5}{12}$ = .8750	380	333	47
9 months.......	$\frac{9.5}{12}$ = .7916	470	372	98
8 months.......	$\frac{8.5}{12}$ = .7083	465	329	136
7 months.......	$\frac{7.5}{12}$ = .6250	601	376	225
6 months......	$\frac{6.5}{12}$ = .5417	731	396	335
5 months......	$\frac{5.5}{12}$ = .4583	864	396	468
4 months......	$\frac{4.5}{12}$ = .3750	1,244	466	778
3 months......	$\frac{3.5}{12}$ = .2917	1,650	481	1,169
2 months......	$\frac{2.5}{12}$ = .2083	2,062	430	1,632
28 to 59 days[1].	$\frac{1.5}{12}$ = .1250	2,714	339	2,375
21 to 27 days..	$\frac{21}{365}+\frac{1}{2}\left(\frac{7}{365}\right)=\frac{49}{730}$ = .0671	707	47	660
14 to 20 days..	$\frac{14}{365}+\frac{1}{2}\left(\frac{7}{365}\right)=\frac{35}{730}$ = .0479	875	42	833
7 to 13 days...	$\frac{7}{365}+\frac{1}{2}\left(\frac{7}{365}\right)=\frac{21}{730}$ = .0288	1,330	38	1,292
6 days.........	$\frac{6}{365}+\frac{1}{2}\left(\frac{1}{365}\right)=\frac{13}{730}$ = .0178	325	6	319
5 days.........	$\frac{5}{365}+\frac{1}{2}\left(\frac{1}{365}\right)=\frac{11}{730}$ = .0151	454	7	447
4 days.........	$\frac{4}{365}+\frac{1}{2}\left(\frac{1}{365}\right)=\frac{9}{730}$ = .0123	611	8	603
3 days.........	$\frac{3}{365}+\frac{1}{2}\left(\frac{1}{365}\right)=\frac{7}{730}$ = .0096	951	9	942
2 days.........	$\frac{2}{365}+\frac{1}{2}\left(\frac{1}{365}\right)=\frac{5}{730}$ = .0068	2,370	16	2,354
1 day..........	$\frac{1}{365}+\frac{1}{2}\left(\frac{1}{365}\right)=\frac{3}{730}$ = .0041	4,090	17	4,073
1 to 23 hours[2].	$\frac{1}{8,760}+\frac{1}{2}\left(\frac{23}{8,760}\right)=\frac{25}{17,520}$ = .0014	12,869	18	12,851
Under 1 hour[2]..	$\frac{1}{2}\times\frac{1}{8,760}=\frac{1}{17,520}$ = .0007≐0	3,330	–	3,330

Separation factors:

f'.......... $\frac{D'}{D'+D''}=\frac{34,982}{39,447}$ = .8868

f".......... $\frac{D''}{D'+D''}=\frac{4,465}{39,447}$ = .1132

[1] Treated as deaths of one month of age. All of the January deaths and one-half of the February deaths of this age were assumed to occur to births of the previous year.

[2] 8.760 represents the number of hours in the year.

Source: Basic data from U.S. National Center for Health Statistics. *Vital Statistics of the United States, 1965*, Vol. II. *Mortality*, Part A, 1967, table 2–8.

single birth cohort may experience deaths at two ages during a year. The data are for children under 10 years and persons 60 to 69 years in Japan in 1962, and mortality rates are calculated according to formulas of type A and B for the former and type B for the latter. (Other procedures, employed in the construction of life tables, are described in the next chapter.) Adapting formula A or (17) for use with population figures, we have:

$$\left(\frac{D_a''+D_{a+1}'}{P_a'}\right)\times 1,000 \qquad (24)$$

where D_a'' and D_{a+1}' represent deaths at age a and deaths at age $a+1$ in a given year, respectively, to the population age a at the beginning of the year, and P_a' represents the population age a at the beginning of the year. It indicates the chance that the population at a given age a at the beginning of a year will die during the following year. The deaths included in the formula relate to two different ages. The 1-year mortality rate for the population aged 2 on January 1, 1962, in Japan is calculated as follows (table 14–10):

$$\left(\frac{D_2''+D_3'}{P_2'}\right)\times 1,000=\left(\frac{1,364+1,227}{1,593,000}\right)\times 1,000=1.6$$

Similarly, the 1-year mortality rate for the population aged 1 is 2.3.

Formula B provides somewhat different mortality rates from formula A, and the results have a somewhat different interpretation. Adapting formula B or (20) for use with population figures, we have:

$$\left(\frac{D_a'}{P_{a-1}'}+\frac{D_a''}{P_a'}\right)\times 1,000 \qquad (25)$$

where D_a' and D_a'' represent the portions of deaths at age a in a particular year associated with different birth cohorts and P_a' and P_{a-1}', the populations in the same age and the next younger age at the beginning of the year. Consider an example based on the data for Japan in table 14–11:

$$\left(\frac{D_2'}{P_1'}+\frac{D_2''}{P_2'}\right)\times 1,000$$

$$\left(\frac{1,515}{1,570,000}+\frac{1,364}{1,593,000}\right)\times 1,000=1.9$$

Such a formula tells the chance of dying at age 2, that is, between one's second birthday and one's third birthday. The chance of dying at age 1, that is, between one's first birthday and one's second birthday, is represented by:

$$\left(\frac{D_1'}{P_0'}+\frac{D_1''}{P_1'}\right)\times 1,000$$

The mortality rate of 1.9 at age 2 compares with a central death rate at age 2 of 1.8 (table 14–11). In general, however, central death rates tend to be higher than mortality rates because, barring net immigration and changes in numbers of births, the midyear population in an age cohort will be smaller than the population at the beginning of the year. This is suggested by the differences between the mortality rates and central death rates for ages 60 to 69 years shown in table 14–11.

Table 14-10.—Calculation of Mortality Rates for the Population Under 10 Years of Age by Formula A, for Japan: 1962

Age on January 1, 1962	Year of birth	Age at death	Deaths 1962 D_a (1)	Population January 1, 1962[1] P'_a (2)	Mortality rate $\dfrac{D'_a + D'_{a+1}}{P'_a} \times 1{,}000$ $[(1) \div (2)] \times 1{,}000 =$ (3)
Under 1 year...	1961	Under 1 and 1 / Under 1.... / 1..........	11,236 / 8,759 / 2,477	1,563,000	7.2
1 year........	1960	1 and 2...... / 1.......... / 2..........	3,640 / 2,125 / 1,515	1,570,000	2.3
2 years........	1959	2 and 3...... / 2.......... / 3..........	2,591 / 1,364 / 1,227	1,593,000	1.6
3 years........	1958	3 and 4...... / 3.......... / 4..........	2,137 / 1,072 / 1,065	1,556,000	1.4
4 years........	1957	4 and 5...... / 4.......... / 5..........	1,653 / 853 / 800	1,515,000	1.1
5 years........	1956	5 and 6...... / 5.......... / 6..........	1,421 / 776 / 645	1,573,000	0.9
6 years........	1955	6 and 7...... / 6.......... / 7..........	1,195 / 624 / 571	1,684,000	0.7
7 years........	1954	7 and 8...... / 7.......... / 8..........	1,088 / 542 / 546	1,717,000	0.6
8 years........	1953	8 and 9...... / 8.......... / 9..........	923 / 463 / 460	1,799,000	0.5
9 years........	1952	9 and 10..... / 9.......... / 10........	918 / 453 / 465	1,881,000	0.5

[1] January 1, 1962, population equals three-fourths of the population estimate for October 1, 1961, plus one-fourth of the population estimate for October 1, 1962.

Source: Basic data from Japan, Ministry of Health and Welfare. *Vital Statistics, 1962, Japan,* Vol. I, table 28; and Japan, Bureau of Statistics. *Population Estimates by Age and Sex as of October 1, 1962,* March 1963, tables 1 and 3.

The rate of 2.3, obtained with formula A as the mortality rate of the population initially 1 year of age, is intermediate to the mortality rates for ages 1 and 2 shown by formula B, 2.9 and 1.9, respectively. In general, the results from formula A fall between the results from formula B for successive ages. This relationship corresponds to the ages of the deaths reflected in each formula.

The considerable deviation of the separation factors from .50–.50 in infancy reflects the fact that infant deaths tend to be concentrated at the earliest ages of infancy, that is, near the time of birth. This concentration results from the predominant role of the endogenous causes of mortality at this age. The greater risk of dying in the first part of an age would also be expected to apply to some of the ages above infancy, but the available data show that this phenomenon tends to disappear quite quickly and is hardly evident once the first year of life has passed. Table 14–11 presents the separation factors for deaths by single ages under 10 implied by the tabulations of deaths by age and year of birth for Japan in 1962. Although there are small fluctuations in the separation factors, they converge rapidly to about .50–.50. Even the separation factors for deaths at ages 1 and 2 are close to these values. At the higher

ages of life also, where mortality rates are rising rapidly, the separation factors implied by the tabulated data for Japan approximate .50–.50 and show no particular trend with increasing age even though such a trend would be expected. Illustrative data for ages 60 to 69 for Japan in 1962 are shown in table 14–11.

There are neither tabulations of death statistics by age and year of birth nor tabulations of deaths by subdivisions of single years of age above infancy for the United States. The separation factors needed to compute mortality rates for ages above infancy in the United States cannot be derived from these sources, therefore. The only study intended to establish the separation factors for ages above infancy in the United States was based on a 10-percent sample of deaths at ages 1 to 4 in 1944, 1945, and 1946.[31] The tabulation of these deaths by year of birth indicated separation factors rather close to .50–.50 for each age. Since a moderate variation in the separation factor has only a slight effect on the mortality rate, it seems best to employ a separation factor of 0.50 for splitting deaths according to birth cohort at the ages above infancy and for calculating 1-year mortality rates at these ages for the United States.

The discussion so far has been confined to mortality rates for single ages in a single year. Mortality rates for longer periods of time and for age groups may also be calculated. Such rates may be constructed from observed data over a number of calendar years and represent a **real cohort,** or be constructed from observed data for a single calendar year and represent a **synthetic cohort.**

For example, it is possible to calculate the risk of dying for a 5-year old in a 5-year period from the appropriate probabilities for this 5-year period. The probabilities at successive ages in successive calendar years would be used. A synthetic probability of this kind could be derived by combining the appropriate single-year-of-age probabilities for a single calendar year. The formula would take the following form:

$$_5q'_5 = 1 - (1 - q'_5)(1 - q'_6)(1 - q'_7)(1 - q'_8)(1 - q'_9) \quad (26)$$

where q'_5 represents the 1-year mortality rate between ages 5 and 6, and $_5q'_5$ the risk of a 5-year old dying in the next five years. Ordinarily, single-year-of-age tabulations of death statistics are not available or are too inaccurate to serve as a basis for the calculation of death rates and mortality rates for single years of age. Adjusted estimates of deaths and death rates in single ages may be derived by various types of interpolation procedures from 5-year age data.

The approximate annual mortality rate for a 5-year age group may be derived as follows:

(1) Set down total deaths in a given age group during the year.

(2) Set down the population in this age group at the middle of the year.

(3) Add one-half of the deaths in step 1 to the population in step 2. (This is the base population for the rate, i.e., the population at the beginning of the year that would be in this age group at the middle of the year.)

(4) Divide the figure in step 1 by the result in step 3.

[31] U.S. National Office of Vital Statistics, *Vital Statistics—Special Reports, Selected Studies,* Vol. 33, No. 7, "Investigation of Separation Factors at Ages 1–4 Based on 10-Percent Mortality Sample," Sept. 6, 1950.

Table 14–11.—Calculation of Separation Factors and Mortality Rates for Ages Under 10 Years and 60 to 69 Years by Formula B, for Japan: 1962

Age	Deaths			Separation factor[1]		Population, January 1, 1962[2]	Mortality rate	Central death rate
	Total D_a	Born in earlier of two possible years D_a''	Born in later of two possible years D_a'	f'' $\dfrac{D_a''}{D_a}$	f' $\dfrac{D_a'}{D_a}$	P_a'	$\left[\dfrac{D_a'}{P_{a-1}'}+\dfrac{D_a''}{P_a'}\right]\times1,000$	$\dfrac{D_a}{P_a}\times1,000$[3]
				$(2)\div(1)=$	$(3)\div(1)=$			
	(1)	(2)	(3)	(4)	(5)	(6)	(7)	(8)
Under 1 year........	42,797	8,759	34,038	.2047	.7953	1,563,000	[4]26.5	27.2
1 year..............	4,602	2,125	2,477	.4618	.5382	1,570,000	2.9	3.0
2 years.............	2,879	1,364	1,515	.4738	.5262	1,593,000	1.9	1.8
3 years.............	2,299	1,072	1,227	.4663	.5337	1,556,000	1.5	1.5
4 years.............	1,918	853	1,065	.4447	.5553	1,515,000	1.2	1.3
5 years.............	1,576	776	800	.4924	.5076	1,573,000	1.0	1.0
6 years.............	1,269	624	645	.4917	.5083	1,684,000	0.8	0.8
7 years.............	1,113	542	571	.4870	.5130	1,717,000	0.7	0.6
8 years.............	1,009	463	546	.4589	.5411	1,799,000	0.6	0.6
9 years.............	913	453	460	.4962	.5038	1,881,000	0.5	0.5
60 years............	11,988	5,911	6,077	.4931	.5069	712,000	16.6	16.6
61 years............	12,271	6,064	6,207	.4942	.5058	657,000	17.9	18.0
62 years............	12,615	6,412	6,203	.5083	.4917	591,000	20.2	20.4
63 years............	13,104	6,732	6,372	.5137	.4863	588,000	22.2	22.9
64 years............	14,139	7,034	7,105	.4975	.5025	551,000	24.9	24.8
65 years............	14,204	6,926	7,278	.4876	.5124	526,000	26.4	27.0
66 years............	15,250	7,686	7,564	.5040	.4960	490,000	30.1	30.3
67 years............	15,760	7,954	7,806	.5047	.4953	447,000	33.7	34.2
68 years............	16,105	8,114	7,991	.5038	.4962	417,000	37.4	38.2
69 years............	16,695	8,540	8,155	.5115	.4885	404,000	40.7	41.7

[1] Implicit in tabulations of deaths by year of birth of decedent.

[2] January 1, 1962, population equals three-fourths of the population estimate for October 1, 1961, plus one-fourth of the population estimate for October 1, 1962.

[3] Based on July 1, 1962, population estimates.

[4] Formula B for infants (under 1 year) is $\dfrac{D_0''^y}{B^y}+\dfrac{D_0''^y}{B^{y-1}}$. Total births numbered 1,589,372 for 1961 and 1,618,616 for 1962.

Source: Basic data from Japan. Ministry of Health and Welfare. *Vital Statistics, 1962, Japan*, Vol. I, table 28; and Japan. Bureau of Statistics. *Population Estimates by Age and Sex as of October 1, 1962*, March 1963, tables 1 and 3.

This procedure disregards the effect of net immigration on deaths and population during the year.

For the probability that a person aged 60 to 64 years of age would die within one year, according to experience in the United States in 1960, the corresponding figures are:

(1) Deaths = 153,444

(2) Population on July 1, 1960 = 7,162,000

(3) Base population = $7,162,000 + \frac{1}{2}(153,444) = 7,238,722$

(4) Rate = $153,444 \div 7,238,722 = .0212$, or 21.2 per 1,000

Cause-specific mortality rates.—Measures giving the chance of dying from a particular cause at a particular age in a year or 5-year period, or of eventually dying from this cause, have been developed, principally in connection with the preparation of life tables by cause of death (ch. 15). Cause-specific probabilities of death can be computed in a manner paralleling the calculation of general probabilities of dying. In the case of general cause-specific probabilities for single ages for single calendar years, the deaths from a particular cause or group of causes at a particular age (a) for a given calendar year may be divided by the population at age a at the middle of the year plus one-half of all deaths at age a during the year. (The base population may alternatively be estimated as the mean of the population at age a and the population at the next younger age ($a-1$) at the beginning of the year, or the mean of the population at age a at the middle of the year and age $a-1$ at the middle of the previous year.) The required data on deaths would normally be obtained by interpolation from 5-year age data. As before, this procedure disregards the net migration occurring during the year of the estimate.

A one-year cause-of-death probability for a 5-year age group may be derived as follows:

(1) Set down (a) total deaths in the given age group, (b) deaths from the particular cause in this age group during the year, and (c) the population in this age group at the middle of the year.

(2) Add one-half of all deaths at these ages (item 1a) to the figure in item 1c. (This is the base population for the rate, i.e., the population at the beginning of the year that would be 60–64 at the middle of the year.)

(3) Divide the figure in item 1b by the result in step 2. For calculating the risk of death from heart disease in the United States at ages 60–64 in 1960, the corresponding figures are:

(1) All deaths, 153,444; deaths due to heart disease, 66,728; and population, July 1, 1960, 7,162,000.

(2) Base population = $7,162,000 + \frac{1}{2}(153,444) = 7,238,722$.

(3) Rate = $66,728 \div 7,238,722 = .00922$, or 922 per 100,000.

Cause-specific probabilities for 5-year time periods, whether for single ages or 5-year age groups, particularly those derived "synthetically" on the basis of data for a single calendar year, are, as suggested earlier, akin to the measures developed in connection with life tables. They are discussed in the next chapter, therefore.

The conventional infant mortality rate, previously described as an approximate measure of the risk of dying in infancy, may be viewed as the sum of a series of rates each of which represents the chance of dying in infancy from a particular cause. For example, the chance of a newborn child dying from a congenital anomaly is measured by $\frac{D_0^c}{B} \times 100,000$ where D_0^c represents infant deaths due to congenital anomalies, and the chance of a newborn child dying from immaturity is measured by $\frac{D_0^i}{B} \times 100,000$, where D_0^i represents infant deaths due to immaturity. In analyzing differences in overall infant mortality rates, it is useful to separate the infant deaths by cause into the two broad classes, endogenous and exogenous, described earlier, and to compute separate rates. The formula for the **endogenous infant mortality rate** is

$$\frac{D_0^{en}}{B} \times 1,000 \qquad (27)$$

and the formula for the **exogenous infant mortality rate** is

$$\frac{D_0^{ex}}{B} \times 1,000 \qquad (28)$$

As suggested above, the endogenous set of causes is related to the genetic makeup of the infant, the circumstances of life in utero, and the conditions of labor. The exogenous set of causes is related to the contact of the infant with the external world. Endogenous infant mortality includes mainly mortality from congenital anomalies, immaturity, birth injuries, postnatal asphyxia, etc. (i.e., conditions which are difficult to prevent or treat in the present state of knowledge). Exogenous infant mortality includes mainly infections and postnatal accidents (i.e., conditions which are relatively preventable or treatable). The endogenous infant mortality rate is estimated at 18.3 in the United States in 1966 and at 14.4 in Norway in 1959. About one-quarter of the infant deaths in these countries are exogenous, whereas in Costa Rica over 60 percent are in this category (table 14-7). As a result, in Costa Rica the exogenous infant mortality rate is well above the endogenous rate. This type of comparative international analysis of the infant mortality rate in terms of endogenous and exogenous causes is useful in determining the extent to which the rate can reasonably be reduced in the present state of knowledge.

The **maternal mortality rate** is a widely used type of cause-specific mortality rate representing approximately the risk of dying as a result of "complications of pregnancy, childbirth, and the puerperium" (Class XI in the Eighth International Classification). This rate is conventionally defined as the number of deaths due to puerperal causes per 10,000 or 100,000 live births. With a constant of 10,000, the formula is:

$$\frac{D^p}{B} \times 10,000 \qquad (29)$$

where D^p represents deaths due to puerperal causes. The maternal mortality rate for Chile in 1960 is, therefore,

$$\frac{789}{274,372} \times 10,000 = 28.8$$

The corresponding figure for Denmark in 1964 is 1.6. These figures encompass the rates for most countries in recent years. In the formula the number of births is employed as an approximation to the number of women exposed to the risk of dying from puerperal causes.

Rates Adjusted for Population Composition. — The level of an observed death rate like that of other observed rates is affected by the demographic composition of the population for which the rate is calculated. The age composition of the population, in particular, is a key factor affecting the level of the crude death rate. For purposes of comparison of death rates over time or from area to area, it is useful to determine the difference between the rates if there were no differences in age composition. It is particularly important to eliminate the effect of the differences in age structure of two populations being compared if one is trying to compare their health conditions. Crude death rates are especially unsatisfactory for this purpose. A crude death rate of a population may be relatively high merely because the population has a large proportion of persons in the older ages, where death rates are high; or it may be relatively low because the population has a large proportion of children and young adults, where death rates are low. The crude death rate of a country may actually rise even though death rates at each age remain stationary, if the population is getting older.

The procedure of adjustment of the crude rates to eliminate from them the effect of differences in population composition with respect to age and other variables is sometimes called **standardization**. Often death rates are adjusted or standardized for both age and sex. Other variables for which death rates may be adjusted or standardized are racial composition, nativity composition, urban-rural composition, etc. We shall confine our discussion largely to adjustment or standardization for age since this is the most important and most common variable for which the adjustment or standardization of death rates is carried out.

It is important to recognize that age-adjusted or age-standardized rates have no direct meaning in themselves. They are meaningful only in comparison with other similarly computed rates. Since they are useful only for comparison, the commonest application of the procedure is to compute such rates for the areas or population groups whose mortality is to be compared and to calculate the relative differences of the resulting rates. The meaningful measure then is a ratio, index, or percent difference between rates similarly adjusted.

A number of methods have been developed for adjusting death rates for age composition or for deriving indexes of age-adjusted rates. We shall consider four principal ones. The measures are the age-adjusted or age-standardized death rate calculated by the direct method, the age-adjusted or age-standardized death rate calculated by the indirect method, the comparative mortality index, and the life table death rate.

Direct standardization. — The simplest and most straightforward measure is the age-adjusted death rate derived by the direct method. For most comparisons, this is the preferred procedure and serves to provide the best basis for determining the relative difference between mortality in two areas or at two dates. In this method, a "standard" population is selected and employed in deriving all the age-adjusted rates in a set to be compared. If the same standard population is employed, as required, all the rates are directly comparable. The formula calls for computing the weighted average of the age-specific death rates in a given area, using as weights the age distribution of the standard population. The formula for direct standardization is:

$$m_1 = \frac{\Sigma m_a P_a}{P} \times 1,000, \qquad \text{or} \qquad \Sigma m_a \frac{P_a}{P} \times 1,000 \qquad (30)$$

where $m_a = \dfrac{d_a}{p_a} =$ age-specific death rate in the given area, P_a represents the standard population at each age, and P or ΣP_a represents the total standard population. (It will be noted that capital letters are used here to identify the characteristics of the standard population and lowercase letters are used to identify the characteristics of the other populations.) Each age-specific rate is multiplied, in effect, by the proportion of the standard population in each age group. (In standardizing a death rate for age and sex jointly, each age-sex-specific death rate is multiplied by the proportion of the total standard population in that age-sex group.) The age-standardized death rate for the standard population is the same as its own crude death rate, since the age-specific death rates for the standard population would be weighted by its own population. The relative mortality of the given area is derived by dividing the age-standardized rate for the area by the crude death rate of the standard population.

Illustrative calculations are shown for a group of countries around 1960 in table 14–12. The population of the United States is employed as a standard population to calculate the age-standardized death rate for Japan (1960), El Salvador (1961), Chile (1960), and Taiwan (1960). The steps in calculating the age-adjusted death rate by the direct method for Japan are as follows:

(1) Record the population in each age group for the United States (standard population).

(2) Record the age-specific death rates for Japan.

(3) Compute the cumulative product of the population figures in step 1 and the death rates in step 2 (1,969,242).

(4) Divide the result in step 3 (1,969,242) by the total population of the United States (179,323,000). The result is 11.0 per 1,000.

The crude death rate in Japan, 7.6 per 1,000 population, falls below the crude death rate in the United States, 9.5, by 20 percent (table 14–12). The adjustment of the crude rate in Japan raised it to 11.0 reflecting the fact that the age composition of Japan's population is more favorable for a low crude death rate than that of the United States. Equating the age distributions changes the relative deficit of the Japan rate to a relative excess of 16 percent. It may be seen that (direct) standardization caused a rise in the death rate in El Salvador, Chile, and Taiwan also. In the case of Taiwan, as well as Japan, the crude rate gives the impression of lower mortality than in the United States; but the difference in the adjusted rates shows that, in fact, the force of mortality is higher in that country than in the United States.

There are a number of possible choices with respect to the selection of the standard to be used in computing the age-adjusted death rate by the direct method. The standard selected may be the age distribution of one of the areas or dates (e.g., earliest, middle, or latest in a series) being compared, or the sum or average of the age distributions for the areas or dates being compared, or an "external" real or theoretical distribution of some sort. Different results for the relative differences between adjusted rates will be obtained depending on the age distribution selected as a standard. In fact, the choice of standard may even affect the direction of the difference between the rates for the populations being compared. Hence, it is desirable to select the standard population carefully. The general rule is to select as a standard an age distribution that is similar to the age distributions of the various populations under study. If the mortality of two populations is being compared, this may best be achieved by using as a standard the (unweighted) average of the two distributions. Since this would apply to time series as well, use of the age distribution of the first year's (or last year's) population in a time series as the standard over a long period is to be avoided. It is not always possible to follow this rule closely since the populations being compared may have quite different age distributions. Yet, the farther apart the age distributions are, the more important it is to make the comparison of their mortality on the basis of adjusted figures. In some cases it may be desirable to forego comparisons of summary measures and compare the schedules of age-specific death rates.

The need for age-adjustment is particularly great in connection with cause-specific death rates. Certain causes of death are concentrated in one or another part of the age distribution. Hence, the level of the observed death rate from these causes is particularly affected by the age distribution of the population. For example, a population with a relatively large proportion of older persons will tend to have a relatively high death rate from heart disease and a population with a relatively small proportion of older persons will tend to have a relatively low death rate from this cause. The calculation of an age-adjusted death rate for a specific cause by the direct method follows the same form as the calculation of a general age-adjusted death rate by the direct method except that age-cause-specific rates are used for the areas under study.

Indirect standardization. – As we have seen, calculation of the age-adjusted death rate by the direct method requires age-specific death rates or deaths by age for the area under study. These may not be available even though a count or estimate of the total number of deaths and an estimate of the crude death rate are at hand. In this case, if counts or estimates of the age distribution of the population are also available, it is possible to adjust the death rate by an indirect method. The formula for indirect standardization is:

$$m_2 = \left(\frac{d}{\Sigma M_a p_a}\right) M \qquad (31)$$

where, for the "standard" population, M_a represents age-specific death rates and M represents the crude death rate; and, for the population under study, d represents the total number of deaths and p_a represents the population at each age. This formula calls for adjusting the crude death rate of the standard population by a factor representing the ratio of the recorded number of deaths to the number expected on the basis of the age-specific deaths rates of the "standard" population and the population by age in the area under study.

The steps in calculating the age-adjusted death rate by the indirect method for Japan in 1961, using the population of the United States as the standard, are as follows (table 14–13):

(1) Set down the age-specific death rates for the United States in 1960 ("standard").

(2) Set down the population by age for Japan in 1960.

(3) Compute the cumulative product of the death rates in step 1 above and the population in step 2 above (617,544). This is the number of deaths expected on the basis of the age-specific death rates in the United States in 1960 and population by age in Japan in 1960.

(4) Divide the result in step 3 (617,544) into the total number of deaths registered in Japan in 1960 (706,599). The result is 1.1442.

(5) Multiply the result in step 4 (1.1442) by the crude death rate of the United States (9.5) to derive the adjusted death rate.

Table 14–12.—Calculation of Age-Standardized Death Rates by the Direct Method, for Several Countries: Around 1960

[Standard population employed is the United States population in 1960. Rates per 1,000 population; calculations were carried out on the basis of rates per person]

Age	Standard population (in thousands) (P_a) United States, 1960	Age-specific death rates (m_a)			
		Japan, 1960	El Salvador, 1961	Chile, 1960	Taiwan, 1960[1]
Under 1 year...............................	4,112	31.3	89.9	150.2	32.4
1 to 4 years...............................	16,209	2.5	16.5	9.8	7.9
5 to 9 years...............................	18,692	0.9	3.3	1.7	1.1
10 to 14 years............................	16,773	0.5	1.5	1.2	0.8
15 to 19 years............................	13,219	1.1	2.4	1.8	1.4
20 to 24 years............................	10,801	1.7	3.5	2.9	2.3
25 to 29 years............................	10,869	1.9	4.0	3.7	2.2
30 to 34 years............................	11,949	2.1	5.0	4.6	2.8
35 to 39 years............................	12,481	2.6	5.5	5.9	3.6
40 to 44 years............................	11,600	3.5	6.8	7.2	5.0
45 to 49 years............................	10,879	5.3	8.4	10.1	6.6
50 to 54 years............................	9,606	8.4	11.5	12.8	10.8
55 to 59 years............................	8,430	13.4	16.0	19.4	17.1
60 to 64 years............................	7,142	21.1	21.6	26.3	25.5
65 to 69 years............................	6,258	34.4	31.9	37.9	41.1
70 to 74 years............................	4,739	57.1	44.7	61.3	63.9
75 to 79 years............................	3,054	94.6	56.5	88.9	99.6
80 to 84 years............................	1,580	146.6	77.3	110.9	} 199.5
85 years and over.........................	929	231.1	131.7	159.1	
(1) Total standard population $= \Sigma P_a = P$......	179,323	(X)	(X)	(X)	(X)
(2) Expected deaths $= \Sigma m_a P_a$............	(X)	1,969,242	2,363,734	2,842,247	2,365,479
(3) Age-adjusted death rate $= \dfrac{\Sigma m_a P_a}{P} = \dfrac{(2)}{(1)}$....	9.5	11.0	13.2	15.8	13.2
(4) Percent difference from United States rate	(X)	+15.8	+38.9	+66.3	+38.9
(5) Crude death rate........................	9.5	7.6	11.3	12.9	6.9
(6) Percent difference from United States rate	(X)	-20.0	+18.9	+35.8	-27.4

X Not applicable.

[1] Data tabulated by year of registration rather than year of occurrence.

Sources: Basic data from United Nations, *Demographic Yearbook* (various dates).

The adjusted rate for Japan using the indirect method is 10.9 and exceeds the rate in the United States by 15 percent, that is, by the factor computed in step 4, minus 1, per 100 (the difference being due to rounding). The level of this rate and the relative excess of the rate over the U.S. rate are rather similar to, but smaller than, the corresponding figures based on the direct method (11.0 and 16 percent). The adjusted rate using the indirect method was far larger than the rate using the direct method for El Salvador and Chile, and moderately larger for Taiwan (table 14–12). In general, the range of variation in the relative mortality of various populations resulting from the use of the indirect method as compared with the direct method or from the use of different standards in the indirect method may be quite wide and, in fact, the indications of relative mortality may go in opposite directions.

We should expect the relative mortality of two areas measured by the direct and indirect methods to be different. Although the rates for the two areas being compared are, in effect, weighted by the same populations in each method, the indirect method weights the age-specific death rates by the population of the area under study and the direct method weights the rates by the standard population.

Comparative mortality index.—The comparative mortality index (CMI) is a measure of relative mortality, usually employed to indicate changes over time in the overall mortality of an area. It uses a shifting pattern of population weights, designated to overcome the problems of prolonged use of a single standard age distribution. The formula is:

$$CMI = \frac{\sum w_a m_a}{\sum w_a M_a} \tag{32}$$

where M_a represents the age-specific death rates in the standard or initial year, m_a represents the age-specific death rates in later years, and

$$w_a = \frac{1}{2}\left(\frac{P_a}{P} + \frac{p_a}{p}\right)$$

where P_a and P are populations of the standard or initial year and p_a and p are populations of later years.

The formula calls for taking a ratio of (1) the weighted sum of age-specific death rates in each year to (2) the similarly weighted sum of age-specific death rates of the initial year. The weights are the average of (a) the proportion of the total population in the age group in the initial year and (b) the corresponding proportion in each later year.

Table 14–13. — Calculation of Age-Standardized Death Rates by the Indirect Method, for Several Countries: A round 1960

[Standard set of age-specific rates are for the United States, 1960. Rates per 1,000 population; calculations were carried out on the basis of death rates per person]

Age	Age-specific death rates (M_a) United States, 1960	Population (in thousands) (P_a)			
		Japan, 1960	El Salvador, 1960	Chile, 1960	Taiwan, 1960[1]
All ages..............................	9.5	93,419	2,511	7,374	10,612
Under 1 year........................	27.0	1,577	97	228	395
1 to 4 years.........................	1.1	6,268	334	876	1,526
5 to 9 years.........................	0.5	9,205	384	981	1,721
10 to 14 years.......................	0.4	11,018	309	836	1,148
15 to 19 years.......................	0.9	9,309	242	725	959
20 to 24 years.......................	1.2	8,318	215	598	764
25 to 29 years.......................	1.3	8,209	173	527	786
30 to 34 years.......................	1.6	7,518	151	507	724
35 to 39 years.......................	2.3	6,038	139	415	618
40 to 44 years.......................	3.7	5,019	112	364	493
45 to 49 years.......................	5.9	4,817	90	324	449
50 to 54 years.......................	9.4	4,201	76	279	342
55 to 59 years.......................	13.8	3,641	51	212	250
60 to 64 years.......................	21.5	2,932	58	183	176
65 to 69 years.......................	31.4	2,160	29	128	115
70 to 74 years.......................	47.2	1,564	21	84	77
75 to 79 years.......................	72.0	955	13	52	46
80 to 84 years.......................	} 147.4 { 117.2	483	9	31	} 24
85 years and over....................	198.6	188	7	22	
(1) Expected deaths = $\sum M_a P_a$.................	(X)	617,544	13,457	44,237	48,062
(2) Registered deaths (d)...................	1,711,982	706,599	28,471	95,486	73,715
(3) Ratio, $\dfrac{\text{Registered deaths}}{\text{Expected deaths}} = \dfrac{d}{\sum M_a P_a} = \dfrac{(2)}{(1)}$...	(X)	1.1442	2.1157	2.1585	1.5337
(4) Age-adjusted death rate = 9.5 x (3).......	9.5	10.9	20.1	20.5	14.6
(5) Percent difference from United States rate	(X)	+14.7	+111.6	+115.8	+53.7

X Not applicable.
[1] Data tabulated by year of registration rather than year of occurrence.

Life table death rate. — The life table death rate, the fourth type of age-adjusted rate considered here, is the most difficult one to derive, requiring the construction of a life table from the observed age-specific death rates for each population examined. It is also the most difficult to interpret. The subject of life tables is treated separately and in detail in the next chapter. We may note here that the assumption of a constant annual number of births and a constant set of age-specific death rates corresponding to the observed age-specific death rates of an area in a given year or period generates a population with an unchanging total size and age distribution, called the "life table stationary population," which has its own crude birth and death rates. The crude death rate of this population is called the "life table death rate." In effect, the life table death rate may be viewed as a result of the weighting of age-specific death rates   by the life table stationary population.

Summary note on age-adjustment. — We have considered a number of ways of calculating age-adjusted death rates. This discussion suggests that there is no perfect method of removing the effects of age composition when the mortality experiences of different populations are being compared. The direct method of standardization employs a common standard population for all the areas or dates being compared, but the results are affected by the choice of standard population and the standard may be unreasonable for widely different populations or over long periods of time. In the indirect method of standardization and in the calculation of the comparative mortality index, the comparability of results is affected by the use of a different standard population for each area or date for which comparative measures are derived. The calculation of life table death rates avoids the problem of the arbitrary choice of a standard population, since life table death rates are generated entirely from the age schedules of age-specific death rates; in effect, however, each set of age-specific rates is weighted by a different population, derived from those very age-specific rates. For a complete analysis of mortality, it is desirable to examine the differences in the individual age-specific rates.

Measures of Pregnancy Wastage

Much loss of potential life occurs as a result of fetal deaths, and there is a close relationship between fetal and neonatal mortality.[32]

---

[32] See United Nations, *Foetal, Infant, and Early Childhood Mortality*, Vol. I, *The Statistics*, Population Studies, Series A, No. 13, 1954, esp. Part III, "Pregnancy Wastage," pp. 12–28; and Earl T. Engle (ed.), *Pregnancy Wastage*, Springfield, Ill., Charles C. Thomas Publisher, 1953, esp. the article by Christopher Tietze, "Introduction to the Statistics of Abortion," pp. 135–145.

Fetal Mortality. — As we have seen, the definition of fetal death complements the definitions of live birth and death. In some countries, however, the definition employed differs from the international recommendations: live-born children dying early in life (e.g., before registration of birth or within 24 hours of birth) may be classed with fetal deaths. A more important problem is the incompleteness and irregularity of reporting of fetal deaths. This limitation applies to the data for most countries of the world. The duration of pregnancy required for registration varies widely. Reporting of early fetal deaths may be seriously incomplete even where required by law. When registration of all fetal deaths is not mandatory, countries differ as to what is to be registered as a late fetal death; 28 weeks is most frequently specified as the minimum period. As a result, international comparability is far greater for the late fetal deaths than for all fetal deaths taken together.

In view of this situation, the United Nations has recommended that fetal deaths be tabulated by period of gestation into four classes — under 20 completed weeks, 20 to 27 completed weeks, 28 to 36 completed weeks, and 37 completed weeks and over (and not stated).[33] It has designated fetal deaths of at least 28 weeks of gestation, combined with fetal deaths of unknown gestational age, as "late" fetal deaths; fetal deaths under 20 weeks' gestation as "early" fetal deaths; and fetal deaths of 20–27 weeks' gestation as "intermediate" fetal deaths. Gestational age reported in months should be allocated to the corresponding intervals in weeks. The data on late fetal deaths are subject to substantial error introduced by incorrect reporting of gestational age. The determination of age is often difficult and comparability of the tabulations is affected by the differences in the skill of the medical attendant in making this determination.

The loss through fetal deaths may be measured by the fetal death ratio or the fetal death rate. The **fetal death ratio** is defined as the number of fetal deaths reported in a year per 1,000 live births in the same year, or

$$\frac{D^f}{B} \times 1,000 \qquad (33)$$

where D^f represents either all fetal deaths or late fetal deaths. Because of the variability in coverage of the early and intermediate fetal deaths mentioned above, it is preferable from the point of view of international comparability to compute the fetal death ratio on the basis of late fetal deaths only.

The **fetal death rate** relates the fetal deaths more closely to the population at risk than the fetal death ratio, i.e., it is more akin to a probability. The formula includes the fetal deaths in the denominator as well as in the numerator:

$$\frac{D^f}{B + D^f} \times 1,000 \qquad (34)$$

The rate can be calculated either with all fetal deaths or late fetal deaths only.

An array of late fetal death ratios and rates for a number of countries around 1960 is as follows:[34]

Country and year	Late fetal death ratio $\dfrac{D^f}{B}$	Late fetal death rate $\dfrac{D^f}{B + D^f}$
Chile (1961)	26.5	25.8
Czechoslovakia (1961)	10.0	9.9
Italy (1962)	23.4	22.8
Japan (1960)	30.8	29.9
Mauritius (1960)	70.5	67.0
Mexico (1961)	17.2	16.9
United States (1960)	12.7	12.5

In spite of the theoretical advantage of the fetal death rate, the fetal death ratio may be considered preferable for international comparisons. The registration of fetal deaths is irregular and the effect of this irregularity is compounded when fetal deaths are included with the births in the base of the fetal death rate. Because of the likelihood that poor registration of fetal deaths will occur in association with poor registration of births and hence that errors in each component will offset one another to some extent, fetal death ratios may sometimes be of satisfactory quality even where the basic data are questionable.

Specific fetal death ratios may be calculated in terms of period of gestation of the fetus or in terms of the age of the mother (requiring data on the age of mother for births and fetal deaths). Other characteristics of principal importance in the analysis of fetal deaths are the marital status of the mother (legitimacy of the fetus), sex of the fetus, the number of children previously born to the mother, and the type of birth (single or plural issue). Other factors include the cause of fetal death, hospitalization, age of father, date of marriage (for legitimate pregnancies), level of education of parents, and the occupational characteristics of the parents.[35]

Perinatal Mortality. — The causes of death in early infancy are believed to be so akin to those accounting for fetal deaths that various measures of mortality combining fetal deaths and deaths of early infancy have been proposed. The combination of these deaths is also intended to eliminate the errors resulting from deliberate and inadvertent misclassification as between fetal deaths, births, and neonatal deaths. The combined risk of dying during the period near parturition (i.e., just before, during, and just after birth) is measured by various so-called perinatal mortality ratios and rates. The formulas differ with respect to the age limits of the infant deaths and the gestational age of fetal deaths to be included, and with respect to whether fetal deaths are included in or excluded from the base of the ratios. Neonatal deaths, deaths under 1 week, or deaths under 3 days, in combination with all fetal deaths or late fetal deaths, are possible ways of operationally defining perinatal deaths. In view of the general lack of tabulated data on infant deaths under 3 days, this coverage is not very useful for international

---

[33] Gestation age or period of gestation is defined as the number of completed weeks which have elapsed between the first day of the last menstrual period and the date of delivery of the fetus. See United Nations, *Principles for a Vital Statistics System*, p. 22.

[34] United Nations, *Demographic Yearbook, 1966*, tables 7 and 9.

[35] See, for example, United Nations, *Foetal, Infant and Early Childhood Mortality*, Vol. II, *Biological, Social and Economic Factors*, Series A, Population Studies, No. 13, Add. 1, 1954; and Ronald Freedman, Lolagene C. Coombs, and Judith Friedman, "Social Correlates of Fetal Mortality," *Milbank Memorial Fund Quarterly*, 44(3):327–344, July 1966.

comparisons. The World Health Organization defines the perinatal period as the period of prenatal existence after viability is reached, the duration of labor, and the early part of extra-uterine life. For computation purposes the lower limit of viability is taken as 28 complete weeks of gestation and the early part of extra-uterine life is taken to be the first 7 days of life.

One formula for the **perinatal mortality ratio** is, then, the number of deaths under 1 week of age and late fetal deaths per 1,000 live births in a year:

$$\frac{D^z + D^f}{B} \times 1,000 \qquad (35)$$

where D^z represents deaths under 1 week, D^f represents late fetal deaths, and B represents births. Another formula is the number of neonatal and late fetal deaths per 1,000 live births in a year:

$$\frac{D^n + D^f}{B} \times 1,000 \qquad (36)$$

where D^n represents neonatal deaths.

The corresponding **perinatal mortality rates** differ by including late fetal deaths in the base of the ratio, thus approximating a probability more closely. Specifically, the perinatal mortality rate is defined as

$$\frac{D^z + D^f}{B + D^f} \times 1,000 \qquad (37)$$

or

$$\frac{D^n + D^f}{B + D^f} \times 1,000 \qquad (38)$$

where the symbols have the same meaning as above.

Illustrative perinatal ratios and rates for several countries around 1960 are as follows: [36]

	Perinatal mortality ratio		Perinatal mortality rate	
Country and year	$\dfrac{D^n + D^f}{B}$	$\dfrac{D^z + D^f}{B}$	$\dfrac{D^n + D^f}{B + D^f}$	$\dfrac{D^z + D^f}{B + D^f}$
Chile (1961).................	61.7	48.6	60.1	47.3
Czechoslovakia (1961).....	22.8	20.2	22.6	20.0
Italy (1962)....................	46.2	39.7	45.2	38.8
Japan (1960).................	47.9	41.4	46.4	40.2
Mauritius (1960)............	102.2	91.8	95.3	87.2
Mexico (1961)...............	43.7	33.4	43.0	32.8
United States (1960)........	31.4	29.4	31.0	29.0

These various measures give essentially the same indications of international differences. The perinatal mortality rate has a small theoretical advantage over the perinatal mortality ratio, but the perinatal mortality ratio is probably more stable for international comparison.

---

[36] United Nations. *Demographic Yearbook, 1966*, tables 7, 9, 12, and 18. and idem. *Demographic Yearbook, 1961*, 1962, table 15.

SUGGESTED READINGS

Antonovsky, Aaron. "Social Class, Life Expectancy, and Overall Mortality." *Milbank Memorial Fund Quarterly* 45(2):31–73, Part I, Apr. 1967.

Arriaga, Eduardo E. "Rural-Urban Mortality in Developing Countries: An Index for Detecting Rural Underregistration." *Demography* 4(1):98–107. Dec. 1967.

Aubenque, Maurice J. "Statement by the Moderator." Session B.3. "Mortality, Morbidity, and Causes of Death." In United Nations, *World Population Conference, 1965* (Belgrade). Vol. I, pp. 151–159. New York, 1966.

Barclay, George W. *Techniques of Population Analysis*. New York, John Wiley and Sons, 1958. Pp. 138–143.

Benjamin, B. *Elements of Vital Statistics*. London, George Allen & Unwin, 1959. Pp. 78–109 and 124–141.

Bourgeois-Pichat, Jean. "Essai sur la mortalité 'biologiqué' de l'homme" (Essay on the "Biological" Mortality of Man). *Population* (Paris), 7(3):381–394. July–Sept. 1952.

Chandra Sekar, C., and Deming, W. E. "On a Method of Estimating Birth and Death Rates and the Extent of Registration." *Journal of the American Statistical Association* 44(245):101–115. Mar. 1949.

Coulter, Elizabeth, and Guralnick, Lillian. "Analysis of Vital Statistics by Census Tract." *Journal of the American Statistical Association* 54(288):730–740. Dec. 1959.

SUGGESTED READINGS – Continued

Cox, Peter R. *Demography*, Chapter 8. Cambridge (Eng.), Cambridge University Press, 1970.

Dorn, Harold F. "Mortality." Chapter 19 in Philip M. Hauser and Otis Dudley Duncan (eds.), *The Study of Population: An Inventory and Appraisal*. Chicago, Ill., University of Chicago Press, 1959. Pp. 437–471.

————. "Underlying and Contributory Causes of Death." In U.S. Public Health Service, *Epidemiological Approaches to the Study of Cancer and Other Chronic Diseases*, edited by William Haenszel. National Cancer Institute Monograph 19, pp. 421–430. Jan. 1966.

El-Badry, M. A. "Higher Female Than Male Mortality in Some Countries of South Asia: A Digest." *Journal of the American Statistical Association* 64(328):1234–1244. Dec. 1969.

Freedman, Ronald, Coombs, Lolagene C., and Friedman, Judith. "Social Correlates of Fetal Mortality." *Milbank Memorial Fund Quarterly* 44(3):327–344. Part 1. July 1966.

Great Britain, Registrar General. *Decennial Supplement, England and Wales, 1951, Occupational Mortality*. Part 1, 1954.

Guralnick, Lillian. Summary of "Multiple Causes of Death, United States, 1955." In United Nations, *World Population Conference, 1965* (Belgrade). Vol. II, p. 452. New York, 1967.

Guralnick, Lillian, and Winter, E. B. "A Note on Cohort Infant Mortality Rates." *Public Health Reports* 80(8):692–694. Aug. 1965.

Hambright, Thea Z. "Comparison of Information on Death Certificates and Matching 1960 Census Records: Age, Marital Status, Race, Nativity and Country of Origin." *Demography* 6(4):413–423. Nov. 1969.

Hamilton, Horace C. "Ecological and Social Factors in Mortality Variation." *Eugenics Quarterly* 2(4):212–223. Dec. 1955.

Hammond, E. I. "Studies in Fetal and Infant Mortality. I. A Methodological Approach to the Definition of Perinatal Mortality." *American Journal of Public Health* 55(7):1012–1023. July 1965.

Herdan, G. "Causes of Excess Male Mortality in Man." *Acta Genetica et Statistica Medica* (Basel), 3(4): 351–376. 1952.

Kannisto, Väinö. "The Value of Certain Refinements of the Infant-Mortality Rate." *Bulletin of the World Health Organization* 16(4):763–782. 1957.

Kilpatrick, S. J. "Occupational Mortality Indices." *Population Studies* (London), 16(2):175–187. Nov. 1962.

Kitagawa, Evelyn M. "Theoretical Considerations in the Selection of a Mortality Index, and Some Empirical Comparisons." *Human Biology* 38(3):293–308. Sept. 1966.

Kitagawa, Evelyn M., and Hauser, Philip M. "Methods Used in a Current Study of Social and Economic Differentials in Mortality." *Emerging Techniques in Population Research*. 1962 Annual Conference, Milbank Memorial Fund, New York, 1963. Pp. 250–266.

————. *Differential Mortality in the United States: A Study in Socioeconomic Epidemiology*. Vital and Health Statistics Monographs. Cambridge, Mass., Harvard University Press, 1973.

Kpedekpo, G. M. K. "Evaluation and Adjustment of Vital Registration Data from the Compulsory Registration Areas of Ghana." *Demography* 5(1):86–92. 1968.

Krueger, Dean E. "New Numerators for Old Denominators– Multiple Causes of Death." In U.S. Public Health Service, *Epidemiological Approaches to the Study of Cancer and Other Chronic Diseases*, edited by William Haenszel. National Cancer Institute Monograph 19, pp. 431–442. Jan. 1966.

Legaré, Jacques M. "Quelques considerations sur les tables de mortalité de génération. Application à l'Angleterre et au Pays de Galles" (Some Considerations Relating to Generation Life Tables. Application to England and Wales). *Population* (Paris), 21(5):915–938. Sept.–Oct. 1966.

Logan, W. P. D. "The Measurement of Infant Mortality." *Population Bulletin of the United Nations*, No. 3. Pp. 30–67. Oct. 1953.

Madigan, Francis C. "Are Sex Mortality Differentials Biologically Caused?" *Milbank Memorial Fund Quarterly* 35(2): 202–223. Apr. 1957.

Martin, W. J. "A Comparison of the Trends of Male and Female Mortality." *Journal of the Royal Statistical Society*. Series A (General), 114(3):287–306. 1951.

Moriyama, Iwao M. "The Eighth Revision of the International Classification of Diseases." *American Journal of Public Health* 56(8):1277–1280. 1966.

Moriyama, I. M., and Guralnick, L. "Occupational and Social Class Differences in Mortality." *Trends and Differentials in Mortality*. 1955 Annual Conference, Milbank Memorial Fund, New York, 1956. Pp. 61–73.

Moriyama, Iwao M., *et al.*, "Inquiry into Diagnostic Evidence Supporting Medical Certifications of Death." *American Journal of Public Health* 48(10):1376–1387. Oct. 1958.

Puffer, Ruth R., and Griffith, G. Wynne. "The Inter-American Investigation of Mortality." United Nations, *World Population Conference, 1965* (Belgrade). Vol. II, pp. 426–432. New York, 1967.

Rosenberg, Harry M. "Recent Developments in Seasonally Adjusting Vital Statistics." *Demography* 3(2):305–318. 1966.

Rosenwaike, Ira. "Seasonal Variation of Deaths in the United States, 1951–60." *Journal of the American Statistical Association* 61(315):706–719. Sept. 1966.

Silcock, H., "The Comparison of Occupational Mortality Rates." *Population Studies* (London), 13(2):183–192. Nov. 1959.

SUGGESTED READINGS – Continued

Sowder, Wilson T. "Why is the Sex Difference in Mortality Increasing?" In U.S. Public Health Service, *Public Health Reports*, Vol. 69, No. 9. Pp. 860–864. Sept. 1954.

Spiegelman, Mortimer, *et al.*, "Problems in the Medical Certification of Causes of Death." *American Journal of Public Health* 48(1):71–80. 1958.

Spiegelman, Mortimer, and Marks, Herbert H. "Empirical Testing of Standards for the Age Adjustment of Death Rates by the Direct Method." *Human Biology* 38(3):280–292. Sept. 1966.

United Nations. *Foetal, Infant, and Early Childhood Mortality*, Vol. I, *The Statistics*. Series A, Population Studies, No. 13, 1954.

————. *Foetal, Infant, and Early Childhood Mortality*, Vol. II, *Biological, Social and Economic Factors*. Series A, Population Studies, No. 13, Add. 1. 1954.

————. *Population Bulletin of the United Nations*, No. 6, *With Special Reference to the Situation and Recent Trends of Mortality in the World*. Esp. pp. 147–201, "Factor Analysis of Sex-Age-Specific Death Rates: A Contribution to the Study of the Dimensions of Mortality." 1962.

U.S. Bureau of the Census. *Vital Statistics-Special Reports*, Vol. 19, No. 21. "Effect of Changing Birth Rates upon Infant Mortality Rates." By I. M. Moriyama and T. N. E. Greville. Nov. 10, 1944. Pp. 401–412.

U.S. National Center for Health Statistics. *Vital and Health Statistics*, Series 2, No. 34. "Comparability of Marital Status, Race, Nativity, and Country of Origin on the Death Certificate and Matching Census Record, United States, May–Aug. 1960." May 1969.

————. *Vital Statistics Rates in the United States, 1940–1960*. By Robert D. Grove and Alice M. Hetzel. PHS Pub. 1677. 1968. Chapter 2.

————. *Vital and Health Statistics*, Series 2, No. 29. "Comparability of Age on the Death Certificate and Matching Census Record, United States, May–Aug. 1960." June 1968.

U.S. National Office of Vital Statistics. *Vital Statistics-Special Reports, Selected Studies*, Vol. 33, No. 10. "Mortality, Occupation, and Socio-Economic Status." By Jean Daric.

Pp. 175–187. Sept. 21, 1951. (Translation of "Mortalité, profession, et situation sociale," *Population* (Paris), 4(4):672–694, Oct.–Dec. 1949.)

————. *Vital Statistics of the United States, 1950*. Vol. I, Chapters 2, 7, and 8. 1954.

Valaoras, V. G. "Refined Rates for Infant and Childhood Mortaltiy." *Population Studies*. (London), 4(3):253–266, December 1950.

Vincent, Paul. "La mortalité des vieillards" (The Mortality of Aged Persons). *Population (Paris), 6(2):181–204, Apr.–June 1951.*

Wolfenden, Hugh H. "On the Theoretical and Practical Considerations Underlying the Direct and Indirect Standardization of Death Rates." *Population Studies* (London), 16(2):188–190. Nov. 1962.

————. *Population Statistics and Their Compilation*. Chicago, Ill., University of Chicago Press, 1954. Chapters IV, VI, IX, XI, and XII.

World Health Organization, *International Classification of Diseases*, 1965 revision, Vol. I, 1967.

Yerushalmy, J. "A Mortality Index for Use in Place of the Age-Adjusted Death Rate." *American Journal of Public Health* 41:907–922. Aug. 1951.

CHAPTER 15

The Life Table

INTRODUCTION

Nature and Use of Life Tables

Some of the measures of mortality that were discussed or used in chapter 14 and earlier chapters are derived from a statistical model known as a life table. A life table is designed essentially to measure mortality, but it is employed by a variety of specialists in a variety of ways. It is used by public health workers, demographers, actuaries, and many others in studies of longevity, fertility, migration, and population growth, as well as in making projections of population size and characteristics and in studies of widowhood, orphanhood, length of married life, length of working life, and length of disability-free life. In the simplest form of a life table, the entire table is generated from age-specific mortality rates and the resulting values are used to measure mortality, survivorship, and life expectation. In other applications the mortality rates in the life table are combined with other demographic data into a more complex model which measures the combined effect of mortality and changes in one or more socioeconomic characteristics, e.g., a table of working life, which combines mortality rates and labor force participation rates and measures their combined effect on working life.

Life tables are, in essence, one form of combining mortality rates of a population at different ages into a single statistical model. They are principally used to measure the level of mortality of the population involved. One of their main advantages over other methods of measuring mortality is that they do not reflect the effects of the age distribution of an actual population and do not require the adoption of a standard population for acceptable comparisons of levels of mortality in different populations. Another is that a life table readily permits making mortality allowances for age cohorts, eliminating the burdensome task of compiling death statistics for age cohorts from annual death statistics by age even when the latter are available.

Types of Life Tables

Life tables differ according to the reference year of the table, the age detail, and number of factors comprehended by the table. We may distinguish two types of life tables according to the reference year of the table—the current or period life table and the generation or cohort life table. The first type of table is based on the experience over a short period of time, such as a year, three years, or an intercensal period, in which mortality has remained substantially the same. Commonly,

the death statistics used for a current life table relate to a period of one to three years, and the population data used relate to the middle of that period (usually close to the date of a census). This type of table, therefore, represents the combined mortality experience by age of the population in a particular short period of time (treated synthetically or viewed cross-sectionally); it does not represent the mortality experience of an actual cohort. Instead, it assumes a hypothetical cohort that is subject to the age-specific death rates observed in the particular period. Therefore, a current life table may be viewed as a "snapshot" of current mortality. It is an excellent summary description of mortality in a year or a short period.

The second type of life table, the generation life table, is based on the mortality rates experienced by a particular birth cohort, e.g., all persons born in the year 1900. According to this type of table, the mortality experience of the persons in the cohort would be observed from their moment of birth through each consecutive age in successive calendar years until all of them die. Obviously, data over a long period of years are needed to complete a single table and it is not possible wholly on the basis of actual data to construct generation tables for cohorts born in this century. This type of table is useful for projections of mortality, for studies of mortality trends, and for the measurement of fertility and reproductivity. An illustration of a generation life table appears in table 15-6.

In general, unless otherwise specified, the term "life table" is used in this chapter to refer to a current life table.

Life tables are also classified into two types—complete (or unabridged) and abridged—according to the length of the age interval in which the data are presented. A complete life table contains data for every single year of age from birth to the last applicable age. An abridged life table, on the other hand, contains data by intervals of 5 or 10 years of age. The simpler abridged life table is usually prepared rather than the more elaborate complete life table. Values for 5 or 10-year intervals are sufficiently accurate for most purposes and the abridged table is less burdensome to prepare. Moreover, it is often more convenient to use. Tables 15-1 and 15-2 are illustrations of complete and abridged life tables, respectively. Occasionally, the basic values from a complete life table are presented only for every fifth age in order to economize on space.

We may also distinguish a standard life table, which is concerned only with the general mortality experience of a cohort by age, from a multiple decrement table, which describes the separate and combined effects of more than one factor. Mortality is always involved. Multiple decrement tables are of several forms. The mortality factor may be applied in terms of component death rates (e.g., by cause of death), or mortality may be combined with changes in one or more socioeconomic characteristic(s) of the population. Some tables of the latter kind describe merely the diminution of an original cohort through these factors (e.g., the attrition of the single population through mortality and marriage), and others describe the effect of accession to and withdrawal from a segment of the original cohort which acquires a given characteristic (e.g., a table of working life which combines mortality rates and labor force participation rates; a table of school life, which combines mortality rates and enrollment rates; or a nuptiality table, which combines mortality rates and data on the prevalence or incidence of marriage). We may appropriately call the latter type of table an increment-decrement table.

Availability of Life Tables

The first life table developed in a logical way, Halley's life table, was published in 1693 and was based on birth and death registration data for the city of Breslau during the years 1687 to 1691. The assumption adopted in the preparation of this table that the population of Breslau had remained stationary (i.e., that the total population and the numbers in each age and sex group did not change over many decades) was not entirely correct and, therefore, the resulting life table could not be regarded as being correct. Other life tables were prepared in the seventeenth and eighteenth centuries on the basis of limited data, but they are subject to necessary simplifying assumptions that render them inexact.

The first scientifically correct life table based on both population and death data classified by age was prepared by Milne and published in 1815. It was based on the mortality experience in two parishes of Carlisle, England, during the period 1779–87. A large number of life tables have been published since then. In the early years, most of these pertained to European countries, particularly Scandinavian countries, but life tables are now available for most countries of the world and every continent is represented.

In the United States, official complete life tables have been prepared since 1900–02 in connection with the decennial censuses of population. These tables are based on the registered deaths in the expanding Death Registration Area (continental United States for the 1929–31 table) in the 3-year period containing the census year. Hence, there is an unbroken decennial series of complete life tables covering this century.[1] There are some tables also for intercensal periods, as for 1901–10, 1920–29, and 1930–39. An annual series of abridged life tables was started in 1945 and has continued to the present. They are now published on a provisional basis for a given year just after the close of the year and on a revised basis in the annual volume on mortality statistics for the year. These annual tables are based on the annual death registration and on postcensal estimates of population.

In addition to national tables, tables have been prepared from time to time for geographic areas varying in size from specific cities to geographic divisions and regions. Sets of life tables for all States have now been published in connection with the censuses of 1940 (whites only), 1950 (nonwhite tables for States in South Region only), and 1960 (color for selected States), i.e., for 1939–41, 1949–51, and 1959–61; and similar tables are planned for 1970.[2] Life tables for geographic divisions or regions corresponding to these State tables have also been published.[3] For the first time, life tables for metropolitan and nonmetropolitan areas as a whole were published for 1959–61.[4] For some periods prior to 1940; separate tables were published for the urban and rural populations.

Special compilations of life tables or analytic studies of life tables were published in connection with the 1910, 1930, and 1940 censuses.[5] The life tables included in these volumes covered the Death Registration States or the United States as a whole, specified individual States and cities, and the urban and rural parts of the Death Registration States in some cases. A general guide to the U.S. life tables for the period 1900 to 1959 has been published.[6]

The United Nations *Demographic Yearbook* for 1966 is the most recent one to contain a comprehensive collection of national life tables. Values of the expectation of life, (1-year) mortality rates, and survivors from the two latest official complete life tables, largely for years in the period 1950–65, are included. These values are shown for single ages up to 5 years and at every fifth age thereafter to 85 years. Similar tables in the 1961, 1957, and 1953 *Yearbooks* carry the life tables back to 1900. In addition, the latest value of the expectation of life is published every year. In the 1966 collection, values for the expectation of life are shown for 115 countries, but the other two life table functions are shown only for 93 or 94 countries. Values for the expectation of life at selected ages are presented for many countries on a current basis in the October–December issue of *Population Index*.

A compilation of life tables for a large number of countries spanning a wide range of time has been published by Keyfitz and Flieger.[7] Life tables for males and females for each country and date included in the compilation were derived by electronic computer on the basis of official data on births, deaths, and population classified by age and sex, where such data were considered to be of satisfactory quality. Only 29 percent of the world's population is covered, however; this population falls mostly in Northern America and Europe; Africa, Asia, and the U.S.S.R. are poorly represented. Arriaga has prepared new life tables for the countries of Latin America

---

[1] U.S. Bureau of the Census, *United States Life Tables, 1930,* 1936; U.S. Bureau of the Census, *United States Life Tables and Actuarial Tables, 1939–41,* by T. N. E. Greville, 1946; U.S. National Office of Vital Statistics, *Vital Statistics-Special Reports, Life Tables for 1949–51,* Vol. 41, No. 1, "United States Life Tables, 1949–51," November 23, 1954; U.S. National Center for Health Statistics, *Life Tables: 1959–61,* Vol. 1, No. 1, "United States Life Tables: 1959–61," December 1964.

[2] U.S. National Office of Vital Statistics, *State and Regional Life Tables, 1939–41,* 1948; U.S. National Office of Vital Statistics, *Vital Statistics—Special Reports,* Vol. 41, Supplement, *State Life Tables: 1949–51,* 1956; and U.S. National Center for Health Statistics, *U.S. Life Tables, 1959–61,* Vol. 2, Nos. 1–26 and Nos. 27–51, June 1966.

[3] U.S. National Office of Vital Statistics, *Vital Statistics—Special Reports, Life Tables for 1949–51,* Vol. 41, No. 4, "Life Tables for the Geographic Divisions of the United States, 1949–51," July 26, 1956; U.S. National Center for Health Statistics, *Life Tables: 1959–61,* Vol. 1, No. 3, "Life Tables for the Geographic Divisions of the United States, 1959–61," May 1965.

[4] U.S. National Center for Health Statistics, *Life Tables: 1959–61,* Vol. 1, No. 5, "Life Tables for Metropolitan and Nonmetropolitan Areas of the United States: 1959–61," December 1967.

[5] U.S. Bureau of the Census, *United States Life Tables, 1890, 1901, 1910, and 1901–10,* by James W. Glover, 1921; idem, *United States Life Tables, 1930,* and idem, *United States Life Tables and Actuarial Tables, 1939–41,* by T. N. E. Greville, 1946.

[6] U.S. National Center for Health Statistics, *Guide to United States Life Tables, 1900–59,* 1963.

[7] Nathan Keyfitz and Wilhelm Flieger, *World Population: An Analysis of Vital Data,* Chicago, University of Chicago Press, 1968.

employing principally census data and stable population theory and techniques.[8]

Sets of model life tables have been published by the United Nations and by Coale and Demeny.[9] These sets of tables correspond to values for life expectancy at birth varying generally by fixed intervals in years. The U.N. tables cover the range from 20 years to 73.9 years, mostly in intervals of 2.5 years. The Coale and Demeny tables relate to four different regions, distinguished by the age pattern of mortality. The tables for females correspond to values for life expectancy at birth ranging from 20 to 77.5 years in intervals of 2.5 years; and companion male tables are presented. These tables are useful for estimating life table functions for countries for which not all the data necessary for the preparation of a life table are available. These tables are discussed in detail in chapter 24. The model life tables for the West Region developed by Coale and Demeny were reproduced by the United Nations in its Manual IV of Manuals on Methods of Estimating Population.[10]

ANATOMY OF THE LIFE TABLE

Life Table Functions

The basic life tables functions—$_nq_x$, l_x, $_nd_x$, $_nL_x$, T_x, and $\overset{\circ}{e}_x$—can be observed in table 15-1. These six columns are generally calculated and published for every life table. However, in some cases, due to limitations of space, some of the columns may be omitted. (For example, the United Nations publishes only q_x, l_x, and $\overset{\circ}{e}_x$ in the *Demographic Yearbook*.) This is done without a significant loss of information since the functions are interrelated and some can be directly calculated from the others. In general, the mortality rate ($_nq_x$) is the basic function in the table, i.e., the initial function from which all other life table functions are derived.

Alternative Interpretations

Life table functions are subject to two different interpretations depending on the interpretation given to the life table as a whole. In the more common interpretation, the life table is viewed as depicting the lifetime mortality experience of a single cohort of newborn babies, who are subject to the age-specific mortality rates on which the table is based. In the second interpretation of the life table, it is viewed as a stationary population resulting from the (unchanging) schedule of age-specific mortality rates shown and a constant annual number of births.

The Life Table as the Mortality Experience of a Cohort. — Under the first interpretation, the life table model conceptually traces a cohort of newborn babies through their entire life under the assumption that they are subject to the current observed schedule of age-specific mortality rates. The cohort

[8] Eduardo E. Arriaga, *New Life Tables for Latin American Populations in the Nineteenth and Twentieth Centuries*, Population Monograph Series No. 3, Institute of International Studies, University of California, Berkeley, California, 1968.

[9] United Nations, *Age and Sex Patterns of Mortality: Model Life Tables for Underdeveloped Countries*, Population Studies, Series A, No. 22, 1955, and United Nations, *Methods for Population Projections by Sex and Age*, Manual III, Manuals on Methods of Estimating Population, Series A Population Studies, No. 25, 1956, esp. pp. 70–81; and A. J. Coale and Paul Demeny, *Regional Model Life Tables and Stable Populations*, Princeton, N.J., Princeton University Press, 1966.

[10] United Nations, *Methods of Estimating Basic Demographic Measures from Incomplete Data*, Series A, Population Studies, No. 42, 1967.

of newborn babies, called the radix of the table, is usually assumed to number 100,000. In this case, the interpretation of the life table functions in an abridged table would be as follows:

x to $x + n$ — The period of life between two exact ages. For instance, "20–25" means the 5-year interval between the 20th and 25th birthdays.

$_nq_x$ — The proportion of the persons in the cohort alive at the beginning of an indicated age interval (x) who will die before reaching the end of that age interval ($x+n$). For example, according to table 15–2, the proportion dying in the age interval 20–25 is 0.00618; that is, out of every 100,000 persons alive and exactly 20 years old, 618 will die before reaching their 25th birthday. In other words, the $_nq_x$ values represent the probability that a person at his xth birthday will die before reaching his $x + n$th birthday.

l_x — The number of persons living at the beginning of the indicated age interval (x) out of the total number of births assumed as the radix of the table. Again, according to table 15–2, out of 100,000 newborn babies, 96,111 persons would survive to exact age 20.

$_nd_x$ — The number of persons who would die within the indicated age interval (x to $x + n$) out of the total number of births assumed in the table. Thus, according to table 15–2, there would be 594 deaths between exact ages 20 and 25 to the initial cohort of 100,000 newborn babies.

$_nL_x$ — The number of person-years that would be lived within the indicated age interval (x to $x + n$) by the cohort of 100,000 births assumed. Thus, according to table 15–2, the 100,000 newborn babies would live 479,098 person-years between exact ages 20 and 25. Of the 96,111 persons who reach age 20, the 95,517 who survive to age 25 would live 5 years each ($95,517 \times 5 = 477,585$ person-years) and the 594 who die would each live varying periods of time less than 5 years, averaging about 2½ years

$$(594 \times 2.55 = 1,513 \text{ person-years}).$$

T_x — The total number of person-years that would be lived after the beginning of the indicated age interval by the cohort of 100,000 births assumed. Thus, according to table 15–2, the 100,000 newborn babies would live 5,053,117 person-years after their 20th birthday.

$\overset{\circ}{e}_x$ — The average remaining lifetime (in years) for a person who survives to the beginning of the indicated age interval. This function is also called the complete expectation of life, or simply, life expectancy. Thus, according to table 15–2 a person who reaches his 20th birthday should expect to live 52.58 years more, on the average.

The interpretation of the functions in a complete life table is the same as in the abridged table except that the q_x, d_x, and L_x values relate to single-age intervals. The l_x, T_x, and $\overset{\circ}{e}_x$ values have the same interpretation as in the abridged table since they are not "interval" values but pertain to exact age x.

The Life Table as a Stationary Population. — An alternate interpretation of the life table is the one associated with the concept of a stationary population. A stationary population is defined as a population whose total number and distribution by age do not change with time. Such a hypothetical population could be obtained if the number of births per year remained constant (usually assumed at 100,000) for a long period of time and each cohort of births experienced the current observed mortality rates throughout life. The annual number of

Table 15–1.— Complete Life Table for the Total Population of the United States: 1959–61

Age interval — Period of life between two exact ages stated in years (x to x+n)	Proportion dying — Proportion of persons alive at beginning of age interval dying during interval ($_nq_x$) (1)	Of 100,000 born alive — Number living at beginning of age interval (l_x) (2)	Of 100,000 born alive — Number dying during age interval ($_nd_x$) (3)	Stationary population — In the age interval ($_nL_x$) (4)	Stationary population — In this and all subsequent age intervals (T_x) (5)	Average remaining lifetime — Average number of years of life remaining at beginning of age interval ($\overset{\circ}{e}_x$) (6)
0–1	.02593	100,000	2,593	97,815	6,989,030	69.89
1–2	.00170	97,407	165	97,324	6,891,215	70.75
2–3	.00104	97,242	101	97,192	6,793,891	69.87
3–4	.00080	97,141	78	97,102	6,696,699	68.94
4–5	.00067	97,063	65	97,031	6,599,597	67.99
5–6	.00059	96,998	57	96,969	6,502,566	67.04
6–7	.00052	96,941	50	96,916	6,405,597	66.08
7–8	.00047	96,891	46	96,868	6,308,681	65.11
8–9	.00043	96,845	42	96,824	6,211,813	64.14
9–10	.00039	96,803	38	96,784	6,114,989	63.17
10–11	.00037	96,765	36	96,747	6,018,205	62.19
11–12	.00037	96,729	36	96,711	5,921,458	61.22
12–13	.00040	96,693	39	96,674	5,824,747	60.24
13–14	.00048	96,654	46	96,630	5,728,073	59.26
14–15	.00059	96,608	57	96,580	5,631,443	58.29
15–16	.00071	96,551	68	96,517	5,534,863	57.33
16–17	.00082	96,483	80	96,443	5,438,346	56.37
17–18	.00093	96,403	89	96,358	5,341,903	55.41
18–19	.00102	96,314	98	96,265	5,245,545	54.46
19–20	.00108	96,216	105	96,163	5,149,280	53.52
20–21	.00115	96,111	110	96,056	5,053,117	52.58
21–22	.00122	96,001	118	95,942	4,957,061	51.64
22–23	.00127	95,883	122	95,822	4,861,119	50.70
23–24	.00128	95,761	123	95,700	4,765,297	49.76
24–25	.00127	95,638	121	95,578	4,669,597	48.83
25–26	.00126	95,517	120	95,456	4,574,019	47.89
26–27	.00125	95,397	120	95,337	4,478,563	46.95
27–28	.00126	95,277	120	95,217	4,383,226	46.00
28–29	.00130	95,157	123	95,095	4,288,009	45.06
29–30	.00136	95,034	129	94,970	4,192,914	44.12
30–31	.00143	94,905	136	94,836	4,097,944	43.18
31–32	.00151	94,769	143	94,698	4,003,108	42.24
32–33	.00160	94,626	151	94,551	3,908,410	41.30
33–34	.00170	94,475	160	94,395	3,813,859	40.37
34–35	.00181	94,315	171	94,229	3,719,464	39.44
35–36	.00194	94,144	183	94,053	3,625,235	38.51
36–37	.00209	93,961	196	93,863	3,531,182	37.58
37–38	.00228	93,765	214	93,658	3,437,319	36.66
38–39	.00249	93,551	232	93,435	3,343,661	35.74
39–40	.00273	93,319	255	93,191	3,250,226	34.83
40–41	.00300	93,064	279	92,925	3,157,035	33.92
41–42	.00330	92,785	306	92,632	3,064,110	33.02
42–43	.00362	92,479	335	92,311	2,971,478	32.13
43–44	.00397	92,144	366	91,961	2,879,167	31.25
44–45	.00435	91,778	400	91,578	2,787,206	30.37
45–46	.00476	91,378	435	91,161	2,695,628	29.50
46–47	.00521	90,943	473	90,707	2,604,467	28.64
47–48	.00573	90,470	519	90,210	2,513,760	27.79
48–49	.00633	89,951	569	89,667	2,423,550	26.94
49–50	.00700	89,382	626	89,069	2,333,883	26.11
50–51	.00774	88,756	687	88,412	2,244,814	25.29
51–52	.00852	88,069	751	87,693	2,156,402	24.49
52–53	.00929	87,318	811	86,913	2,068,709	23.69
53–54	.01005	86,507	870	86,072	1,981,796	22.91
54–55	.01082	85,637	926	85,174	1,895,724	22.14
55–56	.01161	84,711	983	84,220	1,810,550	21.37
56–57	.01249	83,728	1,047	83,204	1,726,330	20.62
57–58	.01352	82,681	1,117	82,123	1,643,126	19.87
58–59	.01473	81,564	1,202	80,962	1,561,003	19.14
59–60	.01611	80,362	1,295	79,715	1,480,041	18.42
60–61	.01761	79,067	1,392	78,371	1,400,326	17.71
61–62	.01917	77,675	1,489	76,930	1,321,955	17.02
62–63	.02082	76,186	1,586	75,393	1,245,025	16.34
63–64	.02252	74,600	1,680	73,760	1,169,632	15.68
64–65	.02431	72,920	1,773	72,033	1,095,872	15.03
65–66	.02622	71,147	1,866	70,214	1,023,839	14.39
66–67	.02828	69,281	1,959	68,302	953,625	13.76
67–68	.03053	67,322	2,055	66,295	885,323	13.15
68–69	.03301	65,267	2,155	64,189	819,028	12.55
69–70	.03573	63,112	2,255	61,985	754,839	11.96
70–71	.03866	60,857	2,352	59,681	692,854	11.38
71–72	.04182	58,505	2,447	57,282	633,173	10.82
72–73	.04530	56,058	2,539	54,788	575,891	10.27
73–74	.04915	53,519	2,631	52,204	521,103	9.74
74–75	.05342	50,888	2,718	49,529	468,899	9.21
75–76	.05799	48,170	2,794	46,773	419,370	8.71
76–77	.06296	45,376	2,857	43,948	372,597	8.21
77–78	.06867	42,519	2,920	41,059	328,649	7.73
78–79	.07535	39,599	2,983	38,108	287,590	7.26
79–80	.08302	36,616	3,040	35,096	249,482	6.81
80–81	.09208	33,576	3,092	32,030	214,386	6.39
81–82	.10219	30,484	3,115	28,926	182,356	5.98
82–83	.11244	27,369	3,078	25,830	153,430	5.61
83–84	.12195	24,291	2,962	22,811	127,600	5.25
84–85	.13067	21,329	2,787	19,935	104,789	4.91
85–86	.14380	18,542	2,666	17,209	84,854	4.58
86–87	.15816	15,876	2,511	14,620	67,645	4.26
87–88	.17355	13,365	2,320	12,205	53,025	3.97
88–89	.19032	11,045	2,102	9,995	40,820	3.70
89–90	.20835	8,943	1,863	8,011	30,825	3.45
90–91	.22709	7,080	1,608	6,276	22,814	3.22
91–92	.24598	5,472	1,346	4,799	16,538	3.02
92–93	.26477	4,126	1,092	3,580	11,739	2.85
93–94	.28284	3,034	858	2,605	8,159	2.69
94–95	.29952	2,176	652	1,849	5,554	2.55
95–96	.31416	1,524	479	1,285	3,705	2.43
96–97	.32915	1,045	344	873	2,420	2.32
97–98	.34450	701	241	580	1,547	2.21
98–99	.36018	460	166	377	967	2.10
99–100	.37616	294	111	239	590	2.01
100–101	.39242	183	72	147	351	1.91
101–102	.40891	111	45	89	204	1.83
102–103	.42562	66	28	52	115	1.75
103–104	.44250	38	17	29	63	1.67
104–105	.45951	21	10	17	34	1.60
105–106	.47662	11	5	8	17	1.53
106–107	.49378	6	3	5	9	1.46
107–108	.51095	3	2	2	4	1.40
108–109	.52810	1	0	1	2	1.35
109–110	.54519	1	1	1	1	1.29

Source: U.S. National Center for Health Statistics, *Life Tables, 1959–61*, Vol. 1, No. 1, "United States Life Tables: 1959–61," December 1964, pp. 8–9.

deaths would thus equal 100,000 also, and there would be no change in the size of the population. In this case, the interpretation of x to x + n, $_nq_x$, and $\overset{\circ}{e}_x$ would be as previously indicated, but for the other life table functions it would be as follows:

l_x — The number of persons who reach the beginning of the age interval each year. Thus, if births remained constant at 100,000 per year, according to table 15–2, there would be 96,111 persons reaching exact age 20 every year.

$_nd_x$ — The number of persons that die each year within the indicated age interval. Thus, if births remained constant at 100,000 per year, according to table 15–2, there would be 594 deaths between exact ages 20 and 25 every year.

$_nL_x$ — The number of persons in the population who at any

Table 15-2.—Abridged Life Table for the Total Population of the United States: 1959-61

Age interval	Proportion dying	Of 100,000 born alive		Stationary population		Average remaining lifetime
Period of life between two exact ages stated in years (x to x + n)	Proportion of persons alive at beginning of age interval dying during interval $(_nq_x)$	Number living at beginning of age interval (l_x)	Number dying during age interval $(_nd_x)$	In the age interval $(_nL_x)$	In this and all subsequent age intervals (T_x)	Average number of years of life remaining at beginning of age interval $(\overset{o}{e}_x)$
	(1)	(2)	(3)	(4)	(5)	(6)
0-1...........................	.02593	100,000	2,593	97,815	6,989,030	69.89
1-5...........................	.00420	97,407	409	388,649	6,891,215	70.75
5-10..........................	.00240	96,998	233	484,361	6,502,566	67.04
10-15.........................	.00221	96,765	214	483,342	6,018,205	62.19
15-20.........................	.00456	96,551	440	481,746	5,534,863	57.33
20-25.........................	.00618	96,111	594	479,098	5,053,117	52.58
25-30.........................	.00641	95,517	612	476,075	4,574,019	47.89
30-35.........................	.00802	94,905	761	472,709	4,097,944	43.18
35-40.........................	.01147	94,144	1,080	468,200	3,625,235	38.51
40-45.........................	.01812	93,064	1,686	461,407	3,157,035	33.92
45-50.........................	.02869	91,378	2,622	450,814	2,695,628	29.50
50-55.........................	.04557	88,756	4,045	434,264	2,244,814	25.29
55-60.........................	.06663	84,711	5,644	410,224	1,810,550	21.37
60-65.........................	.10017	79,067	7,920	376,487	1,400,326	17.71
65-70.........................	.14463	71,147	10,290	330,985	1,023,839	14.39
70-75.........................	.20847	60,857	12,687	273,484	692,854	11.38
75-80.........................	.30297	48,170	14,594	204,984	419,370	8.71
80-85.........................	.44776	33,576	15,034	129,532	214,386	6.39
85 and over...................	1.00000	18,542	18,542	84,854	84,854	4.58

Source: Based on table 15-1.

moment are living within the indicated age interval. Thus, if births remained constant at 100,000 per year, according to table 15-2, there would be 479,098 persons living between exact ages 20 and 25 in the population at any time.

T_x—The number of persons in the population who at any moment are living within the indicated age interval and all higher age intervals. Thus, if births remained at 100,000 per year, according to table 15-2, there would be 5,053,117 persons over exact age 20 living in the population at any time.

Each interpretation has its particular applications. For example, the interpretation of the life table as the history of a cohort is applied in public health studies and mortality analysis, and in the calculation of survival rates for estimating population, net migration, fertility, and reproductivity. The interpretation of the life table as a stationary population is used in the comparative measurement of mortality and in studies of population structure.

Life Span and Life Expectancy

In measuring longevity two concepts should be distinguished—**life span** and **life expectancy**. The first concept tries to establish numerically the extreme limit of age in life, i.e., the maximum age that human beings as a species could reach under optimum conditions. There is no known exact figure for this concept. For purposes of defining the concept more precisely and of excluding rare cases, we might define life span as the age beyond which less than about 0.1 percent of the original cohort lives. We know that very few persons live over a hundred years; but, owing to lack of precision in records, it is not known exactly whether the life span has been increasing, has remained constant, or has declined with time. Life span appears to be about 100 and may not have changed in historical times.

The situation is different with respect to life expectancy, the expected number of years to be lived, on the average. Sufficiently accurate records have been available for some time

for many countries from which estimates have been prepared. These estimates have generally come from a current life table, although in some instances they have been prepared on the basis of death statistics alone or of census data alone. According to this concept, longevity has shown a considerable improvement in modern times in most countries. On the other hand, for many centuries, there was apparently no upward trend in life expectancy and there has been little or no improvement still among some primitive groups (African pygmies, inhabitants of western New Guinea, etc.).

CONSTRUCTION OF CONVENTIONAL LIFE TABLES

General Considerations

The main concern in this section is with methods of constructing life tables where satisfactory data on births, deaths, and population are available. (Model life tables and indirect techniques of life table construction are often employed for statistically underdeveloped countries, as discussed in chapter 24.) Mathematical formulas and demographic procedures are discussed in detail, and techniques for manipulating the mortality data are presented. It should be observed that in every instance underlying these procedures, formulas, and techniques, there is an assumption that the data on births, deaths, and population are fully accurate. It is well known, however, that one of the most important aspects of the preparation of a life table is the testing of the data for possible biases and other errors. Most of the procedures used to check on the accuracy of the data have been discussed in previous chapters and will not be discussed again in the present one. It suffices to indicate that the level of inaccuracy that can be tolerated depends mostly on the intended use of the life table.

One factor that is usually involved in the selection of the method to be used in the construction of a life table is the degree of adherence to the observed data that is desired. Full adherence to the observed values implies that the final life table functions will exhibit all the fluctuations in the observed data, whether these be due to real variations or to errors in the data. On the other hand, overgraduation (i.e., excessive "smoothing" of the data by mathematical methods) would eliminate or reduce true variations in the age pattern of mortality rates. Therefore, it must be decided whether the emphasis in the life table should be on its closeness to the actual data or on its presentation of the underlying mortality picture after fluctuations have been removed. In the statistically developed countries, as the quality and quantity of data have improved with the passage of time, fewer fluctuations have been noticed in the observed mortality rates. As a result, the tendency has been to place stronger emphasis on adherence to the data and less emphasis on eliminating or reducing the fluctuations.

Since the abridged life table is generally sufficient for most purposes of demographic analysis (and since it would be possible, by interpolation, to expand an abridged life table into a complete life table that would satisfy most other needs), our discussion of procedures will be limited to these simpler life tables. Persons interested in the procedures for constructing complete life tables, or for constructing life tables with limited or defective data, will find relevant sources listed among the "Suggested Readings" at the end of this chapter. The former topic is also covered in the full edition of *The Methods and Materials of Demography*.

The Abridged Life Table

Specific Short-Cut Methods.—The most fundamental step in life table construction is one of converting observed age-specific death rates into their corresponding mortality rates, or probabilities of dying. In a complete life table, the basic formula for this transformation is

$$q_x = \frac{2m_x}{2+m_x} \qquad (1)$$

where m_x is the observed death rate at a given age and q_x is the corresponding probability of dying. This formula is based on the assumption that deaths between exact ages x and $x + 1$ occur, on the average, at age $x + \frac{1}{2}$, as, for example, when deaths at age x in a given year are rectangularly distributed by age and time interval. One of the key features of the various short-cut methods described below is the procedure for making this basic transformation. Another difference in the method is in the way the stationary population is derived. We will describe three short-cut methods here—the Reed–Merrell method, the Greville method, and the method of reference to a standard table.

The Reed–Merrell method.—The Reed–Merrell method is one of the most frequently used short-cut procedures for calculating an abridged life table.[11] In this method the mortality rates are read off from a set of standard conversion tables showing the mortality rates associated with various observed central death rates. The standard tables for $_3m_2$, $_5m_x$, and $_{10}m_x$ have been prepared on the assumption that the following exponential equation holds:

$$_nq_x = 1 - e^{-n \cdot _nm_x - an^3 \cdot _nm_x^2} \qquad (2)$$

where n is size of the age interval, $_nm_x$ is the central death rate, a is a constant, and e is the base of the system of natural logarithms. Reed and Merrell found that a value of $a = 0.008$ would produce acceptable results.

The conversion of $_nm_x$'s to $_nq_x$'s by use of the Reed-Merrell tables is usually applied to 5-year or 10-year data, but special

age groups are employed at both ends of the life table. At the younger ages the most frequently used groupings are (1) ages under 1, 1, and 2–4 or (2) ages under 1 and 1–4. Conversion tables for these ages or age groups, as well as for $_5m_x$ and $_{10}m_x$, were worked out by Reed and Merrell; they are reproduced in appendix A.

For example, the death rate for the age group 55–59 ($_5m_{55}$) observed in the United States in 1959, .013970, would be converted to the 5-year mortality rate ($_5q_{55}$), .06765, by using the Reed-Merrell table of values of $_5q_x$ associated with $_5m_x$. We look up .013970 in the $_5m_x$ column, read off the corresponding $_5q_{55}$ value, interpolating as required.

The conversion tables are used to derive $_5q_x$ from $_5m_x$ for all 5-year age groups from 5–9 on. At the higher ages, the mortality rate for the open end group, e.g., 85 years old and over, is evidently equal to one, since the life table ends at the age where there are no more survivors.

Once the mortality rates have been calculated, the construction of the abridged life table continues with the computation of each entry in the survivor column, l_x, and the death column, $_nd_x$, along standard lines, using the formulas

$$l_{x+n} = (1 - _nq_x)l_x \qquad (3)$$

$$_nd_x = l_x - l_{x+n} \qquad (4)$$

All three short-cut methods that are described in this section follow the same procedure in deriving l_x and $_nd_x$.

In the calculation of the next life table function, $_nL_x$, each of the three methods to be discussed follows a different procedure. In the Reed-Merrell method, T_x values are directly determined from the l_x's for ages 10 and over, or 5 and over, by use of the following equations:

$$T_x = -.20833\, l_{x-5} + 2.5\, l_x + .20833\, l_{x+5} + 5 \sum_{\alpha=1}^{\infty} l_{x+5\alpha} \qquad (5)$$

if the age intervals in the table are 5-year intervals, and

$$T_x = 4.16667\, l_x + .83333\, l_{x+10} + 10 \sum_{\alpha=1}^{\infty} l_{x+10\alpha} \qquad (6)$$

if the age intervals in the table are 10-year intervals. These equations are based on the assumption that the area under the l_x curve between any two ordinates is approximated by the area under a parabola through these two ordinates and the preceding and following ordinates (formula 5) or the following ordinate only (formula 6).

For the ages under 10, Reed and Merrell note that L_x may be determined directly from the following linear equations:

$$L_0 = .276\, l_0 + .724\, l_1 \qquad (7)$$

$$L_1 = .410\, l_1 + .590\, l_2 \qquad (8)$$

$$_4L_1 = .034\, l_0 + 1.184\, l_1 + 2.782\, l_5 \qquad (9)$$

$$_3L_2 = -.021\, l_0 + 1.384\, l_2 + 1.637\, l_5 \qquad (10)$$

$$_5L_5 = -.003\, l_0 + 2.242\, l_5 + 2.761\, l_{10} \qquad (11)$$

L_0 should be determined from l_0 and l_1 by use of separation factors appropriate for each situation, however. $_nL_x$ for ages 10 and over may be derived by differencing the T_x's, and e_x° is computed as the ratio of T_x to l_x.

11 Lowell J. Reed and Margaret Merrell, "A Short Method for Constructing an Abridged Life Table," *American Journal of Hygiene*, 30(2):33–62, September 1939; reproduced in U.S. Bureau of the Census, *Handbook of Statistical Methods for Demographers*, 1951, pp. 12–27.

Table 15–3. — Calculation of Abridged Life Table for Males in Rural India, 1957–58, by the Reed-Merrell Method

Age interval (exact ages, x to x + n)	$_nm_x$	$_nq_x$	$_nd_x$	l_x	$\sum\limits_{\alpha=0}^{\infty} l_{x+10\alpha}$	T_x	$_nL_x$	$\overset{\circ}{e}_x$
0 to 1.................	.1980	.154033	15,403	100,000	(^{1})	4,370,894	^{2}89,680	43.71
1 to 5.................	.0426	.157431	13,318	84,597	(^{1})	4,281,214	^{2}301,861	50.61
5 to 15................	.0055	.053734	3,830	71,279	(^{1})	3,979,353	692,386	55.83
15 to 25...............	.0035	.034480	2,326	67,449	362,615	3,286,967	663,161	48.73
25 to 35...............	.0042	.041259	2,687	65,123	295,166	2,623,806	638,501	40.29
35 to 45...............	.0058	.056598	3,534	62,436	230,043	1,985,305	609,698	31.80
45 to 55...............	.0128	.121293	7,144	58,902	167,607	1,375,607	559,479	23.35
55 to 65...............	.0322	.281283	14,559	51,758	108,705	816,128	449,289	15.77
65 to 75...............	.0727	.536649	19,963	37,199	56,947	366,839	267,820	9.86
75 to 85...............	.1700	.855026	14,737	17,236	19,748	99,019	88,466	5.74
85 to 95...............	.3973	.994677	2,486	2,499	2,512	10,553	10,499	4.22
95 and over...........	.9289	1.00000	13	13	13	14	^{3}14	1.08

^{1} Entries not required for the calculations desired.
^{2} Calculated directly from l_0, l_1, and l_5 by use of the formulas $L_0 = .276\, l_0 + .724\, l_1$ and $_4L_1 = .034\, l_0 + 1.184\, l_1 + 2.782\, l_5$.
^{3} Calculated from $l_x \div {}_\infty m_x$.

Source: Observed death rates ($_nm_x$) from India, The Cabinet Secretariat, *The National Sample Survey, Fourteenth Round: July 1958–July 1959*, No. 76, *Fertility and Mortality Rates in India*, 1963, p. 15; and Ajoy Kumar De and Ranjan Kumar Som, "Abridged Life Tables for Rural India, 1957–58," p. 5.

An illustration of the application of the Reed-Merrell method is given in table 15–3, which shows the calculation of an abridged life table for rural India in 10-year age intervals for 1957–58 based on death rates from the National Sample Survey (Fourteenth Round). In this case the conversion tables are employed to obtain $_nq_x$ at all ages and use is made of the tables for m_0, $_4m_1$, and $_{10}m_x$. The steps are as follows:

(1) Read off $_nq_x$ values corresponding to $_nm_x$ values from the appropriate conversion table in appendix A.

(2) Derive the l_x and $_nd_x$ columns by
 (a) Multiplying q_0 (.154033) by the radix l_0 (100,000) to get d_0 (15,403).
 (b) Substract d_0 from l_0 to get l_1 (84,597).
 (c) Continue multiplying the successive values of l_x by the corresponding $_nq_x$ values to get $_nd_x$, and subtracting the successive values of $_nd_x$ from l_x to get l_{x+n}.

(3) Sum the values of l_x from the end of life to age x

$$\left(= \sum_{\alpha=0}^{\infty} l_{x+10\alpha} \right).$$

(4) Substitute these sums and the indicated l_x values in equation (6) to get Tx. For example, T_{25} is obtained from equation (6) as follows:

$$T_{25} = 4.16667\,(65,123) + .83333\,(62,436)$$
$$+ 10\,(230,043) = 2,623,806.$$

(5) Derive $_nL_x$ for the ages under 5 years by use of equations (7) and (9). The separation factors given in equation (7) are believed to be reasonable for rural India, 1958–59.

Greville's method. — A method suggested by T. N. E. Greville converts the observed central death rates to the needed mortality rates by the use of the formula

$$_nq_x = \frac{_nm_x}{\dfrac{1}{n} + {}_nm_x\left[\dfrac{1}{2} + \dfrac{n}{12}\,({}_nm_x - \log_e c)\right]} \quad (12)$$

where c comes from an assumption that the $_nm_x$ values follow

an exponential curve.[12] Empirically, the value of c has been found to be between 1.08 and 1.10. $\log_e c$ could be assumed to be about 0.095 as an intermediate value. Using this method, the observed death rate of .013970 for $_5m_{55}$ (United States, 1959), cited in the preceding section, would also lead to a mortality rate of .06765 as follows:

$$_5q_{55} = \frac{.013970}{1/5 + .013970[\frac{1}{2} + 5/12(.013970 - .095)]}$$

$$_5q_{55} = \frac{.013970}{.206513}$$

$$_5q_{55} = .06765$$

The derivation of $_5q_{55}$ by this method requires several columns in a manual calculation, but it may be programmed for direct calculation by electronic computer on the basis of the $_5m_x$'s and the two constants n and $\log_e c$.

In Greville's method, the central death rates in the life table and the population are assumed to be the same, and the desired value of $_nL_x$ is calculated by the use of

$$_nL_x = \frac{_nd_x}{_nm_x} \quad (13)$$

For the last age interval, that is, the interval with the indefinite upper age limit, the usual approximation for $_\infty L_x$ is

$$_\infty L_x = \frac{l_x}{_\infty m_x} \quad (14)$$

An illustration of the application of the Greville method is presented in table 15–4 for the rural population of India in 1957–58. The basic data are the same as those used above. In this case the value for q_0 is assumed to equal the infant mortality rate observed in the National Sample Survey. The central death rates are transformed into mortality rates by evaluating formula (12), as shown in columns (1) to (4) of the table. The method of calculating the remaining columns in the table is straightforward except for column (7) relating to $_nL_x$. The

---

[12] T. N. E. Greville, "Short Methods of Constructing Abridged Life Tables," *Record of the American Institute of Actuaries*, 32(65):29–42, Part 1, June 1943; reproduced in U.S. Bureau of the Census, *Handbook of Statistical Methods for Demographers*, 1951, pp. 28–34.

Table 15–4. — Calculation of Abridged Life Table for Males in Rural India, 1957–58, by Greville's Method

Age interval (exact ages, x to x + n)	$_n m_x$ (1)	$\frac{1}{2}+\frac{n}{12}\left((1)-.09\right)=$ (2)	$\frac{1}{n}+\left((1)\times(2)\right)=$ (3)	$_n q_x$ (1) ÷ (3) = (4)	$_n d_x$ (5)	l_x (6)	L_x (5) ÷ (1) = (7)	T_x (8)	$\overset{o}{e}_x$ (9)
0 to 1.............	(X)	(X)	(X)	.153200	15,320	100,000	[1]89,736	4,394,579	43.95
1 to 5.............	.0426	.4842	.2706	.157428	13,331	84,680	312,934	4,304,843	50.84
5 to 10............	.0055	.4296	.1024	.053711	3,832	71,349	696,727	3,991,909	55.95
15 to 20...........	.0035	.4279	.1015	.034483	2,328	67,517	665,143	3,295,182	48.81
25 to 35...........	.0042	.4285	.1018	.041257	2,690	65,189	640,476	2,630,039	40.34
35 to 45...........	.0058	.4298	.1025	.056585	3,537	62,499	609,828	1,989,563	31.83
45 to 55...........	.0128	.4357	.1056	.121212	7,147	58,962	558,359	1,379,735	23.40
55 to 65...........	.0322	.4518	.1145	.281223	14,572	51,815	452,547	821,376	15.85
65 to 75...........	.0727	.4856	.1353	.537324	20,012	37,243	275,268	368,829	9.90
75 to 85...........	.1700	.5667	.1963	.866021	14,922	17,231	87,776	93,561	5.43
85 to 95...........	.3973	.7561	.4004	.992258	2,291	2,309	5,766	5,785	2.51
95 and over.......	.9289	(X)	(X)	1.000000	18	18	[2]19	19	1.06

X Not applicable.
[1] Calculated from l_0 and l_1 by use of the formula: $L_0 = .330\, l_0 + .670\, l_1$. *See preceding chapter for a discussion of the derivation of separation factors.*
[2] Calculated from $l_x \div _\infty m_x$.

Source: Observed death rates ($_n m_x$) from India, The Cabinet Secretariat, *The National Sample Survey, Fourteenth Round: July 1958–July 1959*, No. 76, *Fertility and Mortality Rates in India*, 1963, p. 15; and Ajoy Kumar De and Ranjan Kumar Som, "Abridged Life Tables for Rural India, 1957–58," p. 5.

value of L_0 was obtained by weighting l_0 and l_1 with the separation factors .330 and .670, respectively. At the other ages, formula (13) was used, with special formula (14) for the final open-end interval. The resulting value for $\overset{o}{e}_0$ was about the same for the Greville method (43.95 years) as for the Reed-Merrell method (43.71 years).

Method of reference to a standard table. — A third method that is used frequently in routine calculations bases the conversion of the observed $_n m_x$ to the life table $_n q_x$ on the relation that exists in a complete life table between the observed $_n m_x$ and the life table $_n q_x$. Since this method obtains the new table by reference to a standard table, it should only be used when mortality in both tables is of a comparable level.

A simple application of this concept assumes that in each age interval the relation of $_n q_x$ to observed $_n m_x$ shown by the standard table applies to the table under construction.[13] This relation was used in the construction of the annual U.S. abridged tables for the years 1946 to 1953, and the 1939–41 U.S. complete table was used as a standard. In the construction of the annual U.S. life tables by the method of reference to a standard table since 1954, the relation employed was modified and new standard tables (1949–51 or 1959–61) were adopted. The value $_n g_x$ is computed for the standard life table on the basis of the observed central death rate ($_n m_x$) and the mortality rate ($_n q_x$), using the formula

$$_n g_x = \frac{n}{_n q_x} - \frac{1}{_n m_x} \quad (15)$$

where $n - _n g_x$ represents the average number of years lived in the age group by those dying in the age group.[14] The index $_n g_x$ lies between 0 and n. This value is assumed to apply to the new life table at the same age. The required mortality rates in the new life table are computed using the formula

$$_n q_x = \frac{n \cdot _n m_x}{1 + _n g_x \cdot _n m_x} \quad (16)$$

where $_n m_x$ refers to the observed $_n m_x$ values applicable to the new table under construction. In the above example, if the 1949–51 U.S. life table was used as the standard, the value of $_5 g_{55}$ would be equal to 2.2598 and the observed death rate of .013970 for ages 55 to 59 in the United States in 1959 would be converted to a mortality rate of .06771 for 1959 as follows:

$$_5 q_{55} = \frac{5\,(.013970)}{1 + 2.2598\,(.013970)}$$

$$_5 q_{55} = \frac{.069850}{1.031571}$$

$$_5 q_{55} = .067712$$

For the calculation of $_n L_x$ in the abridged table, one may apply the simple relationship $_n L_x \div (l_x + l_{x+n})$ from the standard table.[15] The annual abridged U.S. tables made since 1954 employ another relationship between l_x and $_n L_x$, involving a factor we may designate as $_n G_x$.[16] The value of $_n G_x$, representing the distribution of deaths in the interval x to $x + n$, is obtained from the standard table by the formula

$$_n G_x = \frac{n l_x - _n L_x}{_n d_x} \quad (17)$$

This value is assumed to apply to the new table, and it is used in the formula

$$_n L_x = n l_x - _n G_x \cdot _n d_x \quad (18)$$

to obtain the desired value of $_n L_x$ in the new table. The value for the open-ended interval, $_\infty L_x$, is determined by a special formula. A factor r_x is computed for the standard table as follows:

[13] U.S. National Office of Vital Statistics, *Vital Statistics – Special Reports, Selected Studies*, Vol. 23, No. 11, "U.S. Abridged Life Tables, 1945," April 15, 1947.

[14] U.S. National Center for Health Statistics, *Vital and Health Statistics*, Series 2, No. 4, "Comparison of Two Methods of Constructing Abridged Life Tables by Reference to a 'Standard' Table," Revised March 1966.

[15] U.S. National Office of Vital Statistics, "U.S. Abridged Life Tables, 1945," op. cit.

[16] U.S. National Center for Health Statistics, "Comparison of Two Methods of Constructing Abridged Life Tables by Reference to a 'Standard' Table."

$$r_x = \frac{{}_\infty m_x \cdot {}_\infty L_x}{l_x} \qquad (19)$$

where ${}_\infty m_x$ is the observed central death rate for the open-ended interval. This factor is applied in the new table by using the formula

$$_\infty L_x = \frac{l_x r_x}{{}_\infty m_x} \qquad (20)$$

The method of reference to a standard table is particularly useful and convenient in the construction of an annual series of life tables, since the ${}_n g_x$ and ${}_n G_x$ factors can be calculated once from a complete life table in the initial year and used over and over for each year until a new complete table is prepared.

Comparison of Methods. — A comparison of the mortality rates for the United States in 1959 derived by the three methods described is shown in table 15–5. It may be observed that the results are very similar. Tables 15–3 and 15–4 may be compared for the difference in results when the Reed-Merrell method and the Greville method are applied to rural India in 1957–58. It should be noted that in Greville's method the central death rates in the life table $({}_n m_x^l)$ are required to agree with the central death rates observed in the population $({}_n m_x^p)$. This is not the case in the Reed-Merrell method or in the method of reference to a standard table, however. There is no generally agreed upon requirement that these two sets of central death rates should be equal in numerical value. In fact, it can be maintained that since the distribution of the life table population is different from the distribution of the actual population in any age interval, the two rates should be different. A method has been proposed by Keyfitz to compute a life table that would agree with the observed data after adjustment for the differences in the age distribution of the populations.[17] This method is somewhat complex and is beyond the scope of this book.

GENERATION LIFE TABLES

In previous sections of this chapter, various methods for preparing current life tables have been discussed. In every instance, the basic data for these methods involve a census of the population (or accurate estimates of the population by age and sex), the deaths recorded in the year of the census or in a number of years around the census year, and the births in a few years just prior to the census year. Life tables prepared from these data portray the mortality experience of the population observed during the relatively short period to which the data on deaths apply.

Since life tables are mathematical models that trace a cohort of lives from birth to death according to an assumed series of mortality rates, it would be logical to try to base the life table on the mortality rates experienced by an actual cohort of lives. For this purpose, it is necessary for data to have been collected for many years on an annual basis before a life table could be prepared. For example, a generation life table for the birth cohort of 1890 would employ the observed death or mortality rates for infancy in 1890, for age 1 in 1891, for

Table 15–5. — Comparison of Mortality Rates for the Total United States Population, 1959, Computed by Three Different Abridged Life Table Methods

Age interval (exact ages, x to x + n)	$_n g_x$ Standard[1]	$_n m_x$ Observed[2]	Computed $_n q_x$		
			Reed-Merrell	Greville	Reference to Standard Table
1–5............	18.7253	.001070	.00427	.00427	.00420
5–10...........	17.1188	.000483	.00241	.00241	.00240
10–15..........	−27.7680	.000450	.00225	.00225	.00228
15–20..........	10.2732	.000929	.00464	.00464	.00460
20–25..........	7.5326	.001228	.00612	.00612	.00608
25–30..........	0.1806	.001290	.00643	.00643	.00645
30–35..........	−1.0910	.001655	.00824	.00824	.00829
35–40..........	0.9558	.002316	.01152	.01152	.01155
40–45..........	2.2822	.003654	.01812	.01812	.01812
45–50..........	1.6621	.005737	.02831	.02831	.02841
50–55..........	2.2507	.009145	.04477	.04477	.04480
55–60..........	2.2598	.013970	.06765	.06765	.06771
60–65..........	2.4041	.020773	.09904	.09904	.09892
65–70..........	2.2343	.033315	.15438	.15438	.15503
70–75..........	2.3399	.050037	.22329	.22330	.22396
75–80..........	2.4376	.070893	.30197	.30201	.30224
80–85..........	2.5307	.116898	.45017	.45045	.45105

[1] Computed from the life table for the United States, 1949–51, total population, which is used as the standard table. The $_n g_x$ factors at ages below 40 seem irregular and would be expected to approximate 2.5. However, where death rates are low, the computed $_n q_x$ is relatively insensitive to large variations in $_n g_x$.

[2] Ratio of deaths to population observed in the United States in 1959.

age 2 in 1892, and so on until the latest available rate; the remaining rates would have to be obtained by projection. Table 15—6 presents selected values from a generation life table for the cohort of white females born in 1890 in the United States, prepared by Jacobson.[18]

A sufficient period of time has elapsed since mortality data have been systematically recorded, for a number of generation life tables to have been prepared. For example, Jacobson prepared life tables for white males and females born in the United States in 1840 and every tenth year thereafter through 1960, on the basis of the mortality rates experienced in each year of the life of these cohorts and projected mortality rates.[19] Earlier, Dublin, Lotka, and Spiegelman had developed or described such tables.[20]

Generation life tables can be used to compute generation reproduction rates (ch. 18), to study life expectancy historically (see below), to project mortality into the future (ch. 23), and to make estimates of orphanhood.[21] A series of such tables could represent the development of mortality and life expectancy of real cohorts over time and improve the basis for analyzing the relation between the earlier mortality of a cohort and its later experience. In view of the general tendency for mortality to fall, the expectation of life at birth in a generation

[17] Nathan Keyfitz, "A Life Table That Agrees With the Data" (2 parts), *Journal of the American Statistical Association*, 61(314):305–12, June 1966; and 63(324):1253–68, December 1968.

[18] Paul H. Jacobson, "Cohort Survival for Generations since 1840," *Milbank Memorial Fund Quarterly*, 42(3):36–51, July 1964.
[19] Ibid.
[20] Louis I. Dublin and Mortimer Spiegelman, "Current Versus Generation Life Tables, *Human Biology*, 13:439–458, December 1941; Mortimer Spiegelman, "The Versatility of the Life Table," *American Journal of Public Health*, 47(3):297–304, March 1957; and Louis I. Dublin, A. J. Lotka, and M. Spiegelman, *Length of Life* (rev. ed.), New York, Ronald Press, 1949, esp. pp. 174–182. See also Robert A. M. Case et al., *The Chester Beatty Research Institute Serial Abridged Life Tables, England and Wales, 1841–1960*, Part 1, "Tables, Preface, and Notes," London, Royal Cancer Hospital, 1962, pp. 1–38.
[21] See, for example, Ian Gregory, "Retrospective Estimates of Orphanhood from Generation Life Tables," *Milbank Memorial Fund Quarterly*, 43(3): 323–348, July 1965.

Table 15–6. —Selected Values From the Complete Generation Life Table for the Cohort of White Females Born in the United States in 1890

Calendar year	Age (x)	Mortality rate ($_nq_x$)	Number surviving (l_x)	Average remaining lifetime ($\mathring{e}_x$)
1890	0	147.6	100,000	53.2
1891	1	43.6	85,240	61.3
1895	5	5.8	77,509	63.3
1900	10	2.5	76,025	59.5
1905	15	3.2	75,049	55.2
1910	20	4.2	73,708	51.2
1915	25	4.3	72,086	47.3
1920	30	6.0	69,842	43.7
1925	35	4.8	68,141	39.7
1930	40	5.3	66,434	35.7
1935	45	6.2	64,621	31.6
1940	50	7.6	62,472	27.6
1945	55	9.8	59,862	23.7
1950	60	13.4	56,681	19.9
1955	65	19.7	52,438	16.3
1960	70	28.8	46,535	13.0
1965	[1]75	43.8	39,028	10.1
1970	[1]80	72.9	29,670	7.4
1975	[1]85	119.4	18,420	5.4
1980	[1]90	(NA)	8,702	3.8

NA Not available.
[1] Rates are projected.

Source: Paul H. Jacobson, "Cohort Survival for Generations since 1840," *Milbank Memorial Fund Quarterly*, 42(3), Part 1:39, 46, and 48, July 1964; and unpublished estimates of values of e_x.

life table tends to be higher than in the current table for the starting year of the generation life table. A comparison of life expectations at birth in two such tables reflects the average improvement in mortality for the actual cohort over its lifetime. Conversely, current life tables typically understate the probable life expectancy to be observed in the future for persons at a given age because mortality is expected to continue declining. The generation life table for the cohort of U.S. white females born in 1890, for example, shows a life expectation at birth of 53.2 years, whereas the current life table for 1890 shows a life expectation at birth of 44.5 years.[22] The actual experience of the 1890 cohort added 8.7 years to its life expectation at birth.

The analysis of cohort mortality could serve as an improved basis for projecting mortality. One approach used at present is to trace the mortality experience of a series of cohorts from a group of current life tables prepared in the past for as many years back as is considered reliable. The mortality curves for these cohorts are incomplete to various degrees, but once the cohorts have reached the older ages, the mortality curves can be projected over their remaining life-time in various ways. Currently, the general practice, however, is to project mortality on an age-specific basis rather than on a cohort basis, that is, to analyze the series of rates for the same age group and to project the rates for each age group separately (ch. 23).

As stated, preparation of a generation life table requires compilation of data over a considerable period of time on an annual basis and also the projection of some or many incomplete cohorts. It is, therefore, a burdensome task to prepare many such tables "manually," but the use of the electronic computer should reduce this burden sharply. (The basic "input" for an annual series of generation life tables, for example, is

the matrix of observed central death rates by single ages and single calendar years.) The fact that the basic data for a single table pertain to many different calendar years means that the time reference of the generation life table is somewhat indefinite. The indefiniteness of the time reference is more pronounced for cohort life tables than it is for cohort measures of fertility, for example, which have a more concentrated incidence by age. On the other hand, a current life table describes mortality in a particular period.

One characteristic of the generation life table may be viewed either as an advantage or as a disadvantage, depending on one's interest. A generation life table can involve the combination of mortality rates that are significantly different in nature, owing to the improvement in mortality over a long period of time. For example, the health conditions today are significantly different from those existing at the turn of the century and the corresponding mortality patterns are dissimilar. On the other hand, this type of table does reflect the actual combination of changing health conditions and intracohort influences to which particular cohorts were subject. A current table reflects the influence of a more unitary set of health conditions and a single mortality pattern.

INTERRELATIONS, COMBINATION, AND MANIPULATION OF LIFE TABLE FUNCTIONS

Most demographers are only infrequently faced with the task of constructing a complete, or even an abridged, life table. Many do, however, often have the task of manipulating in various ways life tables or life table functions which are already available. This may involve the calculation of a missing function for a given table, the combination of functions from different tables, or the combination of entire tables.

Interrelations of Functions

The manipulation of life table functions is aided by a review of how these functions are interrelated. All the functions in a life table are dependent on other functions in the table, but usually the q_x function is regarded as being independent of all other functions. A particular q_x value is clearly independent of all other q_x values at either earlier or later ages. A particular l_x, d_x, or L_x value is dependent only on the values of q_x at age x and earlier ages. A particular T_x value is dependent on values of L_x at age x and later ages, and hence on the entire column of q_x values. A particular value of $\mathring{e}_x$ is dependent, in effect, only on the values of q_x at age x and all later ages. Even though $\mathring{e}_x$ is computed from T_x and l_x, which are dependent on earlier values of q_x, the absolute values of the T_x's and l_x's are "washed out" in the calculation of $\mathring{e}_x$. Implicitly, in the derivation of $\mathring{e}_x$, the q_x's for age x and later ages are weighted by the percentage distribution of the L_x's which they generate. Because of the sharp drop from q_0 to q_1, $\mathring{e}_1$ is commonly higher than $\mathring{e}_0$; thereafter, $\mathring{e}_x$ tends to fall steadily.

In some cases values are presented for only some of the life table functions. Some or all of the others may be desired. Let us consider the case where, as in the U.N. *Demographic Yearbook*, we are given 1-year values of q_x, l_x, and $\mathring{e}_x$ at every 5th age and we want $_5L_x$ values for computing 5-year survival rates (see below). We apply the following relationships:

$$\mathring{e}_x = \frac{T_x}{l_x}$$

$$\therefore T_x = \mathring{e}_x l_x$$

$$\text{and } _5L_x = T_x - T_{x+5} \tag{21}$$

[22] Jacobson, op. cit., p. 50.

The $_5q_x$'s corresponding to the l_x values at every fifth age are calculated by $1 - \dfrac{l_{x+5}}{l_x}$, and the $_5d_x$'s corresponding to the l_x values are calculated by $l_x - l_{x+5}$.

Given the same basic information as just noted, we may wish to secure the values for the individual ages of some function. In that case, any of the standard interpolation formulas may be applied. We may interpolate q_x or lx to single ages and obtain single values of d_x as a by-product. Or we may derive T_x at every fifth age as described in the preceding paragraph and then apply one of the interpolation formulas to secure T_x at single ages. Single values for e_x may be secured by direct interpolation or by interpolating the l_x function to single ages and completing the table. The various interpolation procedures are discussed more fully in appendix C at the end of this volume.

Combination of Tables and Functions

First, let us consider the case where life tables for two populations are available and we should like to have a single life table for the combined population. Suppose, for example, we want to derive a single table for the general population from separate life tables for the male and female populations. This can be done by weighting the q_x's from the male and female tables in accordance with the distribution of the population by sex at each age. A new table may now be recomputed from the combined q_x's and an assumed radix of 100,000. If only one of the other functions is wanted, the values from the separate tables may be combined on the basis of population weights in only a few scattered ages, because the shifts from age to age in the distribution of the population by sex could introduce unacceptable fluctuations by age in these functions for the combined population, and the resulting values for different functions may not be consistent with one another. (Note that these functions would not all receive the same population weights, e.g., l_x would be weighted by the population at age x, and T_x and $\overset{\circ}{e}_x$ would be weighted by the population for ages x and over, the weights in each case reflecting the ages covered by the measure.) Only the use of a constant weighting factor over all ages can prevent fluctuations by age in the values of a function. If an entire column for a function is wanted, therefore, it is best to recompute the values for the function by starting with the weighted q_x's.

A different problem is represented by the need for separate male and female life tables whose values bear a realistic absolute relation to one another. The tables for males and females as originally computed do not bear a consistent "additive" relationship to one another since births of boys and girls do not occur in equal numbers, as is implied by the radixes of the separate life tables for males and females. To make the tables additive, it is necessary to inflate the l_x, d_x, L_x, and T_x values of the male life table by the sex ratio at birth in the actual population. Thus, the stationary male population at each age adjusted for absolute comparability with the female stationary population is derived by

$$\frac{B^m}{B^f} \cdot L_x^m \tag{22}$$

and the stationary male population of all ages, adjusted for combination with the total female stationary population is obtained by

$$\frac{B^m}{B^f} \times T_0^m \tag{23}$$

The birth rate (and also the death rate) for the male and female life tables taken in combination is given by

$$\frac{\dfrac{B^m}{B^f} l_0 + l_0}{\dfrac{B^m}{B^f} T_0^m + T_0^f} \tag{24}$$

The ratio of children under five to women of childbearing age in the stationary population, a measure known as the replacement quota (ch. 18) is obtained by

$$\frac{\dfrac{B^m}{B^f} \cdot L_{0-4}^m + L_{0-4}^f}{L_{15-49}^f} \tag{25}$$

We have been describing the procedures by which we may arrive at life table values for a total population consistent in absolute level with the life table values for the separate component male and female populations. Even if the stationary populations are properly combined, and reduced to the level of a single table with a radix of 100,000 by applying the factor

$$\frac{100,000}{\dfrac{B^m}{B^f} 100,000 + 100,000} \tag{26}$$

to the absolute functions in each table, the resulting d_x, l_x, L_x, and T_x would not agree exactly with the values derived directly from the weighted q_x's. Nor will the $\overset{\circ}{e}_x$ or q_x values or the derived survival rates (ratios of L_x) agree. Figures for $\overset{\circ}{e}_0$ consistent with the derivation of a combined table just described may be derived from the reciprocal of the formula given above for the birth rate:

$$\frac{\dfrac{B^m}{B^f} T_0^m + T_0^f}{\dfrac{B^m}{B^f} l_0 + l_0} \tag{27}$$

The q_x's can be obtained as the ratio of d_x to l_x:

$$\frac{\dfrac{B^m}{B^f} d_x^m + d_x^f}{\dfrac{B^m}{B^f} l_x^m + l_x^f} \tag{28}$$

The directly computed table takes account of the actual distribution of the population by sex at each age while the method described links the two tables only by the sex ratio at birth. The differences between these two combined tables tend to be quite small, even inconsequential, however, because the proportions of the sexes approximate equality and have only a small variation by age. On the other hand, the two combined tables for the white and nonwhite population of the United States would show more pronounced differences because the proportions of these groups in the population are quite different and vary more by age. To convert the basic tables for the white and nonwhite populations into additive form, the factor for adjusting the nonwhite table ($B_N \div B_T$) is about .20; and to derive the combined table with a radix of 100,000, the factor is about $100,000 \div 120,000$, or .83.

APPLICATION OF LIFE TABLES IN POPULATION STUDIES

Inasmuch as the measurement of mortality is involved in many types of demographic studies, the life table model and life table techniques as special tools for measuring mortality can be applied in a wide variety of population studies. The life table is a necessary or useful tool in the analysis of fertility, reproductivity, migration, and population structure; in the estimation and projection of population size, structure, and change; and in the analysis of various social and economic characteristics of the population, such as marital status, labor force status, family status, and educational status. In many of these cases, the standard life table functions are combined with probabilities for other types of contingent events (e.g., first-marriage rates, rate of entry into the labor force) in order to obtain new measures.

Mortality Analysis

It is impractical to discuss here all of the many types of analyses of mortality that could be performed on the basis of life table values. The most frequently used procedures involve comparisons between populations of life expectancy at birth and at various ages, of proportions surviving to various ages, and of mortality rates at various ages.

Measurement of the Level of Mortality, Survivorship, and Life Expectancy. — The most common general use of a life table is to measure the level of mortality of a given population. As such, it offers some new measures which eliminate some of the problems involved in the use of existing standard measures. For example, the difference in the crude death rates of two populations, as was indicated in the previous chapter, is affected by the difference in the age composition of the populations. On the other hand, standardized rates have the disadvantage of depending on the particular selection of a standard population. Life tables have the advantage that the overall death rate of the life table (m^l) represents the result of the weighting of the age-specific life table death rates (m^l_x) by a population (L_x) generated by the series of observed death rates themselves. On the other hand, since the weighting scheme involves a variable population distribution from table to table, there is still an issue of **comparability** between the overall life table death rates. As will be described later, **the life table also has the special advantage in providing measures of mortality which automatically are structured in cohort form.** Hence, it eliminates the usually laborious task of compiling death statistics according to age cohorts, even where this is possible. Often insufficient basic data are available for doing this for a particular country and a life table is the only means of making a necessary allowance for the mortality of age cohorts.

The **expectation of life at birth** is the life table function most frequently used as an index of the level of mortality. It also represents a summarization of the whole series of mortality **rates** for all ages combined, as weighted by the life table stationary population. In fact, the reciprocal of the expectation of life, $1/\overset{\circ}{e}_0$, is equivalent to the "crude" death rate, m^l, of the life table population, as can be seen from the following derivation:

$$m^l = \frac{\text{Total number of deaths}}{\text{Total population}} = \frac{l_0}{T_0} = \frac{1}{T_0/l_0} = \frac{1}{\overset{\circ}{e}_0} \quad (29)$$

For example, the death rate in the 1959–61 U.S. life table is calculated from table 15–1 as follows:

$$m^l = \frac{l_0}{T_0} = \frac{100,000}{6,989,030} = .01431, \text{ or} \quad (30)$$

$$m^l = \frac{1}{\overset{\circ}{e}_0} = \frac{1}{69.89} = .01431 \quad (31)$$

The same formulas give the "crude" birth rate, f^l, of the life table population, and the growth rate of this population (r^l) is, of course, zero:

$$f^l = \frac{l_0}{T_0} = \frac{1}{\overset{\circ}{e}_0} = m^l \quad (32)$$

$$\therefore r^l = f^l - m^l = 0 \quad (33)$$

It has been suggested that in some cases, because of the strong effect of the infant mortality rate on the expectation of life at birth, it would be better to use the **expectation of life at age 1** as a comparative measure of the general level of mortality of a population, perhaps in conjunction with the infant mortality rate. Another life table function frequently used is the **expectation of life at age 65.** This value measures mortality at the older ages, the ages where most of the deaths in the developed countries currently occur. (Much of the measured change here may reflect inadequacies in the underlying data.) Other life table values used to measure mortality are the **probability of surviving from birth to age 65,**

$$_{65}p_0 = \frac{l_{65}}{l_0} \quad (34)$$

and the age to which half of the cohort survives, that is, **the median age at death** of the initial cohort assumed in the life table. The median age at death is the age corresponding to l_x value 50,000 in a life table with a radix of 100,000.

The improvement in the mortality record of white males in the United States during this century, as measured by these life table values, is shown in table 15–7. From this table, we could correctly conclude that most of the improvement in mortality has been recorded at ages under 65 and that significant improvement has been recorded during the first year of life. In addition, we could conclude that little progress has been made at any age in the years from 1949–51 to 1966. The variation in these measures is suggested by the figures for several countries shown in table 15–8.

Measures based on the life table have been used in the intensive analysis of vital statistics. Naturally, the type of vital statistics most frequently analyzed with life table measures is mortality, since these measures were originally designed for precisely that application and that application is usually the simplest. The analysis of mortality could cover variations in time and between population groups, by age, sex, color or race, marital status, geographic residence, and other demographic characteristics that may be regarded as having an effect on the overall level of mortality. Table 15–9 shows, for example, the variation in time, by age, of the relative mortality of white males and females in the United States in terms of life table mortality rates. Although mortality of white males has improved tremendously during this century in the United States (table 15–7), the improvement has been greater for white females (table 15–9).

Model life tables may be used to evaluate the quality of mortality data for statistically developed populations or to establish the validity of recorded differences in the patterns of mortality of population groups. The mortality data for a country in earlier historical periods or for some segment of the current population may appear to be so defective as to raise a question whether the corresponding life tables provide an accurate representation of the patterns of mortality actually experienced. Zelnik, for example, compared the age patterns

Table 15–7. — Change in the Mortality of White Males in the United States According to Various Life Table Measures: 1900 to 1966

[Life tables for periods before 1929–31 relate to the Death Registration States]

Measure	Period for the life table								Increase	
	1900-02	1909-11	1919-21	1929-31	1939-41	1949-51	1959-61	1966	1900-02 to 1966	1949-51 to 1966
Expectation of life at birth.........	48.23	50.23	56.34	59.12	62.81	66.31	67.55	67.56	19.33	1.25
Expectation of life at age 1.........	54.61	56.26	60.24	62.04	64.98	67.41	68.34	68.17	13.56	0.76
Expectation of life at age 65........	11.51	11.25	12.21	11.77	12.07	12.75	12.97	12.89	1.38	0.14
Probability of surviving from birth to age 65..............................	.392	.409	.507	.530	.583	.635	.658	.656	.264	.021
Median age at death of initial cohort.	57.2	59.3	65.4	66.4	68.7	70.7	71.4	71.2	14.0	0.5

Source: Various official United States life tables.

Table 15–8. — Mortality of Males According to Various Life Table Measures, for Selected Countries: Around 1960

Country and year	Expectation of life at--			Probability of surviving from birth to age 65 (ℓ_{65})	Median age at death of initial cohort (Md_a^d)
	Birth $(\overset{o}{e}_0)$	Age 1 $(\overset{o}{e}_1)$	Age 65 $(\overset{o}{e}_{65})$		
Argentina, 1959-61........	63.1	66.2	12.9	.597	69.4
Costa Rica, 1962-64.......	61.9	66.8	13.6	.625	70.8
Denmark, 1963-64..........	70.3	70.9	13.5	.735	74.2
El Salvador, 1960-61......	56.6	60.8	14.3	.530	66.9
Guatemala, 1963-65........	48.3	52.5	12.1	.391	57.4
Japan, 1959-60............	65.3	66.6	11.6	.648	70.6
Peru, 1961................	46.9	54.3	11.0	.381	56.5
Sweden, 1961-65...........	71.6	71.8	13.9	.766	75.4
Taiwan, 1959-60...........	61.3	63.2	10.9	.573	67.8
U.S.S.R., 1958-59........	64.4	66.4	14.0	.611	70.4
United Arab Republic, 1960	51.6	56.8	11.9	.492	64.4
United States, 1960.......	66.6	67.6	12.8	.637	70.6

Source: United Nations, *Demographic Yearbook, 1966*, tables 21 and 23; Eduardo E. Arriaga, "New Abridged Life Tables for Peru: 1940, 1950–51, and 1961," *Demography*, 3(1):226, 1966; U.S. National Center for Health Statistics, *Vital Statistics of the United States, 1960*, Vol. II, *Mortality*, Part A, 1963, tables 2-3 and 2-4; Guatemala, Dirección general de estadística, *Tablas de Vida, 1964* (Life tables, 1964), November 1967, p. 21.

Table 15–9. — Mortality Rate of White Females as Percent of the Mortality Rate of White Males in the United States for Selected Ages, As Shown by Various Life Tables: 1900 to 1966

[Life tables for periods before 1929–31 relate to the Death Registration States]

Age (years)	1900-02	1909-11	1919-21	1929-31	1939-41	1949-51	1959-61	1966
0......	83	83	80	80	79	77	76	75
1......	90	92	90	89	89	89	88	85
10.....	90	87	85	77	70	67	67	65
20.....	93	86	101	87	68	45	35	34
30.....	97	91	105	91	79	63	54	54
40.....	88	79	90	78	72	62	57	57
50.....	87	81	91	75	66	55	50	51
60.....	88	84	88	78	67	56	48	44
70.....	91	91	92	84	78	68	58	53
80.....	91	93	95	90	87	82	77	71

Source: Various official United States life tables.

of mortality recorded in the series of official U.S. life tables for whites and Negroes from 1900-02 to 1959-61 with those in the collection of model life tables prepared by Coale and Demeny, seeking to shed light on the quality of the U.S. official life tables for Negroes and on the existence of differences in the patterns of mortality for Negroes and whites.[23]

Analysis of Mortality by Cause of Death. — The comparisons of life expectancy and mortality rates at various ages are generally made on the basis of the total mortality observed at those ages; but, in many instances, the analyses are conducted by causes of death. Life tables have been computed under the assumption that a specific cause of death or group of causes of death is eliminated, that is, that it is impossible to die from

the eliminated cause.[24] Table 15–10 presents an abridged life table for the total population of the United States in 1959–61, prepared on the assumption that the malignant neoplasms (cancer) are eliminated as a cause of death. A series of such life tables considered in comparison with a life table for all causes combined, provides various measures of the relative importance of the different causes of death. Among them is the gain in expectation of life that would occur if a cause of death were eliminated. As an illustration, $\overset{o}{e}_0$ in tables 15–2 and 15–10 may be compared to show that 2.27 (= 72.16−69.89) years of life would be added to life expectancy at birth for the total U.S. population in 1959–61 if cancer were eliminated as a cause of death.[25]

The mathematical procedures followed in the preparation of these tables are somewhat complex and beyond the scope of this book. In general, they require the preparation of a special type of multiple decrement life table, in this case, a life table in which total deaths are subdivided into the different causes or groups of causes of death. This subdivision of total

[23] See Melvin Zelnik, "Age Patterns of Mortality of American Negroes: 1900-02 to 1959-61," *Journal of the American Statistical Association*, 64(326): 433-451, June 1969.

[24] U.S. National Center for Health Statistics, *Life Tables, 1959–61*, Vol. 1, No. 6, "United States Life Tables by Causes of Death, 1959–61," May 1968; and T. N. E. Greville, "Mortality Tables Analyzed by Cause of Death," *The Record*, American Institute of Actuaries, Vol. 37, Part II, No. 76, October 1948.
[25] U.S. National Center for Health Statistics, "United States Life Tables by Causes of Death, 1959–61." For an earlier illustration of the application of this method, see Barnes Woodhall and Seymour Jablon, "Prospects for Further Increase in Average Longevity," *Geriatrics*, 12(10):586–591, October 1957.

Table 15–10. — **Abridged Life Table Eliminating Malignant Neoplasms as a Cause of Death, for the Total Population of the United States: 1959–61**

Age interval (exact ages, x to x + n)	$_nq_x$	l_x	$_nd_x$	$_nL_x$	T_x	$\overset{o}{e}_x$
0–1...........	.02586	100,000	2,586	97,818	7,216,227	72.16
1–5...........	.00378	97,414	368	388,752	7,118,409	73.07
5–10.........	.00202	97,046	196	484,704	6,729,657	69.34
10–15.........	.00190	96,850	184	483,842	6,244,953	64.48
15–20.........	.00417	96,666	403	482,412	5,761,111	59.60
20–25.........	.00573	96,263	551	479,961	5,278,699	54.84
25–30.........	.00569	95,712	544	477,208	4,798,738	50.14
30–35.........	.00681	95,168	649	474,280	4,321,530	45.41
35–40.........	.00934	94,519	883	470,523	3,847,250	40.70
40–45.........	.01431	93,636	1,340	465,058	3,376,727	36.06
40–50.........	.02212	92,296	2,041	456,740	2,911,669	31.55
50–55.........	.03485	90,255	3,146	443,882	2,454,929	27.20
55–60.........	.05117	87,109	4,457	425,063	2,011,047	23.09
60–65.........	.07848	82,652	6,487	397,940	1,585,984	19.19
65–70.........	.11682	76,165	8,897	359,654	1,188,044	15.60
70–75.........	.17534	67,268	11,795	308,077	828,390	12.31
75–80.........	.26597	55,473	14,754	241,641	520,313	9.38
80–85.........	.40981	40,719	16,687	161,657	278,672	6.84
85 and over...	1.00000	24,032	24,032	117,015	117,015	4.87

Source: U.S. National Center for Health Statistics, *Life Tables, 1959–61,* Vol. 1, No. 6, "United States Life Tables by Causes of Death, 1959–61," May 1968, p. 17.

life table deaths by cause is made on the basis of the actual distribution of deaths by cause at each age.

Analysis of Fertility, Reproductivity, and Age Structure

Life table measures and techniques have also been used to analyze other vital phenomena in addition to mortality. The procedures usually involve the combination of mortality with the specific vital rates that are being analyzed. These procedures have particular application in studies of fertility and reproductivity. For example, age-specific fertility rates are combined with survival rates from life tables to calculate the net reproduction rate (ch. 18). The average length of a generation, which is simply the average age of mothers at the birth of their daughters, is another measure based on both fertility rates and life table survival rates.

The life table is an important instrument for the analysis of population dynamics and age structure. Studies of the relation of growth rates, birth rates, and death rates to age structure depend heavily on the use of life tables, particularly as a means of indicating the alternative effect of various levels of mortality on age structure and of developing the stable age distribution corresponding to various levels of fertility. Life tables in themselves indicate the stationary population distribution corresponding to given levels of mortality and a zero growth rate.

Calculation of Life Table Survival Rates. — Survival rates are defined in terms of two ages, and hence two time references, the initial age and date and the terminal age and date. Both age references are equally applicable, but survival rates are more commonly identified symbolically in terms of the initial ages than in terms of the terminal ages. Survival rates express survival from a younger age to an older age, but they can be used to restore deaths to an older population. Survival rates used to reduce a population for deaths are multiplied against the initial population; survival rates used to restore deaths in a reverse calculation are divided into the terminal population.

$$\text{Forward survival: } {_5P_x^t} \cdot {_5s_x^5} = {_5E_{x+5}^{t+5}} \quad (35)$$

$$\text{Reverse survival: } \frac{_5P_x^t}{_5s_x^5} = {_5E_{x-5}^{t-5}} \quad (36)$$

where the elements represent the initial or terminal populations ($_5P_x^t$), the 5-year survival rate ($_5s_x^5$), and the expected later ($_5E_{x+5}^{+5}$) or earlier ($_5E_{X-5}^{t-5}$) populations. Normally, "revival" rates, which directly express "revival" from an older to a younger age, are not used.

The most common form of survival rate employed in population studies is for a 5-year age group and a 5-year time period. The general formula is

$$_5s_x^5 = \frac{_5L_{x+5}}{_5L_x} \quad (37)$$

According to the 1959-61 U.S. life table (table 15-2), the proportion of the population 45 to 49 years of age which will survive 5 years is calculated as follows:

$$_5s_{45}^5 = \frac{434,264}{450,814} = .96329$$

The proportion of the population 75 years and over which will live another 10 years is

$$_\infty s_{75}^{10} = \frac{_\infty L_{85}}{_\infty L_{75}} \quad \text{or} \quad \frac{T_{85}}{T_{75}} \quad (38)$$

$$= \frac{84,854}{419,370} = .20234$$

It will be noted that survival rates for age groups of the population are computed from the L_x values of the life table, the L_x value for the initial age group being given in the denominator of the rate and the L_x value for the terminal age group being given in the numerator of the rate.

A complete life table readily permits calculation of survival rates for single ages for one year or any other time period. A one-year survival rate for a single age is represented by $s_x = \frac{L_{x+1}}{L_x}$. The proportion of the population 64 years of age on a given date which will survive to the same date in the following year, on the basis of the 1959–61 complete U.S. life table (table 15–1), is:

$$s_{64} = \frac{L_{65}}{L_{64}} = \frac{70,214}{72,033} = .97475$$

The proportion of 65-year olds who are expected to live another 10 years is:

$$s_{65}^{10} = \frac{L_{75}}{L_{65}} = \frac{46,773}{70,214} = .66615$$

Survival rates involving birthdays are computed using the l_x values. The proportion of newborn babies who will reach their fifth birthday is

$$\frac{l_5}{l_0} = \frac{96,998}{100,000} = .96998$$

The proportion of 60-year-olds who will reach their 65th birthday is

$$\frac{l_{65}}{l_{60}} = \frac{71,147}{78,371} = .90782$$

The proportion of infants born in a year who will survive to the end of that year (when the cohort is under 1 year of age) is

$$\frac{L_0}{l_0} = \frac{97,815}{100,000} = .97815$$

while the proportion of the newborn infants who will reach their first birthday is

$$\frac{l_1}{l_0} = \frac{97,407}{100,000} = .97407$$

If survival from birth to a given age interval is wanted, then the form of the survival rate is $\frac{_nL_x}{nl_0}$ or, for a 5-year age group, $\frac{_5L_x}{5\,l_0}$. Here n cohorts of 100,000 are subject to death. For survival from birth to age interval 30–34, we have

$$\frac{472,709}{500,000} = .94542.$$

Survival rates involving 5-year age groups may also be computed using values of L_x or l_x for the midpoint age. The survival rate from birth to age interval 30–34 may be approximated by

$$\frac{l_{32.5}}{l_0} = \frac{L_{32}}{l_0} = \frac{94,551}{100,000} = .94551$$

This approximation has little effect on the survival rate, particularly in the ages up through young adulthood.

Survival rates for parts of a calendar year may be calculated by interpolating between the L_x's. A ¾-year survival rate from a complete life table is computed as follows (except at age under 1):

$$\frac{¼L_x + ¾L_{x+1}}{L_x} \qquad (39)$$

For age 45,

$$\frac{¼(91,161) + ¾(90,707)}{91,161} = .99626$$

For age under 1, life tables giving L_x values by months of age or other subdivisions of age under 1 should be used,[26] or special factors may be derived from statistics for infant deaths in a manner similar to the way the separation factors for all infant deaths were derived (ch. 14).

Use of Life Table Survival Rates. — In the use of life table survival rates in population studies, decisions have to be made regarding the selection of the life table and life table survival rates most appropriate for a particular problem. Questions relate both to choice of a life table and to the kinds of survival rates to be calculated. Where a life table is not available for the particular year or period, but for prior and subsequent years or for the initial and terminal years of the period, special survival rates may have to be computed on the basis of the available life tables. Commonly, for example, life tables are available for the census years, but we are interested in measuring population changes or net migration for the intercensal period. Survival rates appropriate for this purpose may be derived by (1) calculating the required survival rates from each of the two tables and (2) averaging them. This assumes that mortality changes occurred evenly over the intercensal period. We should also consider whether they were sufficiently great to justify the additional calculations.

In other cases there may be a question of adjusting for geographic variations or for variations in racial or ethnic composition, urban-rural residence, marital status, educational attainment, or other socioeconomic characteristics. Differences in age-specific death rates for various geographic or ethnic or socioeconomic categories should be examined, when they can be computed, to determine whether life table survival rates for the general population should be adjusted for these differences. For example, although life tables are available for the States of the United States, the convergence of State mortality in each sex-color group to the national level has been so pronounced over the past several decades that, for most purposes, it is not necessary to employ separate life tables for each State in making population projections for States.[27] National life tables for each sex-color group are adequate.

Survival rates for the general population can be adjusted directly for geographic or socioeconomic variations by use of observed central death rates for the various categories, as follows: (1) Take the complement of the general survival rate; (2) calculate an adjustment factor for the particular geographic or socioeconomic category in terms of central death rates, (3) multiply the adjustment factor by the complement of the general survival rate, and (4) take the complement of the result in (3). For example, if s_x^5 is a 5-year survival rate at age x to $x + 4$ and $_5m_x$ is a central death rate at age x to $x + 4$ for the general population, we may develop survival rates for the single population as follows:

(1) $1 - {_5s_x}$ (a type of 5-year mortality rate)

(2) $\dfrac{(_5m_x + {_5m_{x+5}}) \text{ for the single population}}{(_5m_x + {_5m_{x+5}}) \text{ for the general population}}$

(3) $= (1) \times (2)$

(4) $= 1 - (3)$

Sometimes the evidence will not justify making this type of adjustment in the general rates.

Migration may be a troublesome element in making a correct estimate of deaths by means of life table survival rates for populations which are not closed (i.e., affected by migration). Application of survival rates to the initial population (forward procedure) tends to understate deaths for a population with net in-migration and to overstate deaths for a population with net out-migration. A reverse procedure for applying the survival rates has the opposite effect. This problem is considered more fully in chapters 20 and 21.

In many cases official life tables are available but they are based on seriously incomplete statistics on deaths and are not satisfactory for most uses. In these cases and in others where life tables are not available for the country, one may have to decide whether to construct a life table, "borrow" a life table from another country, or employ a model life table. Commonly, the countries which lack life tables simply do not have adequate death statistics for constructing such tables. In some cases one may construct a table by use of population census data and stable population techniques, but it is more practical and convenient to "borrow" another country's table or, preferably, to adopt a model table appropriate to the population under study. The selection of a model table and use of stable population techniques are discussed in chapter 24.

[26] See, for example, U.S. National Office of Vital Statistics, *Vital Statistics— Special Reports, Life Tables for 1949–51*, Vol. 41, No. 3, "U.S. Life Tables for the First Year of Life, 1949–51," September 21, 1955; and U.S. National Center for Health Statistics, "United States Life Tables, 1959–61," p. 8.

[27] Everett S. Lee and Gladys K. Bowles, "Selection and Use of Survival Ratios in Population Studies," *Agricultural Economics Research*, 6(4):120-125, October 1954.

The estimates of survivors will differ depending on the age interval employed in computing the survival rates. (We consider this question apart from the age detail required in the estimates of survivors.) With few exceptions, 5-year survival rates are sufficiently detailed to take account of the important variations in the estimates of survivors. For most calculations 10-year age intervals will be adequate, however. The use of survival rates with a given age interval implies that the actual population has the same distribution by age as the life table population in this interval. For this reason, survival rates for very broad age spans should be avoided except in rough calculations. Hence, the terminal open-ended interval should relate to a relatively limited age span containing only a small percentage of the total population. Terminal group 65 and over is to be avoided except for very "young" populations (e.g., Syria, Costa Rica). On the other hand, for countries with relatively old populations (e.g., France, Great Britain), even 75 and over may be unsatisfactory.

While life table survival rates represent the ratio of l_x or L_x values to one another, the actual level of survival rates is affected only by the q_x's in the age range to which the survival rates apply. Hence, life table survival rates for various populations may properly be combined on the basis of the population distribution at these ages. For example, 5-year survival rates with initial ages 40–44, for the white and nonwhite populations of the United States, would be combined on the basis of the distribution of the population by color at ages 40–44, in order to obtain a survival rate for all races.

National Census Survival Rates. — Another means of allowing for mortality, national census survival rates, employs life table concepts but does not involve actual use of life tables. National census survival rates are particularly applicable in the measurement of internal net migration (ch.21), but they are also used in measuring or evaluating the level of mortality and in constructing life tables, especially for countries lacking adequate vital statistics (ch 24). As noted in chapter 8, they are also useful in evaluating the quality of the census data on the sex and age distribution of the national population. National census survival rates essentially represent the ratio of the population in a given age group in one census to the population in the same age cohort at the previous census. Normally, then, census survival rates pertaining to the children born in the decade are not computed. An illustration of the method of computation, census survival rates for several countries, and a comparison with life table survival rates are presented in table 15–11. National census survival rates measure mortality essentially but they are affected by the relative accuracy of the two census counts employed in deriving them. Underlying their use are certain basic assumptions. They are that the population is a closed one (i.e., that there was no migration during the intercensal period) and that there has been no abnormal influence on mortality (e.g., war). Census survival rates may be rather irregular, often fluctuating up and down throughout the age scale and exceeding unity in some ages. Note particularly the rate for Brazil in table 15–11. The more inconsistent in accuracy are the data from the two censuses, the more erratic the census survival rates are.

In the use of national census survival rates to measure internal net migration for geographic subdivisions of countries, the fact that the rates incorporate both the effect of mortality and relative net census error is considered an advantage. The reasoning is that, since net migration is obtained as a residual and the census survival rates directly reflect the relative accuracy of the census counts in addition to mortality,

the estimates of net migration are unaffected by the errors in the census data. (See ch. 21 for the assumptions involved and their application to the measurement of net migration for small areas.)

The requirement that the population be a closed one is rarely if ever completely met, and it is desirable to try to adjust the census data for intercensal net immigration or net emigration when the necessary data are at hand. For a country with only a small net immigration, this can sometimes be achieved indirectly by basing the census survival rates on the native population. Akers and Siegel have developed rather refined census survival rates for the United States, 1950 to 1960, on the basis of the census data for the native population of the United States, its principal outlying areas, and citizens abroad affiliated with the Federal Government.[28] This broader coverage eliminates from the census survival rates much of the international migration of U.S. natives inasmuch as the population at both ends of the decade is so defined as to include as many American natives as possible wherever they are to be found throughout the world. Separate census survival rates were computed for each sex-color group and for the combined populations of each sex. See table 15–12 for illustrative figures and the corresponding life table survival rates. In this case, census survival rates for the children born during the decade were computed on the basis of births registered during the decade and census data on children under 10 in the 1960 census. National census survival rates for the United States have been used both in the estimation of net migration for States, State economic areas, counties, and metropolitan areas (ch. 21), and in the evaluation of the levels of mortality shown in the official life tables.[29]

Analysis of Socioeconomic Structure and Dynamics

Types of Multiple Decrement Tables. — As mentioned early in this chapter, multiple decrement tables incorporating conventional life table components and using life table techniques have been employed in the analysis of various socioeconomic characteristics of the population.[30] In particular, they have been used in the analysis of the labor force, marital composition, and educational status. These tables may be constructed as multiple decrement tables in which there are two or more forms of exit from the initial cohort, one of which is mortality and the other some change in social or economic status (e.g., exit from the nonworking population, or entry into the labor force; exit from the single population, or marriage). These same tables or extensions of them may then go on to measure the exits from the new status (e.g., retirement from the labor force; divorce or widowhood). Another form of constructing the tables involves the use of prevalence rates in which the life table stationary population is distributed into different statuses according to the prevalence of those statuses in the actual observed population. The working-life table and school-life

[28] U.S. Bureau of the Census, *Current Population Reports*, Series P–23, No. 15, "National Census Survival Rates, by Color and Sex, for 1950 to 1960," by Donald S. Akers and Jacob S. Siegel, July 12, 1965, table 1.

[29] Paul Demeny and Paul Gingrich, "A Reconsideration of Negro-White Mortality Differentials in the United States," *Demography*, 4(2):820–837, 1967. This study used 10-year cumulative (e.g., 25 and over to 35 and over) census survival rates by sex for native whites and Negroes calculated from unadjusted U.S. census data for the 10-year intercensal periods, 1900–1940.

[30] For a discussion of the mathematical theory of multiple decrement tables, see C. W. Jordan, Jr., *Life Contingencies*, Society of Actuaries, Chicago, 1967, chs. 14 and 15.

Table 15–11. — National Census Survival Rates for Males, for Brazil, Greece, Japan, and Mexico

Age--		Brazil			Greece	Japan	Mexico	
At first census	At second census	Population, 1950	Population, 1960	Census survival rate, 1950-60 (2) ÷ (1) =	Census survival rate, 1951-61	Census survival rate, 1955-65	Census survival rate, 1950-1960	Life table survival rate, 1959-1961
		(1)	(2)	(3)	(4)	(5)	(6)	(7)
0-4	10-14	4,235,876	4,287,220	1.01212	.98004	.98812	1.11722	.95132
5-9	15-19	3,560,850	3,445,715	.96767	.93446	.97194	.93228	.98012
10-14	20-24	3,164,704	2,963,688	.93648	.90961	.93366	.87816	.97487
15-19	25-29	2,644,531	2,521,689	.95355	.88514	.95754	.95785	.96278
20-24	30-34	2,384,460	2,254,181	.94536	.95726	.98829	.94595	.95300
25-29	35-39	2,030,312	1,955,652	.96323	.87157	.99262	.97714	.94374
30-34	40-44	1,621,739	1,652,516	1.01898	.89119	.97584	.96428	.93201
35-39	45-49	1,523,976	1,399,968	.91863	.99965	.95908	.81576	.91734
40-44	50-54	1,227,552	1,123,775	.91546	.95415	.93468	.89806	.89831
45-49	55-59	1,018,555	827,679	.81260	.92056	.90398	.75781	.86958
50-54	60-64	810,892	728,135	.89794	.90259	.84234	.91790	.83081
55-59	65-69	549,688	396,121	.72063	.78949	.75814	.77836	.77720
60-64	70-74	473,409	267,198	.56441	.77035	.64314	.60819	.69926
65 and over	75 and over	584,580	262,575	.44917	.47179	.35146	.51352	.39547

Source: Basic data from United Nations, *Demographic Yearbook, 1967*, table 5; United Nations, *Demographic Yearbook, 1960*, table 5; Nathan Keyfitz and Wilhelm Flieger, *World Population: An Analysis of Vital Data*, Chicago, University of Chicago Press, 1968, p. 118, table 2.

Table 15-12. — Comparison of Life Table Survival Rates and National Census Survival Rates for the Native Male and Female Population of the United States: 1950 to 1960

[See text for method of derivation]

Age (years) in--		Census survival rates		Life table survival rates[1]	
1950	1960	Male	Female	Male	Female
Births, 1955-60[2]	Under 5	.94893	.96403	.96716	.97438
Births, 1950-55[2]	5-9	.94765	.96370	.96195	.97016
0-4	10-14	1.01744	1.02292	.99166	.99369
5-9	15-19	.98109	.99586	.99217	.99548
10-14	20-24	.94079	.93331	.98696	.99369
15-19	25-29	.97437	1.00837	.98258	.99151
20-24	30-34	.99994	1.00830	.98094	.98915
25-29	35-39	.99304	.99950	.97733	.98528
30-34	40-44	.98850	.98971	.96766	.97873
35-39	45-49	.95102	.94986	.94963	.96819
40-44	50-54	.91970	.93907	.92037	.95248
45-49	55-59	.90155	.94207	.87794	.93037
50-54	60-64	.82155	.89666	.82099	.89756
55-59	65-69	.81074	.93008	.74756	.84840
60-64	70-74	.72271	.85415	.65619	.77375
65-69	75-79	.56896	.67204	.54278	.66104
70-74	80-84	.41475	.52600	.40468	.50629
75 and over	85 and over	.21519	.27980	.18070	.21782

[1] Average of survival rates from the 1949-51 life table and the 1959-61 life table. Life table survival rates relate to the total population.
[2] Births from April 1, 1955, to March 31, 1960, and from April 1, 1950, to March 31, 1955. The births were not adjusted for underregistration.

Source: Adapted from U.S. Bureau of the Census, *Current Population Reports*, Series P-23, No. 15, "National Census Survival Rates, by Color and Sex, for 1950 to 1960," July 12, 1965, table 2; U.S. National Center For Health Statistics, *Life Tables: 1959-61*, Vol. 1, No. 1, "United States Life Tables: 1959-61," December 1964, table 2; and U.S. National Office of Vital Statistics, *Vital Statistics-Special Reports*, Vol. 41, No. 1, "United States Life Tables: 1949-51," November 23, 1954, table 2.

table described in detail below are of this type, as are the nuptiality tables discussed in chapter 19.

Working-Life Tables. — Working-life tables have been prepared by combining mortality rates with labor force participation rates. Tables of working life are useful in understanding the mechanisms and implications of changes in the labor force.[31] Those which have been constructed have been based on synthetic cohorts and represent the life cycle of economic activity implicit in mortality rates and worker rates for a single year or group of years. These tables provide an indication of the average number of working years to be expected after a given age by all persons or by persons in the labor force attaining that age. In addition, the tables provide information on age-specific rates of accession to, and separation from, the labor force. These measures are useful for studying growth and change in the labor force and related topics, such as estimating lifetime expectations of earnings, estimating replacement needs for industry, and assessing the economic implications of changes in the activity rates and age-structure of the population.

Tables of working life were developed only within the last few decades and complete or abridged tables have now been computed for a number of countries. Complete tables of working life are available for the United States, New Zealand, and Great Britain, for example.[32] Table 15-13 presents selected portions of a complete table of working life. The illustration relates to the male population of the United Arab Republic in 1960.

There are several ways to construct a complete working-life table. The procedure used here distributes the life table stationary population according to the work status of the actual

[31] United Nations, *Methods of Analyzing Census Data on Economic Activities of the Population*, Population Studies, Series A, No. 43, 1968, pp. 19-34; Swee Hock Saw, "Uses of Working Life Tables in Malaya," *Proceedings of the United Nations World Population Conference, 1965*, Vol. IV, p. 336; and Seymour L. Wolfbein and A. J. Jaffe, "Demographic Factors in Labor Force Growth," *American Sociological Review*, 11(4):392-396, August 1946.
[32] S. L. Wolfbein, "The Length of Working Life," *Population Studies*, 3(3):286-294, December 1949; New Zealand, Census and Statistics Department, *New Zealand: Tables of Working Life, 1951*, Special Supplement to November 1955 issue of Monthly Abstract of Statistics; New Zealand Department of Statistics, *New Zealand: Tables of Working Life, 1951*, Special Supplement to February 1957 issue of Monthly Abstract of Statistics; Great Britain, Ministry of Labour, "The Length of Working Life of Males in Great Britain," *Studies in Official Statistics No. 4*, London, HMSO, 1959; and U.S. Department of Labor, Office of Manpower, Automation, and Training, "The Length of Working Life for Males, 1900-60," *Manpower Report No. 8*, by S. H. Garfinkle, July 1963.

population at the same age and assumes that the mortality rates of the general population and the labor force are the same. We can explain the meaning of each column and the method of derivation of the functions of the table by considering each column shown in table 15–13.[33] The special columns lw_x^*, Lw_x^*, Tw_x^*, and $\mathring{e}w_x^*$ represent the same functions as in a standard life table except that these columns refer to working life. This is indicated by the notation w. The functions lw_x^*, ew_x^* and $\mathring{e}_x$ refer to exact ages at each birthday (cols. 5, 7, and 8) while the rest of the functions refer to age intervals between two exact ages (cols. 1 to 4 and 6) or between the midpoint of two age intervals (cols. 9 to 13).

Column (1). w_x.—This column refers to the worker rate or activity rate, or the percentage of the population in the labor force. These rates were obtained for 5-year age groups and taken as central values for these age groups. The values were then interpolated by single years of age and extrapolated beyond age 75 (75 and over was an open interval in the census data). Minor changes were introduced into the central values for the sake of smoothness.

Column (2). L_x.—As in a standard life table, this column refers to the stationary population, or the number of persons who would be living at any age interval out of 100,000 born alive.

Column (3). Lw_x.—This column refers to the stationary male labor force under the prevailing activity rates, or the number of males in the stationary population expected to be in the labor force at each age. It is computed by the formula,

$$Lw_x = L_x \cdot w_x \qquad (40)$$

Column (4). Lw_x^*.—This column represents the number of males in the stationary population who would hypothetically be active if the worker rate at each age under 37 years were the same as at age 37, the maximum worker rate, or

$$Lw_x^* = L_x \cdot w_{37} \qquad (41)$$

This column is required in calculating average number of remaining active years per active survivor at ages under 37 years in order to eliminate the effects of accessions to the active population.

Column (5). lw_x^*.—The number of male survivors at each exact age who would hypothetically be in the labor force if the activity rate at each age under 37 years were the same as at age 37.

$$lw_x^* = \tfrac{1}{2}(Lw_{x-1}^* + Lw_x^*) \qquad (42)$$

Column (6). Tw_x^*.—This column represents the remaining years in the labor force at any age including the hypothetical Lw_x^* values for ages under 37. It may be expressed as follows:

$$Tw_{x_i}^* = \sum_{x=i}^{\infty}(Lw_x^*) \qquad (43)$$

Column (7). $\mathring{e}w_x^*$.—This column refers to the average remaining number of years of active life for males in the labor

force at the given age and is computed from the values of Tw_x^* and the numbers of active survivors, including the hypothetical numbers at ages under 37 (lw_x^*), as follows:

$$\mathring{e}w_x^* = \frac{Tw_x^*}{lw_x^*} \qquad (44)$$

However, for ages 37 and over

$$\mathring{e}w_x^* = \frac{Tw_x}{lw_x} \qquad (45)$$

Column (8). $\mathring{e}_x$.—As in the standard table, this column represents the average number of years of life remaining at the beginning of the given year of age and is represented by the formula

$$\mathring{e}_x = \frac{T_x}{l_x}$$

The underlying functions, T_x and l_x, are not shown since the derivation of $\mathring{e}_x$ was described earlier in this chapter. The difference between $\mathring{e}_x$ and $\mathring{e}w_x^*$ represents the average remaining number of inactive years of life for males in the labor force at any given age.

Column (9). Q_x.—This column represents the mortality rate for males living in the year of age. It is computed as follows:

$$Q_x = \frac{L_x - L_{x+1}}{L_x} \qquad (46)$$

Note that this mortality rate is in terms of the stationary population rather than the survivors at exact ages, as in the computation of a standard life table.

Column (10). $1,000\,A_x$.—This column represents the rate of net accessions to the male labor force between successive years. The rate is derived as the net increase in the stationary labor force per 1,000 persons in the stationary population after allowing for mortality of workers during the year ($Lw_x Q_x$).

$$A_x = \frac{Lw_{x+1} - Lw_x + Lw_x(Q_x)}{L_x} \qquad (47)$$

Column (11). Q_x^s.—This column represents the separation rates from the stationary male labor force due to all causes and was computed as a ratio of the difference between the stationary labor force in successive years to the labor force at age x, or

$$Q_x^s = \frac{Lw_x - Lw_{x+1}}{Lw_x} \qquad (48)$$

For ages 14–37 it was assumed that death was the sole cause of labor force separations and, therefore,

$$Q_{14}^s \text{ to } Q_{37}^s = Q_{14} \text{ to } Q_{37}$$

Column (12). Q_x^d.—This column represents the rates of separation from the labor force due to death under the assumption that the age-specific death rates for males in the labor force are the same as those for all males. Q_x^d is derived from Q_x and Q_x^s as follows:

$$Q_x^d = \frac{Q_x(2 - Q_x^s)}{2 - Q_x} \qquad (49)$$

Table 15–13.—Selected Portions of a Complete Table of Working Life, for the Male Population of the United Arab Republic: 1960

[See text for explanation of symbols]

Age (years) x	Percent of population in labor force	Of 100,000 born alive, number living in year of age		Of 100,000 born alive, number living and in labor force at beginning of year of age	Number of man-years in the labor force remaining in the year of age and later years	Average remaining years of active life for survivors in labor force at beginning of year of age	Complete expectation of life at beginning of year of age	Mortality rate per 1,000 living in year of age	Accessions to the labor force per 1,000 living in year of age	Separations from the labor force per 1,000 in the labor force in year of age		
		In the population	In the labor force							Due to all causes	Due to death	Due to retirement
x	w_x	L_x	Lw_x	Lw_x^{*1}	lw_x^{*}	Tw_x^{*}	$\overset{o}{e}w_x$	$\overset{o}{e}_x$	$1{,}000\,Q_x$	$1{,}000\,A_x$	$1{,}000\,Q_x^{s}$	$1{,}000\,Q_x^{d}$	$1{,}000\,Q_x^{r}$
	(1)	(2)	(3)	(4)	(5)	(6)	(7)	(8)	(9)	(10)	(11)	(12)	(13)
14......	47.2	76,186	35,960	74,738	74,815	3,704,009	49.5	53.1	2.1	80.8	2.1	2.1	–
15......	55.3	76,026	42,042	74,582	74,660	3,629,271	48.6	52.2	2.1	69.9	2.1	2.1	–
16......	62.3	75,866	47,265	74,425	74,504	3,554,689	47.7	51.3	2.1	60.9	2.1	2.1	–
17......	68.4	75,703	51,781	74,265	74,345	3,480,264	46.8	50.4	2.3	48.9	2.3	2.3	–
18......	73.3	75,532	55,365	74,097	74,181	3,405,999	45.9	49.5	2.4	32.9	2.4	2.4	–
19......	77.4	75,355	58,325	73,923	74,010	3,331,902	45.0	48.6	2.4	32.9	2.4	2.4	–
20......	80.7	75,170	60,662	73,442	73,833	3,257,979	44.1	47.7	2.5	30.9	2.5	2.5	–
21......	83.8	74,979	62,832	73,554	73,648	3,184,237	43.2	46.9	2.7	27.9	2.7	2.7	–
22......	86.6	74,780	64,759	73,359	73,457	3,110,683	42.3	46.0	2.7	24.9	2.7	2.7	–
23......	89.1	74,575	66,446	73,158	73,259	3,037,324	41.5	45.1	2.9	21.9	2.9	2.9	–
24......	91.3	74,362	67,893	72,949	73,054	2,964,166	40.6	44.2	2.9	17.9	2.9	2.9	–
32......	97.6	72,377	70,640	71,002	71,140	2,387,015	33.6	37.3	4.0	1.0	4.0	4.0	–
33......	97.7	72,087	70,429	70,717	70,860	2,316,013	32.7	36.4	4.2	1.0	4.2	4.2	–
34......	97.8	71,785	70,206	70,421	70,569	2,245,296	31.8	35.6	4.4	1.0	4.4	4.4	–
35......	97.9	71,469	69,968	70,111	70,266	2,174,875	31.0	34.7	4.6	1.0	4.6	4.6	–
36......	98.0	71,141	69,718	69,789	69,950	2,104,764	30.1	33.9	4.8	1.0	4.8	4.8	–
37......	98.1	70,799	69,454				29.2	33.1	5.0	–	6.0	5.0	1.0
38......	98.0	70,445	69,036				28.4	32.2	5.2	–	6.2	5.2	1.0
39......	97.9	70,079	68,607				27.6	31.4	5.4	–	6.4	5.4	1.0
40......	97.8	69,698	68,165				26.7	30.5	5.7	–	6.8	5.7	1.0
41......	97.7	69,297	67,703				25.9	29.7	6.1	–	7.1	6.1	1.0
60......	89.9	55,475	49,872				11.6	15.1	22.2	–	48.3	21.9	26.4
61......	87.5	54,241	47,461				11.1	14.5	24.3	–	48.8	23.9	24.9
62......	85.3	52,923	45,143				10.6	13.8	26.3	–	51.4	26.0	25.4
63......	83.1	51,532	42,823				10.2	13.1	29.1	–	54.8	28.7	26.1
64......	80.9	50,032	40,476				9.7	12.5	32.9	–	59.2	32.6	26.6
65......	78.7	48,387	38,081				9.3	11.9	36.5	–	63.4	36.0	27.4
66......	76.5	46,622	35,666				8.8	11.3	40.2	–	67.8	39.6	28.2
67......	74.3	44,750	33,249				8.4	10.7	44.2	–	72.5	43.6	28.9
68......	72.1	42,774	30,840				8.0	10.2	48.4	–	77.4	47.7	29.7
69......	69.9	40,706	28,453				7.7	9.6	52.8	–	82.6	52.0	30.6

– Represents zero.

[1] Lw_x^{*} is based on activity rate (w_x) at age 37 and stationary population (L_x) at each age.

Source: Adapted from, United Nations, *Methods of Analyzing Census Data on Economic Activities of the Population*, Population Studies, Series A, No. 43, table A, 1968.

These rates differ from those in column 9 because deaths following retirement during the interval are excluded.

Column (13). Q_x^r.—This column represents the rates of separations from the labor force due to retirement. Q_x^r is obtained by subtracting the rate of separation by death from the total separation rate or

$$Q_x^r = Q_x^s - Q_x^d \qquad (50)$$

Abridged tables of working life are more common than complete tables because data on population and labor force by single years of age are often not available. The abridged form of working-life table differs from the complete form of the table in the same way that the standard complete and abridged tables differ from one another. The rates of accession and separation indicate the probabilities that persons in an age interval will enter or leave the labor force over the next 5-year (or other special) interval. The values for Lw_x, Lw_x^{*}, w_x, Tw_x^{*} refer to the 5-year (or other special) age interval, or the age interval and all later intervals. The 5-year activity

rates are interpolated to exact age x, then applied to l_x of the standard life table to derive lw_x, the number reaching a given age in the labor force. $_5Lw_x$ is then obtained from lw_x by standard short-cut methods. In an alternative procedure $_5Lw_x$ is calculated as the product of $_5w_x$ and $_5L_x$. In general, the major assumptions and methodology present in the construction of single-year tables described also apply to the abridged tables.[34]

[34] For a more complete discussion of the methodology for constructing abridged working-life tables, see U.S. Department of Labor, Bureau of Labor Statistics, *Tables of Working Life*, Bulletin No. 1001, Technical Appendix, August 1950. For abridged tables for various countries, see Saw Swee-Hock, "Malaya: Tables of Male Working Life, 1957," *Journal of the Royal Statistical Society*, Series A (General), 128(2):421–438, 1965; G. M. K. Kpedekpo, "Working Life Tables for Males in Ghana, 1960," *Journal of the American Statistical Association*, 64(325):102–110, March 1969; and Shigemi Kono, "Abridged Working Life Table for Japanese Males: 1930, 1950, and 1955," *Archives of the Population Association of Japan*, The Population Association of Japan, Tokyo, 1963, No. 4. pp. 1–17; and United Nations, University of Chile, Centro Latinoamericano de Demografía, *República de Panamá*, Vol. II, *Proyección de la población económicamente activa, 1950–1975, y Tabla de vida activa masculina para la República y el Distrito de Panamá, 1950*. (Projection of the economically active population, 1950–75, and table of working life for the male population of the republic and the district of Panama, 1950, Part II, by Nivia E. Castro V., Series C, E/CN.CELADE/C.24, Santiago, Chile, 1965.

The construction of tables of working life for females presents some special problems because of the more irregular age pattern of economic activity of women. Such tables may be expanded to take account of labor force participation of women by marital status and presence of children and to provide average number of years of work remaining for women according to these variables. [35]

School-Life Tables. — Another type of multiple decrement life table, the school-life table, combines mortality rates and school enrollment rates to provide estimates of the average number of years of school life for the total population and the enrolled population. [36] Like the working-life table, it is really a type of combined increment-decrement table. The table may be extended to provide age-specific rates of net accession to, and age-specific rates of separation from, the school population, distinguishing separation due to death and separation due to drop-out.

The columns of a school-life table leading to the values of the average expectation of school life are illustrated in table 15–14 for the United States in 1957–59. The average number of years of school life for the total population ($\mathring{e}_{sx}$) is derived by dividing the total number of person-years spent in school by the cohort after a given age (ΣL_{sx} or T_{sx}) by the total number of persons reaching that age (l_x):

---

[35] U.S. Department of Labor, Bureau of Labor Statistics, *Tables of Working Life for Women, 1950*, Bulletin No. 1204, 1957.

[36] For a detailed discussion of tables of school life, see Edward G. Stockwell and Charles B. Nam, "Illustrative Tables of School Life," *Journal of the American Statistical Association*, 58(304):1113–1124, December 1963. See also Mahendra K. Premi and Satis Chandra, Summary of "Tables of School Life for India," in *Population Index*, 34(3):308–309, July–Sept. 1968.

$$\mathring{e}_{sx} = \frac{T_{sx}}{l_x} \tag{51}$$

The values of T_{sx} are derived by summing the L_{sx} column, the stationary school population, and L_{sx} is obtained by multiplying the total stationary population at each age (L_x) by the proportion enrolled (s_x) in each age in the actual population:

$$L_{sx} = L_x s_x \tag{52}$$

The average number of years of school life for enrolled persons ($\mathring{e}'_{sx}$) is derived by dividing the total number of person-years spent in school by the cohort after a given age (ΣL_{sx} or T_{sx}) by the number reaching that age and enrolled in school (l_{sx}):

$$\mathring{e}'_{sx} = \frac{T_{sx}}{l_{sx}} \tag{53}$$

The number reaching a given age and enrolled in school (l_{sx}) is obtained from the L_{sx} column by the formula

$$l_{sx} = \frac{L_{sx-1} + L_{sx}}{2} \tag{54}$$

Table 15–14 also shows, for the United States in 1957–59, net accession rates and separation rates due to deaths for the ages where enrollment rates are rising; and total net separation rates, rates of net separation due to drop-out, and rates of separation due to deaths, for the ages where enrollment rates are falling. The derivation of these figures is described in the source shown in the table.

Table 15–14. — Table of School Life for the Total and Enrolled Populations of the United States: 1957–59

Age (years)	Of 1,000 born alive — Stationary population in age interval L_x (1)	Of 1,000 born alive — Number living at beginning of age interval l_x (2)	Percent of population enrolled in school s_x (3)	Of 100,000 born alive, number living in year of age in the school population L_{sx} (4)	Number of man years remaining in the year of age and later years in the school population T_{sx} (5)	Average number of school years remaining to persons alive at beginning of year of age $\overset{o}{e}_{sx}$ (6)	Number living of 100,000 born alive at beginning of year of age in the school population l_{sx} (7)	Average number of school years remaining to persons alive and enrolled at beginning of year of age $\overset{o}{e}'_{sx}$ (8)	Accessions to school population per 1,000 living in year of age 1,000 A_x (9)	Separations — Due to all causes 1,000 R_x (10)	Separations — Due to death 1,000 R_x^m (11)	Separations — Due to dropping out [1] 1,000 R_x^d (12)
5	96,445	96,482	.62337	55,308	1,327,172	13.76	27,654	13.76	400.1	0.7	0.7	–
6	96,375	96,408	.97389	93,859	1,271,864	13.19	74,584	13.20	19.0	0.6	0.6	–
7	96,313	96,342	.99293	95,632	1,178,005	12.23	94,746	12.43	1.7	0.6	0.6	–
8	96,256	96,283	.99464	95,740	1,082,373	11.24	95,686	11.31	1.7	0.6	0.6	–
9	96,203	96,229	.99634	95,851	986,633	10.25	95,796	10.30	–	0.9	0.5	0.3
10	96,152	96,177	.99602	95,769	890,782	9.26	95,810	9.30	–	0.7	0.5	0.1
11	96,100	96,127	.99590	95,706	795,013	8.27	95,738	8.30	–	2.1	0.6	1.5
12	96,047	96,075	.99436	95,505	699,307	7.28	95,606	7.31	–	2.6	0.6	2.0
13	95,989	96,019	.99237	95,257	603,802	6.29	95,381	6.33	–	6.4	0.7	5.7
14	95,921	95,957	.98668	94,643	508,545	5.30	94,950	5.36	–	31.0	0.8	30.2
15	95,843	95,885	.95684	91,706	413,902	4.32	93,175	4.44	–	76.0	1.0	75.1
16	95,754	95,801	.88496	84,738	322,196	3.36	88,222	3.65	–	170.7	1.1	169.7
17	95,653	95,706	.73471	70,277	237,458	2.48	77,508	3.06	–	418.8	1.4	417.9
18	95,544	95,601	.42747	40,842	167,181	1.75	55,560	3.01	–	314.1	1.5	313.0
19	95,427	95,487	.29358	28,015	126,339	1.32	34,429	3.67	–	215.6	1.5	214.4
20	95,302	95,366	.23058	21,975	98,324	1.03	24,995	3.93	–	282.1	1.6	280.9
21	95,170	95,238	.16576	15,775	76,349	0.80	18,875	4.04	–	364.1	1.8	362.9
22	95,033	95,103	.10556	10,032	60,574	0.64	12,904	4.69	–	200.1	1.7	198.7
23	94,892	94,963	.08457	8,025	50,542	0.53	9,029	5.60	–	101.7	1.6	100.3
24	94,748	94,820	.07609	7,209	42,517	0.45	7,617	5.58	–	48.4	1.6	46.9
25	94,603	94,676	.07251	6,860	35,308	0.37	7,035	5.02	–	161.4	1.6	160.0
26	94,458	94,531	.06091	5,753	28,448	0.30	6,307	4.51	–	150.9	1.7	149.4
27	94,309	94,384	.05180	4,885	22,695	0.24	5,319	4.27	–	75.7	1.7	74.2
28	94,157	94,234	.04795	4,515	17,810	0.19	4,700	3.79	–	156.4	1.9	154.7
29	93,999	94,079	.04052	3,809	13,295	0.14	4,162	3.19	–	168.0	2.0	166.3
30	93,835	93,919	.03377	3,169	9,486	0.10	3,489	2.72	–	201.3	2.1	199.6
31	93,662	93,751	.02702	2,531	6,317	0.07	2,850	2.22	–	251.3	2.3	249.6
32	93,480	93,574	.02027	1,895	3,786	0.04	2,213	1.71	–	334.6	2.5	332.8
33	93,288	93,387	.01352	1,261	1,891	0.02	1,578	1.20	–	500.4	3.2	498.6
34	93,082	93,188	.00677	630	630	0.01	946	0.67	–	1,000.0	3.2	999.2

— Represents zero.

[1] Derived as the difference between R_x and R_x^m; each figure was rounded independently from figures carried to an additional digit.

Source: Edward G. Stockwell and Charles B. Nam, "Illustrative Tables of School Life," *Journal of the American Statistical Association*, 58(304):1114, 1117, 1118, and 1123, December 1963.

SUGGESTED READINGS

Part A. Theory and Construction of Life Tables

Benjamin, B. *Elements of Vital Statistics*. London, George Allen and Unwin. 1959, Pp. 109–122.

Case, Robert A. M., *et al. The Chester Beatty Research Institute Serial Abridged Life Tables, England and Wales, 1841–1960*. Part I, "Tables, Preface, and Notes." London, Royal Cancer Hospital, 1962. Pp. 1–38.

Chiang, Chin Long. *Introduction to Stochastic Processes in Biostatistics*. New York, John Wiley and Sons, 1968. Chapters 9–12.

Derksen, J. B. D. "The Calculation of Mortality-Rates in the Construction of Life Tables. A Mathematical Statistical Study." *Population Studies* (London), 1(4):457–470, Mar. 1948.

Dublin, Louis I., and Spiegelman, Mortimer. "Current Versus Generation Life Tables." *Human Biology* 13(4):439–459. Dec. 1941.

Greville, T. N. E. "Short Methods of Constructing Abridged Life Tables." *Record of the American Institute of Actuaries* 32:29–43. Part I, June 1943.

Jacobson, Paul H. "Cohort Survival for Generations Since 1840." *Milbank Memorial Fund Quarterly* 42(3):36–51. Part 1, July 1964.

Jordan, C. W., Jr. *Life Contingencies*. Chicago, Ill., Society of Actuaries, 1967. Chapters 14 and 15.

Keyfitz, Nathan. "A Life Table that Agrees with the Data." *Journal of the American Statistical Association* 61(314):305–311. June 1966.

————. "A Life Table That Agrees with the Data: II." *Journal of the American Statistical Association* 63(324):1253–1268. Dec. 1968.

————. *Introduction to the Mathematics of Population*. Part I, "The Life Table." Reading, Mass., Addison-Wesley Publishing Co., 1968.

Malaysia, Department of Statistics. *Life Tables for West Malaysia, 1966*. By Lee Jay Cho. Research Paper: No. 2. June 1969.

Reed, Lowell J., and Merrell, Margaret. "A Short Method for Constructing an Abridged Life Table." *American Journal of Hygiene* 30(2):33–62. Sept. 1939. (Reprinted in U.S. Bureau of the Census, *Vital Statistics-Special Reports*, Vol. 9, No. 54, pp. 683–712, June 25, 1940.)

Spiegelman, Mortimer. "Construction of State and Regional Life Tables, 1939–41." *Transactions of the Actuarial Society of America* 49(120):303–327. Part 2, Oct. 1948.

————. *Introduction to Demography* (rev. ed.). Cambridge, Mass., Harvard University Press, 1968. Pp. 116–152.

U.S. Bureau of the Census. *United States Life Tables and Actuarial Tables, 1939–41*. By T. N. E. Greville. 1946.

————. *United States Life Tables, 1890, 1901, 1910, and 1901–1910*. By James W. Glover. 1921.

U.S. National Center for Health Statistics. *Life Tables: 1959–61*. Vol. 1, No. 4. "Methodology of the National, Regional, and State Life Tables for the United States: 1959–61. By T. N. E. Greville. Oct. 1967.

————. *Vital and Health Statistics*, Series 2, No. 4. "Comparison of Two Methods of Constructing Abridged Life Tables by Reference to a 'Standard' Table." By Monroe G. Sirken. Revised, Mar. 1966. Pp. 1–11.

U.S. National Office of Vital Statistics. *Vital Statistics-Special Reports*, Vol. 41, No. 3. "United States Life Tables for the First Year of Life, 1949–51." By Monroe G. Sirken. Sept. 21, 1955. Pp. 56–70.

Wolfenden, H. H. *Population Statistics and Their Compilation* (rev. ed.). Chicago, Ill., University of Chicago Press, 1954. Chapters VI, VII, and VIII.

Part B. Life Tables for Populations with Limited or Defective Data

Abdel-Aty, S. H. "Life-Table Functions for Egypt based on Model Life-Tables and Quasi-Stable Population Theory." *Milbank Memorial Fund Quarterly* 39(2):350–377. Apr. 1961.

Arriaga, Eduardo E. "New Abridged Life Tables for Peru: 1940, 1950–51, and 1961." *Demography* 3(1):218–237. 1966.

————. *New Life Tables for Latin American Populations in the Nineteenth and Twentieth Centuries*. Population Monograph Series No. 3. Institute of International Studies, University of California, Berkeley, Calif., 1968.

Brass, W. "The Construction of Life Tables from Child Survivorship Ratios." *International Population Conference, New York, 1961*. International Union for the Scientific Study of Population, London, 1963. Vol. I, pp. 294–301.

Camisa, Zulma Carmen. *Tabla abreviada de mortalidad de la región Pampeana de la República Argentina, 1946–48*

(Abridged Life Table for the Pampa Region of Argentina, 1946–48). Series C, E/CN.CELADE/C23. United Nations Demographic Center for Latin America, Santiago, Chile, 1964.

Coale, Ansley J., and Demeny, Paul. *Regional Model Life Tables and Stable Populations*. Princeton, N.J., Princeton University Press, 1966.

Demeny, Paul, and Gingrich, Paul. "A Reconsideration of Negro-White Mortality Differentials in the United States." *Demography* 4(2):820–837. 1967.

Kurup, R. S. "A Revision of Model Life Tables." In United Nations, *World Population Conference, 1965* (Belgrade). WPC/WP/314. New York, 1967.

Stolnitz, G. J. *Life Tables from Limited Data: A Demographic Approach*. Princeton, N. J., Princeton University, 1956.

United Nations. *Methods of Using Census Statistics for the Calculation of Life Tables and Other Demographic Measures*. By G. Mortara. Series A, Population Studies, No. 7. 1949.

————. *Age and Sex Patterns of Mortality: Model Life-Tables for Underdeveloped Countries*. Series A, Population Studies, No. 22. 1955.

SUGGESTED READINGS – Continued

————. *Methods for Population Projections by Sex and Age.* Manual III, Manuals on Methods of Estimating Population. Series A, Population Studies, No. 25. Pp. 70–81. 1956.

————. *Methods of Estimating Basic Demographic Measures from Incomplete Data.* Manual IV, Manuals on Methods of Estimating Population. Series A, Population Studies, No. 42. 1967.

Zelnik, Melvin. "Age Patterns of Mortality of American Negroes: 1900–02 to 1959–61." *Journal of the American Statistical Association* 64(326):433–451. June 1969.

Part C. Demographic Applications and Multiple Decrement Tables

Canada, Bureau of Statistics. *Working-Life Tables for Canadian Males.* By Frank T. Denton and Sylvia Ostry. 1969.

Caraballo, Angel Luis. *Tablas de mortalidad y de nupcialidad de Puerto Rico, 1950 y 1960* (Life Tables and Nuptiality Tables for Puerto Rico, 1950 and 1960). Series C, E/ CN.CELADE/C.7, U.N. Demographic Center for Latin America, Santiago, Chile, 1964.

Castro, Nivia E. V. *República de Panamá*, Vol. II, *Proyección de la población economicamente activa, 1950–1975, y Tabla de vida activa masculina para la República y el Distrito de Panamá, 1950* (Projection of the Economically Active Population, 1950–1975, and Table of Working Life for the Male Population of the Republic and the District of Panama, 1950). Part II. Series C, E/CN.CELADE/C.24. U.N. Demographic Center for Latin America, Santiago, Chile, 1965.

Depoid, Pierre. "Tables nouvelles relatives a la population française" (New Tables Relating to the French Population). *Bulletin de la statistique générale de la France* 27(II):269–324. Jan. to Mar. 1938.

Dublin, Louis I., Lotka, Alfred J., and Spiegelman, Mortimer. *Length of Life* (rev. ed.). New York, Ronald Press, 1949.

Grabill, Wilson H. "Attrition Life Tables for the Single Population." *Journal of the American Statistical Association* 40(231):364–375. Sept. 1945.

Great Britain, Ministry of Labour. "The Length of Working Life of Males in Great Britain." *Studies in Official Statistics*, No. 4. London, HMSO, 1959.

Greville, T. N. E. "Mortality Tables Analyzed by Cause of Death." *Record of the American Institute of Actuaries* 37(76):283–294. Part II, Oct. 1948.

Kono, Shigemi. "Abridged Working Life Tables for Japanese Males: 1930, 1950, and 1955." *Archives of the Population Association of Japan.* The Population Association of Japan, Tokyo. No. 4, pp. 1–17. 1963.

Kpedekpo, G. M. K. "Working Life Tables for Males in Ghana, 1960." *Journal of the American Statistical Association* 64(325):102–110. Mar. 1969.

Kumar, Joginder. "Method of Construction of Attrition Life Tables for the Single Population Based on Two Successive Censuses." *Journal of the American Statistical Association* 62(320):1433–1451. Dec. 1967.

Lee, Everett S., and Bowles, Gladys K. "Selection and Use of Survival Ratios in Population Studies." *Agricultural Economics Research* 6(4):120–125. Oct. 1954.

Long, Larry H. "New Estimates of Migration Expectancy in the United States." *Journal of the American Statistical Association* 68(341):37-43. March 1973.

Saveland, Walt, and Glick, Paul C. "First-Marriage Decrement Tables by Color and Sex for the United States in 1958–60." *Demography* 6(3):243–260. Aug. 1969.

Saw, Swee Hock. "Malaya: Tables of Male Working Life, 1957." *Journal of the Royal Statistical Society*, Series A (General), 128(3):421–438. 1965.

————. "Uses of Working Life Tables in Malaya." In United Nations *World Population Conference, 1965* (Belgrade). Vol. IV, p. 336. New York, 1967.

Spiegelman, Mortimer. "The Versatility of the Life Table." *American Journal of Public Health* 47(3):297–304. Mar. 1957.

Stockwell, Edward G., and Nam, Charles B. "Illustrative Tables of School Life." *Journal of the American Statistical Association* 58(304):1113–1124. Dec. 1963.

United Nations. *Methods of Analyzing Census Data on Economic Activities of the Population.* Series A, Population Studies, No. 43. 1968. Pp. 19–34.

U.S. Department of Labor, Bureau of Labor Statistics. *Tables of Working Life for Women, 1950.* By Stuart H. Garfinkle. Bulletin No. 1204. 1957.

————. *Tables of Working Life: Length of Working Life for Men.* Bulletin No. 1001, August 1950. Technical Appendix, pp. 58–74.

U.S. National Center for Health Statistics. *Life Tables: 1959–61.* Vol. 1, No. 6. "U.S. Life Tables by Causes of Death: 1959–61." By Francisco Bayo. May 1968. Pp. 1–14.

Wilbur, George W. "Migration Expectancy in the United States." *Journal of the American Statistical Association* 58(302): 444-453. June 1963.

Wolfbein, Seymour L. "The Length of Working Life." *Population Studies* (London), 3(3):286–294. Dec. 1949.

CHAPTER 16

Natality: Measures Based on Vital Statistics

INTRODUCTION

Concepts

We employ the term "natality" here as a general term representing the role of births in population change and human reproduction. There are also the terms "fertility" and "births." The three terms have alternative and overlapping meanings, and convention has, in fact, often established a particular choice of term to be used in a particular context. The term "fertility," in one sense, is synonymous with natality in referring to the birth factor in population change in the broadest sense. Accordingly, we may speak either of natality statistics, measures, studies, etc., or fertility statistics, measures, studies, etc. In a more restricted sense and more commonly, the term "fertility" refers to the more refined analysis of natality and to certain more or less refined measures of natality. It is because this usage of the term "fertility" is so common that we have preferred to use the less "committed" term "natality" to identify the subjects treated in this and the next chapter.

"Birth statistics" and "birth rates" tend to have a more narrow reference than "natality statistics" and "natality rates" since the former commonly refer to birth statistics *per se* from the registration system and generally exclude those types of natality statistics that are derived from censuses and surveys. Typically, "fertility statistics" come from either source. In this chapter and the next two chapters we have tried to maintain the distinction between birth rates and fertility rates in terms of the source of the data and the complexity of the measures. At the same time, we have tried to recognize the conventions in terminology which have already been established.

"Fertility" refers to actual birth performance, as compared with "fecundity," which refers to the physiological capacity to reproduce.[1] One's fertility is limited by one's fecundity and is usually far below it. Infecund persons are also described as sterile. The term fecundibility refers to a special aspect of fecundity, namely the probability of conceiving measured on a monthly basis.

The terms "natality," "fertility," and "births" may relate to total births, including live births and "stillbirths," but they have come increasingly to refer to live births only. We will use the word "births" to mean "live births" only, unless otherwise specified. We cite again, as given in chapter 3, the statistical definition of a live birth as recommended by the United Nations and the World Health Organization:[2]

"Live birth is the complete expulsion or extraction from its mother of a product of conception, irrespective of the duration of pregnancy, which, after such separation, breathes or shows any other evidence of life, such as beating of the heart, pulsation of the umbilical cord, or definite movement of voluntary muscles, whether or not the umbilical cord has been cut or the placenta is attached; each product of such a birth is considered live-born." According to this definition, the period of gestation, or the state of life or death at the time of registration, are not relevant. The U.S. Public Health Service has recommended for use in the United States the definition of live birth adopted by the United Nations and the World Health Organization.[3] The definition of death, also given in chapter 3, complements that of live birth, since death is "the permanent disappearance of all evidence of life at any time after live birth has taken place." "Fetal deaths" are excluded from both "live births" and "deaths," since fetal death refers to the disappearance of life prior to the explusion or extraction from its mother of a product of conception. Stillbirths, miscarriages, and abortions are types of fetal deaths and, hence, are excluded from live births.

Basic Data for Natality Studies

The basic data for natality studies come from (1) the vital statistics registration system, (2) national censuses, and (3) national sample surveys. In the several countries in which the vital statistics registration system has been incorporated into a national population register system, the latter is then the source of natality data. The first source, the registration system, provides birth statistics principally. The second source, national censuses, provides (a) data on the age composition of the population from which the level of recent fertility can be inferred, (b) direct data on births and fertility, (c) statistics on children by family status of their parents, (d) population data

---

[1] United Nations, *Multilingual Demographic Dictionary*, Population Studies, No. 29, 1958, p. 38. In the Romance languages, the cognate words have opposite meanings, i.e., for example, *fécondité* in French means *fertility* in English. The terms should be paired in terms of their meanings as follows:

English	French	Spanish
Fertility	Fécondité	Fecundidad
Fecundity	Fertilité	Fertilidad

[2] United Nations, *Principles for a Vital Statistics System*, Statistical Papers, Series M, No. 19, August 1953, p. 6; World Health Organization, *Official Records of the World Health Organization*, No. 28, *Third World Health Assembly, Geneva, 8 to 27 May 1950*, December 1950, p. 17.

[3] U.S. National Center for Health Statistics, *Physicians' Handbook on Medical Certification: Death, Fetal Death, Birth*, 1967, Public Health Service Publication No. 593 B.

on fertility-related variables, and (e) population bases for calculating various types of fertility rates. The national sample surveys may provide (a) the same types of data as a census (national data), and (b) additional detailed data permitting a more complete analysis of fertility, including data on special aspects of fertility not amenable to collection in a census and data on the number and timing of births, marriages, and pregnancies.

The types of data and the analytic measures based principally on the birth registration system are sufficiently different from those derived from censuses or surveys to suggest separate detailed treatment of these topics. Therefore, two chapters are devoted to natality in this volume, distinguished in terms of the collection system furnishing the data. The first relates to natality as measured by birth statistics from a registration system and the second to natality as measured by censuses and surveys. A final note on natality data and measures based on national population registers is included in this chapter since the population register system is viewed as closely akin to and as an extension of the registration system.

QUALITY OF BIRTH STATISTICS

Birth statistics suffer from a number of deficiencies very similar to those characteristic of death statistics. We can consider the types of deficiencies under five headings: (1) accuracy of the definition employed and of its application, (2) completeness of registration, (3) accuracy of allocation by place, (4) accuracy of allocation by time, and (5) accuracy of the classification of the births in terms of demographic and socioeconomic characteristics. For the more developed areas, the deficiencies in (1), (2), and (4) have largely been overcome and the concern is primarily with the problems in (3) and (5). For the statistically underdeveloped areas, all of these deficiencies are important, but the problem in (2) looms largest.

Application of Definition

To meet the requirements of the definition recommended by the United Nations, the count of births must include all live-born products of pregnancy and exclude pregnancies not terminating in a live birth, i.e., fetal deaths. One common problem is the failure to register the birth of a child who dies very shortly after birth or who dies before the parents have registered the birth. Depending on national laws and practices, the registration period may vary from a few days to a few years following the date of birth. In some cases newborn infants who die within 24 hours after birth are excluded from the tabulations of live births (e.g., Equatorial Guinea, Northern Zone of Morocco). In others (e.g., Algeria, French Guinea, Taiwan, and Syria) registration of the birth of the infant is not required if the infant dies before the established registration period ends.

Completeness of Registration

As in our discussion of mortality in chapter 14, to which reference should be made, we may distinguish two aspects of the incompleteness of registration of birth statistics: (1) the failure to cover the entire geographic area of a country or all groups in the population, and (2) the failure to register all the vital events in the established registration area. Many less developed countries collect birth statistics only in a designated registration area (e.g., Nigeria, Indonesia, Ghana, Burma, Brazil). Nomadic and indigenous groups (e.g., nomadic Indians in Ecuador, jungle Indians in Venezuela, full-blooded aborigines

in Australia) or particular ethnic or racial groups (Palestinian refugees in Lebanon and Syria, the African population in British Kenya and former Bechuanaland) may be excluded from the groups covered. Special groups (e.g., alien Armed Forces, nonresident foreigners, all foreigners) may be excluded in some cases. Only certain sample registration areas may receive the required effort (e.g., India).

The United Nations finds birth registration to be substantially incomplete for most countries of the world, including nearly all of those considered underdeveloped.[4] Carefully derived estimates of the completeness of registration are rarely available, but estimates are provided to the United Nations by a number of countries and are given in the *Demographic Yearbook*. These estimates are often below 90 percent; even so they probably overstate the actual extent of registration.

Measurement by Match Studies: Experience of the United States. — In the United States the Birth Registration Area was not established until 1915, when it covered only the District of Columbia and 10 States which were mainly in the East, accounting for only 31 percent of the population. Complete geographic coverage of the United States was not achieved until 1933, when the last State joined the registration area. Even so, registration of births in the U.S. Birth Registration Area was not complete. In fact, only 90 percent completeness of registration in a State was a condition for joining the registration area, and the testing procedure was quite crude.

Three systematic national tests of the completeness of birth registration in the United States have been conducted, one in 1940, a second in 1950, and a third in 1969–70. The 1969–70 test was based on a match of listings of children under 5, given over a 9-month period in the National Health Survey-Health Interview Survey sample and in part of the Current Population Survey sample, with birth certificates on file. No attempt was made to determine whether certificates were filed for children who died. The study was intended to provide only national estimates of the completeness of birth registration, by color, in the period 1964–68. The tests of 1940 and 1950 had similar methodological designs; all children under 3 or 4 months of age enumerated in the 1940 and 1950 censuses, respectively, were matched with births registered during the 3 or 4 months preceding the censuses. A special enumeration form was completed for each infant to secure the information needed to locate the birth certificate and to carry out an analysis of the extent of underregistration. The matching study of 1950 was conducted by the U.S. National Office of Vital Statistics, and that of 1969–70 was conducted by the Bureau of the Census in cooperation with the National Center for Health Statistics.

The overall test results indicated a percent completeness of birth registration of 92.5 percent in 1940 and 97.9 percent in 1950. The most impressive differences in each year were found between whites and nonwhites and between births in hospitals and births outside hospitals.

Accuracy of Characteristics of Births

The tabulations of births may suffer from inaccuracies or deficiencies with respect to the geographic allocation of the birth, the allocation to the year of occurrence, and specific demographic and social characteristics of the births and their parents (e.g., age of mother or father, order of birth of child, education of mother or father, etc.).

---

[4] United Nations, *Demographic Yearbook, 1965*, p. 10.

Accuracy of Allocation by Place. – Deficiencies in the geographic allocation of births usually take the form of excessive allocation of births to central cities at the expense of the surrounding suburban and rural areas, and result from the concentration of hospitals and clinics in the central cities and the consequent tendency of mothers to go to the city to have their babies. This bias may occur in those urban agglomerations of either the less industralized or the more industrialized countries where a large percentage of births occur in hospitals and transportation to central city hospitals is relatively easy. It is offset by the availability of hospital facilities outside the central city.

The United Nations recommends that the principal tabulations of birth statistics for geographic areas within countries be made according to place of usual residence.[5] Both types of figures are useful for different purposes. The occurrence data represent the true "service" load, i.e., requirements for maternity services. The residence data are more appropriate for measuring the fertility of the resident population and the relation of the social and economic characteristics of the population to the level of fertility. When the population data from the census are in terms of usual residence, the birth statistics should preferably be tabulated on the same basis, so that the two series will be comparable. Even if a census has been taken on a *de facto* basis, birth statistics by residence are probably more useful than birth statistics by occurrence (e.g., in the measurement of net migration).

On a national level, the United Nations recommends that birth statistics be tabulated on a *de facto* or present-in-area basis.[6]

Accuracy of Allocation by Time. – Many countries tabulate their birth statistics in terms of the year of occurrence of the birth, reassigning the births from the year of registration, as recommended by the United Nations.[7] A considerable number of countries, however, particularly the underdeveloped ones, fail to make this reassignment. The birth statistics published for these countries for any year relate to events which were registered in that year. Inasmuch as the laws requiring early registration are often not observed and are rarely enforced, and the social pressures to register a birth may be few and weak or apply at some late date (e.g., on entering school), many births are not registered for some time, even years, after the birth has occurred. The differences between birth statistics tabulated by year of occurrence and year of registration are minimal in the more developed areas because of the strict requirements for early registration and the general compliance with them. In Belgium and the Netherlands, for example, birth certificates must be filed within 3 days of the birth; in the United States, State requirements vary, but all States require filing within 7 days of the birth; birth certificates have to be filed within 14 days in Japan and within 4 weeks in Norway. Even so, in these countries the births are allocated to the actual year and month of occurrence. (All these countries observe a cut-off date for the births to be included in the national tabulations for each year; in the United States this cut-off date is July 1 of the following year). Because of the considerable delays in registration and the failure to reallocate the births in many countries, the net overstatement or understatement of births in any year may be substantial. Because of the greater pressures to register a death promptly, the difference between the numbers of births on the two bases is likely to be much greater than between the numbers of deaths.

There may also be some error in regard to the year for which a birth is reported if a specific year is regarded as favorable or unfavorable. For example, when the year of the horse starts a new cycle in the Oriental countries, it tends to show low birth rates apparently because of an effort to avoid registration as of that year insofar as possible.[8]

Accuracy of Classification by Socioeconomic Characteristics. – We have noted that errors occur in the tabulations of the demographic and social characteristics of births and in the corresponding tabulations of the population employed to calculate birth and fertility rates. These errors may take the form of differential underregistration of births in certain categories, biases in reporting the characteristics of the births or of the parents of the newborn children, and errors of these kinds in the corresponding populations. We have already alluded to some of these errors in previous chapters. Some further comments will be made later in this chapter as particular characteristics are considered.

GENERAL ASPECTS OF THE ANALYSIS OF NATALITY STATISTICS

Problems in Analysis of Natality Statistics

The analysis of natality is, in several ways, more complicated than the analysis of mortality. Even the statistical definition of birth is more problematic than the definition of death. The differences in the complexity of measurement of natality result from the special and, to some extent, unique characteristics of natality data and of the factors affecting them. These special characteristics give rise to a variety of measures, which may be quite different and which may give inconsistent results.

We may enumerate six such characteristics. First, the entire population is not subject to the risk of having a child. Motherhood is largely restricted to married women of childbearing age but some unmarried women of childbearing age have children. Second, natality may be measured in relation to fathers as well as mothers, or even couples. Two parents, with different demographic, socioeconomic, and other characteristics, are involved in each birth. Third, the event of birth in a sense occurs to both a child and a parent (or parents) and in measuring natality the characteristics of both the child and the parent have to be considered jointly. Death, on the other hand, occurs to an individual only. Fourth, the same adult can have more than one birth in a lifetime and may be more or less continuously exposed to the risk of parenthood even after having had a child. In fact, parenthood may occur twice to the same individual in a single year and, even, in the form of multiple births, twice or more to the same individual at the same hour. Death can occur but once, on the other hand, just as birth can occur but once to the newborn child.

Next, the time period of reference in relation to the population at risk is quite important because of the possibility of large annual fluctuations in fertility and large differences between annual levels of fertility and the levels of fertility performance for couples over a lifetime. In the highly developed countries fertility fluctuates more sharply than mortality, and the current fertility experience of a given age group is more strongly affected by the previous fertility of the cohort than is

[5] United Nations, *Principles for a Vital Statistics System*, Statistical Papers, Series M, No. 19, 1953, p. 18.
[6] Ibid., p. 19.
[7] Ibid., p. 18.
[8] Koya Azumi, "The Mysterious Drop in Japan's Birth Rate," *Transaction*, 5(6):46–48, May 1968.

true for mortality. It is quite important to consider fertility in these countries from the point of view of both annual levels and cumulative levels (for given birth cohorts). In the less developed countries fluctuations in fertility may be relatively small compared to changes in mortality.

Finally, changes in fertility are strongly affected by personal attitudes, preferences, and motivations of couples, and socio-economic factors, as well as physical factors. This makes the interpretation and analysis of changes in fertility a more complex matter than the interpretation and analysis of changes in mortality, which are much less influenced by social and psychological factors.

In at least one respect the handling of birth statistics is somewhat simpler than the handling of death statistics. As a component of change in the estimation of age distributions, births during any period affect only the initial ages of the distribution and, by their nature, define particular "births cohorts" to which deaths and migrants must be assigned. Deaths during the period affect the entire age distribution and, as ordinarily tabulated, must be redistributed by age to correspond to the distribution of the population by age cohorts at the beginning of the period and with the population born during the period.

Factors Important in Analysis

The special characteristics of births and birth statistics just described suggest many of the variables that are important or useful in the measurement and analysis of natality. The variables of prime importance are the age of the child's mother, the age-sex distribution of the population, and the marital status of the mother and of the female population. Fertility variations depend upon the marital status, sex, and age of the parent. Other variables of considerable theoretical importance that reflect sharp differences in fertility are parity of mother (that is, the number of children born to the woman), order of birth of the child, duration of marriage, and interval since previous birth or since marriage; but there are few data on these subjects from the vital statistics tabulations that can be related to population data from a census or survey. Some characteristics of interest in the analysis of natality relate essentially to the births, such as sex, month of occurrence, and place of occurrence in terms of urban-rural residence, metropolitan-nonmetropolitan residence, or size of place. Others relate to the ethnic or socioeconomic characteristics or status of the parents, such as the color, race, or ethnic affiliation of the mother, occupation of the father, educational attainment of the mother or father, religion of mother, and income of the father. Data on these topics provide information on variations in fertility with respect to ethnic characteristics and socioeconomic status but, here too, few data are available on many of them from registration sources.

MEASURES BASED ON BIRTH STATISTICS

There are a great number of measures of natality based on birth statistics. These vary with the aspect of fertility that they are describing, their degree of refinement or elaboration, whether they are summary measures or specific measures, and whether they are measures of fertility *per se* or merely fertility-related measures. As in the case of measures of mortality, our outline of measures of fertility distinguishes, first, observed rates from adjusted rates. The distinction is only approximate. The observed rates are the simpler rates, and are computed directly from actual data in a single brief calculation, whereas the adjusted rates are more complex both with respect to

method of calculation and to interpretation. The latter are often hypothetical representations of the fertility level for a given population group, representing summary measures based on a set of specific fertility rates and various procedures of combining them.

Observed Rates

Crude Birth Rate. — The simplest and commonest measure oı natality is the crude birth rate. The crude birth rate is defined as the number of births in a year per 1,000 midyear population; that is,

$$\frac{B}{P} \times 1,000 \qquad (1)$$

The issues in the choice of a base population are the same as for the crude death rate. As was noted in chapter 14, the midyear population is employed as an approximation of the average population "exposed to risk" during the year or of the number of person-years lived during the year.

An array of crude birth rates based on complete or nearly complete vital statistics registration for a wide range of countries around 1960 (table 16–1) indicates a low of 13.7 for Sweden (1960) and a high of 49 for El Salvador (1961). Rates in recent years in some countries may have been somewhat higher or lower. In fact, reported birth rates for several African countries, based in most cases on sample surveys rather than vital statistics, are higher, even though the reporting of births is described as incomplete. The rates for Mali (1960–61), Malawi (1960, Asiatic population), and Niger (1959–60, African population), for example, are 61, 57, and 52, respectively. Current geographic variation may also encompass reasonably well the range of historical variation. For example, the crude birth rate in the United States appears to have approximated 50 around 1800,[9] and it declined to 15 in 1973. Natural fertility, or fertility uncontrolled in any way, would be higher than has actually been experienced by any country's population. On the basis of an analysis made by Henry of fertility in a number of populations where little or no fertility control is practiced, we may estimate a crude birth rate of about 60 as corresponding to conditions of natural fertility.[10]

In computing the crude birth rate for geographic subdivisions of countries, it is preferable to employ births allocated by place of usual residence, particularly if the population data are distributed in terms of place of usual residence. Satisfactory birth rates by residence may often be computed, however, on the basis of population data tabulated on a *de facto* basis. The use of residence data for births is more important for the smaller geographic units than for the larger ones, since the smaller the area, the more likely the rate is to be affected by the movement of mothers to hospitals to have their babies. Rates by residence more truly reflect the propensity of the population living in an area to have births than rates by occurrence. In computing rates with births by occurrence, some freedom in the choice of type of population is often possible, although use of *de facto* population figures is preferable.

As with the crude death rate, sometimes an annual average crude birth rate covering data for two or three years is computed in order to describe the longer period or to add stability to the rates. As was noted in chapter 14, these annual average

---

[9] Wilson H. Grabill, Clyde V. Kiser, and Pascal K. Whelpton, *The Fertility of American Women*, New York, John Wiley & Sons, 1958, p. 5.
[10] Louis Henry, "La fécondité naturelle. Observation, théorie, résultats," *Population*, 16(4):625–636, Oct.–Dec. 1961.

Table 16–1.—Crude Birth Rates, General Fertility Rates, and Total Fertility Rates, for Various Countries: Around 1960

Country and year	Crude birth rate	General fertility rate	Total fertility rate	Percent difference from the United States		
				Crude birth rate	General fertility rate	Total fertility rate
Chile (1960)............................	35.7	161.2	4,810	+50.6	+36.6	+31.6
Costa Rica (1963)[1].....................	42.7	243.5	7,295	+80.2	+106.4	+99.6
El Salvador (1961)......................	49.4	232.2	6,702	+108.4	+96.8	+83.4
England and Wales (1961)................	17.9	89.4	2,769	−24.5	−24.2	−24.2
Federal Republic of Germany (1961).......	18.3	87.0	2,521	−22.8	−26.3	−31.0
France (1962)..........................	17.7	90.8	2,814	−25.3	−23.1	−23.0
Hungary (1961).........................	14.0	65.5	1,912	−40.9	−44.5	−47.7
Ireland (1961)[1]........................	21.2	116.8	3,791	−10.5	−1.0	+3.7
Israel (1963)..........................	25.0	122.7	3,803	+5.5	+4.0	+4.1
Italy (1961)...........................	18.6	84.0	2,429	−21.5	−28.8	−33.5
Japan (1960)..........................	17.2	70.7	2,006	−27.4	−40.1	−45.1
Mauritius (1962).......................	38.0	197.2	5,863	+60.3	+67.1	+60.5
Panama (1960).........................	40.8	185.2	5,572	+72.2	+56.9	+52.5
Poland (1963)..........................	19.2	89.3	2,696	−19.0	−24.3	−26.2
Puerto Rico (1960).....................	32.3	158.4	4,666	+36.3	+34.2	+27.7
Sweden (1960).........................	13.7	68.1	2,167	−42.2	−42.3	−40.7
Taiwan (1960)[1]........................	44.8	195.8	5,754	+89.0	+65.9	+57.5
United States (1960)...................	23.7	118.0	3,654	(X)	(X)	(X)
Yugoslavia (1961).....................	22.7	101.0	2,783	−4.2	−14.4	−23.8
Average percent difference..............	(X)	(X)	(X)	+41.2	+40.9	+39.8

X Not applicable.

[1] Data tabulated by year of registration rather than year of occurrence.

Source: United Nations, *Demographic Yearbook, 1965*, tables 12, 13, 14, and 17.

rates may be computed in a number of ways, all of which give essentially the same results. The crude rates may be computed for each year and averaged, the annual average number of births may be divided by the annual average population, or the average number of births for the three years may be divided by the midperiod population. The last

$$\frac{1/3(B_1 + B_2 + B_3)}{P_2} \times 1,000 \qquad (2)$$

is the most commonly used procedure.

Like the term "crude death rate," the term "crude birth rate" when unqualified means the crude birth rate for the total population of an area, but one may also speak of the crude birth rate of a particular population group in the area, such as a race, nativity, residence, or occupation group. For example, the crude birth rate for the urban population of Panama in 1960 is represented by:

$$\frac{B_u}{P_u} \times 1,000 = \frac{16,147}{446,213} \times 1,000 = 36.2$$

In this more general sense the principal characteristic of a "crude" birth rate is that all ages and both sexes are represented in the rate.

The birth rate should be adjusted for underregistration of the births and for underenumeration of the population if satisfactory estimates of these errors are available, according to the following formula:

$$\frac{B \div C_b}{P \div C_p} \times 1,000 \qquad (3)$$

where C_b represents the percent completeness of registration of births and C_p represents the percent completeness of the census counts or population estimates.

Official national birth rates for the United States are now based on the uncorrected figures. The U.S. National Center

for Health Statistics has followed this practice since 1959, on the ground that the underregistration of births is rather small and has presumably become smaller than the understatement of the population. It has published an official series of birth rates for the years 1909 to 1959 employing births adjusted for underregistration; during all or most of this period, births were presumed to be less completely reported than the census population.

Monthly Birth Rate.—For births, as for deaths, there is an interest in examining variations over periods shorter than a year. We note only very briefly here the measures of the monthly variation in the birth rate, corresponding to those described for the death rate in chapter 14. The monthly birth rate may be "annualized" by inflating the number of births for a given month by the ratio of the total number of days in the year to the number in the particular month and then dividing by the population for the month. The annualized birth rate in the United States for June 1965, adjusted for underregistration, is obtained as follows:

$$\frac{\frac{365}{30} \times B_{\text{June 1965}}}{P_{\text{June 1965}}} = \frac{\frac{365}{30} \times 311,662}{193,818,000} \times 1,000$$

$$= \frac{3,791,898}{193,818,000} \times 1,000 = 19.6$$

For births, as for deaths, analysis of monthly changes may be carried out on the basis of rates computed for successive 12-month periods ending in each successive month. Rates of this kind for the United States, based on provisional data, are published with a lag of less than two months by the U.S. National Center for Health Statistics.[11] Year-to-year changes

---

[11] U.S. National Center for Health Statistics, *Monthly Vital Statistics Report (Provisional Statistics)*, "Births, Deaths, Marriages, and Divorces, for [Month]."

in the birth rate on a current basis can also be measured by comparing the rates for the same month in successive years. The rates in this comparison do not require adjustment for the number of days in the month or for seasonal variations (except for February in leap year). Annual relative changes can also be measured on a current basis by comparing cumulative rates covering all the months in a year to date with the corresponding period in the preceding year; that is, for example, January to May 1969, and January to May 1968.

General Fertility Rate. — Although the crude birth rate is a valuable measure of natality, particularly in indicating directly the contribution of natality to the growth rate, its analytic utility is extremely limited. This is because it is affected by many factors, particularly the specific composition of a population with respect to age, sex, and other characteristics. Since the age and sex composition of a population has such a strong influence on the level of its crude birth rate, measures of natality that are less affected by differences in age-sex composition from one population group to another are more useful analytically for interarea and intergroup comparisons. A number of such measures have been developed and are variously referred to as specific, general, adjusted, or standardized, and as birth rates, fertility rates, or reproduction rates, depending generally on their degree of complexity or on their particular significance.

The simplest overall age-limited measure is the general fertility rate, defined as the number of births per 1,000 women of childbearing age. It may be represented by:

$$\frac{B}{P_{15-44}^f} \times 1,000 \qquad (4)$$

The total number of births, regardless of age of mother, is employed in the numerator and the female population 15 to 44 years of age is employed in the denominator. Applying the formula to the data for Costa Rica in 1963 (table 16-2), we have:

$$\frac{63,798}{261,963} \times 1,000 = 243.5$$

This figure may be compared with the corresponding figure for the United States in 1963, 108.5. On the basis of this comparison, we can say that fertility in Costa Rica is about twice as great as fertility in the United States. The general fertility rate is sometimes defined in relation to the number of women 15 to 49 years of age but, as noted below, women aged 45–49 contribute relatively few births to the total number. It is preferable, therefore, to avoid dilution of the denominator of the rate with this group when the risk is so low, by excluding it from the population of childbearing age used in the rate. General fertility rates for various countries around 1960 are shown in table 16-1.

Age-Specific Birth Rates. — A set or "schedule" of age-specific birth rates for a given date and population group serves as a basis for a detailed comparison, with corresponding rates for other population groups, that is unaffected by differences between the groups in age-sex composition. A set of rates may consist of the rates for 5-year age groups from 10–14 to 45–49, 15–19 to 45–49, or 15–19 to 40–44. The age classification recommended by the United Nations has 10 categories, under 15 years, quinquennial groups 15–19 to 45–49, a terminal

group 50 and over, and a group of unknown ages;[12] but rates are shown in the *Demographic Yearbook* for the group under 20, 5-year groups to 40–44, and a terminal group 45 and over.

An age-specific birth rate is defined as the number of births to women of a given age group per 1,000 women in that age group:

$$f_a = \frac{B_a}{P_a^f} \times 1,000 \qquad (5)$$

The formula for the birth rate at ages 20–24 is:

$$f_{20-24} = \frac{B_{20-24}}{P_{20-24}^f} \times 1,000$$

Age-specific birth rates as here defined are a type of "central" rate, in somewhat the same sense that age-specific death rates are central rates. The corresponding probabilities are discussed later. The rate for women 20 to 24 years old in Costa Rica (1963) is, therefore (table 16-2):

$$\frac{17,905}{52,929} \times 1,000 = 338.3$$

Table 16-2. — Calculation of General Fertility Rate and Age-Specific Birth Rates, for Costa Rica: 1963

Age of mother	Births		Female population	Rates
	Reported[1]	Unknown distributed[2] (1) x factor[3] =		[(2) ÷ (3)] x 1,000 =
	(1)	(2)	(3)	(4)
15 to 19 years.........	[4]7,893	7,957	65,999	120.6
20 to 24 years.........	17,760	17,905	52,929	338.3
25 to 29 years.........	15,623	15,750	43,523	361.9
30 to 34 years.........	11,315	11,407	38,603	295.5
35 to 39 years.........	7,671	7,733	34,244	225.8
40 to 44 years.........	2,649	2,671	26,665	100.2
45 to 49 years.........	[5]372	375	22,466	16.7
Unknown..............	515	(X)	(X)	(X)
15 to 44 years....	[6]63,798	[6]63,798	261,963	243.5

X Not applicable.
[1] Data tabulated by year of registration rather than year of occurrence.
[2] Births to mothers of unknown age are distributed proportionately among births to mothers of known age.
[3] $\frac{\text{Total births}}{\text{Births to mothers of known age}} = \frac{63,798}{63,283} = 1.00814$
[4] Includes births to mothers under age 15.
[5] Includes births to mothers 50 or over.
[6] All births, regardless of age of mother, registered in 1963 that occurred in 1963 or the previous 7 years.

Source: United Nations, *Demographic Yearbook, 1965*, tables 6 and 13.

Table 16-3 presents schedules of age-specific birth rates for a number of countries around 1960, and figure 16-1 displays these rates graphically. The line graphs approximate a normal curve moderately skewed to the right. The striking fact about these sets of rates and the figure is the great similarity of the age pattern of the rates from one country to another. This similarity is made more evident if the schedules of rates are converted into percent distributions, as is shown in table 16-4. The ranking of the rates and their relative magnitudes over the ages of the reproductive period are the same or approximately the same for most of the 19 countries listed. Clearly,

_____
[12] United Nations, *Principles for a Vital Statistics System*, p. 20.

Table 16–3.—Age-Specific Birth Rates and Median Age of Mother, for Various Countries: Around 1960

[Birth rates are live births per 1,000 women in specified group]

Country and year	Birth rates by age of woman							Median age of mother
	15 to 19 years[1]	20 to 24 years	25 to 29 years	30 to 34 years	35 to 39 years	40 to 44 years	45 to 49 years[2]	
Chile (1960)............................	75.2	213.9	244.2	208.3	143.3	63.0	14.1	28.9
Costa Rica (1963)[3].....................	120.6	338.3	361.9	295.5	225.8	100.2	16.7	28.7
El Salvador (1961)......................	142.0	327.2	320.4	268.0	191.2	70.7	20.8	28.1
France (1962)...........................	23.2	171.4	184.4	108.7	54.6	19.1	1.3	27.4
Hungary (1961)..........................	52.3	154.0	94.9	50.2	23.0	7.6	0.4	24.5
Ireland (1961)[3]........................	9.5	108.2	216.9	209.5	152.2	57.7	4.2	31.1
Israel (1963)...........................	42.9	227.9	229.0	155.6	74.6	25.3	5.6	27.4
Japan (1960)............................	4.3	106.6	181.1	79.7	23.8	5.2	0.4	27.5
Panamá (1960)...........................	138.9	308.1	289.9	198.6	129.1	41.2	8.6	26.9
Sweden (1960)...........................	33.5	128.7	136.7	82.6	38.9	12.2	0.8	27.0
Taiwan (1960)[3].........................	48.4	253.5	333.4	254.9	169.2	78.7	12.7	29.1
United States (1960)....................	90.2	258.1	197.4	112.7	56.2	15.5	0.9	25.4
Yugoslavia (1961).......................	52.1	179.1	154.8	90.9	49.1	25.3	5.6	26.5

[1] Includes births to women under age 15. Base is the female population 15 to 19 years of age.
[2] Includes births to women age 50 and over. Base is the female population 45 to 49 years of age.
[3] Data tabulated by year of registration rather than year of occurrence.

Source: United Nations, *Demographic Yearbook, 1965,* table 14.

the rates for ages 45–49 are of little absolute or relative importance. The rates for ages 40–44 usually fall below those for ages 15–19 and the rates for ages 35–39 usually exceed those for ages 15–19. In general, then, the key fertility ages are 15 to 39; they contribute about 92 to 98 percent of the births. Ages 20 to 29 within this range account for about 50 to 60 percent of

Figure 16–1.—Age-Specific Birth Rates, for Various Countries: Around 1960

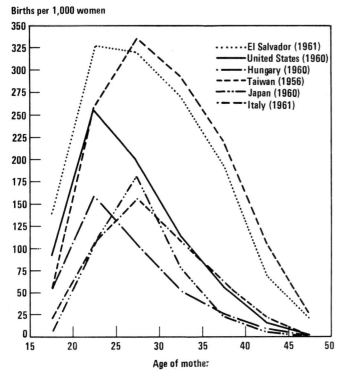

Births per 1,000 women

..........El Salvador (1961)
———United States (1960)
—·—·—Hungary (1960)
— — —Taiwan (1956)
—··—··Japan (1960)
— · —Italy (1961)

Age of mother

NOTE.—Points are plotted at midpoint of age intervals.

Source: See table 16–3.

total births. The older women tend to make their greatest relative contribution in situations where overall fertility is high (e.g., Costa Rica, Taiwan), although there are outstanding exceptions (e.g., Ireland).

Differences in the age pattern of childbearing may also be measured in terms of the median age of childbearing (**median age of mother**) or the mean age of childbearing (**mean age of mother**). Both measures are used but under different circumstances. To eliminate the effect of differences in the age-sex composition of the populations being compared, these measures should be calculated on the basis of age-specific birth rates rather than numbers of births. They may be interpreted then as describing the age pattern of childbearing of a synthetic cohort of women, i.e., a hypothetical group of women who are viewed as having in their lifetime the (fertility) experience recorded in a single calendar year.

Both the median age and the mean age are ordinarily computed from data compiled for 5-year age groups. The mean age of a distribution of birth rates is computed according to the formula:

$$\bar{a} = \frac{\sum_a a f_a}{\sum_a f_a} \qquad (6)$$

where a represents the midpoint of each age interval (17.5, 22.5, etc.) and f_a represents an age-specific birth rate for a 5-year age group. (The formula follows the form of a weighted average of ages, the weights being the age-specific birth rates.)

The considerable similarity of the age distributions of birth rates suggests that the median age would vary only little. For the array of countries shown in table 16–3, the highest and lowest median ages vary by 6.6 years out of a range of about 35 years of reproductive ages, but if the two extreme cases of Hungary (24.5) and Ireland (31.1) are excluded, the difference falls to only 3.7 years. These differences tend to distinguish populations that have their children relatively early in the childbearing period and those that have them relatively later in life. On this basis, childbearing occurs relatively early in Hungary and relatively late in Ireland. These differences may be accounted for either in terms of the differences in the median age at marriage, in the median interval between marriage and each successive birth, or in both.

Fertility Rates Specific for Order of Birth. — Another dimension for analyzing fertility, in addition to the age of the mother, is the order of birth of the child. Order of birth refers to the number of children born alive to the mother, including the present child. The simplest way of analyzing births classified by order of birth consists of calculating the proportional distribution of the births by order. Such a percentage distribution of births by order is much less affected by underregistration of births than are rates and can be computed without use of population data; but at the same time the percentages are affected by such factors as the distribution of the female population by age, marital status, and duration of marriage within the childbearing period. Percentage distributions of births by order for several countries indicate a tremendous variation in the relative frequency of births of different orders in any year: [13]

Birth order	Costa Rica, 1963	United States, 1960	Japan, 1960	United Arab Republic, 1960 [a]
Total............	100.0	100.0	100.0	100.0
First................	16.1	25.6	44.5	10.6
Second............	14.3	24.0	32.6	12.0
Third..............	13.0	18.7	13.8	16.6
Fourth.............	11.2	12.0	5.0	17.0
Fifth................	9.7	6.9	2.2	12.8
Sixth................	8.0	4.0	1.1	10.7
Seventh............	6.7	2.3	0.5	7.3
Eighth and over.	20.8	3.6	0.4	11.5
Unknown..........	0.3	2.9	–	1.6

– Zero or rounds to zero.
[a] Birth registration is incomplete and first births appear to be relatively understated.

General fertility rates may be computed for each order of birth without reference to age of the mother. A **general order-specific fertility rate** is defined as the number of births of a given order per 1,000 women of childbearing age:

$$\frac{B_i}{P^f_{15-44}} \times 1,000 \qquad (7)$$

where B_i represents births of a given order and P^f_{15-44} the number of women 15 to 44 years of age. Thus, the general second-order fertility rate is:

$$\frac{B_2}{P^f_{15-44}} \times 1,000$$

where B_2 refers to births of the second order. This formula is evaluated for Costa Rica in 1963 as follows (table 16–5):

$$\frac{9,114}{261,963} \times 1,000 = 34.8$$

Note that the sum of the general order-specific rates over all

[13] Source of basic data is: United Nations. *Demographic Yearbook, 1965,* table 16.

orders equals the general fertility rate, that is:

$$\sum_{i=1}^{\infty} \frac{B_i}{P^f_{15-44}} = \frac{B}{P^f_{15-44}} \qquad (8)$$

Usually, there is a category of births of unknown order. This category may be treated separately or distributed over the births of known order *pro rata.*

Variations in order-specific birth rates by age may be analyzed in terms of **age-order-specific rates.** A rate of this type is represented by the formula:

$$f^i_a = \frac{B^i_a}{P^f_a} \times 1,000 \qquad (9)$$

where B^i_a represents births of a given order to women of a given age group and p_a^f relates to all the women in a particular age group, without regard to the number of children they have had. This is a type of central rate, as we have defined it. For example, the second-order birth rate for women 20 to 24 years old in Costa Rica (1963) is (table 16–5):

$$\frac{B^2_{20-24}}{P^f_{20-24}} \times 1,000 = \frac{4,310}{52,929} \times 1,000 = 81.4$$

The sum of age-order-specific rates over **all orders** for a particular age group equals the general age-specific birth rate for that age group (the rates being additive because the denominators are the same); that is,

$$\sum_{i=1}^{\infty} \frac{B^i_a}{P^f_a} \times 1,000 = \frac{B_a}{P^f_a} \times 1,000 \qquad (10)$$

The age-order specific rates cannot be added for a particular order over **all ages** to derive the general fertility rate for that order, however (since the denominators of the rates are all different).

In the calculation of age-order specific birth rates, it is desirable to dispose both of births of unknown age and of births of unknown order, both of which ordinarily appear in the basic tabulations. This can be done satisfactorily by (1) prorating the "unknown ages" for each order among the "known ages" for each order, and then (2) prorating the "unknown orders" for each age group among the "known orders" for each age group, or vice versa.

Median ages of childbearing may be computed for each order of birth from age-order-specific birth rates, as shown in table 16–6. These values give an indication of the "typical" age at which mothers have their first, second, etc., child. Differences between values for consecutive orders give a rough idea of the spacing intervals between births: The median age for first births suggests the typical age of starting families; and, by determining the average number of children per woman by other methods, we can identify by interpolation the typical age of completing families. For example, the typical age of starting families in Costa Rica in 1963 is estimated at 21.6 years, the median age of first births. The annual "total fertility rate," discussed below, of 7.3 children corresponds in table 16–6 to a median age of about 32 years, which may be taken as the typical age of completing families in Costa Rica. Each child tends to follow the previous child after an interval of about 1½ to 2 years, with a total average interval of about 10½ years from the birth of the first child to the birth of the last child.

Table 16–4.—Percent Distribution of Age-Specific Birth Rates by Age of Mother, for Various Countries, Around 1960

Country and year	All ages	15 to 19 years[1]	20 to 24 years	25 to 29 years	30 to 34 years	35 to 39 years	40 to 44 years	45 to 49 years
Chile (1960)............................	100.0	7.7	22.3	25.4	21.7	14.9	6.6	1.5
Costa Rica (1963)[2].....................	100.0	8.2	23.2	24.8	20.3	15.5	6.9	1.1
France (1962)...........................	100.0	4.0	30.5	32.8	19.3	9.7	3.4	0.2
Hungary (1961)..........................	100.0	13.7	40.3	24.8	13.1	6.0	2.0	0.1
Ireland (1961)[2]........................	100.0	1.2	14.3	28.6	27.6	20.1	7.6	0.6
Israel (1963)...........................	100.0	5.6	30.0	30.1	20.4	9.8	3.3	0.7
Japan (1960)............................	100.0	1.1	26.6	45.2	19.9	5.9	1.3	0.1
Panamá (1960)...........................	100.0	12.2	27.7	26.1	17.9	11.6	3.7	0.8
Sweden (1960)...........................	100.0	7.7	29.7	31.5	19.1	9.0	2.8	0.2
Taiwan (1956)[2]........................	100.0	3.9	20.4	26.1	22.7	17.0	8.2	1.8
United States (1960)....................	100.0	12.3	35.3	27.0	15.4	7.7	2.1	0.1
Yugoslavia (1961).......................	100.0	9.3	32.2	27.8	16.3	8.8	4.5	1.0

[1] Includes birth to women under age 15.
[2] Births tabulated by year of registration rather than year of occurrence.

Source: United Nations, *Demographic Yearbook, 1965*, tables 14 and 17.

Table 16–5. —Calculation of General Order-Specific Birth Rates and Order-Specific Birth Rates for Women 20 to 24 Years of Age, for Costa Rica: 1963

[Births are tabulated by year of registration rather than years of occurrence]

Birth order	Births			Ages 20 to 24 including part of unknowns[1] $\frac{(1)}{(1)-(3)} \times (2) =$	Rates	
	All ages	Ages 20 to 24 reported	Unknown age		All ages (1) ÷ Female pop. 15 to 44[2] =	Ages 20 to 24 (4) ÷ Female pop. 20 to 24[3] =
	(1)	(2)	(3)	(4)	(5)	(6)
All orders...................	63,798	17,760	515	[4]17,894	243.5	[4]338.1
First......................	10,263	3,754	54	3,774	39.2	71.3
Second.....................	9,114	4,288	46	4,310	34.8	81.4
Third......................	8,264	3,961	61	3,990	31.5	75.4
Fourth.....................	7,132	2,856	64	2,882	27.2	54.5
Fifth......................	6,174	1,631	50	1,644	23.6	31.1
Sixth......................	5,121	713	50	720	19.5	13.6
Seventh....................	4,278	322	41	325	16.3	6.1
Eighth and over............	13,292	196	112	198	50.7	3.7
Unknown....................	160	39	37	51	0.6	1.0
Median order of birth......	(X)	(X)	(X)	(X)	4.1	2.7

X Not applicable.
[1] Births to mothers of unknown age are distributed proportionately among births to mothers of known age.
[2] Female population 15 to 44 years of age = 261,963.
[3] Female population 20 to 24 years of age = 52,929.
[4] Number of births obtained by summation of figures for each order: number and rate differ from figures obtained by distributing unknown ages without regard to order of birth (17,905 and 338.3).

Source: United Nations, *Demographic Yearbook, 1965*, table 16.

Marital Fertility Rates.—Observed fertility rates may be further refined by restricting the base population to married or ever-married women since most births occur within marriage. Use of marital fertility rates of various kinds permits separating the effect of changes in the tendency to marry from the effect of changes in the level of (marital) fertility upon overall fertility.

General and age-specific marital fertility rates.—Alternative forms of general fertility rates that have this refinement, general marital fertility rates and general legitimate fertility rates, employ all live births and legitimate live births, respectively. The **general marital fertility rate** is defined as the total number of live births in a year, regardless of the age and marital status of the mother, per 1,000 married women 15 to 44 years of age at the middle of the year:

$$\frac{B}{P_{fl}^{15-44}} \times 1,000 \qquad (11)$$

where P_{fl}^{15-44} represents married women 15 to 44 years of age. An array of general marital fertility rates for years around 1960 is shown in table 16–7: the rates given here range from 99.6 for Hungary to 671.9 for Panama. The **general legitimate fertility rate** is defined as the number of legitimate live births

Table 16–6. — Age-Order-Specific Birth Rates, for the United States, 1965, and Costa Rica, 1963

Country, year, and order of birth	General fertility rate	Age of mother							Median age of mother
		15 to 19 years[1]	20 to 24 years	25 to 29 years	30 to 34 years	35 to 39 years	40 to 44 years	45 to 49 years[2]	
UNITED STATES (1965)									
All orders[3]......................	96.6	71.3	196.8	162.5	95.0	46.4	12.8	0.8	25.8
First...............................	29.8	52.4	76.0	24.7	7.3	2.8	0.6	–	21.9
Second.............................	23.4	14.9	66.1	40.4	13.3	4.3	0.9	–	24.2
Third...............................	16.6	3.3	33.3	40.2	19.8	7.2	1.5	0.1	27.0
Fourth.............................	10.7	0.6	13.8	26.9	18.5	8.0	1.9	0.1	28.8
Fifth...............................	6.4	0.1	5.2	15.0	13.3	6.8	1.8	0.1	30.3
Sixth...............................	3.8	–	1.7	8.0	8.8	5.1	1.4	0.1	31.6
Seventh............................	2.3	–	0.5	4.1	5.6	3.6	1.1	0.1	32.6
Eighth and over....................	3.7	–	0.2	3.3	8.4	8.5	3.5	0.3	35.1
Median order of birth..............	2.3	1.2	1.8	2.9	3.9	4.6	5.3	6.5	(X)
COSTA RICA (1963)									
All orders[3]......................	243.5	120.6	338.1	361.9	295.5	225.8	100.2	16.7	28.7
First...............................	39.2	70.1	71.3	28.3	10.1	4.6	2.0	0.4	21.6
Second.............................	34.8	32.7	81.4	40.5	15.2	6.7	1.6	0.4	23.5
Third...............................	31.5	12.7	75.4	50.3	22.3	8.8	2.3	0.8	24.9
Fourth.............................	27.2	3.0	54.5	56.2	27.8	12.6	3.1	0.6	26.9
Fifth...............................	23.6	0.6	31.1	59.0	33.4	14.7	4.3	0.7	28.4
Sixth...............................	19.5	0.2	13.6	49.7	36.8	18.7	5.6	0.6	29.9
Seventh............................	16.3	–	6.1	36.6	37.9	20.9	6.2	0.7	31.5
Eighth and over....................	50.7	–	3.7	40.5	111.7	138.6	75.0	12.4	36.3
Median order of birth..............	4.1	1.4	2.7	4.6	6.6	7.7	7.8	7.8	(X)

– Represents zero or rounds to zero.   X Not applicable.
[1] Includes births to mothers under 15 years of age.
[2] Includes births to mothers 50 years of age and over.
[3] Includes births of unknown order. The rates for these births are not shown separately.

Source: U.S. National Center for Health Statistics, *Vital Statistics of the United States, 1965*, Vol. 1, *Natality*, 1967, table 1–10, and United Nations, *Demographic Yearbook, 1965*, tables 6, 16, and 17.

in a year per 1,000 married women 15 to 44 years old at the middle of the year:

$$\frac{B_l}{P_{fl}^{15-44}} \times 1,000 \qquad (12)$$

where B_l represents only legitimate births.
An array of general legitimate fertility rates for years around 1960 is also shown in table 16–7; the rates given here range from 94.1 for Hungary to 245.4 for Ireland.

There are important international differences in the classification of the population by marital status and of births by legitimacy, which have a serious effect on the international comparability of marital fertility rates. The difference between the conventional general fertility rate and the general marital fertility rate is a function of the proportion of women who are married. The difference between the general marital fertility rate and the general legitimate fertility rate is a function of the proportion of births that are legitimate. In those countries, particularly in Latin America, where many persons live in consensual (free or common law) unions (chapter 10), there may be extremely large differences between the general fertility rate and the general marital fertility rate, and between the general marital fertility rate and the general legitimate fertility rate, resulting from the fact that the consensually "married" are not included among the married and their births are not counted as legitimate. The general fertility rate for Panama in 1960 is 185.2; the general marital fertility rate, employing all births in the numerator but only (legally) married women in the denominator, is:

$$\frac{41,258}{61,407} \times 1,000 = 671.9$$

The general legitimate fertility rate, employing only births to (legally) married women in the numerator and only (legally) married women in the denominator, is: [14]

$$\frac{13,808}{61,407} \times 1,000 = 224.9$$

In Panama over half of the couples live in consensual unions. If the general marital fertility rate is based on all "married" women, including the consensually "married," it equals 337.3 and approximates the general legitimate fertility rate more closely.

Age-specific marital birth rates and **age-specific legitimate birth rates** may also be computed for the childbearing ages, that is,

$$\frac{B^a}{P_{fl}^a} \times 1,000 \qquad (13)$$

and

$$\frac{B_L^a}{P_{fl}^a} \times 1,000 \qquad (14)$$

The illegitimacy ratio is conventionally defined as the number

[14] Births of unknown legitimacy are disregarded. They should probably be assigned wholly or largely to the legitimate category.

Table 16-7. —General Marital and Legitimate Fertility Rates and Age-Specific Legitimate Birth Rates, for Various Countries: Around 1960

[General marital fertility rates are total births per 1,000 married women 15 to 44 years of age. General legitimate fertility rates are legitimate live births per 1,000 married women 15 to 44 years of age. Age-specific legitimate birth rates are legitimate live births per 1,000 married women in specified age group]

Country and year	General marital fertility rate	General legitimate fertility rate	Legitimate birth rates by age of woman						
			15 to 19 years	20 to 24 years	25 to 29 years	30 to 34 years	35 to 39 years	40 to 44 years	45 to 49 years
Federal Republic of Germany (1961).....	141.4	133.2	359.8	285.4	210.9	118.6	60.3	23.2	1.2
France (1962).........................	144.9	135.7	544.8	364.6	224.3	122.0	60.5	21.1	1.5
Hungary (1960)........................	99.6	94.1	320.9	228.4	117.1	57.3	27.4	9.5	0.6
Ireland (1961)........................	249.5	245.4	608.6	478.0	392.3	298.6	202.4	77.1	5.8
Japan (1960)..........................	125.6	124.0	319.1	338.1	235.2	91.4	26.9	6.1	0.5
Panama (1960).........................	671.9	224.9	466.8	400.5	301.2	177.0	116.0	39.0	8.8
Poland (1960).........................	157.8	150.8	489.6	336.9	196.6	115.0	65.1	25.4	3.0
Puerto Rico (1960)...................	320.1	221.2	468.0	428.9	264.8	156.4	106.8	31.8	
Sweden (1963)........................	121.8	106.5	509.6	270.0	181.5	99.9	44.0	12.0	0.9
United States (1960).................	170.5	156.3	485.6	354.4	222.3	123.3	61.7	17.4	1.2

Source: United Nations, *Demographic Yearbook, 1965*, various tables, particularly tables 17 and 25.

of illegitimate live births per 1,000 total live births:

$$\frac{B_u}{B_t} \times 1,000 \qquad (15)$$

where B_u represents illegitimate births and B_t total births. This ratio, and illegitimacy ratios in general, are subject to considerable understatement because illegitimate births are more likely to go unregistered than legitimate births and because they may often be reported as legitimate as a result of the social stigma attached to illegitimacy of parenthood. The illegitimacy ratio has an extremely wide range depending on the situation in a given area; it may vary from nearly 0 to 650 or even more:

Country and year	*Illegitimacy ratio*
Greece (1961)...........................	12.2
El Salvador (1961)....................	640.7
Panama (1960)..........................	605.6
United States (1960)................	52.7
Canada (1961)...........................	45.2
Poland (1960)..........................	43.7
Sweden (1960)..........................	112.8
Japan (1960)............................	12.2
Ireland (1961).........................	16.3

As suggested, the illegitimacy ratio in the Latin American countries is particularly affected by the high proportion of consensual unions. The ratio of 606 for Panama counts the offspring of these unions as illegitimate.

The extent of illegitimacy in relation to the "population at risk" is expressed by the general illegitimate fertility rate or, more simply, the **illegitimacy rate,** defined as the total number of illegitimate live births, regardless of age of mother, per 1,000 unmarried (i.e., single, widowed, or divorced) women 15 to 44 years of age:

$$\frac{B_u}{p_{fu}^{15-44}} \qquad (16)$$

where B_u represents illegitimate births and P_{fu}^{15-44} represents

unmarried women 15 to 44 years of age. Illegitimacy rates like illegitimacy ratios vary widely:

Country and year	*Illegitimacy rate*
Greece (1961)...........................	2.2
Panama (1960)..........................	170.7
United States (1960)................	21.8
Canada (1961)...........................	17.9
Sweden (1960)..........................	19.7
Japan (1960)............................	2.0

Duration-specific legitimate fertility rates. — Fertility varies as sharply with the duration of the woman's married life as with the age of the married woman, as suggested by the data in table 16-8. Hence, duration-specific marital fertility data present important possibilities for analysis. Tabulations of births by duration of marriage are rendered more useful for analysis if they are cross-tabulated by age of mother and order of birth.

The simplest measure, the duration-specific legitimate birth ratio, is merely the percent distribution of legitimate births by duration of married life. Such distributions may vary widely, as seen in table 16-8. They may be derived without data on the married female population distributed by duration of married life, which are required for computing duration-specific legitimate fertility rates. A duration-specific legitimate fertility rate is defined as the number of legitimate live births to women married for a specific number of years per 1,000 women 15 to 44 years of age married for this number of years; that is,

$$\frac{B_l^t}{P_{fl}^{15-44^t}} \times 1,000 \qquad (17)$$

where B_l^t represents legitimate live births to women married t years and $P_{fl}^{15-44^t}$ represents women 15 to 44 years of age married t years.

For classifying the births, duration of married life is the number of completed years elapsed between the date of marriage and date of birth of the child. For classifying the population, the duration of married life is the number of completed

Table 16–8. —Percent Distribution of Legitimate Births, and Legitimate Fertility Rates, by Duration of Mother's Married Life, for Selected Countries: Around 1960

[Rates are legitimate live births to women married for specific periods of time per 1,000 married women aged 10 to 49 years or 15 to 44 years with married life of corresponding durations]

Duration of marriage	Percent distribution of legitimate births				Legitimate fertility rates		
	Federal Republic of Germany (1962)	Sweden (1960)	England and Wales (1963)	Venezuela (1963)	Federal Republic of Germany (1962)[1]	Sweden (1960)[2]	England and Wales (1963)[3]
All durations......................	100.0	100.0	100.0	100.0	123.5	80.2	127.2
Under 1 year.........................	9.9	21.1	13.8	0.4	189.6	475.4	350.4
1 year...............................	19.1	11.5	11.8	14.2	364.3	212.9	299.5
2 years..............................	12.9	10.3	11.0	9.6	252.9	189.8	282.6
3 years..............................	10.4	9.8	10.4	9.4	224.9	177.4	271.4
4 years..............................	8.8	8.3	9.1	8.4	202.2	153.9	237.5
5 years..............................	7.3	7.0	7.9	7.8	172.4	131.5	204.5
6 years..............................	6.1	5.7	6.6	6.5	142.3	106.2	170.3
7 years..............................	4.9	4.7	5.5	5.6	118.4	90.1	143.0
8 years..............................	4.0	3.9	4.4	5.3	96.6	75.0	118.1
9 years..............................	3.3	3.3	3.6	3.7	81.0	63.5	100.1
10 to 14 years.......................	10.5	10.0	11.0	16.8	51.5	38.5	62.4
15 to 19 years.......................	2.5	3.5	3.9	7.5	23.5	14.4	26.7
20 years and over....................	0.5	0.7	1.0	3.3	10.5	3.3	11.9
Unknown..............................	-	0.1	-	1.7	-	125.5	-
Median years[4]......................	3.3	3.2	3.8	5.4	4.0	3.1	4.1

— Represents zero or rounds to zero.
[1] Based on married women under 45 years of age.
[2] Based on married women 10 to 49 years of age.
[3] Based on married women 15 to 44 years of age.
[4] In calculating the median, unknowns were excluded from the total. The total for calculating the median birth rate was obtained by adding the rates for durations under 1 year to 9 years, and 5 times the rates for durations 10–14, 15–19, and 20 and over (assumed to represent 20–24).

Source: United Nations, *Demographic Yearbook, 1965*, tables 28 and 29; United Kingdom, General Register Office, *The Registrar General's Statistical Review of England and Wales for the Year 1963, Part II, Tables, Population,* 1965, tables II and JJ.

years elapsed between the date of marriage and the date of the census.

The method of calculating duration of married life from the birth record or census record may introduce a lack of comparability between the rates for different populations. In the absence of information on the exact date of marriage, duration may be based on the difference between the **year** of marriage and the **year** of the census or birth. This procedure introduces an upward bias in the frequencies for all durations except the category under-1-year.[15] On the other hand, the category under-1-year is inflated by births conceived prior to marriage.

Period Adjusted Rates

We have been discussing observed measures of natality, i.e., those measures that are derived directly from tabulations of births in a given year or brief period of years and estimates of midyear or midperiod population. We want to extend the discussion to consider two classes of adjusted rates relating to a given year or brief period of years. The first class of adjusted rates includes various types of summary rates that are adjusted for the demographic composition of the population, particularly its age-sex composition. Comparative fertility analysis often requires other more refined overall measures of the fertility of

a population than the crude birth rate, the general fertility rate, and the general marital fertility rate. The second class covers various types of annual or period rates that have a meaning or structure akin to that of a cohort, i.e., birth probabilities specific for such characteristics of the women as her age, parity, or duration of married life. They are analogous to the mortality rates discussed in chapters 14 and 15. Because an understanding of the calculation of birth probabilities requires an understanding of cohort fertility measures, however, the discussion of birth probabilities will be postponed until after cohort fertility has been considered.

Rates Adjusted for Age-Sex Composition. — As stated earlier, as an overall measure, the crude birth rate in particular is subject to important limitations for analytic studies. Like the crude death rate, it is affected by variations in the demographic composition of the population, particularly its age and sex compostion. The analysis of time trends and of group fertility differences is enhanced by eliminating as completely as possible the effect of differences in the age-sex composition of the populations being compared. This is only partially accomplished by computation of the general fertility rate. Both the crude birth rate and the general fertility rate may be adjusted (or standardized) for the variations in age composition reflected by data for 5-year age groups. In addition, other types of age-sex adjusted measures of fertility, particularly the total fertility rate, may be calculated. As with death rates, birth or fertility rates may be adjusted (or standardized) by either the direct or indirect method. Even though the basic methods of standardization were described in chapter 14, because age-adjustment

---

[15] The category under-1-year is seriously understated because births which occurred within 12 months after marriage but in the next calendar year are classed with the duration one-year rather than in the group under-1-year. The upward bias for all higher categories is important when the number of births is changing rapidly and the duration categories are in single years; otherwise it is negligible.

of birth and fertility rates may differ in detail from age-adjustment of death rates and because rates of a different design may be involved, we consider the calculation of age-adjusted birth and fertility rates separately here. In addition, brief attention is given to adjustment of fertility rates for variables other than age and sex.

Age-sex adjusted birth rate.—We turn first to the calculation of the age-sex adjusted ("crude") birth rate by the direct method. The formula is as follows:

$$\frac{\Sigma f_a P_a^f}{P} \times 1,000, \text{ or } \Sigma f_a \left(\frac{P_a^f}{P}\right) \times 1,000 \quad (18)$$

where f_a equals the age-specific (maternal) birth rates in a particular population, P_a^f equals the number of females by age in the standard population, and P equals the total of the standard population (all ages, both sexes). As may be seen from the version of the formula on the right, the age-specific birth rates are weighted up by the proportions that females of a given age make up of the total population. The use of the overall total population rather than the female population of childbearing age or the total female population is intended to provide an adjusted rate of the approximate magnitude of the crude birth rate.

The calculation of the age-sex adjusted birth rate by the direct method may be illustrated for Costa Rica in 1963, using the population of the United States in 1963 as a standard: [16]

Age of mother (years)	Female population (thous.), United States (1963) P_a^f	Birth rates, Costa Rica (1963) f_a
15–19[a]	7.678	119.6
20–24	6.290	338.3
25–29	5.513	361.9
30–34	5.747	295.5
35–39	6.285	225.8
40–44	6.270	100.2
45–49[b]	5.731	16.7

(1) $\Sigma P_a^f f_a$, expected births: 8.882.704
(2) Age-sex adjusted birth rate = (1) ÷ P
$= \dfrac{8.882.704}{189,417,000^c} =$: 46.9
(3) Percent difference from U.S.[d] = {[(2) ÷ 21.9] − 1.0} × 100: +114
(4) Crude birth rate: 42.7
(5) Percent difference from U.S.[d] = {[(4) ÷ 21.9] − 1.0} × 100: +95

[a] Includes births to women under 15 years of age.
[b] Includes births to women 50 years of age and over.
[c] Total U.S. population in 1963.
[d] U.S. rate = 21.9.

The cumulative product of the age-specific birth rates for Costa Rica and the female population of the United States by age is 8.882.704. Dividing this figure by the total population of the United States, 189,417,000, gives the age-sex adjusted birth rate for Costa Rica, 46.9. This rate may be compared with the age-sex adjusted rate (i.e., the crude birth rate) for the United States in 1963 (21.9) to measure the difference between the two countries in the "underlying" level of fertility. We see that

the adjustment increased the percent difference in fertility between the two countries; the difference was 114 percent when based on adjusted rates but 95 percent when based on crude rates. It is implied that the age-sex composition of Costa Rica's population is less favorable for a high level of fertility than that of the United States. Subtracting from the total difference between the crude rates (95 percent) the difference in age-adjusted fertility rates (114 percent) yields an estimate of the effect of the difference in age composition (−19 percent) on the difference between the crude rates.

The general fertility rate may also be further adjusted for age-sex composition by the direct method by applying the following formula:

$$\frac{\Sigma f_a P_a^f}{\Sigma P_a^f} \times 1,000 \quad (19)$$

This formula is like the formula for the age-sex adjusted birth rate, except that the divisor here is the female population of childbearing age rather than the total population. Hence, age-sex adjusted general fertility rates are of the same order of magnitude as unadjusted general fertility rates. Variations in age-sex adjusted general fertility rates are the same as those in age-sex adjusted birth rates, and in practice the former type of rate is infrequently used.

The crude birth rate may also be adjusted for age-sex composition by the indirect method when age-specific birth rates for the particular population are lacking or defective. The conventional calculation by this method follows the formula:

$$\frac{b}{\Sigma F_a p_a^f} \times F \times 1,000 \quad (20)$$

where the symbols have the following meaning:
For the particular population:
b = total births
p_a^f = population by age
For the "standard" population:
F_a = age-specific birth rates
F = crude birth rate

Since, in computing a series of rates standardized in this fashion, F, the crude birth rate of the "standard" population, is a constant multiplier and its use does not affect the relative magnitudes of the standardized rates being compared, the ratio within formula (20) may be interpreted as the ratio of the fertility level in the particular population to that in the standard population. Formula (20) is equivalent to

$$\frac{\Sigma f_a p_a^f}{\Sigma F_a p_a^f} \times F \times 1,000 \quad (21)$$

This theoretical interpretation is given a practical application in the **United Nations' sex-age adjusted birth rate.**[17] The U.N. sex-age adjusted birth rate is defined as the number of births per 1,000 of a weighted aggregate of the numbers of women in the various five-year age groups from 15 to 44, where the weights (w_a) represent a set of standard age-specific birth rates expressed in rounded whole numbers:

$$\frac{b}{\Sigma F_a p_a^f} \times 1,000 = \frac{b}{\Sigma w_a p_a^f} \times 1,000 \quad (22)$$

This standard set of weights is roughly proportional to the

[16] Data from official national sources.

[17] United Nations, *Methods for Population Projections by Sex and Age*, Manual III, Population Studies, Series A, No. 25, 1956, pp. 42–44.

Table 16–9. — Calculation of United Nations Sex-Age Adjusted Birth Rates, for Various Countries: Around 1960

Age	Population weights, w_a	Female population (in thousands), p_a^f						
		Costa Rica (1963)	El Salvador (1961)	Hungary (1960)	Italy (1961)	Japan (1960)	Taiwan (1960)	United States (1963)
15 to 19 years..................	1	66	125	375	1,861	4,631	465	7,678
20 to 24 years..................	7	53	113	349	2,011	4,193	442	6,290
25 to 29 years..................	7	44	92	368	1,893	4,115	389	5,513
30 to 34 years..................	6	39	78	386	1,940	3,771	340	5,747
35 to 39 years..................	4	34	73	396	1,945	3,275	281	6,285
40 to 44 years..................	1	27	57	239	1,415	2,745	220	6,270
(1) $\sum w_a p_a^f$ (in thousands).........		1,142	2,377	9,533	50,024	101,258	9,666	156,191
(2) Actual births................		57,426	124,871	146,461	929,657	1,606,041	419,442	4,098,020
(3) Sex-age adjusted birth rate = $[(2) \div (1)]$ x 1,000........		50.3	52.5	15.4	18.6	15.9	43.4	26.2
(4) Percent difference from the United States, 1963 = $[(3) \div 26.2) - 1.0]$ x 100..........		+92	+100	−41	−29	−39	+66	(X)

X Not applicable.

Source: United Nations, *Demographic Yearbook, 1965*, tables 6 and 11; *1964*, table 5; and *1961*, table 5; United Nations, *Methods for Population Projections by Sex and Age*, Manual III, Population Studies, Series A, No. 25, p. 42; and U.S. Bureau of the Census, *Current Population Reports*, Series P–25, No. 314, "Estimates of the Population of the United States, by Single Years of Age, Color, and Sex, 1960 to 1964," p. 5.

typical relative fertility rates of the various age groups of the childbearing period, as determined by a review of birth rates by age of mother for various countries. As we have noted, even under widely different fertility conditions, the relative levels of age-specific birth rates are not usually very different. The average percent distribution of age-specific rates for 52 countries, and the population weights derived from this distribution and used for calculating the U.N. rate, are as follows:[18]

Age (years)	Percent distribution	Mean deviation from average	Weights for computation of adjusted birth rates
Total, 15–44......	100.0	(x)	(x)
15–19.................	6.3	±2.7	1
20–24.................	25.3	±3.5	7
25–29.................	27.6	±2.1	7
30–34.................	21.1	±2.1	6
35–39.................	13.4	±2.1	4
40–44.................	6.3	±2.2	1

x Not applicable.

Test calculations show that even a substantial modification of the weights would have only a slight effect on the relative levels of the adjusted birth rates being compared. The absolute values of the weights were so selected that the sum of their products with the corresponding numbers of women in the various age groups (i.e., the denominator of the above ratio) would be of about the same order of magnitude as the total population and the resulting rate would be of the same order of magnitude as a crude birth rate per 1,000 population.

The calculation of the U.N. sex-age adjusted rate for a number of countries is illustrated in table 16–9. For example, for Costa Rica in 1963, we proceed as follows: First, we compute the sum of the products of the standard weights (col. 1) and the number of women in each age group (col. 2). The result, 1,142,000, differs from the total population,

estimated officially at 1,336,274, by about 15 percent. Dividing 1,142,000 into 57,426, the total number of registered births, gives a sex-age adjusted birth rate per 1,000 population of 50.3. Before the adjustment of the crude birth rate, fertility in Costa Rica appeared to be 95 percent greater than in the United States (42.7 ÷ 21.9); following the adjustment, the difference appeared to be only 92 percent (50.3 ÷ 26.2). The result suggests that a difference in age-sex composition accounts for a small part of the difference in crude birth rates.

It may be noted that the results indicated by the U.N. sex-age adjusted birth rate (indirect method) differ from those shown by the conventional age-sex adjusted birth rate (direct method), which gave 114 percent vs. 92 percent for the Costa Rica-United States comparison. Comparison of birth rates adjusted by the direct method and birth rates adjusted by the U.N. (indirect) method for several other countries indicates the following differences:

Country	Adjusted birth rate		Percent difference from U.S. rate	
	Direct[a]	U.N. (indirect)	Direct[a]	U.N. (indirect)
Costa Rica (1963).....	46.9	50.3	+114	+92
El Salvador (1961).....	43.4	52.5	+98	+100
Hungary (1960)........	13.2	15.4	−40	−41
Italy (1961).............	15.3	18.6	−30	−29
Japan (1960)...........	12.4	15.9	−43	−39
Taiwan (1960)..........	37.4	43.4	+71	+66
United States (1963).................	21.9	26.2	....	

[a] U.S. population in 1963 is the standard.

As suggested by the discussion of age-adjustment of death rates in chapter 14, the magnitude of the discrepancy is related indirectly to the general magnitude of the difference in the age composition and, hence, in the fertility of the countries being

_____
[18] Ibid., p. 42.

compared. Moreover, the rates adjusted by the indirect method for different countries are, in effect, based on different population weighting schemes. The results obtained by the direct method of age-adjustment are viewed as more acceptable.

The U.N. sex-age adjusted birth rate is particularly applicable to some statistically underdeveloped countries because only data on the total number of births and the population by age and sex are required for the calculation, a special standard does not have to be selected, and the rate is simpler to compute than the conventional birth rate adjusted by the indirect method.

Total fertility rate.—The total fertility rate (TFR) is another age-sex adjusted measure of fertility which takes account of age detail within the childbearing ages, but it is of a quite different order of magnitude. Once again we begin, in the direct method of computation, with a set of age-specific maternal birth rates for the population under study. In theory, the total fertility rate represents simply the sum of the age-specific birth rates over all ages of the childbearing period. Hence, the age-specific rates are given equal weight and the resulting measure of fertility is of the approximate magnitude of "completed family size," i.e., the total number of children 1,000 women will bear in their lifetime (see below and chapter 18). If the actual rates are for single years of age, then they each receive a weight of one; if the actual rates are for 5-year age groups, as is the usual case, then they each receive a weight of five. In the latter case the calculation is simply carried out by summing the rates and multiplying the sum by 5:

$$5 \sum_{a=15-19}^{a=45-49} f_a \times 1,000 \qquad (23)$$

An array of total fertility rates based on data for 5-year groups was given in table 16–1. The rate for El Salvador, for example, was obtained by summing the age-specific birth rates (per 1,000) shown in table 16–3 and multiplying by 5:

$$5(f_{15-19}+f_{20-24}+f_{25-29}+f_{30-34}+f_{35-39}+f_{40-44}+f_{45-49})$$
$$= 5(142.0+327.2+320.4+268.0+191.2+70.7+20.8)$$
$$= 5(1,340.3)=6,702$$

The rate for ages 15 to 19 includes the few births to women under 15. The total fertility rates shown in table 16–1 vary from 1,912 for Hungary, 1961, to 7,295 for Costa Rica, 1963; that is, the highest fertility shown is over three and a half times as great as the lowest fertility shown.

The total fertility rate is recommended as an easy-to-compute and effective measure of age-sex adjusted fertility for year-to-year or area-to-area comparisons. Because the weighting pattern of the age-specific birth rates is always the same in computing total fertility rates by the direct method and the weighting pattern is a generally reasonable one, there is little question regarding the comparability of a set of total fertility rates computed in this way. The relative level of fertility between two populations, measured by comparing total fertility rates, will be very similar to that shown by age-sex-adjusted birth rates (direct method) since there are roughly the same numbers of women in each of the childbearing ages.

Because the total fertility rate combines the fertility rates over all childbearing ages in a given year, it may also be viewed as representing the completed fertility of a synthetic cohort of women. For example, the total fertility rate of 6,702 for El Salvador in 1961 may be interpreted to mean that a cohort of 1,000 women would have 6,702 children in their lifetime on the average, assuming that they bear children at each age at the rates prevailing in El Salvador in 1961 and that none of the women die before reaching the end of the childbearing period. Under corresponding assumptions, 1,000 Hungarian

women would have only 1,912 children in their lifetime, obviously fewer than the number required to replace their parents.

When only a count of total births is available and a tabulation of births by age of mother is lacking, the indirect method of deriving the total fertility rate may be used. The formula is as follows:

$$\frac{b}{\sum F_a p_a^f} \times 5\sum F_a \times 1,000 \qquad (24)$$

where the symbols have the same meaning as given previously. F_a refers to the "standard" set of age-specific birth rates and, hence, $5\sum F_a$ represents the total fertility rate corresponding to this standard set of rates; b and p_a refer to total births and the population by age for the area or group under study, respectively. In this formula the total fertility rate based on the standard set of age-specific birth rates ($5\sum F_a$) is adjusted upward or downward by the ratio of reported total births for the particular population (b) to the number of births expected in the particular population on the basis of the standard set of age-specific birth rates ($\sum F_a p_a^f$). It will be noted that this formula parallels that for the calculation of the age-sex adjusted ("crude") birth rate by the indirect method:

$$\frac{b}{\sum F_a p_a^f} \times F \times 1,000 \qquad (25)$$

The application of this procedure is illustrated in the following tabular display, where the total fertility rate for Puerto Rico in 1965 is obtained from a set of age-specific rates for 1964: [19]

Age (years)	Age-specific birth rates, 1964 (F_a) (1)	Female population, 1965 (p_a^f) (2)
15–19	104.2	128,700
20–24	257.4	109,900
25–29	191.1	96,800
30–34	117.0	88,600
35–39	88.5	77,900
40–44	41.7	71,400

(1) Expected births, 1965 $=\sum F_a p_a^f$ ...	80,435
(2) Reported births, 1965 $= b$	79,608
(3) Factor $= (2)\div(1)$	.98972
(4) TFR, 1964 $= 5\sum F_a$	3,999.5
(5) TFR, 1965 $=$ TFR, 1964 $\times$ Factor $= (4)\times(3)$	3,958.4

First, the age-specific birth rates in Puerto Rico in 1964 from ages 15–19 to 40–44 are set down (col. 1). The rate for ages 15 to 19 includes the few births to women under 15 and the rate for ages 40 to 44 includes the few births to women 45 and over; that is,

$$\frac{B_{<20}}{P_{15-19}^f} \text{ and } \frac{B_{40+}}{P_{40-44}^f}$$

Next, the total fertility rate for Puerto Rico in 1964 (standard TFR) is calculated as 5 times the sum of the age-specific birth rates in 1964; that is, $5 \times 799.9 = 3,999.5$. The "expected" number of births is obtained as the cumulative product of the rates (on a unit basis) in column (1) and the female population

---

[19] The basic data come from Puerto Rico, Department of Health, *Annual Vital Statistics Report* 1965, San Juan, Puerto Rico, tables 1, 1–A, and 11.

by age for Puerto Rico in 1965 in column (2). This product is 80,435. The reported number of births in 1965 for Puerto Rico is 79,608. Dividing the actual number of births (79,608) by the expected number (80,435) gives the factor (.98972) for adjusting the standard total fertility rate. Accordingly, the estimated total fertility rate for 1965 is .98972 × 3,999.5, or 3,958.4.

This procedure is particularly applicable for updating a time series of total fertility rates. Thus, if the total number of births is known for the past year (1965, in the example above), but births have not yet been tabulated by age of mother, an early estimate of the total fertility rate may be obtained by using the age-specific birth rates of the preceding year (here, 1964).

Separate total fertility rates may readily be derived for significant categories of the population or of births. Three types of examples may be given. The total fertility rate may be calculated for geographic, residence, race, or ethnic group, etc., subdivisions of the population of a country. In these calculations the births and the women both relate to the same geographic, residence, or race, or ethnic group. The resulting figures for various categories of a classification (e.g., white, nonwhite) are independent and cannot be combined. The total fertility rate for each birth order may be obtained by summing the (5-year) age-specific rates for each order over all ages of childbearing and multiplying by 5. Thus, the rate for first births may be obtained by summing first birth rates over all ages ($5\Sigma f_a^1$), the rate for second births may be obtained by summing second birth rates over all ages ($5\Sigma f_a^2$), etc.[20] The sum of order-specific total fertility rates over all orders equals the overall total fertility rate.

Rates Adjusted for Marital Status. – Birth and fertility rates may be adjusted or standardized for other population characteristics in addition to age and sex, such as urban-rural residence, race or ethnic group, marital status, duration of marriage. Standardization for marital status, age, and sex, for example, goes beyond the refinement in general marital or legitimate fertility rates by taking account of legitimate and illegitimate fertility by age within the childbearing period. Crude birth rates and general fertility rates may be adjusted for age, sex, and marital status by an extension of the procedures for direct standardization described earlier. Age-specific legitimate fertility rates (legitimate births per 1,000 married women) and age-specific illegitimate fertility rates (illegitimate births per 1,000 unmarried women) are combined on the basis of a standard distribution by age, sex, and marital status. If the expected births are divided by the total population, an age-sex-marital status adjusted (crude) birth rate is derived:

$$\frac{\sum_a f_{al}P_{al}^f + \sum_a f_{au}P_{au}^f}{P} \quad (26)$$

where the symbols have the meaning assigned in the section above on marital fertility rates. If the expected number is divided by the number of women of childbearing age, we obtain the general fertility rate adjusted for these factors:

$$\frac{\sum_a f_{al}P_{al}^f + \sum_a f_{au}P_{au}^f}{P_{15-44}^f} \quad (27)$$

A change in the crude birth rate, the general fertility rate, and the general marital fertility rate may be considered as resulting jointly from changes in the proportions of married women at each age and from changes in age-sex-specific marital fertility. Comparison of the unadjusted rates and the adjusted rates is useful in identifying the separate contribution of each of these factors to the total change in the unadjusted rates.

Cohort Fertility Measures

We have already considered the calculation and interpretation of schedules of age-specific central birth rates, and of legitimate central birth rates specific for duration of married life, for a particular year. As we have noted, such schedules of birth rates and their sums can be considered as applying to a synthetic cohort. Rates for different calendar years may also be considered in combination to derive various measures describing the fertility of true birth or marriage cohorts of women.

Fertility of Birth Cohorts. – Consider an array of age-specific central birth rates, for a current year and a series of earlier years, for each age 14 to 49 identified in terms of year of birth. Table 16–10 shows a portion of such an array, i.e., age-specific rates for ages 14 to 18 from 1961 to 1965 for the

Table 16–10. – **Central Birth Rates, 1961 to 1965, and Cumulative Fertility Rates, 1962 to 1966, by Age of Mother, for Each Birth Cohort From 1942–43 to 1950–51, for the United States**

[Birth cohort is identified by years shown in margins between the diagonal lines. Central birth rates relate to calendar years indicated. Cumulative rates relate to January 1 of years indicated. Rates are per 1,000 women]

Age	1961	1962	1963	1964	1965	Birth cohort
Current age:			Central rates			
14 years.....	4.0	4.2	4.3	4.5	4.3	1950–51
15 years.....	14.7	11.7	12.5	12.2	12.6	1949–50
16 years.....	38.9	37.7	31.6	33.2	32.6	1948–49
17 years.....	76.2	73.5	74.2	62.4	63.0	1947–48
18 years.....	123.2	123.0	118.8	118.4	99.4	1946–47
	1942–43	1943–44	1944–45	1945–46		

Age	1962	1963	1964	1965	1966	Birth cohort
Exact age:			Cumulative rates			
15 years.....	4.0	4.2	4.3	4.5	4.3	1950–51
16 years.....	(1)	15.7	16.7	16.5	17.1	1949–50
17 years.....	(1)	(1)	47.3	49.9	49.1	1948–49
18 years.....	(1)	(1)	(1)	109.7	112.9	1947–48
19 years.....	(1)	(1)	(1)	(1)	209.1	1946–47
	1942–43	1943–44	1944–45	1945–46		

[1] Cumulative rates cannot be derived on the basis of the annual rates given above.

Source: U.S. National Center for Health Statistics, *Vital Statistics of the United States, 1964*, Vol. I, *Natality*, 1966, tables 1–18 and 1–19; U.S. National Center for Health Statistics, *Vital Statistics of the United States, 1965*, tables 1–17 and 1–18.

[20] Illustrative rates for the United States are presented in: P. K. Whelpton, "Reproduction Rates Adjusted for Age, Parity, Fecundity, and Marriage." *Journal of the American Statistical Association*, 41(236):503–504. December 1946. p. 565.

United States. The age-specific birth rates for successive single ages from age 14 on, in successive years, may now be combined to derive cumulative fertility rates (per 1,000 women) to any given age to any given year for any particular cohort. For example, the cumulative fertility rate for the birth cohort of 1946–47 to exact age 19 on January 1, 1966, is derived as the sum of the following age-specific birth rates: [21]

$$f_{14}^{1961} + f_{15}^{1962} + f_{16}^{1963} + f_{17}^{1964} + f_{18}^{1965}$$

Using the data for the United States given in table 16–10, we have (per 1,000 women):

$$4.0 + 11.7 + 31.6 + 62.4 + 99.4 = 209.1$$

The rates pertaining to any particular cohort fall along diagonal cells in the array. The cumulative fertility rate for the birth cohort of 1947–48 to exact age 18 on January 1, 1966, is derived as the sum of the following age-specific birth rates:

$$f_{14}^{1962} + f_{15}^{1963} + f_{16}^{1964} + f_{17}^{1965}$$

The U.S. figures are:

$$4.2 + 12.5 + 33.2 + 63.0 = 112.9$$

Cumulative and completed fertility rates. — By deriving a set of cumulative fertility rates showing fertility up to each successive age, one can follow the fertility progress of a birth cohort through the childbearing ages. The fertility history of the 1916–17 birth cohort, illustrated with data for the United States, is described as follows:

| | Cumulative fertility rate | | | Age-specific birth rates cumulated through— | |
| | Rate per 1,000 women | Percent of completed rate | Reference date (Jan. 1) | Age | Calendar year |
Age covered					
Up to age 20	244	10.0	1937	19	1936
Up to age 25	891	36.5	1942	24	1941
Up to age 30	1,590	65.1	1947	29	1946
Up to age 35	2,088	85.5	1952	34	1951
Up to age 40	2,362	96.7	1957	39	1956
Up to age 45	2,438	99.8	1962	44	1961
Up to age 50	2,442	100.0	1967	49	1966

If central birth rates over all ages of childbearing for a given birth cohort are combined in the manner described, then the **completed fertility rate** for this cohort is obtained:

$$f_{14}^{y} + f_{15}^{y+1} + f_{16}^{y+2} \ldots + f_{47}^{y+33} + f_{48}^{y+34} + f_{49}^{y+35} \qquad (28)$$

where *y* refers to the calendar year when the cohort is 14 years of age. For example, the completed fertility rate for the cohort born in 1916–17 (exact age of 50 on Jan. 1, 1967) is computed as follows:

$$f_{14}^{1931} + f_{15}^{1932} + f_{16}^{1933} \ldots + f_{47}^{1964} + f_{48}^{1965} + f_{49}^{1966}$$

The completed fertility rate is attained by some age in the late forties, certainly by age 50 and possibly as early as age 45. As is suggested by a comparison of the cumulative rates from age

45 on, very little change occurs after the cohort reaches age 45. The cumulation can, in fact, be terminated with the age-specific birth rate at age 44, or continued to higher ages up through the rate for age 49 if this is possible and convenient.

We have illustrated the procedure for deriving cumulative fertility rates to each age of childbearing for a current or a past year. Cumulative fertility rates for births of a particular order (first, second, etc.) may be derived in a similar manner from an array of age-order specific rates, except that the rates for first births only, second births only, etc., for each birth cohort, are combined. These rates state the total number of first births, second births, etc., that 1,000 women have had by the age and date indicated. Cumulative fertility rates for individual birth orders are illustrated in table 16–11, which shows cumulative rates to exact ages 15 to 24 as of January 1, 1965 (corresponding to the birth cohorts of 1940–41 to 1949–50), for the United States. For example, by this date, the cohort born in 1941–42 and of exact age 23 years, has had 601.9 first births, 339.8 second births, 130.9 third births, etc. (per 1,000 women).

The updating of these cumulative rates to January 1, 1966, is achieved by combining them with the central birth rates, by age and birth order, for 1965. Central birth rates for ages 19 to 24 and cumulative rates by birth order as of January 1, 1966, for ages 20 to 25 are shown in table 16–11. For example, the cohort of 1941–42, which had 601.9 first births (per 1,000 women) by January 1, 1965, when it was at exact age 23, had 62.6 additional first births in 1965, when it was 23 years of age, and hence a cumulative rate of 664.5 first births by January 1, 1966, when it was at exact age 24.

The cumulative or completed fertility rates for several consecutive birth cohorts or ages for a given date may be averaged to represent the cumulative or completed fertility rate for these birth cohorts or ages as a group. The cumulative rates for exact ages 20–24 as of January 1, 1966, shown in table 16–12 were derived by averaging the cumulative rates for ages 20 to 24, respectively, shown in table 16–11.

It should be evident that, because childbearing is spread over more than 30 years of life, the completed fertility rate for any given cohort cannot be assigned to any particular calendar year as a measure of fertility in that year. The rate certainly does not reflect fertility in the year of birth of the cohort, and it tells us little or nothing about fertility in the year the cohort reached childbearing age (age 14, say) or in the year the cohort completed childbearing (age 45, say). Inasmuch as fertility is concentrated at the ages 20 to 29, the completed fertility rate for a particular cohort may be assigned roughly to the decade when the cohort moved through the ages 20 to 29. To simplify our historical analysis, we may choose to assign it arbitrarily to the single calendar year when the cohort was at the mean age of childbearing. Inasmuch as the mean age of childbearing varies from birth cohort to birth cohort for a given population and from population to population, when we are dealing with many cohorts, we may make a rough general assignment of the completed fertility rate to the year when the cohorts were 25 years of age.

Timing of childbearing for cohorts. — Measures of "timing" or "spacing" of births for actual birth cohorts give a more realistic indication of the timing or spacing of births than those calculated from births in a single calendar year. The measures include the mean age of mother (mean age of child bearing), median age of mother, cumulative fertility in percents, and mean and median intervals between births. The median age of mother indicates whether a cohort tends to have its children relatively early or relatively late in the childbearing period,

---

[21] The birth cohort of July 1, 1946, to June 30, 1947, was under 1 (0) year of age (in completed years) on July 1, 1947, 14 on July 1, 1961, 18 on July 1, 1965, and 19 on July 1, 1966. It is assumed to have exact age 19 on January 1, 1966, corresponding to exact age 0 (birth) centered on January 1, 1947.

Table 16–11. — Cumulative Fertility Rates, by Exact Age of Mother and Birth Order, for Women in Each Birth Cohort from 1940–41 to 1949–50, January 1, 1965, and for Women in Each Birth Cohort From 1941–42 to 1945–46, January 1, 1966, for the United States

[Rates are per 1,000 women]

Birth cohort[1]	Age of woman	Total	Birth order								
			1	2	3	4	5	6	7	8 and over
CUMULATIVE FERTILITY RATE BY EXACT AGE, JANUARY 1, 1965											
1949–50	15 years	4.5	4.5	-	-	-	-	-	-	-
1948–49	16 years	16.5	15.4	1.0	0.1	-	-	-	-	-
1947–48	17 years	49.9	44.2	5.2	0.5	-	-	-	-	-
1946–47	18 years	109.7	91.1	16.3	2.1	0.2	-	-	-	-
1945–46	19 years	249.9	190.3	49.4	9.0	1.1	-	-	-	-
1944–45	20 years	416.6	294.1	96.0	22.4	4.1	-	-	-	-
1943–44	21 years	621.9	401.6	161.8	46.1	10.6	1.5	0.2	0.1	-
1942–43	22 years	845.3	496.3	240.3	79.7	22.2	5.3	1.1	0.2	0.2
1941–42	23 years	1,130.0	601.9	339.8	130.9	41.5	11.9	2.9	0.7	0.4
1940–41	24 years	1,427.4	688.9	440.1	195.7	70.3	23.1	6.9	1.8	0.6
CENTRAL BIRTH RATE BY CURRENT AGE, CALENDAR YEAR 1965											
1945–46	19 years	159.2	103.3	41.5	11.7	2.7	-	-	-	-
1944–45	20 years	174.6	94.9	53.2	18.7	5.4	1.9	0.3	0.1	0.1
1943–44	21 years	191.0	85.3	65.6	27.0	9.4	2.8	0.7	0.1	0.1
1942–43	22 years	199.8	75.5	70.2	34.2	13.4	4.7	1.3	0.4	0.1
1941–42	23 years	204.8	62.6	70.7	41.3	19.1	7.5	2.6	0.7	0.3
1940–41	24 years	200.1	49.6	65.0	45.2	23.2	10.6	4.3	1.6	0.6
CUMULATIVE FERTILITY RATE BY EXACT AGE, JANUARY 1, 1966											
1945–46	20 years	409.0	293.6	90.9	20.7	3.8	-	-	-	-
1944–45	21 years	591.2	389.0	149.2	41.1	9.5	1.9	0.3	0.1	0.1
1943–44	22 years	813.7	486.9	227.4	73.1	20.0	4.9	1.0	0.2	0.2
1942–43	23 years	1,045.1	571.8	310.5	113.9	35.6	10.0	2.4	0.6	0.3
1941–42	24 years	1,334.8	664.5	410.5	172.2	60.6	19.4	5.5	1.4	0.7
1940–41	25 years	1,627.5	738.5	505.1	240.9	93.5	33.7	11.2	3.4	1.2

– Represents zero or rounds to zero.
[1] Period from July 1 of initial year to June 30 of terminal year.

Source: U.S. National Center for Health Statistics, *Vital Statistics of the United States, 1965*, Vol. I, *Natality*, 1967, tables 1–17 and 1–18, and *Vital Statistics of the United States, 1964*, Vol. I, *Natality*, 1966, table 19.

Table 16–12. — Cumulative Fertility Rates, by Exact Age of Mother and Birth Order, for Women in Groups of Birth Cohorts From 1911–16 to 1946–51, for the United States: January 1, 1966

[Rates are averages of the rates per 1,000 women for single cohorts]

Birth cohorts[1]	Exact age of mother on January 1, 1966	Total	Birth order							
			1	2	3	4	5	6	7	8 and over
1946–51	15 to 19 years	78.5	64.3	12.1	1.9	0.2	-	-	-	-
1941–46	20 to 24 years	838.8	481.2	237.7	84.2	25.9	7.2	1.8	0.5	0.3
1936–41	25 to 29 years	2,077.8	814.2	626.0	354.5	165.6	70.6	29.0	11.4	6.5
1931–36	30 to 34 years	2,848.2	900.5	781.9	536.3	306.9	159.5	82.2	41.7	39.2
1926–31	35 to 39 years	2,984.7	889.9	778.9	548.4	332.2	187.1	106.6	61.0	80.6
1921–26	40 to 44 years	2,876.5	890.1	747.9	501.1	299.5	172.8	102.6	62.2	100.3
1916–21	45 to 49 years	2,630.5	861.0	687.0	434.6	254.5	147.4	89.4	56.2	100.4
1911–16	50 to 54 years	2,336.7	800.0	600.9	364.6	213.6	127.1	80.4	51.5	98.6

– Represents zero or rounds to zero.
[1] Period from July 1 of initial year to June 30 of terminal year.

Source: Table 16–11 and U.S. National Center for Health Statistics, *Vital Statistics of the United States, 1965*, Vol. I, *Natality*, 1967, table 1–16.

Table 16-13. — Percent Distribution of Women by Parity, by Exact Age, in Groups of Birth Cohorts From 1916-21 to 1946-51, for the United States: January 1, 1966

Birth cohorts [1]	Exact age of woman on January 1, 1966	Total	Parity							
			0	1	2	3	4	5	6	7 or more
1946-51................	15 to 19 years......	100.0	93.6	5.2	1.0	0.2	-	-	-	-
1941-46................	20 to 24 years......	100.0	51.9	24.4	15.4	5.8	1.9	0.5	0.1	0.1
1936-41................	25 to 29 years......	100.0	18.6	18.8	27.2	18.9	9.5	4.2	1.8	1.1
1931-36................	30 to 34 years......	100.0	10.0	11.9	24.6	22.9	14.7	7.7	4.1	4.2
1926-31................	35 to 39 years......	100.0	11.0	11.1	23.1	21.6	14.5	8.1	4.6	6.1
1921-26................	40 to 44 years......	100.0	11.0	14.2	24.7	20.2	12.7	7.0	4.0	6.2
1916-21................	45 to 49 years......	100.0	13.9	17.4	25.2	18.0	10.7	5.8	3.3	5.6

— Represents zero or rounds to zero.
[1] Period from July 1 of initial year to June 30 of terminal year.

Source: Table 16-12.

particularly in comparison with earlier or later cohorts. It is affected both by the timing of marriage, as measured, say, by the median age at first marriage, and by spacing of births within marriage. The basic data for the calculation consist of a distribution of age-specific birth rates for single ages and single calendar years, for a single birth cohort of women; or of age-specific birth rates for 5-year age groups at 5-year time intervals, for a 5-year group of birth cohorts. That is, the median is based on:

$$f_{14}^y, f_{15}^{y+1}, f_{16}^{y+2}, \ldots, f_{47}^{y+33}, f_{48}^{y+34}, f_{49}^{y+35} \quad (29)$$

or

$$f_{15-19}^y, f_{20-24}^{y+5}, f_{25-29}^{y+10}, f_{30-34}^{y+15}, f_{35-39}^{y+20}, f_{40-44}^{y+25} \quad (30)$$

where y represents the calendar year when the cohort was 14 or the cohorts were 15-19 and where the rates for ages 15-19 and 40-44 include any births for the ages under 15 and over 44, respectively. The resulting median age is identified as the median age of births for this particular birth cohort or group of birth cohorts. The terms "median age of births," "median age of childbearing," and "median age of mother" are used interchangeably in this context.

Another general indication of the spacing of a cohort's births is obtained by converting the set of cumulative fertility rates to each successive age for the cohort into percents of the completed fertility rate. An illustration was given in the tabular display at the beginning of this section. A more specific indication of spacing, relating to particular birth intervals, may be obtained from an analysis of order-specific rates for birth cohorts. This type of analysis is parallel to that described earlier in connection with calendar-year data. The median age for first births occurring to the cohort reflects the typical age of initiating childbearing. As before, birth intervals may be approximated by taking the difference between the median ages for successive orders.

Parity distribution of women. — Cumulative fertility rates by live-birth order and exact age of mother, for women in single-year birth cohorts or groups of birth cohorts, at a particular date, provide the basis for deriving the distribution of all women by parity at a particular age or age range (corresponding also to a particular birth cohort or cohorts) at this date. **Parity** refers to the number of children previously born alive to a woman. Zero-parity women are women who have never had a child, 1-parity women are women who have had only one

child, etc. A parity distribution of women can be computed for each age and date (i.e., each birth cohort observed on a particular date) for which cumulative fertility rates have been developed.

Note, first, that the cumulative rate for first births per 1,000 women to any date and age represents the number of women who have had **one or more** children. These women are all different since no women could have had more than one first birth; but it includes women who have had 2, 3, 4, or more children. The percentage of zero-parity or childless women is obtained, therefore, by subtracting the cumulative first birth rate from 1,000 (yielding the number of zero-parity women per 1,000 women) and dividing by 10 (yielding the percentage of zero-parity women). For example, the percentage of zero-parity women aged 40-44 on January 1, 1966 (birth cohorts of 1921-22 to 1925-26) in the United States (11.0 percent in table 16-13) is derived as follows from the data in table 16-12:

$$\frac{1,000 - 890.1}{10} = \frac{109.9}{10} = 11.0 \text{ percent}$$

The proportions of women at individual parities one and over can be obtained by taking the difference between cumulative fertility rates for successive orders (yielding the number of n-parity women per 1,000) and dividing by 10 (yielding the percentage of n-parity women):

Percent at parity n

$$= \frac{(\text{Cumulative rate, order } n - \text{Cumulative rate, order } n+1)}{10}$$

$$(31)$$

For example, the percentage of second-parity women aged 40-44 on January 1, 1966 (birth cohorts of 1921-22 to 1925-26) given in table 16-13 is calculated on the basis of data in table 16-12 as follows:

$$\frac{747.9 - 501.1}{10} = \frac{246.8}{10} = 24.7 \text{ percent}$$

The percentage of women in a terminal (open-ended) parity group is found by dividing the cumulative rate for the lowest parity included in the terminal group by 10. For example, the percentage of 40–44-year-old women in 1966 at seventh and higher parity is found by dividing the cumulative rate for seventh-order births by 10:

$$\frac{62.2}{10} = 6.2 \text{ percent}$$

The percent of zero-parity women at ages 45–49 (13.9 percent in table 16–13) reflects approximately the extent to which the women in a population have been remaining childless, whether because of failure to marry or because of childlessness within marriage (from infecundity or voluntary choice). This percent is different for each new cohort completing the childbearing period and may be higher or lower for cohorts completing childbearing in the future. It represents a maximum estimate of the extent of infecundity for a given cohort. If we subtract from the proportion of zero-parity women at ages 45–49 the proportion of women at these ages who have never married (assuming that all never-married women at these ages are of zero parity), as shown by the census or national sample survey (about 5 percent for U.S. women in 1966), we obtain an estimate of the proportion of married women who are childless. Where voluntary childlessness is extremely small, this adjusted percent (9 percent) serves as a rough estimate of the extent of infecundity of married women.

Birth probabilities.—Birth probabilities represent the probability that a birth will occur over some time period to a woman of a given age, parity, age-parity, parity-birth interval, or duration of marriage, etc., at the beginning of the period. The probabilities are usually expressed as the number of births in a year per 1,000 women in the class at the beginning of the year.

Birth probabilities specific by age indicate the chance that a woman of a given exact age at the beginning of the year will have a birth during the year. An age-specific probability may be represented by $\bar{p}_a$. The value of $\bar{p}_a$ may be approximated by f_a, the central birth rate at age a during the year. For example, the one-year birth probability (per 1,000) for a cohort of women of exact age 20 on January 1, 1965, in the United States is 175 (table 16–11).[22] A similar relation between $\bar{p}_a$ and f_a may be assumed for grouped data; that is, $_n\bar{p}_a$, the probability that a cohort of women of exact ages a to $a+n$ at the beginning of a year, will have a birth during the year, may be approximated by $_nf_a$, the central birth rate for ages a to $a+n$, or $_n\bar{p}_a \doteq {}_nf_a$. For example, the one-year birth probability for the cohort of women of exact age 20–24 on January 1, 1965, in the United States is represented simply by the central birth rate for women 20–24 years of age in 1965, 194.1 births per 1,000 women (computed from table 16–11.

Birth probabilities specific by parity indicate the chances that a woman of a given parity at the beginning of the year will have a child during the year (and attain the next parity). Such rates would only be meaningful for women of childbearing age,

---

[22] Equating the age-specific birth probability for age a and the central birth rate for age a implies that no deaths occur to the women in the half year between the beginning and middle of the year, or between exact age a and completed age a; that is,

$$\bar{p}_a = \frac{B_a}{P_a^f + \frac{1}{2}D_a^f} \doteq \frac{B_a}{P_a^f} = f_a$$

Mortality over a half year tends to be rather low for women in the childbearing ages, especially in countries where mortality is generally low, so that the assumption has only negligible effect.

---

so that normally the probabilities are made specific with respect to both age and parity. **Age-parity-specific birth probabilities** indicate the chances that a woman of a given parity and exact age group at the beginning of a year will have a child during the year (i.e., will attain the next parity before her next birthday). In general, an age-parity-specific birth probability is calculated from central birth rates and cumulative fertility rates by order. The general formula for an age-parity-specific birth probability is:

$$_n\bar{p}_a^i = \frac{_nf_a^{i+1}}{\sum_{\omega_1}^{a+n} {}_nf_a^i - \sum_{\omega_1}^{a+n} {}_nf_a^{i+1}} \times 1,000 \qquad (32)$$

where $_n\bar{p}_a^i$ represents the probability that a cohort of women of exact ages a to $a+n$ and of parity i at the beginning of a year will have an $i+1$th child during the calendar year, and

$_nf_a^{i+1} = i+1$th order birth rate for women of completed age a to $a+n$ years in the calendar year

$\sum_{\omega_1}^{a+n} {}_nf_a^i =$ cumulative fertility rate of order i to beginning of year for a cohort of women of exact ages a to $a+n$

$\sum_{\omega_1}^{a+n} {}_nf_a^{i+1} =$ cumulative fertility rate of order $i+1$ to beginning of year for a cohort of exact ages a to $a+n$.

For example, the birth probability of first parity-women aged 20 to 24 in 1965 in the United States is calculated from the second-order birth rate in 1965 for the cohort aged 20–24 in 1965 (64.9), the cumulative fertility rate for the first-order births to this cohort by the beginning of 1965 (496.6), and the cumulative rate for second-order births by the beginning of 1965 (255.6). Each of these values is derived from table 16–11 by averaging the corresponding rates for the single ages 20 to 24. Each of the cumulative rates includes some women who also had higher order births. The difference between the cumulative rates represents the rate for women who have had births of first order only. Thus, the birth probability of first-parity women aged 20–24 in 1965 in the United States is calculated as follows:

$$\frac{64.9}{496.6 - 255.6} \times 1,000 = 269$$

The birth probability of zero-parity women aged 20 to 24 in 1965 is calculated from the first-order birth rate in 1965 for the cohort 20–24 in 1965 (73.6) and the cumulative fertility rate for first-order births to this cohort by the beginning of 1965 (496.6):

$$\frac{73.6}{1,000 - 496.6} \times 1,000 = 146$$

The calculation for the terminal (open-ended) parity group may be illustrated with the calculation of the probability that seventh and higher parity women of exact age 25–29 on January 1, 1965, will have another (eighth and higher order) child during 1965 (data not shown in tables). The central birth rate for eighth-and-over births in 1965, 3.4, is divided by the cumulative rate to January 1, 1965, for women of exact age 25–29 who have had 7 children, i.e., the difference between the cumulative rate for 7 or more (19.4) and the cumulative rate for 8 or more (7.1):

$$\frac{3.4}{19.4-7.1}\times 1.000=\frac{3.4}{12.3}\times 1.000=276$$

Parity-specific probabilities can also be calculated to refer only to married women. One procedure is to develop probabilities of first marriage among all women by age first, then the probability of first births among zero-parity married women, then the probability of second births among 1-parity women, etc. A substitute for, or addition to, the usual detail by age would be duration of married life or birth interval. Further refinements would, therefore, develop the probabilities separately by the interval in months between marriage and first births, and between births of successive orders. The age variable may be included only in terms of age at marriage in the first marriage probabilities or it may be included throughout the whole set of figures.

Fertility of Marriage Cohorts. — Women who marry in a certain year (i.e., a marriage cohort) share a common experience in relation to the political, social, and economic conditions at the time of marriage and at successive years of married life. This experience may encompass such specific aspects of living as housing conditions, life style, and tastes. Insofar as these factors influence fertility behavior, it is of interest to study the fertility of marriage cohorts. We have already made reference to the relation of marriage duration to fertility (see section "Duration—Specific Legitimate Fertility Rates," above).

A central legitimate birth rate specific for duration of marriage (i.e., for a marriage cohort) for a given calendar year has been defined, in effect, as the number of births occurring to women of a given marriage duration (i.e., a given marriage cohort) in the year per 1,000 married women of that "duration" (i.e., marriage cohort) at the middle of the year:

$$\frac{B_l^{y,t}}{P_{fl}^{y,t}}\times 1.000 \qquad (33)$$

where y defines the calendar year, t the number of years since marriage, and B_l and P_{fl} represent legitimate births and married women, respectively. In order to derive cohort fertility tables for marriage cohorts, we need such rates annually. But annual estimates of the size of a given marriage cohort, which constitute the denominators of the rates, are rarely computed, since census data on marital status by years since marriage, and data on death, divorces, immigration, and emigration by marital status and year of marriage are rarely available. For this reason, the denominator used in computing rates for marriage cohorts is usually the original size of each marriage cohort.[23] Computed in this way, however, duration-specific birth rates tend to be increasingly understated as the duration of marriage grows, since the denominator tends progressively to overstate the current size of the marriage cohort.

---

[23] The marriages corresponding to births in a calendar year should properly be the marriages for 12-month periods, July 1 to June 30. For example, marriages of 1955–56 correspond to women married 5 years on July 1, 1961 and to births in 1961 occurring to women married 5 years.

Paternal Rates

As stated earlier, fertility measures are commonly computed with reference to the female population, but they may equally well be computed with reference to the male population. Accordingly, we distinguish paternal and maternal fertility rates. Analysis of paternal fertility rates provides an opportunity for extending our understanding of differences and changes in fertility, particularly socioeconomic differentials in fertility defined in terms of the characteristics of the father.

The general paternal fertility rate is defined as the number of births, regardless of age of father, per 1,000 males 15 to 54 years of age; that is,

$$GFR_m=\frac{B}{p_{15-54}^m}\times 1.000 \qquad (34)$$

For Panama in 1960, we have:

$$\frac{41,258}{265,979}\times 1.000=155.1$$

The different choice of ages for males from that for females results from the longer span of the reproductive years for males. Because of the use of a broader age group of the population in the base, the paternal rate tends to run somewhat lower than the maternal rate (table 16–14).

Table 16–14. — Comparison of Paternal and Maternal Fertility Measures, for Various Countries: Around 1960

Country and year	Excess of paternal median age (years)	Ratio, paternal to maternal rate	
		General fertility rate	Total fertility rate
Chile (1960).....................	4.5	0.891	1.224
Costa Rica (1963).................	6.1	0.879	1.285
El Salvador (1961)...............	[1]6.8	[2]	[1]1.413
England and Wales (1961)..........	2.6	0.739	0.987
Hungary (1961)...................	4.3	0.806	1.072
Israel (1963)....................	5.2	0.791	1.090
Panamá (1960)....................	6.4	0.837	1.200
Poland (1963)...................	3.6	0.844	1.029
Puerto Rico (1960)...............	5.2	0.907	1.339
United States (1960).............	3.0	0.803	1.047
Yugoslavia (1961)...............	3.7	0.853	1.090
Average.....................	4.7	0.835	1.161

[1] Median age of father and paternal total fertility rate are for 1962.
[2] Not calculated.

Source: Tables 16-1 and 16-3. Other basic data from United Nations, *Demographic Yearbook*, 1965, Tables 14 and 19.

As a result principally of the generally higher age at marriage of males than females and the greater length of the male reproductive age span, the characteristic age pattern of male fertility is different from that of females. In particular, the median age of fathers tends to be a few years older than the median age of mothers.

Unlike the general fertility rate, the total fertility rate for males tends to run somewhat higher than the corresponding rate for females (table 16–14). This is primarily because of the greater length of the reproductive period, the higher age of fathers, and the less stringent physiological limitations on the number of children born to men. The proportions of the sexes have an indirect effect on maternal and paternal total fertility rates, since this factor affects the proportions who marry and, hence, the fertility rate for each sex. Births typically occur at older ages for fathers than for mothers and, therefore, at ages with fewer persons. At these higher ages there is typically a smaller proportion of men than women, and, as a result, a higher proportion of married men.

Whether the measure is the general fertility rate or the total fertility rate, paternal and maternal rates give about the same indications of changes over time for a given area or of differences from area to area. For the 11 countries listed in table 16-14, the male and female total fertility rates have the extremely high correlation coefficient of .99. Paternal fertility rates are generally considered less satisfactory than maternal rates, however, because of the longer and less clearly defined reproductive period, the greater deficiency in the reporting of the characteristics of the father on the birth certificate, and the tendency for children in broken families to remain with the mother (with the consequent effect on the reporting of information regarding the children).

Measures Based on Population Registers

In chapter 2. we considered the population register as a general source of population data comprehending within its scope the functions of both a registration system for births and deaths and a registration system for international and internal migration. In the system as it operates in the Netherlands, for example. a personal card is made out at birth on the basis of notification to the local registrar. and the events subsequently affecting the individual. i.e.. marriage. divorce. change of residence. and death. are entered on the card.[24] Copies of the personal cards with all their particulars on the individual are sent each month to the Central Bureau of Statistics for processing and analysis. Because the population register system of population accounting calls for notification of government authorities by householders, on the occasion of a birth. rather than periodic reporting of births occurring over a particular past period, this basis of collection is more akin to the registration method than to the census or survey method. We consider

here. then. the special problems and potentialities of analysis of natality statistics derived from population registers.

Population registers constitute an extremely valuable source of data for natality analysis. As a self-contained system. the population register system theoretically provides both the births on a residence basis for any period and the resident population at any required date, both nationally and for geographic subdivisions. As operated in some countries. the system ensures correct allocation of births to the municipality of residence of the mother. Registration of births and their coverage in the register. where the population register system is fully developed, is believed by national authorities to be virtually one hundred percent complete.

The system permits calculation of birth rates for small geographic units in a country on a regular monthly and annual basis with extremely short time lags. National estimates of population by age, sex, and marital status provided by the system can be related to births by age and marital status of mother, for the corresponding years, to derive annual age-specific marital fertility rates. In some cases, the register may provide data on other socioeconomic characteristics of the population (e.g., occupation), albeit not regularly up-dated, which may be employed in the calculation of fertility rates for the corresponding socioeconomic groups. Annual age-parity-specific birth probabilities, cumulative fertility rates by order of birth, and parity distributions of women may be developed wholly from the system after it has been operating for a number of years.

The registers may provide data on the mothers or fathers of the newborn children supplementing the information on the individual forms for births. The population registers readily permit the selection of a sample of births or of their parents in a given period. The provincial and local governments can utilize the population registers for the analysis of the local fertility situation; and, if the data are insufficient, they can be extended by field surveys in the area, using the entries in the register as the sampling frame. Finally, the register data may be supplemented by the results of a census or national sample survey which provides information on the socioeconomic characteristics of the population for use in the calculation of various types of vital rates.

---

[24] T. van den Brink. "The Netherlands Population Registers," *Sociologica Neerlandica*, 3(2):32–53, 1966.

SUGGESTED READINGS

Arévalo, Jorge V. *Aplicación a Chile de un método de medición de la fecundidad según el tamaño de la familia* (Application to Chile of a Method of Measuring Fertility According to Size of Family). Series C, No. 17. United Nations Demographic Center for Latin America, Santiago, Chile, 1964.

Blayo, Chantal. "Fécondité des mariages de 1946 à 1964 en France" (Fertility of Marriages in France, 1946–1948). *Population* (Paris), 23(4):649–738. July–Aug. 1968.

Benjamin, B. *Elements of Vital Statistics*, Chapter 4, "Fertility Measures and Trends," pp. 60–76. London, George Allen and Unwin, 1959.

Chandra Sekar, C., and Deming, Edwards W. "On a Method of Estimating Birth and Death Rates and the Extent of Regis-

tration." *Journal of the American Statistical Association*, 44(245):101–115. Mar. 1949.

Coale, Ansley J., and Zelnik, Melvin. *New Estimates of Fertility and Population in the United States*. Princeton, N.J., Princeton University Press, 1963.

Collver, O. Andrew. *Birth Rates in Latin America: New Estimates of Historical Trends and Fluctuations*. Research Series No. 7. Institute of International Studies, University of California, Berkeley, Calif., 1965.

Costa Rica, Dirección General de Estadística y Censos. *Exactitud del registro de nacimientos y algunos análisis demográficos de Costa Rica* (Accuracy of Birth Registration and Selected Demographic Analyses for Costa Rica). By Ricardo Jiménez J. San José, Costa Rica, 1957.

SUGGESTED READINGS – Continued

Cox, Peter R. *Demography,* Chapter 7. Cambridge (Eng.), Cambridge University Press, 1970.

Dandekar, Kumudini. "Intervals between Confinements." *Eugenics Quarterly,* 6(3):180–186. Sept. 1959.

Davis, Kingsley, and Blake, Judith. "Social Structure and Fertility: An Analytical Framework." *Economic Development and Cultural Change,* 4(3):211–235. Apr. 1956.

Easterlin, Richard A. "On the Relation of Economic Factors to Recent and Projected Fertility Changes." *Demography,* 3(1):131–153. 1966.

Gaete-Darbo, Adolfo. "Appraisal of Vital Statistics in Latin America." *Demography and Public Health in Latin America.* Proceedings of the Fortieth Annual Conference of the Milbank Memorial Fund, 1963. *Milbank Memorial Fund Quarterly,* 42(2):86–102. Part 2, Apr. 1964.

Grabill, Wilson H., Kiser, Clyde V., and Whelpton, Pascal K. *The Fertility of American Women.* New York, John Wiley and Sons, 1959. Esp. chapter 9 and appendixes A and B.

Hajnal, John. "The Analysis of Birth Statistics in the Light of the Recent International Recovery of the Birth Rate." *Population Studies* (London), 1(2):137–164. Sept. 1947.

Henry, Louis. *Fécondité des mariages: Nouvelle méthode de mesure* (Fertility of Marriages: A New Method of Measurement). Travaux et documents, Cahier No. 16, Institut national d'études démographiques. Paris, Presses Universitaries de France, 1953.

———. "La fécondité naturelle. Observation, théorie, résultats." *Population* (Paris), 16(4):625–636. Oct.–Dec. 1961.

———. "Some Data on Natural Fertility." *Eugenics Quarterly,* 8(2):81–91. June 1961.

Henry, Louis, and Pressat, Roland. "Évolution de la fécondité en Italie" (The Trend of Fertility in Italy). *Population* (Paris), 10(3):501–528. July–Sept. 1955.

Istituto di statistica, Università di Firenze. *Tavole di fecondità dei matrimoni per l'Italia, 1930–1965* (Tables of the Fertility of Marriages, Italy, 1930–1965). By Massimo Livi Bacci *et al.* Florence (Italy), 1968.

Jacoby, E. G. "A Fertility Analysis of New Zealand Marriage Cohorts." *Population Studies* (London), 12(1):18–39. July 1958.

James, W. H. "Estimates of Fecundability." *Population Studies* (London), 17(1):57–65, July 1963.

Kiser, Clyde V., Grabill, Wilson H., Campbell, Arthur A. *Trends and Variations in Fertility in the United States.* Cambridge, Mass., Harvard University Press, 1968. Esp. chapter 13 and appendixes A and B.

Kpedekpo, G. M. K. "Evaluation and Adjustment of Vital Registration Data from the Compulsory Registration Areas of Ghana," *Demography,* 5(1):86–92. 1968.

Kuczynski, Robert R. *Fertility and Reproduction.–Methods of Measuring the Balance of Births and Deaths.* New York, Falcon Press, 1932.

———. *The Measurement of Population Growth: Methods and Results.* London, Sidgwick and Jackson, 1935.

Kumar, Joginder. "Demographic Analysis of Data on Illegitimate Births." *Social Biology,* 16(2):92–108. June 1969.

Liu, Paul K. C. *The Use of Household Registration Records in Measuring the Fertility Level of Taiwan.* Selected English, Series No. 2, Economic Papers. Academia Sinica, The Institute of Economics, Taipei (Taiwan). Sept. 1967.

Mazur, D. P. "A Demographic Model for Estimating Age-Order Specific Fertility Rates." *Journal of the American Statistical Association,* 58(303):774–788. Sept. 1963.

Parrot, P. "Étude analytique de déclin de la natalité au Canada" (Analysis of the Decline of Fertility in Canada). *Canadian Journal of Public Health,* 57(12):581–585, Dec. 1966.

Potter, Robert G., Jr. "Birth Intervals: Structure and Change." *Population Studies* (London), 17(2):155–166, Nov. 1963.

Pressat, Roland. "Interprétation des variations à court terme du taux de natalité." *Population* (Paris), 24(1):47–56. Jan.–Feb. 1969.

Ryder, Norman B. "Fertility." Chapter 18 in Philip M. Hauser and Otis Dudley Duncan (eds.), *The Study of Population: An Inventory and Appraisal.* Chicago, Ill., University of Chicago Press, 1959. Pp. 400–435.

———. "The Measurement of Fertility Patterns." In Mindel C. Sheps and Jeanne Claire Ridley (eds.), *Public Health and Population Change: Current Research Issues.* Pittsburgh, University of Pittsburgh Press, 1965. Pp. 287–306.

———. "Problems of Trend Determination during a Transition in Fertility." *Milbank Memorial Fund Quarterly,* 34(1):5–21. Jan. 1956.

———. "The Structure and Tempo of Current Fertility." In National Bureau of Economic Research, *Demographic and Economic Changes in Developed Countries.* Princeton, N.J., Princeton University Press, 1960. Pp. 117–136.

Shapiro, S. "Development of Birth Registration and Birth Statistics in the United States." *Population Studies* (London), 4(1):86–111. June 1950.

Spiegelman, Mortimer. "Errors in Census Statistics and Vital Statistics and Their Adjustments." Chapter 3, pp. 44–55 and 77–79, in *Introduction to Demography* (rev. ed.). Cambridge, Mass., Harvard University Press, 1968.

United Kingdom, Royal Commission on Population. "Births, Marriages, and Reproductivity, England and Wales, 1938–47." By John Hajnal. In *Reports and Selected Papers of the Statistics Committee,* Vol. 2, pp. 307–344. London, H. M. Stationary Office, 1950.

SUGGESTED READINGS – Continued

United Nations. *Methods of Appraisal of Quality of Basic Data for Population Estimates*, Chapter II, "The Completeness of Vital Statistics," Manual II. Manuals on Methods of Estimating Population. Series A, Population Studies, No. 23. 1955.

———. *Methods for Population Projections by Sex and Age*. Manual III. Manuals on Methods of Estimating Population. Series A, Population Studies, No. 25. 1956. Pp. 42–44.

———. *Population Bulletin of the United States*, No. 7, *with special reference to conditions and trends of fertility in the world*. 1965.

U.S. National Center for Health Statistics. *Vital and Health Statistics*, Vol. 4, No. 1. "Fertility Measurement." A Report of the United States National Committee on Vital and Health Statistics. Sept. 1965.

———. *Vital Statistics of the United States, 1967*, Vol. I, *Natality*. Section 1, pp. 3–53, and section 3, pp. 1–15. 1969.

———. *Vital and Health Statistics*, Series 21, No. 9. "Seasonal Variation of Births, United States, 1933–63." By Harry M. Rosenberg. May 1966. Pp. 11–17 and 41–42.

U.S. National Office of Vital Statistics. *Vital Statistics of the United States, 1950*, Vol. I, *Analysis and Summary Tables*. 1954. Pp. 108–127.

———. *Vital Statistics-Special Reports*, Vol. 51, No. 1. "Fertility Tables for Birth Cohorts of American Women, Part 1." By Pascal K. Whelpton and Arthur A. Campbell. Jan. 29, 1960.

U.S. National Vital Statistics Division, Public Health Service. *Vital Statistics-Special Reports*, Vol. 4, No. 12. "Matched Record Comparison of Certification and Census Information, United States, 1950." By Joseph Schachter. Mar. 19, 1962.

Vincent, Paul E. "Variations de la fertilité selon l'âge: Méthode de recherche; Aperçu de quelques résultats d'enquête" (Variations in Fecundity according to Age: Research Methods; Summary of Some Research Results). *Bulletin de l'Institut International de Statistique*, 36(2):218–226. 1958.

———. "Recherches sur la fécondité biologique: Étude d'un groupe de familles nombreuses" (Research on Biological Fertility: Study of a Group of Large Families). *Population* (Paris), 16(1):105–112. Jan.–Mar. 1961.

Whelpton, Pascal K. *Cohort Fertility: Native White Women in the United States*. Princeton, N.J., Princeton University Press, 1954.

———. "Reproduction Rates Adjusted for Age, Parity, Fecundity, and Marriage." *Journal of the American Statistical Association*, 41(236):501–516. Dec. 1946.

———. "Why Did the United States Crude Birth Rate Decline During 1957–62?" *Population Index*, 29(2):120–125. Apr. 1963.

Wunsch, Guillaume. *Les Mesures de la Natalité. Quelques Applications à la Belgique* (Measures of Natality. Some Applications to Belgium). Louvain (Belgium), University of Louvain, Départment de Démographie, 1967.

CHAPTER 17

Natality: Measures Based on Censuses and Surveys

Natality data obtained from vital statistics registration systems and the measures derived from them were the subject of the preceding chapter. The present chapter deals with natality data based on national censuses and surveys. As was mentioned in chapter 16, many of the fertility measures derived from birth statistics can be approximated from census data.

COMPARISON WITH MEASURES FROM VITAL STATISTICS

In addition to supplying denominators for vital rates, population censuses or surveys sometimes provide fertility data to be used independently, either as supplements to vital statistics rates or as substitutes for these rates where birth registration does not exist or is inadequate. Fertility statistics from enumerations can be used to evaluate birth statistics from a registration system, and vice versa. Registration statistics present directly the fertility of a particular time period, usually a month or year, whereas census statistics may also show cumulative fertility patterns over a woman's lifetime. Birth registration records usually provide information on a few demographic characteristics of the parents of children born in the relevant period, whereas census inquiries provide data on characteristics of all women or men enumerated who have ever had children and also, and equally important, on those who have never had children.

In general, much more supplementary information of demographic interest can be provided by censuses than by birth records because of the varied kinds of information which are collected in censuses, such as data on socioeconomic characteristics. In addition, there are certain types of information which can be collected in censuses specifically for use with fertility data. Primarily, these are data on the duration of either the present or the first marriage of women, the date of this marriage, or the age at which it took place. This kind of information can add to the meaningfulness of fertility data. If, for example, a woman age 45 reports she has had two live-born children, the implication of this fact for the pattern of fertility is quite different depending on whether she was married at age 28 and has remained married for 17 years, was married at age 19 but has been a widow since she was 25, or was married at age 35 and has remained married for 10 years.

At least two and often three measures of fertility are frequently available from census reports. One is that derived from ordinary age-sex distributions of the population, the ratio of young children to women of childbearing age. The second is based on the classification of women by the number of their own young children living in the same household, derived from data on sex, age, and household relationship. A third and increasingly available measure of fertility is based on the number of children ever born to women. These will be considered in the order named.

CHILD-WOMAN RATIO

Definition

The measure commonly used is the ratio of children under 5 years old to women of childbearing age, referred to as the "general fertility ratio," or the "ratio of children to women," or the "child-woman ratio." This measure is computed by dividing the number of children under 5 years old in the population by the number of women 15 to 49 years old. The childbearing ages in the denominator are sometimes approximated by the ages 20 to 49 or 15 to 44. It must be noted that the children under 5 may have been borne up to 5 years prior to the census date when the women were up to 5 years younger. Although some mothers are left out, they have contributed so few of the children under 5 that the inclusion of younger or older ages would include mostly women who are not "exposed to risk." The computation formula is:

$$\frac{P_{0\text{-}4}}{P_{f\,15\text{-}49}} \times 1{,}000 \qquad (1)$$

where $P_{0\text{-}4}$ is the number of children under 5 years old and $P_{f\,15\text{-}49}$ is the number of women between exact ages 15 and 50. The numerator and denominator for the fertility ratio are drawn from the same universe, the population by age and sex as given by the census. For example, from the Moroccan census of 1963

(a) children under 5 = 2,357,400
(b) females 15 to 49 = 2,826,300

$$\text{ratio}\ \frac{(a)}{(b)} \times 1{,}000 = 834$$

The general fertility ratio may be viewed as either a substitute for the general fertility rate (i.e., the number of births per 1,000 women of childbearing age) or as a measure of effective fertility, which takes into account child mortality. Although the ratio of children 1 year old to the total population in the census could be used as an approximation to the crude birth rate, the measure is seldom used. Children under 5 years old have been traditionally used instead of children under 1 year of age, because:

(a) Age distributions are frequently limited to 5-year groups.

(b) There is proportionately less underenumeration of children under 5 years old than of children under 1 year old.

(c) The ratio is more stable when a larger numerator is used.

Some examples of general fertility ratios for selected countries around 1960 are:

Antigua, 1960	661	Kuwait, 1965	980
Cocos (Keeling)		Mexico, 1960	726
Islands, 1961	835	Norway, 1960	378
Federal Republic of		Tokelau Islands, 1961	852
Germany, 1961	327	Turkey, 1960	700
Gabon (African popula-		United States, 1960	488
tion) 1960–61	398	Uruguay, 1963	390
Ghana, 1960	831		

For a number of these countries, birth registration data are quite inadequate. On the other hand, not all of the selected ratios give a valid indication of recent fertility either. The ratio for Gabon (African population), for example, looks suspiciously low. Either children under 5 were grossly undercounted in the census or child mortality was extremely heavy, or both.

For historical statistics on natality in the United States, we have to rely on this measure. The number of children under 5 was first available for whites in the census of 1830 and for Negroes in that of 1850. Thompson and Whelpton estimated the number of children under 5 for earlier censuses on the basis of the reported number under 10 and then went on to estimate not only the fertility ratio but also the birth rate for whites back to 1800.[1]

Uses and Limitations

It should be clear by now that the great advantage of the general fertility ratio is that it does not require a special question in the census. This ratio is also useful as a means of obtaining fertility statistics for small areas (counties, minor civil divisions, etc.). Since it is often too expensive to tabulate data from special fertility questions or from special codings for such areas, the age-sex distribution, in the absence of birth statistics, may be the only source of fertility information for them. Even for countries with good registration data, birth statistics may not be tabulated for these small areas, for example, because of problems in geographic coding.

The fertility ratio can be computed for certain groups of the population, where the classification of the mother and child by the given characteristic would be the same.

---

[1] Warren S. Thompson and P. K. Whelpton, *Population Trends in the United States,* New York, McGraw-Hill, 1933, p. 263.

For example, a mother and child would be classified the same with respect to residence (political area, urban-rural, farm-nonfarm, etc.), and with respect to race or ethnic group. Separate age distributions for the native of native parentage, the native with mother foreign-born, and the foreign born make it possible to compute fertility ratios for native and foreign-born women separately. Since farm-nonfarm residence is not an item on the birth certificate in the United States, census statistics have been the only source of national information on the fertility of the farm population. The fertility ratio cannot be computed, however, for social and economic groups—unless the children of mothers in a given group are identified. In other words, the ratio for a given group cannot be computed from an ordinary age-sex distribution of the population in that group.

The ratio of children to women has many weaknesses as a measure of fertility. It is directly affected by the underenumeration of young children, which was shown in chapter 8 often to be large and variable from one area to another.

Another factor affecting the interpretation of child-woman ratios is mortality, which affects both the women of childbearing age and the children under 5 who are survivors of births in the preceding 5 years. Since the survival rate is higher for the women, the ratios always understate recent fertility. International comparisons are particularly affected because there is so much variation in infant and child mortality.

The ratio of children under 5 years old to women in the childbearing ages may be regarded as a better measure of "effective" fertility than the general fertility rate (based on births), since children who die early in life do not contribute to the replacement of their mothers in the next generation. From the number of children under 5 years old, the number of births in the preceding five years can be estimated, if allowances are made for underenumeration and for mortality of children and women.

Another inherent defect of the child-woman ratio is that the broad age range used for women does not take account of their age distribution within this range; and fertility is, of course, closely related to age. Age-specific birth rates are at their maximum from ages 20 to 29, and ratios using children under 5, therefore, would be expected to be at their maximum at about ages 22½ to 32½.

Accordingly, it would be desirable to standardize the child-woman ratio for the effects of age composition within the child-bearing years. For direct standardization, age-specific ratios of children under 5 to women, which identify the children according to the ages of their mothers, for the populations under study, are required. These ratios are available normally in terms of own children living in the households of their mothers. If such age-specific ratios are not available for the populations under study, then indirect standardization may be applied, using appropriate age-specific ratios of own children to women or age-specific birth rates for a different but similar population. The method parallels that for the indirect standardization of general fertility rates based on births.

Here we will illustrate the indirect standardization method by using two States of the United States with somewhat different age structures. The standard set of fertility rates is the set of age-specific national ratios of **own** children under 5 per woman. It is assumed here that such age-specific rates would not be available for the subareas of the country. The procedure is outlined in table 17–1 and follows that previously described in chapter 14. The resulting standardized rates are in terms

Table 17-1. — **Number of Children Under 5 Years Old per 1,000 Women 15 to 49 Years Old Standardized for Age of Woman by the Indirect Method, for Pennsylvania and South Carolina: 1960**

[Pattern of age-specific ratios of own children under 5 to women 15 to 49 years old, for the United States, was employed]

Age of woman	United States[1]				Pennsylvania[2]	South Carolina[2]
	Number of women $_P P_a^F$	Own children under 5			Number of women $_1 P_a^F$	Number of women $_2 P_a^F$
		Number	Per woman R_a (2) : (1) =			
	(1)	(2)	(3)		(4)	(5)
Total, 15 to 49 years.....................	41,628,096	19,686,941	(.473)		2,684,501	562,680
15 to 19 years................................	6,588,170	659,635	.100		407,568	107,565
20 to 24 years................................	5,506,512	4,572,435	.830		329,581	80,307
25 to 29 years................................	5,537,428	5,898,737	1.065		333,918	75,907
30 to 34 years................................	6,099,925	4,475,654	.734		397,175	77,563
35 to 39 years................................	6,438,919	2,746,961	.427		427,647	82,047
40 to 44 years................................	5,897,573	1,077,783	.183		410,618	72,803
45 to 49 years................................	5,559,569	255,736	.046		377,994	66,488

a) Ratio of all children under 5 to women 15 to 49, United States[2] = 20,320,901 ÷ 41,600,533 = .488

b) Observed number of children under 5 = ... 1,187,954 ... 294,913

c) Expected number of children under 5 = $\Sigma R_a p_a^F$ = Σ[Col. (3) X Col. (4) or (5)] = ... 1,236,594 ... 266,599

d) Observed number ÷ expected number = (b) ÷ (c) = ... .961 ... 1.106

e) Standardized rate [(d) X .488] = ... .469 ... .540

[1] 25-percent sample statistics.
[2] Complete-count statistics.

Source: Basic data from *U.S. Census of Population: 1960, Women by Children Under 5 Years Old*, PC(2)-3C, 1968, table 1; *U.S. Census of Population: 1960*, Vol. I, *Characteristics of the Population*, Part 40, *Pennsylvania*, 1963, table 16, and Part 42, *South Carolina*, 1963, table 16.

of total children — not own children — under 5, because of the source of the national ratio in step (a). In comparing the ratios for Pennsylvania and South Carolina, the reader should bear in mind what was said in chapter 14 about the use and interpretation of rates or ratios standardized by the indirect method.

Adjustment for Mortality. — So far we have mentioned briefly corrections for net census undercounts and indirect

standardization for age within the childbearing ages. We also mentioned that the general fertility ratio could be adjusted for mortality to make it a closer approximation to the general fertility rate. In other words, from the number of children under 5 in the numerator, we can estimate the births of which they are the survivors. It is also necessary to adjust the number of women 15 to 49 years old to obtain the number of women

Table 17-2. — **Estimated General Fertility Rates Based on Ratios of Children to Women, for Selected Censuses**

Country and census date	Children under 5 years old	Women 15 to 44 years old	Women 20 to 49 years old	Women 17 1/2 to 47 1/2[1] years old 1/2 [(2)+ (3)]=	Ratio of children under 5 per 1,000 women 17 1/2 to 47 1/2[1] [(1)÷(4)] x1,000=	Survival rate for children under 5 (terminal age)	Estimated births (1)÷(6)=	Survival rate for women 17 1/2 to 47 1/2[1] (terminal age)	Estimated women 15 to 44 (4)÷(8)=	Estimated general fertility rate [(7)÷(9)] x1,000=
	(1)	(2)	(3)	(4)	(5)	(6)	(7)	(8)	(9)	(10)
Costa Rica (1963)..,.......	248,947	261,963	218,430	240,197	1,036.4	.93639	265,858	.99178	242,188	1,097.7
El Salvador (1961)........	431,658	537,670	458,851	498,261	866.3	.90961	474,553	.98860	504,007	941.6
Chile (1960)..............	1,104,720	1,617,309	1,414,147	1,515,728	728.8	.86476	1,277,487	.98322	1,541,596	828.7
Taiwan (1964).............	1,987,555	2,371,492	2,077,487	2,224,490	893.5	.93639	2,122,572	.99178	2,242,927	946.3
Sweden (1963).............	521,004	1,531,779	1,476,588	1,504,184	346.4	.98067	531,274	.99729	1,508,271	352.2
Ireland (1961)............	300,790	512,371	484,475	498,423	603.5	.97141	309,643	.99608	500,385	618.8

[1] Exact age limits of the range. Other ages are shown in completed years, as is conventional.

Source: Basic data from United Nations, *Demographic Yearbook, 1965*, table 6; and Ansley J. Coale and Paul Demeny, *Regional Model Life Tables and Stable Populations*, Princeton University Press, Princeton, N.J., 1966.

at the midpoint of the preceding 5-year period. (It is assumed here that there was no net migration of either the children or the women.) The method is illustrated in table 17-2.

The steps are as follows:

Step 1. Enter the enumerated number of children under 5 years old, the enumerated number of women 15 to 44 years old, and the enumerated number of women 20 to 49 years old in cols. (1), (2), and (3), respectively.

Step 2. The estimated number of women 17½ to 47½ years old (exact ages) in col. (4) is obtained by averaging cols. (2) and (3). A more refined method of interpolation could be used.

Step 3. Compute the ratio of (1) to (4) and multiply by 1,000. Enter the result in col. (5). This is a conventional child-woman ratio except for the unusual age range for the childbearing ages. The reason for this choice is that it corresponds to women 15 to 44 years old at the midpoint of the quinquennium, and the general fertility rate is ordinarily the ratio of births to women of that age.

We now need to find the number of births and of women 2½ years younger of which the children and the women at the census date are, respectively, the survivors. In the absence of appropriate life tables for most of these countries, we have recourse to the model life tables developed by Coale and Demeny. For Costa Rica, for example, "model west, mortality level 19" seems appropriate.

Step 4. From birth to ages 0 to 4, the survival rate is given by L_{0-4} divided by 500,000, which from the given model is 0.93639. Enter this survival rate in col. (6).

Step 5. Since we want to use the survival rate in reverse to restore the deaths, we divide it into the number of children to estimate the births and enter the quotient in col. (7), e.g., $\frac{248.947}{.93639} = 265.858$.

Step 6. Applying a survival rate in reverse to the women is more complicated. The survival rate is of the form

$$\frac{L_{f\,17\frac{1}{2} \leftrightarrow 47\frac{1}{2}}}{L_{f\,15-44}}.$$

We first find $L_{f\,20-49}$ and $L_{f\,15-44}$, which for Costa Rica are 2,620,175 and 2,663,975, respectively.

Step 7. The average of these is 2,642,075, which represents $L_{f\,17\frac{1}{2} \leftrightarrow 47\frac{1}{2}}$.

Step 8. The survival rate for Costa Rican women is then .99178 (2,642,075 ÷ 2,663,975). Enter this in column (8).

Step 9. Divide this survival rate into the number of women estimated for col. (4), and enter the result in col. (9).

Step 10. Compute the estimated general fertility rate by dividing col. (7) by col. (9) and multiplying by 1,000. Enter the result in col. (10).

Estimated general fertility rates may be used in the absence of birth statistics to compare fertility levels in various areas. Even when a country, such as Sweden, has adequate birth statistics, we may compute this ratio to permit comparability with other countries.

Comparison With Fertility Measures Based on Births.—The general fertility rate uses a denominator that is much more similar to that of the child–woman ratio than does the crude birth rate. Even after dividing the ratio by 5 to equalize the time periods, variability due to mortality, underenumeration of children, and underregistration of births remains.

Ratio of Children 5 to 9 to Women 20 to 54

By analogy with the ratio using the number of children under 5 years in the numerator, we can compute from census age-sex distributions a ratio of children 5 to 9 for women 20 to 54. This ratio gives us a measure of fertility for an earlier quinquennium. Furthermore, children 5 to 9 years old are often more completely enumerated in a census than children under 5 years old. On the other hand, the cumulative force of mortality (and perhaps net immigration) is greater the older the children are so that, other things being equal, there is a greater understatement of the number of births of which the children are the survivors. Both numerator and denominator are affected but rarely to the same extent so that the canceling of the error is only partial. Table 17-3 contrasts the ratios based on the two age groups of children for several Latin American countries.

In a number of cases, the ratio based on children 5 to 9 is higher than that based on under 5 at the same census, suggesting more nearly complete enumeration of the older children, although effective fertility may also be falling in some cases. In studying international variations in current fertility, differences among the ratios can be computed according to both measures and the results compared to see whether the patterns of differences are reasonably consistent.

Marital Fertility

As explained in chapter 16, **marital fertility** is distinguished from **general fertility,** in that the former includes in the denominator only those women who are or have ever been married. Ideally, illegitimate children should be excluded from the numerator in the measurement of marital fertility, but we cannot identify these children in the census statistics for age groups. In many countries illegitimate births form a sizable proportion of total births because consensual unions are common; thus, limiting the measure of fertility to ever-married women would exclude a significant portion of the population likely to produce children and would adversely affect comparability with other countries.

In countries where most children are legitimate, however, the marital fertility ratio may be based on all children. Here the total number of children under 5 is related to the number

Table 17-3. — **Two Different Child-Woman Ratios for Selected Latin American Countries: Around 1950 and 1960**

Country	Year	Children under 5 per 1,000 women 15 to 49 years old	Children 5 to 9 per 1,000 women 20 to 54 years old
Dominican Republic........	1950	749	733
	1960	769	877
Ecuador.................	1950	705	700
	1962	752	810
El Salvador.............	1950	623	634
	1961	739	771
Honduras................	1950	666	636
	1961	857	883
Nicaragua...............	1950	650	705
	1963	824	937
Venezuela...............	1950	711	672
	1961	804	800

Source: United Nations. *Demographic Yearbook, 1965,* tables 6 and 8; *1964,* table 5; *1962,* table 5; and *1960,* table 5.

of ever-married women of childbearing age instead of all women of childbearing age. Confining the denominator of a fertility ratio to women who have been married is simply an attempt to remove the influence of variations in proportions married when one is comparing the fertility of various groups.

WOMEN OF CHILDBEARING AGE BY NUMBER OF OWN YOUNG CHILDREN

Method of Compilation

From the foregoing discussion, it should be clear that a more useful measure of fertility would be obtained if the young children in a census could be linked to their own mothers since other characteristics of their mothers are known. To do so requires that women of childbearing age on the census schedules be coded according to the number of their own children under a given age in the household. Again, either children under 5 or those 5 to 9 can be used in the numerator. Here, however, age-specific ratios can be computed, for example, the ratio of the number of children under 5 to women 20 to 24 years old.

In most cases it is fairly easy to perform the coding operation because the census enumeration is carried out by households and household members are grouped on the schedules. This procedure would seem to be of great potential value to countries without vital statistics or with poor vital statistics.

This coding procedure has been used extensively in the United States beginning with the census of 1940; but, as far as the authors can discover, it has not been used in the population censuses of other countries. Taiwan, however, has applied this method to the data in its population register and accordingly has been able to show differential fertility ratios in a wealth of socioeconomic detail that is not feasible even for most countries with well-developed vital statistics systems.[2]

Limitations

In most populations, relatively few young children will be omitted from these tabulations because they are not living with their mothers. To measure the loss, one may compare the number of own children living in the same household as their mother with the total number of children of the same age. For the United States, we have the following:

	Own children under 5 as percent of all children under 5	
	1960	1950
White	99.1	98.2
Nonwhite	86.2	85.6

It can be seen that the rates using these data should be quite satisfactory from this standpoint for white women but only fairly adequate for nonwhites (mostly Negroes). Some of the children not living with their mothers were orphans. Others

had living mothers but stayed with other relatives, such as a grandmother, or in institutions. No effort was made to identify own children under 5 in the 1960 and 1950 censuses for women who reported themselves as single, but undoubtedly many single women who had kept their illegitimate children reported themselves as separated or divorced and hence they and their children were included in the tabulation. It is not possible to exclude step-children or adopted children since they are not identified on the census schedule.

Standardized Rates

Since tabulations of women by the number of their own young children of a specified age also give some detail on the age of the women (usually in 5-year groups), it is feasible to standardize the fertility rates for age of women by the direct method, described in chapter 14. Ordinarily, when women in two areas or in two population subgroups are being compared, the national distribution by age of women of childbearing age would be used as the standard.

Experience in the United States

Decennial Census. — The first attempts to develop fertility measures from American data by the method of relating young children to their own mothers were made in the census of 1930.[3] The first extensive use of this technique, however, was made in connection with the 1940 census, at which time fertility ratios were also computed from a sample of the 1910 census schedules.[4]

The basic measures of fertility derived from these statistics were: (1) percent distributions of women by number of children under 5 years old: none, 1, 2, and 3 or more, in most tables; (2) children per 1,000 women. In addition, the supplementary report on *Standardized Fertility Rates and Reproduction Rates* also contained ratios of children per 1,000 women standardized for age, color, urban-rural residence and marital status. Appendix E of this report also describes how estimates of age-specific birth rates can be derived from the age-specific ratios of own children under 5.

Similar tables were also contained in the fertility reports of the 1950 and 1960 censuses.[5] Considerable analysis of the 1950 statistics of this type was contained in the 1950 census monograph by Grabill, Kiser, and Whelpton.[6]

Current Population Survey. — Tabulations of women of childbearing age by number of own children under 5 years old have been made several times from the Current Population Survey — for June 1946, April 1947, April 1952, and annually beginning in 1967.[7] The same kinds of measures and characteristics are

[2] P.K.C. Liu, "Differential Fertility in Taiwan," International Union for the Scientific Study of Population, *Contributed Papers, Sydney Conference,* Sydney (Australia) August 21-25, 1967, pp. 363-370. More recently this method has been used in Korea under the leadership of Lee-Jay Cho. (See citation under "Suggested Readings" at end of chapter).

[3] Frank W. Notestein, "Differential Fertility in the East North Central States," *Milbank Memorial Fund Quarterly,* 16(2):173-191, April 1, 1938; idem, "Intrinsic Factors in Population Growth," *Proceedings of the American Philosophical Society,* 80(4):499-511, February 1939.

[4] *Sixteenth Census of the United States: 1940, Population, Differential Fertility 1940 and 1910, Fertility for States and Large Cities,* 1943; idem, *Standardized Fertility Rates and Reproduction Rates,* 1944, and *Women by Number of Children Under 5 Years Old,* 1945.

[5] *U.S. Census of Population: 1950,* Vol. IV, Special Reports, Pt. 5, ch. C, *Fertility,* 1955; idem, *U.S. Census of Population: 1960, Subject Reports,* PC(2)-3C, *Women by Number of Children Under 5 Years Old,* 1968.

[6] Wilson H. Grabill, Clyde V. Kiser, and Pascal K. Whelpton, *The Fertility of American Women,* New York, John Wiley and Sons, 1958.

[7] U.S. Bureau of the Census. *Current Population Reports, Population Characteristics,* Series P-20, Nos. 8, 18, 27, and 184.

presented as in the decennial reports. For April 1952. moreover. own children under one year old were tabulated by parity and marital status of women 17 to 46 years old.

Factors Important in Analysis

Mainly by way of recapitulation. we may point out that many more types of fertility differentials can be studied when the numerator is own young children rather that all young children of the given age. In the latter case. we are limited to characteristics that are the same for the mother and the child. In the former case. we can compute rates for any characteristic of the woman. her husband. her household. and her dwelling that appear on the census schedule. The 1960 U.S. report on *Women by Number of Children Under 5* shows rates by woman's age at first marriage. age difference between husband and wife. years since first marriage. and total children ever born. for example. Age is a very important determinant of fertility. and the general fertility ratio provides only a moderately good "control" on age. From either datum. a measure or measures of marital fertility can be computed. In the case of total children. we implicitly assume that all the children were borne by the married women: whereas. in the case of own children. there are fewer such erroneous inclusions in the numerator of the rate.

The special fertility report published as part of the 1970 census program, *Women by Number of Children Ever Born,* also contains a wealth of statistics for the analysis of fertility differentials.

CHILDREN EVER BORN

In contrast to the fertility ratios and rates previously discussed in this chapter the number of children ever born requires a special question on the population census schedule. Although the more common practice in a census is to ask the question on children ever born of ever married women. it is sometimes asked of all women. The loss from the total count of children by excluding single women becomes progressively less important with advancing age since. in most countries. many mothers of illegitimate children eventually marry and the children born prior to marriage are also included in the count.

Very often the question on children ever born is accompanied by one or more inquiries on the woman's marriage. such as number of times married. date of marriage. age at marriage. and duration of marriage. The three last items may relate to her first marriage or to her last marriage. In analyzing fertility by age at marriage or duration of marriage. several major groups of women can be distinguished. those women who were married once and are still married at the date of census. those women who were married more than once and are in the current marriage at the date of the census. and women who are widowed or divorced at the date of the census. With separate statistics on children ever born for such groups. the analyst can examine the fertility of women whose marriage has been interrupted. A further refinement would be to compute the total number of years in which women of a given population have been "at risk" of having a child by excluding periods in their adult lives when they were not married.

Definitions

When incorporating a question on number of children ever born in censuses or surveys. both the births to be included and the women of whom the question is to be asked must be specified. The count of children is ordinarily limited to children born alive but stillbirths are sometimes included. There may be a supplementary question on the number of children still living. The two questions in combination provide some limited information on child mortality and on replacement. It is very important to bear in mind that the question on children ever born is not limited to those present in the household. even though that number may provide a fairly good approximation in the case of younger women. Adopted children. foster children. and step-children are excluded. however. The coverage of women in terms of marital status has already been mentioned. In addition. a minimum age may be specified to correspond to the beginning of the childbearing period.

International Recommendations. – For the 1970 censuses, the United Nations has recommended questions on number of children born alive and number of children living.

Information on children ever born may be derived from censuses, sample surveys, or population registers. Some statistics on the female population by age and number of children born alive and by age and number of children living can be found in various issues of the *Demographic Yearbook.* More detailed statistics are given in national publications, of course.

National Practices

Table 17–4 is an example of a table on the distribution of ever married women by number of children ever born (alive) taken from a national publication. The age detail for the woman is adequate, and the categories of number of children are quite detailed. Note the treatment of the various nonresponse categories. Statistics on children ever born tabulated from the Taiwanese population register by a wide variety of social and economic characteristics are given in an article by Liu.[8] Japan has conducted national fertility surveys at irregular intervals.[9] The data on children ever born are tabulated by a number of socioeconomic characteristics and in a number of analytical arrangements (by marriage cohorts, parity and birth order, average interval between births, etc.). Even in very primitive areas of Western New Guinea, a field survey team was able to collect publishable data on children ever born and to compute cohort fertility rates for them.[10]

United States. – Questions on children ever born have been asked in the United States both in the decennial censuses and in national sample surveys. In addition to the official data from the Current Population Survey, data on this subject have been included in a number of important fertility surveys under private auspices, such as the Indianapolis Study of 1941 ("Social and Psychological Factors Affecting Fertility"), the Growth of American Families Studies of 1955 and 1960, and the National Fertility Surveys of 1965 and 1970 (carried out privately with the financial support of the U.S. Public Health Service). Under the name of the National Survey of Family Growth, the survey has now become a regular program of the National Center for Health Statistics (U.S.P.H.S.). It is planned to have a survey every two or three years. The field work for the 1973 survey was conducted by the National Opinion Research Center; a 1975 survey is also scheduled.

---

[8] Liu, op. cit., pp. 363–370.

[9] Japan, Institute of Population Problems, *Summary of the Fertility Surveys in 1940, 1952, 1957, and 1962,* by Hisao Aoki and Eiko Nakano, Research Series No. 177, July 1, 1967.

[10] Netherlands, *Results of the Demographic Research Project: Western New Guinea,* Part 6, *The Progeny of Papuan Women,* 1967.

Table 17–4. — Number of Ever-Married Women 15 Years of Age and Over by Number of Children Ever Born Alive, by Age of Mother, for Thailand: 1960

Number of children ever born alive	Total	Age of mother (years)												Unknown
		15 to 19	20 to 24	25 to 29	30 to 34	35 to 39	40 to 44	45 to 49	50 to 54	55 to 59	60 to 64	65 to 69	70 and over	
Total.........	5,672,390	171,431	739,741	898,893	811,991	651,737	546,462	470,600	400,893	321,913	239,879	160,308	240,603	17,939
None..............	312,240	61,149	89,145	39,463	23,163	18,231	15,701	14,352	13,643	11,656	9,573	6,286	8,661	1,217
1.................	706,252	59,900	264,002	137,682	59,059	37,183	30,135	26,548	23,972	20,397	16,830	11,371	17,370	1,803
2.................	687,659	10,570	193,163	197,329	83,094	44,219	33,842	29,268	26,506	21,827	17,138	11,601	17,576	1,526
3.................	647,785	1,526	82,635	212,698	117,434	55,590	39,398	33,271	29,954	23,951	18,573	12,438	19,071	1,246
4.................	605,656	542	23,368	154,575	149,695	71,978	47,079	39,448	34,548	27,545	20,667	13,826	21,196	1,189
5.................	548,379	115	5,626	75,012	147,652	88,866	55,214	44,928	38,829	30,617	22,446	14,999	22,936	1,139
6.................	476,227	26	2,337	26,244	106,323	95,383	60,601	48,252	40,989	32,691	23,939	15,569	22,895	978
7.................	397,713	11	422	7,662	57,250	86,016	62,363	49,401	41,004	32,692	23,051	15,081	21,834	926
8.................	321,908	2	124	2,862	24,874	63,302	58,955	46,789	38,765	30,602	21,328	13,982	19,603	720
9.................	243,431	-	29	390	9,399	37,102	48,786	41,112	33,177	26,342	18,170	11,866	16,483	573
10..............	171,792	-	-	97	4,270	19,237	35,714	32,439	25,491	19,878	13,745	8,721	11,830	370
11..............	97,422	-	-	27	660	7,950	20,045	20,500	15,853	11,939	8,183	5,064	6,981	220
12..............	74,992	-	-	3	220	4,683	12,211	15,278	12,533	9,921	7,412	4,932	7,564	235
13..............	28,556	-	-	-	67	673	5,051	6,804	5,100	3,761	2,674	1,834	2,518	74
14..............	13,529	-	-	-	1	172	2,710	3,196	2,413	1,708	1,312	797	1,189	31
15..............	7,137	-	-	-	-	39	577	1,635	1,393	1,135	807	585	936	30
16..............	2,687	-	-	-	-	4	179	669	562	420	323	214	305	11
17 or more........	1,826	-	-	-	-	2	100	434	379	274	225	163	234	15
Unknown[1].........	327,199	37,590	78,890	44,849	28,830	21,107	17,801	16,276	15,782	14,557	13,483	10,979	21,421	5,634

– Represents zero.

[1] Including women of unknown marital status.

Source: Thailand, Central Statistical Office, *Thailand, Population Census: 1960, Whole Kingdom*, 1962, table 14, p. 26.

Censuses. — In the United States, the question on children ever born was asked for the first time in the 1890 decennial census, but no tabulations were made from the data of that census. The question was repeated in the decennial censuses of 1900, 1910, and 1940 to 1970. The wording has been as follows:

1890 and 1900: Mother of how many children
Number of these children living
1910: Mother of how many children
Number born
Number living
1940: Number of children ever born (do not include stillbirths)
1950: How many children has she ever borne, not counting stillbirths?
1960: How many babies has she ever had, not counting stillbirths? Do not count her stepchildren or adopted children.
1970: *If this is a girl or woman*—How many babies has she ever had, not counting stillbirths? *Do not count her stepchildren or children she has adopted.*

The question on children ever born was asked of a 5-percent sample in 1940, 3 1/3-percent in 1950, 25-percent in 1960, and 20-percent in 1970. Most of the 1960 tabulations, however, were based on a 5-percent sample. Up until 1970, these questions were confined to women who were or had been married (ever-married women); but in 1970, when self-enumeration is being used extensively, the question is being asked of all females 14 years old and over, including the single. For the same females, questions have been asked on further particulars of the marriage, such as whether married more than once, number of years married to present husband, age at first marriage, and number of years in present marital status.

Table 17–5 is an excerpt from a published table of the 1960 census of the United States and includes various measures, which will be discussed in a later section.

Current surveys. — Questions on the number of children ever born have also been included in the Current Population Surveys for April 1952, April 1954, March 1957, August 1959, March 1962, June 1964, June 1965, and January 1969. The results of these inquiries are published in Series P–20, *Current Population Reports, Population Characteristics*. The statistics are used to study both time trends and differentials in the fertility of socioeconomic groups. In addition to the more conventional presentations, the rates for the same cohort from successive surveys are arranged so as to show the progression of cumulative fertility. (See the subsection below on "Measures.") The large-scale surveys conducted by research foundations, universities, etc., that were mentioned above have collected data on children ever born as part of extensive and intensive interviewing on fertility, which also covered such broad aspects as fertility histories, pregnancy wastage, family limitation practices, and expectations regarding the couple's future number of births.

Uses and Limitations

In contrast to other types of fertility data from censuses, the number of children ever born represents **cumulative** fertility. The number of children ever born per 1,000 women 50 or more years old is also the **completed** fertility rate. It is a short step from this rate to the **gross reproduction rate** described in the next chapter, where only female children ever born are counted. Data on the number of children ever born provide us with in-

Table 17–5. — **Number of Children Ever Born per 1,000 Women 15 Years Old and Over by Marital Status and Age of Woman, for the United States: 1960**

[Based on 5-percent sample]

Age of woman	Total women		Single women	Women ever married								
				Total			Married once, husband present			Other ever married		
	Number of women	Children ever born per 1,000 women		Number of women	Children ever born		Number of women	Children ever born		Number of women	Children ever born	
					Number	Per 1,000 women		Number	Per 1,000 women		Number	Per 1,000 women
Total.........	63,562,641	2,073	10,990,182	52,572,459	131,752,371	2,506	35,019,895	85,529,312	2,442	17,552,564	46,223,059	2,633
15 to 44 years.....	36,068,528	1,746	8,849,881	27,218,647	62,991,470	2,314	21,627,610	50,287,775	2,325	5,591,037	12,703,695	2,272
15 to 19 years...	6,588,191	127	5,527,684	1,060,507	838,598	791	856,131	656,962	767	204,376	181,636	889
20 to 24 years...	5,506,470	1,030	1,572,042	3,934,428	5,670,836	1,441	3,343,945	4,758,163	1,423	590,483	912,673	1,546
25 to 29 years...	5,537,487	2,007	581,377	4,956,110	11,115,714	2,243	4,127,125	9,228,388	2,236	828,985	1,887,326	2,277
30 to 34 years...	6,099,887	2,452	418,699	5,681,188	14,956,928	2,633	4,553,091	12,111,485	2,660	1,128,097	2,845,443	2,522
35 to 39 years...	6,438,939	2,518	390,179	6,048,760	16,212,227	2,680	4,649,516	12,725,155	2,737	1,399,244	3,487,072	2,492
40 to 44 years...	5,897,554	2,407	359,900	5,537,654	14,197,167	2,564	4,097,802	10,807,622	2,637	1,439,852	3,389,545	2,354
45 to 49 years.....	5,559,609	2,247	359,183	5,200,426	12,491,495	2,402	3,652,610	9,031,036	2,472	1,547,816	3,460,459	2,236
50 to 54 years.....	4,927,204	2,179	374,598	4,552,606	10,737,454	2,359	2,983,960	7,240,389	2,426	1,568,646	3,497,065	2,229
55 to 59 years.....	4,405,986	2,290	362,524	4,043,462	10,087,540	2,495	2,410,620	6,117,180	2,538	1,632,842	3,970,360	2,432
60 to 64 years.....	3,718,944	2,503	287,143	3,431,801	9,309,581	2,713	1,794,407	4,920,875	2,742	1,637,394	4,388,706	2,680
65 years and over..	8,882,370	2,942	756,853	8,125,517	26,134,831	3,216	2,550,688	7,932,057	3,110	5,574,829	18,202,774	3,265

Source: *U.S. Census of Population: 1960, Subject Reports, Women by Number of Children Ever Born,* PC(2)–3A, 1964, table 1, p. 1.

formation on the extent of childlessness in a population as well as on the distribution of mothers by number of children. Moreover, the distribution of women by their number of children ever born at a given date in combination with birth statistics by order of birth for the same date, yields parity-specific birth rates.[11]

The number of children ever born as such does not tell us anything about current fertility (except for very young women), because we do not know when each child was born. When the dates of birth of all the children are also listed in the census or survey, chronological information is provided, of course. (See the section below on "Birth rosters.")

The question on children ever born is ordinarily asked of only those women who are or have been married. (In the 1950 census of Puerto Rico, the question was asked also of both single and consensually married women, with a very low nonresponse rate.) When the question is confined to ever-married women, some but not all illegitimate fertility is reported since most women living in common-law marriages probably report themselves as married and even single women with illegitimate children often report themselves as separated, divorced or widowed, particularly, if their children are present in the household.

Women who die or emigrate before the census or survey date are not included in the census or survey statistics, and a record of their fertility is lost. To the extent that they differ from the remaining women with respect to the number of children ever born, the reported fertility for a given cohort of women will be biased. This bias is usually ignored in analysis. It can be examined to some extent, however, by comparing the rates for the same cohort of women at successive censuses. The rates should not decline, of course, if the women who die, emigrate, or immigrate had the same fertility as the other women; and, for women beyond the end of the child-

bearing period, they should neither decline nor increase unless, again, fertility or migration is selective of the less, or the more, fertile women.

Quality of Statistics

The information on the number of children ever born obtained from censuses or surveys may be erroneous because of faulty memory of the woman, especially an older woman who bore her children a long time ago, or because of lack of knowledge on the part of a respondent who may be a person other than the woman involved. Children who died soon after birth are likely to be omitted as are children who left home a long time ago. Another category likely to be omitted are illegitimate children, especially those not kept by the mother. Errors in the count of children ever born include not only underreporting, but also overreporting, although overreporting is much less frequent. An overcount may arise from an inclusion of an adopted child, a stepchild, or a stillbirth when live births are specified.

There is some evidence that the accuracy of reporting the number of children ever born tends to vary inversely with the age of the woman. The following remarks by Brass and Coale although dealing with African countries would seem to have more general reference:

"The number of children ever born is reported with good accuracy by younger women. The events which the young women are asked to recall have happened recently; the total births to each are typically not more than two or three so that the difficulties of counting large numbers in a nonnumerate society do not arise; living children (and a higher proportion of children ever born to younger mothers will survive to the time of the census) will often be present at the interviews, and few will be omitted because they have grown up and left the household."[12]

[11] P. K. Whelpton, *Cohort Fertility: Native White Women In the United States,* Princeton, N.J., Princeton University Press, 1954.

[12] William Brass et. al., *The Demography of Tropical Africa,* Princeton, N.J., Princeton University Press. 1968, p. 91.

Besides these inaccuracies in reporting, one should also consider the failure to report the number of children. Table 16 in the 1963 *Demographic Yearbook* gives the number of "unknowns" in the statistics as published. It is obvious that, for a number of countries, all the nonresponses were allocated prior to publication. In some instances, e.g., the 1957 census of Tanganyika, the nonresponses were allocated uniformly to the category "no children"; in others, they were prorated or were allocated on a probability basis in accordance with various characteristics of the mother. There is some evidence that enumerators are likely to leave a question blank rather than bothering to write in the word "None" or a "0," so that a disproportionate number of the nonresponses may truly be cases of no children ever born. On the whole, nonresponse is not conspicuously bad for this demographic characteristic.

The proportion of women who were or had been married but for whom no information was recorded on the number of children ever born ran fairly high in the 1910, 1940, 1950, and 1960 censuses of the United States. For 1910 and 1940, the overall percentages were 7.6 and 12.7, respectively.

The item on children ever born was among those included in the reinterview study of the 1960 census of the United States.[13] In Study EP–10, women ever married were asked about children who had died or left home and about adoptions and stepchildren; these questions were used to provide more definitive information than that in the census on the number of children actually born to each woman in the sample. The data from the reinterview were compared with 1960 census data for the identical women. Comparisons made indicate that exact agreement on the number of children ever born occurred for 91.9 percent of the women with a report in both surveys; the reinterview count was higher than the census count for 5.4 percent, and lower for 2.8 percent. The total number of children ever born to these women shown by the census was 1.7 percent smaller than that shown by the reinterview.[14]

Particularly in preliterate societies where women who survive the childbearing periods have had both many live births and many other conceptions, the data often suffer from faulty knowledge or memory of the respondent and from misunderstanding of the question. Children who died shortly after birth are particularly likely to be omitted. Some of these errors may be discovered in the more intensive probing of a post-enumeration survey; a list of all pregnancies by type of termination may be made, for example. If birth records are available, a sample of them may be checked against the census questionnaires, comparing the items on order of birth and children ever born for the matched women.

Measures

From the basic statistics on children ever born, various measures can be derived, including percentage distributions and rates. In the percent distribution by children ever born, particular attention attaches to the percent of women who are childless. The commonest rates are the number of children (1) per 1,000 women, (2) per 1,000 mothers, (3) per 1,000 ever-married women, and (4) per 1,000 women, married once, with

husband present in the household. (See table 17–5 for some examples.) These rates can be computed for any age group in the childbearing period. At later ages, they can be used to measure completed lifetime fertility.

If the nonresponses have not been allocated, it is appropriate to omit them from the base of both the percentages and the rates. Consider, for example, the statistics for ever-married women 50 to 54 years old from the 1960 census of Thailand shown in table 17–4. To obtain the base, we would subtract 15,782 from 400,893, getting 385,111 ever-married women of this age who reported on children ever born. Childlessness among these reporting women is thus given by 13,643 divided by 385,111, or 3.5 percent.

General and Marital Fertility Rates. – General fertility refers to the number of children ever born per 1,000 total women, and marital fertility refers to the number of children ever born per 1,000 women ever married. In the nature of the case, marital fertility rates are higher, age for age, than general fertility rates. These differences are large at ages 15 to 19 and 20 to 24 and diminish with advancing age of women. At ages 45 to 49, the differences are relatively small because most women in this age group have been married. General fertility rates, therefore, reflect the influence of proportions married as well as the fertility of married women.

Unless satisfactory data have been collected on the number of children ever borne by single women, it is customary to assume that all single women reported in the census or survey were childless, in order to compute a general fertility rate. In this situation, the same number of children is used in the numerator regardless of whether the denominator of the rate is total women of the given age, ever-married women, or ever-married mothers.

Fertility rates using children ever born may also be standardized by age and other characteristics by usual methods.[15] From table 16 of the 1963 *Demographic Yearbook,* which gives a distribution of women by number of children, by age of woman, the following measures can be computed for women 15 to 44 years old in Jamaica in 1960:

Average number of children per woman = 2.182
Average number of children per mother = 3.385
Proportion childless = 0.3556

The relationship of these measures to each other may be expressed by:

$$\text{Average number of children per mother} = \frac{\text{average number of children per women}}{1 - \text{proportion childless}} \qquad (2)$$

Substituting the measures given above, we have:

$$\frac{2.182}{1 - 0.3556} = 3.386,$$

which agrees, except for rounding, with the number obtained by direct computation.

Suppose we want to standardize the rates for Jamaica according to the age distribution of women in the United States, using all women 15 to 44 for the first and third measures and mothers 15 to 44 for the second measure. Using the age-specific

[13] U.S. Bureau of the Census, *Evaluation and Research Program of the U.S. Censuses of Population and Housing, 1960. Accuracy of Data on Population Characteristics as Measured by Reinterviews,* Series ER 60, No. 4, 1964.

[14] *U.S. Census of Population: 1960, Women by Number of Children Ever Born,* p. XI.

[15] For a large variety of such rates, standardized by age, color, urban-rural residence, and marital status, with an account of the methodology, see: U.S. Bureau of the Census, *Differential Fertility: 1940 and 1910, Standardized Fertility Rates and Reproduction Rates.*

measures derived from the statistics for Jamaica in the *Year-book*, we obtain, by the procedure described in chapter 14 for direct standardization, the following standardized rates for Jamaica:

children per woman $= 2.432$
children per mother $= 3.736$
proportion childless $= 0.326$

A word of warning is in order, here, about the effect of this standardization procedure on the relationship shown in formula (2). Substituting the standardized measures in the formula, we find that

$$\frac{2.432}{1-0.326} = 3.608, \text{ not } 3.736.$$

The discrepancy arises because of the different population standards used and the particular algebraic relationship of the elements (i.e., their nonadditive quality). (This type of standardization problem is not unique to fertility data, of course.) One solution is to obtain two of the three measures independently and then to obtain the third measure from formula (2). For example, the age-standardized number of children per mother could be accepted as 3.608. It is assumed that, after standardization for age (or any other characteristic), we wish to preserve the logical relationship expressed in formula (2).

The marital fertility rate for women of completed fertility is an especially useful measure for comparing the overall level of fertility of two or more populations or a population at different dates. The fertility rates for women still in the childbearing period come closer to representing current fertility. Statistics on young women classified by the number of children ever born, however, can be used to estimate the eventual completed fertility of these women, if assumptions are made as to the future additions to current cumulative fertility. The percent of lifetime fertility completed by each age may be computed from statistics for various previous cohorts and used as a guide for present cohorts.

In the United States, for example, women 30 to 34 years old in 1950 had had 1,871 children per 1,000 women. By 1960 the cumulative fertility of these women had increased to 2,407 children ever born per 1,000 women. The fertility of the women at age 30 to 34 in 1950 was 78 percent of the level they had reached in 1960 when they were 40 to 44 years old. Using this percent as a guide, we can compute the fertility at ages 40 to 44 of women who were 30 to 34 years old in 1965. That cohort had borne 2,778 children per 1,000 by ages 30 to 34.[16] The estimated rate of children ever born to these women at ages 40 to 44 would be 3,562 children. This is a rough estimate. Changes in social and economic conditions during the 10-year period could affect the cumulative fertility of the woman.

Childlessness. — The percentage of childlessness among all women, ever-married women, etc., has already been discussed along with the method of standardizing the percentage or proportion childless. Information on this topic is ordinarily obtained from the question on children ever born, but it could be obtained from a direct question or a birth roster.

In demographic usage childlessness may be defined as the state of never having borne a live child. Under this definition a woman whose only pregnancy terminated in death of the child (abortion or stillbirth) would be reported as childless. A woman whose only child died in infancy, however, would not be reported as childless.

Childlessness is an important subject for consideration since it has a relationship to the replacement of population. If a large proportion of women in a given population are childless, the burden of replacement falls on the remaining women in the childbearing ages of the population. Hence childlessness of women is studied in relation to various social and economic characteristics that seem to be associated with it. For example, Grabill and Glick[17] studied data from several decennial censuses and sample surveys on childlessness of married women with husband present, in relation to such characteristics as age at first marriage, educational attainment of woman, major occupation group of husband, labor force status of woman, and family income.

In the study of childlessness, there has been great interest in the proportion that is voluntary and the proportion that is involuntary. In the absence of medical histories of fecundity impairments, which are available for unrepresentative clinical groups, this information may be sought by retrospective questions in sample surveys. The first Growth of American Families Study collected such data for a representative national sample of women, for example.[18] The American data indicate that voluntary childlessness was relatively rare among young couples in the 1950's, but that it had been much more prevalent in the Great Depression of the 1930's.[19] According to the Indianapolis study in 1941 of women married in 1927 to 1929 and under 30 at marriage, 19 percent were childless and half of these voluntarily. Factors in the subsequent decline of childlessness have been changing attitudes, lower age at marriage, and a decline in gonorrhea, which is sterility-producing. Since fecundity impairments tend to increase with age in adult life, earlier marriage exposes the woman to the risk of conception at a time when her fecundity is relatively high.[20] Some demographers have also been interested in the extent of childlessness (as well as in completed size of family) among populations not practicing contraception. Tietze, for example, reported only 2.4 percent childless among 209 Hutterite (a religious sect) women who had married before age 25, spent at least 20 years with the same husband, and had not used contraception.[21] Similarly, the low percent, 3.1, of childlessness among ever-married women 30 to 44 years old in rural-farm Utah in 1960 would seem to reflect the religious influence of Mormonism. Both genealogical histories and census data may be used to measure or approximate the proportion childless in such population subgroups.

Duration-Specific Rates. — Since duration of marriage has so much influence on the number of children ever born, rates specific for this variable are very useful. Without a control on

[16] Wilson H. Grabill and Maria Davidson, "Recent Trends in Childspacing," *Demography*, 5(1):215, table 1, 1968.

[17] Wilson H. Grabill and Paul C. Glick, "Demographic and Social Aspects of Childlessness: Census Data," *Milbank Memorial Fund Quarterly*, 37(1):1–27, January 1959.

[18] Ronald Freedman, Pascal K. Whelpton, and Arthur A. Campbell, *Family Planning, Sterility, and Population Growth*, New York, McGraw-Hill, 1959.

[19] For this earlier experience, we have the evidence of the Indianapolis study as reported in: Pascal K. Whelpton and Clyde V. Kiser, *Social and Psychological Factors Affecting Fertility*, Vol. 2, New York, Milbank Memorial Fund, 1950, pp. 334–340.

[20] Pascal K. Whelpton, Arthur A. Campbell, and John E. Paterson, *Fertility and Family Planning in the United States*, Princeton, N.J., Princeton University Press, 1966.

[21] Christopher Tietze, "Reproduction Span and Rate of Reproduction Among Hutterite Women," *Fertility and Sterility*, 8(1):89–97, Jan.-Feb. 1957.

Table 17-6.—**Number of Children Ever Born per 1,000 White Women 30 and 40 Years Old, by Years Since First Marriage, for the United States: 1960**

[Based on 5-percent sample]

Years since first marriage	Women 30 years old	Women 40 years old	Years since first marriage	Women 30 years old	Women 40 years old
Less than 1 year.	419	677	14 years.........	3,239	2,254
1 year..........	644	714	15 years.........	3,320	2,319
2 years..........	954	689	16 years.........	3,224	2,418
3 years..........	1,307	927	17 years.........	–	2,447
4 years..........	1,517	967	18 years.........	–	2,496
5 years..........	1,803	1,144	19 years.........	–	2,610
6 years..........	1,999	1,465	20 years.........	–	2,754
7 years..........	2,164	1,450	21 years.........	–	2,890
8 years..........	2,351	1,739	22 years.........	–	3,069
9 years..........	2,477	1,623	23 years.........	–	3,303
10 years.........	2,655	1,912			
11 years.........	2,766	1,929	24 years.........	–	3,299
12 years.........	2,901	2,056	25 years.........	–	3,598
13 years.........	3,064	2,120	26 years.........	–	3,331

– Represents zero.

Source: U.S. Bureau of the Census, *U.S. Census of Population: 1960, Subject Reports, Women by Number of Children Ever Born*, PC(2)-3A, 1964, table 21.

Table 17-7.—**Percent Childless and Number of Children Ever Born per 1,000 Women and per 1,000 Mothers for White Women 35 to 44 and 45 to 54 Years Old, Married More than Once, by Age at Marriage, for the United States: 1960**

Age at marriage	35 to 44 years old			45 to 54 years old		
	Percent childless	Children ever born		Percent childless	Children ever born	
		Per 1,000 women	Per 1,000 mothers		Per 1,000 women	Per 1,000 mothers
14 to 17 years....	9.7	2,895	3,206	12.9	2,719	3,122
18 and 19 years...	12.5	2,519	2,879	16.9	2,272	2,733
20 and 21 years...	16.1	2,248	2,679	23.0	1,935	2,514
22 to 24 years....	21.1	1,978	2,508	28.9	1,666	2,344
25 and 26 years...	26.8	1,805	2,466	35.9	1,466	2,286
27 to 29 years....	30.2	1,663	2,382	41.7	1,274	2,183
30 to 44 years....	33.8	1,607	2,427	49.5	1,154	2,287

Source: *U.S. Census of Population: 1960, Subject Reports, Women by Number of Children Ever Born*, PC(2)-3A, 1964, table 18.

current age of woman, however, they tend to reflect mainly the influence of age. Table 17-6 compares women 30 years old with those 40 years old in the United States in 1960, giving the number of children ever born per 1,000 white women by years since first marriage. At both ages, the rate tends to increase with duration of marriage. By subtraction, we can approximate both the age at first marriage and the year of first marriage. Comparing 30-year-old women with 40-year-old women, married the same number of years, we observe the combined influence of the difference in age at marriage (about 10 years) and of the secular conditions to which the two groups of women were exposed during their lives. Forty-year-old women married 20 years were married at about the same age as 30-year-old women married 10 years, but they have been exposed to 10 more years of married life. Since most children are born before the mother is 30, however, the superiority in children ever born per 1,000—2,754 over 2,655 is not very great. In conclusion, these considerations suggest that arranging the data in terms of birth cohorts or marriage cohorts will permit more satisfactory analysis.

Rates Specific for Age at Marriage.—Rates specific for age at (first) marriage may receive similar treatment to those specific for duration of marriage. In the report of the 1960 census of the United States on *Women by Number of Children Ever Born*, the percent childless, the rate per 1,000 women, and the rate per 1,000 mothers are shown by age at first marriage by detailed marital status for women 35 to 44 years old and 45 to 54 years old at the census. Table 17-7 shows such statistics for white women married more than once. With these statistics, we have, in effect, two 10-year cohorts. Within each cohort, we can study the effect of age at first marriage. We can also compare the fertility of the two cohorts for women married at the same age. Since even the 35-year-old women have had most of their children already, the longer duration of marriage is only a minor source of bias in comparing the two age groups, or cohorts. In societies where the childbearing ages tend to be more prolonged, it would be better to compare somewhat older age groups.

Rates Specific for Year of Birth or Year of Marriage.—Just as there are rates of children ever born specific for age of woman, duration of marriage, and age at marriage, rates are also published by year of birth (of woman) or year of marriage. For a given census, of course, year of birth is just another way of expressing age and year of marriage is another way of expressing duration of marriage.

Parity-Specific Rates.—In chapter 16, we defined parity-specific rates in terms of the number of births of order n per 1,000 women of parity $n - 1$. Such measures can be approximated from data that are compiled completely from censuses or surveys. The number of children ever born is equal to the woman's parity, of course. From the census or survey data, we can also determine whether or not the woman had a child under 1 year old living with her. From these two pieces of information, we can estimate the woman's parity 1 year before the census date. The number of children under 1 approximates the number of births in the preceding year. (Adjustments can be made for mortality and for the proportion of children under 1 not living with their mothers.) From the 1950 census of the United States, the data were coded and tabulated so as to yield the rate of first births to women of zero parity in 1949, of second births to women of first parity, etc. The rates were shown by age and other characteristics of the woman.[22] These approximate rates per 1,000 ever-married white women look like this:

Age	First births to zero parity women	Second births to first parity women	Third births to second parity women	Fourth births to third parity women
15 to 19 years.....	261	214	179	(X)
20 to 24 years.....	258	230	175	183
25 to 29 years.....	188	188	133	139
30 to 34 years.....	102	124	80	91
35 to 39 years.....	37	56	40	51
40 to 44 years.....	12	12	12	18

X Not applicable.

Birth and Marriage Cohorts.—From the discussion of the measures in the two preceding subsections, we may surmise that the analysis of fertility trends can be made more incisive if

---

[22] *U.S. Census of Population: 1950, Fertility*, tables 30 and 31.

the statistics are arranged explicitly in terms of birth or marriage cohorts. It is logical, for example, with birth cohorts, to study changes in cumulative fertility with advancing age, and with marriage cohorts, to study changes with increasing duration of marriage. Such cohort analysis cannot be conducted, however, from the statistics on children ever born from a single census or survey. We need to relate the rates for the same cohort from successive censuses or surveys. Information on when each child was born can be obtained from birth rosters, which are the subject of the next section, in order to show the progressive fertility over time of a given cohort.

In the case of women of presumably completed fertility, however, we can make genuine cohort comparisons. If we assume that no females marry before age 15 and that childbearing is completed by age 49, then we can subtract 30 years from the census date to estimate the latest year of marriage that corresponds to completed fertility.

Factors Important in Analysis.—We have already mentioned the importance of marital status, age at marriage, and duration of marriage. These factors provide some rough indication of the number of years of married life during which the woman has been exposed to the risk of having a live birth. Sometimes tabulations are confined to women, "married once, husband present," who represent a group of women with relatively continuous exposure since marriage. Another approach is to take explicit account of the effects of marital dissolution and of remarriage. Lauriat has attempted to make quantitative estimates of these effects using statistics from the 1960 census of the United States.[23] Of course, not all women have been at risk from the date of a marriage to its dissolution. During part of that time, the husband may have been away in the armed forces, for example. Very intensive surveys would be required to estimate fertility per month of marital exposure.

Davis and Blake have discussed intermediate variables through which social factors must operate to influence the level of fertility.[24] Complete or partial sterility during specified periods, frequency of coitus, and prolonged separation within marriage are among the relevant factors.

The number of children ever born has been cross-classified with various social and economic factors in reports based on national censuses and population registers as a means of studying fertility differentials. Particularly in the United States, the study of these relationships has been carried much further in surveys conducted by private organizations. Here the interest has not been restricted to relationships with fertility *per se* but has included mediating variables such as the acceptance, practice, and effectiveness of contraception, as well. The Indianapolis Study, the two studies of the Growth of American Families, and the National Fertility Survey, all previously mentioned, as well as the Princeton "Third Child Study"[25] have made important methodological and analytical contributions here.

Status (as reflected in educational attainment, occupation, or income), farm background, wife's labor force participation, race, and religion have all been found to be highly correlated with various measures of fertility. Demographers and other social scientists have been less successful in finding significant associations with psychological factors when social and economic factors are held constant. Some of them feel, however, that the problem lies in their failure to discover meaningful and practicable measures of underlying psychological factors.

BIRTH ROSTERS

A birth roster is a list of children born to a woman with their birth dates. Thus, a birth roster for a woman who has borne four children includes birth dates of her first, second, third and fourth children. The children to be listed are identical with those who would be counted for the number of children ever born (q.v.). A roster may also include information on the sex of each child, on whether it is living at home or elsewhere, on whether it is deceased, and on the woman's date of marriage (or marriages).

There may be complete or truncated birth rosters. Complete rosters include the birth dates of all children ever born to a woman. Truncated birth rosters are often used to save space on the schedule, or money, or both. Truncated rosters may, for example, include birth dates of the first four children and the last (of fifth or higher order) or the first four and the next to last and the last, or some other combination.

Sources

Birth rosters are usually collected from national surveys, or censuses, but they could be compiled from a population register. An early example of birth rosters obtained from a large-scale survey is that of the 1946 Family Census in Great Britain.[26] Other examples are the Current Population Survey of August 1959 and of June 1965 in the United States.[27]

Approximations from Household Rosters

This method makes use of rosters of own children living in the same household with their mother at the time of a census or a survey. The children are listed in order by age or date of birth. Such birth rosters are deficient to the extent that they include only own children at home, whereas the birth dates of other children born to the woman who are either absent or deceased have to be estimated. The total number of children ever born also has to be known.

Such birth rosters were constructed from the household rosters in the 1960 census of the United States, with the birth dates of all missing children having been allocated.[28] In order to allocate children absent from home, the number of absent children had to be determined by subtracting the number of own children present at home from the number of children ever born to the woman. Next, allocations were made of the birth order of absent children. Then, allocations were made of the spacing of each absent child in relation to a prior child or to the mother's marriage date to obtain the birth date of the child.

[23] Patience Lauriat, "The Effect of Marital Dissolution on Fertility," *Journal of Marriage and the Family*, 31(3):484–493, August 1969.

[24] Kingsley Davis and Judith Blake, "Social Structure and Fertility: An Analytic Framework," *Economic Development and Cultural Change*, 4:211–235, 1955–56.

[25] Charles F. Westoff et al., *Family Growth in Metropolitan America*, Princeton, N.J., Princeton University Press, 1961.

[26] D. V. Glass and E. Grebenik, *The Trend and Pattern of Fertility in Great Britain, A Report on the Family Census of 1946*, London, Royal Commission on Population, Pt. I: *Report* and Pt. II: *Tables*, 1954.

[27] U.S. Bureau of the Census, *Current Population Reports*, Series P–20, No. 108, "Marriage, Fertility, and Childspacing: August 1959," July 12, 1961; idem, No. 186, "Marriage, Fertility, and Childspacing: June 1965," Aug. 6, 1969.

[28] *U.S. Census of Population: 1960, Subject Reports, Childspacing*, PC(2)–3B, 1968.

Table 17–8. — Number of Children Ever Born per 1,000 White Women by Specified Duration of Marriage and by Marriage Cohort, for the United States: 1959

Marriage cohort	Duration of marriage in completed years								
	1 year	2 years	3 years	4 years	5 years	7 years	10 years	15 years	20 years
1950–1954......................	348	692	1,016	1,319	1,596	2,069	-	-	-
1945–1949......................	329	654	932	1,190	1,427	1,833	2,284	-	-
1940–1944......................	244	529	760	971	1,178	1,549	1,952	2,381	-
1935–1939......................	255	527	746	949	1,140	1,469	1,833	2,229	2,434
1930–1934......................	242	510	732	933	1,096	1,368	1,713	2,118	2,325
1925–1929......................	274	560	790	994	1,180	1,470	1,787	2,162	2,375
1920–1924......................	300	594	849	1,081	1,280	1,614	1,982	2,381	2,594
1910–1919......................	338	665	939	1,183	1,413	1,796	2,239	2,745	3,016
1900–1909......................	315	654	955	1,226	1,468	1,916	2,463	3,168	3,600

— Represents zero.

Source: Derived from U.S. Bureau of the Census, *Current Population Reports*, Series P–20, No. 108, "Marriage, Fertility, and Childspacing: August 1959," July 12, 1961, table 16.

Uses and Measures

In discussing children ever born, we mentioned several types of measures that could be derived from birth rosters. These were characteristically completed fertility rates at each year of age for a birth cohort or for each year of married life for a marriage cohort. We should also consider the application to child spacing, that is the measurement of the time interval between date of marriage and the date of birth of a given order or between successive births.

Cohort Measures.—We can compare not only the cumulative fertility of a birth cohort after successive ages but also that of different birth cohorts at the same age.

Table 17–8 presents data for American cohorts by duration of marriage. Subtracting successive entries along the rows of this table, one can obtain duration-specific (not cumulative) fertility rates for successive years or periods of married life. Comparing two cohorts, one cohort married in the depression (1930–34) and the other cohort married just after the end of the Second World War (1945–49), we may notice the difference in family building patterns. The marriage cohort of 1930–34 had a birth rate of 242 per 1,000 women in its first year of marriage but a rate of only 163 (1,096 minus 933) per 1,000 women in the fifth year of marriage, whereas the 1945–49 marriage cohort had an initial birth rate of 329 per 1,000 women and a rate of 237 (1,427 minus 1,190) per 1,000 women in the fifth year. The cohort married in the more prosperous years was continuing its advantage, year by year, over that married in the depression as far as we are able to compute rates by single years of duration of marriage.

Such duration-specific fertility rates are helpful in studying the way in which families are built up over the period of married life and in distinguishing between changes that affect the completed family size and those changes that leave the completed size unchanged but that influence the speed with which families are completed.

Parity-Specific Rates.—In the discussion of measures based on the number of children ever born, we showed how parity-specific fertility rates could be approximated when the number of the woman's own children under 1 is known. The same kind of information is available from birth rosters. In fact, the dates of births permit the estimation of such rates for a series of years, on a retrospective basis, and not just for the year preceding the census. From the Current Population Survey of August 1959, the U.S. Bureau of the Census has published parity-specific rates for 5-year periods from 1930–34 to

1955–59.[29] The women surviving to 1959 may be a biased sample, with respect to fertility, of all women from the same cohorts who were alive in these past years. Approximate rates can be estimated from the ages of own children still living in the household. The U.S. Bureau of the Census labels these "presumed births of order '$N+1$' per 1,000 women of parity 'N'."[30]

Childspacing.—As we have mentioned, childspacing can be measured from birth rosters. It can also be studied from vital records where the birth certificate gives not only the order of and date of birth but also the date of marriage. It is very rare that we also have the date of the preceding birth in vital records, however.

Childspacing is measured by the length of the interval between first marriage and first birth and also between successive births. The unit of measurement of childspacing is preferably the month or the quarter. The year is a rather gross unit for this purpose. The statistics used for summarizing the length of interval can be the arithmetic mean or the median.

Let us consider, first, measurement from census or survey statistics. Table 17–9 is based on data of the Current Population Survey of August 1959 for the United States and gives intervals between marriage and first birth for various marriage cohorts. The median interval is also shown between marriage and second and third births, respectively. Both the distribution by number of months and the median number are given. To obtain the median length of interval between first and second births, we subtract the median length of interval between marriage and first birth from the median length of interval between marriage and second birth. The data in the following illustration are for the marriage cohort 1935–39.

62.2 months − 21.1 months = 41.1, the median interval between first and second births

Similarly, 96.9 months − 62.2 months = 34.7, the median interval between second and third births

Limitations and Quality

The remarks made concerning the limitations of statistics on children ever born when used to measure various aspects of fertility, and on their accuracy, tend to apply to birth rosters

---

[29] *Current Population Reports*, Series P–20, No. 108, "Marriage, Fertility, and Childspacing: August 1959," table 28.
[30] *Current Population Reports*, Series P–20, No. 184, "Women by Number of Own Children Under 5 Years Old: 1968 and 1967," June 16, 1969, tables 4 and 5.

Table 17-9. — Number of Children Ever Born by Order of Birth per 1,000 Women Ever Married, by Succssive Intervals Since First Marriage, for White Women Who First Married in 1900 to 1939, for the United States: August 1959

[Noninstitutional population]

Order of birth and interval since first marriage	Year of first marriage of white woman					
	1935 to 1939	1930 to 1934	1925 to 1929	1920 to 1924	1910 to 1919	1900 to 1909
FIRST CHILDREN EVER BORN PER 1,000 WOMEN						
6 months...................	51	54	44	48	55	45
7 months...................	68	69	56	59	69	57
8 months...................	86	90	81	83	89	74
9 months...................	118	123	122	127	139	133
10 months..................	160	166	173	189	203	205
12 months..................	233	225	256	285	308	298
14 months..................	286	278	311	348	375	368
16 months..................	328	323	357	395	431	415
18 months..................	371	365	394	433	472	466
20 months..................	415	398	439	473	511	510
22 months..................	448	430	476	507	547	550
24 months..................	473	465	504	538	589	596
30 months..................	545	531	574	610	662	675
36 months..................	594	589	623	656	700	716
48 months..................	661	662	687	714	753	774
60 months..................	715	708	733	749	793	801
84 months..................	774	753	778	783	827	834
120 months.................	816	792	806	809	849	855
180 months.................	840	822	826	828	864	887
240 months.................	847	828	834	834	865	891
241 months and over........	850	831	839	838	869	898
MEDIAN INTERVAL BETWEEN MARRIAGE OF MOTHER AND BIRTH OF ... CHILD						
First child..............	21.1	21.6	19.6	17.8	16.7	17.8
Second child.............	62.2	62.1	56.7	49.6	48.2	48.5
Third child..............	96.9	100.6	90.1	80.0	79.6	79.2

Source: U.S. Bureau of the Census, *Current Population Reports*, Series P-20, No. 108, "Marriage, Fertility, and Childspacing: August 1959," July 12, 1961, table 16.

as well. On the one hand, the listing of each birth probably produces a more accurate count of children ever born; on the other hand, dating the birth of each child is undoubtedly difficult for some respondents.

EXPECTED AND IDEAL SIZE OF FAMILY

In a fairly large number of sample surveys in the United States, Puerto Rico, and various European and Asian countries, attitudinal questions have been included, addressed to the wife and sometimes to the husband, regarding the total number of children expected, the number regarded as ideal, etc. The gap between the number of children a couple would like to have and the number it expects to have may indicate the potential for family planning programs on the basis of present attitudes. Attempts have been made to use statistics on the expected size of completed family as a guide in projecting the future fertility of the cohorts concerned. One of the earliest uses of such questions was in a fertility survey carried out in Puerto Rico in 1947-48 by the Office of Population Research,

Princeton University, and the Social Science Research Center, University of Puerto Rico.[31]

The Special Demographic Survey following the 1966 census of South Korea asked for the ideal numbers of sons and of daughters. The two studies on the Growth of the American Family, the National Fertility Survey, and The Third Child Study—all previously mentioned—also included questions of this character.

Some studies attempt to distinguish among the ideal, the expected, the intended, and the wanted number or some selection of these. For some further particulars, see contributions by Pressat and by Freedman and Coombs.[32]

PATERNAL FERTILITY

Several fertility measures such as number of own children under 5 years old, number of children ever born, and reproduction rates have been computed from census and survey data for men as well as for women. In theory, all of the measures described in this chapter—or some analogue taking account of biological differences between the sexes—could be computed if the appropriate data were collected. Studies in fertility seldom use fertility rates for men, however. There are several reasons for preferring maternal fertility rates. In the first place, the reproductive span for women has well defined limits, whereas that for men is rather vague. Male fecundity declines rather gradually within the older ages. It is much easier to interview women because women are more likely to be at home, and there is a greater likelihood that the children will be living with the mother than the father in the case of broken marriages, consensual marriages, and illegitimacy.

There may be advantages, however, in employing paternal rates for particular purposes. Tietze, for example, pointed out that differential fertility between occupational groups can be studied better in terms of paternal rates.[33] When maternal rates are used, women are classified by occupation of husband; the rates are confined to married women, husband present; and it is impossible to obtain general fertility rates, which would reflect differences in proportions married at each age among occupational classes.

In some societies, certain socioeconomic factors such as labor force status or literacy are more useful in studying fertility differentials for women than for men because practically all men are in the labor force or are literate but the proportions are more moderate for women. In most societies, too, the average age at marriage is later for men and the relative age-specific profiles of fertility rates are affected by this factor.

[31] Paul K. Hatt, *Backgrounds of Human Fertility in Puerto Rico*, Princeton, N.J., Princeton University Press, 1952.

[32] R. Pressat, "The Ideal and the Actual Number of Children," in *Studies on Fertility and Social Mobility*, Egon Szabady, ed., Budapest, Hungarian Academy of Sciences, 1964, pp. 45–56; R. Freedman and L. C. Coombs, "Expected Family Size and Family Growth Patterns: A Longitudinal Study," in: *World Views of Population Problems*, Egon Szabady, ed., Budapest, Hungarian Academy of Sciences, 1968, pp. 83–95.

[33] Christopher Tietze, "The Measurement of Differential Reproduction by Paternity Rates," *Eugenics Review*, 30(2):101–107, July 1938; idem, "Differential Reproduction in England," *Milbank Memorial Fund Quarterly*, 17(3):293, July 1939; idem, "Differential Reproduction in the United States: Paternity Rates for Occupational Classes Among the Urban White Population," *American Journal of Sociology*, 49(3):242–247, November 1943.

SUGGESTED READINGS

Cho, Lee-Jay, Grabill, Wilson H., and Bogue, Donald J. *Differential Fertility in the United States*. Chicago, Community and Family Study Center, University of Chicago, 1970.

Cho, Lee-Jay. "The Own-Children Approach to Fertility Estimation: An Elaboration." International Union for the Scientific Study of Population, International Population Conference, Lige, 1973. Vol. 2, pp. 263–279.

Freedman, R., and Coombs, L. C. "Expected Family Size and Family Growth Patterns: A Longitudinal Study." *World Views of Population Problems*. Egon Szabady, ed. Budapest, Hungarian Academy of Sciences, 1968. Pp. 83–95.

Freedman, Ronald, Whelpton, Pascal K., and Campbell, Arthur A. *Family Planning, Sterility, and Population Growth*. New York, McGraw-Hill, 1959. Pp. 10–16; 431–443; 478–480.

Glass, D. V., and Grebenik, E., *The Trend and Pattern of Fertility in Great Britain, A Report on the Family Census of 1946*. Part I: *Report* and Part II: *Tables* London, Royal Commission on Population, 1954.

Grabill, Wilson H., and Davidson, Maria. "Recent Trends in Childspacing." *Demography*, 5(1):212–225. 1968.

Grabill, Wilson H., and Glick, Paul C. "Demographic and Social Aspects of Childlessness: Census Data." *Milbank Memorial Fund Quarterly*, 37(1):1–27. January 1959.

Grabill, Wilson H., Kiser, Clyde V., and Whelpton, P. K. *The Fertility of American Women*. New York, John Wiley & Sons, 1958. Especially pp. 41–50 and 400–419.

Henry, Louis. "Etude statistique de l'espacement des naissances." *Population* (Paris) 6(3):425–444. July–September 1951.

Japan. Institute of Population Problems. *Summary of the Fertility Surveys in 1940, 1952, 1957, and 1962*. By Hisao Aoki and Eiko Nakano. Research Series No. 177. July 1, 1967.

Kiser, Clyde V., Grabill, Wilson H., and Campbell, Arthur A. *Trends and Variations in Fertility in the United States*. Cambridge, Mass., Harvard University Press, 1968. Especially pp. 297–310.

Lauriat, Patience. "The Effect of Marital Dissolution on Fertility." *Journal of Marriage and the Family*, 31(3):484–493. August 1969.

Liu, P. K. C., "Differential Fertility in Taiwan." International Union for the Scientific Study of Population. *Contributed Papers. Sydney Conference*. Sydney (Australia). August 21 to 25, 1967. Pp. 363–370.

Malaysia. Department of Statistics. *Estimates of Fertility for West Malaysia (1957–1967)*, Kuala Lumpur, June 1969, Research Paper No. 3, by Lee Jay Cho.

Netherlands. *Results of the Demographic Research Project Western New Guinea*. Part G, *The Progeny of Papuan Women*. 1967.

Pressat, R. "The Ideal and the Actual Number of Children." *Studies on Fertility and Social Mobility*. Egon Szabady, ed. Budapest, Hungarian Academy of Sciences, 1964. Pp. 45–56.

Ryder, Norman B., and Westoff, Charles F. "Relationships among Intended, Expected, Desired, and Ideal Family Size: United States, 1965." *Population Research*. Center for Population Research, National Institute of Child Health and Human Development, March 1969.

Thompson, Warren S., and Whelpton, P. K. *Population Trends in the United States*. New York, McGraw-Hill, 1933. Pp. 262–291.

Tietze, Christopher. "Differential Reproduction in the United States: Paternity Rates for Occupational Classes among the Urban White Population." *American Journal of Sociology*, 49(3):242–247. November 1943.

United Nations. *Demographic Yearbook, 1965*. Pp. 22–24; 230–261.

———. *Handbook of Population Census Methods*. Vol. III, *Demographic and Social Characteristics of the Population*. Studies in Methods. Series F, No. 5, rev. 1. 1959. Pp. 33–44.

U.S. Bureau of the Census. *Current Population Reports*. Series P–20, No. 46. "Fertility of the Population: April 1952." Dec. 31, 1953.

———. *Current Population Reports*. Series P–20, No. 108. "Marriage, Fertility, and Childspacing: August 1954." July 12, 1961.

———. *Current Population Reports*. Series P–20, No. 184. "Women by Number of Own Children under 5 Years Old: 1968 and 1967." June 16, 1969.

———. *Current Population Reports*. Series P–20, No. 186. "Marriage, Fertility, and Childspacing: June 1965." Aug. 6, 1969.

———. *Ratio of Children to Women: 1920*. By Warren S. Thompson. Census Monograph XI. 1931. Especially pp. 15–17.

———. *Sixteenth Census of the United States: 1940. Population. Differential Fertility, 1940 and 1910. Standardized Fertility Rates and Reproduction Rates*. 1944.

———. *U.S. Census of Population: 1950. Special Reports*. Series P–E, No. 5C, *Fertility*. 1955.

———. *U.S. Census of Population: 1960. Subject Reports. Women by Number of Children Ever Born*, PC(2)–3A. 1964.

———. *U.S. Census of Population: 1960. Subject Reports. Childspacing*, PC(2)–3B. 1968.

SUGGESTED READINGS — Continued

————. *U.S. Census of Population: 1960. Subject Reports. Women by Number of Children Under 5 Years Old*, PC(2)-3C. 1968.

Westoff, Charles F., et al. *Family Growth in Metropolitan America*. Princeton, N.J., Princeton University Press, 1961.

Whelpton, Pascal K., Campbell, Arthur A., and Patterson, John E. *Fertility and Family Planning in the United States*. Princeton, N.J., Princeton University Press, 1966. Especially pp. 1–31; 416–419.

Whelpton, Pascal K., and Kiser, Clyde V. *Social and Psychological Factors Affecting Fertility*. Vol. 2. New York, Millbank Memorial Fund, 1950. Pp. 334–340.

CHAPTER 18

Reproductivity

GENERAL CONCEPT

The study of reproductivity is concerned with the extent to which a group is replacing its own numbers by natural processes. Measures of reproductivity or population replacement are thus measures of natural increase expressed in terms of a generation rather than a year or other brief period of time. The group may be a true or a synthetic birth cohort, or a true or a synthetic marriage cohort. The analysis of reproductivity has led mathematical demographers to the concept of the stable population and of its vital measures. Originally viewed as indicating the inherent or ultimate results of current fertility and mortality, these concepts have been found to have much wider applications in the estimation of current vital measures and population structure and in many types of demographic analysis.

Certain measures on an annual, instead of a generation basis, are sometimes considered simple measures of reproductivity. Among these are the crude rate of natural increase and the vital index. Natural increase was defined in chapter 13 as the difference between the number of births and the number of deaths. The **crude rate of natural increase** is thus the (algebraic) excess of births over deaths per 1,000 of the population or the difference between the crude birth rate and the crude death rate.

This rate can be expressed as

$$r_n = \frac{B-D}{P} \times 1,000 \qquad (1)$$

$$= b - d \qquad (2)$$

where r_n is the rate of natural increase, B = births during a calendar year, D = deaths during a calendar year, P = midyear population, b = the birth rate, and d = the death rate.

The **vital index**, a term coined by Raymond Pearl, is the ratio of the number of births to the number of deaths, times 100. In other words,

$$VI = \frac{B}{D} \times 100 \qquad (3)$$

One of the few virtues of this index is that it can be computed for an area for which postcensal population estimates are not available. It does indicate the extent to which the force of natality exceeds that of mortality at a given time.

The rate of natural increase is the most direct indication of how rapidly a given population actually grew during a given year as the result of vital processes. If births exceed deaths, there is growth and the rate is positive. If deaths outnumber births, the population fails to increase during the year, and the rate is negative. The crude rate of natural increase, like its two components, is influenced by the current age structure. If, for example, a relatively large proportion of a population is in adolescence and early adulthood, its population will tend to have a relatively high birth rate and a relatively low death rate; these will result in a high crude rate of natural increase. If, however, a relatively small proportion of the population is within these ages, its rate of natural increase tends to be relatively low. Since the age composition of a population at a given time is determined by the previous trend and fluctuations of its birth and death rates and by its migration history during the same span, any measure of reproductivity that fails to take account of the age composition is an inadequate measure of the long-term replacement tendencies in that population. For example, after a devastating war, the sex ratio of the population in the reproductive ages may be abnormally low so that the crude rate of natural increase will be affected for some years thereafter.

The actual experience of a generation may be observed over a period of some 30 years, or the experience may be synthesized from current data. For many years, reproduction rates were computed only from current vital statistics because it was thought that the fertility of women in, say, 30 successive years of age in the childbearing period in a single calendar year could be combined to approximate the fertility and reproductivity of an actual cohort of women passing through the childbearing period. In the late 1940's, demographers started to realize that these synthetic rates were frequently not reliable measures of reproductivity and began to observe actual cohorts.

CONVENTIONAL REPRODUCTION RATES

Conventional reproduction rates are measures of lifetime reproductivity that are based on the experience of a single year or other short period. Reproductivity is the net force of fertility and mortality, and the net reproduction rate measures this force. We also have to consider, however, two related measures that represent limiting cases in which mortality is assumed to be nil until the end of the childbearing period. These are the gross reproduction rate and the total fertility rate.

313

Total Fertility Rate

As the reader may recall from chapter 16, the total fertility rate is the sum of the age-specific birth rates of women over their reproductive span, as observed in a given year. It is expressed as

$$TFR = \sum_{\omega_1}^{\omega_2} \frac{B_x}{P_x} \times 1,000 \qquad (4)$$

where B_x is the number of live births registered during the year to mothers of age x, where "x" represents an interval of one year of age, and P_x is the midyear population of women of the same age. The age-specific fertility rates are cumulated from $x = \omega_1$ to $x = \omega_2$, where ω_1 to ω_2 may be 10 to 54, 15 to 44, or whatever range is available and appropriate.

In usual practice, the total fertility rate is calculated by a shorter method. The specific birth rates are calculated for 5-year age groups and thus the subscript i represents 5-year intervals such as 15 to 19, 20 to 24, and 45 to 49 years. The expression for total fertility, using seven 5-year age groups, for example is then

$$TFR = 5 \sum_{i=1}^{7} \frac{B_i}{P_i} \times 1,000 \qquad (5)$$

where B_i is the number of live births registered during the year to mothers of age group i, i is an interval of 5 years, and P_i is the mid-year population of women of the same age. The purpose of the factor 5 in formula (5) is to apply the average rate for the age group to five successive single years so that the sum of the age-specific rates will be commensurate with that in (4).

When, for convenience, we use seven 5-year age groups in our calculations, the age limits will not encompass the full childbearing range. Births to younger women should then be allocated to the youngest age group shown and those to older women to the oldest age group shown. For example, any births registered for women aged 52 would be allocated to age 49, or to the age group 45 to 49. Likewise, births for which the age of mother was not reported should be allocated over the childbearing ages.

The total fertility rate states the number of births 1,000 women would have if they experienced a given set of age-specific birth rates throughout their reproductive span. A rate of 5,602 for Peru in 1961 (table 18–1), for example, means that if a hypothetical group of 1,000 women were to have the same birth rates at each single year of age as were observed in Peru in 1961, they would have a total of 5,062 children by the time they reached the end of the reproductive period, taken as 49 years old, assuming all of them survived to that age. The procedure for computing the total fertility rate is shown in the table.

Reproduction Rates

The conventional reproduction rate (called *taux classique de reproduction* by French demographers) measures the replacement of the female population only, whereas the total fertility rate involves births of both sexes. The net reproduction rate was devised by the German statistician, Richard Böckh in 1884.[1] It was vigorously advocated by his pupil, Robert R. Kuczynski, in many publications, his most definitive discussion

Table 18–1.—**Computation of Total Fertility Rate, for Peru: 1961**

Age	Number of women (1)	Number of births (2)	Births per 1,000 women [(2)÷(1)] x1,000= (3)
15 to 19 years..................	480,162	[1]38,537	80.3
20 to 24 years..................	428,007	100,210	234.1
25 to 29 years..................	381,755	94,573	247.7
30 to 34 years..................	308,741	62,822	203.5
35 to 39 years..................	279,640	43,192	154.5
40 to 44 years..................	211,469	14,745	69.7
45 to 49 years..................	187,480	[2]4,240	22.6
Summation..............		[3]358,318	1,012.4
Total fertility rate 5 x Σ col. (3).............			5,062.0

[1] Includes births to women under 15 years of age.
[2] Includes births to women 50 years of age and over.
[3] Includes 4,973 births to mothers of unknown age that have been prorated to the age groups shown.

Source: United Nations, *Demographic Yearbook, 1965*, tables 6 and 13.

having been given in *The Measurement of Population Growth*.[2] Kuczynski himself computed many such reproduction rates and compiled those that had been computed by others. As will be described later, we are indebted to Lotka for the first rigorous mathematical analysis of these rates as measures of reproductivity and of their place in a whole set of measures relating to intrinsic population change and the stable population.[3]

Gross Reproduction Rate

The gross reproduction rate (GRR) is a special case of the total fertility rate. Whereas the total fertility rate measures the total number of children a cohort of women will have, the gross reproduction rate measures the number of daughters it will have. Thus, a total fertility rate can be converted to a gross reproduction rate simply by multiplying it by the proportion of the total births that were female births in the calendar year or years for which it is computed. This conversion is not mathematically exact, but it is a close approximation to the gross reproduction rate we should obtain if births at each age of woman were tabulated by the sex of the child. As was mentioned in chapter 7, there is a slight variation of the sex ratio at birth with the age of the mother. The gross reproduction rate also assumes that all females survive to the end of the childbearing period. The formula for the gross reproduction rate is:

$$GRR = \frac{B^f}{B^t} \sum_{\omega_1}^{\omega_2} \frac{B_x}{P_x} \cdot k \qquad (6)$$

where B_x is the number of infants born to mothers of age x, P_x is the number of women of age x in the midyear population, ω_1 and ω_2 are, respectively, the lower and upper limits of the childbearing period, the fraction B^f/B^t is the proportion of total births that are female, and k is a constant—unity (1), or 100, or 1,000.

If data are available on the number of *female* births by age of mother, the GRR can be calculated directly without the B^f/B^t adjustment as follows:

$$GRR = \sum_{\omega_1}^{\omega_2} \frac{B_x^f}{P_x} \cdot k \qquad (6a)$$

---

[1] In *Statistisches Jahrbuch der Stadt Berlin, 1884*, pp. 30–34, and "Die statistische Messung der ehelichen Fruchtbarkeit," *Bulletin de l'Institut international de statistique*, Vol. V First Section, 1890, pp. 165–166.

[2] New York, Oxford University Press, 1936.

[3] Louis I. Dublin and Alfred J. Lotka, "On the True Rate of Natural Increase." *Journal of the American Statistical Association*, 20(151):305–339, September 1925.

where B_x^f is the number of *female* infants born to mothers of age x.

If the computation uses 5-year age groups, (6) becomes

$$\text{GRR} = 5 \frac{B^f}{B^t} \sum_{i=1}^{7} \frac{B_i}{P_i} \cdot k \qquad (7)$$

The gross reproduction rate is such an important demographic measure that we should realize that it may be viewed and defined several ways, namely, as (1) an age-standardized fertility rate with each age given a weight of one, (2) the average number of daughters that a group of females starting life together would bear if all the initial group of females survived the childbearing period, (3) the ratio between the number of females in one generation at, say, age 15 and the number of their daughters at the same age, if there were no mortality during the childbearing period, or (4) the ratio between female births in two successive generations assuming no deaths before the end of the childbearing period. The last three definitions are phrased as if the measure applied to a real cohort, or pair of cohorts; but any of these interpretations can be used regardless of whether the GRR was computed from synthethic cohort data or data compiled over the lifetime of a real cohort.

Our initial definition of reproductivity stated that it is the natural increase of a population over a generation and hence implied that it was the resultant of mortality and fertility over this period. The gross reproduction rate thus represents a limiting case of this concept where mortality through the end of the childbearing period is zero.

Net Reproduction Rate

The net reproduction rate (NRR) is a measure of the number of daughters that a cohort of newborn girl babies will bear during their lifetime assuming a fixed schedule of age-specific fertility rates and a fixed set of mortality rates. Thus, the net reproduction rate is a measure of the extent to which a cohort of newly born girls will replace themselves under given schedules of age-specific fertility and mortality. Some girls will die before attaining the age of reproduction, others will die during the reproductive span, and others will live to complete their reproductive life. Expressed in symbols we have

$$\text{NRR} = \frac{B^f}{B^t} \sum_{\omega_1}^{\omega_2} \frac{B_x}{P_x} \cdot \frac{L_x}{l_o} \qquad (8)$$

where $B_x / P_x = f_x$ the age-specific fertility rate at age x and L_x / l_o is a life-table survival rate, with l_o being the radix of the life table or 100,000. Again when 5-year age groups are used, we have

$$\text{NRR} = 5 \frac{B^f}{B^t} \sum_{i=1}^{7} \frac{B_i}{P_i} \cdot \frac{{}_5 L_x}{l_o}$$

$$= \frac{B^f}{B^t} \sum_{i=1}^{7} \frac{B_i}{P_i} \cdot \frac{{}_5 L_x}{100,000} \qquad (9)$$

If the survivors in each age interval are assumed to bear daughters as specified by a current schedule of age-specific birth rates, then to obtain the net reproduction rate one multiplies each value ${}_n L_x$ by the corresponding age-specific birth rate for baby girls, sums the cross-products, and divides by 100,000.

A rate of 1.00 (or 100 or 1,000, depending on the value of k)

means exact replacement, a rate above unity indicates that the population is more than replacing itself, and a rate below unity means that the population is not replacing itself. The parallel interpretations to those given for the gross rate are: (2) the average number of daughters borne by a group of females starting life together, (3) the ratio between the number of females in one generation at, say, age 15, and the number of their daughters at the same age, and (4) the ratio between female births in two successive generations. As previously indicated the net reproduction rate is frequently abbreviated NRR, but the usual symbol in mathematical demography is Lotka's R_o. The rationale of R_o is made clear in the next section.

The procedure for computing the gross and net reproduction rates is shown in table 18-2. From statistics on births and population in the *Demographic Yearbook,* we have computed age-specific birth rates for Peru, 1961. For convenience, the age-specific rates are computed from births of both sexes, so we use the symbol ${}_5 f_x^T$. The transition to female births only is made later, in a single step, by applying the proportion of all births that are female. Also for convenience, we have grouped our data in 5-year age groups. In this form the figures in column (2) do not look like survival rates because the survival rates have been cumulated over 5 years. We could have used instead the female survival rate to the midpoint of each age interval $\frac{l_{17.5}}{l_o}$, $\frac{l_{22.5}}{l_o}$, etc. For example, the survival rate to age 17.5 is $\frac{77,672}{100,000} = 0.77672$. In this approach, the sum of column (3) would need to be multiplied by 5 to produce the net reproduction rate. From the female life table for 1961, we read off, or sum, the values of L_x for x taken successively at

Table 18-2.—**Computation of Gross and Net Reproduction Rates, for Peru: 1961**

Age of mother	Age-specific birth rate per female $_5 f_x^T$ (1)	$\frac{_5 L_x^F}{\ell_{\bar{o}}^F}$ (2)	$_5 f_x^T \cdot \frac{_5 L_x^F}{\ell_o^F}$ (1) x (2)= (3)
15 to 19 years	.08026	3.88727	.31199
20 to 24 years	.23413	3.79787	.88920
25 to 29 years	.24773	3.68824	.91369
30 to 34 years	.20348	3.57334	.72710
35 to 39 years	.15446	3.45368	.53346
40 to 44 years	.06973	3.32597	.23192
45 to 49 years	.02262	3.17970	.07192
Summation	1.01241	(X)	3.67928

$$\text{GRR} = 5 \frac{B^F}{B^T} \Sigma_5 f_x^T = 5(.48439)(1.01241) = 2.45201$$

$$\text{NRR, or } R_o = 5 \frac{B^F}{B^T} \Sigma_5 f_x^T \cdot \frac{_5 L_x^F}{5\ell_o^F}$$

$$= \frac{B^F}{B^T} \cdot \Sigma_5 f_x^T \frac{_5 L_x^F}{\ell_o^F}$$

$$= (.48439)(3.67928) = 1.78221$$

X Not applicable.

Source: Table 18-1 and Eduardo E. Arriaga, "New Abridged Life Tables for Peru: 1940, 1950-51, and 1961," *Demography,* 3(1):226, 1966; and United Nations, *Demographic Yearbook, 1965,* table 22.

ages 15, 20, 25, . . . 45. Since these are for 100,000 women, we divide them by 100,000 to obtain values per woman (the 3.88727 years for the age group 15 to 19 is equivalent to 388,727 years lived by a cohort of females who number 100,000 at birth).

The specific steps are as follows:

Step 1. Compute the age-specific birth rate per female, and enter it in column (1).

Step 2. From an appropriate life table, find the number of years lived by females in the stationary population for each interval of age. In an abridged life table, $_5L_x$ will be given directly. (In a life table by single years of age, $_5L_x$ can be obtained by summing five successive values of L_x or by subtracting T_{x+5} from T_x. See chapter 15.)

Step 3. Divide $_5L_x$ by 100,000, and enter the result in column (2).

Step 4. Multiply each entry of column (2) by the corresponding entry of column (1) and enter the result in column (3).

Step 5. To obtain the gross reproduction rate, multiply the sum of the entries in column (1) by the proportion of births that are female (.48439) and then by the factor 5.

Step 6. To obtain the net reproduction rate, apply the factor .48439 to the sum of column (3). These rates are expressed per woman, but they could also have been per 100 women or per 1,000 women.

It is important to keep in mind that, in the conventional reproduction rates, age-specific fertility and mortality rates all refer to the same fixed period of time, say a given calendar year or the average of several years. Moreover, if a life table is not available for exactly the same year as that of the fertility schedule, approximate reproduction rates for that year may be obtained by using a life table for an adjacent year or years. There is certainly no need to insist on identity of the two time references if the basic population, fertility, and mortality statistics are not very accurate.

These reproduction rates are defined in terms of the female population, but analogous rates can also be computed for both sexes combined or for the male population only. (See section on "Reproduction of the Male Population" below.) Reproductivity is usually studied in terms of mothers and daughters because the fecund period for females is shorter than it is for males and because characteristics (age, etc.) are much more likely to be known for the mothers of illegitimate babies than for their fathers.

The ratio of the net reproduction rate to the gross reproduction rate is called the **reproduction-survival ratio.** It is the proportion of potential reproductivity that survives the effects of mortality.[4]

THE STABLE POPULATION: ITS VITAL RATES AND OTHER CHARACTERISTICS

The assumption of the constancy of age-specific birth rates and death rates for an indefinite period, which is made in the calculation of the gross reproduction rate and the net reproduction rate, defines the basic conditions of a general theoretical

model. Lotka developed a model that defined the age composition implicit in the given vital rates and expressed reproductivity on an annual as well as on a generation basis. This model is designated the "stable population" model. Lotka demonstrated in 1907 that, if a population is subject to a fixed schedule of age-specific fertility rates and a fixed schedule of age-specific mortality rates for an indefinte period of time, and, if meanwhile there is no migration, ultimately the age composition of the population would assume a fixed characteristic distribution.[5] Coale has investigated the time required for a population with a given age structure and age-specific fertility and mortality to approach its ultimate stable form.[6]

In 1925 Lotka proved that a closed population with constant age-specific fertility and mortality schedules would eventually have a constant rate of natural increase.[7] Lotka called this rate the **true rate of natural increase**. It has also been called the **intrinsic rate of natural increase** or the **implicit rate of natural increase**. (The **life table** or **stationary population** can be viewed as a stable population with a natural increase of zero.)

Intrinsic Rate of Natural Increase

Lotka computed the true rate of natural increase by solving the equation,

$$\int_0^\infty e^{-rx} f(x) p(x) dx = 1 \tag{10}$$

where $p(x)$ is the probability of surviving from birth to age x, r is the intrinsic rate of increase per head per annum, and $f(x)$ is the number of live female births per annum to each woman of age x. $p(x)$ is the L_x of the life table divided by l_o. Since $f(x)$ is zero outside the childbearing period, we could substitute the limits ω_1 and ω_2, or approximately 15 to 49, as the limits of the definite integral.

In practice, a very close approximation to the real root of equation (10) is given by the quadratic equation,

$$\frac{1}{2} \beta r^2 + \alpha r - \log_e R_o = 0 \tag{11}$$

where

$$\alpha = \frac{R_1}{R_o} \tag{12}$$

$$\beta = \alpha^2 - \frac{R_2}{R_o} = \left(\frac{R_1}{R_o}\right)^2 - \frac{R_2}{R_o} \tag{13}$$

R_o is the net reproduction rate, and R_1 and R_2 are the first and second moments of the curve representing the age schedule of net reproductivity. The general equation for these moments is given by

$$R_n = \int_0^\infty x^n f(x) p(x) dx \tag{14}$$

The intrinsic measures that are presented below as relatively exact employ at most the second moments. Fortunately, the

---

[4] Frank Lorimer and Frederick Osborn, *Dynamics of Population*, New York, MacMillan Company, 1934, p. 351.

[5] Alfred J. Lotka, "Relation Between Birth Rates and Death Rates," *Science*, New Series 26(653):21–22, July 5, 1907.
[6] Ansley J. Coale, "Convergence of a Human Population to a Stable Form," *Journal of the American Statistical Association*, 63(322):395–435, June 1968.
[7] Dublin and Lotka, "On the True Rate of Natural Increase." See their Appendix, pp. 329–339, for derivations.

terms involving the higher moments converge quite rapidly toward zero.

Solving the quadratic equation (11) for r, we obtain, for the positive radical, which corresponds to the real root,

$$r = \frac{-\alpha + \sqrt{\alpha^2 + 2\beta \log_e R_0}}{\beta} \qquad (15)$$

and substituting for α and β in terms of R_0, R_1, and R_2, Glass [8] obtains

$$r = \frac{\frac{R_1}{R_0} - \sqrt{\left(\frac{R_1}{R_0}\right)^2 - 2\left[\frac{R_2}{R_0} - \left(\frac{R_1}{R_0}\right)^2\right]\log_e R_0}}{\frac{R_2}{R_0} - \left(\frac{R_1}{R_0}\right)^2} \qquad (16)$$

As we shall see from the computations for table 18–3, this form enables us to evaluate r, since R_0 was derived in table 18–2, and R_1 and R_2 can also be derived from the same basic statistics.

Mean Length of Generation

The mean length of generation is defined as the mean age of mothers at the birth of their daughters. Since the stable population is growing at the annual rate r, compounded continuously, and the net reproduction rate R_0 is its rate of growth in one generation, T years, we may write

$$R_0 = e^{rT} \qquad (17)$$

$$T = \frac{1}{r}\log_e R_0 \qquad (18)$$

Since from (11),

$$\log_e R_0 = \frac{1}{2}\beta r^2 + \alpha r \qquad (19)$$

$$T = \alpha + \frac{1}{2}\beta r \qquad (20)$$

Since every year the stable population is $(1 + r)$ times larger than the year before, at the end of a generation, it will be larger by a factor equal to the net reproduction rate. Of course, if r is negative, R_0 will be less than one.

Intrinsic Birth Rate

The true or intrinsic birth rate is the birth rate that would eventually be reached in a population subject to fixed fertility and mortality schedules. Thus, it is the birth rate of the stable population. The intrinsic birth rate may be expressed by the equation,

$$b = \frac{1}{\int_0^x e^{-rx}p(x)\,dx} \qquad (21)$$

where b is the birthrate per head per annum. This transcendental equation can be solved directly for b in a manner similar to that used for solving (10) for r; or, if we first determine the arithmetic value of r, we can substitute it in (21) and evaluate the definite integral.

A satisfactory approximation to equation (21) may be obtained from,

$$b = \frac{1}{\sum_0^x e^{-r(x+1/2)} \cdot \frac{L_x}{l_0}} \qquad (22)$$

Intrinsic Death Rate

The true, or intrinsic, death rate is the death rate that would eventually be reached in a population subject to fixed fertility and mortality schedules, or the death rate of the stable population. Obviously, the intrinsic death rate is equal to the difference between the intrinsic birth rate and the intrinsic rate of increase,

$$d = b - r \qquad (23)$$

Stable Age Distribution

The proportion of females within the age interval x to $x + dx$ may be expressed by,

$$c(x) = be^{-rx}p(x) \qquad (24)$$

We see that $c(x)$ can be found when b and r have been determined.

Ordinarily these computations will have been performed for the female population only. We may also desire to know the corresponding vital measures of the stable male population, or of the population of both sexes combined. We need first to know the sex ratio at birth which fortunately is very nearly constant from year to year. According to Lotka, the sex ratio of the stable population of all ages is

$$\frac{N_m}{N_f} = \frac{B_m}{B_f} \cdot \frac{\bar{L}_m + rL_m'}{\bar{L}_f + rL_f'} \qquad (25)$$

where $\frac{B_m}{B_f}$ is the sex ratio at birth, $\bar{L}$ is the mean length of life in the life table, and L' the first moment of the function L_x. The subscript denotes the sex. The birth rate of both sexes combined then becomes,

$$b_{m+f} = \frac{b_f \cdot \left(1 + \frac{B_m}{B_f}\right)}{\frac{N_m + N_f}{N_f}} \qquad (26)$$

Lotka stated that the rate of natural increase, r, must be the same for both males and females. R_0 is not the same for males and females in actual populations, however, because T is greater for males than for females. The greater male mean length of generation in turn stems from later average age at marriage and a longer fecund period. Hence, in the human species, there have probably been more female generations than male generations. Mathematical demographers have not yet been able to arrive at a complete system of logical relationships between the measures for males and females.

Computation of Measures

Intrinsic Rate of Natural Increase. — Let us now illustrate how r, the intrinsic rate of natural increase, can be computed from (16). We do so in table 18–3 using the statistics from the Peruvian census of 1961, previously applied in tables 18–1 and 18–2.

[8] D. V. Glass, *Population Policies and Movements*, Oxford Press, 1940, p. 407.

Step 1. Compute age-specific birth rates per woman and multiply each age-specific birth rate by .48439 (the proportion of all births that are female) to obtain birth rates of daughters (col. 1). In this particular case, we can begin with the rates in column (3) of table 18–1 and multiply them by .48439 and divide by 1,000.

Step 2. Enter the pivotal values of each of the age intervals (col. 2).

Step 3. From the appropriate life table, record the number of years lived by females in the stationary population for each 5-year age interval and divide by 100,000 (col. 3).

Step 4. Multiply each entry of column (1) by the corresponding entry of column (3). The result shows the expected female births to the female stationary population (col. 4).

Step 5. Multiply each entry of column (2) by the corresponding entry of column (4) and enter the result in column (5).

Step 6. Multiply each entry of column (2) by the corresponding entry of column (5) and enter the result in column (6).

Step 7. Sum columns (1), (4), (5), and (6). (Note that the gross reproduction rate is equal to the sum of column (1) multiplied by 5.) R_1 is the sum of column (5) and R_2, that of column (6).

Step 8. Obtain R_o, R_1, and R_2 from the sums of columns (4), (5), and (6), respectively.

Step 9. Compute α and β from the expressions

$$\alpha = \frac{R_1}{R_o} \quad \text{and} \quad \beta = \alpha^2 - \frac{R_2}{R_o}$$

Step 10. Compute r from formula (16). With the natural logarithm of 1.78224 equal to 0.57787, this equation reduces to $r = 0.020146$.

Mean Length of Generation. — We have from (20), $T = \alpha + \frac{1}{2}\beta r$. Using the values previously computed from the 1961 census of Peru,

$$T = 29.2182 + \frac{1}{2}(-53.0132 \times 0.02015)$$

$$= 29.2182 + \frac{1}{2}(-1.0682) = 28.6841.$$

Intrinsic Birth and Death Rates and Stable Age Distribution. — Having calculated the intrinsic rate of natural increase (see table 18–3), we can proceed to compute the intrinsic birth rate, the intrinsic death rate, and the stable age distribution of

Table 18–3.—Calculation of Intrinsic Rate of Increase (r), Peru: 1961

Age of mother x to x + 5	Annual births of daughters per female $(_5F_x)$	Pivotal (central) age (x + 2 1/2)	$\frac{_5L_x^F}{\ell_0}$	Products		
				Zero moment (R_o)	First moment (R_1)	Second moment (R_2)
				(1) x (3)=	(2) x (4)=	(2) x (5)=
	(1)	(2)	(3)	(4)	(5)	(6)
15 to 19 years....................	.03888	17.5	3.88727	.15114	2.64495	46.28663
20 to 24 years....................	.11341	22.5	3.79787	.43072	9.69120	218.05200
25 to 29 years....................	.12000	27.5	3.68824	.44259	12.17123	334.70883
30 to 34 years....................	.09856	32.5	3.57334	.35219	11.44618	372.00085
35 to 39 years....................	.07482	37.5	3.45368	.25840	9.69000	363.37500
40 to 44 years....................	.03378	42.5	3.32597	.11235	4.77488	202.93240
45 to 49 years....................	.01096	47.5	3.17970	.03485	1.65538	78.63055
Summation....................	.49041	(X)	(X)	1.78224	52.07382	1,615.98626

GRR = 5 Σ col. (1) = 5x.49041 = 2.45205

NRR = R_0 = Σ col. (4) = 1.78224

R_1 = Σ col. (5) = 52.07382

R_2 = Σ col. (6) = 1,615.98626

$\alpha = \dfrac{R_1}{R_0}$ = 29.2182

$\beta = \alpha^2 - \dfrac{R_2}{R_0}$ = 853.7032 - 906.7164 = -53.0132

$\log_e R_0$ = .57787

r = .02015

X Not applicable.

Source: Table 18–2.

The values to be entered are

$$r = \frac{\dfrac{52.07382}{1.78224} - \sqrt{\left(\dfrac{52.07382}{1.78224}\right)^2 - 2\left[\left(\dfrac{1,615.98626}{1.78224}\right) - \left(\dfrac{52.07382}{1.78224}\right)^2\right]\log_e 1.78224}}{\dfrac{1,615.98626}{1.78224} - \left(\dfrac{52.07382}{1.78224}\right)^2} = 0.02015$$

the female population. The computations are set forth in table 18–4.

Step 1. Set down the midpoint of the age intervals (col. 1).

Step 2. Multiply each midpoint of column (1) by 0.02015, the value of r. Enter the result in column (2).

Step 3. Compute the value $e^{-r(x+2.5)}$ using: U.S. National Bureau of Standards, *Tables of the Exponential Function e^x*, Applied Mathematics Series 14, 1951.

Step 4. From the female life table, record the average number of years one woman would live during each age interval. This is obtained by dividing $_5L_x$ for the appropriate interval of the female stationary population by 100,000. Record the results in column (4).

Step 5. Multiply each entry in column (4) by the corresponding entry in column (3). Record the products in column (5). Sum these entries and record the result at the bottom of column (5).

Step 6. From the corresponding male life table, multiply the average number of years each male would live in the interval by the sex ratio at birth and record the product in column (6).

Step 7. Multiply each entry in column (6) by the corresponding entry in column (3), and record the products in column (7). Sum these entries, and record the result at the bottom of column (7).

Step 8. Compute the intrinsic birth and death rates. The total number of female person-years is the total of column (5), or 27.78557. There is 1 female birth. (The radix of the female life table was 100,000 births, but we have changed it to 1 in the calculations shown in table 18–4.) Therefore, to obtain the female birth rate per head, we use the fraction $\frac{1}{27.78557}$, which is 0.03599. The intrinsic birth rate per 1,000 of the female population—the conventional form—is, therefore, 35.99. The birth rate for the entire stable population (both sexes) is

$$\frac{1 + 1.06444}{27.78557 + 27.92017} = \frac{2.06444}{55.70574} = .03706,$$

or 37.06 per 1,000. The intrinsic death rate equals the intrinsic birth rate minus the intrinsic rate of natural increase, or $37.06 - 20.15$, which equals 16.91 per 1,000.

Step 9. To compute the stable population by age and sex, we add the sums of columns (5) and (7), obtaining 55.70574, which is the total stable population derivative of both sexes. To express the distribution per 100,000 of the total population, divide 100,000 by this total and multiply each value in columns (5) and (7) by this constant quotient. Post the results in columns (8) and (9), respectively.

The formula (24), $c(x) = be^{-rx}p(x)$, gives the age distribution per unit for a given sex. Since the number of female births is 1, then the female birth rate is $\frac{1}{\Sigma e^{-rx}p(x)}$ or, in our table, the reciprocal of the total of the entries in column (5). In the actual computations in our worksheet, the value of b does not appear explicitly.

The results obtained give a stable population consisting of 100,000 persons with the appropriate numbers of males and females in accordance with the sex ratio at birth. Often the distribution is wanted for the female (or male) population only. Then we use the sum of column (5)—or column (7)—alone rather than the total of the two sums. For example, for females under 5 years old,

$$\frac{4.02712}{27.78557} \times 100,000 = 14,494$$

Table 18–4. — Computation of the Intrinsic Birth and Death Rates, and the Age Distribution of the Stable Population, for Peru: 1961

Age interval (x to x+5)	Midpoint (x+2.5)	r(x+2.5) or .02015(x+2.5) .02015x(1)=	e^{-r(x+2.5)} or 1/e^{r(x+2.5)}	$\frac{_5L_x^F}{\ell_o^F}$	Female stable population derivative (3)x(4)=	$\frac{_5L_x^m}{\ell_o^m}$. (sex ratio at birth, 1.06444)	Male stable population derivative (3)x(6)=	Stable population Female (5)x$\frac{100,000}{\Sigma(5)+\Sigma(7)}$=	Stable population Male (7)x$\frac{100,000}{\Sigma(5)+\Sigma(7)}$=
	(1)	(2)	(3)	(4)	(5)	(6)	(7)	(8)	(9)
0 to 4 years.........	2.5	0.05038	.95087	4.23519	4.02712	4.39801	4.18194	7,229	7,507
5 to 9 years.........	7.5	0.15113	.85974	3.99952	3.43855	4.12774	3.54878	6,173	6,371
10 to 14 years.......	12.5	0.25188	.77734	3.94482	3.06647	4.05691	3.15360	5,505	5,661
15 to 19 years.......	17.5	0.35263	.70284	3.88727	2.73213	3.98527	2.80101	4,905	5,028
20 to 24 years.......	22.5	0.45338	.63548	3.79787	2.41347	3.87702	2.46377	4,333	4,423
25 to 29 years.......	27.5	0.55413	.57457	3.68824	2.11915	3.74901	2.15407	3,804	3,867
30 to 34 years.......	32.5	0.65488	.51950	3.57334	1.85635	3.61841	1.87976	3,332	3,374
35 to 39 years.......	37.5	0.75563	.46971	3.45368	1.62223	3.47974	1.63447	2,912	2,934
40 to 44 years.......	42.5	0.85638	.42470	3.32597	1.41254	3.32180	1.41077	2,536	2,533
45 to 49 years.......	47.5	0.95713	.38399	3.17970	1.22097	3.13065	1.20214	2,192	2,158
50 to 54 years.......	52.5	1.05788	.34719	2.99918	1.04129	2.89337	1.00455	1,869	1,803
55 to 59 years.......	57.5	1.15863	.31392	2.76999	.86956	2.59856	0.81574	1,561	1,464
60 to 64 years.......	62.5	1.25938	.28383	2.46504	.70087	2.23554	.0.63451	1,258	1,139
65 to 69 years.......	67.5	1.36013	.25663	2.07434	.53234	1.79986	0.46190	956	829
70 to 74 years.......	72.5	1.46088	.23203	1.58196	.36706	1.30649	0.30314	659	544
75 to 79 years.......	77.5	1.56163	.20979	1.03649	.21745	0.81137	0.17022	390	306
80 to 84 years.......	82.5	1.66238	.18969	0.54295	.10299	0.39886	0.07566	185	136
85 years and over....	90.0	1.81350	.16308	0.27610	.04503	0.14805	0.02414	81	43
Summation........	(X)	(X)	(X)	(X)	27.78557	(X)	27.92017	49,880	50,120

$b_F = .03599$ $d_F = .01584$ $b_M = .03812$ $d_M = .01797$

$b_T = .03706$ $d_T = .01691$ $r = .02015$

X Not applicable.

Source: Table 18–3 and Eduardo E. Arriaga, "New Abridged Life Tables for Peru: 1940, 1950–51, and 1961," *Demography*, 3(1):226, 1966.

Some Approximations

The amount of calculation required for some of these measures can be reduced appreciably by using short-cut formulas described by Coale. These appear in two separate articles. We will describe first his earlier proposal for approximating the intrinsic rate of natural increase.[9] Coale's calculations involve two relationships:

$$NRR = e^{rT} \tag{27}$$

$$\frac{NRR}{GRR} \cong \frac{l_T}{l_0} \tag{28}$$

where NRR is the net reproduction rate, GRR is the gross reproduction rate, r is the intrinsic rate of increase, T is the mean length of female generation, and $\frac{l_T}{l_0}$ is the female probability of surviving to age T.

The intrinsic rate of natural increase can be estimated by determining T from the approximate relationship (28), thus,

$$l_T \cong l_0 \frac{NRR}{GRR},$$

and calculating r from (27), thus

$$\log NRR = rT \log e \tag{29}$$

$$r = \frac{\log NRR}{T \log e} \tag{30}$$

or, using natural logarithms, we obtain,

$$r = \frac{\log_e NRR}{T} \tag{31}$$

Because r and T are reciprocally related, the percent error in estimating r varies inversely with that in estimating T. Coale tested the accuracy of the approximation by comparing the results it yields with the results of exact computations of T for 17 different countries. The root mean square error was about 1.8 percent of the average value of T. (Coale also gives an approximation, first suggested by Lotka, for b, the intrinsic birth rate.[10])

Let us use this approximation to estimate r and compare the result with the closer estimate of table 18-3. We take the values of NRR and GRR from that table.

$$l_T = \frac{1.78224}{2.45205} \times 100,000 = 72,684.$$

Consulting the Peruvian life table for 1961 and interpolating, we find that this value of l_T corresponds to a T of 29.881 years. Substituting in (31) we have

$$r = \frac{\log_e 1.78224}{29.881} = \frac{0.57788}{29.881} = .01934,$$

which may be compared with the .02015 found by the more exact method.

The short-cut method is not very close in this case, being off by about 4 percent. There are other possible approximations, however. Since $T = \alpha + \frac{1}{2}\beta r$, it can be seen that the second term can be ignored when r is small. Then

$$T \cong \alpha = \frac{R_1}{R_0} \tag{32}$$

From table 18-3, $T = \frac{52.07382}{1.78224} = 29.218$ years.

$$\frac{R_1}{R_0} = \frac{\Sigma x f(x) p(x)}{\Sigma f(x) p(x)},$$

$f(x)$ being the fertility rate at age x and $p(x)$ the survival rate to age x. This can be seen as the weighted age (x) in which the weights are the rates of age-specific survivors of births.

We may then substitute this new estimate of T in the equation $R_0 = e^{rT}$ obtaining,

$$r = \frac{0.57788}{29.218} = .01978.$$

This is a closer approximation to .02015.

Coale has a later article in which he gives still another way of approximating r.[11] This method involves using 29 years as an estimate of T and getting r_1, a first approximation to the value of r, which is then adjusted according to the extent to which

$$\int_{\omega 1}^{\omega 2} e^{-r_1 x} f(x) p(x) dx$$

fails to equal 1.

The excess of the estimate from 1 is designated δ and the adjusted value of r becomes

$$r_1 + \frac{\delta}{29 - \frac{\delta}{r_1}} \tag{33}$$

Coale regards this approximation as better than that based on (28).

AVAILABILITY OF THESE MEASURES

The *Demographic Yearbooks* of the United Nations for 1954 and 1965 have included female gross and net reproduction rates. The more recent of these yearbooks presents annual values, insofar as they are available, for the period 1930 to 1964, for 132 countries, distributed by continents as follows:

Africa	41
America, North	18
America, South	11
Asia	26
Europe	31
Oceania	4
U.S.S.R.	1

Many of these were computed at the United Nations for its *Population Bulletin*, No. 7 (1965), which dealt with conditions and trends of fertility in the world. A fair number of countries

[9] Ansley J. Coale, "The Calculation of Approximate Intrinsic Rates," *Population Index*, 21(2):94–97, April 1955.

[10] Ibid., p. 96.

[11] Ansley J. Coale, "A New Method for Calculating Lotka's r—the Intrinsic Rate of Growth in a Stable Population," *Population Studies* (London), 11(1):92–94, July 1957.

publish such rates in their statistical yearbook or elsewhere. Net reproduction rates are shown less frequently than gross rates.

The April–June issue of *Population Index* presents not only the reproduction rates for countries of the world but also the mean length of generation ("mean age of fertility"), and the intrinsic rates of birth, death, and natural increase compared with the corresponding crude rates. The intrinsic rates are based on model stable populations, and the required data are the gross reproduction rate, the mean age of the fertility schedule, and the expectation of life at birth.

Using several standard computer programs, Keyfitz and Flieger have produced a book containing tables of intrinsic vital rates and related measures for most of the countries of the world, including many without adequate vital statistics.[12] Many other demographic measures are contained in this report.

REPRODUCTIVITY ESTIMATED FROM CENSUS STATISTICS

In the absence of adequate birth statistics, attempts have been made to compute measures of reproductivity wholly from census statistics or from census statistics and life tables. The general approach is similar to that in estimating fertility measures from census data described in chapter 17.

In the United States, reproduction rates, like other fertility measures, were first based on census data, for these reasons: (a) the failure of the Birth Registration Area to cover the entire United States until 1933; (b) the lack of national statistics on resident births until 1937; (c) the fact that fertility data on some characteristics (such as farm-nonfarm residence) are available from the census but not from the registration system. The last of these reasons is still an important one. On the other hand, we must recognize that undercounting of young children in the United States is now more serious than underregistration of births, and we have less information on the geographic variations of the former than of the latter.

The methods described in this section are applicable to many countries with good age-sex statistics from censuses but with inadequate birth statistics. It is often desirable, however, to make some kind of correction for undercounting young children, or otherwise to adjust the age-sex distribution, when measuring reproductivity from census data.

Measures Based on Age-Sex Distributions — The Replacement Index

The **replacement index** as currently used is computed as follows: "Given an age distribution of actual female population and a female life table for the same period, the replacment index has, as its numerator, the ratio of children under age five years to females at the reproductive ages in the actual population and, as its denominator, the corresponding ratio in the stationary or life table population."[13] Thus defined, this measure relates to reproductivity in the five years preceding the observed age distribution, but we may exercise a choice in the age groups for children and women just as we do in the case of the total fertility ratio.

In his 1920 census monograph, Thompson devised several types of measures that he called "replacement indexes."[14] Of these, his **permanent replacement index** corresponded to the index just described. Net replacement indexes similar to those defined by Thompson were computed for many countries of the world as part of the computer program devised by Keyfitz and Flieger.[15]

It remained for Lotka to give a generalized mathematical treatment of replacement indexes and to show how a replacement index is related to the intrinsic rate of natural increase.[16] He defined the index J in the same way as Thompson's permanent replacement index, namely, as the quotient of two ratios, the numerator, the ratio of children under a given age to women in the childbearing ages in the actual population and the denominator, the same ratio in the life table population. Thus, in integral notation

$$J = \frac{\int_p^q c(a)da}{\int_u^v c(a)da} \div \frac{b_0 \int_p^q p(a)da}{b_0 \int_u^v p(a)da} \qquad (34)$$

where $c(a)$ is the coefficient of age distribution in the actual population and $p(a)$ is the probability at birth of surviving to age a; p and q are the lower and upper limits, respectively, of the junior age group of both sexes, and u and v are the lower and upper limits, respectively, of the age groups of women in the reproductive ages. The symbol b_0 denotes the birth rate per head in the life table population in which

$$c(a) = b_0 p(a) \qquad (35)$$

In the more conventional demographic notation, this is

$$J = \frac{P_{0-4}}{P_{15-49}^f} \div \frac{L_{0-4}^f + \frac{B_m}{B_f} L_{0-4}^m}{L_{15-49}^f} \qquad (36)$$

The replacement index, then, gives us a method for approximating the net reproduction rate in the absence of vital statistics. Table 18-5 illustrates the computation of the J-Index using the 1961 statistics for Peru that were used in tables 18-3 and 18-4. We have used the simpler statistical notation instead of Lotka's integral equation.

$$r' = \frac{\log_e J}{\alpha_2 - \alpha_1} \qquad (37)$$

where α_2 and α_1 are defined in table 18-5.

The values of J (2.1205) and r (.0256) are seen to be considerably higher than those of R_0 (1.7822) and r (.0201) in table 18-3. The facts that the 1961 population of Peru was not of stable form and that R_0 was not very close to unity affect the interpretation of J as equivalent to R_0.

[12] Nathan Keyfitz and Wilhelm Flieger, *World Population: An Analysis of Vital Data*, Chicago, University of Chicago Press, 1968.

[13] Mortimer Spiegelman, *Introduction to Demography*, rev. ed., Cambridge, Mass., Harvard University Press, 1968, p. 285.

[14] Warren S. Thompson, *Ratio of Children to Women: 1920*, Census Monograph XI, Washington, U.S. Bureau of the Census, 1931, pp. 157–174.
[15] Op. cit.
[16] Alfred J. Lotka, "The Geographic Distribution of Intrinsic Natural Increase in the United States, and an Examination of the Relation Between Several Measures of Net Reproductivity," *Journal of the American Statistical Association*, 31(194):273–294, June 1936.

Table 18-5.—**Computation of Replacement Index (J) and Substitute Intrinsic Rate of Natural Increase, for Peru: 1961**

Steps	Values
1. Ratio, children 0–4 to women 15–49 years: census population	
a. P_{0-4}^{T}..	1,671,526
b. P_{15-49}^{F}..	2,275,249
c. Ratio (a) ÷ (b).............................	.73466
2. Ratio, children 0–4 to women 15–49 years: life table population	
a. L_{0-4}^{M}..	413,636
b. L_{0-4}^{F}..	423,519
c. L_{15-49}^{F}..	2,490,607
d. $\dfrac{1.06220 \ (a) + (b)}{(c)}$..	.34645
3. Replacement index (J) = (1,c) ÷ (2,d)...............	2.12054
4. Rate of increase $r' = \dfrac{\log_e J}{\alpha_2 - \alpha_1}$	
a. $\log_e J$..	0.75167
b. α_1 = median age of children 0–4 in life table population[1]..	2.48
c. α_2 = median age of women 15–49 in life table population[2]..	31.84
d. $\alpha_2 - \alpha_1$..	29.36
e. r' = (a) ÷ (d)..	.0256

[1] The mean age for each sex was based on values for L_0 and L_{1-4} in the abridged life table for Peru, and midpoints of 0.5 and 3.0, respectively, were assumed for these age groups. The resulting mean ages for males and females were then weighted in proportion to the assumed distributions of births by sex in Peru.

[2] Computed on the basis of the life table population in 5-year age groups and the midpoints of these age intervals.

Source: United Nations, *Demographic Yearbook, 1965*, table 6, and Eduardo E. Arriaga, "New Abridged Life Tables for Peru: 1940, 1950–51, and 1961," *Demography*, 3(1):226, 1966.

Reproduction Rate Estimated From Women's Own Children Under 5

Up to this point in our discussion of the use of census data to approximate net reproductivity, we have dealt only with the data obtainable from age-sex distributions. Since it was not until the 1940 census of the United States that data were tabulated and published on the distribution of women by the number of their own children under 5 years old in the household, it is not surprising that no attention was paid by demographers to the computation of reproduction rates from such data. In 1942 Grabill developed a method for calculating gross and net reproduction rates from data on own children under 5 years of women of childbearing age, which was used in a special report of the 1940 census and elsewhere to estimate reproduction rates, intrinsic rates, etc.[17] This method is most completely described, however, in a recent article by Grabill and Cho.[18]

[17] U.S. Bureau of the Census, *Sixteenth Census of the United States: 1940, Population, Differential Fertility: 1940 and 1910, Standardized Fertility Rates and Reproduction Rates*, 1944; U.S. Bureau of the Census, *Current Population Reports*, Series P-20, No. 8, "Differential Fertility: June 1946," December 31, 1947.

[18] Wilson H. Grabill and Lee Jay Cho, "Methodology for the Measurement of Current Fertility from Population Data on Young Children," *Demography*, 2:50–73, 1965.

They begin by estimating the age-specific birth rates corresponding to the ratios of own children to women by age. Let B_x^f denote the number of daughters born to women of age x and P_x^f the number of women of age x; then

$$F_x = \frac{B_x^f}{P_x^f} \tag{38}$$

is the age-specific birth rate.

Let s_x be the life table probability of survival from birth to age x, or the probability that daughters will live to their mother's age. Thus, using life table probabilities, one can expect that a census or survey taken i years after the birth of the daughters will show $B_x^f s_i$ surviving daughters of age i. Similarly, the survivorship among women "x" to age "$x+i$" would be $P_x^f \left(\dfrac{s_{x+i}}{s_x} \right)$. The ratio of surviving daughters born to surviving women would be

$$R_{x+i} = \frac{B_x^f s_i}{P_x^f \left(\dfrac{s_{x+i}}{s_x} \right)} \tag{39}$$

Since $F_x = \dfrac{B_x^f}{P_x^f}$, substituting $F_x P_x^f$ for B_x^f in (39), we get

$$R_{x+i} = \frac{F_x s_x s_i}{s_{x+i}} \tag{40}$$

and

$$F_x = R_{x+i} \left[\frac{s_{x+i}}{s_x s_i} \right] \tag{41}$$

Replacing the F_x values in the conventional formulas for gross and net reproduction rates with the mathematically identical values $R_{x+i} \left[\dfrac{s_{x+i}}{s_x s_i} \right]$ yields

$$GRR = \sum R_{x+i} \cdot \frac{s_{x+i}}{s_x s_i} = \frac{1}{s_i} \sum R_{x+i} \cdot \frac{s_{x+i}}{s_x} \tag{42}$$

since s_i is a constant

$$NRR = \left[\sum R_{x+i} \left(\frac{s_{x+i}}{s_x s_i} \right) \cdot s_x \right] = \frac{1}{s_i} \sum R_{x+i} \cdot s_{x+i} \tag{43}$$

The summation in (42) and (43) is to be taken over all ages for which $R_{x+i} > 0$.

Using a life table notation and daughters under 5 years old, we may write

$$NRR = \frac{1}{\sum\limits_{0}^{4} L_x} \cdot \sum R_A^{0-4} \cdot L_A \tag{44}$$

where $\sum\limits_{0}^{4} L_x$ is a female life table stationary population under 5 years old expressed on a unit-radix basis, A is the age of women at the census date, R_A^{0-4} is the ratio of daughters under 5 years old to women aged A, and L_A is the female life table stationary population age A, expressed on a unit-radix basis.

Using a specific survival rate notation, $_5s_A$, or $_5s'_x$ if we understand x to cover the childbearing ages, we have

$$\text{NRR} = \frac{1}{_5s_0} \cdot \sum {}_5R_x \cdot {}_5s'_x \qquad (45)$$

Similarly, the gross reproduction rate is

$$\text{GRR} = \frac{500,000}{\sum\limits_0^4 L_x} \sum R_A^{0-4} \cdot \frac{L_A}{L_{A-2.5}} \qquad (46)$$

It is assumed in expression (46) that women age A at the census date were, on the average, 2½ years younger at the time their daughters under 5 years old were born. This assumption is not very reasonable for females near the beginning or end of the childbearing period, but the errors tend to cancel so that the overall effect on the GRR is slight.

Again, in our survival rate notation, we have

$$\text{GRR} = \frac{1}{s_{0-4}} \sum \frac{_5R_x \cdot {}_5s'_x}{_5s''_x} \qquad (47)$$

where $_5s'_x$ is the survival rate of women from birth to ages x to $x+5$, and $_5s''_x$ is the survival rate to ages $x-2.5$ to $x+2.5$.

One important consideration in the use of age-specific ratios of own children to compute reproduction rates has not yet been brought out, however, namely, that "own children" fail to account for all children under 5 years (or under 1, etc.). This problem was discussed in chapter 17. The obvious remedy is to make a correction on the basis of the total number of children in the population. This may be an overall correction, not taking account of variations with age of mother since these do not have much effect on a reproduction rate, which is summed over all ages.

The availability of statistics forced us to draw our illustration from the United States. We wish to emphasize, however, that countries without adequate birth statistics could profitably use this method if they would code the mother's own children on the census schedule.

In table 18-6, we have translated the L_A's, etc., into survival rates for convenience of manipulation. In summing columns (4) and (6), we would ordinarily multiply by five since we are dealing with 5-year age groups of women, and the rates are averages for the 5 years. On the other hand, the children under 5 in the numerator of column (1) are survivors of births over a 5-year period, and they need to be reduced to annual births. Hence, the 5's cancel.

This method can be applied to a country like Peru for which there are no census data on own children by age of mother. The application requires a form of indirect standardization. We know the total number of girls under 5 and the number of women by age for Peru. If we use a schedule of age-specific rates of own female children under 5 per 1,000 women for the United States, or some part thereof, we can compute the expected number of female children under 5 in Peru, and divide this into the total number of girls under 5 to obtain an adjustment factor.[19] This factor is then applied to the U.S.

Table 18-6.—Computation of Net and Gross Reproduction Rates from Census Ratios of Own Children to Women, for the Rural-Farm Population of the United States: 1960

Age of women	Own children under 5 per 1,000 women $_5R_x$ (1)	Adjusted ratio[1] $_5R'_x$ col. (1) x factor= (2)	Survival rate $_5L_x \div 500,000 = {}_5s'_x$ (3)	$_5R'_x \cdot {}_5s'_x$ (2)x(3)= (4)	Survival rate $_5L_{x-2.5} \div 500,000 = {}_5s''_x$ (5)	$_5R'_x \cdot \frac{_5s'_x}{_5s''_x}$ (4)÷(5)= (6)
15 to 19 years...	69	35	.97077	34	.97186	35
20 to 24 years...	963	489	.96770	473	.96923	488
25 to 29 years...	1,188	604	.96423	582	.96596	603
30 to 34 years...	845	429	.96010	412	.96217	428
35 to 39 years...	529	269	.95435	257	.95723	268
40 to 44 years...	260	132	.94588	125	.95011	132
45 to 49 years...	72	37	.93319	35	.93954	37
50 to 54 years...	18	9	.91386	8	.92353	9
Summation....				1,926		2,000

Survival rate for children (from life table) $= \frac{_5L_0}{_5\ell_0} = \frac{489,135}{500,000} = 0.9783$

$\text{NRR} = \frac{1,926}{.9783} = 1,969$ or 1.969 per woman

$\text{GRR} = \frac{2,000}{.9783} = 2,044$ or 2.044 per woman

[1] Adjustment factor

$= \frac{\text{Total female children under 5}}{\text{Own children under 5 of women 15 to 54}} = \frac{652,589}{1,284,276} = .508138$

Source: Basic data from *U.S. Census of Population: 1960, Subject Reports*, "Women by Children Under 5 Years Old," PC(2)-3C, 1968, p. 3; U.S. National Center For Health Statistics, *Life Tables: 1959–1961*, Vol. I, No. 5, "Life Tables For Metropolitan and Nonmetropolitan Areas of the United States: 1959–1961," December 1967, table 8.

schedule to obtain an estimated schedule of rates for Peru. The following steps would then proceed as before with the use of an appropriate Peruvian life table. Grabill and Cho illustrate this procedure using two States, one of which is assumed to lack statistics on own children.[20]

A worksheet like table 18-6 does not yield estimates of annual age-specific birth rates as a by-product. Grabill and Cho give two procedures for making such estimates, however.[21]

With the aid of life table survival rates, age-specific ratios of children under 5 to women can be adjusted to restore deaths among children and women. The figures thus adjusted become equivalent to birth rates cumulated over a 5-year period for women aged 2½ years younger than at the end of the period. In order to derive birth rates for conventional 5-year age groups, one has to apply either Sprague's interpolation formula or some other technique. The reader is referred to the article for the details of the procedures and for a table of multipliers that can be used for this purpose.

Children Ever Born

From the observed or estimated sex ratio at birth and the number of children ever born, one may compute the number of daughters ever born by a given group of women. For

---

[19] The U.S. schedule happens to be the only one readily available. The level is certainly much inferior to that of the Peruvian rates. We are, in effect, assuming that the patterns or profiles are the same. As the United Nations has pointed out in connection with sex-age adjusted birth rates, "The use of a standard system of weights for this purpose is justified by the observation that, even

under greatly differing conditions of fertility, the relative levels of age-specific rates for women in the age groups from 15–19 to 40–44 are not very different." (Manual III, *Methods for Population Projections by Sex and Age*, Series A, Population Studies No. 25, 1956, p. 42.)

[20] Grabill and Cho, op. cit., p. 59.
[21] Ibid., pp. 58–69.

women who have completed the childbearing period, say those 50 years old and over, the average number of daughters ever born represents an estimate of the generation gross reproduction rate for these cohorts of women. It is obviously an upward-biased estimate, however, because those women who died before attaining the given age certainly averaged fewer children than those who survived. If we compare two age groups that are both beyond the reproductive period (for example, 50 to 54 and 55 to 59), the difference between them in the average number of children (daughters) ever born is also affected by selective mortality, although here the direction of the selection is uncertain because of our lack of knowledge about the relationship between fertility and mortality.

MISINTERPRETATIONS AND SHORTCOMINGS OF CONVENTIONAL MEASURES

Interpretation of Conventional Rates

During the 1920's and 1930's, Dublin, Lotka, Lorimer, Osborn, Notestein, and other prominent demographers stated explicitly that the net reproduction rate and other measures of the stable population merely described what would happen if the fertility and mortality schedules of a given period continued unchanged sufficiently long in a closed population. These measures did not represent a description of what was happening or forecasts of what would eventually have happened to the given population. It was recognized that both fertility and mortality were indeed changing.

Since the general trend in fertility had been downward for many decades, however, the reproduction rates came to be regarded as conservative indicators of how far reproductivity would eventually fall. It was frequently stated that the current levels of the crude rate of natural increase were due to transitory favorable age distributions (resulting from past births, deaths, and migration) and could not be maintained. When the net reproduction rate of the United States fell below unity in the 1930's, some demographers wrote that our population was no longer replacing itself.

Then, after the end of the war, a sudden, very fundamental change took place. The sharp upturn of age-specific fertility rates led some demographers to challenge the inevitability of a decline in the population and at the same time the interpretation of the classical reproduction rates was profoundly modified and their utility was seen to be much less than had been thought.[22]

Let us try to see how the misinterpretation of the earlier reproduction rates came about. One of the ingredients in reproduction rates is a schedule of age-specific fertility rates for a given year or other short period of time. This schedule is treated as if it represented the performance of a cohort of women as they pass through the successive ages of the childbearing period. Likewise the mortality schedule is treated as if it could be used to describe the proportion of women in a given cohort who survive from birth to successive ages of the childbearing period. The conventional life table is the same kind of "synthetic" cohort. The usefulness of these "synthetic" measures depends, in large degree, upon the extent to which they describe the fertility, mortality, or reproductivity of some past or future period, or, from another viewpoint, upon whether actual cohorts have had this experience or could have had this experience.

It is obvious that the conventional rates assume that the fertility rate at one age is independent of the rates at earlier ages for the same group of women. Suppose in a given year, t, women aged 25 had a fertility rate f_{25}^t and those aged 35 had a fertility rate f_{35}^t. Then 10 years later would f_{35}^{t+10} for the younger cohort be likely to equal f_{35}^t for the older cohort, if at prior ages they had had very different fertility rates? One cohort may have lived through ages 20 to 24 in a depression, the other in prosperity; (or one in wartime, the other in peacetime). Their distribution by age at marriage may have been very different. The proportions marrying by different ages may also have been different, perhaps because there was a great shortage of marriageable men arising from war losses. Marriages and births are believed to be postponed or advanced because of the economic and psychological conditions prevailing. In the 1930's and 1940's, most of the reproduction rates that were available to demographers were for the interwar period for countries that were belligerents in World War I and had suffered from the great depression of the 1930's. In fact, relatively few reproduction rates then available applied to populations that had spent their reproductive lives in periods of stability or gradual social change. More generally, there was a tendency to extrapolate short-run conditions into long-run consequences.

Other Limitations of Conventional Reproduction Rates

The conventional reproduction rates may also imply impossible values by order of birth, and corresponding male and female rates may be very different. Some of the shortcomings have been remedied by refinements that are described in the next major section.[23]

Impossible Reproduction Rates by Order of Birth. — Reproduction rates can be computed separately for each order of birth. Working with American fertility data for the years during World War II, Whelpton noticed some logical absurdities.[24] When he added the first birth rates over all ages from 15 through 49 for native white women in 1942, he obtained a rate of 1,084 as illustrated in table 18-7. In other words, it was implied that 1,000 women living through the childbearing period would have 1,084 first births! In view of the existence of some involuntary sterility and spinsterhood, even 1,000 first births would be impossible for a real cohort.

The explanation of this paradox lies in the facts that many women in their thirties had their first child in 1942 after postponing marriage and childbearing during the depression and many younger women were marrying and beginning childbearing relatively early as the result of the psychology

---

[22] Harold F. Dorn, "Pitfalls in Population Forecasts and Projections," *Journal of the American Statistical Association*, 45(251):311–334, September 1950.

[23] For additional critiques of the conventional rates see the following: J. Hajnal, "Births, Marriages, and Reproductivity, England and Wales, 1938–1947," in *Papers of the Royal Commission on Population*, Vol. II, *Reports and Selected Papers of the Statistics Committee*, London, H.M. Stationery Office, 1950, pp. 303–400; idem, "Some Comments on Mr. Karmel's Paper 'The Relations Between Male and Female Reproduction Rates,' " *Population Studies*, 2(3):354–360, December 1948; P. H. Karmel, "A Rejoinder to Hajnal's Comments," loc. cit., pp. 361–372; idem, "The Relations Between Male and Female Nuptiality in a Stable Population," *Population Studies*, 1(4):353–387, March 1948; idem, "The Relations Between Male and Female Reproduction Rates," *Population Studies* (London) 1(3):249–274, December 1947. Hannes Hyrenius, "La mesure de la reproduction et de l'accroissement naturel" (The measurement of reproduction and of natural increase), *Population*, 3(2):271–292, April–June 1948; George J. Stolnitz and Norman B. Ryder, "Recent Discussion of the Net Reproduction Rate," *Population Index*, 15(2):114–128, April 1949; T. J. Woofter, "The Relation of the Net Reproduction Rate to Other Fertility Measures," *Journal of the American Statistical Association*, 44(248):501–517, December 1949.

[24] P. K. Whelpton, "Reproduction Rates Adjusted for Age, Parity, Fecundity, and Marriage," *Journal of the American Statistical Association*, 41(236):501–516, December 1946.

Table 18–7.—**First Births Per 1,000 Native White Women by 5-year Age Periods, United States: 1940 to 1944**

[Rates are adjusted for incomplete registration in accordance with data of the Division of Vital Statistics based on the test period December 1, 1939, tc April 1, 1940]

Age period	1940	1941	1942	1943	1944
15 to 19 years[1]	35.09	37.88	42.41	42.47	36.70
20 to 24 years	65.88	75.01	90.49	81.00	72.60
25 to 29 years	40.87	45.99	55.38	48.24	39.31
30 to 34 years	16.41	18.03	21.17	19.88	17.10
35 to 39 years	4.82	5.34	6.37	6.74	6.68
40 to 44 years	.81	.88	.95	1.09	1.23
45 to 49 years[2]	.06	.05	.05	.06	.06
Total	163.94	183.18	216.82	199.48	173.68
Total x 5	819.70	915.90	1,084.10	997.40	868.40

[1] Includes the few births to women younger than 15.
[2] Includes the few births to women 50 or older.

Source: P. K. Whelpton, "Reproduction Rates Adjusted for Age, Parity, Fecundity, and Marriage," *Journal of the American Statistical Association,* 41(236):503, December 1946, table 1.

of the prosperous wartime period. This combination of events could not occur in the lifetime of an actual cohort of women. This analysis brings out clearly the fact that fertility at one age is not independent of fertility at earlier ages. The absurdity of the rate of first births in 1942 cast doubt on the conventional interpretation of reproduction rates, covering all orders of birth, for other calendar years.

Inconsistent Male and Female Reproduction Rates.— Although reproduction rates are usually computed only for the female population, rates for the male population can also be computed. The calculation of reproduction rates for males and females has shown that they are not compatible, however.

Kuczynski, for example, calculated a net reproduction rate of only 977 per 1,000 French females in 1920–23 but one of 1,194 per 1,000 males.[25] He attributed the inconsistency to a lack of balance between the sexes in the reproductive ages, arising from deaths of French men in World War I. A lack of balance could also arise in some countries from the greater international migration of men than of women. In monogamous societies, persons of the sex in short supply in the reproductive ages would be expected to have the higher reproduction rate because a greater proportion of them would be able to find marital partners. It should also be mentioned that **apparent** values of the conventional reproduction rates are affected by underenumeration, misstatements of age, underregistration of births, and errors in coverage and age-reporting of deaths. Differences in the calculated rates for the two sexes are affected by only the sex differentials in the quality of the basic data, however.

Differences between male and female reproduction rates may lead to different conclusions regarding population replacement. Could the female net reproduction rates below unity, especially prevalent in Western Europe during the 1930's, validly have been interpreted as foreshadowing failure of the population to reproduce itself when corresponding male reproduction rates were well above unity? Part of the rise in German fertility during the early years of Nazi rule may have reflected an increasingly favorable balance of the sexes as men too

young to have fought in World War I attained marriageable age. Hence, some of the increase in female reproduction rates appeared to represent the passing of a temporary shortage of husbands rather than any fundamental change in the fertility of married couples. On the other hand, war losses of men in World War I were not the only cause of declining reproduction rates during the interwar period. This is suggested by the fact that reproductivity also declined in the neutral Scandinavian countries.

Thorough analyses of the theoretical relationships between male and female reproduction rates were made independently but about the same time by Vincent and Karmel.[26] (Reproduction rates for the male population are discussed in a separate section of this chapter.) Vincent's analysis was limited to net reproduction rates.

We have discussed this problem of consistent male and female reproduction rates in the context of the conventional (period or synthetic) reproduction rates. It is necessary to conclude, however, by pointing out that these inconsistencies do not disappear when the improvements to be discussed in the next section—including rates for real cohorts—are introduced.

NEW AND IMPROVED MEASURES OF REPRODUCTIVITY

Much of the criticism of the conventional reproduction rates that has just been summarized is mingled with suggestions for improving them. In the United States, the attempts first took the form of continuing to use period data but taking account of additional demographic factors.

Period Rates Specific for Parity, Fecundity, and Marriage

Such an attempt was made by Whelpton, shortly after the end of World War II.[27] As mentioned in the preceding section of this chapter, Whelpton showed in this article that the conventional reproduction rates could lead to absurd results when dissected into their components by order of birth. This conclusion led him to the computation of rates that were specific for parity (and, of course, age).

Birth statistics by order of birth and by age of mother were readily available, for the United States; and there were also population data giving the distribution of women by age and parity for selected periods. Whelpton used the distribution for 1940 and 1910 published in the 1940 census reports. To obtain such distributions for intermediate and later years, a large amount of computation and some fairly broad assumptions were obviously required.

Starting with a hypothetical cohort of women, he allowed for mortality and parity by subtracting deaths and successive numbers who bore a first child, second child, etc. Whelpton's allowance for "fecundity and marriage" represents an attempt to confine the specific rates to women "at risk" in the actuarial sense. On the basis of the meager evidence then available, Whelpton assumed that 10 percent of women were "sterile" (involuntarily childless) and that an additional 10 percent

[25] Robert R. Kuczynski, *Fertility and Reproduction*, New York, Falcon Press, 1932, pp. 36–38.

[26] Paul Vincent, "De la mesure du taux intrinsèque d'accroissement naturel dans les populations monogames" (On the measurement of the intrinsic rate of natural increase in monogamous populations, *Population* (Paris), 1(4):699–712, October-December 1946; Karmel, "The Relations Between Male and Female Reproduction Rates."

[27] Whelpton, "Reproduction Rates Adjusted for Age, Parity, Fecundity, and Marriage."

could not marry before the end of the childbearing period. This method of allowing for marital status is more limited than the computation of nuptial reproduction rates, which introduces marital status as another specific characteristic in the rates.

For the years from 1920 to 1944, Whelpton found that net reproduction rates adjusted for age-parity-marriage-fecundity were lower than the conventional reproduction rates for every year except 1921, for which the rates were the same. The differences, however, were not large. For example, the net reproduction rates (per 100) for 1942 were as follows:

Age adjusted.. 116
Age-parity-fecundity-marriage adjusted.......... 112

Generation Reproduction Rates

So far, the improved measures of reproductivity that we have described have simply been annual, or period, reproduction rates computed from fertility rates specific for other demographic variables as well as age. In taking up generation reproduction rates, we are passing to measures of reproductivity that are based on the fertility and mortality experience of an actual cohort of women during its reproductive years and not on the experience of one calendar year—or other short period—cumulated over all the reproductive ages. With this very important difference, the generation gross and net reproduction rates are defined just as they are in the case of the conventional, or "classical," reproduction rates.

Generation rates can be derived in several ways. We will consider first generation rates that can be obtained from vital statistics and then rates that can be obtained from census or survey statistics on children ever born. In both cases, more attention has been paid to gross rates than to net rates.

Generation Rates From Vital Statistics.—A direct method of computing a net generation reproduction rate would be to start with female births of a given year, say 1900. The female births to **native** women of the same cohort would be added throughout its childbearing period. Thus, we would add the births in 1915 to girls 15 years old, the births in 1916 to girls 16 years old, the births in 1917 to girls 17 years old, and so on through the births in 1949 to women 49 years old. Mortality of the mothers (earlier generation) is introduced at each successive age so that no survival rates are required. The sum of these successive female births divided by the initial size of the cohort of mothers is the net reproduction rate for the cohort of 1900. This method would not be appropriate for countries of heavy emigration. Earlier in the chapter, we discussed approximate methods of computing conventional rates from census data in the absence of adequate birth statistics. Given adequate vital statistics, we now have shown that it is possible to compute generation reproduction rates without population data from censuses.

Generation reproduction rates were first computed by Pierre Depoid [28] for France. The first American generation rates were computed by Woofter. [29]

Depoid's method.—Depoid calculated the conventional age-specific fertility rates (15 to 19 through 45 to 49 years) for 5-year cohorts beginning with 1826–1830 and ending with

1900–1905. Multiplying the sum of these fertility rates (daughters only) for a cohort by 5 gives the gross reproduction rate. For the same cohorts, he computed the probability of survival from birth to various ages up to 50 for females. The summed products of the fertility rates and the survival rates gave the generation net reproduction rate. All reproduction rates were for women who had completed their childbearing period.

Table 18-8 is adapted from Depoid's article. Note that this table juxtaposes rates for cohorts born in a given period against the "synthetic" rates of the same period. Females born in a given period were not at the beginning of their childbearing period until about 15 years later, at the average age of childbearing until about 30 years later, and at the end thereof until about 50 years later. Any of these lags would seem more reasonable than the comparisons implied by the table, and one can indeed introduce the lags by reading the table along the appropriate diagonals. For example, the net generation rate of 0.95 for 1826–30 could be compared with the period rate of 0.98 for 1856–60. But, as we said in chapter 17, there is no one-to-one correspondence between the fertility of a birth cohort and that of any particular short period. In any case, note that the fluctuations in the period rates are much greater than those in the generation rates.

Woofter's method.—To obtain a generation gross reproduction rate, Woofter computed the number of daughters surviving to the exact ages of mothers at the time when their daughters were born. More precisely, the age-specific fertility rates used were those of the calendar years when the cohort of women attained specific ages. Woofter originally named the cohorts according to the calendar year when they were 15—not according to the calendar years of their birth. Thus, they were identified by the year when they entered the childbearing period. For his "1915" cohort, for example, he cumulated the age-specific fertility rate of 15-year-old females in 1915, the rate of 16-year-olds in 1916, and so on through the rate of 44-year-olds in 1944. The sum is his **generation gross reproduction rate.** Applying to each cohort of daughters the mortality rates appropriate to the calendar years through which they must

Table 18-8.—Generation and Conventional Reproduction Rates per Woman, for France: 1826 to 1905

Birth period of cohort	Generation reproduction rates		Rates based on fertility and mortality of the period	
	Gross	Net	Gross	Net
1826–1830.................	1.66	0.95	1.92	1.05
1831–1835.................	1.66	0.94	1.86	1.03
1836–1840.................	1.66	0.97	1.79	1.05
1841–1845.................	1.65	0.98	1.77	1.04
1846–1850.................	1.63	0.97	1.69	0.99
1851–1855.................	1.60	0.95	1.65	0.96
1856–1860.................	1.54	0.93	1.68	0.98
1861–1865.................	1.49	0.91	1.71	1.03
1866–1870.................	1.40	0.87	1.71	1.02
1871–1875.................	1.31	0.82	1.68	1.00
1876–1880.................	1.28	0.83	1.69	1.08
1881–1885.................	1.16	0.78	1.65	1.05
1886–1890.................	1.09	0.74	1.53	0.99
1891–1895.................	1.01	0.70	1.45	0.97
1896–1900.................	0.97	0.69	1.41	0.98
1901–1905.................	1.03	0.76	1.37	0.98

Source: P. Depoid, "Reproduction nette en Europe depuis l'origine de statistiques de l'état civil," Etudes démographiques, No. 1, Statistique générale de la France, Imprimerie nationale, 1941, p. 34.

[28] P. Depoid, "Reproduction nette en Europe depuis l'origine de statistiques de l'état civil," *Etudes démographiques,* No. 1, Statistique générale de la France, Imprimerie nationale, 1941.
[29] T. J. Woofter, "Completed Generation Reproduction Rates," *Human Biology,* 19(3):133–153, September 1947.

Table 18-9. — Conventional Reproduction Rates in 1923-38 and Generation Reproduction Rates for Women Attaining Age 28 in 1923-38: United States

[Rates per 100 native white women]

| Calendar year | Women born in | Reproduction rates for— | | | |
| | | Calendar year | | Generation | |
		Gross	Net	Gross	Net
1923	1895	145	124	139	122
1924	1896	145	124	136	120
1925	1897	140	121	133	117
1926	1898	136	118	129	114
1927	1899	133	116	126	112
1928	1900	126	111	122	109
1929	1901	119	105	119	107
1930	1902	121	107	116	108
1931	1903	115	102	114	104
1932	1904	111	98	113	103
1933	1905	104	93	112	103
1934	1906	107	97	112	103
1935	1907	106	96	111	103
1936	1908	104	95	111	103
1937	1909	105	96	111	104
1938	1910	108	99	111	104

Source: T. J. Woofter, "Generation Reproduction Rates," U.S. National Office of Vital Statistics, *Vital Statistics, Special Reports, Selected Studies,* Vol. 33, No. 4, 1950, table 4, p. 68.

survive and summing the products of the appropriate fertility and survival rates, Woofter obtained the **generation net reproduction rate.**

In table 18-9, Woofter's generation reproduction rates for women by year of **birth** are compared with conventional rates of 28 years later, i.e., at the time when the cohort was near its average age of childbearing. Generations that exhibited fertility well below average in some periods made up the deficit by fertility well above average in other periods.

Woofter's method of computing generation reproduction rates mixes two generations. In other words, the mortality used for the computation of his generation reproduction rate is not that to which a single generation of women has been exposed but is made up, in varying proportions, of the mortality to which the mothers have been exposed and the mortality to which their daughters have been exposed. This method requires the projection of mortality rates into future years even when we are concerned with fertility that was completed in the recent past.[30]

Whelpton's method.—The preparation of cumulative and completed fertility rates for cohorts was discussed in chapter 16. When such complete fertility rates are multiplied by the proportion of births that are female, we have the generation gross reproduction rate. In the United States, Whelpton had done pioneering work on data assembly and adjustment and on methodology.[31] This work was extended by Whelpton and Campbell for the National Office of Vital Statistics. Their report describes the methodology that is still followed at the National Center for Health Statistics (the successor agency).[32]

(Most of the complications in the methodology stem from inadequacies in the underlying vital statistics and population statistics, particularly for earlier years.) Generation gross reproduction rates for more recent years can be readily computed from completed fertility rates for annual birth cohorts of women by current age of woman and live-birth order as given in the annual volumes of *Vital Statistics of the United States,* Vol. I, *Natality.*

It is surprising that so little attention has been paid to the derivation of generation net reproduction rates from these materials. Ideally, one would develop a generation life table for the same cohort of females, or, what amounts to the same thing, apply each year an annual survival rate that comes from the life table of that year. This procedure would involve a great deal of work but probably less than is devoted to estimating cumulative fertility. A short-cut would be to use a life table applying to the date at which the cohort had completed roughly half of its childbearing. This age tends to be fairly constant.

Whelpton paid some attention to this problem in the mid-fifties. Appendix chapter V in *Cohort Fertility* was entitled "Net Reproduction Tables for Hypothetical Cohorts Based on Age-Specific Birth Rates, and on Age-Parity Specific Birth Rates Adjusted for Spinsterhood and Sterility." A few years later he computed net reproduction rates for the U.S. cohorts of 1901–05 and 1906–10. These were 905 and 874, respectively, indicating a failure of these particular cohorts to replace themselves.[33] These values were lower than any period values of the NRR that had been observed. Among recent European writers, Pressat devotes some space in his text to methods of computing both net and gross reproduction rates for actual cohorts.[34]

Generation Rates From Children Ever Born.— It has already been mentioned that the derivation of the gross reproduction rate for children ever born to women of completed fertility is quite direct. For example, the number of children ever born per 1,000 women 45 to 49 years old as reported in Chile in 1960 was 3,621. Multiplying this rate by the proportion of female births in 1960, 0.49166, gives 1,780. (This proportion is very nearly constant over time.) This is a generation gross reproduction rate, albeit one that has been affected by any selective mortality and net immigration among the women of this cohort. To approximate the generation net reproduction rate, we multiply by the survival rate $\frac{L_{27}}{l_o}$ from a Chilean life table for 1940, when this cohort was roughly at its average age of childbearing. Numerically, 1,780 times 0.63623 gives 1,115, or NRR = 1.115.

Another approximate method has been utilized in a report based on statistics of the Current Population Survey (United States).[35] This approximation is based on a "replacement quota." This quota is computed as follows:

(1) From the sex ratio at birth, compute the total number of births corresponding to 1,000 female births; e.g., for a sex ratio of 1,050, this would be 2,050. This is the number

[30] Alfred J. Lotka, "Critique de certains indices de reproductivité," (Critique of some suggested measures of reproductivity), unpublished paper presented to the International Union for the Scientific Study of Population, Geneva, September 1949.

[31] Pascal K. Whelpton, *Cohort Fertility,* Princeton, N.J., Princeton University Press, 1954.

[32] U.S. National Office of Vital Statistics, *Vital Statistics, Special Reports, Selected Studies,* Vol. 51, No. 1, Part 1, "Fertility Tables for Birth Cohorts of American Women," by Pascal K. Whelpton and Arthur A. Campbell, January 29, 1960.

[33] Wilson H. Grabill, Clyde V. Kiser, and Pascal K. Whelpton, *The Fertility of American Women,* New York, John Wiley & Sons, 1958, p. 331.

[34] Roland Pressat, *Principes d'analyse: cours d'analyse démographique de l'Institut de démographie de l'Université de Paris* (Principles of analysis: Course in demographic analysis of the Institute of Demography of the University of Paris), Editions de l'Institut national d'études démographiques, Paris, 1966, pp. 51–59 and 113–123.

[35] U.S. Bureau of the Census, *Current Population Reports,* Series P-20, No. 147, "Fertility of the Population: June 1964 and March 1962," January 5, 1966, p. 6.

of births required for a cohort of 1,000 women to replace itself if all of them survived through the childbearing period.

(2) Divide this "gross" quota by the survival rate of women to the average age of childbearing. The quotient is the replacement quota.

In the report cited, the average age was taken as 27 and the survival rate as .963 yielding a quota of 2,130 births. The number of children ever born reported by women 45 to 49 years old in 1964 was 2,437. This is 1.144 times the quota, so that we might speak of an approximate generation net reproduction rate of 1.144 for these women. The survival rate used was based on a current life table; and, again, as a refinement, a life table corresponding to the year when the cohort was 27 years old could be used instead. This would be about 20 years earlier, or 1944.

INCOMPLETE REPRODUCTIVITY OF COHORTS

Very little of recent fertility is reflected in the reproductivity of cohorts that have completely passed through the childbearing age. It is usually this recent fertility that is of greatest interest, however, since we wish to get some clue concerning the effect of recent events upon the eventual size of completed families. Much can be learned by comparing the cumulative reproductivity of cohorts still in the midst of the childbearing period with that of earlier cohorts when these latter were at the same age.

We may then want to extrapolate, by some method or other, the net reproductivity of the cohort to the end of its childbearing period. Shortly after they began studying cohort reproductivity, both Woofter and Whelpton addressed themselves to this problem.[36] Since that time, demographers have studied the problem of extrapolating the incomplete fertility of cohorts; but they have paid very little attention to adjusting the cumulative fertility to take account of future changes in mortality. This situation may reflect the fact that, in the United States and other western countries, survival rates are high and fairly stable at the ages concerned. Moreover, in making population projections, demographers customarily treat future fertility and future mortality separately and are not explicitly concerned with reproductivity.

NUPTIAL REPRODUCTION RATES

To the analysis of the possibilities latent in constant age-specific fertility and mortality rates Wicksell, Charles, and other demographers have added constant age-specific first marriage rates.[37] The resultants are called the nuptial gross reproduction rate and the nuptial net reproduction rate. The nuptial gross reproduction rate is obtained by applying current marital fertility rates to the proportions of married women at each age that would result from current marriage rates. Charles defines this rate as "the number of girls who would be born on the average to each woman passing though the childbearing period if the specific fertility rates of single and married women and the marriage rate at each age for a given year were all to remain constant."[38] The nuptial net reproduction rate

is the number of girls who would be born on the average to a birth cohort of females if, in their lifetime, they were subject at each age to the specific fertility rates of single and ever-married women, the marriage rate, and the mortality for a given year.

An age-specific fertility rate may be viewed as a weighted average of rates specific for additional demographic variables such as duration of marriage and parity.[39] The conventional reproduction rates are then implicitly a function of the existing composition by marital status, etc. Consequently, conventional net reproduction rates do not effectively remove the influence of the current population's demographic history. Moreover, net rates adjusted for duration of marriage, parity, etc., may be strikingly different from unadjusted ones.

In the United States, the necessary data on first marriage by age of bride are still lacking for States that are not in the Marriage Registration Area. (Volume III of *Vital Statistics of the United States, 1969* presents statistics for an MRA consisting of 39 states and 5 other registration areas.) Some kinds of nuptial reproduction rates can be computed from statistics on marital status by age instead of marriages by age at marriage, however.

Spiegelman[40] expresses the nuptial reproduction rate by the formula

$$R_o = \frac{1}{{}^f l_o} \sum_{\omega_1}^{\omega_2} {}^f i'_x \, {}^f n_x \, {}^f l_x + \frac{1}{{}^f l_o} \sum_{\omega_1}^{\omega_2} {}^f i''_x (1 - {}^f n_x) {}^f l_x, \qquad (48)$$

where ${}^f l_x$ is the survival rate of women to age x, ${}^f n_x$ is the corresponding age-specific proportion of married women in the current population, and ${}^f i'_x$ and ${}^f i''_x$ are the fertility of married and unmarried women, respectively, at age x. The two terms represent legitimate and illegitimate reproductivity, respectively. In this formulation, the conventional net reproduction rate has simply been divided into these two components without any effect on the total.

A nuptial reproduction rate may also be computed by the use of a hypothetical standard population of women by marital status, age, and duration of marriage. The standard population is developed by life table techniques ("multiple decrement" procedure) from an initial group of 100,000 single women who are then reduced by death rates, marriage rates, and rates of dissolution of marriage, in a given period (see chapter 15).[41] The standard population gives the number of women at each age who are unmarried and, further, the number of married women at each age distributed by duration. We apply age-specific birth rates of unmarried women to the unmarried women of the standard population, and birth rates of married

[36] T. J. Woofter, "The Relation of the Net Reproduction Rate to Other Fertility Measures"; P. K. Whelpton, "Cohort Analysis of Fertility," *American Sociological Review*, 14(6):735–749, December 1949.
[37] S. D. Wicksell, "Nuptiality, Fertility, and Reproductivity": *Skandinavisk Aktuarietidskrift*, 14(3):125–157, 1931; Enid Charles, "Differential Fertility in Scotland, 1911–1931," *Transactions of the Royal Society of Edinburgh*, 59(3):673–686, 1938–1939.
[38] Op. cit., p. 681.

[39] Stolnitz and Ryder, "Recent Discussion of the Net Reproduction Rate," pp. 116–118.
[40] Mortimer Spiegelman, *Introduction to Demography*, p. 286.
[41] For examples of this general method as applied to data for Australia, the United States, and Germany, respectively, see John Hajnal, "The Analysis of Birth Statistics in the Light of the Recent International Recovery of the Birth-Rate," *Population Studies*, 1(2):137–164, September 1947; P. K. Whelpton, "Reproduction Rates Adjusted for Age, Parity, Fecundity, and Marriage," *Journal of the American Statistical Association*, 41 (236):501–516, December 1946; Germany, Statistisches Reichsamt, "Die Lebensbilanz des deutschen Volkes im Jahre 1933," *Neue Beiträge zum deutschen Bevölkerungs-problem*, (The vital balance of the German people in the year 1933, new contributions to the German population problem) Supplement to *Wirtschaft und Statistik*, No. 15, 1935, pp. 53–82.

women at each age and marriage duration to the married women of the standard population. The resulting cumulative product represents the total number of births that would be born to a cohort of women with the fertility rates of the country and year under examination and the marriage and mortality rates of the standard population.

REPRODUCTIVITY OF MARRIAGE COHORTS

Another approach to reproductivity is to deal with cohorts of marriages, i.e., with the subsequent fertility and mortality of couples marrying in a given year. If this analysis is to be brought to bear on population replacement, however, one must go on to consider the proportion of each sex that eventually marries, age at marriage, dissolution of marriage by death or divorce, remarriage, and illegitimate fertility. Most of the computations of reproductivity for marriage cohorts have been confined to gross reproduction rates and have not been extended to net reproduction rates.

Using data collected in the 1946 Family Census in Great Britain, Glass and Grebenik estimated the net fertility of specific cohorts of marriage as follows:[42]

(1) They constructed generation life tables for males and females.

(2) For three marriage periods, they distributed marriages by age of bride and jointly by age of groom, using 5- and 10-year age groups respectively.

(3) For these joint age groups, they computed joint survival rates from the generation life tables.

(4) These joint survival rates were applied to the corresponding duration-specific fertility rates from the Family Census tabulations.

(5) These results were weighted by the age distribution of marriages for the corresponding cohort, so as to obtain the overall net fertility of that marriage cohort.

These calculations were regarded as sort of an exercise by the authors, and they regarded their estimates of generation replacement rates on a birth-cohort basis as being more useful. "To go beyond these estimates of net marital fertility and to apply the concept of replacement, means introducing the missing factors into the calculations—in this case, premarital mortality, the probability of first marriage, the chances of dissolution of marriage by divorce as well as death, the likelihood of remarriage, and the contribution of illegitimacy."[43] Even though Glass and Grebenik did not then undertake these further calculations, their work called for very detailed tabulations, exceedingly complex calculations, and many assumptions. This major reliance on date of marriage rather than on the mother's date of birth arose, in part, from the fact that the latter item was not included on the birth certificate in England and Wales until 1938. Nonetheless, taking account of these factors regarding marriage has enriched the analysis of fertility trends and may have provided a better basis for projecting fertility into the future.

The use of marriage cohorts is one way of avoiding inconsistent male and female reproduction rates. These inconsistencies were discussed earlier, and the next section deals explicitly with male reproductivity.

REPRODUCTION OF THE MALE POPULATION

Although both mother and father are involved in every birth, it has been customary to disregard this biological fact and to measure fertility and reproduction primarily for females and only secondarily for males and the total population. Paternal rates could be used as easily as maternal rates, however, if adequate data were collected. Reproduction rates calculated directly for males do occur in the literature.[44]

A male reproduction rate, in terms of the number of sons sired by a "synthetic cohort" of males starting life together, was first published by Kuczynski.[45] The first male reproduction rates based on American data were published by Myers.[46]

Paternal reproduction rates are particularly useful in measuring the replacement of various social and economic groups defined by a characteristic of the father, such as occupation. To obtain the necessary fertility data by such characteristics of the father, census or survey data on men by number of children are usually needed.

In a previous section the conventional replacement indexes, using all children and women, have been discussed. The numerator in formula (36) could have been restricted to girls. Here we will illustrate formulas for replacement indexes for males and for both sexes combined. For males, then, the formula may be written:

$$\text{RI}_m = \frac{P^m_{0-4}}{P^m_{15-54}} \div \frac{L^m_{0-4}}{L^m_{15-54}} \qquad (49)$$

where the two ratios are for the actual and life table male populations, respectively. Similarly, for both sexes, we may write

$$\text{RI}_t = \frac{P^t_{0-4}}{P^t_{15-54}} \div \frac{L^t_{0-4}}{L^t_{15-54}} \qquad (50)$$

These indexes are easy to compute since they call only for the age distribution and a life table; they are especially useful for countries where adequate birth registration does not exist.

RELATIONSHIPS BETWEEN VITAL RATES AND AGE STRUCTURE IN ACTUAL AND STABLE POPULATIONS

The stable population model has been used, (1) to assist in analyzing implications of vital rates, (2) to show the manner in which population would grow over time under specified conditions, (3) to estimate vital rates of populations of statistically underdeveloped countries,[47] and (4) to evaluate, estimate, or correct age distributions of such countries. The first two uses of the stable population model are connected with its use in projections of populations for developed countries

---

[42] D. V. Glass and E. Grebenik, "The Trend and Pattern of Fertility in Great Britain," in *Papers of the Royal Commission on Population*, Vol. VI, Part I, *Report*, 1954, Appendix 2 to chapter 6, pp. 136–137.

[43] Ibid., p. 278.

[44] W. A. B. Hopkin and J. Hajnal, "Analysis of the Births in England and Wales, 1939, By Father's Occupation," (2 parts), *Population Studies* (London), 1(2):187–203, September 1947, and 1(3):275–300, December 1947; Robert R. Kuczynski, *Fertility and Reproduction*, New York, Falcon Press, 1932, chapters 3 and 5; Christopher Tietze, "Differential Reproduction in England," *Milbank Memorial Fund Quarterly*, 17(3):288–293, July 1939.

[45] Robert R. Kuczynski, op. cit., pp. 36–38.

[46] Robert J. Myers, "The Validity and Significance of Male Net Reproduction Rates," *Journal of the American Statistical Association*, 36(214):275–282, June 1941.

[47] United Nations, *The Aging of Population and Its Economic and Social Implications*, Series A, Population Studies, No. 26, 1956, pp. 25–31; United Nations, *Methods of Estimating Basic Demographic Measures from Incomplete Data*, Manual IV, Series A, Population Studies, No. 42, 1967; United Nations, *The Concept of a Stable Population: Application to the Study of Populations of Countries with Incomplete Demographic Statistics*, Series A, Population Studies, No. 39, 1968.

and also with studies of past trends designed to measure the relative contribution of the components of population growth to total growth, especially by age.

One of the most useful kinds of demographic analysis is that of the relationship between changes in the components of population change and changes in age-sex structure. This analysis has been both historical and theoretical. The empirical, historical analysis has reached some surprising conclusions, upsetting, for example, intuitive beliefs concerning the relative importance of fertility and mortality in the aging of western populations since the turn of the century. The basic approach is to "hold constant" one component (fertility, mortality, migration) at a time and to use the actual historical values of the other components. The contribution of each component to changes in age structure can thus be. isolated.

An analysis of changes in the American population by Hermalin reached conclusions that seem to apply to many other western countries.[48] Hermalin's article contains a good bibliography of similar studies.[49] He extended and refined the work of Valaoras and like him considered future as well as past changes.

Hermalin's analysis established that:

"1. The measure of mortality that determines the effect of changes in mortality on age composition and growth is the relative change in survival rates, rather than the relative change in the corresponding mortality rates.

"2. Improvements in mortality can be made which have no effect at all on the age distribution. Specifically, a proportional increase in the probability of surviving a fixed number of years, which was the same at all ages, would leave the age distribution unaffected though it would increase the growth rate.

"3. The effect of mortality improvement on the growth rate varies with the initial level of mortality as well as the magnitude of the change. A change in expectation of life at birth from 30 to 40 years, for example, will produce a higher growth rate than a change from 60 to 70 years, for a given level of fertility."[50]

Furthermore, it was found that:

(1) Immigration has led to a younger population.

(2) Changes in fertility have led to a marked aging of the population.

(3) Whether declining mortality rates have served to age or rejuvenate the population depends on the measure one wishes to employ. On balance, however, declining mortality from 1900 to 1960 has led to a somewhat younger population in the United States. Coale had shown earlier that the reason why declining mortality had had so little effect on age structure lay in the U-shaped or J-shaped age pattern of historical improvements in survival rates.[51] The reduction of mortality

has had a considerable effect on population growth, however; the 1960 population is estimated to have been about one-third larger as the result of improvements in survival rates since 1900. In the rest of this century, on the other hand, further improvements in mortality could have only a slight effect on population growth and fertility will be the chief determinant.

The contributions of these factors to the increase of the population 65 years old and over in the United States was examined by Hornseth.[52] Hornseth was interested in the absolute size of the elderly population rather than in the proportion of elderly persons in the total population.

He concluded that the most important element in the sharp increase in the aged population during the past 50 years had been the rapid increase in births in the last half of the nineteenth century and that the large immigration in the first quarter of the century and the reduction in mortality ranked much lower and in that order.

We may also start with a stable population and vary one of the two factors, expectation of life and gross reproduction rate, the other remaining constant.[53] This is a simpler approach than projecting age-specific mortality and fertility rates. The underlying logic is that the expectation of life characterizes a particular life table (and hence its age-specific mortality rates) and that the gross reproduction rate is largely independent of the particular profile of age-specific fertility rates so long as their total is known. Having changed $\mathring{e}_o$ and/or the GRR then, we can determine the new stable population that will eventually result.

An article by Keyfitz, too recent to have been cited by Hermalin, makes a number of important theoretical contributions.[54] We recall that Coale had shown that the effect of declines in mortality rates upon age structure depends on the age-pattern of these declines. Keyfitz's analysis suggested that ". . . a *neutral* change in mortality may be defined as one which is either constant at all ages or, without being constant, has an incidence such that the age distribution of the population, or at least its mean age, is unaffected. If the incidence of improved mortality is on balance at younger ages, then the age distribution becomes younger, and vice versa."[55] He derived an index, or set of weights, that tells us whether the change is neutral, and, if not, whether it falls on the younger or the older side of neutrality. Keyfitz also gives expressions for decomposing changes in the intrinsic rate of natural increase and the net reproduction rate into component factors.

SOME CONCLUSIONS

The reader who has persisted through the involved discussion in this chapter on reproductivity may have joined the authors in reaching several conclusions.

1. The conventional reproduction rates may be readily computed if the necessary in-put statistics are available, but they must be interpreted with considerable caution.

2. In areas of fertility decline, the answer to whether population is reproducing itself requires statistics for real cohorts.

[48] Albert I. Hermalin, "The Effect of Changes in Mortality Rates on Population Growth and Age Distribution in the United States," *Milbank Memorial Fund Quarterly*, 44(4):451–469, October 1966, Part 1.

[49] Several of these are: Ansley J. Coale, "How the Age Distribution of a Human Population is Determined," *Cold Spring Harbor Symposia on Quantitative Biology*, 1957, Vol. 22, pp. 83–89; United Nations, "The Cause of the Ageing of Populations: Declining Mortality or Declining Fertility?", *Population Bulletin*, No. 4, pp. 30–38, December 1954; Vasilios G. Valaoras, "Patterns of Aging of Human Populations," *The Social and Biological Challenge of Our Aging Population, Proceedings of the Eastern States Health Education Conference, March 31–April 1, 1949*, New York, Columbia University Press, 1950, pp. 67–85.

[50] Hermalin, op. cit., pp. 451–452.

[51] Ansley J. Coale, "The Effect of Declines in Mortality on Age Distribution," in *Trends and Differentials in Mortality*, New York, Milbank Memorial Fund, 1956, pp. 125–132.

[52] Richard A. Hornseth, "Factors on the Size of the Population 65 Years and Older," unpublished paper summarized in *Population Index*, 19(3):181–182, July 1953.

[53] United Nations, *The Aging of Populations and Its Economic and Social Implications*, pp. 25–31.

[54] Nathan Keyfitz, "Changing Vital Rates and Age Distributions," *Population Studies* (London), 22(2):235–251, July 1968.

[55] Ibid., p. 237.

3. Despite the application of considerable ingenuity and rigorous analysis, mathematical demographers have not been able to develop a consistent set of equations describing the reproductivity of the male and female populations.

4. Nonetheless, stable population concepts have been found to have many useful applications beyond those described by Lotka. For example, in underdeveloped countries with an actual population that is roughly stable or quasi-stable (constant fertility but declining mortality), the model can be used to approximate the age structure and vital rates.

5. Empirical approaches, such as cohort fertility tables, usually require a great deal of computation, including estimates and adjustments because of gaps and shortcomings in the data. Hence, for some time to come, such tables may be produced by only a few specialists with adequate resources and experience.

6. The electronic computer, however, has opened up new possibilities for processing detailed input on reproductivity and for deriving a wide variety of measures relating to replacement and to the stable population. (Unfortunately, these procedures are so complex that it is difficult to illustrate them in a textbook.)

SUGGESTED READINGS

Charles, Enid. "Differential Fertility in Scotland, 1911–1931." *Transactions of the Royal Society of Edinburgh*, 59(3):673–686, 1938–1939.

Coale, Ansley J. "A New Method for Calculating Lotka's *r* – The Intrinsic Rate of Growth in a Stable Population." *Population Studies* (London), 11(1):92–94, July 1957.

————. "The Calculation of Approximate Intrinsic Rates." *Population Index*, 21(2):94–97, April 1955.

————. "Convergence of a Human Population to Stable Form." *Journal of the American Statistical Association*, 63(322):395–435, June 1968.

————. "The Effect of Declines in Mortality on Age Distribution" in *Trends and Differentials in Mortality*, New York, Milbank Memorial Fund, 1956. Pp. 125–132.

————. "How the Age Distribution of a Human Population is Determined." *Cold Spring Harbor Symposia on Quantitative Biology, 1957*, Vol. 22, pp. 83–89.

Coale, Ansley J., and Tye, C. Y. "The Significance of Age-Patterns of Fertility in High Fertility Populations." *Milbank Memorial Fund Quarterly*, 39(4):631–646, October 1961.

Depoid, P. "Reproduction nette en Europe depuis l'origine de statistiques de l'état civile." *Etudes démographiques*, No. 1. Statistique générale de la France, Imprimerie nationale. 1941.

Dublin, Louis I., and Lotka, Alfred J. "On the True Rate of Natural Increase." *Journal of the American Statistical Association*, 20(150):305–339, September 1925.

Dorn, Harold F. "Pitfalls in Population Forecasts and Projections." *Journal of the American Statistical Association*, 45(251):320–322, September 1950.

Germany. Statistisches Reichsamt. "Die Lebensbilanz des deutschen Volkes im Jahre 1933." *Neue Beiträge zum deutschen Bevölkerungs-problem*, Supplement to *Wirtschaft und Statistik*, No. 15, 1935. Pp. 53–82.

Glass, D. V. *Population Policies and Movements*, New York, Oxford University Press, 1940. Pp. 383–415.

Glass, D. V., and Grebenik, E. "The Trend and Pattern of Fertility in Great Britain" in *Papers of the Royal Commission on Population*, Vol. VI, Part I, *Report*, 1954, Appendix 2 to Chapter 6, pp. 136–137.

Grabill, Wilson H., and Cho, Lee Jay. "Methodology for the Measurement of Current Fertility from Population Data on Young Children." *Demography*, 2:50–73, 1965.

Hajnal, John. "The Analysis of Birth Statistics in the Light of the Recent International Recovery of the Birth-Rate." *Population Studies*, 1(2):137–164, September 1947.

————. "Births, Marriages, and Reproductivity, England and Wales, 1938–1947" in *Papers of the Royal Commission on Population*, Vol. II, *Reports and Selected Papers of the Statistics Committee*, London, H. M. Stationery Office, 1950. Pp. 303–400.

————. "Some Comments on Mr. Karmel's Paper 'The Relations between Male and Female Reproduction Rates'." *Population Studies* (London), 2(3):354–360, December 1948.

————. "The Study of Fertility and Reproductivity: A Survey of Thirty Years." Pp. 11–37 in *Thirty Years of Research in Human Fertility: Retrospect and Prospect*, Papers presented at the 1958 Annual Conference of the Milbank Memorial Fund, October 22–23, 1958, Part II, New York, Milbank Memorial Fund, 1959.

Hermalin, Albert I. "The Effect of Changes in Mortality Rates on Population Growth and Age Distribution in the United States." *Milbank Memorial Fund Quarterly*, 44(4):451–469, October 1966, Part I.

Hopkin, W. A. B., and Hajnal, J. "Analysis of the Births in England and Wales, 1939, by Father's Occupation." (2 parts) *Population Studies* (London), 1(2):187–203, September 1947, and 1(3):275–300, December 1947.

Hyrenius, Hannes. "La mesure de la reproduction et de l'accroissement naturel." *Population* (Paris), 3(2):271–292, April–June 1948.

Karmel, P. H. "An Analysis of the Sources and Magnitudes of Inconsistencies Between Male and Female Net Reproduction Rates in Actual Populations." *Population Studies* (London), 2(2):240–273, September 1948.

————. "A Rejoinder to Hajnal's Comments." *Population Studies* (London), 2(3):361–372, December 1948.

————. "The Relations Between Male and Female Nuptiality

in a Stable Population." *Population Studies* (London), 1(4):353–387, March 1948.

———. "The Relations Between Male and Female Reproduction Rates." *Population Studies* (London), 1(3):249–274, December 1947.

Keyfitz, Nathan. "The Intrinsic Rate of Natural Increase and the Dominant Root of the Projection Matrix." *Population Studies* (London), 18(3):293-308, March 1965.

———. "Changing Vital Rates and Age Distributions." *Population Studies* (London), 22(2):235-251, July 1968.

Keyfitz, Nathan and Flieger, Wilhelm. *World Population: An Analysis of Vital Data*, Chicago, University of Chicago Press, 1968.

Kuczynski, Robert R. *Fertility and Reproduction*, New York, Falcon Press, 1932.

———. *The Measurement of Population Growth*, New York, Oxford University Press, 1936.

Lorimer, Frank. "Dynamics of Age Structure in a Population with Initially High Fertility and Mortality" in United Nations. *Population Bulletin*, No. 1, December 1951. Pp. 31–41.

Lotka, Alfred J. "The Geographic Distribution of Intrinsic Natural Increase in the United States, and an Examination of the Relation Between Several Measures of Net Reproductivity." *Journal of the American Statistical Association*, 31(194):273–294, June 1936.

———. "Relation Between Birth Rates and Death Rates." *Science*, New Series, 26(653):21–22, July 5, 1907.

Myers, Robert J. "The Validity and Significance of Male Net Reproduction Rates." *Journal of the American Statistical Association*, 36(214):275–282, June 1941.

Pressat, Roland. *Principes d'analyse: cours d'analyse démographique de l'Institut de démographie de l'Université de Paris*. Editions de l'Institut national d'études démographiques, Paris, 1966. Pp. 51–59 and 113–123.

Spiegelman, Mortimer. *Introduction to Demography*, rev. ed., Cambridge, Mass., Harvard University Press, 1968. Pp. 283–292.

Stolnitz, George J., and Ryder, Norman B. "Recent Discussion of the Net Reproduction Rate." *Population Index*, 15(2): 114–128, April 1949.

Thompson, Warren S. *Ratio of Children to Women: 1920*. Census Monograph XI, Washington, U.S. Bureau of the Census, 1931. Pp. 157–174.

Tietze, Christopher. "Differential Reproduction in England." *Milbank Memorial Fund Quarterly*, 17(3):288–293, July 1939.

United Kingdom. Royal Commission on Population. *Report*,

London, H. M. Stationery Office, 1949. Pp. 60–63; 241–258.

United Nations. *The Aging of Population and Its Economic and Social Implications*, Series A, Population Studies, No. 26, 1956. Pp. 25–31.

———. "The Cause of the Aging of Populations: Declining Mortality or Declining Fertility?" *Population Bulletin*, No. 4, pp. 30–38, December 1954.

———. *The Concept of a Stable Population: Application to the Study of Populations of Countries with Incomplete Demographic Statistics*. Series A, Population Studies, No. 39, 1968.

———. *Demographic Yearbook, 1965*. Pp. 35–36; 605–617.

———. *Methods of Estimating Basic Demographic Measures from Incomplete Data*, Manual IV, Series A, Population Studies, No. 42, 1967.

U.S. Bureau of the Census. *Current Population Reports*. Series P–20, No. 147. "Fertility of the Population: June 1964 and March 1962." January 5, 1966, especially p. 6.

———. *Sixteenth Census of the United States: 1940. Population. Differential Fertility: 1940 and 1910. Standardized Fertility Rates and Reproduction Rates*. 1944. Pp. 3–5; 39–40.

U.S. National Office of Vital Statistics. *Vital Statistics. Special Reports. Selected Studies*. Vol. 51, Part 1. "Fertility Tables for Birth Cohorts of American Women" by Pascal K. Whelpton and Arthur A. Campbell, January 29, 1960.

Valaoras, Vasilios G. "Patterns of Aging of Human Populations." *The Social and Biological Challenge of Our Aging Population, Proceedings of the Eastern States Health Education Conference, March 31–April 1, 1949*. New York, Columbia University Press, 1950. Pp. 67–85.

Vincent, Paul. "De la mesure du taux intrinsèque d'accroissement naturel dans les populations monogames." *Population* (Paris), 1(4):699–712, October–December 1946.

Whelpton, P. K. *Cohort Fertility: Native White Women in the United States*, Princeton, New Jersey, Princeton University Press, 1954. Especially pp. 480–492.

———. "Reproduction Rates Adjusted for Age, Parity, Fecundity, and Marriage." *Journal of the American Statistical Association*, 41(236):501–516, December 1946.

Wicksell, S. D. "Nuptiality, Fertility, and Reproductivity." *Skandainavisk Aktuarietidskrift*, 14(3):125–157, 1931.

Wolfenden, Hugh H. *Population Statistics and Their Compilation*, Published for the Society of Actuaries. University of Chicago Press, 1954. Pp. 220–224.

Woofter, T. J. "Completed Generation Reproduction Rates." *Human Biology*, 19(3):133–153, September 1947.

———. "The Relation of the Net Reproduction Rate to Other Fertility Measures." *Journal of the American Statistical Association*, 44(248):501–517, December 1949.

CHAPTER 19

Marriage and Divorce

INTRODUCTION

The role of marital status in demographic analysis was treated in chapter 10. Marital status provides a static representation of the population of an area with respect to its marital composition. This chapter focuses on the dynamic aspect of marriage, i.e., the nature of moves from one marital status to another and the correlates thereof. The material for chapter 10 was drawn mainly from censuses and surveys, whereas this chapter draws on material derived from vital statistics systems; however, censuses and surveys are discussed to some extent in this chapter as they sometimes include questions about vital events occurring in the past and can provide estimates of the number and rate of occurrence of such events.

Marriage and divorce rates directly measure changes in population composition characteristics rather than changes in population size as do the other rates treated in this section on population dynamics—fertility, mortality, and migration. But if the change in population size due to births is considered as a function of the broader process of one generation's reproducing another—a process that in most societies takes place largely through formation of families that produce offspring—then the rates of family formation and dissolution are appropriate for consideration under a section on population dynamics.

Despite the fact that nuptiality statistics have important demographic, economic, and social implications, they have received relatively little attention compared to natality and mortality statistics. The study of **nuptiality** according to the *Multilingual Demographic Dictionary* ". . . deals with the frequency of marriages . . .; with the characteristics of persons, united in marriage; and with the dissolution of such unions." [1] The term thus bears roughly the same relationship to marriage and divorce as natality (or fertility) does to birth and mortality does to death.

DEFINITIONS OF TERMS

The United Nations recommends that marriage and divorce statistics be collected by the national registration method. The paragraphs below present their recommended definitions of marriage, divorce, and related events.

Marriage entails a change from any other marital status to the status of "married." In accordance with the recommendation made in *Principles for a Vital Statistics System*, marriage

refers to "the legal union of persons of opposite sex. The legality of the union may be established by civil, religious, or other means as recognized by the laws of each country." [2] There are basically two types of marriage: **first marriage,** which refers to persons moving from the marital status "single" to the status "married," and **remarriage,** which includes persons moving from the "divorced" or "widowed" status to the "married" status.

The concept of **consensual union** (sometimes referred to as stable *de facto* union) refers to the establishment of a marital union without recorded legal sanction. In many countries, particularly those in Central and South America, consensual unions constitute a large proportion of all marriages.

The legal marriage contract can be broken in three ways: (1) by a divorce decree, (2) by annulment, and (3) by death of one of the spouses.

Divorce is defined as ". . . a final legal dissolution of a marriage, that is, the separation of husband and wife by a judicial decree which confers on the parties the right to civil and/or religious remarriage, according to the laws of each country." [3]

Annulment is another of the ways in which the marriage contract can be dissolved. An annulment makes the contract void from its inception, whereas a divorce makes the contract void as of the date of the decree. It is defined as "the invalidation or voiding of a marriage by a competent authority, according to the laws of each country, which confers on the parties the status of never having been married to each other." [4] Although most countries provide for annulment, the compilation of annulment statistics is by no means universal.

An annulment, like a divorce, is a legal decree voiding the marriage contract. Annulment, however, is distinct from divorce in that the former is granted because of some condition existing at the time of the contract, whereas the latter is granted because of a condition arising during the contract. As a result, annulment confers on the parties the status they held before the annulled marriage (usually "single"), whereas divorce confers the status of "divorced."

Becoming **widowed** is a third way in which the legal contract of marriage may be dissolved. Upon the death of one of the spouses, the surviving spouse experiences the event of widowhood. The United Nations does not recommend that direct information on widowhood be collected through the vital

---

[1] United Nations, *Multilingual Demographic Dictionary: English Section,* 1958, p. 31.

[2] United Nations, *Principles for a Vital Statistics System,* Statistical Papers, Series M, No. 19, 1953, p. 7.
[3] Ibid.
[4] Ibid.

statistics system, even though the characteristics of the spouse of the decedent would be of interest.

National Practices

Marriages. — For demographic purposes, it is necessary to define a marriage even though it is recognized that the concept of marriage is basically legal, ceremonial, or religious. Although most countries have officially adopted the recommended United Nations definition of marriage for statistical purposes, the national definition of marriage ultimately depends upon the law governing the civil contract or upon the tribal or customary rules governing the union. In some populations, unions by mutual consent without ceremonial or legal rites constitute formal, legal marriage contracts that are in no way different from those sanctioned by religious or other customary rites. "In some countries, Moslem law may allow the male to have a plurality of wives while under another law, in the same country, monogamy may be stipulated." [5]

The definition of marriage used by the National Center for Health Statistics of the United States follows the recommended definition of the United Nations. Although the requirements for obtaining a marriage license vary widely among the States, there are usually minimum age and health requirements and prohibitions regarding marriage within certain degrees of consanguinity. Variations in law and regulations are likely to have an effect on the comparability of marriage statistics among the States; for example, States with lenient marriage laws have higher marriage rates than States with stricter marriage laws. In many instances a religious ceremony is performed. Whether the ceremony be religious or civil, however, the marriage is not recorded in the vital statistics record system unless it has been registered with the proper authority.

Divorces. — The recommended U.N. definition limits divorces to final decrees that confer the right to remarry.

Divorce statistics are subject to the same kinds of legalistic considerations as marriage statistics. In fact, they are subject to even wider variability among countries. The situation concerning the dissolution of marriage ranges from no legal provisions whatsoever in predominantly Catholic countries (Argentina, Ireland, Italy, the Philippines, etc.) to the granting of divorce in response to a statement of intent by Moslem men. The incidence of divorce tends to vary inversely with the degree of difficulty in obtaining a decree. Annulment is frequently an alternative to divorce in countries where divorce decrees are not granted.

Annulments. — The vehicle frequently used for dissolving marriages in areas with strict divorce laws is annulment. In countries or States where divorce is readily obtainable, the number of annulments tends to be relatively small. In countries where divorce is not readily obtainable, such as Italy, annulments are sometimes substituted for divorce and, as a result, the frequencies are considerably higher. In the United States, in States like Nevada, where divorce is a relatively simple procedure, annulments are seldom granted; but in New York State, which until recently had only a few grounds for divorce, annulments have accounted for a large proportion of marriage dissolutions.

MARRIAGES

Sources of Data

Data on marriage are obtained directly from national vital statistics registration systems or, less often, from continuous population registers. In addition, data on marital status available from censuses and sample surveys in most countries may be used to estimate the number of marriages and marriage rates. This indirect method represents the only source in countries where census data are satisfactory but registration data are lacking or inadequate. Only a few of the economically underdeveloped countries in Africa, Asia, and the Americas publish registration statistics on marriages that the United Nations considers relatively complete in coverage.

International. — Perhaps the best single source of international marriage statistics is the *Demographic Yearbook* of the United Nations. Two tables are shown annually — one on marriages and crude marriage rates for the last several years and the other on age of bride and age of groom (not cross-classified) for the latest year available.

Many countries publish yearly marriage statistics in statistical yearbooks. Tabulations peculiar to the individual country are often found in these yearbooks along with tabulations of the form found in the U.N. *Demographic Yearbook.*

United States. — It was mentioned in chapters 2 and 3 that marriage and divorce statistics have lagged behind mortality and natality statistics in the United States. Even now the Marriage Registration Area (MRA) is still incomplete.

The National Center for Health Statistics (NCHS) is the agency responsible for collecting, tabulating, and publishing marriage data at the national level. Two methods are used by NCHS to collect statistics on marriages (and on divorces and annulments). Total numbers for States and counties are complete counts provided by State and local officials in all parts of the United States. [6] Data on the characteristics of these events and the persons involved in them are based on samples of marriage records from States and other registration areas in the MRA.

NCHS obtains State and county data on total numbers of marriages from those States with central files, whether or not they are in the Marriage Registration Area. For the remainder of the States, these data are collected either from State officials or from county officials. It should be noted, however, that in some instances the marriage data relate to licenses issued rather than ceremonies performed; and, for several areas, the numbers are estimated by NCHS in order to arrive at national totals. Further, statistics are compiled on the basis of place of occurrence rather than place of usual residence of either bride or groom.

NCHS publishes marriage data in three forms: (1) the *Monthly Vital Statistics Report* (provisional monthly State figures, with an annual supplement giving advance final figures; monthly national totals are adjusted to approximate expected final totals), (2) the annual volume on *Marriage and Divorce* in *Vital Statistics of the United States,* and (3) the *Marriage Statistics Analysis* report in Series 21.

There are two types of annual supplements to the monthly series. One gives an annual summary of the provisional statistics for births, deaths, marriages, and divorces. The other contains advance final figures, separately, for the major vital events.

In addition to the conventional item on marital status, decennial censuses and surveys in the United States have included such topics as whether married more than once, age at first marriage, and duration of marriage. There was a subject report on *Age at First Marriage* from the 1960 census, for example. Some of the data included in this report are age at

---

[5] United Nations, *Handbook of Vital Statistics Methods,* p. 60.

[6] U.S. National Center for Health Statistics, *Vital Statistics of the United States,* Vol. III, *Marriage and Divorce,* p. 4–3.

first marriage, duration of marriage, times married, and difference between ages of husband and wife, with cross-classifications by race, ethnic origin, education, income, and occupation.

The Current Population Survey has also been used to collect retrospective data on marriages.

Uses and Limitations

Statistics on marriage are obviously of interest to demographers and others for the light they shed on family formation, family composition, and fertility. Analysis proceeds by studying marriages as a component in the growth of the married population, which in turn is related to many other demographic, social, and economic phenomena. The trend in the number of marriages has implications for housing requirements and community services. Private industry is interested in marriage and in marital status as factors in labor supply and market analysis. The female labor supply is particularly affected by changes in marital status. The size of the unmarried population by age and that of the married population give, respectively, the population "at risk" of marriage and divorce. Marriage statistics provide guidance for housing programs, community planning, and the development of enterprises concerned with providing for the needs of newly-married couples, e.g., life and health insurance, real estate enterprises, firms producing or selling furniture and clothing and businesses that produce or sell infant clothing, food, and pharmaceuticals.

In making population projections by the component method, age-specific marriage rates can be used to refine the projections of both fertility and mortality. Forecasting of marriages and hence of marital status is necessary for the setting up of annuity systems, social security, old-age benefits, and the like.[7] The ages of the bride and groom have implications for the couple's fertility, the probability of divorce or widowhood in a specified period of subsequent years, and the consumption of certain goods and services. Differentials in marriage rates in terms of religious or ethnic group, socioeconomic class, urban and rural residence, etc., may be studied from the standpoints of both their genesis and their consequences.

Marriage statistics are subject to the general limitations of vital statistics that have been discussed in chapters 14 and 16. In addition, international comparability is greatly affected by the variations in the national forms of marriage and, to a lesser extent, in statistical definitions.

Perhaps the most serious limitation arises in connection with statistics for countries where consensual unions are prevalent. For example, the crude marriage rate for El Salvador in 1966 was only 3.3 per 1,000, whereas for European countries it ranged from 5.8 (Ireland) to 9 or 10. Characteristics of marriages, such as age at marriage and previous marital status, are also affected by the omission of consensual marriages. This limitation is particularly prominent in many South and Central American countries.

In some areas, including parts of the United States, the statistics relate to licenses issued rather than to ceremonies performed; and, accordingly, they may overstate the actual number of legal marriages.

Still another consideration is the time reference. In about 38 percent of the 148 countries collecting data on marriage and divorce in 1966, the information was collected by date of registration rather than by date of occurrence. In many of these countries, however, the difference is very slight. It is only in

countries like Japan where registration is not required to legalize the marriage that the delay is often appreciable.

Another limitation involves the distinction between place of registration and place of residence of bride and groom. For legal and other reasons, couples are often married at a place other than their usual place of residence. If marriage data are intended to indicate the number of newly formed family or subfamily units in the area under consideration, then this distinction may be considered a limitation, as not all marriages registered in an area have equal relevance for the area. Along these lines a problem also arises for the geographic assignment of marriages where the bride and groom had different residences before marriage. In the United States, this limitation cannot be eliminated until all States are in the MRA so that allocation back to State of usual residence can be made.

Quality of the Statistics

The quality of statistics on marriage varies widely, with a major factor being variations in the completeness of registration. Part of the incompleteness is a result of the lack of uniformity and universality in national legislation. In some countries, for example, there is no national provision for compulsory registration; in others the established "registration area" comprises only part of the country; and in still other countries certain segments of the population may be excluded from the statistics.

In the United States, States that are accepted for the Marriage Registration Area agree to tests of the completeness of accuracy of marriage registration in cooperation with the National Center for Health Statistics. To date about nine States have conducted checks of records filed in State offices against the original records in county or local offices. About 99 percent were found in all cases.

Measures

There are many measures of nuptiality that are based primarily on vital statistics. These measures vary according to their degree of refinement, whether they are summary measures or specific measures, and the aspect of marriage described.

Crude Marriage Rate. – The index most commonly used to describe the incidence of marriage is the crude marriage rate. If M is the total number of marriages among residents in an area during the year, and P is the average number of persons

Table 19-1. – **Crude Marriage Rates for Selected Countries: 1965**

[Rates per 1,000 of the population]

Country	Rate	Country	Rate
Cape Verde Islands	3.8	Taiwan	7.4
Mauritius	5.4	Japan	*9.7
Barbados	3.9	Czechoslovakia	*7.9
Bermuda	8.8	West Germany	8.3
Canada	7.4	Gibraltar	19.9
Mexico	6.8	Ireland	5.9
Puerto Rico	*10.2	United Kingdom	7.7
United States	*9.2	Australia	8.2
Chile	7.6	U.S.S.R.	8.7

*Provisional.

Source: United Nations, *Demographic Yearbook, 1966*, table 24.

living in the area during the year, then the formula for the crude marriage rate μ is:

$$\mu = \frac{M}{P} \times 1,000 \tag{1}$$

Crude marriage rates from selected countries (table 19–1) indicate a low of 3.8 for the Cape Verde Islands and a high of 19.9 for Gibraltar. Some of the factors that contribute to variation in the crude marriage rate are: (1) the proportion of the population of marriageable age, (2) the proportion of these who have previously married, (3) the area's economic situation, (4) provisions for the dissolution of marriage contracts, and (5) customs pertaining to remarriage. The very high rate for Gibraltar results mainly from the exclusion from the base of the armed forces stationed there.

General Marriage Rate. — Although the crude marriage rate is useful as a gross measure of the relative frequency of marriages in the same area over a relatively short period of time and for crude international comparisons, it does not take account of the variation between areas in "marriageable" population, which may be derived from the marital status and the age distribution of the population. An initial refinement consists of restricting the population to persons of marriageable age and may be represented by:

$$\frac{M}{P_{15+}} \times 1,000 \tag{2}$$

The total number of marriages in the year is again employed in the numerator whereas the population 15 years of age and older is now employed in the denominator.

Further refinements involve taking account of the marital-status and sex composition along with the age composition. The general marriage rate for the total marriageable population may be represented by:

$$\frac{M}{P_{15+}^{un}} \times 1,000 \tag{3}$$

where the total number of marriages is still employed in the numerator but now the unmarried population 15 years of age and older is employed in the denominator. (The unmarried population is the sum of the single, divorced, and widowed populations, and, often, of the consensually married population as well.)

The corresponding ratios for males and females will **differ** from each other in accordance with the level of the sex ratio in the marriageable ages. For females, for example, we may write

$$\frac{M}{P_{15+,f}^{un}} \times 1,000 \tag{4}$$

Such a rate is roughly twice the order of magnitude of the rates defined in formula (3).

The general marriage rate for the marriageable population provides a better measure of the nuptiality level among different populations and also a better index to changes over time in the same population. It has the disadvantage, however, of requiring a population base that is frequently not available except for census years.

Age-Sex Specific Rates. — General marriage rates that take partial account of the marital status and age distribution of a population allow certain comparisons to be made between countries, but internal variations in the age composition within the population of marriageable age are not taken into account by these rates. In order to eliminate the effects of differing age-sex compositions, age-sex specific rates may be computed. Since marriage involves the union of two persons of the opposite sex but usually not of the same age, the rates, to be meaningful, must be both age and sex specific. Marriage rates for males are published and analyzed more frequently than are fertility rates for males.

An age-sex specific marriage rate is defined as the number of marriages to persons of a given age-sex group per 1,000 persons in the age-sex group: [8]

$$\mu_a = \frac{M_a^{m \, or \, f}}{P_a^{m \, or \, f}} \times 1,000 \tag{5}$$

The formula for the marriage rate of males 25 to 29 years old is, then,

$$\frac{M_{25-29}^m}{P_{25-29}^m} \times 1,000 \tag{6}$$

Order-Specific Rates. — The demographic and the sociological analysis of marriage often calls for a distinction between first marriages and remarriages. The two types tend to have very different characteristics, for example, average age. More generally, we are interested in the order of the marriage. Although in some instances it is possible to compute the rate for marriages of a specific higher order (2nd, 3rd, etc.), the usual procedure is to compute a remarriage rate that includes all marriages other than first marriages. An order-specific general marriage rate is defined as the number of marriages of a given order i during a year per 1,000 persons 15 years old and over, or per 1,000 persons 15 years old and over at order $i-1$, at the middle of the year:

$$\frac{M_i}{P_{i-1}^{15+}} \times 1,000 \tag{7}$$

The first marriage rate may then be represented as:

$$\frac{M_1}{P_s^{15+}} \times 1,000 \tag{8}$$

where M_1 refers to marriages of the first order, and P_s^{15+} refers to the single population 15 years old and over. Similarly, the remarriage rate may be represented as:

$$\frac{M_{2+}}{P_{w+d}} \times 1,000 \tag{9}$$

where P_{w+d} is the widowed and divorced population.

These rates are frequently made specific for age and sex. The formula for the first-marriage rate among females 20 to 24 years old is, then,

$$\frac{M_{20-24}^{f,1}}{P_{20-24}^{f,s}} \times 1,000 \tag{10}$$

where $M_{20-24}^{f,1}$ represents the number of first marriages of females aged 20–24, and $P_{20-24}^{f,s}$ refers to the number of single females at age 20–24.

---

[8] Throughout this section we have used the subscript "a" to represent age. Strictly speaking, it should be used only to represent a single year of age. The adaptation of the formula to the conventional statistics in 5-year age groups is straightforward, however.

The remarriage rate for females at age (or age group) a may be represented by the formula:

$$\frac{M_a^{f,\,2+}}{P_a^{f,\,w+d}} \times 1,000 \qquad (11)$$

Remarriage rates can be prepared by parallel methods for only widowed (or only divorced) persons by limiting the numerator to widowed, or divorced) persons remarrying and the denominator to persons who are currently widowed (or divorced).

At any age, remarriage rates tend to be higher for males than for females. There is also considerable variation in remarriage rates among countries. Much of the variation is due to three factors (1) the incidence of divorce, (2) differential mortality (particularly relevant at older ages), and (3) norms and laws pertaining to remarriage.

Standardized Rates. — Comparative marriage analysis, like comparative fertility or mortality analysis, may make use of standardized rates. Again, as compared with rates specific for age and other factors (urban-rural residence, race or ethnic group, marital status, etc.), standardized rates have the advantages and the disadvantages of summary measures. Although marriage rates, like other vital rates, may be standardized either by the direct or indirect method, depending on the availability of the data and other considerations, only standardization by the direct method is discussed in this chapter.

Age-sex standardized marriage rates. — Standardized rates may approximate the level of the crude marriage rate or of the general marriage rate, depending on whether the total population or the population of marriageable age is used. Rates may be standardized by age, etc., for the male and female populations separately; or they may be standardized simultaneously by age and sex for the population of both sexes combined.

The equivalent of a crude rate standardized for age only and for females is given by:

$$\sum_{a=15}^{\infty} \frac{\mu_a^f P_a^f}{P} \times 1,000 \qquad (12)$$

where μ_a^f equals the age-specific female marriage (nuptiality) rates in a particular population, P_a^f equals the number of females by age in the standard population, and P equals the total of the standard population. The weights, $\dfrac{P_a^f}{P}$, are the proportions of the total population made up of females of a given age. The use of the total population in the denominator provides an adjusted rate of a magnitude similar to that of the crude marriage rate. The denominator of the weights could be the total female population; but, in that case, the standardized rate would be about twice the level of the crude rate.

The adjusted marriage rate may also be computed so that it is of a magnitude similar to that of a general marriage rate. Let us consider a standardized rate for males of marriageable age. We may use the formula:

$$\frac{\sum\limits_{a=15}^{\infty} \mu_a^m P_a^m}{\sum\limits_{a=15}^{\infty} P_a^m} \times 1,000 \qquad (13)$$

where $\sum\limits_{a=15}^{\infty} P_a^m$ equals the number of males fifteen years of age and over in the standard population.

A further refinement would be to confine the rates to the unmarried population, thereby more closely approximating the population at risk of the event. This refinement results in a rate of a magnitude similar to the unstandardized rate for the unmarried population of marriageable age for a given sex.

$$\frac{\sum\limits_{a=15}^{\infty} \mu_a^{m,\,un} P_a^{m,\,un}}{\sum\limits_{a=15}^{\infty} P_a^{m,\,un}} \times 1,000 \qquad (14)$$

where $\mu_a^{m,\,un}$ refers to an age-specific nuptiality rate for unmarried males in a particular population, and $P_a^{m,\,un}$ refers to the number of unmarried males at age a in the standard population.

Age-standardized first marriage rates. — A still further possible refinement of the adjusted general marriage rate is to compute it separately for first marriages and remarriages. The standardized first-marriage rate for males is:

$$\frac{\sum\limits_{a=15}^{\infty} \mu_a^{m,\,1} P_a^{m,\,s}}{\sum\limits_{a=15}^{\infty} P_a^{m,\,s}} \times 1,000 \qquad (15)$$

where the superscript "1" denotes a first-marriage rate and the superscript "s," the single population, each at the given age.

As an example, the adjusted first-marriage rates for males in the United States (1960) and Israel (1966) were computed (table 19-2). Again, the rate for the United States is higher then the corresponding rate for Israel, 93 and 77, respectively, when the population of England and Wales is used as the standard.

Total Marriage Rate. — By analogy with the total fertility rate, TFR (see chapter 16), we may define the total marriage rate, TMR. The total fertility rate is a measure of completed family size, i.e., the total number of children 1,000 women will bear in a lifetime. The TMR indicates the total number of marriages in a synthetic cohort passing through life together. The age-specific rates are given equal weights. We have then

$$TMR = \sum_{a=15}^{\infty} \frac{M_a^m}{P_a^m} \times 1,000 \qquad (16)$$

where M_a^m is the **number** of marriages of males aged a.

Similarly, the total first marriage rate, $TFMR$, is obtained by summing age-specific first marriage rates and indicates the number of males or females who will ever marry per 1,000 males or females in a population. The total first marriage rate for females of marriageable age is computed by summing the age-specific nuptiality rates for females:

$$(TFMR)^f = \sum_{a=15}^{\infty} \frac{M_a^{f,\,1}}{P_a^f} \times 1,000 \qquad (17)$$

and for males by summing the age-specific nuptiality rates for males:

$$(TFMR)^m = \sum_{a=15}^{\infty} \frac{M_a^{m,\,1}}{P_a^m} \times 1,000 \qquad (18)$$

Table 19–2. — Age-Standardized First Marriage Rates for Males in Israel (Jews Only), 1966, and the United States, 1960, With the Single Population of England and Wales, 1961, as the Standard Population

Age	Male, age-specific first marriage rate		Single male population England and Wales (1961)	Expected first marriages	
	Israel (1966)	United States (1960)		Israel (1966) (1) x (3) =	United States (1960) (2) x (3) =
	(1)	(2)	(3)	(4)	(5)
Total.....................................	(X)	(X)	4,314,183	330,616	401,262
15 to 19 years........................	.006	.031	1,604,910	9,629	49,752
20 to 24 years........................	.116	.210	989,661	114,801	207,829
25 to 29 years........................	.249	.190	425,643	105,985	80,872
30 to 34 years........................	.189	.111	262,656	49,642	29,155
35 to 44 years........................	.098	.060	373,918	36,644	22,435
45 to 64 years........................	.026	.019	498,549	12,962	9,472
65 years and over....................	.006	.011	158,846	953	1,747

Computation of age-standardized rates for adult population, using formula (14)

for Israel (1966) $\frac{330,616}{4,314,183} \times 1,000 = 77$

for United States (1960) $\frac{401,262}{4,314,183} \times 1,000 = 93$

X Not applicable.

Source of basic data: Official national reports.

where $M_a^{f,1}$ and $M_a^{m,1}$ equal the number of first marriages for females and males, respectively, at the specified ages, and P_a^f and P_a^m equal the female and male population at the same ages.

By subtraction of the *TFMR* from the *TMR*, we obtain a total remarriage rate, or we may obtain this rate directly. In all three types of rates, the population in the denominator is the total population in the age group, not the single or the widowed plus the divorced population.

The total marriage rate is a summary rate for a synthetic cohort. Since the rates for all ages (or age groups) are given equal weights, it does not take account of the mortality to which a true cohort would be subject. The marriage rates at the advanced ages are so low, however, that they contribute relatively little to the overall rate. Note also that the age-specific rates are based on the total males (or females) in the age group rather than on the number unmarried, which is the true number at risk.

Rates on a Probability Basis

Marriage probabilities represent the chance that a marriage will occur over a specified period of time to a person of a given age, sex, and marital status at the beginning of the period. The probabilities are usually expressed as the number of marriages in a year per 1,000 persons in the class at the beginning of the year.

Marriage probabilities specific for age indicate the chances that a person of a given age will marry during the period, usually the year. The value of $\bar{\mu}_a$, an age-specific probability, may be approximated by μ_a, the central marriage rate at age a during the year. Marriage rates may be converted into marriage probabilities for persons at exact years of age using the formula:

$$\bar{\mu}_a = \frac{2\mu_a}{2 + \mu_a} \quad (19)$$

(See ch. 15 for analogous transformations of mortality rates.)

Marriage probabilities by marital status indicate the chances that a man or woman of a given marital status at the beginning of the year will marry during the year. These marriage probabilities are usually also made specific with respect to age so that age-marital-status-specific marriage probabilities indicate the chances that a person of a given marital status and exact age at the beginning of the year will marry during the year.

It is also possible to derive marriage probabilities using only census data. This can be done by estimating the number of marriages in a year as the difference in the proportion single at two adjoining ages, and relating this difference to the proportion single at the younger age. For example, the marriage probability for single females 20 years of age is computed from the following ratio:

$$\frac{S_{20}^f - S_{21}^f}{S_{20}^f} \quad (20)$$

where S_{21}^f represents the proportion single for the female population 21 years of age at last birthday and S_{20}^f represents the proportion single for the female population 20 years of age. Since these proportions are derived directly from census statistics treated as a synthetic cohort, the exact ages concerned are "offset" by a half year, e.g., we have the probability of marrying in a 1-year period from exact age 20.5 to exact age 21.5. Such a series of probabilities can be converted to conventional 1-year age intervals by interpolation.

Nuptiality Tables

Perhaps the most refined device for investigating marriage patterns is the nuptiality table. There are basically two types of nuptiality tables: (1) the gross nuptiality table, and (2) the net nuptiality table, a type of double-decrement table. The techniques of constructing the gross nuptiality table, which assumes that no person dies before passing through the marriageable ages, are the same as those used in life table construction. Construction of a gross nuptiality table consists of computing the age-specific probabilities of marrying for single males or females and decreasing the population by these rates to obtain the "survivors," that is, those remaining single at each age. From this table, one can determine what proportion of a cohort of single males or females would be married at various ages assuming that the marriage rates used in constructing the table continued to prevail.

Net nuptiality tables take into account mortality as well as marriage. This type of table indicates the pace at which a group of single persons is decreased annually by marriage and death, and it also gives the probability of a single person marrying at each year of age according to the current nuptiality and mortality rates. This table also provides information on the average age at marriage, a measure of the proportion of single persons who remain single at each age, and conversely,

Table 19-3. —Nuptiality Table for White Females in the United States: 1958 to 1960

Age x	Of 1,000 alive and single at start of age x		Of 100,000 born alive					Stationary population		
	Number marrying at age x $1,000 n_x$	Number dying at age x $1,000 q_x$	Alive and single at start of age x ℓ'_x	Deaths at age x while single d'_x	First marriages			At age x L'_x	At age x and all later ages T'_x	Years of single life remaining at start of age x $\overset{\circ}{e}'_x$
					At age x v'_x	At age x and all later ages N'_x	Percent at age x and all later ages $\% N'_x$			
	(1)	(2)	(3)	(4)	(5)	(6)	(7)	(8)	(9)	(10)
0 years	-	19.64	100,000	1,964	-	94,937	94.9	98,319	2,157,010	21.57
14 years	11.67	0.36	97,411	35	1,137	94,937	97.5	96,948	789,410	8.10
15 years	30.72	0.41	96,239	39	2,956	93,800	97.5	94,942	692,462	7.20
16 years	63.61	0.47	93,244	42	5,930	90,844	97.4	90,532	597,520	6.41
17 years	109.19	0.51	87,272	42	9,527	84,914	97.3	82,839	506,988	5.81
18 years	184.99	0.54	77,703	38	14,370	75,387	97.0	70,660	424,149	5.46
19 years	211.68	0.55	63,295	31	13,395	61,017	96.4	56,466	353,489	5.58
20 years	232.86	0.56	49,869	25	11,609	47,622	95.5	43,907	297,023	5.96
21 years	259.88	0.58	38,235	19	9,934	36,013	94.2	33,093	253,116	6.62
22 years	270.09	0.60	28,282	15	7,636	26,079	92.2	24,250	220,023	7.78
23 years	241.27	0.62	20,631	11	4,976	18,443	89.4	17,955	195,773	9.49
24 years	209.19	0.63	15,644	9	3,272	13,467	86.1	13,894	177,818	11.37
25 years	190.91	0.65	12,363	7	2,359	10,195	82.5	11,112	163,924	13.26
26 years	163.44	0.68	9,997	6	1,633	7,836	78.4	9,126	152,812	15.29
27 years	135.07	0.71	8,358	6	1,129	6,203	74.2	7,763	143,686	17.19
28 years	133.74	0.74	7,223	5	966	5,074	70.2	6,721	135,923	18.82
29 years	118.05	0.79	6,252	5	738	4,108	65.7	5,862	129,202	20.67
30 years	94.03	0.85	5,509	4	518	3,370	61.2	5,235	123,340	22.39
31 years	85.32	0.91	4,987	4	425	2,852	57.2	4,765	118,105	23.68
32 years	72.74	0.97	4,558	4	331	2,427	53.2	4,385	113,340	24.87
33 years	68.27	1.05	4,223	4	288	2,096	49.6	4,072	108,955	25.80
34 years	56.59	1.13	3,931	4	222	1,808	46.0	3,814	104,883	26.68
35 years	48.51	1.22	3,705	4	180	1,586	42.8	3,610	101,069	27.28
36 years	43.59	1.33	3,521	5	153	1,406	39.9	3,440	97,459	27.68
37 years	40.47	1.45	3,363	5	136	1,253	37.3	3,291	94,019	27.96
38 years	37.40	1.58	3,222	5	120	1,117	34.7	3,158	90,728	28.16
39 years	34.45	1.74	3,097	5	107	997	32.2	3,040	87,570	28.28
40 years	31.66	1.90	2,985	6	94	890	29.8	2,934	84,530	28.32
41 years	29.02	2.09	2,885	6	84	796	27.6	2,839	81,596	28.28
42 years	26.60	2.29	2,795	6	74	712	25.5	2,754	78,757	28.18
43 years	24.38	2.52	2,715	7	66	638	23.5	2,678	76,003	27.99
44 years	22.37	2.76	2,642	7	59	572	21.6	2,608	73,325	27.75
45 years	20.28	3.03	2,576	8	52	513	19.9	2,546	70,717	27.45
46 years	18.86	3.31	2,516	8	47	461	18.3	2,488	68,171	27.09
47 years	17.68	3.62	2,461	9	43	414	16.8	2,435	65,683	26.69
48 years	16.48	3.96	2,409	9	40	371	15.4	2,384	63,248	26.25
49 years	15.30	4.32	2,360	10	36	331	14.0	2,337	60,864	25.79
50 years	14.15	4.73	2,314	11	33	295	12.7	2,292	58,527	25.29
51 years	13.00	5.17	2,270	12	29	262	11.5	2,249	56,235	24.77
52 years	11.89	5.60	2,229	12	26	233	10.5	2,210	53,986	24.22
53 years	10.85	6.01	2,191	13	24	207	9.4	2,172	51,776	23.63
54 years	9.84	6.42	2,154	14	21	183	8.5	2,136	49,604	23.03
55 years	8.88	6.87	2,119	14	19	162	7.6	2,102	47,468	22.40
56 years	7.99	7.40	2,086	15	17	143	6.9	2,070	45,366	21.75
57 years	7.18	8.05	2,054	16	15	126	6.1	2,038	43,296	21.08
58 years	6.50	8.86	2,023	18	13	111	5.5	2,008	41,258	20.39
59 years	5.90	9.81	1,992	19	12	98	4.9	1,977	39,250	19.70
60 years	5.40	10.88	1,961	21	11	86	4.4	1,945	37,273	19.01
61 years	4.96	12.03	1,929	23	10	75	3.9	1,913	35,328	18.31
62 years	4.58	13.25	1,896	25	9	65	3.4	1,879	33,415	17.62
63 years	4.32	14.54	1,862	27	8	56	3.0	1,845	31,536	16.94
64 years	4.13	15.92	1,827	29	7	48	2.6	1,809	29,691	16.25
65 years and over	22.89	1000.00	1,791	1,750	41	41	2.3	27,882	27,882	15.57

— Represents zero.

Source: Adapted from, Walt Saveland and Paul C. Glick, "First-Marriage Decrement Tables by Color and Sex for the United States in 1958-60," *Demography*, 6(3):247, August 1969.

of the proportion of men and women in the cohort who will eventually marry.

These two types of nuptiality tables also provide nuptiality rates analogous to the first-order gross reproduction rate and net reproduction rate. The gross nuptiality rate represents the proportion of men and women who, according to the current marriage rates, would marry if none died before passing through the marriageable ages, whereas the net nuptiality rate also takes into account the prevailing mortality rates. It should be emphasized that both tables reflect only first marriages; remarriages are not taken into account.

Table 19-3 is a nuptiality table for white females in the United States for the period 1958-60 from an article by Saveland and Glick. The original article contained tables for nonwhite females, white males, and nonwhite males, as well.[9]

Details for constructing this table are discussed in the original article. Here we will limit ourselves to a description of the columns:

Column 1 (1,000 $_n x$) represents the annual number marrying per 1,000 alive and single at the beginning of each year of age. These are marriage probabilities and may be obtained from any of the methods mentioned earlier.

Column 2 (1,000 q_x) represents the annual number who were single at the beginning of each year of age and who died during the year. These rates are from the *United States Life Tables: 1959-61*. In the absence of a life table for the single population, it was assumed that the life table for the total population of corresponding sex and color could be applied.

Column 3 (l'_x) represents the number of single persons living at the beginning of each year out of 100,000 born alive. These persons are sometimes known as "net nuptiality" survivors since they "escape" both death and marriage. This column is, therefore, analogous to the l_x column in a standard life table.

Column 4 (d'_x) and column 5 (v'_x) are together analogous to the d_x column in a standard life table, i.e., they show the number of losses by death and marriage from the single population at each age as the cohort passes through life.

Column 6 (N'_x), which has no counterpart in the standard life table, represents the number of single persons expected to marry during the remainder of their lives. This column is also needed to compute the percentage of single persons who will ever marry (col. 7).

Column 7 (%N'_x) represents the percentage of single persons living at each age who will ever marry during the remainder of their lives. This column, like column (6), has no counterpart in the standard life table.

Column 8 (L'_x) represents the number of years lived in any 1 year of age by the survivors of 100,000 births as they pass through each year of age. This column is analogous to the L_x column in a standard life table. Although each person who passes through the age lived 1 year in the given year of life, not all of the cohort lived throughout the year of age. As a result, the years lived in any 1 year of age by the entire cohort is intermediate between the number of survivors at the beginning and end of each year of age. This column might alternatively be described as the single population by age that would arise from a constant flow of 100,000 births per year, with no migration and no change in the mortality and nuptiality rates.

Column 9 (T'_x) consists of the figures in column (8) cumulated upward. The top value indicates that, under the

prevailing mortality and nuptiality conditions, the survivors of 100,000 births would live an aggregate of 2,157,010 years as single females before the last survivor died or married.

Column 10 ($\mathring{e}'_x$) which corresponds to the $\mathring{e}'_x$ column in a standard life table, represents the average number of years remaining before death or marriage.

Remarriage tables for widowed and divorced persons may also be computed. The basic inputs and calculations for these tables are similar to those in the nuptiality table for single persons with the primary difference being that the radix of the former is comprised only of widowed and divorced persons. Ideally, the mortality rates should be specific for widowed and divorced persons too; but, as in the case of the single just above, mortality rates for all marital statuses combined may be used if the necessary statistics are lacking. Remarriage rates have been computed by Jacobson separately for widowed and and divorced persons.[10]

Age at Marriage and Remarriage

One of the most important demographic characteristics of a marriage is the age of the bride and the groom. Age at marriage is closely related to fertility (especially the size of completed families), the duration of married life, and the stability of marriage. Age at marriage is inversely related to fertility so long as marriage occurs within the childbearing period and not before it. For a constant duration of marriage, women marrying at the youngest ages (again, within the childbearing period) tend to have the highest fertility. In modern western societies, very early marriage tends to be associated with higher divorce and separation rates.

Distribution of Age at Marriage and Remarriage. — The percentage distribution of marriages by age of bride or age of groom is often used for descriptive and comparative purposes. By comparing distributions for different countries or distributions for the same country at different dates, the researcher is able to observe the variations in age at marriage. Regardless of the country, most marriages occur between the ages 15 and 30; however, within these ages, the distribution may vary considerably (table 19-4). For example, whereas 37 percent of the

Table 19-4. — **Percent Distribution of Marriages by Age of Bride, for Selected Countries: 1956 or 1957**

Age	United States (1956)	El Salvador (1957)	Japan (1956)	Hungary (1957)	Netherlands (1956)	England and Wales, (1956)
Total........	100.0	100.0	100.0	100.0	100.0	100.0
Under 15 years...	0.3	2.2	(Z)	0.2	(NA)	(NA)
15 to 19 years...	37.4	35.9	6.7	33.1	8.7	20.3
20 to 24 years...	32.5	27.6	61.0	35.9	47.1	47.7
25 to 29 years...	10.2	14.1	25.0	11.8	28.0	14.1
30 to 34 years...	5.8	8.3	4.6	6.6	7.5	5.9
35 to 39 years...	4.2	4.7	1.5	4.0	2.9	3.5
40 to 44 years...	3.1	3.0	0.6	2.6	2.0	2.6
45 to 49 years...	2.4	2.0	0.3	2.3	1.4	2.1
50 to 54 years...	1.5	1.1	0.2	1.5	1.0	1.5
55 to 59 years...	1.1	0.6	0.1	1.0	0.7	1.0
60 years and over	1.5	0.4	0.1	0.9	0.7	1.4
Median..........	21.9	22.2	23.6	22.3	24.4	23.1

Z Less than 0.05.
NA Not available.

Source: United Nations, *Demographic Yearbook, 1958*, table 18.

---

[9] Walt Saveland and Paul C. Glick, "First-Marriage Decrement Tables, by Color and Sex for the United States in 1958-60," *Demography* 6(3):243-260, August 1969.

[10] See, Paul H. Jacobson, *American Marriage and Divorce*, New York, Rinehart & Company, Inc., 1959, pp. 83-86, tables 38-41.

Table 19–5. — Percent Distribution of First Marriages and Remarriages by Age of Bride, for Selected Countries: 1956 or 1957

Age	First marriages			Remarriages		
	United States (1956)	El Salvador (1957)	East Germany (1956)	United States (1956)	El Salvador (1957)	East Germany (1956)
Total........	100.0	100.0	100.0	100.0	100.0	100.0
Under 15 years...	0.4	2.2	-	-	-	-
15 to 19 years...	44.4	36.4	23.7	3.1	3.0	0.3
20 to 24 years...	39.2	27.9	54.5	14.7	7.5	6.5
25 to 29 years...	8.9	14.0	13.7	16.7	25.6	13.2
30 to 34 years...	3.3	8.1	4.7	15.2	16.5	14.2
35 to 39 years...	1.7	4.7	1.6	13.4	10.5	15.5
40 to 44 years...	0.9	2.9	0.6	11.1	13.5	18.5
45 to 49 years...	0.6	1.8	0.5	9.0	12.0	15.5
50 to 54 years...	0.3	1.0	0.3	6.1	5.3	8.6
55 to 59 years...	0.2	0.6	0.2	4.4	3.0	4.6
60 years and over	0.2	0.4	0.2	6.3	3.0	3.0
Median.........	20.7	22.0	22.4	29.6	34.3	35.7

— Represents zero or rounds to zero.

Source: United Nations, *Demographic Yearbook, 1958,* table 22.

Table 19–6. — Computation of Median Age at First Marriage by Sex, for the United States, 1960, and Israel, 1966

Age	Number of marriages			
	United States		Israel	
	Bride[1]	Groom[1]	Bride	Groom
Total..............	1,012,405	1,025,035	16,404	16,212
Under 18 years............	153,411	18,745	1,712	64
18 years..................	180,272	64,669	2,054	184
19 years..................	148,702	91,066	2,760	471
20 years..................	111,130	93,782	2,662	1,111
21 years..................	107,188	144,965	2,202	1,762
22 years..................	81,197	132,818	1,453	1,955
23 years..................	48,647	98,236	954	1,820
24 years..................	31,316	71,470	631	1,639
25 years..................	26,149	57,313	429	1,411
26 years..................	17,831	44,571	326	1,179
27 years..................	15,497	32,972	259	1,005
28 years..................	13,290	23,067	179	757
29 years..................	10,522	21,585	159	554
30 to 34 years............	31,842	67,394	350	1,478
35 to 39 years............	15,507	26,906	147	455
40 to 49 years............	12,319	24,337	80	291
50 years and over.........	7,585	11,139	47	76
1/2 Total..............	506,203	512,518	8,202	8,106
Median age.................	20.2	22.9	20.6	24.5

Median $= L + \frac{n_1}{n_2} (i)$, where:

L = lower limit of median class
n_1 = number of frequencies to be covered in median class to reach middle item
n_2 = number of frequencies in median class
i = width of median class

Example: For brides in U.S. (col. 1)

Median $= 20.0 + \frac{23,818}{111,130} (1)$
$= 20.2$

[1] Excludes unknowns.

Source: U.S. Public Health Service, National Vital Statistics Division, *Vital Statistics of the United States, 1960,* Vol. III, *Marriage and Divorce,* 1964, table 2–6; Israel, Central Bureau of Statistics, *Statistical Abstract of Israel, 1968.* No. 19. Jerusalem. September 1968. table C–8.

total marriages in the United States in 1956 involved females 15 to 19 years old, the corresponding figure for Japan was only 7 percent. The distribution of remarriages is obviously shifted more to the older ages than is that of first marriages; and, furthermore, the dispersion (interquartile range, standard deviation, etc.) is greater for remarriages (table 19–5).

Although the percentage distribution according to age at marriage is a useful and convenient tool, provided only a few broad age groups are used, it becomes awkward and cumbersome when analyzing more than a small number of distributions. As a result, an average or other summary index is generally used when comparing the age at marriage for a large number of countries.

Average Age at Marriage. — The average age at first marriage (median or mean) is a useful summary measure for comparing the distributions of various areas or the distributions of first marriages and remarriages.

There are several alternative methods of calculating the average age at marriage from vital statistics. (Methods using census data were discussed in ch. 10.) These methods can be described in terms of four general characteristics: (1) the time reference, (2) whether the statistics refer to a real cohort or a synthetic cohort, (3) whether the statistics are based on absolute numbers or rates, and (4) the order of marriage, i.e., first, second, etc. For illustrative purposes, only first marriages are considered here.

Since such a large proportion of first marriages occur within the span of a few years, it is desirable to compute the median from a distribution by single years of age – at least in and around this span. (See table 19–6 for the computation of the median age at first marriage for the United States, 1960, and Israel, 1966.) These medians are for a specific year.

A second method involves following members of a birth cohort through the marriageable ages. This method is analogous to the cohort method of determining completed family size or a generation life table. Since not all the members of a cohort will eventually marry, it would be necessary to trace the cohort to extinction to determine the total number of first marriages, which is needed for the calculation. In practice, it would be found that, beyond a certain age, there are few additional first marriages annually. At some arbitrary age, then, the cumulated

number of marriages would closely approximate the total number and the median (less so the mean) would have only a slight downward bias. The result of this cohort method does not apply to any particular year, of course, any more than a generation reproduction rate does. This method is feasible in a country with a long historical series of vital statistics or a long-established population register.

It is also possible to compute the mean and median ages at first marriage from the nuptiality table. Applying the standard formula, the mean may be computed from the information in column (5) and the median most readily from the information in column 6 of table 19–3. (Note, however, that frequencies are cumulated **upward** in col. (6).)

Age of Bride by Age of Groom. — Another useful way of analyzing age at marriage is through the distribution of age of bride by age of groom. This distribution provides information on the differential age at marriage of bride and groom at each age. Comparisons of the median age of bride with the median age of groom will indicate that men marry at older ages, on the average, than women but will measure the average difference between the ages of brides and grooms only approximately. The average difference may be estimated more appropriately from a cross-classification of the ages. A table such as table 19–7, which cross-classifies age of groom and age of bride in 5-year age groups for marriages in Israel in 1966, would better serve this purpose if the statistics were tabulated in terms of single years of age. From this table the median age of groom is 25.2 years and the median age of bride, 21.7

Table 19–7. — Age of Bride by Age of Groom for Marriages in Israel (Jews Only): 1966

Age of groom	Total marriages	Age of bride											
		Under 20	20 to 24 years	25 to 29 years	30 to 34 years	35 to 39 years	40 to 44 years	45 to 49 years	50 to 54 years	55 to 59 years	60 to 64 years	65 years and over	Un-known
Total..............	18,599	6,551	8,128	1,672	627	410	293	228	227	189	154	108	12
Under 20 years.........	723	582	127	10	3	-	-	-	-	-	-	-	1
20 to 24 years.........	8,362	3,998	4,124	210	17	7	3	1	-	-	1	-	1
25 to 29 years.........	5,179	1,644	2,821	616	75	16	2	2	-	-	-	-	3
30 to 34 years.........	1,823	269	831	486	167	51	11	1	1	-	1	1	4
35 to 39 years.........	744	40	170	229	180	80	39	6	-	-	-	-	-
40 to 44 years.........	465	6	37	78	116	128	71	21	6	-	1	-	1
45 to 49 years.........	269	-	10	26	44	75	51	44	17	2	-	-	-
50 to 54 years.........	250	1	5	7	14	32	61	53	56	16	4	1	-
55 to 59 years.........	233	1	1	4	10	10	35	51	64	41	10	6	-
60 to 64 years.........	187	-	-	2	1	6	8	28	41	58	34	9	-
65 years and over.......	348	3	-	2	-	4	12	19	42	72	103	90	1
Unknown...............	16	7	2	2	-	1	-	2	-	-	-	1	1

— Represents zero.

Source: Israel, Central Bureau of Statistics, *Statistical Abstract of Israel, 1968*, No. 19, Jerusalem, September 1968, table C–10.

years, giving a difference of 3.5 years, whereas the median difference between the ages of brides and grooms is 4.2 years.

Duration of Marriage

Duration of marriage is useful in the analysis of fertility, fetal mortality, and divorce. Duration of marriage is defined as the time elapsed between the exact date of marriage and the date of the other event, for example, the live birth, fetal death, or the divorce. Date of marriage is a first priority item on the divorce record (see below), but mother's date of marriage is less often available from the birth and fetal death certificates.

Information on duration of marriage is also sometimes obtained from censuses and sample surveys. There may be a direct question on duration, or it may be derived by subtraction (1) of date of marriage from date of census or (2) of age at marriage from age at census. If a choice must be made between a question on date of first marriage and date of current marriage, the former is preferred because of its greater relevance in the analysis of fertility. When duration is obtained by subtraction, the finer the detail in the two components, the less error there will be in the estimate of duration.

Factors Important in Analysis

Most of the demographic and socioeconomic characteristics discussed in this book may be used in the analysis of marital status, either as control variables or in the study of interrelationships. Thus, marriage rates can be made specific for age, sex, race or ethnic group, country of origin, previous marital status, migration status, occupation, educational attainment, religion, etc. The sex ratio in the population of marriageable age will also influence the marriage rate. Marriage may also be studied in relation to subsequent fertility, mortality, and migration. In comparing one area with another, one should take account of differences in laws and customs. Marriage statistics may be subjected to the conventional methods of analysis of time series to discover trends, patterns of seasonality, and the effects of the business cycle, and of nonperiodic events (wars, famines, pestilences, changes in laws, etc.).

Race and ethnicity are also important factors in the study of family formation. Differences exist between race and ethnic groups in terms of such characteristics as average age at

marriage, crude marriage rate, age-specific marriage rates, and the proportion who ever marry. In the United States in 1960, for example, the nonwhite population had an older mean age at marriage than the white population, a lower marriage rate for the females 18 to 44 years of age, a higher first marriage rate at ages 14 to 17 and 30 and over, and a higher remarriage rate at ages 45 and over.

Marriage patterns are also likely to vary with nativity and country of origin. Such differentials in the median age at first marriage in the United States in 1960 are presented for females in table 19–8 , which shows that the youngest median age at first marriage tends to be for natives of native parentage (persons with both parents born in the United States). The foreign-born population has the oldest median age at first marriage, whereas the native population of foreign or mixed parentage has a median age at first marriage intermediate between the others.

Religion is another important factor for the analysis of marriage patterns. Religion influences age at marriage and the proportion who never marry, for example, celibate priests and

Table 19–8 — Median Age at First Marriage for the White Female Population 25 to 79 Years Old, by Nativity, Country of Origin, and Age, for the United States: 1960

Nativity, parentage, and country of origin	Total, 25 to 79 years	25 to 34 years	35 to 44 years	45 to 54 years	55 to 64 years	65 to 79 years
Total...............	21.4	20.3	21.5	21.9	21.8	22.3
Native of native parentage.	20.9	20.0	21.0	21.5	21.5	21.9
Native of foreign or mixed parentage.................	22.4	21.2	22.6	22.9	22.7	23.3
Parent(s) born in--						
Ireland.................	23.9	22.0	23.5	24.6	24.2	24.9
Germany.................	22.5	21.0	22.0	22.6	22.6	23.1
Italy...................	22.4	21.6	23.1	22.9	21.6	21.5
Foreign born.............	22.6	21.9	23.3	22.9	22.5	22.6
Born in--						
Ireland.................	26.4	24.3	26.6	26.1	26.7	26.8
Germany.................	23.7	22.7	23.8	23.8	24.6	23.6
Italy...................	21.6	21.9	23.6	21.3	21.2	21.5

Source: *U.S. Census of Population, 1960, Subject Report*, PC (2)–4D, "Age at First Marriage," 1966, table 6.

nuns. Earlier, we pointed out that Hinduism discourages the remarriage of widows. Religious beliefs are also related to the existence of polygamy or polyandry in a society.

Background Factors. – Since the balance of the sexes has an important influence on the marriage rate, the sex ratio of the population in the marriageable ages is a factor in short-run fluctuations in the marriage rate. In general, the number of marriages is greater if the sexes at the marrying ages are equal than if either sex markedly outnumbers the other. In many countries, women tend to marry men a few years older than themselves. Immediately following World War II, there was a very sharp rise in birth cohorts in some of the former belligerent countries, and as these cohorts reached the marriageable ages, women found that there were not enough men in the "appropriate" age group, since such men were members of earlier and much smaller cohorts. This phenomenon has been called the "marriage squeeze." [11]

Laws and customs influence the marriage rate and the proportion of the population that eventually marries. Age at marriage is limited or at least affected by the minimum age imposed by law. This may be either an absolute minimum or a minimum below which the young person may not marry without parental consent. (Young persons who marry below the legal minimum age frequently falsify their ages.) The ages within which school attendance is compulsory and the facilities for higher education also affect the age at marriage. Furthermore, the extent to which a country has a subsistence economy or a highly developed market economy and the degree of industrialization and urbanization influence the prevalence of unpaid family work, the participation of women in the labor force as gainful workers, and the occupations pursued by women, which in turn affect the marriage patterns.

Other cultural factors also exert an influence on marriage patterns. For example, the role and status of women in society have an effect on the incidence of marriage and an even greater effect on its stability. The less restrictive confinement of women to the home, their increasing participation in the labor force both before and after marriage, their increasing ability to be self-supporting, and changes in the attitudes regarding bachelorhood, spinsterhood, and remarriage have contributed to changes in the marriage rates of various countries. Related aspects are the importance placed by the society upon being a parent and the ideals held by members of the society as to family size. Marriage patterns are thus related to prospective fertility. In child marriages in India, cohabitation usually took place only some years after the formal ceremony.

Countries with high mortality also tend to be countries in which most persons enter relatively early into some form of marital union. The nature of this relationship is complex, however. The two phenomena may merely have common causes; and it is not at all evident, for example that in traditional societies with high mortality, men tend to marry early in order to be assured of male offspring to carry on the family name and to perform the necessary memorial rituals.

Migration, like fertility amd mortality, is an important factor in the understanding of marriage patterns in certain countries. Where migration occurs before marriage and is sex-selective, the sex ratio will not be in equilibrium in either the area of origin or the area of destination so that the marriage rate, other

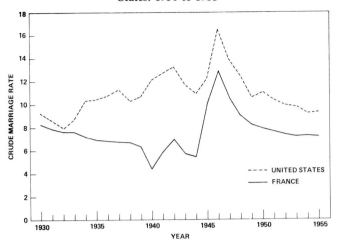

Figure 19–1. – **Crude Marriage Rates for France and the United States: 1930 to 1955**

Source: United Nations, *Demographic Yearbook, 1958*, table 16.

things being equal, will be depressed in both areas. For example, in western countries, rural-urban migration tends to be selective of single females. Hence, many bachelors are left behind in agricultural areas.

The impact of war on marriage patterns depends upon such factors as its nature, intensity, and duration, and upon whether young men are sent out of the country for a long period of time. The outbreak of war often leads to a rush to the altar in anticipation of the imminent separation. Figure 19–1 graphs the crude marriage rate for France and the United States over the period 1930 to 1955, which included World War II. Even though France entered that conflict first, the rate began its rise earlier in the United States—part of the rise in the United States may represent recovery from the low level of the Great Depression. In both countries, the rate dipped after 1942, rose sharply through 1945 and 1946 at end of the war, and then dropped off again.

Analysis of Time Series. – Standard methods may be used to extract the secular trend and the cyclical, seasonal, and random movements. The type of marriage rate chosen for this type of analysis could be any of those previously discussed in this chapter.

The marriage rate tends to vary positively with the business cycle. If an index of the economic conditions of a country and the crude marriage rate for a number of years are juxtaposed, one finds that they vary concomitantly, i.e., when the economic conditions are good, the marriage rate rises and vice versa. Some indexes of the economic situation in the United States that have been used are business failures, unemployment, industrial production, and real personal per capita income. Regardless of the index used, a close parallel has been found between economic conditions and the marriage rate.

One way to measure the association between marriage rates and economic conditions is to fit trend lines to the marriage rate and to the economic indicator, measure the deviations of each from its own trend, and compute the correlation coefficient between the two sets of deviations. [12]

Marriage rates are often computed on a monthly basis.

[11] The term was first used in Paul C. Glick, David M. Heer, and John C. Beresford, "Family Formation and Family Composition: Trends and Prospects," in Marvin B. Sussman, *Sourcebook in Marriage and the Family*, Houghton-Mifflin Company, 1963, pp. 30–40.

[12] Dorothy Swaine Thomas, *Social Aspects of the Business Cycle*, New York, Gordon and Breach Science Publishers (first published 1927 by Alfred A. Knopf), pp. 79–93; 173–179.

Consequently, we can determine the seasonality in marriage. Factors such as religious customs, climate, date of the harvest, and a number of other national or regional conditions affect the choice of month of marriage. From the unadjusted general marriage rates for the United States in table 19–9, it is apparent that June, August, and September are the preferred months for marriage. The last column indicates the pattern of seasonality. A factor value above 1 indicates that the month is favored; a value below 1, that it is avoided. These factors are derived from a computerized seasonal adjustment program developed by the U.S. Bureau of the Census.[13] This program takes account not only of seasonality *per se* but also of the number of days in the month, monthly variations in the number of specified days of the week (Saturday tends to be favored), and year-to-year variations in the dates of Lent (during which marriages are avoided by members of some religious denominations). The adjustment factors when divided into the corresponding unadjusted general marriage rates produce the seasonally adjusted rates in the table. Such adjustments help the demographer to interpret shortrun changes in marriage upon the receipt of data for part of a year.

DIVORCE

The dissolution of marriages has demographic and socioeconomic implications as does the contracting of marriages, but naturally the effect is reversed. Thus, divorces represent withdrawals from the married population and, therefore, tend to diminish the population to which births are likely to occur and to change the number, composition, needs, and functions of households.

Sources of Data

Since both marriage and divorce statistics are collected via essentially the same systems, there are many similarities in the characteristics of these two sets of data. Hence, in this section emphasis is placed on the characteristics peculiar to divorce statistics.

Divorce data, like marriage data, may be obtained from three sources: (1) national vital statistics registration systems, (2) continuous population registers, and (3) censuses and sample surveys. Most countries collect divorce data through national registration systems; however, in countries with continuous population registers such as Sweden and the Netherlands, divorce data are obtained from this source. Censuses and sample surveys provide information on marital status, which may be used to gauge the general trend of the number of divorces and the divorce rate in countries where census data are satisfactory but registration data are inadequate. (Changes from one date to another in the divorced population represent the number of divorces as reduced by death and remarriage.)

The United Nations *Demographic Yearbook* is the most complete source of international statistics on divorce. The number of divorces and crude divorce rates are published for the latest available year in each yearbook.

In addition, many countries publish yearly divorce statistics in statistical yearbooks or other publications. Tabulations peculiar to the individual country are often found in these yearbooks along with more recent tabulations of the form found in the United Nations *Demographic Yearbook*.

---

[13] U.S. Census Bureau Technical Paper No. 15, November 1965, *The X-11 Variant of the Census Method II Seasonal Adjustment Program;* Harry M. Rosenberg, "Seasonal Adjustment of Vital Statistics by Electronic Computer," *Public Health Reports* 80(3):201–208, March 1965.

Table 19–9. — General Marriage Rates, Seasonally Adjusted and Unadjusted, and Adjustment Factors, for the United States: 1968

Month	Number of marriages	Rate per 1,000 women, 15 to 44 years old		Adjustment factor
		Unadjusted	Seasonally adjusted	
January..............	119,000	34.6	46.1	0.751
February.............	137,000	42.6	50.0	0.852
March................	133,000	38.6	53.4	0.723
April................	149,000	44.6	50.5	0.883
May..................	155,000	44.9	47.6	0.943
June.................	258,000	77.1	48.3	1.596
July.................	185,000	53.4	51.6	1.035
August...............	223,000	64.3	55.4	1.161
September............	195,000	58.0	50.1	1.158
October..............	156,000	44.8	47.0	0.953
November.............	163,000	48.3	50.6	0.955
December.............	187,000	53.6	54.9	0.975

Source: U.S. National Center for Health Statistics, *Monthly Vital Statistics Report, Provisional Statistics,* Vol. 17, No. 12, "Births, Marriages, Divorces, and Deaths for 1968," March 12, 1969, table 7.

In the United States, the National Center for Health Statistics (NCHS), is the agency responsible for divorce statistics. Its Divorce Registration Area (DRA) comprises 28 States, Puerto Rico, and the Virgin Islands, which account for 55 percent of the estimated number of divorces in the United States. This area is thus appreciably smaller than the Marriage Registration Area. As with marriage statistics, NCHS publishes divorce data in three forms: (1) the *Monthly Vital Statistics Report* (provisional monthly State figures, with an annual supplement giving advance final figures, and monthly national totals adjusted to approximate expected final totals), (2) the annual volume on marriage and divorce in *Vital Statistics of the United States,* and (3) the report *Divorce Statistics Analysis* in Series 21.

Uses and Limitations

Much of what was said in the corresponding section for "Marriages" also applies here. Statistics on divorce are useful in the analysis of the change in the number of married couples, families, and unrelated individuals. Divorces are, of course, a component in the net change of some social units, but a social unit does not necessarily disappear or a new one (or ones) appear because of the divorce. Thus, a divorced person may remarry almost immediately (thus helping to form a new married couple), set up his or her own housekeeping unit (thus forming a new household), return to his paternal family (no change except in family size), become a lodger in a private household, or move into some kind of group quarters.

Divorce statistics have indirect application in fertility analysis. For a married person, divorce represents at least a temporary end to his or her period of exposure to the risk of (legitimate) fertility. Refined measures of maternal fertility may take account of the number of years of married life within the childbearing period. Age-specific or duration-of-marriage-specific fertility rates differ for first marriages and for remarriages. The level and trend of divorce also have implications for future mortality and migration. Age-for-age, death rates are relatively high for divorced persons; and recently divorced persons have an obviously high propensity to move.

Sociologists are interested in divorce as an index of marital stability, as well as in the living arrangements and emotional adjustment of the divorced parties, the care of the children, and the effects upon personal income and labor force participation. Divorced women often find it necessary to enter or re-enter the labor force in order to support themselves. A divorced man is often burdened with a double responsibility, support of the divorced wife and her children along with the support of himself and perhaps a second family.

Some of these aspects are also of interest to economists, of course. Moreover, marketing analysts are concerned with the effects of divorce upon the number, composition, needs, and economic resources of spending units.

The length of married life when terminated by divorce has important demographic and sociological implications, as it measures the lifetime of these families or subfamilies as units. Further, when combined with age at marriage, it provides a means for investigating the extent to which early age at marriage is conducive to marriage instability.

Divorce statistics are beset by many of the same kinds of limitations as marriage statistics. Legalistic considerations present in divorce data may make them less meaningful for the demographer. Moreover, laws regulating divorce differ even more widely among countries than laws regulating marriage. These variations affect international comparability and, to a lesser extent, comparability among the political units of the same country if, as in the States of the United States, these have different divorce laws.

Societal norms and mores also contribute to variations in the frequency of divorce. One often finds that even in countries where divorce is recognized, its incidence is a reflection of the religious convictions of the population. In countries where the population is largely Roman Catholic, divorces generally refer to the dissolution of civil contracts or to marriage among non-Catholics.

Those husbands and wives for whom divorce is not permitted may nonetheless live apart. Separation (legal or otherwise), annulment, and desertion are alternative forms. (Desertion has been called the poor man's divorce.) From a sociological or demographic point of view, then, divorce statistics may not be indicative of the actual frequency of marital disruption, as divorces reflect the extent to which the population can afford divorce or the rigidity of the divorce laws rather than how often marital disruption actually occurs.

Quality of the Statistics

Again, since both marriage and divorce data are usually collected by means of the same system, many of the quality considerations present in marriage data are also present in divorce data, e.g., fragmentary coverage—both geographic and ethnic. Divorce is defined as a legal act, which is always registered in the court and frequently in the civil register as well. When the statistics are based on court records, incompleteness comes about only through clerical errors. When the statistics are based on entries in the civil register, delays in registering or failure to register may also occur.

Measures

Crude Divorce Rate.—The simplest and most common measure of marriage dissolution by divorce is the crude divorce rate. This rate is expressed as the number of divorces per 1,000 of the total population, or

$$d = \frac{D}{P} \times 1,000 \qquad (21)$$

Table 19–10. — Crude Divorce Rate for Selected Countries: 1965

[Rate per 1,000 of the population]

Country	Crude divorce rate	Country	Crude divorce rate
Mauritius	0.17	Japan	*0.78
United Arab Republic	*2.16	Jordan	1.02
Canada	*0.46	Austria	1.16
Costa Rica	0.13	Czechoslovakia	*1.32
El Salvador	0.23	West Germany	0.93
Mexico	0.58	Hungary	2.01
Puerto Rico	*3.04	Australia	0.75
Venezuela	0.26	U.S.S.R.	1.60
Taiwan	0.39	United States	*2.48

* Provisional.

Source: United Nations, *Demographic Yearbook, 1966*, table 26, and *1967*, table 32.

The crude divorce rate for Sweden (1960), for example, is:

$$\frac{8,958}{7,485,615} \times 1,000 = 1.2$$

The corresponding figure for the United States is 2.2. The highest rate shown in table 19–10 for selected countries in 1965 was 3.04 for Puerto Rico, and the highest rate shown for that year in the 1967 *Yearbook* was 5.69 for the Virgin Islands of the United States. In both cases, divorces granted to persons who were demographically (if not legally) nonresidents were responsible for the high rate. In some countries, the divorce rate approaches zero because there is no provision for divorce.

General Divorce Rate.—By analogy with the general marriage rate, we may define the general divorce rate as the number of divorces per 1,000 persons of "divorceable" age, or

$$d_{15+} = \frac{D}{P_{15+}} \times 1,000 \qquad (22)$$

A further refinement is to confine the denominator to the population 15 years old and over of a single sex, $\frac{D}{P^m_{15+}}$ or $\frac{D}{P^f_{15+}}$.

Divorce Rates for Married Persons.—The United Nations uses the term "modified crude divorce rate" to denote the divorce rate based on the number of married couples.[14] In intercensal years, it is difficult to obtain a good estimate of the denominator in many countries. Furthermore, there are usually differences between the numbers of males and females reported as "married." The denominator may be obtained by averaging these numbers or by using one or the other. (The United States uses the number of married females.) The formula in such a case would be:

$$d_{mar} = \frac{D}{P^{m\,(or\,f)}_{mar}} \qquad (23)$$

The divorce rate for the married population (both sexes, males, or females) corresponds to the marriage rate for the unmarried population in that the denominator is approximately the population at risk.

To smooth out the possible wide annual fluctuations in this rate, the number of divorces may be averaged for 3 years centering on the census.

[14] United Nations, *Handbook of Vital Statistics Methods*, p. 195.

Age-Sex Specific Rates. — The formula for an age-sex-specific divorce rate is similar to that for other age-sex-specific vital rates, i.e.,

$$d_a^{m(or\ f)} = \frac{D_a^{m(or\ f)}}{P_a^{m(or\ f)}} \times 1,000 \qquad (24)$$

where $D_a^{m(or\ f)}$ refers to the number of divorces to either males or females in the specified age group and $P_a^{m(or\ f)}$ refers to the corresponding male or female population in the specified age group.

Age-sex-specific rates may also be confined to the married population in the denominator, but such refined rates are even more likely to be feasible only in and around the census year. The expression for a rate of this type for males would be:

$$\frac{D_a^m}{P_a^{m,\ mar}} \times 1,000 \qquad (25)$$

Divorce Rates Specific for Duration of Marriage. — Another rate sometimes desired in the analysis of divorce is the divorce rate specific for duration of marriage. This rate may be represented by the formula:

$$\frac{D_i}{P_i} \times 1,000 \qquad (26)$$

where D_i refers to the number of divorces that occurred to a specific marriage-duration group (i) of a population in a given area during a given year and P_i refers to the midyear married population of the same specific marriage-duration group in the given area during the same year.

This is a meaningful type of rate for the demographer; but, again, the denominator may be hard to come by, particularly in postcensal years.

Order-Specific Rates. — Divorce rates may be made specific also for such variables as number of children involved in the divorce, age of husband or wife at time of marriage, marriage order, occupation, education, etc. For example, divorce rates may be computed separately for first marriage and for remarriages to study differentials by order of marriage. Such rates, in turn, may be refined by making them specific for age, duration of marriage, etc.

Standardized Rates. — Most of the age-standardized or age-adjusted marriage rates have their counterpart divorce rates. Age-standardized or age-adjusted divorce rates are usually computed separately for males and females. Ages at marriage and hence ages at divorce tend to be different for the sexes. For example, the age adjustment to the crude divorce rate would be:

$$\sum_{a=15}^{\infty} \frac{d_a^f P_a^f}{P} \qquad (27)$$

where d_a^f is the female divorce rate at age a and $\frac{P_a^f}{P}$, the proportion of the total standard population in each age group of females.

The adjusted divorce rate may also be computed so that its magnitude is comparable to that of the general divorce rate. The equivalent of (14) for the general marriage rate would be:

$$\frac{\sum_{a=15}^{\infty} d_a^m P_a^m}{\sum_{a=15}^{\infty} P_a^m} \times 1,000 \qquad (28)$$

where the superscript m denotes males. The next refinement is a standardized rate confined to the married population just as (15) in the case of marriage was confined to the unmarried population of marriageable age. The general procedure is to restrict the denominator to married persons (male or female) 15 years of age and older and may be represented by the formula (for females):

$$\frac{\sum_{a=15}^{\infty} d_a^f P_a^{f,\ mar}}{\sum_{a=15}^{\infty} P_a^{f,\ mar}} \times 1,000 \qquad (29)$$

where $P_a^{f,\ mar}$ equals the number of married women at age a in the standard population. The lower age limit, 15, is hardly necessary in this case.

It would also be appropriate to define the **total divorce rate,** analogous to the total marriage rate or the total fertility rate, by summing the age-sex-specific divorce rates without weighting, thus

$$TDR = \sum_{a=15}^{\infty} d_a^{m\ (or\ f)} \qquad (30)$$

Again, the denominators of the age-specific rates should be the total population, not the married population, of that age. Furthermore, just as we have defined the total first marriage rate, we may define the **total first divorce rate.** For d_a in (24) we would substitute an age-specific rate with first divorces in the numerator and the total population of the given age in the denominator.

Rates on a Probability Basis. — Probabilities of divorce may be computed with respect to either age of wife (or husband) or duration of marriage. They are usually expressed as the number of divorces in a given year at a given age of wife (or husband) per thousand wives (or husbands) reaching that age at the beginning of the year, or as the number of divorces in a given year in a given marriage-duration class per thousand marriages reaching that duration class at the beginning of the year. Probabilities of divorce are computed like probabilities of marriage from central divorce rates. (Probabilities of widowhood may be defined and calculated in an analogous manner.)

Marriage Dissolution Tables. — Probabilities of divorce by age of wife (or husband) may be combined with probabilities of death of husband or wife by age of wife (or husband) for computing one type of marriage dissolution table. Another type of marriage dissolution table may be prepared from probabilities of divorce and death of either husband or wife by duration of marriage. Both of these may be considered types of nuptiality tables broadly defined and are calculated like the nuptiality tables previously discussed. They provide such information as the probability that a marriage will eventually end in divorce or death and the average number of years of married life remaining to the couple before dissolution by divorce or death.

Analysis of Divorce by Marriage Cohorts. — The general methods of cohort analysis are described in Appendix C; and they have already been applied to some extent in our earlier chapters on fertility. Cohorts are exposed to wars, prosperity, depression, etc., at different stages of their life cycle. A cohort may tend to postpone divorce for financial reasons during a depression and then to have unusually heavy divorce rates in the first years of returning prosperity. Likewise, divorces may be below average in wartime, when, because of separation, there is less marital friction, and then rise in frequency after the end of the war. Hence, fluctuations in the "total divorce rate" (see above) may be less for real than for synthetic cohorts, just as in the case of fertility.

Factors Important in Analysis

Many factors important for the analysis of marriage are also important for divorce analysis. Even laws and customs that affect the ease of marriage (particularly at young ages) and of remarriage may also affect the incidence of divorce. Divorce rates reflect the stringency of divorce laws, the extent to which a population can afford divorce, the frequency of marital discord, and the ways in which couples deal with it. Further, the divorce laws determine to a large extent the distribution by legal grounds for the decree because the legally permissible grounds for divorce vary according to the jurisdiction involved. For example, the most common grounds for divorce in Australia, Switzerland, and the United States are desertion, "alienation," and cruelty, respectively. As a result, changes in laws relating to divorce must be considered when analyzing changes in the incidence of divorce as well as in the legal grounds for divorce.

Divorces, like marriages, are affected by a variety of demographic factors (region, size of place, age, race, national origin, race, etc.) and of background factors (economic conditions, war and peace, and the legal factors just mentioned).

In the United States, the annual volumes on *Marriage and Divorce* give the age of husband and of wife at time of decree for divorces and annulments in each State of the Divorce Registration Area. No rates are shown there, but estimated age-sex-specific national rates per 1,000 of the population are

Table 19-11. — **Estimated Number and Rate of Divorces and Annulments, by Age of Husband and Wife at Decree, for the United States: 1965**

Age at decree	Number of divorces	Rate per 1,000 of the population in each age-sex group
HUSBAND		
Total	479,000	10.8
Under 20 years	6,700	27.1
20 to 24 years	81,000	28.8
25 to 29 years	93,900	21.7
30 to 34 years	72,300	15.7
35 to 44 years	120,200	11.7
45 to 54 years	70,900	7.5
55 to 64 years	24,900	3.7
65 years and over	9,100	1.6
WIFE		
Total	479,000	10.6
Under 20 years	31,600	30.6
20 to 24 years	114,500	26.0
25 to 29 years	87,700	17.7
30 to 34 years	63,700	12.6
35 to 44 years	106,800	9.7
45 to 54 years	55,600	6.0
55 to 64 years	15,300	2.6
65 years and over	3,800	1.0

Source: U.S. National Center for Health Statistics, *Vital and Health Statistics,* Series 21, Number 17, "Divorce Statistics Analysis, United States, 1964 and 1965," October 1969, table 8.

Table 19-12. — **Percent Distribution of Divorces by Couple's Number of Children Under 18, Israel (Jews Only): Selected Years, 1955 to 1966**

Number of children	1966	1965	1964	1960	1955
Total	100.0	100.0	100.0	100.0	100.0
Childless	56.6	55.6	54.8	57.2	59.1
One child	24.7	25.3	26.0	25.0	28.0
Two children	12.1	12.7	13.8	12.2	9.3
Three children	3.4	3.6	3.0	3.5	2.2
Four children	1.6	1.8	1.4	1.2	0.8
Five children	1.2	0.7	0.7	0.7	0.3
Six or more children	0.4	0.3	0.3	0.2	0.3

Source: Israel, Central Bureau of Statistics, *Statistical Abstract of Israel, 1968,* No. 19, Jerusalem, September 1968, table C-19.

available elsewhere (see table 19-11). The modal rates are for husbands 20 to 24 years old and for wives under 20.

In view of the fact that the divorce rate is rising in many countries, it is important to determine if childless couples, couples with children, or both are responsible for the increase. In Israel, it appears that a larger proportion of divorces was granted to couples with two or more children in 1966 than in 1955 (table 19-12). From the experience in Israel and the United States, it appears that children are becoming much less a deterrent to divorce than previously believed. In the United States during a similar period of time, the average number of children per divorce decree increased from 0.90 in 1954 to 1.34 in 1967.[15] A factor that is not controlled is the change in the average number of own children under age 18 still living per husband-wife family; in the United States, the average rose from 1.26 children in 1954 to 1.47 in 1965 and then declined to 1.39 in 1967. We do not know whether the divorce rates specific for number of own minor children were also rising during this period.

The importance of duration of marriage as a factor in divorce was mentioned previously. Statistics on duration of marriage can be used to measure potential loss of fertility as well as to provide some indication of family stability. In general, the relationship is one where the likelihood of divorce decreases with increasing duration of marriage after the first years of marriage.

The previous marital status of husband and wife has an important influence on the likelihood of divorce. More specifically, there is some evidence that divorced persons who remarry have a higher probability of subsequent divorce than do other married persons, other things being equal. The preferred method of examining the association between divorce and marriage order would be through order-specific divorce rates. These rates, however, are seldom computed because population bases are not available. An indirect method that may be used to suggest the relative likelihood of divorce among persons of different numbers of times married is to compare a distribution of divorces by marriage order of husband and wife with a similar distribution of marriages performed. If a pattern were found indicating that certain marriage-order categories are consistently higher among persons who were divorced than among those who were recently married, it would suggest that persons belonging to these categories have a higher likelihood of divorce than those belonging to the others.

[15] U.S. National Center for Health Statistics, *Monthly Vital Statistics Report, Final Statistics,* Vol. 18, No. 1, Supplement, "Divorce Statistics, 1967," April 16, 1969, table 3.

SUGGESTED READINGS

Agarwala, S. N. *Some Problems of India's Population.* Bombay, Vova & Co., 1966. Pp. 85–92.

Akers, Donald S. "On Measuring the Marriage Squeeze." *Demography,* 4(2):907–924.1, 1967.

Carter, Hugh, and Glick, Paul C. *Marriage and Divorce: A Social and Economic Study.* Cambridge, Mass., Harvard University Press, 1970.

Carter, Hugh, Glick, Paul C., and Lewit, Sarah. "Some Demographic Characteristics of Recently Married Persons, Comparisons of Registration Data and Sample Survey Data." *American Sociological Review,* 20(2):165–172, April 1955.

Charles, Enid. "The Nuptiality Problem With Special Reference to Canadian Marriage Statistics." *Canadian Journal of Economics and Political Science,* 7(3):447–477, August 1941.

Glick, Paul C. "The Life Cycle of the Family." *Marriage and Family Living,* 17(1):3–9, February 1955.

Glick, Paul C., Heer, David M., and Beresford, John C. "Family Formation and Family Composition: Trends and Prospects." *Sourcebook in Marriage and the Family.* Marvin B. Sussman, ed. Boston, Houghton-Mifflin Co., 1963. Pp. 30–40.

Jacobson, Paul H. *American Marriage and Divorce.* New York, Rinehart & Co., 1959.

Jamison, Ellen Linna, and Akers, Donald S. "An Analysis of the Difference between Marriage Statistics from Registration and Those from Censuses and Surveys." *Demography,* 5(1):460–474, 1968.

Karmel, P. H. "The Relations between Male and Female Nuptiality in a Stable Population." *Population Studies* (London), 1(4):353–387, March 1948.

Kuczynski, R. R. "The Analysis of Vital Statistics. I. Marriage Statistics." *Economica,* New Series, 5(18):138–163, May 1938.

Rowntree, Griselda, and Carrier, Norman H. "The Resort to Divorce in England and Wales, 1858–1957." *Population Studies* (London), 11(3):188–233, March 1958.

Saveland, Walt, and Glick, Paul C. "First-Marriage Decrement Tables by Color and Sex for the United States in 1958–60." *Demography,* 6(3):243–260, August, 1969.

Thomas, Dorothy Swaine. *Social Aspects of the Business Cycle.* New York, Gordon and Breach Science Publishers (first published by A. A. Knopf in 1927). Pp. 79–93; 173–179.

United Nations. *Demographic Yearbook, 1968.* Pp. 13–14; 31–38; 446–783.

————. *Handbook of Vital Statistics Methods.* Studies in Methods. Series F, No. 7. Especially pp. 60–63; 76–87; 118–119; 178–179; 193–195.

U.S. Bureau of the Census. *U.S. Census of Population: 1960. Subject Reports. Age at First Marriage.* PC(2)–4D. 1966.

————. *U.S. Census of Population: 1960. Subject Reports. Marital Status,* PC(2)–4E. 1966.

United States, National Center for Health Statistics. *Vital and Health Statistics.* Series 21, No. 2. "Demographic Characteristics of Persons Married Between January 1955 and June 1958." By Carl E. Ortmeyer. April 1965.

————. National Center for Health Statistics. *Vital and Health Statistics.* Series 21, No. 13. "Divorce Statistics Analysis: United States, 1963." October 1967.

————. National Center for Health Statistics. *Vital and Health Statistics.* Series 21, No. 17. "Divorce Statistics Analysis: United States, 1964 and 1965." By Alexander A. Plateris. October 1969.

————. National Center for Health Statistics. *Vital and Health Statistics.* Series 21, No. 16. "Marriage Statistics Analysis: United States, 1963." By Carl E. Ortmeyer and Russell P. Kuhn. September 1968.

————. National Center for Health Statistics. *Vital Statistics of the United States.* Vol. III, *Marriage and Divorce, 1966.* 1969.

CHAPTER 20

International Migration

INTRODUCTION

Uses

Migration is the third basic factor affecting change in the population of an area; the other two factors, births and deaths, have been treated in earlier chapters. The importance of migration in affecting the growth and decline of populations and in modifying the demographic characteristics of the areas of origin and the areas of destination has long been recognized.

Migration is an important element in the growth of the population and the labor force of an area. A knowledge of the number and characteristics of persons entering or leaving an area is required together with census data and vital statistics in order to analyze the changes in the structure of the population and labor force of an area. The measurement and analysis of migration are important in the preparation of population estimates and projections for a nation or parts of a nation. Data on the sex, age, citizenship, mother tongue, duration of residence, occupation, etc., of the immigrant facilitate an understanding of the nature and magnitude of the problem of social and cultural assimilation that often results in areas with heavy immigration.

The sociologist is concerned with the social and psychological effects of migration upon the migrant and upon the populations of the receiving and sending areas and the acculturation and adjustment of migrant populations. The economist has been interested in the relation of migration to the business cycle, the supply of skilled and unskilled labor, the growth of industry, and the occupational and employment status of the migrant. The legislator and political scientist are concerned with the formulation of policies and laws regarding immigration and, to a lesser extent, internal migration, and the enfranchisement and voting behavior of migrants.

Definition

Migration is a form of geographic or spatial mobility involving a change of usual residence between clearly defined geographic units. Some changes of residence, however, are temporary and do not involve changes in usual residence; these are usually excluded from "migration." They include brief excursions for visiting, vacation, or business, even across national boundaries. Other changes in residence, although permanent, are short-distance movements and, hence, are also excluded from "migration." In practice, such short-distance movements affect the scope of internal but not international migration. Thus, the term migration has in general usage been restricted to relatively permanent changes in residence between specifically designated political or statistical areas, or between type-of-residence areas.

For demographic purposes, two broad types of migration are identified, international migration and internal migration. The former refers to movement across national boundaries. It is designated as emigration from the standpoint of the nation from which the movement occurs and as immigration from that of the receiving nation. The term internal migration refers to migration within the boundaries of a given nation.

The distinction between international and internal migration is not always clear because non-self-governing territories have some but not all of the characteristics of independent states. Depending on the purpose of the statistics, such movements as those between the occupied zones of postwar Germany, between Puerto Rico and the United States, or between Metropolitan France and the former Algeria, might be classified as either internal or international migration. A historical series for a country may include an area for part of the series which later becomes independent, and it may not be possible to reconstruct the figures from internal to international migration. For example, the migration statistics of the United Kingdom include the Irish Republic prior to April 1, 1923.[1]

The sources of data, the types of data available, and the techniques of estimation and analysis are sufficiently different for international and internal migration, however, to warrant separate treatment of these two types of migration. Therefore, two chapters are devoted to migration in this volume, the first to international migration and the second to internal migration.

Types of International Migration

International migratory movements may be variously classified as temporary or permanent movements, movements of individuals and families or movements of whole nations or tribes, movements of citizens or aliens, voluntary or forced movements, peaceful or nonpeaceful movements, movements of civilians or military personnel, and movements for work, study, or other purposes.[2] A common basis of classification of immigration statistics, important because of its relation to the collection systems often employed, is the mode of travel or type of entry or departure point (i.e., sea, air, or land).

The more permanent movements are generally made as a result of racial, ethnic, religious, political, or economic

---

[1] Great Britain, *The Registrar General's Statistical Review of England and Wales, the Year 1963*, Part II, *Tables, Population*, 1965, p. vi and table S.
[2] For a general typology of migration see William Petersen, *Population*, 2nd ed., Collier-MacMillan Limited, Toronto, 1969, pp. 289–300, and *idem*, "A General Typology of Migration" in Charles B. Nam (ed.), *Population and Society*, Houghton-Mifflin Company, pp. 288–297.

pressures, or a combination of these, in the area of emigration, and corresponding attractive influences in the area of immigration. Of the large number of persons who enter or leave any country in a given period, only a portion are true immigrants or emigrants. At the present time, much of the international mobility of labor, although of economic significance, is of a temporary nature. This movement would not properly be included in the statistics of (permanent) migration but in those of temporary movements; yet there is strong interest in such figures. This is particularly true on the European continent, where daily, weekly, or seasonal movements over national boundaries occur on a considerable scale. On the American continent, a similar case may be seen in the seasonal traffic across the border between Mexico and the United States and the daily commuting between Canada and the United States (e.g., between Detroit and Windsor) from home to factory or office.

Conquest, invasion, colonization, forced population transfers, and refugee movements represent types of mass movements. Conquest and invasion illustrate types of nonpeaceful, mass movements of a tribe or nation (e.g., the invasion of England by the Normans in the eleventh century, the invasion of Korea and Manchuria by the Japanese in the twentieth century). The racial and nationalistic ideologies of countries have been major factors in forced migrations of tribes or nations (e.g., the importation of Negro slaves into the Western Hemisphere during the eighteenth century and the transfer of millions to slave labor camps in Germany during World War II). War and other political upheavals have resulted in numerous forced population transfers and refugee movements (e.g., the exchange of 18 million persons between India and Pakistan after the partition of India in 1947, the exchange of population between Greece and Turkey in the early nineteen twenties, and the movement of Arab refugees in Palestine after the establishment of the state of Israel in 1948). The shipment of troops under military orders may be considered a type of forced migration.

Colonization is a movement of a tribe or nation, or of individuals and families, for the purpose of settling in a relatively uninhabited area discovered or conquered by the "mother" country (e.g., the movement to the Western Hemisphere by Europeans during the Colonial period, the movement to Australia by the British in the early part of the twentieth century). Another important type of international movement associated with colonization is the movement related to the coolie-contract system, which was theoretically voluntary and often led to permanent settlement. It flourished in the nineteenth century on the initiative of the great colonial powers, especially Great Britain, France, and the United States. Britain recruited Indians for labor on plantations and in mines in Burma, Ceylon, Fiji, East Africa, and various Caribbean Islands; France recruited Indochinese for New Caledonia; the United States recruited Orientals and Filipinos for Hawaii, and Mexicans for the Far West; etc.

COLLECTION SYSTEMS

General Sources

Data on international migration may be derived from a variety of sources.[3] We may distinguish six classes of migration data corresponding to these several sources:

1. Statistics collected on the occasion of movement of people across international borders, mostly as a by-product of the administrative operations of frontier control.

2. "Passenger statistics" obtained from lists of passengers on sea or air transport manifests.

3. Statistics of passports and of applications for passports, visas, work permits, etc.

4. Statistics obtained in connection with population registers.

5. Statistics obtained in censuses or periodic national surveys on the basis of inquiries regarding previous residence, place of birth, or citizenship.

6. Statistics collected in special or periodic inquiries regarding migration, previous and present residence, or citizenship, such as a registration of aliens or a count of citizens overseas.

In addition, estimates of total net migration or net migration of particular groups (foreign-born, aliens, civilian citizens, armed forces) for particular past periods, may be made on the basis of these or other statistics. For example, estimates of net migration of citizens of the country, including armed forces, may be made in some cases on the basis of counts or estimates of the citizens overseas. The various ways of estimating net migration by indirect methods will be treated in a subsequent section; in this section we propose to treat the direct sources of migration statistics, describe the kinds of statistics provided, and consider their relation to one another.

Frontier Control Data and Similar Sources

General Aspects. — We consider here the first four of the sources of migration statistics listed above. Frontier control data are the most important source for the direct measurement of migration, if not the most frequently available of the various sources, and hence, this source is given most detailed consideration in this chapter. Statistics from the operations of frontier control relate to any of several systems of collecting migration statistics at the point of actual movement across international borders. These collection systems may distinguish "land frontier control statistics," which relate particularly to movement across land borders, from "port control statistics," which relate particularly to movement into and out of a country via its airports and seaports. The collection of information at land frontiers is much more difficult than at ports. This is due in large part to the heavier traffic, with only a small proportion of all travelers being classified as migrants. Declaration forms are not commonly used at land borders and, hence, types of travelers are not distinguished. Another system employs coupons detachable at points of departure and arrival from special identity documents issued to migrants by their own governments. It covers both land and port movement.

Collection of migration statistics on the basis of travel documents or special forms requires that the agency responsible for the operation have agents at the frontier points authorized to request from international travelers the statistical information desired. Passports and work permits carried by travelers crossing the frontiers may be used to facilitate obtaining data on some classes of travelers.

Normally a distinction is made between permanent and temporary "migrants," on the one hand, and "frontier traffic," on the other. Persons residing in the frontier areas may make frequent moves across the border and employ special simplified travel documents (frontier crossing cards). Frontier traffic is then usually omitted from the principal tabulations on migration.

The types of administrative organization for collecting and

---

[3] We are concerned here principally with sources of data relating to the volume of immigration and emigration. Vital statistics provide some information on international migration through questions on country of birth, citizenship, etc., of parents of infants born in a particular year.

compiling immigration statistics vary from that where the data are collected and published by a single immigration agency to that where several administrative agencies are involved in the collection of the migration data, with publication by each of them, one of them, or a central statistical office. The organization adopted varies with the situation in each country. Questionnaires may be required from travelers and migrants by such departments as immigration, police, customs, exchange control, or public health. The information obtained on these forms by the various authorities participating in frontier control activities may or may not be used for statistical purposes. It is still a major problem in many countries to eliminate duplication between the items of information collected and published by the different agencies and to insure that these items cover the field without leaving serious gaps.

Instead of or in addition to statistics based on various travel documents or special report forms, the country may collect "passenger statistics," or "statistics of sea and transport manifests." Tabulations of "passenger statistics" are based on counts of names on copies of passenger lists furnished by steamship companies and airlines, or statistical returns compiled from them by the transportation companies. Migrants cannot usually be distinguished from other travelers (unless the ship on which they are traveling is specifically an "emigrant ship"), and the classification of passengers in terms of previous country of usual residence or intended country of usual residence is not precise. Emigrants do not necessarily embark from their country of last usual residence or disembark at their country of intended usual residence. Often, transport manifests are checked in the frontier control operations against the identity papers of the travelers. This type of collection system is not applicable to land frontier movement.

International Recommendations. — The United Nations considers frontier control data and, where available, population register data most satisfactory for the measurement of international migration. It recommends that national governments collect and tabulate total arrivals (immigrants and other arrivals) and departures (emigrants and other departures) and subdivide these totals into several defined categories of arrivals and departures. Such categories are intended to aid in the interpretation of the migration statistics for a given country from year to year, in making valid comparisons from country to country, and in using the statistics in conjunction with other demographic data.

Availability of Data. — Statistics of international migration are now available for a fair number of countries. For the most part, those collecting such data normally collect only the data they need for their own administrative purposes. For many countries detailed statistics on migration are scattered through the publications of several national collecting and processing agencies. To facilitate use of these data, the United Nations has assembled and published a bibliography of statistics on international travelers and migrants covering 24 selected countries.[4] It would be preferable, however, as the United Nations recommends, for each country to present its migration statistics in a single, easily available publication.

International compilations of migration data for the post-World-War-II period are available in various issues of the United Nations' *Demographic Yearbook,* with the current frequency being biennial (even years). Statistics for the period 1918–1947 are available in another United Nations publication;[5] and data on the

economic characteristics of long-term migrants were published for selected countries in a source-book published in 1958.[6]

Collection of Frontier Control Data in the United States. — Immigration data for the United States are available or may be developed from several sources. As indicated in chapter 2, the principal source of data on immigration for the United States is the tabulations of the Immigration and Naturalization Service, U.S. Department of Justice, resulting from the administrative operations of frontier control. One collection system secures tabulations of aliens on the basis of visas or other documents surrendered; these cover "documented" movements. The Immigration and Naturalization Service also compiles data on passengers on air and seagoing vessels, "border crossers," and crewmen. In addition, the decennial census reports, as well as the Current Population Survey of the U.S. Census Bureau, contain limited direct information on the volume of immigration and other data which serve as a basis for making estimates of net immigration for intercensal periods. Several other Federal agencies compile statistics of incidental use in the measurement of international movements. Finally, limited comparative or supplementary data for the United States may be obtained from the reports on immigration or the censuses of various foreign countries. We will consider each of these sources of immigration statistics for the United States under the appropriate heading identifying a particular source, beginning here with the "frontier control" data of the U.S. Immigration and Naturalization Service.

History of collection of migration data. — Official records of immigration to the United States have been kept by a Federal agency since 1820; official records of emigration have been kept only since 1908. The statistics were compiled by the Department of State, the Department of Labor, and other Departments before this work was shifted from the Department of Labor to the Justice Department in 1944, where it is now located.

Since 1820 the official immigration statistics have changed considerably in completeness and in the basis of reporting. Some of the more important changes may be noted. Reports for Pacific ports were not included until 1850. Entries of Canadians and Mexicans over the land borders were first reported in 1906. Until 1904 only third-class passengers were counted as immigrants; first- and second-class passengers were omitted. The current series on "immigrant aliens admitted" (i.e., aliens admitted for permanent residence in the United States) began in 1892 (except 1895–1897); earlier, the figures related to "immigrant aliens arrived" or "alien passengers."

The immigration statistics of the Immigration and Naturalization Service are now published in the *Annual Report of the Immigration and Naturalization Service* and the *I and N Reporter.* The immigration office has issued an annual report on immigration each year since it was established in 1820, except for the years from 1933 to 1942, when the report appeared only in abbreviated form or was not published at all. Although these reports have been designed primarily to describe the administrative operations of the immigration office, an extensive body of statistical data on immigration has been included in them. The annual report is currently published with a lag of 8 to 12 months following the close of the year (July 1 to June 30) to which the figures relate. The *I and N Reporter* or a predecessor has been published monthly or quarterly since 1943.

[4] United Nations, *Analytical Bibliography of Statistics on International Migration Statistics, 1925–1950,* Population Studies, Series A, No. 24, 1955.
[5] United Nations, *Sex and Age of International Migrants for Selected Countries, 1918–1947,* Population Studies, Series A, No. 11, 1953.
[6] United Nations, *Economic Characteristics of International Migrants, Statistics for Selected Countries, 1918–1954,* Population Studies, Series A, No. 12, 1958.

Principal collection systems. — The data on migration of the Immigration and Naturalization Service do not fit any simple classification scheme and, in fact, because of the complexity and variety of the data, more than one classification scheme is required to present them. [7] We may identify two principal collection systems and a few subsidiary and supplementary ones. The first is confined to aliens and is based on visa forms surrendered by aliens at ports of entry. We will refer to the resulting data as "admission statistics." This system covers only a small part of the movement across the United States borders. The second principal collection system is more inclusive than the first and, in general, covers all persons arriving at and departing from American ports of entry. We shall refer to these data as "arrival statistics." From a demographic point of view, however, these data have serious limitations not shared by the first classification. This system covers three subsidiary groups: passengers arriving or departing principally by sea or air, land border crossers, and crewmen. Some supplementary groups are covered by other means. The types of statistics compiled and their precise definitions vary from one period to another. The description given applies to the current situation.

In general, "admission statistics" covers (1) aliens admitted to the United States as "immigrant aliens admitted," (2) aliens departing from the United States as "emigrant aliens departed," (3) aliens admitted as "nonimmigrant aliens admitted," and (4) aliens departing as "nonemigrant aliens departed." Immigrant aliens are nonresident aliens admitted to the United States for permanent residence (or with the declared intention of residing here permanently) or persons residing in the United States as nonimmigrants, refugees, or "parolees" who acquired permanent residence through adjustment of their status.

Emigrant aliens are resident aliens departing from the United States for a permanent residence abroad (or with the declared intention of residing permanently abroad). Statistics on emigrant aliens were discontinued as of July 1, 1957, when persons departing were no longer inspected.

The basic classes of alien migrants considered are supplemented by two additional classes of alien admissions or departures — nonimmigrant aliens admitted and nonemigrant aliens departed. In general, nonimmigrant aliens are nonresident aliens admitted to the United States for a temporary period [8] or resident aliens returning to an established residence in the United States after a temporary stay abroad (i.e., an absence of more than 12 months). On the basis of the experience of 1966–67, numerically the most important group of nonimmigrant aliens is the class "temporary visitors for pleasure." Other numerically important groups are "returning residents," "temporary visitors for business," "transit aliens," "temporary workers and industrial trainees," and "students." Also included among nonimmigrant aliens are "foreign government officials," "exchange aliens," and members of international organizations.

"Nonemigrant aliens departed" are nonresident aliens departing after a temporary stay in the United States [9] or resident aliens departing for a temporary stay abroad (i.e., for more than 12 months). Data on nonemigrant aliens were tabulated up to July 1, 1956; such figures are not available since

Table 20–1. — **Aliens and Citizens Admitted at United States Ports of Entry: 1966–67**

[Each entry of same person counted separately]

Class	Total	Aliens	Citizens
Total.................	206,837,454	120,196,406	86,641,048
Border crossers[1].........	195,143,536	114,630,122	80,513,414
Canadian...............	67,265,449	37,044,010	30,221,439
Mexican...............	127,878,087	77,586,112	50,291,975
Crewmen...............	3,046,559	2,036,877	1,009,682
Others admitted..........	8,647,359	[2]3,529,407	[3]5,117,952

[1] Partially estimated.
[2] Includes immigrants, documented nonimmigrants, aliens with multiple entry documents other than border crossers and crewmen, and aliens returning from Canada or Mexico after extended visits.
[3] Includes all citizens arriving by sea and air and citizens returning from Canada or Mexico after extended visits.

Source: U.S. Immigration and Naturalization Service, *Annual Report of the Immigration and Naturalization Service, 1967,* 1968, table 3.

that date. The classes of arrival and the classes of departure do not correspond to each other completely because the intended length of stay as declared does not always correspond to the actual length of stay. Thus, persons who are admitted as nonimmigrant aliens for a temporary stay but remain longer than a year are classified as emigrant aliens on departure, and aliens who are admitted for permanent residence but decide to depart within a year are classified as nonemigrant aliens on departure.

The second collection system provides "arrival" and "departure" statistics and may be viewed as having three distinct components. The first component covers principally arrivals and departures by sea and air; the second covers "border crossers," i.e., persons who cross over frequently to or from Canada or Mexico; and the third covers crewmen. The statistics are classified by citizenship. The first component may also be designated as "passenger" statistics; the count is derived from lists of names on passenger manifests prepared by the airlines and steamship companies.

The second component, "border crossers," represents principally a count, made by immigration inspectors at established points of entry, of persons entering the United States over its land borders with Canada and Mexico. It is a count of crossings; hence, the same persons may be counted more than once.

The figures on border crossers and crewmen are combined with the figures on "others admitted" in table 20–1 to obtain the total number of arrivals in the United States in fiscal year 1966–67. "Others admitted" includes immigrants, documented nonimmigrants, aliens with multiple entry documents other than "border crossers" and crewmen, citizens arriving by sea and air, and aliens and citizens returning from Canada and Mexico after extended visits.

Some statistical information is secured for several other groups of persons who cross the borders of the United States in addition to the group of migrants already enumerated. Some account should be taken of these in any assessment of the impact of immigration on the population, particularly on the *de facto* population: [10]

1. Contract laborers, particularly Mexican. The program of importing unskilled agricultural and industrial laborers was

---

To supplement the description of the categories of migrants given in the *Annual Report of the Immigration and Naturalization Service,* the reader may consult: U.S. Immigration and Naturalization Service, *I and N Reporter,* "International Migration Statistics as Related to the United States," by Gertrude D. Krichefsky, Part I, July 1964, and Part II, October 1964.

[8] More than three days if admission is at the Mexican border and more than six months if admission is at the Canadian border.

[9] See footnote 8.

[10] For a more detailed listing and description of the classes of alien migrants, see E. P. Hutchinson and Ernest Rubin, "Estimating the Resident Alien Population of the United States," *Journal of the American Statistical Association* 42(239):392–398, September 1947.

set up during World War II; its scope was gradually reduced until it was eliminated in the mid-sixties.

2. Aliens paroled into the United States. From time to time, special legislation allows political refugees to enter and remain in the United States outside the requirements of the Immigration Act. On this basis refugees from Hungary after the revolution in 1956 and refugees from the Communist regime in Cuba in the 1960's were granted asylum in the United States.

3. Arrivals from and departures to the outlying areas of the United States. In the basic tabulations on admissions, the United States and its outlying areas are treated as a unit. Data on movement between the United States and Puerto Rico are currently available, however, in the form of passenger statistics compiled by the Puerto Rico Planning Board.

4. U.S. military personnel. Direct data are not available but their number may be estimated from data on the number of U.S. military personnel overseas given in census reports and reports of the U.S. Department of Defense.

5. Illegal entrants and unrecorded departures.

6. Aliens deported from the United States or departing voluntarily under deportation proceedings. During 1966-67, 162,000 deportable aliens were located. About half were aliens who had entered illegally and about half became deportable after violating the status for which they were admitted.

All of the movement across a country's borders, however temporary, except daily commuting, should be considered demographically significant in relation to a *de facto* count of the population. The groups of migrants who would be considered consistent with a *de jure* count of the population would be much more restricted. In the case of the United States, these include members of the armed forces who are transferred into and out of the United States, all "immigrant aliens admitted" and "emigrant aliens departed;" certain classes of "nonimmigrant aliens admitted" and "nonemigrant aliens departed," such as students, resident aliens arriving and departing, some temporary visitors for business, and temporary workers and industrial trainees; "refugees" and "parolees" who enter under special legislation and may later have their status adjusted to that of permanent residence; and citizens who change their usual residence (movement to or from outlying areas and foreign countries). In view of the discontinuance of data collection for certain of these categories (emigrant aliens departed and nonemigrant aliens departed) and the volatility of the figures for passenger movement (citizens arrived and departed, aliens departed), additions through immigration to the *de jure* population may reasonably be restricted to "immigrant aliens" and allowances for citizens based principally on sources other than passenger statistics, as is now the practice at the U.S. Census Bureau.[11]

Quality of Statistics. — The quality of data on international migration based on frontier control operations is generally much poorer than that of census counts or birth and death statistics. Such data tend to suffer from serious problems of completeness and international comparability. There are several reasons for the poor quality of the data. First, there are many forms of international movement, and they are not easy to define or classify. Second, classification based on duration of stay or purpose of migration depends on statements of intentions, and the actual movements may not correspond to these statements of intentions. Next, the mere counting of persons on the move is extremely difficult, especially when a country has a very long boundary which is poorly patrolled. It is certain that many international migrants enter or leave a country unrecorded under these conditions. Controls over departures are usually less strict than over arrivals so that statistics of emigration are more difficult to collect and less accurate than statistics of immigration. This type of problem is illustrated by unrecorded movement over the Mexican border to the United States. Between 1950 and 1964, 3.9 million aliens were apprehended for being in the United States illegally; about three-fourths of these were Mexicans apprehended between 1950 and 1955 as "wetbacks." However, most of those apprehended had been in the United States only for a matter of weeks.[12]

International comparability of migration statistics is also seriously impeded by the complexity and diversity of definitions and classification systems used in different countries.

Data From Population Registers

A fully developed system of national population accounting would cover movement into and out of a country as well as births, deaths, and internal movements. Under this system international movement, including arrivals into the country and departures from it, is simply a special type of change of residence which must be reported to the local registrar. Reporting of change of residence is generally exempted from declaration if the duration of absence is short. At present, this is a relatively uncommon source of migration statistics. Very few countries in the world, perhaps less than ten, have a system of continuous population registration from which it is possible to derive satisfactory immigration and emigration statistics. These include Belgium, Czechoslovakia, Denmark, Federal Republic of Germany, Israel, Japan, the Netherlands, Sweden, and Taiwan.

Census or Survey Data

Residence Abroad at a Previous Date. — Censuses or national sample surveys may, in effect, provide information on immigration during a fixed period prior to the census or survey date (chapter 9). From a theoretical point of view, the type of immigration data obtained in a census or survey differs in several respects from data compiled in connection with the administrative operations of frontier control. The latter represent counts of arrivals and departures during a given period. The former represent a classification of the population living in the country at a particular date according to residence inside or outside the country at some previous specified date. The census data on migration cover only persons who were alive both at the census date and at the previous specified date. Hence, the number of "immigrants" reported in the census or survey is deficient as a count of immigrants during the period between the previous date and the census date by the number of children born abroad during this period who immigrated into the country, and by the number of immigrants who died (in the country of immigration) or returned to the country of origin during the period. Even though the census or survey figures are affected by some departures, they cannot properly be viewed as estimates of net immigration because they fail

[11] For other discussions of the U.S. immigration statistics, see E. P. Hutchinson, "Notes on Immigration Statistics of the United States," *Journal of the American Statistical Association* 53(284): 963-1.025, December 1958; idem, *Our Statistics of International Migration: "Comparability and Completeness for Demographic Use,"* Report to the Committee on Population Statistics of the Population Association of America, July 1965; and Donald S. Akers, "Immigration Data and National Population Estimates for the United States," *Demography* 4(1):262-272, 1967.

[12] Akers, "Immigration Data and National Population Estimates for the United States," p. 268.

to allow for the departure of persons who were living in the country prior to the migration period. Censuses or surveys cannot readily provide any separate information on emigration for a country.

Data on prior residence abroad are available only for the brief periods before each census or survey in which a question on "previous residence" is asked. On the other hand, the census or survey data on immigration are likely to be comprehensive with respect to the types of migrants included: e.g., aliens and citizens, civilians and military personnel, persons having permanent residence status and refugees, etc. In addition, the data are more consistent with the general body of census or survey data from which they are drawn than "frontier-control" data and certain demographic characteristics of the "immigrants" (especially age and sex) may be readily tabulated.

In spite of the simplicity of the question and its value in providing data on the volume of immigration and the character-istics of immigrants, data of this kind are available for few countries. They are illustrated in table 20-2. Among these few countries is the United States. Data relating to the volume of immigration and the characteristics of immigrants for the United States, based on a question relating to previous resi-dence, is given in the reports on internal migration derived from the decennial census and the Current Population Survey (*Current Population Reports*, Series P-20). A category of "persons abroad" is shown in the published tables as one of three major categories of migration status (in effect, non-migrant, internal migrant, and immigrant). This category refers to persons living in the United States at the census or survey date who reported that their place of residence at a specified previous date was in an outlying area (not a State) of the United States or in a foreign country.

Nativity. – Census data on nativity, particularly on the foreign born, serve both as direct indications of the volume and characteristics of immigrants and as a basis for estimating them (chap. 9). Data on the foreign born are especially valuable for measuring migration when "frontier-control" data on migra-tion are lacking, are of poor or questionable quality, or are irregularly compiled. Some important kinds of classifications may not be available in the "regular" immigration tabulations, but may appear in the census data. Hence, measurement of

the volume of immigration of certain groups or of certain characteristics of immigrants may be possible only from census data. Where similar material is available from both sources, the census data may aid in validating the indications of the "regular" immigration data in spite of the difficulties of this comparison. For most countries, including the United States, the body of the census information relating to the foreign-born population is more extensive and detailed, more comparable internally, and of better quality than the immigration data collected at time of arrival.

A classification of the U.S. population by nativity has been made since 1850 in connection with the "State or country of birth" question on the census schedule. The number of sched-ule inquiries relating to the foreign-born population increased with each succeeding census after 1850 until the peak in 1920, when there were questions, among others, on country of birth, mother tongue, and year of immigration. In each year since 1850, the foreign born were tabulated at least by age and sex.

Volume of immigration. – Census tabulations of the foreign-born population provide information on "net lifetime immigra-tion" of the foreign born, i.e., net immigration over the lifetime of the present population. Census data on the foreign born do not provide an indication of the volume of immigration during any particular past period of time. Foreign-born persons who returned to live abroad or who died prior to the census date are excluded. At the older ages both of these factors may have important impact in reducing the numbers of foreign born far below the numbers of immigrants who may have originally entered the country. Surviving immigrants are counted only once even though they may have moved to the country in question more than once in a lifetime.

In sum, the census figures provide information specifically on the balance of immigration and emigration of foreign-born persons during the last century, diminished by the number of deaths of immigrants in the country prior to the census date. Thus,

$$P_F = (I_F - E_F) - D_F \qquad (1)$$

$$I_F - E_F = P_F + D_F \qquad (2)$$

where P_F refers to the foreign-born population, I_F and E_F refer to immigrants and emigrants born abroad, respectively, and

Table 20–2. — Number of Persons Abroad at a Previous Fixed Date as Reported in the Censuses of Population, for Various Countries: Around 1960

Country and date of census	Previous date of reference	Census data			Reported immigration during period
		Number abroad	Total population	Percent abroad	
Canada, June 1, 1961	June 1, 1956	469,915	15,302,621	3.1	[1]744,088
England and Wales, April 23, 1961	April 23, 1960	[2]342,750	46,166,830	0.7	(NA)
Scotland, April 23, 1961	April 23, 1960	[2]47,070	5,166,440	0.9	(NA)
United States:					
April 1, 1940	April 1, 1935	[3]359,499	132,164,569	0.3	[4]299,272
April 1, 1950	April 1, 1949	505,265	151,325,798	0.3	[4]254,456
April 1, 1960	April 1, 1955	2,002,822	179,323,175	1.1	[4]1,423,451

NA Not available.

[1] Immigrant arrivals only.

[2] Includes persons living elsewhere in the British Isles.

[3] A figure from the 1940 census which is more nearly comparable with the "registration" statistics on immigration is 226,911, for the foreign-born population abroad in 1935.

[4] Immigrant aliens admitted only.

Source of basic data: Official national reports.

D_F refers to deaths of foreign-born persons in the country. It should be particularly noted that the quantity $I_F - E_F$ does not represent net migration in the usual sense, since the emigrants here consist solely of former immigrants ("return migrants"), and movement of native persons is entirely excluded.

Demographic characteristics of immigrants. — Much information regarding the geographic and residence distribution and the demographic characteristics of immigrants can be obtained from census data on the nativity of the population (chapter 9). The characteristics of the surviving immigrants are reflected in the current composition of the foreign-born population with respect to age, sex, country of birth, mother tongue, occupation, educational attainment, etc. Some of these characteristics (e.g., place of residence, occupation) undergo changes after the date of arrival, others (e.g., country of birth, mother tongue) should not change at all, and still others change in measurable ways (age) or may change little or not at all (educational attainment); but the distributions are affected by differences in the mortality of immigrants in different categories. Hence, the adequacy of the foreign-born data in reflecting the characteristics of immigrants varies with the characteristic.

The distinctive geographic distribution of immigrants and the influence of immigration on the geographic distribution of the general population may be inferred from a comparison of the geographic distribution of the native, foreign born, and total populations of a country.

Differences in the fertility level of immigrants and the non-immigrant population may be ascertained from differences between the foreign born and the native population in the general fertility rate or in the average number of children ever born.

For studies of the impact of immigration on a country's social and economic structure, census data on the foreign born, especially by age and sex, are particularly valuable. Census data giving detailed tabulations of the foreign-born population by area of present residence and country of birth permit the calculation of indexes of the relative concentration of various ethnic groups among the immigrants in various parts of a country.[13] Census data on the detailed occupation and country of birth of the foreign born provide the basis for measuring the tendency of various ethnic groups among the immigrants to concentrate in certain lines of work. Census data on the proportion of foreign born in the labor force, by occupation and industry, are also useful in measuring the minimal impact of immigration on a country's labor force and economy over a broad period of time.[14]

Studies of the changes in the social and economic status of immigrants and of the cultural assimilation of immigrants involve comparisons of the characteristics of the immigrants and the general population and observation of changes in the immigrants over time. The use of census data is necessary or preferable in such studies. Census data must also be employed where the immigrants are to be classified according to their status with respect to characteristics that necessarily change or that may change after immigration (e.g., citizenship, employment status, occupation, language spoken), or where the immigration data are not tabulated in terms of the characteristic

(e.g., educational attainment). Greater comparability in the classification is probably achieved by use of census data even where these conditions do not apply.[15]

Timing of immigration. — There is specific interest in the timing of immigration. A knowledge solely of the number and characteristics of immigrants over the indeterminate past is of limited practical value in migration analysis. How long the immigrant has been in the host country is relevant to his assimilation, including his opportunities for acquiring citizenship there or learning the language of the host country. We should like to know the number and characteristics of immigrants at least for intercensal periods. Inferences regarding intercensal trends in the volume of migration or the characteristics of migrants cannot safely be made directly from a series of census figures for the total number of foreign-born persons or from a series of derived measures based on statistics for the foreign born.

Tabulations by year of immigration would indicate directly the volume of immigration (for surviving immigrants) for particular past periods. Such statistics would be useful in refining migration analyses based on date or place of birth. Each immigrant would be asked the date of arrival in the area of present residence or the date of departure from the place of birth. This method requires an additional question relating to date of migration on the questionnaire — a question which is a relatively difficult one. The data may be tabulated in terms of year or **period of immigration** or **duration of residence** in the country of immigration.

A question on year of immigration or number of years of residence of the foreign-born population was asked in the decennial censuses of the United States from 1890 to 1930[16] and was restored to the schedule in 1970.

Miscellaneous Other Sources

Other sources of immigration data are given brief treatment here since they are relatively infrequent. These include special surveys, registrations of aliens, tabulations of permits for work abroad, and tabulations of passports and visas issued.

The occasional special surveys may involve inquiries relating to the nativity, previous residence, or citizenship of the resident population. Surveys may be taken of the country's citizens living abroad. The figures obtained from such a survey would represent net lifetime emigration of a country's citizens including the natural increase of these emigrants but excluding former citizens who had become naturalized in the countries of emigration.

Occasional, periodic, or continuous special registrations of aliens may be carried out. In several countries (e.g., United

[13] E. P. Hutchinson, *Immigrants and Their Children, 1850–1950*, 1950 Census Monograph, New York, John Wiley & Sons, 1956, table 14.

[14] Hutchinson, op. cit., chapter 5. See also A. Ross Eckler and Jack Zlotnick, "Immigration and the Labor Force." *The Annals of the American Academy of Political and Social Science*, 262:94 and 100, March 1949.

[15] Examples of the use of census tabulations on the foreign born in the study of cultural and ethnic assimilation of immigrants are given in: Alma F. Taeuber and Karl E. Taeuber, "Recent Immigration and Studies of Ethnic Assimilation," *Demography* 4(2):798–808, 1967; and C. V. Kiser, "Cultural Pluralism." *Annals of the American Academy of Political and Social Science* 262:117–130, March 1949. A description of the statistical material available both in the United States and abroad for studying the cultural assimilation of immigrants and examples of the uses to which this material has been put are given in: International Union for the Scientific Study of Population, *Cultural Assimilation of Immigrants* (papers presented at General Assembly at Geneva, 1949). Supplement to *Population Studies* (London), March 1950. See especially the article by Max Lacroix and Edith Adams, "Statistics for Studying the Cultural Assimilation of Migrants." pp. 69–108.

[16] In the 1890 and 1900 censuses the question was on number of years of residence of the foreign born in the United States. The tabulation in 1890 was limited to alien males 21 years old and over.

States, Japan, and New Zealand) aliens are required to register annually. The count of alien registrants in a single registration may be interpreted as representing lifetime net surviving alien immigrants, i.e., total immigration of aliens, less deaths, emigration, and naturalizations of aliens prior to the census date. Obviously, such data tell us nothing about the timing of migration without tabulations by year of entry (or duration of residence) or age. Typically, country of nationality is obtained rather than country of last permanent residence. A registration of aliens was conducted in the United States in 1940 by the Immigration and Naturalization Service in accordance with the requirements of the Alien Registration Act of 1940, and an annual registration has been conducted since 1951 in accordance with the requirements of the Internal Security Act of 1950 and the Immigration and Nationality Act of 1952. Each alien, regardless of date of immigration or length of stay, is required to register in January of each year and provide information regarding his address, sex, date of birth, country of birth, citizenship, date of entry into the United States, permanent or temporary status, and current occupation. Tabulations of aliens who report under the Alien Address Program are shown in the *Annual Report of the Immigration and Naturalization Service* and the *I and N Reporter*. They distinguish permanent residence from nonpermanent residence and show numbers of aliens by nationality and State of current residence

Foreign workers may have the obligation of securing a work permit, as in most European countries. The resulting statistics of emigration are weakened by the fact that not all permits are used and the procedure followed in renewing permits may be ill-adapted to the recording of international movements. The statistics of alien identity cards also have this defect.

Counts of passports and visas issued by a particular country relate to citizens and aliens, respectively. Not all passports and visas are actually used, or, if used, there is no way of knowing exactly when the traveler departs or returns and how many trips he makes. Tabulation of the number of travelers listed on a passport or visa is required to determine the possible number of travelers involved. Although passports and visas have severe limitations for measuring the actual volume of international migration, they may be useful in evaluating data or estimates from other sources.

COMBINATION, EVALUATION, AND ESTIMATION OF INTERNATIONAL MIGRATION STATISTICS

Combination and Comparison of International Migration Statistics

As we have suggested, adequate data on immigration, emigration, or net migration are often not available for a country even though a number of direct sources of data bearing on international migration for that country may exist. One method of expanding the quantity of information on migration for a given country is to refer to the migration statistics for other countries furnishing or receiving migrants from the country under study. This approach depends on the fact that any international movement involves two countries— the country of immigration and the country of emigration—and therefore may be reported by two countries. Since overlapping as well as complementary information may be available, this approach also serves as a basis of evaluation of migration data.

Let us imagine two arrays presenting statistics of migration between all the pairs of countries in the world. One such array would give the statistics reported by the country of immigration and the second would give the statistics reported by the country of emigration. If immigration and emigration were completely and consistently reported for all the countries of the world, the two arrays would agree or nearly agree. For example, the total immigration reported for a given country would coincide with the emigration reported by all the other countries which furnish emigrants to the country in question. Likewise, the total emigration reported for a given country would coincide with the immigration received by all other countries from the country under consideration.

In practice, of course, migration statistics are lacking for many countries and the statistics available are often incomplete and inconsistent. The lack of comparability of the migration statistics from one country to another is a major problem. The types of movement counted as migration and the categories of persons classed as migrants will differ from country to country, partly because there are many different ways of defining a migrant and the categories of migration and because there are many different systems for collecting the data. A small part of the difference in the counts of migrants reported by different countries for a given year may result from the fact that there is a gap in time between departure from one country and arrival in another. Hence, births and deaths may occur en route, and some travelers may change their destination or not be admitted to a country. There is also a problem in the identification of the country of origin as well as destination. Although the statistics usually relate to country of last permanent residence, in some cases they relate to country of birth, country of last previous residence, or country of citizenship—which may, of course, differ from one another.

In spite of the limitations mentioned, the binational reporting of international migration provides a useful, if not always fully adequate, basis for evaluating the accuracy of migration data for a country or for filling in missing data, particularly on emigration.

The possibility of combining and comparing statistics from different countries applies to census statistics as well as to statistics of frontier control. Suppose we are trying to estimate the net total movement to or from a country or a particular stream and its counterstream. We can approximate the true "net lifetime immigration" for a country somewhat more closely, for example, if we try to "balance" the census count of the foreign-born population in the country with the census count of persons in other countries who had been born in the country in question. In practice, it is usually possible to identify at least those few countries that are principal destinations of the emigrants of some country or that are the principal origins of the immigrants of that country. Many national censuses identify a long list of countries of origin for their foreign born. If the principal movement from a country is to one or two countries only, examination of the census reports of only these few countries is required to measure net migration for the country of emigration. For example, 617,056 persons living in the United States in 1960 were born in Puerto Rico and 50,910 persons living in Puerto Rico in 1960 were born in the United States; these figures indicate a "net surviving emigration" of 566,146 persons from Puerto Rico to the United States up to 1960.

With census statistics as with regular immigration statistics, there is a possibly serious problem of comparability resulting from differences in the nature of the census questions, the type of census (*de facto* or *de jure*), and the dates of the two censuses.

Estimation and Evaluation of Net Migration

In view of the lack of adequate statistics on migration for many countries and many periods, it is often necessary to estimate the volume of migration. Although there is interest in separate figures on immigration and emigration, the available methods do not permit making adequate estimates of immigration or emigration separately; only the balance of migration can be satisfactorily estimated by these methods. For the most part, the same methods are useful in evaluating the reported migration data as for deriving alternative estimates of net migration. Accordingly, we treat the procedures for the estimation and evaluation of net migration jointly in the present section.

Intercensal Component Method for the Total Population.— Approximate estimates of net immigration can often be derived for intercensal periods by use of census data on the total population. The general formula for estimating the total volume of net immigration in an intercensal period involves a rearrangement of the elements of the standard intercensal component equation:

$$I - E = (P_1 - P_0) - (B - D) \qquad (3)$$

Net immigration is derived as a residual; as previously stated, estimates of immigration and emigration cannot be obtained separately. The method simply involves subtracting an estimate of natural increase $(B - D)$ during the period from the net change in population during the period $(P_1 - P_0)$. If the data used to arrive at the estimate are exact, an exact estimate of the balance of all in-movements and out-movements is obtained. Because, however, the census counts and vital statistics as recorded are subject to unknown degrees of error, the residual estimates of net immigration may be in substantial error. The relative error in net immigration may be considerable when the amount of migration is small.

Estimates of intercensal net migration for several countries

are derived by the residual method in table 20-3. For Jamaica during the 1943-60 period, population increase amounted to 1,610,000 minus 1,237,000, or 373,000, and natural increase amounted to 568,000; hence, the estimated net migration was 373,000 minus 568,000, or −195,000.

Intercensal Cohort-Component Method.— The cohort-component method is applicable to the estimation of net migration, by age, for the total population and for segments of the population which are fixed (e.g., sex, race, and country of birth). This procedure involves the calculation of estimates by age groups on the basis of separate allowances for the components of population change. In preparing estimates by age, the compilation of death statistics by age cohorts to allow for the mortality component is so laborious, even where the basic statistics on death are available, that survival rates are normlly used instead. The survival rates may be life table survival rates or so-called national census survival rates.

One formula for this purpose covering age cohorts other than those born during the intercensal period is:

$$I_a - E_a = P_a^1 - s P_{a-t}^0 \qquad (4)$$

where I_a and E_a represent immigrants and emigrants in a cohort defined by age at the end of the period, P_a^1 the population at this age in the second census, P_{a-t}^0 the population t years younger at the first census, and s the survival rate for this age group for an intercensal period of t years (that is, s is a simplified representation of $_n s_{a-t}^t$). For the new-born cohorts the formula is:

$$I_a - E_a = P_a^1 - s B \qquad (5)$$

where B represents the births which occurred in the intercensal period.

Table 20-4 illustrates the procedure for estimating intercensal net migration of males between 1950 and 1960 by age for Puerto Rico by use of life table survival rates. Ten-year survival rates for 5-year age groups are first computed from the 1954-56 life table for Puerto Rico by the formula $\frac{_5L_{x-10}}{_5L_x}$.[17] These rates (col. 3) are applied to the population at the first census (col. 1) to derive an estimate of the expected survivors (allowing for deaths but not for net migration) 10 years older at the later census date (col. 4). The difference (col. 5) between the population at the second census (col. 2) and the expected population (col. 4) is an estimate of net migration.

The above calculations represent an application of the conventional survival-rate procedure. One defect of this method is that it has a tendency to understate or overstate the number of (implied) deaths during the intercensal period. In an "emigration" country, the initial population in an age cohort overstates the average population exposed to risk during the following intercensal period and the terminal population understates the average population exposed to risk, because some persons emigrate. A more satisfactory estimate of (implied) deaths and net migration may be made, therefore (table 20-4), by (1) "aging" the initial population to the date of the second census (col. 4); (2) calculating the corresponding "forward" estimate

Table 20-3.— Calculation of the Estimated Intercensal Net Migration by the Residual Method, for Puerto Rico, Jamaica, and Australia

[Numbers in thousands]

Item	Puerto Rico, April 1, 1950 to 1960	Jamaica, Jan. 4, 1943, to April 7, 1960	Australia, June 30, 1954 to 1961
(1) Population, first census.......	2,211	1,237	8,987
(2) Population, second census......	2,350	1,610	10,508
(3) Net change, (2)-(1)...........	139	373	1,522
(4) Births........................	811	(NA)	1,544
(5) Deaths........................	180	(NA)	601
(6) Natural increase, (4)-(5)......	631	568	944
Estimated net immigration:			
(7) Residual estimate, (3)-(6)...	−493	−195	578
(8) Direct count................	−452	(NA)	585
(9) Recorded arrivals...........	(NA)	(NA)	1,767
(10) Recorded departures.........	(NA)	(NA)	1,182
(11) Difference, (7)-(8).........	−41	(NA)	−7

NA Not available.

Source: U.S. Bureau of the Census, *Current Population Reports*, Series P-25, No. 336, "Estimates of the Population of Puerto Rico and Other Outlying Areas: 1950 to 1965," April 26, 1966, table 3; K. Tekse, "Comentarios sobre la creciente fecundidad de la población de Jamaica," *Estadística*, 25 (96/97): 548-564, September-December 1967; and Australia, Commonwealth Bureau of Census and Statistics, *Demography, 1964*, Bulletin No. 82, p. 12.

---

[17] The $_5L_x$ values can be derived from the life table functions in the U.N. *Demographic Yearbook, 1965*, by the formulas

$$T_x = e_x^0 \, l_x \quad \text{and} \quad _5L_x = T_x - T_{x-5}$$

Table 20–4. — Calculation of Estimates of Net Migration of Males, by Age Cohorts, by the Life-Table Survival-Rate Method, for Puerto Rico: 1950–60

Age in		Population		10-year life-table survival rate [1]	Forward method		Reverse method		Average estimate of net migration
1950	1960	Census, April 1, 1950	Census, April 1, 1960		Survivors	Net migration	"Younged" population	Net migration	$\left[(5)+(7)\right] \div 2 =$
		(1)	(2)	(3)	(1) x (3) = (4)	(2) - (4) = (5)	(2) ÷ (3) = (6)	(6) - (1) = (7)	(8)
All ages............	All ages...........	1,110,946	1,162,764	(X)	1,421,306	-258,542	1,247,507	-265,856	-262,201
Births, 1955-60........	Under 5................	[2]196,140	179,619	.93747	183,875	-4,256	191,600	-4,540	-4,398
Births, 1950-55........	5-9....................	[3]206,277	165,930	.92024	189,824	-23,894	180,312	-25,965	-24,930
Under 5................	10-14.................	185,014	162,244	.97658	180,681	-18,437	166,135	-18,879	-18,658
5-9....................	15-19.................	161,446	122,602	.99004	159,838	-37,236	123,835	-37,611	-37,424
10-14.................	20-24.................	138,696	79,792	.98717	136,917	-57,125	80,829	-57,867	-57,496
15-19.................	25-29.................	108,984	61,971	.97984	106,787	-44,816	63,246	-45,738	-45,277
20-24.................	30-34.................	91,269	58,723	.97361	88,860	-30,137	60,315	-30,954	-30,546
25-29.................	35-39.................	76,528	61,592	.96898	74,154	-12,562	63,564	-12,964	-12,763
30-34.................	40-44.................	66,769	53,087	.96148	64,197	-11,110	55,214	-11,555	-11,333
35-39.................	45-49.................	67,324	53,781	.95196	64,090	-10,309	56,495	-10,829	-10,569
40-44.................	50-54.................	47,745	39,832	.94135	44,945	-5,113	42,314	-5,431	-5,272
45-49.................	55-59.................	39,893	34,404	.92389	36,857	-2,453	37,238	-2,655	-2,554
50-54.................	60-64.................	36,548	29,095	.88796	32,453	-3,358	32,766	-3,782	-3,570
55-59.................	65-69.................	24,692	24,525	.81677	20,168	+4,357	30,027	+5,335	+4,846
60-64.................	70-74.................	25,636	16,370	.71369	18,296	-1,926	22,937	-2,699	-2,313
65-69.................	75-79.................	16,270	10,275	.61203	9,958	+317	16,788	+518	+418
70-74.................	80-84.................	10,679	4,636	.52046	5,558	-922	8,908	-1,771	-1,347
75 and over..........	85 and over..........	13,453	4,286	.28604	3,848	+438	14,984	+1,531	+985

X Not applicable.
[1] Derived from the 1954–56 life table for Puerto Rico, as given in United Nations, *Demographic Yearbook, 1966*, tables 21 and 23.
[2] Equals three-fourths of the births in 1955 plus the births in 1956, 1957, 1958, 1959, and one-fourth of the births in 1960.
[3] Equals three-fourths of the births in 1950 plus the births in 1951, 1952, 1953, 1954, and one-fourth of the births in 1955.

Source of basic data: U.S. Bureau of the Census, official reports.

of net migration (col. 5); (3) "younging" the terminal population to the date of the first census by dividing the terminal population by the survival rate (col. 2 ÷ col. 3 = col. 6); and (4) calculating the corresponding "reverse" estimate of net migration by subtracting the initial population (col. 1) from the "younged" population (col. 6); and (5) averaging the two estimates of net migration in steps (5) and (7) (col. 8). Now the formulas are: [18]

$$\text{Forward estimate: } M_1 = (I_a - E_a)_1 = P_a^1 - sP_{a-t}^0 \qquad (6)$$

$$\text{Reverse estimate: } M_2 = (I_a - E_a)_2 = \frac{P_a^1}{s} - P_{a-t}^0 \qquad (7)$$

$$\text{Average estimate: } M_3 = \frac{M_1 + M_2}{2} \qquad (8)$$

The calculation of deaths to the age cohorts born during the period requires special treatment. Exposure is less than the full intercensal period (10 years in the example of table 20–4). The survival rate for the cohort under 5 years of age at the end of the decade is $\frac{L_{0-4}}{5l_0}$ and the survival rate for the cohort 5 to 9 is $\frac{L_{5-9}}{5l_0}$. The forward and reverse survival procedures are then applied in the same way as for the older cohorts, the births in the two 5-year periods being taken as the initial population for the first two age groups.

Intercensal Component Method for the Foreign-Born Population. — Intercensal changes in the volume of immigration are obscured in a historical series on the number of foreign-born persons. Combined use of such data at successive censuses in some form may serve as a basis for estimating intercensal net immigration.

The intercensal change in foreign-born population understates the net immigration of foreign-born persons during an intercensal period by the number of deaths of foreign-born persons in the area during the period. Thus, solving the intercensal component equation for the foreign-born population for $I_F - E_F$, we have

$$P_F^1 = P_F^0 + I_F - E_F - D_F \qquad (9)$$

$$I_F - E_F = P_F^1 - P_F^0 + D_F \qquad (10)$$

where P_F^0 and P_F^1, represent the foreign-born population at the first and second censuses, respectively, I_F and E_F, persons born abroad entering and leaving the area during the intercensal period, respectively, and D_F, the number of deaths of foreign-born persons in the area during the intercensal period. The factor of births does not figure, of course, among the components of change of the foreign-born population. As suggested earlier, total (gross) immigration of the foreign born cannot be estimated by this residual procedure, only the net numbers of foreign-born persons arriving.

The number of deaths may be quite large and have a considerable effect on the estimate of net immigration, particularly if large numbers of immigrants arrived in preceding decades who would now be of advanced age. The difference in the number of foreign-born persons may even suggest net emigration when net immigration actually occurred. The figure for the

---

[18] The logic and procedure of forward, reverse, and average survival calculations in the estimation of net migration are considered further in chapter 21.

Table 20–5. – Calculation of Estimates of Net Immigration of Foreign-Born Males, by Age Cohorts, by the Life-Table Survival-Rate Method, for the United States: 1950–60

Age in 1950	Age in 1960	Foreign-born population Census, April 1, 1950 [1] (1)	Census, April 1, 1960 (2)	10-year life-table survival rate [2] (3)	Forward method Survivors (1) x (3) = (4)	Net immigration (2) - (4) = (5)	Reverse method "Younged" population (2) ÷ (3) = (6)	Net immigration (6) - (1) = (7)	Averaged net immigration [(5) + (7)] ÷ 2 = (8)
All ages	All ages	5,315,139	4,760,432	(X)	3,822,909	[3]801,983	6,231,822	916,683	927,410
(X)	Under 5	(X)	51,442	.99199	(X)	(X)	51,857	51,857	[4]51,857
(X)	5-9	(X)	84,098	.99770	(X)	(X)	84,292	84,292	[4]84,292
Under 5	10-14	33,039	119,935	.99290	32,804	87,131	120,793	87,754	87,443
5-9	15-19	32,718	98,724	.99274	32,480	66,244	99,446	66,728	66,486
10-14	20-24	36,141	137,519	.98797	35,706	101,813	139,193	103,052	102,433
15-19	25-29	50,845	171,310	.98428	50,046	121,264	174,046	123,201	122,233
20-24	30-34	95,065	204,769	.98331	93,478	111,291	208,245	113,180	112,236
25-29	35-39	167,913	262,716	.98041	164,624	98,092	267,965	100,052	99,072
30-34	40-44	158,928	219,307	.97064	154,262	65,045	225,941	67,013	66,029
35-39	45-49	265,836	315,526	.95438	253,709	61,817	330,608	64,772	63,295
40-44	50-54	411,033	422,367	.92699	381,023	41,344	455,633	44,600	42,972
45-49	55-59	563,782	541,383	.88465	498,750	42,633	611,974	48,192	45,413
50-54	60-64	653,704	542,483	.82810	541,332	1,151	655,094	1,390	1,271
55-59	65-69	722,473	566,414	.75427	544,940	21,474	750,943	28,470	24,972
60-64	70-74	734,631	504,663	.66068	485,356	19,307	763,854	29,223	24,265
65-69	75-79	595,398	306,872	.54331	323,486	-16,614	564,819	-30,579	-23,597
70-74	80-84	384,988	141,429	.40438	155,681	-14,252	349,743	-35,245	-24,749
75 and over	85 and over	408,645	69,475	.18410	75,232	-5,757	377,376	-31,269	-18,513

X Not applicable.

[1] Includes data for Alaska and Hawaii.

[2] Average of the 1950 and 1960 life tables for the white population of the United States. Special survival rates are required for immigrants born during the decade; see text for explanation.

[3] Excludes net immigration of children under 10 years of age in 1960

[4] Reverse estimate only.

Source of basic data: U.S. Bureau of the Census, official reports.

United States for the 1950–60 decade (– 693,000), for example, suggests a substantial net emigration, whereas frontier control data show a substantial net immigration of aliens during this period (2,238,000). When formula (10) is used to estimate net migration to the United States, 1950–60 (i.e., allowing for the deaths of the foreign born), the resulting figures show a substantial net immigration of foreign-born persons:

Central death rates applied to midperiod
    population... 1,736,000
Life-table survival-rate method...................... 1,805,000
Census survival-rate method......................... 1,450,000

A discussion of each of these estimates is presented in the following paragraphs.

The allowance for the deaths of the foreign-born population during the intercensal period can be made by use of statistics of deaths or by estimating deaths on the basis of death rates or survival rates. Illustrations will be given here of several procedures. The calculations are quite simple when only an estimate of total net immigration is wanted and both population and deaths are tabulated by nativity. As formula (10) shows, it is merely necessary to take the difference between the counts of the foreign-born population in the two censuses and add the number of deaths of the foreign born. (This method corresponds to the "vital statistics" method of estimating net migration described in chapter 21.)

In the event that statistics of deaths of foreign-born persons are lacking, as is the common situation, the number may be estimated by applying appropriate central death rates to the midperiod foreign-born population. This procedure has been worked out for the United States for 1950–60. The specific steps consist of (1) cumulatively multiplying central age-specific death rates for the general population by the estimated foreign-born population by age for 1955 (i.e., an average of the foreign-born population by age in 1950 and 1960), to determine the average annual number of deaths of foreign-born persons, and (2) multiplying the result by 10, the number of years in the period. The estimate of total deaths of the foreign-born population of the United States between 1950 and 1960 obtained in this way is 2,428,000. The resulting estimate of net immigration of foreign-born persons following formula (10), is

$$(9,738,000 - 10,431,000) + 2,428,000 = 1,736,000.$$

Estimates of net immigration of the foreign born by age cohorts are obtained by use of a survival-rate procedure. Either life-table survival rates or census survival rates may be employed. The life-table survival-rate procedure is illustrated in table 20–5, which develops estimates of net immigration of the foreign-born male population of the United States by age from 1950 to 1960. The steps are similar to those described earlier when the estimates of net immigration were derived on the basis of the total population (table 20–4). The foreign-born male population in 1950 (col. 1) is aged 10 years by appropriate survival rates (col. 3) to 1960 (col. 4). (The survival rates used are averages of 10-year survival rates for white males from the U.S. life tables for 1950 and 1960.) The survivors 10 years old and over in column (4) are then subtracted from the foreign-born population 10 years old and over in 1960 (col. 2) to obtain the "forward" estimate of net immigration in column (5).

The procedure just described is the conventional one but it

does not provide an estimate of net immigration for the new-born children. A reverse survival-rate procedure provides an alternate estimate for ages 10 and over (col. 7) and makes possible the calculation of estimates of net immigration of children born during the intercensal period. The two sets of estimates of net immigration are then averaged (col. 8) to derive the final estimates of net immigration of foreign-born males to the United States, 1950–60 ([(col. 5) + (col. 7)] ÷ 2). To reduce the calculations, estimates for the children born during the period derived by the reverse method may be combined with estimates for the other age groups derived by the forward method. The result in this case is 938,000 as compared with the 927,000 obtained by the average method. Note again how little difference results from the variations in survival-rate procedure. Once again the calculations relate to age cohorts, so that age groups 10 years apart are paired in the two censuses; all the calculations for the same age cohort are shown on the same line of the table.

The estimation of deaths for the age cohorts born abroad in the intercensal period (i.e., under 5 and 5 to 9 years of age at end of period) requires special treatment since these groups are exposed to the risk of death in the United States for much less than 10 years as a result both of recent birth abroad and of the staggered timing of immigration. For example, the population under 5 years old in 1960 has been at risk for only $1\frac{1}{4}$ years on the average ($2\frac{1}{2} \times \frac{1}{2}$ years), and the population 5 to 9 years old in 1960 has been at risk for only $3\frac{3}{4}$ years on the average ($7\frac{1}{2} \times \frac{1}{2}$ years). The first factors (i.e., $2\frac{1}{2}$ and $7\frac{1}{2}$) represent the average period of time between birth and the 1960 census, and the second factor (i.e., $\frac{1}{2}$) reflects the fact that the migrants entered the country or departed at various times during the period between birth and the census, with an assumed average residence in the area of one-half the period. Therefore, the survival rates

for the populations under 5 and 5 to 9, respectively, in **exact** ages, are as follows:

$$\frac{L_{0-5}}{1.25 \; l_0 + L_{0-3.75}} \text{ and } \frac{L_{5-10}}{L_{1.25-6.25}}.$$

As was stated above, alternative estimates of net immigration of the foreign-born population employing the survival-rate procedure may be derived by use of national census survival rates instead of life-table survival rates. It may be recalled from chapter 15 that census survival rates represent the ratios of the population in a given age at one census to the population in the same age cohort at an earlier census and are often computed for the native population so as to exclude the effects of international migration from the rates. They are intended to represent the combined effect of mortality and the relative change from one census to the next in the rate of net census error for a given age cohort. Their presumed advantage is that the estimates of net migration resulting from their use may be more accurate than when life-table survival rates are used, since errors in census enumeration are incorporated in the census survival rates rather than in the estimates of net migration, as in the case when life-table survival rates are used.

The steps are the same as with life-table survival rates, once the census survival rates have been calculated. The cohorts born during the intercensal period once again require special treatment. For the computations relating to the U.S. foreign-born population, as before, the populations under 5 and 5 to 9 years old in 1960 have to be "younged" to the estimated date of arrival in the United States, not to the date of birth; and only a reverse estimate of net migration is possible. A comparison of the estimates of net immigration of the foreign born into the United States by age and sex, for 1950–60, derived by the

Table 20–6. — Comparison of Estimates of Net Immigration by Age Cohorts and Sex, Based on the Census Survival-Rate Method, the Life-Table Survival-Rate Method, and Visa Data, for the United States: 1950–60

[Figures relate to the resident population of the United States, including Alaska and Hawaii]

Age of cohort in 1960 (years)	Estimates of net immigration						Percent difference from census survival-rate method[2]			
	Based on census survival-rate method		Based on life-table survival-rate method		Based on visa data[1]		Life-table survival-rate method		Visa data	
	Male	Female	Male	Female	Male	Female	Male $\left[\frac{(3)-(1)}{(1)}\right] \times 100 =$	Female $\left[\frac{(4)-(2)}{(2)}\right] \times 100 =$	Male $\left[\frac{(5)-(1)}{(1)}\right] \times 100 =$	Female $\left[\frac{(6)-(2)}{(2)}\right] \times 100 =$
	(1)	(2)	(3)	(4)	(5)	(6)	(7)	(8)	(9)	(10)
All ages.........	741,986	707,529	927,410	877,329	1,013,200	1,225,100	+25.0	+24.0	+36.6	+73.2
Under 5..............	52,562	51,075	51,857	50,741	30,100	27,600	-1.3	-0.7	-42.7	-46.0
5–14................	171,365	164,989	171,735	166,290	169,200	164,100	+0.2	+0.8	-1.3	-0.5
15–24...............	172,516	210,373	168,919	210,156	160,300	229,700	-2.1	-0.1	-7.1	+9.2
25–34...............	231,396	309,813	234,469	316,318	247,800	379,400	+1.3	+2.1	+7.1	+22.5
35–44...............	158,031	160,205	165,101	165,882	200,600	204,800	+4.5	+3.5	+26.9	+27.8
45–54...............	105,353	83,616	106,267	73,496	125,800	119,100	+0.9	-12.1	+19.4	+42.4
55–64...............	29,921	6,098	46,684	12,067	59,500	68,900	+56.0	+97.9	+98.9	+1,029.9
65 and over.........	-179,158	-278,640	-17,622	-117,621	20,000	31,400	+90.2	+57.8	+111.2	+111.3

[1] Net immigration of civilian aliens. Source: U.S. Bureau of the Census, *Current Population Reports*, Series P–25, No. 310.

[2] A minus sign indicates that the estimate of net immigration or net emigration based on the life-table survival-rate method or the visa data is less than the estimate based on the census survival-rate method. A positive sign indicates that the estimate of net immigration or net emigration based on the life-table survival-rate method or the visa data is greater than the estimate based on the census survival-rate method, or that the former methods give an estimate of net immigration whereas the latter methods give an estimate of net emigration.

Source of basic data: Official national reports.

life-table survival-rate procedure and the census survival-rate procedure, is given in table 20-6. The percent differences are small for most ages but become rather large among older persons. Table 20-6 also shows a comparison of estimates of the net migration of the foreign born based on the census survival-rate method and estimates based on visa data, or immigration statistics. The differences are quite large for females and for most adult ages. It is apparent that the U.S. censuses and the visa data reflect quite different levels of immigration.

The general procedure just described may be extended to derive estimates of net immigration, by age and sex, according to race, country of birth, mother tongue, or other stable ethnic characteristic, or combination of them. Table 20-7 presents the calculation of estimates of the net immigration of males born in Italy to the United States between 1950 and 1960,

based on U.S. census statistics on country of birth. A (forward) estimate of net immigration to the United States of males born in Italy is made by carrying the male population of the United States born in Italy, for 1950, forward to 1960, by the use of 10-year life-table survival rates, and comparing the survivors with the corresponding 1960 census figures. Reverse estimates of net immigration of children under 10 are prepared by "younging" the 1960 census figures back to the average date of immigration. The overall estimate of net immigration of Italians to the United States, for males (114,000) and females (66,000) combined, derived by this method, is 180,000. This estimate may be compared with the count of 187,000 immigrants coming to the United States in this decade born in Italy, as shown by visa statistics on immigration.

Table 20-8 illustrates the calculation of estimates of the

Table 20-7. — Calculation of Estimates of Net Immigration to the United States of Males Born in Italy, by Age Cohorts, by the Life-Table Survival-Rate Method: 1950-60

Age in		U.S. males born in Italy		10-year life table survival rates[2]	Survivors	Net migration
1950	1960	Census, April 1, 1950[1]	Census, April 1, 1960		(1) x (3) =	(2) - (4) =
		(1)	(2)	(3)	(4)	(5)
All ages...............	All ages...............	794,650	683,697	(X)	578,678	113,680
(X).....................	Under 5.................	(X)	4,547	.99199	[3]4,584	[3]4,584
(X).....................	5-9.....................	(X)	4,022	.99770	[3]4,031	[3]4,031
Under 5.................	10-14...................	1,170	8,918	.99290	1,162	7,756
5-14....................	15-24...................	3,285	23,083	.99036	3,253	19,830
15-24...................	25-34...................	15,090	37,068	.98379	14,845	22,223
25-34...................	35-44...................	33,990	50,795	.97554	33,159	17,636
35-44...................	45-54...................	80,490	92,538	.94081	75,726	16,812
45-54...................	55-64...................	202,395	181,431	.85705	173,463	7,968
55-64...................	65-74...................	265,245	199,282	.71014	188,361	10,921
65-74...................	75-84...................	149,425	73,446	.48235	72,075	1,371
75 and over............	85 and over............	43,560	8,567	.18410	8,019	548

X Not applicable.
[1] White population born in Italy.
[2] Average of the 1950 and 1960 life tables for the white male population of the United States. Special survival rates are required for immigrants born during the decade, as explained in connection with table 20-5.
[3] Reverse estimate.

Source of basic data: U.S. Bureau of the Census, official reports.

Table 20-8. — Calculation of Estimates of Net Migration of Males from Puerto Rico to the United States, for Selected Age Cohorts, Using Place-of-Birth Data: 1950-60

[Reverse survival method]

Age in		Excess of U.S. population born in Puerto Rico over Puerto Rican population born in the United States		10-year life table survival rates[1]	"Younged" population	Net migration
1950	1960	Censuses, April 1, 1950	Censuses, April 1, 1960		(2) ÷ (3) =	(4) - (1) =
		(1)	(2)	(3)	(4)	(5)
All ages...............	All ages...............	101,272	280,304	(X)	292,700	191,428
(X).....................	Under 5.................	(X)	2,873	.93747	3,065	3,065
(X).....................	5-9.....................	(X)	15,394	.92024	16,728	16,728
Under 5.................	10-14...................	2,325	27,047	.97658	27,696	25,371
5-9.....................	15-19...................	5,863	28,496	.99004	28,783	22,920
10-14...................	20-24...................	7,685	36,792	.98717	37,270	29,585
15-19...................	25-29...................	7,833	39,534	.97984	40,347	32,514

X Not applicable.
[1] Derived from 1954-56 life table for the total population of Puerto Rico. Special survival rates are required for migrants born during the decade, as explained in connection with table 20-5.

Source of basic data: Official national reports.

Table 20–9. — Net Immigration, by Components, for the United States: July 1, 1960–68

[Minus sign (−) denotes outward movement from the United States]

Year (July 1 to June 30)	Net immigration	Net civilian immigration						Net movement of armed forces
		Total	Immigrant aliens	Puerto Ricans	Citizens from abroad	Other[1]	
1967–1968.........................	380,975	448,327	355,344	18,681	29,166	45,136	−67,352
1966–1967.........................	175,989	454,701	361,972	34,174	43,291	15,264	−278,712
1965–1966.........................	239,776	449,463	323,040	30,089	62,234	34,100	−209,687
1964–1965.........................	321,943	346,082	296,697	10,758	33,508	5,119	−24,139
1963–1964.........................	376,004	363,792	292,248	4,366	48,446	18,732	12,212
1962–1963.........................	384,802	378,388	306,260	4,798	50,930	16,400	6,414
1961–1962.........................	315,949	384,526	283,763	11,363	13,933	75,467	−68,577
1960–1961.........................	396,921	402,597	271,344	13,762	53,164	64,327	−5,676

[1] Conditional entrants, including Cuban parolees and nonimmigrants, Hong Kong Chinese, net addition of refugees, and Cuban parolees who came in as alien immigrants.

Source: U.S. Bureau of the Census, unpublished records.

net movement of population between two areas on the basis of place-of-birth data from the census of each area. The illustration relates to the movement of the male population in the younger age cohorts between the United States and Puerto Rico in the 1950–60 decade. The basic procedure is the same as in the preceding illustration, but here the movements in both directions are taken account of. The net movement of United States-born males between the United States and Puerto Rico (who are mostly of Puerto Rican ancestry) offsets, in part, the net movement of Puerto Rico-born males between the two areas. Here, by way of illustration, the reverse-survival procedure alone is used to estimate net migration at all ages. The resulting estimate of total net migration of Puerto Ricans to the United States, for males and females combined, is $191,000 + 188,000$, or 379,000, as compared with the count of the net movement of passengers to the United States obtained from the Puerto Rican government, which is 452,000.

Changes in Alien Population. — The same types of procedures as described in the previous section may be employed in connection with census counts of aliens, where available. The change in the number of aliens between two dates is affected by naturalizations as well as by net immigration and deaths:

$$P_A^1 = P_A^0 + I_A - E_A - D_A - N \qquad (11)$$

$$I_A - E_A = (P_A^1 - P_A^0) + D_A + N \qquad (12)$$

where A refers to aliens and N refers to naturalizations. Hence the net immigration of aliens is estimated as the sum of the net change in the number of aliens, deaths of aliens, and naturalizations.

Data on Nationals Abroad. — Estimates of the net migration of nationals or a particular segment of them (e.g., employees of the national government and their dependents; armed forces) during some period, complementing the estimates for the foreign born or aliens, may be derived by a similar method from statistics or estimates of these groups living outside the country at the beginning and end of the period. The formula may be arrived at by solving the component equation relating to this special population for the net migration component:

$$M_0 = [0_1 - 0_0] - [B_0 - D_0] \qquad (13)$$

where 0_0 and 0_1 represent the country's overseas population at the initial and terminal date in the period, respectively, B_0 births to the country's overseas population, D_0 deaths to the country's overseas population, and M_0 net movement overseas. The sign of the result is then changed so that it will refer to movement vis-a-vis the country in question, not the country's population abroad. Net movement into the country of its nationals from overseas equals $-M_0$, therefore.

United States Procedure of Estimating Net Migration

The estimates of net civilian immigration to the United States currently employed by the U.S. Census Bureau in its population estimates illustrate the composite use of several of the sources and methods that have been described. The statistics of net civilian immigration for the United States since 1950 cover the following categories, data for which are summarized in table 20–9 for 1960 to 1968:

1. Immigrant aliens
2. Emigrant aliens through June 1957
3. Net admissions of aliens for temporary residence (nonimmigrant aliens less nonemigrant aliens) through June 1956
4. Net arrivals from Puerto Rico
5. Net arrivals of civilian citizen passengers through March 1960
6. Net immigration since April 1960 of civilian citizens affiliated with the U.S. Government
7. Refugee escapees (covered by the Act of July 14, 1960) and refugees from Cuba and Hong Kong.

Categories 1, 2, 3, and 7 represent "frontier control" statistics compiled by the U.S. Immigration and Naturalization Service. Beginning in July 1957 collection of statistics on emigrant aliens was discontinued, and the Census Bureau discontinued making an allowance for this category. Categories 4 and 5 are based on passenger statistics compiled by the Planning Board of the Commonwealth of Puerto Rico and the U.S. Immigration and Naturalization Service, respectively. Category 6 represents estimates derived from the data on overseas population provided by the U.S. Defense Department and the U.S. Civil Service Commission.

TECHNIQUES OF ANALYSIS

Few specialized techniques have been devised for the analysis of international migration. Many of the techniques of analysis used in other demographic fields, especially for internal migration, apply equally to international migration. We will, accordingly, touch upon this subject only briefly in this chapter, leaving the more complete discussion to the chapter on internal migration.

General Aspects of Migration Analysis and Demographic Factors Important in Analysis

Analysis of international migration presents certain complexities arising from the special characteristics that distinguish it from, say, mortality analysis and natality analysis. The conceptual difficulties in defining a migrant are much greater than in defining a birth or a death. These difficulties are associated with the variety of definitions in a given country and between countries and with the variety of collection systems, and result in serious problems of international comparability. Unlike the events of birth or death, migration necessarily involves two areas, the area of origin and the area of destination. This fact leads to interest in particular migration streams and counterstreams as well as in total migration to an area, to the need to analyze jointly immigration, emigration, and net migration, and to difficulties in formulating measures of analysis, particularly rates. Formulation of migration rates involves practical and theoretical problems. Immigration cannot be viewed as a risk to which the members of the receiving population are subject in the same sense that this population is subject to the risk of death or bearing children, since the migrants come from outside the population. On the other hand, one can view emigration as a risk associated with the population of the area of origin. Analysis in terms of net migration, which is often necessary because of the lack of adequate statistics on immigration and emigration, presents special arithmetic and logical problems since net migration may be a positive or negative quantity and may differ considerably in magnitude from the gross volume of movement into or out of the country.

Unlike death, but like bearing a child, the event of migration may occur repeatedly to the same person or may not occur at all to an individual in his lifetime. Hence, there is need to differentiate migration (number of moves) from migrants (number of persons who have moved) and to specify precisely the period for which migration status (migrant, nonmigrant) is defined. The time period must be chosen carefully because it introduces one of the main complications in defining migration rates.

Analysis of international migration involves some of the same factors found to be important in other demographic analysis as well as other factors especially pertinent to this subject. Analysis of international migration streams calls for data on country of last permanent residence of immigrants and country of next permanent residence of emigrants. Country of birth or citizenship may serve as a substitute for previous permanent residence or as a supplement to it. For both administrative and demographic uses, it is important to have data on citizenship, type of migration (e.g., permanent, temporary, and "commuters" or "border crossers"), and the legal basis of entry of aliens.

Tabulations on type of migration and legal basis of entry may also provide general or detailed information on causes of migration, that is, whether the migrant intends to settle permanently, study, tour, take a temporary job, etc. There is interest in the demographic, social, and economic characteristics of migrants. In recommending a program to improve international migration statistics, the U.N. Population Commission at its third Session in May 1948 urged the "provision of statistics most relevant to the study of demographic trends and their relation to economic and social factors, including statistics on age, sex, marital condition, family size, occupation and wages of migrants."[19] The variation of rates of migration by age and sex may be quite pronounced and is important in its effect on the composition of the population of the sending and receiving countries. It is also of value to have data on major occupation group and industrial classification, particularly in connection with an analysis of the impact of migration on the sending and receiving country's economy. Data on the mother tongue, ethnic origin or race, education, marital status, and family composition of the immigrants are useful in the analysis of the assimilation of immigrants. Tabulations by state or province (political subdivision of country) of intended future residence for immigrants and by state of last permanent residence for emigrants are useful for making current estimates of population for such subdivisions.

Some analysts of international migration also take into account various nondemographic factors, such as changes in legislation regarding immigration and emigration, programs to assist immigrants, modes and costs of travel, economic conditions at origin and destination, wars, persecutions of minorities, business cycles, adequacy of harvests, and many other social and economic factors that are not necessarily measured by the population census or immigration statistics.

Net Migration, Gross Migration, and Migration Ratios

Movement into a country, previously referred to as immigration, may also be referred to as **gross immigration**. Similarly, movement out of a country, previously referred to as emigration, may also be referred to as **gross emigration**. The balance of gross immigration and gross emigration for a given country is referred to as **net immigration** or **net emigration**, depending on whether immigration or emigration is larger. A series of figures on gross immigration or emigration for several countries constituting some broader geographic area, say a continent, cannot be combined by addition to obtain the corresponding figures on gross immigration or emigration for the broader area, since some of the movement may have occurred between the constituent countries and hence would balance out in the measurement of immigration to, or emigration from, the area as a whole.

Net migration figures, however, can be added together to obtain totals for broader areas or subtracted from one another to obtain figures for constituent areas. The latter relation may be seen either by an algebraic or a graphic presentation. If we consider the movement affecting two countries, then we may diagram this movement as follows:

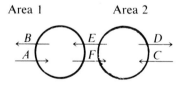

Thus, we have the following relationships:

(1) = Immigration to areas 1 and 2 as a unit, excluding movement between areas 1 and $2 = A + C$.

(2) = Emigration from areas 1 and 2 as a unit, excluding movement between areas 1 and $2 = B + D$.

(3) = (1) − (2) = net immigration to areas 1 and 2 as a unit, calculated directly = $(A + C) - (B + D)$.

(4) = Net immigration, area 1, taken separately = $(A + E) - (B + F)$.

[19] United Nations, *Problems of Migration Statistics*, Series A, Population Studies, No. 5, 1958, p. 2.

(5) = Net immigration, area 2, taken separately = $(C+F) - (D+E)$.

(6) = (4) + (5) = Net immigration, areas 1 and 2 as a unit, calculated by summing net migration for the component areas = $(A+C) - (B+D)$.

Note that the results in (3) and (6) are the same. They can be extended to cover n areas, in fact the entire world. At the global level we know that only births and deaths affect population growth. If comparable international migration data were available for every country for the same period—a purely hypothetical situation at present—the sum of the net immigration or net emigration figures for the various countries would therefore have to be zero.

It should be evident that any one of many very different amounts of immigration and emigration may underlie any given net figure. A net figure of zero may represent the balance of two equally large migration currents or the absence of all movement. There is interest, then, in the basic immigration and emigration figures even when the net movement is known. In analyzing the differences between the total movement and the net movement, a useful concept is that of **gross migration**, the sum of immigration and emigration. This figure may also be called **migration turnover.** It is intended to represent the total movement across the borders of an area during a period. For example, the 68,000 net immigration into Australia in 1961 represented the balance of an immigration of 128,000 and an emigration of 59,000. In total, 187,000 (i.e., 128,000 plus 59,000) moves occurred across the Australian borders in that year (table 20–10).

Various types of ratios may be computed to indicate the relative magnitude of immigration (I), emigration (E), net migration $(I-E$, or $M)$, and gross migration $(I+E)$, to or from a country.

$$\frac{\text{Emigration}}{\text{Immigration}} = \frac{E}{I} \tag{14}$$

$$\frac{\text{Net immigration}}{\text{Immigration}} = \frac{I-E}{I} \text{ where } I > E \tag{15}$$

$$\frac{\text{Net emigration}}{\text{Emigration}} = \frac{E-I}{E} \text{ where } E > I \tag{16}$$

$$\frac{\text{Immigration}}{\text{Gross migration}} = \frac{I}{I+E} \tag{17}$$

$$\frac{\text{Emigration}}{\text{Gross migration}} = \frac{E}{I+E} \tag{18}$$

$$\frac{\text{Net migration}}{\text{Gross migration}} = \frac{I-E}{I+E} \tag{19}$$

The ratio to be selected for some particular analytic study depends on the type of analysis being made and the migration characteristics of the area under study. For an area characterized by immigration, the ratio of net immigration to gross immigration may be used to measure the proportion of (gross) immigration which is effectively added to the population, i.e., the proportion which is uncompensated by emigration. Similarly, for an area characterized by emigration, the ratio of net emigration to (gross) emigration may be used to measure the proportion of the (gross) emigration which is effectively lost, i.e., the proportion which is uncompensated by immigration. Table 20–10 illustrates the computation of these measures on the basis of data for the United States for the 5-year periods 1910–15 to 1950–55 and on the basis of data for selected foreign countries for years around 1960.

Another measure, the ratio of net migration to gross migration, or net migration to migration turnover (formula 19), is

Table 20–10. — Amounts and Ratios of Net and Gross Migration, for the United States, 1910 to 1955, and for Selected Countries, Around 1960

Country and year or period	Immigration	Emigration	Net migration	Gross migration	Ratio, net migration to immigration or emigration[1]	Ratio, net migration to gross migration
			(1) − (2) =	(1) + (2) =	(3) ÷ (1) or (3) ÷ (2) =	(3) ÷ (4) =
	(1)	(2)	(3)	(4)	(5)	(6)
Australia (1961)	127,586	59,147	68,439	186,733	.5364	.3665
Colombia (1961)	6,335	18,833	−12,498	25,168	(−).6636	(−).4966
Cyprus (1960)	1,006	14,589	−13,583	15,595	(−).9310	(−).8710
Federal Republic of Germany (1961)	732,684	301,601	431,083	1,034,285	.5884	.4168
Israel (1960)	23,644	7,206	16,438	30,850	.6952	.5328
Italy (1961)	27,700	57,526	−29,826	85,226	(−).5185	(−).3500
Sweden (1960)	26,143	15,138	11,005	41,281	.4210	.2666
United Kingdom (1961)	83,724	90,959	−7,235	174,683	(−).0795	(−).0414
United States:[2]						
1910–1915	4,459,831	1,444,530	3,015,301	5,904,361	.6761	.5107
1915–1920	1,275,980	702,464	573,516	1,978,444	.4495	.2899
1920–1925	2,638,913	697,397	1,941,516	3,336,310	.7357	.5819
1925–1930	1,468,296	347,679	1,120,617	1,815,975	.7632	.6171
1930–1935	220,209	323,863	−103,654	544,072	(−).3201	(−).1905
1935–1940	308,222	135,875	172,347	444,097	.5592	.3881
1940–1945	170,952	42,696	128,256	213,648	.7502	.6003
1945–1950	864,087	113,703	750,384	977,790	.8684	.7674
1950–1955	1,087,638	134,220	953,418	1,221,858	.8766	.7803

[1] Ratio of net immigration to (gross) immigration and of net emigration to (gross) emigration.
[2] Aliens only.

Source: Cols. (1) and (2) from, U.S. Bureau of the Census. *Historical Statistics of the United States, Colonial Times to 1957,* 1960, pp. 62 and 64; and United Nations. *Demographic Yearbook, 1962,* tables 27 and 28.

a measure of migration effectiveness. It measures the relative difference between the effective addition or loss through migration and the overall gross movement. The ratio varies from zero to one, the higher the ratio the fewer the moves required to produce a given net gain or loss in population for a particular country.

The logic of the interpretation of migration ratios where negative values appear in the numerator or the denominator, or where the numerator or denominator is very small, should be considered. A negative sign may simply be taken to indicate that emigration exceeds immigration. Apart from this fact, the negative signs in the ratios shown in table 20–10 may be disregarded. For example, the migration effectiveness ratio for Australia is about the same as for Italy, even though the signs differ. Extremely large ratios resulting from very small denominators must ordinarily be interpreted with special care, but we have designed the ratios in table 20–10 to avoid this difficulty. Extremely small ratios resulting from very small numerators simply indicate that the effective addition or loss is small in relation to gross migration or migration turnover.

Migration Rates

Only limited use has been made of migration rates in the analysis of international migration or national population growth. In fact, no particular set of rates has yet become standard. Theoretically, the analogues of some of the types of rates used in natality or mortality analysis could be employed here. The logical difficulties of determining the form of migration rates and of interpreting them are probably greater for the reasons suggested earlier. The analytic measures used could also follow the form of those used in internal migration analysis since many of the problems of analysis are the same (chap. 21).

Several crude rates may be constructed on the basis of separate figures on immigration and emigration. These rates represent the amount of immigration, emigration, net migration, or gross migration, per 1,000 of the midyear population of a country and may be symbolized as follows:

$$\text{Crude immigration rate} = \frac{I}{P} \times 1,000 \qquad (20)$$

$$\text{Crude emigration rate} = \frac{E}{P} \times 1,000 \qquad (21)$$

Crude net migration (i.e., net immigration or net emigration)
$$\text{rate} = \frac{I - E}{P} \times 1,000 \qquad (22)$$

$$\text{Crude gross migration rate} = \frac{I + E}{P} \times 1,000 \qquad (23)$$

We may illustrate the application of the various formulas by computing the rates for Australia in 1961. The number of immigrants during 1961 was 127,586, the number of emigrants during that year, 59,147, and the estimated population on July 1, 1961, 10,508,186. Substituting the first and third values in the formula for the crude immigration rate, we have (127,586 ÷ 10,508,186) × 1,000, or 12.1. The crude emigration rate is (59,147 ÷ 10,508,186) × 1,000 or 5.6. The difference between the crude immigration rate and the crude emigration rate equals the crude net immigration rate (6.5), and the sum of the two rates equals the crude gross migration rate (17.8). The net migration rate may be either a crude immigration rate or a crude emigration rate. The gross migration rate is a measure of the relative magnitude of migration turnover and the population which it affects. Additional illustrative computations are given in table 20–11.

Table 20–11. — Various Rates of Migration, for the United States, 1910 to 1955, and for Selected Countries, Around 1960

Country and year or period	Immigration	Population[1]	Immigration rate $[(1) \div (2)] \times 1,000 =$	Emigration	Emigration rate $[(4) \div (2)] \times 1,000 =$	Net migration rate $\left[\frac{(1) - (4)}{(2)}\right] \times 1,000 =$	Gross migration rate $\left[\frac{(1) + (4)}{(2)}\right] \times 1,000 =$
	(1)	(2)	(3)	(4)	(5)	(6)	(7)
Australia (1961)	127,586	10,508,186	12.1	59,147	5.6	6.5	17.8
Colombia (1961)	6,335	14,443,000	0.4	18,833	1.3	-0.9	1.7
Cyprus (1960)	1,006	577,615	1.7	14,589	25.3	-23.5	27.0
Federal Republic of Germany (1961)	732,684	53,977,418	13.6	301,601	5.6	8.0	19.2
Israel (1960)	23,644	2,116,996	11.2	7,206	3.4	7.8	14.6
Italy (1961)	27,700	50,623,569	0.5	57,526	1.1	-0.6	1.7
Sweden (1960)	26,143	7,459,129	3.5	15,138	2.0	1.5	5.5
United Kingdom (1961)	83,724	46,269,000	1.8	90,959	2.0	-0.2	3.8
United States:[2]							
1910–1915	4,459,831	96,280,000	46.3	1,444,530	15.0	31.3	61.3
1915–1920	1,275,980	103,238,000	12.4	702,464	6.8	5.6	19.2
1920–1925	2,638,913	110,998,000	23.8	697,397	6.3	17.5	30.1
1925–1930	1,468,296	119,772,000	12.3	347,679	2.9	9.4	15.2
1930–1935	220,209	125,320,000	1.8	323,863	2.6	-0.8	4.3
1935–1940	308,222	129,465,000	2.4	135,875	1.0	1.3	3.4
1940–1945	170,952	135,800,000	1.3	42,696	0.3	0.9	1.6
1945–1950	864,087	145,379,000	5.9	113,703	0.8	5.2	6.7
1950–1955	1,087,638	158,332,000	6.9	134,220	0.8	6.0	7.7
1955–1960	1,427,841	172,630,000	8.3	(NA)	(NA)	(NA)	(NA)

NA Not available.
[1] Estimated midperiod population.
[2] Migration of civilian aliens. Population relates to conterminous United States, and includes all U.S. armed forces, both at home and abroad after 1930. Amounts and rates of migration are for 5-year periods.

Source: Cols. (1) and (4) from, U.S. Bureau of the Census, *Historical Statistics of the United States, Colonial Times to 1957*, 1960, pp. 62 and 64; and United Nations, *Demographic Yearbook, 1962*, tables 27 and 28.

Various kinds of specific rates may also be computed. The rates may be specific for age, sex, race, or other characteristics of the migrants. An age-specific net migration rate is computed as the amount of net migration (net immigration or net emigration) at a given age per 1,000 of the midyear population at this age:

$$\frac{I_a - E_a}{P_a} \times 1,000 \qquad (24)$$

where I_a and E_a represent immigration and emigration at age a, respectively, and P_a the midyear population at age a.

The most appropriate general base for the calculation of rates describing the relative frequency of migration for a country during a period is the population of that area at the middle of the period. This is particularly the case where the migration data are frontier control data, or visa data, which cover all movements into and out of the country during the period. The midperiod population represents here the average population at "risk" of sending out emigrants or receiving immigrants during the migration period. The midperiod population can serve as a common base for the calculation of rates of immigration, emigration, net immigration, and net emigration.

When the migration data come from a census or survey, they are ordinarily restricted to the cohorts of persons living both at the beginning and at the end of the migration period (i.e., excluding immigrants who were born, died, or emigrated during the period). In this case use of a midperiod population may be less appropriate and less convenient, particularly for migration periods of more than one year, and usually practical substitutes are used. Since population data corresponding to these immigration data are readily available in the census or survey, it is common to use the census population as a base. Census or survey data on the number of persons resident in the country who are living abroad at a particular previous date (e.g., t years earlier) are typically expressed as a percent of the census or survey population t years old and over. These rates may loosely be interpreted as the rates of immigration for the area during the period or, more exactly, the proportion of immigrants in the population at the census or survey date.

The current census population may also be used as a base where rates of net immigration are computed from census data on nativity. As may be recalled, data on the foreign born represent lifetime immigration excluding immigrants who subsequently died or emigrated. For example, the rate of lifetime net immigration may be computed from the data on nativity in a single census as the percent (or per thousand) which the population living in the country but born outside the country constitutes of the total population living in the country.

The various types of rates we have been discussing can all be computed on an age-specific basis provided the migration data and the population data have been tabulated by age. Some are central rates in the same sense as crude death rates and age-specific death rates; others are reverse cohort rates, as when the terminal population in the same cohort as the migrants is used as a base. They are not "true" rates or probabilities of migration; i.e., they do not represent the chance that a person observed at some date will migrate into or out of an area during a specified subsequent period. Normally, probabilities are expressed for relatively restricted categories of the population, such as an age group, but we can extend the term loosely here to relate to general populations. If probabilities of migration are to be computed, it is not always apparent what population should be used in the denominator. The population in a country at the beginning of the specified migration period plus the

immigrants during the period represents the approximate population exposed to the risk of losing members through emigration during the subsequent period. In an age-specific rate for a one-year period, this population can be approximated by the midperiod population at some age plus one-half the deaths at that age during the year. An annual age-specific probability of emigration between exact age a and $a + 1$ may be represented by

$$\epsilon_a = \frac{E_a}{P_a + \frac{1}{2}D_a} \qquad (25)$$

where E_a represents emigrants at a given age during the year, P_a the midyear population at age a, and D_a deaths during the year at age a. The immigrants at age a who arrive during the year and who are at risk of emigration during the year are already included in the midperiod population figure (P_a).

Probabilities of immigration cannot be based on the initial population of an area. Immigration to an area during a period is not a "risk" to which the population of the area at the beginning of the period is subject since the immigrants are not a part of this initial population. The number of immigrants entering a country is logically related to the population at the beginning of the period in all the other countries of the world, or at least those countries where the immigrants originated; this is the population at risk of losing these persons through emigration to the country in question. An immigration rate computed in this manner would have only theoretical interest and would be of little or no practical value in analyzing population growth in a particular country. Inasmuch as probabilities of immigration cannot be computed, the substitute procedure is employed of computing the "probability" that an area will receive immigrants; this "probability" is based on the population of the area at the end of the migration period.

We can consider computing probabilities of **net** immigration or **net** emigration by employing the fiction that they are gross immigration or gross emigration rates, respectively. Net emigration can reasonably be related to the population at the beginning of the migration period, but, again, there is no simple, logical base for computing a net immigration rate. It would also be confusing to have a different base for different rates in a time series or in an array of rates for various countries. Under these circumstances, since we must abandon our effort to compute actual probabilities of net migration, the midpoint population seems like the best compromise. Where this choice is inconvenient, as is often the case, we would fall back on the census or survey population, which is normally the terminal population.

Immigration and Population Growth

We consider next certain aspects of the measurement of immigration in relation to the other components of population growth and to the overall population growth.

Immigration as a Component of Population Growth. — It may be desired to express the relative importance of migration as a component of national population growth during a period in terms of the percent which each component of population change contributes to the total increase or decrease during the period. For this purpose, since some components are positive (births and immigration) and others negative (deaths and emigration), it is best to combine them so that logically related items in the distribution have a common sign for the calculation of the percents. We can compute the percentages which net immigration (+) and natural increase (+) constitute of the total increase, but not the percentages which the individual components constitute of total increase.

The importance of net immigration or net emigration in relation to population growth for a country may be more sensitively measured by the ratio of net immigration or net emigration to natural increase during the period. These ratios express the amount of net migration as a "percent" of the amount of natural increase:

$$\frac{M}{B-D} \times 100 \qquad (26)$$

Other measures of the relation of migration to population change may be developed on the basis of the concepts of **migration turnover** (i.e., the sum of immigration and emigration), **natural turnover** (i.e., births plus deaths), and **population turnover** (i.e., the sum of the four components). These values may be related to one another and to the total population, in order to measure the magnitude of the basic demographic changes which a population experiences and has to deal with over a year or longer period.

Total Contribution of Migration to Population Change. — It is sometimes desired to measure not simply net immigration or emigration during a period for a country but the total effective contribution of immigration or emigration to the country's population growth or loss during the period. Estimation of the "net population change attributable to migration" involves adjusting net immigration or emigration, as reported, to allow for the natural increase of the migrants, or involves estimating net migration in a special way so as to incorporate its own natural increase. In general, net immigration or net emigration during a period must be reduced for the deaths of the migrants and increased for the births occurring to them during the period; for all older age cohorts only the factor of deaths is involved. The estimate of population loss attributable to emigration must include the natural increase of the emigrants after emigration since this natural increase as well as the emigrants themselves were lost to the population. Inasmuch as all the descendents of the migrants during a period represent gains or losses due to migration during the period, births occurring to the (native) children of the migrants will also have to be taken into account in a long period of observation. All of these adjustments must be estimated since statistics on the deaths of migrants and births to migrant women are not normally available.

An estimate of the net population gain due to net immigration or net loss due to net emigration for one or more intercensal periods may be derived by (1) aging to a later census date the initial census population and estimated births occurring to the initial population during the intercensal period, and (2) subtracting the resulting estimates of expected survivors from the later census counts. The aging of the population may be accomplished by means of life-table survival rates or national census survival rates. Death statistics should not be used since deaths occurring to the initial population are required, not deaths occurring to the average resident population. The procedure for estimating net gain or loss due to migration is illustrated by the material in tables 20–12 and 20–13 relating to selected age cohorts of the male population of Canada for the period 1951–61.

For the age cohorts already born by the initial census date the calculations can be simpler than when one is estimating actual net immigration or emigration by age as a residual. It is necessary neither to compile death statistics by age cohorts nor to estimate deaths from survival rates by a special "aver-

age" procedure. The conventional (forward) survival procedure suffices. On the other hand, the calculation of births occurring to the initial population presents a special problem (table 20–13). In the present case this was done by (1) computing general fertility rates for each year of the decade for the general population (cols. 1, 2, and 3) and (2) applying these rates (col. 3) to the survivors of the initial female population for each year (col. 4). The births to the initial female population are shown in column (5). In step (1) above, annual intercensal estimates of the population by age and sex are required; for Canada the official estimates were employed. In step (2), annual figures for the expected female survivors 15 to 44 years of age were derived by "aging" the census population in 1951 to 1956 and 1961 by use of life-table survival rates and then interpolating the figures for 1951, 1956, and 1961 linearly to individual years.

The resulting births were then "aged" to the end of the decade, like the initial population, using life-table survival rates. The estimated net gain due to net immigration is then obtained as the difference between the 1961 census population and the surviving population. The results of our calculations for Canada, 1951 to 1961, may be summarized as follows:

	Male	Female	Total
(1) Census population, June 1, 1951	7,088,873	6,920,556	14,009,429
(2) Births to initial population	2,171,472	2,053,701	4,225,173
(3) Survivors, June 1, 1961, assuming no net immigration	8,501,340	8,414,786	16,916,126
(4) Census population, June 1, 1961	9,218,893	9,019,354	18,238,247
(5) Estimated net gain due to net immigration	717,553	604,568	1,322,121

An alternative form for presenting the components of the estimate of the total net gain due to net immigration is as follows:

(1) Net immigration estimated as residual		1,082,904
(a) Pop., June 1, 1951	14,009,429	
(b) Pop., June 1, 1961	18,238,247	
(c) Births as reported	4,468,909	
(d) Deaths as reported	1,322,995	
(2) Births to net migrants		243,736
(a) Births as reported	4,468,909	
(b) Births to initial population	4,225,173	
(3) Deaths of net migrants		4,519
(a) Deaths as reported	1,322,995	
(b) Deaths of initial population and of its births	1,318,476	
(4) = (1) + (2) − (3) Net gain due to net immigration		1,322,121
(5) = (1a) − (1b) = Net population gain		4,228,818
(6) = (4) ÷ (5) Proportion of net population gain due to net immigration		.313

Table 20–12.—**Calculation of Estimates of the Net Gain in the Male Population Due to Net Immigration, for Selected Age Cohorts, for Canada: 1951–61**

Age in 1951	Age in 1961	Census population, June 1, 1951 (1)	10-year life-table survival rate[1] (2)	Survivors assuming no net migration (1) x (2) = (3)	Census population, June 1, 1961 (4)	Estimated net gain due to net immigration (4) − (3) = (5)
All ages	All ages	7,088,873	(X)	8,501,340	9,218,893	717,553
Births, 1956–61	Under 5	[2]1,118,692	.96163	1,075,768	1,157,091	81,323
Births, 1951–56	5–9	[2]1,052,780	.95509	1,005,500	1,060,840	55,340
Under 5	10–14	879,063	.98964	869,956	948,160	78,204
5–9	15–19	713,873	.99156	707,848	729,035	21,187
10–14	20–24	575,122	.98762	568,002	587,139	19,137

X Not applicable.

[1] Average of survival rates derived from the 1951 and 1961 life tables of Canada.

[2] Excludes births to immigrants. Derived from the births of both sexes given in table 20–13 by applying the proportion male among all births in each 5-year period to births occurring to women present at the initial date; i.e., calculation for the period 1956–61 = .513880 × 2,176,951 = 1,118,692, and calculation for the period 1951–56 = .513997 × 2,048,222 = 1,052,780. These figures are not included in the total of all ages.

Source of basic data: Canada, Dominion Bureau of Statistics, official reports.

Table 20–13.—**Calculation of Estimates of Births to the Immigrant Population, for Canada: June 1, 1951–61**

Year	Estimated female population, 15-44[1] (1)	Reported births[2] (2)	General fertility rate $[(2) \div (1)] \times 1,000 =$ (3)	Expected female population, 15-44[3] (4)	Births to initial female population $[(3) \times (4)] \div 1,000 =$ (5)
1951	[4]3,103,807	381,092	122.78	3,103,807	381,085
1952	3,171,600	403,599	127.25	3,126,669	397,869
1953	3,228,600	417,884	129.43	3,149,531	407,644
1954	3,291,500	436,198	132.52	3,172,393	420,406
1955	3,349,700	442,937	132.23	3,195,255	422,509
1956	3,405,100	450,739	132.37	3,218,119	425,982
1957	3,496,600	469,093	134.16	3,253,029	436,426
1958	3,567,800	470,118	131.77	3,287,939	433,252
1959	3,616,800	479,275	132.51	3,322,849	440,311
1960	3,670,300	478,551	130.38	3,357,759	437,785
1961	[4]3,721,651	475,700	127.82	3,392,667	433,651
June 1, 1951—June 1, 1956	(X)	2,110,732	(X)	(X)	2,048,222
June 1, 1956—June 1, 1961	(X)	2,358,177	(X)	(X)	2,176,951

Estimate of births to immigrant female population:
Reported births (col. 2) . . . 4,468,909
Births to initial female population (col. 5) . . . 4,225,173
Difference . . . 243,736

X Not applicable.
[1] Estimates as of June 1.　[2] Calendar-year births.
[3] Population expected in each year without allowance for net immigration, calculated by "surviving" the 1951 census population by age to 1956 and 1961 and interpolating for the other years.
[4] Enumerated population.

Source: Cols. (1) and (2) from official national reports.

Births to net migrants are obtained as the difference between births as reported and births as estimated for the initial population. Deaths of net migrants may be derived as the difference between (1) the reported deaths and (2) the deaths occurring to the initial population and its births. The latter figure is the difference between (a) the initial population and its births and (b) their survivors.

Another procedure of estimating the net gain or loss in population due to international migration consists of applying life table survival rates to the reported net migration figures, distributed by age and sex, for a period (e.g., 5 or 10 years) to obtain estimates of surviving "net migrants" at the end of the period, and then of applying general fertility rates, age-adjusted birth rates, or age-specific birth rates to "net migrant" women by age (estimated for the middate of the period) to obtain estimates of births for the whole period. In each case, the required rates must ordinarily be borrowed, possibly from the general population. Since the migrants enter or leave at various dates in the estimate period, it may be assumed that exposure to death or birth is for one-half of the period; that is, the migrants are aged for only one-half of the estimate period and the births for women entering in the estimate period occur at one-half the rate of the general population. The births are then aged to the end of the estimate period and added to the survivors of the net migrants.

The "net gain attributable to migration" is of special interest in connection with the analysis of projected population changes and the development of population policy. The proportion of total projected population growth attributable to migration may be derived as the difference between a series of projections assuming no immigration after some current date and a second series allowing for net migration, say, by a fixed annual amount in future years.[20] For example, the cumulative additions to the population of the United States between 1966 and 1990 resulting from immigration, by components of change, have been projected as follows. About 9.6 million "net migrants" (400,000 per year for 24 years) are assumed to arrive in the United States; but as a result of a natural increase of 4.0 million (4.6 million births less 0.6 million deaths) among the "net immigrants," the net addition to the population amounts to 13.6 million. By 1990 some of the births due to immigration are grandchildren of immigrants.

---

[20] U.S. Bureau of the Census, *Current Population Reports*, Series P-25, No. 381, "Projections of the Population of the United States, by Age, Sex, and Color to 1990, With Extensions of Population by Age and Sex to 2015," December 1967, esp. pp. 41-44.

Graphic Techniques

In addition to the standard types of charts, including various types of maps, a few special graphic techniques can be employed in the description and analysis of international migration. One technique, for example, would be to use maps containing arrows of varying width to indicate the volume and direction of migration between areas. On such maps or charts the width of each arrow would be directly proportional to the volume of migration; and the length and position of the arrows would identify the areas of origin and destination.

A somewhat different type of chart, the population pyramid, described in detail in chapter 8, has a special application to immigration analysis. A pyramid may serve to depict, albeit roughly, the historical sequence of the various waves of immigration into an area and their relative numerical importance. Heavy immigration at some era in a population's history will often be reflected prominently in the contour of the pyramid. A protrusion of the bars at the upper ages suggests immigration many decades earlier, since international migration tends to occur in the ages of youth. The ethnic identity of the migration waves may be reflected in the pyramids if the various ethnic groups are shown separately or if separate pyramids are drawn for each group. Pyramids for the total foreign-born population of Hawaii, and for the major racial groups among the foreign born, in 1950, would make evident the succession of the waves of foreign immigration to Hawaii over the last three-quarters of a century—Chinese, Japanese, Filipinos, and Caucasians—and the predominance of males among the immigrants, as is illustrated partially by figure 20-1.

Figure 20–1. — Population Pyramids of the Total Foreign-Born Population and of the Foreign-Born Filipino and Japanese Populations of Hawaii: 1950

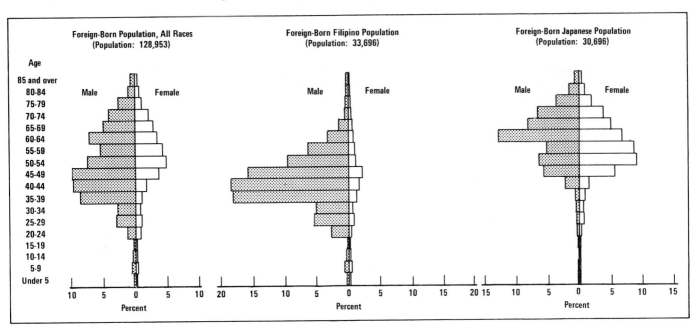

Source: Basic data from *U.S. Census of Population: 1950*, Vol. II, Parts 51-54, *Territories and Possessions*, 1953, p. 52-35.

SUGGESTED READINGS

Part A. Methodological and Analytical Studies

Akers, Donald S. "Immigration Data and National Population Estimates for the United States." *Demography* 4(1):262–272. 1967.

Borrie, W. D. "Statement by the Moderator." Meeting B.9, "International Migration as Related to Economic and Demographic Problems of Developing Countries." In United Nations, *World Population Conference, 1965* (Belgrade), Vol. I, *Summary Report,* pp. 138–150. 1966.

Germani, Gino. "Mass Immigration and Modernization in Argentina." *Studies in Comparative International Development* 2(11):165–182. 1966.

Hutchinson, E. P. "Notes on Immigration Statistics of the United States." *Journal of the American Statistical Association* 53(284):963–1025. Dec. 1958.

International Union for the Scientific Study of Population. *Cultural Assimilation of Immigrants* (papers presented at the General Assembly in Geneva, 1949). Supplement to *Population Studies* (London), March 1950.

Krichefsky, Gertrude D. "International Migration Statistics as Related to the United States." In U.S. Immigration and Naturalization Service, *I and N Reporter,* Part 1, July 1964, and Part 2, Oct. 1964.

Kuznets, Simon, and Rubin, Ernest. *Immigration and the Foreign Born.* Occasional Paper 46, New York, National Bureau of Economic Research, 1954.

Petersen, William. "A General Typology of Migration." In Charles B. Nam (ed.), *Population and Society.* New York, Houghton-Mifflin Company, 1968. Pp. 288–297.

Price, Charles A. "Some Problems of International Migration Statistics—An Australian Case-Study." *Population Studies* (London), 19(1):17–27. July 1965.

Recchini de Lattes, Zulma L. "Demographic Consequences of International Migratory Movements in the Argentine Republic, 1870–1960." In United Nations, *World Population Conference, 1965* (Belgrade), Vol. IV, pp. 211–215. 1967.

Rubin, Ernest. "The Demography of Immigration to the United States." *Annals of the American Academy of Political and Social Science* 367:15–22. Sept. 1966.

Tabah, Leon, and Cataldi, Alberto. "Effects d'une immigration dans quelques populations modèles" (Effects of immigration on some model populations). *Population* (Paris), 18(4):683–696. Oct.–Dec. 1963.

Taeuber, Alma F., and Taeuber, Karl E. "Recent Immigration and Studies of Ethnic Assimmilation." *Demography* 4(2):798–808. 1967.

Thomas, Brinley. "International Migration." In Philip M. Hauser and Otis Dudley Duncan (eds.), *The Study of Population: An Inventory and Appraisal.* Chicago, Ill., University of Chicago Press, 1959. Pp. 510–543.

Union Panamericana, Organizacion de los Estados Americanos. *La Inmigración en América Latina* (Immigration in Latin America). By Fernando Bastos de Avila, S. J. *Revista Interamericana de Ciencias Sociales,* Vol. 3, Special No. 1964.

United Nations. *International Migration Statistics.* Series M, Statistical Papers, No. 20. 1953.

———. *International Research on Migration.* Miscellaneous Studies, No. 18. 1953.

———. *Problems of Migration Statistics.* Series A, Population Studies, No. 5. Nov. 1949.

———. *World Population Conference, 1965* (Belgrade). Vol. IV, *Migration, Urbanization, and Economic Development.* Meeting B.9 Papers, "International Migration as Related to Economic and Demographic Problems of Developing Countries." 1967. Pp. 191–249.

Part B. Illustrative Studies of International Migration Trends Including Statistics, and Statistical Compilations

Appleyard, R. T. *British Emigration to Australia.* Toronto, University of Toronto Press, 1964.

Borrie, W. D. *Trends and Patterns in International Migration since 1945.* Background paper B.9/14/E/474, United Nations, World Population Conference, 1965 (Belgrade).

Canada, Dominion Bureau of Statistics. *1961 Census of Canada.* Series 1.2, *Population.* Bulletin 1.2–8, *Citizenship and Immigration,* 1963.

Cook, Robert C. (ed.). "World Migration, 1946–55." *Population Bulletin* 13(5):77–95. Aug. 1957.

Kirk, Dudley. *Europe's Population in the Interwar Years.* League of Nations. Princeton, N.J., Princeton University Press, 1946.

———. "Major Migrations Since World War II." *In Selected Studies of Migration Since World War II.* New York, Milbank Memorial Fund, 1958. Pp. 11–29.

Kulischer, Eugene M. *Europe on the Move: War and Population Changes, 1917–47.* New York, Columbia University Press, 1948.

Neiva, Artur Hehl. "International Migrations Affecting Latin America." *Milbank Memorial Fund Quarterly* 43(4):119–135. Part 2, Oct. 1965.

Pankhurst, K. V. "Migration between Canada and the United States." *Annals of the American Academy of Political and Social Science* 367:53–62. Sept. 1966.

Schechtman, Joseph B. *Population Transfers in Asia.* New York, Hallsby Press, 1949.

SUGGESTED READINGS — Continued

———. *The Refugee in the World: Displacement and Integration.* New York, Barnes, 1963.

Taft, Donald R., and Robbins, Richard. "History and Statistics of World Migration." Chapter 2 in *International Migrations.* New York, Ronald Press Company, 1955. Pp. 36–41.

United Nations. *Economic Characteristics of International Migrants: Statistics for Selected Countries, 1918–1954.* Series A, Population Studies, No. 12, 1958.

———. *Sex and Age of International Migrants: Statistics for 1918–1947.* Series A, Population Studies, No. 11, 1953.

U.S. Immigration and Naturalization Service. *Annual Report of the Immigration and Naturalization Service.* [Annual publication]

Willcox, Walter F. (ed.) *International Migrations.* 2 vols., New York, National Bureau of Economic Research, 1929 and 1931.

Woytinsky, W. S., and Woytinsky, E. S. *World Population and Production.* New York, Twentieth Century Fund, 1953. Chapter 3, pp. 66–87.

Zubrzycki, Jerzy. *Immigrants in Australia.* 2 vols., Melbourne, Melbourne University Press, 1960.

CHAPTER 21

Internal Migration and Short-Distance Mobility

The mobility of a people within its national borders did not receive as early attention as international migration, but it has been the subject of increasing study in recent decades. Of all the components of population change, internal mobility has been the most difficult to measure. Unless the country in question is so fortunate, demographically speaking, as to have a good continuous population register, it has not been easy to obtain accurate measures of net or gross movement. Nonetheless, both indirect and direct methods of measurement have gradually been introduced and refined so that there is a variety of approaches from which to choose.

The demographer's interests have included (1) migration as a factor in population change and hence in measures or estimates of migration for use in making current population estimates and population projections; (2) migration as usually the primary factor in population redistribution among geographic or type-of-residence areas; (3) differentials in short-distance mobility and migration and the selectivity of those two types of movement. The aspects that are of interest to other social scientists, planning officials, etc., have been described by Shryock.[1]

In the next section we will define some of the more common mobility terms and concepts. In that section and, indeed, in many other parts of this chapter we have drawn heavily on a recent United Nations manual, *Methods of Measuring Internal Migration*.[2]

CONCEPTS OF MOBILITY

"Mobility" as used in demography usually refers to spatial, physical, or geographic movement whereas in sociology it usually refers to a change in status, e.g., of occupation. The two forms may be distinguished by calling them geographic mobility and social mobility, respectively. This chapter and the preceding one deal only with the former except when changes in residence are related to changes in status.

Moreover, not all geographic movements qualify as migration. First, we require a change in usual place of residence. Thus, commuting, or the diurnal movement (German *Pendelwanderung*) between home and workplace, home and school, etc., does not qualify as migration. The distinction between international migration and internal migration was made in chapter 20. But not all changes of usual residence within a

country are customarily regarded as internal migration either. A distinction is made between short-distance or local movers, on the one hand, and migrants, on the other hand.

It is just here that our operational definitions have to become arbitrary. Some theorists would define migration as a change of environment. There are many different aspects of one's environment, however, and there are questions as to which aspects are most demographically relevant. Even if there were consensus on this point, there would usually be difficulty in measuring such changes in censuses or surveys. A more mechanical but practical approach would be to require a move over a minimum distance. Even this has its difficulties, however, either because respondents may make fairly gross errors in reporting the distance of their moves or because the coding of distance on the basis of a comparison of place of origin and place of destination is a complex and expensive operation—although it could be handled in a sufficiently large electronic computer. Recording distance alone, furthermore, does not tell us a great deal about a move.

The usual procedure, then, is to define a migrant as a mover who changed the political area of his usual residence. This may be the primary, secondary, or even the tertiary division in the country. (See ch. 5 for these types of divisions.) The name of the specific political area of prior residence is usually also recorded. With this information, migrants defined as movers between different areas at a given political level can be characterized according to whether or not the migration was also between higher levels of political (or statistical) areas. For example, if migrants in India are defined as movers between different districts, we can also distinguish interstate migrants and interregional migrants (the latter being defined as migrants between two natural regions).

In this approach, namely the definition of a migrant as a mover between two political or administrative areas, the concept of a change in environment or milieu has been retained somewhat, albeit crudely. One political area or region may differ in a number of cultural traits from another. For example, the states of India are distinguished primarily by the fact that their inhabitants speak different languages.

Nonetheless, we have to admit that such a definition of a migrant remains arbitrary. Some movers within a political area conform more closely to the theoretical conception of migration, and move shorter distances, than some movers across area boundaries. This type of problem in analytical classification arises whenever either cutting scores or arbitrary areas are set up, and it appears to be unavoidable. A definition of "migrant" in terms of a minimal distance moved would also be arbitrary unless there were some natural break in the

---

[1] Henry S. Shryock, Jr., *Population Mobility Within the United States*, Community and Family Study Center, University of Chicago, 1964, pp. 1–5.
[2] United Nations, *Methods of Measuring Internal Migration*, Manuals on Methods of Estimating Population, Manual VI, 1970.

continuous distribution of moves. It has, indeed, been suggested that a migrant be defined as a mover within a labor market area,[3] and accordingly the minimum distance might be set at the point at which commuting to work becomes so time-consuming and expensive as to require the substitution of a change of residence.

Political or administrative units are not usually delineated in terms of a grid that yields rectangular units of equal area. The effect of the size and shape of political units upon the measurement of migration has been discussed by Lee and by Wendel.[4] Since countries, and even regions of the same country, differ with respect to the average area and other geometric characteristics of the geographic units used in their definition of migration, van den Brink contends that there cannot be meaningful international comparisons of migration rates. Larry Long adds that the only really comparable mobility rates are those including all changes of usual residence (address) in the numerator since these are independent of the country's geographic subdivisions.[5]

In measuring mobility or in defining who is a mover and who is a nonmover, the time-period also has to be specified. The time-period can be either variable or fixed. Examples of variable periods are the period since birth (which yields lifetime mobility) and the period since the last move. Examples of fixed periods are one year, five years, and ten years. If the mobility period coincides with the last intercensal period, the resulting migration statistics may be more useful in measuring the components of population change (see ch. 13) or in studying the consistency of the population and migration statistics. Too long a period results in more nonresponse and reporting errors, however, and omits a substantial proportion of the population— those born and those dying during the mobility period—from the mobility statistics. In addition to the date of the last census, mobility questions have also referred to dates of historic significance, such as the beginning or end of a war (Current Population Survey of the United States, 1946) or a political coup (Special Demographic Survey of the Republic of Korea, 1966).

Mobility data are usually obtained from questions that compare current residence with residence at a prior date, those persons who have made a specified type of change in residence being classified as "migrants." These data yield a classification of the population by **mobility status**. An example of such a classification, using American geographic units is as follows:

Same house (nonmover)
Different house in the same county (intracounty movers)
Migrants (intercounty movers)
   Different county, same state (intrastate migrants)
   Interstate migrants
      Between contiguous states
      Between noncontiguous states
Movers from abroad

In such a classification, only those persons whose residences at the beginning and end of the period were different are counted as movers. Movers who died during the period are omitted from the classification altogether, and those who returned during the period to their residence at the beginning of the period are classified as nonmovers. Furthermore, only one move per person can be counted during the period. From survey questions about mobility histories, on the other hand, the total number of moves of any type can be determined for a specified period, but again only for persons who survive the period. To obtain a count of all moves, including those of subsequent decedents, data would need to be extracted from a continuous population register, or survey respondents would need to be asked to report on persons no longer in the household.

Definitions

From this general discussion of concepts, we may derive some basic definitions of terms. It should be emphasized, however, that although the present definitions are mostly supported by the consensus of users, terminology in the field of population mobility is not yet as well standardized as that in natality or mortality. The terms given here are of fairly general applicability (for example, they are mostly applicable to both variable-period and fixed-period mobility); but, in using them, one should indicate the time period unless that is clear from the context.

Mobility or migration period (or interval). — The period to which the question on previous residence applies. In fixed-period migration, the period may be defined by either specifying a prior date (beginning of the period) or specifying the length of the interval (5 years, etc.).

Mobility status. — A classification of the population into major categories on the basis of comparison of residences at two dates. (See illustration above.)

Mover. — A person who moved from one address (house or apartment) to another.

Short-distance or local mover. — A person who moved only within a political or administrative area.

Migrant. — A person who moved from one political area to another. The number of **nonmigrants** is equal to the number of nonmovers plus the number of short-distance movers.

Mover from abroad. — An "immigrant" or another kind of person who has moved from outside the country. (See discussion in ch. 20.)

Area of origin (departure). — The area from which a migrant moves.

Area of destination (arrival). — The area to which a migrant moves. (With some sources of migration data, there have been intervening residences that are not recorded as origins or destinations.)

In-migrant. — "Every move is an out-migration with respect to the area of origin and an in-migration with respect to the area of destination. Every migrant is an out-migrant with respect to the area of departure and an in-migrant with respect to the area of arrival. An in-migrant is thus a person who enters a migration-defining area by crossing its boundary from some point outside the area, but within the same country. He is to be distinguished from an 'immigrant' who is an international migrant entering the area from a place outside the country."[6]

Out-migrant. — "An out-migrant is a person who departs from a migration-defining area by crossing its boundary to a

[3] John B. Lansing and Eva Mueller, *The Geographic Mobility of Labor,* Ann Arbor (Michigan), Survey Research Center, University of Michigan, 1967, p. 7.

[4] Everett S. Lee et al., *Methodological Considerations and Reference Tables,* Vol. I of : Simon Kuznets and Dorothy Swaine Thomas, eds., *Population Redistribution and Economic Growth, United States, 1870–1950,* Philadelphia, American Philosophical Society, 1957, pp. 10–12; Bertil Wendel, "Regional Aspects of Internal Migration and Mobility in Sweden, 1946–50," in *Migration in Sweden: A Symposium,* David Hannerberg, Torsten Hägerstrand, and Bruno Odeving, eds. (Lund Studies in Geography, Ser. B. Human Geography No. 13), Lund, The Royal University of Lund, 1957, pp. 18–20.

[5] "Residential Mobility: International Comparisons," unpublished paper read to Population Association of America, Atlantic City, N.J., April 11, 1969.

[6] United Nations, *Methods of Measuring Internal Migration.*

point outside it, but within the same country. He is to be distinguished from an 'emigrant' who is an international migrant, departing to another country by crossing an international boundary."[7]

Net migration.—The balance between in-migration and out-migration. According to the direction of the balance, it may be characterized as **net in-migration** or **net out-migration.** In a column of net migration figures, whether the net flow is in or out is indicated by a plus (+) or minus (−) sign, respectively.

Gross migration.—Either in-migration or out-migration is **gross migration.** Their sum is sometimes referred to as the **turnover** for an area. As was stated in chapter 20 regarding international migration, so here the number of in-migrants (or out-migrants) is not additive when a set of secondary divisions of a country is combined into a set of primary divisions. For example, the sum of in-migrants to all the counties of a State is not equal to the in-migrants to the State because migrants from one county to another within the State are not in-migrants to the State. Net migration is additive, however. (We seek to avoid the term "net migrant" as far as possible because it is not reasonable to think of a group of people as being net migrants.)

Lifetime migration.—Migration that has occurred between birth and the time of the census or survey. A **lifetime migrant** is one whose current area of residence is different from his area of birth, regardless of intervening migrations. Lifetime migration for an area may be either gross or net, and the terms lifetime in-migrant and lifetime out-migrant are also used.

Migration stream.—A group of migrants having a common origin and destination in a given migration period. Although strictly speaking a stream represents the movement between two geographic areas, by analogy, it may also be used to describe the movement between two type-of-residence areas, such as from nonmetropolitan to metropolitan areas, where neither the origin nor the destination represents contiguous territory. The movement in the opposite direction to a stream is called its **counterstream.** Thus, if the stream is from Area A to Area B, the counterstream is that from B to A during the same period. The concepts of stream and counterstream were used first by E. G. Ravenstein for describing rather heavily unidirectional flows, like those between rural areas and towns in the nineteenth century. In the general sense, we can think of a counterstream as the smaller of the two movements. The two are often of nearly equal size and indeed may exchange rankings from time to time. Eldridge refers to the stream in the "prevailing" direction as the "dominant stream" and to the counterstream as the "reverse stream."[8]

The difference between a stream and its counterstream is the **net stream,** or **net interchange,** between the two areas. Similarly, the sum of the stream and the counterstream is called the **gross interchange** between the two areas. In order to show which area of the pair received the net in-migration, some convention must be adopted in terms of plus and minus signs. Such a convention is illustrated in table 21-1.

Return migration.—A return migrant is a person who moved back to the area where he formerly resided. Not all return migration is identified in the usual sources of migration data. It is necessary to know the origin and destination of individual migrants for at least two migration periods.

Table 21-1.—Net and Gross Interchange of Population Between Each Pair of Regions of the United States: 1953–54 to 1957–58, Average

Pair of regions	Net migration			Gross population interchange		
	Number[1] (thousands)	Percent of population		Number (thousands)	Percent of combined populations	Percent of all interregional migrants
		In first region	In second region			
NORTHEAST:						
North Central...	−28	0.07	0.06	184	0.20	6.8
South..........	+64	0.15	0.13	488	0.54	17.9
West..........	−42	0.10	0.19	188	0.29	6.9
NORTH CENTRAL:						
South..........	+22	0.05	0.04	846	0.87	31.1
West..........	−107	0.22	0.48	459	0.65	16.8
SOUTH:						
West..........	−58	0.12	0.26	560	0.78	20.6

[1] Plus sign (+) indicates net in-migration to first-named region; minus sign (−) indicates net out-migration from first-named region and net in-migration to second-named region.

Source: Adapted from, Henry S. Shryock, Jr., *Population Mobility Within the United States,* 1964, table 8.12, p. 249.

Bases of Rates

For the most part, definitions of mobility that involve derived figures are discussed in the several subsections on "Measures Used in Analysis." Common to population–based mobility rates derived from various sources, however, is the question of the proper base to use. This question is discussed at some length by Hamilton, by Thomlinson, and in the U. N. Manual VI.[9]

We seek a measure of the rate of mobility, or the ratio of the number of movers in an interval of time to the population at risk during that interval, or

$$m = \frac{M}{P} k \ \ldots \ldots \tag{1}$$

where m = the mobility rate

M = the number of movers

P = the population at risk

k = a constant, such as 100 or 1,000

For mobility and movers, we may substitute the more specific terms, migration and migrants.

For the country as a whole, this rate measures the overall level of mobility or migration. In the case of migration, three analogous rates are those of in-migration, out-migration, and net migration, which are given respectively by,

$$m_i = \frac{I}{P} k \ \ldots \ldots \tag{2}$$

$$m_o = \frac{O}{P} k \ \ldots \ldots \tag{3}$$

$$m_n = \frac{I-O}{P} k \ \ldots \ldots \tag{4}$$

[7] Ibid.

[8] Hope T. Eldridge, "Primary, Secondary, and Return Migration in the United States, 1955–60," *Demography,* 2:445, 1965.

[9] C. Horace Hamilton, "Practical and Mathematical Considerations in the Formulation and Selection of Migration Rates," *Demography,* 2:429–443, 1965; Ralph Thomlinson, "The Determination of a Base Population for Computing Migration Rates," *Milbank Memorial Fund Quarterly,* 40(3):356–366, July 1962.

where I and O are the number of in-migrants and out-migrants, respectively.

For the discussion of a suitable time reference for these base populations, see Chapter 20. If the migration interval is short, say a year or less, the initial, final, or average population all yield about the same rates. For a 5-year period, however, there can be considerable difference.

The next choice is, "The population of what area?". This logical problem does not arise in the case of the national rates (although it would in the case of immigration). For out-migration from an area, the population at risk is clearly that of the area itself. For in-migration, however, the population at risk is that of the balance of the country. This base is practically never used, but Shryock gives some illustrations of such in-migration rates.[10] Mainly as a practical matter, then, in-migration rates are also based on the population of the area of destination. Net migration rates, of course, must be the difference of in- and out-migration rates that have the same population base.

SOURCES OF DATA AND STATISTICS

The most common source of data on population mobility is a census or sample survey of households. Another source is the continuous population register. From what has been said in chapter 2 about the relative frequency of these two sources of demographic data, it will be recognized that the former two are potential sources for a much greater number of countries than the latter. Miscellaneous sources include the records of national social security systems, some of the other partial population registers described in chapter 2 (compulsory military service, aliens, etc.), public utility records, and city directory listings. The last two types have local rather than national coverage.

Censuses and Surveys

Mobility data from censuses and surveys are of two broad types. The first type consists of those derived from direct questions about mobility or about prior residence—place of birth, place of residence at a fixed past date, duration of residence, last prior residence, mobility history, number of moves, etc. The second type consists of estimates of net migration derived from (1) total counts of population or population by age and sex, at two censuses and (2) natural increase (births minus deaths) or intercensal survival rates, which are derived in turn from (a) life tables or (b) comparison of the age distributions of successive censuses. These are called estimates by the "residual method." The difference between the total change in population during an intercensal period and the change due to natural increase is imputed to net migration. This difference also usually includes net immigration from abroad, however.

Population Registers

As was described in chapters 2 and 3, a universal, continuous population register requires that a person who moves transfer his record from one local registry office to another. From these changes of residence, migration statistics are compiled. Such statistics have been published for many years by the Scandinavian countries and a few other western European countries and more recently by eastern European countries as well.

As far as we are able to ascertain the countries now publishing statistics of internal migration on the basis of their population registers are the following:

Denmark, Finland, Iceland, Norway, Sweden
Belgium, Netherlands, West Germany
Czechoslovakia, Hungary, Poland, U.S.S.R.
Italy, Spain
Israel
Japan, Singapore, Taiwan

In the European registers, a migrant is one who changes his commune (or the like) of usual residence. Hungary, Israel, Singapore, and Taiwan appear to be the only countries publishing registration statistics of total moves.

Sweden.—The country with the longest existing series is Sweden. Although its continuous population register was established in the eighteenth century, the annual migration statistics first seem to have attained adequate national coverage and accuracy about 1895.[11]

Miscellaneous Sources

Some of the "partial" population registers described in chapter 2 (registers of a population subgroup) have been, or could be, used to estimate the migration of that subgroup. By various assumptions, migration rates for the subgroup could be extended to the general population; or the data may simply be used to describe and analyze migration differentials among classes within the subgroup. A few examples will be given.

In the United States, the national social insurance scheme that provides pensions to retired workers and their surviving dependents is called Old Age and Survivors Insurance and is operated by the Social Security Administration. A one-percent continuous work history sample gives the age, sex, and place of employment of workers covered by the scheme. By comparison of places of employment of identical workers at yearly intervals, migration can be approximated. The chief shortcomings of the procedure are the occasional lack of correspondence between area of employment and area of residence, incompleteness of coverage (for example, in the past some industries have not been covered), and sampling error. These data are better suited, therefore, for making estimates for large areas like States. A pioneering study with this material for two States was carried out by Bogue, who tried to measure job mobility as well as geographic mobility.[12]

Croze, of France's Institut national de la statistique et des études économiques (INSEE), estimated the net migration for the period 1950–52 by departments, using electoral lists.[13] Registers of electors (voters) are maintained and up-dated locally, but changes in the commune of registration are reported to INSEE.

In the United States, commercial firms produce "city directories" that list the names and addresses of adults living in most of the sizable cities. Some of these are published

[10] Shryock, *Population Mobility Within the United States*, pp. 31–33.

[11] Svend Riemer, "Notes on Methods of Analysis of Swedish Migration Data," Appendix C3 in: Dorothy Swaine Thomas, *Research Memorandum on Migration Differentials*, New York, Social Science Research Council, Bulletin 43, 1938, p. 400.

[12] Donald J. Bogue, *A Methodological Study of Migration and Labor Mobility in Michigan and Ohio in 1947*, Scripps Foundation Studies in Population Distribution, No. 4, Miami, Ohio, Scripps Foundation for Research in Population Problems, June 1952.

[13] Marcel Croze, "Un instrument d'étude des migrations intérieures: les migrations d'électeurs" (An instrument for studying internal migration: the migration of electors), *Population* (Paris), 11(2):235–260, April–June 1956.

Table 21–2.— Comparison of Estimates of the Net Migration Rate, 1946–53, by (A) National Growth Rate Method and (B) Vital Statistics Method, for Ceylon, by Districts

District	Population 1953 (1)	Population 1946 (2)	Population change, 1946-1953 Amount (3)	Population change, 1946-1953 Rate (4)	Estimated net migration — National growth rate method — Based on population change (5)	Estimated net migration — National growth rate method — Based on natural increase (6)	Vital statistics method — Natural increase, 1946-1953 (7)	Vital statistics method — Net migration, 1946-1953 Amount (8)	Vital statistics method — Net migration, 1946-1953 Rate (9)
CEYLON, TOTAL................	8,097,905	6,657,239	1,440,666	21.6	(X)	(X)	1,354,841	+85,825	+1.3
Colombo.........................	1,708,726	1,420,232	288,494	20.3	-1.3	-0.1	246,396	+42,098	+3.0
Kalutara.........................	523,550	456,572	66,978	14.7	-6.9	-5.7	78,138	-11,160	-2.4
Kandy...........................	840,382	711,449	128,933	18.1	-3.5	-2.3	153,022	-24,089	-3.4
Matale..........................	201,049	155,720	45,329	29.1	+7.5	+8.7	37,793	+7,536	+4.8
Nuwara Eliya....................	325,254	268,121	57,133	21.3	-0.3	+0.9	59,999	-2,866	-1.1
Galle...........................	524,369	459,785	64,584	14.0	-7.6	-6.4	80,329	-15,745	-3.4
Matara..........................	413,431	351,947	61,484	17.5	-4.1	-2.9	77,596	-16,112	-4.6
Hambantota......................	191,508	149,686	41,822	27.9	+6.3	+7.5	38,675	+3,147	+2.1
Jaffna..........................	491,849	424,788	67,061	15.8	-5.8	-4.6	74,150	-7,089	-1.7
Mannar..........................	43,689	31,538	12,151	38.5	+16.9	+18.1	6,325	+5,826	+18.5
Vavuniya........................	35,112	23,246	11,866	51.0	+29.4	+30.6	6,773	+5,093	+21.9
Batticaloa......................	270,493	203,186	67,307	33.1	+11.5	+12.7	40,425	+26,882	+13.2
Trincomalee.....................	83,917	75,926	7,991	10.5	-11.1	-9.9	12,602	-4,611	-6.1
Kurunegala......................	626,336	485,042	141,294	29.1	+7.5	+8.7	117,047	+24,247	+5.0
Puttalam........................	58,830	43,083	15,747	36.6	+15.0	+16.2	11,251	+4,496	+10.4
Chilaw..........................	170,072	139,764	30,308	21.7	+0.1	+1.3	27,567	+2,741	+2.0
Anuradhapura....................	229,282	139,534	89,748	64.3	+42.7	+43.9	39,503	+50,245	+36.0
Badulla.........................	466,896	372,238	94,658	25.4	+3.8	+5.0	84,900	+9,758	+2.6
Ratnapura.......................	421,555	343,620	77,935	22.7	+1.1	+2.3	78,806	-871	-0.3
Kegalla.........................	471,605	401,762	69,843	17.4	-4.2	-3.0	83,544	-13,701	-3.4

Col. (3) = (1)-(2); col. (4) = $\frac{(3)}{(2)}$ x 100; col. (5) = district rate in (4) minus national rate; col. (6) = district rate in (4) minus $\left[\frac{1,354,841}{6,657,239} \times 100 = 20.4\right]$, the national rate of natural increase; col. (8) = (3) minus (7); col. (9) = $\frac{(8)}{(2)}$ x 100.

X Not applicable.

Source: Adapted from, S. Vamathevan, *Internal Migration in Ceylon: 1946–53*, Monograph No. 13, Ceylon Department of Census and Statistics, 1961, pp. 64–65.

annually. Comparisons of the names in the alphabetical listings of successive directories, with allowances for death and the attainment of adulthood (usually age 18 in the directories) identifies persons who have entered or left the area. The classic study of this type was made of Norristown, Pennsylvania, by Goldstein.[14] He gives a detailed account of the validity of the data and of his methodology. By studying migration histories over three successive decades, he was able to establish, for example, that out-migration rates of in-migrants in the preceding decade were higher than those of previous nonmigrants.

INTERNATIONAL RECOMMENDATIONS AND NATIONAL PRACTICES

It has been stated that several different types of questions relating to internal migration have been included in population censuses. Reflecting its long history of inclusion in many national censuses, "place of birth" is found on the United Nations' list of recommended topics for the 1970 censuses. The list of suggested topics also includes "place of previous residence" and "duration of residence"—with only an oblique recognition of fixed-period migration. "Duration of residence" is included in the basic regional program of ECAFE and IASI and "Place of previous residence" in that of ECE and IASI.[15]

[14] Sidney Goldstein, *Patterns of Mobility, 1910–1950: The Norristown Study*, Philadelphia, University of Pennsylvania Press, 1958, pp. 68–122.

[15] United Nations, *Principles and Recommendations for the 1970 Population Censuses*, Statistical Papers, Series M, No. 44, p. 151.

MEASUREMENT OF MOBILITY

Indirect Estimates of Net Migration

Since the indirect estimates do not require special questions and may be computed from population counts by age and sex, or even from population totals, it seems logical to discuss them first. The indirect estimates to be discussed in this section are (1) the national growth rate method, and (2) the residual method, comprising (a) the vital statistics method and (b) the survival rate method. The use of place-of-birth statistics is sometimes considered an indirect method; but, since it requires a special question and since there are some direct measures derivable from the data, it will be treated in the next major section.

National Growth Rate Method.—This is a crude but fairly commonly used method. Let P_T^0 and P_T' be the national population at the beginning and end of the intercensal period, respectively. Let $P_1^0, P_2^0, P_3^0, P_i^0, \ldots$ be the populations of the geographic subdivisions at the beginning of the period and $P_1', P_2', \ldots$ their populations at the end of the period.

Then the estimated migration rate, m_i, for area i is given by

$$m_i = \left[\frac{P_i' - P_i^0}{P_i^0} - \frac{P_T' - P_T^0}{P_T^0}\right]k \qquad (5)$$

This rate is customarily multiplied by a constant, such as 100 or 1,000. In table 21–2, 100 is used. Thus, for a geographic division, a rate of growth greater than the national average is interpreted as net in-migration, and a rate less than the national average as net out-migration. The same procedure can be applied to specific age-sex groups.

The method is illustrated for Ceylon in table 21–2, columns (1) through (5). The results are shown in column (5) and the steps are outlined at the bottom of the table.

Figures on the estimated amount of net migration by the National Growth Rate Method can be computed by applying the difference rates in column (5) to the population at the beginning of the intercensal period in column (2). To avoid too much rounding, however, the rates need to be carried out to more places than are shown in the table. For example,

Proportionate change in population, 1946–53:
(a) Ceylon.. .216406
(b) Colombo District...................................... .203132
(c) = (b) − (a) Deficit below national proportion.....−.013274
(d) = 1,420,232 × (−.013274) = Estimated out-migration for Colombo District .... −18,852

This method yields an estimate of the rate of internal migration for each geographic subdivision on the assumption that rates of natural increase and of net immigration from abroad are the same for all parts of the country. It requires no vital statistics.

A variant of the method, described in the paper by the Bombay Centre is defined as follows: "If the total counts of the population of an area are available from two censuses, a rough indication of the extent of net migration may be obtained by comparing the rate of growth of this area with the rate of natural growth of the nation. This method assumes that the rate of natural increase is the same throughout the country." [16]

$$\text{Here, } m_i = \left[\frac{P_i' - P_i^o}{P_i^o} - \frac{B_T - D_T}{P_T^o} \right] k, \qquad (6)$$

where the last term is the national rate of natural increase. Column (6) in table 21–2 gives the net total migration rate using this method, since this formula yields an estimate of the total rate of net migration, including international migration, rather than the net internal rate. Therefore, these estimates are systematically a little higher algebraically than those in column (5).

Where net immigration is negligible, formulas (5) and (6) yield virtually the same results. In the case of Ceylon, net immigration, according to records, was 62,556, or 1.0 percent of the 1946 population, a marginally important figure.

When either net immigration or natural increase is not available, it can be estimated by subtraction of its counterpart from total increase. For Ceylon, net immigration estimated by subtracting natural increase from total increase is 85,825, somewhat greater than the recorded figure. One danger of estimating either natural increase or net immigration by subtraction of its counterpart from total increase is that there may be considerable underregistration or that one census may be more nearly complete than the other. The exercises in this subsection relate to the national growth rate method *per se* and not to the evaluation of the accuracy of the statistics used.

The estimates in column (6) may be compared with those in column (9), which are obtained by the "vital statistics" variant of the residual method, since both purport to relate to **total** net

migration. (That method is discussed in the next subsection.) If the country had reasonably complete vital statistics and if the two successive censuses were of about the same amount of completeness, then these estimates should be more accurate than those in column (6). Although there are a few differences of sign, the two sets of rates are reasonably similar. The average difference disregarding the sign of the differences is 3.2 percentage points.

Residual Methods – Vital Statistics Method. – As previously mentioned, this method requires virtually complete registration of both births and deaths during the intercensal period. If, however, the extent of underregistration has been estimated, the corrected vital statistics should be used. The births and deaths should also be by date of occurrence (not date of registration) and ordinarily by place of residence (not place of occurrence). Even in the case where only *de facto* population counts are available from the censuses, the mixture of *de facto* population and births and deaths by residence should usually yield more meaningful estimates of net migration, i.e., those approximating net changes of usual residence, than would using births and deaths classified by place of occurrence.

The vital statistics method is almost always used to estimate net migration for the total population of an area, i.e., for all age-sex groups combined. (This method was discussed in ch. 20 as a means of estimating net international migration. The "intercensal component method" is a means of estimating the residual, i.e., net migration.) The simplest equation is a form of the "balancing equation":

$$M = (P' - P^o) - (B - D), \qquad (7)$$

or net migration equals population change during the intercensal period minus natural increase during that period.

The result is the difference between the total number of persons moving into an area during a given intercensal period and the number moving out. This estimate of net migration reflects both the in-migrants and out-migrants who died before the second census.[17] The net migration obtained for a given area is that with respect to all other areas and thus represents net international migration combined with net internal migration. This method may be used to estimate net migration for a sex, race, or nativity group or a group defined by any other characteristic that is invariant over time, provided that the population and vital statistics are published by the characteristic. The main issue concerning the method is not the theoretical validity of formula (8), which represents what Siegel and Hamilton call "exact net migration," but rather the effect of errors in the terms on the right-hand side of the equation upon the accuracy of the estimate. A further issue is the accuracy of this method as applied to actual statistics in comparison with that of other methods, such as the survival rate method. Most of the voluminous methodological literature on the residual method deals with these topics.[18]

The straightforward calculation of estimates of net migration

[16]Demographic Training and Research Center, Bombay, "Internal Migration in Some Countries of the East," in International Union for the Scientific Study of Population, *International Population Conference, New York, 1961*, Vol. I, London, 1963, p. 421.

[17]Jacob S. Siegel and C. Horace Hamilton, "Some Considerations in the Use of the Residual Method of Estimating Net Migration," *Journal of the American Statistical Association*, 47(259):480–483, September 1952; Shryock, *Population Mobility Within the United States*, pp. 15–16.
[18]Hope T. Eldridge, "Vital Statistics Versus Census Survival Ratios for Estimating Net Intercensal Migration," Section VII in, *Net Intercensal Migration for States and Geographic Divisions of the United States, 1950–60: Methodological and Substantive Aspects*, Analytical and Technical Reports, No. 5, Philadelphia, Population Studies Center, University of Pennsylvania, 1965; C. Horace Hamilton, "Effect of Census Errors on the Measurement of Net Migration," *Demography*, 3(2):393–415, 1966; Siegel and Hamilton, op. cit.; Leroy O. Stone, "Evaluating the Relative Accuracy and Significance of Net Migration Estimates," *Demography*, 4(1):310–330, 1967.

by the vital statistics method can be illustrated by referring back to table 21–2. The pertinent columns are (3) and (7) to (9). Column (3) minus column (7) gives column (8); and column (9) is column (8) ÷ column (2), since the rate shown here is based on the population at the beginning of the period.

Changes in area boundaries. — One source of error in these estimates of net migration is a change in the boundaries of the geographic area or areas in question. Unless the population figures and those on natural increase can be adjusted to represent a constant area (ordinarily the present rather than the original area), the estimates will reflect the population in the transferred territory as well as net migration (see ch. 13). If the territory transferred is an entire administrative unit of some sort (for example, a commune that is transferred from one province to another), the requisite statistics will usually be readily available. Often, however, the transferred territory does not follow any previous legal boundaries. In that case, rough estimates will have to be made, for example, on the basis of new and old maps, using the land areas involved as a rough indicator of the proportions of an old administrative unit's population and vital events that are to be attributed to the new units. The transfer will usually occur during the intercensal period. Vital events for following years will often be on the new geographic basis and will require no adjustment. There may be a lag in the geographic assignment of vital events, however; and, in any case, if the transfer occurred during the year and the vital statistics are available for whole years only, a special proration must be made for the year of transfer.

Exclusion of international migration. — As previously mentioned, any variant of the residual method includes in its estimate of net migration net immigration from abroad—unless a specific attempt is made to remove that component. One approach is to confine the computations to the native population. In the case of the vital statistics method, this approach requires that the native population be available separately from the two censuses and that deaths also be tabulated by nativity. The inputs and method of this approach may be illustrated as follows:

(1) Native population of state at second census
(2) Native population of state at first census
(3) Total births, in intercensal period
(4) Deaths of natives, in intercensal period

Estimated net migration of native population to the state from other states, 1950–60, is then given by (1) − (2) − (3) + (4).

As was explained in chapter 9, "native" means born in the country, not just born in the state. Births are all native by definition, and babies born during the intercensal period are included in the native population counted at the end of the period.

The other approach is to try to allow directly for net immigration from abroad by subtraction. For this purpose, of course, one must know not only the total immigration but also the area (state, province, etc.) of intended residence of the immigrants and the area of last residence of the emigrants. It is necessary to assume that immigrants went to their announced area of destination and stayed there until the next census and that emigrants were counted at the preceding census in the area given as that of last residence.

Effects of errors in census and vital statistics. — Since the estimates of net migration represent a residual obtained by subtraction, relatively moderate errors in population counts or in statistics of births or deaths produce much larger percentage errors in the migration estimates. Fortunately, these errors are sometimes offsetting. For example, the population

counts at the two successive censuses may represent about the same amount of underenumeration. The assessment of these errors is discussed in the sources listed in footnote 18, among others. For the same percentage error, errors in the population statistics have more effect on the estimates of net migration than do errors in the vital statistics.

Estimates by age. — Although the vital statistics method is adaptable to making estimates of net migration by age, it is rarely used for this purpose, mainly because of the large amount of work involved both in data compilation and in calculations. Now that electronic computers are so widely available, however, it should be much more feasible to produce estimates in this detail. A fairly detailed account and evaluation of this method is given elsewhere by Hamilton.[19]

Residual Methods—Survival Rate Method. — The survival rate method is the commonly used variant in statistically underdeveloped countries because it does not require accurate vital statistics. It is also frequently used in statistically developed countries as well, partly because it yields estimates of net migration by age and sex without nearly as much labor as is involved in the use of deaths by age.

This method as applied to the estimation of net international migration was discussed in chapter 20. The basic formula for estimating net migration is

$$M_{x+t} = P^t_{x+t} - sP^o_x, \tag{8}$$

where x is an age or age group
t is the interval in years between censuses
P^o_x is the population aged x at the first census
P^t_{x+t} is the population at the next census at age $x + t$
s is the survival rate

Thus, the second term on the right is again the expected population at the second census had there been no net migration.

Two main types of survival rates are used, those from life tables and those from censuses. The former are derived from a life table, if possible, for the same geographic area and time period as the estimate of net migration applies to. Usually, however, for a county or city without an appropriate life table, one could use a recent life table for its province, for its region, or for the national population, making the choice on the basis of available information as to comparative mortality. For a city, for example, the life table for the national urban population might be most appropriate. The census survival rate represents the ratio of the numbers in the same national cohort at successive censuses. The objective is to approximate a closed population. This method requires no life table, or no vital statistics, and has the advantage of eliminating the effects of some of the errors in the population statistics.

Unlike the vital statistics method, neither variant of the survival rate method measures net migration exactly even when there are no errors in the underlying population and vital statistics. Formula (9) estimates the number of net migrants who were living at the end of the period. This difference arises because the implicit number of deaths is not necessarily equivalent to the number of deaths to residents of the given area during the given period, that is, deaths to nonmigrants plus deaths to in-migrants.

[19] C. Horace Hamilton, "The Vital Statistics Method of Estimating Net Migration by Age Cohorts," *Demography*, 4(2):464–478, 1967.

Table 21–3. — **Worksheet Showing the Procedure for Estimating Net Migration by Age, by Forward Life Table Survival Rate Method, for Males, Greater Bombay: 1941–51**

Age		Population, 1941	Ten-year life table survival rate	Population, 1951	Expected survivors, 1951	Net migration, 1941-1951
1941	1951				(1) x (2) =	(3) – (4) =
		(1)	(2)	(3)	(4)	(5)
0–4	10–14	77,135	.9087	132,870	70,093	+62,777
5–9	15–19	85,434	.9573	170,227	81,786	+88,441
10–14	20–24	79,185	.9471	263,971	74,996	+188,975
15–19	25–29	82,603	.9308	253,964	76,887	+177,077
20–24	30–34	126,247	.9223	195,373	116,438	+78,935
25–29	35–39	155,344	.9161	151,259	142,311	+8,948
30–34	40–44	138,843	.9047	118,383	125,611	–7,228
35–39	45–49	109,356	.8850	76,421	96,780	–20,359
40–44	50–54	81,626	.8548	65,897	69,774	–3,877
45–49	55–59	47,062	.8122	32,265	38,224	–5,959
50–54	60–64	36,908	.7535	22,248	27,810	–5,562
55–59	65–69	15,134	.6726	9,655	10,179	–524
60 and over	70 and over	25,094	.3866	10,100	9,701	+399
Total, all ages	Total, 10 and over	1,059,971	(X)	1,502,633	[1] 940,590	+562,043

[1] Obtained by summation.
X Not applicable.

Source: United Nations, *Methods of Measuring Internal Migration*, table C.2.

There are several ways of applying the survival rates. In the ordinary method, called the **forward survival rate method,** the estimate of net migration is obtained as in (9). Another way that has been proposed is called the **reverse survival rate method.** Here the survival rates are divided into the number in the age group at the end of the intercensal period, thus:

$$M' = \frac{P^t_{x+10}}{s} - P^o_x \qquad (9)$$

The two methods always give different results but are linked by the relationship. $M' = \frac{M}{s}$.

As mentioned in chapter 20 (see section "Intercensal Cohort-Component Method"), the results of the forward and reverse methods may also be averaged, $M'' = \frac{M + M'}{2}$. The average implies a more meaningful assumption regarding the timing of net migration than does either the forward or reverse method. The amount of difference between the migration estimates from the forward and reverse methods depends on the amount of net migration and on the level of the survival rate. The percentage difference is a function of only the survival rate. The relationships among these estimates and their rationales are also discussed at length by Siegel and Hamilton and in the U.N. Manual VI. In this book the term "survival rate method" when used unqualified refers to the forward method.

Life table survival rate method. — A more specific expression may be substituted at this point for the survival rate, *s*. If we express the survival rate for a 5-year age group and 10-year period in life table notation, namely,

$$s_x = {}_5L_{x+10} \div {}_5L_x, \qquad (10)$$

equations (8) and (9) may be adapted accordingly. As an example,

$$M_{20-24} = P^{10}_{20-24} - \left[\frac{{}_5L_{20}}{{}_5L_{10}} \times P^0_{10-14} \right]$$

If there is an open end interval, say 75 years and over, in the census age distribution, then $s_{70} = T_{75} \div T_{70}$.

In using actual life tables, the demographer would prefer to have one covering the full intercensal period. If, instead, there are available only tables centering on the two census dates, survival rates can be computed from both life tables and the results averaged so as to give a better representation of the conditions prevailing over the decade. In chapter 20, table 20–4 illustrated all three techniques (forward, reverse, and average) of applying survival rates, as well as their application to cohorts born during the intercensal period, for Puerto Rico. The same procedures would be used for a geographic subdivision of a country and, in that case, the estimated residual would represent mostly internal (e.g., interstate) migration.

Table 21–3, which is for greater Bombay, derives its survival rates from the United Nations Model Life Table, $\overset{o}{e}_o = 45$,[20] since the death statistics for Bombay are not sufficiently complete for construction of a life table. The Indian census figures have been adjusted for area changes.

When the intercensal period is not 5 or 10 years, there are some complications. The survival rates will have to be calculated for the number of years in the intercensal period, and this calculation will require extra work if only an abridged life table is available. Methods of interpolation in this situation are described in chapter 15. Furthermore, the survivors will appear in unconventional 5-year age groups. They should be redistributed into conventional 5-year age groups so that they may be compared with the age groups of the second census. Estimates of net migration obtained by subtraction will then be in terms of conventional age groups.

Even if there were no external migration, estimated net migration over all the geographic subdivisions could not be expected to add to zero at each age group. There are errors of coverage and age reporting in the input data, and the life table will only approximate deaths to internal (interarea) migrants.

---

[20] United Nations, *Methods for Population Projections by Sex and Age,* Manuals on Methods of Estimating Population, Manual III. Series A. Population Studies, No. 25, 1956, pp. 80–81.

As a final step, then, the net migration figures for each age-sex group should be adjusted to add to zero (or to the net external migration). Moreover, as in the case of the vital statistics method, it will often be desirable to smooth the reported age distribution in the census and to make corrections for other types of gross errors.

It will be noted that table 20-4 included the net migration of babies born during the intercensal period, whereas table 21-3 does not. Applications of the survival rate method do frequently omit the cohorts born during the intercensal period even when adequate statistics on registered births are available. The present authors recommend the more comprehensive figures, however, in the interests of a more nearly complete estimate of net migration and of greater comparability with the vital statistics method. As described in chapter 15, the survival rates for children born during the intercensal period are of a different form from those for the older ages. Babies born during the first quinquennium of a 10-year intercensal period will be 5 to 9 years old and those born during the second quinquennium will be under 5 years old. Births can be represented by the radix, l_0, of the life table so that

$$s_{5-9} = \frac{{}_5L_5}{5l_0} \quad \text{and} \quad s_{0-4} = \frac{{}_5L_0}{5l_0}. \quad (11)$$

Census survival rate method. — A census survival rate is the ratio of the population aged $x + n$ at the second census to that aged x at the first census, where the censuses are taken n years apart. Thus,

$$s_x^n = \frac{P_{x+n}^{t+n}}{P_x^t} \quad (12)$$

Here t is the date of the first census. We want a rate that reflects mortality but not migration. Hence, census survival rates have to be based on national population statistics; and, if there is appreciable external migration, it is preferable to base them on the native population as counted in the two national censuses. Having secured survival rates based on a closed population, however, it is permissible to apply them to the total population figures for local areas so as to include the net migration of the former immigrants in the estimates. (In so doing, it is assumed that the foreign-born has had the same survival rates as the native population.) The census survival rates measure mortality plus relative coverage and reporting errors in the two censuses. The confounding of the two effects is actually an advantage because it is then unnecessary to correct for the disturbing influence of the errors in the population data and because these errors are, in effect, largely excluded from the estimates of net migration. There are two very important assumptions with this method that need to be kept in mind. They are (1) that true survival rates are the same for the geographic subdivisions as for the nation and (2) that the pattern of relative errors in the census age data is also the same.

The first assumption, of course, is also made when we use a life table for a larger area containing the area in question or when we use a model life table. The second assumption specifically means that the relative change in the percent completeness of coverage for a particular age cohort between the two censuses is the same for the country as a whole and for each area for which net migration is being estimated.

Because of coverage and age reporting errors in the census, or because of net immigration from abroad, a survival rate will sometimes exceed unity. This is an impossible value, of course, as far as survival itself is concerned; but, for the purpose of estimating net migration, this is the value of the rate that should be used. An intercensal survival rate of 1.019725 (as given in table 21-4) for the cohort under 5 years old at the first census probably indicates that young children are characteristically less completely enumerated in the census than are older children. This fact has to be allowed for when estimating the expected population 10 to 14 years old 10 years later. U.N. Manual VI treats the problem of estimating net migration of children born during the intercensal period when adequate birth statistics are not available. ". . . the following approximate method, which uses area-specific child-woman ratios, derived from the second census, may be applied. If the ratios of children aged 0-4 to women aged 15-44 and of children aged 5-9 to women aged 20-49 are denoted by CWR_0 and CWR_5 respectively, then estimates of net migration for the age groups 0-4 (denoted by Net ${}_5M_{0,i}$) and 5-9 (denoted by Net ${}_5M_{5,i}$) are given by:

$$\text{Net } {}_5M_{0,i} = \frac{1}{4}CWR_0 \cdot \text{Net } {}_{30}M_{15,i}^{(f)} \quad (13)$$

$$\text{Net } {}_5M_{5,i} = \frac{3}{4}CWR_5 \cdot \text{Net } {}_{30}M_{20,i}^{(f)} \quad (14)$$

where Net ${}_{30}M_{15,i}^{(f)}$ and Net ${}_{30}M_{20,i}^{(f)}$ are the area estimates of net migration for females aged 15-44 and 20-49 respectively. If we assume that the flow of migration was even and fertility ratios constant, then one-fourth of the younger and three-fourths of the older children would have been born before their mothers migrated." These proportions are derived as follows: The children under 5 years old at the census were born, on the average, 2.5 years earlier; only ¼ of their mothers' migration occurred after that date. The children 5 to 9 years old at the census were born, on the average, 7.5 years earlier; ¾ of their mothers' migration occurred after that date.

Considerable methodological discussion of intercensal survival rates with tables for the United States are contained in a report by the U.S. Bureau of the Census.[21] This publication also contains a fairly complete bibliography on the subject. Rates are based on both the total population and the native population. Table 21-4 applies the former rates to the 1950 population of Franklin County, Ohio. The last column represents an adjustment (by the plus–minus procedure described in Appendix C) of the preliminary estimates of net migration by age to the results of the vital statistics method for all ages combined.

Estimates like those in table 21-4 have been prepared for all American counties in the 1950-60 decade by the U.S. Department of Agriculture in cooperation with several other agencies.[22]

If we add the estimates at a given age for all the States, we obtain totals approximating zero. This is always the case whatever the nature of error in the age data or in the survival rates and is one of the features that distinguishes the Census Survival Rate (CSR) method from the Life Table Survival Rate

[21] *Current Population Reports*, Series P–23, No. 15, "National Census Survival Rates, by Color and Sex, for 1950 to 1960," July 12, 1965.

[22] U.S. Department of Agriculture, Economic Research Service, *Net Migration of the Population, 1950–60, by Age, Sex, and Color*, Vol. I, *States, Counties, Economic Areas, and Metropolitan Areas* (in 6 parts), May 1965; Vol. II, *Analytical Groupings of Counties*, November 1965.

Table 21-4. — Estimating Net Migration by Age Cohort, for Franklin County, Ohio, by Use of National Census Survival Rates, for 1950–60

[Forward survival procedure]

Age of cohort		Census population, 1950	Births	National census survival rates[1]	Expected population, 1960 [(1) or (2)] x (3)=	Census population, 1960	Net migration	
Age in 1950 (or birth date)	Age in 1960						Preliminary estimate (5) – (4) =	Adjusted estimate (6) adjusted =
		(1)	(2)	(3)	(4)	(5)	(6)	(7)
Total, all ages........	Total, all ages.......	503,410	(X)	(X)	608,900	682,962	74,062	75,756
Born Apr. 1, 1955 to 1960..	Under 5....................	(X)	[2]89,091	[3].947031	84,372	85,464	1,092	1,115
Born Apr. 1, 1950 to 1955..	5-9......................	(X)	[2]69,628	[3].939721	65,431	69,635	4,204	4,293
Under 5....................	10-14....................	52,007	(X)	1.019725	53,033	58,043	5,010	5,117
5-9........................	15-19....................	36,585	(X)	.987918	36,143	47,602	11,459	11,703
10-14......................	20-24....................	29,954	(X)	.963345	28,856	52,454	23,598	24,101
15-19......................	25-29....................	31,972	(X)	.989183	31,626	50,587	18,961	19,364
20-24......................	30-34....................	46,644	(X)	.999510	46,621	50,959	4,338	4,430
25-29......................	35-39....................	47,657	(X)	.992511	47,300	49,523	2,223	2,270
30-34......................	40-44....................	40,390	(X)	.986082	39,828	42,132	2,304	2,353
35-39......................	45-49....................	37,138	(X)	.948344	35,220	37,007	1,787	1,825
40-44......................	50-54....................	34,329	(X)	.925976	31,788	33,201	1,413	1,443
45-49......................	55-59....................	31,067	(X)	.916818	28,483	28,849	366	374
50-54......................	60-64....................	28,138	(X)	.853888	24,027	23,803	-224	-219
55-59......................	65-69....................	24,693	(X)	.859476	21,223	19,812	-1,411	-1,381
60-64......................	70-74....................	20,467	(X)	.776462	15,892	15,074	-818	-801
65-69......................	75-79....................	17,102	(X)	.608023	10,398	10,084	-314	-307
70-74......................	80-84....................	11,629	(X)	.461477	5,367	5,456	89	91
75 and over...............	85 and over.............	13,638	(X)	.241403	3,292	3,277	-15	-15

X Not applicable.
[1] Based on total population rather than native population.
[2] Adjusted for underregistration.
[3] Based on adjusted births.

Source: U.S. Bureau of the Census, *Current Population Reports,* Series P-23, No. 15, "National Census Survival Rates, by Color and Sex, for 1950 to 1960," July 12, 1965, table A.

method.[23] The census survival rates computed from national statistics for the total population reflect both mortality and net immigration from abroad. Hence, the estimates of net migration represent internal migration plus any excess or deficit of the area's rate of net immigration relative to the national rate. Furthermore, the estimates summed over all areas must balance to zero for any age-sex group.

The assumption that mortality, or survival, levels are nearly equal throughout the various geographic areas of the country deserves scrutiny particularly in countries where mortality is high. Where mortality is high, or where there is much regional variation in mortality, or where net migration rates are low, some adjustment for differences in survival rates is necessary. Available information on mortality differences between the geographic subdivisions and the nation may be used to adjust the census survival rates. External evidence on mortality from vital statistics or sample surveys is best for this purpose. The methodology of making adjustments is described at length in the U.N. Manual. In brief, the ratio of the regional life–table survival rate to the corresponding national survival rate must be computed and applied to the national census survival rates as an adjustment factor.

Table 21-5 is our own illustration of such calculations for the Province of Quebec, Canada, over the period 1951–56 using an actual provincial life table. Unfortunately, age distributions of the native population were not available from both censuses so the national census survival rates were derived for the total population despite the occurrence of

considerable immigration from abroad. As in a number of our other illustrations, then, the calculated values are not to be taken as realistic. Instead of hypothesizing regional differences in survival rates, as Zachariah did for Korea, we compared the actual differences between the life-table survival rates for the Province of Quebec and those for all Canada and used those ratios to adjust the national census survival rates.

Sources of error. — Estimates by the survival rate methods like those by the vital statistics method are affected by changes in area boundaries and reflect international migration to some extent unless specific allowances are made for these phenomena. As has been shown, the effects of errors in the components of the balancing equation are somewhat different, however. In the vital statistics variant of the residual method, it is the relative errors in the population figures at the two censuses for the area in question that concern us. (If the error for an age-sex group at the first census is different from that at the second census, the difference will be included in the estimated net migration.) In the survival rate variant, on the other hand, it is the applicability to the given geographic area of the relative national errors at the two censuses that is in question. Similarly, for mortality, in the first case, we are concerned with the completeness of death registration (and perhaps also with the accuracy of reported ages at death); whereas, in the second case, we are concerned with the applicability of the survival rates used to the area in question. Stated in different terms, the key to the accuracy of the census survival rate method in a given case lies in the applicability of the assumptions stated in the opening paragraph of this section. The sources of error in the survival

[23] United Nations, *Methods of Measuring Internal Migration.*

Table 21-5. — Estimated Net Migration of Female Population of Quebec Province, Canada, by Age: 1951–56

Age (years) 1951	Age (years) 1956	Population of Canada 1951 (1)	Population of Canada 1956 (2)	National census survival rate (2)÷(1) = (3)	Quebec life-table survival rate[1] (4)	National life-table survival rate[2] (5)	Ratio of Quebec survival rate to national rate (4)÷(5) = (6)	Adjusted census survival rate (3) x (6) = (7)	Population of Quebec 1951 enumerated (8)	Population of Quebec 1956 expected (7) x (8) = (9)	Population of Quebec 1956 enumerated (10)	Estimated net migration for Quebec, 1951–1956 (10)−(9) = (11)
All ages....	5 and over..	6,920,556	6,957,184						2,033,554	2,054,117	2,017,608	−36,509
Under 5.......	5–9..........	843,046	887,101	1.05226	0.99362	0.99479	0.99882	1.05102	265,559	279,108	273,989	−5,119
5–9..........	10–14........	683,952	702,562	1.02721	0.99731	0.99756	0.99975	1.02695	227,157	233,279	229,358	−3,921
10–14........	15–19........	555,661	575,666	1.03600	0.99700	0.99723	0.99977	1.03576	177,528	183,876	184,960	+1,084
15–19........	20–24........	525,792	561,931	1.06873	0.99601	0.99631	0.99970	1.06841	169,736	181,348	179,360	−1,988
20–24........	25–29........	551,106	592,301	1.07475	0.99493	0.99554	0.99939	1.07409	176,403	189,473	181,892	−7,581
25–29........	30–34........	578,403	613,750	1.06111	0.99334	0.99451	0.99882	1.05986	171,660	181,936	177,118	−4,818
30–34........	35–39........	530,177	558,622	1.05365	0.99097	0.99257	0.99839	1.05195	152,840	160,780	156,251	−4,529
35–39........	40–44........	495,562	502,784	1.01457	0.98692	0.98905	0.99785	1.01239	139,163	140,887	139,635	−1,252
40–44........	45–49........	422,767	422,988	1.00052	0.98078	0.98321	0.99753	0.99805	121,301	121,064	118,116	−2,948
45–49........	50–54........	356,971	351,215	0.98388	0.97190	0.97494	0.99688	0.98081	100,615	98,684	97,207	−1,477
50–54........	55–59........	322,195	307,271	0.95368	0.95620	0.96212	0.99385	0.94781	86,090	81,597	80,107	−1,490
55–59........	60–64........	278,126	259,265	0.93219	0.93059	0.94086	0.98908	0.92201	69,095	63,706	63,499	−207
60–64........	65–69........	241,828	226,562	0.93687	0.89372	0.90849	0.98374	0.92164	57,777	53,250	52,287	−963
65–69........	70–74........	205,421	183,218	0.89191	0.83751	0.85674	0.97755	0.87189	46,387	40,444	39,397	−1,047
70–74........	75–79........	154,674	113,948	0.73670	0.74536	0.76872	0.96961	0.71431	33,945	24,247	24,005	−242
75–79........	80–84........	94,261	61,460	0.65202	0.61449	0.64086	0.95885	0.62519	21,019	13,141	13,019	−122
80 and over...	85 and over...	80,614	36,540	0.45327	0.37978	0.40760	0.93175	0.42233	17,279	7,297	7,408	+111

[1] Average of survival rates computed from provincial life tables for 1950–52 and 1955–57.
[2] Average of survival rates computed from national life tables for 1950–52 and 1955–57.

Source: Canada, Dominion Bureau of Statistics, official reports.

rate methods and their importance have been discussed at length by Hamilton, Price, Stone, and Wunsch, and in U.N. Manual VI, among others.[24]

Comparative Results From Different Methods. — Estimates of net migration by age and sex made by the survival rate method are sometimes adjusted to add to an estimate for the total population made by the vital statistics method. It was demonstrated by Siegel and Hamilton that the latter gives a theoretically exact measure of net migration whereas the former does not.[25] Which method gives more accurate estimates in an actual situation depends upon a host of empirical considerations and has been debated for particular situations.[26] On the basis of American data, Eldridge, Hamilton, and Tarver all found that the forward census survival rate method tends to give lower (algebraically) estimates of net migration than the vital statistics method. The authors of U.N. Manual VI state that, "Since the number of deaths is likely to be larger in the larger of the two components of net migration (in-migration and out-migration), CSR estimates obtained by the forward method will generally be smaller than those obtained

by the VS method." [27] They conclude, however, that it is difficult to make a general statement regarding the relative accuracy of the two methods for the net migration of all ages combined.

Uses and Limitations. — In summary, the residual method cannot be used to estimate gross in-migration or out-migration or migration streams. The migration period must be the intercensal period. In any of its variations, this method can be used to estimate net migration for a fixed area, for a group defined by an unchanging characteristic (sex, race, nativity, etc.) or for a group defined by a characteristic that changes in a fixed way with time (age).

The residual method cannot ordinarily be used for social and economic groups, mainly because the corresponding characteristics (marital status, occupation, income, etc.) change frequently and unpredictably during the intercensal period. In most countries, education changes so seldom for adults, however, that net migration by educational attainment could probably be estimated fairly well for the adult population. Such estimates have indeed been made by Hamilton for the Southern region of the United States, over the intercensal decade 1940–50, by age and farm-nonfarm residence, and for the national farm and nonfarm populations by color and sex.[28]

Rural-urban migration is such an important element in internal migration, particularly in the developing countries, that there is a great interest in measuring it by some means. As with other methods, the residual method has many pitfalls

[24] C. Horace Hamilton, "Effect of Census Errors on the Measurement of Net Migration," *Demography*, 3(2):393–415, 1966; Daniel O. Price, "Examination of Two Sources of Error in the Estimation of Net Internal Migration," *Journal of the American Statistical Association*, 50(271):689–700, September 1955; Leroy O. Stone, op. cit.; United Nations, *Methods of Measuring Internal Migration*, 1970; G. Wunsch, "Le calcul des soldes migratoires par la methode de la 'population attendu'–caracteristiques et evaluation des biais" (The calculation of net migration by the method of 'expected population'–characteristics and evaluation of bias), *Population et famille* (Brussels), 18:49–62, 1969; K. C. Zachariah, "A Note on the Census Survival Ratio Method of Estimating Net Migration," *Journal of the American Statistical Association*, 57(297):175–183, March 1962.
[25] Siegel and Hamilton, op. cit., pp. 483, 487.
[26] Hope T. Eldridge, op. cit., pp. 82–99; C. Horace Hamilton, "The Vital Statistics Method of Estimating Net Migration by Age Cohorts," pp. 464–465 and pp. 474–478; James D. Tarver, "Evaluation of Census Survival Rates in Estimating Intercensal State Net Migration," *Journal of the American Statistical Association*, 57(300):841–862, December 1962.

[27] United Nations, *Methods of Measuring Internal Migration*.
[28] C. Horace Hamilton, "Educational Selectivity of Net Migration From the South," *Social Forces*, 38(1):33–42, October 1959; idem, "Educational Selectivity of Migration From Farm to Urban and to Other Nonfarm Communities" in *Mobility and Mental Health*, Mildred B. Kantor, ed., Springfield (Illinois), Charles C Thomas, 1965, pp. 166–195.

Table 21-6. — Place of Birth by Place of Enumeration, for Gabon: 1960 to 1961

[From 10-percent sample used in second stage of census]

Place of enumeration	Place of birth											
	Woleu N'Tem	Estuaire	Ogooué Maritime	Moyen Ogooué	Ogooué Ivindo	Ogooué Lolo	Haut Ogooué	N'Gounie	Nyanga	Total Gabon	Abroad	Grand total[1]
MALE												
Woleu N'Tem..................	33,827	103	-	10	150	-	31	26	47	34,193	771	34,965
Estuaire.....................	2,505	15,260	440	584	1,289	1,411	1,189	3,871	1,687	28,236	3,265	31,523
Ogooué Maritime..............	127	567	10,575	1,303	68	1,352	1,049	2,892	1,316	19,249	1,883	21,137
Moyen Ogooué.................	92	68	417	11,773	367	890	213	2,442	716	16,976	228	17,213
Ogooué Ivindo...............	90	45	11	45	15,739	11	11	11	-	15,966	249	16,214
Ogooué Lolo..................	-	19	19	-	-	15,350	77	10	19	15,494	38	15,590
Haut Ogooué..................	-	-	-	-	-	98	17,392	-	-	17,491	517	18,032
N'Gounie.....................	37	356	198	230	16	528	22	31,765	812	33,964	594	34,558
Nyanga.......................	-	-	49	82	-	-	-	197	14,776	15,104	361	15,465
Total...................	36,678	16,418	11,709	14,027	17,629	19,640	19,984	41,214	19,373	196,673	7,906	204,698
FEMALE												
Woleu N'Tem..................	40,783	423	21	41	412	-	16	57	155	41,907	1,440	43,346
Estuaire.....................	2,394	17,428	385	704	781	928	485	2,501	1,213	26,821	2,695	29,552
Ogooué Maritime..............	64	612	12,037	1,320	50	942	458	2,938	868	19,289	1,063	20,366
Moyen Ogooué.................	29	139	502	12,230	292	647	150	2,309	384	16,680	164	16,844
Ogooué Ivindo...............	23	34	-	79	18,420	11	34	68	11	18,680	260	18,940
Ogooué Lolo..................	-	29	29	10	48	20,496	192	48	19	20,870	38	21,072
Haut Ogooué..................	-	-	-	-	148	172	23,173	-	-	23,493	615	24,157
N'Gounie.....................	26	205	215	293	42	581	44	40,981	1,199	43,587	498	44,084
Nyanga.......................	-	66	66	115	-	-	-	640	20,024	20,910	279	21,205
Total...................	43,319	18,936	13,255	14,792	20,193	23,777	24,552	49,542	23,873	232,237	7,052	239,567
TOTAL												
Woleu N'Tem..................	74,609	526	21	52	562	-	47	83	201	76,100	2,211	78,311
Estuaire.....................	4,899	32,688	825	1,288	2,070	2,339	1,674	6,373	2,900	55,057	5,960	61,075
Ogooué Maritime..............	191	1,180	22,612	2,623	118	2,294	1,506	5,830	2,184	38,538	2,946	41,503
Moyen Ogooué.................	121	207	918	24,003	659	1,537	362	4,751	1,100	33,656	392	34,057
Ogooué Ivindo...............	113	79	11	124	34,160	23	45	79	11	34,646	509	35,154
Ogooué Lolo..................	-	48	48	10	48	35,846	269	58	38	36,365	77	36,662
Haut Ogooué..................	-	-	-	-	148	271	40,565	-	-	40,984	1,132	42,189
N'Gounie.....................	62	561	413	523	58	1,109	66	72,747	2,011	77,550	1,092	78,642
Nyanga.......................	-	66	115	197	-	-	-	836	34,801	36,014	640	36,670
Total...................	79,995	35,355	24,963	28,820	37,823	43,419	44,534	90,757	43,246	428,910	14,959	444,264

— Represents zero.
[1] Including place of birth not reported.

Source: Gabon, Service de statistique, and France, I.N.S.E.E., *Recensement et enquête démographiques, 1960–61* (Population census and survey, 1960–61), Paris, p. 116.

in this application—all of those previously catalogued plus some new possibilities for error as well. Other things being equal, the method works best when the urban and rural areas are defined in terms of whole administrative units, such as communes, and changes in classification are rarely made. Here one may obtain constant territories over the intercensal period by reassigning whole communes, localities, etc., that have been shifted from the rural to the urban classification, or vice versa. When, however, the reclassification of territory involves annexations and retrocessions or a radically different set of boundaries for the units in question, there are serious problems in adjusting the statistics, which may be insurmountable. U.N. Manual VI describes some of the devices that can be used to handle these difficulties.[29]

As previously stated, census errors in classification as well as in coverage will be reflected as errors in estimated net migration. Furthermore, death statistics are not available separately for all the areas or groups in question, and life tables for larger populations will be inappropriate in varying degrees.

Place of Birth

The traditional item that represents a direct question relating to migration is "place of birth." This item has long been included in national censuses, and it is occasionally found in sample surveys. The first national census to contain such an item was that of England and Wales in 1841. As was explained in chapter 9, two kinds of specificity are usually called for—in the case of the foreign born, the country of birth; in that of the native population, the geographic subdivision (state, province, etc.) or often also the secondary type of subdivision, such as the district in India or Pakistan. "Native" may be defined sometimes to include persons born in outlying territories belonging to the country—territories that are not covered by the census in question. Usually those natives are separately identified, however.

---

[29] United Nations, op. cit.

Table 21-7. — Place of Birth by State of Residence in 1955 and 1960, for the Population 5 Years Old and Over, by Age, for the United States

[Based on 25-percent sample]

Age	Total population, 5 years old and over	Native								Foreign born		
		Living in--						Not living in State of birth in 1960, abroad in 1955, or State not reported	Other [1]	Living in--		Abroad in 1955 or State of 1955 residence not reported
		State of birth in 1955 and 1960	State of birth in 1960, but not 1955	State of birth in 1960, abroad in 1955, or State not reported	Same State in 1955 and 1960, but born in a different State	State of birth in 1955, but not 1960	Different States in 1955 and 1960, both different from State of birth			Same State in 1955 and 1960	Different State in 1955 and 1960	
Total............	159,003,811	98,550,799	2,291,746	697,479	30,865,143	6,675,068	4,423,613	860,798	5,002,960	8,058,749	494,344	1,083,112
5 to 9 years........	18,659,145	14,981,710	282,828	52,815	1,132,531	1,148,551	364,743	63,037	467,072	56,283	13,763	95,812
10 to 14 years.....	16,815,970	13,121,219	226,814	40,778	1,748,290	638,172	349,772	54,316	399,567	142,949	16,544	77,549
15 to 19 years.....	13,287,439	9,675,074	151,441	42,316	1,660,737	813,929	300,097	44,839	396,107	111,880	13,983	77,036
20 to 24 years.....	10,803,169	6,479,182	236,217	106,568	1,330,033	1,322,050	459,554	101,401	460,799	115,662	25,129	166,574
25 to 29 years.....	10,870,386	6,071,176	434,015	186,929	1,798,258	769,775	626,803	182,612	413,514	169,157	38,919	179,228
30 to 34 years.....	11,951,709	6,944,311	268,110	66,788	2,588,043	541,822	554,281	112,348	397,190	288,220	49,365	141,231
35 to 44 years.....	24,075,532	13,817,546	326,432	78,501	6,353,658	686,666	872,233	168,025	743,242	785,014	78,165	166,050
45 to 54 years.....	20,625,380	11,650,123	170,172	48,731	5,764,261	351,664	436,459	67,405	645,861	1,335,680	63,132	91,892
55 to 64 years.....	15,707,844	8,123,603	101,349	32,404	4,269,590	207,602	240,001	33,053	541,299	2,032,504	75,722	50,717
65 to 74 years.....	10,847,899	5,219,741	67,426	23,303	2,797,667	139,827	158,553	19,515	326,511	1,987,270	83,794	24,292
75 years and over...	5,359,338	2,467,114	26,942	18,346	1,422,075	55,010	61,117	14,247	211,798	1,034,130	35,828	12,731
Median age.........	33.8	29.1	27.9	27.8	43.1	22.8	31.0	29.6	34.6	60.0	46.8	28.5

[1] Includes persons born in outlying areas or born abroad or at sea of American parents, as well as native persons for whom State of birth was not reported.

Source: Abstracted from, *U.S. Census of Population, 1960*, PC(2)-2D, *Lifetime and Recent Migration*, table 1.

Several examples will be given of the treatment of data on internal migration from the question on birthplace in national population censuses. The most detailed and basic statistics, from which various summary statistics are derived, are given in the cross-classification of residence at birth by residence at the time of the census. An example of such a cross-classification for Gabon is reproduced as table 21-6. The geographic areas shown here are the country's primary divisions. Some countries (e.g., Costa Rica, India, Japan) show such cross-classifications, or consolidations thereof, for secondary divisions.[30] Changes of residence in these cross-classified statistics may be viewed as representing migration streams between the time of birth and the time of enumeration.

Frequently the statistics on streams are consolidated into categories like "living in given state, born in different state" and "born in given state, living in different state," which may be viewed as representing lifetime in-migrants and lifetime out-migrants, respectively. Characteristics (age and social and economic characteristics) of lifetime migrants, are usually shown in terms of these consolidated categories (whether or not born in area of enumeration), because the full detail would be very space-consuming; but they are sometimes shown for migration streams.[31]

The United States. — Data on the State of birth of the native population have been collected at every census beginning with that of 1850. A cross-classification of each State of residence at the time of the census with each State, territory, and possession at time of birth has been shown in full detail. In addition, the State of birth has been shown in some reports for the urban, rural-nonfarm, and rural-farm parts of States and for individual cities of varying minimal size (e.g., 50,000 in 1930 and 100,000 in 1940). The census inquiry does not, of course, provide information on urban-rural residence, or city of residence, at birth.

A special feature of the 1960 migration reports was a very detailed cross-classification of division of present (1960) residence, by division of 1955 residence, by division of birth, by age, sex, and color, with some tables containing a further cross-classification by educational attainment.[32] These statistics provide one way of studying **longitudinal migration,** i.e., an abstract of the person's lifetime migration history. From them, return migration and progressive (or in Eldridge's nomenclature, "secondary") migration can be measured. A simple summarization of these statistics is shown in table 21-7. "Progressive" migrants are those in the seventh column, "Different States in 1955 and 1960, both different from State of birth."

Uses and Limitations. — We are concerned at this point not with the accuracy or quality of the statistics on place of birth of the native population but with their pertinence to internal migration. On the assumption that the statistics are accurate, what would they actually measure, and how useful are these measures to the demographer?

Unlike the estimates of migration by the residual method, which are limited to net movements, place-of-birth data can

[30] Costa Rica, *Dirección general de estadística y censos, El area metropolitana de San José según los censos de 1963 y 1964*, December 1967, table 31, pp. 77-78; India, Office of the Registrar General, *Census of India, 1961*, Vol. IV Bihar, Part II-D, *Migration Tables*, 1965, pp. 11-197 (not all State reports give this detail); Irene B. Taeuber, *The Population of Japan*, Princeton (New Jersey), Princeton University Press, 1958, pp. 126-129 and 139-141.
[31] See, for example, India, Office of the Registrar General, *Census of India, 1951*, Volume 1, *India*, Part II-A, *Demographic Tables*, 1955, pp. 260-297. This long table gives the "livelihood classes" of interstate streams.
[32] *U.S. Census of Population: 1960, Subject Reports, Lifetime and Recent Migration*, PC(2)-2D, 1963.

represent in-migrants, out-migrants, and specific streams. For instance, table 21–6 indicates that there were an estimated 1,687 surviving males born in the Nyanga region of Gabon but living in Estuaire at the time of the census. The statistics often tell us nothing about intrastate migration; and, even when secondary subdivisions are specified in the recording of birthplace, intra-area mobility (short-distance movement) is not covered. Moreover, the statistics do not take account of intermediate migrations between the time of birth and the time of the census; and persons who have returned to live in their area of birth appear as nonmigrants. In sum, these statistics do not indicate the total number of persons who have moved from the area in which they were born to other areas, or to any specific area, during any given period of time.

The question of time reference is deserving of particular attention. The statistics from one census not tabulated by age tell us nothing about when the move occurred. With a tabulation by age, the only specification is that given by age itself, for example, a migrant 35 years old must have moved within the 35 years preceding the census. Thus, the older the migrant the less we know about the date of the move and the more likely that he made intervening moves between other areas of the same class. Even statistics tabulated by age for two successive censuses are not fully adequate for measuring migration in an intercensal period although they do greatly enhance the value of the data. The ways of exploiting this extra information, some of which are fairly complex, are discussed below.

By definition, the statistics on state, province, etc., of birth relate to the native population of the country. Hence, the internal migration of the foreign-born population subsequent to its immigration is not included.

Quality of the Statistics. — There is little quantitative evidence of the accuracy with which birthplace is reported. The fact that the item does not change for the individual should make it easy for him to remember it; but since, for everybody except young children, it relates to a more remote date than does the migration question regarding residence at a fixed past date, there should be more lack of knowledge on the part of respondents other than the person himself.

Statistics on place of birth are subject to the types of errors of reporting and data processing that affect the generality of demographic characteristics; and, in addition, they have some sources of error that are *sui generis*. These include uncertainties about area boundaries at the time of birth and about the reporting of birthplace for babies who were not born at the usual residence of their parents.

There have been few attempts to measure the gross or net effects of all these sources of error by such methods as re-interviews or matching studies of a sample of the original records. Since, for purposes of measuring internal migration, we are interested in the birthplace of the native population only, our first concern is with the accuracy of the classification of the population as native or foreign born. In the United States, the Post-Enumeration Survey of the 1950 census indicated a high degree of accuracy in the classification of the population as native or foreign born. Only 0.6 percent of the enumerated population was misclassified in the census. In the reporting of State of birth, about 3 percent of those reporting gave a different State in the census and in the P.E.S. The net effect of these differences was small.

Boundary changes. — From the standpoint of measuring internal migration, it would be ideal if birthplace were reported in terms of present boundaries. (Otherwise, a person who lived

in a part of State *A* that was transferred to State *B* is automatically classified as a migrant whether he moved or not.) Rarely, however, are instructions provided in the census on this point.[33] This is a problem with other migration questions as well, but the chances are greater that a boundary change may have occurred if lifetimes are being considered. In the United States, West Virginia was detached from Virginia and became a separate State in 1863. It is evident from the subsequent statistics for many decades thereafter that some respondents born in West Virginia before 1863 gave Virginia as their birthplace whereas others gave West Virginia. In the compilation made at the University of Pennsylvania, therefore, these two States are combined as one birthplace for 1870.[34]

Births away from parents' usual residence. — In modernized countries, increasing proportions of births take place at hospitals rather than in the home. The two places may be located in different areas, thus introducing some ambiguity in the question on birthplace. From what was said earlier about the desirability of measuring changes in usual residence rather than *de facto* residence, it is clear that our preference is for the location of the parents' usual residence, rather than that of the hospital. Since most hospitals are located in urban areas, a bias would be introduced toward urban birthplaces unless the parents' usual residence were reported. When the home and hospital are located within the same tabulation area, the birthplace statistics are not affected, of course.

The U.N. Manual VI points out a related problem in certain countries where births occur under more traditional auspices. In India, for example, it is customary for a woman to return to her father's household to bear the first child and often the second and subsequent children. This custom gives rise to some spurious migration as measured from place-of-birth statistics.

Sampling error and bias. — In a few countries, the item on place of birth has been collected in the census from a fairly large sample of the population. In these cases the statistics will be affected by both sampling error and sampling bias

In the intended 10-percent sampling taken as part of the 1960–61 census of Gabon, there was an overall shortfall so that, at the national level, the reported numbers had to be multiplied by 11.6 instead of 10. Among the regions, the inflation factor ranged from 9.5 to 24.6. Thus, there was obviously sampling bias. In the 1950 census of the United States, the migration items were collected for a 20-percent sample of persons and, in the 1960 census, for a 25-percent sample of households. Each time there was a slight under-sampling, particularly of such persons as unrelated individuals. For example, in 1950, the ratio of the complete count of the native population to the inflated sample count was 1.0029 for the United States as a whole.

Nonresponse. — Most countries show separately the number of persons who did not reply to the question on birthplace or

[33] The following extract from the text of a census report for Ghana is an interesting exception: " . . ., the definition of regional boundaries should have been the one existing at Census date. The specification of region was expected to be a difficult one because of many reasons, the main ones being the changes in boundaries of regions in past years and the occurrence of the same name of locality in more than one region. To assist the enumerator, a list of traditional areas formerly called 'states' with the names of the paramount chiefs (customary heads of traditional areas), classified by the existing seven regions was given in the Enumerator's Manual. Many persons especially from rural areas could indicate the traditional area in which they were born or the name of its 'present' paramount chief rather than the region." (Ghana, Census Office, *1960 Population Census of Ghana*, Vol. III, *Demographic Characteristics*, 1964, p. ix.
[34] Lee et al., p. 295.

who gave uncodable information. For example, from table 21-6, the number of nonresponses in the Gabon census can be obtained by subtraction as 395, or 0.1 percent of the total population. Similarly low percentages of nonresponse were published officially for the earlier censuses of the United States, but these low percentages according to Lee et al., reflect considerable but unmeasured assignment of a State of birth on the basis of other information about the individual or members of his family in the course of editing the schedules.[35]

Measures Used in Analysis. — In the first major section of this chapter, there was a discussion of the appropriate bases of migration rates. In the case of place-of-birth statistics, there are appropriate situations for using either the population at origin or the population at destination. In either case, however, the population at the time of the census must be used. This decision is imposed by the fact that the population at risk does not have a fixed birthdate.

Rates. — Several kinds of rates could be computed for the areas of Gabon in table 21-6, for example.

(a) **Interregional migration rate**

$$m_r = \frac{\Sigma N_{ij} - \Sigma N_{i=j}}{N} \times 100 \qquad (15)$$

where N is the total native population; subscript i refers to the region of enumeration and subscript j to the region of birth; and N_{ij} is the number of natives living in region i and born in region j, including those living in the region of birth ($i=j$). Thus, $\Sigma N_{ij} = N$.

$\Sigma N_{i=j}$ is the number of natives living in the region of birth.

$N_{i \neq j}$ is an interregional migration stream, which we may call M_{ij}.

This leads us to an alternative expression of the rate:

$$m_r = \frac{\Sigma M_{ij}}{N} \times 100 \qquad (16)$$

Formula (15) represents the shorter of the two procedures and is described as follows:

Step 1. In the block for both sexes, sum along the leading diagonal. This total of 372,031 represents nonmigrants, from the regional standpoint, for Gabon.

Step 2. Subtract the result of Step 1 from the total native population.

$$423,910 - 372,031 = 56,879 \text{ interregional migrants.}$$

Step 3. Divide the result of Step 2 by the total native population and multiply by 100.

$$\frac{56,879}{428,910} \times 100 = 13.3$$

(b) **In-migration rate for a region**

$$m_1^I = \frac{\Sigma M_{1j}}{N_1} \times 100 \qquad (17)$$

where M_{1j} is the migrants living in region 1 who were born in region j; and N_1 is the native population enumerated in region 1.

[35] Ibid., p. 60.

Note that $\Sigma M_{1j} = \Sigma N_{1j} - N_{11} = N_1 - N_{11}$

Step 1. For a given region, say Woleu N'Tem, subtract the nonmigrants from the total native population living in the area at the time of the census to obtain the number of in-migrants

$$76,100 - 74,609 = 1,491.$$

Step 2. Divide the result of Step 1 by the total native population enumerated in the region and multiply by 100.

$$\frac{1,491}{76,100} \times 100 = 2.0.$$

(c) **Out-migration rate for a region**

$$m_1^O = \frac{\Sigma M_{i1}}{N_1'} \times 100 \qquad (18)$$

where ΣM_{i1} is the migrants from region 1 to the ith region and N_1' is the total population born in region 1. Again,

$$\Sigma M_{i1} = \Sigma N_{i1} - N_{11} = N_1' - N_{11}$$

Step 1. For Woleu N'Tem, subtract the nonmigrants from the total population born in the region to obtain the total number of out-migrants

$$79,995 - 74,609 = 5,386.$$

Step 2. Divide the result of Step 1 by the total population born in the region.

$$\frac{5,386}{79,995} \times 100 = 6.7$$

(d) **Net migration rate for a region**

$$m_1^n = \frac{\Sigma M_{1j} + \Sigma M_{i1}}{N_1} \times 100 \qquad (19)$$

Step 1. Subtract the out-migrants of (c) from the in-migrants of (b)

$$1,491 - 5,386 = -3,895.$$

Step 2. Divide the result of Step 1 by the total native population living in Woleu N'Tem to obtain the net out-migration rate from that region.

$$\frac{-3,895}{76,100} \times 100 = -5.1$$

If we had computed the out-migration rate using the native population living in the region as the base, we could have subtracted the out-migration rate from the in-migration rate.

$$\frac{5,386}{76,100} \times 100 = 7.1.$$

$$2.0 - 7.1 = -5.1.$$

(e) **Turnover rate**

$$m_1^T = \frac{\Sigma M_{1j} + \Sigma M_{i1}}{N_1} \times 100 = m_1^I + m_1^O \qquad (20)$$

(where the term m_1^O is based on N_1 and not N_1')
For Woleu N'Tem

$$\frac{1,491 + 5,386}{76,100} \times 100 = \frac{6,877}{76,100} \times 100 = 9.0.$$

The turnover rate does not carry a sign. The difference between the rate computed directly and the same rate computed by adding the in- and out-migration rates is due to rounding $(2.0 + 7.1 = 9.1)$.

Birth-Residence Index.—This index is simply the net gain or loss for an area through interarea migration. In other words, it is the net effect of lifetime migration upon the surviving population. The formula for Area 1 may be written as,

$$BR_1 = \Sigma M_{1j} - \Sigma M_{i1} = I_1 - O_1 \ldots \quad (21)$$

This is obviously the numerator of the net migration rate (19). Thus, the sum of the Birth-Residence Indexes taken over all areas of the country must be equal to zero, or

$$\Sigma BR_i = 0 \quad (22)$$

"A particular net gain or loss may be the resultant of qualitatively different patterns of in-migration and out-migration over time. For example, a zero balance may have arisen from (1) zero balances at every decade in the past, (2) net in-migration in recent decades balanced by heavier net out-migration in earlier decades, (3) net out-migration in recent decades balanced by heavier net in-migration in earlier decades, or (4) more complex patterns. A migration of a given size in a recent decade has the effect ordinarily of a migration of larger size at an earlier decade because of the different proportion of survivors from migrants of the two decades, but the analysis is further complicated by differences in the age distributions of the migrants in the two decades and by changing mortality conditions." [36]

Intercensal change in the Birth-Residence Index.—This measure "is defined as the difference in the birth-residence indices of two consecutive census periods for a given state (or group of states) and is intended to approximate the net migration during the latter decade. This measure is thus designed to relate state of birth data to a fixed period as opposed to the many different lifetimes represented by a surviving population.

"How accurate is this approximation? Since the basic data are confined to the native population, changes in the birth-residence index are produced by only internal migration and deaths (aside from errors in the census data themselves)." [37]

As Thornthwaite points out, "A birth-residence surplus in one state and the coincidental birth-residence deficit in another state may both be reduced by the death of people who migrated into the first state from the second. It is probable that when a large birth-residence surplus in a state begins to shrink, the shrinkage is due at least in part to deaths of the earlier immigrants, and the shrinkage of some of the large deficits may likewise be attributed to the same cause." [38]

"It does not follow, however, that the decennial change in the birth-residence index is always reduced arithmetically by mortality. The decennial approximation is to net migration, and mortality may have more effect upon the smaller gross component (say, in-migration) than upon the larger gross component (say, out-migration)." [39]

More detailed discussion of the failure of the index to measure intercensal migration is given by Bogue, Lee and his co-authors, and Shryock.[40] Strictly speaking we should not speak of the "error" in this measure because the defect arises not primarily from errors in the census data but rather from the only approximate validity of the measure. In other words, place-of-birth statistics are not really designed to measure migration in a fixed period of time no matter how they are manipulated, and at best the demographer can obtain only approximations to what he would like to have. Furthermore, we do not know the direction of the bias although sometimes it can be inferred.

The measure we have just defined is

$$BR_2 - BR_1 = (I_2 - I_1) - (O_2 - O_1) \ldots \quad (23)$$

where the subscripts 2 and 1 indicate the first and second censuses, respectively.

Gross intercensal interchange of population.—The sum of the absolute values of the change during an intercensal period in the number of nonresident natives of an area and the change in the number of resident natives of other states has been termed the "gross intercensal interchange of population." This may be viewed as a measure of gross interstate migration, or of population turnover, for a given state. This measure will almost certainly be too low because four out of five of the "error" terms for both in-migrants and out-migrants are errors of omission.[41] The formula may be written as,

$$GIIP = |I_2 - I_1| + |O_2 - O_1| \ldots \ldots \quad (24)$$

We have not complicated the formula by introducing a second type of subscript ($I_{i,t}$, etc.), but it is to be understood as applying to a particular State or province.

Refined measurement of intercensal migration.—Since statistics on place of birth are often the only available statistics relating to gross internal migration and migration streams in a country, it is important to consider how, despite their shortcoming, they may be refined to serve the demographer's interests. When we move from the data on lifetime migration from a single census to estimates made by differencing figures from successive censuses, we are moving from direct to indirect measurement of migration. First, it should be pointed out that biased estimates can be obtained of intercensal in-migration, out-migration, or a migration stream and not just of net migration (the Intercensal Change in the Birth-Residence Index). The first step as before is to subtract the figure for the earlier census from the corresponding figure from the later census. For example, according to censuses of India, there were 10,301 persons living in Orissa State in 1961 who were born in Uttar Pradesh as against 2,329 in 1951. By subtraction we get a biased estimate of 7,972 migrants from Uttar Pradesh to Orissa between 1951 and 1961.

The next step is to remove part of the bias. All the adjustments proposed for this purpose are allowances for intercensal mortality. No adjustments have been proposed to allow for return migration or progressive migration. Migration of these types during the intercensal period are not included in the Change Index. Eldridge and Kim have published discussions of these adjustments, which are also covered in U.N. Manual VI.[42]

To allow for intercensal mortality, formula (23) may be modified as follows:

$$BR_2 - BR_1 = (I_2 - S^I I_1) - (O_2 - S^O O_1) \quad (25)$$

[36] Shryock, *Population Mobility Within the United States*, p. 19.

[37] Ibid., p. 20.

[38] C. Warren Thornthwaite, *Internal Migration in the United States*, Study of Population Redistribution, Philadelphia, University of Pennsylvania Press, 1934, p. 9.

[39] Shryock, op. cit., p. 20.

[40] Donald J. Bogue, *Methods of Studying Internal Migration*, Part III, "Indirect Methods of Measuring Migration Streams: Place-of-Birth Statistics," technical paper prepared for a regional seminar on Population in Central and South America, held in Rio de Janeiro, Brazil, December 1955, pp. 44–57; Lee et al., op cit., pp. 57–64; Shryock, *Population Mobility Within the United States*, pp. 20–21 and 431–434.

[41] Shryock, *Population Mobility Within the United States*, pp. 431–434.

[42] Hope T. Eldridge and Yun Kim, *The Estimation of Intercensal Migration from Birth-Residence Statistics: A Study of Data for the United States, 1950 and 1960*, Analytical and Technical Reports, No. 7, Philadelphia, Population Studies Center, University of Pennsylvania, February 1968; United Nations, *Methods of Measuring Internal Migration*.

Table 21–8. – Net Migration to New England by Division of Birth, Native White Males: 1950–60
(Procedure 2)

Division of birth	Native white males in conterminous U.S.			Native white males in New England			
	Total enumerated in 1950	10 years old and over enumerated in 1960	10-year survival rate 1950-60	Total enumerated in 1950	10 years old and over in 1960		Net change due to migration 1950-1960
					Expected	Enumerated	
			$(2) \div (1)=$		$(4) \times (3)=$		$(6) - (5)=$
	(1)	(2)	(3)	(4)	(5)	(6)	(7)
New England........................	4,018,516	3,696,112	0.919770	3,448,223	3,171,572	2,984,526	-187,046
Middle Atlantic....................	12,526,609	11,505,221	0.918463	223,158	204,962	264,743	+59,781
East North Central................	13,070,675	11,914,402	0.911537	46,661	42,533	63,772	+21,239
West North Central................	7,882,937	7,145,528	0.906455	20,915	18,959	28,311	+9,352
South Atlantic....................	7,373,563	6,766,652	0.917691	34,110	31,302	45,401	+14,099
East South Central................	5,183,050	4,677,577	0.902476	10,759	9,710	15,270	+5,560
West South Central................	6,015,384	5,640,579	0.937692	10,293	9,652	15,132	+5,480
Mountain...........................	1,980,217	1,894,899	0.956915	6,083	5,821	7,856	+2,035
Pacific...........................	3,186,973	3,074,806	0.964805	10,833·	10,452	13,988	+3,536
All divisions.....................	61,237,924	56,315,776	0.919623	3,811,035	3,504,963	3,438,999	-65,964

Source: Adapted from, United Nations, *Methods of Measuring Internal Migration*, tables B.4 and B.5.

where S^1 and S^0 are the intercensal survival rates for the lifetime in-migrants and out-migrants as counted at the earlier census, respectively. The two terms of (25) give net migration among persons born outside the area and inside the area, respectively. (Any of the formulas in this section may be taken to apply to a birth cohort as well as to the total native population.)

The most obvious type of survival rate to use is an intercensal survival rate. We ought to have, then, the place-of-birth data tabulated by age at both censuses.

U.N. Manual VI gives procedures for three situations, namely, where place of birth has been tabulated by age for neither of two successive censuses, for one but not the other, and for both. In the first situation, it is possible to use only

an overall intercensal or life-table survival rate $\left(\frac{T_{10}}{T_0}\right)$, thus

presuming that migrants had the age distributions and age-specific mortality rates of the general (life table) population. Unrealistic as this assumption is, this adjustment is probably better than not allowing for mortality at all.

When age is cross-tabulated at only one census that census is very likely to be the latest one, e.g., the 1962 census of France. This circumstance is what is preferred because we can then compute the intercensal survival rate for a period of k years by getting the ratio of the population k years old and over at the second census to that of all ages at the first census. Since the population born in a given state or province is a closed population just like the total native population of the country (neglecting emigration of natives abroad), we can compute a forward intercensal survival rate for any tabulated area of birth.

Drawing on materials assembled or computed by Eldridge and by Zachariah, we may illustrate the steps for this procedure. Table 21–8 gives the computation of the intercensal survival rates for the population born in each area, here the geographic divisions of the conterminous United States, and shows how these survival rates are applied so as to obtain the net change due to migration for one of these divisions, namely, New England. An underlying assumption here is that persons born

in a given area will have the same survival rates regardless of their subsequent migration history. Thus, the expected number of native white males 10 years old and over living in New England and born in a given division is determined by the survival rate of their division of birth. As a byproduct, we find that the estimated net intercensal out-migration for native white males of New England – 65,964 – is the algebraic sum of two components. These are a net intercensal out-migration of 187,046 for those born in New England and living there in 1950 and a net intercensal in-migration of 121,082 for those born in other divisions.

These components need not necessarily have signs of minus and plus, respectively. There may be an indicated net in-migration of what Eldridge and Kim call the "in-born" and a net out-migration of what they call the "out-born". In fact, they actually found a few cases of these types for other divisions over the 1950–60 decade.[43] Assuming that the intercensal survival rates used are applicable, the first type could be interpreted as an excess of return migrants among natives of the division over recent out-migrants, and the second, as an excess of out-migrants among former in-migrants over recent in-migrants. A great advantage of carrying out the estimation in this disaggregated fashion is that we can come much closer to approximating the gross migration during the decade, even though, as Eldridge and Kim demonstrate by their detailed analysis, uncontrolled factors still remain that leave these refined estimates of gross migration still below their likely true values.

The third procedure – that for two censuses with State of birth tabulated by age – is similar in its strategy to the second, but the computations are much more detailed. The computations are carried out separately for each age cohort.

Comparisons with other methods of measuring migration. – Comparisons of estimates of net migration from place-of-birth statistics with those from the residual method have been made by Vamathevan and by Lee and his associates.[44] Eldridge and Kim have compared results from the present source with those

---

[43] Eldridge and Kim, op. cit., p. 32. Their table 3 is based on the U.N.'s "third procedure", but the principle is the same.

[44] S. Vamathevan, *Internal Migration in Ceylon: 1946-53*, Monograph No. 13, Ceylon Department of Census and Statistics, 1961, pp. 10-12; Lee et. al., op. cit., pp. 65-95.

from the question on residence at a fixed past date, which is described in the next subsection. In making such comparisons, one must bear in mind that these different sources and methods do not define migration in exactly the same way.

From comparisons with direct measures of fixed-period migration and from theoretical considerations, Eldridge and Kim concluded that the estimates of gross intercensal migration (migration streams, etc.) from state-of-birth data are seriously too low even when corrections are made for mortality during the intercensal period.[45] These indirect estimates are much less biased, however, for net migration including net interchanges between pairs of areas.

Residence at a Fixed Past Date

Fixed-period migration can be obtained by a question of the form, "Where did you live on December 31, 1965?" Often, however, the question is in several parts, with the parts specifying the levels of area detail required. For example, the format used in the 1970 Census of the United States was as follows:

19a. **Did he live in this house on April 1, 1965?** *If in college or Armed Forces in April 1965, report place of residence there.*

  ○ Born April 1965 or later ⎫
  ○ Yes, this house .... ⎬ *Skip to 20*
  ○ No, different house ⎭

b. **Where did he live on April 1, 1965?**

  (1) State, foreign country,
     U.S. possession, etc. _____

  (2) County _____

  (3) Inside the limits of a city, town, village, etc.?
     ○ Yes   ○ No

  (4) If "Yes," name of city,
     town, village, etc. _____

In many ways this is the best single item on population mobility. It counts the migrants over a definite period of time within the surviving population and provides gross as well as net migration statistics. These gross statistics do not, however, include all of the moves during the period or even all of the persons who have moved during the period. A wide range of useful measures can be derived from the absolute figures; and useful analysis of geographic patterns, time changes, differentials, etc., can be based on these measures. These considerations are discussed further in the subsection on "Uses and Limitations."

Despite the weak recognition of this kind of mobility question in international recommendations, it is being used in an increasing number of censuses and sample surveys. These differ somewhat with respect to geographic detail called for, length of mobility period, and character of move (permanent or temporary change of residence).

National Practices—Censuses.—For national censuses, table 21–9 summarizes the countries inquiring about mobility within a fixed period.

The United States census of 1940 was the first to include a

[45] Eldridge and Kim, op. cit., pp. 64–65.

Table 21–9. — National Censuses With Question About Fixed-Period Migration

Country	Date	Migration interval (years)	Sampling rate (percent)	Area detail on schedule
AMERICA:				
Canada[1]	1961	[2]5	20	Same or different house; locality; municipality; farm or nonfarm
United States	1940	5	100	Same or different house; locality; county; farm or nonfarm
	1950	1	20	Same or different house; county; farm or nonfarm
	1960	5	25	Same or different house; locality; county
	1970	5	[3]15	Same or different house; locality; county
Venezuela	1961[4]	1	100	Municipio
ASIA:				
Japan	1960	1	100	Same or different house; same or different locality in same prefecture; prefecture
Thailand[5]	1960	5	100	Province (Changwad)
EUROPE:				
France	1962	8+	[6]5	Commune; département
Greece[7]	1961	5+	[6]2	Municipality; commune
Hungary	1960	11	100	County (megye) and independent city
Portugal	1960	1	100	Different house; district; concelho
United Kingdom[8]				
England and Wales	1961	1	10	Full address
Scotland	1961	1	10	Full address

[1] Persons 15 years old and over in sample households (excluding group quarters).
[2] Intercensal period.
[3] State of previous residence was obtained for a 20-percent sample.
[4] If moved within last year, previous place of residence.
[5] Persons who migrated to province of present residence since date 5 years before census.
[6] Collected on 100-percent basis but tabulated on a sample basis.
[7] Excluded persons living in group quarters and visitors in private households.
[8] Excluded members of armed forces living in barracks or aboard ship.

Source: Schedules on file at Statistical Office of the United Nations and national census reports.

question of this type. The 1970 question was essentially the same as that of 1960.

National Practices—Surveys.—The inclusion of fixed-period migration questions in national sample surveys is also becoming more prevalent. So far, only those for Turkey and the United States include fixed-place migration questions on a regular basis, however.

United States.—Again, the United States seems to have been the first country to exploit the potentialities of a national survey for studying fixed-period migration. In the Current Population Survey of March 1945, near the end of World War II, there were questions about place of residence in December 1941, when the United States entered that conflict. After experimenting with a variety of migration intervals, the U.S. Bureau of the Census settled on a one-year interval for its survey of April 1948 and has included virtually identical questions through 1971. Beginning in 1972, the question has referred to residence at the last census date so that there is now a mobility interval of increasing length.

At the present time, the characteristics that are cross–classified with mobility status and for which mobility rates are shown comprise: race, sex, age, educational attainment, marital status, number of own children, relationship to head of household, employment status, major occupation group, and income in preceding year. Other characteristics have been cross–classified in occasional years, for example, date of first marriage, duration of marital status, size and type of family, full versus part–time employment, weeks unemployed and number of different jobs held in preceding year, employment status, occupation and industry at beginning of mobility period, number of weeks worked in preceding year, and tenure of home.

Most of the statistics are shown only for the Nation as a whole; but reduced detail is shown for:

    (a) 4 regions (including interregional streams)
    (b) type of present residence, i.e.
        Metropolitan
           In central cities
           Outside central cities
        Nonmetropolitan

Uses and Limitations. The question on place of usual residence at a fixed date is one of the most effective means of measuring migration. The measures discussed below illustrate some of the specific uses. Like other migration questions in censuses, it is not very well suited to a *de facto* census.

So-called gross migration from the fixed-period question is partly net because circular migrants and migrants who died during the migration period are not counted as migrants, multiple migrants are counted only once, and children born after the reference date are not included in the statistics. With regard to the last of these categories, the planned tabulations from the French census of 1962 call for assigning the 1954 residence of the mother to children born during the migration (intercensal) period, and the same convention was employed in a few Current Population Surveys in the United States. [46]

Choice of Mobility Period. — In choosing the reference date for the mobility question, we may find that considerations of usefulness and of accuracy are in conflict to some extent. From the standpoint of demographic analysis, the date of the previous census has many advantages, since the components of population change can then readily be studied for the intercensal period. If the intercensal period is 10 years rather than 5 years, however, errors of memory and lack of knowledge may tend to be excessive in reporting prior residence. Anthropologists have found that other events are remembered relatively well if they tied in with some event of historical significance, such as the outbreak of a war or the achievement of national independence; but the irregular time intervals thus defined do not lend themselves very well to time series or to demographic analysis in general.

The longer the migration interval, the more children will be omitted from the coverage of the question and the less will characteristics such as age and marital status recorded in the census correspond to those at the time of the move. On the other hand, a very short period, 1-year or less, may not yield enough migrants for detailed analysis of migration streams. Furthermore, one must consider the representativeness of the 1-year period or to, a lesser extent, any period shorter than the intercensal period. There was considerable

evidence that the April 1949 to April 1950 period used in the 1950 census of the United States was somewhat unrepresentative of both the intercensal decade and the postwar years with respect to rate of mobility and much more so with respect to net migration streams and to the selectivity of migration. [47] The U.S. Bureau of the Census, as well as most experts in the field, has decided that, on balance, a 5-year mobility period is optimal for a census; but this preference does not gainsay the great usefulness of a 1-year period in an annual sample survey.

Average Annual Movers. — Another problem arises in the computation of average annual numbers or rates. If the number of moves has been compiled from a population register over a period of years, it is quite appropriate to obtain such averages by dividing the number of moves by the number of years. When, however, the number of movers is defined by the number of persons whose residence at the beginning of a fixed period was different from that at the census or survey date, then division produces an estimate with a downward bias and the longer the interval the greater the bias.

The reasons for this bias are as follows:

"1. A larger proportion of movers will have died over the longer period than over the shorter period. Emigration from the country has the same effect on the statistics as death.

2. A mover has a greater opportunity to return to his original residence over a longer period than over a shorter period. Hence, the proportion of persons per year who will appear to be movers will be smaller for the longer period.

3. Since a mover is counted as such only once no matter how many times he moves, we should expect to find the proportion of movers in any one of five 1-year periods to be larger than one-fifth the proportion over a 5-year period-given a constant proportion of movers per year.

4. When children born after the base date are not covered by the mobility questions, fewer movers are counted for the longer period. Children born during the mobility interval are then excluded from the base population. The effect on the over-all rate also depends on the age-specific mobility rates. Young children do tend to be more mobile than the average for the population of all ages." [48]

Nonetheless, averages computed for periods of unequal length can be used with appropriate caution in the analysis. If the apparent average was greater for the longer period than for the shorter period, then we can infer that the true difference was in the same direction and at least as large.

This bias also affects annual estimates computed from the Intercensal Change in the Birth-Residence Index and estimates of net migration made by the survival rate method. In the case of the latter method, the only disturbing factor is mortality. Multiple moves are not a factor.

Quality of the Data and Statistics. — The question on residence at a fixed past date is subject to essentially the same types of errors of reporting and tabulation in censuses and surveys as are other demographic, social, and economic characteristics. In addition, the reporting of this item is affected

[46] France, Institut national de la statistique et des études économiques, *Recensement de la population de 1962, Programme de tableaux,* 3rd Part, "Migrations internes; migrations: lieu de naissance – lieu de résidence," etc. Internal migration; migration: place of birth – place of residence," etc., Document No. 727, April 26, 1963, p. 1; and Shryock, *Population Mobility Within the United States,* p. 24.

[47] Shryock, *Population Mobility Within the United States,* pp. 410 *et passim.*
[48] Ibid., p. 25.

by the types of errors that are peculiar to reporting past events and their placement in time.

Sample surveys.—Some indications of the level of sampling error from national sample surveys are available. (It is desirable, of course, that this kind of information be made available in all reports giving statistics on internal migration that were collected in sample surveys.)

In the United States, information on mobility status in the year 1949–50 was obtained from both the census and the CPS, and the distributions obtained from these two sources are compared in table 21–10.

Among persons in households interviewed in the Current Population Survey, the proportion failing to answer the question on residence one year ago is quite low—of the order of a fraction of one percent. These omissions are now supplied by allocations in the computer. Of households eligible for interview, however, about 5 percent are not interviewed at all in an average month. The members of these households are also nonrespondents on the migration questions, of course. Inflating the sample data to control totals by age, sex, etc., gives these nonrespondents the same characteristics as those persons in the specific age-sex group who reported, although actually they may have had a somewhat different distribution on such a characteristic as mobility. Thus, the published statistics do not show any "unknowns" on migration.

Censuses.—As table 21–9 indicated, several of the fixed-period migration questions in national censuses were asked on a sample basis.

The percent of nonresponse on fixed-period mobility in the last few censuses of the United States is as follows:

1940... 0.9
1950... 1.7
1960... 1.6
1970... 5.2

Population mobility was one of the items included in the post-enumeration survey of both the 1950 and 1960 censuses of the United States. The 1960 findings may be briefly summarized as follows:

"Analysis of the CES (Content Evaluation Study) data on mobility status indicates that, in general, the tendency is for the census to overestimate the more stable elements in the society and to underestimate the more mobile. The census showed fewer people living in a different State or abroad 5 years earlier than actually were living in a different State or abroad, and conversely, more people living in the same State, same county, and same house 5 years earlier than the resurvey

revealed to be the case. One measure of the extent of the discrepancy is the distribution of interstate migrants in the resurvey by their classification in the census enumeration. Of those classified as interstate migrants in the CES, 85.6 percent were classified the same in the census, and the remainder were classified as having lived in the same State 5 years earlier (13.5 percent) or abroad (0.9 percent)." [49]

A recurring difficulty in both censuses and sample surveys of the United States has been the biased reports of the urban-rural origin of movers. This difficulty arises partly from the rather complex urban-rural definition now in use and partly from a strong tendency for persons living outside a city but in its suburbs to give the city as their residence. This tendency leads to an overstatement of out-migration from urban areas and an understatement of that from rural areas. As a result, the United States had had to discontinue the direct measurement of rural-urban migration. In countries where the areas classified as urban are communes, etc., with relatively permanent boundaries, this problem may not be so acute.

Measures Used in Analysis.—Many of the measures used in analyzing fixed-period mobility are identical or similar to those used in the kinds of mobility data that were discussed previously. There are also similar problems regarding the choice of population base, annual averages, etc.

Mobility or migration status.—Among the simplest types of derived figures is the percentage distribution by mobility status. The following distribution is for the male population of Canada in the 1956–1961 period: [50]

Mobility status	Percent of total population
Total population 5 years old and over.........	100.0
Nonmovers...	54.9
Movers within Canada.................................	42.1
Within same municipality...........................	25.1
Between different municipalities....................	16.9
Within same province.............................	13.3
Between different provinces.....................	3.5
Contiguous provinces..........................	1.8
Noncontiguous provinces......................	1.6
Place of 1956 residence not stated................	0.2
Immigrants from abroad................................	3.1

For some purposes (e.g., comparisons of two states in the same country), it would be better to base the percent distribution on the total excluding the unknowns. In the case of Canada, the unknowns shown are only partial unknowns since the nonresponses had been partly allocated. The percentages for various types of mobility can be regarded as mobility rates per 100 of the resident population.

In a distribution for a particular geographic subdivision, some of the figures also represent in-migration rates; for example, the category "different county," in the distribution by mobility status for a county. Of necessity, such a distribution excludes out-migrants.

Propensity rates.—The "status" rates in the upper half of

Table 21–10.—**Percent Distribution of Persons 1 Year Old and Over in the United States by Mobility Status: 1950 Census and March 1950 Current Population Survey**

Mobility status	Census	CPS
Total reporting mobility status...	100.0	100.0
Same house as in 1950...................	82.6	80.9
Different house in the U.S. (movers)....	17.1	18.7
Intracounty movers......................	11.4	13.1
Intercounty migrants....................	5.7	5.6
Intrastate migrants..................	3.0	3.0
Interstate migrants..................	2.7	2.6
Abroad in 1949..........................	0.4	0.3

Source: Henry S. Shryock, Jr., *Population Mobility Within the United States*, p. 58.

[49] *U.S. Census of Population: 1960. Mobility for State and State Economic Areas*, PC(2)–2B, 1963, p. XIV.

[50] Canada, Dominion Bureau of Statistics, *1961 Census of Canada*, Series 4.1, *Population Sample*, Bulletin 4.1–9, *General Characteristics, Migrant and Non-Migrant Population*, 1965, table I–1.

Table 21–11.—Status and Propensity Migration Rates, by Type of Migration, for the United States, by Regions: 1935–40

[See text for explanation]

Type of rate and type of migration	North-east	North Central	South	West
STATUS RATES				
Population 5 years old and over reporting..............thous..	33,304	36,736	37,254	12,696
Intercounty (migrants)...........	8.0	12.5	14.3	24.8
Intrastate....................	4.8	7.9	8.9	11.2
Interstate....................	3.2	4.6	5.4	13.5
PROPENSITY RATES				
Estimated 1935 population...thous..	35,554	39,412	39,428	12,637
Intercounty....................	7.9	12.8	14.2	17.9
Intrastate....................	4.4	7.4	8.5	11.3
Interstate....................	3.4	5.5	5.8	6.6
Residue.......................	6.4	5.7	4.9	7.2

Source: Henry S. Shryock, Jr., *Population Mobility Within the United States*, table 6.2, p. 124, and 6.3, p. 126.

table 21–11 do not represent the propensity of the region's population to migrate, i.e., move across county lines, because they include in-migrants to the region but exclude out-migrants, who were there at the beginning of the migration period. In the case of fixed-period migration, the population at the beginning, rather than end, of the period more nearly represents the population "at risk." This population, usually, is available or can be estimated.

The choice of a population base for the rates is significant not so much for its effect on the size of the base as for its effect on the number of migrants. Persons who are excluded or included to approximate the population at the beginning of the period have a 100-percent migration rate, and their number must be deducted from or added to the number of migrants as well as the base. In table 21–11 we have an independent intercensal estimate of the population by regions in 1935. From the resident migrants in each region, we wish to remove the in-migrants from outside the region and restore the out-migrants from the region in order to include just those migrants who lived in the region in 1935. We may do so by subtracting the region's net migration, 1935 to 1940, from the total number of migrants enumerated in the region in 1935. Then the numerator will contain only persons who are also included in the base population shown in the lower portion of the table.

The procedure for deriving the "propensity" rates for the West, shown in the lower portion of table 21–11, is as follows:

Basic data—
(1) Resident migrants, 1940............................. 3,145
(2) Net interregional migrants, 1935–40.............. 887
(3) Resident interstate migrants, 1940................. 1,720
(4) Estimated population, 1935......................... 12,637
(5) Resident nonmigrants, 1940........................ 9,471
Calculations—
(6) Estimated migrants, 1935–40, who lived in the West in 1935 (1) minus (2)...................... 2,258
(7) Estimated interstate migrants, 1935–40 who lived in the West in 1935 (3) minus (2)........ 833
(8) Migration rate (6) ÷ (4)............................. 17.9
(9) Interstate migration rate (7) ÷ (4).................. 6.6

It will be noted in the above calculations that the 1935 population is not equal to the sum of nonmigrants and migrants $(12,637 - [9,471 + 2,258] = 908)$. This difference of 908,000 persons consists of deaths, emigrants, persons not reporting migration status, and net error in the data.

The number of intrastate migrants does not have to be adjusted at the region, division, or State level but would need to be adjusted for net migration if the propensity rate were being computed for a county or State economic area. The "at risk" migration rate of 17.9 for the West, 1935–40, is markedly lower than 24.8, the corresponding "status" rate. For 1949–50, a 1-year period, the difference was relatively slight—8.4 versus 8.8.

In- out- and net migration rates.—In-, out-, and net migration rates may be expressed by formulas analogous to those shown above for place-of-birth statistics. Again, it is possible to base the rate on the population at the beginning or end of the period or on the midperiod population.

From the 5-percent sample of the 1962 census of the population of France, we have the following statistics on migration between 1954 and 1962 for the Alsace region: [51]

Nonmigrants............................. 1,210,140
In-migrants............................. 57,840
Out-migrants........................... 52,940

The total population in 1962 was 1,318,070, and that in 1954 (with a slight correction for changes in the *de jure* definition), was 1,226,719. The average population may be estimated at 1,272,395. Corresponding rates are then:

Numerator	Rate		
	1962 base	1954 base	Average base
In 57,840.............................	4.4	4.7	4.5
Out 52,940.............................	4.0	4.3	4.2
Net 57,840 − 52,940 = 4,900.......	+0.4	+0.4	+0.4
Turnover 57,840 + 52,940 = 110,780	8.4	9.0	8.7

Rates involving migration streams.—From the 1-percent sample of the 1960 census of Japan, we have the following information on usual residence one year ago (1959): [52]

Population of Hokkaido, 1960.................. 4,961,100
Population of Aomori-ken, 1960................ 1,383,900
Migrants, 1959–60, Hokkaido to Aomori.... 3,000
Migrants, 1959–60, Aomori to Hokkaido.... 16,600

The last two lines of data represent migration streams. From these we can compute partial migration rates. A **partial migration rate** is defined as the number of migrants to an area from a particular origin, or from an area to a particular destination, per 1,000 (or per 100, etc.) of the population at either origin or destination. [53] The **partial out-migration rate** can be expressed as,

$$m_{ij}^o = \frac{M_{ij}}{P_i} \times 1,000 \qquad (26)$$

---

[51] France, Institut national de la statistique et des études économiques, *Recensement générale de la population de 1962, Résultats du sondage au 1/20ᵉ. Population—Ménages—Logements—Immeubles, Récapitulation pour la France entière* (General census of population, 1962; results from the 5-percent sample; Population, households, dwellings, other living quarters; Summary for France as a whole), 1964, p. 25.
[52] Japan, Bureau of Statistics, *1960 Population Census of Japan*, Vol. 2, *One Percent Sample Tabulation*, Part 2, *Migration*, 1962, table 1.
[53] See Shryock, *Population Mobility Within the United States*, pp. 33–34.

where M_{ij} is the stream from area i to area j. Taking P_1 as the population of Hokkaido and P_2 as that of Aomori, we have, for the stream from the former to the latter, the partial out-migration rate

$$m_{12}^0 = \frac{3,000}{4,961,100} \times 1,000 = 0.6.$$

The **partial in-migration rate** can likewise be expressed as,

$$m_{ji}^I = \frac{M_{ji}}{P_i} \times 1,000 \qquad (27)$$

where M_{ji} is the stream to area i from area j. For in-migrants to Hokkaido from Aomori,

$$m_{21}^I = \frac{M_{21}}{P_1} = \frac{16,600}{4,961,100} \times 1,000 = 3.3$$

As we have said, the population at risk for an in-migration rate is actually the population at origin, however. Therefore, the partial in-migration rate to Hokkaido from Aomori could well be expressed as the partial out-migration rate from Aomori to Hokkaido, or $\frac{16,600}{1,383,900} \times 1,000 = 12.0$.

The **gross rate of population interchange** may be defined as,

$$GRI_{i \leftrightarrow j} = \frac{M_{ij} + M_{ji}}{P_i + P_j} \qquad (28)$$

which, in our illustration is

$$\frac{3,000 + 16,600}{4,961,100 + 1,383,900} \times 1,000 = 3.1.$$

The net rate of population interchange is then,

$$NRI_{\leftrightarrow j} = \frac{M_{ij} - M_{ji}}{P_i + P_j} \qquad (29)$$

Viewed from the standpoint of Hokkaido, this is

$$\frac{+13,600}{6,345,000} \times 1,000 = +2.1$$

whereas from the standpoint of Aomori, it would be -2.1. For parallelism with the usual complete and partial rates, (28) and (29) might also be computed with only P_1 in the base. The series of partial rates relative to all other areas in the country would then be additive to the turnover rate and net migration rate, respectively, for Area 1 (Hokkaido).

Effectiveness of migration.—The ratio of net migration to turnover may be considered a measure of the "effectiveness" of internal migration. The higher the ratios for a set of areas the fewer the moves that are required to effect a given amount of population redistribution among them. There are patently other important aspects of the effectiveness of migration that are not comprehended in this measure.[54]

This ratio ranges from 0 to 100. We may also measure the effectiveness of a stream and its counterstream in this fashion.

Thus, the effectiveness of the migration between Hokkaido and Aomori, in both directions, would be $\frac{13,600}{19,600} \times 100 = 69.4$. This is a relatively high ratio. Often, the counterstream is as large as the stream, so there is little net migration and a low ratio of effectiveness.

We may also be interested in a summary measure of the effectiveness of migration for all the states, provinces, or other primary geographic divisions of a country. Here, we define

$$\text{Effectiveness Index} = \frac{\Sigma \text{ Net in-migrants}}{\text{Total interarea migrants}} \times 100 \qquad (30)$$

For example, this index for the United States in the periods 1949–50 and 1935–40, respectively, was 8 and 21. Another method of making a summary comparison would be in terms of the median State ratio for each period.[55]

Migration Preference Index.—This measure was first suggested by Bachi.[56] This index is defined as the ratio (times a constant) of the actual to the expected number of migrants in a stream when the expected number is directly proportionate to both the population at origin and the population at destination. "This measure indicates, then, whether streams are greater or smaller than would be expected from considerations of population size alone. It does not include any assumption about the expected effect of distance, but we can compare measures for different streams in the light of our knowledge of contiguity and of mileage and, in fact, in the light of knowledge that we may have about the relative attractiveness of the areas."[57]

There are several ways in which this index can be computed. The most useful one, however, relates to interarea migrants only, i.e., the Preference Index for a State relates to interstate migration only. To compute the index, we assume that interstate migrants are distributed proportionately to the population at origin and the population at destination. Take the national rate of interarea (interprovincial, interregional, etc.) migration. Assume that it is uniformly the out-migration rate for all areas of the given class. Compute the expected total number of out-migrants from a given area to all destinations. Distribute these among the other areas in proportion to their population, to obtain the expected number of migrants in each stream. "The Preference Index is then given by

$$P.I. = \frac{M_{OD}(\Sigma P_i - P_O)}{m P_O P_D} \times 100. \qquad (31)$$

where M_{OD} = actual number of migrants from O to D
$\quad P_O$ = population at origin, O
$\quad P_D$ = population at destination, D
$\quad \Sigma P_i$ = national population
$\quad m$ = proportion of interarea migrants in the national population

This index may vary from 0 to ∞. The total number of expected in-migrants can be obtained by summing the expected numbers

[54] Shryock, *Population Mobility Within the United States*, p. 285.

[55] For further discussion of this measure, see ibid., pp. 285–294.
[56] Roberto Bachi, "Statistical Analysis of Geographic Series" in *Bulletin de l'Institut international de statistique*, 36(2):234–235. (Proceedings of the 30th meeting of the Institute, Stockholm, 1957.) See also Shryock, *Population Mobility Within the United States*, pp. 267–284.
[57] Shryock, *Population Mobility Within the United States*, p. 267.

in all the in-migration streams, but the indices themselves are not additive." [58]

Table 21-12 gives the Preference Indexes for the four regions of Thailand for the migration period, 1955 to 1960; and table 21-13 shows the underlying computations. The population figures used were those at the census date. Had they been available, the 1955 population figures would have been used. The table shows, for example, that the stream from the North to the Central region is most preferred and that from the Northeast to the South is least preferred. Bangkok, the capital, is located in the Central region.

Distance of move. — A very different way of reducing the data is to compute the distance (in kilometers or miles) between the points of origin and destination. Actually instead of points (addresses), it would be practical to start with statistics on origin and destination grouped in tabulation areas, and estimate the distance between the centers of the areas of origin and destination. (Such a procedure was used in the 1961 Census of Great Britain.)

Duration of Residence and Last Prior Residence

The typical question on duration of residence has been of the form, "How long has he been living in this area?" (i.e., the area of usual residence). The logical companion question and that recommended by the United Nations is one concerning the name of the previous area of residence; but in national censuses most countries asking the question on duration of residence have been content to ask only for place of birth. There are two ways of defining a migrant from such data:

(1) a person who had moved into the area at any time in the past and was still resident there. This category would include primary, secondary, and return migrants in Eldridge's terminology. By this definition, the number of migrants would exceed that of lifetime migrants.

(2) a person who had moved into the area since a given date — 1 year ago, 2 years ago, etc. Again this might be a person who had migrated only once since his birth, a secondary (or progressive) migrant, or a return migrant. The areas referred to above are those areas, such as municipalities, counties, etc., that are "migration-defining" for the particular country.

National Practices. — Table 21-14 summarizes the inclusion of the item on duration of residence in national censuses taken around 1960. This item was especially popular in the Americas. A number of countries (Guatemala, Panama, Argentina, Brazil, Chile, Colombia, Ecuador, Paraguay, Uruguay, Belgium, and Yugoslavia) also asked for previous area of residence.

Table 21-12. — Preference Indexes for Interregional Migration Streams, Thailand: 1955-60

Region of destination	Total in-migrants	Region of origin			
		Central	Northeast	North	South
Total out-migrants....	(X)	112	117	78	61
Central.................	152	(X)	58	182	139
Northeast..............	40	150	(X)	109	50
North..................	127	158	24	(X)	20
South..................	76	148	12	13	(X)

X Not applicable.

Source: Based on data from, Thailand, Central Statistical Office, *Thailand Population Census, 1960, Whole Kingdom, 1962,* table 6, p. 16.

---

[58] Ibid., p. 268.

Table 21-13. — Illustrative Example of Computation of Preference Index for Interregional Migration, Thailand: 1955-60

[Worksheet for table 21-12]

Step 1 Proportion migrating between regions:

Population 5 years old and over reporting.....	22,015,400
Interregional migrants......................	214,273
Proportion.................................	.0097329

Step 2 Regional distribution of population, and expected out-migrants by region

1960	Population 5 years old and over reporting (1)	Population in other regions (2)	Proportion of outside population in each other region (3)	Expected number of out-migrants (col. 1 x .0097329) (4)
Central........	6,979,798	15,035,602	NE .499359 N .316244 S .184397	67,934
Northeast......	7,508,165	14,507,235	C .481125 N .327762 S .191113	73,076
North..........	4,754,919	17,260,481	C .404380 NE .434992 S .160628	46,279
South..........	2,772,518	19,242,882	C .362721 NE .390179 N .247100	26,984

Step 3 Distribution of out-migrants by destination

Residence in 1960 / 1955	Expected (col. 4 x col. 3 of step 2)			
	Central	Northeast	North	South
Central.........	-	33,923	21,484	12,527
Northeast.......	35,159	-	23,952	13,966
North..........	18,714	20,131	-	7,434
South..........	9,788	10,529	6,668	-

Residence in 1960 / 1955	Actual			
	Central	Northeast	North	South
Central.........	-	19,648	39,170	17,379
Northeast.......	52,605	-	26,002	6,998
North..........	29,507	4,896	-	1,482
South..........	14,487	1,252	847	-

Step 4 Preference Indexes of migration streams. Based on step 3; (actual ÷ expected) x 100

Destination / Origin	Central	Northeast	North	South
Central.........	-	58	182	139
Northeast.......	150	-	109	50
North..........	158	24	-	20
South..........	148	12	13	-

Step 5 Preference Indexes of total in-migrants and out-migrants (Totals derived from step 3)

Region	Total in-migrants			Total out-migrants		
	Expected	Actual	Preference index	Expected	Actual	Preference index
Central........	63,661	96,599	152	67,934	76,197	112
Northeast......	64,582	25,796	40	73,076	85,605	117
North..........	52,103	66,019	127	46,279	35,885	78
South..........	33,927	25,859	76	26,984	16,586	61

In the United States, a question on year moved to present address was included in the censuses of 1960 and 1970. The 1970 format is as follows:

18.	When did this person move into this house (or apartment)?
	Fill circle for date of last move.
	○ 1969 or 70 ○ 1965 or 66 ○ 1949 or earlier
	○ 1968 ○ 1960 to 64 ○ Always lived in
	○ 1967 ■ 1950 to 59 this house or apartment

This question had been asked earlier in the Current Population Survey, in the supplements of 1952 and 1959.[59] Since, however, the year of the move into the county of residence has not been determined, this item has not yielded any migration statistics for the United States.

Uses and Limitations. — If only the last previous place of residence has been obtained or tabulated, the resulting statistics, like those from the item on place of birth, have an indefinite time reference. On the other hand, these statistics describe direct moves whereas the place-of-birth statistics may conceal intervening moves.

The chief virtue of the question on duration of residence (or on year of last move) is that it gives the distribution of lifetime movers by recency of latest move. If the previous place of residence has also been ascertained, then the time is also fixed for streams of migration. Such statistics describe the in-migrants now living in an area, but they do not produce a very useful time series. Since only the latest move is recorded, the number of moves in the earlier years will be seriously understated because of multiple moves and deaths. Origin-destination tabulations for the most recent migration interval will yield data approximating those from the fixed-period item for the same interval; the shorter the interval, the closer the approximation. Furthermore, percentage distributions by duration of residence, enable us to distinguish those parts of the country to which migration has been relatively recent. Thus, U.N. Manual VI points out that 29 percent of lifetime migrants to Amazonas State in Peru had arrived in the 12 months preceding the 1960 census, whereas the national average for interstate lifetime migrants was only 16 percent.[60]

A number of countries have cross-tabulated duration of residence by place of birth. The United States has tabulated year moved into present home by mobility status in the 1955–60 period.[61]

Table 21–15 shows for Rôndonia State, Brazil, in 1960, the distribution of in-migrants by duration of residence and state of last previous residence. From such statistics on migration streams, in-, out-, and net migration can be computed, as well as the corresponding rates using the current population as the base. As with statistics on place of birth, it is not possible to compute "propensity" rates (using as a base the origin population at a fixed past date), however.

Quality of the Statistics. — The statistics on duration of residence and area of previous residence are subject to the usual types of reporting errors that have been discussed for other migration questions. In addition, there is the likelihood of preferences for round numbers in the reporting of duration in

Table 21–14. — National Censuses With Question About Duration of Residence: 1955 to 1964

Country	Date	Area of reference
AFRICA:		
Madagascar	1956	City
Morocco	1960	Habitual residence
Togo	1960	Commune
AMERICA, NORTH:		
Antigua	1960	Parish, town, island
Bahama	1963	Town
British Honduras	1960	District
Cayman Islands	1960	Parish, town, island
Costa Rica	1963	Cantón
Guatemala	1964	Municipio
Jamaica	1960	Parish
Montserrat	1960	Parish, town, island
British Virgin Islands	1960	Parish, town, island
St. Christopher	1960	Parish, town, island
Nevis	1960	Parish, town, island
Aguilla, Turks, and Caicos Is.	1960	Parish, town, island
Panama	1960	Place
Trinidad and Tobago	1960	Present residence
AMERICA, SOUTH:		
Argentina	1960	Habitual residence
Brazil	1960	Municipality
Colombia	1964	Municipio
Ecuador	1962	Locality
Paraguay	1962	District
Peru	1961	Province
Uruguay	1963	Locality
Venezuela	1961	Municipio
ASIA:		
India	1961	District
Israel	1961	Locality
EUROPE:		
Belgium	1961	Commune
Yugoslavia	1961	Locality
United Kingdom		
England and Wales	1961	Full address
Scotland	1961	Full address

Source: National census questionnaires on file in Statistical Office of United Nations and national census reports.

years. For example, Zachariah reports that, in the 1961 Census of India, in-migrants to Greater Bombay showed considerable heaping on 10 and 15 years.[62] Nonresponse, however, was slight—nine per thousand total migrants.

Measures Used in Analysis. — Not many additional measures have been proposed for this subject. Table 21–15 gives the percentage distribution by duration of residence and the median years. A measure of dispersion, like the interquartile range, could also be computed. There is a high positive correlation between duration of residence and age, and tables giving a cross-classification by age are a prerequisite for adequate analysis. When the same tabulation is available for successive censuses, there are opportunities for more complex methods and measures. The U.N. manual carries a dummy table illustrating the computation for an intercensal period.[63]

OTHER ITEMS FROM SURVEYS

Almost all of the additional mobility questions that have been asked have been included in surveys rather than censuses. The most general type is the mobility or migration history—a roster of previous usual residences with the dates of moves.

---

[59] U.S. Bureau of the Census, *Current Population Reports*, Series P–20, Nos. 47 and 104.

[60] United Nations, *Methods of Measuring Internal Migration.*

[61] *U.S. Census of Population: 1960, Subject Reports, Mobility for States and State Economic Areas*, PC(2)–2B, 1963, table 4.

[62] United Nations, *Methods of Measuring Internal Migration.*

[63] United Nations, *Methods of Measuring Internal Migration,* table B.15.

Table 21-15. — Percent Distribution by Duration of Residence of In-migrants to Rôndonia State, Brazil, by State of Last Previous Residence: 1960

[Percent and median not shown where base is less than 100]

State of last residence	All in-migrants	Reporting duration of residence									Median duration
		Number	Percent distribution by duration of residence								
			Less than 1 year	1 year	2 years	3 years	4 years	5 years	6 to 10 years	11 years and over	
Total......................	31,665	31,605	10.5	5.9	9.0	7.7	4.7	7.7	23.4	31.2	7.0
Acre........................	2,588	2,583	12.0	6.9	9.0	7.6	5.4	7.9	23.8	27.4	8.5
Amazonas...................	10,850	10,833	12.2	8.3	9.1	9.0	5.1	8.1	23.5	24.8	5.8
Roraima....................	218	217	4.6	2.8	6.5	11.1	2.8	11.5	49.8	11.1	7.1
Pará.......................	3,776	3,764	8.3	3.6	6.8	6.2	3.4	9.1	31.0	31.6	8.0
Amapá......................	92	92	-	-	-	-	-	-	-	-	-
Maranhão...................	1,157	1,155	17.0	4.8	8.7	6.5	5.1	7.8	17.1	33.0	6.0
Piauí......................	437	436	5.0	2.8	6.7	11.5	5.7	7.1	28.0	33.3	8.0
Ceará......................	6,198	6,189	3.8	5.5	13.8	9.3	5.9	9.4	24.1	28.3	6.5
Rio Grande do Norte........	859	859	0.9	1.6	7.9	4.9	6.6	4.0	26.9	47.1	10.5
Paraíba....................	714	714	1.8	2.7	9.9	7.1	1.5	4.8	25.1	47.1	10.4
Pernambuco.................	394	393	3.1	3.8	6.1	6.1	2.8	3.6	24.4	50.1	11.0
Alagoas....................	112	112	1.8	4.5	6.2	2.7	1.8	8.9	21.4	52.7	+11
Fernando de Noronha........	7	7	-	-	-	-	-	-	-	-	-
Sergipe....................	95	95	-	-	-	-	-	-	-	-	-
Bahia......................	218	218	8.7	8.3	2.8	2.8	4.6	8.7	16.1	48.2	10.4
Minas Gerais...............	136	136	66.9	5.1	6.6	1.5	-	0.7	10.3	8.8	0.7
Serra dos Aimorés..........	3	3	-	-	-	-	-	-	-	-	-
Espírito Santo.............	20	20	-	-	-	-	-	-	-	-	-
Rio de Janeiro.............	62	62	-	-	-	-	-	-	-	-	-
Guanabara..................	173	173	32.4	12.7	9.8	3.5	-	3.5	17.3	20.8	2.5
São Paulo..................	350	350	82.9	2.9	3.7	0.9	1.1	1.7	4.3	2.6	0.6
Paraná.....................	53	53	-	-	-	-	-	-	-	-	-
Santa Catarina.............	23	23	-	-	-	-	-	-	-	-	-
Rio Grande do Sul..........	25	25	-	-	-	-	-	-	-	-	-
Mato Grosso................	2,963	2,951	9.4	3.1	4.2	4.1	2.4	4.1	15.6	57.1	+11
Goiás......................	100	100	39.0	2.0	10.0	8.0	4.0	2.0	16.0	19.0	2.9
Distrito Federal...........	42	42	-	-	-	-	-	-	-	-	-

- Represents zero or rounds to zero.
Source: Adapted from, Brazil, Serviço nacional de recenseamento, *Censo demográfico de 1960*, Série regional, Vol. I, Book 1, Part 1, Rondônia – Roraima – Amapá. 1967, pp. 38-39.

Such a history may be paralleled by a similar listing of all jobs for persons of working age. From it may be summarized the number of lifetime moves or the number of moves in a given period; but such information is sometimes obtained by a simple, direct question. Likewise, instead of determining occupation, employment status, marital status, etc., at a fixed-past date from a more or less complete history, a direct question may be employed referring to the past date that is central to the study.

Other topics include temporary versus permanent migration; number of household members who moved together; month of migration; reasons for moving; migration plans or intentions; and the cost of moving.

Migration Histories

Migration histories are not limited to national sample surveys but may come from surveys of specific areas. Since a relatively small proportion of the population may be casual drifters who change residence very frequently, the surveys do not usually attempt to record all former residences (or all moves) but only the last "k" residences and perhaps also the area of residence at birth. Furthermore, the exact address is usually not recorded but only the area of residence, and there may be other restrictions on the information recorded. Thus, for example, the May

1958 supplement to the Current Population Survey recorded three previous residences; and, if these did not account for all previous residences, the area of birth was also requested. A minimum stay of 1 year was required for an area (other than that of birth) to be recorded in the list of previous residences. Moves within a county were not recorded unless they involved a change in type of residence, e.g., from farm to nonfarm. Furthermore, migration connected with service in the armed forces was not recorded. [64]

This 1958 survey was sponsored by the U.S. Public Health Service. Their most definitive report on it from the standpoint of migration analysis is: Karl E. Taeuber, Leonard Chiazze, Jr., and William Haenszel, *Migration in the United States: An Analysis of Residence Histories*, Public Health Monograph, No. 77, U.S. Department of Health, Education, and Welfare, 1968.

The retrospective migration data when arrayed by year of move and age at time of move can be analyzed for birth cohorts. Thus, in analyzing the 1958 data from the Current Population Survey, K. Taeuber assigns each move to a specific cohort

---

[64] U.S. Bureau of the Census, *Current Population Reports*, Series P-23, No. 25, "Lifetime Migration Histories of the American People," March 8, 1968, p. 1.

Table 21-16. – Cohort Age Periods for Analysis of Statistics From U.S. Current Population Survey of May 1958

Age in 1958 (years)	Age at moving (years)				
	0-17	18-24	25-34	35-44	45 and over
18-24..........	(1933-58)	(1951-58)	(X)	(X)	(X)
25-34..........	(1923-51)	(1941-57)	(1948-58)	(X)	(X)
35-44..........	(1913-41)	(1931-47)	(1938-57)	(1948-58)	(X)
45-54..........	(1903-31)	(1921-37)	(1928-47)	(1938-57)	(1948-58)
55-64..........	(1893-21)	(1911-27)	(1918-37)	(1928-47)	(1938-58)
65 and over....	(To 1893)	(To 1917)	(To 1927)	(To 1937)	(To 1958)

X Not applicable.

Source: Karl E. Taeuber, "Cohort Migration," *Demography*, 3(2):421, 1966, table 1. © 1967 by the Population Association of America.

and to the age group at which the move occurred.[65] For each of the cells in table 21–16 he tabulates size of place of residence at origin and destination, and additional available characteristics can be introduced as controls. "One way to handle these data is to percentage each row across and regard each table as a stochastic transition matrix. A variety of methods for handling social mobility data as transition matrices has been elaborated recently."[66]

Status at Prior Date

To obtain a person's current characteristics, the demographer may make use of questions that are a regular part of a general purpose survey. To obtain the characteristics at a past date, extra questions must be added. Fortunately, this action is being taken in a growing number of surveys. Classifications according to status prior to migration form the logical bases for measuring differential propensities to migrate. This kind of information also indicates important temporal sequences and answers such questions as, "Did the unemployment of a given worker precede or follow his move?" These temporal sequences, in turn, yield some insights into causal relationships.

The greatest defect of these data on past status is the additional response error that tends to be introduced as more distant dates are referred to. There is some evidence from recent tests in the United States, for example, that changes in status as measured by retrospective questions are considerably underestimated. Comparisons of current survey reports with reports of matched persons in a decennial census 6 years earlier indicated that the number of shifts in occupation was considerably understated. As a corollary, the retrospective description of jobs 6 years earlier must have been subject to a good deal of response error. Presumably, errors of recall would be fewer for a 1-year period, however.

Reasons for Moving

Questions on reasons for moving are among the more popular items in recent sample surveys on internal migration. Some of the sample surveys containing a question or questions on this topic have already been mentioned. These questions represent an attempt to determine motivation by asking migrants why they moved. This approach is quite different from trying to tease inferences on the causes of migration from data on migration differentials or on the comparative characteristics of sending and receiving areas. In the studies ex-

amined, there were no attempts to provide a "control group" of nonmigrants by seeking to measure the prevalence among them of the conditions cited by migrants as reasons for moving. Thus, we cannot say, for example, whether unsatisfactory housing conditions are more prevalent among migrants in a given period than among those who did not migrate. On the other hand, such studies probably do measure the subjective importance of the condition cited as a reason for leaving.

There has been little standardization of categories or reasons among the various surveys that have investigated this topic. Although there has been some repetition of categories used in earlier surveys, the topic seems to be essentially in its exploratory stage with meaningful categories and classifications being developed partly by trial and error. The respondent is often allowed to give more than one reason so that the sum of reasons given may exceed the number of persons reporting. There is frequently an attempt, either in the questions themselves or in the tabular classification of the replies, to distinguish job-related from other (personal or social) reasons. "Push-pull" theories of migration are seldom tested explicitly by asking the respondent to compare his attitudes towards his areas of origin and destination. The main problems of measurement for this topic seem to be (1) choice of meaningful universes in the coverage of the survey and of subuniverses in the tabulations; (2) choice of a reasonable number of predesignated reasons that are mutually exclusive; and (3) choice of analytically relevant classifications of reasons. Write-in entries for other reasons provide a chance to correct during the office coding for poor choices of categories in (2).

The distribution of reasons obviously differs considerably between short-distance movers and migrants, with job-related reasons being relatively more frequent among the latter. In the March 1963 Current Population Survey of the United States, none of the reasons shown was given exclusively by one type of mover, however, as is shown in table 21–17.

In analyzing the 1946 data from the Current Population Survey, Shryock first divided the migrants into those "who made the basic decisions either for themselves alone or for their families" and "those whose migration was merely derivative from a decision made by the head of the family."[67] This dichotomization was made on the basis of the reason itself. Thus, those who migrated to "join head of family" or "moved with head" were classified as "secondary migrants" and those giving other reasons were classed as "primary migrants." The percentage of migrants giving a primary reason was as follows:[68]

Age	*Male*	*Female*
Under 14 years................	4.6	3.3
14 to 24 years................	71.6	36.2
25 to 44 years................	98.3	23.2
45 years and over............	95.5	43.0

Miscellaneous Topics

Direct questions on the number of moves were asked in the U.S. Current Population Survey of October 1946 (number of moves as a civilian in the preceding 14 months, i.e., since the end of World War II) and in the Special Demographic Survey of 1966 in the Republic of Korea. Table 21–18 is extracted from statistics in the report of the latter.

In the investigation of internal migration carried out by the National Sample Survey of India during the 1950's, migrations

[65] Karl E. Taeuber, "Cohort Migration," *Demography*, 3(2):420–421, 1966.
[66] Ibid., p. 421.

[67] Shryock, *Population Mobility Within the United States*, p. 404.
[68] Ibid., p. 405.

Table 21–17.—Male Movers 18 to 64 Years Old, by Reason for Move and Type of Mobility, for the United States: March 1963

[Numbers in thousands]

Reason for moving	Intracounty movers				Migrants			
	All reasons	One reason only	Two or more reasons	Primary reason[1]	All reasons	One reason only	Two or more reasons	Primary reason[1]
Total persons.....................	6,292	5,754	538	6,292	3,519	2,974	545	3,519
All reasons......................	6,857	5,754	1,103	6,292	4,101	2,974	1,127	3,519
Related to job......................	794	488	306	780	2,374	1,838	536	2,287
To take a job...................	188	126	62	190	964	772	192	966
To look for work...............	65	42	23	66	394	269	125	389
Job transfer...................	29	25	4	28	297	254	43	268
Commuting and Armed Forces...........	512	295	217	496	719	543	176	664
Easier commuting............	459	264	195	(NA)	272	174	98	(NA)
Enter or leave Armed Forces..........	53	31	22	(NA)	447	369	78	(NA)
Not related to job..................	6,044	5,247	797	5,512	1,709	1,118	591	1,232
Housing......................	4,127	3,704	423	3,895	461	324	137	362
Better housing...............	3,783	3,398	385	(NA)	435	304	131	(NA)
Forced move..................	344	306	38	(NA)	26	20	6	(NA)
Family status..................	1,304	1,081	223	1,143	664	419	245	480
Change in marital status...........	751	659	92	(NA)	165	133	32	(NA)
Join or move with family...........	553	422	131	(NA)	499	286	213	(NA)
Other.......................	614	462	152	485	583	375	208	398
Health......................	77	47	30	(NA)	116	71	45	(NA)
All other reasons...................	537	415	122	(NA)	467	304	163	(NA)
Not reported.....................	19	19	(NA)	(NA)	18	18	(NA)	(NA)
PERCENT DISTRIBUTION								
All reasons[2].....................	100.0	100.0	100.0	100.0	100.0	100.0	100.0	100.0
Related to job......................	11.6	8.5	27.7	12.4	58.1	62.2	47.6	65.0
To take a job...................	2.7	2.2	5.6	3.0	23.6	26.1	17.0	27.3
To look for work...............	1.0	0.7	2.1	1.0	9.6	9.1	11.1	11.1
Job transfer...................	0.4	0.4	0.4	0.4	7.3	8.6	3.8	7.6
Commuting and Armed Forces...........	7.5	5.1	19.7	7.9	17.6	18.4	15.6	18.9
Easier commuting............	6.7	4.6	17.7	(NA)	6.7	5.9	8.7	(NA)
Enter or leave Armed Forces..........	0.8	0.5	2.0	(NA)	10.9	12.5	6.9	(NA)
Not related to job..................	88.1	91.5	72.3	87.6	41.9	37.8	52.4	35.0
Housing......................	60.4	64.6	38.3	61.9	11.3	11.0	12.2	10.3
Better housing...............	55.3	59.3	34.9	(NA)	10.7	10.3	11.6	(NA)
Forced move..................	5.0	5.3	3.4	(NA)	0.6	0.7	0.5	(NA)
Family status..................	19.1	18.8	20.2	18.2	16.3	14.2	21.7	13.6
Change in marital status...........	11.0	11.5	8.3	(NA)	4.0	4.5	2.8	(NA)
Join or move with family...........	8.1	7.4	11.9	(NA)	12.2	9.7	18.9	(NA)
Other.......................	9.0	8.1	13.8	7.7	14.3	12.7	18.5	11.3
Health......................	1.1	0.8	2.7	(NA)	2.8	2.4	4.0	(NA)
All other reasons...................	7.9	7.3	11.1	(NA)	11.5	10.3	14.5	(NA)

NA Not available.

[1] Persons reporting more than one reason were assigned a single reason on the basis of the order in which reasons appear in the stub. At the same time the detail was consolidated into the subtotals indicated.

[2] Based on reported cases.

Source: U.S. Bureau of the Census, "Reasons for Moving: March 1962 to March 1963," *Current Population Reports*, Series P–20, No. 154, August 22, 1966, table E.

were classified as permanent or temporary. When the migrant reported that he was more or less permanently settled in the area to which he migrated, the case was treated as one of permanent migration. On the other hand, a sojourn might continue for years without permanent settlement. In such cases, the migration was considered to be of temporary nature.[69] Thus, this distinction was not made strictly on the basis of length of stay.

Among the miscellaneous topics that have been included in surveys are questions on expectations, intentions, or desires with respect to moving. This topic is even more attitudinal in nature than that of motivations of past moves. Validation of a question on intentions or expectations may be carried out by checking at the end of the reference period to determine whether the respondent has indeed left his original address. In the case of migration (as opposed to merely moving), it would also be necessary to ascertain whither the person had moved (to establish whether he had also left the county, commune, etc.). It may be fairly expensive to secure this additional information from other members of the household or neighbors, change-of-address files maintained at the post office, etc.

FACTORS IMPORTANT IN ANALYSIS

Geographic Areas

In most countries, short-distance mobility and migration rates vary considerably from one part of the country to another,

[69] India, Directorate of National Sample Survey, *The National Sample Survey, Ninth, Eleventh, Twelfth, and Thirteenth Rounds, May 1955–May 1958*.

Table 21–18. — Percent Distribution by Number of Moves in the Period 1961–66 of Urban Male Migrants, by Age, for the Republic of Korea

Age	All movers	1	2	3	4	5	6 or more	Unknown
Total...........................	100.0	50.2	23.1	16.2	4.7	3.5	1.1	1.2
5 to 9 years......................	100.0	46.4	25.6	16.0	4.1	6.1	1.0	0.7
10 to 14 years....................	100.0	46.3	23.1	19.4	4.5	5.6	0.7	0.4
15 to 19 years....................	100.0	58.3	19.9	10.7	6.6	1.8	0.7	1.8
20 to 24 years....................	100.0	49.1	25.5	12.7	4.8	1.2	2.4	4.2
25 to 29 years....................	100.0	51.9	22.9	18.4	3.8	1.9	0.4	0.8
30 to 34 years....................	100.0	54.1	23.7	14.6	3.2	2.4	0.8	1.2
35 to 39 years....................	100.0	44.7	22.8	17.9	7.3	3.3	4.1	-
40 to 44 years....................	100.0	44.4	23.1	21.3	3.7	6.5	-	0.9
45 to 49 years....................	100.0	40.9	22.7	21.2	9.1	4.5	1.5	-
50 to 54 years....................	100.0	47.7	15.9	25.0	4.5	-	2.3	4.5
55 to 59 years....................	100.0	66.7	17.8	13.3	2.2	-	-	-
60 to 64 years....................	100.0	52.4	28.6	4.8	4.8	4.8	4.8	-
65 to 69 years....................	100.0	44.4	22.2	11.1	-	11.1	-	11.1
70 years and over.................	100.0	44.4	33.3	22.2	-	-	-	-

— Represents zero or rounds to zero.

Source: E. H. Choe and J. S. Park, *Some Findings from the Special Demographic Survey*, Seoul (Korea), Seoul National University, and Bureau of Statistics, April 1969, p. 302.

reflecting such factors as differential age-sex structure, the rate of natural increase (which may be so high as to induce pressure of population on resources), economic opportunities, and climatic and other amenities.

Because of differences in the average size and arrangement of the migration–defining areas, overall migration rates are never comparable among countries. Moreover, the same applies to comparisons among the states, provinces, etc., of a single country. The effects of size of area, shape of area, spatial distribution of population within the area, etc., upon various types of migration rates has been described by Everett S. Lee and others.[70]

Type-of-Residence Areas

One reason that different geographic areas have different migration rates is their differences with respect to urban-rural or metropolitan-nonmetropolitan character, composition by size of place, etc. Differentials with respect to type-of-residence are often quite large. This generalization holds particularly true of net migration rates, since the most rural areas are usually the ones of heaviest migratory loss.

Age and Sex

"Mobility rates by age tend to reflect the several phases of the family cycle. They are high at the ages at which children leave home to find jobs, marry, and set up their own households. Additional mobility occurs in response to increased family size and increased income. Mobility rates decline with age because of the now strengthened ties of the household head to his residence, his job, and his community. Changes in marital status upset this equilibrium and contribute to additional mobility since the event of divorce or separation involves a change of residence of one of the marriage partners. Widowhood may also lead to a change of residence, especially if the previous living arrangements now become unsatisfactory."[71]

Table 21–19 illustrates the effects of these processes upon

the age profile of Swedish migration rates. Although Jamaican rates are lower than the Swedish rates, their profiles are fairly similar despite the great differences in the national cultures and the fact that the Swedish statistics were based on the population register and the Jamaican statistics, on the census data regarding duration of residence (less than 1 year).[72] The Jamaican rates do peak at younger ages, probably reflecting earlier entry into the labor market and earlier marriage. Sex differences in mobility rates are much less pronounced than age differences in most areas; but, in examining differentials among social and economic groups, it is highly desirable to cross-classify the

Table 21–19. — Migration Rate per 1,000 of the Midyear Male and Female Population, for Sweden: 1966

Age	Male	Female
All ages.....................	66.15	67.66
Under 5 years.....................	112.61	111.17
5 to 9 years......................	64.36	63.97
10 years..........................	44.34	43.48
15 years..........................	54.48	113.82
20 years..........................	152.51	213.28
25 years..........................	164.41	144.78
30 years..........................	106.05	82.46
35 years..........................	69.20	54.93
40 years..........................	48.88	39.40
45 years..........................	37.76	31.05
50 years..........................	27.68	24.74
55 years..........................	21.22	21.18
60 years..........................	18.06	19.94
65 years..........................	20.36	18.44
70 years..........................	15.01	14.43
75 years..........................	13.10	13.63
80 years..........................	13.87	13.70
85 years..........................	12.46	16.05
90 years..........................	11.54	12.87
95 years and over.................	15.03	16.13

Source: Adapted from, Sweden, Statistiska Centralbyrån *Folkmängdens Förändringar, 1966* (Population changes), Stockholm, 1967, p. 102.

[70]Lee *et al.*, op. cit., pp. 10–12. See also Bertil Wendel, "Regional Aspects of Internal Migration and Mobility in Sweden, 1946-50," in *Migration in Sweden;* David Hannerberg, Torsten Hägerstrand, and Bruno Odeving, eds., Lund Studies in Geography, Series B, Human Geography, No. 13, Royal University of Lund (Sweden), 1957; and Gunnar Kulldorff, *Migration Probabilities*, Lund Studies in Geography, Series B, Human Geography, No. 14, Royal University of Lund (Sweden), 1955.

[71] U.S. Bureau of the Census, *Current Population Reports*, Series P–20, No.

171, "Mobility of the Population of the United States: March 1966 to March 1967," April 30, 1968, p. 1.

[72] Jamaica, Department of Statistics, *Internal Migration in Jamaica*, by Kalman Tekse, April 1967, p. 17.

data by both age and sex. (Such cross-classification is much less necessary in the study of population redistribution and of migration streams.)

Economic and Social Factors

Chapter VI, "Economic and Social Factors Affecting Migration" in the United Nations publication *The Determinants and Consequences of Population Trends,* summarizes the literature on that subject through the early 1950's.[73] An up-dated edition of this report has recently been published (see Suggested Readings).

MIGRATION DIFFERENTIALS

Migrants tend to have different personal, social, and economic characteristics from those of nonmigrants. In other words, migrants are far from constituting a representative

sample of the national population. Moreover, the characteristics of movers tend to vary with the distance spanned, short-distance movers differ on the average from migrants, and so on. Migrants or short-distance movers may be compared with the total national population (or with the nonmovers). In-migrants may be compared with the total population, or nonmigrants, in the area of destination, out-migrants with the population in the area of origin, and a migration stream with the population at either origin or destination. The importance of holding constant age and sex in studying migration differentials in terms of social and economic conditions has already been stressed. A discussion of the more significant types of migration differentials, the methods of measuring them, and the state of scientific knowledge about them at the time was contained in a 1938 bulletin of the Social Science Research Council by Dorothy Swaine Thomas and her collaborators.[74] This publication examined differentials in age, sex, family status, physical health,

Figure 21–1. —Net Streams Between Divisions, Native Whites 10 Years Old and Over, Conterminous United States: 1950 to 1960

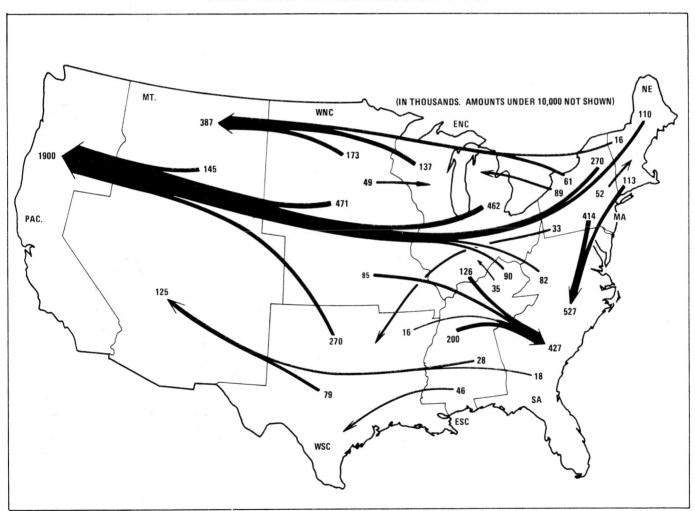

Source: Hope T. Eldridge and Yun Kim, *The Estimation of Intercensal Migration from Birth-Residence Statistics: A Study of Data for the United States, 1950 and 1960,* Philadelphia, University of Pennsylvania, Population Studies Center, Analytical and Technical Reports, No. 7, February 1968, figure 7, p. 62.

[73] Population Studies, Series A, No. 17, 1953, pp. 98–133.

[74] *Research Memorandum on Migration Differentials,* New York, Social Science Research Council, Bulletin 43, 1938.

Figure 21-2. — **Net and Gross Interchange of Nonwhite Migrants Between Each Pair of Regions, for the United States: 1935 to 1940**

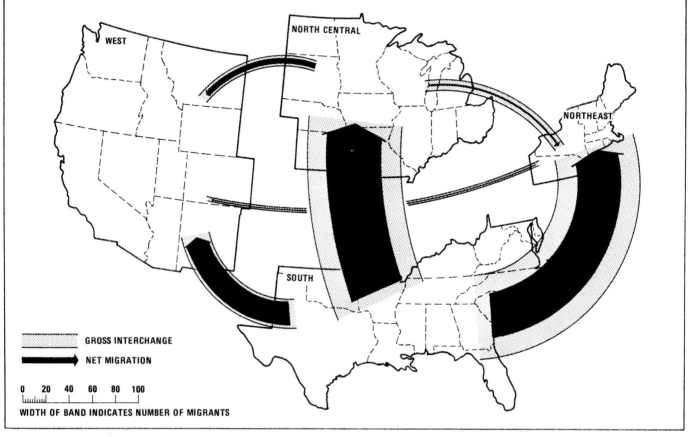

Source: Henry S. Shryock, Jr., *Population Mobility Within the United States,* Community and Family Study Center, University of Chicago, 1964, figure 8.3, p. 242

mental health, intelligence, occupation, and motivation and assimilation.

Selectivity of Migration

The selectivity of migration is measured by comparing the migrants with the population from which they are drawn, i.e., at the national level, with the total population or, at the local level, that at the area of origin. A very skillful analysis of the selectivity of mobility on the basis of retrospective data from a sample survey is given in the chapter on "Geographical and Social Mobility" in the book on *The American Occupational Structure* by Blau and Duncan.[75]

POSSIBILITIES OF LONGITUDINAL ANALYSIS

Survey data offer a number of possibilities for the longitudinal analysis of internal migration. Some of these methods are not readily possible, or even possible at all, from multipurpose complete censuses that are taken at intervals of 5 or 10 years. Nonetheless, Eldridge has shown how estimates of net migration by the residual method can be used to study

the migratory behavior of cohorts.[76] A sufficiently long time series on place of birth or on fixed-period migration could be analyzed in a similar fashion. There is considerable interest in knowing the migration histories, or parts of them, for persons, families, or cohorts.[77] From this information, one could compile statistics on total number of migrations since birth or during a fixed period of time, the extent of migration by stages and of return migration, the effect of earlier migration on the probability of migration in a subsequent period, and so on.

There are several ways in which data of these sorts can be produced from surveys. These include (a) tabulation of age cohorts from successive surveys, (b) matching cases between successive surveys, and (c) the collection of migration histories. (In an earlier section on "Migration Histories," we described the methods of cohort analysis that have been applied to such data.) In all of these approaches, the products parallel those that could be produced from a continuous population register. On the whole, the figures from surveys will ordinarily not be as accurate as those from registers, either because of absence of the continuous feature (at best, one has a series of closely spaced "snapshots") or because of the greater difficulty of

[75] Peter M. Blau and Otis Dudley Duncan, *The American Occupational Structure,* New York, John Wiley & Sons, 1967, pp. 243–275.

[76] Hope T. Eldridge, "A Cohort Approach to the Analysis of Migration Differentials," *Demography,* 1(1):212–219, 1964.
[77] Henry S. Shryock, Jr. and Elizabeth A. Larmon, "Some Longitudinal Data on Internal Migration," *Demography,* 2:581–583, 1965.

controlling on incomplete and erroneous reporting and other sources of bias. The retrospective data of lifetime migration histories are especially subject to bias—errors of recall, lack of knowledge, attrition of cohorts over time, etc. On the other hand, surveys here as elsewhere have the potentialities of collecting a rich variety of social and economic characteristics for the study of differentials (*vide* the book by Lansing and Mueller), whereas registers are limited to a relatively few standard items.

GRAPHIC PRESENTATION

Most of the graphic forms used in other branches of demography can be used to portray migration statistics. Migration, because of its definitional association with geographic areas, is especially well suited to statistical mapping.

One fairly common technique, mentioned in the preceding chapter, is the use of flow charts on a map to portray the number and proportion of migrants in a given stream. Figure 21-1 gives an example (here the width of the arrow is proportionate to the change in the Birth–Residence Index). Net and gross interchange can also be shown on the same map, as in Figue 21–2. Some other proposals for graphic treatment of this subject are given by Bachi.[78]

---

[78] Roberto Bachi, "Analysis of Geographical Data on Internal Migration," in: United Nations, World Population Conference, 1965, Vol. IV, *Migration, Urbanization, Economic Development*, 1967, pp. 475–482.

SUGGESTED READINGS

Bachi, Roberto. "Analysis of Geographical Data on Internal Migration." In: United Nations. *World Population Conference, 1965* (Belgrade), Vol. IV. 1967. Pp. 475–479.

Bogue, Donald J. "Internal Migration." In: *The Study of Population*. Philip M. Hauser and Otis Dudley Duncan, eds. Chicago, University of Chicago Press, 1959. Pp. 486–509.

———. *A Methodological Study of Migration and Labor Mobility in Michigan and Ohio in 1947*. Scripps Foundation Studies in Population Distribution, No. 4. Miami, Ohio. Scripps Foundation for Research in Population Problems, June 1952.

———. *Methods of Studying Internal Migration*. Technical Paper Prepared for a Regional Seminar on Population in Central and South America, held in Rio de Janeiro, Brazil, December 1955.

Bogue, Donald J., Shryock, Henry S., Jr., and Hoermann, Siegfried A. *Subregional Migration in the United States, 1935–40*. Vol. I. *Streams of Migration between Subregions*. Scripps Foundation Studies in Population Distribution No. 5. Oxford, Ohio. Scripps Foundation for Population Research, 1957. Pp. 1–36.

Canada. Dominion Bureau of Statistics. *Special Labor Force Studies*, No. 4. "Geographic Mobility in Canada: October 1964–October 1965." By May Nickson, April 1967.

Croze, Marcel. "Un instrument d'étude des migrations intérieures: les migrations d'électeurs." *Population* (Paris), 11(2):235–260, April–June 1956.

Das Gupta, Ajit. "Types and Measures of Internal Migration." In International Union for the Scientific Study of Population. *International Population Conference*, Vienna, 1959. Pp. 619–624.

Demographic Training and Research Centre, Bombay. "Internal Migration in Some Countries of the East." International Union for the Scientific Study of Population. *International Population Conference, 1961*. Vol. I. London, 1963. Pp. 420–427.

Eldridge, Hope T. "A Cohort Approach to the Analysis of Migration Differentials." *Demography*, 1(1):212–219, 1964.

———. *Net Intercensal Migration for States and Geographic Divisions of the United States, 1950–60: Methodological and Substantive Aspects*. Analytical and Technical Reports No. 5. Philadelphia, Population Studies Center, University of Pennsylvania, 1965.

———. "Primary, Secondary, and Return Migration in the United States, 1955–60." *Demography*, 2:444–455, 1965.

Eldridge, Hope T., and Kim, Yun. *The Estimation of Intercensal Migration from Birth-Residence Statistics: A Study of Data for the United States, 1950 and 1960*. Analytical and Technical Reports, No. 7. Philadelphia, Population Studies Center, University of Pennsylvania, February 1968.

Elizaga, Juan C. "A Study of Migration to Greater Santiago (Chile)." *Demography*, 3(2):352–377, 1966.

———. *Medición del volumen y de las características de las migraciones interiores*. Santiago, Chile. Centro latino-americano de demografía. Series A, No. 38, 1965. Pp. 6–14.

Friedlander, D., and Roshier, R. J. "A Study of Internal Migration in England and Wales." Part II. "Recent Migrants—Their Movements and Characteristics." *Population Studies* (London), 20(1):45–59, July 1966.

Goldstein, Sidney. *Patterns of Mobility, 1910–50: The Norristown Study*. Philadelphia, University of Pennsylvania Press, 1958. Pp. 68–122.

Hamilton, C. Horace. "Educational Selectivity of Net Migration from the South." *Social Forces*, 38(1):33–42, October 1959.

———. "Effect of Census Errors on the Measurement of Net Migration." *Demography*, 32(2):393–415, 1966.

———. "Practical and Mathematical Considerations in the Formulation and Selection of Migration Rates." *Demography*, 2:429–443, 1965.

———. "The Vital Statistics Method of Estimating Net Migration by Age Cohorts." *Demography*, 4(2):464–478, 1967.

SUGGESTED READINGS – Continued

Hashmi, Sultan S., Khan, Masihur Rahman, and Krótki, Karol J. *The People of Karachi.* Pakistan Institute of Development Economics, Statistical Papers No. 4, Karachi, June 1964.

India. Directorate of National Sample Survey. *The National Sample Survey, Ninth, Eleventh, Twelfth, and Thirteenth Rounds: May 1955–May 1958,* Number 53. "Tables with Notes on Internal Migration." Delhi, 1962.

Kulldorff, Gunnar. *Migration Probabilities.* Lund Series in Geography. Series B, Human Geography, No. 14. Royal University of Lund (Sweden), 1955.

Lansing, John B. and Mueller, Eva. *The Geographic Mobility of Labor.* Ann Arbor, Michigan, Survey Research Center, University of Michigan, 1967. Pp. 3–10.

Lee, Everett S. et al. *Methodological Considerations and Reference Tables.* Vol. I of: Simon Kuznets and Dorothy Swaine Thomas. *Population Redistribution and Economic Growth, United States, 1870–1950.* Philadelphia, American Philosophical Society, 1957. Pp. 1–106.

Miller, Ann R. *Net Intercensal Migration to Large Urban Areas of the United States: 1930–40, 1940–50, 1950–60.* Analytical and Technical Reports No. 4. Philadelphia, Population Studies Center, University of Pennsylvania, May 1964.

Neymark, Ejnar E. "Migration Differentials in Education, Intelligence, and Social Background: Analysis of a Cohort of Swedish Males." *Bulletin of the International Statistical Institute* (Proceedings of the 34th Session, Ottawa, 1963), 40(1):350–379. Toronto, 1964.

Pourcher, Guy, "Un essai d' analyse par cohorte de la mobilité géographique et professionnelle." *Population* (Paris), 21(2):357–378, March–April 1956.

Price, Daniel O. "Examination of Two Sources of Error in the Estimation of Net Internal Migration." *Journal of the American Statistical Association,* 50(271):689–700, September, 1955.

Saben, Samuel. "Geographic Mobility and Employment Status, March 1962–March 1963." *Special Labor Force Report,* No. 44. U.S. Bureau of Labor Statistics, August 1964.

Shryock, Henry S., Jr. *Population Mobility Within the United States.* Chicago, Community and Family Study Center. University of Chicago, 1964. Pp. 1–62.

Shryock, Henry S., Jr., and Larmon, Elizabeth A., "Some Longitudinal Data on Internal Migration." *Demography,* 2:581–583, 1965.

Siegel, Jacob S., and Hamilton, C. Horace. "Some Considerations in the Use of the Residual Method of Estimating Net Migration." *Journal of the American Statistical Association,* 47(259):475–500, September 1952.

Stone, Leroy O. "Evaluating the Relative Accuracy and Significance of Net Migration Estimates." *Demography,* 4(1):310–330, 1967.

Taeuber, Karl E. "Cohort Migration." *Demography,* 3(2): 416–422, 1966.

Taeuber, Karl E., Chiazze, Leonard, Jr., and Haenszel, William. *Migration in the United States: An Analysis of Residence Histories,* Public Health Monograph No. 77. U.S. Public Health Service, 1968. Pp. 1–13.

Tarver, James D. "Evaluation of Census Survival Rates in Estimating Intercensal State Net Migration." *Journal of the American Statistical Association,* 57(300):841–862, December, 1962.

Thomas, Dorothy Swaine. "Age and Economic Differentials in Interstate Migration." *Population Index,* 24(4):313–325, October 1958.

———. *Research Memorandum on Migration Differentials.* New York, Social Science Research Council, Bulletin 43, No. 38.

Thomlinson, Ralph. "The Determination of a Base Population for Computing Migration Rates." *Milbank Memorial Fund Quarterly,* 40(3):356–366, July 1962.

Thornthwaite, C. Warren. *Internal Migration in the United States.* Philadelphia, University of Pennsylvania Press, 1934.

United Nations. *Birthplace Tabulations in Recent Censuses.* Studies of Census Methods, No. 7. March 1949.

———. *The Determinants and Consequences of Population Trends.* Series A, Population Studies, No. 50, Vol. I, 1973.

———. *Handbook of Population Census Methods,* Vol. III, *Demographic and Social Characteristics of the Population.* Studies in Methods, Series F, No. 5, Rev. 1. 1959. Pp. 12-16.

———. *Methods of Measuring Internal Migration.* Manuals on Methods of Estimating Population, Manual VI, 1970.

———. *National Programmes on Analysis of Population Census Data as an Aid to Planning and Policy-Making.* Series A, Population Studies, No. 36. 1964. Pp. 17–19.

———. *World Population Conference, Belgrade, 1965.* Vol. I, *Summary Report.* 1966. Statements by the Moderator (D. J. Bogue) and by the Rapporteur (K. C. Zachariah), "Internal Migration, with Special Reference to Rural-Urban Movement." Meeting A.3. Pp. 162–168.

U.S. Bureau of the Census. *Current Population Reports,* Series P-20, No. 154. "Reasons for Moving: March 1962 to March 1963." August 22, 1966.

———. *Current Population Reports,* Series P-23, No. 7. "Components of Population Change, 1950 to 1960, for Counties, Standard Metropolitan Statistical Areas, State Economic Areas, and Economic Subregions." November 1962.

SUGGESTED READINGS – Continued

————. *Current Population Reports,* Series P–23, No. 15. "National Census Survival Rates, by Color and Sex, for 1950 to 1960." July 12, 1965.

————. *Current Population Reports,* Series P–23, No. 25, "Lifetime Migration Histories of the American People." March 8, 1968.

————. *U.S. Census of Population: 1960. Subject Reports, Lifetime and Recent Migration,* PC(2)–2D. 1963.

U.S. Department of Agriculture. Economic Research Service. *Net Migration of the Population, 1950–60, by Age, Sex, and Color,* Vol. I. *States, Counties, Economic Areas, and Metropolitan Areas* (in 6 parts). May 1965; Vol. II, *Analytical Groupings of Counties.* November 1965.

Vaidyanathan, K. E. "A Comparison of the CSR Estimates of Net Migration, 1950–60 and the Census Estimates, 1955–60 for the United States." *Social Forces,* 48(2):233–243, December 1969.

Vamathevan, S. *Internal Migration in Ceylon, 1946–53.* Monograph No. 13. Ceylon, Department of Census and Statistics, 1961.

Wendel, Bertil. "Regional Aspects of Internal Migration and Mobility in Sweden, 1946–50." In: David Hannerberg, Torsten Hägerstrand, and Bruno Odeving, editors. *Migration in Sweden: A Symposium* (Lund Studies in Geography. Series B, Human Geography, No. 13), Lund, Royal University of Lund, 1957. Pp. 18–20.

Wunsch, G. "Le calcul des soldes migratoires par la méthode de la 'population attendue' – caractéristiques et évaluation des biais." *Population et famille* (Brussels), 18:49–62, 1969.

Zachariah, K. C. *A Historical Study of Internal Migration in the Indian Sub-continent, 1901–31.* New York, Asia Publishing House, 1964.

————. "A Note on the Census Survival Ratio Method of Estimating Net Migration." *Journal of the American Statistical Association,* 57(297):175–183, March 1962.

Zitter, Meyer and Nagy, Elisabeth S. "Use of Social Security's Continuous Work History Sample for Measurement of Net Migration by Geographic Area." Washington, American Statistical Association. *Proceedings of the Social Statistics Section,* 1969. Pp. 235–260.

CHAPTER 22

Population Estimates

INTRODUCTION

The Nature and Use of Population Estimates

At the present time, the most complete and reliable source of information on the population of countries and their geographic subdivisions is the census based on a house-to-house enumeration. Population is constantly changing, sometimes quite rapidly, however, and hence statistics for every tenth or even fifth year are not adequate for most current purposes. Public health officials, market research analysts, public and private planners, and others need current data to plan and evaluate programs. Although state or provincial, and even local governments, sometimes conduct special censuses of their areas, these sparse data do not by any means meet public needs. Moreover, the method of complete enumeration is expensive, laborious, and time-consuming, and is not applicable to past and future dates.

In order to meet the need for basic population figures more fully, a wide variety of estimating techniques, including the use of sample surveys, has been developed. Like the census, sample surveys are rather expensive and cannot provide data for past or future dates. Nonsurvey or analytic techniques involving the use of vital statistics data, immigration data, other data symptomatic of population change, and mathematical methods, however, are relatively inexpensive to apply and can be used to prepare estimates for past and future dates as well as for current dates. This chapter will confine itself largely to a discussion of these nonsurvey or analytical methods, and the term "estimates" will be used herein to refer mainly to approximations arrived at by these methods.

Types of Population Estimates

Estimates viewed broadly can be divided into several types on the basis of their time reference and basis of derivation. These types, which pose different methodological problems and, on the average, achieve different levels of reliability, are: (1) **intercensal estimates**, which relate to a date intermediate to two censuses and take the results of these censuses into account; (2) **postcensal estimates**, which related to a past or current date following a census and take that census and possibly earlier censuses into account, but not later censuses; and (3) **projections**, which relate to dates following the last census, usually future dates, for which no current reports are available.[1] Both postcensal estimates and projections can be

regarded as extrapolations, and intercensal estimates as interpolations, taking these terms in their more general sense.

Estimates vary in several other respects. They may be made for a variety of types of areas, such as a whole country, the major geographic subdivisions of a country, or broad classes of areas within the country (e.g., urban and rural areas, city-size classes). Estimates may be made of the total population of an area or of particular classes of the population in the area distinguished by age, sex, color, race, nativity, family and marital status, educational attainment, employment status, etc. An important aspect of the type of population estimates to be made relates to the definition of population employed for the estimate. Estimates, like census counts, can vary as to whether they refer to the *de jure* (usual resident) population or the *de facto* (physically present) population. Countries will tend to employ the same type of population in their estimates as in their last census. Another dimension of the problem of definition relates to coverage of armed forces, both at home and abroad, and coverage of nationals abroad.

National and International Programs of Population Estimates

In planning a national population estimates program, the responsible agency in the national government determines which estimates it will make according to the demand or need for various kinds of figures, the availability and quality of basic data, the amount of work necessary to produce the estimates, and the resources (i.e., funds, personnel, and time) available. On this basis, as a general guide, we suggest that estimates of the total population and of the population classified according to age and sex, for a nation, are most important; then come estimates for its primary political subdivisions, first total population, then age and sex.

Estimates of total population for secondary geographic subdivisions (e.g., counties in the United States) would be of next importance in a national program. National estimates of the population classified by ethnic composition, marital status, educational attainment, literacy, employment status, broad occupation and industry groups, etc., are important, but normally, apart from data in censuses or population registers, they are most satisfactorily obtained from continuing or periodic national sample surveys. These characteristics present special problems of estimation because they may change during the lifetime of individuals (e.g., a change from married to divorced) and the requisite data on the components of change are not available or cannot be estimated satisfactorily.

The best sources for the population estimates prepared by national governments are the publications of the governments themselves. The figures characteristically appear in national

---

[1] Additional types of estimates may be identified to complete the picture: (a) Precensal estimates, which relate to a date preceding a census and take account of that census and possibly later censuses but not earlier censuses; and (b) acensal estimates, which are not based on census data at all.

statistical yearbooks, but they also often appear in special reports. The estimates usually fail to include adjustments for deficits in census coverage or other census errors, and may lack comparability with estimates from other countries because of differences in the categories of the population represented. The scope of national population estimates programs varies enormously with respect to the resources devoted to it, the frequency and detail of the estimates, and the kind of publication, including the extent to which the methodology is explained and the results are analyzed. As our discussion of the methodology will show, there are wide differences in the methods employed and the quality of the results, depending on the resources and data available.

United Nations Program. — Each U.N. *Demographic Yearbook* (table 1 in the *Yearbook* for 1948 through 1964 and table 2 in later *Yearbooks*) presents for each country of the world, sovereign and nonsovereign, estimates of population with about a one-year lag.

Other estimates published regularly in the *Yearbook* include an annual table showing aggregates of population for the world, continents, and regions, both at decennial intervals starting with 1920 or 1930 and for the current year; estimates of the total population by age and sex for selected countries; and estimates for the total population of capital cities and of each city that had 100,000 or more inhabitants according to the latest available data.

Generally, the estimates displayed in the *Yearbook* are official figures that are consistent with the results of national censuses or sample surveys taken in the period. Thus, they have been revised by the national government or by the United Nations on the basis of a census or survey where discontinuities appeared to exist, so as to form a consistent series. They refer to July 1 of the estimate year, but may have been computed by the United Nations as the mean of two year-end official estimates. When an acceptable official estimate of population is not available for a given year, the United Nations prepares its own estimate. These estimates may take into account available information on the reliability of census and survey results and data on natural increase and net migration, so as to produce estimates for various postcensal and intercensal years comparable to one another and to the figures for the census date. The methodology and quality of the estimates are indicated by a type-of-estimate code accompanying each estimate. The code is composed of four parts, identifying the nature of the basic data, their recency, the nature of the adjustment since the base date, and the quality of this adjustment.

United States. — The official population estimates for the United States prepared by the U.S. Bureau of the Census relate to the population on the basis of usual residence rather than the *de facto* (actually present) population, just as the census counts do. As a result of the tremendous increase in the size of the armed forces during and following World War II, three types of population estimates are regularly prepared for the United States as a whole: the total population residing in the United States (that is, the population as usually defined in the decennial census); the total population including armed forces overseas (that is, the total population resident in the United States plus armed forces of the United States serving overseas); and the civilian population (that is, the total resident population minus armed forces serving in the United States). A fourth type of estimate, presented only occasionally, is the most inclusive and covers, in addition to the armed forces overseas, civilian citizens of the United States employed overseas by the U.S.

Government and dependents of civilian and military overseas employees of the U.S. Government. Since in previous decennial censuses individuals were assigned geographically according to their usual place of residence and since the armed forces overseas were not allocated to a residence in the United States, only the first and third types of population estimates have been prepared for subdivisions of the United States.

The most important source of population estimates for the United States is the U.S. Bureau of the Census. This agency regularly publishes a wide variety of population estimates in *Current Population Reports, Population Estimates and Projections,* Series P–25.[2] The content of these brief reports provides an indication of the scope of the U.S. Census Bureau's basic estimates program at the present time. For the United States as a whole, postcensal and intercensal estimates of total population are published at monthly intervals. Three types of population figures are given for each date, as defined above. For States, postcensal and intercensal estimates of total population are published for each midyear date. Two types of population figures — total resident and civilian resident — are given for each date. Estimates of the total population for standard metropolitan statistical areas are now made on an annual basis. In recognition of the widespread need for small–area population estimates of uniform quality from State to State, the Census Bureau has developed a cooperative program with the States for the preparation and publication of county population estimates. In 1969, the Bureau established Series P–26 (*Federal–State Cooperative Program for Population Estimates*) to serve as a vehicle for postcensal county estimates produced under this program.

The following types of estimates of population characteristics appear in the P–25 Series of Census Bureau reports: For the United States as a whole, postcensal and intercensal estimates of population classified by 5-year age groups and single years of age, sex, and color are available for July 1 of each year. Three types of population figures are given. (Unpublished postcensal estimates for single years of age are available for each month.) For States, postcensal estimates for special broad age groups of functional importance (e.g., under 5, 5–17, 18–44, 45–64, 65 and over, 14 and over, 21 and over) are available on an annual basis. As noted in chapter 2, national estimates of households, families, marital groups, and school enrollment, based on the Current Population Survey, appear in *Current Population Reports,* Series P–20, at annual intervals, and estimates of labor force and employment appear in the Bureau of Labor Statistics' *Employment and Earnings* at monthly intervals. Estimates of households for States comparable with the national survey-based figures, are derived by analytic techniques and are published annually.

In addition to the publication of actual estimates, the Census Bureau's program of population estimates includes the occasional publication of reports presenting detailed descriptions of one or another method of making State or county population estimates for use by local technicians.[3]

Population estimates for States, counties, and cities are also published by many State and local government agencies and by private organizations. Population estimates for a State and its counties may appear in the reports of the State Health Department, a State Planning Agency, the Business or Social Research Bureau at the State University, or the Budget Office

---

[2] Predecessor series were titled *Population-Special Reports* and *Population.*
[3] For example, U.S. Bureau of the Census, *Current Population Reports,* Series P–25, No. 339, "Methods of Population Estimation: Part 1, Illustrative Procedure of the Census Bureau's Component Method II," June 6, 1966.

or equivalent.[4] The reports of many local planning commissions contain current estimates for the local area.

METHODOLOGY

General Considerations

Choice of Data and Method. — The most important factor determining the choice of the method to be used in preparing a given population estimate is the type and quality of data available for this purpose. If, for example, an estimate of the total population of an area is wanted and the only relevant information at hand is the total size of the population at two or more census dates, then a purely mathematical or graphic approach may have to be used. The level of accuracy required and the amount of time, funds, and trained personnel available are other important considerations in determining the choice of method.

The data on which a population estimate may be based can be divided roughly into two categories: (1) "Direct" data and (2) "indirect" data. We are concerned here with the data for the base date and data for the period between the base date and the estimate date. The classification depends on the specific kind of data and their use in a given method. Illustrative of the first type of data are (a) those obtained from censuses, population registers, and special compulsory or quasi-compulsory registrations,[5] and (b) recorded data on the components of population change (i.e., statistics on births, deaths, and migration) when these data are used to measure these phenomena themselves. Examples of the second type of data, indirect data, are school enrollment and school census data; statistics on gas and electric meter installations; employment statistics; statistics on voter registration; birth and death statistics when used to measure total population change directly rather than to measure natural increase; and statistics on housing construction, conversion, and demolition. It should be apparent that data of a given type may be direct for one kind of estimate and indirect for another and that there is no rigid dividing line between the two classes of data. Data on registrations for military service represent indirect data if they are being employed to estimate the total male population as such and direct data if they are being employed to estimate the male population of registration age directly. Both direct and indirect data may be used in combination in preparing a given population estimate. To complete an estimate the available direct and indirect data may have to be manipulated on the basis of hypotheses or assumptions. These hypotheses or assumptions may involve the use of a mathematical formula or its equivalent, such as a graph. Estimation by use of assumptions or a mathematical formula is required to make effective use of indirect data. If the analyst lacks reliable data of both the direct and indirect types, he is forced to the use of mathematical devices as the principal procedure, i.e.; he must make some assumptions as to the trend of population change following the base date and express these in terms of some mathematical formula or graphic device.

The usefulness of indirect data for population estimation depends on the extent to which factors other than population size and distribution influence them. Changes in the number of children attending school may result from changes in the laws relating to attendance and in their enforcement and in the availability of school facilities, as well as from changes in the number of children of school age. Employment, housing construction, and the number of public utility customers change with economic conditions as well as with population and households. The number of deaths varies not only with the size of the population but with the "force of mortality," which sometimes shows sharp fluctuations, for example, as the result of an epidemic. It is apparent that the usefulness of indirect data as indicators of population change will vary with the particular situation and that many of them will be of little or no value in preparing estimates for underdeveloped areas.

In general, the data to be used are to be carefully evaluated according to the requirements set forth in previous chapters. The coverage of the latest census is especially important. An understanding of definitions and collection procedures may be important in a particular case. A spurious population change between two dates would be indicated if population estimates based on school data employed statistics on school enrollment as of the end of one school year and statistics on average daily attendance in another school year. The method used in collecting the data may give important indications as to the consistency of a series and the likelihood of over- or undercounting. For example, a house-to-house survey made by trained enumerators to ascertain the number of children of school age will yield a more accurate count than a public request to all parents to report the number of their children of school age to the school authorities.

Some Estimating Principles. — At the risk of excessive generalization, we have set down below some principles of population estimation which may serve as rough guides in spite of their many limitations, including numerous exceptions.

1. More accurate estimates can generally be made for an entire country than for the geographic subdivisions of the country. The national population is much more likely to be a closed population than is that of a subdivision of the country. Moreover, when there is immigration, it is likely to be registered for administrative reasons while internal migration will go unrecorded. In general, more direct data, data of better quality, and more information on how to adjust these data for deficiencies, are available for the larger areas, particularly for entire countries, than for the smaller areas. Furthermore, the size of small populations may fluctuate widely, with the result that accurate estimation is extremely difficult or impossible. Depressed economic opportunities in one region of a country would have little discernible effect on the size of the population of the country but a particular state or province in this region might be sharply affected. The shutting down of a single factory would have little or no effect on the size of the population of a state, but it might cause the population of a small county in the state to be reduced sharply. It is usually advisable, therefore, to adjust estimates for geographic subareas to agree with an independently estimated area total; e.g., estimates for provinces should be adjusted to the national total.

2. More accurate estimates can generally be made for the total population of an area than for the demographic characteristics of the population of the area. Fewer data and data of poorer quality are usually available for making estimates of the population of a given area classified by age, color, sex, and

[4]For example, U.S. Bureau of the Census, *Current Population Reports*, Series P-25, No. 454, "Inventory of State and Local Agencies Preparing Population Estimates—Survey of 1969," December 31, 1970.

[5] When population figures are obtained rather directly from registers or registrations without considerable adjustment to include or exclude particular groups, then the data can be considered as being virtually in a class with a census count. When the original data must be subjected to considerable adjustment, we may consider the results as estimates.

other characteristics than of the total population of the area. It is usually advisable, therefore, to adjust estimates for such classes to the area total for the characteristic; e.g., estimates for age classes in the population of a province should be adjusted to the estimated total of all ages for the province.

3. In general, assuming that the available data are of good quality, direct data are to be preferred to indirect data from the standpoint of the accuracy of the resulting estimates, either in preparing separate estimates or in elaborating a single estimate. The more nearly the basic data approximate an exact count of the population being estimated or reflect actual change in that population since some base date (when the population figure is closely known), and the less adjustment or manipulation of the data required, the smaller the error to be expected in the resulting estimate. In actual practice, allowing that the direct or indirect data may in fact be defective, the choice may be determined by the accuracy, completeness, internal consistency, and recency of the data. In measuring population change, use of data that reflect actual population change (i.e., direct data, such as births, deaths, etc.) and of methods whose steps parallel actual demographic processes (e.g., aging) may be expected, on the average, to produce more accurate estimates than the use of data and methods which are indirect. For example, if the population under 5 years of age is being estimated, a highly reliable procedure is to trace the "demographic development" of the group from birth, making allowance for deaths and migration.

Again assuming that the available data are of good quality, this principle suggests, first, the use of direct data before use of indirect data in the preparation of a particular estimate and the use of both direct and indirect data before mathematical methods are resorted to as a main procedure. The principle suggests, secondly, the desirability of employing a "cohort" approach, where possible, since such procedures by their very nature follow actual demographic changes. The most common application of a cohort approach is in the preparation of estimates of age groups.

4. An estimate may be checked by comparing it with another estimate derived by an equally accurate, or more accurate, method using different data and assumptions. Two or more independent estimates based in whole or part on different data or different methods, each considered highly accurate, can sometimes be worked out. If the estimates differ considerably from one another, doubt is cast on both; if they are quite similar, one may have greater confidence in each. Often, the best estimate is a weighted average of such alternative estimates.

5. Since the making of assumptions is a precarious matter, from the standpoint of accuracy mathematical methods are to be avoided when possible. However, as a mathematical method, interpolation is almost invariably more reliable than extrapolation. In interpolation the two census figures usually establish narrow limits on the possible size of the intercensal estimate. The greater accuracy of interpolation than of extrapolation suggests that fairly accurate estimates for years between censuses may be obtained by less exact methods; here rapid mathematical methods can often serve.

6. Finally, the quality of the base data, the quality of the data used to allow for change since the base date, and the period of time which has elapsed since the base date all have a major effect on the accuracy of the final estimate. It is reasonable to assume that the poorer the quality of the data and the longer the estimating period, the less reliable are the resulting population estimates.

National Estimates

We come now to a discussion of the methodological techniques used in preparing specific types of population estimates. Our general plan is to consider estimates of national population first and then estimates of the geographic subdivisions of countries. Under each of these headings we will first consider estimates for the total population and then estimates for the two most basic characteristics, age and sex. In presenting the techniques that are applicable for a given area, techniques using direct data solely will be described first, then those depending both on indirect data and direct data or on indirect data only, and finally, those involving principally mathematical assumptions. Techniques for making intercensal estimates of a given type will be described along with the corresponding postcensal estimates, since intercensal estimates often involve first the preparation of postcensal estimates and then the additional step of "tying in" the postcensal estimates with the following census.

Total Population. — Several methods for making postcensal estimates of the total population of a country are available, each applicable under different circumstances.

Component method. — First, a simple component method may be used when a satisfactory census count and satisfactory data on births, deaths, and migration are available. The method consists essentially of adding natural increase and net immigration for the period since the last census to the latest census count or the latest previous estimate. The basic estimating equation is as follows:

$$P_1 = P_0 + B - D + I - E \qquad (1)$$

where the symbols have the same meaning as previously. For example, the estimated population of Japan for October 1, 1967, is obtained as follows from the estimate for October 1, 1966: [6]

Estimated population, Oct. 1, 1966.........	99,053,735
For Oct. 1966 to Sept. 1967	
Live births..	1,866,361
Deaths...	673,627
Natural increase................................	1,192,734
Entries...	491,148
Exits..	494,906
Entries minus exits............................	−3,758
Net population increase.........................	1,188,976
Estimated population, Oct. 1, 1967..............	100,242,711

If an estimate including the country's armed forces overseas is required, P_0 should include the armed forces overseas and D should include the military deaths overseas. If an estimate of the resident population of a country is required, one procedure is to carry the resident population at the census date forward by adding resident births, subtracting resident deaths (only), and adding net immigration including movements

[6] Japan, Office of the Prime Minister, Bureau of Statistics, *Population Estimates*, Series No. 32, "Population Estimates by Age and Sex as of October 1, 1967," March 1968, p. 16.

Table 22-1. — Estimation of the Resident Population of the United States, for July 1, 1960 to 1965

Date	Resident population[1]	Births	Resident deaths	Change in the following period					
				Net migration					
				Resident	Civilian	Net military movement overseas[2]			
						Total	Net change in Armed Forces overseas	Death to Armed Forces overseas	Net recruitment of Armed Forces abroad
				$(5)-(6)=$		$(7)+(8)-(9)=$			
	(1)	(2)	(3)	(4)	(5)	(6)	(7)	(8)	(9)
April 1, 1960	[3]179,323,175	1,004,260	407,622	72,010	80,645	8,635	8,281	354	–
July 1, 1960	179,991,823	4,349,693	1,681,362	396,921	402,597	5,676	6,760	1,260	2,344
July 1, 1961	183,057,075	4,259,490	1,742,514	315,949	384,526	68,577	67,395	1,110	-72
July 1, 1962	185,890,000	4,185,492	1,802,462	384,704	378,388	-6,316	-7,511	1,146	-49
July 1, 1963	188,657,734	4,119,104	1,780,406	376,004	363,792	-12,212	-11,086	1,194	2,320
July 1, 1964	191,372,436	3,939,918	1,819,660	321,943	346,082	24,139	29,360	1,187	6,408
July 1, 1965	193,814,637	(X)	(X)	(X)	(X)	(X)	(X)	(X)	(X)

– Represents zero.　　X Not applicable.

[1] Equals the resident population at the beginning of the period plus births minus resident deaths plus resident net migration during the period (i.e., col. (1) + col. (2) – col. (3) + col. (4) on the previous line).

[2] Signs are to be reversed when these data are combined with civilian net migration.

[3] Census count.

of the armed forces into and out of the country. Another is to subtract the armed forces overseas on the estimate date from an estimate of population including armed forces overseas. Table 22-1 illustrates these computations for the period April 1, 1960, to July 1, 1965, for the United States.

Where effective population registers are in operation, they may be used in preparing current estimates of total population. The nature and functioning of population registers have been described in previous chapters. The data from the register are employed merely to update the count from the last census rather than to provide the current estimates directly. Typically the information from the register at the census date is evaluated on the basis of the census returns and the register is adjusted to agree with the census.

Use of national sample surveys and registrations. — It is preferable, when possible, to prepare postcensal estimates on the basis of census counts and direct data on postcensal changes from registration systems or a population register. Adequate postcensal estimates may also be derived by updating the results of a national sample survey or a national registration to the estimate date, on the basis of the balance of births, deaths, and migration. No country has actually employed the procedure with a national sample survey. This situation arises because the countries with adequate data on births, deaths, and migration or with population registers are typically those which have taken a recent complete census.

From time to time, a special national registration may be taken which may serve as a basis for an estimate of national population or for evaluating an estimate of national population derived by other methods. For example, such a use was made of registrations for food ration books in 1942 and 1943 to derive estimates of the U.S. population.

Estimates based on extremely limited data. — For many statistically underdeveloped countries, the data available for making current population estimates are extremely limited. Following the classification of the United Nations, we may distinguish the types of situations for which estimates have to be derived from extremely limited data, in terms of the base figures available, as follows: (1) Estimates based on one census only, (2) estimates based on incomplete censuses, (3) estimates based on noncensal estimates, and (4) estimates based on conjectures, or "reasoned guesses." (In general, in these cases, data on population change since the base date are lacking.)[7]

Estimates based on incomplete censuses include estimates based on censuses covering, say, less than half the population or those conducted over an extended period of time (e.g., more than 12 months); estimates based on partial sample surveys; and estimates based on counts of selected groups in the population, such as those covered in agricultural, school, or other censuses, and those listed in various types of special registers, such as lists of taxpayers or voters. Noncensal estimates include estimates made from counts of dwellings (such as huts or tents) which did not list the members of the population individually, and estimates based on combinations of local administrative reports of persons living in an area. Where the whole population or an important part of it has not been counted, population estimates have to be based on "conjectures." A conjectural estimate is one based on numerical data not relating to the population itself. Conjectural procedures include the conversion of inhabited land area, the returns of the salt tax, or total production or consumption of a staple commodity, to population, by applying a factor representing the ratio of population to the unit of measurement.

Standard mathematical formulas are not directly usable in countries where only one census was taken, where the enumeration is only partial or a combination of partial counts, or where the available data do not relate to the population itself. Population estimates can be made under these circumstances only by the use of rather arbitrary assumptions concerning the "estimating ratio" (i.e., the relation of the total population to the unit of measurement or the "indicator" data) and concerning rates of population growth.

With even a single census, the estimating situation is vastly better than without a complete census. In the former case, one has a firm base to which an estimate of postcensal change can be added, or can usually compute a firm estimating ratio

---

[7] United Nations, *Methods of Estimating Total Population for Current Dates*, Manuals on Methods of Estimating Population, Manual I, Series A, Population Studies, No. 10, 1952.

by which the total population can be estimated from the indicator data. With a partial census, sample survey, or registration of individuals, the error in the total population figure is compounded of errors in (1) the partial count, (2) the estimating ratio or the adjustment factor used in arriving at the total population on the base date, and (3) the estimate of change since the base date. Conjectural estimates are subject to a very wide margin of error. The rank order of the accuracy of the types of estimates based on extremely limited data corresponds, in general, to the order of the listing of these types of estimates at the beginning of this section.

The need to evaluate and correct the basic data for population estimates is all the more important in the case of those statistically underdeveloped countries which have little experience in census taking or systematic collection of vital statistics (e.g., countries with only one or no censuses). Reference should be made to the discussion in chapters 4, 8, 14, 16, and 20, and especially to U.N. Manual II,[8] for procedures of appraising the accuracy of the basic data for population estimates. An evaluation of such data, and their correction where necessary, are essential steps in making reliable population estimates and in determining the limits of confidence of the estimates made.

Assumed rates of growth. — For countries lacking current data on population changes — and this includes most countries in the world — the figure for the base year is typically updated by use of an assumed rate of population increase. This is, of course, a form of mathematical extrapolation, specifically geometric extrapolation.

The rate of growth assumed for the postcensal period may be the average annual rate of growth in the previous intercensal period, an extrapolation of the rates for the two previous intercensal periods, or a rate assumed when only one or no census was taken. (Linear extrapolation or polynomial formulas of higher degree are not usually employed for extrapolating the population.) This method of updating the latest census figure or other base figure implies, of course, that the population has been growing at a more or less constant rate since the base date. The specific steps to be taken for projecting a population by use of an assumed rate of increase may be illustrated with data for Mexico. If we assume that the average annual growth rate between June 6, 1950, and June 8, 1960 (i.e., the last intercensal period), continued to 1968, the estimated population of Mexico on July 1, 1968, may be determined as follows: From the general formula for population growth at a constant rate,

$$\frac{P_1}{P_0} = e^{rt} \qquad (2)$$

we write

$\frac{P_{60}}{P_{50}} = e^{r(10.01)}$, where $t = 10.01$ (i.e., the number of years in the intercensal period)

or

$$\frac{34,923,129}{25,791,017} = e^{10.01r}$$

$$1.35408111 = e^{10.01r}$$

[8] United Nations, *Methods of Appraisal of Quality of Basic Data for Population Estimates,* Manuals on Methods of Estimating Population, Manual II, Population Studies, Series A, No. 23, 1955. See also United Nations, Manual I, op. cit., pp. 4–9.

Then, from the tables for the function e^x, we establish that $10.01r = .303123$, or $r = .030282$ (see chapter 13 for further explanation). Solving for the population of July 1, 1968:

$$P_{68} = P_{60} e^{(.030282)(8.06)}$$

where $t = 8.06$ (i.e., the number of years in the postcensal period)

$$P_{68} = 34,923,129 \, e^{.24407292}$$

From the table for the function e^x, $e^{.24407292} = 1.27643741$ and

$$P_{68} = 34,923,129 \times 1.27643741, \text{ or } 44,577,188.$$

Where three censuses are available, one may compute average annual rates of increase for each intercensal period and, if reasonable, extrapolate the two intercensal rates to the postcensal period. The average annual rate may be extrapolated linearly to the midpoint of the postcensal period and may then be assumed to represent the average rate for this period. We may illustrate this procedure by deriving population estimates for Mexico for 1968 on the basis of the 1940–50 and 1950–60 growth rates. The annual average growth rates for 1940–50 and 1950–60 derived from formula (2) are .026514 and .030282. Linearly extrapolating these rates to 1964 we have

Middate (rounded)	r	Length of period (years)
April 15, 1945..........	$r_1 = .026514$	$t_1 = 10.13$
June 1, 1955............	$r_2 = .030282$	$t_2 = 9.04$
June 15, 1964...........	$r_3 = .033645$	

$$r_3 = r_2 + \frac{t_2}{t_1}(r_2 - r_1) \qquad (3)$$

$$r_3 = .030282 + \frac{9.04}{10.13}(.030282 - .026514) = .033645$$

Hence, .033645 is entered on the line for June 15, 1964, as r_3. Solving the equation for the population in 1968 with this r, we have

$$P_{68} = P_{60} e^{(.033645)(8.06)}$$

$$P_{68} = (34,923,129)(e^{.271179})$$

$$P_{68} = (34,923,129)(1.31150981)$$

$$P_{68} = 45,802,026$$

If annual population estimates are required, the extrapolated rate can be determined for each year of the period.

Intercensal estimates. — Intercensal estimates present the additional problem of allowing for the difference between the "expected" number at the later census date (P_1') and the actual number enumerated (P_1) — the so-called error of closure. This difference represents the balance of errors in the elements of the estimating equation (including the population counts from the earlier and later censuses).

Assuming that the census counts are to be maintained without change, the error of closure may be scaled back to zero at the earlier census date according to any of several procedures. One simple arithmetic device assumes that the adjustment for the error of closure is purely a function of time elapsed since the first census; hence, the correction for each year is derived by interpolating linearly between zero at the earlier census date and the error of closure assigned to the later census date. These interpolated corrections may then be combined with the original postcensal population estimates. Table 22-2 illustrates this procedure in connection with the preparation of intercensal estimates for the United States from 1950 to 1960.

The error of closure may also be distributed over the intercensal period in proportion to the postcensal population, total population change, or one or more of the components of change.

If three or more censuses are taken, the annual average growth rates for the two intercensal periods may be determined and variable growth rates for each year of the second intercensal period may be estimated by interpolation or extrapolation. The two average annual rates can be assigned to the middate of each intercensal period, then they can be interpolated and extrapolated linearly to derive values for July 1 of each year of the period. For example, to derive population estimates for Tunisia in 1957-65, we first determine the average annual growth rates for the intercensal periods November 1946 to February 1956 and February 1956 to May 1966, assign the rates to June 1951 and March 1961, and then compute interpolated and extrapolated rates for each year 1956-57 to 1964-65. The 10-year average annual growth rates are .021539 and .011967, respectively. The rate for 1957-58 (Jan. 1958), for example, is derived by interpolating between .021539 for June 1951 and .011967 for March 1961:

$$r_{58} = r_{51} + \frac{6.54}{9.74}(r_{61} - r_{51})$$

$$r_{58} = .021539 + \frac{6.54}{9.74}(.011967 - .021539)$$

$$r_{58} = .021539 - .006428$$

$$r_{58} = .015111$$

Table 22-2. — Calculation of Annual Intercensal Estimates of the Resident Population of the United States: July 1, 1950 to 1959

Date	Postcensal population estimates (1)	Intercensal adjustment[1] (2)	Intercensal population estimates (1) + (2)= (3)
April 1, 1950..............	[2]151,325,798	–	[2]151,325,798
July 1, 1950...............	151,868,270	-69	151,868,201
July 1, 1951...............	153,982,789	-345	153,982,444
July 1, 1952...............	156,393,937	-620	156,393,317
July 1, 1953...............	158,957,174	-896	158,956,278
July 1, 1954...............	161,884,943	-1,172	161,883,771
July 1, 1955...............	165,070,640	-1,447	165,069,193
July 1, 1956...............	168,090,071	-1,723	168,088,348
July 1, 1957...............	171,188,616	-1,999	171,186,617
July 1, 1958...............	174,151,529	-2,275	174,149,254
July 1, 1959...............	177,137,611	-2,550	177,135,061
April 1, 1960..............	179,325,932	-2,757	[2]179,323,175

– Represents zero.
[1] The adjustment for each year is derived by interpolating linearly between zero at the earlier census date and the error of closure, −2,757, which is assigned to the later census date.
[2] Census count.

The series of estimated rates for 1956-66 is then applied sequentially to the population in 1956, to obtain annual estimates of population between 1956 and 1966. The estimate for July 1, 1958, is computed from the census figure for February 1956 as follows:

$$P_{58} = P_{56}(e^{r_1 t_1})(e^{r_2 t_2})(e^{r_3 t_3})$$

where $r_1 =$ rate for July 1955-56
$r_2 =$ rate for July 1956-57
$r_3 =$ rate for July 1957-58
and $t =$ number of years since the previous estimate ($t_1 = .42$ for Feb.–July 1, 1956; $t_2 = 1$ for July 1956-57; and $t_3 = 1$ for July 1957-58).
Hence,

$$P_{58} = 3,943,273 \ e^{.016802(.42)} \cdot e^{.016094(1)} \cdot e^{.015111(1)}$$

$$P_{58} = 3,943,273 \ (e^{.007057}) (e^{.016094}) (e^{.015111})$$

$$P_{58} = 3,943,273 \ (1.0070820)(1.0162242)(1.0152258)$$

$$P_{58} = 4,097,075$$

Age-Sex Composition. – Let us consider next the procedures that can be used to prepare postcensal estimates of the age, sex, and race composition of the population of a country. These include sample surveys, population registers, complete or partial registrations, the cohort-component method, mathematical methods, stable population analysis, and various combinations of these methods. The method typically employed where the basic data are available and a population register is lacking is the specific variation of the component method called the cohort–component, or cohort–survival, method.

Cohort-component method. – The basic estimating equation for the cohort-component method is similar to that for the component method as applied to the total population, except that the component equation must be evaluated for each age group and the birth component is included only at the very youngest ages. Several methodological problems are thus introduced that merit special attention. A special problem in making age estimates relates to the determination of the number of deaths and migrants that belong to a particular cohort. This ordinarily involves both subdividing the reported data on deaths and migration into single years of age, and then further interpolating the data to obtain those segments of the deaths and migrants falling into each age cohort.

Let us consider the simplest case. Suppose that estimates of the population of Japan by single years of age on October 1, 1967, are desired, given estimates of this kind for October 1, 1966 (table 22-3). The estimates for 1966 were based on the census counts of October 1, 1965. The basic equations, representing estimates for single years of age over a 1-year period are: [9]

For the population under 1: $P_0^{t+1} = B - f_0 D_0 + \frac{1}{2} M_0$ (4)

For the population aged 1:

$$P_1^{t+1} = P_0^t - [(1-f_0)D_0 + \frac{1}{2}D_1] + \frac{1}{2}(M_0 + M_1) \quad (5)$$

_____
[9] Infant migrants would tend to be very few. A separation factor of $\frac{1}{2}$ for infant migrants is shown even though it does not allow for a possible lag after birth in the tendency to migrate.

For the population at all higher ages:

$$P_{a+1}^{t+1} = P_a^t - \tfrac{1}{2}(D_a + D_{a+1}) + \tfrac{1}{2}(M_a + M_{a+1}) \text{ or} \qquad (6)$$

$$P_a^t - [\tfrac{1}{2}(D_a - M_a) + \tfrac{1}{2}(D_{a+1} - M_{a+1})]$$

All the elements in each equation refer to the same cohort and the capital letters refer, of course, to the population at the beginning (P^t) or end (P^{t+1}) of the year or to the components of change during the year. The subscript o refers to infancy and the subscript a refers to age. The element f_0 refers to the separation factor for infant deaths.

1. First, the base population must be set down in single years of age according to ages on October 1, 1966 (col. 1). These figures are presumed to be available. Next, the population should be adjusted to its ages on the estimate date, but since the population is given in single ages and is to be carried forward a whole year, the adjusted population is simply the same as the initial population assigned to the next higher age. It is not necessary, therefore, to re-post the figures on the next line; this will be taken care of, in effect, when the columns are combined.

2. Births during the 12-month period October 1, 1966, to September 30, 1967, are set down at the head of column 1. This group corresponds to the population which will be under 1 year of age on the estimate date.

3. Next, we need estimates of the number of deaths occurring to each age cohort during the year October 1, 1966, to September 30, 1967. Death statistics are conventionally tabulated in 5-year age groups (except for single years under 5). The available statistics for Japan are preliminary data for 5-year age groups for the year October 1–September 30 (col. 2). This normally involves two steps, interpolation of death statistics to single ages and redistributing deaths according to age cohorts. It is necessary, then, to estimate deaths by single years of age from the 5-year totals. Rectangular subdivision of

Table 22–3. — Estimation of the Population of Japan, for Selected Ages: October 1, 1967

| Age | Population, Oct. 1, 1966 [1] (1) | Deaths, Oct. 1, 1966-1967 [1] (D) (2) | Net migrants, Oct. 1, 1966-1967 [1] (M) (3) | Deaths minus net migrants | | Population, Oct. 1, 1967 (1) adj. - (5)= (6) |
				By age at time of occurrence [2] (2) - (3)= (4)	By age on Oct. 1, 1967 [3] (5)	
Total........................	[4]99,053,735	673,627	−7,777	681,420	681,420	100,238,676
Births, Oct. 1, 1966-1967.............	1,866,361					
0 to 4 years....................	7,999,569	36,291	−1,985	38,276	37,608	8,262,567
Under 1 year...................	1,430,286	28,592	−844	29,436	[5]23,296	1,843,065
1 year.........................	1,735,998	2,995	−552	3,547	[6,7]7,914	1,422,372
2 years........................	1,655,726	1,962	−333	2,295	2,921	1,733,077
3 years........................	1,611,804	1,483	−178	1,661	1,978	1,653,748
4 years........................	1,565,755	1,259	−78	1,337	1,499	1,610,305
5 to 9 years...................	7,788,552	4,016	−139	4,155	4,429	7,835,838
5 years........................	1,549,887			849	1,093	1,564,662
6 years........................	1,566,849			853	851	1,549,036
7 years........................	1,598,571			842	848	1,566,001
8 years........................	1,559,205			820	831	1,597,740
9 years........................	1,514,040			792	806	1,558,399
10 to 14 years.................	8,732,995	3,233	−423	3,656	3,771	8,316,860
10 years.......................	1,606,246			1,066	929	1,513,111
11 years.......................	1,695,257			809	938	1,605,308
12 years.......................	1,692,488			646	728	1,694,529
13 years.......................	1,812,600			569	608	1,691,880
14 years.......................	1,926,404			566	568	1,812,032
55 to 59 years.................	4,165,786	45,865	−170	46,035	44,578	4,245,067
55 years.......................	878,738			7,994	7,683	910,297
56 years.......................	857,972			8,663	8,329	870,409
57 years.......................	836,593			9 282	8,973	848,999
58 years.......................	798,362			9,808	9,545	827,048
59 years.......................	794,121			10,288	10,048	788,314
60 to 64 years.................	3,330,516	60,311	−19	60,330	58,635	3,415,025
60 years.......................	667,720			10,787	10,538	783,583
61 years.......................	687,873			11,263	11,025	656,695
62 years.......................	651,556			11,880	11,572	676,301
63 years.......................	672,390			12,718	12,299	639,257
64 years.......................	650,977			13,683	13,201	659,189
65 to 69 years.................	2,669,247	80,429	−196	80,625	78,706	2,781,336
65 years.......................	617,668			14,578	14,131	636,846
66 years.......................	573,330			15,413	14,996	602,672
67 years.......................	511,332			16,199	15,806	557,524
68 years.......................	506,735			16,911	16,555	494,777
69 years.......................	460,182			17,524	17,218	489,517
70 years and over..............	3,750,885	315,120	−2,075	317,195	325,957	3,866,892

[1] Source: Japan, Office of Prime Minister, *Population Estimates Series*, No. 32, "Population Estimates by Age and Sex as of October 1, 1967," March 1968.
[2] Single years of age 5 and over obtained by osculatory interpolation of 5-year age groups (Sprague multipliers).
[3] $1/2(D_a - M_a) + 1/2(D_{a-1} - M_{a-1})$ where D refers to deaths, M to net migrants, and a to age.
[4] Total excludes births during the period, Oct. 1, 1966, to Oct. 1, 1967.
[5] Obtained by applying formula: $0.8D_0 - 0.5M_0 = 0.8(28,592) - 0.5(-844) = 23,296$.
[6] Obtained by applying formula: $0.2D_0 + 0.5D_1 - 0.5M_0 - 0.5M_1 = 0.2(28,592) + 0.5(2,995) - 0.5(-844) - 0.5(-522) = 7,914$.

the totals is not too satisfactory because of the marked changes in the number of deaths from one 5-year age group to another, particularly at the higher ages. The totals for 5-year age groups may be satisfactorily subdivided by osculatory interpolation. An alternative technique would consist of (1) multiplying the central death rates for single years implicit in some unabridged life table (e.g., 1959–61 Japanese life table) by the population figures in single years of age at the base date and then (2) adjusting the resulting deaths to the established total for each 5–year age group. In this illustration deaths by single years of age for ages 5 and over were obtained by osculatory interpolation (Sprague Multipliers).[10]

4. The next step logically is to determine the deaths in each cohort that must be subtracted from the population. It is evident that the deaths classified in terms of age at time of death do not represent the deaths experienced by any single age cohort or group of age cohorts. Some of the deaths occurred to persons attaining that age during the year and, hence, to persons who are not in the cohort aged x at the beginning of the year; it is not correct, therefore, to subtract deaths at age x from the population aged x at the beginning of the year. Nor is it correct to subtract deaths at age x from the population aged $x-1$ at the beginning of the year because some of the deaths occurred to persons who were already aged x at the beginning of the year. It is necessary, therefore, to separate deaths at each single year of age into the two groups corresponding to the two population cohorts with which the deaths are associated before they can be subtracted from the population.

In the present case, where we are estimating over a 12-month period, the following procedure may be followed: Above infancy, the distribution of the deaths at a particular year of age, within the year of age and over the estimate period is so nearly rectangular that a separation factor of .50 may be used for subdividing the deaths; that is, one-half of the deaths at a given age are estimated to occur to the population at that age at the beginning of the year and one-half to the population at the next younger age at the beginning of the year. It should be noted that a rectangular assumption disregards any available information regarding monthly changes in the number of deaths.

For infancy, the distribution of deaths within the year of age is sufficiently uneven to require the use of special separation factors. To subdivide the deaths by cohorts, the Japan Bureau of Statistics uses proportions derived from tabulations of deaths by year of birth. On this basis a separation factor of .80 for deaths of infants in 1960 is obtained. Hence, the number of deaths corresponding to the cohort under 1 on October 1967 is $.80 D_o$ (where D_o represents infant deaths between October 1966 and September 1967); to the cohort aged 1, $.20 D_0 + .50 D_1$; to the cohort aged 2, $.50 D_1 + .50 D_2$; etc. The calculations for adjusting the deaths at the earliest ages are shown at the foot of table 22–3.

5. The international migration component requires the same general treatment as the mortality component: Determining the age distribution for the estimate period in age groups, subdividing the data into single years of age, and then redistributing the figures by age to obtain migrants corresponding to population age cohorts. As in the case of deaths, osculatory interpolation may be used to divide the age data by single years; and a rectangular assumption may be employed

throughout to redistribute net migrants according to population age cohorts.

6. Inasmuch as the same operations are performed on deaths and net migrants and the same adjustment factors are applied to both components at ages 2 and over, these calculations may be performed on both components in a single step: First take the difference between deaths and net migrants, or column 2 minus column 3, and then apply the interpolation and adjustment procedures to the difference (col. 4). Thus, the entry in column 5 of table 22–3 for age 4 was derived by taking one-half the entry in column 4 at age 3 and one-half the entry in column 4 at age 4. For the ages under 1 and 1, the deaths and net migrants are adjusted separately because different separation factors are employed.

7. The final estimates (col. 6) are obtained by subtracting the difference between deaths and net migrants, adjusted to terminal ages (col. 5), from the initial population one year younger (col. 1). For example, the estimated population aged 4 on October 1, 1967 (1,610,305) was derived by subtracting adjusted net deaths 4 years of age (1,499) from the initial population 3 years of age (1,611,804).

A set of annual midyear postcensal estimates by age normally has to be built up from the census counts by single years of age, which may require adjustment for various types of reporting errors, such as underenumeration and age misreporting, particularly age heaping. The census figures would usually also have to be carried forward to the middate of the first postcensal year.

Adjustments for underenumeration and age misreporting in age groups may be available in the form of estimates of net census undercounts, that is, the combination of these two types of errors. A number of courses can be taken with respect to the use of these estimates. The census counts may be employed unchanged, without adjustment, on the ground that the postcensal estimates by age should be comparable with the official census counts or that the correction factors are subject to too much error to be used with confidence. An alternative is to apply the corrections and to carry the corrected population forward. If the corrections are reasonably valid, the resulting population estimates are more accurate reflections of the numbers at each age and the estimates of change by age groups since the last census are more accurate than with the uncorrected numbers. Another possibility is to "inflate" the census counts, carry these corrected figures forward by age cohorts to an older age, then "deflate" the results to census level by use of the corrections at the last census applicable at the older age group. Note in the following formula that the "deflation" factor (c_{a+5}) differs from the "inflation" factor (c_a) because a different age group is involved at the different dates:

$$_5P_{a+5} = \frac{(c_a \cdot _5P_a) - D + M}{c_{a+5}} \qquad (7)$$

The resulting estimates are at a level comparable to the official census counts and should be related to these in measuring changes by age in the postcensal period.

Often annual estimates of the population in the conventional 5-year age groups are all that is desired. Even so, it is probably more efficient to carry out most of the calculations in single years of age because of the changing identity of the cohorts in the estimate for any 5-year age group.

Limited cohort-component method. — The cohort-component method may be applied in a more limited way than described

Table 22–4 — Estimation of the Population of the Philippines, by Age, for 1965 and 1962, by the Survival-Rate Method

Age in		Census population, Feb. 15, 1960 [1]		Survival rate [2]		Survivors (both sexes), Feb. 15, 1965 $[(1)\times(3)]+[(2)\times(4)]=$	Estimated population, Feb. 15, 1965 (5) adjusted [3] =	Estimated population, July 1, 1962	
1960	1965	Male	Female	Male	Female			Initial [4]	Adjusted [5] .99717281x (7)=
		(1)	(2)	(3)	(4)	(5)	(6)	(7)	(8)
All ages.........	All ages........	[6]13,662,869	[6]13,424,816	(X)	(X)	29,428,978	31,489,265	29,178,436	29,095,943
Births, 1960–65 [7]	Under 5 years.......	2,104,067	1,909,308	.87196	.88895	3,531,942	5,077,159	(X)	(X)
Under 5 years..........	5 to 9 years........	2,354,038	2,218,377	.95941	.96125	4,390,902	4,648,438	4,812,168	4,798,563
5 to 9 years..........	10 to 14 years......	2,254,566	2,114,832	.98620	.98603	4,308,741	4,360,338	4,501,942	4,489,214
10 to 14 years.......	15 to 19 years......	1,765,992	1,669,435	.98536	.98523	3,384,915	3,425,449	3,874,760	3,863,805
15 to 19 years.......	20 to 24 years......	1,384,759	1,429,547	.97870	.97998	2,756,191	2,789,196	3,104,599	3,095,822
20 to 24 years.......	25 to 29 years......	1,194,182	1,264,441	.97392	.97572	2,396,778	2,425,479	2,615,645	2,608,250
25 to 29 years.......	30 to 34 years......	952,368	1,000,981	.97094	.97236	1,898,006	1,920,735	2,177,611	2,171,454
30 to 34 years.......	35 to 39 years......	764,978	791,473	.96591	.96864	1,505,552	1,523,581	1,729,486	1,724,596
35 to 39 years.......	40 to 44 years......	702,568	725,906	.95792	.96426	1,372,966	1,389,407	1,473,650	1,469,484
40 to 44 years.......	45 to 49 years......	546,393	552,585	.94643	.95810	1,046,554	1,059,086	1,236,932	1,233,435
45 to 49 years.......	50 to 54 years......	524,638	508,045	.92904	.94636	·968,203	979,797	1,045,224	1,042,269
50 to 54 years.......	55 to 59 years......	365,354	344,745	.90354	.92742	649,835	657,617	838,206	835,836
55 to 59 years.......	60 to 64 years......	252,394	235,536	.86569	.89661	429,679	434,824	568,531	566,924
60 to 64 years.......	65 to 69 years......	231,786	199,118	.81117	.84916	357,101	361,377	432,766	431,542
65 to 69 years.......	70 to 74 years......	112,702	113,126	.73552	.77830	170,941	172,988	290,214	289,394
70 to 74 years.......	75 to 79 years......	106,799	102,141	.63147	.67648	136,537	138,172	191,863	191,321
75 years and over.....	80 years and over...	149,352	154,528	.39303	.42345	124,135	125,622	284,839	284,034

X Not applicable.

[1] Source: United Nations, *Demographic Yearbook, 1962*, table 5.

[2] The model life table employed here corresponds to "Model West, mortality level 15," in Ansley J. Coale and Paul Demeny, *Regional Model Life Tables and Stable Populations*, Princeton, N.J., Princeton University Press, 1966.

[3] The difference (2,060,287) between the independent estimate of the total population (31,489,265) and the total number of survivors (29,428,978) was distributed among the survivors as follows: 75 percent to ages 0–4, 12.5 percent to ages 5–9, and 12.5 percent to ages 10 and over.

[4] Obtained by linear interpolation: $.525P_{1960}+.475P_{1965}$.

[5] Factor equals ratio of estimated total population obtained by geometric extrapolation to total obtained in col. (7): $\frac{29,095,943}{29,178,436}=.99717281$.

[6] Births are excluded.

[7] The sex ratio of births in 1963 was applied to total births for 1964 and 1965 to obtain male and female births for these years. Five-year totals equal 87.5 percent of births in 1960, all births in 1961–64, and 12.5 percent of births in 1965.

above, that is, in less detailed or precise form, at less frequent intervals, or in combination with mathematical or other procedures not employing components. We describe below one variation that is particularly applicable to a country for which good death statistics by age are lacking and which has only a negligible volume of net immigration. Statistics or estimates of births for the postcensal estimate period and an estimate of total population on the estimate date are required in addition to data on the age distribution of the population at the last census. Typically, in the underdeveloped countries reported births would be grossly underregistered and, hence, some allowance for underregistration should be made. The method consists simply of carrying forward the population by 5-year age groups, as enumerated at the last census, for 5 years by the use of life-table survival rates and then adjusting the surviving population by age proportionately to the independent estimate of the total population. Annual estimates of population by age may then be secured by interpolating linearly to each calendar year between the census counts and the estimates 5 years later by age. This interpolation may be applied to the absolute numbers or the percent distribution by age, but in each case the interpolated figures should be tied in with the preestablished total population figure.

This method of deriving estimates at quinquennial and annual intervals is illustrated with data for the Philippine Islands in table 22–4. Census counts by age and sex for February 15, 1960, as enumerated (cols. 1 and 2) are carried forward to February 15, 1965 (col. 5) by use of survival rates from a model life table (cols. 3 and 4). The procedure for selecting a model life table for use in making population estimates is described in chapter 24; the selection here of model West, level 15, of the Coale-Demeny volume on regional model life tables is partly an arbitrary one.[11] The survival calculations are carried out separately for males and females because the survival rates come separately for each sex, but only the survivors for both sexes combined need to be recorded. The total number of survivors is 29,429,000, but the independent estimate for February 15, 1965, obtained by assuming a continuation of the 1950–60 growth rate (3.0 percent) amounts to 31,489,000. This discrepancy may reflect the failure to allow for net immigration, understatement of births, and overestimation of deaths. Net immigration is believed to be negligible. Given a fixed or rising growth rate at the approximate level of 3 percent, the number of births is clearly grossly understated, possibly by as much as one third, or over 1½ million; this appears to be confirmed by more elaborate analysis.[12] Instead of adjusting the distribution of survivors *pro rata* to the assigned total in 1965, therefore, even in a rough calculation it

[11] Ansley J. Coale and Paul Demeny, *Regional Model Life Tables and Stable Populations*, Princeton (N.J.), Princeton University Press, 1966, p. 16.

[12] Frank Lorimer, "Analysis and Projections of the Population of the Philippines," *First Conference on Population, 1965*, Proceedings of the Conference sponsored by the Population Institute of the University of the Philippines, Quezon City, University of Philippines Press, 1966, pp. 200–314, esp. 227–231.

would seem preferable to assign most of the discrepancy to the population under 5 in 1965 as a way of allowing for the underregistration of births. We have arbitrarily assigned $\frac{3}{4}$ of the discrepancy to the population under 5, $\frac{1}{8}$ to the population 5 to 9, and $\frac{1}{8}$ to the population 5 and over (col. 6). This type of adjustment implies a more reasonable progression of numbers in ages under 5, 5–9, and 10–14 in 1965 and of births in the 15 years preceding 1965. (Some adjustment of the 1960 census counts may be desirable to assure consistency with the adjusted numbers for 1965 before the next step is carried out; but, since we are merely illustrating the major steps, this step has been omitted.)

Table 22–4 goes on to illustrate the calculation of estimates of population by age for July 1, 1962. Having established the age distribution in 1965 by a cohort method, relatively simple mathematical procedures will give a close approximation to consistent cohort estimates for prior dates. First approximations are obtained by linear interpolation of the absolute numbers for the same age groups in 1960 and 1965:

$$P_a^t = m_0 P_a^0 + m_1 P_a^1 \qquad (8)$$

where m_o and m_1 represent the interpolation multipliers. The interpolation multipliers are based on the following fractions:

$$\frac{2.375}{5} = .475 \text{ and } \frac{2.625}{5} = .525$$

These multipliers are the proportions of the (5–year) intercensal period before (.475) and after (.525) the estimate date, applied in reverse order. Substituting in equation (8):

$$P_a^{1962} = .525 \, P_a^{1960} + .475 \, P_a^{1965}$$

These results are then adjusted *pro rata* to the total for July 1, 1962 (29,095,943), derived on the basis of the assumption of a constant growth rate (3.0 percent).

Interpolation of age groups by cohorts to July 1, 1962, for

example between ages 15–19 in 1960 and ages 20–24 in 1965, or between ages 15–24 and ages 20–29, would be undesirable for several reasons. Cohort interpolation of 5- or 10-year age groups to some intermediate date within a 5-year time period would initially produce estimates in "odd" 5-year age groups which would then require redistribution into the conventional ages. Such calculations would be more numerous, and not necessarily more exact or consistent with the initial figures, than the calculations for linear interpolation. Under these circumstances, it is preferable to interpolate between figures for the same 5- or 10-year age groups; this procedure does not require the redistribution of the interpolated figure by age.

Mathematical methods. — A number of mathematical procedures may be used in preparing postcensal estimates of national population by age group (see appendix C). These procedures give relatively rough results and are to be used only when lack of adequate data or information on changes due to births and deaths, and other considerations, make it impractical to use some type of cohort–component method.

Two types of mathematical procedures, both employing a single independently determined estimate of the current total population are illustrated in table 22–5. Estimates have been made for broad age groups for Panama on July 1, 1968, by (1) linear extrapolation of the absolute census counts for 1950 and 1960 and (2) linear extrapolation of the percent of the population in each age group at the two censuses. In the first procedure, the census counts are extrapolated to July 1, 1968, by use of the multipliers -0.756 for the 1950 figure and 1.756 for the 1960 figure (col. 3).

The preliminary estimates at each age derived by linear extrapolation receive a further proportional adjustment to the assigned figure for the total population. Such a figure was calculated by geometric extrapolation (1,338,101). The factor for *pro rata* adjustment represents the ratio of the assigned total to the total for Panama calculated by linear extrapolation. The preliminary figures receive an adjustment of about 4.6 percent (col. 4).

In the second procedure, the calculation of estimates by the method of linear extrapolation of the percents, the specific steps are as follows: First, the percent distributions by age

Table 22–5. — Estimation of the Population of Panama, by Age, for July 1, 1968, by Linear Extrapolation of Numbers and Percents

Age	Census population		Linear extrapolation of numbers, July 1, 1968		Census percent distribution		Linear extrapolation of percents, July 1, 1968	
	Dec. 10, 1950[1]	Dec. 11, 1960[2]	Initial estimates[3]	Adjusted estimates[4] 1.0455098x(3)=	Dec. 10, 1950	Dec. 11, 1960	Percents[3]	Population estimates 1,338,101x(7)=
	(1)	(2)	(3)	(4)	(5)	(6)	(7)	(8)
All ages....................	805,285	1,075,541	1,279,855	1,338,101	100.000	100.000	100.000	1,338,101
Under 5 years....................	131,168	181,939	220,322	230,349	16.288	16.916	17.391	232,709
5 to 14 years....................	205,706	285,445	345,728	361,462	25.545	26.540	27.291	365,181
15 to 24 years....................	148,784	197,895	235,023	245,719	18.476	18.400	18.343	245,448
25 to 34 years....................	118,949	141,155	157,943	165,131	14.771	13.124	11.879	158,953
35 to 44 years....................	85,609	109,417	127,416	133,215	10.631	10.173	9.827	131,495
45 to 54 years....................	54,390	75,958	92,263	96,462	6.754	7.062	7.295	97,614
55 to 64 years....................	35,203	45,838	53,878	56,330	4.371	4.262	4.180	55,933
65 years and over.................	25,476	37,894	47,282	49,434	3.164	3.523	3.794	50,768

[1] Source: United Nations, *Demographic Yearbook, 1960*, table 5.

[2] Source: United Nations, *Demographic Yearbook, 1962*, table 5.

[3] Multipliers are −.756 for the 1950 population and 1.756 for the 1960 population.

[4] Factor equals $\frac{1,338,101}{1,279,855} = 1.0455098$. Total of all ages was obtained independently by geometric extrapolation.

in 1950 and 1960 are computed (cols. 5 and 6). Second, estimates of this distribution on July 1, 1968 are derived by linear extrapolation, employing the same multipliers as for linear extrapolation of the absolute census counts (col. 7). The extrapolated percents will automatically add to 100.0 percent for all Panama as required. The extrapolated percents are then multiplied by the previously assigned total population for July 1, 1968 (1,338,101) to secure the population estimates for age groups on that date.

The limitations characteristic of mathematical extrapolation should be borne in mind in using this method to make postcensal estimates of population by age. They apply with particular force in connection with age estimates. The assumption that the direction and average amount of population change by age in a past intercensal period will continue for another intercensal period is subject to considerable error. We should, therefore, expect such methods to be less accurate than the cohort-component method, even where mortality and fertility by age have to be estimated.

Intercensal estimates. — Less refined methods are satisfactory for the calculation of intercensal estimates; but, as we have noted, to make full use of the available data, we have the special problem of dealing with the error of closure. Here, the adjustment for the error of closure must be made age by age. Intercensal adjustments may be made as a part of, or a supplement to, a cohort-survival procedure and involve either the calculation of forward-and-reverse-survival estimates, purely mathematical calculations, or an "inflation-deflation" procedure (see below). If a mathematical method has been used to prepare the "preliminary" intercensal estimates, the adjustment for error of closure would be wholly incorporated in it and further adjustment would be unnecessary. Not all of the difference between the census count for an age group and the count for the same group (cohort) at the later census can be accounted for by errors in the available estimates of net change due to deaths and net migration (and births for youngest age groups). Part of the discrepancy is a consequence of the difference between the net undercounts in the two censuses for the age groups in the same cohort. This is evident from the sharp variation in the residuals by age shown in table 8-5 relating to intercensal changes for the United States, 1950–60. This irregularity cannot reasonably be attributed entirely to errors in the independent estimates of net change.

Several alternative procedures for handling the error of closure in connection with adjusting postcensal estimates by age made by the cohort-survival method may be considered first. These methods produce estimates which are in one way or another comparable with the census counts.

The forward-reverse survival procedure of making intercensal estimates of the age distribution of a population is illustrated by the calculation of such estimates for Panama for July 1, 1955 (table 22–6). The procedure involves the calculation of two preliminary estimates, one by aging the first census (cols. 1–6) and the second by "younging" the second census (cols. 7–12), and then averaging the two estimates (cols. 13–14). First, the December 1950 census population was aged to December 1955 by use of one of the U.N. model life tables (average of mortality levels 60 and 65).[13] It was assumed that net immigration equalled or approximated zero. The "forward" estimates of the population for July 1, 1955 (col. 6) were then derived by linear interpolation, at each age, between the census counts for

December 1950 and the survivors in December 1955. The equation employed for this purpose, expressing the calculations in terms of multipliers, is:

$$P_a^{7/55} = \frac{0.44}{5.00} P_a^{12/50} + \frac{4.56}{5.00} P_a^{12/55}$$

$$P_a^{7/55} = .088 P_a^{12/50} + .912 P_a^{12/55}$$

A second set of preliminary estimates for July 1, 1955, was prepared by the "reverse" procedure. The December 1960 census counts were "younged" to December 1955 by use of one of the U.N. model life tables (average of mortality levels 65 and 70). Estimates of population for July 1955 were then made by interpolating linearly between these estimates for December 1955 and the census counts for 1950 at each age. The equation expressing the latter calculation in terms of multipliers is identical to that given above in connection with the forward estimates of the population for July 1, 1955.

The differences between the forward and reverse estimates are principally a reflection of differences in census net undercounts for a given cohort at the two censuses, but they also reflect any net immigration during the intercensal period. The final estimates are derived by (1) averaging the forward and reverse estimates, with weights in reverse relation to the time lapse from the census dates (col. 13), and (2) forcing the results *pro rata* to the independent estimate of the total population on July 1, 1955, derived by geometric interpolation of the census counts for 1950 and 1960 (col. 14). The preliminary estimates for 1955 are averaged with weights of .544 for the forward estimate and .456 for the reverse estimate. The factor for adjusting the weighted estimates to the assigned total is 918,582 ÷ 905,344, or 1.01462207. Annual estimates by age consistent with these middecade intercensal estimates may be derived by (1) linear interpolation between the census figures and the middecade estimates and (2) *pro rata* adjustment to annual national totals.

Availability of postcensal estimates in single years of age permit the use of a simple mathematical adjustment for the error of closure to derive highly accurate and consistent intercensal estimates. The error of closure may be interpolated linearly along age cohorts.

Other procedures may be wholly mathematical, designed to avoid the direct handling of the error-of-closure problem and to simplify the calculations. The census counts themselves may be interpolated by age cohorts to the middle of an intercensal decade or to each year of the decade. Further mathematical redistribution of the initial estimates will then be necessary. The interpolation between censuses may be a linear interpolation between the same 5-year age groups. Percent distributions by age may be preferable in the latter case. Such estimates may be unreasonable when the figures for the same age cohorts are examined.

Some of the alternative procedures for handling the error of closure within the framework of the cohort-survival method are illustrated by the procedures employed by the U.S. Bureau of the Census in preparing its intercensal estimates for the last three decades. The procedure applied for the 1930–40 decade illustrates one type of forward-reverse survival procedure.[14] In this decade net immigration was near zero. The method consisted of (1) aging the 1930 census figures by single

[13] United Nations, *Methods for Population Projections by Sex and Age,* Manual III, Population Studies, Series A, No. 25, pp. 78–79.

[14] U.S. Bureau of the Census, *Current Population Reports,* Series P–25, No. 114, "Estimates of the Population of the United States, by Age, Color, and Sex: 1900 to 1940," April 27, 1955.

Table 22-6. — Calculation of Intercensal Estimates of the Population of Panama, by Age, for July 1, 1955, by the Forward-Reverse Survival-Rate Method

Forward method

Age in 1950	Age in 1955	Census population, Dec. 10, 1950 Male (1)	Female (2)	Survival rate[1] Male (3)	Female (4)	Survivors, Dec. 10, 1955 (both sexes) $[(1) \times (3)] + [(2) \times (4)] =$ (5)	Preliminary population estimates, July 1, 1955[2] (6)	Weighted population estimates, July 1, 1955 Initial[8] (13)
All ages	All ages	[3]409,589	[3]395,694	(X)	(X)	895,493	887,555	905,344
Births, 1950-55[4]	Under 5 years	77,759	74,842	.84810	.86665	130,809	130,841	144,229
Under 5 years	5 to 9 years	66,288	64,882	.94814	.95001	124,489	123,631	126,495
5 to 9 years	10 to 14 years	58,051	56,686	.98452	.98400	112,931	110,998	109,282
10 to 14 years	15 to 19 years	46,511	44,457	.98378	.98309	89,462	88,434	89,743
15 to 19 years	20 to 24 years	37,908	39,877	.97564	.97618	75,912	75,480	76,170
20 to 24 years	25 to 29 years	35,140	35,859	.97066	.97145	68,944	68,557	67,890
25 to 29 years	30 to 34 years	32,456	32,089	.96929	.96958	62,572	61,853	61,003
30 to 34 years	35 to 39 years	28,080	26,324	.96649	.96770	52,613	52,389	52,285
35 to 39 years	40 to 44 years	26,133	23,931	.96042	.96458	48,182	47,070	45,987
40 to 44 years	45 to 49 years	19,099	16,446	.94947	.95811	33,891	33,474	33,582
45 to 49 years	50 to 54 years	15,021	14,134	.93262	.94651	27,387	27,198	26,826
50 to 54 years	55 to 59 years	13,226	12,009	.90827	.92852	23,163	22,646	22,950
55 to 59 years	60 to 64 years	9,155	8,133	.87225	.89899	15,297	[5]15,527	16,009
60 years and over	65 years and over	22,521	20,867	.67779	.69854	29,841	[5]29,457	32,893

Reverse method

Age in 1960	Age in 1955	Census population, Dec. 11, 1960 Male (7)	Female (8)	Survival rate[6] Male (9)	Female (10)	Younged population, Dec. 11, 1955 (both sexes) $[(7) \div (9)] + [(8) \div (10)] =$ (11)	Preliminary population estimates, July 1, 1955[7] (12)	Weighted population estimates, July 1, 1955 Final[9] $1.01462207 \times (13) =$ (14)
All ages	All ages	545,774	529,767	(X)	(X)	938,269	926,566	918,582
Under 5 years	(X)	91,868	90,071	.86298	.88105	(X)	(X)	(X)
5 to 9 years	Under 5 years	79,251	76,598	.95509	.95719	163,002	160,201	146,337
10 to 14 years	5 to 9 years	65,961	63,635	.98660	.98630	131,376	129,912	128,345
15 to 19 years	10 to 14 years	52,804	54,431	.98576	.98541	108,804	107,234	110,880
20 to 24 years	15 to 19 years	45,026	45,634	.97850	.97937	92,610	91,305	91,055
25 to 29 years	20 to 24 years	37,792	37,818	.97414	.97529	77,571	76,993	77,284
30 to 34 years	25 to 29 years	33,366	32,179	.97302	.97367	67,340	67,094	68,883
35 to 39 years	30 to 34 years	30,063	28,724	.97053	.97196	60,529	59,990	61,895
40 to 44 years	35 to 39 years	26,656	23,974	.96499	.96899	52,364	52,162	53,050
45 to 49 years	40 to 44 years	23,073	20,618	.95489	.96281	45,577	44,694	46,659
50 to 54 years	45 to 49 years	17,199	15,068	.93889	.95182	34,149	33,710	34,073
55 to 59 years	50 to 54 years	12,502	11,999	.91546	.93480	26,493	26,382	27,218
60 to 64 years	55 to 59 years	11,054	10,283	.88068	.90673	23,893	23,312	23,286
65 to 69 years	60 to 64 years	7,153	6,737	.82907	.86060	[5]16,456	16,584	16,243
70 years and over	65 years and over	12,006	11,998	.61896	.64134	[5]38,105	36,993	33,374

X Not applicable.

[1] Average of survival rates derived from U.N. model life tables with mortality levels 60 and 65.

[2] Derived by linear interpolation between the Dec. 10, 1950, and Dec. 10, 1955, population of the *same* age. Weights are .088 for P_{1950} and .912 for P_{1955}.

[3] Excludes births 1950-55.

[4] Births by sex, Dec. 10, 1950, to Dec. 10, 1955, equal 4.17 percent of births in 1950, plus 1951-54 plus 95.83 percent of births in 1955.

[5] Census population aged 60-64 in 1950 is 17,915 and census population aged 65 and over in 1950 is 25,473.

[6] Average of survival rates derived from U. N. model life tables with mortality levels 65 and 70.

[7] Derived by linear interpolation between groups of the *same* age on Dec. 10, 1950, and Dec. 11, 1955; weights are .088 for P_{1950} and .912 for P_{1955}.

[8] Weighted average of preliminary estimates. Weights equal .544 for the forward method and .456 for the reverse method.

[9] Adjusted to independent estimate for the total population, 918,582. Factor equals $\frac{918,582}{905,344} = 1.01462207$.

ages (smoothed) to each midyear date of the subsequent decade, 1930 to 1939, and to April 1, 1940, by life-table survival rates, (2) "younging" the 1940 census figures for single ages (smoothed) to each earlier midyear date back to 1930 by life-table survival rates, (3) averaging the forward and reverse estimates for each age and year, with weights inversely proportional to the time lapse from the census dates (for example, for July 1, 1932, 77.5 percent of the forward estimate and 22.5 percent of the reverse estimate), and (4) forcing these average estimates *pro rata* to the assigned intercensal estimates for all ages for each year.

In the research for preparing the intercensal estimates for the United States for 1940-50, a few procedures, such as those described earlier, were tested before the final procedure, the "inflation-deflation" procedure, was selected.[15] The inflation-deflation procedure combines use of postcensal estimates by the cohort-component method, cohort adjustment for the error of closure in single ages, and allowance for net census

[15] U.S. Bureau of the Census, *Current Population Reports*, Series P-25, No. 98, "Estimates of the Population of the United States and of the Components of Change, by Age, Color, and Sex: 1940 to 1950," August 13, 1954.

undercounts. The cohort method ideally requires a base free of net undercounts. It appeared desirable, therefore, to construct a tentative set of estimates of a "true" population by age on April 1, 1940, i.e., approximately corrected for net undercounts, to which the estimate of net cohort change could be added. This calculation provided "true" intercensal estimates by age and, when continued to April 1, 1950, provided a "true" population in each year from 1940 to 1950. The difference between the "true" population in 1940 and 1950 and the census counts represented tentative estimates of net undercounts by age in these years. In order to maintain the same age pattern of "net undercounts" in the intercensal estimates as in the census counts, tentative estimates of "net undercounts" by age for intercensal dates were obtained by interpolation between the estimated amounts of net undercount at the census dates. These estimates were then subtracted from the "true" population estimates, to provide intercensal estimates comparable with the census levels. The "deflated" population by age for any year was adjusted to agree with previously established estimates of the total population for that year.

The procedure employed for the 1950-60 decade was essentially the same as for the 1940-50 decade, but there were several differences.[16] The calculations were carried out in terms of single years of age; carefully prepared estimates of net census undercounts for 1950 were employed and consistent estimates of net census undercounts for 1960 were derived in the process of preparing postcensal estimates for the 1950-60 decade; and ratios of census counts to the "true" population in 1950 and 1960 were computed and interpolated (linearly) between 1950 and 1960 to each estimate date. In the final step, the postcensal estimates of "true" population were deflated to a level comparable to the two censuses by multiplying them by the interpolated ratios of uncorrected estimates to corrected population.

If mathematical interpolation is employed to prepare intercensal estimates for age groups, the error of closure is auto-

matically taken care of in the principal calculation and a separate adjustment for this error is unnecessary. Two types of mathematical procedures, both employing a single estimate of the current total population, are illustrated in table 22-7. Estimates have been made for broad age groups for Panama on July 1, 1955, on the basis of the 1950 and 1960 census counts, by (1) linear interpolation of the absolute census counts, and (2) linear interpolation of the percents of the population in each age group at the two censuses. In the first procedure, the preliminary estimates require a *pro rata* adjustment to the assigned total population obtained directly by geometric interpolation. In the second procedure, the interpolated percents will automatically add to 100 percent for all Panama.

The specific steps in the calculation of estimates by the method of linear interpolation of the percents are as follows: First, the percent distributions by age in 1950 and 1960 are computed. Second, estimates of this distribution on July 1, 1955, are derived by linear interpolation, employing 0.544 as the multiplier for the 1950 percent and 0.456 as the multiplier of the 1960 percent at each age. The interpolated percents are then multiplied by the assigned total population for July 1, 1955 (918,582) obtained by geometric interpolation of the census counts, to secure the population estimates for age groups on that date.

In the derivation of annual historical series, in the absence of data on components of change, polynomials of various degrees may usefully be fitted to census data. For example, the annual estimates of the U.S. population by age and sex for 1900 to 1919 were derived by 5-point (4th-degree) polynomial interpolation of figures for identical age groups from five successive censuses.[17] For the 1900-10 decade, the 1880 to 1920 censuses were used; annual estimates for 1910 to 1919 were based on the 1890 to 1930 censuses. A fourth-degree polynomial equation yields a flexible curve capable of reflecting a constantly changing trend as determined by the relation of one census count to another. There is an advantage in em-

Table 22-7. — Calculation of Intercensal Estimates of the Population of Panama, by Age, for July 1, 1955, by Linear Interpolation of Numbers and Percents

Age	Census population		Linear interpolation of numbers, July 1, 1955		Census percent distribution		Linear interpolation of percents, July 1, 1955	
	December 10, 1950[1]	December 11, 1960[1]	Initial estimates[2]	Adjusted estimates[3] .989294815x(3)=	December 10, 1950	December 11, 1960	Percents[2]	Adjusted estimates 918,582x(7)=
	(1)	(2)	(3)	(4)	(5)	(6)	(7)	(8)
All ages......................	805,285	1,075,541	928,522	918,582	100.000	100.000	100.000	918,582
Under 5 years...................	131,168	181,939	154,320	152,668	16.288	16.916	16.574	152,246
5 to 14 years...................	205,706	285,445	242,066	239,476	25.545	26.540	26.000	238,831
15 to 24 years..................	148,784	197,895	171,179	169,346	18.476	18.400	18.441	169,396
25 to 34 years..................	118,949	141,155	129,075	127,693	14.771	13.124	14.020	128,785
35 to 44 years..................	85,609	109,417	96,465	95,432	10.631	10.173	10.422	95,735
45 to 54 years..................	54,390	75,958	64,225	63,537	6.754	7.062	6.894	63,327
55 to 64 years..................	35,203	45,838	40,053	39,624	4.371	4.262	4.321	39,692
65 years and over...............	25,476	37,894	31,139	30,806	3.164	3.523	3.328	30,570

[1] See table 10.

[2] Multipliers are .544 for the 1950 population and .456 for the 1960 population.

[3] Adjusted to independent estimate of the total population, 918,582; factor equals $\frac{918,582}{928,522} = .989294815$.

[16] U.S. Bureau of the Census, *Current Population Reports*, Series P-25, No. 310, "Estimates of the Population of the United States and Components of Change, by Age, Color, and Sex: 1950 to 1960," June 30, 1965.

[17] U.S. Bureau of the Census, *Current Population Reports*, Series P-25, No. 114, p. 1.

ploying data for the interpolation lying on both sides beyond the period to be interpolated.

Geographic Subdivisions of Countries

Total Population. — The preparation of estimates of the total population of geographic subdivisions of a country, such as states, provinces, counties, and cities, generally requires a somewhat different method from estimates for a country as a whole because fewer data are available and these are of poorer quality. Although data on births and deaths for such area units are available for many countries on a regular basis, direct information on the volume of migration into and out of these areas is not ordinarily available. Though a component method may still be used, therefore, the net migration component must be estimated indirectly. Techniques which employ estimates of net migration directly are applicable in some cases.

Individual states or provinces sometimes take their own censuses or the national government may take state or provincial censuses. If the state census figures are adjusted for any differences between them and the national census figures in the definition of the population covered, they can serve as additional base points in the preparation of estimates for other years.

Population registers and registrations. — Data from population registers, registration data, and city directories may be used to develop population estimates for small geographic areas for occasional dates or on a regular basis. According to the general review of the use of continuous population registers conducted by the United Nations in the early 1960's, there is evidence that population registers were used in the preparation of local population estimates for about one-third of the 46 countries studied.

A registration must be compulsory (e.g., a military registration) or quasi-compulsory (i.e., voluntary but supported by strong pressures to participate, as, for example, registration for food ration books), so as to ensure reasonably complete coverage of the population. The registration data may have to be adjusted to include certain segments of the population not required to register and to exclude others not in the population for which estimates are being prepared. Military registration data are of limited usefulness for making population estimates, primarily because they usually cover only a narrow range of the age distribution and are not very comparable in coverage with census data.

Delayed registrations create a special problem. Since they may be numerous, it is desirable to include registrations for a short period following the initial registration date. On the other hand, it is hazardous to use a count of registrants for a date far removed from the date of the initial registration because the registration lists may not be adjusted to exclude persons who died or left the area after the initial registration date or to include persons who migrated into the area during this period.

A city directory may be used to estimate directly the adult population of many cities in the United States; these estimates may then be supplemented with estimates of the child population obtained in another manner. The general procedure is to count the names listed in the directory. City directories are compiled by private directory firms on the basis of a house-to-house enumeration and are available on a current basis for most large cities. They customarily list alphabetically, according to surnames, all persons 21 years old and over and employed persons 18 to 20 years old residing in a given city or a

city and its surrounding area. For city estimates, persons living outside the city, identified by the address listed, must be excluded from the count. Because a complete counting job is quite laborious and costly, sampling methods may be used. The estimating procedure may be improved by modifying the sampling ratio to include an adjustment factor representing the ratio of the census count of the covered population to the directory count for the census year. The method and quality of directory enumeration vary from area to area and from year to year; hence, before a city directory is used for estimating population, the directory company should be consulted regarding these matters.

Component methods: Direct data on migration. — Estimates of regional and local population may be prepared for current dates by a direct component method if satisfactory data on births, deaths, and migration are available. Recall the basic relationship

$$P_e = P_c + B - D + M \qquad (9)$$

where P_e is the estimated total population on the estimate date, P_c is the census count, and B, D, and M represent the components of change in the postcensal period. For statistically developed countries, the required birth and death statistics or estimates are available with only a brief time lag. The availability of adequate data on migration for local areas for current years, particularly on a continuing basis, presents a special problem. Migration data may be secured for geographic subdivisions of a country on a current basis from continuing national sample surveys, surveys on internal migration, population registers or registrations, special tabulations from appropriate administrative records such as tax returns, records of a family allowance system or social security system, etc.

Typically, the migration data from sample surveys fail to include those migrants who died during the estimate period. The deaths of in-migrants must be included in the data on in-migrants just as they are included in the reported death statistics. The adjustment of the migrant data for deaths can be effected by a simple estimating procedure. One such procedure is illustrated later in this chapter.

The inclusion of a migration (residence at fixed previous date) question on the census schedule makes possible the preparation of estimates for a specific precensal date by a component method involving direct measurement of the migration component. The general formula is:

$$P^{t-x} = P^t + D^x - B^x - M^x \qquad (10a)$$

or

$$P^{t-x} = P_c^t + D_c^x - M_c^x \qquad (10b)$$

where P^{t-x} is the estimated total population at a particular date x years before the census (i.e., 1 year, 5 years, etc.), P^t is the census population, D^x is the number of deaths in the period of x years between the estimate date and the census date, B^x is the number of births during the period, and M^x is the number of (net) migrants during the period. The formula may take two forms. The elements may relate to all ages (10a) or to the cohorts x years of age and over at the census date (10b). In the former case P^t and P^{t-x} relate to the total population at the census date and the estimate date, respectively, and D^x, B^x, and M^x relate to total deaths, births, and (net) migrants, respectively. In the latter case, P^{t-x} is the estimated total population at the estimated date, P_c^t is the census population aged x and over (e.g., 1 and over, 5 and over, etc.), D_c^x is the number of deaths to the cohorts aged x and over on the census date, and M_c^x is the number of (net) migrants affecting these cohorts.

The two alternative procedures may be illustrated by the computations for estimating the population of the State of Connecticut on April 1, 1955. The specific steps in the first procedure are as follows:

1. Total population on April 1, 1960................ 2,535,234
2. Births between April 1, 1955, and March 30, 1960.. 271,995
3. Deaths between April 1, 1955, and March 30, 1960.. 113,606
4. Net migration between April 1, 1955, and March 30, 1960.
   a. Net internal migration of persons aged 5 years old and over alive on April 1, 1960 (population living in other States in 1955)... *a* 35,827
   b. Net internal migration of children under 5 alive on April 1, 1960 (native population born in other States)............................. 18,781
   c. Adjustment for deaths after migration during period (see below)................................ 114
   d. Adjustment for overseas migration............ 31,044
   e. Net migration between April 1, 1955, and March 30, 1960 ($a+b+c+d$)................. 85,766
5. Estimated population on April 1, 1955 ($1-2+3-4e$).. 2,291,079

a This figure does not include an allowance for persons who failed to report migration status in the census.

Data on in-movement from overseas for the population 5 years old and over are available in the 1960 census of the United States, but corresponding data on out-movement overseas are not available. A figure on in-movement only would tend to overstate the net movement. The net movement may be viewed as composed of net movement of aliens, net movement of Puerto Ricans, net movement of other civilian citizens, and net movement of military personnel. The last three groups are not easily measurable or are of negligible volume. Hence, only the movement of alien immigrants is included here. The portion of all alien immigrants from the frontier-control statistics allocated to Connecticut has been determined on the basis of the distribution of the foreign-born population by States in 1960.

The adjustment for deaths of migrants was determined by applying life table survival rates to the net surviving migrants to Connecticut, 1955–60, distributed by age. Specifically, 5-year survival rates were divided into the (net surviving) migrants to determine the hypothetical "total" number of net migrants in 1955. The difference between the total number of net migrants and the number of surviving migrants represents the adjustment for deaths over the whole period (227); one-half of this number may be taken to represent the adjustment for deaths of net migrants after migration (114). The calculations are shown in table 22–8.

The specific steps in the second procedure are as follows:

1. Population 5 years old and over on April 1, 1960... 2,257,004
2. Deaths between April 1, 1955, and April 1, 1960 (cohorts 5 years old and over on April 1, 1960):

Table 22–8. — **Estimation of Deaths of Net Migrants, for Connecticut, Between April 1, 1955, and April 1, 1960**

Age in		Surviving net migrants[1] in 1960	Survival rate[2]	Cohort of net migrants in 1955
1955	1960	(1)	(2)	$(1) \div (2) =$ (3)
All ages......	All ages.......	54,608	(X)	54,835
(X)................	0 to 4 years.......	[3]18,781	.97293	19,304
(X)................	5 years and over...	35,827	(X)	35,531
0 to 9 years....	5 to 14 years.....	11,761	.99678	11,799
10 to 19 years....	15 to 24 years...	1,857	.99560	1,865
20 to 29 years....	25 to 34 years...	15,774	.99331	15,880
30 to 39 years....	35 to 44 years...	7,067	.98799	7,153
40 to 49 years....	45 to 54 years...	2,288	.97025	2,358
50 to 59 years....	55 to 64 years...	−255	.93158	−274
60 to 69 years....	65 to 74 years...	−2,394	.85441	−2,802
70 years and over.	75 years and over	−271	.60528	−448

Adjustment for deaths to migrants $= 1/2 \left[(3) - (1) \right]$
    All ages   = 114
    5 and over   = −148

X Not applicable.
[1] Includes only in-migrants and out-migrants coming from or going to another State; source, *U.S. Census of Population: 1960, Subject Reports, Mobility for States and State Economic Areas*, 1963, table 24.
[2] Calculated from: U.S. National Center for Health Statistics, *United States Life Tables: 1959–61*, Vol. 1, No. 1, 1964, table 1.
[3] Source: *U.S. Census of Population: 1960, Subject Reports, State of Birth*, 1963, tables 25 and 31.

   a. Total deaths (all ages)............................. 113,606
   b. Deaths of children born during this period... 6,399
   c. ($= a-b$) Deaths to cohorts 5 years old and over on April 1, 1960............................ 107,207
3. Net migration between April 1, 1955, and April 1, 1960 (cohorts 5 years old and over on April 1, 1960):
   a. Net internal migration of persons aged 5 years old and over alive on April 1, 1960 (census)... 35,827
   b. Adjustment for deaths (see below)............ −148
   c. Net migration from overseas.................... 28,989
   d. Net migration of cohorts 5 years old and over on April 1, 1960 ($a+b+c$)............... 64,668
4. Estimated population on April 1, 1955 ($1+2c-3d$)... 2,299,543

In this procedure it is unnecessary to allow for births and for deaths and net migration in the cohort under age 5 in 1960. As in the first procedure described, the only component of the overseas movement for which allowance was made is alien immigration. The estimate of alien immigration for Connecticut was made in the same way as in the first procedure, except that the figure was adjusted to exclude the movement of children under 5. The adjustment for deaths of migrants in the second procedure differs from that in the first because the deaths of the migrants under 5 are omitted. The calculation of deaths in the cohorts under 5 in 1960 is required only to derive by subtraction an estimate of deaths in the cohorts 5 years and over (table 22–9). That estimate was obtained by applying appropriate separation factors to the reported numbers of deaths aged under 1 and 1–4 in each year,

Table 22–9. — Estimation of Deaths Between April 1, 1955, and April 1, 1960, of Children Born During this Period, for Connecticut

Age	Recorded or estimated deaths[1]					Percent distribution of life table deaths (d_x), 1959-61[2]
	1955-56	1956-57	1957-58	1958-59	1959-60	
Total deaths under 1 year..................	1,134	1,193	1,260	1,319	1,251	(X)
1 to 4 years....................................	(X)	158	168	187	174	100.00
1 year.......................................	(X)	64	68	75	70	40.34
2 years......................................	(X)	(X)	41	46	43	24.69
3 years......................................	(X)	(X)	(X)	36	33	19.07
4 years......................................	(X)	(X)	(X)	(X)	28	15.90
Deaths to cohort under 5 on April 1, 1960......	[3].843 $D_0 =$	$D_0 + 1/2\ D_1 =$	$D_{0-1} + 1/2\ D_2 =$	$D_{0-2} + 1/2\ D_3 =$	$D_{0-3} + 1/2\ D_4 =$	
	956	1,225	1,349	1,458	1,411	(X)

Total deaths to cohort under 5 = 6,399

X Not applicable.

[1] April to April years. Deaths under 1 obtained from monthly data. Deaths 1–4 interpolated from calendar-year data. Single years of age obtained by applying percent distribution of life-table deaths for single ages 1–4 to the group total 1–4. See U.S. National Center for Health Statistics, *Vital Statistics of the United States, Mortality,* annual volumes 1955 to 1960.

[2] Derived from 1959–61 life table for the total population of the United States. See U.S. National Center for Health Statistics, *United States Life Tables: 1959–61,* Vol. 1, No. 1, 1964, table 1.

[3] Separation factor $= \dfrac{l_0 - L_0}{d_0}$, based on the 1959–61 life table for the total population of the United States.

1955 to 1960. The ages of deaths affecting the population cohorts under 5 on April 1, 1960, may be roughly identified in the following sketch:

The "birth" years extend from April of one year to March of the following year. All deaths under 4 in "1959," under 3 in "1958," under 2 in "1957," and under 1 in "1956" are included, but only fractions of the deaths aged 4 in "1959," aged 3 in "1958," aged 2 in "1957," aged 1 in "1956," and aged under 1 in "1955." The first steps are to redistribute the reported calendar year figures to an April–March basis (e.g., on a rectangular assumption) and then to subdivide deaths at ages 1–4 into single ages (e.g., using the distribution of deaths, or d_x, aged 1–4 in a recent U.S. life table). Employing .843 as a separation factor for infant deaths [18] and .5 for all older ages, we have:

$$.843\ D_0^{55} + D_0^{56} + D_0^{57} + D_0^{58} + D_0^{59}$$
$$+ .5\ D_1^{56} + D_1^{57} + D_1^{58} + D_1^{59}$$
$$+ .5\ D_2^{57} + D_2^{58} + D_2^{59}$$
$$+ .5\ D_3^{58} + D_3^{59}$$
$$+ .5\ D_4^{59}$$

where the deaths of a given age relate to the 12 months beginning in April of the year shown.

If life-table survival rates are employed in combination with the terminal population, instead of actual death statistics, to allow for mortality between 1955 and 1960, then the number of deaths during the period tends to be overstated because the terminal population would contain all the "net migrants," who were in the area only part of the time. At the same time, as we know, the reported net migration is understated by the deaths of the "net migrants." These two types of "errors" are additive and, hence, it is desirable to avoid the use of life tables in this type of reverse calculation.

A set of estimates of the population of the States of the United States for 1955 were prepared by use of census data on internal migration for the period 1955–60; the report gives a full description of the methodology.[19] If population estimates for States as a set are being prepared, the State distribution of each component, particularly net migration, should be adjusted to agree with the independently established national total for that component. For example, the State distribution of the number of net internal migrants would be adjusted to zero, the required national figure on net internal migration, and the figures on net movement from abroad would be adjusted to agree with the best available figure on net immigration to the country for this period. More simply, internal

[18] Derived from the 1959–61 U.S. unabridged life table on the basis of the formula $\dfrac{l_0 - L_0}{d_0}$.

[19] U.S. Bureau of the Census, *Current Population Reports,* Series P–25, No. 304, "Revised Estimates of the Population of States and Components of Population Change: 1950 to 1960," April 8, 1965.

migration and net immigration combined may be adjusted to the national total on net immigration.

Component methods: Use of symptomatic series to measure migration. — In the absence of actual data on internal migration, we may estimate this component by use of "symptomatic" data. Let us assume that we want annual population estimates for some class of geographic area, such as states or counties. To serve this purpose, the symptomatic data on internal migration must be available on a continuing current basis, must relate to a substantial segment of the population, must be internally comparable from year to year, and must fluctuate principally in response to changes in population.

Many series of administrative data may be considered for this purpose, for example, data on school enrollment, social security account-holders or beneficiaries, tax returns or population covered by tax returns, or families under the family allowance systems. A specific way in which such data can be used to measure migration will become apparent from the description given below of the use of school enrollment data in making current estimates of State and local population in the United States. School data in various forms — school census, school enrollment, and school attendance data — are well adapted to the preparation of annual population estimates for areas within the United States.

Theoretically, school census data with carefully defined age limits would be most useful for estimating purposes; but, in the few States where school censuses are taken, the quality of enumeration is usually quite poor and varies appreciably from year to year. School enrollment data, even if only by grade, are generally more dependable than school census data for measuring year-to-year population changes; enrollment data by age are preferable in principle to data by grade but are not usually available. Since the school series serves to measure changes in the population of school age, its coverage must be restricted to those ages where attendance is virtually complete, i.e., the compulsory school ages, or to the grades attended, for the most part, by children of compulsory school age. If grade data alone are available, only the elementary grades (excluding kindergarten but including any special and ungraded classes on the elementary level) should be included; high school enrollment data are unsatisfactory since many children drop out of high school and the drop-out rates vary from year to year. The age or grade coverage must be the same from year to year, and the figures must relate to the same date in the school year.

Two methods of estimating the civilian population that employ school enrollment data to estimate civilian net migration into an area, designated as the U.S. Census Bureau's Component Methods I and II, will be described in general terms here.[20] Each method takes direct account of natural increase and the net loss to the armed forces (inductions and enlistments less separations). The first method employing school data to estimate net migration, the Census Bureau's **Component Method I,** is fairly simple to apply. The procedure rests on the basic assumption that the migration rate of school-age children of a local area may be estimated as the difference between the percent change in the population of school age in the area and the corresponding figure for the United States;

the latter figure is presumed to represent for each area the effect of change due to all factors except internal migration. The migration rate of the total population of the local area is then assumed to be the same as the migration rate of the school-age population and is applied to the total population of the area at the census date and one-half of the births in the postcensal period, to derive the estimate of net migration.

The assumption in Component Method I that the trend of fertility in the local area during the postcensal period is the same as that in the country as a whole is subject to question. Changes in fertility vary notably from area to area, even in the short run; for example, the percentage change in the number of births between 1946–52 and 1953–59 (corresponding to the cohorts 6 to 13 years of age in January 1960 and 1967) **differs substantially in** a number of States from the corresponding figure for the United States. The other major assumption, the equivalence of the rate of net migration of school-age children and the total population, is only a rough rule to follow; areas vary greatly in the age pattern of migration. Because of the more realistic nature of its assumptions and its more logical approach in measuring migration, Method II would be expected to yield more accurate results. This is borne out, in general, by actual tests, as shown later.

The U.S. Census Bureau's **Component Method II** calls, first, for estimating net migration for the cohort of school-age children by comparing a current estimate of school-age children with the expected number (excluding migration) derived from the last census, next converting the number to a migration rate, and then converting this rate to a rate for the whole population. More specifically, the net migration component is estimated as follows: (1) First, enrollment in elementary grades 2 to 8 at the estimate date is adjusted to approximate the population of elementary school age ($7\frac{1}{2} \leftrightarrow 15\frac{1}{2}$ years)[21] on the basis of the relative size of these two groups at the last census (relating local school enrollment data to census counts in each case); (2) next, the "expected" population (assuming no net migration) of elementary school age on the estimate date is computed by "surviving" the population in the same cohort at the time of the last census (including, if necessary, births following the census) to this date; (3) then, net migration of children of school age is estimated as the difference between the "actual" population of school age (item 1) and the "expected" population of school age (item 2); (4) next, net migration of school-age children (item 3) is converted into a migration rate by dividing by the population in the same age cohort at the time of the last census (including, if necessary, one-half the natural increase during the postcensal period; (5) next, the migration rate of school-age children is adjusted to represent the migration rate of the total population on the basis of national gross migration experience as estimated in the Current Population Survey for the same postcensal period; (6) finally, total net migration is obtained by applying the migration rate obtained in step (5) to the total population at the last census plus one-half the natural increase during the subsequent postcensal period.

The principal steps in the calculation of the estimate of the population of Middlesex County, New Jersey, for July 1, 1963, are as follows:

1. Total resident population on April 1, 1960...   433,856
2. Military personnel residing in the county on April 1, 1960......................................   722

---

[20] A more detailed description of the first method mentioned is given in: U.S. Bureau of the Census, *Current Population Reports,* Series P–25, No. 20, "Illustrative Examples of Two Methods of Estimating the Current Population of Small Areas," May 6, 1949. For a detailed illustrative example of the application of the second method mentioned, see idem, *Current Population Reports,* Series P–25. No. 339.

[21] Arrow denotes exact limits of age range.

3. Civilian resident population on April 1, 1960 (item 1 – item 2).. 433,134

4. Births, by residence, adjusted for underregistration, April 1, 1960, to July 1, 1963......... 33,459

5. Deaths, by residence, April 1, 1960, to July 1, 1963.. 10,988

6. Elementary enrollment, April 1, 1960, obtained by linear interpolation between October figures for 1959 and 1960................... 67,402

7. Elementary enrollment, July 1, 1963, obtained by linear interpolation of October figures for 1962 and 1963......................... 79,292

8. Estimated population of elementary school age on July 1, 1963
   a. Population 7.25 to 15.24 years old on April 1, 1960....................................... 66,562
   b. Number enrolled on April 1, 1960............. 67,402
   c. Ratio (item 8a ÷ item 8b)....................... .9875
   d. Number enrolled on July 1, 1963.............. 79,292
   e. Estimated population 7.50 to 15.49 on July 1, 1963 (item 8c × item 8d)............... 78,301

9. Expected population of elementary school age, assuming no migration, July 1, 1963
   a. Cohort on April 1, 1960, 4.25 to 12.24 years of age................................. 76,362
   b. Survival rate.. .99860
   c. Survivors 7.50 to 15.49 years of age on July 1, 1963....................................... 76,255

10. Net change in population in this cohort due to migration (item 8e – item 9c)................... +2,046

11. Migration rate for this cohort, April 1, 1960, to July 1, 1963 (item 10 ÷ item 9a)............. +.0268

12. Migration rate for the population of all ages, April 1, 1960, to July 1, 1963 (item 11 × 1.1775, the intercounty migration ratio, see below).. +.0316

13. Estimated total net civilian migration, April 1, 1960, to July 1, 1963
   a. Civilian noninstitutional population of April 1, 1960.. 430,645
   b. One-half of the births during the period (1/2 × item 4).................................... 16,730
   c. One-half of the deaths during the period (1/2 × item 5).................................... 5,494
   d. One-half of the net movement of civilians into the armed forces during the period (1/2 × item 14).................................... 176
   e. Population base for computing estimate of net migration of all ages: (item 13a + item 13b – item 13c – item 13d)...................... 441,705
   f. Net civilian migration at all ages (item 13e × item 12) =.. +13,958

14. Net movement between the civilian and military population, April 1, 1960, to July 1, 1963 (based on official data for New Jersey).. 352

15. Civilian resident population, July 1, 1963 (item 3 + item 4 – item 5 + item 13f – item 14)... 469,211

16. Resident armed forces on July 1, 1963......... 431

17. Total resident population on July 1, 1963 (item 15 + item 16)............................... 469,642

As stated, the ratio of the migration rate for the total population to the rate for the school-age children is derived from national data on interstate or intercounty migration shown by the Current Population Survey. Because migration rates change, though slightly, from year to year, and the ages of "migration exposure" over the postcensal period, for the school-age cohort, are determined by the length of the estimating period, the ratio of migration rates changes also as the estimating period increases.

Two important assumptions are made in this procedure: (1) That there has been no change since the last census in the ratio of the population of elementary school age to the number enrolled in the elementary grades, and (2) that the ratio of the net migration rate of the total population to the migration rate of the school-age population for a given postcensal period, for a given local area, corresponds to that for gross interstate or intercounty migrants in the United States for the same period. The validity of both of these assumptions can be examined on the basis of census data or intercensal estimates. Change in the ratio of the population of elementary school age to enrollment in grades 2 to 8 can be examined for several preceding censuses, but in view of the very high proportion of children attending school, this assumption would give rise to relatively little error. Moreover, the error is, in general, reduced by a *pro rata* adjustment of the initial estimates of school-age population for a set of local areas (e.g., States) to the independent estimate for the parent area (e.g., United States).

Geographic variations in the ratio of the migration rate of the total population to the migration rate of the elementary school-age population, reflecting area differences in the age composition of migrants, are far more important. For the period 1955–60, the State ratios averaged 1.12 rather than 1.00, the ratio shown for all interstate migrants in the United States in this period; and in about three-fourths of the States, the ratio fell outside the range 1.00–1.24. The movement of certain groups in the population are poorly represented by the movements of school-age children – e.g., single men and women, new migrants to suburbs, institutional population, and college students. On the other hand, the civilian population of an area with many families headed by military personnel may be overestimated since their school-age children will be included in the figure on school enrollment. The assumption may give rise to rather large errors in the population estimates, therefore. This error is, in general, reduced by a *pro rata* adjustment of the initial estimates of net migration or population to an independent estimate of net migration or population for the parent area.

Numerous other variations in the use of school data in connection with a component method of estimating the population of geographic subdivisions of countries may be considered. We note only two of these to illustrate the variety of possibilities. In the **grade-progression method,** annual net migration of school-age children is determined by comparing the number of children enrolled in, say, grades 2 to 7 in one year with the number enrolled in grades 3 to 8 in the following year. In the **age-progression method,** the number of children enrolled in school aged, say, 7 to 13 years is compared with the number enrolled aged 8 to 14 in the following year, to measure annual net migration of school-age children. Factors other than migration play only a small part in this year-to-year change, but allowance may be made for them. The other steps in the school-progression methods are modeled along the lines of Component Method II.

Other types of administrative data which have proved useful or may prove useful in estimating net migration or even gross migration for the geographic parts of countries include data from the family allowance system, the social security system, tax collection system, and the automobile registration system, or data resulting from some other type of registration. The development of the data from these systems to yield estimates of net migration for, say, the principal and secondary political units of a country as a set may be quite complex and involve modifications of the basic reporting form, the need for special tabulations, difficult problems of assigning residence, etc. Reports of movements of families in receipt of family allowances under a national family allowance system have been infrequently used although family allowance systems are widespread. Canada's Dominion Bureau of Statistics prepared its provincial population estimates for the years 1956–67 by use of such data.[22] The U.S. Census Bureau has been exploring the possibility of using a sample of the records of the U.S. Social Security Administration relating to the work histories of covered workers to measure interstate migration.[23] The problems that have appeared arise from the differences between State of employment and State of residence, particularly around metropolitan areas, high sampling errors for a 1-percent sample, omission of certain classes of workers and of all nonworkers, and errors and inconsistencies from year to year in the reporting of place of employment. In addition, there is the problem, characteristic of most of the methods using symptomatic statistics, of converting the migration rate for the symptomatic series (i.e., working-age population) to represent the migration rate of the total population. The possibility is also being explored of using Federal income tax returns to derive estimates of the population of each county in the United States.

Use of symptomatic series to estimate total population: Censal-ratio and ratio-correlation methods.—We come next to several methods involving the use of symptomatic series to estimate total population, or total population change, directly rather than merely the component of migration. Two types of such methods may be noted: censal-ratio methods and ratio-correlation methods. The former include any of a number of methods which project the ratio of a symptomatic series to the total population of an area to estimate its population. The latter include those methods which employ a regression equation with a number of types of symptomatic data as independent variables to estimate population change.

The **censal-ratio methods** consist more specifically of (1) computing the ratio of the symptomatic data to the total population at the census date, (2) extrapolating the ratio to the estimate date, and (3) dividing the estimated ratio into the value from the symptomatic series for the estimate date. We may symbolically represent these steps as follows:

$$\text{(a)} \quad r_o = \frac{S_o}{P_o} \tag{11}$$

$$\text{(b)} \quad r_t = \varphi r_o = \varphi \left(\frac{S_o}{P_o}\right) = \left(\frac{S}{P}\right)_t \tag{12}$$

$$\text{(c)} \quad S_t \div \left(\frac{S}{P}\right)_t = S_t \times \left(\frac{P}{S}\right)_t = P_t \tag{13}$$

where S_o/P_o, or r_o, is the ratio at the census date, computed from the separate figures for the symptomatic series (S_o) and the population (P_o); φ is the factor by which r_o is extrapolated to the estimate date; $\left(\frac{S}{P}\right)_t$, or r_t, is the extrapolated ratio for the estimate date; and S_t is the reported current level of the symptomatic series. If the symptomatic data are to be useful, accurate and comparable data must be available at frequent intervals, including the census date, and the number of annual cases of the "event" should be high in relation to population size. It is also necessary for the ratio to be fairly stable or to change in a regular fashion if it is to be accurately projected from the census date to the estimate date. The necessity for great predictability and accuracy in the ratio must be stressed because a given percentage error in the ratio will result in a corresponding percentage error in the population estimate.

Many series of data have been considered useful as "symptomatic" series. In the industralized countries the list includes school enrollment or school census data, number of electric, gas, or water meter installations or customers, volume of bank receipts, volume of retail trade, number of building permits issued, number of residential postal "drops" (residential units where mail is deposited), voting registration, welfare recipients, auto registration, birth statistics, death statistics, and tax returns. The series of symptomatic data available in nonindustralized countries are relatively few and most of those just mentioned are not among them. Such countries may have current data on school enrollment, poll taxes, salt rations or the weight of salt sold by the state monopoly, and loads of "nightsoil" collected.

Some of the series are clearly not well adapted to the direct measurement of population change (as in the censal ratio methods) though they may be useful as ingredients in other methods (as the ratio-correlation methods) or in evaluating estimates prepared by use of other data. This is particularly true of the economic indexes such as volume of bank receipts and volume of retail trade. Changes in the buying power of the population and limitations on the availability of goods and services preclude a perfect or near-perfect correlation between population change and economic change, and thus preclude the possibility of measuring one accurately in terms of the other. These series may fluctuate sharply in response to factors other than population change.

Each of the three principal series of data employed in applying the U.S. Census Bureau's Component Method II—birth statistics, death statistics, and school enrollment data—is a possible choice in connection with the direct estimation of the total population of an area by the ratio method. For example, the estimated crude birth rate or crude death rate may be used in combination with the number of births or deaths in a given year, respectively, to derive an estimate of the population from each series. Similarly, the estimated ratio of school enrollment to population for the local area at the last census date may be used in combination with current data on school enrollment. In effect, this procedure assumes that postcensal population changes are reflected in the changes in school enrollment. Another possibility is to extrapolate the ratio on the basis of the percent change in the national enrollment rate during this period.

A single censal-ratio estimate is not ordinarily used to estimate the total population of an area. Bogue has described in detail a censal-ratio procedure for estimating the population in postcensal years employing both crude birth and death

[22] Canada, *Canada Yearbook,* 1966, p. 206.

[23] U.S. Bureau of the Census, *Current Population Reports,* Series P–23, No. 31, "Use of Social Security's Continuous Work History Sample for Population Estimation," April 10, 1970.

rates—which he called the **vital rates method**.[24] This method extends the design of the simple ratio method by computing two intermediate estimates—one based on birth rates, the other based on death rates—which are then averaged to derive a single composite estimate. Bogue's suggested procedure for estimating the birth and death rates at current dates takes account of the postcensal changes in these rates in some broader area for which the current rates are known or readily ascertainable. Because the vital rates method cannot satisfactorily measure changes in the number of military personnel stationed in an area and these numbers vary sharply from area to area, it is preferable to use the civilian population for the principal calculations. The main steps in the method, illustrated with data and estimates for Dallas, Texas, Standard Metropolitan Statistical Area for July 1, 1967, are as follows:

1. Total resident population, April 1, 1960...... 1,119,410
2. Armed forces, April 1, 1960..................... 1,598
3. Civilian resident population, April 1, 1960 (item 1 — item 2)................................. 1,117,812
4. Births, 1959–60 average, SMSA................ 28,385
5. Birth rate, 1959–60 average, SMSA (item 4 ÷ item 3)................................. 25.39
6. Birth rate, 1959–60, State (civilian population base)............................ 26.49
7. Birth rate, 1967, State (civilian population base)............................. 19.52
8. Estimated birth rate, 1967, SMSA $\left(\dfrac{\text{item 5}}{\text{item 6}}\right.$ $\times$ item 7$\Big)$............................. 18.71
9. Births, 1967, SMSA............................. 27,835
10. Estimated population A (item 9 ÷ item 8)..... 1,487,707
11. Deaths, 1959–60 average, SMSA.............. 8,369
12. Death rate, 1959–60 average, SMSA (item 11 ÷ item 3)............................ 7.49
13. Death rate, 1959–60, State (civilian population base)........................... 8.03
14. Death rate, 1967, State (civilian population base)............................. 8.09
15. Estimated death rate, 1967, SMSA $\left(\dfrac{\text{item 12}}{\text{item 13}}\right.$ $\times$ item 14$\Big)$............................. 7.55
16. Deaths, 1967, SMSA............................. 10,470
17. Estimated population B (item 16 ÷ item 15)... 1,386,755
18. Average of two estimates of civilian population $\dfrac{(\text{item 10} + \text{item 17})}{2}$...................... 1,437,231
19. Armed forces, July 1, 1967..................... 1,738
20. Estimated total resident population, July 1, 1967 (item 18 + item 19)........................ 1,438,969

The reliability of this method depends principally on the correctness of the assumption that the birth rates and death rates of local areas vary in the same general manner as the rates of the larger areas which contain them. Although national and regional changes in fertility and mortality are generally reflected in the States and smaller areas, extreme deviations from the broader pattern of change sometimes occur, particularly for the birth rate. The averaging process may partly offset opposite biases characteristic of the birth-rate estimate and the death-rate estimate. If one population estimate is too low

as a result of an overestimate of the birth rate, the other estimate is likely to be too high as a result of an underestimate of the death rate, since an age distribution which favors a high birth rate also favors a low death rate. One of the objections occasionally given to use of the vital rates method of making population estimates is that the resulting estimates cannot properly be used to compute birth and death rates. Since both the birth rate and the death rate are used in combination and the population estimate is different from a figure based on a single rate, this objection would appear to be only partially valid at most.

Schmitt has proposed the combination and weighting of several censal-ratio estimates based on different indicators for making estimates of county population in the United States. His preferred method weights three types of censal-ratio estimates by the reciprocals of the coefficients of variation of the censal ratios, as measured by intercounty variations for one State at the census date (1940).[25] These censal-ratio estimates were based on voter registration, school enrollment, and births; they had the lowest coefficients of variation in his study out of six types of censal-ratio estimates. The increase in the accuracy achieved by this method results in part from the fact that the censal ratios for subareas were assumed to change postcensally by a constant percentage.

Some indicators reflect changes in the number of housing units or households more directly than changes in the number of individuals. These indicators include the number of building permits issued by local governments (for new construction, demolitions, and alterations), electric, gas, or water utility connections or customers, residential postal delivery "drops," "certificates of occupancy," etc. In the United States, a housing-unit method is often used to make population estimates for cities and metropolitan counties, the areas for which such data are available on a current basis. This method rests on the assumption that changes in the number of occupied housing units essentially reflect changes in population. The estimate of change in the number of housing units in an area in the postcensal period is derived from various kinds of data reflecting new residential construction and losses in the housing inventory, e.g., permits for building or demolition, or changes in residential electric utility connections. Changes in the population are affected also by change in the vacancy rate, the average number of persons occupying a housing unit, and the size of the population living in group quarters (e.g., hotels, rooming houses, college dormitories, institutions, etc.). It is desirable, therefore, to try to take into account changes in these factors between the census date and the estimate date, possibly on the basis of changes in the country as a whole, for which data may be available. The basic formula for estimating the current population in households (occupied housing units), without allowance for change in average size of household, is:

$$(H_o + U) \times \frac{P_o}{H_o} \tag{14}$$

where P_o = population in households on the last census date,
H_o = occupied housing units on the last census date,
U = net increase in occupied housing units between census date and estimate date, and
$\dfrac{P_o}{H_o}$ = average population per occupied housing unit at the last census.

[24] Donald J. Bogue, "A Technique for Making Extensive Population Estimates," *Journal of the American Statistical Association*, 45(250):149–163, June 1950.

[25] Robert C. Schmitt, "Short-Cut Methods of Estimating County Population," *Journal of the American Statistical Association* 47(258):232–238, June 1952.

The term U encompasses any change in the vacancy rate or in the use of units as group quarters. If building-permit data are employed to measure postcensal changes in occupied housing units, an occupancy rate will have to be applied; if utility-connection data are employed for this purpose, this adjustment is not necessary, but there is the problem of distinguishing residential from nonresidential units.

Another more elaborate approach to the use of symptomatic data to estimate the total population of an area directly, designated the **ratio-correlation method,** or regression method, involves mathematically relating changes in several indicator series to population changes by a multiple regression equation. More specifically, a multiple regression equation is derived to express the relationship between (1) the change over the previous intercensal period in an area's share of the total for the parent area for several symptomatic series and (2) the change in an area's share of the population of the parent area. The types of symptomatic data which have been used for this purpose in the applications of the method in the United States are births, deaths, elementary school enrollment, tax returns filed, motor vehicle registrations, employment, voting registration or votes cast, bank deposits, and sales taxes. Other series are, of course, applicable. This method has been described by Schmitt, Goldberg, and others.[26]

The method can be employed to make estimates for either the primary or secondary political divisions of a country. Let us illustrate the use of the method in preparing estimates of the population of the States of the United States for some date in the 1960's on the basis of relationships recorded between 1950 and 1960. The dependent variable (X_o) in the regression equation represents the ratio of a State's share of the national total population in 1960 to its share in 1950, that is,

$$\frac{\text{Percent of total U.S. population in State } i, 1960}{\text{Percent of total U.S. population in State } i, 1950} = \frac{\left(\frac{P_s}{P_{us}}\right)^{1960}}{\left(\frac{P_s}{P_{us}}\right)^{1950}}$$

The independent variables (X_1, X_2, etc.), for example births are expressed in a corresponding manner,

$$\frac{\text{Percent of total U.S. births in State } i, 1960}{\text{Percent of total U.S. births in State } i, 1950} = \frac{\left(\frac{B_s}{B_{us}}\right)^{1960}}{\left(\frac{B_s}{B_{us}}\right)^{1950}}$$

or in general,

$$\frac{\left(\frac{X_s^j}{X_{us}^j}\right)^{1960}}{\left(\frac{X_s^j}{X_{us}^j}\right)^{1950}} \tag{15}$$

The correlation coefficients between the independent variables and the dependent variable are as follows:

Variable	Symbol	Correlation coefficient
Births	X_1	.95
Deaths	X_2	.92
Elementary school enrollment	X_3	.93
Personal tax returns	X_4	.73
Motor vehicle registration	X_5	.81
Nonagricultural employment	X_6	.87
Votes cast in Congressional elections	X_7	.25
Building permits	X_8	.69

The multiple correlation coefficient for the first six variables, $R_{0.123456}$, is .987. The corresponding regression equation is
$$X_{0.123456} = .06 + .30X_1 + 14X_2 + .22X_3 + .08X_4 + .07X_5 + .12X_6.[27]$$

Estimates for 1968 (July) would be prepared by substituting in the equation appropriate data for the 1960-68 period. For example, the value X_1 for a given State in 1968 would be computed as:

$$\frac{\text{Percent of U.S. births in State } i, 1968}{\text{Percent of U.S. births in State } i, 1960}$$

Values for other independent variables would be derived in similar fashion. When the equation is solved for each State, the results represent estimates of the following form:

$$\frac{\text{Percent of U.S. population in State } i, 1968}{\text{Percent of U.S. population in State } i, 1960}$$

The ratio so computed for each State is applied to each State's percentage of the U.S. population in 1960, as shown by the 1960 census, to arrive at its estimated percentage of the U.S. population in 1968. The 1968 percentages for all States are summed and adjusted to add to 100 percent. These percentages are then applied to the estimated resident population for the United States for July 1, 1968, yielding an estimate of the resident population of each State on July 1, 1968. If population estimates for counties are to be prepared, the parent area would then be the State, the various percents would represent the county's share of the State total, and a current estimate of the total population of the State and current indicator data for the counties are needed.

The success of the ratio-correlation method depends upon the accuracy of the underlying assumption that the observed statistical relationship between the independent and dependent variables in the past intercensal period will persist in the current postcensal period. The adequacy of this assumption is dependent on the size of the multiple correlation and the number of variables used; with the use of electronic computers and the increasing availability of new series of data, many combinations of variables can be included. High multiple correlation coefficients for two past intercensal periods would suggest that the degree of association of the variables is not changing very rapidly. In such a case, the regression based on the last intercensal period should be applicable to the current postcensal period. Furthermore, it is assumed that deficiencies in coverage in the basic data series will remain constant, or change very little, in the present period.

Composite method.—In general, a characteristic of the various methods of making local population estimates that we have described is that they are single methods, not combinations of methods, which measure change in the total population of an area directly rather than in terms of its principal segments, such as geographic or age segments. The

[26] Robert C. Schmitt and Albert H. Crosetti, "Accuracy of Ratio-Correlation Method for Estimating Postcensal Population," *Land Economics,* 30(3):279–280, August 1954; and David Goldberg, Allan Feldt, and J. William Smit, *Michigan Population Studies,* No. 1, "Estimates of Population Change in Michigan: 1950–60," Ann Arbor, Michigan, University of Michigan, 1960, pp. 36–39.

[27] U.S. Bureau of the Census, *Current Population Reports,* Series P–25, No. 436, "Estimates of the Population of States: July 1, 1968 and 1969," January 7, 1970, p. 5.

various indicators and methods are often. more appropriate for one age group than another in the population, however. Furthermore, the uncertainties in the way of projecting crude censal ratios, resulting from changes in the age and sex composition of the population and in the force of vital and other trends, suggest the desirability of employing the censal ratio method and indicator data specific for age groups. We can also consider various weighted combinations of the basic methods we have described as useful additions to our store of methods.

In the "composite method," independent estimates are prepared for the population in several age groups or age-sex groups, using methods and basic data considered most appropriate for each age group.[28] An estimate is then derived for the population as a whole by summing the independently derived estimates for each age group. Many alternative combinations are possible. For example, the number of deaths of persons 45 years old and over by age may be used to estimate the population in this age range; the number of births may be used to estimate the number of females in the childbearing ages (18 to 44 years), which, in turn, may be used to estimate the number of males in this age range; school enrollment data may be used to estimate the population of school ages (5 through 17 years old); and the number of births in the previous 5-year period, in conjunction with school enrollment data, may be used to estimate the population under 5 years of age. The ratio method or the component method are used for one age group or another as appropriate.

The principal steps in applying this version of the composite method are as follows:

1. Population 45 years old and over:

a. Compute age-sex-color specific death rates by 10-year age groups for the census year, say 1960, on the basis of the census population, starting with the population 45 to 54 years and ending with 75 years old and over, for the country and each area;

b. Compute the corresponding death rates for the country for the 12-month period centered on the estimate date;

c. Prepare estimates of the age-sex-color specific death rates for each area for the 12-month period centered on the estimate date, on the assumption that the change in the death rate for each area from the census year was the same as for the country as a whole;

d. Compute the estimated population for each area on the estimate date in each age-sex group by dividing the number of deaths for each group in the year centered on the estimate date by its current specific death rate as obtained above.

In the smaller areas, when deaths are distributed by age and sex, there may be extremely small numbers of deaths in some categories. The thinness of these data makes their use as bases for estimates by this technique very questionable. Consequently, in very small areas the procedure is modified so that estimates are prepared for the age group 45 years old and over as a whole.

2. Population 18 to 44 years of age: Estimates of the number of females 18 to 44 years old are first developed by the censal ratio method in a manner corresponding to steps (a) through (d) above using data on the number of births and the number of females 18 to 44 years of age. As with the death rate 45 and over, this method assumes that the change in age-specific birth rates for the local area was the same as for the country as a whole. Then, the ratio of the number of civilian males to females at the last census in each area in this age range, adjusted for the change in this ratio for the country as a whole between the census date and the estimate date, is used to arrive at an estimate of the number of civilian males in the area. The number of civilian males, an estimate of military personnel, and the number of females are summed to yield an estimate of the total population 18 to 44 years of age.

3. Population under 18 years of age: The estimated population in this age group may be developed by a component procedure similar to that described for Component Method II above:

a. Obtain the census population in the cohort that would be under 18 years old on the estimate date;

b. Add births for the postcensal period;

c. Subtract deaths for the cohort for the same period; and

d. Add an estimate of net migration.

The calculations may be carried out for two age groups separately, under 5 and 5 to 17 years. Estimates of net migration for the group under 18 years of age are obtained from the migration rate of the school-age population derived earlier as part of the Component Method II procedure. The factor used to convert the migration rate for the school-age population to the rate for the population under 18 years of age may be based on national ratios of interstate or intercounty migrants under 18 years of age to interstate or intercounty migrants in the school ages, for the postcensal period.

The details of the composite method may be varied depending on kind and quality of data available. The procedures of Component Method II may be applied to ages 18 to 44 as well as ages under 5 and 5 to 17. On the other hand, ages under 5 and 5 to 17 may be estimated by a censal-ratio procedure, using births and school enrollment as symptomatic data.

Averaging of methods. — The averaging of methods may be employed as a basis for improving the accuracy of population estimates for geographic subdivisions of countries. The methods to be averaged should employ different indicators or essentially different procedures and assumptions. Averaging may affect the accuracy of population estimates in two ways. It may reduce the risk of an extreme error and it may partly offset opposite biases characteristic of the two types of estimates being averaged. The methods to be averaged may be selected subjectively or on the basis of various quantitative indications given by studies of the accuracy of the methods (see below). For example, two methods which have relatively low average errors but which have opposite biases may be considered good candidates for averaging. The existence of opposite biases is indicated by a low or negative correlation between the percent errors for the geographic units in a distribution (e.g., States) according to the two methods. The methods to be averaged may be given the same weights or different weights. These too may be determined subjectively or quantitatively on the basis of evaluation studies.[29]

Intercensal estimates. — Annual intercensal population estimates for geographic subdivisions may also be prepared by

[28] Donald J. Bogue and Beverly Duncan, "A Composite Method for Estimating Postcensal Population of Small Areas by Age, Sex, and Color," reproduced in U.S. National Office of Vital Statistics, *Vital Statistics-Special Reports* 47(6), August 24, 1959.

[29] See, for example, the use of coefficients of variation of censal ratios by Schmitt, op. cit.

the component method if satisfactory data on births, deaths, and migration are available. In making annual intercensal estimates, however, there is the problem of allowing for the error of closure. The error of closure for each subarea may be distributed over the intercensal period in proportion to one of the components of change for the subarea (e.g., net migration), the total population change, or the population itself. In addition, the intercensal estimates of change due to each component, population change as a whole, or the population itself for all subareas in each year must be adjusted to agree with the corresponding intercensal figures established for the whole country.

Even if a country has the required birth and death statistics for making intercensal estimates for geographic subdivisions, it often lacks the required migration data or even data symptomatic of migratory changes. If this is the case, intercensal estimates of population for regions may be developed on the basis of mathematically derived estimates of net migration. This procedure involves, first, estimating intercensal net migration (including the error of closure) for each geographic unit as a residual by the vital statistics method described in chapter 21; second, allocating an equal portion of the intercensal net migration to each year of the intercensal period; third, adjusting the preliminary estimates of net migration or population to the independently established estimates of net immigration or population (adjusted for error of closure) for the entire country in each year of the intercensal period.

As with previous methods for intercensal estimates of geographic subdivisions, preliminary estimates of population for such areas should be adjusted to agree with the independently established intercensal estimates of the national population. One method is to adjust the geographic distribution of each component, particularly net migration, to agree with the established national estimate for that component (e.g., net immigration). A simpler method is to adjust the preliminary

estimates of the population of the subareas to the established estimates of the national population without reference to the individual components.

Mathematical methods. — In the censal ratio methods and the ratio-correlation methods, use of mathematical relationships is an incidental aspect; these methods involve essentially the use of a symptomatic series for the local area presumably reflecting population changes in the area. A simpler method is to estimate the total population of the area for postcensal dates by fitting some form of mathematical curve directly to the census figures for the area. In some cases very limited use may be made of current data.

The mathematical procedures described in Appendix C are useful in estimating the population of the geographic subdivisions of the underdeveloped countries in the absence of current data. The mathematical curves most conveniently and commonly employed for this purpose are straight lines and geometric curves, although a great variety of other forms may be employed (e.g., higher degree polynomials, reverse geometric curves, and more complex exponentials such as the logistic). Geometric extrapolation is more logical than linear extrapolation in making population estimates at the national level in underdeveloped areas, where birth rates may remain nearly constant at a high level and death rates are constant or declining, but linear extrapolation may be more appropriate for some of the geographic subdivisions of these same countries because growth rates may be falling as a result of out-imigration. The subareas of the same country could reasonably be treated differently, therefore, taking account of information, however limited, identifying the areas of in- and out-migration.

The steps for estimating the population of the geographic parts of countries by linear and geometric extrapolation are illustrated on the basis of calculations for the Provinces of Panama for July 1, 1968, using census data for 1950 and 1960 (table 22-10). For linear extrapolation, the preliminary popu-

Table 22–10. — Estimation of the Population of the Provinces of Panama for July 1, 1968, by Linear and Geometric Extrapolation of Numbers and by Linear Extrapolation of the Percent Distribution

Provinces	Census population		Linear extrapolation of numbers, July 1, 1968		Geometric extrapolation of numbers, July 1, 1968		Census percent distribution			Linear extrapolation of percents, July 1, 1968
	Dec. 10, 1950[1]	Dec. 11, 1960[2]	Initial estimates[3]	Adjusted estimates[4] 1.045509842 x (3)=	Initial estimates	Adjusted estimates[5] .993346270 x (5)=	Dec. 10, 1950	Dec. 11, 1960	Percents[3]	Population estimates 1,338,101 x (9)=
	(1)	(2)	(3)	(4)	(5)	(6)	(7)	(8)	(9)	(10)
Panama................	805,285	1,075,541	1,279,855	1,338,101	1,347,064	1,338,101	100.000	100.000	100.000	1,338,101
Bocas del Toro..........	22,392	32,600	40,317	42,152	43,305	43,017	2.781	3.031	3.220	43,087
Chiriquí...............	138,136	188,350	226,312	236,611	238,105	236,520	17.154	17.512	17.783	237,955
Coclé.................	73,103	93,156	108,316	113,245	111,892	111,148	9.078	8.661	8.346	111,678
Colón................	90,144	105,416	116,962	122,285	118,657	117,867	11.194	9.801	8.748	117,057
Darién...............	14,660	19,715	23,537	24,608	24,664	24,500	1.820	1.833	1.843	24,661
Herrera..............	50,095	61,672	70,424	73,629	72,169	71,689	6.221	5.734	5.366	71,802
Los Santos...........	61,422	70,554	77,458	80,983	78,349	77,828	7.627	6.560	5.753	76,981
Panamá..............	248,335	372,393	466,181	487,398	505,860	502,494	30.838	34.624	37.486	501,601
Veraguas.............	106,998	131,685	150,348	157,190	154,063	153,038	13.287	12.244	11.455	153,279

[1] Source: United Nations, *Demographic Yearbook, 1952*, table 5.

[2] Source: United Nations, *Demographic Yearbook, 1962*, table 8.

[3] Multipliers are −.756 for the 1950 population and 1.756 for the 1960 population.

[4] Adjusted to independent estimate of the total population, 1,338,101; factor equals $\frac{1,338,101}{1,279,855} = 1.045509842$.

[5] Adjusted to independent estimate of the total population, 1,338,101; factor equals $\frac{1,338,101}{1,347,064} = .993346270$.

lation estimates are derived by taking $\frac{7.56}{10.00}$, or 75.6 percent, of the 1950–60 increase for each Province and adding the resulting estimate of the postcensal increase to the 1960 census count. These preliminary figures are then adjusted proportionately to the total for Panama for July 1, 1968, derived previously by geometric extrapolation. In the geometric method, the 1950–60 average annual rate of growth is calculated for each Province by use of either the formula for annual compounding or the formula for continuous compounding and then applied sequentially to the 1960 population. Here, too, the results are adjusted to the preassigned total for Panama for July 1, 1968. It may be observed that the adjustment process tends to produce similar results regardless of the method of extrapolation. In principle, all mathematical methods involve adjustment to independent estimates for the parent area; in practice, no adjustment is necessary if linear extrapolation is used for both the parent area and its subdivisions.

The extrapolation process may be applied to percents representing the proportion that the population of an area constitutes of the population of some more inclusive area for which annual totals are available. In the illustrative calculations shown in table 22–10, the populations of the Provinces of Panama are estimated for July 1, 1968, by linear extrapolation of percents. (A simpler and generally rougher procedure would use merely the percentage distribution of the population by Provinces in the 1960 census.)

Another primarily mathematical method of estimating subnational population which also takes limited account of current data is the so-called **apportionment method.**[30] This method essentially involves distributing the postcensal increase in population in the parent area (e.g., a country) among the component parts of that area (e.g., states) in accordance with the distribution of the intercensal increase in the parent area among the component parts. If a subarea lost population in the intercensal period, its population size as indicated by the latest census is assumed to remain unchanged to the postcensal date. Except for the special treatment of areas that lost population, the apportionment method is approximately equivalent to arithmetic extrapolation when the results by the latter method are adjusted *pro rata* to the independent estimate of the total population of the parent area. Like the other mathematical methods, this method requires a current total for the parent area.

The limitations characteristic of mathematical extrapolation in general, as described earlier, should be borne in mind in the present connection. On *a priori* grounds we should expect such methods to be, on the average, less accurate than the various component methods and other methods using symptomatic data, and, in fact, the validation tests made so far confirm this expectation (see below). In the absence of any current data, postcensal estimates made by mathematical methods (like other methods) are methodologically in the same class as projections made by such methods, even though they relate to past rather than future dates. These methods usually employ the assumption that population change in the postcensal period proceeded in the same direction and at the same pace as in the preceding intercensal period or periods. A comparison of intercensal changes for counties and even States in the United States for 1940–1950 and for 1950–1960 will readily reveal numerous cases inconsistent with this assumption and will emphasize the need for use of symptomatic data pertaining to the specific areas for which estimates are being prepared. In the case of the apportionment method, the assumption that the postcensal growth in local areas bears a fixed relation to the earlier growth in the area, as indicated by the change in the parent area, is an assumption that may readily be challenged with census data. The method is logically inapplicable if the postcensal change in the parent area differs in direction from the intercensal change in that area. On the other hand, there is usually a high correlation between growth trends for a given area in two successive intercensal periods and growth trends of broader areas are often reflected in their geographic subdivisions. Mathematical methods have certain advantages. They are simple to apply, take relatively little time, and require only readily available data.

The same general procedures as described above for postcensal estimates apply in developing intercensal estimates. In this case the population of each geographic subdivision is interpolated to the estimate date on the basis of the two census figures, and the figures for all the subdivisions are adjusted *pro rata* to the figure for the parent area (e.g., the entire country). As with the linear extrapolations shown in Table 22–10, the interpolation process may be applied either to numbers or to percents representing the proportion that the population of an area constitutes of the population of some more inclusive area for which annual totals are available.

As the evaluation studies show, the mathematical methods are not very reliable for making postcensal population estimates for geographic subdivisions of countries, although they may be necessary in the absence of symptomatic data for the postcensal period. Their reliability is quite different when they are used to prepare intercensal estimates, however. Because the general direction and average rate of population growth are established for the intercensal period, rather reliable annual estimates may readily be prepared by mathematical methods, particularly the method of interpolating the proportional distribution of the population of the parent area.

Age-Sex Composition. – Several methods of developing estimates of the age-sex composition of the population of geographic areas have, in effect, been set forth in the preceding section relating to estimates of the total population of such areas. We may recall that component or censal ratio estimates for different age-sex groups are used to derive estimates of total population by the composite method. The range of symptomatic data employed for one age group or another may be extended to cover many other series than those ordinarily employed in the composite method, however. Age-sex estimates may be prepared not only by means of a component method or a censal ratio method but also by a purely mathematical method, at one extreme, and by use of population registers, at the other.

The component estimates may be based either on actual statistics on the age-sex composition of migrants or on symptomatic data for migrants. In Canada considerable use is made of data on interprovincial migration by age and sex from the Family Allowance System, which maintains records on the movements of families in receipt of family allowances and the ages of persons moving. In the United States school enrollment data may be used as an indicator of population change due to migration both at the school ages and at certain other ages, that is, in variations of Component Method II or the composite method.

The annual estimates of population for States for broad age

---

[30] This method was used in the first part of this century by the U.S. Bureau of the Census in estimating State populations.

groups prepared by the U.S. Bureau of the Census incorporate estimates of net civilian migration for every age group except 65 and over derived from estimates of migration rates for school-age children. The ratios of the national interstate migration rates for four broad age groups to the rates for the school-age population aged $7\frac{1}{2} \longleftrightarrow 15\frac{1}{2}$ for the period April 1960 to July 1968 are as follows: [31]

Age on estimate date	Ratio
Under 5 years	0.519
5 to 17 years	0.953
18 to 44 years	1.373
45 to 64 years	0.439

It is assumed that these same ratios apply equally to all States for a given period and, for a given State, that the direction of the net migration is the same for all the broad age groups under 65. These factors will, of course, vary with the length of the estimating period. The initial migration estimates derived for the age groups under 65 in each State are adjusted to be consistent with State in-migration totals for all ages independently calculated as described earlier. These adjusted estimates of net migration for ages under 65 are then employed to derive preliminary estimates of population for States by age.

The last previous and present methods employed by the U.S. Census Bureau to estimate the interstate migration of persons 65 and over illustrate the use of two other types of administrative data as demographic indicators. The first are data obtained from the Social Security Administration relating to the interstate movement of "aged beneficiaries," for each State for various years. The reported number of beneficiaries who migrated was adjusted for differences between States in insurance coverage, and the State estimates of net migration were then adjusted to tie in with the level of interstate migration of persons 65 and over for the United States for the whole postcensal period, as estimated from the Current Population Survey.

"Medicare" statistics now provide an improved basis for estimating the population 65 years old and over in each of the States; only a few small classes among the aged are excluded from coverage and there is a very high rate of participation of the population in this age group in the program. The number of persons enrolled under the basic provisions of the program closely approximates the total aged population in the United States for current years as estimated by extending the census counts.[32] The estimates of State population 65 years and over now used represent basically the number of persons covered by the "Medicare" program.

Similarly, censal ratio methods employing a variety of symptomatic data may be used to derive estimates of age-sex groups for geographic areas. The principal indicators available for this purpose in the United States are death statistics, birth statistics, and school enrollment data. In fact, the entire age distribution may be covered fairly well by these types of symptomatic data: Births for population under 5 and for females 18 to 44, deaths for age groups 45 and over, and school enrollment for ages 5 to 17. Other possible indicators include data on social security account holders by age, which may be useful in developing estimates of the population of working age in each State, at least of the male population. The principal area of uncertainty is the estimate of postcensal change in the age-specific death rate or proportion of an age group enrolled in school or covered by social security.

Another possibility, a kind of limited component method, is to employ the changes in age cohorts during the previous intercensal period in combination with some current data. Specifically, the intercensal rates of change in age cohorts may be adapted to represent the rates of change for age cohorts in the postcensal period. Statistics or estimates of births and a current estimate of total population are then required to complete the estimates.

Purely mathematical methods have very limited application in the preparation of postcensal estimates of age-sex groups for geographic areas. It is possible, of course, to apply the age distribution at the last census date to a preassigned current total for the area, or to extrapolate the last two census age distributions to the current date and apply the extrapolated distribution to the current total. The resulting estimates are subject to such large errors, however, that these methods are viewed as having little value.

In general, intercensal estimates involve an adjustment for the intercensal error of closure at each age and for inconsistencies with annual estimates of the total population for each area. In a cohort-component method, we would compute annual preliminary estimates of population by age and sex for each geographic unit, i.e., provinces, with allowance for net migration, determine the error of closure for each age cohort, distribute the error of closure over the expected population by age cohorts, and then adjust the figures jointly to the previously estimated area totals over all ages and to the national estimates for each age-sex group. A cruder procedure involves calculation of the expected population by use of life tables or death statistics and then allowing for net migration by cohorts as part of the adjustment for error of closure. A procedure of this general type was employed in preparing intercensal age-sex estimates for the Provinces of Canada, 1931–41.[33] Mathematical interpolation is a convenient and simple method of making intercensal estimates, given annual estimates of the total population of each area to which the estimates for age groups may be adjusted. Alternatively, the percent of the total population of the area in each age group may be interpolated linearly and applied to the estimates of the total population of the area at the estimate date.

EVALUATION OF METHODS

Design of Evaluation Studies

We have discussed a variety of methods of making population estimates for countries and their geographic subdivisions, by age and sex, with relatively little explicit concern for their relative accuracy as indicated by evaluation studies. Any consideration of the accuracy of methods depends on the availability of an adequate standard against which to judge the methods and the establishment of criteria of accuracy. We may consider as general criteria of the accuracy of postcensal estimates one or more of the following: Consistency of the estimate with a previous census count, agreement with a subsequent census count, or agreement with another type of

[31] U.S. Bureau of the Census, *Current Population Reports*, Series P-25, No. 437, "Estimates of the Population of States, by Age, 1968, With Provisional Estimates for July 1, 1969," January 16, 1970.

[32] See U.S. Social Security Administration, *Health Insurance Enrollment under Social Security*, July 1, 1966; and U.S. Bureau of the Census, *Current Population Reports*, Series P-25, No. 437, "Estimates of the Population of States, by Age, 1960 to 1968," January 16, 1970, pp. 3-4.

[33] Canada, Dominion Bureau of Statistics, *Population 1921-1966; Revised Estimates of Population by Sex and Age, Canada and the Provinces*, June 1968.

independent estimate. Thus, the standard may be a census count, a figure from a population register or registration of some sort, an estimate from a sample "census," or other independent estimate. There are different definitions of census comparability too, particularly for estimates of age composition.

There is often an implication in the use of the census as a standard that the earlier and later censuses are completely consistent, but such consistency is hardly to be taken for granted. The criterion used may be agreement with the next census count, even though the actual estimate may have been based on the previous census count, as is the case with most methods. When estimates comparable to the last census are sought, the same methods of estimation may be employed, but the standard for evaluating the estimates is less evident, particularly estimates of age groups. Estimates of age groups may be judged for their adequacy by their comparability either with the same age cohort or with the same age group at the last census.

One practical device for securing a "rough and ready" standard for an intercensal year is to interpolate between census counts to the date of the estimate. These estimates can then be employed to evaluate postcensal estimates. Annual standards for evaluating postcensal estimates can also be developed by adjusting the postcensal estimates for each year, carried forward to the date of the second census, for the intercensal error of closure.

Securing an adequate standard to represent the "true" population is often difficult or impossible. A special registration or a population register may provide more correct figures than the census. Moreover, many of the methods for evaluating census data can be extended to the evaluation of population estimates, particularly national estimates. Because of the lack of a standard to judge the absolute validity of most estimates, and because of the implicit character of most estimates both as comparable to the last census and as anticipating the results of the next census, the common practice is to employ the later census as a standard for evaluating postcensal estimates.

National Estimates

When a postcensal estimate of the total population of a country is evaluated by comparison with the census count and when the estimate has been made by a component method, we have labelled the difference between the postcensal estimate and the later census count as the error of closure. The ratio of the error of closure to the census count as standard may then be taken to reflect the percent of error in the postcensal estimate carried over a whole intercensal period in relation to the level of the census count. We can then examine the intercensal errors of closure for various countries to judge the accuracy of the postcensal estimates in relation to their census counts. For example, the decennial percent deviations of the U.S. population estimates in 1950 and 1960 from the census counts were both less than 0.2 percent and the quinquennial percent deviation from the census count for Japan in 1960 was 0.5 percent.

The estimates of national population based on some population registers are viewed as having at least as great accuracy as the census counts; but they are, in fact, closely akin to the estimates prepared by the component method. Commonly, at a census date the register is reviewed and corrected on the basis of the census results, and then the national total from the register, validated by the previous census, is kept up to date by use of the component data from the register. The

discrepancies between census counts and totals from population registers are quite small in most countries having registers.

For the statistically less developed countries, the data on components are often much too inadequate to serve as a firm basis of preparing postcensal estimates; where such data are used, the postcensal estimates are subject to considerable error. Whether this method or mathematical extrapolation is employed, it is probable also that in these countries the census counts are fairly incomplete and less comparable in coverage with one another. Some underdeveloped areas have relatively good censuses but poor vital statistics, however (e.g., South Korea).

The errors in postcensal estimates by age, as measured by intercensal errors of closure, tend to be considerably greater than the error in the total for an area, particularly because of the effect of errors in reporting age in the two censuses. The intercensal errors represent in part the effect of carrying reporting errors at one age in the first census forward to a later age at the second census. The structuring of the calculations by age cohorts results in good comparability of age estimates with the previous census figures for the same birth cohorts but poor comparability with figures for the same age groups in either the first or second census.

Estimates for Geographic Subdivisions

A large and growing number of studies have been carried out, particularly in the United States, designed to evaluate various methods of preparing estimates of population for the geographic subdivisions of countries. The U.S. Bureau of the Census has carried out an extensive program of developing and evaluating postcensal estimates for States and local areas over the past three decades, and several important studies have been conducted by other investigators. These studies typically employ census counts as the standard by which the estimates are judged.

First, we may consider evaluation studies relating to the total population of States. Systematic comparisons have been made by the U.S. Bureau of the Census between the census counts for States for 1940, 1950, and 1960 and the corresponding estimates derived by several different basic methods and selected averages of these basic methods.[34] The methods are: Component Methods I and II, the composite method, the vital rates method, the ratio-correlation method, two mathematical methods (arithmetic and geometric extrapolation), and various averages of Component Method II, vital rates method, the composite method, and the ratio-correlation method. The principal measures of accuracy used have been the mean percent deviation of the estimates from the census counts over all States for each method, the errors being taken without regard to sign ($|x|$); the number of State deviations of 5 or 10 percent or more; and the number of positive deviations for States. The first measure describes average error, the second

[34] Meyer Zitter and Henry S. Shryock, Jr., "Accuracy of Methods of Preparing Postcensal Population Estimates for States and Local Areas," *Demography* 1(1):227–241, 1964; Jacob S. Siegel, "Status of Research on Methods of Estimating State and Local Population," in *Proceedings of the Social Statistics Section, American Statistical Association*, Washington, D.C., 1960, pp. 172–79; Henry S. Shryock, Jr., "Development of Postcensal Population Estimates for Local Areas," with discussion by John N. Webb and Ormond C. Corry, in National Bureau of Economic Research, "Studies in Income and Wealth," Vol. XXI, *Regional Income*, Princeton, Princeton University Press, 1957, pp. 377–99; Jacob S. Siegel, Henry S. Shryock, Jr., and Benjamin Greenberg, "Accuracy of Postcensal Estimates of Population for States and Cities," *American Sociological Review*, 19(4):440–46, August 1954.

Table 22–11. — **Summary of Percentage Deviations From Census Counts, of State Estimates for 1960 and 1950 Prepared by Various Methods, for the United States**

[Excludes Alaska, Hawaii, and District of Columbia]

Summary measures	Component method II (X_1)	Vital rates method (X_2)	Composite method (X_3)	Ratio-correlation method (X_4)	Component method I (X_5)	Arithmetic method (X_6)	Geometric method (X_7)	Averages		
								(X_1,X_2)	(X_1,X_4)	(X_3,X_4)
_1960_										
Average percent deviation..........	2.00	2.37	2.07	2.75	6.85	5.65	5.23	1.58	1.49	1.84
Deviations of 10 percent or more...	-	-	-	1	13	8	7	-	-	-
Deviations of 5 percent or more....	3	6	3	8	27	18	17	2	1	4
Positive deviations................	28	24	31	20	21	25	21	26	25	27
_1950_										
Average percent deviation..........	3.16	4.42	2.53	(NA)	5.88	6.27	6.07	3.54	(NA)	(NA)
Deviations of 10 percent or more...	1	4	-	(NA)	10	10	10	-	(NA)	(NA)
Deviations of 5 percent or more....	8	19	3	(NA)	22	23	22	15	(NA)	(NA)
Positive deviations................	25	22	25	(NA)	28	29	27	25	(NA)	(NA)

— Represents zero.

NA Not available.

Source: Meyer Zitter and Henry S. Shryock, Jr., "Accuracy of Methods of Preparing Postcensal Population Estimates for States and Local Areas," *Demography*, 1(1):229, 1964.

the frequency of extreme error, and the third, the presence or absence of an upward or downward bias in the method.

The results of these tests show that, in terms of the several different criteria, all the symptomatic methods except Component Method I appear to be more accurate for State estimates than the mathematical methods (table 22–11). Among the basic methods, Component Method II and the composite method have the lower average errors in 1960 — 2.0 percent and 2.1 percent. Of the basic symptomatic methods, only the ratio-correlation method had any deviations of 10 percent or more (one case); with Component Method II and the composite method no States had deviations of 10 percent or more and only three States had deviations of 5 percent or more. There was evidence of directional bias for most of the methods tested. In the case of Component Method II, it resulted presumably because adjustments were not made at several computational stages to national totals. As a result of improvements in data and in procedures as well as greater regularity of population growth during the 1950's, the methods, in general (the ratio-correlation method was not tested in 1950), showed increases in accuracy between 1950 and 1960. Studies show only a very weak correlation between the percentage errors of the 1950 and 1960 State estimates calculated by Component Method II; this finding suggests that a compensatory system of adjustments from one census period cannot safely be applied to the next census period.

As suggested earlier, combining estimates for each State derived by two of the better methods results in even greater accuracy. For example, the mean State error for the average of Component Method II and the ratio-correlation method was 1.5 percent in 1960, and with this average method there was only one case with an error exceeding 5 percent and none with an error exceeding 10 percent. A relatively low mean error for States was obtained also by averaging Component Method II and the vital rates method (1.6 percent). The relatively low average errors for Component Method II, the composite

method, the vital rates method, and the ratio-correlation method in 1960, and the opposite biases of Component Method II and the composite method, on the one hand, and the vital rates method and the ratio-correlation method, on the other, suggest the kinds of combinations of methods which may be effectively averaged.

We turn now to evaluation of estimates of metropolitan areas and their constituent counties. The U.S. Bureau of the Census carried out test calculations on Component Method II and the vital rates method for 46 SMSA's having a total of 132 counties.[35] The test results showed Component Method II to be less accurate than the vital rates method, both for SMSA's as a whole and for their constituent counties. Moreover, Component Method II had a pronounced downward bias. With both methods, mean percentage errors were higher for individual metropolitan counties than for SMSA's as a whole. Greater accuracy is achieved both with Method II and the vital rates method for slowly and moderately growing counties than for rapidly growing counties and counties losing population. Averaging Method II and the vital rates method reduces the mean errors below those of each of the two basic methods (except for moderately growing counties).

A number of other studies relating to counties, usually the whole set of counties in a State, have been conducted by other investigators. Goldberg and Balakrishnan tested four methods (Method II, vital rates, composite, and ratio-correlation) of estimating the population of the counties of Michigan in 1958.[36] The ratio-correlation method proved best among these methods but averaging some of them tended to reduce the errors even further.[37] A study by Rosenberg tested the effect of stratifying counties by demographic characteristics in applying the

---

[35] Zitter and Shryock, op. cit.

[36] David Goldberg and T. R. Balakrishnan, *A Partial Evaluation of Four Estimating Techniques*, Michigan Population Studies No. 2, Ann Arbor, Michigan, University of Michigan, 1961.

[37] Other studies of the accuracy of the ratio-correlation method are described in: Robert C. Schmitt and Albert H. Crosetti, op. cit; and David Goldberg, V. R. Rao, and N. K. Namboodiri, "A Test of the Accuracy of Ratio-Correlation Estimates," *Land Economics*, 40(1):100–102, Feb. 1964.

Table 22-12.—Summary of 1960 Tests for Counties of the United States, Showing Average Percent Deviation From Census Counts of Estimates Prepared by Various Estimating Techniques

Method	All counties		Large counties[1]		Medium-size counties[2]		Small counties[3]	
	Number of areas	Percent deviation	Number of areas	Percent deviation	Number of areas	Percent deviation	Number of areas	Percent deviation
(1) Component Method II....................	1,101	7.5	87	3.7	292	6.1	722	8.5
(2) Vital rates method....................	993	6.8	81	4.3	245	4.6	568	7.9
(3) Composite method......................	402	4.5	64	2.3	163	4.1	175	5.9
(4) Ratio-correlation method..............	182	4.8	15	2.0	31	4.4	37	4.5
(5) Average of (1) and (2)..............	894	5.1	81	3.1	245	3.7	568	6.0
(6) Average of (1) and (3)..............	402	4.2	64	2.2	163	3.6	175	5.5
(7) Average of (1) and (4)..............	83	3.9	15	2.4	31	3.8	37	4.1
(8) Average of (1), (2), and (4).........	83	3.5	15	2.3	31	3.7	37	3.2

[1] 1960 population greater than 100,000.
[2] 1960 population between 25,000 and 100,000.
[3] 1960 population less than 25,000.

Source: M. Zitter. D. E. Starsinic, and D. L. Word, "Accuracy of Methods of Preparing Postcensal Population Estimates for Counties: A Summary Compilation of Recent Studies," unpublished paper presented at the Annual Meeting of the Population Association of America, Boston, April 18–20, 1968.

ratio-correlation method.[38] When he applied the method to the 88 counties of Ohio in 1960 by grouping the counties according to such characteristics as percent urban and percent employed in agriculture, he found measurable improvement in the estimates. Pursell has tested the effect of combining stratification and use of "dummy" variables to represent county type in the correlation method. Dummy variables represent such non-quantitative county characteristics as location or transportation network. The average percent error in 1960 for the 55 counties of West Virginia was sharply reduced when dummy variables and stratification were used in addition to the conventional indicators.[39]

Hillery found, in testing Method II for the counties of Kentucky, that the use of different factors for converting the migration rate of school-age population to the migration rate for all ages for different areas had little effect on average accuracy although the estimates for individual counties were often changed appreciably.[40] Poole tested Component Method II and the vital rates method on the counties of six midwestern States.[41] The same methods were tested in four other States by the Study Group on Postcensal Estimates of the Public Health Conference on Records and Statistics.[42] In both of the last two studies, the composite method was also included for a few States. There were other tests by Jeffries and Hiller (Minnesota counties, 1960), Lee (Iowa counties, 1960), Hamilton (North Carolina counties, 1965 and 1966), and the U.S. Bureau of the Census (39 selected SMSA's for 1964 to 1966).

Altogether the studies mentioned, including the tests for 46 SMSA's conducted by Zitter and Shryock (but excluding the tests by Rosenberg and Pursell), covered 14 States and 1,200 counties. The results of these studies have been summarized at the U.S. Bureau of the Census (table 22-12).[43] (There is some question regarding the uniformity of application of methods and consistency of the basic data employed in the various studies.) These tests indicate that the accuracy of any particular method varies considerably from State to State and from county to county over an extremely wide range. Component Method II showed the largest mean error (7.5 percent) followed closely by the vital rates method (6.8 percent); the composite method had a lower mean error (4.5 percent). Ratio-correlation had the lowest error (4.8 percent) in the two States where it was tested. Averaging of the better methods tends to reduce mean errors below that of each of the separate methods. For example, the average of Component Method II and the vital rates method had a mean error of 5.1 percent. The lowest mean error was secured when Component Method II, vital rates, and ratio-correlation were averaged (3.5 percent). Generally, single methods gave large errors in comparison with averages of these methods. The test results also indicate a tendency for percent error to vary inversely with population size. Errors were larger for the more rapidly growing areas than the less rapidly growing areas, but areas with population losses had the highest average error. There is no firm evidence that errors in estimates vary with the length of the estimating period.

In order to broaden the range of information relating to the adequacy of various estimating methods for subnational areas in the United States, the U.S. Bureau of the Census has sponsored a Federal–State Cooperative program in the preparation and evaluation of county estimates for 1970.[44] It is expected that the outcome of this testing program will provide a wider body of research materials for selecting methods of estimating local population in the decade of the 1970's.

[38] Harry Rosenberg, "Improving Current Population Estimates through Stratification," *Land Economics*, 44(3):331–338, August 1968.

[39] Donald E. Pursell, "Improving Population Estimates with the Use of Dummy Variables," *Demography*, 7(1):87–91, Feb. 1970.

[40] George A. Hillery, Jr., *Post-censal Population Estimates for Small Areas: An Evaluation of Selected Techniques*, KAES RS-17, Lexington, Ky., University of Kentucky, Agricultural Experimental Station, September 1962.

[41] Richard W. Poole *et al.*, *An Evaluation of Alternative Techniques for Estimating County Population in a Six-State Area*, Stillwater, Oklahoma, Economic Research Series No. 3, Oklahoma State University, 1965.

[42] U.S. National Center for Health Statistics, *Preliminary Report of the Study Group on Postcensal Population Estimates*, The Public Health Conference on Records and Statistics, June 11, 1962.

[43] M. Zitter, D. E. Starsinic, and D. L. Word, "Accuracy of Methods of Preparing Postcensal Population Estimates for Counties: A Summary Compilation of Recent Studies," unpublished paper presented at the Annual Meeting of Population Association of America, Boston, Mass., April 18–20, 1968.

[44] U.S. Bureau of the Census, *Current Population Reports*, Series P-26, No. 21, "Federal-State Cooperative Program for Local Population Estimates: Test Results—April 1, 1970," April 1973.

SUGGESTED READINGS

Part A. Estimates for Countries

Bharier, Julian. "A Note on the Population of Iran, 1900–1966." *Population Studies* (London), 22(2):273–279. July 1968.

Ganiage, Jean. "La population de la Tunisie vers 1860. Essai d'évaluation d'après les registres fiscaux." *Population* (Paris), 21(5):857–886. Sept.–Oct. 1966.

Grauman, John V. "Population Estimates and Projections." Chapter 23 in Philip M. Hauser and Otis Dudley Duncan (eds.), *The Study of Population: An Inventory and Appraisal*. Chicago, Ill., University of Chicago Press, 1959, Pp. 544–575.

Japan, Office of the Prime Minister, Bureau of Statistics. *Population Estimates*, Series No. 32. "Population Estimates by Age and Sex as of October 1, 1967." Mar. 1968.

Lorimer, Frank W. "Analysis and Projections of the Population of the Philippines." In *First Conference on Population, 1965*. Proceedings of the Conference sponsored by the Population Institute of the University of the Philippines. Quezon City, University of Philippines Press, 1966. Pp. 200–287.

Muhsam, H. V. "Population Estimates Based on Census Enumeration and Coverage Check." *Population Studies* (London), 13(3):278–281. Mar. 1960.

United Nations. *Methodology and Evaluation of Population Registers and Similar Systems*. Series F, Studies in Methods, No. 15. 1969.

———. *Methods of Appraisal of Quality of Basic Data for Population Estimates*. Manual II, Manuals on Methods of Estimating Population. Series A, Population Studies, No. 23. 1955.

———. *Demographic Yearbook, 1964*. Pp. 12–14.

———. *Methods of Estimating Total Population for Current Dates*, Manual I, Manuals on Methods of Estimating Population. Series A, Population Studies, No. 10. 1952.

U.S. Bureau of the Census. *Current Population Reports*, Series P–25, No. 310. "Estimates of the Population of the United States and Components of Change, by Age, Color, and Sex: 1950 to 1960." June 30, 1965.

Wolfenden, Hugh H. *Population Statistics and Their Compilation* (rev. ed.). Chicago, Ill., University of Chicago Press, 1954. Pp. 76–88.

Part B. Estimates for Subdivisions of Countries

Atchley, Robert C. "A Shortcut Method for Estimating the Population of Metropolitan Areas," *Journal of the Institute of Planners*, 34:259–62, 1968.

Bogue, Donald J. "A Technique for Making Extensive Population Estimates." *Journal of the American Statistical Association*, 45(250):149–163. June 1950.

Ericksen, Eugene P. "A Method for Combining Sample Survey Data and Symptomatic Indicators to Obtain Population Estimates for Local Areas." *Demography*, 10 (2):137–160, May 1973.

———. "A Regression Method for Estimating Population Changes of Local Areas." *Journal of the American Statistical Association,* 69 (348):867–875, December 1974.

Goldberg, David, and Balakrishnan, T. R. *A Partial Evaluation of Four Estimation Techniques*. Michigan Population Studies, No. 2. Ann Arbor, Mich., Department of Sociology, University of Michigan, 1961.

Goldberg, David, Feldt, Allan, and Smit, J. William. *Estimates of Population Change in Michigan: 1950–1960*. Michigan Population Studies, No. 1. Ann Arbor, Mich., Department of Sociology, University of Michigan, 1960.

Goldberg, David, Rao, V. R., and Namboodiri, N. K. "A Test of the Accuracy of Ratio Correlation Population Estimates." *Land Economics*, 40(1):100–104, Feb. 1964.

Harewood, J. "Deriving Current Estimates of the Population of Trinidad and Tobago by Administrative Areas and by Age." In International Union for the Scientific Study of Population, *Sydney Conference, Contributed Papers*, Sydney, Australia, 1967. Pp. 937–948.

Hillery, George A., Jr. *Postcensal Population Estimates for Small Areas: An Evaluation of Selected Techniques*. Agricultural Experiment Station, KAES RS-17. Lexington, Ky., University of Kentucky, Sept. 1962.

Japan, Office of the Prime Minister, Bureau of Statistics. *Population Estimates*, Series No. 35. "Population Estimates by Prefecture, as of October 1, 1961 to 1964 (rev.)," Mar. 1969. Appendix 1.

———. *Population Estimates*, Series No. 26. "Population Estimates by Prefecture, as of October 1, 1956 to 1959 (rev.)." Sept. 1964.

Malaysia, Department of Statistics. *Estimates of Population for West Malaysia*. By Lee Jay Cho. Research Paper No. 1. Kuala Lumpur, 1967.

Poole, Richard W., *et al. An Evaluation of Alternative Techniques for Estimating County Population in a Six-State Area*. College of Business, Economic Research Series No. 3. Stillwater, Okla., Oklahoma State University, 1966.

Pursell, Donald E. "Improving Population Estimates with the Use of Dummy Variables." *Demography*, 7(1):87–91. Feb. 1970.

Rosenberg, Harry. "Improving Current Population Estimates Through Stratification," *Land Economics*, 44(3):331–338. Aug. 1968.

Schmitt, Robert C. "Short-Cut Methods of Estimating County Population." *Journal of the American Statistical Association*, 47(258):232–238. June 1952.

Schmitt Robert C., and Crosetti, Albert H. "Accuracy of the Ratio-Correlation Method for Estimating Postcensal Population." *Land Economics*, 30(3):279–280. Aug. 1954.

Schmitt, Robert C., and Grier, John M. "A Method of Estimating the Population of Minor Civil Divisions." *Rural Sociology*, 31(2):355–361. June 1966.

SUGGESTED READINGS – Continued

Schneider, J. R. L. "Note on the Accuracy of Local Population Estimates." *Population Studies* (London), 8(2):148–150. Nov. 1954.

Shryock, Henry S. "Development of Postcensal Population Estimates for Local Areas." In National Bureau of Economic Research, *Studies in Income and Wealth*, Vol. XXI, *Regional Income*. Princeton, N.J., Princeton University Press, 1957. Pp. 377–399.

Siegel, Jacob S. "Status of Research on Methods of Estimating State and Local Population." *Proceedings of the Social Statistics Section, 1960*. Washington, D.C., American Statistical Association, 1960. Pp. 172–179.

Siegel, Jacob S., Shryock, Henry S., and Greenberg, Benjamin. "Accuracy of Postcensal Estimates of Population for States and Cities." *American Sociological Review*, 19(4):440–446. Aug. 1954.

Silcock, H. "Precision in Population Estimates." *Population Studies* (London), 8(2):140–147. Nov. 1954.

Starsinic, Donald E., and Zitter, Meyer. "Accuracy of the Housing Unit Method in Preparing Population Estimates for Cities." *Demography*, 5(1):475–484. 1968.

U.S. Bureau of the Census. *Current Population Reports*, Series P–25, No. 432. "Estimates of the Population of 100 Large Metropolitan Areas: 1967 and 1968." Oct. 3, 1969.

————. *Current Population Reports*, Series P–25, No. 454. "Inventory of State and Local Agencies Preparing Population Estimates: Survey of 1969." Dec. 1970.

————. *Current Population Reports*, Series P–25, No. 339. "Methods of Population Estimation: Part I, Illustrative Procedure of the Census Bureau's Component Method II." June 6, 1966.

————. *Current Population Reports*, Series P–25, No. 436. "Estimates of the Population of States: July 1, 1968 and 1969." Jan. 7, 1970.

————. *Current Population Reports*, Series P–25, No. 437. "Estimates of the Population of States, by Age, 1968, With Provisional Estimates for July 1, 1969." Jan. 16, 1970.

————. *Current Population Reports*, Series P–25, No. 304. "Revised Estimates of the Population of States and Components of Population Change: 1950 to 1960." Apr. 8, 1965.

U.S. National Center for Health Statistics. *Vital Statistics – Special Reports*, Vol. 47, No. 6. "A Composite Method for Estimating Postcensal Population of Small Areas, by Age, Sex, and Color." By Donald J. Bogue and Beverly Duncan. Aug. 24, 1959.

Zitter, Meyer, and Shryock, Henry S. "Accuracy of Methods of Preparing Postcensal Population Estimates for States and Local Areas." *Demography*, 1(1):227–241. 1964.

CHAPTER 23

Population Projections

INTRODUCTION

The Nature and Types of Population Projections

The distinction was made between projections and current estimates in the previous chapter. As may be recalled, current estimates make use of actual postcensal data from the recent past in the form of vital statistics, tabulations from population registers, or statistics that are merely correlated with population change. Where there are no such data, the current estimate reduces methodologically to a short-range projection; but even here account could be taken of qualitative information concerning, for example, a natural disaster, war, famine, epidemic, or mass migration. Conventionally, projections into the future make no attempt to speculate about such possibilities because they are essentially unforeseeable.

Although we think of projections as applying to future population, the projection may also be into the past. There is sometimes interest in historical figures for dates preceding the first census. One way of making such a "precensal" estimate is to project the population backward by essentially the same techniques that are used for making forward projections.

The distinction should be made between projections and forecasts.[1] When the author or the subsequent user of a projection is willing to describe it as indicating the most likely population at a given date, then he has made a forecast. At the other extreme, a model worked out to illustrate certain analytical relationships, on assumptions that are described as highly unlikely, would not be regarded as constituting a forecast of future population growth. It is apparent that, in this usage, all forecasts are projections but not all projections are forecasts.

Population projections, then, are essentially concerned with future growth. They may be prepared for the total population of nations, their principal geographic subdivisions, or specific localities within them. Projections may also be prepared for residence classes, such as urban and rural population and size-of-locality classes. The principal characteristics for which projections need to be made are age and sex. Projections may also be made for various social and economic subgroups of the population and for other demographic aggregates. The most frequently required and produced are projections of the population in terms of (1) educational characteristics (i.e., enrollment and attainment), (2) economic characteristics (i.e., economically active population, employment distributed by occupation or industry), (3) social aggregates like households and families.

Uses of Population Projections

The principal uses of population projections and other demographic projections relate to government or private planning. Demographic projections may be used directly or as the basis for preparing other more specialized types of projections. These include, for example, projections of the expected number of retirements from the labor force in a given period, and of the required number of teachers or classrooms, housing units, medical personnel and facilities, etc. The users include national and local governments, business firms, labor unions, university research centers, and social service organizations. A fairly detailed discussion of these uses and users is given in a recent publication of the United Nations.[2] The less advanced countries of the world have recognized the necessity of making concrete, comprehensive plans for achieving specific goals of public policy related to accelerating their social and economic development. A first step in planning is to study relevant aspects of the population and economy both at the present time and in the recent past. As the United Nations notes, "such study provides a basis for projections representing plausible future courses of development under the assumption that future conditions will evolve in an orderly manner from those of the present and past."[3] In addition to the uses in the field of planning, there are important uses in demographic analysis and related types of scientific studies.

PROGRAMS OF POPULATION PROJECTIONS

International Programs

Some of the international agencies have both compiled and published national population projections and made their own projections for the world and regions. They have provided leadership in many aspects of concepts and methodology and have published a number of instructional manuals.

The United Nations has published two comprehensive sets

---

[1] John V. Grauman, "Population Estimates and Projections" in Philip M. Hauser and Otis Dudley Duncan (ed.), *The Study of Population*, Chicago, University of Chicago Press, 1959, pp. 544–575; and Irving Siegel, "Technological Change and Long-Run Forecasting," *Journal of Business*, 26(3):146, July 1953.

[2] United Nations, *General Principles for National Programmes of Population Projections as Aids to Development Planning*, Series A, Population Studies, No. 38, 1965.
[3] Ibid, p. 2.

NOTE.—This chapter makes substantial use of Background Paper WPC/WP/454, *Projections of the Total Population and of Age-Sex Structure*, by Henry S. Shryock, Jr., and Background Paper WPC/WP/494, *Projections of Urban and Rural Population and Other Socio-Economic Characteristics*, by Jacob S. Siegel, both prepared for the United Nations World Population Conference, Belgrade, 1965.

of population projections for the countries of the world, one in 1958 and another in 1966. The later report presented projections of the total population of the world, major areas, and regions for 1960 to 2000, and projections of the total population of each country for 1960 to 1980.[4] These projections take into account the results of censuses taken in many countries in 1960 and 1961. The 1958 report also presented projections of the future population of the world, regions, and countries, but on the basis of the censuses taken around 1950.[5] For various dates between 1954 and 1959, the United Nations also published rather detailed projections by sex and age groups for certain regions and countries, particularly in Central and South America and Asia.[6]

The national projections prepared by national agencies have been assembled in unified reports issued by international organizations on various occasions, and these sets of projections have followed common guidelines in varying degrees. The national projections made according to a common set of assumptions and methodology by a single agency or office are much more likely to be comparable, however. Even here, however, comparability is affected by differences in the availability and quality of the basic national data. In fact, a uniform methodology tends toward the least common denominator of what is available. The dilemma then is whether to have more regard to uniformity or to the inclusion of those individual projections that are noteworthy for sophistication of assumptions, data, and methodology. The most recent study of world population prospects by the United Nations attempts to embrace the best features of both approaches.[7] The population models developed at the United Nations also provide a unifying framework for assumptions concerning future trends in the components of population change. Underlying these models, there seems to be a theory of demographic transition, although Grauman and others have characterized the demographic transition as one of the discredited laws of population growth.[8] In a very general way, there does seem to be such a phenomenon although demographic changes in a given country may not conform very closely to any of the models.

National Programs

Many countries have recently produced their first official population projections, often as a basis for their formal plan of economic development. Other countries have previously produced one or more sets of projections and their program involves updating the figures from time to time, taking account of the rapidity with which the demographic situation is changing and particularly the suspected occurrence of a turning point in previous trends. Some countries change their projections at frequent intervals if they are repeatedly getting out of line with actual demographic developments.

The projections differ from one country to another as to the method employed. This method may often involve the separate projection of the components of migration, births, and deaths, particularly where adequate vital statistics are available. For a given country alternative projections may be derived with different assumptions as to migration, or fertility, or mortality, or a combination of these components.

Table 23–1 gives a summary of projections for Pakistan according to three assumptions regarding future fertility after 1960– (I) constant fertility; (II) constant fertility until 1970, then 30 percent linear decline to 1985, constant thereafter; (III) constant fertility until 1965, then 50 percent linear decline to 1985, constant thereafter.

The pioneering work of Warren S. Thompson and P. K. Whelpton of the Scripps Foundation in developing the component method has had an enormous influence on subsequent work around the world. Their projections for the United States were adopted as the official U.S. Government projections in the 1930's and early 1940's; however, by the midforties the method and the official responsibility for preparing projections were taken over by the Bureau of the Census. The Scripps Foundation returned to this field in the late fifties and sixties when Whelpton and his associates developed projections of the U.S. population on the basis of fertility "expectations" data obtained from the Growth of American Families Studies. In an earlier generation, the logistic curve, promoted by Raymond Pearl and Lowell J. Reed of Johns Hopkins University, was a tool widely used by demographers and others in the projection of future population growth, both in the United States and abroad.

GENERAL ISSUES AND PRINCIPLES

The Framework of Assumptions

Basic Assumptions.—It is almost commonplace to preface published projections with a statement that it is assumed that the area will not be visited by a war or a natural disaster. Furthermore, no attempt is made ordinarily to allow for future economic fluctuations of a cyclical nature. Of course, the age-sex structure of the population, the marital status and cumulative fertility of its women, and other measures of its current demographic status reflect the impact of such events in the past. The reports presenting the projections do sometimes state that conditions of nearly full-employment are expected to continue or that the gross national product is expected to increase at a given percentage per annum—even though relatively little is known about how variations in these economic phenomena affect population growth.

Number of Series and Combination of Assumptions.—Because of the evident uncertainties regarding future population changes, it is desirable to present more than one series of projections, in the case of both national projections and subnational projections. These may represent combinations of multiple assumptions concerning one or more of the components of population change. For example, there may be high, medium, and low assumptions for mortality and fertility. The high assumption may represent constant mortality or fertility; or all assumptions may call for decline and the variations may relate only to the date of onset or the rate of decline. In our era, increases in these components are regarded as very unlikely in many countries.

In preparing alternate projection series, care should be taken to avoid combinations of unlikely assumptions. Thus, one would not assume that expectation of life at birth would rise to 70 and

[4] United Nations, *World Population Prospects as Assessed in 1963*, Series A, Population Studies, No. 41, 1966.

[5] United Nations, *The Future Growth of World Population*, Series A, Population Studies, No. 28, 1958. Earlier sets of world population projections were published by the United Nations in "The Past and Future Growth of World Population—A Long-Range View," *Population Bulletin*, No. 1, 1952, and "Framework for Future Population Estimates, 1950–80, by World Regions," *Proceedings of the World Population Conference, 1954* (Rome), Vol. III, 1955.

[6] United Nations, *The Population of Central America (including Mexico), 1950–80*, Series A, Population Studies, No. 16, 1954; *The Population of South America, 1950–80*, Series A, Population Studies, No. 21, 1955; *The Population of South-East Asia (including Ceylon and China: Taiwan), 1950–80*, Series A, Population Studies, No. 30, 1959; and *The Population of Asia and the Far East, 1950–80*, Series A, Population Studies, No. 31, 1959.

[7] United Nations, *World Population Prospects as Assessed in 1963*, pp. 44–45.

[8] Grauman, op. cit., p. 551.

Table 23–1.—Projections of the Population of Pakistan by Sex According to Three Assumptions Regarding Fertility: 1965 to 2000

[In thousands. Midyear population]

Sex and assumption	1960	1965	1970	1975	1980	1985	1990	1995	2000
BOTH SEXES									
Assumption I..................	97,720	113,948	134,939	163,027	200,009	247,254	305,767	378,718	470,546
Assumption II.................	97,720	113,948	134,939	159,696	188,328	219,903	258,704	305,141	359,413
Assumption III................	97,720	113,948	131,667	151,582	172,984	193,810	218,515	246,471	276,874
MALE									
Assumption I..................	50,455	58,881	69,652	83,963	102,746	126,728	156,476	193,618	240,421
Assumption II.................	50,455	58,881	69,652	82,261	96,764	112,714	132,359	155,915	183,478
Assumption III................	50,455	58,881	67,981	78,107	88,906	99,246	111,770	125,858	141,197
FEMALE									
Assumption I..................	47,265	55,067	65,287	79,064	97,263	120,525	149,291	185,100	230,125
Assumption II.................	47,265	55,067	65,287	77,437	91,564	107,189	126,345	149,226	175,935
Assumption III................	47,265	55,067	63,686	73,475	84,078	94,464	106,745	120,613	135,676

Source: Adapted from, Pakistan Institute of Development Economics, *Population Projections for Pakistan: 1960–2000*, by Lee L. Bean, Masihur Rahman Khan, and A. Razzaque Rukanuddin, Karachi, January 1968, table V, p. 28.

continue at that high level while fertility remained constant, say at a total fertility rate of 6.0. Such combinations have never been observed in demographic history; and they would imply remarkable inconsistencies with regard to modernizing developments in the fields of medicine, public health, economic development, and attitudes toward marriage and the family.

One series may be designated as "most probable" or as a "best judgment" series. This usage indicates that an effort is being made to "forecast" and not just to "project." The medium or central series is ordinarily taken to be the most probable series. If only three series are prepared, the distribution of the high and low series about the medium series does not have to be symmetrical, but the usefulness of the projections is enhanced if the series are at least roughly symmetrical. The range from the highest to the lowest of the "reasonable" series presented may be regarded as a rough indicator of the degree of uncertainty regarding future population change. It is doubtful whether the range should be interpreted in a probability sense, however.

Principal and Analytic Series.—The series we have just described may or may not all represent reasonable possibilities. Assumptions or combinations of assumptions regarded as very unlikely of attainment may be presented for their analytical value or to set an upper or lower limit within which population growth is virtually certain to remain. Series I of the projections made for the United States in 1958 represents an example of the latter type.[9] Another example is provided by the projection series for South Korea that was based on the assumption of the continuation of current mortality and fertility levels. The report noted that the assumption is not realistic, but it was included because it provided a useful benchmark against which to measure the effects of declining mortality.[10] The difference between the series allowing for declining

mortality and the model analytic series assuming constant mortality would represent the effect of declining mortality.

As suggested above, in these and other examples, certain series may be characterized as reasonable possibilities and others as unreasonable possibilities. A more explicit way of distinguishing the two kinds of projections is to include some or all of the former as "principal series" in the regular tables of the published report and to place the latter in an appendix as "analytic series," including any series that were elaborated simply to serve as a benchmark for the other series. An appendix may also include reasonable series which represent intermediate assumptions, series derived by a method that still may be at an experimental stage, etc.

Length of Projection Period

Projections may extend for varying numbers of years into the future depending upon the type of area in question, the needs to be served, the conception of the problem by the analyst, and the available resources. Projections made nearly a century ahead by the U.S. Social Security Administration, as a basis for the long-range cost estimates of the social security system of the United States, represent one extreme.[11] The projections for Great Britain made for the Royal Commission on Population also extended a century ahead, covering the period 1947 to 2047. Projections of just a few years used for short-term economic analysis represent the other extreme. Long-term projections (over 25 years) are needed in connection with the development of water and forestry resources, major transportation and recreational facilities, and planning for provision of food, whereas only middle-range projections (10 to 25 years) are required for planning educational and medical facilities and services, and housing needs. Relatively simple methods may be used both for short-range projections (under 10 years) and for the extensions of middle-range projections beyond 25 years. For projections up to 25 years ahead the more

---

[9] U.S. Bureau of the Census, *Current Population Reports*, Series P-25, No. 187, "Illustrative Projections of the Population of the United States, by Age and Sex: 1960 to 1980," Nov. 10, 1958.

[10] Republic of Korea, Economic Planning Board, *The New Population Projections for Korea: 1960–2000*, Seoul, 1964, p. 39.

[11] U.S. Social Security Administration, Office of the Actuary, "United States Population Projections for OASDHI Cost Estimates," by Francisco Bayo, *Actuarial Study No. 62*, December 1966.

elaborate methods are usually employed and are recommended. Differences in the accuracy of national and subnational projections suggest that, in general, the latter be extended for substantially fewer years into the future (see below).

Frequency and Nature of Revision

There is, in general, a real problem as to how often an organization should revise its projections and how fundamental the changes should be. Primary factors pertinent to these decisions include: (1) The extent to which the present projections are out of line with current estimates, (2) the mere passage of time, say a few years, during which at least some of the series of projections must deviate from the current estimates, (3) the availability of additional data, not only for more recent periods but also (more refined data) for earlier years, (4) advances in methodology. Secondary factors include (1) the availability of resources and (2) the need to meet an established publication schedule. The revision may consist of merely "splicing in" adjusted projections for the next few years or of making a small absolute or ratio adjustment of all the original projections, in order to make them consistent with newly available current data. In such cases the basic assumptions are not otherwise modified. In other cases, the most recent data may suggest the need for major modifications in the basic assumptions and hence a recalculation of all the projections.

Detail of Presentation

National reports on population projections will differ greatly in the detail shown, but they should be fully informative as to how the projections were made. In addition to the principal series, the report may usefully present a number of supplementary series, including additional combinations of the basic assumptions and analytic series. Single ages or broad age groups and annual or quinquennial figures may be presented, depending on the quality and need for the figures. When there are many series, the age or chronological detail may be abridged for those series that are not considered the principal series. The text should fully discuss the choice of assumptions, the methodology, the resulting projections, and even their implications. It is useful also to present the relevant information on the past trends of the components of population change which were taken into account in choosing the assumptions, the implied birth and death rates at future dates, and other demographic summary measures.

METHODOLOGY

Some Methodological Principles

From the above discussion of principles, several conclusions, or indeed recommendations, regarding methodology emerge. The choice of methods, although affected by the other considerations mentioned, should be determined primarily by the data available. The availability of adequate statistics on the age-sex distribution of the population and of adequate statistics on age-sex-specific fertility and mortality ordinarily calls for a cohort-component method. Using theoretical models, the cohort-component method may also be employed without benefit of recorded age detail for fertility and mortality. Even if there are some defects in the basic data, however, it may often be advisable to adjust the data rather than to have recourse to a more theoretical model. More time may then need to be spent on evaluating and adjusting the basic data than on computing the projections.

The trend toward refinement of the methodology of population projections has led to increased specificity of the assumptions and the estimation categories on the ground that more accurate assumptions can be made for those more specific categories. The impetus toward specificity derives in part from the thesis that the quality and usefulness of projections will be enhanced by employing a method whose steps parallel the order of events in real life. This principle leads to the use of methods based on birth or marriage cohorts and specific rates for the frequency of events affecting these cohorts as they progress through life. In its fully developed form we have a population model.

In order to take advantage of the most up-to-date knowledge of recent events, the benchmark of the projections should be the latest postcensal estimates (including sample survey estimates) or, failing that, the latest census counts. There would ordinarily be only one set of population figures at the benchmark date; however, if the current estimates are subject to considerable error, including sampling error, a range of estimates may be employed. For later dates, there should be at least two principal projection series. Moreover, the various series in a single set of projections should conform to some general system of projections, permitting easy comparison of the assumptions and results. Finally, the assumptions of each principal series should be mutually consistent and should not represent an unreasonable combination of developments.

Classification of Methods

The methods of making projections may be classified in a number of different ways. One possible way is to distinguish those that can be applied independently to any type of area from those that are dependent on or require antecedent projections for other areas. The latter class of methods includes (1) summing of projections, as in the case of adding country projections to secure regional or world population totals, (2) obtaining a projection for one area on the basis of changes in some other similar area or more inclusive area for which a projection is already available. The latter procedure covers a wide range of techniques; e.g., distributing the projected population of an area among its subdivisions taking account of the past proportionate distribution (ratio methods) and projecting the components of population change for an area on the basis of changes in these components for a similar area or more inclusive area for which such projections are already available. Most typically, the independent methods are applied in the case of national populations, but they could be employed to project the population of any type of area from the world total to a particular locality. These methods include mathematical methods and component methods applied at any geographic level, but particularly at the national level. Even here the results over a number of areas may be adjusted to population figures for a parent area. In fact, it is a principle of population projections to extend the degree of interdependence where possible to assume consistency of assumptions and results for the various areas for which projections are being prepared.

Another typology of the methods distinguishes among mathematical and ratio methods, various methods using a series of indicators of population change, component methods, and combinations of these. The mathematical methods employ relatively simple mathematical equations to describe the nature of future population change of an area, taken as a whole. The component methods involve the separate projection of mortality, fertility, and net migration. The methods using indicators of population change embrace a variety of procedures, some estimating total population directly, others estimating net migration only, which is then combined with a separate estimate of natural increase. The indicators (e.g., employment, per capita income, etc.) may be used to project population or net migration by a regression or a ratio procedure.

The description of methods will begin with those methods applicable to national populations. It is convenient to consider the projection of the age-sex distribution of national populations at the same time because the totals are often obtained by summation of projections for age groups. Projections for the principal geographic subdivisions of countries are then considered. Again, projections of the age-sex distribution for such areas are considered at the same time. Finally, we will consider briefly the methodology for projections for certain social and economic subgroups of the population, including urban and rural groups, the enrolled population, the labor force, and social aggregates like households and families.

National Projections of Total Population and Age-Sex Composition

Methods of preparing national projections of total population fall very simply into two classes, mathematical methods and component methods. As is indicated later, the component method is usually the method of choice, but there are many variations of the component method and results may differ substantially from one variation to another. Mathematical and component methods may also be employed as principal methods of preparing projections of age-sex distribution for countries. They may be combined with ratio procedures as subsidiary elements in the method; as, for example, when the percent distribution of a population by age is projected by a mathematical formula. The component method is also the method of choice for preparing projections of age-sex distribution; this is so even when very limited historical data on components or component data of doubtful quality are available. In the latter situation, these data must first be corrected before the actual projections are calculated.

Mathematical Methods. — Mathematical methods are simplest conceptually, usually require relatively little time to apply, and historically antedated the use of component methods. The mathematical methods involve application of some mathematical formula directly to the total population from one or more censuses, to derive projections of total population. The mathematical equations useful for preparing projections correspond to those described in chapter 13 for measuring population change. Among the most commonly used are the geometric curve, with (1a) annual or (1b) continuous compounding, and (2a or 2b) the logistic curve.

$$P_t = P_o(1+r)^t \qquad (1a)$$

or

$$P_t = P_o e^{rt} \qquad (1b)$$

$$P_t = \frac{\frac{1}{a}}{1 + \frac{b}{a e^{-rt}}} \qquad (2a)$$

or

$$P_t = \frac{K}{1 + e^{a+bt}} \qquad (2b)$$

where r is the growth rate, t is the number of years, a and b

are constants, and e is the base of the natural system of logarithms. In (2), the constant $1/a$ or K is the upper asymptote.

For making a population projection, one may apply equation (1a) or (1b) either by using the latest intercensal rate of change, the average rate over a longer period, or an arbitrary rate, or by fitting a curve by the method of least squares to a series of census totals. Since growth rates are likely to change in the long term, these formulas are recommended for use only in making short-term projections.

Fitting a logistic curve is a more complicated procedure and requires a greater number of observations covering a longer period. At the same time, it is useful for projection over a greater period of years than the simple geometric procedures, particularly if the past series has reached the point of inflexion. This form of projection overcomes a principal weakness in the use of the geometric rate, namely, the possibility of obtaining extremely large population figures after a short period.

Mathematical methods are now much less frequently employed to estimate the population of countries than formerly, even though they have by no means been abandoned.[12] Component methods have been displacing the mathematical methods.

Component Methods. — These methods involve the separate projection of mortality, fertility, immigration, and emigration. As in preparing population estimates, the last two components are usually combined in the form of net migration; and even this combined component is often omitted, in effect, by use of an assumption of "no net migration." Since the method may be easily applied, and is conventionally applied, by age-sex groups, projections of age-sex structure are usually obtained directly by this method; and total population is commonly obtained by combining the projections for age-sex groups. The component method makes explicit the assumptions regarding the components of population growth and hence can give one considerable insight into the way population changes. Valuable by-products, such as various measures of fertility and mortality, as well as estimates of the effect of alternative levels of fertility, mortality, or immigration on population growth, may also be obtained. In its simpler form the component method is mechanistic in many ways. In its more complex forms, however, the component method becomes a rather elaborate model of population growth.

Use of crude rates. — In principle, crude birth and death rates may be employed as measures in projecting the numbers of births and deaths. They are rarely used in a direct fashion for this purpose, however, since they are strongly affected by the structure of the population, particularly its age-sex structure. Crude rates have been employed occasionally in otherwise elaborate studies of the future on the ground that methodological refinement is unjustified by the state of our knowledge.[13] Assumed absolute numbers have not been used for projecting births and deaths, but such an assumption is sometimes made for the migration component.

Cohort-component methods. — In these methods, the computations are carried out separately for age-sex groups on the basis of separate allowances for components. Specifically, one starts with the population distributed by age and sex at

[12] Luigi Amoroso, "Le quattro fasi di espansione demografica" (The four phases of demographic development), *Rivista di politica economica*, 48(11): 1168–1175, November 1958; and H. C. Plessing, "Om den Logistiske Kurve og dens Anvendelse i Praksis" (On the logistic curve, its use and practice), *Erhversekonomisk Tidsskrift* (Copenhagen), 26(3):205–231, 1962.
[13] Marion Clawson *et al*, *Land for the Future*, printed for Resources for the Future, Inc., by the John Hopkins University Press, Baltimore, 1960.

the base date, applies assumed survival rates and age-sex specific fertility rates or birth probabilities, and makes allowances for net migration by age and sex, if desired. The base population should be the latest dependable postcensal estimates of the national population distributed by age and sex. In the event that current estimates are not available or calculable, the data from the last census can be used. The age groups usually have 5-year class intervals.

Projections of Mortality. — In projecting the component of mortality, two measures of mortality may be employed, age-specific death rates or age-specific survival rates; hence, our discussion will be concerned principally with the different ways of projecting rates of these types.

Projections of past trends. — Where there are adequate mortality statistics for the country itself, we may start with the country's own age-sex-specific death rates or survival rates for the most recent year or period. The survival rates are usually calculated for 5-year age groups and 5-year time intervals and are applied to the population distributed in 5-year age groups at one date to obtain the number of survivors 5 years older 5 years later.

$$_5s_x^5 = {}_5P_x \cdot \frac{_5L_{x+5}}{_5L_x} \qquad (3)$$

Projections for single years of age or individual calendar years may be obtained by interpolation from the grouped data at quinquennial intervals. Once single ages have been derived at 5-year intervals, annual projections may be obtained by interpolation along single-age cohorts. If an electronic computer is available, however, the preferred procedure is to carry out the survival calculations directly in single ages for individual calendar years.

The various procedures of projecting mortality commonly involve a different and progressive set of survival rates for each succeeding 5-year period or other time unit over which the population is projected. Hence, each age cohort is carried forward, in effect, by a type of generation life table although generation life tables are not actually computed.

Projected death rates for a given country may be derived on the basis of a number of general procedures or assumptions. These are: (1) Maintaining the latest observed death rates, (2) extrapolating past trends in the country's own death rates in some fashion, (3) applying standard percentage decreases in death rates depending on the level of the death rate at each successive date, and (4) establishing target rates for a distant future date and securing rates for intermediate dates by some form of interpolation.

The first alternative would be applicable only where (a) mortality had reached a standstill and there were no signs of further progress or (b) the implications of other more realistic assumptions were being evaluated against this "base." The series of death rates may be extrapolated either graphically or by fitting an appropriate mathematical curve. The extrapolations under (2), (3), and (4) should all provide for improvement in mortality, with rare exceptions, and, if death rates are already low, the improvement should be at a decreasing rate. The extrapolations should be carried out independently for each age-sex group; and the same or different method of extrapolation may be used for different age-sex groups.

The risk of arriving at unreasonably low levels of mortality in the projection period when direct extrapolation is employed suggests that a limit should be set to the improvement

assumed to occur. Accordingly, after a given future date, the rates at every age may be assumed not to change any farther. The target rates set to avoid unreasonable levels of future mortality may be reached in different years for different age-sex groups. They may be derived in a number of different ways: (1) Use of the rates already attained in some advanced geographic subdivision of the country; (2) use of the rates already attained in another, more advanced country, somewhat similar to the given country in certain socioeconomic features but having better public health organization and lower death rates;[14] (3) analysis of age-specific death rates in terms of components, such as principal causes of death, for which judgmental projections could more confidently be made; and (4) determination of the lower biological limit for mortality at each age on the basis of present knowledge, and of the date by which this limit will be attained.

The third procedure for deriving target rates has been employed in a number of important studies in the United States.[15] It may be explained on the basis of the particular projection procedure described in Actuarial Study No. 62 of the U.S. Social Security Administration. Past trends in age-sex specific death rates were analyzed in terms of 10 cause-of-death classes, and medical research in progress was evaluated from the standpoint of the prospects of "breakthroughs" that would importantly affect the level of these rates.

Mortality models. — Mortality models, representing generalized schemes for projecting mortality applicable to a group of countries, are generally employed when satisfactory statistics are not available regarding the current level and trend of mortality for a country or group of countries. They are also used occasionally when such data are available. Mortality models may relate to the pattern by which current death rates are reduced over successive time units in the projection period, or to the way in which the terminal levels of mortality are determined, or to both; or they may establish specific levels of terminal mortality.

An early generalized scheme of projecting mortality was the method developed by Coale at Princeton University in the 1940's.[16] His method was based on an historical analysis of age-specific death rates for populations of European origin. This analysis showed that the rate of decline of these death rates tended to vary directly with their level, without regard to country or year. He derived a set of curves for age-specific mortality rates $(_5q_x)$ by (1) separating the observed rates at each age into four segments on the basis of the level of the rates, (2) fitting straight lines by least squares to the three segments with the higher rates, (3) fitting an exponential curve to the segment with the lowest rates (the form of curve being chosen to prevent the rates from becoming negative when extrapolated), and (4) smoothing the junctures between segments.

[14] See, for example, J. J. Paes Morais and A. Costa Leal, "A evolucão demográfica nacional e o desenvolvimento económico (National population trends and economic development), *Revista de economia* (Lisbon), 11(4):149-167, December 1958.

[15] U.S. Bureau of the Census, *Forecasts of the Population of the United States, 1945-75,* by P. K. Whelpton *et al.,* 1947, pp. 8-13; U.S. Social Security Administration, Office of the Actuary, "Illustrative United States Population Projections," by T. N. E. Greville. *Actuarial Study No. 46,* May 1957, pp. 9-16; and idem, "United States Population Projections for OASDI Cost Estimates," by Francisco Bayo, *Actuarial Study No. 62,* December 1966, pp. 10-16. *Actuarial Study No. 46* was the basis of the projections of mortality incorporated in the population projections published by the U.S. Bureau of the Census in *Current Population Reports,* Series P-25, Nos. 286 and 381.

[16] Frank W. Notestein, *et. al., The Future Population of Europe and the Soviet Union: Population Projections, 1940-70,* Geneva, League of Nations, 1944, pp. 183-189.

The Princeton method was based wholly, or in large part, on pre-World-War-II mortality experience. In general, the decline of mortality has accelerated in the post-World-War-II period. Using postwar data, Campbell derived new constants for Coale's height-slope relationships and introduced certain "ultimate" levels of mortality at each age.[17] His principal equation is:

$$m_{x.t} = (m_{x.o} - m_{x.\infty})e^{-b_x t} + m_{x.\infty} \qquad (4)$$

where $m_{x.o}$ is a central death rate in the base year (last observed rate) at age x, $m_{x.t}$ is a central death rate t years after the base year at age x, $m_{x.\infty}$ is the ultimate central death rate at age x, b_x is a constant specific for age x, to be determined, and t represents the number of years for which the projection is desired.

The constant b_x is a quantity representing the slope of a straight line which expresses absolute annual decline in mortality as a function of the level of mortality. The method of deriving b_x involves first setting the level of m_∞ independently for each age. Then a value of b is determined for each observation (each country) from the equation,

$$b = \frac{m_t'}{m_t - m_\infty} \qquad (5)$$

which is derived from equation (4). The value m_t' is the first derivative of m_t and represents the average annual decline for any death rate (m_t).

The hypothetical low death rates, $m_{x.\infty}$, chosen do not greatly affect the projections of population for most age groups as long as the death rates assumed are below present levels and reasonable. Campbell derived the low death rates by (1) compiling a set of lowest observed age-specific death rates, (2) smoothing the observed lows graphically, (3) calculating maximum possible percentage reductions below the smoothed observed lows (ranging up to 50 percent for ages 20–24 years), and (4) reducing the smoothed observed lows by these percentages to obtain a set of hypothetical lows. Such mortality rates are intended only for use in making population projections. They are not intended to yield independently valid projections of future mortality rates.

The formulas are of doubtful utility in extrapolating death rates in countries whose postwar mortality experience has diverged widely from the average international trend. Similarly, they are not applicable to countries having mortality rates much higher than any of those used to estimate the constants of the projection formulas. The formulas may become obsolete as medical progress leads to improved mortality beyond the limits of the minimum mortality rates used in the projection formula.

Projections of mortality could also be based on the research findings at France's Institut national d'études démographiques regarding the separate trends of deaths from endogenous and exogenous causes and the biological "limit" of the death rate at each age.[18]

Without specifying either height-slope relationships or ultimate levels of mortality, the United Nations has developed a set of model life tables for projecting mortality.[19] They are based on analysis of 158 life tables representing the experience in the first half of the 20th century of 50 countries. The calculation of the model tables rests on the principles that the schedule of age-specific mortality rates is fairly well defined if the expectation of life at birth is known, and that expectation of life tends to increase by a fairly uniform annual amount until a moderately high level of life expectancy is reached.

The model tables were designed on the assumption that an annual gain of 0.5 year in expectation of life at birth will occur whenever the expectation is less than 55 years. At older ages, the gain increases, then declines. Twenty-four model tables (males and females for each model) are presented, reflecting differences of 2.5 years in expectation of life at birth (up to 55 years) over each 5-year time period and covering a time span of 115 years. The expectation of life for both sexes combined ranges from 20 to 73.9 years. Once a particular model table is selected to represent the current level of mortality or once the current level is established, the tables to be used for future mortality allowances are selected in rank order from the set of model tables. (See illustration below for Costa Rica in table 23-4). To use the system of model life tables, the assumptions with respect to mortality have to be expressed in terms of $\mathring{e}_o$ for both sexes combined. This $\mathring{e}_o$ identifies a particular model life table which describes a whole age-pattern of mortality conditions.

Another system of model life tables, the Coale-Demeny model life tables, distinguishes regional variations in mortality patterns; four sets of life tables are given, with expectation values of 20.0 to 77.5 years, at intervals of 2.5 years.[20] The selection of tables was not designed to provide a progression of values directly applicable to projections, as were the U.N. tables. Once some assumptions regarding the improvement in mortality in a country have been developed, however, the Coale-Demeny model tables can effectively be employed to implement the assumptions.

These systems of model tables may be used to project the mortality of countries for which only limited data on mortality are available. They may also be used to project mortality in developed countries under a unified system of projections. The use of a common system of mortality projections should add to the international comparability of the projections even though the procedure may be less refined than is possible for some countries.

Projections of Fertility.—We may distinguish three types of methods of projecting fertility as part of a cohort-component method of projecting population, namely, the period-fertility method, the cohort-fertility method, and the marriage-parity-interval progression method. Other methods of projecting fertility may be viewed as elaborations or combinations of these basic methods.

Period-fertility method.—In the period-fertility method, or the **age-specific birth rate method,** assumptions concerning future fertility are stated in terms of calendar-year or period age-specific birth rates, period sex-age adjusted birth rates, period total fertility rates, or period gross reproduction rates. The prototype procedure consists essentially of (1) calculating and analyzing a time series of birth rates for each (5-year) age group of woman; (2) holding the most recent rates constant

[17] U.S. Bureau of the Census, *International Population Reports*, Series P-91, No. 5, "A Method of Projecting Mortality Rates Based on Postwar International Experience," by Arthur A. Campbell, 1958, pp. 3–16.

[18] John V. Grauman. op. cit., p. 571; Jean Bourgeois-Pichat, "Essai sur la mortalité 'biologique' de l'homme" (Essay on the "biological" mortality of man), *Population* (Paris), 7(3):381–394, July–Sept. 1952.

[19] United Nations, *Age and Sex Patterns of Mortality: Model Life Tables for Underdeveloped Countries*, Series A, Population Studies, No. 22, 1955; and idem, *Methods for Population Projections by Sex and Age*, Series A, Population Studies, No. 25, Manual III, Methods of Estimating Population, 1956.

[20] Ansley J. Coale and Paul Demeny, *Regional Model Life Tables and Stable Populations*, Princeton, N.J., Princeton University Press, 1966.

over the projection period or projecting the rates on the basis of various assumptions and techniques to some future year; and then (3) applying the projected schedule of rates to the projected female population by 5-year age groups for a given future year to determine the corresponding number of births. The characteristic feature of the method is that the trend analysis is in terms of rates for a given age group. If the calculations are done manually, it is most convenient (1) to obtain the projected rates for the end of each quinquennium, (2) multiply the rates cumulatively by the number of women by age previously calculated to have survived to the end of the quinquennium, (3) average the births for the first and last years of each quinquennium, and (4) multiply by five (to secure births for a 5-year period).

This procedure may be abbreviated, with little change in the results, by projecting the period total fertility rate or the period gross reproduction rate instead of the schedule of age-specific birth rates. Applying any single reasonable set of age-specific rates (e.g., the rates for the current year) to the future numbers of women of childbearing age, with an adjustment to the projected total fertility rate (or gross reproduction rate), produces very nearly the same number of births as applying a projected schedule of age-specific birth rates. Variations in the pattern of age-specific birth rates tend to have little effect on the corresponding numbers of births, given the overall reproduction rate, so that the total fertility rate (or the gross reproduction rate) is a useful substitute for making projections of births by the age-specific birth rate method. This variation permits the application of the period-fertility method to areas where age-specific birth rates are lacking and where only the total number of births is known. The initial age-specific birth rates and total fertility rate can be estimated by the indirect method from (1) the total number of births, (2) the female population of childbearing age, and (3) an available set of age-specific birth rates for a similar area (ch. 16).

Among the several variations we have mentioned so far, we will illustrate first the use of the total fertility rate to project births, given current age-specific birth rates. Let us consider an example using the total fertility rate to project births for Costa Rica from 1965 to 1985 (table 23–2, **Method B**). We employ for this purpose the basic data and some of the assumptions of an official Costa Rican study, including the estimates of the population by age and sex and the recorded age-specific birth rates for 1965.[21] First, the total fertility rate is projected on some basis. In the present case, following the procedure of the official study, we made three assumptions regarding the future course of fertility, representing high, medium, and low alternatives. The high series assumes a continuation of the total fertility rate of 1965, the medium series a decline in the rate of 5 percent each 5 years, and the low series a decline in the rate of 10 percent each 5 years (table 23–3).

First, the current schedule of age-specific birth rates is assumed initially for all future dates and series. Next, these assumed rates are applied to the projected female population of childbearing age in each year (see below for method of derivation) to obtain the "expected" number of total births for that year (for all series). For example, under all assumptions the expected number of births in 1970 is derived by taking the cumulative product of the age-specific birth rates of 1965 and

the projected female population of childbearing age in 1970 (table 23–2). This product is 75,344. Next, this number is multiplied by the ratio of the projected total fertility rate (*TFR*) in 1970 for a particular series to the total fertility rate for 1965, corresponding to the percent change in the rate for that series for 1965–70. For the medium series, for example, this ratio is .95 and the projected number of births in 1970 is 75,344 × .95, or 71,577. The projected number of births for 1965–70 is obtained by inflating by 5 (for 5 years) the average number of births for the period [$\frac{5}{2}(62,821 + 71,577) = 335,995$]. The projected number of female births is then derived by applying as assumed proportion female among births; e.g.,

$$335,995 \times .49 = 164,638$$

Similarly, to derive the projected number of births in 1975, under the medium series the expected number of births in 1975 is adjusted by the ratio of the projected *TFR* in 1975 under the medium series to the current *TFR* (this is, .9025). Note that, after 15 years of projecting births, it is necessary to begin carrying forward to older ages the female births which were projected in the first future 5-year period, so as to derive the female population of reproductive age; for example, the births of 1965–70 will become 15 to 19 years of age in 1985.

The calculations for the high and low series of births are presented in table 23–3. Since the high series assumes no change in the *TFR*, it corresponds to the expected numbers of births computed in table 23–2, the adjustment factors being 1.00 in each case. The low series of births in 1970 (67,810) was derived by adjusting the expected number of births in 1970 (75,344) by the ratio of the low series *TFR* for 1970 (5,886.5) to the *TFR* for 1965 (6,540.5) —.90; the low series of births in 1975 (75,088) was derived by adjusting the expected number of births in 1975 (92,701) by the ratio of the low series *TFR* for 1975 (5,297.9) to the *TFR* for 1965 (6,540.5) —.81; etc.

Use of the total fertility rate in this way is equivalent precisely to assuming that the age-specific birth rates change uniformly by age by the same percentage as the total fertility rate (table 23–2, **Method A**). Even if a different pattern of change in fertility by age is expected, the summary procedure is acceptable because of the negligible effect of variations in the age pattern of fertility on the number of births, once the total fertility rate is fixed. It may be recognized that the procedure of using the total fertility rate to project the number of births, just described, involves the same logic as the calculation of this rate by the indirect method, albeit the order of calculation is different. In the former case, we are given the population by sex and age, a standard set of assumed age-specific birth rates, and the total fertility rate, and we are asked to calculate the corresponding number of births. In the latter case, we are given the population by sex and age, a standard set of assumed age-specific birth rates, and the total number of births, and we are asked to calculate the corresponding total fertility rate. The procedure and logic with gross reproduction rates are essentially the same; there is no advantage in the use of this rate and either measure allows for a change in the sex ratio of births with equal convenience.

The medium series of births, derived as described above, was employed in completing a set of projections of the female population of Costa Rica by 5-year age groups at 5-year time intervals to 1985 (table 23–4.) For this purpose, we have adopted the current population estimates given in the Costa Rican report cited above. The base population in 1965 (derived from

[21] Costa Rica. Dirección general de estadística y censos, Proyección de la problación de Costa Rica por sexo y grupos de edad, 1965–90" (Projections of the population of Costa Rica by sex and age, 1965–90), by Ricardo Jiménez J. *et al.*, Serie Demográfica No. 5, *Revista de estudios y estadísticas*, San José, No. 8, October 1967.

Table 23-2. — Calculation of the Projected Medium Number of Female Births in Costa Rica by the Period-Fertility Method, for 1965 to 1985

[Methods A, B, and C represent three variations in the application of the period-fertility method]

Age of mother (years)	Age-specific birth rates (f_a)					Population (p_a)[2]					Population weights (W_a)
	1965[1]	1970	1975	1980	1985	1965	1970	1975	1980	1985	
15 to 19.....................	110.9	105.4	100.1	95.1	90.4	72,385	92,466	116,881	142,217	157,262	1
20 to 24.....................	301.9	286.8	272.5	258.9	245.9	58,115	71,914	92,022	116,472	141,847	7
25 to 29.....................	319.6	303.6	288.4	274.0	260.3	48,275	57,650	71,483	91,617	116,076	7
30 to 34.....................	258.1	245.2	232.9	221.3	210.2	42,325	47,836	57,241	71,090	91,214	6
35 to 39.....................	212.4	201.8	191.7	182.1	173.0	36,875	41,872	47,420	56,835	70,671	4
40 to 44.....................	89.3	84.8	80.6	76.5	72.7	29,815	36,381	41,395	46,960	56,352	1
45 to 49.....................	15.9	15.1	14.3	13.6	12.9	25,200	29,260	35,781	40,786	46,340	-
Total fertility rate $= 5 \Sigma f_a$	6,540.5	6,213.5	5,902.8	5,607.7	5,327.3						

	Births (B)				
	1965	1970	1975	1980	1985

Method A:

1. Expected births, year $y = \Sigma f_a^y p_a^y$: 62,821 | 71,579 | 83,661 | 99,011 | 115,405

2. Expected female births, 5 year period[3] $= \left[\left(B^y + B^{y+5}\right) \div 2\right] \times 5 \times .49$: 164,640 | 190,169 | 223,773 | 262,660

Method B:

3. Expected births, year $y = \Sigma F_a^{65} p_a^y$: 62,821 | 75,344 | 92,701 | 115,478 | 141,684

4. Projected medium births $= (3) \times \dfrac{TFR^y}{TFR^{65}}$: 62,821 | 71,577 | 83,663 | 99,009 | 115,403

5. Projected medium female births, 5 year period[3] $= \left[\left(B^y + B^{y+5}\right) \div 2\right] \times 5 \times .49$: 164,638 | 190,169 | 223,773 | 262,655

Method C:

6. Weighted population $= \Sigma W_a\, p_a^y$: 1,248,380 | 1,490,299 | 1,835,937 | 2,299,680 | 2,849,043

7. Sex-age adjusted birth rate, year $y = s.a.a.b.r.^{y-5} \times .95$: .5032 | .4780 | .4541 | .4314 | .4098

8. Projected medium births $=$ weighted pop. $\times$ s.a.a.b.r. $= (6) \times (7)$: 62,818 | 71,236 | 83,370 | 99,208 | 116,754

9. Projected medium female births, 5 year period[3] $= \left[\left(B^y + B^{y+5}\right) \div 2\right] \times 5 \times .49$: 164,216 | 189,392 | 223,658 | 264,553

— Represents zero.

[1] Source: Costa Rica, Dirección general de estadística y censos, "Proyección de la población de Costa Rica por sexo y grupos de edad, 1965–90" (Projections of the population of Costa Rica by sex and age, 1965–90), by Ricardo Jiménez J, *et al.*, Serie Demográfica No. 5, *Revista de estudios y estadísticas*, San José, No. 8, October 1967, table 34.

[2] See table 23-4.

[3] These births relate to the 5-year period from July 1 of initial year to June 30 of terminal year.

the census of 1963) was adjusted for underenumeration of children under 10 years old and for age heaping in the range 10 years old and over. As in that report, we have assumed that there would be no immigration and that mortality would show a moderate gradual improvement after 1965. The projected survival rates were based on the observed death rates for 1963; they followed the assumption that expectation of life at birth would increase from 1.5 to 2.5 years over each 5-year time period, corresponding to the model life tables in U.N. Manual III.[22] The observed rates in 1963 corresponded most closely, on the average, to the rates in U.N. model life table 90 (or $\mathring{e}_0 = 65.8$). Therefore, for projection periods 1965–70, 1970–75, 1975–80, and 1980–85, we used model tables 95, 100, 105, and 110, corresponding to life expectation values of 68.2, 70.2, 71.7, and 73.0 years, respectively.

Five-year survival rates for each 5-year time period corresponding to each required mortality level were selected from the model tables and applied to the female population. The 1965–70 factors were applied to the 1965 population, by sex, to derive the projected population 5 years old and over in 1970, then the 1970–75 factors were applied to this population to derive the projected population 10 and over in 1975, etc. The projected births were developed on the basis of the projections of female population of childbearing age, as described earlier (Method B, medium series, in table 23-2). The births in 1965–70 were then carried forward to ages 0–4 in 1970 by use of the appropriate survival rate, the survivors 0–4 in 1970 were carried forward to ages 5–9 in 1975, etc.

A still simpler procedure for projecting births which is applicable when only total births and an age-sex distribution are available employs the U.N. sex-age adjusted birth rate (ch. 16). In this procedure we do not have to assume any particular pattern of age-specific rates. We have applied the procedure in table 23-2 (**Method C**) to Costa Rica, using the same general assumptions as before but without making use of the available age-specific birth rates. This procedure represents still another application of the logic of indirect standardization.

[22] United Nations, *Methods for Population Projections by Sex and Age*, table V.

Table 23-3. — Calculation of Alternative Projections of Female Births in Costa Rica by the Period-Fertility Method, for 1965 to 1985

Series	1965	1970	1975	1980	1985
HIGH SERIES					
1. Total fertility rate, year $y = 1.00$ TFR^{y-5}	6,540.5	6,540.5	6,540.5	6,540.5	6,540.5
2. Expected births $= \Sigma F_a^{65} p_a^y$	62,821	75,344	92,701	115,478	141,684
3. Projected female births, 5-year period[1]	169,252	205,855	255,019	315,023	
MEDIUM SERIES					
4. Total fertility rate, year $y = .95$ TFR^{y-5}	6,540.5	6,213.5	5,902.8	·5,607.7	5,327.3
5. Projected births $= (2) \times \frac{(4)}{(1)}$	62,821	71,577	83,663	99,009	115,403
6. Projected female births, 5-year period[1]	164,638	190,169	223,773	262,655	
LOW SERIES					
7. Total fertility rate, year $y = .90$ TFR^{y-5}	6,540.5	5,886.5	5,297.9	4,768.1	4,291.3
8. Projected births $= (2) \times \frac{(7)}{(1)}$	62,821	67,810	75,088	84,185	92,961
9. Projected female births, 5-year period[1]	160,023	175,050	195,109	217,004	

[1] $[(B^y + B^{y+5}) \div 2] \times 5 \times.49$. These births relate to the 5-year period from July 1 of initial year to June 30 of terminal year.

Use of the sex-age adjusted birth rate was recommended by the United Nations as a simple means of preparing comparable projections for many countries.[23]

For purposes of the population projections for Costa Rica, we begin by determining the U.N. sex-age adjusted birth rate for a current year (50.32). We then project the rate on the same assumptions as we projected the total fertility rate. For example, the sex-age adjusted birth rate under the medium assumption in 1970 is $.95 \times 50.32$, or 47.80. We next calculate the weighted aggregate of women of childbearing age in each projection year and multiply the projected rate by the weighted aggregate:

$$B = s.a.a.b.r. \times W_a p_a^f \qquad (6)$$

For the medium series in 1970, we have:

$$B = .0478 \times 1,490,299$$

$$B = 71,236$$

It may be observed that in the present case the differences among the projections of births derived on the basis of the total fertility rate and the U.N. sex-age adjusted birth rate are very small for all years (table 23-2).

Some of the methods of extrapolating fertility rates parallel those of extrapolating mortality rates. Straight lines or curves may be fitted to determine the trend of the rates in the past, a rate of change may be assumed for each future quinquennium, a target schedule of rates may be assumed for the end of the projection period (or beyond) and intermediate values calculated by interpolation, and so on. Knowledge from a wide variety of sources can be brought to bear in making these assumptions. These include the trends in other countries, fertility differentials (e.g., urban-rural, geographic, ethnic) within the country in question, current practices and attitudes about family limitation as obtained from sample surveys or clinic records, national programs and plans in the field of family limitation, and information or theory concerning the relationship between fertility, on the one hand, and marriage patterns, family systems, sexual practices, infant mortality, survival to childbearing age, desired and expected numbers of children, employment of women outside the home, and economic conditions, on the other hand. As in the case of mortality, studies at the Institut national d'études démographiques in Paris are helping to provide information about the biological limits on fertility.[24]

Cohort-fertility method. — The period-fertility method is very simple to apply operationally, but it does not always yield reasonable levels of implied family size. Furthermore, there is no logical basis for projecting the trend of annual fertility and the levels assumed for various dates in the projection period are extremely arbitrary. The cohort-fertility approach tries to overcome these weaknesses of the period-fertility approach.

In the cohort-fertility approach, the fertility assumptions are directly formulated in terms of the completed fertility of real cohorts of women, so that unreasonable or unlikely assumptions concerning completed family size may be avoided. In fact, projections of births derived by other procedures should be evaluated, in part, in terms of the implied completed family size. The cohort-fertility approach makes possible the use of additional information and relationships, such as information on the expressed expectations of women regarding completed family size that have been obtained in national or local sample surveys, and historical evidence regarding the relation of cumulative fertility or mean age of childbearing to completed fertility size, etc.

The cohort-fertility method may be applied in different

[23] United Nations, *Methods for Population Projections by Sex and Age,* pp. 41-50.

[24] Louis Henry, "Fécondité et famille" (Fertility and family, *Population,* 12(3):413-444, July-September 1957.

Table 23–4. — Calculation of Projections of the Female Population of Costa Rica, by Age, According to the Medium Fertility Assumption, for 1970 to 1985

[Projections assume no net migration. Heavy lines identify a 5-year age cohort]

Age (years)	Estimated population, July 1, 1965[1] (1)	Survival rate, 1965-70[2] (2)	Projected population, July 1, 1970 (1) x (2)= (3)	Survival rate, 1970-75[2] (4)	Projected population, July 1, 1975 (3) x (4)= (5)	Survival rate, 1975-80[2] (6)	Projected population, July 1, 1980 (5) x (6)= (7)	Survival rate, 1980-85[2] (8)	Projected population, July 1, 1985 (7) x (8)= (9)
All ages[3]	742,080	(X)	884,782	(X)	1,053,262	(X)	1,254,427	(X)	1,492,527
Births[4]	164,638	.9660	190,169	.9744	223,773	.9801	262,655	.9838	(X)
Under 5	144,520	.9892	159,040	.9925	185,301	.9948	219,320	.9963	258,400
5 to 9	117,715	.9962	142,959	.9972	157,847	.9981	184,337	.9986	218,509
10 to 14	92,875	.9956	117,268	.9967	142,559	.9976	157,547	.9982	184,079
15 to 19	72,385	.9935	92,466	.9952	116,881	.9965	142,217	.9974	157,263
20 to 24	58,115	.9920	71,914	.9940	92,022	.9956	116,472	.9966	141,847
25 to 29	48,275	.9909	57,650	.9929	71,483	.9945	91,617	.9956	116,076
30 to 34	42,325	.9893	47,836	.9913	57,241	.9929	71,090	.9941	91,214
35 to 39	36,875	.9866	41,872	.9886	47,420	.9903	56,835	.9915	70,671
40 to 44	29,815	.9814	36,381	.9835	41,395	.9853	46,960	.9868	56,352
45 to 49	25,200	.9731	29,260	.9754	35,781	.9775	40,786	.9792	46,340
50 to 54	21,275	.9609	24,522	.9637	28,540	.9661	34,976	.9680	39,938
55 to 59	16,065	.9402	20,443	.9437	23,632	.9466	27,572	.9491	33,857
60 to 64	12,520	.9036	15,104	.9083	19,292	.9122	22,370	.9161	26,169
65 to 69	9,205	.8429	11,313	.8490	13,719	.8549	17,598	.8609	20,493
70 to 74	6,190	.7525	7,759	.7602	9,605	.7677	11,728	.7764	15,150
75 to 79	4,085	.6323	4,658	.6410	5,898	.6493	7,374	.6592	9,106
80 and over	4,640	.3781	[5]4,337	.3828	[5]4,646	.3869	[5]5,628	.3913	[5]7,063

X Not applicable.

[1] Source: Costa Rica, Dirección general de estadística y censos, "Proyección de la población de Costa Rica por sexo y grupos de edad, 1965–1990" (Projections of the population of Costa Rica by sex and age, 1965–1990), by Ricardo Jiménez J. et al., Serie Demográfica No. 5, Revista de estudios y estadísticas, San José, No. 8, October 1967, table 34.

[2] Selected from table V of: United Nations, Methods for Population Projections by Sex and Age, Series A. Population Studies, No. 25, 1956. See text for explanation.

[3] Total excludes births.

[4] See table 23-3, medium series.

[5] Derived from populations and survival rates on same and previous lines: $[(p_{75-79}^{y-5} \times s_{75-79}) + (p_{80+} \times s_{80+})]$.

ways. We describe the variation employed by the U.S. Census Bureau in its 1967 projections. The steps are: [25]

1. Selecting "terminal" levels of completed fertility for the annual birth cohorts of women which will reach childbearing age in the future (table 23-5). Four alternatives, corresponding to the four series of population projections planned, were selected. They are as follows: Series A, 3,350; Series B, 3,100; Series C, 2,775; Series D, 2,450. In arriving at these figures, account was taken of "expectations" data from various national fertility surveys and of historical trends regarding the variations in completed family size.

2. Assigning age patterns, representing the distribution of births by age of mother over the childbearing span, to the terminal levels of completed fertility in step (1) (table 23-5). Four patterns corresponding to each of the four terminal levels of completed fertility were selected. They are based on the age patterns of fertility of various historical cohorts which had completed fertility rates similar to those set for the four series.

3. Developing assumptions on the completed fertility of each annual birth cohort of women which had already reached childbearing age by the base year of the projections, consistent with cumulative fertility to date for each cohort and in line with the assumed distribution by age and the assumed completed fertility rates for the cohorts which will reach childbearing age in future years determined in steps (1) and (2) (table 23-6). The completed fertility rates for these older cohorts would not agree with those for the younger cohorts in step (1) because of the fertility cumulated to date.

4. Calculating the cumulative fertility rates for each cohort to each age of childbearing, and the implied age-specific birth rates, in each future year (table 23-7).

5. Applying the age-specific rates in step (4) to projections of the female population of childbearing age to obtain the number of births for each year.

The cohort-fertility rates are first extended to the most recent year for which the total number of births is known and then tied in with the overall level of fertility estimated for that year. This is accomplished by projecting the available age-specific birth rates according to various assumptions to the current year, selecting that series which most closely yields the reported number, and then adjusting the rates to yield exactly the reported number of births (cf. indirect method of standardization, ch. 16). In the procedure followed by the U.S. Bureau of the Census, cumulative fertility data available from the National Center of Health Statistics to January 1, 1965, were projected to January 1, 1966, and were then forced to be in line with the estimated number of births for 1965. The cumulative rates for the various cohorts for January 1, 1966, were then extended to the end of the childbearing period on the basis of previously established assumptions regarding their completed fertility and the pattern of age-specific birth rates. Similar patterns were used to distribute the assumed completed fertility of new cohorts entering the childbearing ages after 1965 over the childbearing period.

[25] U.S. Bureau of the Census, Current Population Reports, Series P-25, No. 381, "Projections of the Population of the United States, by Age, Sex, and Color to 1990, With Extensions of Population by Age and Sex to 2015," by Jacob S. Siegel, December 1967, pp. 18–32 and appendix A.

Table 23-5.—**Summary of Age-Specific Birth Rates Used in Distributing Terminal Completed Cohort Fertility Rates by Age, According to Four Projection Series, for the United States**

[Rates are based on the female population adjusted for net census undercounts. These rates apply exactly to cohorts born after July 1, 1954 and reaching age 14 after July 1, 1968]

Age of woman (years)	Series A	Series B	Series C	Series D
BIRTH RATES				
15 to 19[1]	87.4	77.0	64.5	51.8
20 to 24	233.7	203.6	168.0	131.8
25 to 29	180.9	167.9	150.8	133.8
30 to 34	102.5	103.1	101.6	100.4
35 to 39	50.7	53.0	54.3	55.9
40 to 44	14.0	14.6	15.0	15.4
45 to 49[2]	0.8	0.9	0.9	0.9
Completed fertility rate[3]	3,350.0	3,100.0	2,775.0	2,450.0
Median age of mother	25.3	25.8	26.4	27.2
PERCENT OF TOTAL				
15 to 19	13.0	12.4	11.6	10.6
20 to 24	34.9	32.8	30.3	26.9
25 to 29	27.0	27.1	27.2	27.3
30 to 34	15.3	16.6	18.3	20.5
35 to 39	7.6	8.5	9.8	11.4
40 to 44	2.1	2.4	2.7	3.1
45 to 49	0.1	0.1	0.2	0.2
Completed fertility rate	100.0	100.0	100.0	100.0

[1] Includes births to women under 15 years of age.
[2] Includes births to women 50 years old and over.
[3] Sum of age-specific birth rates over all childbearing ages for cohorts born after July 1, 1954 and reaching age 14 after July 1, 1968.

Source: U.S. Bureau of the Census, *Current Population Reports*, Series P-25, No. 381, "Projections of the Population of the United States, by Age, Sex, and Color to 1990, With Extensions of Population by Age and Sex to 2015," December 18, 1967, table S.

In preparing these projections limited account was taken of information on the expectations regarding completed family size reported by representative national samples of married couples included in the Growth of American Families (GAF) Studies of 1955 and 1960 and in the University of Michigan national sample surveys of 1962, 1963, and 1964.[26]

The cohort method of projecting fertility may be extended to take into account other variables in addition to age, particularly marital status. Like the general cohort method, this method employs cumulative fertility rates for birth cohorts of women and assumed levels of completed fertility. In this more elaborate method, however, the rates relate only to married women, and hence assumptions must be made about the proportion of women in each birth cohort who will have married by various later ages and about the fertility rates for ever-married women cumulated to these ages. The cumulative fertility experience of each birth cohort to date for ever-married women is taken into account, of course.

---

[26] The former studies were carried out by the Scripps Foundation for Research in Population Problems, Miami University, and the Survey Research Center, University of Michigan, and the latter studies were carried out by the Population Studies Center and the Survey Research Center, University of Michigan. The methods and results of the 1955 and 1960 surveys are described in: Ronald Freedman, Pascal K. Whelpton, and Arthur A. Campbell, *Family Planning, Sterility, and Population Growth*, New York, McGraw Hill, 1959; and Pascal K. Whelpton, Arthur A. Campbell, and John E. Patterson, *Fertility and Family Planning in the United States*, Princeton, N.J., Princeton University Press, 1966. The design and results of the 1962-64 studies are described in: Ronald Freedman and Larry Bumpass, "Fertility Expectations in the United States: 1962-64," *Population Index*, 32(2):181-197, April 1966.

Table 23-6.—**Estimated and Assumed Completed Fertility Rates, for 5-Year Birth Cohorts of Women, According to Four Projections Series, for the United States: Birth Years, 1900-05 to 1960-65**

[Average number of children born by end of childbearing period per 1,000 women. Rates below the heavy line are projections. Completed fertility rates for birth periods 1950-55 and later correspond approximately (1950-55) or exactly (1955-60 and later) to the "terminal" rates]

Birth period of women[1]	Age on July 1, 1965 (years)	Cumulative fertility rate to Jan. 1, 1966	Completed fertility rate			
			Series A	Series B	Series C	Series D
1900-1905	60 to 64	2,421	2,421	2,421	2,421	2,421
1905-1910	55 to 59	2,273	2,273	2,273	2,273	2,273
1910-1915	50 to 54	2,310	2,310	2,310	2,310	2,310
1915-1920	45 to 49	2,553	2,553	2,553	2,553	2,553
1920-1925	40 to 44	2,844	2,865	2,865	2,863	2,863
1925-1930	35 to 39	2,978	3,133	3,122	3,117	3,115
1930-1935	30 to 34	2,913	3,383	3,372	3,366	3,357
1935-1940	25 to 29	2,284	3,368	3,346	3,322	3,295
1940-1945	20 to 24	1,084	3,305	3,111	2,971	2,883
1945-1950	15 to 19	157	3,320	3,087	2,778	2,504
1950-1955	10 to 14	3	3,347	3,098	2,775	2,451
1955-1960	5 to 9	–	3,350	3,100	2,775	2,450
1960-1965	Under 5	–	3,350	3,100	2,775	2,450
1965 and later	([2])	–	3,350	3,100	2,775	2,450

– Represents zero.
[1] Period extends from July 1 of initial year to June 30 of terminal year.
[2] Born after July 1, 1965.

Source: U.S. Bureau of the Census, *Current Population Reports*, Series P-25, No. 381, "Projections of the Population of the United States, by Age, Sex, and Color to 1990, With Extensions of Population by Age and Sex to 2015," December 18, 1967, table R.

Applications of this method were made by Whelpton and his associates, who based their projections of completed fertility of ever-married women on the fertility expectations reported by married women in the 1955 and 1960 GAF Studies.[27] In their second study, they developed three series of population projections—high, medium, and low—based on different assumptions regarding the percent of women who will ever marry, the "average size of family" (i.e., completed fertility rate for ever-married women), and the distribution of birth rates by age of mother for ever-married women over the childbearing span. (The mortality assumptions were essentially the same as those used in the Census Bureau projections, except that the computations were carried out for 5-year age groups by 5-year time periods. Net immigration was assumed to be 300,000 per year, distributed by age and sex like the immigrant aliens in the period 1957 to 1962.) The specific steps followed in developing the three series of projections of births are as follows:

1. Projections were made of the proportion of women who will have married by specified ages in groups of birth cohorts (tables 23-8 and 23-9). The three series imply corresponding changes in median age at marriage. For example, the medium series implies a small increase (less than 1 year) in the median age at marriage.

2. Next, projections were made of the cumulative fertility rates of ever-married women up to specified ages in groups of cohorts, including the final ages of fertility. The cumulative marital fertility rates for each cohort group were projected to the end of the childbearing ages on the basis of the average expectations regarding the size of completed families reported

---

[27] Freedman, Whelpton, and Campbell, op. cit., chapter 11; and Whelpton, Campbell, and Patterson, op. cit., chapter 10.

Table 23–7. — Estimates and Projections of Cumulative and Completed Fertility Rates, by Birth Cohort of Women, According to All Series and Series C, for the United States: Birth Years, 1900–01 to 1939–40

[Rates represent cumulative live births per 1,000 women up to age indicated. Rates below heavy lines are based, in whole or part, on age-specific birth rates projected for years after 1966]

Birth year of mother	Up to age 20	Up to age 25	Up to age 30	Up to age 35	Up to age 40	Completed fertility rate	Median age of mother
				All series			
1900-1901........................	287	1,112	1,780	2,195	2,430	2,511	25.9
1901-1902........................	310	1,115	1,750	2,135	2,355	2,435	25.7
1902-1903........................	314	1,113	1,737	2,121	2,344	2,426	25.7
1903-1904........................	315	1,091	1,697	2,083	2,310	2,392	25.8
1904-1905........................	314	1,053	1,646	2,026	2,261	2,343	25.9
1905-1906........................	311	1,028	1,602	1,981	2,226	2,306	25.9
1906-1907........................	307	1,004	1,564	1,946	2,202	2,282	26.1
1907-1908........................	303	982	1,539	1,934	2,201	2,279	26.3
1908-1909........................	290	933	1,478	1,888	2,154	2,230	26.5
1909-1910........................	281	916	1,476	1,910	2,188	2,268	26.8
1910-1911........................	279	897	1,467	1,926	2,205	2,285	27.1
1911-1912........................	274	888	1,472	1,951	2,222	2,303	27.2
1912-1913........................	264	879	1,487	1,979	2,246	2,328	27.3
1913-1914........................	248	868	1,499	1,988	2,253	2,334	27.4
1914-1915........................	240	856	1,498	1,986	2,252	2,331	27.4
1915-1916........................	243	871	1,538	2,035	2,307	2,387	27.3
1916-1917........................	244	891	1,590	2,088	2,362	2,442	27.2
1917-1918........................	248	930	1,649	2,155	2,442	2,522	27.3
1918-1919........................	263	1,003	1,742	2,269	2,560	2,640	27.2
1919-1920........................	273	1,034	1,827	2,391	2,695	2,777	27.2
				Series C			
1920-1921........................	269	1,016	1,841	2,406	2,701	2,778	27.0
1921-1922........................	273	1,049	1,886	2,457	2,747	2,822	27.0
1922-1923........................	281	1,087	1,924	2,500	2,782	2,854	26.9
1923-1924........................	289	1,126	1,980	2,558	2,834	2,903	26.8
1924-1925........................	280	1,166	2,048	2,622	2,893	2,959	26.7
1925-1926........................	266	1,205	2,105	2,676	2,937	3,002	26.5
1926-1927........................	275	1,253	2,162	2,724	2,971	3,036	26.3
1927-1928........................	320	1,322	2,256	2,817	3,058	3,124	26.2
1928-1929........................	351	1,376	2,315	2,867	3,100	3,166	26.0
1929-1930........................	377	1,449	2,415	2,966	3,191	3,258	25.8
1930-1931........................	384	1,480	2,445	2,974	3,197	3,264	25.7
1931-1932........................	407	1,550	2,526	3,034	3,257	3,324	25.5
1932-1933........................	426	1,633	2,619	3,112	3,337	3,404	25.3
1933-1934........................	438	1,675	2,649	3,126	3,352	3,420	25.2
1934-1935........................	448	1,704	2,658	3,120	3,348	3,416	25.0
1935-1936........................	454	1,729	2,655	3,110	3,339	3,408	24.9
1936-1937........................	464	1,730	2,615	3,069	3,300	3,369	24.8
1937-1938........................	479	1,735	2,576	3,032	3,264	3,333	24.7
1938-1939........................	481	1,714	2,516	2,973	3,207	3,276	24.7
1939-1940........................	487	1,691	2,462	2,920	3,155	3,225	24.6

Source: U.S. Bureau of the Census, *Current Population Reports*, Series P–25, No. 381, "Projections of the Population of the United States, by Age, Sex, and Color to 1990, With Extensions of Population by Age and Sex to 2015," Dec. 18, 1967, table A–1.

in the GAF Studies. The cumulative figures up to specified ages also had to be determined. For example, for cohorts entering the marriageable ages about 1980–85, 74.0 percent of the cumulative births for ever-married women would have occurred by age 30 in the medium series, as compared with 59.0 percent for the cohorts entering the marriageable ages in the early 1930's.

3. The cumulative fertility rates for ever-married women in groups of birth cohorts were then converted into rates for **all** women by multiplying the cumulative fertility rate for ever-married women in a given age group at a given date (item 2 above) by the proportion of women who have married by that age (item 1 above):

$$\frac{B_{cum}}{P_t^f} = \frac{P_{em}^f}{P_t^f} \times \frac{B_{cum}}{P_{em}^f} \qquad (7)$$

4. The rise from one age group to another in the cumulative fertility rate of all women, for a given group of birth cohorts (item 3), was used to derive the number of births added per 1,000 women in a given initial age group during each 5-year interval.

5. The projections of births per 1,000 women in a given initial age group in a 5-year period (item 4) were then applied cumulatively to the female population by age at the beginning of the interval to obtain the number of births occurring to the women during the period.

Another demographic variable which may be taken into account in the application of the cohort-fertility method is the order of birth of the child. Completed fertility rates for a given birth cohort and birth order would represent the number of births of a given order per 1,000 women or ever-married women at the end of childbearing. The available cohort fertility tables for the United States include order of birth as a variable, but not marital status.[28] It would be possible, however, to

[28] See, for example, U.S. National Center for Health Statistics, *Vital Statistics of the United States, 1966*, Vol. I, Natality, 1968.

Table 23–8. — **Estimated and Projected (Medium Series) Cumulative Marriage and Fertility Rates for Cohorts of Women Born in 1930 to 1935, by Successive Ages, for the United States**

[These cohorts reach 15 to 19 years of age in 1950 and 45 to 49 years of age in 1980. Percents and rates below the horizontal line are projections]

Age (years)	Cumulative percent ever married (1)	Number of births per 1,000 women ever married (2)	Number of births per 1,000 total women (1) x (2)= (3)
15–19....................	17.3	729	126
20–24....................	69.8	1,397	975
25–29....................	89.3	2,390	2,134
30–34....................	92.5	3,075	2,844
35–39....................	94.0	3,350	3,149
40–44....................	95.0	3,440	3,268
45–49....................	95.5	3,450	3,295

Source: U.S. Bureau of the Census, *Current Population Reports,* Series P–25, No. 286, "Projections of the Population of the United States, by Age and Sex: 1964 to 1985, With Extensions to 2010," July 1964, table D–2.

extend the cohort-fertility method to include both order of birth and marital status. The expectations data from the fertility surveys provide a basis for making assumptions regarding the fertility rates for first births, second births, etc., at the end of childbearing. This approach requires assumptions not only about the completed fertility of each cohort and the timing of their future births, but also about the future age at marriage, or percents ever-married by age, and order-specific marital fertility. These additional assumptions, like the basic ones required by the simpler variation, are subject to considerable uncertainty and may contribute to an increase in the "error" of the projections. On the other hand, the explicit introduction of the additional variables relating to marriage

Table 23–9. — **Estimated and Projected (Medium Series) Cumulative Marriage and Fertility Rates Up to Ages 45 to 49, for 5-year Birth Cohorts of Women, for the United States: Birth Years, 1900–05 to 1950–55**

[Percents and rates below the horizontal line are projections]

Birth period of woman [1]	Year in which cohorts reach-- Ages 15 to 19 (1)	Year in which cohorts reach-- Ages 45 to 49 (2)	Percent ever married (3)	Births per 1,000 women ever married (4)	Births per 1,000 total women (3) x (4)= (5)
1900–1905.........	1920	1950	92.2	2,625	2,420
1905–1910.........	1925	1955	92.4	2,458	2,271
1910–1915........	1930	1960	93.3	2,481	2,315
1915–1920........	1935	1965	95.0	2,700	2,565
1920–1925...	1940	1970	96.5	3,000	2,895
1925–1930..	1945	1975	96.0	3,300	3,168
1930–1935........	1950	1980	95.5	3,450	3,295
1935–1940........	1955	1985	95.0	3,300	3,135
1940–1945........	1960	1990	94.5	3,200	3,024
1945–1950........	1965	1995	94.5	3,100	2,930
1950–1955........	1970	2000	94.0	3,000	2,820
1955 or later....	1975 or later	2005 or later	94.0	3,000	2,820

[1] Period extends from July 1 of initial year to June 30 of terminal year.

Source: U.S. Bureau of the Census, *Current Population Reports,* Series P–25, No. 286, "Projections of the Population of the United States, by Age and Sex: 1964 to 1985, With Extensions to 2010," July 1964, table D–1.

and order of birth assures reasonable implicit assumptions in these components of the overall fertility rates.

In spite of the apparently superior logic of the cohort-fertility approach to projections of births as compared with the period-fertility method, it has a number of weaknesses. First, there are serious difficulties in determining the future level of completed fertility of each cohort; yet, the results depend heavily on these assumptions. Expressed birth expectations may be unreliable because of changing circumstances, particularly for young women who have recently married and who have not begun childbearing. Making assumptions about the level of completed fertility is particularly hazardous for cohorts which have not yet married or even entered the childbearing ages, for whom expectations data are not available.

There are problems also in developing realistic assumptions regarding the timing pattern of the various cohorts—the other important determinant of the annual number of future births. (Changes in timing affect period rates even when the size of the completed family does not change.) The past may not be a good guide because of prospective changes in the socioeconomic milieu, and the survey data on birth expectations now available can tell us little or nothing about future annual changes in fertility, particularly after a few years of the projection period have passed.[29] Mathematical procedures for projecting the fertility of cohorts which have already started childbearing may be useful for these cohorts but cannot serve as a satisfactory basis of projecting the total fertility and timing pattern of cohorts which have not yet entered childbearing.[30] Accordingly, research has turned to an alternative method, the marriage-parity-interval progression method, which involves a greater disaggregation of the data and a more realistic structuring of the components.

Projections of Net Immigration. — For many countries international migration has long been of negligible magnitude, and for others it has dwindled away to a trickle as compared to the heavy flows of the past. The passing of the frontiers of settlement and the enactment of very rigid national laws with respect to either immigration or emigration, or both, have contributed to this reduction of migration flows in relation to the corresponding national population totals, and there is little likelihood of a resumption of heavy flows. The volume of net immigration will often be too small, therefore, to justify separate treatment of this component in population projections. Frequently, also, the mere lack of adequate historical information on the volume of migration dictates an assumption of no net migration. In making projections of national population for many, and possibly most, countries of the world, therefore, it is satisfactory to ignore this component, i.e., to assume that the net balance will be nil.

In a fair number of countries, however, net migration is a major factor in population change. Moreover, even though the volume of net migration may be small, population projections by age may be strongly affected because of the special sex–age composition of the migrants. It is desirable, therefore, to allow for net migration in some cases.

[29] For a general criticism of expectations data as a basis for projections of births, see Norman B. Ryder and Charles F. Westoff, "The Trend of Expected Parity in the United States: 1955, 1960, and 1965," *Population Index* 33(2):153–168, April–June, 1967; and Jacob S. Siegel and Donald S. Akers, "Some Aspects of the Use of Birth Expectations Data from Sample Surveys for Population Projections," *Demography* 6(2):101–115, May 1969.

[30] Donald S. Akers, "Cohort Fertility vs. Parity Projection as Methods of Projecting Births, *Demography,* 2:414–428, 1965.

The allowance for international migration may be an arbitrary rounded amount, roughly in line with recent experience, which is assumed to remain constant for the duration of the projection period. If the calculations are done by age and sex, a constant age-sex distribution may also be assumed. Alternatively, a constant rate of migration may be assumed or the amount or rate of net migration may be projected on the basis of the recent trend. The age-sex distribution of migrants is unlikely to resemble the population of the sending or receiving country, young adult males tending to be overrepresented among migrants. Therefore, the age-sex distribution is best taken from an available distribution for recent migrants.

To simplify the computations, the migrants may be added to or subtracted from the population at the end of the year of entry or departure or, less precisely, the quinquennium of entry or departure. This simple assumption eliminates the need of allowing for deaths or births of the migrants during the period of entry or departure (the average period of exposure to the risk of birth and death being one-half of each projection period). At the same time, the age distribution of the migrants should be adjusted to reflect the change in age between the date of migration and the end of the migration period. Following the period of arrival, the same birth rates and death rates may be applied to the immigrants as to the general population. Any difference in the level of fertility or mortality of the general population and the immigrant population would have very little effect on the future population. These simplifying assumptions eliminate the need for special calculations of fertility rates and survival rates, and special adjustments of the data on migrants.

The four principal series of projections published for the United States in 1967 each employed the same assumption of a constant annual net immigration of 400,000 from July 1, 1966, to July 1, 1990.[31] This number corresponds roughly to the volume of net immigration in the middle sixties. For the age and sex composition of net civilian immigration, the same allowance was used for each year in each series of population projections, corresponding to the distribution in 1961–64. The future additions to the population resulting from an annual net immigration of 400,000 persons between 1966 and 1990 may be determined by comparing projections assuming this amount of net immigration with projections assuming no migration, under the same fertility and mortality assumptions.

The most recent national projections prepared at the United Nations illustrate the projection of a changing amount of migration. For a few countries (in Europe, the Americas, and Oceania), assumed annual amounts of net immigration for the initial period (1960–80) were projected linearly to zero between 1980 and 2000.[32]

Because it is often impossible to arrive at a satisfactory measure of past trends in the volume and age-sex distribution of net migration and because it is extremely difficult in most cases to predict the future course of migration with any confidence, model allowances and model tables may be employed to allow for this component. These may either be incorporated into the principal population projections, or they may be computed and displayed separately from those projections. The model can be used to measure the effect of any given annual amount of migration upon the population figures for the

various sex-age groups. A model may be applicable to a group of countries in a particular geographic region, although it may be based on the statistics of recent immigration and emigration for a single country. Other models, suited to different situations, can be developed by the same methods. The U.N. projections for Latin America were accompanied by a model which could be used to estimate the effect of any given annual amount of immigration and emigration upon the sex-age distribution for various countries in Latin America.[33]

Normally, the two components of international migration, immigration and emigration, are projected in combination as net migration. Consideration may be given to projecting immigration and emigration separately when separate trends in these components can be identified and net migration is sufficiently large. Even when there is an overall balance of roughly zero net migration for a population, there may be substantial gross migration and, under these circumstances, the contribution of immigration and emigration to the gross migration may vary from one age to another. If the age-sex distributions of immigrants and emigrants differ considerably, there may be reason to project, and evaluate the effect of, immigration and emigration separately.

Subnational Projections of Total Population and Age-Sex Composition

In turning from the methodology of national projections to the methodology of projections for geographic subdivisions, we must become concerned with the outlook for internal migration. Although the assumption that future international migration will be nil or negligible is justified for many countries, internal migration (particularly rural-urban migration) involves a large proportion of the people in almost all countries.

The problem of making population projections is rather different for political units, such as States and counties, and for subdivisions that represent statistical classifications, such as urban-rural areas and areas classified by size. As was noted in chapter 6, the former are subject to relatively few boundary changes, whereas the latter are frequently reclassified after the results of each census become available. Changes in population resulting from boundary changes and revised statistical classifications constitute, in effect, another major component along with migration which must be considered in making projections for these areas. We will consider here only the methods of projecting population for relatively stable, political areas not affected by boundary changes.

The range of methods applicable to subnational projections is more extensive than for national projections. This situation results from the fact that both independent methods and methods dependent on the projection for another area (either another geographic subdivision or, more commonly, the country as a whole) may conveniently be employed in the calculation, from the fact that the methodology must consider internal migration as an additional factor, and from the fact that many types of data are relevant and often available. Although internal migration is often an important factor in local population growth, and it must be taken account of in projections, the allowance for this factor does not have to be explicit. The various methods for subnational projections include mathematical methods and ratio methods; cohort-component

[31] U.S. Bureau of the Census, *Current Population Reports*, Series P–25, No. 381, p. 42.
[32] United Nations, *World Population Prospects as Assessed in 1963*, pp. 46–47.

[33] United Nations, *The Population of South America, 1950–1980*, Annex C; and United Nations, *Methods for Population Projections by Sex and Age*, tables 47–50, pp. 64–67.

methods; methods using economic analysis, particularly correlation with "indicators," i.e., variables whose changes more or less reflect changes in population; and combinations of these methods. The various analytic methods attempt to take explicit account of one or more of the demographic or socioeconomic variables with which change in the geographic distribution of the population is related. Once again, because some methods derive the projections of total population by combining projections for age groups and because others may be applicable for deriving projections of both total population and age-sex structure directly, it is convenient to consider the projection of total population and age-sex structure at the same time.

Mathematical and Ratio Methods. — We consider first those methods which employ a minimum of independent data, assumptions, and variables for making the projections. Even when additional statistics, including estimates of past net migration, can be developed, one does not have to use the more complex, analytic methods, especially if projections must be prepared for a large number of geographic subdivisions and rough figures are satisfactory. In the mathematical methods, as previously noted, typically the series of total population figures for past years for the area are directly extended to future years by use of some mathematical formula, without benefit of other related series of projections. In the ratio procedures, typically the series of total population figures for past years are extended by mathematical formula as a ratio of the population for some larger area for which population projections are already available.

Mathematical methods. — The forms of mathematical curves useful for making projections of local population are very much the same as those useful for making national projections, as noted earlier, and will not be elaborated further here. We should note, however, that special caution is necessary in the use of mathematical curves for projection purposes. A mathematical curve may project an unreasonably large figure for a local area in relation to a prior projection for the parent area and, in extreme cases, may project a larger figure, even with the same mathematical curve. None of the curves, except possibly the logistic, is applicable over a long period because they imply either unlimited growth or indefinite continuation of the same growth pattern.

Ratio methods. — The ratio method of projecting total population is peculiar to projections for geographic subdivisions. In this method the percentage distribution of the parent population (e.g., a country) among the geographic subdivisions (e.g., States) is observed for one or more past dates, projected to future dates, and applied to an independently derived projection of the parent population. For the projection of the percents, the percent distribution may be held constant at the last observed level or may be modified in some way to take account of the past trend. Although, in principle, the chief factor affecting the redistribution of population is internal migration and this component of change is quite likely to be influenced by economic developments, there is no direct practical way of taking account of plans or expectations concerning the economic development of regions in the ratio method. The percent distribution may be assumed to approach a stable condition after a great number of years on the ground that differences in fertility and mortality will have disappeared and net migration will have fallen off to zero for each area. There are many ways of accomplishing a shift to stability in this distribution.

One procedure for calculating projections by the ratio method may be illustrated for the provinces of Canada (table 23-10). These projections assume that the average annual rate of change in the percents observed between June 1, 1961, and June 1, 1969, for each area will fall off linearly to zero by the year 2,025.[34] First, the percent distributions of the national population by provinces in 1961 and 1969 are calculated (cols. 1 and 2). Second, the average annual rates of change between 1961 and 1969 are derived (col. 3). They are assigned to the middate of the period, Dec. 1965. These rates of change are then extrapolated according to the basic assumption noted above (cols. 4 to 9 and cols. 13 to 17). The population percents of June 1969 are then extrapolated to 1975 (col. 10) and 1980 (col. 18) according to the annual percent changes previously determined for each projection year. These percents do not automatically add to 100.0 and require some adjustment (cols. 11 and 19). Finally, in columns 12 and 20 the absolute figures are obtained by multiplying the percents by an independently derived projection of the total population of the country.[35]

Another variation of the ratio method employs an assumed relationship between the area's own growth rate and that of another area for which a projection is already available. In this case the other area may be a broader area or a coordinate geographic subdivision. The method is especially applicable when (a) adequate population and vital statistics are lacking for a given area, (b) data are lacking for part of an area's population, for example the tribal or nomadic population, or (c) the relatively small population of the area or the limited time available make it impractical to construct an independent projection.

The ratio method in either form may also be used, of course, when a single geographic subdivision requires a population projection. Moreover, it is readily capable of extension to a hierarchy of subdivisions. For example, having been used for states or provinces, it may next be used for the counties or communes of which these primary subdivisions are constituted; in this case, the projections for the primary subdivisions are employed as totals to which the secondary set of percentage distributions is applied.

The ratio method may be readily adapted to derive projections of the age-sex distribution of geographic subdivisions. Two types of ratio techniques are considered here for projecting the age-sex distribution of the primary divisions of a country. Both require prior projections of the national population by age and sex and of the total population of each area. The first employs the ratio of the percent of the area's population in each age-sex group to the corresponding percent for the national population at the last census.[36] It is then assumed that the ratios of the percents will not change, will change according to the intercensal "trend," or will change according to some other principle, as for example, that the ratios will approach unity by some distant date. The projected ratios are then applied to the projections of the percent distribution of the population by age for the country as a whole to obtain projections of the percent distribution by age for each area. Each of these percent distributions must then be proportionately adjusted to 100 percent for all ages before being applied to projections of the total population of each area. Proportionate

---

[34] All calculations are actually carried out in terms of proportions rather than percents (i.e., on a unit basis).

[35] An alternative briefer method of projecting the ratios is illustrated in: Helen R. White, J. S. Siegel, and Beatrice M. Rosen, "Short Cuts in Computing Ratio Projections of Population," *Agricultural Economics Research,* 5(1):5–11, January 1953.

[36] For an application of this method, see Margaret J. Hagood and J. S. Siegel, "Projections of the Regional Distribution of the Population of the United States to 1975," *Agricultural Economics Research,* 3(2):41–52, April 1951.

Table 23-10. — Projection of the Population of the Provinces of Canada by the Ratio Method from 1969 to 1975 and 1980

Province	Percent distribution of population[1]		Average annual rate of change in percent[2]							Preliminary percent distribution, June 1, 1975
			Estimated, 1961-69	Projected						
	Census, June 1, 1961	Estimates, June 1, 1969		1969-70	1970-71	1971-72	1972-73	1973-74	1974-75	
	(1)	(2)	(3)	(4)	(5)	(6)	(7)	(8)	(9)	(10)
Canada.......................	100.000	100.000	(X)	(X)	(X)	(X)	(X)	(X)	(X)	100.166
Newfoundland................	2.510	2.441	-0.349	-0.322	-0.316	-0.310	-0.304	-0.298	-0.292	2.399
Prince Edward Island........	0.574	0.522	-1.188	-1.093	-1.072	-1.051	-1.030	-1.009	-0.988	0.489
Nova Scotia..................	4.041	3.623	-1.365	-1.257	-1.233	-1.209	-1.185	-1.161	-1.137	3.371
New Brunswick...............	3.278	2.968	-1.241	-1.142	-1.120	-1.098	-1.076	-1.054	-1.032	2.780
Quebec......................	28.836	28.413	-0.185	-0.171	-0.168	-0.165	-0.162	-0.159	-0.156	28.074
Ontario.....................	34.192	35.383	+0.428	+0.396	+0.389	+0.382	+0.375	+0.368	+0.361	36.241
Manitoba....................	5.054	4.648	-1.046	-0.965	-0.947	-0.929	-0.911	-0.893	-0.875	4.399
Saskatchewan................	5.073	4.553	-1.351	-1.247	-1.224	-1.201	-1.178	-1.155	-1.132	4.239
Alberta.....................	7.303	7.412	+0.185	+0.171	+0.168	+0.165	+0.162	+0.159	+0.156	7.502
British Columbia............	8.932	9.814	+1.178	+1.088	+1.068	+1.048	+1.028	+1.008	+0.988	10.437
Yukon and Northwest Territories...	0.206	0.223	+0.991	+0.914	+0.897	+0.880	+0.863	+0.846	+0.829	0.235

Province	Projected population, June 1, 1975		Projected average annual rate of change in percent[2]					Projected population, June 1, 1980		
	Adjusted percent distribution	Number	1975-76	1976-77	1977-78	1978-79	1979-80	Percent distribution		Number
								Preliminary	Adjusted	
	(11)	(12)	(13)	(14)	(15)	(16)	(17)	(18)	(19)	(20)
Canada.......................	100.000	[3]23,256,025	(X)	(X)	(X)	(X)	(X)	100.162	100.000	[3]25,250,634
Newfoundland................	2.395	556,982	-0.286	-0.280	-0.274	-0.268	-0.262	2.360	2.356	594,905
Prince Edward Island........	0.488	113,489	-0.967	-0.946	-0.925	-0.904	-0.883	0.464	0.463	116,910
Nova Scotia..................	3.365	782,565	-1.113	-1.089	-1.065	-1.041	-1.017	3.187	3.182	803,475
New Brunswick...............	2.775	645,355	-1.010	-0.988	-0.966	-0.944	-0.922	2.645	2.641	666,869
Quebec......................	28.027	6,517,966	-0.153	-0.150	-0.147	-0.144	-0.141	27.831	27.786	7,016,141
Ontario.....................	36.181	8,414,262	+0.354	+0.347	+0.340	+0.333	+0.326	36.764	36.704	9,267,994
Manitoba....................	4.392	1,021,405	-0.857	-0.839	-0.821	-0.803	-0.785	4.214	4.207	1,062,294
Saskatchewan................	4.232	984,195	-1.109	-1.086	-1.063	-1.040	-1.017	4.011	4.005	1,011,288
Alberta.....................	7.490	1,741,876	+0.153	+0.150	+0.147	+0.144	+0.141	7.544	7.532	1,901,878
British Columbia............	10.420	2,423,278	+0.968	+0.948	+0.928	+0.908	+0.888	10.897	10.879	2,747,016
Yukon and Northwest Territories...	0.235	54,652	+0.812	+0.795	+0.778	+0.761	+0.744	0.245	0.245	61,864

X Not applicable.

[1] Sources of basic data: Canada, Dominion Bureau of Statistics, *Census of Canada, 1961*, Vol. I (Part 1), *Population: Geographical Distribution*, 1962; and idem, "Estimated Population by Sex and Age Group, for Canada and Provinces," June 1, 1969, Dec. 1969.

[2] Rate for 1961-69 derived from the percent distributions for 1961 and 1969. Rates for 1969-70 to 1979-80 derived by linear interpolation between the rate of change for the period 1961-69 (with July 1, 1965 as the assumed central date) and the rate for the period 2020-25, which is assumed to be zero in every case.

[3] Derived by adjusting the projections given in, Wolfgang M. Illing, *Population, Family, Household and Labor Force Growth to 1980*, Staff Study No. 19, prepared for the Economic Council of Canada, Sept. 1967, table 2-7 (series with medium immigration and medium fertility).

adjustments of the figures for all areas at each age to the national totals at that age, and of the age-sex figures for each area to the total of all ages for the area, follow in order. This cycle of adjustments is repeated until full, or nearly full, agreement with marginal totals is obtained (known as the method of iterative proportions or two-way raking).

Two-way raking is also applied in the second ratio procedure for projecting the age-sex distributions for a set of subnational areas. Here, the male population and the female population at each age in the parent population at the first projection date are distributed by geographic areas according to the regional distribution at the last census, and the results over all ages are then adjusted proportionately to the independent projection of the total population of each area. The resulting figures are again subjected to the same adjustment cycle until complete reconciliation with assigned marginal totals has been achieved. These final results are then utilized as a basis for deriving the projections at the next projection date by the same procedure, and so on.

These two ratio techniques may be effectively merged into a single procedure. Consider the case where data on age, sex, and geographic subdivisions are available for a single census and it is assumed that the sub-area-to-total-area ratios of the percents for each age group will remain unchanged from one projection date to the next. The final figures for the first projection year, after raking is completed, may then be employed as a basis for calculating the ratios of the percents to be used in the second projection year, and so on.[37]

In sum, the mathematical and ratio methods of projecting regional population generally have the advantage of simplicity of computation, but they do not take into explicit account, and hence provide little or no information with respect to, the demographic or socioeconomic correlates or components of the indicated trends. Nevertheless, the results may usefully serve

---

[37] A step-by-step description of this procedure, including applications to the countries of Central America, is given in: United Nations, Economic Commission for Latin America, *Human Resources of Central America, Panama, and Mexico, 1950-80, in Relation to Some Aspects of Economic Development*, by Louis J. Ducoff, 1960, app. B and pp. 41-45.

as approximations to the future geographic distribution of the population of an area, particularly when comparable figures for a set of constituent areas are desired quickly and with little investment of resources.

Cohort-Component Method. — The cohort-component method is now the most widely used of the analytic methods for preparing regional projections. The general outline of the method is the same as indicated for national projections, but the component of internal migration must be incorporated into the procedure. This method is especially appropriate if projections by age and sex are wanted in addition to totals. Carrying out component projections on an age-specific basis is recommended even when projections of only the total population are sought, because of the added specificity of the assumptions and the provision of data on the age-sex distribution of the population as a by-product. The cohort-component method may be applied with various degrees of refinement and complexity, from a variation in which mortality and migration are handled jointly and a single assumption is made for this joint component, to one in which migration is treated as three components, namely, net immigration, gross in-migration, and gross out-migration, and several assumptions are made with respect to each component of change. In applying the cohort-component method, the choice of a specific procedure is suggested in large part by the type and quality of data available and the resources for developing the projections.

Sufficient testing has not been carried out to support the conclusion that the analytic methods, including the cohort-component method, provide more accurate or realistic projections than the ratio and mathematical methods, but they may be preferred for a number of reasons, quite apart from the relative accuracy of the methods. As in the case of national projections, they may be expected to provide more meaningful and useful results for subnational areas since they attempt to take explicit account of the components of change and of available knowledge regarding the course of these components. In addition, direct use can be made of national projections of mortality, fertility, and net migration from abroad. From the standpoint of accuracy, however, the choice of method may be less important than the choice of assumptions, particularly those relating to individual components. Like any set of projections based on recent past trends, projections using the cohort-component method will fail to predict a shift away from these trends during the projection period and, hence, may differ from actual developments.

Use of census cohort-change rates. — In the simplest form of the cohort-component method the components of net migration and mortality are treated as a unit in the projections for the cohorts already alive at the base date, and a single assumption is made that the rates remain unchanged. In general, the method involves carrying forward the latest census or current population by age and sex to future years by use of census cohort-change rates (sometimes called migration-survival ratios), representing the ratio of the number of persons enumerated in a given age group in one census to the number of persons in the same birth cohort enumerated in a previous census. Census cohort-change rates include, in addition to mortality and net migration, the effect of relative errors of enumeration between successive censuses. In the formula of Hamilton and Perry cohort-change rates for a given area between the two censuses are assumed to remain constant in future years: [38]

[38] C. Horace Hamilton and Josef Perry, "A Short Method for Projecting Population by Age From One Decennial Census to Another," *Social Forces* 41(2):163–170, Dec. 1962.

$$_nP_{a+k}^{t+k} = \frac{_nP_{a+k}^t}{_nP_a^{t-k}} \, _nP_a^t \tag{8}$$

where P is population, a is the initial age of the age interval at the second census, n is the size of the age interval, t is the year of the second census, and k is the intercensal interval in years (a multiple of n, usually 5 or 10 years). The formula does not directly allow for the calculation of projections of children in the cohorts born during the intercensal period. In the Hamilton-Perry procedure births during the projection period are estimated by holding constant recent age-specific birth rates. In our illustration below, we hold constant the last observed child-women ratio. Although formula (8) assumes no change from the intercensal period to the projection period in age-specific death rates, migration rates, and patterns of net census error, alternative assumptions are possible.

Censuses are taken most commonly at 10-year time intervals and show abridged age detail for geographic subdivisions of the country. In this case, it is most convenient to prepare projections at 10-year time intervals for a combination of 5- and 10-year age groups (depending on the original age detail). An illustration, relating to Sivas Province in Turkey, employs census data for 1955 and 1965, to derive projections for the female population in 5- and 10-year age groups for 1975 (table 23-11). The age data shown for 1955 and 1965 represent the maximum common detail available in these censuses, and projections for the same age groups were prepared. Census data are also available for 1960 but were disregarded for this illustration. No adjustments were made for errors in coverage or age reporting in the 1955 and 1965 censuses. To adapt to the age detail available, we have calculated 10-year census cohort-change rates for 5-year age groups up to 20–24 and for 10-year age groups 25 and over; for example,

$$P_{20\text{-}24}^{1975} = \frac{P_{20\text{-}24}^{1965}}{P_{10\text{-}14}^{1955}} \cdot P_{10\text{-}14}^{1965}$$

$$P_{25\text{-}34}^{1975} = \frac{P_{25\text{-}34}^{1965}}{P_{15\text{-}24}^{1955}} \cdot P_{15\text{-}24}^{1965}$$

$$P_{65+}^{1975} = \frac{P_{65+}^{1965}}{P_{55+}^{1955}} \cdot P_{55+}^{1965}$$

For the ages under 10 in 1975, we have employed the following ratios of children to women:

$$P_{0\text{-}4}^{1975} = \frac{P_{0\text{-}4}^{1965}}{P_{15\text{-}54}^{1965}} \cdot P_{15\text{-}54}^{1975}$$

$$P_{5\text{-}9}^{1975} = \frac{P_{5\text{-}9}^{1965}}{P_{20\text{-}54}^{1965}} \cdot P_{20\text{-}54}^{1975}$$

Where the censuses are five years apart and 5-year age data are available for the geographic units, projections can conveniently be made at 5-year time intervals for 5-year age groups by census cohort-change rates, except for young children. The projections for children under 5 in the first projection year can be derived by assuming that the ratio of children to women of childbearing age in that year and later years will be the same as shown by the latest census.

Direct estimates of net migration. — A more refined procedure, which we may call the **cohort migration-survival method,** treats migration as a separate component. In the most common variation, births, deaths, and net migration (combining

Table 23-11.—**Projection of the Female Population of Sivas Province, Turkey, by Age, by the Census Cohort-Change Method (10-Year Intercensal Period), from 1965 to 1975**

Age (a) (years)	Census, 1955[1] (1)	Census, 1965[1] (2)	Ratio $(2)_a \div (1)_{a-10} =$ (3)	Projected population, 1975 $(3)_a \times (2)_{a-10} =$ (4)
Total..........	300,383	354,809	(X)	418,741
0-4...............	54,095	59,569	(X)	[2]71,992
5-9...............	41,990	55,957	(X)	[2]66,156
10-14.............	27,566	41,585	.7687	45,791
15-19.............	25,757	29,291	.6976	39,036
20-24.............	28,216	25,455	.9234	38,400
25-34.............	47,127	55,516	1.0286	56,312
35-44.............	28,831	35,614	.7557	41,953
45-54.............	23,040	22,028	.7640	27,209
55-64.............	14,446	18,347	.7963	17,541
65 and over.......	9,315	11,446	[3].4317	[4]14,351

X Not applicable.
[1] Sources: Turkey, General Statistical Office, *Census of Population, Oct. 23, 1955, Population of Turkey*, Vol. 67, table 16; idem, *Census of Population, Oct. 24, 1965, Social and Economic Characteristics of the Population*, table 12. Ages not reported have been prorated.
[2] Calculated by special formula; see text.
[3] Derived by: $11,446 \div (14,446 + 9,315)$.
[4] Derived by: $.4817 (18,347 + 11,446)$.

net immigration and net internal migration) are the components directly manipulated.

A variety of choices are available for making assumptions regarding future migration for geographic subdivisions. The migration assumptions may be expressed in terms of amounts or rates of net migration. The amounts or rates assumed should be based on actual experience in some recent period in the specific local area; they may be held constant or projected according to some formula. Gradual change to half or some other proportion of the previous intercensal amounts or rates by some future date is a type of alternative assumption. To simplify the expression and application of the assumptions, rounded amounts or rates may be employed. Estimates of net migration by age for a current period, for use in population projections for local areas, may be computed in various ways, as described in chapter 21. Either actual death statistics, life table survival rates, or national census survival rates may be employed to make the allowance for mortality in a residual procedure. The "forward" formula is well adapted to the needs of projections. The estimated amounts of net migration may then be converted to rates for the purpose of projections by dividing them by the "expected" population, that is, the survivors, at the end of the intercensal period, of the initial population.

Procedures for deriving projections of population by the cohort migration-survival method using census survival rates are illustrated for the female population of Sivas Province, Turkey. Projections of the population to 1975 by 5-year or 10-year age groups, based on the census data for 1955 and 1965 (disregarding the 1960 census) and conforming to the age data available from the 1965 census, are presented in table 23-12. The steps are as follows:

A. Derivation of net migration rates by age, 1955-65:

1. Calculate national census survival rates for the female population of Turkey, 1955-1965 (col. 2).

2. Apply these rates (col. 2) to the census population of Sivas Province in 1955 (col. 1) to derive the expected population 10 years older in 1965 (col. 3). (No allowance was made for regional variation in mortality levels.)

3. Subtract the survivors (col. 3) from the census population of Sivas in 1965 (col. 4) to derive the estimated net migration by age cohorts, 1955-65 (col. 5).

4. Derive the corresponding rates of net migration, 1955-65 (col. 6), by dividing the estimates of net migration (col. 5) by the expected population (col. 3). (These rates have an unusual age pattern in comparison with the age patterns of net migration rates observed elsewhere but have been accepted for the present use without modification.)

B. Projection of the 1965 population to 1975:

5. Calculate 10-year life-table survival rates from model table South No. 20 (col. 7), chosen on the basis of an analysis of national census survival rates for Turkey and reflecting some future improvement.[39]

6. "Survive" the 1965 census population (col. 4) to 1975 (col. 8) by use of the life-table survival rates (col. 7).

7. Calculate net migration for the 1965-75 period (col. 9) by multiplying the migration rates (col. 6) by the survivors in 1975 (col. 8). This procedure assumes that net migration rates will remain unchanged.

8. Add net migrants (col. 9) to the expected population (col. 8) to derive population projections for 1975 (col. 10).

Inasmuch as the birth statistics of Turkey are inadequate, we have developed our projection of the population under 10 years of age in 1975 on a different basis. We have simply assumed that the ratio of children under 10 to women 15 to 54 years old in 1975 is the same as that observed in 1965. This procedure encompasses the effect of fertility, mortality, and net migration in determining the size of these juvenile age cohorts.

Other assumptions of net migration should be employed to prepare additional series of population projections. We have assumed above that net migration rates were the same as observed in 1955-65. Possible additional assumptions are: (1) Net migration rates, 1955-65, reduced 25 percent; (2) net migration rates, 1955-65, increased by 25 percent; etc.

We could have employed national census survival rates in projecting the population of Sivas Province instead of life table survival rates. As we developed the projections above, net census error in one age group is carried down to the next age group over a projection period. Use of national census survival rates would help to keep the net census errors of the Sivas data in their original (1965) ages and, hence, increase the accuracy of the projected changes by age. This procedure imposes the pattern of change in net census errors by age cohorts between 1955 and 1965 for Turkey on the census data for Sivas Province. Implicitly, mortality in Turkey as a whole at the level of 1955-65 is assumed for 1965-75, without change. Note that, ideally, we would adjust the 1965 census population of Sivas for net census errors by age and then carry the adjusted population forward by projected life table survival rates and net migration rates; a basis for adjusting the population is lacking, however.

A comparison of the 1975 projections given in tables 23-11 and 23-12 reveals substantial differences at many ages. Differences in population projections for the same date reflect mainly differences in the underlying assumptions on net migration.

[39] Ansley J. Coale and Paul Demeny, *op. cit.* See also Paul Demeny and Frederic C. Shorter, *Estimating Turkish Mortality, Fertility, and Age Structure*, Publication No. 218, Faculty of Economics, University of Istanbul, Istanbul, 1968, pp. 8-28.

Table 23–12. — **Projection of the Female Population of Sivas Province, Turkey, by Age, by the Migration-Survival Method (10-Year Intercensal Period), from 1965 to 1975**

Age (a) (years) in 1955	in 1965	in 1975	Population of Sivas Province, census of 1955[1] (1)	10-year census survival rates, Turkey, 1955 to 1965 (2)	Expected survivors, 1965 (1)x(2)= (3)	Population of Sivas Province, census of 1965[1] (4)	Net migration, 1955 to 1965 (4)-(3)= (5)	10-year net migration rate (5)÷(3)= (6)	10-year life-table survival rate[2] (7)	Expected survivors, 1975 (4)x(7)= (8)	Projected net migration, 1965 to 1975 $(6)_a \times (8)_{a+10}$= (9)	Projected population, 1975 (8)+(9)= (10)
Total........	Total.......	Total.......	300,383	(X)	(X)	354,809	(X)	(X)	(X)	(X)	(X)	432,132
(X)............	(X)............	0-9..........	(X)	(X)	(X)	(X)	(X)	(X)	(X)	(X)	(X)	[3]143,345
(X)............	0-4............	10-14.........	(X)	(X)	(X)	59,569	(X)	(X)	.9798	58,366	-11,837	46,529
(X)............	5-9............	15-19.........	(X)	(X)	(X)	55,957	(X)	(X)	.9926	55,543	-11,231	44,312
0-4............	10-14..........	20-24.........	54,095	.9643	52,164	41,585	-10,579	-.2028	.9910	41,211	-4,632	36,579
5-9............	15-19..........	25-34.........	41,990	.8744	36,716	29,291	-7,424	-.2022	} .9867	54,019	-854	53,165
10-14..........	20-24..........	25-34.........	27,566	1.0403	28,677	25,455	-3,222	-.1124				
15-19..........	25-34..........	35-44.........	53,973	1.0451	56,407	55,516	-891	-.0158	.9809	54,456	-9,748	44,708
25-34..........	35-44..........	45-54.........	47,127	.9205	43,380	35,614	-7,766	-.1790	.9678	34,467	-4,894	29,573
35-44..........	45-54..........	55-64.........	28,831	.8905	25,674	22,028	-3,646	-.1420	.9331	20,554	-2,695	17,859
45-54..........	55-64..........	65-74.........	23,040	.9165	21,116	18,347	-2,769	-.1311	.8285	15,200	[4]-3,785	[4]16,062
55 and over....	65 and over....	75 and over...	23,761	.5952	14,143	11,446	-2,697	-.1907	.4060	4,647 }		

X Not applicable.

[1] Ages not reported have been prorated.

[2] Computed from model life table for region South, level 20, given in, Ansley J. Coale and Paul Demeny, *Regional Model Life Tables and Stable Populations,* Princeton, N.J., Princeton University Press, 1966, p. 675. .

[3] Calculated by use of the formula:

$$\frac{P^{1965 \text{ (census)}}_{0-9}}{P^{1965 \text{ (census)}}_{15-54}} \times P^{1975 \text{ (projected)}}_{15-54}$$

[4] Ages 65 and over.

Projections of population are often wanted at 5-year time intervals. A procedure for deriving projections of Sivas' population to 1970 is presented in table 23–13. It employs 10-year net migration rates estimated from the 1955 and 1965 censuses but requires a conversion of these 10-year rates to 5-year rates. (For purposes of this illustration also, we omit use of the 1960 census.) We assume that the 5-year net migration rates as estimated on the basis of 1955–65 population changes will continue unchanged for 1965–70. Having derived estimates of expected survivors in 1965, and of net migration for 1955–65 for 5-year age groups from 10 to 24 and 10-year age groups above age 25 (table 23–12, cols. 3 and 5), we can proceed as follows (table 23–13) :

1. Subdivide the expected survivors in 1965 and the net migration estimates, 1955 to 1965, into 5-year age groups above age 25 (cols. 1 and 2). In the present illustration Sprague's osculatory multipliers [40] were employed for all categories except net migration 25 to 34 years. For the latter category, negative results were obtained by osculatory interpolation and Newton's interpolation formula for halving a group [41] was employed in its place.

2. Combine the population in adjacent ages (col. 3) and the estimates of net migration in adjacent ages (col. 4). The purpose of this step is to take advantage of the fact that the net migration of each 5-year age group is reflected in two of the cohorts employed in the calculations. For example, net migration for terminal ages 30–34 is represented by the two shaded areas in the following sketch:

3. Divide the net migration in column (4) by the population in column (3) to derive a hypothetical 10-year migration rate for 5-year age groups (col. 5).

4. Divide the results in column (5) by 2 to derive 5-year migration rates for 5-year age groups (col. 6).

5. Compute the expected female population of Sivas in 1970 (col. 9) by applying life-table survival rates (col. 8) to the census population in 1965 (col. 7).

6. Calculate the projected net migration for 1965–70 (col. 10) by applying the rates in column (6) to the expected survivors in column (9). In every case the figures are matched in terms of terminal ages.

7. Combine the projected net migration in column (10) with the expected survivors in column (9) to derive the projected population for 1970.

[40] For a discussion of the mathematical theory of osculatory interpolation, see T. N. E. Greville, "The General Theory of Osculatory Interpolation," *Transactions of the Actuarial Society of America,* 45 (112), Pt. II: 202-265, October 1944. See also Thomas B. Sprague, "Explanation of a New Formula for Interpolation," *Journal of the Institute of Actuaries* 22:270, 1880-81.

[41] Margaret J. Hagood, *Statistics for Sociologists,* New York, Reynal and Hitchcock, Inc., pp. 760-763, 1941.

Table 23–13.—**Projection of the Female Population of Sivas Province, Turkey, by Age, by the Migration-Survival Method (10-Year Intercensal Period), from 1965 to 1970**

Age (a) (years) in 1965	in 1970	Expected survivors, Sivas Province, 1965[1] (1)	Net migration, 1955 to 1965[1] (2)	$(1)_a +$ $(1)_{a+5}=$ (3)	$(2)_a +$ $(2)_{a+5}=$ (4)	10-year net migration rate $(4)\div(3)=$ (5)	5-year net migration rate $(5)\div2=$ (6)	Population of Sivas Province, census of 1965[2] (7)	5-year life-table survival rate[3] (8)	Expected survivors, 1970 $(7)\times(8)=$ (9)	Projected net migration, 1965 to 1970 $(9)_{a+5}\times(6)_a=$ (10)	Projected population, 1970 $(9)+(10)=$ (11)
Total........	Total........	(X)	(X)	(X)	(X)	(X)	(X)	354,809	(X)	(X)	(X)	386,525
(X)............	0–4............	(X)	(X)	(X)	(X)	(X)	(X)	(X)	(X)	(X)	(X)	[4]65,957
0–4............	5–9............	(X)	(X)	(X)	(X)	(X)	(X)	59,569	.98339	58,580	–5,940	52,640
5–9............	10–14............	(X)	(X)	(X)	(X)	(X)	[5]–.1014	55,957	.99630	55,750	–5,647	50,103
10–14............	15–19............	52,164	–10,579	88,880	–18,003	–.2026	–.1013	41,585	.99625	41,429	–3,372	38,057
15–19............	20–24............	36,716	–7,424	65,393	–10,646	–.1628	–.0814	29,291	.99469	29,135	–961	28,174
20–24............	25–29............	28,677	–3,222	58,241	–3,848	–.0661	–.0330	25,455	.99327	25,284	–200	25,084
25–29............	30–34............	29,564	[6]–626	56,407	–891	–.0158	–.0079	28,517	.99211	28,292	–1,067	27,225
30–34............	35–39............	26,843	[6]–265	50,655	–3,819	–.0754	–.0377	26,999	.99069	26,748	–2,394	24,354
35–39............	40–44............	23,812	–3,554	43,380	–7,766	–.1790	–.0895	20,258	.98821	20,019	–1,954	18,065
40–44............	45–49............	19,568	–4,212	34,214	–6,681	–.1953	–.0976	15,356	.98443	15,117	–1,073	14,044
45–49............	50–54............	14,646	–2,469	25,674	–3,646	–.1420	–.0710	12,177	.97796	11,909	–551	11,358
50–54............	55–59............	11,028	–1,177	20,638	–1,912	–.0926	–.0463	9,851	.96781	9,534	–625	8,909
55–59............	60–64............	9,610	–735	21,116	–2,769	–.1311	–.0656	8,874	.94993	8,430	–768	7,662
60–64............	65–69............	11,506	–2,034	[7]18,578	[7]–3,383	–.1821	–.0911	9,473	.91659	[8]16,464	[8]–1,571	[8]14,893
65 and over....	70 and over....	14,143	–2,697	[9]14,143	[9]–2,697	–.1907	[10]–.0954	11,446	.67984			

X Not applicable.

[1] Age groups 25–29 to 60–64 were interpolated from 10-year age groups (cols. 3 and 5 of table 23–12) using Sprague's osculatory multipliers.

[2] Ages not reported were prorated. Age groups 25–29 to 60–64 were interpolated from 10-year age groups (col. 4 of table 23–12) using Sprague's osculatory multipliers.

[3] Model life table for region South, level 20, in, Ansley J. Coale and Paul Demeny, *Regional Model Life Tables and Stable Populations*, Princeton, N.J., Princeton University Press, 1966, p. 675.

[4] Calculated by use of the formula:

$$\frac{P_{0-4}^{1965 \text{ (census)}}}{P_{15-49}^{1965 \text{ (census)}}} \times P_{15-49}^{1970 \text{ (projected)}}$$

[5] Net migration rate for ages 0–4 (initial) to 5–9 (terminal) assumed to equal one-half the net migration rate (–.2028) for ages 0–4 (initial) to 10–14 (terminal). See table 23–12, col. 6.

[6] Computed by applying Newton's interpolation formula for halving a group.

[7] Value for ages 60–64 plus one-half value for ages 65 and over in col. 1 or col. 2.

[8] Ages 65 and over.

[9] Same as value for ages 65 and over in col. 1 or col. 2.

[10] Net migration rate for ages 60 and over (initial) to 65 and over (terminal) assumed to equal one-half the net migration rate (–.1907) for ages 55 and over (initial) to 65 and over (terminal).

It will be noted in table 23–13 that the conversion of the data to 5-year time intervals was carried out after the estimates of 10-year net migration were derived since the latter should be calculated from unadjusted census data (assuming the use of national census survival rates). The derivation of 5-year amounts and rates of net migration for 5-year age groups, from 10-year amounts and rates for 5-year and 10-year age groups, may be accomplished in different ways. For example, we could interpolate our 10-year data to 5-year data at a different point in the calculations, as by converting the 10-year net migration rates directly to 5-year rates. This approach might improve on the procedure described earlier, which does not, in effect, handle the calculation of the denominator of the 5-year rates very satisfactorily. One could simply compute population projections for 10-year age groups 10 years ahead and derive other required figures by interpolation.

Even if a country has good current data on internal migration, it is desirable for the responsible agency to prepare a special series of projections assuming no further net migration among the geographic subdivisions. In this way user agencies can make their own assumptions to serve special needs. Calculation of a series assuming no net migration among the geographic subdivisions of a country in future years permits an evaluation of the contribution of net migration to projected population change as reflected in the various series with net migration. In sum, a no-migration series may have considerable value to planners even though it may be quite unrealistic in itself.[42]

Projection of births and deaths.—We have already mentioned one or more ways of projecting births and deaths in preparing subnational population projections by the component method. We may consider this problem more generally.

Birth rates by age and survival rates (or age-specific death rates) may be projected independently, or by comparing the rates for the local area and its parent area for one or more past dates and applying the current or extrapolated ratio of these rates to previously available projections of the rates for the parent area. The parent area will usually be the country as a whole. In addition to historical analysis, an analysis of regional

[42] See, for example: Sweden, Statistika Centralbyrån, *Befolkningsprojektion för kommunblocken till 1970, 1975, 1980 och 1985* (Population projections for cooperating communes to 1970, 1975, 1980, and 1985), Stockholm, 1969; idem, *Sveriges beräknade framtida folkmängd*, Part 2, *Underlag för lokala prognoser 1965, 1970 och 1975* (Estimated future population of Sweden, Pt. 2, Basis for local projections, 1965, 1970, and 1975), Statistiska Maddelanden B 1963 (4), Stockholm, p. 5; France, Institut de la statistique et des études économiques, "Perspectives d'évolution naturelle de la population par departement" (Projections of natural increase of population by department), *Etudes Statistiques, Supplement Trimestriel du Bulletin Mensuel de Statistique* (Paris), No. 4, Oct.–Dec. 1957, pp. 63–64; and J. R. L. Schneider, "Local population projections in England and Wales," *Population Studies* (London), 10(1):98, July 1956.

differences for other countries at various stages of economic development should prove useful in this regard.

Calculation of the number of births for each 5-year projection period by use of age-specific birth rates involves a special step. We have to interpolate the female population of childbearing age, by age, to the middle of each projection period, apply the projected age-specific birth rates cumulatively to this population to derive the average number of births in the period, and multiply this number by five to obtain the estimate for the 5-year period. Alternatively, births may be calculated for the first and last year of each 5-year projection period, added together, and then inflated by a factor of 2.5. The projected number of births and deaths over all subdivisions of an (parent) area should, as a final step, be adjusted to the projected total of births and deaths for the area.

Because birth and death statistics by age are not always available for small geographic areas and the volume of computations in the procedure just described is quite large, it is desirable or necessary to try to abbreviate the procedure. Fertility may be projected by an indirect method, such as one employing a single schedule of age-specific birth rates and projected total fertility rates, or sex-age adjusted birth rates. Mortality may also be projected by an indirect method, involving a single set of age-specific death rates but with some "control" figures at the local level. When mortality varies only moderately among the local areas, the same mortality rates may be assigned to several areas or even to all areas in a country. The projections for geographic subdivisions can be made in terms of the ratio of crude birth rates and crude death rates for the local areas to the corresponding figures for the country as a whole, provided there have been no radical changes in the age composition of local areas and carefully prepared national projections of births and deaths are available. [43]

Because of the possibility of substantial differences between the fertility of regions, consideration must be given to the question whether migrants will be assigned the fertility rates of their area of origin or the fertility rates of their area of destination during the period of their arrival and subsequently.

Separate projection of in- and out-migration. — More meaningful and possibly more realistic projections of internal migration may be derived by projecting in- and out-migration separately when the appropriate data are available, rather than net migration. This is because the variation in the relation of in-migration to out-migration, i.e., net migration, may be expected to be much greater than the variation in the corresponding gross in- or out-migration. Furthermore, total in-migration between geographic units in a country is in a sense dependent upon total out-migration from these units and projections for these totals can be made equal to one another very simply when in- and out-migration are projected separately. On the other hand, use of net migration rates often results in serious imbalances between total net in-migration and net out-migration. One general approach for projecting in- and out-migration separately employs age-specific out-migration rates to derive figures for out-migrants for each area and the distribution of the resulting total number of out-migrants among the

several areas to derive estimates of in-migrants. This general approach was used in preparing projections of the population of States of the United States published in 1967, metropolitan areas of the United States in 1969, and communes of Sweden in 1969. [44]

Other approaches involving the separate projection of in- and out-migration for projecting regional population are based on the development of a model of interregional migration or a national demographic model incorporating regional changes. The possibility of developing a model of internal migration for use in projecting population has been reviewed by H. Ter Heide. [45] He concluded that, although there has been substantial progress in the development of a model which describes past migration, the problems of employing this model for purposes of population projections are almost insurmountable, if only because of the general lack of projections of the several independent variables on which the model depends. Ter Heide was concerned with models which incorporate economic, social, and psychological factors. On the other hand, Rogers has proceeded to develop "descriptive" models of interregional migration for use in projecting population; his models are less analytic and he has dealt so far only with the situation where the interregional migration rates are assumed to remain constant. [46]

Methods Taking Account of Economic Variables. — It is generally believed that internal movements are significantly affected by differential economic opportunities and that any drastic changes in the economic advantages of one area over another will have substantial impact on the future size of migration streams and even on the direction of net movement. Some research is being directed toward these relationships. [47] A number of methods have been developed which employ economic variables directly in the context of a ratio, component, or correlation method, or a combination of these methods. These methods take account of economic prospects quantitatively by basing the projections of population on projections of employment, per capita income, production, land use, or other economic variables. We consider the methods in two groups, those methods involving correlation with economic indicators and other methods using economic analysis.

Correlation with economic indicators. — Regression analysis may be employed to project the total population directly or to project the net migration component only (natural increase being projected separately in the latter procedure). An example of the use of this method is given by the projections of the population of States of the United States made at the Stanford Research Institute. [48] Projections of net migration for States were derived from a regression equation relating net migration and average per capita income. The fitted equation was as follows:

[43] Illustrations of the various procedures noted here are given in: U.S. Bureau of the Census, *Current Population Reports,* Series P–25, No. 375, "Revised Projections of the Population of States, 1970 to 1985," October 3, 1967; Schneider, op. cit., pp. 105–111; France, Institut national de la statistique et des études économiques, "Perspectives d'évolution naturelle de la population par department," pp. 63–64; Bureau of the Census *Current Population Reports,* Series P–25, No. 160, "Illustrative Projections of the Population, by States: 1960, 1965, and 1970," August 9, 1957; and Sweden, Statistiska Centralbyrån, *Sveriges Beräknade Framtida Folkmängd,* Pt. 2.

[44] U.S. Bureau of the Census, *Current Population Reports,* Series P–25, No. 375, "Revised Projections of the Population of States, 1970 to 1985," Oct. 3, 1967; idem, *Current Population Reports,* Series P–25, No. 415, "Projections of the Population of Metropolitan Areas: 1975," Jan. 31, 1969; and Sweden, Statistika Centralbyrån, *Befolkningsprojektion för kommunblocken till 1970, 1975, 1980 och 1985.*

[45] H. Ter Heide, "Migration Models and Their Significance for Population Forecasts," *Milbank Memorial Fund Quarterly,* 41(1):56–76, Jan. 1963.

[46] Andrei Rogers, "A Markovian Policy Model of Interregional Migration," *Regional Science Association Papers,* 17:205–224, 1966; and idem, "The Multiregional Matrix Growth Operator and the Stable Interregional Age Structure," *Demography,* 3(2):537–544, 1966.

[47] Ira S. Lowry, *Migration and Metropolitan Growth: Two Analytical Models,* San Francisco, Calif., Chandler Publishing Company, 1966.

[48] Pietro Balestra and W. Koteswara Rao, *Basic Economic Projections: United States Population, 1965–80,* Stanford Research Institute, Menlo Park, Calif., 1964, esp. pp. 37–44.

$$Y_c = 38.94255 + .402863X$$

in which Y_c stands for the net migration rate, 1950–60 (i.e., net migration for the decade as a percent of the 1950 population) and X stands for the percent change in per capita personal income for the decade expressed as a percent of the corresponding U.S. figure. Per capita income had been previously projected to 1970, and 1980 on the basis of the converging trend in per capita income observed in the 1950–60 decade. This procedure assumes that net migration for States is more closely correlated with per capita income than is total population change, that labor tends to move toward areas of higher per capita income, and that the systematic influence of income on migration flows will continue in the future.

The basic regression shows that a unit change in the per capita income of a State (in percentage form) causes a change of 0.4 in the percentage of net migration of that particular State. This factor was applied to the change in per capita income from 1960 to 1970 to derive "unadjusted" estimates of net migration in the 1960–70 decade. The estimates were then adjusted on the basis of the fact that the sum of the unadjusted figures over all States was far in excess of the expected total interstate migration for the United States computed separately. The adjustment procedure assumes different adjustment coefficients for different groups of similar States. An illustration of these steps is given for two States:

	Arkansas	Ohio
(1) Migration rate, 1950–60 (percent)......	−22.5	+5.7
(2) Percent change in per capita income, 1960–70 (independently projected)..	+7.2	−2.4
(3) Effect of income change = (2) × 0.4.....	+2.88	−.96
(4) Unadjusted migration rate, 1960–70 (percent) = (1) + (3).....................	−19.6	+4.7
(5) 1960 population (in thousands).........	1,786.3	9,706.4
(6) Unadjusted net migration, 1960–70 (in thousands) = (4) × (5).................	−350	+456
(7) Adjusted migration rate, 1960–70 (percent) [a]..	−19.6	+2.9
(8) Adjusted net migration, 1960–70 (in thousands) = (5) × (7)...................	−350	+281

[a] See source for explanation of adjustment in net migration rates.

Other methods using economic analysis. — Other methods of projecting the population of geographic subdivisions using economic analysis may involve an intensive study of the economic prospects for each area. One approach involves separate consideration of several main branches of the economy, proceeding from national to local employment in these branches, then to total employment in the area, and finally to total population. The local projections of employment in various branches of the economy may be made as a proportion of the corresponding national projections.

Another class of methods employs a limited type of component procedure which depends on prior projections of employment or labor force. In the simpler application, the projections of employment or labor force, and then net migration, are made directly for the population of all ages; in the more elaborate application, the projections of employment or labor force, and then net migration, are made by age groups. The method sequentially calculates employment or labor force, net migration of the labor force, net migration of the total population, and finally, the total population, combining the net migration with the expected population allowing for births and deaths. A specific example is provided by the population projections made by the Oregon State government for the State as a whole and its state economic areas, by age and sex, 1964.[49]

The more intensive procedures are difficult to apply since they require considerable data and involve the problem of demographic and economic interdependence. Careful study of the economy of an area, involving measurement of the future requirements for workers and of the degree to which these future requirements can be filled by the available population, on the one hand, and by net in-migration, on the other, is needed. This analysis would require some prior assumption as to the amount of net in-migration and prior projections of the size of the expected population. Initial assumptions would then have to be modified on the basis of what the initial projections imply as to labor deficits or surpluses. Even if the simplest method involving economic analysis is employed, the calculations become voluminous when age and sex detail is included and projections have to be prepared for a large number of areas. Under these circumstances, it becomes desirable or even necessary to carry out the work by electronic computer.

EVALUATION OF PROJECTIONS

Design of Evaluation Studies

As in the case of estimates, the evaluation of projections requires some standard by which to judge their quality. The possibilities of evaluating a set of projections are limited because current estimates or census counts for many years subsequent to the base date are needed. In the practical situation, this would permit evaluation of projections only after a long period of time has elapsed; by this date the methodology in current use may have changed and, hence, would not be encompassed by the evaluation. However, there is a broader issue. The concept of "accuracy" becomes less meaningful where several series of projections are offered as reasonable possibilities and, particularly, where none is designated as a "forecast." In this case would we judge the accuracy of the "medium" series, the one described as "most likely," each series individually, or some of them, recognizing that the projections were not offered as predictions?

When a particular projection has been designed as a forecast or prediction, it seems perfectly appropriate to measure its accuracy by a subsequent comparison with a census count or current estimate. We also believe that those projections which fit our definition of principal series (i.e., which claim to incorporate realistic assumptions and which are offered as reasonable possibilities of future population size) are also proper candidates for evaluation, especially the medium series in a set. One may reasonably compare each projection in a set of such projections with the population actually recorded to indicate how it deviates from the current figure. Such comparisons, expressed in terms of percent differences, have in fact, often been made.[50] They are also of value in selecting a series of projections for later use.

To achieve a limited evaluation of a set of projections, one may compare them with a revised set of projections made in

[49] Oregon State Board of Census, *Population Bulletin,* Release No. P–10, "Population Forecast, State of Oregon and Economic Areas: 1960–85," by Richard B. Halley and Morton Paglin, Portland, Oreg., April 1964.

[50] For illustrations see references given in footnotes 53 and 56.

subsequent years. Such a comparison suggests the probable direction of errors of the longer-term projections, as well as the actual error of the short-term figures which are compared with current estimates or revised short-term projections.

Attention may profitably be focused on the accuracy of the projection of net change and its components (births, deaths, and net migration) rather than on the population projection itself since these are the elements actually projected and since the analyst knew the initial population to begin with. The percentage error of the projected change will be very much larger than the percentage error in the projected population. Comparison of the actual components of change with the projected figures is particularly valuable since it provides insight into the reasonableness of the various assumptions and shows how the overall population projections may have benefited from compensating errors. A supplementary approach is to consider the errors in projections by age, identifying separately, if possible, the age group born since the base date of the projections. This comparison will not only indicate differences in the relative accuracy of projections of age groups, but will also indirectly provide some insight into the relative contribution of births and the other components to the total error.

Keyfitz has suggested further that, since the analyst ordinarily knows the current growth rate or other measures of current change, such as age-specific birth rates and death rates, relative success should be measured by the degree to which the analyst anticipates the deviation from the change resulting from the current rates of growth.[51] By that criterion the percentage errors will tend to be even larger for a given series of projections.

If the analyst's success in making projections for different dates and areas is being judged comparatively, an important standardizing factor is the length of the projection period. One would expect errors to be greater when the projection period is longer. The size of the area and its rate of growth would also be factors characterizing the area which might systematically affect the accuracy of the projections.

A possible further basis for evaluating projections is in terms of the range from the highest to the lowest series in a set of principal projections. The width of the range from the highest to the lowest projections depends on the regularity of past demographic trends, knowledge regarding past trends, ability to measure them accurately, and finally, the analyst's judgment of the likely course of future change. The range is, in a sense, a reflection of the analyst's confidence in the medium series of projections. As the range widens, he is indicating that he has less and less confidence that the medium figures will correspond to the actual figures.

The real difficulty here is the assumption that one range covers the same confidence interval as another range. A few demographers have considered the problem of developing probabilistic measures of the accuracy of projections analogous to the sampling error of estimates derived from sample surveys. The issue has been examined most extensively by Muhsam and Sykes.[52]

National Projections

Several systematic studies of the accuracy of national projections have been made but these do not usually distinguish the method of projection, the components of error, or the length of the projection period.[53] The period-fertility variation of the component method, the method most widely used to project national population where vital statistics are available, has often been found to produce unsatisfactory results. Typically, the greatest source of error has been in the projections of births. Where several series have been projected, the range has usually been quite wide and yet has occasionally failed to encompass the actual population. The prospects for improving the accuracy of national population forecasts are not great although they may have considerable value as analytic tools.[54]

The United Nations has systematically compared the projections for the regions and countries of the world published in 1966 with those published in 1958.[55] The differences for regions reflect reestimation of the present size of the population in each area and of current fertility and mortality levels, and a change in the assumptions used in the projections. The result of the new calculations for most of the world, in comparison with the earlier ones, is an initial acceleration of population growth and a subsequent deceleration. Judging the earlier projections on the basis of the current estimates for 1960 and the revised projections, the earlier medium projections for 1960 and 2000 were too low in some areas and too high in others by considerable percentages. The indicated percent "errors" in the projections for major world regions for various dates are shown in table 23-14.

The projections of U.S. population made during the 1930's and 1940's by the Scripps Foundation and the U.S. Bureau of the Census proved to be consistently too low, primarily because of understatement of future births. Although the range of the projections made during the 1950's and 1960's still encompassed the actual figures after several years, the projections were generally too low or too high, depending on whether the actual trend of fertility was up or down at the time the projections were being prepared.

Geographic Subdivisions

Projections for states, provinces, localities, etc., are subject to much greater average error than those for whole countries (i.e., for a given length of projection period). The greater inaccuracy results partly from the added uncertainties of internal migration and partly from the fact that errors (or deviations in general) tend to vary inversely with population size.

Siegel's review in 1954 of the various U.S. tests of subnational projections concluded that no one method uniformly gave the best results and that in many cases a simple mathematical method was as accurate as a component method using

[51] Nathan Keyfitz, "La proyección y la predicción en demografía: Un revisión del estado de este arte" (Projection and prediction in demography: A review of the state of the art), *Conferencia Regional Latinoamericana de Población, 1970, Actas,* Mexico City, El Colegio de Mexico, 1972.

[52] H. V. Muhsam, "The Utilization of Alternative Population Forecasts in Planning," *Bulletin of the Research Council of Israel,* 5(2-3):133-146, March-June 1956; idem, "The Use of Cost Functions in Making Assumptions for Population Forecasts," in United Nations, *World Population Conference, 1965* (Belgrade), Vol. III, pp. 23-26; and Z. M. Sykes, "Some Stochastic Versions of the Matrix Model for Population Dynamics," *Journal of the American Statistical Association,* 64(325):111-130, March 1969.

[53] Joseph S. Davis, *The Population Upsurge in the United States,* War-Peace Pamphlets No. 12, Food Research Institute, Stanford University, December 1949; Harold F. Dorn, "Pitfalls in Population Forecasts and Projections," *Journal of the American Statistical Association,* 45(251):311-334, September 1950; Henry S. Shryock, Jr., "Accuracy of Population Projections for the United States," *Estadistica,* 12(45):587-598, December 1954; Robert J. Myers, "Comparison of Population Projections with Actual Data," in United Nations, *World Population Conference, 1954* (Rome), pp. 101-111; and Cézar A. Peláez, "The Degree of Success Achieved in the Population Projections for Latin America Made since 1950. Sources of Error. Data and Studies Needed in Order to Improve the Basis for Calculating Projections," in United Nations, *World Population Conference, 1965* (Belgrade), pp. 27-33. John V. Grauman, "Success and Failure in Population Forecasts of the 1950's: A General Appraisal," in United Nations, *World Population Conference, 1965* (Belgrade), pp. 10-14.

[54] John Hajnal, "The Prospects for Population Forecasts," *Journal of the American Statistical Association,* 50(270):309-322, June 1955.

[55] United Nations, *World Population Prospects as Assessed in 1963,* Series A, Population Studies, No. 41, 1966, pp. 15-17.

Table 23-14. — Comparison of United Nations Medium Projections for Major World Regions Published in 1966 and Corresponding Projections Published in 1957

[Percents represent the deviation of projections published in 1957 from the current estimates or revised projections as a percent of the current estimates or revised projections]

Country	Population, 1960 (millions)		Percent difference				
	1957	1966	1960	1970	1980	1990	2000
World..............................	[1]2,910	2,998	-2.9	-3.1	-2.5	-0.9	+2.4
East Asia..........................	796	794	+0.3	+6.2	+15.7	+27.5	+44.0
South Asia.........................	827	865	-4.4	-8.5	-10.4	-10.3	-7.2
Europe.............................	424	425	-0.2	+0.7	+3.3	+5.6	+7.8
U.S.S.R............................	215	214	+0.5	+3.3	+6.8	+7.3	+7.4
Africa.............................	235	273	-13.9	-19.7	-25.8	-30.2	-32.7
Northern America...................	197	199	-1.0	-0.9	-3.1	-7.5	-11.9
Latin America......................	206	212	-2.8	-6.4	-7.9	-8.4	-7.2
Oceania............................	16.3	15.7	+3.8	+3.7	-0.4	-4.8	-8.2

[1] Rounded to nearest 10 million.

Source: Adapted from United Nations, *World Population Prospects as Assessed in 1963*, Series A, Population Studies, No. 41, 1966, tables 4.3 and 4.4.

age-specific rates.[56] The more elaborate methods, such as the cohort-component methods, would still be preferred for their analytic value, however.

He concluded further that the "error" rate tends to vary directly with the rate of population growth and with the length of the projection period. After 20 years no method any longer provided accurate forecasts. The longer the projection period the greater the likelihood of unforseen developments which can cause the actual population to fall outside the range projected. Similarly, population trends are less regular for small populations than large ones. Accordingly, projections for subnational areas should be carried out for fewer years than projections for countries as a whole. These findings and the practical need for consistency with current estimates suggest the need for frequent revision of the projections for geographic areas.

PROJECTIONS OF SOCIOECONOMIC CHARACTERISTIS—INTRODUCTION

The remainder of this chapter reviews the current status of the methodology of projections of selected "socioeconomic" characteristics of population. Projections for two classes of socioeconomic characteristics are discussed here. The first class relates to type-of-residence categories, in particular the classification into urban and rural residence. The second class relates to selected characteristics of the population conventionally considered socioeconomic, such as household or family status, school enrollment, and economic activity.

Projections of the population characteristics noted above are essential for national planning. The implementation of development programs requires information on constituent groups in the population; and, hence, the preparation of projections of total population and age-sex structure may be viewed as only antecedent to the preparation of projections of socioeconomic characteristics. The importance of these types

of projections in the development and implementation of national policies and plans has come to be realized on a broad scale by national governments. The need for them has been clearly identified by the participants in various international conferences, including U.N. regional seminars and conferences and the meetings of the U.N. Population and Statistical Commissions.[57]

Applications

The preparation of projections of urban and rural population, households, enrollment, labor force, and other socioeconomic characteristics must be recognized as only a preliminary step in the development of a wide range of projections needed for the evaluation and solution of specific problems of public planning and administration. The types of projections listed can be extended to represent various types of applied projections, such as projections of requirements for housing units; supply and demand for workers in various occupational categories; requirements for hospitals, hospital beds, and medical personnel; demand for schools, classrooms, and teachers; requirements for food and other consumer goods, etc. Calculations of this kind require some assumptions for the future regarding the relation between requirements (e.g., number of teachers) and population, consumers, or consuming units (e.g., number of students). Although current ratios may be maintained, more commonly some "improvement" may be sought. Even current ratios may be viewed as representing an inadequate standard or norm for the ratio of "supply" to "demand." Moreover, the so-called norms may have to be constituted of elements which reflect the differences in the cultural and socioeconomic situation in each country or sector of the country. Thus, the types of housing units, the specific composition of diets, the size of school classes and structures, etc., will vary among countries and among areas within countries according to the available supply of materials, local preferences, and variations in needs.

---

[56] Jacob S. Siegel, "Some Aspects of the Methodology of Population Forecasts for Geographic Subdivisions of Countries," in United Nations, *World Population Conference, 1954* (Rome). See also Helen R. White, "Empirical Study of the Accuracy of Selected Methods of Projecting State Populations" *Journal of the American Statistical Association*, 49(267):480–498, September 1954; Jacob S. Siegel, "Forecasting the Population of Small Areas," *Land Economics*, 29(1):72–87, February 1953; and Robert C. Schmitt and Albert H. Crosetti, "Accuracy of the Ratio Method for Forecasting City Population," *Land Economics*, 27(4):346–348, November 1951.

[57] See, for example: United Nations, *Seminar on Population Problems in Africa*, Cairo, United Arabic Republic, Oct. 29–Nov. 10, 1962 (E/CN.14/186-E/CN.9 CONF. 3/1); idem, *Seminar on Evaluation and Utilization of Census Data in Latin America*, Santiago, Chile, Nov. 30–Dec. 18, 1959 (ST/TAO/SER.C/46-E/CN.9/CONF. 1/1 Rev. 1); and idem, *Seminar on Evaluation and Utilization of Population Census Data in Asia and the Far East*, Bombay, India, June 20–July 8, 1960 (ST/TAO/SER.C/47-E/CN.9/CONF. 2/1).

Methods of Projection

In general, the methods and problems of projections of socioeconomic characteristics differ from those for the total population. In part, these differences result from the fact that these characteristics typically may change during the lifetime of an individual and additional components have to be estimated and manipulated. The array of methods ranges from simple mathematical procedures to very elaborate cohort-component methods using sequential probabilities.

The age-specific "participation" rate procedure has predominated in the published studies among the variety of procedures which have been applied or proposed for preparing projections of the various socioeconomic characteristics, particularly the characteristics other than urban and rural distribution. In this procedure assumed proportions of the population in each age group having the characteristic, such as the proportion enrolled or the proportion in the labor force, etc., are applied to projections of the population at these ages to derive the projections of these characteristics. The age-specific proportions or rates may be held constant or they may be projected in any of a number of ways. The trend of observed rates at the same age may be projected according to some mathematical formula. For example, the observed rates may be extended on the assumption of a straight-line or a simple exponential trend. More elaborate mathematical formulas have been employed, involving three or more observations and various curve-fitting procedures (e.g., least squares). Alternatively, an analytic procedure for projecting the rates may be employed. Specifically, projections of the rates may be based on historical analysis of the changes in the rates in relation to changes in various other related socioeconomic characteristics, or current regional or international analysis may identify correlated socioeconomic characteristics.

In the specific approaches cited above, the projected levels follow from the general method of projection, but the procedure may involve initially setting the terminal levels of the rates and deriving the rates for the intermediate years by interpolation. Methods vary in the basis for setting or selecting the target rates and the method of interpolation. The terminal rates may be borrowed from another country viewed as advanced or borrowed from an advanced region or population group in the same country.

Cohort methods have also been introduced in various forms. The projection of the age-specific rates may be carried out on a cohort basis: that is, ratios of rates in successive ages and years in some recent period may be held constant or projected in some fashion. This requires the basic data for two recent dates, preferably 5 years apart to match the rates for 5-year age groups: with data for three dates, two sets of ratios may be computed and projected. With the development of additional types of data and techniques, it has become possible to apply the cohort-component method directly in the projection of socioeconomic characteristics. This method involves projecting the population (usually distributed by age), subdivided into those having the characteristic and those lacking the characteristic, to successive future dates on the basis of probabilities of attaining or losing the characteristic. The development of tables of working life, school life, nuptiality, etc., representing extensions of the concept of the life table to these special areas and providing probabilities of labor force entry and withdrawal, school entry and withdrawal, and marriage, widowhood, and divorce, etc., and the development of interest in elaborate demographic-economic models, have

given impetus to the cohort method of analysis and projection of socioeconomic characteristics. In fact, in recent years, much research has been devoted to the development of national demographic-economic models, involving integration of projections of population and several related socioeconomic characteristics.

As in the case of projections of total population, those of the socioeconomic characteristics have been increasingly viewed as indications of probable future developments resulting from various stated assumptions rather than as predictions of future population. Accordingly, several series of projections are ordinarily prepared on the basis of alternative assumptions varying particularly for the level of the characteristic in question; these would show the greatest variation and potentially greatest impact on the results. It is useful also to include a series "holding present rates constant" and to carry out a component analysis of prospective changes in terms of the demographic (i.e., age and sex) and socioeconomic factors (e.g., school-enrollment rates or worker rates).

Accuracy of Projections

In projections of socioeconomic characteristics, variations and trends in the relevant socioeconomic variable (school enrollment, economic activity, household status, urban-rural residence, etc.) as well as in fertility, mortality, and migration have to be taken into account. The accuracy of such projections is particularly affected by the age range to which the characteristic applies. For example, intermediate-term projections of enrollment (e.g., 5 to 15 years ahead) tend to be subject to considerably greater error than projections of labor force because the former depend upon projections of births over the next decade whereas the latter depend only upon projections of cohorts already alive at the base date of the projections.

Regional projections are subject to a much greater margin of error than national projections because of the greater uncertainty with respect to future population changes in small areas and the additional highly variable factor of internal migration. In regional projections of socioeconomic characteristics, regional variations and trends in the relevant socioeconomic variable as well as in fertility, mortality, and internal migration have to be considered.

POPULATION BY TYPE OF RESIDENCE

Introduction

Projections of the urban and rural sectors by age and sex may serve as the basis for designing realistic national plans relating to urban and rural development, particularly in the less developed areas. Furthermore, they may be useful in preparing, and in appraising the demographic implications of, projections of national population by age and sex, which are basic ingredients in a wide variety of projections needed for developing and implementing national plans.

The period covered by the projections which have been made in recent decades varies, extending occasionally more than 25 years ahead. The projections of urban and rural population for Jamaica prepared by Roberts in 1963 extended only to 1970, but those for Finland published by the Central Statistical Office in 1963 extended to 1990.[55] Projections of the urban

[55] G. W. Roberts, "Provisional Assessment of Growth of the Kingston–St. Andrew Area, 1960–70," *Social and Economic Studies* (Jamaica), 12(4):432–441. Dec. 1963; Finland. Central Statistical Office, "Alueellinen väestoennuste vuoteen 1990 saakka," by Tor Hartman. *Tilastokatsauksia* (Helsinki), 38(10):45–49. 1963.

and metropolitan population of the United States prepared by Pickard extend from 1960 to 2000.[59] Such long-term projections have been prepared in spite of the fact that projections of urban-rural population for periods longer than about 15 years are subject to extremely large errors and are quite speculative. Available evidence indicates that projections for geographic areas within a country become seriously unrealistic after a period of this duration, and urban-rural projections may be subject to even greater uncertainty.

The methods and problems of projections by type of residence are similar in many respects to those for geographic areas within countries, both being affected by internal variations in natural increase and by internal migration. Projections of population by type of residence involve, in addition, however, special problems of the availability of adequate basic data, both in terms of quantity, quality, and comparability over time, and the need to take account of additional factors of change (e.g., reclassification of areas).

Several methods of projecting urban and rural population call for comparable census data on urban and rural population, possibly by age and sex, for at least two censuses. Not only may the geographic limits of the urban territory change from census to census because of reclassification of areas and annexations, but basic definitions may be modified. It has been found impossible, therefore, in many cases to construct the urban population on a comparable basis for a series of censuses or to achieve a realistic definition in even a single census. Alternative definitions designed to improve the comparability of the data from one census to another for projection purposes divide the population into one or more size-of-locality intervals, say at 2,000, 10,000, or 20,000 population, or identify a specific list of places or class of places. Another criterion has been to separate the population into that residing in the principal urban agglomerations and the remainder, and, possibly, further to subdivide the latter into places above and below 10,000 or 20,000. Variations within the urban population may be so marked that a division into large cities (strictly urban) and smaller towns (semiurban) may usefully be made.

Methodology

The methods of projecting the urban and rural population may be classified into the mathematical methods, the ratio methods, and the analytic methods. The methods are generally viewed as varying in reliability and analytic usefulness in the order indicated.

Mathematical and Ratio Methods. — Simple mathematical devices for projecting the total urban and rural population have often been employed. The average annual rates of change in the rural population may show a pattern of sufficient regularity in the past to provide a basis for projection. The rural population often has, or may be assumed to have, a steady, low or negligible rate of growth, whereas the urban population may grow rapidly, absorbing all or nearly all the increase of the national population. Given projections of the total population, it becomes a simple task to derive projections of the urban and rural population on this basis.[60] Alternatively, the relation between the urban or rural and total rate of growth may be

projected.[61] Other mathematical curves may be employed in projecting the urban or rural population of a country. For example, Davis used a logistic curve to project the urban population of India to the year 2000.[62]

Projections of total urban and rural populations may be derived by a ratio method, for example, by projecting the trend of the percentage urban in the national total on some mathematical or other basis and applying the projected proportions to the available national totals for future years.[63] For example, the average annual rate of change in the proportion urban observed in one or more previous intercensal periods may be assumed to remain constant, or to avoid the implication of indefinite increase in the proportion, one may assume that the rate would tend gradually to zero by some terminal date (i.e., that the proportion would eventually reach stability). The United Nations used a logistic curve to project the percent urban for the populations of the major world regions to 1980, setting a limit of 70 percent for any region.[64]

Two ratio techniques of projecting the age-sex distribution of urban-rural population may be described. Both require prior projections of the national population by age and sex and of the total urban and rural population. The first employs the ratio of the percent of the urban (rural) population in each age-sex group to the corresponding percent for the total population at the last census.[65] It is then assumed that the ratios of the percents will not change, will change according to the intercensal "trend," or will change in some other manner, as for example, that the ratios will approach unity by some distant date. The projected ratios are then applied to the projections of the percent distribution of the population by age for the country as a whole to obtain projections of the percent distribution by age for the urban (rural) population. The percent distributions for the urban and rural populations must then be proportionately adjusted to 100 percent for all ages before being applied to projections of the total urban and rural populations, respectively. Proportionate adjustments of the urban and rural figures at each age to the national totals at that age, and of the age-sex figures for the urban and rural populations to the urban and rural totals, respectively, follow in order ("two-way raking"). This cycle of adjustments is repeated until full agreement with rim totals is obtained (app. C).

Two-way raking is also applied in the second ratio procedure, designated the Registrar General's Square Table Method. Here, the total male population and the total female population

[59] Jerome P. Pickard. *Dimensions of Metropolitanism*, Research Monograph 14. Urban Land Institute, Washington, D.C., 1967.

[60] See, for example: United Nations, Economic Commission for Latin America, *Some Aspects of Population Growth in Colombia* (E/CN.12/618), November 1962, Santiago, Chile, esp. p. 27; César A. Peláez, *Proyección de la población urbana y rural de la República de Panamá*, Vol. I, *Proyecciones de población*

1950–1980 (Projection of the urban and rural population of the Republic of Panama, Vol. I, *Projections of population, 1950–80*) (E/CN.CELADE/C.19), Latin American Demographic Center, Santiago, Chile, 1964; and Berto Cataldi D., *La situación demográfica del Uruguay en 1957 y proyecciones a 1982 (The demographic situation of Uruguay in 1957 and projections to 1982)* (E/CN.CELADE/C.15), Latin American Demographic Center, Santiago, Chile, 1964.

[61] See, for example, National Council of Applied Economic Research (New Delhi, India), *Long-Term Projections of Demand for and Supply of Selected Agricultural Commodities, 1960–61 to 1975–76*, Commercial Printing Press Limited, Bombay, 1962, especially pp. 28–30.

[62] Kingsley Davis, "Urbanization in India: Past and Future," in: *India's Urban Future*, Roy Turner, ed., University of California Press, 1962, pp. 3–26.

[63] See U.S. Senate, Select Committee on National Water Resources, "Population Projections and Economic Assumptions," *Water Resources Activities in the United States*, Committee Print No. 5, 1960, pp. 27–35; and Margaret J. Hagood and Jacob S. Siegel, "Projections of the Regional Distribution of the Population of the United States to 1975," *Agricultural Economic Research*, 3(2):41–52, April 1951.

[64] United Nations, "World Urbanization Trends, 1920–60" (an interim report on work in progress), Working Paper No. 6, Inter-Regional Seminar on Development Policies and Planning in Relation to Urbanization, University of Pittsburgh, Pittsburgh, Pa., Oct. 24–Nov. 7, 1966, pp. 37–40.

[65] United Nations and Government of the Philippines, *Population Growth and Manpower in the Philippines*, United Nations, Series A, Population Studies, No. 32, 1960, Appendix C and pp. 50–51.

at each age at the first projection date are distributed by urban and rural residence according to the urban-rural distribution at the last census, and the results over all ages are then adjusted proportionately to the previously estimated total urban and rural population. The resulting figures are again subjected to the same adjustment cycle until complete reconciliation with assigned marginal totals has been achieved. These final results are then utilized as a basis for deriving the projections at the next projection date by the same procedure and so on.[66]

These two ratio techniques may be effectively merged into a single procedure. Consider the case where data on age, sex, and urban-rural residence are available for only one census and it is assumed that the rural-to-total or urban-to-total ratio of the percents in each age group will remain unchanged. The final figures for the first projection year, after raking is completed, may then be employed as a basis of calculating the ratios of the percents to be initially used in the second projection year, and so on.[67]

The mathematical and ratio methods have the advantage of simplicity of computation. On the other hand, relatively mechanical methods of this kind do not take into account the demographic or socioeconomic correlates or components of the population changes, and hence provide no information on these variables. Nevertheless, the results can usefully serve both as approximations of the future urban-rural distribution and as bases for improving projections of socioeconomic characteristics.

Analytic Methods. — The so-called analytic methods attempt to take explicit account of one or more of the demographic or socioeconomic variables to which change in population by type of residence is related. They may be expected to provide more meaningful and useful results, therefore. There is no evidence that such projections will be more accurate, however.

These methods are widely used in regional projections, but to date they have received scant consideration in projecting the population by type of residence. For this discussion of analytic methods applicable to urban and rural populations, we select the following: ratio method based on agricultural labor force; the cohort-component method; the area-component procedure; and use of migration and population models.

The **ratio method based on agricultural labor force** involves the relationship between the proportion of the labor force in agriculture and the proportion of the population in rural areas, by age and sex. (The relationship between the proportion of the population dependent on agriculture and the proportion of the population in rural areas, and between the proportion of the labor force in nonagricultural industries and the proportion of the population in urban areas, or male population only, are also applicable, the former being the most satisfactory alternative.) Prior projections of total population by age and sex and of labor force by age and sex, in agriculture and in nonagricultural industries, are required. Methods of developing the latter

projections are discussed in a later section of this chapter. The absolute or percent difference between the proportion of the labor force in agriculture and the proportion of the population rural can be used to convert the projected proportion in the agricultural labor force to a projection of the proportion rural, as illustrated in two U.N. studies.[68] This difference should tend to show a considerable regularity or even stability. Projections of the urban population are derived by subtraction of the rural population from the total population. The procedure can be applied to estimate the age distribution as well as the total population.

Growth of the urban and metropolitan populations may be viewed as consisting of (1) the overall growth of national population leading to increased density and reclassification of areas; (2) balance of fertility and mortality in the area; and (3) internal migration from rural or nonmetropolitan areas. The **cohort-component method**, which is widely used for regional projections, involves carrying forward the current urban and rural populations (or alternative areas) by age and sex, to future dates, by allowing separately for fertility, mortality, migration, and changes due to reclassification.[69] The method, therefore, poses the difficult problem of securing recent data, by age and sex, on the above components. A simpler cohort method involves carrying forward the current urban and rural populations by age and sex to future years, by assuming that the cohort-change rates of a previous intercensal period will remain unchanged; i.e., that age-specific vital rates, migration rates, rates of urban-rural reclassification, and patterns of error in census data will be the same in future decades as in the past intercensal period.[70]

Data on mortality by type of residence are usually not available. Urban-rural variations in mortality may be reflected in regional variations shown by the registration system or a sample survey. An indication of the existing urban-rural differences in mortality may possibly be secured from data for the principal city and for the remainder of the country, or for the principal political subdivisions grouped as primarily urban or rural. Although interarea differences in mortality exist, they are usually not large enough to have any substantial impact on the projections. The projection of the mortality rates for residence areas to future dates may be done in proportion to prior projections of overall mortality.[71] Whether urban and rural mortality will converge with the passage of time, and what the rate of convergence will be, are matters of judgment. The choice of rates would be aided by the availability of model urban-rural life tables related to various levels of overall mortality.

The estimation of current differences in urban and rural fertility is somewhat more important for component projections than differences in mortality, but the task may be easier. Even if the registration system does not provide reliable indications

[66] United Nations, Economic Commission for Latin America, *Application of the Registrar-General's Square Table to a Population Projection with Respect to Urban and Rural Population by Groups of Sex and Age*, Santiago, Chile, Jan. 25, 1961. Applications of the method are given in: Peláez, op. cit.; Cataldi, op. cit.; and Jorge Somoza and Luis Llano, *Proyección de la población de Bolivia (Projection of the population of Bolivia)*, E/CN.CELADE/C.9), Latin American Demographic Center, Santiago, Chile, 1963.

[67] Applications of this procedure to the countries of Central America are illustrated in: United Nations Economic Commission for Latin America, *Human Resources of Central America, Panama, and Mexico, 1950–80, in Relation to Some Aspects of Economic Development*, by Louis J. Ducoff, 1960, Appendix B and pp. 37–45.

[68] See, for example, United Nations and Government of the Philippines, op. cit. In this study the projection is based on the percent difference between these proportions obtained in a single observation. See also Ducoff, op. cit., for an application of this method to the countries of Central America. Here the absolute differences in the proportions in the last census are employed for each country.

[69] James D. Tarver, *A Component Method of Estimating and Projecting State and Subdivisional Populations*, Miscellaneous Publications MP–54, Oklahoma State University, Agriculture Experiment Station, 1959.

[70] See C. Horace Hamilton and Josef Perry, "A Short Method for Projecting Population by Age from One Decennial Census to Another," *Social Forces*, 41(2):163–170, Dec. 1962.

[71] See, for example, U.S. Bureau of the Census, *Current Population Reports*, Series P–25, No. 415, "Projection of the Population of Metropolitan Areas: 1975." Jan. 31, 1969.

of the differences, the census may be a valuable source. Census ratios of children to women represent a ready basis for measuring the differences in the level of recent urban and rural fertility, although such ratios are affected by other factors, including differential net undercounts of children and women and migration of families between urban and rural areas after a child is born.

Projection of urban and rural fertility presents a more difficult task than for mortality since the range of effective uncertainty is greater. Once again, urban and rural rates may be extended in relation to the fertility of the whole country. In addition to historical analysis, an analysis of urban-rural differences for other countries at various stages of economic development should prove useful in this regard.

The most difficult and critical problem is the measurement of the components of net migration and of area reclassification. Information on the component of net migration may possibly be provided directly by the census or by a sample survey. Census data on place of birth, or place of residence at some previous date, of the population classified by current urban-rural residence, age, and sex, are occasionally tabulated in terms of prior urban-rural residence. A more general practical possibility is to obtain estimates of net migration as a residual by removing natural change from total change in the urban and rural populations, by age cohorts, between the last two censuses.[72] There are a number of difficult conceptual and technical problems in applying this residual method. Such a residual ordinarily would represent a combination of net migration and reclassification of population from rural to urban. The effects of reclassification can be excluded if "urban" is defined not by a size criterion but as referring to a specified list of areas such as "county seats." On the other hand, it may be useful to measure the two elements in combination in the past so that joint allowance for them can be made in the future. "Census survival rates" may serve as a useful tool for estimating past intercensal migration between urban and rural areas by the residual method. Census survival rates are ordinarily based on national data by age and sex from two censuses and theoretically would eliminate from the total change by age cohorts, for urban and rural areas, the parts due to mortality and changing net undercounts in the two censuses, leaving estimates of net migration, combined with the population reclassified, for each sector (ch. 21).

The component method can easily grossly under- or overestimate the prospective urban population while over- or underestimating the prospective rural population, depending on the assumptions made with respect to the various components. Furthermore, only infrequently can it be manipulated to take explicit account of future reclassification of areas from rural to urban, the effect of which may vary considerably from one period to another. The method has severe limitations if comparable data by residence are not available from two previous censuses for estimating net migration, and it cannot allow for any future change in the definition of urban and rural areas.

In the **area-component method**, projections are prepared for individual urban agglomerations, individual urban districts, or urban parts of provinces or regions; these are then combined to represent the total urban population or a specific segment

of it.[73] The entire country may be divided into "economic areas"—small, economically integrated areas akin to metropolitan areas or type-of-farming areas—which may be identified as primarily urban or rural and for which separate population projections are prepared.[74] To project the population of each area, various methods, including particularly the cohort-component method (or its simpler version based on local intercensal cohort-change rates) and various types of economic analysis, may be used.

Other approaches to be considered for projecting urban-rural population (or metropolitan-nonmetropolitan population) are based on the development of a model of migration to urban and metropolitan areas, or of a **demographic-economic model** to predict population growth in those areas.

The differences in the economic condition and structure of urban and rural areas, including differences in job opportunities and per capita income, basically determine the flow of migrants from one type of area to the other. It seems useful, therefore, to consider a method of projecting the total or working-age population of urban and rural areas, or rural-urban migration, that takes explicit account of the difference in the relative economic status of urban and rural areas. Studies are needed to determine the predictive value of various economic and other indicators of urban-rural migration. Regression techniques applied to spatial units (e.g., urban-rural parts of each principal subdivision) may be useful here. The demographic trend in the area of origin (usually rural areas) may also be quite important.

One indicator of the number of persons who will migrate from farms, a major factor in rural population changes for many countries, is the replacement ratio, the excess of the number of persons reaching working age over the number leaving through death or retirement. Similarly, the relative change between the population aged 20 to 29 at one census date and the expected population 20 to 29 at a subsequent date (allowing only for death since the preceding census), or the ratio of the population 20 to 29 to the population 10 to 19 in the same census, may be used to reflect the change in the potential supply of farm labor.

Grauman has suggested a model of urban-rural population change which would, in effect, incorporate migration into a system of stable population models.[75]

Projections for Regions

The value of urban-rural projections for planning uses is enhanced if figures are developed for each major geographic division of the country. Projections of urban-rural population by geographic subdivisions, even by age and sex, have been prepared from time to time.[76] Such projections may be derived by various simple mathematical and ratio methods which may

[72] See Juan C. Elizaga, *Tasas de migración rural-urbana por edad* (*Urban-rural migration rates by age*), (E/CN.CELADE/A.7), Latin American Demographic Center, Santiago, Chile, 1963.

[73] See, for example: Pickard, op. cit., and Ira S. Lowry, *Metropolitan Populations to 1985: Trial Projections*, Rand Corporation, Santa Monica, Calif., 1964.

[74] Such a project was undertaken in the Netherlands by the Committee for Regional Planning Forecasts, Government Physical Planning Service. Two reports were issued, one in 1952 and a second in 1959. See also L. H. J. Angenot, "Regional Population Projecting in the Netherlands," in United Nations, *Proceedings of the World Population Conference, Rome 1954*, Vol. III, New York, pp. 1–10.

[75] John V. Grauman, "Development of a Model of Rural-Urban Population Change, with Relevance to Latin America," *International Population Conference, New York, 1961*, Vol. I, International Union for the Scientific Study of Population, London, 1963, pp. 448–457.

[76] See, for example: United Nations, Economic Commission for Latin America, *Some Aspects of Population Growth in Colombia*.

require projections of national population by age and sex, the total population of regions, and national urban and rural population. For example, the total urban and rural population of each region may be projected on the basis of the growth rate in the previous intercensal period; the results are then adjusted to the national urban and rural totals and to the total population of each region. Alternatively, the ratio method may be employed to distribute projections of total urban and rural population by regions in the first projection year on the basis of the census counts, and then the square table method may be applied to effect agreement with rim totals, etc. More generally, any of the methods applicable for projecting the urban and rural population nationally may be adapted for regional projections, although, on the whole, mathematical and ratio methods may be most conveniently used for this purpose.

HOUSEHOLDS AND FAMILIES

Introduction

Projections of households and families are required for many uses, particularly those which depend on information regarding future numbers of consumer units. A principal use of projections of households is in deriving projections of housing needs, but they are also needed to anticipate demand for other consumer goods and services required by households and families as a unit.

Because of the doubling-up of families, married couples, and individuals within households (so-called subfamilies and secondary families, and individuals), the inadequacies of the present supply of housing, and the uncertainties with respect to future changes in household relationships, it is desirable to distinguish projections of the "actual" (probable) and "potential" (desirable) number of households and families and their characteristics. The line between these two types of projections is not a precise one. In addition to demographic factors, the number of households will depend on the extent to which the housing supply and costs, income levels, and local practices permit doubled-up individuals and family units to spread out into separate units. The different procedures for projecting the number of households and families may vary according to whether they represent essentially extensions of past trends, reproducing the current characteristics of households and the current pattern with respect to the density of occupancy of housing units by families and individuals, or whether various norms relating to the size and composition of households are introduced on the basis of assumptions regarding more favorable conditions of the supply and cost of housing, family income, and other factors.[77] If, as a result of a shortage of housing, inability to afford the available housing, or the custom of living in joint families, many families or individuals live in the households of others, "actual" projections of

households would be derived by maintaining or nearly maintaining the present population-to-household ratios. These might also represent a kind of minimal projection of the number of households or families to be expected. The potential number would assume such highly favorable conditions as to allow a separate household or dwelling unit for most or all "individuals" and nuclear family units—much more favorable conditions than ordinarily exist. The impact of this distinction is far greater in the case of the economically underdeveloped countries than in the case of the economically developed ones.

Methodology

Methods at several levels of refinement or elaboration are considered here, as for other types of projections, without our being able to specify the precise differences in the accuracy of the results. The crude methods depend on the overall level of households and fail to take into account many of the factors which influence the growth of the number of households or to provide projections of household characteristics. The simplest procedure is to assume that the number of households would increase at the same rate as the total population; that is, that the current proportion of the population living in private households and the average size of households would remain unchanged. Since household heads are usually adults, the change in the population 18 years of age and over should be a better basis for making rough projections of the number of households than the total population.

The more refined methods of projection take into account the composition of the population by age, sex, marital status, relation to the household head, and other variables which have an important effect on changes in the number and characteristics of households.

Guides for preparation of household projections were published by the United Nations in connection with its studies of the measurement of housing needs.[78] Since the detailed methods of household projections cannot be used for most countries because the necessary data are lacking, the United Nations guides have suggested the use of the simpler methods to fill the gap.

Component Methods. — One class of methods develops projections of the total number of households in terms of components; these components may be the principal types of households, the components of change in households or families, or both. Only limited use of data on the age composition of the population is made. The projections of households published by the Economic Council of Canada in 1967 illustrate this type of method.[79]

Use of Age-Specific Headship Rates. — A basic element in the most common class of methods is the use of projections of population distributed by age and sex. In these methods the pertinent variables are projected and applied by age and sex; hence, the effects of changes in age and sex distribution are fully taken into account. The general procedure consists of applying to the population, projected by age and sex, various estimating ratios which are related to marital, household, and family status, such as the proportion of the population in each marital category and the proportion who are household heads

[77] H. V. Muhsam, "Estimates and Projections of Numbers and Characteristics of Families and Households in Relation to Housing Requirements" (E/CN.9/CONF.1/L.15), document prepared for United Nations Seminar on Evaluation and Utilization of Population Census Data in Latin America, Santiago, Chile, Nov. 30–Dec. 18, 1959; and J. S. Siegel, "Demographic Information Required for Housing Programs with Special Reference to Latin America" (ST/ECLA/CONF.9/L.12), Demographic Center for Latin America, Santiago, Chile, document prepared for United Nations Seminar on Housing Statistics and Programs, Sept. 2–25, 1962, Copenhagen, Denmark. (See also the Spanish version of the latter document, "Informaciones demográficas para la formulación de programas de vivienda con especial referencia a América Latina," *Estadística*, 21(79):227–281, June 1963.)

[78] United Nations, Economic Commission for Europe, *Techniques of Surveying a Country's Housing Situation, Including Estimating of Current and Future Housing Requirements* (ST/ECE/HOU/6), Geneva, 1962; and United Nations, Statistical Office, *Methods of Estimating Housing Needs* (E/CN.11/ASTAT/HSP/L.4), 1963.

[79] Wolfgang M. Illing, *Population, Family, Household, and Labor Force Growth to 1980*, Staff Study No. 19, Economic Council of Canada, Sept. 1967.

("headship rates"). The total number of households is obtained by summing the number of heads obtained by age and sex. This procedure requires only projections of adult population by age and sex at the projection dates although projections of total population are required to compute the projected average size of household.

When census data on households or household relationships are lacking, projections of households may be based on projections of the population classified by marital status only, in addition to age and sex. Future distributions by marital status may be derived by holding constant the age-specific proportions at the last census or survey date or by extrapolation of past trends in these proportions. The proportions in each marital class in each age may be projected on an age basis [80] or on a cohort basis.[81]

As suggested, with the use of information on marital status alone, one tends to develop projections of the potential number of households rather than of the "actual" number. More realistic projections of households are usually obtained when census data on the proportions of household heads in each marital class are employed. Even so, the assumptions involving data on household heads may be designed to represent either the actual future number of households or the potential number. To derive the potential number, the proportion of heads are so selected as to represent all the clusters of individuals who would occupy separate housing units under relatively favorable conditions—e.g., nuclear families, most individuals, some groups of unrelated persons, etc. It is particularly appropriate to hold headship rates of this kind constant at the recent observed level.[82]

The usual procedure is to project the "actual," or probable, number of households on the basis of census age-specific headship rates. As suggested earlier, use of age-specific headship rates permits taking account both of changes in the proportion of heads by age and sex and changes in the size and age-sex distribution of the population. The proportions may be employed without use of data on marital status, if such data are lacking or in order to simplify the procedure. On the other hand, if data on marital status by age and sex and on household heads by age, sex, and marital status are available, it is preferable to take advantage of these data, i.e., to apply age-sex-marital-status specific headship rates to projections of population by marital status, age, and sex. (If data on heads by age and sex or by age, sex, and marital status are lacking, schedules of headship rates by age may be "borrowed" from another country. The model schedules should first be tested against the population by age from the census for the area under study and adjusted so as to yield the total number of households shown by the census, if possible.)

In a still more elaborate form of the age-specific rate method, data on marital status, by age and sex, and data on various categories of family and household status, by age and sex of head, are combined, and the pertinent age-specific proportions and rates are projected or held constant as seems appropriate, to provide projections of married couples and of households and families by type. Depending on the data available and the procedure employed, one could obtain, for example, the number of households by age and sex of head; the number of households headed by families, by type of family (e.g., husband-wife, other male head, female head); the number of households headed by individuals, by sex; the number of (secondary) families who live with other (primary) families in the same household; the number of married couples and nuclear families, by whether they are living in their own households or in the households of others, etc. With such a design, not only are projections of "actual" households and families obtained but so also are the elements for a variety of projections of potential households.

This type of projection was prepared by the U.S. Bureau of the Census in 1963.[83]

Multiple series have been recognized as useful here, as for other types of demographic projections, in order to take account of the uncertainty in the components entering into the projection design, to serve the different needs of projections, and to permit analysis of the components of change. For areas of low mortality and little net immigration, the areas for which projections of households have mostly been prepared, a single series of population projections suffices since the possible variation in future fertility does not affect the population projections entering into the calculation of households for the first 15 to 20 years. On the other hand, the possible future variation in the proportion of the population in different marital and relationship categories, especially the proportion of single persons and the proportion of married couples having their own households, would seem to indicate the need for more than one series of assumptions in projecting these categories.

In addition to the principal series of household projections, it is advantageous to develop a series with constant headship rates even if it is unrealistic, so as to permit analysis of future changes in the number of households in terms of the part due to the size and age-sex distribution of the population and the part due to assumed change in propensities for household formation and dissolution. The additional detail on household and family units which is provided in the more elaborate projections may permit the user to derive several additional series representing a range of possibilities for potential households.

Extension of the Cohort Method.—We have already alluded to the use of the cohort approach in projecting the percentages in each marital category when making projections of marital status. The cohort approach may be extended to include the projection of the percentages of heads, when data on heads of households by age are available for a series of census dates. This general procedure is directly applicable to series like percent married or percent heads which are essentially cumulative

[80] See United Nations, Economic Commission for Europe, *Techniques of Surveying a Country's Housing Situation.*

[81] See, for example: Roland Pressat, "Un essai de perspectives de ménages" (Illustrative household projections), *International Population Conference, Vienna, 1959,* International Union for the Scientific Study of Population, pp. 112–121; U.S. Bureau of the Census, *Current Population Reports,* Series P–20, No. 90, "Illustrative Projections of the Number of Households and Families, 1960 to 1980," Dec. 29, 1958; and W. S. Hocking, "A Method of Forecasting the Future Composition of the Population of Great Britain by Marital Status," *Population Studies* (London), 12(2):131–148, Nov. 1958.

[82] See, for example: William Steigenga, "Family Structure, Age Composition, and Housing Needs," *International Population Conference, New York, 1961,* International Union for the Scientific Study of Population, Vol. 1, pp. 243–250; and Gerard Calot, "Perspectives du nombre des ménages de 1954 a 1976" (Projections of the number of households from 1954 to 1976), *Etudes statistiques Supplement trimestriel du Bulletin mensuel de statistique* (Paris), 12(2):149–15?. April–June 1961.

[83] U.S. Bureau of the Census, *Current Population Reports,* Series P–20, No. 123, "Interim Revised Projections of the Number of Households and Families: 1965 to 1980," April 11, 1963. See also Paul C. Glick, *American Families,* New York, John Wiley and Sons, 1957, pp. 164–191; and U. S. Bureau of the Census, *Current Population Reports,* Series P–20, No. 90.

by age. It has the advantage that it may contribute to the avoidance of proportions single, married, or heads which imply unreasonable rates of marriage or household formation.

An important shortcoming of the procedures for preparing projections of households described so far is their failure to provide information on gross household formation and dissolution. Much of the practical interest in short-term projections of households, particularly annual projections, centers on gross changes rather than net changes. (For example, changes in the market for household durable goods reflect the new households formed, not the net change in households, which is determined in part by households dissolved by death, divorce, etc.) This type of data would be provided by a cohort method involving estimation of the components of change in households and paralleling more closely the stages in the life cycle of the household—formation of the household, change in its composition, dissolution through death or divorce, etc.

Projections by Number of Members

Largely because of the importance of the size of dwelling units as an element in housing programs, some projections of households and families have included detail by size or number of members. The simplest procedure is to distribute the total number of projected households by size in accordance with the size distribution of all households in the last census. A more refined procedure would employ separate size distributions by age and sex of head, marital status of head, etc.[84] This procedure takes advantage of the close relation between the age of the head and the size of the household. It may also be possible to take account of such additional elements as past trends in size distribution and the prospective fertility trends assumed for the general population.

Regional Projections

A simple method of projecting the number of households for subnational areas involves the assumption that the number of households will increase at the same rate as the population between the base date and the projection date, i.e., that the average size of household will remain unchanged. It is usually more reasonable, however, to project the average size of household in the area than to hold it constant.

The more complex and refined procedures involve the application of age-specific headship rates to the population projected by age and sex. (If such rates are known only for the country as a whole, they can be "borrowed" for use in each political subdivision.) The projections of households for the States of the United States published in 1968 were prepared by applying age-specific headship rates and age-specific proportions relating to household status and relationship to the projected population by age, and adjusting the resulting preliminary projections in each category to the corresponding national figures.[85] If the appropriate data on both marital status and relationship are available, projections of the distribution of the population at each age by marital status can be prepared first and, then, the projected marital classes can be subdivided into heads and nonheads. An application of this procedure was made by Walkden for areas within Great Britain.[86]

In sum, once projections of total population or population by age and sex have been prepared, the possible procedures for projecting households for principal political subdivisions or urban-rural areas parallel those described for the country as a whole. The need for alternative series is even greater for regional projections of households than for national projections of households in view of the greater number of highly variable factors involved, particularly internal migration.

SCHOOL ENROLLMENT

Introduction

Projections of the number of children who will be enrolled in school are needed to formulate educational policies and plan educational programs and, specifically, to plan for needed schools, classrooms, and teachers. Since almost all children in the appropriate ages for elementary school attendance should be attending school, projections of the total population of elementary school age, in relation to the number expected to attend, are also useful in determining needs, particularly in the less developed areas. In addition, projections of school enrollment rates can be used in preparing projections of labor force participation rates since the two sets of rates are inversely related, especially at certain age groups.

Projections of the educational attainment of a population are also needed for national planning. The present and prospective educational attainment of a population influences many aspects of its development. Social and economic relationships in a population change in many ways as its educational level changes. In underdeveloped countries, it is important to set national priorities which, among other things, take into account particularly the extent of illiteracy and the problems created or intensified by it. There has been greater emphasis by national governments on projections of school enrollment than on projections of educational level or of illiteracy. This situation might have resulted from the fact that enrollment presents more urgent problems, i.e., the need for schools, teachers, and classroom materials.

Methodology

This discussion of the methods of making projections of school enrollment covers the two principal methods, the age-specific rate method and the cohort method. A simple component method can also be employed (i.e., using new entrants, dropouts, deaths, etc.) but is not usually practical. In addition to total enrollment, the age and grade of enrollees is considered. A case study of methods of enrollment projections applied to New Zealand has been published by UNESCO.[87]

Age-Specific Enrollment Rate Method.—The most widely used procedure of preparing projections of enrollment has been to employ assumed age-sex-specific enrollment rates (or proportions of the population enrolled in school at each age) in combination with the projected population by age and sex. The assumption relating to enrollment rates may be quite simple. Some studies have, in fact, assumed the continuation of

---

[84] H. V. Muhsam. "Population Data and Analyses Needed in Assessing Present and Future Housing Requirements" (E/CN.9/CONF.2/L.10), paper prepared for the United Nations Seminar on the Evaluation and Utilization of Population Census Data in Asia and the Far East, Bombay, 1960.

[85] U.S. Bureau of the Census. *Current Population Reports,* Series P-25, No. 387. "Projections of the Number of Households, by States, July 1, 1970 and 1975," Feb. 20. 1968.

[86] A. H. Walkden, "The Estimation of Future Numbers of Private Households in England and Wales," *Population Studies* (London), 15(2):174–186, Nov. 1961.

[87] UNESCO, *Methods of School Enrolment projection,* by E. G. Jacoby, Educational Studies and Documents No. 32, Paris, 1959.

current or recent enrollment rates,[88] or included a series with this assumption.[89] Commonly the series on enrollment has been based on enrollment rates which have been projected. Past trends in the rates may be assumed to continue as observed or in a modified fashion. In those instances where the enrollment rates approach 100 percent and extrapolation using a growth rate could cause these enrollment rates to exceed 100, nonenrollment rates, or the complements of the enrollment rates, may be projected; the final rates cannot then exceed 100.0 percent.

Alternatively, the rates in some "advanced" section of a country may be selected as terminal values for the country as a whole, or the current rates from a relatively developed country may be borrowed for use as terminal values for a relatively undeveloped country.

Current legal requirements and practices with respect to the ages or grades of school attendance have ordinarily been incorporated into projections of school-age population and school enrollment. The possibility of developing projections of enrollment which incorporate new norms representing more inclusive age and grade spans, and even hypothetical enrollment rates for the base year, needs to be considered if provision is to be made in the future for upgrading the current level of school services.

For the economically developed countries, where nearly all children of statutory school age are enrolled in school, projected trends in the number of children of statutory school age give an approximate indication of the expected trend in the number of pupils enrolled in the corresponding grades. On the other hand, because only a portion of the population of high school age is enrolled in high school and this percentage is changing, changes in the population of high school age do not adequately reflect changes in the numbers enrolled in high school. For the less developed countries, the need to take careful account of enrollment trends is imperative because of the often considerable gap between enrollment and population in the school ages.

The special need for alternative series on enrollment arising from the heavy dependence of such projections on projections of births should be recognized. At the same time it is desirable to take account of the possible variation in future enrollment rates by incorporating alternative sets of rates in different series of enrollment projections. The desirability of calculating a series assuming a continuation of recent enrollment rates for analytic purposes, in combination with the principal series, should also be recognized. This approach lends itself readily to analysis of the total projected change in enrollment in terms of a demographic component and an enrollment-rate component. The difference between the series assuming constant enrollment rates and the series assuming variable rates represents the change in enrollment which would result from the change in enrollment rates. This valuable type of analysis has not often been carried out.[90] Typically, in the developed countries, where enrollment rates are high, the bulk of the prospective increase is attributable to population changes, but in the less developed areas changes in enrollment rates may play a significant part.

The availability of alternative sources of data has also served

as a basis for alternative series. Projections based on the administrative statistics of the school system may be combined with projections based on census data or data from a national survey.[91]

Enrollment rates have also been projected on the basis of analytic studies relating changes in enrollment rates to changes in other socioeconomic variables. Socioeconomic factors which may affect rates of school enrollment include urban-rural residence, labor force status, marital and family status, and military service. Changes in socioeconomic composition may be allowed for by separate projection of rates for the component categories. For example, the Economic Commission for Latin America projected urban and rural rates of enrollment separately in its 1962 study of Colombia.[92] Among the special factors proposed as a basis for projecting college enrollment is the educational attainment of the fathers.[93] Some data on intergenerational changes in education are available in the United States for testing this approach.[94] A second approach would employ survey data relating to the college plans of youths, particularly high school seniors. Again, some pertinent data are available in the United States.[95]

When projections of school enrollment are made by the enrollment-rate method, they rarely contain detail on grade or school level. To obtain such figures, one procedure is to prepare projections by age by the enrollment-rate method first, and then to distribute the projected total enrollment at each age by grade or school level on the basis of recent census, survey, or administrative data (holding the distribution constant or extrapolating it).[96] Alternatively, projections of enrollment for broad school levels at each age may be calculated by the use of age-level enrollment rates (i.e., in relation to total population at each age) and total enrollment at each age may then be derived by summation.[97] It is useful to match this total with the total enrolled at each age computed directly by projection of the overall age-specific enrollment rate.

Cohort Method. – As suggested, projection of age-specific enrollment rates on a period basis in the framework of an enrollment-rate method has been the predominant method of making projections of school enrollment. For the highly developed countries at the college level, and for the less developed countries at all school levels, the possible variation in enrollment rates renders projection of enrollment rates a task fraught with great uncertainty. Lines of further investigation include analysis of enrollment rates on a cohort basis and possible application of a cohort-component method or a grade-cohort method of projection. The cohort-component method is particularly useful for deriving enrollment projections by age as well as for providing separate information on entries into and withdrawals from school by age. Here, one begins with a distribution of persons enrolled by age and carries this population forward by use of age-specific rates of net school accession and net school withdrawal, separate rates possibly being

[88] National Planning Council of Ceylon, The Planning Secretariat, *The Ten-Year Plan*, Colombo, Ceylon, 1959, pp. 466–472.

[89] See, for example, Hector Gutierrez R., "Proyección de la población escolar de Chile, 1957–82" (Projection of the school population of Chile, 1957–82), (E/CN.9/CONF.1/L.22), contribution to the United Nations Seminar on the Evaluation and Utilization of Population Census Data in Latin America, Santiago, Chile, Nov. 30–Dec. 18, 1959.

[90] But, see, for example, U.S. Bureau of the Census, *Current Population Reports*, Series P-25, No. 232, "Illustrative Projections to 1980 of School and College Enrollment in the United States," June 22, 1961.

[91] Gutierrez, op. cit.

[92] United Nations, Economic Commission for Latin America, *Some Aspects of Population Growth in Colombia*.

[93] Louis H. Conger, "College and University Enrollment Projections," *Economics of Higher Education*, U.S. Department of Health, Education, and Welfare, 1962.

[94] See, for example, U.S. Bureau of the Census, *Current Population Reports*. Series P-20, No. 132, "Educational Changes in a Generation: March 1962," Sept. 22, 1964.

[95] See, for example, U.S. Bureau of the Census and U.S. Department of Agriculture, Series Census-ERS (P-27), No. 32, "Factors Related to College Attendance of Farm and Nonfarm High School Graduates: 1960." June 1962.

[96] See, for example, U.S. Bureau of the Census, *Current Population Reports*, Series p. 25, No. 365, "Revised Projections of School and College Enrollment in the United States to 1985," May 5, 1967.

[97] See, for example, Gutierrez, op. cit.

applied for dropouts and deaths. A set of net accession rates and net withdrawal rates may be derived by taking the relative difference between enrollment rates at successive ages given in the census. A net accession rate is obtained for the ages where enrollment rates are increasing and a net withdrawal rate is obtained where enrollment rates are decreasing. Rates of net accession are applied to the population not-enrolled to derive new enrollees, who are added to the enrolled population, and rates of net withdrawal are applied to the enrolled population to derive dropouts and deaths, who are removed from the enrolled population.

The cohort method may also be employed to secure enrollment projections by grade.[98] In this procedure the number of enrolled persons by grade is carried forward to each subsequent calendar year by use of projected grade-retention rates or grade-progression rates, representing the proportion of children in a given grade who will advance to the next grade in the course of a year. A historical series of grade-retention rates may be developed on the basis of survey data or data from administrative records of the school system and then projected forward on an annual basis.

As with age-specific enrollment rates, in projecting grade-retention rates, the regular methods of extrapolation can be used, but if the rates are very high (i.e., near 100 percent) and the trend is one of rapid increase, conventional methods can produce values in excess of 100 percent. In this case an asymptotic equation which prevents getting a projected value of 100 percent or more can be used, or the complements of the grade retention rates (i.e., grade-dropout rates) can be extrapolated on a geometric basis.

Normative Projections and Applications

When enrollment rates are held constant or projected according to past trends, the projections of enrollment tend to represent the number of pupils who may actually be served, excluding, as in the past, those who do not attend because of lack of classrooms or teachers or for other reasons such as financial inability of parents. As suggested earlier, however, future enrollment rates may be designed to represent normative or target levels rather than actual or expected levels. The goals may represent the level of enrollment to be achieved by some future date as a consequence of implementing a particular national economic development plan.

Norms may also be set which would reflect not only future requirements but current requirements and deficits as well. The actual current ratio of pupils per teacher or classroom may represent a concession to a gross shortage of teachers or classrooms, reflected in part-time attendance, failure to attend, or crowded classrooms.

ECONOMICALLY ACTIVE POPULATION

Role of Projections

The widespread concern about economic planning and development both in the industrial countries and in the less developed countries has led to a considerable increase in interest and effort in preparing projections of economic characteristics of the population in recent decades. Projections of the economically active population are needed to give an indication of the number and characteristics of the workers who will be available for employment in future years so that appropriate plans and policies can be made. Information is needed not only regarding the total labor force but also regarding its composition by age, sex, occupational classes, industrial classes, and educational level. Labor force projections are also needed to indicate the number of jobs which the economy must make available and as a point of departure in making projections of the economy as a whole. Labor force projections must be matched by projections of manpower requirements for effective national economic planning, however.

The methodology of labor force projections has, in general, tended to assume that the rates of labor force participation or economic activity are independent of the general state of the economy, although in fact the number of persons seeking work does depend on the demand for labor, the availability of jobs, and on the need for a job within the family. The interaction between worker rates and the general state of the economy is recognized but often excluded from the projections, particularly for the developed countries, by an assumption of "full employment" or a large demand for labor. Since economic planning regards full employment as a desirable goal, this assumption would produce a principal type of labor force projection needed for such economic planning. Here, in fact, the important question may be the size of the prospective manpower shortage. The same practice of projecting worker rates without taking account of the general state of the economy and prospective change in it occurs in areas where full employment is far from being achieved. For less developed countries, worker rates may include considerable unemployment and underemployment, and the important question may be the size of the prospective manpower surplus. Projections of labor force have sometimes been modified to take account of the prospective demand for labor (see below).

The predictive character of labor force projections is limited for another reason. The very use of projections of the labor force in the formulation of future economic policy may render such projections invalid. After taking labor force projections into account, the national plan may call for greater production through higher participation rates and may, for example, exert pressure on housewives to work, with a consequent rise in participation rates.

Methodology

Population of Working Age. — The simplest approach to the measurement of prospective changes in the labor force has been to employ projections of the population of working age, by age and sex, without further adjustment. In this case projections of the population in an appropriate age range, such as 15 to 69, by age and sex, are used to represent the population of working age. The selection of the age range varies on the basis of differences in practices from country to country with respect to the age at which persons begin to join the labor force or to retire in large numbers. Persons typically begin working at a later age and retire at an earlier age in the economically advanced countries than in the less advanced countries.

Because worker rates are quite high for males 25 to 54 years of age, changes in estimates and projections of the male population in this range reflect rather well the changes in the economically active male population in the same ages. However, worker rates for males outside this range and for females are much lower and tend to change more sharply over time. Projections of working-age population are, therefore, inadequate to measure the changes in the active population in these groups

[98] UNESCO, *Estimating Future School Enrollment in Developing Countries, A Manual of Methodology*, by Bangnee Alfred Liu, Series A, Population Studies, No. 40, 1966, pp. 13–14.

and in the total active population. For similar reasons, projections of working-age population involving these groups represent even less adequately the level of the economically active population at any future date.

Use of Labor Force Participation Rates. — Preferably actual labor force projections by age and sex should be prepared to measure prospective labor force changes. A simple age-specific rate method may be employed; proportions of the population in the labor force by age and sex (i.e., labor force participation rates or activity rates) are assumed for future dates and applied to projections of the population of working age (e.g., 14 years and over) by age and sex. In the simplest form of the method, these activity rates are held constant at the level observed in the last census or some recent survey. Such projections take account only of expected shifts in the future size, age, and sex composition of the population, but not in activity rates.

The projections may allow explicitly for expected future changes in activity rates. The rates may be projected on the basis of past trends for each age-sex group or of judgment regarding the factors which could influence future change in the rates at each age. The past trend may be defined in terms of two or more prior observations, and some mathematical formula may be used to project the rates.

For many ages of males, say from 25 to 54, the particular projection assumption is not important because nearly all males are in the labor force and a "projection" of rates may not be considered necessary. The crucial assumptions are those relating to the changes in the rates for males at the fringe ages of economic activity and for female at all ages. Accordingly, it is desirable to give special attention to the projection of these rates, taking account particularly of the relation between worker rates and various socioeconomic variables affecting labor force participation.[99] Since the variables of marital status and the presence or absence of young children define quite different levels of economic activity for female workers, it is especially useful to consider rates specific for these variables separately in developing projections of worker rates for women.[100] Average age at marriage and of childbearing and number of children previously born may also be taken into account directly or indirectly. Consideration may be given to the availability of public services and facilities, such as day-care nurseries, which could affect the activity rates for women with children. Other factors which may be allowed for, explicitly or implicitly, include school attendance and trends in age at retirement.

Other general factors affecting labor force participation include trends in urban-rural distribution and internal migration, agricultural and nonagricultural employment, the general level of economic development and unemployment, the general level of health, etc. The relationship between these factors and the worker rates should be intensively investigated before

they are used as a basis for projections. In his projections of the labor force of India (1963), Tilak gave particular attention to expected changes in the general level of health.[101]

A comparison of worker rates in a number of countries classified according to their degree of social and economic development has proved useful in developing assumptions for projecting national worker rates, particularly for the youngest and the oldest working ages. The United Nations has compiled and analyzed the data for 100 countries.[102] In projecting the labor force for the youngest and oldest working ages of males in the Philippines the United Nations and the Government of the Philippines employed the correlation between labor force and industrialization (i.e., percent of the labor force in agriculture), for some 30 countries, in combination with expected trends in industrialization.[103] The correlation between the level of economic development (industrialization) and the labor force for geographic areas within a country may also serve as a guide to possible future changes in the national labor force, given certain assumed levels of future industrialization. Ducoff employed this approach in his projections of the agricultural and nonagricultural labor force of the countries of Central America.[104]

The prospective change in the urban and rural distribution or regional distribution may be taken into account explicitly as variables in projections of the economically active population. Current participation rates for the urban and rural sectors or for geographic subdivisions may be held constant for the projection period or weighted together on the basis of independent projections for these areas, to derive projections of overall labor force by age and sex.[105] Worker rates for urban and rural areas or for geographic regions may also be projected, for use with prior projections of the urban and rural populations or regional population by age and sex.[106]

Inasmuch as the labor force projections usually extend only a few decades ahead, only one series of projections of the working-age population is usually necessary, particularly for the countries with low or moderate death rates. An exception may be made where migration is an important factor in national growth.[107] It is commonly more important to allow for possible variation in worker rates than in population size and age-sex composition. Alternative series with different worker rates may be based on different assumptions with regard to the factors influencing these rates.

[99] See, for example: Gertrude Bancroft, *The American Labor Force, Its Growth and Changing Composition*, John Wiley and Sons, Inc., New York, 1958; John D. Durand, *The Labor Force in the United States, 1890–1960*, Social Science Research Council, New York, 1948; and United Nations, "Studies of Factors Affecting the Size, Composition, and Growth of the Labor Force," Chapter II in *Methods of Analyzing Census Data on Economic Activities of the Population*, Series A, Population Studies, No. 43, 1968.

[100] See, for example, the following reports of the United States Department of Labor: *Population and Labor Force Projections for the United States, 1960 to 1975*, Bulletin No. 1242, 1959; "Interim Revised Projections of United States Labor Force, 1965–75," *Special Labor Force Report* No. 24, 1962; "Labor Force Projections for 1970–80," by Sophia Cooper and Denis F. Johnston, *Monthly Labor Review*, 88(2):129–140, Feb. 1965 (same as *Special Labor Force Report* No. 49).

[101] V. R. K. Tilak, "The Future Manpower Situation in India, 1961–76," *International Labour Review*, 80(5):434–446, May 1963.

[102] United Nations, *Demographic Aspects of Manpower*, Report I, *Sex and Age Patterns of Participation in Economic Activities*, Series A, Population Studies, No. 33, 1961.

[103] United Nations and Government of the Philippines, op. cit.

[104] Ducoff, op cit., Ch. IV.

[105] See, for example: U.S. Bureau of the Census, *International Population Statistics Reports*, Series P-90, No. 16, "The Labor Force of Bulgaria," by Zora Prochazka, 1962; idem, Series P-90, No. 14, "The Labor Force of Rumania," by Samuel Baum, 1961; and idem, Series P-90, No. 20, "The Labor Force of Poland," by Zora Prochazka and Jerry W. Combs, Jr., 1964. See also United Nations and Government of the Philippines, op. cit. (female labor force only); and T. Chellaswani, "Population Trends and Labour Force in India: 1951–66," *Population Review* (Madras), 2(2):42–48, July 1958.

[106] See, for example: Eduardo Arriaga, *Venezuela, Proyeccion de la población economicamente activa, 1950–75 (Venezuela, Projection of the economically active population, 1950–75)*, Series C, No. 26, Latin American Demographic Center, Santiago, Chile, 1965

[107] See, for example: U.S. Bureau of the Census, *International Population Statistics Reports*, Series P-90, No. 11, "The Labor Force of the Soviet Zone of Germany and the Soviet Sector of Berlin," by Samuel Baum and Jerry W. Combs, Jr., 1959; and Wolfgang M. Illing et al., *Population, Family, Household, and Labour Force Growth to 1980*, Staff Study No. 19, Economic Council of Canada, Sept. 1967.

The calculation of alternative projections of the economically active population by the age-specific rate method readily permits a component analysis of the prospective changes in the number of active persons in terms of the demographic component and the activity-rate component. A series of projections based on constant activity rates indicates directly the component of future change in the labor force due to changing population size and age-sex distribution. These projections may then be compared with projections based on projected activity rates, to obtain the component due to prospective changes in activity rates.

Cohort Methods. — The possible application of cohort methods to projections of labor force has received scant consideration and actual applications are few. First, activity rates may be projected on a cohort basis rather than on a period basis. A cohort technique of projecting worker rates was employed by the U.S. Bureau of the Census for the projections of the "normal" labor force made in the early forties.[108] In these projections, worker rates for 5-year age groups based on data from a number of decennial censuses were first interpolated to the middle of each decade, differences between rates in successive 5-year age groups at successive 5-year intervals were taken, and the trend of these differences was extrapolated to future periods. This procedure is not directly applicable to the youngest and oldest age groups. Cohort ratios of the worker rates rather than cohort differences may be used instead. Furthermore, the worker rates for age cohorts may be projected by analysis of the pattern of rates for each successive cohort, with possible measurement of the changing parameters (e.g., maxima, "spacing" characteristics) of the labor force cohorts. The French INSEE appears to have adopted a cohort method of this general type for projecting the economically active population engaged in agriculture in France from 1956 to 1971.[109] These variations of the cohort method permit taking into account explicitly the previous history of labor force participation of each cohort; this procedure may be particularly useful in projecting the rates for women, whose participation rates at the older ages may be importantly affected by their participation rates at the younger ages.

A more characteristic and elaborate form of the cohort method consists of carrying forward the economically active population by age and sex to future dates by use of probabilities of net entry and probabilities of net withdrawal through death or retirement. The probabilities of net entry are applied to the inactive population to determine accessions for the first year, and the probabilities of net withdrawal, separately for retirement and death if possible, are applied to the active population to determine separations for the year. The estimated new entrants are then subtracted from the inactive population and added to the active population; similarly, separations due to retirement are subtracted from the active population and added to the inactive population. Rates of net entry and of net withdrawal due to retirement may be derived from the relative change in activity rates at consecutive ages as given

by census data. Until about age 35, these changes are positive and the rates are considered rates of net entry. At the older ages, the changes are negative and the rates are considered rates of net withdrawal due to retirement and death. The rates can then be adjusted to exclude the effect of mortality. This variation of the cohort method has the virtue of providing, as valuable by-products, the annual number of net entrants into, and net withdrawals from, the labor force, by age. Such information is of specific use in national economic planning, both for the utilization of new workers and for management of retirement and other programs for older ages.

The development of tables of working life may give further impetus to use of the cohort method of projecting the labor force. These show for the year of the table the rates of net accessions and the rates of net separation through net retirement and death, by age, along with the activity rates on which the entry rates and withdrawal rates were based. (See section on "Tables of Working Life," ch. 15.)

Equilibration of Supply of and Demand for Workers

If, as a number of studies have shown, labor supply is responsive to changes in the demand for labor,[110] further refinement of projections of the economically active population may be secured, particularly in "full-employment" economies, by taking account of historical and prospective information on the demand for labor. For example, separate projections may be made of the supply of labor on the basis of general activity rates, on the one hand, and of the demand for labor on the basis of expected employment in various industrial branches,[111] on the other; and these projections of supply and demand may then be reconciled.

Labor Force by Industry, Educational Level, and Region

To serve national planning needs, it is desirable to extend the projections of the labor force to include its composition by occupation and industry, its level of skill, and its distribution by urban-rural residence and by region. So far such projections have been made rather infrequently in spite of the recognized need for them.

Major Industry Groups. — There is more interest in projections for industry groups of the labor force than for occupation groups. Continuation of past trends has been a common basis for the projections of major industry groups. The Economic Commission for Latin America published projections for three broad industrial categories—primary, secondary, and tertiary—for Colombia from 1950 to 1981.[112] Mathematical extrapolation, involving a constant rate of growth, was employed for the first two of these categories, and the remaining national population growth was assumed to be absorbed by tertiary industry.

[108] U.S. Bureau of the Census, *Population-Special Reports*, Series P-44, No. 12, "Normal Growth of the Labor Force in the United States: 1940 to 1950," June 12, 1944.

[109] Institut national de statistique et des études économiques, "La population agricole française: Structure actuelle et évolution" (The agricultural population of France: Present and prospective composition), by M. Febvay, *Etudes et conjoncture* (Paris), 11th year, Aug. 1956, pp. 707–739.

[110] See Alfred Tella, "The Relation of Labor Force to Employment," *Industrial and Labor Relations Review*, April 1964, pp. 454–469; and U.S. Department of Labor (Cooper and Johnston), op. cit., particularly pp. 138–140.

[111] See, for example: J. E. Morton, *On Manpower Forecasting*, Methods for Manpower Analysis No. 2, W. E. Upjohn Institute for Employment Research, Kalamazoo, Mich., Sept. 1968, esp. pp. 26–27; U.S. Bureau of Labor Statistics, *The Forecasting of Manpower Requirements*, Report No. 248, prepared for the U.S. Agency for International Development, April 1963; and H. Goldstein, "Projections of Manpower Requirements and Supply," *Industrial Relations*, 5(3):17–27, May 1966.

[112] United Nations, Economic Commission for Latin America, *Some Aspects of Population Growth in Colombia*, p. 109.

Other studies have given more attention to intermediate variables. In the Philippines manpower study, projections of employment in four industry groups were developed on the basis of projected national income and income per worker in each industry group.[113] Ducoff's projections of the agricultural-nonagricultural sectors of the labor force took account of the relation between the level of the agricultural labor force and economic development as indicated by regional data.[114]

Economic Activity and Educational Level. – In order to secure projections of the labor force classified by level of skill, one may project labor force and educational attainment jointly. The amount of education of the prospective labor force will determine its suitability for the jobs which may have to be filled to carry out the tasks of the future economy. Educational requirements will differ depending on the technological advancement of the economy. The U.S. Department of Labor has made some rough projections of the expected level of schooling of the future labor force on the basis of analysis of past trends.[115] Tilak has sketched a methodology for assessing the relationship between the number of future new entrants into the labor force and the projected demand for manpower by level of education. He attempted then to bring about an adjustment between supply and demand for specific categories of manpower.[116] Jones and Gingrich used projected school-attendance rates of various countries to estimate the activity rates at ages 10 to 24 and then used the resulting activity rates to estimate the educational attainment of the labor force in the corresponding age cohorts at subsequent years; regression equations were employed to measure these relationships.[117]

Regional Distribution. – Projections of labor force or employment for geographic areas within a country, prepared particularly under local auspices, have become quite common in a number of the industrialized countries. Projection of employment by major industry may, in fact, be a first step in the projection of the total population for geographic areas. A procedure sometimes used in this case is to project the ratio of regional to national employment by major industry groups and apply the resultant ratios to independent projections of national employment by these major industry groups. Activity rates have also been applied to regional population projections by age and sex.

CONCLUDING NOTE

The preparation of projections of various socioeconomic characteristics of the population and their use in the development and implementation of national plans have grown considerably during recent decades. We are also seeing increasing use of the more analytic approaches to the making of projections, i.e., ones which take account of specific variables affecting the structure of the population. We have noted, for example, a tendency to prepare national projections of socioeconomic characteristics on the basis of separate projections of these characteristics for the urban and rural sectors of the country; this has permitted making allowance for both the differences in the characteristics of the two sectors and expected shifts in the urban-rural redistribution of the population. The preparation of adequate projections has been hampered in many countries, however, by the lack of a reliable body of demographic information for the present and past. Progress in projections has been delayed also by the lack of evaluation studies indicating the relative reliability of alternative methods.

For the most part, the various types of projections for a given area have been prepared independently of one another on the basis of independent assumptions.[118] For example, the projections of marital status and fertility implied by the general population projections have not directly been employed in the projections of families by size. There has been emerging, however, a recognition of the desirability and even the necessity of integrating and balancing the projections of the total population (or their underlying assumptions) and the projections of various socioeconomic characteristics of the population with one another into a single consistent system of demographic projections; and of integrating the latter system with other economic and social projections, such as projections of production, consumption, health services, educational services, etc., so as to arrive at an overall consistent national plan. This logical integration of assumptions would involve study of the relationship of many pertinent variables, such as the relationship of fertility, labor force participation, and school enrollment; marital status, fertility, and household formation; enrollment trends and trends in educational level, etc. Research in a few countries has already been initiated into the development of a national "population model" or a model of the national economy, including population and households as variables.[119] The advent and use of the electronic computer render feasible such broad integration of projections and the preparation of national population models.

SUGGESTED READINGS

Part A. General Studies

Grauman, John V. "Population Estimates and Projections." Chapter 23 in Philip M. Hauser and Otis Dudley Duncan (eds.), *The Study of Population: An Inventory and Appraisal.* Chicago, University of Chicago Press, 1959. Pp. 544–575.

Hajnal, John. "The Prospects for Population Forecasts." *Journal of the American Statistical Association,* 50(270): 309–322. June 1955.

Keyfitz, Nathan. "La proyección y la predicción en demografía: Una revisión del estado de este arte" (Projection and Prediction in Demography. A Review of the State of the Art).

---

[113] United Nations and Government of the Philippines, op. cit.

[114] Ducoff, op. cit.

[115] U.S. Department of Labor, *Projections of the Labor Force of the United States,* testimony by Harold Goldstein presented before the U.S. Senate Subcommittee on Labor and Public Welfare, Sept. 26, 1963, pp. 21–23.

[116] Tilak, op. cit.

[117] Gavin Jones and Paul Gingrich, "The Effects of Differing Trends in Fer-

tility and of Educational Advance on the Growth, Quality, and Turnover of the Labor Force," *Demography,* 5(1):226–248, 1968.

[118] See, for example, U.S. Bureau of the Census, *Current Population Reports,* Series P-25, No. 388, "Summary of Demographic Projections," by Donald S Akers, March 14, 1968.

[119] See, for example, Guy H. Orcutt et al, *Microanalysis of Socioeconomic Systems: A Simulation Study,* New York, Harper and Brothers, 1961.

SUGGESTED READINGS—Continued

Conferencia Regional Latinoamericana de Población, 1970 (Mexico City). International Union for the Scientific Study of Population and Colegio de México, Mexico City (in press).

Muhsam, H. V. "The Utilization of Alternative Population Forecasts in Planning." *Bulletin of the Research Council of Israel,* 5(2–3):133–146. March–June 1956.

United Nations. *General Principles for National Programmes of Population Projections as Aids to Development Planning.* Series A, Population Studies, No. 38. 1965.

Widén, Lars. *Methodology in Population Projection: A Method Study Applied to Conditions in Sweden.* Reports 9. Gothenburg (Sweden), Demographic Institute, University of Gothenburg, 1969.

Part B. Projections for Countries

Adams, Edith, and Menon, P. Sankar. "Types of Data and Studies Needed to Improve the Basis for Population Projections in Tropical Africa." In United Nations, *World Population Conference, 1965* (Belgrade). Vol. III, pp. 1–5. 1967.

Bean, Lee L., Khan, Masihur Rahman, and Rukanuddin, A. Razzaque. *Population Projections for Pakistan, 1960–2000.* Karachi, Pakistan Institute of Development Economics, 1968.

Camisa, Zulma C. "Argentina. Proyección de la población por sexo y edad, 1960–1980" (Argentina. Projection of the Population by Sex and Age, 1960–1980). Series C, No. 62. Latin American Demographic Center, Santiago, Chile, 1965.

Cox, Peter R. *Demography,* chapters 14 and 15. Cambridge (Eng.), Cambridge University Press, 1970.

Das Gupta, A., and Sen Gupta, S. "Population Projections for Thailand and a Study of the Elements and Criteria." In United Nations, *World Population Conference, 1965* (Belgrade). Vol. II, pp. 13–16. 1967.

Dorn, Harold F. "Pitfalls in Population Forecasts and Projections." *Journal of the American Statistical Association,* 45(251):311–334. Sept. 1950.

Grauman, John V. "Success and Failure in Population Forecasts of the 1950's: A General Appraisal," In United Nations, *World Population Conference, 1965* (Belgrade). Vol. III, pp. 10–14. 1967.

Illing, Wolfgang M. *Population, Family, Household and Labour Force Growth to 1980.* Staff Study No. 19, prepared for the Economic Council of Canada. Chapter 2, "Population," by Yoshiko Kasahara. Ottawa, 1967.

Israel, Central Bureau of Statistics. *Projections of the Population of Israel (1955–1970).* By Benjamin Gil. Special Series No. 69. April 1958.

Keyfitz, Nathan. "The Population Projection as a Matrix Operator." *Demography,* 1(1):56–73. 1964.

Kono, Shigemi. "Forecasts in Some Asian Areas During Recent Years: Criticism and Suggestions." In United

Nations, *World Population Conference, 1965* (Belgrade). Vol. III, pp. 15–18. 1967.

Lorimer, Frank W. "Analysis and Projections of the Population of the Philippines." In *First Conference on Population, 1965.* Proceedings of the Conference sponsored by the Population Institute of the University of the Philippines. Quezon City (Philippine Islands), University of Philippines Press, 1966. Pp. 287–314.

Mortara, Giorgio. "Ancora a proposito di prospettive demografiche" (Another Look at Demographic Projections). *Giornale degli Economisti e Annali di Economia* (Padua), 21(5–6):330–335. May–June 1962.

Myers, Robert J. "Comparison of Population Projections with Actual Data." In United Nations, *World Population Conference, 1954* (Rome). Vol. III, pp. 101–111. 1955.

Notestein, Frank W., *et al. The Future Population of Europe and the Soviet Union,* Appendixes I and II. Geneva, League of Nations, 1944.

Pearl, Raymond, Reed, Lowell J., and Kish, Joseph F. "The Logistic Curve and the Census Count of 1940." *Science,* 92(2395):486–488. Nov. 22, 1940.

Peláez, César A. "The Degree of Success Achieved in the Population Projections for Latin America Made Since 1950. Sources of Error. Data and Studies Needed in Order to Improve the Basis for Calculating Projections." In United Nations, *World Population Conference, 1965* (Belgrade). Vol. III, pp. 27–33. 1967.

Reed, Lowell J. "Population Growth and Forecasts." *Annals of the American Academy of Political and Social Science,* 188:159–166, Nov. 1936.

Romaniuk, A. "Projection Basis for Populations of Tropical Africa: A General Discussion." In United Nations, *World Population Conference, 1965* (Belgrade). Vol. III, pp. 40–43. 1967.

Shryock, Henry S. "Accuracy of Population Projections for the United States." *Estadística,* 12(45):587–597. Dec. 1954.

Somoza, Jorge, and Llano, Luis. "Proyección de la población de Bolivia" (Population Projections for Bolivia). E/CN. CELADE/C.9. United Nations Demographic Center for Latin America, Santiago, Chile. 1963.

Stukel, A. "Population Projections, Canada, 1961–1991." Pp. 269–316 in T. M. Brown (ed.), *Canadian Economic Growth.* Royal Commission on Health Services, Ottawa, 1964.

Sykes, Z. M. "Some Stochastic Versions of the Matrix Model for Population Dynamics." *Journal of American Statistical Association,* 64(325):111–130. March 1969.

Taeuber, Irene B. "The Development of Population Predictions in Europe and the Americas." *Estadística,* 2:323–346. 1944.

United Kingdom, Royal Commission on Population, *Reports*

SUGGESTED READINGS—Continued

and Selected Papers of the Statistics Committee. Papers of the Royal Commission on Population, Vol. II. London, H. M. Stationery Office, 1950.

United Nations. *Methods for Population Projections by Sex and Age.* Manual III, Manuals on Methods of Estimating Population. Series A, Population Studies, No. 25. 1956.

_____. *World Population Prospects as Assessed in 1968,* Series A, Population Studies, No. 53, 1973.

——. Chapter III and Appendix B in *Population Growth and Manpower in the Philippines: A Joint Study of the United Nations and the Government of the Philippines.* Series A, Population Studies, No. 32. 1960.

U.S. Bureau of the Census. *Current Population Reports,* Series P-25, No. 381. "Projections of the Population of the United States, by Age, Sex, and Color to 1990, with Extensions of Population by Age and Sex to 2015." By Jacob S. Siegel. Dec. 18, 1967.

——. "Projections of the Population of Pakistan, by Age and Sex: 1965-1986." By James W. Brackett and Donald S. Akers. June 1965.

Visaria, Pravin M. "Population Projections for Countries of Middle South Asia During the 1950's." In United Nations, *World Population Conference, 1965* (Belgrade), Vol. III, pp. 47-50. 1967.

Wolfenden, H. H. *Population Statistics and Their Compilation* (rev. ed.). Chicago, University of Chicago Press, 1954. Pp. 88-94.

Yugoslavia. "Forecast of the Population of Yugoslavia for the Period 1961-1981." *Yugoslav Survey* (Belgrade), No. 16. Jan.-March 1964.

Part B(1). Projection of Mortality

Basavarajappa, K. G. "The Significance of Differences in Patterns of Mortality for Projections of Population." In International Union for the Scientific Study of Population, *Sydney Conference: Contributed Papers,* Sydney, Australia, 1967. Pp. 521-527.

Beard, R. E. "A Theory of Mortality Based on Actuarial, Biological, and Medical Considerations." In *International Population Conference, 1961* (New York). International Union for the Scientific Study of Population, London, 1963. Vol. I, pp. 611-625.

Pollard, A. H. "Methods of Forecasting Mortality Using Australian Data." *Journal of the Institute of Actuaries,* 75:151-170. 1949.

Spiegelman, Mortimer. "Mortality Projections and Theories." Chapter 6 in *Introduction to Demography* (rev. ed.). Cambridge, Mass., Harvard University Press, 1968. Pp. 153-170.

Tarver, James D. "Projections of Mortality in the United States to 1970." *Milbank Memorial Fund Quarterly,* 37(2):132-143. April 1959.

U.S. Bureau of the Census. *International Population Reports,* Series P-91, No. 5. "A Method of Projecting Mortality Rates Based on Postwar International Experience." By Arthur A. Campbell. 1958.

U.S. Social Security Administration, Office of the Actuary. *United States Population Projections for OASDHI Cost Estimates.* By Francisco Bayo. Actuarial Study, No. 62. Dec. 1966.

Part B(2). Projection of Fertility

Akers, Donald S. "Cohort Fertility Versus Parity Progression as Methods of Projecting Births." *Demography,* 2:414-428. 1965.

Beshers, James M. "Birth Projections with Cohort Models." *Demography,* 2:593-599. 1965.

Bumpass, Larry and Westoff, Charles F. "The Prediction of Completed Fertility." *Demography,* 6(4):445-454. Nov. 1969.

Canada, Dominion Bureau of Statistics. *Projection of Incomplete Cohort Fertility for Canada by Means of the Gompertz Function.* By A. Romaniuk and S. M. Tanny. Analytical and Technical Memorandum No. 1. March 1969.

Freedman, R. and Coombs, L. C. "Expected Family Size and Family Growth Patterns: A Longitudinal Study." In Egon Szabady (ed.), *World Views of Population Problems.* Budapest, Akadémiai Kiadó, 1968. Pp. 83-95.

Freedman, Ronald, Coombs, Lolagene C., and Bumpass, Larry. "Stability and Change in Expectations about Family Life: A Longitudinal Study," *Demography,* 2:250-275. 1965.

Freedman, Ronald, and Bumpass, Larry. "Fertility Expectations in the United States; 1962-64," *Population Index,* 32(2):181-197. April 1966.

Freedman, Ronald, Whelpton, Pascal K., and Campbell, Arthur A. *Family Planning, Sterility, and Population Growth,* Chapter 7 and Appendix F. New York, McGraw-Hill Book Company, 1959.

Ryder, Norman B., and Westoff, Charles F. "The Trend of Expected Parity in the United States: 1955, 1960, and 1965." *Population Index,* 33(2):153-168. April-June 1967.

Siegel, Jacob S., and Akers, Donald S. "Some Aspects of the Use of Birth Expectations Data from Sample Surveys for Population Projections." *Demography,* 6(2):101-115. May 1969.

Whelpton, P. K. "Cohort Analysis and Fertility Projections." *Emerging Techniques in Population Research.* New York, Milbank Memorial Fund, 1963. Pp. 39-64.

Whelpton, P. K., Campbell, A. A., and Patterson, John E. *Fertility and Family Planning in the United States,* chapters 1, 2, and 10. Princeton, N.J., Princeton University Press, 1966.

SUGGESTED READINGS—Continued

Part C. Projections for Geographic Subdivisions of Countries

Balestra, Pietro, and Rao, N. Koteswara. *Basic Economic Projections*. Menlo Park, Calif., Stanford Research Institute, 1964.

Bendiksen, Bjørnulf. "Schematized Local Projections in Connection With a Population Census." In United Nations, *World Population Conference, 1965* (Belgrade). Vol. III, pp. 6–9. 1967.

Czamanski, Stanislaw. "A Method of Forecasting Metropolitan Growth by Means of Distributed Lags Analysis." *Journal of Regional Science*, 6(1):35–49. Summer 1965.

Greenberg, Michael R., Krueckeberg, Donald A., and Mautner, Richard. *Long–range Population Projections for Minor Civil Divisions: Computer Programs and User's Manual*. Center for Urban Policy Research, Rutgers University, New Brunswick, New Jersey. May 1973.

Hagood, Margaret Jarman, and Siegel, Jacob S. "Projections of the Regional Distribution of the Population of the United States to 1975." *Agricultural Economic Research*, 3(2):41–52. April 1951.

Hamilton, C. Horace, and Perry, Josef. "A Short Method for Projecting Population by Age from One Decennial Census to Another." *Social Forces*, 41(2):163–170. Dec. 1962.

Hollmann, Walter P. "Population Projections for Counties and Metropolitan Statistical Areas in California." *Proceedings of the Social Statistics Section, 1967*. Washington, D.C., American Statistical Association, 1968.

Isard, Walter F., *et al. Methods of Regional Analysis: An Introduction to Regional Science*. New York, John Wiley and Sons, 1960. Chapters 2 and 3.

Loomer, Harlin G. "Accuracy of the Ratio Method for Forecasting City Population: A Reply." *Land Economics*, 28(2):180–183. May 1952.

Lowry, Ira S. *Metropolitan Populations to 1985: Trial Projections*. Memorandum RM–4125–RC. Santa Monica, Calif., The Rand Corporation, Sept. 1964.

Morrison, Peter A. *Demographic Information for Cities: A Manual for Estimating and Projecting Local Population Characteristics*. R–618–HUD. RAND, Santa Monica, California. June 1971.

National Planning Association, Center for Economic Projections. "Economic and Demographic Projections for Two Hundred and Twenty-Four Metropolitan Areas." *Regional Economic Projection Series*, Report No. 67–R–1. Washington, D.C., National Planning Association, 1967.

Pobedina, A. F. "The Use of Electronic Computers for Population Projections." In United Nations, *World Population Conference, 1965* (Belgrade). Vol. III, pp. 34–39. 1967.

Pickard, Jerome P. *Metropolitanization of the United States*, Appendix B. Research Monograph 2. Washington, D.C., Urban Land Institute, 1959.

————. *Appendixes to Dimensions of Metropolitanism*. Research Monograph 14A. Washington, D.C., Urban Land Institute, 1967.

Schmitt, R. C., and Crosetti, A. H. "Short-Cut Methods of Forecasting City Population." *Journal of Marketing*, 17:417–424, April 1953.

————. "Accuracy of the Ratio Method for Forecasting City Population." *Land Economics*, 27(4):346–348. Nov. 1951.

Schneider, J. R. L. "Local Population Projections in England and Wales." *Population Studies* 10(1):95–114. July 1956.

Siegel, Jacob S. "Some Aspects of the Methodology of Population Forecasts for Geographic Subdivisions of Countries." In United Nations, *World Population Conference, 1954* (Rome). Vol. III, pp. 113–133. 1955.

————. "Forecasting the Population of Small Areas." *Land Economics*, 29(1):72–87. Feb. 1953.

Sweden, National Central Bureau of Statistics. *Befolkningsprojektion for kommunblocken till 1970, 1975, 1980 och 1985* (Population Projections for Cooperating Communes to 1970, 1975, 1980, and 1985). Stockholm. 1969.

Tarver, James D. "A Component Method of Estimating and Projecting State and Subdivisional Populations." Miscellaneous Publication MP–54. Stillwater, Okla., Oklahoma State University, Dec. 1959.

————. "Predicting Migration." *Social Forces*, 39(3):207–213. March 1961.

Tarver, James D., and Black, Therel R. *Making County Population Projections—A Detailed Explanation of a Three-Component Method, Illustrated by Reference to Utah Counties*. Technical Bulletin 459. Logan, Utah, Utah Agricultural Experiment Station, June 1966.

Ter Heide, H. "Migration Models and their Significance for Population Forecasts." *Milbank Memorial Fund Quarterly*, 41(1):56–76. Jan. 1963.

U.S. Bureau of the Census. *Current Population Reports*, Series P–25, No. 375, "Revised Projections of the Population of States: 1970 to 1985." Oct. 3, 1967.

————. *Current Population Reports*, Series P–25, No. 415, "Projections of the Population of Metropolitan Areas: 1975." Jan. 31, 1969.

U.S. Department of Commerce, Office of Field Service and Office of Industry and Commerce. *Better Population Forecasting for Areas and Communities: A Guidebook for Those Who Make or Use Population Projections*. By Van Beuren Stanbery. Domestic Commerce Series, No. 32. Sept. 1952.

White, Helen R. "Empirical Study of the Accuracy of Selected Methods of Projecting State Populations." *Journal of the American Statistical Association*, 49(207):480–498. Sept. 1954.

SUGGESTED READINGS – Continued

Part D. Projections for Various Socioeconomic Characteristics

France, Institut National de la Statistique et des Études Économiques. "Projections démographiques pour la France." By G. Calot *et al. Les Collections de l'INSEE.* Series D, No. 6, Demographie et emploi. March 1970.

France, Institut National de la Statistique et des Études Économiques. "Perspectives d'évolution de la population de la France, population totale, population active et scolaire, ménages." Extrait de *Études Statistiques,* No. 3, July–Sept. 1964.

Illing, Wolfgang M. *Population, Family, Household and Labour Force Growth to 1980.* Staff Study No. 19, prepared for the Economic Council of Canada. Ottawa, Sept. 1967.

United Nations. *General Principles for National Programmes of Population Projections as Aids to Development Planning.* Series A, Population Studies, No. 38. 1965.

———. "Selected Bibliography on Methods of Projecting the School-Age Population, the Economically Active Population, the Urban and Rural Populations and the Number and Size of Households." United Nations, Seminar on Evaluation and Utilization of Population Census Data in Asia and the Far East, Bombay, India, 1960. E/CN.9/CONF.2/L.11. April 7, 1960.

U.S. Bureau of the Census. *Current Population Reports,* Series P–25, No. 388. "Summary of Demographic Projections." By Donald S. Akers. March 14, 1968.

Part E. Projections of Population by Type of Residence

Angenot, L. H. J. "Regional Population Projecting in the Netherlands." In United Nations, *Proceedings of the World Population Conference, 1954* (Rome), Vol. III, pp. 1–10. 1955.

Cataldi D., Alberto. "La situación demográfica del Uruguay en 1957 y proyecciones a 1982," E/CN.CELADE/C.15. Santiago, Chile, United Nations Demographic Center for Latin America, 1964.

Chasteland, J. C., Amani, M., and Puech, O. A. "La population de l'Iran, perspectives d'evolution de 1956 a 1986." Institut d'Études et de Recherches Sociales, Université de Téhéran, Téhéran, 1966.

Davis, Kingsley. "Conceptual Problems of Urban Population in Developing Countries." In United Nations, *World Population Conference, 1965* (Belgrade), Vol. III, pp. 61–65. 1967.

———. "Urbanization in India: Past and Future." In Roy Turner (ed.), *India's Urban Future.* Berkeley, Calif., University of California Press, 1962. Pp. 3–26.

Etherington, D. M. "Projected Changes in Urban and Rural Population in Kenya and the Implications for Development Policy." *East African Economic Review* (Nairobi), 1 (New Series No. 2):65–83. June 1965.

Finland, Central Statistical Office. "Aluellinen vaestoennuste vuoteen 1990 saakka." By Tor Hartman. *Tilastokatsauksia* (Helsinki), 38(10):45–49. 1963.

Grauman, John V. "Development of a Model of Rural-Urban Population Change, with Relevance to Latin America." In *International Population Conference, 1961* (New York), Vol. I, pp. 448–457. London, International Union for the Scientific Study of Population, 1963. (Published also as "Desarrollo de un modelo para el cambio de población de rural a urbana referido a Latinoamérica,"*Estadística,* 20(75): 242-254, June 1962.)

Hodge, Gerald. "Do Villages Grow?–Some Perspectives and Predictions." *Rural Sociology,* 31:183–196. 1966.

Peláez, César A. "Proyección de la población urbana y rural de la República de Panamá," E/CN.CELADE/C.19. Santiago, Chile, United Nations Demographic Center for Latin America, 1964.

Pickard, Jerome P. *Appendixes to Dimensions of Metropolitanism.* Research Monograph 14A. Washington, D.C., Urban Land Institute, 1967.

Roberts, G. W. "Provisional Assessment of Growth of the Kingston-St. Andrew Area, 1960–70." *Social and Economic Studies* (Jamaica), 12(4):432–441. Dec. 1963.

Seklani, Mahmoud. "Villes et campagnes en Tunisie. Évaluations et previsions." *Population* (Paris), 15(3):485–512. June–July 1960.

Somoza, Jorge, and Llano, Luis. "Proyección de la población de Bolivia." E/CN.CELADE/C.9. Santiago, Chile, United Nations Demographic Center for Latin America, 1963.

United Nations, Economic Commission for Latin America. *Human Resources of Central America, Panama and Mexico, 1950–1980, in Relation to Some Aspects of Economic Development.* By Louis J. Ducoff. 1960. Appendix B and pp. 37–45.

———. *Some Aspects of Population Growth in Colombia.* E/CN.12/618. Santiago, Chile. Nov. 1962.

United Nations. Chapter IV in *Population Growth and Manpower in the Philippines: A Joint Study by the United Nations and the Government of the Philippines.* Series A, Population Studies, No. 32. 1960.

———. "World Urbanization Trends, 1920–1960." Working Paper No. 6, Agenda Item No. 2. Inter-Regional Seminar on Development Policies and Planning in Relation to Urbanization, University of Pittsburg, Pittsburg, Pa. 1966. Pp. 37–40.

U.S. Bureau of the Census. *Current Population Reports,* P–25, No. 415. "Projections of the Population of Metropolitan Areas: 1975." Jan. 31, 1969.

U.S. National Resources Board. *Estimates of Future Population by States.* By Warren S. Thompson and P. K. Whelpton. Dec. 1934.

Part F. Projections of Households and Families

Brown, S. P. "Analysis of a Hypothetical Stationary Popula-

SUGGESTED READINGS – Continued

tion by Family Units." *Population Studies* (London), 4(4):380–394. March 1951.

Cabello, Octavio. "Estimación de necesidades de viviendas en América Latina." *Proceedings of the International Statistical Institute, 1969* (London). London, International Statistical Institute, 1969.

Calot, Gérard. "Perspectives du nombre des ménages de 1954 a 1976." *Études statistiques* (Paris), 12(2):149–159. April–June 1961. Supplement trimestriel du Bulletin mensuel de statistique.

Douša, J. "Problémy zjišt'ování perspektivní skladby domácností" (Research problems in the projections of household composition). *Statistický Obzor* (Prague), 39(12):536–544. 1959.

Garcia, Agustin. "Panamá. Estimación de las necesidades de vivienda, 1950–1980." Series C, No. 27. Santiago, Chile, United Nations Demographic Center for Latin America, 1965. Pp. 45–100.

Glass, Ruth, and Davidson, F. G. "Household Structure and Housing Needs." *Population Studies* (London), 4(4):395–420. March 1951.

Glick, Paul C. *American Families*, Appendix C, "A Technical Note on Household and Family Projections." New York, John Wiley & Sons, 1957. Pp. 226–232.

Hocking, W. S. "A Method of Forecasting the Future Composition of the Population of Great Britain by Marital Status." *Population Studies* (London), 12(2):131–148. Nov. 1958.

Kono, Shigemi. "Household Projections for Japan, 1960 to 1975." *Jinko Mondai Kenkyu* (Tokyo), 83:1–13, July 1961.

Morales V., Julio. "Estimación de las necesidades de viviendas en Chile, 1952–1982." E/CN.CELADE/C.20. Santiago, Chile, United Nations Demographic Center for Latin America, 1964.

Muhsam, H. V. "Population Data and Analysis Needed in Assessing Present and Future Housing Requirements." E/CN.9/CONF.2/L.10. Paper prepared for the United Nations Seminar on the Evaluation and Utilization of Population Census Data in Asia and the Far East, Bombay, 1960. (Also published as "Datos de población y análisis necesarios para evaluar la demanda presente y futura de viviendas." *Estadística*, 21(79):301–322, June 1963.)

Parke, Robert, Jr., "The Choice of Assumptions in Household and Family Projections." In United Nations, *World Population Conference, 1965* (Belgrade), Vol. III, pp. 78–82. 1967.

Parke, Robert, Jr., and Grymes, Robert O. "New Household Projections for the United States." *Demography*, 4(2):442–452. 1967.

Pressat, Roland. "Un essai de perspectives de ménages." In *International Population Conference, 1959* (Vienna). Vienna, International Union for the Scientific Study of Population, 1959. Pp. 112–121.

Siegel, Jacob S. "Demographic Information Required for Housing Programmes with Special Reference to Latin America." ST/ECLA/CONF.9/L.12. Santiago, Chile, United Nations Demographic Center for Latin America, 1962. Document prepared for United Nations Seminar on Housing Statistics and Programs, Copenhagen, Denmark, 1962. (Also published as "Informaciones demográficas para la formulación de programas de vivienda con especial referencia a América Latina," *Estadística*, 21(79):260–267, June 1963.)

———. "Desarrollo y exactitud de proyecciones de población y de hogares censales en los Estados Unidos" (Development and Accuracy of Projections of Population and Households in the United States), prepared for the Conferencia Regional Latinoamericana de Población, Mexico City, 1970, conference sponsored by IUSSP and Colegio de México.

Steigenga, Willem. "Family Structure, Age Composition, and Housing Needs." In *International Population Conference, 1961* (New York). Vol. I, pp. 243–250. London, International Union for the Scientific Study of Population, 1963.

United Nations. "Proposed Methods of Estimating Housing Needs." Statistical Commission, Eleventh Session. E/CN.3/274. Jan. 20, 1960. Pp. 30–41.

———. *Methods of Projecting Households and Families.* Manual VII, Manuals on Methods of Estimating Population. Series A, Population Studies, No. 54. 1973.

U.S. Bureau of the Census. *Current Population Reports*, Series P–25, No. 394. "Projections of the Number of Households and Families, 1967 to 1985." June 6, 1968.

———. *Current Population Reports*, Series P–25, No. 387. "Projections of the Number of Households, by States, July 1, 1970 and 1975." Feb. 20, 1968.

Walkden, A. H. "The Estimation of Future Numbers of Private Households in England and Wales." *Population Studies* (London), 15(2):174–186. Nov. 1961.

Widén, Lars. *Methodology in Population Projection: A Method Study Applied to Conditions in Sweden.* Reports 9. Gothenberg (Sweden), Demographic Institute, University of Gothenberg, 1969.

Part G. Projections of Educational Characteristics

Amine-Zadeh, F. "Impact de la croissance démographique en Iran sur les effectifs de la population scolaire entre 1966 et 1986." In *Contributed Papers, Sydney Conference*, Sydney, Australia, International Union for the Scientific Study of Population, Aug. 1967. Pp. 201–209.

Ceylon, Department of Census and Statistics. *Projections of the School-Going Population of Ceylon, 1961–1981.* By T. Nadarajah. 1965.

Ceylon, National Planning Council, The Planning Secretariat. *The Ten-Year Plan.* Colombo, Ceylon. 1959. Pp. 466–472.

Choudhari, R. E., and Ramachandran, K. V. "Projections of Primary and Secondary School Populations for Maharashtra State, 1951–1981." *Artha Vijñana* (Poona), 4(2):98–116. June 1962.

SUGGESTED READINGS — Continued

Folger, John K. *Some Methods for Projecting School and College Enrollments*. Atlanta, Georgia, Southern Regional Education Board, 1954.

Goldberg, David. *College Enrollment Potential in Michigan, 1960–1975*. Michigan Population Studies No. 3. Ann Arbor, Mich., Department of Sociology, University of Michigan, 1961.

Gutierrez R., Hector. "Proyección de la población escolar de Chile, 1957–1982." E/CN.9/CONF.1/L.22. Paper contributed to the United Nations Seminar on the Evaluation and Utilization of Population Census Data in Latin America, Santiago, Chile, 1959.

Im, T. B., and Ramachandran, K. V. "Future School Population in the Republic of Korea, 1960 to 1975." Bombay, India, United Nations Demographic Training and Research Center, 1962.

Lee, Eun Sul. *College Enrollments and Projections in North Carolina*. Raleigh, N.C., North Carolina Board of Higher Education, May 1968.

New Zealand, House of Representatives. *School and University Enrolment Projections for the Years 1967–80*. By A. E. Kinsella, Minister of Education. 1967.

Peláez César A. "Panamá. Estimación de la matrícula escolar y de las necesidades de maestros y escuelas, 1960–1980." Series C, No. 27. Santiago, Chile, United Nations Demographic Center for Latin American, 1965. Pp. 1–44.

Pratt, William F. "Illustrative Projections of School Enrollment." In *First Conference on Population, 1965*. Proceedings of the Conference Sponsored by the Population Institute of the University of the Philippines. Quezon City, University of the Philippines, 1966. Pp. 389–422.

Pressat, Roland. "Croissance des effectifs scolaires et besoins en maîtres." *Population* (Paris), 13(1):9–38 (Part 1), Jan.–March 1958, and 13(2):193–214 (Part 2), April–June 1958.

Ruiz Carlos, Haroldo. "Proyección de la población escolar, estimación de la probable matrícula, y de las necessidades de maestros y salas de clase en la enseñanza primaria de Guatemala, 1960–1980." B.61.2/5.1. Paper submitted to the Conference on Education and Social and Economic Development in Latin America, Santiago, Chile, 1962.

Schmid, Calvin F., and Miller, Vincent A. *Enrollment Forecasts: State of Washington: 1967 to 1975*. Seattle, Wash., Washington State Census Board, 1967.

UNESCO. *Methods of School Enrolment Projection*. By E. G. Jacoby. Educational Studies and Documents, No. 32. 1959.

United Nations/UNESCO. *Estimating Future School Enrollment in Developing Countries: A Manual of Methodology*. By Bangnee Alfred Liu. Series A, Population Studies, No. 40. 1966.

U.S. Bureau of the Census. *Current Population Reports, Series P–25, No. 201. "Projections of the School-Age Population, by States, 1959 to 1963." June 1, 1959.

———. *Current Population Reports, Series P–25, No. 365. "Revised Projections of School and College Enrollment in the United States to 1985." May 5, 1967.

———. *Current Population Reports, Series P–25, No. 390. "Projections of Educational Attainment, 1970 to 1985." March 29, 1968.

U.S. Office of Education. *Projections of Educational Statistics to 1977–78*. By Kenneth A. Simon and Marie G. Fullam. 1968 Edition. 1969. Esp. chapter 1, "Enrollment."

Yun, Kim, Lagamon, Lucila J., and Ramachandran, K. V. "Illustrative Projections of School Populations in the Philippines, 1960–1980." *Philippine Statistician*, 9(4):133–158. Dec. 1960.

Part H. Projections of Economically Active Population

Arriaga, Eduardo. "Venezuela. Proyección de la población económicamente activa, 1950–1975." Series C, No. 26. Santiago, Chile, United Nations Demographic Center for Latin America, 1965.

Bose, S. R. "Labour Force and Employment in Pakistan, 1961–86: a Preliminary Analysis." *Pakistan Development Review* (Karachi), 3(3):371–398. Autumn 1963.

Bourcier de Carbon, Philippe. "Précision sur les perspectives de population active. Application au Mexique." *Population* (Paris), 25(1):77–96. Jan.–Feb. 1970.

Ceylon, National Planning Council, Planning Secretariat. *Manpower Resources of Ceylon, 1956–1981*. By R. M. Sundrum, V. R. Rao, and S. Selvaratnam, Colombo, Ceylon. 1959.

De Wolff, P. "Employment Forecasting by Professions." In United Nations, *World Population Conference, 1965* (Belgrade), Vol. III, pp. 109–114. 1967.

Dumard, Jean, and Merle, Monique. "Une nouvelle méthode de prévision de l'emploi: le modèle Hermès. Application au secteur du commerce." *Population* (Paris), 25(Special Number):43–70. Feb. 1970.

Easterlin, Richard A. "Recent and Projected Labor Force Growth in the Light of Longer-Term Experience." Chapter 6 in *Population, Labor Force, and Long Swings in Economic Growth*. New York, National Bureau of Economic Research, 1968. Pp. 141–164.

Elizaga, Juan C. "Proyección de la población masculina económicamente activa de Chile, 1950–1969." E/CN.CELADE/A.11. Santiago, Chile, United Nations Demographic Center for Latin America, 1963.

Federici, N. "Prospettive di evoluzione delle forze di lavoro femminili." *Statistica* (Bologna), 23(3). July–Sept. 1963.

Finland, Central Statistical Office. "Aktuvinen väestö vuosina 1957–71." By A. Tunkelo. In *Tilastokatsauksia* (Helsinki), 32(4):38–40. 1957.

SUGGESTED READINGS – Continued

Finland, National Planning Bureau. *Population Development and Labor Force Resources in Finland during 1950–1970.* By A. Strommer. Series A.5. Helsinki, 1959.

Fourastié, J. "La prévision de l'emploi en France." *Revue du Travail* (Brussels), 64(1–2). Jan.–Feb. 1963.

Gnanasekaran, K. S. "Labour Force Projections for India, 1951–1976." *Artha Vijñana* (Poona) 2. March 1960.

Goldstein, H. "Projections of Manpower Requirements and Supply." *Industrial Relations*, 5(3):17–27. May 1966.

Hama, H. "An Estimate of the Future Labour Force in Japan: 1960–1970." Japan, Institute of Population Problems. *Annual Reports of the Institute of Population Problems*, No. 7, Tokyo, 1962.

Im, T. B., and Ramachandran, K. V. "Labor Force Projections for the Republic of Korea, 1960–1975." Bombay, India, United Nations Demographic Training and Research Center, 1961.

Jones, G. W. "A Projection of the Labour Force in Malaya." In *Contributed Papers, Sydney Conference.* Sydney, Australia, International Union for the Scientific Study of Population, Aug. 1967. Pp. 125–136.

Jones, Gavin, and Gingrich, Paul. "The Effects of Differing Trends in Fertility and of Educational Advance on the Growth, Quality, and Turnover of the Labor Force." *Demography*, 5(1):226–248. 1968.

Morton, J. E. *On Manpower Forecasting: Methods for Manpower Analysis*, No. 2. Kalamazoo, Mich., W. E. Upjohn Institute for Employment Research. Sept. 1968.

Netherlands, Central Bureau of Statistics. "Berekeningen omtrent de toekomstige loop van de nederlandse bevolking." *Maandschrift*, 56(7):657–661. July 1961.

Novacco, Nino. "Prévision pour l'année 1975 sur la population italienne selon la qualification professionnelle et l'instruction." *Population* (Paris), 16(3):463–472. July–Sept. 1961.

Organization for Economic Co-Operation and Development. *Demographic Trends, 1965–1980, in Western Europe and North America.* Paris, 1966.

Organization for European Co-Operation and Development. *Employment Forecasting.* Paris, 1963.

Pressat, Roland. "Vues prospectives sur la population active par départment de 1960 à 1970. *Population* (Paris), 16(3):401–426. July–Sept. 1961.

SVIMEZ. "Popolazione e forze di lavoro. Prospettive demografiche fino al 2000 per l'Italia meridionale, Sicilia, Sardegna, Mezzogiorno e Italia." By G. De Meo. Rome, 1952.

Swerdloff, Sol. "How Good Were Manpower Projections for the 1960's." *Monthly Labor Review*, 92(11):17–22, Nov. 1969.

Tabah, Leon. "Représéntations matricielles de perspectives de population active." *Population* (Paris), 23(3):437–476. May–June 1968.

Tilak, V. R. K. "The Future Manpower Situation in India, 1961–76." *International Labor Review* (Geneva), 80(5):435–446. May 1963.

Tilak, V. R. K., and Panitchpakdi, Prom. "Population and Labour Force Growth in Thailand." In *Contributed Papers, Sydney Conference.* Sydney, Australia, International Union for the Scientific Study of Population, August 1967. Pp. 137–143.

United Nations. *Human Resources of Central America, Panama and Mexico, 1950–1980, in Relation to Some Aspects of Economic Development.* By Louis J. Ducoff. 1960. Chapters III (Part B) and IV, and appendixes B and C.

————. *Methods of Analysing Census Data on Economic Activities of the Population.* Series A, Population Studies, No. 43. 1968.

————. Chapter VI in *Population Growth and Manpower in the Philippines: A Joint Study by the United Nations and the Government of the Philippines.* Series A, Population Studies, No. 32. 1960.

U.S. Agency for International Development. Part III, "The Labor Force," in *Demographic Techniques for Manpower Planning in Developing Countries.* Prepared by the U.S. Department of Labor. 1963. Pp. 211–221.

U.S. Bureau of the Census. *International Population Statistics Reports*, Series P–90, No. 20. "The Labor Force of Poland," pp. 32–38. By Zora Prochazka and Jerry W. Combs, Jr. 1964.

U.S. Department of Labor. *The Forecasting of Manpower Requirements.* BLS Report No. 248. Prepared for the U.S. Agency for International Development. April 1963.

————. "Labor Force Projections by State, 1970 and 1980." By Denis F. Johnston and George R. Methee. *Special Labor Force Report*, No. 74. (Same as *Monthly Labor Review*, 89(10):1098–1104. Oct. 1966.)

————. "The U.S. Labor Force: Projections to 1985." By Sophia C. Travis. *Special Labor Force Report*, 119. (Reprinted from *Monthly Labor Review*, 93(5):3–12. May 1970, with supplemental tables.)

————. *Manpower Projections: An Appraisal and Plan of Action.* Report of the Working Group on Manpower Projections to the President's Committee on Manpower. Aug. 1967.

Vimont, Claude. "Perspectives nouvelles des recherches sur les prévisions d'emploi." *Population*, 25(Special):19–41, Feb. 1970.

————. "Comparaison des prévisions d'emploi par profession, du Ve plan et de l'évolution réele de l'emploi de 1962 à 1968." *Population* (Paris), 25(Special):9–18. Feb. 1970.

CHAPTER 24

Some Methods of Estimation for Statistically Underdeveloped Areas

INTRODUCTION

Deficiencies of Demographic Statistics

An adequate description of the basic demographic processes of fertility and mortality (and hence population change) requires the combined use of two types of information. One type of information concerns stocks: the number of persons at a given date, suitably classified by various characteristics, geographic area, and so on. The second type of information concerns flows: events, notably births and deaths, occurring to, or originating from, the population during some specified time period such as a calendar year, and thereby changing its size, composition, and geographic distribution.

Traditionally these two types of data have been obtained from two separate statistical sources, the stocks from periodic population censuses and the flows from a system of vital registration. In any modern state, a combination of economic and political requirements leads almost inevitably both to the introduction of periodic census taking as a matter of routine and to the setting up of an organization for recording the legal facts of life and death. It follows that virtually without exception the economically more advanced countries are endowed with a system sufficient to describe the fundamental demographic processes that are taking place in their territory.

Outside the circle of the more advanced countries, however, such a system seldom exists; indeed it can be said that the large majority of the world's population lives in countries that are underdeveloped not only economically but from the point of view of demographic statistics as well. Since the latter deficiency is simply a facet of the broader problem of underdevelopment, it is likely that its elimination will be brought about almost as a concomitant of general social and economic advance. But, given the historically unprecedented configurations of demographic parameters in today's underdeveloped countries, the very effort to speed up economic and social advance requires an increased amount of information on demographic matters that the existing statistical sources and the existing analytical expertise have been increasingly unable to satisfy. Hence there exists an urgent need to develop and disseminate methods for a better utilization of the statistical data that are already at hand in these countries, as well as a need to increase the supply of pertinent demographic information to be generated in the coming years.

Approaches to Obtaining Demographic Information

The "Classical" System as a Model. — It is often assumed that the best way to serve the objective of increasing the supply of basic demographic information in the statistically underdeveloped countries is to bring the performance of censuses and of the deficient vital registration system up to the level now prevailing in the more developed countries; in other words to adopt what may be called the "classical" system as soon as possible. There are, however, at least two fundamental reasons that make such an assumption highly questionable.

First, for many statistically underdeveloped countries, the development of a comprehensive vital registration system is either unobtainable in the foreseeable future (because of technical constraints and/or because of the attitudes of the population), or an early attainment of such a system would require an expenditure of resources on a scale that would be impossible to justify from an economic standpoint. The validity of this proposition is supported, if only impressionistically, by many well-known facts. Thus, by and large, it is not in the interest of the population voluntarily to cooperate with the registrar even though such cooperation may be elicited by intensive educational programs and strict enforcement of the registration laws. Similarly, the most casual observations or checks are sufficient to establish the presence of gross errors even in vital registration systems that have been in existence for many decades. Western experience confirms that the attainment of full or nearly full coverage of vital events is a long and gradual process. Accordingly, although the objective of a comprehensive vital registration system should be pursued, there are strong incentives to look for investments with a quicker payoff in producing the needed demographic information.

Second, even where both censuses and the vital registration are reasonably accurate, these two sources seldom meet all needs of developing countries for demographic information. The fact that the primary objectives of obtaining either set of records are legal and administrative, and the demographic data are a by-product, constitutes a limitation. Owing to this fact, there is always a strong and understandable resistance to the inclusion of demographic "frills," so defined from the viewpoint of legal and administrative needs. Such resistance is further justified by the necessarily comprehensive coverage of these records, which calls for brevity in the questionnaires utilized, both for reasons of economy and for considerations having to do with the possible negative reactions of the affected public if too detailed questions are asked. Finally, a major limitation originates from the difficulty brought about by the combined use of two sets of records typically generated by two separate data-collecting agencies. An example will suffice to suggest the nature of the problems. The calculation of a simple measure of mortality (e.g., the crude death rate) is effectively accomplished by relating the total number of deaths in a single year (a flow) to the stock of the total population enumerated at the midpoint of that year or estimated for that date. To differentiate such a rate by sex is also easily accomplished, since a classification by sex will be routinely available in both sets of records and the criterion for such classification is unambiguously defined. But a differentiation by age may be far more problematical as the biases affecting the two sets of data may differ and, hence, may reinforce rather than offset each other to some extent. Such problems multiply rapidly as we move toward the calculation of more specialized types of rates (e.g., by marital status, occupation, race, or religion) where the classifying criteria are subject to varying interpretations, assuming, of course, that the necessary classifications are available in both sets of records.

New Approaches. — The difficulties of developing a system based on a combination of censuses and conventional vital registration (with respect to costs and organizational constraints), and the limitations inherent in such a system (with respect to accuracy and coverage) even when fully developed, explain the fact that efforts to remedy the lack of demographic statistics during the last 15 years or so have been increasingly directed toward the development of alternative sources of demographic information. Two general approaches have emerged from these efforts: one relying on special sample surveys and census-type information alone, and another relying primarily on sample registration systems, usually in combination with related sample surveys. (The advantages and disadvantages of sample surveys vis-a-vis complete censuses were discussed in ch. 3.) Complementing and to a large extent guiding these advances was the development of new techniques for the analysis of existing "unorthodox" data aimed at producing estimates that were previously obtained only through the classical combination of censuses and vital statistics. The main purpose of this chapter is to give an account of these developments.

Scope of the chapter. — The organization of this chapter and the relative length of its subdivisions reflect the emphasis found in the contemporary literature. Recent developments in the analysis of existing data have been concentrated almost entirely on methods using census or census-type data as inputs. The simple explanation for this state of affairs is that, in statistically less developed countries, such data are far more abundant, and generally of a much higher quality, than are vital statistics. In this respect the situation in present-day underdeveloped countries is in an interesting and felicitous contrast to the typical case prevailing in the early statistical history of Western nations where the development of vital registration preceded the introduction of periodic censuses. The problem early demographers faced was the estimation of rates from vital data only, a famous example being the astronomer Halley's attempt to construct a life table from statistics of deaths alone. The task of making estimates from census data in the absence of vital statistics is fortunately a more rewarding one.

First of all, a cross-sectional view of the population as observed by a census reflects the cumulative results of past demographic flows and offers a base for estimating such flows, particularly if more than one census has been taken,

separated by not too wide intervals. Many underdeveloped countries today meet this requirement.

Secondly, a census-type operation need not be limited to a still photograph of the situation at a given moment — to employ a somewhat abused metaphor. Suitable questions in a census can elicit direct information on past events with or without a specified reference period for such events, thus supplying a substitute for data traditionally provided by vital registration only.

Not surprisingly, recent proposals for improved data sources are aimed at utilizing precisely this capability of a cross-sectional survey to generate flow-type data. To some degree the scope of the traditional census can be extended for these purposes, and such extension has been recommended and carried out with increasing frequency. However, since costs of a census increase somewhat as the questionnaire is expanded and, furthermore, since obtaining retrospective information of reasonable quality requires an intensity of field work and a quality of field personnel that is seldom obtainable for a full census in a less developed country, attention has increasingly shifted towards sample surveys of a much more limited size as a substitute for, or more typically as a complement to, the census as conventionally defined. This trend was reinforced by the realization that many of the characteristics required can be estimated with sufficient accuracy from sample surveys of a relatively modest size and, furthermore, that any loss owing to less than full coverage, notably a loss in geographic detail, is amply compensated for by the flexibility of the sample surveys, permitting repetitiveness of the operation and data of better quality, greater timeliness, and richness of contents.

A common characteristic of the innovation aimed at remedying the lack of traditional vital registration data through cross-sectional surveys is that it relies heavily on various analytical techniques, notably demographic models, for transforming the unconventional information into familiar types of data, for determining general characteristics from particular pieces of evidence, and for filling gaps in the existing information.

This latter feature is also present, though to a less pronounced degree, in the second type of methodological innovation, which aims at obtaining conventional types of vital registration data but on a sample basis rather than through initial coverage of all parts of a country. Estimates are then prepared using such information in conjunction with traditional census data or, more often, with data obtained from special cross-sectional sample surveys where the sampling units for the surveys coincide with those for the vital registration. These methods, however, often go far beyond merely replacing full-coverage vital statistics by information derived from a sample. One notable feature is that they incorporate techniques insuring a higher level of control over omissions by matching records from different sources, and by comparing conventional flow/stock rates with rates derived from survey data alone as referred to above. The refinement of such techniques, which at present are still in the experimental stage, promises to provide a kit of demographic tools that would supply not only almost all information that can be obtained from the conventional methods but that in some respects would be superior to the latter. Unfortunately, this increased power is typically obtained at a price of a considerable increase in costs and in organizational complexities when compared to the more modest requirements of estimates based on survey data alone.

The two main new approaches to generating data — surveys

and sample registration—do not exhaust the full range of possible sources of information on demographic processes that may be considered as alternatives to the traditional census and vital registration system. Continuous population registers, longitudinal panel studies, or intensive observation of small populations utilizing all available tools for recording demographic facts (as is customary, for instance, in intensive anthropological field work) are additional examples of non-conventional approaches to obtaining demographic data. But, because of their cost, their overly specialized nature, or because of their narrow range of applicability, they are of limited interest as possible solutions for a pervasive lack of adequate demographic statistics. Accordingly, the discussion that follows is organized under two main headings. One examines methods based on cross-sectional data, either from censuses or from surveys, or from both, whereas the other focuses on methods utilizing data collected in sample registration areas, typically linked with census-type recording systems in various ways. Preceding these two sections is a discussion of some general demographic models that are utilized extensively in applying both of the two unconventional approaches.

TOOLS OF ESTIMATION: SOME DEMOGRAPHIC MODELS

Demographic models are generalizations about a demographic phenomenon or process. In the present context, we are concerned with such models as tools of estimation. In this role the use of models becomes particularly important when the available data are limited and defective. While it is seldom justified to have recourse to such models when relevant demographic data are easily available, they are indispensable in checking and adjusting data, in filling gaps in the available records, and in deriving reliable estimates from fragmentary pieces of evidence, each of which may be defective if taken in isolation. Two types of models of particular importance—model life tables and model stable populations—will be singled out in this section for separate discussion because of their pervasive use and general importance in methods of demographic estimation for statistically underdeveloped areas.

Model Life Tables

The history of demographic analysis records numerous examples of attempts to formulate generalizations about the age pattern of human mortality.[1] A simple example is the construction of life tables from observed data, where methods of graduation, interpolation, and extrapolation often modify the raw measures greatly, substituting for them an idealized model of reality. These techniques are justifiable only if the basic data are essentially reliable and the analyst's task is merely to remove the effects of random deviations from the true underlying values or to obtain estimates for age intervals different from those according to which the basic data are grouped.

However, in statistically underdeveloped countries information on age-specific mortality may be deficient to such an extent that a reliable direct description of the pattern of mortality is not, or is hardly, feasible. In such cases, when an estimate of the true life table is desired, the solution is often to use the best available evidence to select some actual life table reflecting reliably recorded mortality experience in some other population, such as in a neighboring country. A generalization of this approach is to construct a set of model

life tables based on a wider range of recorded data. Probably the best known such set consists of the tables published by the United Nations.

The United Nations Model Life Tables. — It has been observed that the level of mortality in any given age group can be closely predicted if the level of mortality in an adjacent age group is known. The United Nations set of model life tables, first published in 1955[2] and made available in a more elaborate form in 1956,[3] were constructed from parabolic regression equations indicating the relationships between adjacent pairs of life table $_nq_x$ values as observed in 158 life tables collected from a wide selection of countries and representing different periods of time. Thus, starting from a specified level of infant mortality, q_0, a value for $_4q_1$ can be determined. From that calculated value of $_4q_1$ a value of $_5q_5$ is estimated, which in turn serves as an estimator of $_5q_{10}$, and so on, until the life table is completed. By repeating this procedure starting from various specified levels of q_0, a system of model life tables is obtained spanning the entire range of human mortality experience.

The U.N. model life tables have proven very useful in a wide range of practical applications, notably in preparing population projections with a specified pattern of mortality change. The construction of these tables, however, has been subjected to various criticisms. Apart from the statistical bias introduced by the chain-like use of a series of regression equations to construct the life table, the criticism mainly concerns two points.

First, it may be argued that the collection of life tables that went into calculating the regression equations is not sufficiently representative of the whole range of reliably recorded human mortality experience. Moreover, it includes some tables that in turn incorporate not only a great deal of actuarial manipulation but also the outright use of models. Thus, in particular, the life tables for India have a heavy influence on the pattern of mortality at low levels of $\overset{\circ}{e}_0$ shown in the U.N. model tables. But the childhood mortality values of these Indian tables are essentially extrapolations from mortality at more advanced ages and cannot be considered as indicative of reliably recorded experience. Second, the suggestion implicit in the U.N. model life tables that a single parameter (such as q_0, or any other life table value) can determine all other life table values with sufficient precision is clearly dependent on the particular use of the life table so calculated. Although high mortality in one age group does tend to imply high mortality in all other age groups as well, the detailed age patterns of human mortality do display substantial variation. To assume away the existence of such variation may be legitimate in some uses but unacceptable in others.

Regional Model Life Tables. — Regional tables, published in 1966 by Coale and Demeny[4] and reproduced in part by the United Nations in 1967,[5] consists of four sets of model life tables, labeled "West," "East," "North," and "South." Each set contains 24 tables, calculated for males and females separately, with equal spacing of the values of the expectation of life at birth for females, ranging from an $\overset{\circ}{e}_0$ of 20 years (labeled

[1] For a brief survey, see Ansley J. Coale and Paul Demeny, *Regional Model Life Tables and Stable Populations*, Princeton, New Jersey, Princeton University Press, 1966, ch. 1.

[2] United Nations, *Age and Sex Patterns of Mortality*, Series A, Population Studies, No. 22, 1955.
[3] United Nations, *Methods for Population Projections by Sex and Age*, Series A, Population Studies, No. 25, 1956.
[4] Op. cit.
[5] United Nations, *Methods of Estimating Demographic Measures From Incomplete Data*, Manual IV, Manuals on Methods of Estimating Population, Series A, Population Studies, No. 42, 1967.

as level 1) to an $\dot{e}_0$ of 77.5 years (labeled as level 24). The mortality level shown in the male tables differs from the mortality level of the female tables with which they are paired; this difference reflects the typical relationship between male and female mortality occurring in a particular population. Among the four sets of tables, the "East," "North," and "South" models are based on relatively small collections of tables in which the age pattern of mortality showed substantial and systematic deviations from the average world patterns calculated from a set of over 300 life tables (each for males and females separately) reflecting actually recorded mortality experience in various countries. These sets of model life tables are clearly of a regional character. The "East" tables are based mainly on Central European experience, whereas the "North" and "South" tables were derived from life tables of Scandinavian and South European countries, respectively.

The "West" tables, on the other hand, are representative of a broad residual group. This model set was based on some 125 life tables from over 20 countries, including Canada, the United States, Australia, New Zealand, South Africa, Israel, Japan, and Taiwan, as well as a number of countries from Western Europe. The mortality experience in these countries did not show the systematic deviations from mean world experience found in the other three groups.

The differences among the age (and sex) patterns of mortality in the four regional models are slight in some respects and pronounced in others. These differences also vary in character as one moves from higher to lower levels of mortality. Thus, no simple rules can summarize the extent to which the use of one set, in preference to another, will affect the outcome in any particular application. In general, the use of the "East," "North," and "South" models is recommended only if there is some evidence suggesting that the mortality in the population for which the model is to be used has some of the peculiarities characterizing these three models. Otherwise the use of the "West" model is to be preferred.

In almost all instances when the use of model life tables is necessary the analyst has, of course, little or no reliable information concerning the true pattern of mortality. Although the rule just given favors the use of the "West" model, it should always be remembered that in such instances there is no assurance that the true pattern is well represented by that model. Lacking substantive evidence, there exists no sound basis for deciding where a particular country's mortality fits within the range represented by the four regional models. Indeed it is certain that even the four regional models fail to span the range covered by mortality patterns in contemporary situations. But the existence of the four model patterns should make the analyst sensitive to the fact that some of his estimates may be substantially affected by the choice of a model pattern.

Consider for instance the problem of estimating infant mortality from given values of the expectation of life at age 5. What are the levels of q_0 for females in the four regional model life tables and in the U.N. set, assuming an $\dot{e}_5$ of 45, 55, and 65 years, respectively? We locate those tables that bracket the required values for $\dot{e}_5$, find the corresponding values of q_0, and then determine by interpolation the values of q_0 corresponding exactly to the required value of $\dot{e}_5$. The answer is supplied in the following figures:

Set of model tables	$\dot{e}_5 = 45.0$	$\dot{e}_5 = 55.0$	$\dot{e}_5 = 65.0$
"West"	.234	.126	.048
"East"	.331	.178	.070
"North"	.200	.112	.048
"South"	.238	.154	.092
United Nations	.210	.140	.058

Obviously these figures indicate that any attempt to estimate infant mortality from an observed or estimated value of $\dot{e}_5$ is subject to the risk of considerable error. It should also be noted that insensitivity of a particular estimate to a choice between two or more model sets cannot be interpreted to indicate similar insensitivity in other applications. For instance, given $\dot{e}_5 = 45$, the "West" and the "South" tables result in closely similar values for q_0. However, another measure of early childhood mortality, $_5q_0$, if derived from the same model tables ($\dot{e}_5 = 45$), yields quite different figures: .357 if the "West" tables are used and .439 if the "South" tables are applied. Lacking some reliable indication as to the true pattern of mortality, the analyst should obviously seek to avoid making estimates that are strongly dependent on the nature of the assumption concerning that pattern. Failing this, at least some rough indication of the magnitude of the error that may affect the estimate owing to the choice of a wrong model is an essential requirement.

Comments concerning the sensitivity of various methods to the assumption on the age pattern of mortality will be made in this chapter whenever appropriate. These comments are based on applying alternative regional models to the same problem and comparing the results. The analyst who has access to the sets of regional tables is encouraged to follow similar procedures routinely. In making estimates that are not overly sensitive to the choice of model pattern, however, the "West" set may be recommended with the greatest confidence. In most applications this set and the U.N. model life tables are close substitutes. Excerpts from the "West" tables only are reproduced in appendix B; a calculated set of stable populations also excerpted in this appendix is based on the "West" model tables. The use of a regional model serves as a constant warning that the model describes only a certain type of experience, without claiming generality.

Quite apart from the question of the reliability of the age pattern of mortality, the user of these tables should also be aware of the fact that the reliability of the tables as representations of real, recorded experience varies considerably with the level of mortality. Generally speaking, the tables are most reliable in a broad middle range of mortality. At very low expectations of life, the recorded experience is virtually nil, hence the models should be considered as somewhat tentative extrapolations. Similar caution should be exercised in attributing significance to minor details of the age pattern shown in the models representing very low levels of mortality.

Table B-1 of appendix B presents abridged life tables for females only from the "West" model at five different levels of mortality, namely levels 9, 11, 13, 15, and 17. The corresponding table for males at level 9 is also shown. Table B-2 gives a more detailed description of mortality under age 5, in terms of the function l_x, for 12 mortality levels, levels 1, 3, . . . 21, and 23, separately for males and females. In view of the fact that estimates of l_x for $x = 1, 2, . . . 5$ are often obtainable only for the two sexes together, this table also gives values for both sexes, assuming a sex ratio at birth of 1.05. Although the entire set of "West" model life tables is published in the volume by Coale and Demeny on *Regional Model Life Tables and Stable Populations,* tables at alternate levels of mortality provide a sufficient density of information that values at intermediate mortality levels can be obtained by interpolation. Simple linear interpolation can be expected to give sufficient precision in all applications. The following examples illustrate the method of calculating various life table values not directly available from the tabulations given in the appendix. Underlying the calculations in all examples is the assumption that the age pattern of mortality is well described by the "West" model.

Example 1. What is the proportion surviving from birth to age 27 among females assuming that the expectation of life at birth is 49.2 years? To find the answer, we have to interpolate between level 11 ($\mathring{e}_0 = 45$) and level 13 ($\mathring{e}_0 = 50$). Also, since abridged life tables do not contain information on l_{27}/l_0, we have to interpolate between l_{25}/l_0 and l_{30}/l_0. From appendix table B-1, we have the following figures (taking l_0 as equal to 100,000, as usual):

	$\mathring{e}_0 = 45.0$	$\mathring{e}_0 = 50.0$
l_{25}	69,022	74,769
l_{30}	66,224	72,326

Two procedures may now be followed. Either first we calculate both l_{25} and l_{30} for $\mathring{e}_0 = 49.2$ and then interpolate between l_{25} and l_{30} to obtain l_{27}, or first we calculate l_{27} for both $\mathring{e}_0 = 45.0$ and $e_0 = 50.0$ and then interpolate between $\mathring{e}_0 = 45.0$ and $\mathring{e}_0 = 50.0$ to obtain $e_0^\circ = 49.2$. Often there is no reason to prefer one sequence of the necessary calculations to another. In general, however, the more convenient interpolations should be performed first. At this point a short numerical illustration of the method of linear interpolation may be useful.

To arrive at $\mathring{e}_0 = 49.2$ from $\mathring{e}_0 = 45.0$ requires travelling "up" a fraction of the distance from 45 to 50. The proper fraction is:

$$\frac{49.2 - 45.0}{50.0 - 45.0} = \frac{4.2}{5} = .84$$

Alternatively, if we start from $\mathring{e}_0 = 50$, a fraction of the total distance,

$$\frac{50.0 - 49.2}{50.0 - 45.0} = \frac{0.8}{5} = .16$$

should be travelled "down" toward $\mathring{e}_0 = 45$. The sum of the two fractions (.84 and .16) is of course 1. This suggests that it is convenient to express the intermediate figure — 49.2 — as a weighted average of the values between which we interpolate. The larger fraction is used to weight the bracketing value to which the intermediate figure is closer, whereas the smaller fraction is used to weight the bracketing value at the other end of the interval:

$$49.2 = (45.0 \times .16) + (50.0 \times .84)$$

The same weighting factors should be used for obtaining l_{25} and l_{30} for $\mathring{e}_0 = 49.2$

$$l_{25} = (69,022 \times .16) + (74,769 \times .84) = 73,849$$

$$l_{30} = (66,224 \times .16) + (72,326 \times .84) = 71,350$$

To get l_{27} from l_{25} and l_{30}, we interpolate by weighting l_{25} with $0.6 \left(= 1 - \frac{27-25}{30-25}\right)$ and l_{30} with 0.4 $(= 1 - 0.6)$. Note that $27 = (25 \times 0.6) + (30 \times 0.4)$. Specifically for $\mathring{e}_0 = 49.2$:

$$l_{27} = 73,849 \times 0.6 + 71,350 \times 0.4 = 72,849$$

Example 2. What is the value of $\mathring{e}_{65}$ at mortality level 9, for males and females combined? In general, a simple arithmetic mean of the male and female $\mathring{e}_{65}$ values will give a good approximation. A more exact answer would call for calculating a value of T_{65} for males and females together and dividing this figure by the corresponding value of l_{65}. Assuming a sex ratio at birth of 1.05, the calculation is as follows:

	T_{65}	l_{65}
(1) Females, level 9	308,507	29,527
(2) Males, level 9	229,910	24,006
(3) Males, level 9, adjusted: line (2) × 1.05	241,406	25,206
(4) Males adjusted + females, level 9: line (1) + line (3)	549,913	54,733

The answer is then $T_{65}/l_{65} = 549,913/54,733 = 10.05$

Example 3. The value of $_2q_0$ (proportion dying under age 2) is estimated as .270 for the sexes combined. What is the implied level of mortality? Proportions surviving to age 2 out of 100,000 births (males and females combined) are tabulated in appendix table B-2. In the present example, l_2 is $(1-.270) \times 100,000 = 73,000$. This figure is bracketed by levels 7 and 9 in table B-2:

Level	l_2
7	71,112
9	75,813

The interpolation requires calculating the fraction

$$\frac{73,000 - 71,112}{75,813 - 71,112} = \frac{1,888}{4,701} = .402$$

The level of mortality corresponding to $l_2 = 73,000$ is then $7 \times (1 - .402) + 9 \times (.402) = 7 \times .598 + 9 \times .402 = 7.80$. Any life table parameters corresponding to level 7.8 can now be obtained by applying the same weights to the appropriate values in level 7 and level 9 life tables.

Model Stable Populations

In demographic analysis, it is sometimes of interest to inquire about the ultimate consequences of indefinitely maintaining in the future some specified current behavior. For instance we may ask: What would be the ultimate level of vital rates in a closed population if current age-specific death rates and fertility rates remained fixed? As was explained in chapter 18, the answer requires the calculation of stable values — the so-called intrinsic rates — to which current vital rates would converge if the current conditions of fertility and mortality indeed remained constant.[6]

Although the theory of stable population as well as computational routines required to determine stable population parameters had been worked out decades ago, it was realized only recently that constancy of fertility and mortality schedules for a wide range of human populations has been, to a close approximation, a fact of life in the indefinite past. But then the state of these populations as observed in the present must be close to the stable state. This discovery has provided us with a powerful tool in estimating population characteristics for populations where demographic statistics are deficient or erroneous and where the assumption of stability is a realistic one.

The essence of the stable estimating procedure consists of two basic steps: (1) on the basis of the available evidence relating to a given population, a stable population is constructed; and (2) the various parameters of the stable population are assigned to the actual population as estimates of the corresponding parameters in that population. The power of the technique lies in the fact that step (1) can be made with confidence, often even on the basis of fragmentary data, whereas step (2) yields a whole series of sophisticated measures for which no direct information exists at all in the actual population. The weaknesses of the method are twofold. First, the actual situation may be poorly approximated by a stable model. The reasons for this can be substantial migratory movements, particularly if sex-selective; substantial, if temporary, deviations from constant fertility and mortality in the past (e.g., owing to epidemics, wars, or other unusual conditions), even if both fertility and mortality have been following horizontal trends; and systematic changes in the level of fertility or mortality. Second, even if the true situation is close to a stable state, the available data may be too fragmentary or biased to permit the derivation of the appropriate stable population. When the quality of the basic recorded data is poor, such data will appear to be consistent with too broad a range of stable populations to be of much value for the analyst.

Despite these defects the method, or modifications thereof, has proven effective under many circumstances and a significant portion of our current knowledge on world demographic trends and characteristics comes directly from the applications of stable population analysis. Although it is to be hoped that improvements in the volume and quality of demographic data in the statistically underdeveloped countries will tend to reduce the need for stable population techniques in the future, these techniques at present are still indispensable.

The two steps in preparing stable estimates that were indicated above can be made somewhat mechanically by following detailed rules and examples set forth at later portions of this chapter. A full understanding, however, of the logic underlying the method that would alone permit its creative application in unusual situations requires familiarity with the essence of the stable model, as well as familiarity with its computation, even though in practice the analyst might routinely use a precomputed set of stable populations such as given in the Coale and Demeny volume or U.N. Manual IV and illustrated in appendix table B–3. A method of deriving a stable population distribution when the fixed mortality and fertility schedules are known was described in Chapter 18.

The Stable Population Model. — Consider a female population in which the number of life births during 1970 was B_{1970}. Suppose that during the indefinite past this population (a) has been closed to migration; (b) the number of births has been growing at a constant annual rate r; and (c) mortality, as described by a life table, has been constant. What will be the size of the population at each age recorded by such a census?

Evidently the number of children under 1 year old at the end of 1970 will be the survivors of the birth cohort of 1970; the 1-year olds will be the survivors of the births in 1969; and in general the number of x-year olds will be the survivors of births that have occurred x years before 1970. (Ages are expressed as exact age at last birthday.) Thus, to give an answer to the question posed above requires (1) the calculation of the number of births in each of the 100 or so years preceding 1970 and (2) the calculations of the survivors to the end of 1970.

Since in 1970 there were B_{1970} births and since this number was growing annually at the rate of r, we have

Births in 1970 $= B_{1970}$

Births in 1969 $= B_{1970}/(1+r)$ or, more exactly, $B_{1970}e^{-r}$

Births in 1968 $= B_{1970}/(1+r)^2$ or $B_{1970}e^{-2r}$

Births in 1932 $= B_{1970}/(1+r)^{38}$ or $B_{1970}e^{-38r}$

Births in 1970$-x = B_{1970}/(1+r)^x$ or $B_{1970}e^{-rx}$

Of the two alternative expressions on the right hand side of the above equations, the one using the exponential function will be retained in the following illustrations as it is computationally more convenient and also corresponds better to the continuous nature of population growth than the formula assuming yearly compounding. Of course, for values of r within the range of human experience, particularly when the absolute value of r is small, the two formulas yield results that are numerically very close to each other.

The population alive at the end of 1970 can be now easily calculated:

Persons at age 0 $= B_{1970}L_0/l_0$

Persons at age 1 $= B_{1970}e^{-r}L_1/l_0$

Persons at age 2 $= B_{1970}e^{-2r}L_2/l_0$

Persons at age $x = B_{1970}e^{-rx}l_x/l_0$

What will be the rate of growth of this population? Consider first the rate of growth at an arbitrarily selected age, for instance, those aged 38. The number of 38-year-olds at the end of 1970 (P_{38}^{1970}) was obtained by calculating first the size of the birth cohort of 1932 and multiplying that number by a factor indicating survival from birth to age 38.

$$P_{38}^{1970} = B_{1932}L_{38}/l_0 = B_{1970}e^{-38r}L_{38}/l_0 \qquad (1)$$

The same procedure applies in calculating P_{38} for some other date, such as for the end of 1969:

$$P_{38}^{1969} = B_{1931}L_{38}/l_0 = B_{1970}e^{-39r}L_{38}/l_0 \qquad (2)$$

---

[6] Since such an assumption appears to be unrealistic under most circumstances, it is sometimes charged that the exercise in question is at best of an academic interest. But such a judgment is based on the misinterpretation of the stable measures. Just as a speedometer reading of 60 miles an hour is a measure of the **current** speed, and the implied prediction (that the vehicle will be 60 miles away if **current** speed is maintained for an hour) is of secondary importance, so the calculated stable population and its various parameters are of primary interest as reflections of a **current** situation.

Dividing the right-hand side of equation (2) into equation (1) gives r as the rate of growth of the age group 38 during the calendar year 1970. The same reasoning holds for any other age group or any other time interval. In other words, in a closed population where mortality has been constant and where births are growing at a fixed annual rate, r, each age group, and consequently the population as a whole, is also growing at the same annual rate. It follows that in such a population the size of each age group relative to any other age group, or to the population as a whole, remains constant; in other words that the age distribution is "stable."

In the previous example, the population at age x at the end of 1970 was:

$$P_x^{1970} = B_{1970-x} L_x/l_0 = B_{1970} e^{-rx} L_x/l_0 \qquad (3)$$

The total population P is obtained by summing the number of people at each age group from age 0 to the highest age ω:

$$P^{1970} = \sum_{x=0}^{\omega} B_{1970} e^{-rx} L_x/l_0 = B_{1970} \sum_{x=0}^{\omega} e^{-rx} L_x/l_0 \qquad (4)$$

By dividing the right-hand side of equation (4) into equation (3), we obtain the expression for the proportion at age x in a stable population:

$$c_x = \frac{e^{-rx} L_x/l_0}{\sum e^{-rx} L_x/l_0} \qquad (5)$$

Note that the expression no longer contains a dated quantity, such as B_{1970}. Given the constancy of r and of the life table, the age distribution remains fixed over time.

What will be the birth rate in such a stable population? To obtain the birth rate for 1970, we must divide B_{1970} by the population at mid-1970. As before, we may obtain the total population by summing up individual age groups. The number of those aged x at mid-1970 will be, according to equation (3)

$$P_x^{\text{mid-1970}} = B_{1970} e^{-r(x+1/2)} L_x/l_0 \qquad (6)$$

Note that the expression $B_{1970} e^{-r(x+1/2)}$ gives the number of births during yearly intervals, going backwards from the mid-point of 1970, in terms of the number of births during the calendar year 1970 and the annual rate of growth. Thus, for $x=0$, the expression yields the number of births from mid-1969 to mid-1970. For $x=1$ the number of births calculated are those that took place between mid-1968 and mid-1969, etc.

Total population at mid-1970 is calculated as

$$P^{\text{mid-1970}} = \sum_{x=0}^{\omega} P_x^{\text{mid-1970}} = \sum_{x=0}^{\omega} B_{1970} e^{-r(x+1/2)} L_x/l_0 \qquad (7)$$

and the birth rate b is given by

$$b = \frac{B_{1970}}{P^{\text{mid-1970}}} = \frac{B_{1970}}{\sum B_{1970} e^{-r(x+1/2)} L_x/l_0} = \frac{1}{\sum e^{-r(x+1/2)} L_x/l_0} \qquad (8)$$

Again, the expression does not contain quantities with a time-subscript; in a stable population, the birth rate is constant. As the rate of growth of the population was shown to be also constant, the death rate d is constant as well:

$$d = b - r \qquad (9)$$

The same calculations can also be carried out if the initial stable conditions specify a life table and a constant set of age-specific fertility rates, f_x, as such a combination implies a rate of growth. Clearly, the number of births in 1970 is the cumulative product of the age-specific fertility rates and the number of women over all the childbearing ages from ω_1 to ω_2 (roughly from ages 15 to 49):

$$B_{1970} = \sum_{x=\omega_1}^{\omega_2} P_x^{\text{mid-1970}} f_x$$

$$B_{1970} = \sum_{x=\omega_1}^{\omega_2} B_{1970} e^{-r(x+1/2)} f_x L_x/l_0$$

$$1 = \sum_{x=\omega_1}^{\omega_2} e^{-r(x+1/2)} f_x L_x/l_0 \qquad (10)$$

It can be shown that equation (10) has a unique solution for r. In other words, a set of fixed age-specific fertility rates and a life table determine a unique stable population, with unique vital rates and with a unique age distribution.

On the other hand, specification of a stable growth rate and a fixed life table is not sufficient to determine the series of age-specific fertility rates since for any fixed growth rate and life table an arbitrarily large number of f_x schedules can be constructed that would satisfy equation (10). The determination of the f_x schedule and of such measures of the stable population as the gross reproduction rate $(GRR = \Sigma f_x)$ or the net reproduction rate $(NRR = \Sigma f_x L_x/l_0)$ requires a specification of the age pattern of fertility as well as the rate of growth and the life table. Suppose that such an age pattern of fertility is described by a fertility schedule f_x^*, so that the true fertility schedule f_x is simply a multiple k of f_x^* at each age: $f_x = k f_k^*$. If we know r, the life table schedule L_x/l_0, and the f_x^* schedule, equation (10) permits calculation of k:

$$k = \frac{1}{\sum_{x=\omega_1}^{\omega_2} e^{-r(x+1/2)} f_x^* L_x/l_0} \qquad (11)$$

Once k is known, the true fertility schedule $k f_x^*$, and consequently such summary stable indexes of fertility and reproduction as the GRR and the NRR, are easily calculated.

Regional Stable Population Models. — Despite the essential simplicity of the stable population model, typical applications tend to require onerous calculations if attempted without access to a precomputed set of model stable populations. Suppose, for instance, that we should like to know the birth rate in a male stable population with a given proportion of persons under age 10 and with a given death rate. As there exists no convenient analytical expression from which such a birth rate could be directly calculated, the only feasible approach is to calculate a trial stable population, to observe its proportion under 10 and its death rate and, using the difference of the observed values and the desired values, to calculate another stable population close to the desired one. This procedure usually will have to be repeated several times before we can reach the stable population with exactly the desired proportion under 10 and with the desired death rate. Since actual populations are never exactly stable and since observed parameter values are often distorted by reporting errors, in making estimates the analyst should carry out a whole series of such calculations to explore the range of various estimates

Table 24-1. — **Calculation of the Net Reproduction Rate (NRR) in a Model Stable Population ("West" Females, $e_0 = 45.0$, $r = .025$) Assuming a Mean Age at Maternity of 29 Years**

[Implied GRR is 3.03]

Age (x to x + n)	Expected person years (in birth cohort of 100,000) $(\overset{o}{e}_0 = 45.0)$	Fertility schedule assuming GRR=1.0 and $\bar{m} = 29$[1]	Expected births (2) x (3)=
(1)	(2)	(3)	(4)
15 to 19 years.....	363,207	.018	6,538
20 to 24 years.....	351,543	.042	14,765
25 to 29 years.....	338,115	.056	18,934
30 to 34 years.....	323,525	.044	14,235
35 to 39 years.....	307,872	.028	8,620
40 to 44 years.....	291,388	.010	2,914
45 to 49 years.....	273,969	.002	548
Sum...........			66,554

$$NRR = \frac{66,554 \times 3.03}{100,000} = 2.02$$

[1]Calculated by authors.

resulting from different combinations of observed parameters. The need for such exploration tends to make the needed computations prohibitively voluminous.

The foregoing considerations suggest that a flexible and efficient application of stable techniques presupposes the existence of a tabulated network of stable populations spanning the entire feasible range of mortality and fertility experience. Such a network may be most conveniently computed by combining a series of model life tables from lowest to highest levels of mortality with a broad range of growth rates. A tabulation of this type is illustrated in appendix table B-3. Table B-3 presents a series of stable populations excerpted from the volume, *Regional Model Life Tables and Stable Populations*, cited above. The tabulations printed in the appendix tables are all from the "West" family. They were obtained by combining various mortality levels in the "West" model life tables shown in appendix table B-1 (selected tables only) with 13 evenly spaced values of the rate of increase ranging from $r = -.010$ to $r = .050$ (whole array shown only for levels 9 and 11). (Such tables are computed separately for females and males.) For each stable population, table B-3 gives the proportionate age distribution, the cumulated age distribution (up to age 65), and various stable parameters, such as the rates of birth and death, and gross reproduction rates associated with four values of the mean age at maternity. The detailed characteristics of the life table underlying some of the stable populations can be established by referring to the model life tables in table B-1. United Nations Manual IV of its Manuals on Methods of Estimating Population presents model stable populations for 12 levels of mortality representing every second level of mortality out of the 24 presented in the Coale and Demeny volume.[7] The network of the model stable populations in this manual is sufficiently detailed so that we may obtain with adequate precision intermediate stable populations (or single parameter values) through linear interpolation.

The method of calculating stable population parameters not directly available in appendix table B-3 is illustrated in the following examples:

[7] United Nations, *Methods of Estimating Basic Demographic Measures From Incomplete Data*, op. cit.

Example 1. Calculate the birth rate (b), the proportion under age 35 ($C(35)$), and the gross reproduction rate assuming a mean age at maternity of 28.2 years — $GRR(28.2)$ — in a "West" female stable population with $\overset{o}{e}_0 = 48.7$ and $r = .0263$.

The sought-for stable population is bracketed by the four stable populations tabulated for level 11 and level 13 ($\overset{o}{e}_0 = 45.0$ and 50.0, respectively) at $r = .025$ and $r = .030$. First, we interpolate between $r = .025$ and $r = .030$ in appendix table B-3 (levels 11 and 13, females) to obtain cols. (1) and (2) below, both representing stable populations with $r = .0263$. Second, we interpolate between cols. (1) and (2) to obtain a stable population (shown in col. (3)) having both $r = .0263$ and level 12.48 ($\overset{o}{e}_0 = 48.7$). The gross reproduction rates are calculated in this exercise for $\bar{m} = 27$ and $\bar{m} = 29$. To obtain GRR (28.2), a final interpolation is required between $GRR(27)$ and $GRR(29)$. The result is given in the bottom line of col. (3):

Parameter	Level 11 $\overset{o}{e}_0 = 45.0$ $r = .0263$ (1)	Level 13 $\overset{o}{e}_0 = 50.0$ .0263 (2)	Level 12.48 $\overset{o}{e}_0 = 48.7$ .0263 (3)
b......................	.0456	.0421	.0430
$C(35)$..................	.7710	.7580	.7614
$GRR(27)$..............	2.94	2.71	2.77
$GRR(29)$..............	3.14	2.89	2.96
$GRR(28.2)$...........	(3.06)	(2.82)	2.88

Other parameter values of the same stable population are to be obtained by similar interpolations. The reader may check his understanding by calculating values for $C(10)$ — the proportion under age 10 — and l_2/l_0. (The answers are .3110 and .8399, respectively.)

Example 2. What is the sex ratio in the stable population defined by "West" level 9 mortality and a growth rate of .020, and what is the birth rate for the sexes combined assuming that the sex ratio at birth is 1.05? From table B-3 we have: $b_{female} = .0433$ and $b_{male} = .0456$. For 1,000 current female births (the choice for the size of the current birth cohort is arbitrary and does not affect the final results), the female population is $\frac{1,000}{.0433} = 23,095$. For 1,000 male births, the male population is $\frac{1,000}{.0456} = 21,930$. Since there are 1.05 male births for each female birth, the male population must be inflated accordingly. The population sex ratio is then $\frac{1.05 \times 21,930}{23,095} = \frac{23,027}{23,095} = .997$. The birth rate for both sexes is calculated by dividing the sum of the male and female births by the sum of the male and female population: $\frac{1,050 + 1,000}{23,027 + 23,095} = \frac{2,050}{46,122} = .0444$, or 44.4 per 1,000. (Note that a simple arithmetic mean of the male and female birth rates would give a very close approximation in most instances.)

Example 3. What is the net reproduction rate in a "West" female stable population with an $\overset{o}{e}_0 = 45.0$ and $r = .025$, if we assume that the mean age at maternity is 29 years? The needed calculation is summarized in table 24–1. The expected person-years to be lived in the childbearing ages by an original cohort of 100,000 women shown in col. (2) is taken from appendix table B-1 (females, level 11). The calculation in cols. (3) and (4) indicates that with a fertility schedule assuming $GRR(29) = 1.00$, 66,554 female children would be born to the 100,000 women by the end of the childbearing ages. However, appendix table B-3 shows that the actual GRR in the stable population defined above equals 3.03. Thus, the actual number of female children born to 100,000 women will be 66,554 x 3.03 = 201,659, or 2.02 per woman. A good estimate of the NRR can be obtained directly, and much more easily, by using the approximation: NRR = GRR l_m/i_o. In this instance $m = 29$ and $l_{29}/l_0 = .6678$ (by interpolating between l_{25} and l_{30} in the appropriate life table). Thus, NRR = 3.03 x .6678 = 2.02.

METHODS OF ESTIMATION BASED ON DATA FROM CENSUSES AND SURVEYS

Suppose that in a given country vital statistics are non-existent or are so grossly deficient as to be useless for purposes of demographic estimation. Suppose, furthermore, that in such a country the population has been enumerated in one or more censuses or one or more cross-sectional demographic surveys have been taken. These assumptions correspond to reality in a large majority of the statistically underdeveloped areas in the contemporary world. This section seeks to answer two broad questions that demographers will be asked when the above-stated conditions prevail.

First, given the available census and survey data, what are the methods by which measures of demographic flows (such as birth and death rates, gross reproduction rates, life tables, etc.) may be derived? Clearly, no general answer can be given to this question since the tools the analyst may use depend on the exact nature of the available data. Therefore, in discussing the answer, it will be necessary not only to distinguish between various problems of estimation according to the demographic quantity to be measured (e.g., birth rate, death rate, etc.), but also to consider various situations that differ with respect to the existing data base.

Second, given the existing tool-kit of demographers, what is the best advice the analyst can give to the census and survey taker as to the kinds of data to be collected? Again, no generally correct answer can be given since the answer will necessarily depend on a weighing of the needs of the users of the final estimates against the costs and efficiency of data-collecting in specific situations. A fairly general ranking of the various pieces of data by order of importance can be suggested with some confidence, however.

Estimation of Vital Rates From Two Consecutive Censuses

Method Based on Observed Intercensal Growth Rate and Census Survival Rates.—Consider the information presented in table 24-2 on the population of Fiji. It summarizes perhaps the most basic cross-tabulation that is likely to be available in any census—that by age and sex. The table gives such data for two consecutive censuses, those of 1956 and 1966.[8] Since, during the intercensal years, the population of Fiji was essentially closed to migration, a comparison of the total female population figures yields the natural growth rate per annum (r) for the intercensal period t (females only):

$$\frac{233,874}{167,064} = 1.3999 = e^{rt} \qquad r = \frac{\log_e 1.3999}{t}$$

Given that the censuses were taken 10 years apart ($t = 10$), this calculation yields $r = .0336$ (i.e., an annual rate of growth of 3.36 percent). Actually the official census dates were separated by a distance of 14 days less than 10 years. With $t = 9.962$ years the calculation yields $r = .0338$. In view of the likely errors in the counts, however, we have used the simpler computation for present purposes.

Estimation of the level of mortality.—Relying upon the data in table 24-2 alone, can we say anything about the absolute magnitudes of the crude birth rate and the crude death rate during the 1956-66 period that produced a growth rate of .0336? Since the censuses were taken 10 years apart, it is easy to identify identical 5-year birth cohorts enumerated in 1956 and in 1966. Those aged 0-4 in 1956 should appear as aged 10-14 in 1966, for instance. In a closed population, the depletion of each cohort will reflect the effect of mortality on that cohort. But total deaths during the period cannot be calculated, since the population under age 10 in 1966 was not yet

Table 24-2. — Population of Fiji by Age and Sex as Enumerated: 1956 and 1966

Age (x to x+ n)	Population				Proportionate age distribution			
	Female		Male		Female		Male	
	1956	1966	1956	1966	1956	1966	1956	1966
(1)	(2)	(3)	(4)	(5)	(6)	(7)	(8)	(9)
Total, all ages.................	167,064	233,874	178,254	242,607	1.0000	1.0000	1.0000	1.0000
Under 5 years......................	30,917	40,545	31,668	41,918	.1851	.1734	.1777	.1728
5 to 9 years.......................	25,795	37,951	26,679	39,372	.1544	.1623	.1497	.1623
10 to 14 years.....................	21,642	31,061	22,452	31,892	.1295	.1328	.1260	.1315
15 to 19 years.....................	18,162	25,438	18,312	25,473	.1087	.1088	.1027	.1050
20 to 24 years.....................	14,776	21,128	15,133	20,677	.0884	.0903	.0849	.0852
25 to 29 years.....................	12,105	17,247	12,667	17,060	.0725	.0737	.0711	.0703
30 to 34 years.....................	9,757	13,399	10,167	14,054	.0584	.0573	.0570	.0579
35 to 39 years.....................	8,749	11,455	9,299	11,957	.0524	.0490	.0522	.0493
40 to 44 years.....................	6,775	9,093	7,775	9,675	.0406	.0389	.0436	.0399
45 to 49 years.....................	4,898	8,072	5,960	8,819	.0293	.0345	.0334	.0364
50 to 54 years.....................	3,733	6,205	4,285	7,091	.0223	.0265	.0240	.0292
55 to 59 years.....................	2,875	4,147	3,467	4,689	.0172	.0177	.0194	.0193
60 to 64 years.....................	2,635	3,111	3,709	3,527	.0158	.0133	.0208	.0145
65 to 69 years.....................	1,671	1,981	2,679	2,372	.0100	.0085	.0150	.0098
70 to 74 years.....................	1,247	1,344	1,914	1,666	.0075	.0057	.0108	.0069
75 to 79 years.....................	636	782	966	1,107	.0038	.0033	.0054	.0046
80 years and over..................	691	915	1,122	1,258	.0041	.0039	.0063	.0052

Source: Official census reports.

---

[8] The population for which age was not stated is omitted from the tabulation as it represented a negligible fraction (about 1/20th of 1 percent) of the total.

alive at the time of the 1956 census. Thus, comparison of two consecutive census counts gives only a very partial picture of the number of deaths in infancy and during the early childhood ages that have occurred in the intercensal period. A measure of the mortality at roughly age 5 and over is certainly implicit, however, in the data of table 24–2. If that mortality is calculated, an extrapolation to ages 0–4 may successfully complete the task of estimating the level of overall mortality.

The first problem, then, is to derive the level of mortality over age 5. One possibility is to calculate 10-year census survival rates and construct a life table from such rates. Under the typical conditions of age misreporting and differential under-enumeration that prevail in countries with inadequate statistics, such a procedure is almost always unworkable. In the instance of Fiji, the female census survival rate from age 0–4 in 1956 to 10–14 in 1966 is 1.005 – an obvious impossibility in the absence of migration. The other survival rates remain under unity, hence are theoretically possible; but they show a marked fluctuation that is patently spurious in the light of reliably recorded mortality experience elsewhere. To cope with this problem, one may wish to smooth the age distributions at the census dates to reduce the effect of age misreporting (and possibly correct them for net omissions) prior to calculating the census survival rates. Alternatively, a smoothing of the highly erratic individual census survival rates may be attempted. There exists, unfortunately, no methodology that would result in a satisfactory solution in either case. A third solution, however, eliminates the need for drastically arbitrary assumptions and permits the extracting of the information on mortality over age 5 implicit in two consecutive censuses in a relatively satisfactory manner. The essence of the solution consists of calculating mortality levels that correspond to reported "cumulative survival rates" (e.g., proportions surviving from age 0 and over at one census to age 10 and over at a census taken 10 years later; from age 5 and over to age 15 and over, etc). Such mortality levels will generally show a reasonably high level of

consistency, hence they permit the estimating of a single mortality level (e.g., by selecting the median of the series) with some confidence. The calculation assumes that a set of model life tables is available. The computational routine is illustrated in tables 24–3 to 24–6 for the female population of Fiji.

Column (2) of table 24–3 reproduces the female age distribution as reported in 1956. This population is projected to 1966 in cols. (7) to (10) under various assumptions as to the level of mortality during 1956–66. The 10-year survival rates corresponding to the various specified mortality levels are calculated from the West female model life tables in the usual fashion, i.e., as $_5L_{x+10}/_5L_x$. Thus, for mortality level 11, the proportion surviving from age 15–19 to age 25–29 is

$$338,115/363,207 = .9309, \text{ etc.}$$

The mortality levels in this calculation should be selected in such a manner that, when the projected populations are cumulated (so as to show numbers at age 10 and over, 15 and over, 20 and over, etc., the cumulated figures will bracket the corresponding reported population figures in 1966, as in the case in table 24–4. The selection may involve a simple trial and error procedure, starting from an arbitrary mortality assumption first and successively modifying that assumption in the light of the calculated results. Naturally, the computation is likely to be much simpler if the first hypothetical mortality level by which the projection is performed is judiciously chosen.

It is seen from table 24–4 that with mortality levels of 13 and 15 the female population at age 20 and over in 1966 (projected from 1956) would have been 98,218 and 99,833, respectively (3rd line, cols. 4 and 5). By linear interpolation we calculate that the actually reported 1966 population of 98,879 would have resulted from a mortality level of 13.82. Similar results for other ages x and over up to $x = 50$ are shown in

Table 24–3. — Projection of Female Population of Fiji From 1956 to 1966 Assuming Various Levels of Mortality

Age (x to x + n)	Population, 1956	10-year survival rates in "West" female model life tables at various levels of mortality [1]				Projected population in 1966 assuming various levels of mortality			
		Level 11 ($\overset{o}{e}_0$=45.0)	Level 13 ($\overset{o}{e}_0$=50.0)	Level 15 ($\overset{o}{e}_0$=55.0)	Level 17 ($\overset{o}{e}_0$=60.0)	Level 11 (2) x (3)=	Level 13 (2) x (4)=	Level 15 (2) x (5)=	Level 17 (2) x (6)=
(1)	(2)	(3)	(4)	(5)	(6)	(7)	(8)	(9)	(10)
Under 5 years....................	30,917	.9063	.9273	.9478	.9649				
5 to 9 years.....................	25,795	.9516	.9618	.9715	.9798	28,020	28,669	29,303	29,832
10 to 14 years...................	21,642	.9439	.9552	.9655	.9752	24,547	24,810	25,060	25,274
15 to 19 years...................	18,162	.9309	.9444	.9562	.9680	20,428	20,672	20,895	21,105
20 to 24 years...................	14,776	.9203	.9355	.9488	.9621	16,907	17,152	17,377	17,581
25 to 29 years...................	12,105	.9106	.9273	.9419	.9563	13,598	13,823	14,019	14,216
30 to 34 years...................	9,757	.9007	.9186	.9340	.9492	11,023	11,225	11,402	11,576
35 to 39 years...................	8,749	.8899	.9083	.9239	.9391	8,788	8,963	9,113	9,261
40 to 44 years...................	6,775	.8713	.8986	.9067	.9225				
45 to 49 years...................	4,898	.8377	.8597	.8777	.8956	7,786	7,947	8,083	8,216
50 to 54 years...................	3,733	.7836	.8101	.8315	.8530	5,903	6,034	6,143	6,250
55 to 59 years...................	2,875	.7045	.7361	.7614	.7871	4,103	4,211	4,299	4,387
60 to 64 years...................	2,636	.5987	.6335	.6609	.6894	2,925	3,024	3,104	3,184
65 to 69 years...................	1,671	.4644	.4992	.5265	.5553	2,025	2,116	2,189	2,263
70 and over.....................	2,574	[2].1951	[2].2128	[2].2286	[2].2463	2,856	3,051	3,209	3,379
Total.....................	167,064					148,909	151,697	154,196	156,524

[1] $_5L_{x+10}/_5L_x$.
[2] T_{80}/T_{70}.

Table 24-4.—Female Population of Fiji at Age _x_ and Over in 1966 as Reported by the Census of 1966 and as Projected From 1956 Assuming Various Levels of Mortality in "West" Female Model Life Tables

Age (x and over)	Census population	Projected population assuming various mortality levels			
		Level 11	Level 13	Level 15	Level 17
(1)	(2)	(3)	(4)	(5)	(6)
10 years and over...	155,378	148,909	151,697	154,196	156,524
15 years and over...	124,317	120,878	123,028	124,893	126,692
20 years and over...	98,879	96,342	98,218	99,833	101,418
25 years and over...	77,751	75,914	77,546	78,938	80,313
30 years and over...	60,504	59,007	60,394	61,561	62,732
35 years and over...	47,105	45,409	46,571	47,542	48,516
40 years and over...	35,650	34,386	35,346	36,140	36,940
45 years and over...	26,557	25,598	26,383	27,027	27,679
50 years and over...	18,485	17,814	18,436	18,944	19,463

table 24-5, both in terms of mortality level and in terms of the corresponding expectation of life at age 5. (At values over 50 years of age, the results are very sensitive to age misreporting and should be ignored.) They show a fairly high level of consistency and are well described by the median value. Thus, the best available estimate of $\mathring{e}_5$ may be taken as 57 years.

Calculation of the crude death rate.—The estimate of $\mathring{e}_5$ is of considerable interest for its own sake, but of itself is not sufficient for the calculation of the crude death rate. For that purpose we also need estimates of the age-specific death rates, including the death rate under age 5. Lacking other information, we may simply assign the $_nm_x$ values "found" in the median level (13.77) model life table; these values are obtained by interpolation of values at levels 13 and 15. The death rate under age five ($_5m_0$) is calculated from the model life tables as

$$\frac{l_0 - l_5}{L_0 + {}_4L_1}$$

In combination with the average population by age between 1956 and 1966, these death rates now permit calculation of the absolute total number of deaths per annum during the intercensal period. This calculation is shown in col. 4 of table 24-6. The crude death rate for females is then obtained as the ratio of the number of calculated deaths per annum to the mean population during the decade, or, in this instance, as $(2,696.5)/(200,470) = .0135$ per head, or 13.5 per 1,000 of the population.

Table 24-5.—Levels of Mortality of Fiji Females and Corresponding Expectations of Life at Age 5 Derived From Proportions Surviving to Age _x_ and Over in 1966 From Age _x_-10 and Over 10 Years Earlier

Age (x and over)	Level of mortality	Expectation of life at age 5
(1)	(2)	(3)
10 years and over............	16.02	60.21
15 years and over............	14.38	57.82
20 years and over............	13.82	57.02
25 years and over............	13.29	56.28
30 years and over............	13.19	56.13
35 years and over............	14.10	57.42
40 years and over............	13.77	56.95
45 years and over............	13.54	56.63
50 years and over............	13.19	56.13
Median......................	13.77	56.95

Calculation of the crude birth rate.—The female crude birth rate now can be calculated as the sum of the observed rate of increase plus the death rate as estimated from census survival rates:

$$b_f = .0336 + .0135 = .0471$$

An exactly analogous but independent calculation may be followed with respect to the male population shown in table 24-2. This calculation is left to the reader as an exercise. It results in an estimated male death rate of .0135 and a male birth rate of .0443. For the total population of Fiji during the 1956-66 period, we then have the following combined estimates:

Rate of natural increase.........	.0322
Death rate..........................	.0135
Birth rate...........................	.0457

Validity of estimates based on observed growth rate and census survival rates.—In preparing any estimate, the analyst should be able to indicate to what extent his estimates are insensitive, or "robust," (1) to the various assumptions incorporated in the estimating procedure and (2) to the possible errors in the data themselves. The most convenient way to shed light on these problems is to calculate multiple sets of estimates based on alternative assumptions in applying the estimating procedure and alternative hypotheses as to the accuracy of the basic data. Such procedures, guided by the knowledge of special local circumstances, should be routinely followed so as to arrive at qualified results. Only brief comments can be made in the present context concerning this topic.

The major questionable assumption incorporated in the calculations based on census survival rates (apart from any assumption concerning completeness of data) has to do with the model life table pattern. What are the consequences of selecting West model life tables in preference to other models? The answer may be given in two parts. First, the estimates of mortality at age 5 and over, whether expressed in terms of $\mathring{e}_5$ or of some other measure, are largely insensitive to the model pattern chosen. In other words, even the

Table 24-6.—Calculation of the Crude Death Rate for the Female Population of Fiji in the Period 1956-66 Corresponding to the Recorded Age Distribution and to a Mortality Level Estimated From Cumulated Census Survival Rates

Age	Mean population, 1956-1966	Death rate (Level=13.77, "West", females)	Mean annual deaths, 1956-1966 (2) x (3)=
(1)	(2)	(3)	(4)
Total...............	200,470	[1].0135	[2]2,696.5
Under 5 years..........	35,731	.0367	1,311.3
5 to 9 years...........	31,873	.0039	124.3
10 to 14 years.........	26,352	.0031	81.7
15 to 19 years.........	21,800	.0042	91.6
20 to 24 years.........	17,952	.0054	96.9
25 to 29 years.........	14,676	.0061	89.5
30 to 34 years.........	11,578	.0070	81.0
35 to 39 years.........	10,102	.0078	78.8
40 to 44 years.........	7,934	.0088	69.8
45 to 49 years.........	6,485	.0104	67.4
50 to 54 years.........	4,969	.0141	70.1
55 to 59 years.........	3,511	.0190	66.7
60 to 64 years.........	2,873	.0288	82.7
65 to 69 years.........	1,826	.0423	77.2
70 years and over.......	2,808	.1095	307.5

[1] (3) = (4)/(2).
[2] By summation.

selection of sharply differing life table patterns will leave the estimate of mortality over age 5 largely unaffected. Second, the opposite is true when mortality estimated for age 5 and over is extrapolated to ages under 5. As was shown above in connection with the discussion of model life tables, the record indicates that historically very different levels of early childhood mortality are associated with a fixed level of mortality at higher ages. If the life table pattern chosen underestimates the level of infant mortality, for example, the estimated total death rate will be lower than warranted and there will be an error of the same absolute magnitude and direction in the estimated birth rate. Thus, the basic weakness of the method is that there exists no way to extract information on the entire mortality pattern from reported age distributions alone.

Concerning errors in the basic data, the method of cumulative census survival rates is relatively insensitive to errors of age misreporting. No such statement can be made, however, as to the effect of differential net under- or over-enumeration of the population in two consecutive censuses. Such errors affect, of course, the observed intercensal growth rate; the closer the two censuses and the larger the difference between the relative completeness of the two censuses, the more serious is the error. A two percent relative underenumeration in the 1966 Fiji census, for instance, would have caused an underestimation of the value of $\mathring{e}_5$ in the intercensal period by more than 2 years.

Stable Population Estimates From Observed Age Distribution and Intercensal Growth Rate. — The discussion of the stable population model earlier in this chapter indicated that, if an observed configuration of population parameters closely resembles those of a particular stable population, then various other parameters of that stable population may be accepted as estimates of the corresponding parameters of the observed population. Ideally, derivation of such stable population estimates should be attempted only if the existence of approximate stability can be verified by direct evidence, such as through the observed stability of the age distribution in consecutive censuses and through observed stability of the rate of growth in successive census intervals. Sundry indications of unchanging fertility behavior and stability of mortality conditions may present additional and important, if impressionistic, evidence in support of the hypothesis that a population is approximately in the stable state. (When the recorded data from censuses and the registration system are deemed accurate, then there is no need to apply stable analysis.)

Such requirements for "proofs" of stability are strictly speaking not satisfied in the instance of the population of Fiji. Comparison of the 1956 and 1966 censuses shows noticeable (if relatively minor) changes in the reported age distribution (cf. cols. 6 to 9 in table 24–2). There is evidence of an acceleration of the rate of population growth and of a decline of mortality. Earlier migratory movements still exert an influence on the age and sex composition. Furthermore, the population is not homogeneous with respect to the major demographic characteristics but is composed of racial groups with some markedly different demographic traits.

Similar objections to the validity of applying stable population analysis can be made in almost all countries. Yet, the technique of stable analysis proves useful under a wide range of circumstances provided that such analysis contains a reasonably full exploration of the conflicting evidence presented by the observed data. Indeed, if the conditions of stability are not fulfilled, this fact will be shown by the very results of the stable population analysis itself.

In the instance of Fiji, the overwhelmingly young age distribution does show that no sustained and substantial decline of fertility took place prior to 1966. The examination of census survival rates strongly suggests the presence of age misreporting and probably some omission of children under age 5. Variations in the observed age distributions thus may reflect differential occurrence of such errors of enumeration in the censuses of 1956 and 1966. It is also known that an orderly decline of mortality has only a relatively minor effect on the age distribution and, furthermore, that a population that was initially stable but has undergone a decline of mortality is always closely approximated by a stable population having the current growth rate and the current life table of the actual population. Finally, lack of homogeneity biases only slightly the estimates for the total population.

These considerations justify an attempt to derive estimates of the vital rates for Fiji by the stable technique. Again, the only data used in the analysis are those contained in table 24–2. To determine stable populations corresponding to the recorded data, we combine the observed intercensal rate of growth with measures of the observed age distribution in 1966. Naturally a large number of indexes of age distribution can be constructed, defining a more or less broad spectrum of stable populations, all characterized by the observed intercensal r. It is important to select the indexes of age distribution in such a manner that they will be least affected by errors of age reporting, but at the same time the indexes should reflect a broad range of observations. Various considerations and experience suggest the selection of proportions from the cumulative age distribution—proportions under age 5, under age 10, under age x in general, denoted as $C(5)$, $C(10)$, $C(x)$—for this purpose. Going beyond $C(45)$ in the analysis is not recommended.

Computational routine. — The technique is illustrated in table 24–7 for the female population of Fiji. Column (2) shows the observed values of $C(5)$, $C(10)$, ..., $C(45)$. When each of these nine indexes of the age distribution is combined with the same observed growth rate of .0336, they define nine different populations within the tabulated network of a set of model stable populations. To locate these populations, first construct model populations having a growth rate of .0336 (by interpolating between columns with $r = .030$ and $r = .035$) for various levels of mortality. The mortality levels are to be selected in such a manner that the various observed $C(x)$ values are bracketed between the corresponding $C(x)$ values in the models. The results of this part of the calculation are presented in cols. (3) to (7). Note that the parameters to be estimated (the birth rate, the death rate, etc.) are shown along with the $C(x)$ values.

The second part of the calculation is best illustrated by a numerical example. What is the birth rate in a "West" female stable population having an r of .0336 and a $C(15)$ of .4685? The models shown in cols. (3) and (4) already have the requisite r, but the values for $C(15)$ are .4780 and .4677, respectively. Appropriate linear interpolation, as described earlier in this chapter, between these two bracketing populations yields the population with the required $C(15)$. When the same interpolation factors used in this calculation are applied to the birth rates (.0493 and .0460) they yield the birth rate (.0463: third entry in col. 8) in the stable population with $r = .0336$ and $C(15) = .4685$. Exactly analogous procedures result in various estimated parameter values (cols. 8 to 14) corresponding to all $C(x)$ values shown in col. (2).

Validity of the stable estimates. — Interpretation of the results of table 24–7, if unaided by additional outside information or

Table 24–7.—Stable Population Estimates of Fertility and Mortality Based on the Age Distribution of the Female Population of Fiji as Reported in the Census of 1966 and on the Observed Intercensal Growth Rate

$$[r = .0336]$$

Age x	Proportionate population up to age x C(x)	Values of C(x) and of various parameters in female stable populations with r = .0336 and levels of mortality as indicated					Parameter values in stable populations with C(x) as shown in col. (2) and r = .0336						
		Level 13	Level 15	Level 17	Level 19	Level 21	Birth rate (b)	Death rate (d)	Level of mortality	$\overset{o}{e}_0$	$\overset{o}{e}_5$	GRR(29)	GRR(31)
(1)	(2)	(3)	(4)	(5)	(6)	(7)	(8)	(9)	(10)	(11)	(12)	(13)	(14)
5 years..............	.1734	.1944	.1881	.1824	.1775	.1732	.0391	.0055	20.91	69.77	67.62	2.76	2.96
10 years.............	.3357	.3495	.3406	.3324			.0443	.0107	16.20	57.99	60.48	3.14	3.39
15 years.............	.4685	.4780	.4677				.0463	.0127	14.84	54.61	58.48	3.29	3.55
20 years.............	.5773	.5846	.5735				.0471	.0135	14.32	53.29	57.73	3.36	3.63
25 years.............	.6676	.6723	.6613				.0479	.0143	13.85	52.14	57.07	3.42	3.70
30 years.............	.7413	.7442	.7336				.0484	.0148	13.55	51.37	56.64	3.46	3.74
35 years.............	.7986	.8029	.7931				.0479	.0143	13.88	52.20	57.11	3.42	3.69
40 years.............	.8476	.8505	.8418				.0482	.0146	13.67	51.67	56.81	3.44	3.72
45 years.............	.8865	.8891	.8815				.0482	.0146	13.68	51.71	56.83	3.44	3.72
Birth rate (b)......		.0493	.0460	.0432	.0409	.0390							
Death rate (d)......		.0157	.0124	.0096	.0073	.0054							
$\overset{o}{e}_0$		50.0	55.0	60.0	65.0	70.0							
$\overset{o}{e}_5$		55.86	58.70	61.67	64.68	67.77							
GRR (29)............		3.53	3.27	3.06	2.89	2.75							
GRR (31)............		3.82	3.53	3.30	3.11	2.95							

knowledge of special national conditions, should be necessarily cautious. The remarkably consistent estimates of the birth rate associated with $C(20)$ to $C(45)$ tend to confirm the existence of a sustained high level of fertility in the past. However, accepting the precise magnitudes of the birth rates (about .047 to .048 per head in this instance) and of other parameters would imply that a high degree of confidence is attached to the appropriateness of the West model life tables for describing the true mortality pattern of Fiji and to the precision of the observed intercensal growth rate. For these reasons the conformity of the results to those obtained from census survival rates must not be construed as evidence of reliability; the two methods are sensitive to similar biases.

The estimates associated with $C(5)$ to $C(15)$ show relatively lower levels of fertility (and mortality). This could be interpreted as evidence of fertility decline during the decade or so prior to 1966, or as a consequence of a tendency to exaggerate the ages of children in census reports or to omit children in the census, particularly in the youngest ages. Analysis of census survival rates suggests that the last explanation is correct, or at least dominant.

These comments underline the need for reliable age reports in stable population analysis. Alternatively, the analyst should endeavor to obtain all information that is helpful in determining the most reliable segments of the age distribution or in isolating particular errors that affect age reports. Such information may lead either to the selection of particular indexes (e.g., $C(10)$ or $C(35)$) of the age distribution as preferable to others or to an adjustment of the age distribution prior to stable analysis. In general, the former procedure is to be preferred. In the absence of particular reasons for preferring one index of the age distribution to another, and in the absence of very marked fluctuations in the estimates produced by various $C(x)$ values, the median of the series should be selected as the single best estimate. In the present instance, this rule favors the stable population associated with $C(35)$, i.e., an estimated birth rate of .0479, a death rate of .0143, and a mortality level of 13.9 ($\overset{o}{e}_5 = 57.1$).

The application of similar procedures to the male population is straightforward, and it is left to the reader as an exercise. The appropriate calculations yield a male birth rate of .0457, a death rate of .0149, and an $\overset{o}{e}_5$ value of 55.5. Thus, for the total population of Fiji (both sexes), we obtain the following stable estimates:

Rate of natural increase.........	.0322
Death rate..........................	.0146
Birth rate...........................	.0468

It should be noted that knowledge of local conditions often suggests that the estimates for one sex are more accurate than for the other sex. For instance, when ages are inadequately known and have to be estimated by enumerators, female age distributions are typically more amenable to correct interpretation than male age distributions. Under such circumstances it is preferable to derive the estimate for males, and for the total population, from stable analysis of the female population only.

Calculation of the gross reproduction rate by stable analysis.—In the preceding example (cf. table 24–7, two alternative values of the GRR were obtained corresponding to a mean age of the fertility schedule ($\bar{m}$) of 29 and of 31 years. The median estimates were 3.42 and 3.69, respectively.

To arrive at a single estimate of the GRR, a prior estimate of the true value of $\bar{m}$ is necessary. When age-specific fertility rates are not available, as is likely to be the case, a rough estimate can be made by various methods. The largely self-explanatory calculation shown in table 24–8 illustrates one such method. Its application is based on the assumption that fertility outside marriage is negligible and that within marriage little or no contraception is practiced. A standard age pattern of "natural" fertility (i.e., without contraception) shown in col. (3) can then be combined with the proportions of married women as reported in the census to give the likely pattern (but of course not the level) of age-specific fertility rates (col. 4). The mean of this schedule is calculated as 30.6 years. The stable estimate of the gross reproduction rate for Fiji can now be calculated as 3.64 by interpolation between GRR ($\bar{m} = 29$) and GRR ($\bar{m} = 31$) to GRR ($\bar{m} = 30.6$). If the sex ratio at birth is assumed to be 1.05, the total fertility rate (TFR) may be given as $3.64 \times 2.05 = 7.46$. The approximate nature of such calculations should always be kept in mind. Alternative estimates of $\bar{m}$ run as low as 29.2 years, giving $GRR = 3.45$ and $TFR = 7.07$.

Quasi-stable estimates.—When an actual population is reasonably approximated by the stable model, it is often described as being in the "quasi-stable" state, and the estimates derived from the model are referred to as "quasi-stable" estimates. Two meanings of the term, as used in the literature, should be distinguished, however.

In one sense the expression is intended merely as a reminder that the correspondence between the model and the actual population is imperfect, generally both for inadequately fulfilled stability conditions and for distortions in the data. In this loose interpretation, all estimates described above as "stable" should be considered "quasi-stable." Although use of the term is a matter of definition, it is preferable to refer to such estimates simply as stable estimates. It should always be understood that such estimates are subject to biases owing to deviations of the actual population from the stable model used and owing to erroneous measurements.

In a second, more precise meaning, the term "quasi-stable" is applied only to populations that were initially stable (as always, only in the sense of a close approximation) but that have undergone a process of destabilization generated by an orderly and sustained decline of mortality. The impact of such a decline on the age distribution and on the values of other parameters has been analyzed, and the biases in estimating population parameters for such a population on the assumption of strict stability have been investigated. The findings of this research permit the analyst to make appropriate numerical adjustments on the stable estimates, taking into account the decline of mortality. The final estimates are then called quasi-stable in the more narrow, technical sense of the term.

Table 24-8.—Calculation of the Mean of the Fertility Schedule ($\bar{m}$) for the Female Population of Fiji From Proportions Married as Reported in the 1966 Census and From a Standard Age Pattern of Fertility Rates Reflecting "Natural" Fertility

Age (x to x + n)	Proportions married, females, Fiji, 1966[1]	Age pattern of natural fertility rates	Calculated age pattern of fertility rates, Fiji (2) x (3)=	Midpoint of age interval	Calculation (4) x (5)=
(1)	(2)	(3)	(4)	(5)	(6)
15 to 19 years.....	0.166	[2]1.084	0.1799	17.5	3.15
20 to 24 years.....	0.672	1.000	0.6720	22.5	15.12
25 to 29 years.....	0.872	0.935	0.8153	27.5	22.42
30 to 34 years.....	0.910	0.853	0.7762	32.5	25.23
35 to 39 years.....	0.911	0.685	0.6240	37.5	23.40
40 to 44 years.....	0.883	0.349	0.3082	42.5	13.10
45 to 49 years.....	0.830	0.051	0.0423	47.5	2.01
Total, 15 to 49 years.........			3.4179		104.43

$$\bar{m} = \frac{104.43}{3.4179} = 30.55$$

[1] Source: Official census reports.
[2] Estimated as $1.2 - 0.7$ (proportion married at age 15-19) = $1.2 - 0.7(.166) = 1.084$.

The special interest attached to populations where mortality has declined while fertility remained stable is, of course, due to the prevalence of this condition in many contemporary populations. Unfortunately, proper quantitative adjustments for quasi-stability require information on the duration and rapidity of the mortality decline that is seldom available in the desired form and detail. Hence, the parameters of the mortality decline itself have to be estimated from often fragmentary pieces of evidence. For discussions and illustrations of the estimating methods that have been worked out, the reader is referred to the specialized literature.[9] It may be observed, however, that the general effect of mortality change is to introduce a downward bias into the stable estimates of the birth rate (and of the GRR) when such estimates are obtained from observed intercensal growth rates and from $C(x)$'s for values of x of 20 years and over. If the decline of mortality was very rapid, and if the decline lasted for several decades, this downward bias can be quite pronounced—e.g., the stable estimate associated with $C(35)$ may be .040 when the true (quasi-stable) birth rate is in fact .044. On the other hand, estimates derived from $C(10)$ or $C(15)$ are likely to be only slightly affected by declining mortality.

Estimates Based on the Reverse Survival Technique.—When two consecutive censuses supplying age-sex distributions but no other demographic information are at hand and the population is not a stable one, the analyst may wish to apply the familiar reverse survival technique to estimate the crude birth rate in the 5- or 10-year period preceding the second census. The technique requires the construction of an appropriate life table by which the population is projected "backwards" by 5 or 10 years. A by-product of such a reverse projection is the absolute number of births during the 5 or 10 years prior to the census. These quantities are obtained by multiplying the populations under 5 and under 10 years old by the factors $5l_0/{_5}L_0$ and $10l_0/{_{10}}L_0$, respectively. Dividing the average annual number of births during the given interval by the total population calculated at the midpoint of the interval yields the estimate of the birth rate.

Insofar as the life table is estimated from census survival rates (hence infant and early childhood mortality is essentially an extrapolation), the method is subject to a source of uncertainty that plagues the previously discussed estimates as well. Since in making reverse survival estimates, no assumption of stability is made, the method may appear attractive under many situations where the stable conditions do not obtain. It can be shown, however, that regardless of the existence of stability or the lack of it, the results of reverse survival estimates and the stable estimates obtained from the population under age 5 or 10 are essentially identical. In fact, using a set of tabulated model stable populations is a quick and efficient way to obtain reverse survival estimates whether the population is stable or not. Thus, the first two lines of the example set forth in table 24-7 illustrate the results of a reverse survival analysis for Fiji. This example calls attention to a weakness of the method. By necessity the estimate of the birth rate is obtained primarily from the numbers under 5 and under 10 reported in the second census. But because of frequent undercounting of the age group under 5 years old in censuses, the correctness of these measures of the age distribution is often

[9] United Nations, *Methods of Estimating Demographic Measures From Incomplete Data*, op. cit.

especially suspect. To avoid the pitfall of interpreting the level of fertility in the light of the recorded numbers in childhood ages only, the analyst again may find it advantageous to examine a full set of computations such as is summarized in table 24–7, even if the stability conditions are very inadequately fulfilled.

Estimates Based On Data Collected in a Single Census or Survey

Fertility Measures Derived From a Recorded Age Distribution. – The most conspicuous common feature of the methods discussed in the preceding sections was their reliance on the analysis of the age distribution recorded in the most recent census. The existence, however, of an earlier enumeration permitting the calculation of an intercensal growth rate or of census survival rates represents an essential prerequisite for the successful application of these methods. The question may then be raised whether it is possible to derive the general magnitudes of fertility, mortality, and growth from a single recorded age distribution alone. The answer to this question must be essentially negative. If age and sex distributions have been tabulated for numerous geographical subdivisions of the country, however, and are available for other sub-populations as well, such as racial or linguistic groups, internal comparisons and checks may reveal sufficient consistency of age reporting so that various measures of the age distributions may be accepted as reliable. Since past fertility is the dominant factor determining the shape of the age distribution and, in particular, proportions in the youngest ages depend upon recent fertility, a rough estimate of the level of the birth rate may be obtained by an examination of the age structure. Alternative assumptions concerning the level of mortality (e.g., contrasting a plausible low and a plausible high hypothesis for early childhood mortality) might show, for instance, that either assumption leads to a birth rate of over 40 per thousand. In most cases it will be true, however, that the uncertainty as to the level of mortality removes even qualitative precision from the estimate. Experimentation in the hands of a skilled analyst will normally result in determining nontrivial thresholds or ceilings for the birth rate, but the method will be unable to differentiate between, for example, "very high" or just "moderately high" fertility.

The explicit introduction of assumptions as to probable mortality levels may be omitted if only comparative estimates of fertility levels are aimed for. Thus, some indexes of the age distribution that are, *ceteris paribus*, highly correlated with fertility levels may show sufficient regional contrasts as to warrant valid conclusions concerning differential fertility. The most commonly used index of age distribution for such purposes is the ratio of children under age 5 to the number of women in the childbearing ages (usually defined as ages 15 to 49), the child-woman ratio discussed in chapter 17. It may be noted that general indexes of age distribution (e.g., proportions under age 5, 10, or 15) may perform as well or better for this purpose and present a smaller temptation for being interpreted as a measure of fertility as such. The interest in differential patterns shown by such indexes is, of course, greatly strengthened if analysis (easily performed by means of tabulated model stable populations) can show that mortality differentials or plausible directions and magnitudes leave such patterns at least qualitatively unaffected.

Information on the age distribution alone is entirely insufficient to support meaningful estimates with respect to either absolute values or to differentials of growth rates and of mortality rates.

Estimates of Fertility From Retrospective Reports on Child-bearing. – By tradition, censuses have been used primarily to record a cross-sectional view of the state of a population at a given moment of time. As we have seen, however, inquiries concerning past demographic events experienced by individuals can also be introduced in a census or survey. When vital registration is deficient, the recording and analysis of such events may be especially rewarding. With respect to fertility, two types of questions have appeared with increasing frequency in recent censuses and surveys. One type of question concerns the number of births that have occurred during a specified time, usually a year, preceding the survey. Another question inquires about the number of children ever born by each woman up to the time of the inquiry.

A priori reasoning as well as experience suggests, however, that reports of women on past births may be subject to serious biases. As to the question on births during a specified period, a chief problem lies in the difficulty on the part of the respondents of locating a past event on a time scale, especially when no written record of that event is available. Thus, the average length of time covered by the reports may span more or less than the intended 12-month period, often by a margin of several months. Accordingly, age-specific fertility rates, and hence total fertility, calculated from such statistics, may under- or over-estimate fertility.

With respect to children ever born, the possibility of over-reporting is rather remote. On the other hand, understatement of the true number appears to be common, owing to various factors such as memory failure and omission of dead children or children who have already left the home. By their very nature, these biases are likely to affect particularly women of older age and of higher parity. This circumstance diminishes the value of precisely the sort of information that is of the greatest interest for analysis, namely the number of children ever born by the end of the childbearing age. (Clearly, under conditions of approximately constant fertility – and ignoring differential mortality and migration – the average number of children ever born to women of about age 50 would automatically supply an estimate of the current total fertility rate.)

Methods that would permit the evaluation and correction of distortions in retrospective fertility reports would, therefore, greatly enhance the usefulness of such information. A technique worked out by Brass to serve that purpose is illustrated below through data taken from the Turkish Demographic Survey of 1965–66.[10] The data relate to a sample area covering a rural region, thus also demonstrating the possibilities of retrospective fertility reports to generate information for sub-populations, as well as for a country as a whole.

Column (3) of table 24–9 shows age-specific fertility rates calculated from 12-month retrospective reports. (Since women in a given 5-year age group at the time of the survey were on the average half a year younger at the time the births occurred, the age-specific fertility rates actually relate to the unconventional age groups bounded by exact ages 14.5 and 19.5; 19.5 and 24.5; etc.) Cumulation of these rates, shown in col. (4) (0.605 children up to age 19.5; 2.390 children up to age 24.5; etc.), when carried to the end of the childbearing period, gives an estimate of the current (1965–66) total fertility rate of the population in question, assuming, of course, that the reports are correct. The estimated total fertility rate from these "current" reports exceeds considerably the average number of

---

[10] William Brass et al., *The Demography of Tropical Africa*, Princeton, New Jersey, Princeton University Press, 1968, pp. 89–104 and 140–142.

children ever born reported by women at the end of the childbearing period, as shown in col. (7). This comparison is, however, inconclusive as to the validity of the "current" fertility reports; the number of children ever born is often underreported by older women or, alternatively, past fertility may have been genuinely lower than current fertility.

These considerations suggest that it would be of interest also to compare cumulated current fertility with corresponding reports on children ever born at younger age groups. But apart from the age at the end of the childbearing period, cols. (4) and (7) are not directly comparable. Column (4) shows cumulated fertility up to ages 19.5; 24.5, . . . etc., whereas in col. (7) cumulative fertility relates roughly to the midpoint of the age groups shown in col. (1); i.e., up to 17.5, 22.5, . . . etc. Adjusting to these ages the cumulated "current" fertility shown in col. (4) by linear interpolation is a possibility, but assuming an even distribution of fertility within each age interval is clearly unrealistic. To eliminate an unnecessary source of bias, a more sophisticated adjustment is performed in cols. (5) and (6) with the help of table B-4 of appendix B, which gives a set of adjustment factors ω_i to obtain values of cumulated "current" fertility (F_i) directly comparable to average numbers of children ever born (P_i) calculated for the conventional 5-year age groups of women.

The correction factors reflect the curvature of an underlying set of model age-specific fertility patterns. The appropriate model, hence the appropriate correction factor, is selected by either of two summary measures of the age-specific fertility rate: the mean of the fertility schedule, $\bar{m}$ (in this instance calculated from col. (3) as 29.4), or the steepness of the take-off of the fertility curve measured by the ratio f_1/f_2 (in this instance .340, also from col. (3)). The two adjustments give very nearly identical results—hence only the adjustment through factors selected on the basis of $\bar{m}$ is illustrated in table 24-9 .

The values of P_i and F_i are compared in col. (8). The values calculated for ages 15 to 20 are always highly uncertain and best ignored. The rest of the P_i/F_i figures show the familiar declining tendency, in all probability reflecting progressive forgetting of offspring with advancing age. The interpretation of the P_i/F_i ratios between ages 20 and 35 is more problematical. Ordinarily, it could be assumed that reports on children

ever born to women aged 20 to 24 tend to be reliable (by definition these reports are not affected by the problem of time-reference error, and forgetting children at such an age is highly unlikely). Hence, any discrepancy between the value of P_2/F_2 and the expected value of 1 reflects a period-reference error in the "current" fertility reports. Since there is no reason to expect that such time-reference errors are related to the age of the respondent, the correction factor P_2/F_2 could be used to adjust upwards the entire series of reported current fertility rates.

In the present instance, however, the uncertainty inherent in the relatively small number of events on which the rates for age 20 to 24 are based and, in particular, the marked plateau of the P_i/F_i ratios between 25 and 35 strongly suggest that comparison of numbers of children ever born for young women and cumulated current fertility tends to confirm the latter by disproving the assumption about the presence of substantial time-reference errors in this particular instance. Column (9) demonstrates the mechanics of adjustment through inflating the reported "current" age-specific fertility rates by the factor P_3/F_3. Obviously the effect in the present case is trivial. The chief conclusion that emerges is that the retrospective fertility reports strongly support an estimate of a total fertility rate of 7.5, i.e., of a gross reproduction rate of nearly 3.7.

Estimates of Mortality From Retrospective Reports on Deaths. — On the analogy of estimating fertility from survey reports on childbearing during a specified period prior to the survey, it is tempting to try to derive mortality estimates from survey data on deaths obtained from retrospective reports. Indeed, a number of censuses as well as surveys have experimented with such questions. Unfortunately, the experience thus far has been entirely negative. In retrospective birth reports, the rate of omission or the degree of distortion with respect to the length of the reference period may be reasonably assumed to be insensitive to the age of the reporting women, hence the pattern of fertility shown by such reports can be accepted as approximately correct. In contrast, any assumption of uniformity of errors in reported deaths with respect to age at death appears to be patently false. Differential completeness is generated by the fact that the importance of death to the survivors varies with the personal attributes of the deceased,

Table 24–9. — Estimation of Total Fertility From Survey Reports on Births During a 12-Month Period Preceding the Survey for Rural Region I of Turkey in 1966

Age of woman at time of survey	Age interval (i)	Average number of births in 12 months preceding the survey, per woman[1] (f_i)	Cumulative fertility to the beginning of interval $i-1$ ($5\sum_{j=0} f_j$)	Multiplying factors for estimating average fertility[2] ($\bar{m} = 29.4$) w_i	Estimated average cumulative fertility ($F_i = (4) + w_i f_i$) $(4) + [(5) \times (3)] =$	Average number of children ever born per woman (P_i)	$\dfrac{P_i}{F_i}$ (7) / (6) =	Adjusted age-specific fertility rates $f_i' = f_i \times \dfrac{P_3}{F_3}$ (3) x 1.022 =
(1)	(2)	(3)	(4)	(5)	(6)	(7)	(8)	(9)
15 to 19 years..............	1	0.121		1.716	0.208	0.301	1.447	0.124
20 to 24 years..............	2	0.357	0.605	2.798	1.604	1.842	1.148	0.365
25 to 29 years..............	3	0.338	2.390	2.993	3.401	3.475	1.022	0.345
30 to 34 years..............	4	0.280	4.080	3.103	4.949	4.956	1.001	0.286
35 to 39 years..............	5	0.228	5.480	3.224	6.215	5.916	0.952	0.233
40 to 44 years..............	6	0.118	6.620	3.458	7.028	6.370	0.906	0.121
45 to 49 years..............	7	0.058	7.210	4.224	7.455	6.498	0.872	0.059
Total, 15 to 49 years...		1.500	7.500					

[1] For age intervals one-half year younger than shown in col. (1); i.e., in exact ages, 14.5–19.5, 19.5–24.5, etc.

[2] From appendix table **B-4**.

Source: Adapted from Turkey, Ministry of Health and Social Welfare, *Vital Statistics from the Turkish Demographic Survey:* 1965–66, Ankara, 1967.

and such attributes are highly correlated with age. Also, although retrospective birth reports are supplied by a well-defined group—women in the childbearing age—directly connected with the event of birth and subject to low mortality, no such logical respondent category exists with respect to past deaths. Thus, retrospective death reports often contain not only errors of omission and of reference period, but also of multiple reporting of the same event.

Furthermore, unlike the case of retrospective birth reports, no technique exists by which the average degree of the erroneous lengthening or shortening of the reference period can be estimated. Hence, no correction is possible for reference-period errors. The assumption that the reference-period error is of the same magnitude as the one calculated for fertility by the method described earlier is unacceptable, since the distortion in the perception of time elapsed is likely to be different for the two events.

Estimates of Infant and Child Mortality From Proportions Living Among Children Ever Born.—When a census or survey collects data from women past age 15 concerning the number of children still living as well as children ever born, the ratio of these two statistics—i.e., the proportion of children surviving, or rather its complement to unity, the proportion of dead children—supplies an index of child mortality during the precensal period. Such an index, by definition, contains no reference-period error; and, apart from minor biases, such as those originating from a possible correlation between child mortality and mortality of women in the childbearing ages, it is likely to be affected only by underreporting insofar as the degree of such underreporting differs in the numerator and in the denominator of the index. (The underreporting would also tend to vary inversely with the recency of the births concerned.) It is almost always certain that proportionate underreporting affects the number of children ever born more than the number reported as surviving, since children already dead at the time of the survey are more likely to be omitted than children who are still alive. Thus, even if the magnitude of the bias in the value of the proportion dead is unknown, its direction is unambiguously defined—the index gives a minimum estimate of mortality.

Reported proportions dead (when controlled for age of the reporting women) supply a measure directly usable for purposes of roughly describing patterns of differential mortality. But the usefulness of the index as a measure of mortality is obviously limited unless it can be interpreted in terms of conventional mortality indexes.

A method developed by Brass that permits such a translation is illustrated in table 24–10 by data for a rural region of Turkey.[11] Columns (3) and (4) present the raw statistics of children ever born (P) and children surviving (S) by age of women. (Note that in the survey the questions were asked of ever-married women only. Their reports were transformed to estimates relating to all women by assuming that single women have had no children and that the rates of children ever born and surviving among ever-married women who did not report were the same as among those reporting.) Calculated proportions dead are shown in col. (5). Clearly these proportions reflect the chances of dying from the moment of birth to some age x (in standard life table symbolism, $_xq_0$), where the value of x is an average determined by the lengths of time elapsed during which births to women of various age groups were

exposed to the risk of dying. If the age pattern of fertility and the age pattern of the risk of dying are known, or can be estimated, the value of x can be calculated. When such calculations are performed for typical age patterns of fertility and mortality, the value of x is found to be very close to 1 for proportions dead reported by women 15 to 19 years old, very close to 2 for reports by women 20 to 24 years old, etc. (See col. 7.) Thus, for example, proportions dead reported by women 25 to 29 years old supply an estimate of $_3q_0$, the probability of dying between birth (age zero) and age 3. It can be demonstrated that such estimates are robust to known variations in the pattern of infant and child mortality. Very early or very late childbearing does affect the exact value of x, however. If childbearing is especially early, the true x is larger, and the converse is true if childbearing is very late. To obtain $_xq_0$ estimates for the desired round values of x shown in col. (7), it is, therefore, desirable to correct reported values of proportions of dead children to take into account the age pattern of fertility prevailing in the population in question.

Multipliers that perform the needed correction are tabulated in table B–5 in appendix B. The multipliers are to be selected on the basis of one or more of three alternative indexes of the age pattern of fertility shown in the bottom three lines of table B–5. These three indexes are (1) the ratio, P_1/P_2, where P_1 and P_2 are the average number of children ever born reported by women aged 15 to 19 and 20 to 24, respectively; (2) the mean of the fertility schedule $\bar{m}$; and (3) the median of the fertility schedule $\bar{m}'$. The first of these indexes was used for determining correction factors in the example shown in table 24–10. The value of P_1/P_2 is determined from col. (3) of the table (as .301/1.842). The multipliers, obtained through linear interpolation from table B–5, are given in col. (6) of table 24–10. The products of cols. (5) and (6) give the final estimates for $_xq_0$, shown in col. (8). In this instance (as in most instances), results obtained by utilizing multiplying factors selected on the basis of indexes of the estimated fertility function ($\bar{m}$ or $\bar{m}'$) differ only trivially from these estimates.

The $_xq_0$ values given in col. (8) of table 24–10 may be expressed in terms of mortality levels through locating model life tables (usually by interpolation) having the same $_xq_0$ values. Apart from the estimate of $_1q_0$ derived from reports of women aged 15 to 19 years old (which is often affected by grave biases and is best ignored), the estimates show an impressive consistency, suggesting a mortality level of roughly 10 (in terms of the "West" model life tables), i.e., an expectation of life at birth in the neighborhood of 40 years. Naturally, a translation of child mortality into $\mathring{e}_0$ values implies an extrapolation to adult mortality using a particular model life table. The validity of such an operation should be, if possible, corroborated by additional evidence. Alternatively, various model life table patterns should be used to gauge the sensitivity of the estimate to plausible variations in the age patterns of mortality.

Consistency of the estimates obtained from women of various ages is not sufficient to establish the proposition that the model used is correct or that mortality has remained unchanged. It should be remembered that the various estimates refer to various time periods prior to the census, depending on the age of the women reporting. The older the reporting women, the longer the period represented. Thus, for $_2q_0$ the reference period is roughly 4 to 5 years; for $_3q_0$ it is 6 to 8 years prior to the census. For women over 30, the period is of course much longer. Accordingly, consistent estimates for mortality levels may be a fortuitous outcome of two biases pulling in opposite directions: increasing underestimation of mortality

---

[11] Brass et al., op. cit., pp. 104–114.

Table 24–10. — **Estimation of Values of $_xq_0$ (Proportions Dead by Age x) From Survey Reports on Children Ever Born and Children Surviving for Rural Region I of Turkey in 1966**

Age of woman	Age interval (i)	Average number of children ever born per woman[1] (P_i)	Average number of children surviving per woman[1] (S_i)	Proportion of children dead $(1 - \frac{S_i}{P_i})$	Multipliers for col. (5)[2] $(P_1/P_2 = .163)$	Age x	Proportion dead by age x $(x^q{}_0)$
(1)	(2)	(3)	(4)	(5)	(6)	(7)	(8)
15 to 19 years........................	1	0.301	0.236	.216	1.020	1	.220
20 to 24 years........................	2	1.842	1.374	.254	1.032	2	.262
25 to 29 years........................	3	3.475	2.568	.261	1.006	3	.263
30 to 34 years........................	4	4.956	3.608	.272	1.011	5	.275
35 to 39 years........................	5	5.916	4.215	.288	1.021	10	.294
40 to 44 years........................	6	6.370	4.392	.311	0.999	15	.311
45 to 49 years........................	7	6.498	4.407	.322	0.998	20	.321

[1] Source same as table 24-9.
[2] From appendix table B-5.

from retrospective reports of older women and relatively higher child mortality in earlier years. The retrospective reporting may be subject to errors of memory and to errors arising from an aversion to mentioning dead children, especially those just recently deceased.

Reports of younger women, on the other hand, are likely to contain only minor errors due to memory failure, since the events reported are recent and parity is low. These reasons single out the estimates of $_2q_0$ and $_3q_0$ as the most reliable, as well as most interesting. In view of the fact that these estimates are to be regarded as minimum estimates of mortality (as dead children are more likely to go unreported), it is notable that for typical underdeveloped countries such estimates tend to indicate higher levels of child mortality than estimates based on vital registration. To obtain more precise estimates of child mortality is of significance not only because of the interest in measuring child mortality itself but also because of the use of such estimates as a tool in estimating fertility from a reported age distribution.

Estimates of Fertility From Child Mortality and Age Distribution. – It was pointed out earlier that a recorded age distribution reflects past processes of fertility and mortality. In particular, an accurate count of persons in childhood permits reconstruction of births in recent years, provided that a satisfactory correction for child mortality can be accomplished. Estimates of child mortality obtained by the method just described from census or survey reports supply the data needed for such a correction.

As before, interpretation of a reported age distribution may be attempted on the assumption of stability; or, if there is no proof that the population is in fact stable, or if there exists positive evidence that the population is not stable, the analyst may choose to rely on estimates based on reverse survival alone. Since stable analysis is also a simple way of making reverse survival calculations, the method is illustrated for that assumption only. Column (2) of table 24–11 shows the reported female cumulative age distribution for the same rural region in Turkey for which mortality estimates in table 24–10 were obtained. Accepting the value of $_3q_0$ (= .263) as the most reliable estimate of the $_xq_0$ values estimated, the "West" model tables (cf. appendix table B–1) indicate a mortality level of 9 for the two sexes combined. Since information on sex differentials in Turkish child mortality is lacking, the simplest procedure is to assume that the sex differences incorporated in the "West" model prevail also in Turkey, and to apply level 9 mortality for both males and females. If the true sex pattern of

mortality differs from that shown by the model – as is likely to be the case in this instance – this procedure will give biased estimates for both males and females (such as in estimating the male and female birth rates), but the biases will be equal in size and different in direction; hence in the merged (average) estimates for both sexes they will cancel out. However, only the calculation for the female population is illustrated here.

Columns (3) to (7) show various parameter values in "West" female stable populations that share the characteristic of having a mortality level of 9 and that have proportions under age 5, 10, ... 45 as shown in col. (2). The parameter values are obtained by linear interpolation from appendix table B–3; as the estimated $_3q_0$ corresponds to a round mortality level, only one set of interpolations is necessary.

Examination of col. (3) of table 24–11 shows a tendency toward lower birth rates when estimates are derived from increasingly larger segments of the cumulated age distribution. Explanations consistent with this finding include gradually falling fertility in the decades prior to the survey or distortion due to falling mortality (i.e., to quasi-stability in the strict sense). Neither of these explanations is convincing under the present circumstances. In fact, evidence indicates that the Turkish population is poorly described by the stable model. In this instance the effect of internal migration is a further disturbing element.

Thus, a cautious analysis should rely on the reverse survival technique only, as summarized in the indexes derived from proportions under age 5 and 10, and to a lesser extent, under age 15. These parameters, in combination with the estimated $_3q_0$, suggest a crude birth rate between .047 and .050 per head, a crude death rate between .023 and .027, and a growth rate of about .024. The reliability of these estimates would be increased, and the range of uncertainty narrowed, if some additional information on age reporting were also available. Differences between the estimates of the crude birth rate derived from $C(5)$ and $C(10)$, for instance, may be explained by exaggeration of age of children under 5, by differential omission of infants, or by falling fertility. Elimination of some of these possibilities on the basis of local evidence or confirmation of one of the interpretations as the correct one would be most helpful for the analyst.

Estimates of Fertility and Mortality Through Reconstruction of Pregnancy Histories. – Attempts to obtain information on past flows of vital events through retrospective reports in a census or survey can be logically extended beyond the relatively simple goals of recording the number of children ever

Table 24–11. — Stable Population Estimates of Fertility and Mortality Based on the Age Distribution of the Female Population and on a Level of Mortality Derived From Reported Child Survival Rates for Rural Region I of Turkey in 1966

Age x	Proportionate population cumulated up to age x[1] C(x)	Values of various parameters in female stable populations with C(x) shown in col. (2) and with mortality level of 9				
		Birth rate	Death rate	Rate of natural increase	Gross reproduction rate	
					$\overline{m} = 29$	$\overline{m} = 31$
(1)	(2)	(3)	(4)	(5)	(6)	(7)
5 years	.1724	.0466	.0231	.0235	3.17	3.39
10 years	.3314	.0503	.0266	.0237	3.49	3.76
15 years	.4457	.0486	.0250	.0236	3.34	3.58
20 years	.5365	.0466	.0231	.0235	3.17	3.39
25 years	.6037	.0434	.0201	.0233	2.92	3.10
30 years	.6739	.0426	.0191	.0235	2.85	3.03
35 years	.7479	.0437	.0204	.0233	2.95	3.13
40 years	.8070	.0444	.0210	.0234	3.00	3.19
45 years	.8438	.0424	.0191	.0233	2.84	3.01

[1] Source: Same as table 24–9.

born and surviving or the number of children born (or dead) during a specified period, such as 12 months preceding the survey. A small but substantial step in this direction may be to ask about vital events during a 24-month period, as well as a 12-month period, prior to the survey. Even in reports for such recent periods, however, survey experience has shown that the respondent often makes errors in placing the event in the correct time interval. Ideally, tabulation of such data would give a crude birth rate, as well as various age-specific fertility rates, for 2 consecutive years.

Increasing the detail of such questions may conceivably lead to the establishment of a full pregnancy history for each woman past age 15, specifically to the recording of the timing of each conception and its outcome—fetal death, live birth, or death.[12] If such records were accurate, a highly refined description of past fertility could be obtained, at least up to the point—perhaps 20 or 30 years before the date of the survey—where the effects of increasingly scarce survivors in the older ages and the correlation of differential fertility and mortality become strong enough to destroy the representativeness of retrospective reports.

Many persons, however, particularly in largely illiterate populations, are unfamiliar with the calendar or the concept of chronological age, and are unable to recall past events correctly and to locate these events with some precision on a time scale. These facts make the collection of usable pregnancy histories not only costly but also an exceedingly difficult enterprise. Under many circumstances, in fact, even the most careful field work will fail to elicit the sought-for information. If the attempt is made, nevertheless, there is some danger that the results will primarily reflect the judgment of enumerators on what is "normal" (e.g., with respect to birth intervals) rather than the actual situation. Naturally even under such conditions various by-products of the pregnancy history—such as more reliable figures on children ever born and surviving—may still be highly useful and justify the extra costs.

A less ambitious effort to establish dated records of fertility performance of women with respect to children alive at the time of the census may be attempted through the so-called "own children" technique. If in a population almost all young children live with their mothers, i.e., if the extent of adoption (including de facto adoption) is limited, household schedules obtained in a census can be used to record the number of live own children by age for each mother even without asking any direct questions on fertility. If children are fully counted and their age records (as well as their mothers') are accurate, age-specific birth rates for some 10 years prior to the survey can be calculated on a year-by-year basis for any suitable subgroup of women. Naturally, an allowance for mortality is necessary, requiring an estimate of infant and child mortality. As is the case with reverse-survival estimates, a further problem with this method is that young children tend to be omitted from the census altogether or placed in the wrong age group. When these distortions are mild or controllable, the "own children" technique may be a useful adjunct to the tool kit of the demographer. The development of this technique by Grabill and his applications of it in the United States were described in chapter 17. Cho has applied this technique to an Asian population.[13]

Data Requirements for Estimation in Censuses and Surveys

The analysis of an existing body of census or survey data is largely circumscribed by prior decisions that cannot be modified by the analyst, notably by decisions with respect to questionnaire content, coding, tabulation, and publication. In the light of the present stage of development of analytical techniques surveyed above, it can be said that the contents of past surveys and their form of presentation often severely limit the possibilities of applying some of the more powerful methods of estimating demographic measures from survey data. Such limitations may reflect a need to minimize the costs of a census or survey. It should be observed, however, that below a certain minimum package of information the analytical possibilities become so restricted that for some purposes, such as estimating vital rates, the data may no longer be useful at all.

On the other hand, deficiencies in survey content and in its form of presentation often arise simply from a lack of coordination between producers and future users of the data. With better coordination such deficiencies could be easily avoided. Although the nature of information necessary for applying

[12] Donald J. Bogue and Elizabeth J. Bogue, "The Pregnancy History Approach to Measurement of Fertility Change," 1967 Proceedings of the Social Statistics Section, American Statistical Association, pp. 212–231.

[13] Malaysia, Department of Statistics, Estimates of Fertility for West Malaysia (1957–67), Kuala Lumpur, June 1969, Research Paper No. 3, by Lee Jay Cho.

the analytical techniques described above is implicit in that description, a brief review of the data requirements may be useful at this point.

In setting such requirements, obviously no general rules are possible. It should be stressed, however, that when the reliability of the basic data is demonstrably weak, or is at least open to suspicion, it is highly desirable that the same measures be estimated on the basis of several methods. The same principle suggests that statistics should be collected and tabulated in sufficient detail not only to permit the application of various alternative methods, but also to facilitate checks within the methods themselves. For example, calculation of the birth rate from the age distribution should be based, if possible, on separate estimates of the male and female birth rates derived by means of sex-differentiated estimates of child mortality. The inconsistencies that are inevitably found when such procedures are applied will help the analyst to isolate points of weakness and of strength in the data and hence to arrive at more reliable estimates.

Some flexibility and ranking of priorities are nevertheless called for even within what might be called a minimum program. A basic "menu" of tabulations determined by the data needs of basic techniques for the estimation of vital rates may be set forth as follows:

Symbol	Tabulation
A	Population by age, sex, and marital status.
B	Women by age; and total number of children born alive, for each age group of women.
C	Women by age; and total number of children living, for each age group of women.
BX	Women by parity and by age.
BB	Women by age; and total number of children born alive, for each age group of women, by sex.
CC	Women by age; and total number of children living, for each age group of women, by sex.
D	Number of women who have had a live birth during the 12 months preceding the census, by age.
DX	Women by length of time that has elapsed since the birth of their last live-born child, by age; separately for currently married women and other women.

In all these tables, age classifications are assumed to be based on standard 5-year age groups—for all ages in tabulation A, and at least for ages 15 to 49 in the other tabulations. In tabulation BX, parities at least up through parity 7 should not be grouped. In tabulation DX, column headings might be: "no live birth ever"; 0-2 months; 3-5 months; . . . 15-17 months; 18-23 months; 24-29 months; 30 months or more.

From the above list, variants of a minimum tabulation program may be selected. The number of meaningful combinations is limited by needs of the various methods for joint tabulations. Tabulations BB, CC, and DX account for tabulations, B, C, and D; given the former the latter do not constitute separate tabulations. Tabulation BX accounts for tabulation B, only if parities are given in full detail, which is seldom the case. (In practice, tabulation B would be independently tabulated rather than obtained from BX.) The five basic variants of a minimum tabulation program for obtaining estimates of vital rates from a census are as follows:

I	II	III	IV	V
A	A	A	A	A
B	B	BB	BB	BB
C	C	CC	CC	CC
	BX	BX	BX	BX
			D	DX

Although the analytical possibilities would differ appreciably depending on which of the five specific programs indicated in the table has been carried out, the feasibility of deriving an accurate fertility estimate in each instance from the same source (e.g., from reported age distribution plus child mortality) gives underlying unity to the various approaches that would be adopted in the analysis. To have less than variant I, which is suggested here as a *minimum minimorum,* would drastically curtail the field of maneuver for the analyst. On the other hand, to go beyond the program suggested in variant V (e.g., by introducing age at marriage as a variable or, preferably, by preparing parity-specific tabulations of tabulation DX) would certainly be desirable but would involve a major step towards greater complexity and cost.

METHODS OF ESTIMATION FROM SAMPLE REGISTRATION AREAS

The techniques relying upon census or survey data described in the preceding section represent the cheapest, quickest, and most flexible approach towards generating estimates of vital rates in statistically underdeveloped countries. But, although estimates obtained by such techniques may be perfectly adequate for some purposes—for charting the basic parameters of the population problem in a given country, for describing patterns of differential demographic behavior, or for the formulation of general population policy—in other uses their precision is less than would be desirable. For measuring the effects of a family planning program, for instance (particularly in the initial stages when the effects are small), survey data typically serve as too crude an instrument. Similarly, the effective management of a public health program would require detailed data on age at death by sex in combination with various other characteristics, notably cause of death. Such data cannot be reliably obtained from a survey, even if it is repeated at regular intervals.

Thus, the logic of the demand for detailed demographic information leads to the proposition that in many applications reliance on survey and census data alone is merely a stopgap measure until such data may be combined with information obtained from a continuous vital registration system. But, as was pointed out earlier, in less developed countries the building up of a reliable vital registration system is both an expensive and a necessarily protracted process.

Sample Registration

A solution for this dilemma is to substitute a sample registration scheme for the standard system of comprehensive registration. With such a sample, it may be possible to make arrangements assuring a far higher level of accuracy than could be the case for the population as a whole. This can be done by a variety of administrative devices designed to compensate for the lack of motivation on the part of the populace to register vital events. Even the official registrars themselves are often inadequately motivated. The relatively small size of a sample, for instance, makes it possible—even within the confines of a limited budget—to select better registrars; to provide them with more thorough training, better supervision, and greater remuneration; perhaps to employ them on a full-time basis; to facilitate registration through organization of a continuous house-to-house canvass and through employing a network of informants who have particularly easy access to relevant local information; and so on. Within the sample these improvements may be effected through upgrading the existing deficient vital registration system or through the introduction of an entirely new system, possibly organized under a different agency from the one responsible for the general vital registration. The optimal mixture of the various methods for promoting more complete coverage and the specific administrative arrangements for the scheme will necessarily vary depending on local circumstances.

If the sample is scientifically designed, it can be taken as a representation of the entire population, and vital rates observed in the sample may be used to estimate vital rates for the entire population within the limits of quantifiable errors.

Taken in isolation, the statistics supplied by sample registration, even when of a high quality, are not adequate for a full description of demographic processes. For some purposes, however, incompleteness of vital registration may not be important when identification of a trend alone suffices. If the degree of incompleteness can be taken as roughly uniform over time, even grossly incomplete data may reliably reveal a fall or a rise in the birth rate. Moreover, careful examination of flows alone—sometimes referred to as numerator analysis— may reveal processes normally described by indexes based on both stock-and-flow-type data. For instance, shifts in the distribution of reported birth order of children may serve as an approximate index of changes in reproductive behavior. The obvious advantage of such measures is their simplicity of calculation and their lack of dependence on stock-type data. It remains true, however, that more sophisticated measures of fertility and mortality do require not only data from continuous registration but also stock-type data obtainable from a census or survey. For practical purposes, therefore, registration on a sample basis must generally be combined with periodic surveys based on a corresponding sample.

Dual Systems Based on Sample Registration Areas and Surveys

It follows from the proposition just made that the introduction of sample registration will require a dual system of measurement, which can be used also to obtain all information necessary for applying the techniques developed for the analysis of survey data that were discussed in the preceding section. Thus, the two approaches of estimation based on survey data, on the one hand, and sample registration, on the other, are not competitive alternatives. Rather, the latter subsumes the former and may be considered a powerful extension of techniques using survey data only. Obviously this extension can only be achieved at the cost of a substantial increase in the resources invested in the operation of the system.

It will be recalled that estimates of vital rates from survey data alone can be generated by obtaining direct information on events (e.g., births) during a specified time period from a cross-sectional investigation. But in a dual system, in addition to such information, the same events are also observed through continuous observation, i.e., through sample registration. Thus, dual systems provide a possibility for comparing numbers of births and deaths obtained by alternative means. If the independence of the two approaches is scrupulously maintained—e.g., by assigning the two tasks to two different organizations and by preventing their cooperation through suitable administrative controls—such comparisons will permit the evaluation of the quality of the system and possibly the correction of any deficiencies revealed.

The capacity of the analyst to make appropriate corrections will naturally depend upon the nature of the comparisons between the results obtained from the two systems. Obviously,

a simple comparison of the total number of births, for instance, would not be particularly illuminating. If a discrepancy were found, the reasons for the discrepancy may not be identifiable from the summary comparison. Similarly, if the two systems give essentially the same result, this circumstance alone cannot be interpreted as a confirmation of the validity of the estimates, since both systems may be affected by biases of identical magnitudes and directions, even if originating from different sources.

If comparisons are performed for progressively smaller units (e.g., by comparing numbers of births reported by the two systems in small territorial subdivisions of the total sample), the pattern of the discrepancies found between the numbers of events registered by the two systems may turn out to be quite uneven, thus indicating the location and possible source of underlying weaknesses in the data. At any event, it is most likely that, by diminishing the size of the units compared, increasingly large discrepancies between the two sets of data will be revealed. Hence, the more detailed such comparisons are, the better the picture of the errors affecting the two systems will emerge.

Chandra Sekar and Deming's Method.—The logic of these considerations suggests that, other things being equal, the best results will be obtained if comparisons are performed at the level of the smallest unit, i.e., for individuals or for individual events. When the results of two systems—such as a sample survey and sample vital registration—are matched on such a level, it is possible to obtain a numerical estimate of the degree of completeness of both systems and hence to estimate the true total number of events, on the basis of assumptions described below. Case-by-case matching of data from a registration system and a survey was employed in connection with the 1940 and 1950 Censuses of the United States and the Current Population Survey in 1969–70 (to measure completeness of birth registration, or of both infant underenumeration and birth underregistration), but the technique of estimating the total number of events was refined and tested by Chandra Sekar (now Chandrasekaran) and Deming.[14]

The essential features of the Chandra Sekar-Deming procedure may be summarized as follows (using as an example statistics of births). Suppose that births are recorded for a given year in a sample vital registration system and in a corresponding sample survey (conducted at the end of the year) in which a question on births during the 12-month period preceding the survey is asked. Suppose, furthermore, that the two sets of birth records so obtained are matched event by event. From the matching procedure, the following classification of these events may be obtained:

C = number of events recorded in both registration and survey
N_1 = events recorded only by registration
N_2 = events recorded only by the survey
X = events missed by both systems

The classification may be represented in the following schematic table:

---

[14] C. Chandra Sekar and W. Edwards Deming, "On a Method of Estimating Birth and Death Rates and the Extent of Registration," *Journal of the American Statistical Association*, 44(245):101–115, March 1949. See also Ansley J. Coale, "The Design of an Experimental Procedure for Obtaining Accurate Vital Statistics," pp. 372–375 in *International Population Conference, New York, 1961*, Vol. II, London, International Union for the Scientific Study of Population, 1963.

	In registration system	Not in registration system	Total
In sample survey	C	N_2	S
Not in sample survey	N_1	X^*	
Total	R		N^*

*Estimate.

An estimate of the total number of events N may be then obtained as

$$N = C + N_1 + N_2 + X$$

In turn, X is estimated as $N_1 N_2 / C$, thus giving the expression for the estimated total number of events as

$$N = C + N_1 + N_2 + \frac{N_1 N_2}{C} \qquad (12)$$

which may also be written as

$$N = \frac{(C + N_1)\,(C + N_2)}{C} = \frac{RS}{C} \qquad (13)$$

where R denotes the number of events recorded by registration and S denotes the number of events recorded by the survey.

From rearranging the last formula, it can be seen that the Chandra Sekar-Deming formula estimates the completeness of the coverage of the registration system as the match rate of the survey, and estimates the completeness of the coverage of the survey as the match rate of the registration: [15]

$$\frac{R}{N} = \frac{C}{S} \qquad \text{and} \qquad \frac{S}{N} = \frac{C}{R}$$

In reviewing estimates based on a single system (survey), it has been emphasized that comparisons of estimates based on alternative estimating procedures constitute essential checks on the quality of the results obtained. It is the great merit of the dual systems of estimating vital events that such checking is a built-in feature of the estimating procedure. Unfortunately, the simplicity of the estimating formulas conceals a number of difficulties in the practical application of the method. The nature of the major problems will be discussed below briefly.

Suppose that in a dual system the number of births in a given year and in a given geographic area is found to be 1,200 when recorded by birth registration, whereas the number registered in a retrospective survey is 1,300. Suppose also that subsequent individual matching of the births recorded in the two systems is successful in 900 instances. Using the notation given above, we have:

$$R = 1,200$$
$$S = 1,300$$
$$C = 900$$
$$N_1 = 300$$
$$N_2 = 400$$

[15] Cf. William Seltzer, "Some Results from Asian Population Growth Studies," *Population Studies* (London), 23(3):395–406, Nov. 1969.

Applying the Chandra Sekar-Deming technique, we estimate the number of births missed by both registration and survey as

$$X = \frac{300 \times 400}{900} = 133$$

Hence the estimated total number of births is obtained as

$$N = 900 + 300 + 400 + 133 = 1,733$$

Inserting these figures in our schematic table, we have:

	In registration system	Not in registration system	Total
In sample survey	900	400	1,300
Not in sample survey	300	133*	433*
Total	1,200	533*	1,733*

*Estimate.

In other words, the completeness of the registration of births is estimated as 69.2 percent $\left(\frac{1,200}{1,733}\text{ or }\frac{900}{1,300}\right)$ and the completeness of the listing of births in the survey is estimated as 75.0 percent $\left(\frac{1,300}{1,733}\text{ or }\frac{900}{1,200}\right)$.

The validity of the above estimates will depend, however, on the fulfillment of the following main conditions:

(1) The matching procedure successfully identifies all true matches and, conversely, only true matches are identified as matches.

(2) All events identified in either of the two systems are true events, i.e., occurred in the population under investigation and in the appropriate time period.

(3) The two systems are independent, i.e., the probability of an event being omitted from one system is not related to the chance of the event being omitted from the other system.

Practical and General Considerations. — Deviations from the above conditions may seriously affect the accuracy of the estimates derived from the method. Yet these conditions are quite stringent and the degree of deviation from their fulfillment is at best difficult to ascertain. First of all, the notion of what constitutes a proper match is ambiguous in almost all practical applications. Events may be described through listing a variety of alternatives. In the instance of births, for example, the statistics may record the address of the head of the household in which the event has occurred, the name, date of birth, and sex of the newborn, the age and the parity of the mother, and related items. In general, the more stringent the definition for a match, i.e., the larger the number of the attributes that must coincide in order to establish a "true" match, the smaller will be the estimate of C, and the larger will be N_1 and N_2 and, consequently, the larger will be the estimate of N. There is a danger that overly stringent matching criteria, apart from making the matching process especially laborious will result in inflated estimates of the true number of births. On the other hand, an estimate of N based on loose matching criteria may yield an underestimate.

The task of finding the golden mean between such extremes is difficult. The obvious solution – that of investigating every suspected match in detail – is limited by cost considerations. Usually simple rules will have to be imposed, but the fact that such rules necessarily must take into account the peculiarities of the specific situation makes generalizations about them difficult. If, for instance, addresses are nonexistent or ambiguous, or if names have many variations or are shared by many people, the power of these otherwise most useful matching criteria is greatly diminished or at least the possibility of mechanizing the matching operation is eliminated. Uncertainty about dates and ages makes matching by these particulars unrealistic; but, here again, the middle road between too stringent and too loose requirements is hard to establish. It should be emphasized, however, that the very act of matching and the problems revealed by adopting alternative matching criteria will provide the analyst with valuable insights into the quality of the data, and hence make their interpretations more nuanced.

When reasonable matching criteria are applied, it can be accepted without further investigation that matched records of events are correct unless both are "out of scope." This generalization does not apply to entries recorded in only one of the two systems. A failure to match a registered event may mean either that the event in question was erroneously omitted in the other system or that it was erroneously included in the first system. Such erroneous inclusions may originate, for instance, from errors of time reference in a retrospective survey. Insofar as no correction is possible or is carried out for such errors, the validity of the method will be affected. The above formulas imply that nonmatches are investigated and false entries are eliminated from the statistics. Various devices, notably the use of overlapping reference periods in consecutive surveys may lessen the need for such investigations but only at the expense of carrying out a prior matching procedure for events reported for the overlapping survey periods themselves.

Possible false entries in one of the two reporting systems may also make their appearance because of in-migration to or out-migration from the area covered. The effects of such migrations are often particularly strong on the phenomena under observation. Thus, deaths and births in a sample population may commonly occur in a hospital outside the sample area, or the sample area may contain a hospital attracting outsiders. Similarly, many women often return to their parents' home for the birth of a baby. Accordingly, if survey and registration data are based on a *de facto* definition of the population (which would be the simplest solution from an administrative viewpoint), there is a definite risk that the results will show false discrepancies and that total births and deaths will be overestimated. This risk can be reduced or eliminated if the same rules are applied in both collection systems. (If many events occur in hospitals and a *de facto* approach is used, then hospitals should be sampled separately in order to reduce sampling error.) By adopting a *de jure* concept instead, the method can solve this problem, but only at the cost of following up residents moving out of the area and keeping track of events affecting temporary residents. The technical difficulties involved in such a solution are likely to be formidable.

Finally, the above formulas provide no correction for the presumably not uncommon situation where certain events tend to be omitted from both systems for the same reasons. A simple application of the estimating formulas will then give an estimate of the completeness of coverage which is biased upward. If, for example, both systems missed the same 20 percent of births (e.g., all illegitimate births) but both included all other births, the method would erroneously indicate full coverage for both systems.

Chandrasekaran and Deming have shown that this bias can be reduced by subdividing the population into various homogeneous subgroups (e.g., geographic subdivisions; births by sex, or by survival status of the infant one month after birth; deaths by age at death, etc.), preparing estimates for each subgroup separately, and then obtaining a total estimate as the sum of the estimates for the various subgroups. Although the theoretical value of this procedure is clear and experimentation with it is to be recommended, its practical usefulness remains to be demonstrated.

Another manifestation of lack of independence between the two systems may be that the general quality of each is influenced by the existence of the other. Although such influences are typically positive and hence would normally be welcome, they do make it difficult to derive conclusions as to the completeness of coverage that can be generalized for areas where only one system is in operation. It is by no means certain, however, that a dual registration system will tend to improve over time if maintained for a given area for a longer time period. Beyond the general difficulty of sustaining a complicated and demanding system at a high level of efficiency, it will be particularly hard to avoid the deleterious effects of a possible collusion between the officials responsible for the operation of the two systems. Such collusion will naturally tend to be established as soon as it is understood that the quality of the work done by the registrar and the survey takers can be evaluated by observing changes in the match rates achieved in the survey and the vital registration.

Experience with dual systems of vital records described above has been building up rapidly during the last decade in Africa, Asia, and Latin America. This procedure underlies the Pakistan Growth Estimation Study, for example.[16] Seltzer has summarized the results from a number of population growth estimation studies conducted in Asia since 1945.[17] Although the underlying principles have generally been the same, a wide variety of specific attempts have been made seeking to minimize the biases just mentioned. Thus, attempts using sample registration differ not only in the size of the sample and in the design of the sampling scheme, but even more with respect to the length of the reference period and to the peculiarities of the field operation, with respect to the frequency and scope of the periodic surveys, with respect to the registrars' mode of operation, and, in particular, with respect to the existence, detail, and quality of the matching operation and of the investigation of the validity of nonmatches. Experience indicates that the measurement of mortality remains very difficult even by means of dual systems. For fertility, however, such systems represent a major advance toward more reliable measurement procedures in statistically underdeveloped countries.

[16] Pakistan Institute of Development Economics, *Report of the Population Growth Estimation Experiment: Description and Some Results for 1962-63*, Karachi, Pakistan, December 1968. A summary of the study appears in W. Parker Mauldin, "Estimating Rates of Population Growth," *Family Planning and Population Programs: A Review of World Developments*, Bernard Berelson, ed., University of Chicago Press, 1966, pp. 642-647.
[17] Seltzer, op. cit.

SUGGESTED READINGS

Agrawal, B. L. "Sample Registration in India." *Population Studies* (London), 23(3):379–394. November 1969.

Ahmed, Nazir, and Krótki, Karol J. "Simultaneous Estimations of Population Growth—the Pakistan Experiment." *Pakistan Development Review,* 3(1):37–65. Spring 1963.

Alauddin Chowdhury, A. K. M., Aziz, K. M. A., and Mosley, Wiley H. *Demographic Studies in Rural East Pakistan: Second Year, May, 1967–April, 1968,* Dacca, Pakistan—SEATO Cholera Research Laboratory. June 1969.

Arriaga, Eduardo E. "Rural-Urban Mortality in Developing Countries: An Index for Detecting Rural Underregistration." *Demography,* 4(1):98–107. 1967.

Blacker, J. G. C., and Martin, C. J. "Old and New Methods of Compiling Vital Statistics in East Africa." In *International Population Conference, New York, 1961* (International Union for the Scientific Study of Population), Vol. I, pp. 355–362.

Bogue, Donald J., and Bogue, Elizabeth J. "The Pregnancy History Approach to Measurement of Fertility Change." *Proceedings of the Social Statistics Section, 1967.* Washington, American Statistical Association. Pp. 212–231.

Bourgeois-Pichat, Jean. "Utilisation de la notion stable pour mesurer la mortalité et la fecondité des populations des pays sous-développés." *Bulletin de l'Institut international de statistique,* 36(2):94–121. (Proceedings of the 30th Session, Stockholm, 1957). Uppsala, 1958.

Brass, William. "The Construction of Life Tables From Child Ratios." In *International Population Conference, New York, 1961* (International Union for the Scientific Study of Population), Vol. I, pp. 294–301.

Brass, William, and Coale, Ansley J. "Methods of Analysis and Estimation." In William Brass *et al., The Demography of Tropical Africa,* Princeton, Princeton University Press, 1968. Chapter 3, pp. 88–150.

Carmen, Arretx G., and Somoza, Jorge L. "Survey Methods, Based on Periodically Repeated Interviews, Aimed at Determining Demographic Rates." *Demography,* 2:289–301. 1965.

Cavanaugh, J. A. "Sample Vital Registration Experiment." In *International Population Conference, New York, 1961* (International Union for the Scientific Study of Population), Vol. II, pp. 363–371.

Chandrasekaran, C. "Fertility Indices from Limited Data." *International Population Conference, Ottawa, 1963* (International Union for the Scientific Study of Population), Liège 1964. Pp. 91–105.

Chandra Sekar, C., and Deming, W. Edwards. "On a Method of Estimating Birth and Death Rates and the Extent of Registration." *Journal of the American Statistical Association,* 44(245):101–115. March 1949.

Clairin, Rémy. "Les données susceptibles d'être utilisées pour une évaluation du mouvement de la population en Afrique au sud du Sahara." In *International Population Conference, Ottawa, 1963* (International Union for the Scientific Study of Population), Liège, 1964. Pp. 107–120.

Coale, Ansley J. "The Design of an Experimental Procedure for Obtaining Accurate Vital Statistics." In *International Population Conference, New York, 1961,* Vol. II, pp. 372–375.

———. "Estimates of Various Demographic Measures through the Quasi-Stable Age Distribution." In *Emerging Techniques in Population Research,* New York, Milbank Memorial Fund, 1963. Pp. 175–193.

Coale, Ansley J., and Demeny, Paul. *Regional Model Life Tables and Stable Populations.* Princeton, New Jersey, Princeton University Press, 1966.

Coale, Ansley J., and Hoover, Edgar M. *Population Growth and Economic Development in Low-Income Countries.* Princeton, New Jersey, Princeton University Press, 1958. Pp. 337–374.

Cooke, Dorothy S. "Population Growth Estimation Experiment in Pakistan." *Statistical Reporter* (U.S. Bureau of the Budget), May 1969. Pp. 173–176.

Demeny, Paul. "Estimation of Vital Rates for Populations in the Process of Destabilization." *Demography,* 2:516–530. 1965.

———. "A Minimum Program for the Estimation of Basic Fertility Measures from Censuses of Population in Asian Countries with Inadequate Demographic Statistics." In *Contributed Papers, Sydney Conference,* Sydney, Australia 1967, International Union for the Scientific Study of Population. Pp. 818–825.

Demeny, Paul, and Shorter, Frederic C. *Estimating Turkish Mortality, Fertility, and Age Structure: Application of Some New Techniques,* Publication No. 218, Faculty of Economics, University of Istanbul, Istanbul, 1968.

El-Badry, M. A. "Some Demographic Measurements for Egypt Based on the Stability of Census Age Distributions." *Milbank Memorial Fund Quarterly,* 33(3):268–305. July 1955.

El-Badry, M. A., and Chandrasekaran, C. "Some Methods for Obtaining Vital Statistics in India." *International Population Conference, New York, 1961* (International Union for the Scientific Study of Population), Vol. II, pp. 377–386.

Goldberg, David, and Adlakha, Arjun. "Infant Mortality Estimates Based on Small Surveys on the Ankara Area." Chapter 7 in *Turkish Demography: Proceedings of a Conference, Izmir, February 21–24, 1968.* Hacettepe University, Publication No. 7, Ankara, Turkey, 1969. Pp. 133–145.

Grabill, Wilson H., and Cho, Lee Jay. "Methodology for the Measurement of Current Fertility from Population Data on Young Children." *Demography,* 2:50–73. 1965.

Holzer, Jerzy. "Estimate of the Age Structure of Ghana's Population. An Application of the Stable Population Model." In *Contributed Papers, Sydney Conference,* Sydney, Aus-

SUGGESTED READINGS – Continued

tralia, 1967, International Union for the Scientific Study of Population. Pp. 838–849.

India, Office of the Registrar General. *Sample Registration in India: Report on Pilot Studies in Urban Areas: 1964–67,* 1969.

Indian Statistical Institute. "The Use of the National Sample Survey in the Estimation of Current Birth and Death Rates in India." *International Population Conference, New York, 1961* (International Union for the Scientific Study of Population), Vol. II, pp. 395–402.

Jabine, Thomas B. "Collection of Data on Population Growth and Fertility Within the Framework of the Atlantida Program." *Proceedings of the Interregional Workshop on the Methodology of Demographic Sample Surveys at Copenhagen, September 24– October 3, 1969,* forthcoming.

Jabine, Thomas B., and Bershad, Max A. "Some Comments on the Chandrasekaran-Deming Technique for the Measurement of Population Change." Presented at the CENTO Symposium on Demographic Statistics, Karachi, Pakistan, November 5–12, 1968.

Krótki, Karol J. "Estimating Population Size and Growth from Inadequate Data." *International Social Science Journal,* 17(2):246–258. 1965.

———. "First Report on the Population Growth Experiment." In International Union for the Scientific Study of Population, *International Population Conference, Ottawa, 1963,* Liège, International Union for the Scientific Study of Population, 1964. Pp. 159–173.

———. "The Problem of Estimating Vital Rates in Pakistan." In *World Population Conference, 1965* (Belgrade), United Nations, 1966. Pp. 154–158.

Lauriat, Patience. "Field Experience in Estimating Population Growth." *Demography,* 4(1):228–243. 1967.

Liberia. Department of Planning and Economic Affairs. *Liberian Population Growth Survey. Handbook 1969.*

Marks, Eli S., Seltzer, William, and Krotki, Karol J. *Population Growth Estimation: A Handbook of Vital Statistics Measurement,* New York, The Population Council, 1974.

Mauldin, W. Parker. "Estimating Rates of Population Growth." In *Family Planning and Population Programs,* Proceedings of the International Conference on Family Planning Programs, Geneva, August 1965, Chicago, University of Chicago Press, 1966. Pp. 635–653.

Muhsam, Helmut V. "Moderator's Introductory Statement." In *International Population Conference: Ottawa, 1963* (International Union for the Scientific Study of Population). Liège, 1964. Pp. 25–46.

Myburgh, C. A. L. "Estimating the Fertility and Mortality of African Populations from the Total Number of Children Ever Born and the Number of these Still Living." *Population Studies* (London), 10(2):193–206. November 1956.

Pakistan Institute of Development Economics. *Report of the Population Growth Estimation Experiment,* Karachi, December 1968.

Rele, J. R. "Estimation of Reproduction Rates for Asian Countries from Census Data." In *Contributed Papers, Sydney Conference,* Sydney, Australia, 1967, International Union for the Scientific Study of Population. Pp. 929–934.

———. *Fertility Analyses through Extension of Stable Population Concepts,* Berkeley, California, International Population and Urban Research, University of California, 1967.

Roberts, G. W. "Improving Vital Statistics in the West Indies." *International Population Conference, New York, 1961* (International Union for the Scientific Study of Population), Vol. II, pp. 420–426.

Romaniuk, A. "Estimation of the Birth Rate for the Congo Through Nonconventional Techniques." *Demography,* 4(2):688–709. 1967.

Sabagh, Georges, and Scott, Christopher. "A Comparison of Different Survey Techniques For Obtaining Vital Data in a Developing Country." *Demography,* 4(2):759–772. 1967.

Seltzer, William. "Some Results from Asian Population Studies." *Population Studies* (London), 23(3)395–406. Nov. 1969.

Som, Ranjan K. "On Some Techniques of Demographic Analysis of Special Relevance to the Asian Countries." In *Contributed Papers, Sydney Conference,* Sydney, Australia, 1967, International Union for the Scientific Study of Population. Pp. 807–813.

Srivastava, M. L. "The Relationships Between Fertility and Mortality Characteristics in Stable Female Populations." *Eugenics Quarterly,* 14(3): 171–180. September 1967.

———. "Selection of Model Life Tables and Stable Populations." In *Contributed Papers, Sydney Conference,* Sydney, Australia, 1967, International Union for the Scientific Study of Population. Pp. 904–911.

United Nations. *Age and Sex Patterns of Mortality. Model Life Tables for Underdeveloped Countries,* Series A, Population Studies, No. 22. 1955.

———. *Guanabara Demographic Pilot Survey: A Joint Project of the United Nations and the Government of Brazil,* Series A, Population Studies, No. 35. 1964.

———. *Methods of Estimating Basic Demographic Measures from Incomplete Data,* Series A, Population Studies, No. 42. 1967.

———. *Methods of Using Census Statistics for the Calculation of Life Tables and Other Demographic Measures,* by Giorgio Mortara, Series A, Population Studies, No. 7. November 1949.

———. *The Mysore Population Study,* Series A, Population Studies No. 34, 1961. Pp. 21–52, 221–404.

United States. National Center for Health Statistics. *Vital and Health Statistics,* Series 2, No. 32. "Methods for Measuring Population Change: A Systems Analysis Summary" by Forrest E. Linder. March 1969.

Wells, H. Bradley and Agrawal, B. L. "Sample Registration in India." *Demography,* 4(1):374–387. 1967.

Zelnik, Melvin, and Khan, Masihur Rahman. "An Estimate of the Birth Rate in East and West Pakistan." *Pakistan Development Review,* 5(1):64–93. Spring 1965.

APPENDIXES

APPENDIX A

Reference Tables for Constructing an Abridged Life Table by the Reed-Merrell Method

Table A-1.— Values of q_0 Associated With m_0 by the Equation:

$$q_0 = 1 - e^{-m_0(.9539 - .5509 m_0)}$$

m_0	q_0	Δ	m_0	q_0	Δ	m_0	q_0	Δ	m_0	q_0	Δ
		.000			.000			.000			.000
.000	.000 000	953	.050	.045 262	857	.100	.085 960	770	.150	.122 510	691
.001	.000 953	951	.051	.046 119	856	.101	.086 730	769	.151	.123 201	690
.002	.001 904	949	.052	.046 975	853	.102	.087 499	767	.152	.123 891	689
.003	.002 853	947	.053	.047 828	851	.103	.088 266	766	.153	.124 580	686
.004	.003 800	944	.054	.048 679	850	.104	.089 032	764	.154	.125 266	685
.005	.004 744	943	.055	.049 529	849	.105	.089 796	762	.155	.125 951	684
.006	.005 687	941	.056	.050 378	846	.106	.090 558	760	.156	.126 635	683
.007	.006 628	939	.057	.051 224	845	.107	.091 318	759	.157	.127 318	680
.008	.007 567	937	.058	.052 069	842	.108	.092 077	757	.158	.127 998	679
.009	.008 504	935	.059	.052 911	842	.109	.092 834	756	.159	.128 677	678
.010	.009 439	933	.060	.053 753	839	.110	.093 590	754	.160	.129 355	677
.011	.010 372	931	.061	.054 592	837	.111	.094 344	752	.161	.130 032	674
.012	.011 303	929	.062	.055 429	835	.112	.095 096	751	.162	.130 706	673
.013	.012 232	928	.063	.056 264	834	.113	.095 847	749	.163	.131 379	672
.014	.013 160	925	.064	.057 098	832	.114	.096 596	747	.164	.132 051	671
.015	.014 085	923	.065	.057 930	831	.115	.097 343	746	.165	.132 722	669
.016	.015 008	921	.066	.058 761	828	.116	.098 089	745	.166	.133 391	667
.017	.015 929	919	.067	.059 589	827	.117	.098 834	742	.167	.134 058	666
.018	.016 848	918	.068	.060 416	825	.118	.099 576	741	.168	.134 724	665
.019	.017 766	915	.069	.061 241	823	.119	.100 317	739	.169	.135 389	663
.020	.018 681	913	.070	.062 064	821	.120	.101 056	739	.170	.136 052	661
.021	.019 594	912	.071	.062 885	820	.121	.101 795	736	.171	.136 713	660
.022	.020 506	910	.072	.063 705	818	.122	.102 531	734	.172	.137 373	659
.023	.021 416	907	.073	.064 523	817	.123	.103 265	733	.173	.138 032	657
.024	.022 323	906	.074	.065 340	814	.124	.103 998	732	.174	.138 689	656
.025	.023 229	904	.075	.066 154	813	.125	.104 730	730	.175	.139 345	654
.026	.024 133	902	.076	.066 967	811	.126	.105 460	728	.176	.139 999	653
.027	.025 035	900	.077	.067 778	809	.127	.106 188	727	.177	.140 652	651
.028	.025 935	898	.078	.068 587	808	.128	.106 915	725	.178	.141 303	651
.029	.026 833	897	.079	.069 395	805	.129	.107 640	724	.179	.141 954	648
.030	.027 730	894	.080	.070 200	805	.130	.108 364	722	.180	.142 602	647
.031	.028 624	892	.081	.071 005	802	.131	.109 086	720	.181	.143 249	646
.032	.029 516	891	.082	.071 807	801	.132	.109 806	719	.182	.143 895	644
.033	.030 407	889	.083	.072 608	799	.133	.110 525	717	.183	.144 539	643
.034	.031 296	886	.084	.073 407	797	.134	.111 242	716	.184	.145 182	641
.035	.032 182	885	.085	.074 204	795	.135	.111 958	714	.185	.145 823	640
.036	.033 067	884	.086	.074 999	794	.136	.112 672	713	.186	.146 463	639
.037	.033 951	881	.087	.075 793	793	.137	.113 385	711	.187	.147 102	637
.038	.034 832	879	.088	.076 586	790	.138	.114 096	710	.188	.147 739	636
.039	.035 711	878	.089	.077 376	789	.139	.114 806	708	.189	.148 375	634
.040	.036 589	875	.090	.078 165	787	.140	.115 514	706	.190	.149 009	633
.041	.037 464	874	.091	.078 952	786	.141	.116 220	705	.191	.149 642	631
.042	.038 338	872	.092	.079 738	783	.142	.116 925	704	.192	.150 273	630
.043	.039 210	870	.093	.080 521	782	.143	.117 629	702	.193	.150 903	629
.044	.040 080	868	.094	.081 303	780	.144	.118 331	700	.194	.151 532	628
.045	.040 948	866	.095	.082 083	779	.145	.119 031	699	.195	.152 160	625
.046	.041 814	865	.096	.082 862	777	.146	.119 730	697	.196	.152 785	625
.047	.042 679	863	.097	.083 639	776	.147	.120 427	696	.197	.153 410	623
.048	.043 542	861	.098	.084 415	773	.148	.121 123	694	.198	.154 033	622
.049	.044 403	859	.099	.085 188	772	.149	.121 817	693	.199	.154 655	620
.050	.045 262	857	.100	.085 960	770	.150	.122 510	691	.200	.155 275	

NOTE.—The tables in this appendix were extracted from, Lowell J. Reed and Margaret Merrell, "A Short Method for Constructing an Abridged Life Table," *American Journal of Hygiene*, 30(2):52–61, September 1939. Copyright, 1939, by the *American Journal of Hygiene* (now the *American Journal of Epidemiology*), The Johns Hopkins University.

Table A-2. — Values of q_1 Associated With m_1 by the Equation:

$$q_1 = 1 - e^{-m \cdot (.9510 - 1.921 m_1)}$$

m_1	q_1	Δ	m_1	q_1	Δ
		.000			.000
.000	.000 000	949	.050	.041 847	725
.001	.000 949	944	.051	.042 572	721
.002	.001 893	939	.052	.045 293	717
.003	.002 832	934	.053	.044 010	712
.004	.003 766	930	.054	.044 722	708
.005	.004 696	925	.055	.045 430	704
.006	.005 621	920	.056	.046 134	699
.007	.006 541	916	.057	.046 833	696
.008	.007 457	911	.058	.047 529	692
.009	.008 368	907	.059	.048 221	687
.010	.009 275	902	.060	.048 908	683
.011	.010 177	897	.061	.049 591	679
.012	.011 074	893	.062	.050 270	675
.013	.011 967	887	.063	.050 945	671
.014	.012 854	884	.064	.051 616	666
.015	.013 738	878	.065	.052 282	663
.016	.014 616	875	.066	.052 945	659
.017	.015 491	869	.067	.053 604	654
.018	.016 360	866	.068	.054 258	650
.019	.017 226	860	.069	.054 908	647
.020	.018 086	856	.070	.055 555	641
.021	.018 942	852	.071	.056 196	639
.022	.019 794	847	.072	.056 835	634
.023	.020 641	842	.073	.057 469	630
.024	.021 483	838	.074	.058 099	626
.025	.022 321	834	.075	.058 725	622
.026	.023 155	829	.076	.059 347	617
.027	.023 984	825	.077	.059 964	614
.028	.024 809	820	.078	.060 578	610
.029	.025 629	816	.079	.061 188	606
.030	.026 445	812	.080	.061 794	602
.031	.027 257	807	.081	.062 396	598
.032	.028 064	802	.082	.062 994	594
.033	.028 866	799	.083	.063 588	590
.034	.029 665	793	.084	.064 178	585
.035	.030 458	790	.085	.064 763	583
.036	.031 248	785	.086	.065 346	578
.037	.032 033	781	.087	.065 924	574
.038	.032 814	776	.088	.066 498	570
.039	.033 590	772	.089	.067 068	566
.040	.034 362	768	.090	.067 634	562
.041	.035 130	763	.091	.068 196	559
.042	.035 893	760	.092	.068 755	554
.043	.036 653	755	.093	.069 309	551
.044	.037 408	750	.094	.069 860	547
.045	.038 158	746	.095	.070 407	542
.046	.038 904	742	.096	.070 949	540
.047	.039 646	738	.097	.071 489	534
.048	.040 384	733	.098	.072 023	532
.049	.041 117	730	.099	.072 555	527
.050	.041 847	725	.100	.073 082	

Table A-3. — Values of $_3q_2$ Associated With $_3m_2$ by the Equation:

$$_3q_2 = 1 - e^{-3\,_3m_2 - .008(3)^3\,_3m_2^2}$$

$_3m_2$	$_3q_2$	Δ	$_3m_2$	$_3q_2$	Δ
		.00			.00
.000	.000 000	2 996	.010	.029 576	2 911
.001	.002 996	2 987	.011	.032 487	2 903
.002	.005 983	2 979	.012	.035 390	2 895
.003	.008 962	2 970	.013	.038 285	2 886
.004	.011 932	2 962	.014	.041 171	2 878
.005	.014 894	2 953	.015	.044 049	2 870
.006	.017 847	2 945	.016	.046 919	2 862
.007	.020 792	2 936	.017	.049 781	2 854
.008	.023 728	2 928	.018	.052 635	2 845
.009	.026 656	2 920	.019	.055 480	2 837
.010	.029 576	2 911	.020	.058 317	

Table A-4. — Values of $_4q_1$ Associated With $_4m_1$ by the Equation:

$$_4q_1 = 1 - e^{-4\,_4m_1(.9806 - 2.079\,_4m_1)}$$

$_4m_1$	$_4q_1$	Δ	$_4m_1$	$_4q_1$	Δ
		.00			.00
.000	.000 000	3 906	.020	.072 370	3 316
.001	.003 906	3 875	.021	.075 686	3 289
.002	.007 781	3 843	.022	.078 975	3 262
.003	.011 624	3 812	.023	.082 237	3 235
.004	.015 436	3 781	.024	.085 472	3 209
.005	.019 217	3 750	.025	.088 681	3 183
.006	.022 967	3 720	.026	.091 864	3 156
.007	.026 687	3 689	.027	.095 020	3 131
.008	.030 376	3 659	.028	.098 151	3 104
.009	.034 035	3 630	.029	.101 255	3 079
.010	.037 665	3 600	.030	.104 334	3 054
.011	.041 265	3 570	.031	.107 388	3 028
.012	.044 835	3 542	.032	.110 416	3 003
.013	.048 377	3 512	.033	.113 419	2 979
.014	.051 889	3 484	.034	.116 398	2 954
.015	.055 373	3 455	.035	.119 352	2 929
.016	.058 828	3 427	.036	.122 281	2 905
.017	.062 255	3 399	.037	.125 186	2 882
.018	.065 654	3 372	.038	.128 068	2 856
.019	.069 026	3 344	.039	.130 924	2 834
.020	.072 370	3 316	.040	.133 758	

Table A–5. — Values of $_5q_x$ Associated With $_5m_x$ by the Equation:

$$_5q_x = 1 - e^{-5_5m_x - .008(5)^3{}_5m_x^2}$$

$_5m_x$	$_5q_x$	Δ	$_5m_x$	$_5q_x$	Δ	$_5m_x$	$_5q_x$	Δ
		.00			.00			.00
.000	.000 000	4 989	.050	.223 144	3 952	.100	.399 504	3 116
.001	.004 989	4 965	.051	.227 096	3 935	.101	.402 620	3 100
.002	.009 954	4 943	.052	.231 031	3 915	.102	.405 720	3 085
.003	.014 897	4 920	.053	.234 946	3 897	.103	.408 805	3 070
.004	.019 817	4 897	.054	.238 843	3 879	.104	.411 875	3 056
.005	.024 714	4 876	.055	.242 722	3 861	.105	.414 931	3 041
.006	.029 590	4 852	.056	.246 583	3 842	.106	.417 972	3 026
.007	.034 442	4 830	.057	.250 425	3 824	.107	.420 998	3 011
.008	.039 272	4 808	.058	.254 249	3 807	.108	.424 009	2 998
.009	.044 080	4 786	.059	.258 056	3 788	.109	.427 007	2 982
.010	.048 866	4 763	.060	.261 844	3 770	.110	.429 989	2 969
.011	.053 629	4 742	.061	.265 614	3 753	.111	.432 958	2 953
.012	.058 371	4 720	.062	.269 367	3 735	.112	.435 911	2 940
.013	.063 091	4 698	.063	.273 102	3 717	.113	.438 851	2 926
.014	.067 789	4 676	.064	.276 819	3 700	.114	.441 777	2 911
.015	.072 465	4 655	.065	.280 519	3 682	.115	.444 688	2 897
.016	.077 120	4 633	.066	.284 201	3 665	.116	.447 585	2 883
.017	.081 753	4 612	.067	.287 866	3 647	.117	.450 468	2 870
.018	.086 365	4 590	.068	.291 513	3 630	.118	.453 338	2 855
.019	.090 955	4 570	.069	.295 143	3 613	.119	.456 193	2 842
.020	.095 525	4 547	.070	.298 756	3 596	.120	.459 035	2 827
.021	.100 072	4 527	.071	.302 352	3 579	.121	.461 862	2 815
.022	.104 599	4 506	.072	.305 931	3 562	.122	.464 677	2 800
.023	.109 105	4 485	.073	.309 493	3 545	.123	.467 477	2 787
.024	.113 590	4 464	.074	.313 038	3 528	.124	.470 264	2 773
.025	.118 054	4 444	.075	.316 566	3 511	.125	.473 037	2 760
.026	.122 498	4 423	.076	.320 077	3 495	.126	.475 797	2 746
.027	.126 921	4 402	.077	.323 572	3 478	.127	.478 543	2 733
.028	.131 323	4 382	.078	.327 050	3 461	.128	.481 276	2 720
.029	.135 705	4 361	.079	.330 511	3 445	.129	.483 996	2 707
.030	.140 066	4 341	.080	.333 956	3 429	.130	.486 703	2 693
.031	.144 407	4 321	.081	.337 385	3 412	.131	.489 396	2 680
.032	.148 728	4 301	.082	.340 797	3 396	.132	.492 076	2 667
.033	.153 029	4 281	.083	.344 193	3 380	.133	.494 743	2 655
.034	.157 310	4 261	.084	.347 573	3 364	.134	.497 398	2 641
.035	.161 571	4 241	.085	.350 937	3 347	.135	.500 039	2 628
.036	.165 812	4 221	.086	.354 284	3 332	.136	.502 667	2 616
.037	.170 033	4 201	.087	.357 616	3 316	.137	.505 283	2 603
.038	.174 234	4 182	.088	.360 932	3 300	.138	.507 886	2 590
.039	.178 416	4 162	.089	.364 232	3 284	.139	.510 476	2 577
.040	.182 578	4 143	.090	.367 516	3 268	.140	.513 053	2 565
.041	.186 721	4 123	.091	.370 784	3 253	.141	.515 618	2 552
.042	.190 844	4 104	.092	.374 037	3 237	.142	.518 170	2 540
.043	.194 948	4 085	.093	.377 274	3 222	.143	.520 710	2 527
.044	.199 033	4 066	.094	.380 496	3 206	.144	.523 237	2 515
.045	.203 099	4 047	.095	.383 702	3 191	.145	.525 752	2 503
.046	.207 146	4 028	.096	.386 893	3 176	.146	.528 255	2 490
.047	.211 174	4 008	.097	.390 069	3 160	.147	.530 745	2 478
.048	.215 182	3 990	.098	.393 229	3 145	.148	.533 223	2 466
.049	.219 172	3 972	.099	.396 374	3 130	.149	.535 689	2 454
.050	.223 144	3 952	.100	.399 504	3 116	.150	.538 143	2 442

Table A–5. — Values of $_5q_x$ Associated With $_5m_x$ by the Equation:

$$_5q_x = 1 - e^{-5_5m_x - .008(5)^3 {}_5m_x^2} - \text{Continued}$$

$_5m_x$	$_5q_x$	Δ	$_5m_x$	$_5q_x$	Δ	$_5m_x$	$_5q_x$	Δ
		.00			.00			.00
.150	.538 143	2 442	.200	.646 545	1 904	.250	.730 854	1 476
.151	.540 585	2 430	.201	.648 449	1 894	.251	.732 330	1 469
.152	.543 015	2 418	.202	.650 343	1 885	.252	.733 799	1 462
.153	.545 433	2 406	.203	.652 228	1 876	.253	.735 261	1 453
.154	.547 839	2 394	.204	.654 104	1 866	.254	.736 714	1 447
.155	.550 233	2 382	.205	.655 970	1 856	.255	.738 161	1 439
.156	.552 615	2 371	.206	.657 826	1 847	.256	.739 600	1 432
.157	.554 986	2 359	.207	.659 673	1 838	.257	.741 032	1 424
.158	.557 345	2 347	.208	.661 511	1 829	.258	.742 456	1 417
.159	.559 692	2 336	.209	.663 340	1 819	.259	.743 873	1 409
.160	.562 028	2 324	.210	.665 159	1 810	.260	.745 282	1 403
.161	.564 352	2 313	.211	.666 969	1 802	.261	.746 685	1 395
.162	.566 665	2 301	.212	.668 771	1 792	.262	.748 080	1 388
.163	.568 966	2 290	.213	.670 563	1 783	.263	.749 468	1 381
.164	.571 256	2 279	.214	.672 346	1 774	.264	.750 849	1 374
.165	.573 535	2 267	.215	.674 120	1 765	.265	.752 223	1 366
.166	.575 802	2 257	.216	.675 885	1 756	.266	.753 589	1 360
.167	.578 059	2 245	.217	.677 641	1 747	.267	.754 949	1 353
.168	.580 304	2 234	.218	.679 388	1 739	.268	.756 302	1 345
.169	.582 538	2 223	.219	.681 127	1 729	.269	.757 647	1 339
.170	.584 761	2 211	.220	.682 856	1 721	.270	.758 986	1 332
.171	.586 972	2 201	.221	.684 577	1 712	.271	.760 318	1 325
.172	.589 173	2 190	.222	.686 289	1 704	.272	.761 643	1 318
.173	.591 363	2 180	.223	.687 993	1 695	.273	.762 961	1 311
.174	.593 543	2 168	.224	.689 688	1 686	.274	.764 272	1 304
.175	.595 711	2 157	.225	.691 374	1 678	.275	.765 576	1 298
.176	.597 868	2 147	.226	.693 052	1 669	.276	.766 874	1 291
.177	.600 015	2 137	.227	.694 721	1 661	.277	.768 165	1 284
.178	.602 152	2 125	.228	.696 382	1 652	.278	.769 449	1 278
.179	.604 277	2 115	.229	.698 034	1 644	.279	.770 727	1 271
.180	.606 392	2 105	.230	.699 678	1 636	.280	.771 998	1 264
.181	.608 497	2 094	.231	.701 314	1 627	.281	.773 262	1 258
.182	.610 591	2 083	.232	.702 941	1 619	.282	.774 520	1 251
.183	.612 674	2 073	.233	.704 560	1 611	.283	.775 771	1 245
.184	.614 747	2 063	.234	.706 171	1 602	.284	.777 016	1 239
.185	.616 810	2 053	.235	.707 773	1 595	.285	.778 255	1 231
.186	.618 863	2 042	.236	.709 368	1 586	.286	.779 486	1 226
.187	.620 905	2 032	.237	.710 954	1 578	.287	.780 712	1 219
.188	.622 937	2 022	.238	.712 532	1 570	.288	.781 931	1 213
.189	.624 959	2 012	.239	.714 102	1 562	.289	.783 144	1 206
.190	.626 971	2 002	.240	.715 664	1 555	.290	.784 350	1 201
.191	.628 973	1 992	.241	.717 219	1 546	.291	.785 551	1 193
.192	.630 965	1 982	.242	.718 765	1 538	.292	.786 744	1 188
.193	.632 947	1 972	.243	.720 303	1 531	.293	.787 932	1 182
.194	.634 919	1 962	.244	.721 834	1 522	.294	.789 114	1 175
.195	.636 881	1 952	.245	.723 356	1 515	.295	.790 289	1 169
.196	.638 833	1 943	.246	.724 871	1 507	.296	.791 458	1 163
.197	.640 776	1 933	.247	.726 378	1 500	.297	.792 621	1 157
.198	.642 709	1 923	.248	.727 878	1 492	.298	.793 778	1 151
.199	.644 632	1 913	.249	.729 370	1 484	.299	.794 929	1 145
.200	.646 545	1 904	.250	.730 854	1 476	.300	.796 074	1 139

Table A–5. — Values of $_5q_x$ Associated With $_5m_x$ by the Equation:

$$_5q_x = 1 - e^{-5\,_5m_x - .008(5)^3\,_5m_x^2} - \text{Continued}$$

$_5m_x$	$_5q_x$	Δ	$_5m_x$	$_5q_x$	Δ	$_5m_x$	$_5q_x$	Δ
		.00			.000			.000
.300	.796 074	1 139	.350	.846 261	874	.400	.884 675	667
.301	.797 213	1 133	.351	.847 135	869	.401	.885 342	663
.302	.798 346	1 128	.352	.848 004	865	.402	.886 005	660
.303	.799 474	1 121	.353	.848 869	860	.403	.886 665	656
.304	.800 595	1 115	.354	.849 729	856	.404	.887 321	653
.305	.801 710	1 110	.355	.850 585	850	.405	.887 974	649
.306	.802 820	1 103	.356	.851 435	847	.406	.888 623	646
.307	.803 923	1 098	.357	.852 282	842	.407	.889 269	642
.308	.805 021	1 092	.358	.853 124	837	.408	.889 911	638
.309	.806 113	1 087	.359	.853 961	833	.409	.890 549	635
.310	.807 200	1 080	.360	.854 794	828	.410	.891 184	632
.311	.808 280	1 075	.361	.855 622	824	.411	.891 816	628
.312	.809 355	1 070	.362	.856 446	819	.412	.892 444	625
.313	.810 425	1 063	.363	.857 265	816	.413	.893 069	621
.314	.811 488	1 059	.364	.858 081	810	.414	.893 690	618
.315	.812 547	1 052	.365	.858 891	807	.415	.894 308	614
.316	.813 599	1 047	.366	.859 698	802	.416	.894 922	612
.317	.814 646	1 042	.367	.860 500	798	.417	.895 534	607
.318	.815 688	1 036	.368	.861 298	793	.418	.896 141	605
.319	.816 724	1 030	.369	.862 091	789	.419	.896 746	601
.320	.817 754	1 026	.370	.862 880	785	.420	.897 347	598
.321	.818 780	1 019	.371	.863 665	781	.421	.897 945	594
.322	.819 799	1 015	.372	.864 446	776	.422	.898 539	592
.323	.820 814	1 009	.373	.865 222	773	.423	.899 131	588
.324	.821 823	1 003	.374	.865 995	768	.424	.899 719	585
.325	.822 826	0 999	.375	.866 763	764	.425	.900 304	581
.326	.823 825	0 993	.376	.867 527	760	.426	.900 885	579
.327	.824 818	0 988	.377	.868 287	756	.427	.901 464	575
.328	.825 806	0 982	.378	.869 043	751	.428	.902 039	572
.329	.826 788	0 978	.379	.869 794	748	.429	.902 611	569
.330	.827 766	0 972	.380	.870 542	744	.430	.903 180	566
.331	.828 738	0 967	.381	.871 286	739	.431	.903 746	562
.332	.829 705	0 962	.382	.872 025	736	.432	.904 308	560
.333	.830 667	0 957	.383	.872 761	732	.433	.904 868	556
.334	.831 624	0 952	.384	.873 493	728	.434	.905 424	554
.335	.832 576	0 947	.385	.874 221	723	.435	.905 978	550
.336	.833 523	0 941	.386	.874 944	720	.436	.906 528	548
.337	.834 464	0 937	.387	.875 664	716	.437	.907 076	544
.338	.835 401	0 932	.388	.876 380	712	.438	.907 620	541
.339	.836 333	0 927	.389	.877 092	708	.439	.908 161	539
.340	.837 260	0 922	.390	.877 800	705	.440	.908 700	535
.341	.838 182	0 917	.391	.878 505	700	.441	.909 235	532
.342	.839 099	0 912	.392	.879 205	697	.442	.909 767	530
.343	.840 011	0 907	.393	.879 902	693	.443	.910 297	526
.344	.840 918	0 903	.394	.880 595	689	.444	.910 823	524
.345	.841 821	0 897	.395	.881 284	686	.445	.911 347	521
.346	.842 718	0 893	.396	.881 970	682	.446	.911 868	518
.347	.843 611	0 888	.397	.882 652	678	.447	.912 386	514
.348	.844 499	0 884	.398	.883 330	674	.448	.912 900	513
.349	.845 383	0 878	.399	.884 004	671	.449	.913 413	509
.350	.843 261	0 874	.400	.884 675	667	.450	.913 922	

Table A–6.— Values of $_{10}q_x$ Associated With $_{10}m_x$ by the Equation:

$$_{10}q_x = 1 - e^{-10\,_{10}m_x - .008(10)^3\,_{10}m_x^2}$$

$_{10}m_x$	$_{10}q_x$	Δ	$_{10}m_x$	$_{10}q_x$	Δ	$_{10}m_x$	$_{10}q_x$	Δ
		.00			.00			.00
.000	.000 000	9 959	.050	.405 479	6 391	.100	.660 404	3 920
.001	.009 959	9 874	.051	.411 870	6 332	.101	.664 324	3 879
.002	.019 833	9 792	.052	.418 202	6 273	.102	.668 203	3 840
.003	.029 625	9 709	.053	.424 475	6 214	.103	.672 043	3 800
.004	.039 334	9 627	.054	.430 689	6 156	.104	.675 843	3 762
.005	.048 961	9 546	.055	.436 845	6 098	.105	.679 605	3 723
.006	.058 507	9 465	.056	.442 943	6 041	.106	.683 328	3 685
.007	.067 972	9 384	.057	.448 984	5 985	.107	.687 013	3 646
.008	.077 356	9 305	.058	.454 969	5 928	.108	.690 659	3 610
.009	.086 661	9 225	.059	.460 897	5 872	.109	.694 269	3 571
.010	.095 886	9 147	.060	.466 769	5 816	.110	.697 840	3 535
.011	.105 033	9 068	.061	.472 585	5 762	.111	.701 375	3 499
.012	.114 101	8 990	.062	.478 347	5 706	.112	.704 874	3 462
.013	.123 091	8 913	.063	.484 053	5 653	.113	.708 336	3 426
.014	.132 004	8 836	.064	.489 706	5 598	.114	.711 762	3 390
.015	.140 840	8 760	.065	.495 304	5 546	.115	.715 152	3 355
.016	.149 600	8 684	.066	.500 850	5 492	.116	.718 507	3 320
.017	.158 284	8 608	.067	.506 342	5 439	.117	.721 827	3 285
.018	.166 892	8 534	.068	.511 781	5 388	.118	.725 112	3 251
.019	.175 426	8 459	.069	.517 169	5 335	.119	.728 363	3 216
.020	.183 885	8 386	.070	.522 504	5 284	.120	.731 579	3 183
.021	.192 271	8 312	.071	.527 788	5 234	.121	.734 762	3 149
.022	.200 583	8 239	.072	.533 022	5 182	.122	.737 911	3 116
.023	.208 822	8 167	.073	.538 204	5 132	.123	.741 027	3 083
.024	.216 989	8 095	.074	.543 336	5 083	.124	.744 110	3 050
.025	.225 084	8 023	.075	.548 419	5 033	.125	.747 160	3 018
.026	.233 107	7 953	.076	.553 452	4 984	.126	.750 178	2 986
.027	.241 060	7 882	.077	.558 436	4 935	.127	.753 164	2 954
.028	.248 942	7 812	.078	.563 371	4 887	.128	.756 118	2 923
.029	.256 754	7 743	.079	.568 258	4 840	.129	.759 041	2 891
.030	.264 497	7 673	.080	.573 098	4 791	.130	.761 932	2 861
.031	.272 170	7 606	.081	.577 889	4 745	.131	.764 793	2 830
.032	.279 776	7 536	.082	.582 634	4 698	.132	.767 623	2 799
.033	.287 312	7 470	.083	.587 332	4 652	.133	.770 422	2 769
.034	.294 782	7 402	.084	.591 984	4 605	.134	.773 191	2 740
.035	.302 184	7 336	.085	.596 589	4 560	.135	.775 931	2 710
.036	.309 520	7 269	.086	.601 149	4 515	.136	.778 641	2 680
.037	.316 789	7 204	.087	.605 664	4 470	.137	.781 321	2 652
.038	.323 993	7 139	.088	.610 134	4 425	.138	.783 973	2 623
.039	.331 132	7 073	.089	.614 559	4 382	.139	.786 596	2 594
.040	.338 205	7 010	.090	.618 941	4 337	.140	.789 190	2 567
.041	.345 215	6 946	.091	.623 278	4 294	.141	.791 757	2 538
.042	.352 161	6 881	.092	.627 572	4 251	.142	.794 295	2 511
.043	.359 042	6 820	.093	.631 823	4 209	.143	.796 806	2 483
.044	.365 862	6 756	.094	.636 032	4 166	.144	.799 289	2 456
.045	.372 618	6 695	.095	.640 198	4 123	.145	.801 745	2 429
.046	.379 313	6 633	.096	.644 321	4 083	.146	.804 174	2 402
.047	.385 946	6 572	.097	.648 404	4 041	.147	.806 576	2 376
.048	.392 518	6 511	.098	.652 445	4 000	.148	.808 952	2 350
.049	.399 029	6 450	.099	.656 445	3 959	.149	.811 302	2 324
.050	.405 479	6 391	.100	.660 404	3 920	.150	.813 626	2 298

Table A–6. — Values of $_{10}q_x$ Associated With $_{10}m_x$ by the Equation:

$$_{10}q_x = 1 - e^{-10_{10}m_x - .008(10)^3 {}_{10}m_x^2} \quad \text{— Continued}$$

$_{10}m_x$	$_{10}q_x$	Δ	$_{10}m_x$	$_{10}q_x$	Δ	$_{10}m_x$	$_{10}q_x$	Δ
		.00			.00			.000
.150	.813 626	2 298	.200	.901 726	1 290	.250	.950 213	693
.151	.815 924	2 273	.201	.903 016	1 274	.251	.950 906	685
.152	.818 197	2 248	.202	.904 290	1 259	.252	.951 589	675
.153	.820 445	2 222	.203	.905 549	1 244	.253	.952 264	666
.154	.822 667	2 198	.204	.906 793	1 228	.254	.952 930	658
.155	.824 865	2 174	.205	.908 021	1 215	.255	.953 588	649
.156	.827 039	2 149	.206	.909 236	1 199	.256	.954 237	641
.157	.829 188	2 125	.207	.910 435	1 185	.257	.954 878	633
.158	.831 313	2 102	.208	.911 620	1 171	.258	.955 511	624
.159	.833 415	2 078	.209	.912 791	1 157	.259	.956 135	617
.160	.835 493	2 054	.210	.913 948	1 142	.260	.956 752	608
.161	.837 547	2 032	.211	.915 090	1 129	.261	.957 360	601
.162	.839 579	2 008	.212	.916 219	1 115	.262	.957 961	593
.163	.841 587	1 986	.213	.917 334	1 102	.263	.958 554	585
.164	.843 573	1 964	.214	.918 436	1 088	.264	.959 139	577
.165	.845 537	1 941	.215	.919 524	1 075	.265	.959 716	570
.166	.847 478	1 920	.216	.920 599	1 062	.266	.960 286	562
.167	.849 398	1 897	.217	.921 661	1 049	.267	.960 848	555
.168	.851 295	1 876	.218	.922 710	1 036	.268	.961 403	548
.169	.853 171	1 855	.219	.923 746	1 024	.269	.961 951	541
.170	.855 026	1 833	.220	.924 770	1 011	.270	.962 492	534
.171	.856 859	1 813	.221	.925 781	0 998	.271	.963 026	526
.172	.858 672	1 792	.222	.926 779	0 986	.272	.963 552	520
.173	.860 464	1 771	.223	.927 765	0 974	.273	.964 072	513
.174	.862 235	1 751	.224	.928 739	0 962	.274	.964 585	506
.175	.863 986	1 731	.225	.929 701	0 950	.275	.965 091	499
.176	.865 717	1 711	.226	.930 651	0 939	.276	.965 590	493
.177	.867 428	1 692	.227	.931 590	0 926	.277	.966 083	486
.178	.869 120	1 672	.228	.932 516	0 916	.278	.966 569	480
.179	.870 792	1 652	.229	.933 432	0 904	.279	.967 049	473
.180	.872 444	1 633	.230	.934 336	0 892	.280	.967 522	467
.181	.874 077	1 615	.231	.935 228	0 882	.281	.967 989	461
.182	.875 692	1 596	.232	.936 110	0 871	.282	.968 450	455
.183	.877 288	1 577	.233	.936 981	0 859	.283	.968 905	449
.184	.878 865	1 559	.234	.937 840	0 849	.284	.969 354	443
.185	.880 424	1 540	.235	.938 689	0 839	.285	.969 797	436
.186	.881 964	1 523	.236	.939 528	0 827	.286	.970 233	431
.187	.883 487	1 505	.237	.940 355	0 818	.287	.970 664	426
.188	.884 992	1 487	.238	.941 173	0 807	.288	.971 090	419
.189	.886 479	1 470	.239	.941 980	0 797	.289	.971 509	414
.190	.887 949	1 452	.240	.942 777	0 787	.290	.971 923	408
.191	.889 401	1 436	.241	.943 564	0 777	.291	.972 331	403
.192	.890 837	1 418	.242	.944 341	0 767	.292	.972 734	397
.193	.892 255	1 402	.243	.945 108	0 758	.293	.973 131	392
.194	.893 657	1 386	.244	.945 866	0 748	.294	.973 523	387
.195	.895 043	1 368	.245	.946 614	0 738	.295	.973 910	381
.196	.896 411	1 353	.246	.947 352	0 729	.296	.974 291	377
.197	.897 764	1 337	.247	.948 081	0 720	.297	.974 668	371
.198	.899 101	1 320	.248	.948 801	0 710	.298	.975 039	366
.199	.900 421	1 305	.249	.949 511	0 702	.299	.975 405	361
.200	.901 726	1 290	.250	.950 213	0 693	.300	.975 766	356

Table A–6. — Values of $_{10}q_x$ Associated With $_{10}m_x$ by the Equation:

$$_{10}q_x = 1 - e^{-10_{10}m_x - .008(10)^3 \,_{10}m_x^2} \quad \text{—Continued}$$

$_{10}m_x$	$_{10}q_x$	$\triangle$	$_{10}m_x$	$_{10}q_x$	$\triangle$	$_{10}m_x$	$_{10}q_x$	$\triangle$
		.000			.000			.0000
.300	.975 766	356	.350	.988 667	175	.400	.994 908	82
.301	.976 122	352	.351	.988 842	173	.401	.994 990	82
.302	.976 474	346	.352	.989 015	171	.402	.995 072	80
.303	.976 820	342	.353	.989 186	168	.403	.995 152	80
.304	.977 162	337	.354	.989 354	165	.404	.995 232	77
.305	.977 499	333	.355	.989 519	163	.405	.995 309	77
.306	.977 832	328	.356	.989 682	161	.406	.995 386	76
.307	.978 160	323	.357	.989 843	158	.407	.995 462	74
.308	.978 483	319	.358	.990 001	157	.408	.995 536	73
.309	.978 802	315	.359	.990 158	153	.409	.995 609	72
.310	.979 117	310	.360	.990 311	152	.410	.995 681	71
.311	.979 427	306	.361	.990 463	149	.411	.995 752	70
.312	.979 733	302	.362	.990 612	147	.412	.995 822	69
.313	.980 035	297	.363	.990 759	145	.413	.995 891	68
.314	.980 332	294	.364	.990 904	143	.414	.995 959	66
.315	.980 626	289	.365	.991 047	141	.415	.996 025	66
.316	.980 915	285	.366	.991 188	139	.416	.996 091	64
.317	.981 200	282	.367	.991 327	136	.417	.996 155	64
.318	.981 482	277	.368	.991 463	135	.418	.996 219	63
.319	.981 759	274	.369	.991 598	133	.419	.996 282	61
.320	.982 033	269	.370	.991 731	130	.420	.996 343	61
.321	.982 302	266	.371	.991 861	129	.421	.996 404	60
.322	.982 568	263	.372	.991 990	127	.422	.996 464	58
.323	.982 831	258	.373	.992 117	125	.423	.996 522	58
.324	.983 089	255	.374	.992 242	123	.424	.996 580	57
.325	.983 344	252	.375	.992 365	121	.425	.996 637	56
.326	.983 596	248	.376	.992 486	120	.426	.996 693	56
.327	.983 844	244	.377	.992 606	117	.427	.996 749	54
.328	.984 088	241	.378	.992 723	116	.428	.996 803	53
.329	.984 329	237	.379	.992 839	114	.429	.996 856	53
.330	.984 566	234	.380	.992 953	113	.430	.996 909	52
.331	.984 800	231	.381	.993 066	111	.431	.996 961	50
.332	.985 031	228	.382	.993 177	109	.432	.997 011	51
.333	.985 259	224	.383	.993 286	107	.433	.997 062	49
.334	.985 483	221	.384	.993 393	106	.434	.997 111	49
.335	.985 704	218	.385	.993 499	104	.435	.997 160	47
.336	.985 922	215	.386	.993 603	103	.436	.997 207	47
.337	.986 137	212	.387	.993 706	101	.437	.997 254	47
.338	.986 349	209	.388	.993 807	100	.438	.997 301	45
.339	.986 558	206	.389	.993 907	098	.439	.997 346	45
.340	.986 764	203	.390	.994 005	096	.440	.997 391	44
.341	.986 967	200	.391	.994 101	096	.441	.997 435	43
.342	.987 167	197	.392	.994 197	093	.442	.997 478	43
.343	.987 364	194	.393	.994 290	093	.443	.997 521	42
.344	.987 558	192	.394	.994 383	090	.444	.997 563	42
.345	.987 750	188	.395	.994 473	090	.445	.997 605	40
.346	.987 938	186	.396	.994 563	088	.446	.997 645	40
.347	.988 124	184	.397	.994 651	087	.447	.997 685	40
.348	.988 308	180	.398	.994 738	085	.448	.997 725	38
.349	.988 488	179	.399	.994 823	085	.449	.997 763	39
.350	.988 667	175	.400	.994 908	082	.450	.997 802	

Selected "West" Model Life Tables and Stable Population Tables, and Related Reference Tables

Table B–1.—Selected "West" Model Life Tables Arranged by Level of Mortality

Age interval (exact ages x to x + n)	l_x	$_nm_x$	$_nq_x$	$_nL_x$	$\frac{_5L_{x+5}}{_5L_x}$	T_x	$\overset{o}{e}_x$
			LEVEL 9 Female				
0-1	100,000	.2010	.1777	88,447	[1].7835	4,000,000	40.00
1-5	82,226	.0320	.1179	303,316	[2].9100	3,911,553	47.57
5-10	72,530	.0069	.0338	356,520	.9698	3,608,237	49.75
10-15	70,078	.0054	.0264	345,762	.9694	3,251,718	46.40
15-20	68,227	.0071	.0350	335,172	.9606	2,905,956	42.59
20-25	65,842	.0090	.0440	321,964	.9533	2,570,784	39.04
25-30	62,944	.0102	.0495	306,933	.9474	2,248,820	35.73
30-35	59,829	.0115	.0559	290,781	.9412	1,941,886	32.46
35-40	56,483	.0128	.0618	273,690	.9355	1,651,105	29.23
40-45	52,993	.0139	.0673	256,043	.9291	1,377,415	25.99
45-50	49,424	.0155	.0747	237,894	.9144	1,121,372	22.69
50-55	45,733	.0205	.0975	217,525	.8891	883,478	19.32
55-60	41,277	.0268	.1257	193,410	.8481	665,953	16.13
60-65	36,087	.0400	.1818	164,037	.7895	472,543	13.09
65-70	29,527	.0560	.2457	129,500	.7100	308,507	10.45
70-75	22,272	.0845	.3488	91,943	.6006	179,007	8.04
75-80	14,505	.1253	.4772	55,221	[3].3657	87,064	6.00
80 and over	7,584	.2382	1.0000	31,843		31,843	4.20
			LEVEL 9 Male				
0-1	100,000	.2408	.2074	86,106	[1].7567	3,730,053	37.30
1-5	79,263	.0321	.1183	292,227	[2].9088	3,643,947	45.97
5-10	69,888	.0065	.0322	343,818	.9722	3,351,720	47.96
10-15	67,639	.0047	.0233	334,258	.9722	3,007,902	44.47
15-20	66,064	.0066	.0324	324,977	.9610	2,673,644	40.47
20-25	63,926	.0094	.0459	312,305	.9517	2,348,667	36.74
25-30	60,995	.0104	.0508	297,226	.9454	2,036,363	33.39
30-35	57,895	.0121	.0585	281,009	.9365	1,739,137	30.04
35-40	54,509	.0142	.0688	263,172	.9240	1,458,128	26.75
40-45	50,760	.0175	.0837	243,182	.9088	1,194,955	23.54
45-50	46,512	.0209	.0994	220,999	.8872	951,774	20.46
50-55	41,887	.0273	.1277	196,068	.8572	730,775	17.45
55-60	36,540	.0348	.1602	168,066	.8135	534,706	14.63
60-65	30,686	.0489	.2177	136,730	.7510	366,640	11.95
65-70	24,006	.0676	.2891	102,678	.6693	229,910	9.58
70-75	17,066	.0967	.3893	68,718	.5598	127,231	7.46
75-80	10,421	.1418	.5234	38,471	[3].3425	58,514	5.62
80 and over	4,967	.2478	1.0000	20,043		20,043	4.04

[1] Proportion surviving from birth to 0–4 years of age, $_5L_0/_5l_0$.
[2] $_5L_5/_5L_0$.
[3] T_{80}/T_{75}.

NOTE.—The tables in this appendix were reproduced from, United Nations, *Methods of Estimating Basic Demographic Measures from Incomplete Data*, Population Studies, No. 42, 1967, pp. 85–89, 93, 99–105, 111, 124, and 125. Tables B–1 and B–3 were originally published in Ansley J. Coale and Paul Demeny, *Regional Model Life Tables and Stable Populations*, pp. 10, 12, 14, 16, 18, 42, 46, 50, 54, 58, 62, 66, and 138 (copyright © 1966 by Princeton University Press.) Reprinted by permission of Princeton University Press.

Table B–1. — Selected "West" Model Life Tables Arranged by Level of Mortality — Continued

Age interval (exact ages x to x + n)	l_x	$_nm_x$	$_nq_x$	$_nL_x$	$\dfrac{_5L_{x+5}}{_5L_x}$	T_x	$\overset{\circ}{e}_x$
				LEVEL 11 Female			
0-1.......................................	100,000	.1615	.1461	90,502	[1].8219	4,500,000	45.00
1-5.......................................	85,388	.0250	.0937	320,442	[2].9288	4,409,498	51.64
5-10......................................	77,389	.0055	.0272	381,683	.9758	4,089,056	52.84
10-15.....................................	75,285	.0043	.0212	372,430	.9752	3,707,373	49.24
15-20.....................................	73,687	.0058	.0284	363,207	.9679	3,334,942	45.26
20-25.....................................	71,596	.0073	.0360	351,543	.9618	2,971,735	41.51
25-30.....................................	69,022	.0083	.0405	338,115	.9569	2,620,192	37.96
30-35.....................................	66,224	.0094	.0459	323,525	.9516	2,282,077	34.46
35-40.....................................	63,186	.0105	.0510	307,872	.9465	1,958,552	31.00
40-45.....................................	59,963	.0116	.0562	291,388	.9402	1,650,680	27.53
45-50.....................................	56,592	.0131	.0636	273,969	.9267	1,359,291	24.02
50-55.....................................	52,996	.0175	.0837	253,884	.9039	1,085,322	20.48
55-60.....................................	48,558	.0232	.1095	229,493	.8669	831,439	17.12
60-65.....................................	43,239	.0347	.1596	198,946	.8127	601,946	13.92
65-70.....................................	36,339	.0495	.2203	161,682	.7367	403,000	11.09
70-75.....................................	28,333	.0758	.3184	119,112	.6304	241,319	8.52
75-80.....................................	19,311	.1144	.4448	75,083	[3].3856	122,207	6.33
80 and over...............................	10,722	.2275	1.0000	47,125		47,124	4.40
				LEVEL 13 Female			
0-1.......................................	100,000	.1282	.1183	92,310	[1].8566	5,000,000	50.00
1-5.......................................	88,169	.0188	.0717	335,996	[2].9453	4,907,690	55.66
5-10......................................	81,848	.0043	.0214	404,871	.9810	4,571,694	55.86
10-15.....................................	80,100	.0034	.0166	397,178	.9804	4,166,823	52.02
15-20.....................................	78,771	.0046	.0226	389,403	.9743	3,769,645	47.86
20-25.....................................	76,990	.0059	.0289	379,397	.9693	3,380,242	43.90
25-30.....................................	74,769	.0066	.0327	367,736	.9652	3,000,845	40.13
30-35.....................................	72,326	.0076	.0370	354,934	.9608	2,633,107	36.41
35-40.....................................	69,647	.0085	.0415	341,009	.9561	2,278,173	32.71
40-45.....................................	66,756	.0095	.0464	326,031	.9500	1,937,165	29.02
45-50.....................................	63,656	.0111	.0538	309,724	.9375	1,611,134	25.31
50-55.....................................	60,234	.0149	.0717	290,374	.9170	1,301,410	21.61
55-60.....................................	55,916	.0200	.0953	266,259	.8834	1,011,036	18.08
60-65.....................................	50,587	.0301	.1401	235,225	.8332	744,777	14.72
65-70.....................................	43,503	.0439	.1980	195,982	.7603	509,552	11.71
70-75.....................................	34,890	.0683	.2918	149,002	.6566	313,570	8.99
75-80.....................................	24,711	.1051	.4163	97,836	[3].4055	164,568	6.66
80 and over...............................	14,424	.2162	1.0000	66,732		66,732	4.63
				LEVEL 15 Female			
0-1.......................................	100,000	.0996	.0934	93,745	[1].8889	5,500,000	55.00
1-5.......................................	90,661	.0129	.0500	350,729	[2].9613	5,406,255	59.63
5-10......................................	86,127	.0032	.0157	427,251	.9860	5,055,527	58.70
10-15.....................................	84,773	.0025	.0122	421,284	.9852	4,628,276	54.60
15-20.....................................	83,740	.0035	.0174	415,061	.9800	4,206,992	50.24
20-25.....................................	82,284	.0046	.0227	406,751	.9757	3,791,931	46.08
25-30.....................................	80,416	.0053	.0259	396,874	.9724	3,385,180	42.10
30-35.....................................	78,333	.0060	.0294	385,907	.9686	2,988,306	38.15
35-40.....................................	76,029	.0068	.0334	373,805	.9643	2,602,399	34.23
40-45.....................................	73,493	.0078	.0382	360,445	.9581	2,228,595	30.32
45-50.....................................	70,686	.0094	.0458	345,343	.9464	1,868,150	26.43
50-55.....................................	67,452	.0128	.0619	326,821	.9274	1,522,807	22.58
55-60.....................................	63,276	.0175	.0840	303,101	.8966	1,195,986	18.90
60-65.....................................	57,964	.0266	.1246	271,765	.8492	892,885	15.40
65-70.....................................	50,742	.0398	.1808	230,771	.7783	621,120	12.24
70-75.....................................	41,567	.0629	.2716	179,610	.6765	390,349	9.39
75-80.....................................	30,277	.0984	.3948	121,501	[3].4235	210,740	6.96
80 and over...............................	18,323	.2053	1.0000	89,239		89,238	4.87

[1] Proportion surviving from birth to 0-4 years of age, $_5L_0/5l_0$.
[2] $_5L_5/_5L_0$.
[3] T_{80}/T_{75}.

Table B-1. — Selected "West" Model Life Tables Arranged by Level of Mortality — Continued

Age interval (exact ages x to x + n)	l_x	$_nm_x$	$_nq_x$	$_nL_x$	$\frac{_5L_{x+5}}{_5L_x}$	T_x	$\overset{o}{e}_x$
				LEVEL 17 Female			
0-1...........................	100,000	.0745	.0707	94,785	[1].9171	6,000,000	60.00
1-5...........................	92,934	.0085	.0332	363,755	[2].9744	5,905,215	63.54
5-10..........................	89,854	.0022	.0110	446,805	.9902	5,541,460	61.67
10-15.........................	88,868	.0017	.0085	442,445	.9895	5,094,655	57.33
15-20.........................	88,110	.0025	.0125	437,800	.9855	4,652,210	52.80
20-25.........................	87,010	.0033	.0165	431,463	.9822	4,214,410	48.44
25-30.........................	85,575	.0039	.0191	423,798	.9795	3,782,947	44.21
30-35.........................	83,944	.0044	.0219	415,125	.9763	3,359,149	40.02
35-40.........................	82,106	.0052	.0255	405,299	.9722	2,944,024	35.86
40-45.........................	80,014	.0061	.0302	394,022	.9660	2,538,725	31.73
45-50.........................	77,595	.0077	.0379	380,623	.9550	2,144,704	27.64
50-55.........................	74,655	.0180	.0523	363,504	.9378	1,764,080	23.63
55-60.........................	70,747	.0151	.0727	340,876	.9097	1,400,576	19.80
60-65.........................	65,603	.0231	.1093	310,088	.8652	1,059,700	16.15
65-70.........................	58,432	.0356	.1633	268,300	.7968	749,612	12.83
70-75.........................	48,888	.0574	.2508	213,785	.6969	481,312	9.85
75-80.........................	36,626	.0917	.3729	148,989	[3].4431	267,528	7.30
80 and over...................	22,970	.1938	1.0000	118,539		118,538	5.16

[1] Proportion surviving from bith to 0-4 years of age, $_5L_0/5l_0$.
[2] $_5L_5/_5L_0$.
[3] T_{80}/T_{75}.

Table B-2. — Values of the Function l_x for $x = 1, 2, 3,$ and 5 in "West" Model Life Tables at Various Levels of Mortality, for Females, Males, and Both Sexes

[$l_0 = 100,000$. The l_x values for both sexes assume that the sex ratio at birth is 1.05]

(l_0 = 100,000)

Level	l_1	l_2	l_3	l_5	Level	l_1	l_2	l_3	l_5
	Female					Male--Continued			
1....................	63,483	55,000	51,199	46,883	13....................	86,058	82,912	81,534	79,961
3....................	69,481	61,829	58,399	54,506	15....................	88,864	86,523	85,498	84,327
5....................	74,427	67,671	64,643	61,205	17....................	91,379	89,790	89,056	88,184
7....................	78,614	72,765	70,145	67,169	19....................	93,713	92,796	92,338	91,774
9....................	82,226	77,271	75,051	72,530	21....................	95,909	95,508	95,285	94,989
11....................	85,388	81,300	79,468	77,389	23....................	97,856	97,719	97,636	97,521
13....................	88,169	84,939	83,492	81,848		Both sexes			
15....................	90,661	88,364	87,324	86,127					
17....................	92,934	91,419	90,709	89,854	1....................	60,722	52,597	48,996	44,897
19....................	95,006	94,143	93,724	93,201	3....................	67,118	59,709	56,425	52,688
21....................	96,907	96,559	96,385	96,160	5....................	72,392	65,798	62,877	59,551
23....................	98,484	98,377	98,321	98,248	7....................	76,857	71,112	68,567	65,670
	Male				9....................	80,709	75,813	73,646	71,177
					11....................	84,080	80,019	78,220	76,173
1....................	58,093	50,308	46,898	43,005	13....................	87,088	83,901	82,489	80,881
3....................	64,868	57,690	54,546	50,957	15....................	89,740	87,421	86,389	85,205
5....................	70,454	64,015	61,195	57,976	17....................	92,137	90,584	89,862	88,999
7....................	75,183	69,537	67,064	64,242	19....................	94,144	93,453	93,011	92,455
9....................	79,263	74,425	72,307	69,888	21....................	96,396	96,020	95,822	95,560
11....................	82,835	78,800	77,032	75,015	23....................	98,162	98,040	97,970	97,876

Table B-3. — Selected "West" Model Stable Populations Arranged by Level of Mortality and Annual Rate of Increase

LEVEL 9
Female (e_0 = 40.0 years)

Age and parameter	−.010	−.005	.000	.005	.010	.015	.020	.025	.030	.035	.040	.045	.050
						Annual rate of increase							
Age interval:					Proportion in age interval								
Under 1.........	.0158	.0188	.0221	.0257	.0295	.0336	.0379	.0424	.0471	.0518	.0567	.0617	.0667
1-4.............	.0557	.0653	.0758	.0870	.0988	.1111	.1238	.1367	.1498	.1629	.1760	.1890	.2018
5-9.............	.0684	.0786	.0891	.1000	.1111	.1221	.1330	.1436	.1538	.1636	.1728	.1814	.1894
10-14...........	.0698	.0781	.0864	.0946	.1024	.1098	.1167	.1229	.1284	.1332	.1372	.1405	.1431
15-19...........	.0711	.0776	.0838	.0894	.0945	.0988	.1024	.1051	.1071	.1084	.1089	.1087	.1080
20-24...........	.0718	.0765	.0805	.0838	.0863	.0880	.0890	.0891	.0886	.0874	.0856	.0834	.0808
25-29...........	.0720	.0747	.0767	.0779	.0783	.0779	.0767	.0750	.0727	.0699	.0668	.0635	.0600
30-34...........	.0717	.0726	.0727	.0720	.0705	.0684	.0658	.0627	.0593	.0556	.0518	.0480	.0443
35-39...........	.0709	.0701	.0684	.0661	.0632	.0598	.0560	.0521	.0480	.0439	.0400	.0361	.0324
40-44...........	.0698	.0672	.0640	.0603	.0562	.0519	.0474	.0430	.0387	.0345	.0306	.0270	.0236
45-49...........	.0681	.0640	.0595	.0546	.0497	.0447	.0399	.0352	.0309	.0269	.0233	.0200	.0171
50-54...........	.0655	.0600	.0544	.0487	.0432	.0379	.0330	.0284	.0243	.0207	.0174	.0146	.0122
55-59...........	.0612	.0547	.0484	.0423	.0365	.0313	.0265	.0223	.0186	.0154	.0127	.0104	.0084
60-64...........	.0546	.0476	.0410	.0350	.0295	.0246	.0204	.0167	.0136	.0110	.0088	.0070	.0056
65-69...........	.0453	.0385	.0324	.0269	.0221	.0180	.0145	.0116	.0092	.0073	.0057	.0044	.0034
70-74...........	.0338	.0280	.0230	.0186	.0150	.0119	.0093	.0073	.0056	.0043	.0033	.0025	.0019
75-79...........	.0213	.0173	.0138	.0109	.0085	.0066	.0051	.0039	.0029	.0022	.0016	.0012	.0009
80 and over....	.0131	.0103	.0080	.0061	.0046	.0035	.0026	.0019	.0014	.0010	.0007	.0005	.0004
Age:					Proportion under given age								
1...............	.0158	.0188	.0221	.0257	.0295	.0336	.0379	.0424	.0471	.0518	.0567	.0617	.0667
5...............	.0715	.0842	.0979	.1127	.1284	.1448	.1617	.1791	.1968	.2147	.2327	.2506	.2685
10..............	.1400	.1627	.1871	.2127	.2394	.2668	.2947	.3227	.3507	.3783	.4055	.4321	.4579
15..............	.2097	.2408	.2735	.3073	.3419	.3767	.4114	.4456	.4791	.5115	.5427	.5725	.6010
20..............	.2809	.3185	.3573	.3968	.4363	.4755	.5137	.5507	.5862	.6198	.6516	.6813	.7090
25..............	.3527	.3949	.4378	.4806	.5227	.5635	.6027	.6399	.6747	.7072	.7372	.7647	.7898
30..............	.4247	.4697	.5145	.5585	.6009	.6414	.6794	.7148	.7474	.7771	.8040	.8282	.8498
35..............	.4963	.5423	.5872	.6305	.6715	.7098	.7452	.7775	.8067	.8328	.8559	.8762	.8940
40..............	.5673	.6124	.6556	.6966	.7346	.7696	.8012	.8296	.8547	.8767	.8958	.9123	.9265
45..............	.6370	.6796	.7197	.7568	.7908	.8214	.8487	.8726	.8933	.9112	.9264	.9393	.9501
50..............	.7052	.7436	.7791	.8115	.8405	.8662	.8885	.9078	.9243	.9381	.9497	.9593	.9672
55..............	.7707	.8036	.8335	.8602	.8837	.9041	.9215	.9363	.9486	.9588	.9671	.9739	.9794
60..............	.8319	.8583	.8819	.9025	.9202	.9354	.9481	.9586	.9672	.9742	.9798	.9843	.9878
65..............	.8865	.9059	.9229	.9374	.9497	.9600	.9684	.9753	.9808	.9852	.9886	.9913	.9934
					Parameter of stable populations								
Birth rate......	.0178	.0212	.0250	.0291	.0336	.0383	.0433	.0486	.0540	.0597	.0654	.0713	.0773
Death rate......	.0278	.0262	.0250	.0241	.0236	.0233	.0233	.0236	.0240	.0247	.0254	.0263	.0273
GRR (27)........	1.24	1.42	1.63	1.86	2.12	2.41	2.75	3.12	3.55	4.02	4.56	5.17	5.85
GRR (29)........	1.25	1.44	1.66	1.91	2.20	2.53	2.91	3.34	3.83	4.38	5.01	5.73	6.54
GRR (31)........	1.25	1.46	1.70	1.97	2.30	2.67	3.09	3.58	4.15	4.79	5.54	6.39	7.36
GRR (33)........	1.25	1.47	1.73	2.04	2.40	2.81	3.30	3.86	4.52	5.28	6.17	7.20	8.39
Average age.....	36.2	33.9	31.6	29.5	27.4	25.5	23.7	22.0	20.5	19.1	17.8	16.7	15.6
Births/population 15-44..	.042	.048	.056	.065	.075	.086	.099	.114	.130	.149	.170	.194	.221

Table B-3.—Selected "West" Model Stable Populations Arranged by Level of Mortality and Annual Rate of Increase—Continued

LEVEL 9
Male (e^o_o = 37.3 years)

Age and parameter	Annual rate of increase												
	−.010	−.005	.000	.005	.010	.015	.020	.025	.030	.035	.040	.045	.050
Age interval:													
Proportion in age interval													
Under 1	.0168	.0198	.0231	.0267	.0305	.0346	.0389	.0433	.0480	.0527	.0576	.0625	.0675
1–4	.0583	.0680	.0783	.0894	.1010	.1131	.1255	.1382	.1510	.1639	.1768	.1895	.2022
5–9	.0718	.0818	.0922	.1028	.1136	.1244	.1350	.1453	.1552	.1647	.1737	.1821	.1899
10–14	.0733	.0815	.0896	.0975	.1051	.1122	.1187	.1247	.1299	.1344	.1383	.1414	.1438
15–19	.0750	.0812	.0871	.0925	.0972	.1012	.1045	.1070	.1087	.1097	.1101	.1098	.1089
20–24	.0757	.0801	.0837	.0867	.0888	.0902	.0908	.0907	.0899	.0885	.0866	.0842	.0815
25–29	.0758	.0781	.0797	.0804	.0804	.0797	.0782	.0762	.0737	.0707	.0675	.0640	.0604
30–34	.0753	.0757	.0753	.0742	.0723	.0699	.0669	.0636	.0599	.0561	.0522	.0483	.0445
35–39	.0741	.0727	.0706	.0678	.0644	.0607	.0567	.0525	.0483	.0441	.0400	.0361	.0324
40–44	.0720	.0689	.0652	.0611	.0566	.0520	.0474	.0428	.0384	.0342	.0303	.0267	.0233
45–49	.0688	.0642	.0592	.0541	.0490	.0439	.0390	.0344	.0301	.0261	.0225	.0193	.0165
50–54	.0642	.0584	.0526	.0468	.0413	.0361	.0313	.0269	.0230	.0194	.0164	.0137	.0114
55–59	.0578	.0513	.0451	.0392	.0337	.0287	.0243	.0203	.0169	.0140	.0115	.0094	.0076
60–64	.0495	.0428	.0367	.0311	.0261	.0217	.0179	.0146	.0119	.0096	.0077	.0061	.0048
65–69	.0390	.0330	.0275	.0228	.0186	.0151	.0121	.0097	.0077	.0060	.0047	.0037	.0028
70–74	.0275	.0226	.0184	.0149	.0119	.0094	.0074	.0057	.0044	.0034	.0026	.0020	.0015
75–79	.0162	.0130	.0103	.0081	.0063	.0049	.0037	.0028	.0021	.0016	.0012	.0009	.0006
80 and over	.0089	.0070	.0054	.0041	.0031	.0023	.0017	.0013	.0009	.0007	.0005	.0003	.0002
Age:													
Proportion under given age													
1	.0168	.0198	.0231	.0267	.0305	.0346	.0389	.0433	.0480	.0527	.0576	.0625	.0675
5	.0751	.0877	.1014	.1161	.1315	.1477	.1644	.1815	.1990	.2166	.2343	.2520	.2697
10	.1468	.1695	.1936	.2189	.2452	.2721	.2994	.3268	.3542	.3813	.4080	.4341	.4596
15	.2202	.2510	.2832	.3164	.3502	.3843	.4181	.4515	.4841	.5158	.5463	.5755	.6034
20	.2951	.3322	.3703	.4089	.4474	.4855	.5226	.5585	.5928	.6255	.6563	.6853	.7123
25	.3709	.4123	.4541	.4956	.5363	.5767	.6134	.6492	.6827	.7140	.7429	.7695	.7938
30	.4466	.4904	.5338	.5760	.6167	.6553	.6916	.7253	.7564	.7847	.8104	.8335	.8542
35	.5219	.5661	.6091	.6502	.6890	.7252	.7585	.7889	.8163	.8409	.8626	.8818	.8986
40	.5961	.6389	.6796	.7179	.7534	.7859	.8152	.8414	.8646	.8850	.9027	.9180	.9311
45	.6681	.7078	.7448	.7790	.8101	.8379	.8626	.8843	.9031	.9192	.9330	.9446	.9544
50	.7369	.7719	.8041	.8331	.8590	.8818	.9016	.9186	.9331	.9453	.9555	.9640	.9709
55	.8011	.8303	.8566	.8800	.9003	.9179	.9329	.9455	.9561	.9648	.9719	.9777	.9824
60	.8589	.8817	.9017	.9191	.9340	.9466	.9572	.9659	.9730	.9788	.9834	.9871	.9900
65	.9084	.9245	.9384	.9502	.9601	.9683	.9751	.9805	.9849	.9883	.9910	.9932	.9948
Parameter of stable populations													
Birth rate	.0194	.0229	.0268	.0311	.0356	.0405	.0456	.0510	.0566	.0623	.0682	.0742	.0804
Death rate	.0294	.0279	.0268	.0261	.0256	.0255	.0256	.0260	.0266	.0273	.0282	.0292	.0304
GRR (27)	1.29	1.47	1.68	1.92	2.19	2.49	2.84	3.22	3.66	4.16	4.71	5.34	6.04
GRR (29)	1.29	1.49	1.72	1.98	2.28	2.62	3.01	3.45	3.96	4.53	5.18	5.92	6.76
GRR (31)	1.29	1.51	1.76	2.05	2.38	2.76	3.20	3.71	4.29	4.96	5.73	6.61	7.62
GRR (33)	1.30	1.53	1.80	2.12	2.49	2.92	3.42	4.00	4.68	5.47	6.39	7.45	8.69
Average age	34.7	32.5	30.4	28.4	26.5	24.7	23.0	21.5	20.0	18.7	17.5	16.4	15.4
Births/population 15–44	.043	.050	.058	.067	.077	.089	.103	.118	.135	.154	.176	.201	.229

Table B–3.—Selected "West" Model Stable Populations Arranged by Level of Mortality and Annual Rate of Increase—Continued

LEVEL 11
Female ($\overset{o}{e}_0$ = 45.0 years)

Age and parameter	Annual rate of increase												
	−.010	−.005	.000	.005	.010	.015	.020	.025	.030	.035	.040	.045	.050
Age interval:													
						Proportion in age interval							
Under 1	.0142	.0170	.0201	.0235	.0272	.0311	.0352	.0396	.0440	.0487	.0534	.0582	.0631
1–4	.0516	.0610	.0712	.0822	.0939	.1061	.1187	.1316	.1447	.1579	.1711	.1842	.1971
5–9	.0643	.0743	.0848	.0957	.1069	.1181	.1292	.1401	.1506	.1606	.1702	.1792	.1875
10–14	.0660	.0743	.0828	.0911	.0992	.1069	.1141	.1206	.1265	.1316	.1360	.1396	.1425
15–19	.0676	.0743	.0807	.0867	.0920	.0967	.1006	.1038	.1061	.1077	.1086	.1087	.1082
20–24	.0688	.0738	.0781	.0818	.0847	.0868	.0881	.0887	.0884	.0875	.0860	.0840	.0816
25–29	.0696	.0727	.0751	.0767	.0775	.0775	.0767	.0753	.0732	.0707	.0677	.0645	.0611
30–34	.0700	.0714	.0719	.0716	.0706	.0688	.0664	.0635	.0603	.0568	.0531	.0493	.0455
35–39	.0700	.0696	.0684	.0665	.0639	.0607	.0572	.0534	.0494	.0453	.0413	.0375	.0337
40–44	.0697	.0676	.0648	.0614	.0575	.0533	.0490	.0446	.0402	.0360	.0320	.0283	.0249
45–49	.0689	.0651	.0609	.0563	.0514	.0465	.0417	.0370	.0326	.0284	.0247	.0213	.0182
50–54	.0671	.0619	.0564	.0508	.0453	.0400	.0349	.0302	.0260	.0221	.0187	.0157	.0131
55–59	.0637	.0574	.0510	.0448	.0390	.0335	.0286	.0241	.0202	.0168	.0138	.0114	.0093
60–64	.0581	.0510	.0442	.0379	.0321	.0270	.0224	.0185	.0151	.0122	.0098	.0079	.0062
65–69	.0496	.0425	.0359	.0300	.0248	.0203	.0165	.0132	.0105	.0083	.0065	.0051	.0040
70–74	.0384	.0321	.0265	.0216	.0174	.0139	.0110	.0086	.0067	.0052	.0039	.0030	.0023
75–79	.0255	.0207	.0167	.0133	.0104	.0081	.0063	.0048	.0036	.0027	.0020	.0015	.0011
80 and over	.0170	.0134	.0105	.0081	.0062	.0047	.0035	.0026	.0019	.0014	.0010	.0007	.0005
Age:													
						Proportion under given age							
1	.0142	.0170	.0201	.0235	.0272	.0311	.0352	.0396	.0440	.0487	.0534	.0582	.0631
5	.0658	.0780	.0913	.1057	.1210	.1371	.1539	.1711	.1887	.2065	.2245	.2424	.2602
10	.1301	.1523	.1761	.2014	.2279	.2552	.2831	.3112	.3393	.3672	.3947	.4215	.4477
15	.1961	.2266	.2589	.2926	.3271	.3621	.3971	.4318	.4658	.4988	.5306	.5611	.5902
20	.2637	.3009	.3396	.3792	.4191	.4588	.4978	.5356	.5719	.6065	.6392	.6698	.6984
25	.3325	.3747	.4177	.4610	.5038	.5457	.5859	.6242	.6604	.6940	.7252	.7539	.7800
30	.4021	.4474	.4929	.5377	.5814	.6231	.6626	.6995	.7336	.7647	.7930	.8184	.8411
35	.4721	.5188	.5648	.6094	.6519	.6919	.7290	.7630	.7938	.8215	.8460	.8677	.8867
40	.5421	.5884	.6332	.6758	.7158	.7527	.7862	.8164	.8432	.8668	.8874	.9052	.9204
45	.6117	.6559	.6979	.7372	.7733	.8060	.8352	.8610	.8835	.9028	.9194	.9335	.9453
50	.6806	.7211	.7588	.7934	.8247	.8525	.8769	.8980	.9160	.9313	.9441	.9547	.9635
55	.7476	.7829	.8152	.8443	.8700	.8925	.9118	.9282	.9420	.9534	.9628	.9705	.9766
60	.8114	.8403	.8662	.8891	.9090	.9260	.9404	.9523	.9622	.9702	.9766	.9818	.9859
65	.8695	.8913	.9104	.9270	.9411	.9530	.9628	.9708	.9772	.9824	.9865	.9897	.9921
						Parameter of stable populations							
Birth rate	.0156	.0188	.0222	.0260	.0302	.0346	.0393	.0443	.0494	.0547	.0602	.0658	.0715
Death rate	.0256	.0238	.0222	.0210	.0202	.0196	.0193	.0193	.0194	.0197	.0202	.0208	.0215
GRR (27)	1.13	1.29	1.48	1.69	1.92	2.19	2.50	2.84	3.23	3.66	4.16	4.71	5.33
GRR (29)	1.13	1.30	1.50	1.73	2.00	2.30	2.64	3.03	3.47	3.98	4.55	5.20	5.94
GRR (31)	1.12	1.31	1.53	1.78	2.07	2.41	2.79	3.24	3.75	4.34	5.01	5.78	6.68
GRR (33)	1.12	1.32	1.56	1.83	2.15	2.53	2.97	3.47	4.07	4.76	5.56	6.48	7.56
Average age	37.6	35.2	32.9	30.6	28.5	26.4	24.6	22.8	21.2	19.7	18.4	17.2	16.1
Births/population 15–44	.038	.044	.051	.059	.068	.078	.090	.103	.118	.135	.155	.177	.201

Table B–3.—Selected "West" Model Stable Populations Arranged by Level of Mortality and Annual Rate of Increase—Continued

Age and parameter	LEVEL 13 Female (e^0_o = 50.0 years) Annual rate of increase			LEVEL 15 Female (e^0_o = 55.0 years) Annual rate of increase			LEVEL 17 Female (e^0_o = 60.0 years) Annual rate of increase			LEVEL 19 Female (e^0_o = 65.0 years) Annual rate of increase			LEVEL 21 Female (e^0_o = 70.0 years) Annual rate of increase		
	.025	.030	.035	.025	.030	.035	.025	.030	.035	.025	.030	.035	.025	.030	.035
Age interval *(Proportion in age interval)*															
Under 1	.0372	.0415	.0460	.0350	.0392	.0436	.0331	.0372	.0415	.0316	.0356	.0398	.0304	.0343	.0385
1–4	.1271	.1402	.1534	.1230	.1362	.1494	.1194	.1325	.1459	.1161	.1293	.1426	.1131	.1262	.1397
5–9	.1368	.1476	.1579	.1339	.1449	.1555	.1311	.1422	.1530	.1283	.1397	.1507	.1257	.1373	.1485
10–14	.1184	.1246	.1301	.1165	.1230	.1287	.1145	.1212	.1272	.1126	.1195	.1257	.1106	.1178	.1242
15–19	.1025	.1052	.1070	.1013	.1043	.1065	.1000	.1033	.1057	.0987	.1022	.1048	.0973	.1010	.1039
20–24	.0881	.0882	.0875	.0876	.0880	.0876	.0870	.0876	.0874	.0862	.0871	.0872	.0854	.0865	.0868
25–29	.0754	.0736	.0712	.0755	.0739	.0717	.0754	.0740	.0721	.0752	.0741	.0723	.0748	.0739	.0723
30–34	.0642	.0611	.0577	.0648	.0618	.0586	.0652	.0624	.0593	.0654	.0628	.0598	.0655	.0631	.0602
35–39	.0544	.0505	.0466	.0554	.0516	.0476	.0562	.0525	.0486	.0568	.0532	.0494	.0572	.0537	.0500
40–44	.0459	.0416	.0374	.0471	.0428	.0385	.0482	.0439	.0396	.0491	.0448	.0406	.0498	.0456	.0414
45–49	.0385	.0340	.0298	.0398	.0353	.0310	.0411	.0365	.0322	.0422	.0376	.0332	.0431	.0385	.0341
50–54	.0319	.0274	.0234	.0333	.0287	.0246	.0346	.0300	.0258	.0358	.0312	.0268	.0369	.0322	.0278
55–59	.0258	.0217	.0180	.0272	.0229	.0192	.0286	.0242	.0203	.0300	.0254	.0214	.0312	.0266	.0224
60–64	.0201	.0165	.0134	.0215	.0177	.0144	.0230	.0190	.0155	.0244	.0202	.0165	.0258	.0214	.0176
65–69	.0148	.0118	.0094	.0161	.0129	.0103	.0176	.0141	.0113	.0190	.0153	.0122	.0204	.0165	.0132
70–74	.0099	.0077	.0060	.0111	.0087	.0067	.0123	.0097	.0075	.0137	.0108	.0084	.0151	.0119	.0093
75–79	.0057	.0044	.0033	.0066	.0050	.0038	.0076	.0058	.0044	.0087	.0066	.0051	.0098	.0076	.0058
80 and over	.0034	.0025	.0018	.0041	.0031	.0022	.0051	.0038	.0028	.0064	.0047	.0035	.0079	.0058	.0043
Age *(Proportion under given age)*															
1	.0372	.0415	.0460	.0350	.0392	.0436	.0331	.0372	.0415	.0316	.0356	.0398	.0304	.0343	.0385
5	.1642	.1817	.1994	.1580	.1754	.1930	.1525	.1698	.1873	.1477	.1649	.1824	.1434	.1606	.1781
10	.3010	.3293	.3574	.2920	.3203	.3485	.2836	.3120	.3404	.2760	.3046	.3331	.2692	.2978	.3266
15	.4195	.4539	.4874	.4085	.4433	.4772	.3981	.4333	.4676	.3886	.4241	.4589	.3798	.4156	.4508
20	.5219	.5591	.5945	.5099	.5476	.5837	.4981	.5365	.5733	.4873	.5262	.5637	.4770	.5166	.5547
25	.6101	.6472	.6820	.5975	.6356	.6713	.5851	.6241	.6607	.5735	.6133	.6509	.5624	.6031	.6415
30	.6854	.7208	.7533	.6729	.7095	.7430	.6605	.6981	.7328	.6487	.6874	.7231	.6373	.6770	.7139
35	.7496	.7819	.8110	.7377	.7713	.8016	.7257	.7606	.7921	.7141	.7502	.7830	.7028	.7401	.7741
40	.8041	.8325	.8575	.7931	.8228	.8492	.7818	.8130	.8407	.7709	.8034	.8323	.7599	.7939	.8241
45	.8500	.8741	.8949	.8402	.8656	.8877	.8300	.8569	.8803	.8199	.8482	.8729	.8097	.8395	.8655
50	.8885	.9081	.9247	.8800	.9009	.9187	.8711	.8934	.9125	.8621	.8858	.9061	.8528	.8780	.8996
55	.9203	.9355	.9481	.9132	.9296	.9433	.9057	.9234	.9382	.8979	.9170	.9330	.8898	.9102	.9275
60	.9461	.9572	.9662	.9405	.9526	.9625	.9344	.9476	.9585	.9279	.9424	.9543	.9210	.9368	.9498
65	.9662	.9736	.9796	.9620	.9703	.9769	.9574	.9666	.9740	.9523	.9626	.9709	.9468	.9582	.9674
Parameter of stable populations															
Birth rate	.0408	.0457	.0507	.0378	.0425	.0473	.0354	.0399	.0445	.0333	.0377	.0422	.0316	.0358	.0402
Death rate	.0158	.0157	.0157	.0128	.0125	.0123	.0104	.0099	.0095	.0083	.0077	.0072	.0066	.0058	.0052
GRR (27)	2.62	2.98	3.38	2.43	2.76	3.14	2.28	2.59	2.95	2.16	2.45	2.79	2.06	2.34	2.66
GRR (29)	2.78	3.19	3.66	2.58	2.96	3.39	2.41	2.77	3.17	2.28	2.61	3.00	2.17	2.49	2.85
GRR (31)	2.96	3.43	3.97	2.74	3.17	3.67	2.55	2.96	3.43	2.41	2.79	3.23	2.28	2.65	3.07
GRR (33)	3.17	3.71	4.34	2.92	3.42	4.00	2.71	3.18	3.72	2.55	2.99	3.50	2.41	2.83	3.31
Average age	23.5	21.8	20.3	24.2	22.4	20.8	24.8	23.0	21.3	25.4	23.5	21.8	26.0	24.1	22.3
Births/population 15–44	.095	.109	.124	.088	.101	.115	.082	.094	.103	.077	.089	.102	.073	.084	.097

Table B–4. — Table for Estimating Cumulated Fertility From Age-Specific Fertility Rates Calculated From Survey Reports on Births During a 12-Month Period Preceding the Survey

[Multiplying factors w_i for estimating the average value over 5-year age groups of cumulated fertility (F_i) according to the formula:

$$F_i = 5 \sum_{j=0}^{i-1} f_j + w_i f_i$$

when $f_0 = 0$, f_1 = age-specific fertility rate for ages 14.5 to 19.5, f_2 = rate for ages 19.5 to 24.5, etc.]

Age interval (1)	Exact limits of age interval	Multiplying factors w_i for values of f_1/f_2 and $\overline{m}$ as indicated in lower part of table							
1	15–20	1.120	1.310	1.615	1.950	2.305	2.640	2.925	3.170
2	20–25	2.555	2.690	2.780	2.840	2.890	2.925	2.960	2.985
3	25–30	2.925	2.960	2.985	3.010	3.035	3.055	3.075	3.095
4	30–35	3.055	3.075	3.095	3.120	3.140	3.165	3.190	3.215
5	35–40	3.165	3.190	3.215	3.245	3.285	3.325	3.375	3.435
6	40–45	3.325	3.375	3.435	3.510	3.610	3.740	3.915	4.150
7	45–50	3.640	3.895	4.150	4.395	4.630	4.840	4.985	5.000
	f_1/f_2	.036	.113	.213	.330	.460	.605	.764	.939
	$\overline{m}$	31.7	30.7	29.7	28.7	27.7	26.7	25.7	24.7

Table B–5. — Table for Estimating Mortality From Child Survivorship Rates

[Multiplying factors for estimating the proportion of children born alive who die by age $a - q(a)$ — from the proportion dead among children ever born reported by women classified in 5-year age intervals]

Mortality measure estimated (1)	Exact limits of age interval of women (2)	Multiplying factors to obtain q(a) shown in col. 1 from proportion of children reported as dead by women of ages specified in col. 2; for values of P_1/P_2, $\overline{m}$, and $\overline{m}'$ as specified in lower part of table							
q(1)	15–20	0.859	0.890	0.928	0.977	1.041	1.129	1.254	1.425
q(2)	20–25	0.938	0.959	0.983	1.010	1.043	1.082	1.129	1.188
q(3)	25–30	0.948	0.962	0.978	0.994	1.012	1.033	1.055	1.081
q(5)	30–35	0.961	0.975	0.988	1.002	1.016	1.031	1.046	1.063
q(10)	35–40	0.966	0.982	0.996	1.011	1.026	1.040	1.054	1.069
q(15)	40–45	0.938	0.955	0.971	0.988	1.004	1.021	1.037	1.052
q(20)	45–50	0.937	0.953	0.969	0.986	1.003	1.021	1.039	1.057
q(25)	50–55	0.949	0.966	0.983	1.001	1.019	1.036	1.054	1.072
q(30)	55–60	0.951	0.968	0.985	1.002	1.020	1.039	1.058	1.076
q(35)	60–65	0.949	0.965	0.982	0.999	1.016	1.034	1.052	1.070
	P_1/P_2	0.387	0.330	0.268	0.205	0.143	0.090	0.045	0.014
	$\overline{m}'$	24.7	25.7	26.7	27.7	28.7	29.7	30.7	31.7
	$\overline{m}$	24.2	25.2	26.2	27.2	28.2	29.2	30.2	31.2

APPENDIX C

Selected General Methods

This chapter treats a variety of topics concerned with methods of measurement and analysis. None are strictly demographic; yet all are applicable to many fields of demography. The methods and basic concepts presented are ones which the demographer will find useful regardless of his special subject interest, but they are especially pertinent for work in population estimates and projections.

INTERPOLATION OF POINT DATA

Introduction

Some Definitions. — **Interpolation** is narrowly defined as the art of inferring intermediate values in a given series of data by use of a mathematical formula or a graphic procedure. **Extrapolation** is the art of inferring values that go beyond the given series of data by use of a mathematical formula or a graphic procedure. Many of the techniques used for interpolation are suitable also for extrapolation; hence, the term interpolation is often used to refer to both types of inference. Broadly considered, interpolation encompasses not only mathematical and graphic devices for estimating intermediate or external values in a series (e.g., annual population figures from decennial figures, survivors in a life table for single ages from survivors at every fifth age) but also mathematical and graphic devices for subdividing grouped data into component parts (e.g., figures for single years of age from data for 5-year age groups) and for inferring rates for subgroups from rates for broad groups (e.g., birth rates by duration of marriage). Typically, these devices reproduce, or are consistent with, the given values. In this case we say that the fit is exact. In other cases, modified interpolation formulas are used and the interpolated series does not pass through the original values or maintain the original group totals. Then, we say that the fit is approximate.

Interpolation is, in a sense, a form of estimation, but normally "interpolation" relates only to those forms of estimation that involve the direct application of mathematical or graphic devices to observed data. Sometimes, however, it is used loosely to include some forms of estimation involving a simple use of some external series of data suggestive of the pattern or trend in the range of interpolation. A principal type of "interpolation" of this kind is interpolation by prorating. We discuss this method later in this chapter. Interpolation as a means of making population estimates is considered more directly and fully in chapter 22.

Though there is frequent need for interpolated estimates in demographic work, the degree of precision required by the user in practice or actually supported by the data is often too low to justify anything more than the application of the more simple forms of interpolation. Indeed, for some purposes demographic data could satisfactorily be interpolated by running a smooth line by hand through a set of plotted points. For others, complex methods of interpolation are essential. Sometimes, however, where highly complex methods of interpolation appear necessary, the problem may be that the initial data are too defective or the number and spacing of the observations are inadequate.

Interpolation by mathematical formula has the quality of imputing a regularity or smoothness to the given series of data or even imposing these characteristics on the data. The regularity imputed or imposed may be unreal. There are often true fluctuations in population growth or in the age distribution due to past variations in births, deaths, and migration, especially if there have been wars, epidemics, population transfers, and refugee movements, etc. Interpolation may usefully serve to adjust defective data, even though some real fluctuations are removed, or to eliminate abnormalities from a series, such as those due to war, when the underlying pattern or trend is wanted.

We turn first to a consideration of the methods of interpolating "point" values in a series, such as a time series; and among these we consider first those methods that can be employed to fit the given data exactly. These methods include graphic interpolation, polynomial interpolation, use of some types of exponential functions, osculatory interpolation, and use of spline functions.

Graphic Interpolation

For many purposes it may suffice to plot a series of given data on a large-scale graph, draw a free-hand curve through the plotted points, and then read off the intermediate points from the graph as needed.[1] Mechanical aids, such as a French curve or a spline and weights, are sometimes used to draw the line through the plotted points. However satisfactory from the viewpoint of general accuracy, this procedure does not satisfy several purely technical and aesthetic requirements that those working with large amounts of demographic data seek, and it can take more time and work than some arithmetic methods. One requirement is reproducibility. Different workers will produce different results with the hand-and-eye technique, even

---

[1] Morton D. Miller, *Elements of Graduation,* Actuarial Monographs. No. 1, The Actuarial Society of America, American Institute of Actuaries. 1946, ch. 2.

though the differences may not be large. The second requirement is computability. The easy access nowadays to desk calculators and electronic computers generates the demand for computable interpolation procedures which can take advantage of the speed and economy of these machines. Another disadvantage of graphic interpolation is that it may be difficult to read off more than two or three significant figures from the graph.

If the series of data show a nonlinear trend when plotted on an arithmetical scale, it is sometimes helpful to replot the data in a "transformed" manner that yields a nearly linear trend, and then interpolate. Thus, the use of a logarithmic scale for one or both axes of the graph may sometimes be helpful in narrowing the range of judgement for drawing a free-hand curve. An example of the use of a semilogarithmic scale was given in figure 13–2. Other possibilities include the plotting of the reciprical or the square root of either or both coordinates of the given points. There is room for ingenuity in devising methods of graphic interpolation.

Polynomial Interpolation

General Form of Equation. — Polynomial interpolation is interpolation where the series is assumed to conform to an equation of the general type, $y = A + Bx + Cx^2 + Dx^3$ . . . More or fewer terms may be used. As already pointed out in chapter 13, the equation $y = A + Bx$ is a straight line, or linear equation, which can be passed through any two given points. The equation $y = A + Bx + Cx^2$ is a quadratic, or parabola, which can be passed through any three given points. The equation $y = A + Bx + Cx^2 + Dx^3$ is a cubic, which can be passed through any four given points. More generally, a polynomial equation of the nth degree can be passed through $n + 1$ given points. Although one has decided to fit a polynomial of higher degree to the observed data and a polynomial of the nth degree will give an exact fit for $n + 1$ observations, one must still decide upon how many observations to use. The choice of the degree of the polynomial would depend on the nature of the data to be interpolated. Usually, the simplest equation that describes the data reasonably well and gives a smooth series is the one wanted. The criterion of a smooth fit of the given data normally requires use of a higher-degree equation than a straight line. Greater smoothness would normally be achieved by employing at least two observations before, and two observations after, the point of interpolation. This would seem to call for at least 4-point interpolation by a third-degree polynomial, where possible, but often 3-point or even 2-point interpolation will give about the same results.

In what follows, we will need symbols for given points. The symbol $f(a)$ means the value of the function when x equals a, $f(b)$ the value of the function when x equals b, etc. Hence, the symbol $f(a)$ will be the observed value of the y ordinate for the abscissa $x = a$, the symbol $f(b)$ will be the observed value of y for $x = b$, and so on. The symbol $f(x)$ will be the desired interpolated value of y for $x = x$.

Methods of Application. — Polynomials that pass through the given data may be fitted in several different ways operationally while producing the identical results. One is by general solution of the polynomial equation and derivation of the values of the constants A, B, C, D, etc.; another is by use of interpolation coefficients; still another is by sequential linear interpolation; and a fourth is by use of "differences". Usually in polynomial interpolation by exact fit, the method

involving the general solution of the polynomial equation is not employed since it is too cumbersome.

Waring's formula. — The formulas for polynomial interpolation can be set forth in the form of linear compounds, that is, as the sum of the products of certain coefficients or multipliers and certain given values. The Waring formula, also known as the Lagrange Formula or the Waring-Lagrange formula, is used to derive the multipliers to interpolate for the $f(x)$ value corresponding to a given x value. The Waring formula for interpolating between four points by a polynomial, i.e., for fitting a cubic, is as follows:

$$f(x) = f(a)\frac{(x-b)(x-c)(x-d)}{(a-b)(a-c)(a-d)} + f(b)\frac{(x-a)(x-c)(x-d)}{(b-a)(b-c)(b-d)}$$

$$(1)$$

$$+ f(c)\frac{(x-a)(x-b)(x-d)}{(c-a)(c-b)(c-d)} + f(d)\frac{(x-a)(x-b)(x-c)}{(d-a)(d-b)(d-c)}$$

This is equivalent to the polynomial $y = A + Bx + Cx^2 + Dx^3$ passing through the four points $f(a)$, $f(b)$, $f(c)$, and $f(d)$ to derive $f(x)$. The points do not have to be equally spaced. By the formula, a particular value of $f(x)$ can be obtained from given values of $f(a)$, $f(b)$, $f(c)$, and $f(d)$.

The formula is especially suitable for computing the coefficients or "multipliers" to be applied to the $f(a)$, $f(b)$, $f(c)$, and $f(d)$ values to obtain $f(x)$. These multipliers may be used again and again so long as a y value on the same x abscissa is being sought and there are four given points spaced in the same way. In this way, for example, the same multipliers may be used for all the age-sex groups in a distribution or for all the States in a country to secure interpolated values at the same date. The multipliers for any particular interpolation formula add to 1.00.

Similarly, the formula

$$f(x) = f(a)\frac{(x-b)(x-c)}{(a-b)(a-c)} + f(b)\frac{(x-a)(x-c)}{(b-a)(b-c)}$$

$$+ f(c)\frac{(x-a)(x-b)}{(c-a)(c-b)} \qquad (2)$$

is equivalent to the polynomial $y = A + Bx + Cx^2$, a parabola passing through three points $f(a)$, $f(b)$, and $f(c)$. This is Waring's 3-point formula. By this formula, $f(x)$ can be obtained from given values of $f(a)$, $f(b)$, and $f(c)$. Extension to more points or fewer points should be obvious from an inspection of formulas (1) and (2).

Suppose the population in 1950 is P_a, the population in 1960 is P_b, the population in 1970 is P_c, and it is desired to use 3-point interpolation to estimate the population in 1965. Then,

Population in 1965 = $f(1965)$

$$= P_a\frac{(1965-1960)(1965-1970)}{(1950-1960)(1950-1970)}$$

$$+ P_b\frac{(1965-1950)(1965-1970)}{(1960-1950)(1960-1970)}$$

$$+ P_c\frac{(1965-1950)(1965-1960)}{(1970-1950)(1970-1960)}$$

$$= P_a \left(\frac{5}{-10}\right)\left(\frac{-5}{-20}\right) + P_b \left(\frac{15}{10}\right)\left(\frac{-5}{-10}\right) + P_c \left(\frac{15}{20}\right)\left(\frac{5}{10}\right)$$

$$= -.125P_a + .750P_b + .375P_c$$

The reader should note that the computational work could have been simplified if, instead of using numbers (dates) like 1950, 1960, 1965, and 1970, x (representing 1965) had been taken as "0" and the other dates as -3, -1, and $+1$ for 1950, 1960 and 1970, respectively. (The simplified recodes come by noting that 1950 is three units of 5 years each before 1965, 1960 is one unit of 5 years before 1965, and 1970 is one unit of 5 years past 1965.) Accordingly,

$$f_x = f(o) = P_a \frac{(1)(-1)}{(-2)(-4)} + P_b \frac{(3)(-1)}{(2)(-2)} + P_c \frac{(3)(1)}{(4)(2)}$$

$$= -.125P_a + .750P_b + .375P_c$$

The a, b, c, etc., values may be recoded in any desired way so long as they maintain the same relative values. Thus, they can be multiplied or divided by a constant and the differences between them can be divided by a constant without any effect on the results.

Aitken's iterative procedure. — Aitken's iterative procedure is a system of successive linear interpolations equivalent to interpolation by a polynomial of any desired degree.[2] It is especially suitable for use with desk calculators.

Waring's two-point formula:

$$f(x) = f(a)\frac{(x-b)}{(a-b)} + f(b)\frac{(x-a)}{(b-a)}$$

can be rewritten as

$$f(x) = \frac{f(a)(b-x) - f(b)(a-x)}{(b-x) - (a-x)} \quad (3)$$

which is an expression that will appear in Aitken's procedure as outlined below. Aitken's system is set up in the following basic format for interpolation between four given points for the value of $f(x)$:

Given ordinates	Computational stages			Proportionate parts
	(1)	(2)	(3)	
$f(a)$				$(a-x)$
$f(b)$	$f(x; a, b)$			$(b-x)$
$f(c)$	$f(x; a, c)$	$f(x; a, b, c)$		$(c-x)$
$f(d)$	$f(x; a, d)$	$f(x; a, b, d)$	$f(x; a, b, c, d)$	$(d-x)$

Only the first two lines would be used for 2-point or linear interpolation, and there would be just one computational stage. The first three lines and two computational stages would be used for 3-point interpolation. Additional lines and computational stages are used as required for more points. As many

points as desired can be used. The first column, "given ordinates," symbolizes the given data, i.e., the four observations. The "proportionate parts" in the extreme right-hand column are differences between the given abscissa and the one for which the interpolation is wanted. The abscissa values may be transformed into simplest terms in order to simplify the calculations, as in the case of the Waring formula.

The entries in computational stage (1) are each calculated by computing diagonal cross-products, differencing them, and dividing by the difference between the proportionate parts, as follows:

$$f(x; a, b) = \frac{f(a)(b-x) - f(b)(a-x)}{(b-x) - (a-x)} \quad (4)$$

$$f(x; a, c) = \frac{f(a)(c-x) - f(c)(a-x)}{(c-x) - (a-x)} \quad (5)$$

$$f(x; a, d) = \frac{f(a)(d-x) - f(d)(a-x)}{(d-x) - (a-x)} \quad (6)$$

Each of the expressions $f(x; a, b)$, $f(x; a, c)$, $f(x; a, d)$, etc. is an estimate of $f(x)$ obtained by linear interpolation or extrapolation of $f(a)$ and one of the subsequent $f(b)$, $f(c)$, or $f(d)$ values.

The general process of successive linear interpolations is repeated for computational stage (2), but this time we use the results of computational stage (1) and their associated diagonal multipliers. Thus,

$$f(x; a, b, c) = \frac{f(x; a, b)(c-x) - f(x; a, c)(b-x)}{(c-x) - (b-x)} \quad (7)$$

$$f(x; a, b, d) = \frac{f(x; a, b)(d-x) - f(x; a, d)(b-x)}{(d-x) - (b-x)} \quad (8)$$

Suppose, for example, we want the population in 1935, given data on population in 1920, 1930, 1940, and 1950. The calculations are summarized as follows:

Date	Population	Computational stages			Proportionate parts
		(1)	(2)	(3)	
1920	16,321				1920−1935 = −15
1930	30,567	37,690			1930−1935 = −5
1940	52,108	43,161	40,426		1940−1935 = +5
1950	87,724	52,022	41,273	40,002	1950−1935 = +15

The successive computations for stage (1) are:

$$\frac{(16,321)(-5) - (30,567)(-15)}{(-5) - (-15)} = \frac{-81,605 + 458,505}{-5 + 15}$$

$$= \frac{376,900}{10} = 37,690$$

$$\frac{(16,321)(+5) - (52,108)(-15)}{(+5) - (-15)} = \frac{+81,605 + 781,620}{+5 + 15}$$

$$= \frac{863,225}{20} = 43,161$$

[2] A. C. Aitken, "On Interpolation by Iteration of Proportionate Parts, Without the Use of Differences," *Proceedings of the Edinburgh Mathematical Society*, Series 2, 3:56–76, 1932.

$$\frac{(16,321)(+15)-(87,724)(-15)}{(+15)-(-15)}=\frac{+244,815+1,315,860}{+15+15}$$

$$=\frac{1,560,675}{30}=52,022$$

The successive computations for stage (2) are:

$$\frac{(37,690)(+5)-(43,161)(-5)}{(+5)-(-5)}=\frac{+188,450+215,805}{10}$$

$$=\frac{404,255}{10}=40,426$$

$$\frac{(37,690)(+15)-(52,022)(-5)}{(+15)-(-5)}=\frac{565,350+260,110}{20}$$

$$=\frac{825,460}{20}=41,273$$

The computation for stage (3) is:

$$\frac{(40,426)(+15)-(41,273)(+5)}{(+15)-(+5)}=\frac{606,390-206,365}{15-5}$$

$$=\frac{400,025}{10}=40,002$$

Our final interpolated figure for 1935, 40,002, is the result of computational stage (3).

The given observations need not be equally spaced as they are in the example. Also, their order of arrangement in the table can be mixed, that is, they do not have to follow a prescribed order.

This interpretation of the results assumes, however, that the given data are not too widely spaced for a reliable result. If the given data are widely spaced and have a high degree of curvature in the region of interpolation, none of the interpolation procedures, including Aitken's, will yield a reliable result; observations closer to the desired abscissa must be used.

In interpolations of demographic data, the computational stages should probably stop at about the point where it is clear that there is no longer a clear convergence from stage to stage.

Aitken's iterative procedure involves a relatively large amount of work to arrive at a single result if many observations are used, and the same amount of work must be repeated each time another interpolation is carried out. It is efficient to use this procedure, therefore, when only a few interpolations at most are required. In constrast, it is more efficient to use the Waring formula when many interpolations are being carried out for the same abscissa or x-value, especially ones based on relatively few observations. Under these circumstances, the coefficients, once derived, can be used over and over again.

Osculatory Interpolation

One of the chief difficulties met in adjusting rough data by the usual (single polynomial) interpolation formulas, as described above, is that, at points where two interpolation curves meet, there are sudden breaks in the values of the first-order differences. Various methods have been employed to effect a smooth junction of the interpolations made for one range of data with the interpolations made for the next (adjacent) range. Osculatory interpolation is a method which accomplishes that purpose. It involves combining two overlapping polynomials into one equation. One of the polynomials begins sooner and ends sooner than the other, and the interpolations are limited to the overlapping parts. The second of the two polynomials in the first range then becomes the first polynomial in the second range. The use of one polynomial in common for each pair of successive ranges permits a continuous welding of results from range to range. The two overlapping polynomials are generally forced to have specified conditions in common at the beginning and at the end of the range in which interpolation is desired. The specified conditions may include a common ordinate, a common tangent (slope), and/or a common radius of curvature,[3] usually accomplished by making the first derivative or the first two derivatives equal for the two polynomials.

Illustrative Formulas. — Although osculatory interpolation encompasses a wide variety of possible equations, only a few have seen much use. We consider here specifically Sprague's fifth-difference equation, Karup-King's third-difference equation, and Beers' six-term formulas.

Sprague's fifth-difference equation. — The fifth-difference equation developed by Sprague is expressed in terms of leading differences.[4] The equation is based on two polynomials of the fourth degree, forced to have a common ordinate, a common tangent, and a common radius of curvature at y_{n+2} and at y_{n+3}:

$$y_{n+2+x}=y_n+\frac{x+2}{1!}\Delta y_n+\frac{(x+2)(x+1)}{2!}\Delta^2 y_n$$

$$+\frac{(x+2)(x+1)x}{3!}\Delta^3 y_n+\frac{(x+2)(x+1)x(x-1)}{4!}\Delta^4 y_n$$

$$+\frac{x^3(x-1)(5x-7)}{4!}\Delta^5 y_n \qquad (9)$$

Six given observations, designated y_n, y_{n+1}, y_{n+2}, y_{n+3}, y_{n+4}, and y_{n+5}, are involved in the leading differences, Δy_n, ..., $\Delta^5 y_n$. In the formula, n denotes any integral number, including 0, and x denotes any fraction less than unity. Thus, interpolation is to be limited to a middle range, from abscissa $n+2$ to abscissa $n+3$, or to "midpanel" interpolation. The six given observations must be equally spaced along the abscissa. Other procedures exist or can be developed for use with unevenly spaced observations and also for interpolation in other than a middle range, but midpanel formulas for use with equally spaced observations cover most situations.

Karup-King's third-difference equation. — As another example of an osculatory interpolation equation, we present a third-difference equation based on two overlapping polynomials of the second degree, with ordinates, tangents, and radius of curvature forced to be common to both polynomials at the abscissas $n+1$ and $n+2$. The equation is designed to interpolate between the abscissas $n+1$ and $n+2$; that is, it is limited to midpanel interpolation. The four given points, y_n, y_{n+1}, y_{n+2}, y_{n+3}, must be equally spaced. The formula is again expressed in terms of leading differences:

$$y_{n+1+x}=y_n+\frac{(x+1)}{1!}\Delta y_n+\frac{(x+1)x}{2!}\Delta^2 y_n$$

$$+\frac{x^3(x-1)(3-2x)}{2!}\Delta^3 y_n \qquad (10)$$

[3] The radius of curvature of a line at some point is equivalent to the radius of a circle whose curvature, or rate of change in direction, is the same as the curvature of the line at that point.
[4] U.S. Bureau of the Census, *United States Life Tables, 1890, 1901, 1910, and 1901–10*, by James W. Glover, 1921, pp. 344–348; Thomas Bond Sprague, "Explanation of a New Formula for Interpolation," *Journal of the Institute of Actuaries*, 22:270, 1880–81.

If only a common tangent is required but not a common radius of curvature, the corresponding equation would be:

$$y_{n+1+x} = y_n + \frac{(x+1)}{1!}\Delta y_n + \frac{(x+1)x}{2!}\Delta^2 y_n + \frac{x^2(x-1)}{2!}\Delta^3 y_n \quad (11)$$

The last equation is the Karup-King osculatory interpolation formula.

Beers' six-term ordinary and modified formulas. — In most interpolation work the interest is in the interpolated points themselves, and a procedure which yields smooth trends in terms of the interpolated points is logically sounder than one which forces a specified number of derivatives to be equal at junction points. The two overlapping curves can be fitted in a manner that minimizes the squares of a certain order of differences within the interpolation range. Beers did this by a minimization of fifth differences for a six-term formula.[5] The resulting formulas generally yield smoother results than are possible from the usual osculatory interpolation formulas.

"Ordinary" osculatory equations, such as the Beers' six-term formula mentioned, reproduce the given values. The requirement that the given values be reproduced sometimes causes undesirable undulations in the interpolated results. "Modified" equations relax the requirement that the given values be reproduced and yield smoother interpolated results than would otherwise be possible. The Beers' six-term modified formula is an example of a formula which combines interpolation with some smoothing or graduation of the given values.[6] It minimizes the fourth differences of the interpolated results. This formula is recommended for use when smoothness of results is more important than maintenance of the given values. In the next section on "use of multipliers," procedures are described for applying both the ordinary and the modified interpolation formulas. The analyst has to decide for himself whether he wishes to maintain the original data unchanged at a cost of less smoothness for the interpolated results or prefers results that are smoother and only approximate the original data.

Use of Multipliers. — The actual application of the equations given above takes a different form from that shown. The formulas for osculatory interpolation can be expressed in **linear compound form,** that is, in terms of coefficients or multipliers that are applied to the given data. An interpolated value can then be readily computed by multiplying the given data by the corresponding coefficients and by accumulating the products. In this way the analyst has only to select the method of interpolation and to know how to use the multipliers; he does not need to be familiar with the formula itself or with the mathematical derivation of the multipliers. In effect, then, carrying out the interpolation becomes a purely clerical operation.

This appendix presents selected sets of multipliers for point interpolation. The sets presented (see tables C–4 to C–7 at the end of the text of this appendix) are based on four different formulas:

(1) Karup-King Third-Difference Formula
(2) Sprague Fifth-Difference Formula
(3) Beers Six-Term Ordinary Formula
(4) Beers Six-Term Modified Formula

The Karup-King formula is applied to four points, the Sprague formula to six points (for midpanel interpolation), and the Beers formulas to six points (for midpanel interpolation). For all formulas the given points must be equally spaced, and the given values are maintained in the interpolation for all formulas except Beers' modified formula.

We may illustrate the application of the multipliers by interpolating to single ages between l_{45} and l_{50} in the 1959–61 U.S. life table by the use of the Karup-King formula. For these interpolations four points are used. The general form of the equation is simply:

$$N_{2+x} = m_1 N_{1.0} + m_2 N_{2.0} + m_3 N_{3.0} + m_4 N_{4.0} \quad (12)$$

where x is a fraction between 0 and 1; $N_{1.0}$, $N_{2.0}$, $N_{3.0}$, and $N_{4.0}$ represent four given values; and m_1, m_2, m_3, and m_4 are the four multipliers associated with the four given points. In this case, if we wish l_{48}, a value 0.6 of the way from l_{45} to l_{50}, we have:

$$N_{2.6} = m_1 N_{40} + m_2 N_{45} + m_3 N_{50} + m_4 N_{55}$$

or

$$l_{48} = m_1 l_{40} + m_2 l_{45} + m_3 l_{50} + m_4 l_{55}$$

We have the following table of multipliers for interpolating the middle interval.

| | Coefficients to be applied to: | | | |
Interpolated point	$N_{1.0}$	$N_{2.0}$	$N_{3.0}$	$N_{4.0}$
$N_{2.0}$	.000	1.000	.000	.000
$N_{2.2}$	−.064	+.912	+.168	−.016
$N_{2.4}$	−.072	+.696	+.424	−.048
$N_{2.6}$	−.048	+.424	+.696	−.072
$N_{2.8}$	−.016	+.168	+.912	−.064

Selecting the multipliers for $N_{2.6}$ from this table and the values of l_x from table 15–2, we have then

$$l_{48} = -.048(93,064) + .424(91,378) + .696(88,756) \\ - .072(84,711) = 89,952$$

Similarly we derive the following values for l_{46}, l_{47}, and l_{49}:

l_x	Computed by Karup-King formula	Published
l_{45}	[a] 91,378	91,378
l_{46}	90,936	90,943
l_{47}	90,465	90,470
l_{48}	89,952	89,951
l_{49}	89,386	89,382
l_{50}	[a] 88,756	88,756

[a] Given points.

[5] Henry S. Beers, "Six-Term Formulas for Routine Actuarial Interpolation," *The Record of the American Institute of Actuaries,* 33, Pt. II (68):245–260, November 1944; and idem, "Discussion of Papers Presented in the Record, No. 68: 'Six-Term Formulas for Routine Actuarial Interpolation', by Henry S. Beers," *The Record of the American Institute of Actuaries,* 24, Pt. I (69):59–60, June 1945.
[6] Henry S. Beers, "Modified-Interpolation Formulas that minimize fourth Differences," *The Record of the American Institute of Actuaries,* 34, Pt. I (69):14–20, June 1945.

Exponential Functions

Exponential functions are another class of mathematical equations useful in interpolation and extrapolation of series of data. As we saw in chapter 13, this class of curves is important in connection with the measurement and analysis of population growth. Exponential equations are used for many other demographic purposes. Our discussion here is intended to describe the types of exponential functions and note their general relationship to one another.

An exponential function is one in which one or more of the variables is expressed as a power of some parameter or constant in the formula. Thus, $y = A^x$ is an exponential function because x is a power of the parameter A. In contrast, $y = x^A$ is not an exponential function because the variable x is not a power of a parameter. Exponential functions take many forms, as indicated below. One general form of an exponential function, the power function, is:

$$y = AB^x \tag{13}$$

With modifications this general equation lends itself to many uses. Power functions are also known as **growth curves.**

Geometric Curve. — The simple geometric curve is a special case of the power function. In the geometric curve the given y values form a geometric progression while the corresponding x values form an arithmetic progression. The curve can be fitted exactly through two points. If there are more than two observations, the simple geometric curve may be fitted approximately by various methods (see below). An example of the simple geometric curve is the "compound interest" curve. This curve commonly takes either of two forms — **annual compounding** or **continuous compounding.** For a quantity (population) compounded annually the formula is:

$$y = A(1 + B)^x \tag{14}$$

and for a quantity (population) compounded continuously, the formula is:

$$y = Ae^{Bx} \tag{15}$$

where A is the initial amount
  x is the period of time over which growth occurs
  B is the growth rate per unit of time
  e is the base of the system of natural logarithms
  y is the amount (population) at time x.

The "compound interest law" has several applications in demographic analysis. For example, it is the basis for Lotka's equations for a stable population, discussed in chapters 18 and 24. As noted there, stable population analysis is valuable in understanding the implications of birth and death rates at a particular date for changes in the age distribution of a population. Lotka began with life table proportions of persons surviving from birth to age x, expressed on a unit basis. For instance, in a life table showing 96,368 persons reaching age 20 out of 100,000 born alive, the proportion surviving to age 20 would be $P(20) \div P(0) = 96,368 \div 100,000 = .96368$. This amounts to putting the radix of the life table at 1.0000 (or unity). To convert the life table $P(x)$ proportions into proportions for a stable population growing or decreasing at a constant rate r, the compound interest formula is used in a reverse fashion. The life table $P(x)$ values are taken as the "present values", or y values, and r is taken as the growth rate per year, or B, in the compound interest formula $y = Ae^{Bx}$. Then, A becomes the proportion that persons x years of age would constitute of the

births (1.0000 birth per year in the stable population). There would be a different A value for each age x, so we may call it $A(x)$. The formula can now be written in the form:

$$P(x) = A(x)e^{rx} \tag{16}$$

$$A(x) = \frac{P(x)}{e^{rx}} \quad \text{or} \quad A(x) = P(x)e^{-rx} \tag{17}$$

This is the equation for computing the proportion $A(x)$ of the population at age x in a stable population growing (or declining) at a constant rate r, from a life table series of proportions $P(x)$.

Other Growth Curves. — As we noted above, there are many possible modifications of the general exponential equation $y = AB^x$ in addition to the simple geometric curve with annual or continuous compounding. These growth curves do not usually fit the given data exactly; hence, they also belong under the heading "Curve fitting."

The equation

$$y = K + AB^x \tag{18}$$

is a **modified exponential equation** which yields an ascending asymptotic curve when A is negative and B is a fractional value between 0 and 1. It describes a series in which the absolute growth in the y values decreases by a constant proportion. When $x = 0$, $y = K - A$. As x increases, y approaches K as an upper limit. In other variations A is positive and B is between 0 and 1 or greater than 1.

More commonly used than the modified exponential equation just described is the **Gompertz curve,** the equation of which is:

$$y = KA^{B^x} \tag{19}$$

which reduces to the equivalent logarithmic form:

$$\log y = \log K + (\log A)B^x \tag{20}$$

The Gompertz curve is exactly like the modified exponential curve above except that it is the increase in the logarithms of the y values which decreases by a constant proportion. The Gompertz curve fits many types of growth data much better than the modified exponential curve.

Another type of growth curve which has the same general shape as the Gompertz curve is the **logistic curve,** also known as the **Pearl-Reed curve.**[7] The logistic curve has the general equation

$$\frac{1}{y} = K + AB^x, \tag{21}$$

or, when fitted by the method of selected points,

$$y = \frac{K}{1 + e^{A + Bx}} \tag{22}$$

The reader is cautioned that no matter how well an asymptotic growth curve fits observed data, projections that go beyond the observations will not necessarily be realized. No empirically fitted curve can magically anticipate future changes when these are dependent on circumstances that

---

[7] Further information regarding the fitting of growth curves may be found in many general statistics texts. See, for example, Frederick E. Croxton. Dudley J. Cowden, and Sidney Klein. *Applied General Statistics*, 3rd edition, New York. Prentice-Hall. 1967, pp. 256–282.

are beyond the ken of the curve. It is often easy to fit a variety of modified logistic curves to the same observations in a manner that will yield very different projections.

Curve Fitting

Although a series of demographic data may not be subject to any mathematical law, the data may follow a typical trend or pattern that can be represented empirically by some mathematical equation. Curve fitting consists of finding a suitable equation to represent that trend or pattern. Curves to be fitted might be polynomials, osculatory equations, exponential equations, trigonometric equations (useful for data that have periodic fluctuations or seasonal patterns), or still other curves. The aim may be to fit a curve to the data in an approximate fashion, in which case crude methods, such as graphic methods or moving averages, may be suitable, or to fit a curve by a more sophisticated method, as by the method of moments or by the method of least squares. Whether or not the fitted curve is suitable for interpolation or extrapolation would depend on the nature of the given data, on the choice of curve, and on the goodness of fit. Probably demographers most often fit straight lines or polynomials of second or third degree.

Method of Least Squares.—Curves are commonly fitted by the method of least squares or by the use of moments. We consider first the method of least squares and illustrate it by fitting a second-degree polynomial ($y = A + Bx + Cx^2$) to a time series of data on median family income in the United States for selected dates from 1947 to 1966.[8] The method of least squares minimizes the sum of the squares of the differences between the observed or given points and the points calculated from the fitted curve. That is:

$$\sum w(Y_{\text{obs.}} - Y_{\text{calc.}})^2 = \text{min.} \qquad (23)$$

where $Y_{\text{obs.}}$ is the observed value of y at each given point, $Y_{\text{calc.}}$ is the corresponding value from the fitted curve, and w is a weight (1 for each observed value if all are assumed to be of equal precision, as in the present example). Three "normal" equations have to be solved. The normal equations in general form are:

$$\text{I.} \quad \sum y = AN + B\sum x + C\sum x^2$$

$$\text{II.} \quad \sum xy = A\sum x + B\sum x^2 + C\sum x^3$$

$$\text{III.} \quad \sum x^2 y = A\sum x^2 + B\sum x^3 + C\sum x^4$$

The origin ($x = 0$) is arbitrarily taken at the year 1957, or roughly midway in the range of dates considered. We now develop the following data needed to solve the normal equations:

Year	Median family income $y_{\text{obs.}}$	^{a} x	x^2	x^3	x^4
1947..........	$3,031	−10	100	−1,000	10,000
1950..........	3,319	−7	49	−343	2,401
1955..........	4,421	−2	4	−8	16
1960..........	5,620	3	9	27	81
1964..........	6,569	7	49	343	2,401
1965..........	6,957	8	64	512	4,096
1966..........	7,436	9	81	729	6,561

^{a} Year minus 1957.

$$N = 7 \text{ (observations)} \qquad \sum x = 8$$

$$\sum y = 37,353 \qquad \sum x^2 = 356$$

$$\sum xy = 123,038 \qquad \sum x^3 = 260$$

$$\sum x^2 y = 1,903,440 \qquad \sum x^4 = 25,556$$

These values are now inserted in the normal equations as required:

$$\text{I.} \quad 37,353 = 7A + 8B + 356C$$
$$\text{II.} \quad 123,038 = 8A + 356B + 260C$$
$$\text{III.} \quad 1,903,440 = 356A + 260B + 25,556C$$

Solution of the three normal equations for A, B, and C by any of a number of methods yields:[9]

$$A = 4,808.8; B = 233.81; C = 5.115$$

The desired equation is, therefore:

$$y = 4,808.8 + 233.81x + 5.115x^2$$

The computed values of y are shown in figure C-1.

Figure C-1. — **Fitting a Second-Degree Polynomial by Least Squares to Median Family Income in the United States, 1947 to 1966**

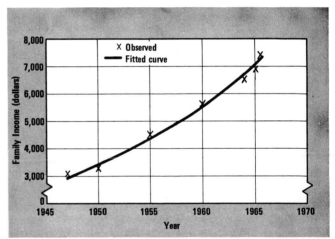

	Median family income (dollars)		
Year	Observed	Fitted curve[1]	Difference
1947	3,031	2,982	− 49
1950	3,319	3,423	+104
1955	4,421	4,362	− 59
1960	5,620	5,556	− 64
1964	6,569	6,696	+127
1965	6,957	7,007	+ 50
1966	7,436	7,327	−109
		Total =	-

[1] $y = 4,808.8 - 233.81x - 5.115x^2$ (origin at year 1957).

Source of observed data: U.S. Bureau of the Census, *Statistical Abstract of the United States, 1968*, p. 324.

---

[8] Further information regarding the fitting of polynomials by least squares is given in many general statistics texts. See, for example, Croxton, Cowden, and Klein, op. cit., pp. 263–274.

[9] The Doolittle method for solution is recommended when there are three or more unknowns. The Doolittle method is described in many textbooks on elementary statistics. See, for example, Frederick E. Croxton and Dudley J. Cowden, *Applied General Statistics*, 2nd edition, New York, Prentice-Hall, 1955, p. 431.

If the equation were also used to project a value for 1967, the result would be $7,658. The observed value for 1967 (not used in the curve fitting) is $7,974. The projection falls short of the observed value, therefore, by $316, or 4 percent. This percent difference is greater than for any year within the range of the observed data and is noted as an example of the hazards of extrapolation.

We have skipped over illustrating the procedure for fitting a straight-line by the method of least squares since the illustration just given encompasses the basic steps. Only two normal equations have to be solved:

$$\text{I.} \quad \sum y = AN + B \sum x$$

$$\text{II.} \quad \sum xy = A \sum x + B \sum x^2$$

Hence, we need to compute only $\sum y$, $\sum xy$, $\sum x$, and $\sum x^2$. These have the same values as given above. The normal equations which we solve for A and B then become:

$$\text{I.} \quad 37,353 = 7A + 8B$$

$$\text{II.} \quad 123,038 = 8A + 356B$$

and the desired equation is $y = 5,071.40 + 231.65x$.

The pattern of the normal equations is evident. For fitting a third-degree polynominal by least squares, the normal equations are:

$$\text{I.} \quad \sum y \ = AN \quad + B \sum x \ + C \sum x^2 + D \sum x^3$$

$$\text{II.} \quad \sum xy = A \sum x \ + B \sum x^2 + C \sum x^3 + D \sum x^4$$

$$\text{III.} \quad \sum x^2 y = A \sum x^2 + B \sum x^3 + C \sum x^4 + D \sum x^5$$

$$\text{IV.} \quad \sum x^3 y = A \sum x^3 + B \sum x^4 + C \sum x^5 + D \sum x^6$$

Readers who wish a full explanation of the principles of least squares, and a general solution that can be applied to almost any kind of equation, may wish to consult Deming, *The Statistical Adjustment of Data.*[10]

Curves Based on Moments About the Mean. — Many kinds of population data, such as age-specific birth rates and distributions of families by income, exhibit skewed bell-shaped patterns of distribution which can be fitted reasonably well by formulas involving the use of moments. Recall the pattern of birth rates: the birth rates are low at the young ages of childbearing, rise to a peak at some central age, and then taper off at the later ages of childbearing (ch. 16). The pattern may resemble a normal curve somewhat, but it is not symmetrical; the pattern may be flatter (platykurtic) or more peaked (leptokurtic) than a normal curve, and may differ in other respects. English statisticians have developed whole systems of curves to fit to skewed bell-shaped patterns by modifying the equation for a normal curve, which is, of course, perfectly symmetrical, and they have provided criteria for selecting the proper equation automatically and for computing the constants or parameters of the equation. The derivation of the formulas involves the use of moments about the mean. Although the fitting of such a curve usually involves the use of moments, a least squares method of fitting is also possible.

The two main systems of curves based on moments are (1) the Gram-Charlier curves and (2) the Pearsonian curves. They can take varied forms besides a bell-shaped distribution, such as an S-shaped, J-shaped, or U-shaped pattern. The Gram-Charlier curves will fit anything that the Pearsonian curves will fit, have a more logical basis for their derivation, and may be easier to fit because thay make use of data on derivatives of the normal curve that are available in published tables.[11]

By way of illustration, we show the results of fitting a Pearsonian Type I curve by the method of moments to child-women ratios for whites in the United States in 1960:

Age of woman (years)	Own children under 5 years old per 1,000 women	
	Observed ($y_{obs.}$)	Computed from fitted curve ($y^{calc.}$)
15 to 19....................	94	69
20 to 24....................	826	841
25 to 29....................	1,074	1,044
30 to 34....................	730	775
35 to 39....................	417	416
40 to 44....................	175	166
45 to 49....................	41	47
50 to 54....................	10	10

The Pearsonian Type I curve, with origin at the mode, has the form:

$$y_{\text{calc.}} = y_0 \left(1 + \frac{x}{a_1} \right)^{m_1} \left(1 - \frac{x}{a_2} \right)^{m_2} \qquad (24)$$

where $\dfrac{m_1}{a_1} = \dfrac{m_2}{a_2}$ and y_0, a_1, and a_2 are defined as in the following sketch:

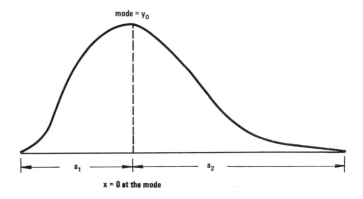

---

[10] W. Edwards Deming, *The Statistical Adjustment of Data*, New York, John Wiley & Sons, 1943, chapters 2 and 9.

[11] A good explanation of the Gram-Charlier system may be found in Thornton C. Fry, *Probability and Its Engineering Uses*, New York, D. Van Nostrand Company, 1928. A good explanation of the Pearsonian system appears in W. P. Elderton, *Frequency Curves and Correlation*, 4th edition, London, Cambridge University Press, 1953.

The parameters y_0, a_1 and a_2, and m_1 and m_2 are functions of the first four moments about the mean.

Although they have not been discussed in this section, it should be noted that there are many other types of curves which can be used in curve fitting, such as the Poisson distribution, the chi-square distribution, and the binomial distribution.

INTERPOLATION OF GROUPED DATA

Introduction

We have been concerned in the preceding part of this chapter with interpolation and curve fitting as applied to point data. We turn now to a consideration of interpolation as applied to grouped or "area" data. Interpolation of grouped data may serve any of several purposes. The most common purpose probably is the estimation of data in finer detail than is available in published data, as for estimating numbers of persons in single years of age from published data for 5-year age groups. Another purpose may be the smoothing or graduation of data that are available in fine detail, as by combining data given in single years of age into data for 5-year age groups, which are then interpolated to obtain smoothed estimates of data by single years of age.

It should be noted that the methods to be described have one thing in common. They assume that the pattern of distribution of grouped data is a valid indication of the pattern of the distribution within groups. There are some kinds of demographic data where the distribution within groups is known to have a special pattern that is not reflected by grouped data; in such instances the methods described here may not apply. For example, in the United States it is common for persons to work either 40 or 48 hours a week—a fact which is not evident from broad groupings of hours worked.

"Interpolation" by Prorating

Sometimes the best estimates of the subdivisions of grouped data come not from elaborate mathematical techniques but rather from simple prorating. In this procedure, a relative distribution taken from some other similar group that has satisfactory detailed information is used to split up a known total for a given group. Such a procedure depends for its accuracy on how well the distribution of the former group represents conditions in the latter group.

For example, a series of annual birth statistics for years in which persons now 25 to 29 years of age were born may be a useful basis for prorating the number of persons 25 to 29 years of age so as to obtain estimates of the population by single years of age. It is not necessary to allow for deaths since the birth period or for net migration, if it is thought that the **relative** distribution of the annual births is still a reasonably good indicator of how the population is distributed by age within the 25-to-29 year age group. Furthermore, the birth registration need not have been complete so long as the percent completeness of birth registration was reasonably similar from year to year. Interpolations obtained in this manner may be superior to interpolations from a mathematical equation that involves an assumption of a smooth flow of events from age to age.

As another example of a type of problem where prorating may be superior, consider the task of securing the percent married for single ages from the percent married for a 5-year age group, say for ages 15 to 19. A suggested procedure is to (1) take the percent married for age group 15 to 19 in the given population, (2) multiply that percent by 5 to obtain an approximate value for the sum of the percentages for each of the five single years of age, and then (3) estimate the percentages for single years of age by prorating this total according to known single-year-of-age percentages from some other population. In the United States the decennial census provides data on marital status by single years of age, but the data from the Current Population Survey are usually tabulated only by broad age groups; the census data may be a good basis, therefore, for splitting up the current survey data. Additional applications of prorating procedures are discussed in the section on "Adjustment of distributions to marginal totals."

Use of a Rectangular Assumption

The simplest and perhaps the most commonly used method of subdividing grouped data employs the assumption that the data are rectangularly distributed within the interval to be subdivided. This assumption is that the values of the parts are all equal. They are derived, then, by dividing the total for the interval by the number of parts desired. A rectangular assumption is a useful basis for deriving rough estimates of detailed categories under many different circumstances in demographic studies. For example, it may be employed to derive life table d_x's in single years of age from $_5d_x$'s over an age range such as 20 to 39 (and hence, to derive single l_x's in this range) or to derive the central d_x in each 5-year interval over a much wider range of years. We illustrate with data based on the 1959–61 U.S. life table for the total population:

$x+2$	Published $_5d_x$	d_{x+2} Rectangular assumption $_5d_x \div 5$	d_{x+2} Published value
17....................	440	88	89
27....................	612	122	120
37....................	1,080	216	214
47....................	2,622	524	519
57....................	5,644	1,129	1,117
67....................	7,920	1,584	1,586

Births or deaths on a calendar-year basis may be shifted to a "fiscal-year" (i.e., July-to-June) basis by simply assuming that one-half the births or deaths of each year occurs in the first or second half of the year. Even if there is a pronounced seasonal variation and a sharp trend up or down through the two years, the 12-month estimate may be quite adequate because the excess or deficit in the estimate for the first half of the period may be largely offset by the deficit or excess in the estimate for the second half of the period.

Graphic Interpolation

Graphs are useful for deriving rough estimates for subdivisions of grouped data as well as for estimating values in a point series. They are especially useful when the grouped data are unevenly spaced as, for example, in subdividing data for combinations of 1-, 5-, and 10-year age groups into single ages. One procedure for graphic interpolation is to draw smooth lines through the histograms for the grouped data, as the sketches below indicate:

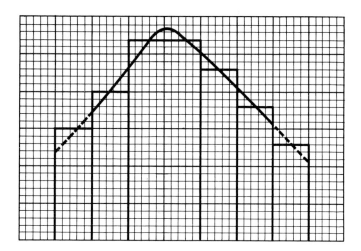

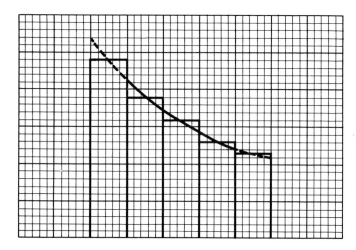

When drawing the curve through the tops of the histograms, an effort should be made to equalize the area under the curve and the area above the curve that falls above and below (respectively) the tops of the histograms. Once the curve is drawn, the interpolated results are determined by reading off the values of the ordinates along the curve that are at desired abscissas. Another procedure is to cumulate the grouped data and plot the resulting point values at the appropriate ordinates. With age data, the age references change as follows:

Age of grouped data (years)	Age of cumulated data	
	Tabulated	Plotted at—
Under 5	Under 5	5.0
5–9	Under 10	10.0
10–14	Under 15	15.0
15–19	Under 20	20.0
20–24	Under 25	25.0

The accumulated values are plotted at 5.0, 10.0, 15.0, etc., and the required values for a given age are obtained by reading off values of the ordinates along the curve and taking the differences between these ordinate values. (For further explanation, see the discussion below on the "cumulation-differencing" procedure.)

Sometimes graphs are of special help in interpreting and solving an interpolation problem. Suppose, for example, one wishes to compute separation factors for apportioning annual birth data by age of mother into the births occurring to mothers who would be of given ages at some date during the period. Specifically, we may wish to determine the number of births in a 12-month period ending on April 1, 1960, which occurred to mothers who are "x" years of age on April 1, 1960. A diagram of the type shown as figure C–2 may be helpful.

Figure C–2. – **Diagram for Determining the Proportions of Births in 1959 and 1960 That Occurred in the 12-Month Period Preceding April 1, 1960, to Mothers x Years Old on April 1, 1960**

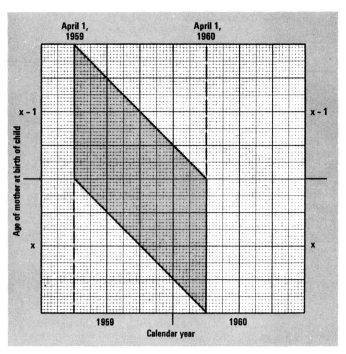

If the births are assumed to be uniformly distributed by date of occurrence and age of mother, then the desired separation factors can be figured from the proportionate parts of the shaded areas. The procedure consists of estimating directly the triangular area included in the shaded area for each age and year; or estimating the rectangular area from April 1, 1959, to December 31, 1959 (¾ year), and from January 1, 1960, to March 31, 1960 (¼ year), and subtracting from it the triangular area which is not included in the shaded area:

Proportion of $B_{1960}^x = [¼ − ½(¼ × ¼)] B_{1960}^x = {}^{7}/_{32} B_{1960}^x$

Proportion of $B_{1960}^{x-1} = [½(¼ × ¼)] B_{1960}^{x-1} = {}^{1}/_{32} B_{1960}^{x-1}$

Proportion of $B_{1959}^x = [½(¾ × ¾)] B_{1959}^x = {}^{9}/_{32} B_{1959}^x$

Proportion of $B_{1959}^{x-1} = [¾ − ½(¾ × ¾)] B_{1959}^{x-1} = {}^{15}/_{32} B_{1959}^{x-1}$

The width of the interval from January 1 to March 31, 1960, is ¼ year and the height of the January 1, 1960, vertical line is ¾ year for the shaded area relating to age x and ¼ year for the shaded area relating to age $x-1$. Hence, the shaded areas for 1960 are computed at $7/32$ of 1960 births to mothers x years old at childbirth plus $1/32$ of 1960 births to mothers $x-1$ years old at childbirth. In a similar manner, the desired proportions for the interval between April 1 and December 31, 1959, are $9/32$ of 1959 births to mothers x years old at childbirth plus $15/32$ of 1959 births to mothers $x-1$ years old.

Midpoint and Cumulation-Differencing Methods

One general procedure for the interpolation of grouped data may, for want of a better term, be called the "midpoint" approach. Another, usually more reliable approach, involves a cumulation and differencing calculation. Polynomial interpolation, particularly in the form of Aitken's procedure, is then ordinarily combined with one of these approaches to obtain the final results. Both methods will be briefly explained in terms of illustrative examples.

Midpoint Method Using Data on Percentages. — Suppose one has data on the proportion of women ever married for age groups 14–17, 18–19, 20–24, and 25–29 and wants an estimate of the proportion ever married for women 20 years of age. The lower limit of age group 14–17 is the 14th birthday (exact age 14.0) and the upper limit of that age group is the 18th birthday (exact age 18.0); the midpoint of age group 14–17 is, therefore, $(14.0+18.0)/2$, or 16.0. In a similar manner the midpoints of the other age groups in this example may be determined to be, respectively, 19.0 for age group 18–19, 22.5 for age group 20–24, and 27.5 for age group 25–29. The midpoint of the desired year of age 20 is 20.5. By equating the given percentages ever married with the corresponding midpoints of age groups and by interpolating among the former, we can obtain an estimate of the percentage ever married corresponding to the midpoint of age 20, i.e., for age 20.

In the following example of midpoint interpolation, Aitken's iterative procedure is applied to marital data for the United States in 1960:

Age group (years)	Midpoint age	Percent ever married	Computations (1)	(2)	(3)	Proportionate parts[a]
14-17...	16.0	5.4	......	......	......	−4.5
18-19...	19.0	32.4	45.9	......	......	−1.5
20-24...	22.5	71.6	51.2	48.2	......	2.0
25-29...	27.5	89.5	38.3	44.6	49.7	7.0

[a] Difference between desired midpoint age 20.5 and the midpoint for each age group.

The last figure in column (3) of the computations is the desired result: 49.7 percent. Volume I of the U.S. Census of Population for 1960 gives 54.0 as the percent ever married for women age 20; so that the estimate of 49.7 is actually too low. As will be seen later, a much more accurate result will be obtained by the cumulation-differencing process.

Cumulation-Differencing Method Using Data on Absolute Numbers. — The cumulation-differencing method has a sounder theoretical basis for interpolation of groups than the midpoint approach just described. This is because, in fact, group averages seldom apply exactly to the midpoints of groups, as

is assumed in the midpoint approach; whereas in the cumulation-differencing approach the observed data are associated with the precise points to which they actually apply.

Consider the numbers of women in the age groups 14-17, 18-19, 20-24, and 25-29 for the United States in 1960, as before. Tentatively, for illustrative purposes, assume that there are K girls under age 14. (The number K will drop out as the work progresses, so that its value does not matter; it is used here simply to help clarify the exposition.) The upper limit of the age range under 14 is 14.0. The number of girls aged 14 to 17 plus K (the girls under 14) is then the cumulated number under 18 years old; the upper limit of that age range is 18.0. The population 18 to 19 years old, plus the population 14 to 17 years old, plus K is the cumulated number under 20 years old; the upper limit of that age range is 20.0. Continuing this process, one obtains the cumulated numbers at exact ages 14.0, 18.0, 20.0, 25.0, and 30.0. The cumulated data represent the "ogive" transformation of the original data for groups into data for specific points along the age scale. The transformed data are thus associated with precise points of age.

Interpolation of the transformed data can now be performed by any appropriate method but must be done twice — once for the upper limit of the subgroup for which interpolation is desired and once for the lower limit. Thus, to estimate the population in age 20 from the data for age groups 14-17, 18-19, 20-24, and 25-29, one estimates the population under age 20 and then estimates the population under age 21. The difference between the two estimates will be the population between the 20th and 21st birthdays. Since K is common to both the population under 20 and the population under 21, the subtraction causes K to vanish. This means that K can be taken as zero (instead of some other arbitrary number), thereby simplifying the operation:

Age group (years)	Upper limit of age group	Number of women — In age group	Cumulated from youngest group	Aitken's procedure — Computations for age 21.0 (upper limit) (1)	(2)	(3)	Proportionate parts[a]
14-17...	18.0	5,516	5,516	....	....	....	−3.0
18-19...	20.0	2.417	7,933	9,142	....	....	−1.0
20-24...	25.0	5,520	13,453	8,918	9,097	....	4.0
25-29...	30.0	5,537	18,990	8,885	9,116	9,082	9.0

[a] Difference between desired upper limit 21.0 and upper limits of given age groups.

The figure of 9,082 in column (3) is the interpolated estimate of the number of women cumulated to age 21.0. We also need the cumulated number to age 20.0, but that is already given as 7,933. The desired estimate of the population in age 20 is therefore the difference (1,149) between the 9,082 cumulated to age 21.0 and the 7,933 cumulated to age 20.0.

Cumulation-Differencing Method Using Data on Percentages. — In applying the cumulation-differencing method to percentage data, the percentages require weighting by the class intervals associated with them as a first step. The work then proceeds in the same manner as for absolute numbers. The following example uses the same data as those used in the example of interpolation of percentages by the midpoint method, but we interpolate for the value of the percents ever married cumulated to age 21.0:

Age group (years)	Upper limit of age group	Number of single years in age group (1)	Percent ever married (2)	Estimated sums of percentages for single ages		Computations			Proportionate parts [a]
				Within a group (3) = (1) × (2)	Cumulated (4) = Σ(3)	(5)	(6)	(7)	
14–17......	18.0	4	5.4	21.6	21.6	....	....	....	−3.0
18–19......	20.0	2	32.4	64.8	86.4	118.8	....	....	−1.0
20–24......	25.0	5	71.6	358.0	444.4	202.8	135.6	....	4.0
25–29......	30.0	5	89.5	447.5	891.9	239.2	130.8	139.4	9.0

[a] Difference between desired upper limit of 21.0 and upper limits of given age groups.

The figure of 139.4 in column (7) is the interpolated estimate of percentages cumulated to age 21.0. We also need the cumulated figure for age 20.0; it is already available (86.4) in column (4). The desired estimate of the percent ever married for women in age 20 is the difference (53.0) between the figures for the upper (139.4) and lower (86.4) limits of the year of age. Note that this estimate is much closer to the reported figure (54.0) than that obtained by the midpoint approach.

The above examples employ Aitken's iterative procedure. If one has many interpolations to make for the same spacings of the abscissas (ages in this case), it might save time to use Waring's formula to derive interpolation multipliers which can be used for all the interpolations.

Osculatory Interpolation

Both Aitken's procedure and Waring's procedure use a single curve (i.e., polynomial) and, as mentioned earlier, this circumstance can give rise to a lack of smoothness in the junction between interpolated results when passing from one group to another. Osculatory interpolation or other smooth-junction procedures are often preferred for interpolating demographic data, therefore.

Tables of Selected Sets of Multipliers. — As we noted earlier, formulas for interpolation can be expressed in linear compound form, that is, in terms of coefficients or multipliers that are applied to the given data. Appendix tables C–4 to C–8 present selected sets of multipliers for "area" interpolation, i.e., for subdivision of grouped data. These sets of multipliers are based on five different formulas:

1. Karup-King Third-Difference Formula
2. Sprague Fifth-Difference Formula
3. Beers Six-Term Ordinary Formula
4. Beers Six-Term Modified Formula
5. Grabill's Weighted Moving Average of Sprague Coefficients

Sets of multipliers based on all five formulas are given for subdividing intervals into fifths, that being the most common need. These are suitable for subdividing age data given in 5-year groups into single years of age. For the first two formulas, sets of multipliers are also presented for subdividing grouped data into tenths and halves. They may be used for subdividing data for 10-year age groups into single years of age and into 5-year age groups.

The multipliers can be manipulated in various ways (e.g., used in combination) to meet special needs. For example, one set might be used to split 10-year groups into 5-year groups and then another set used to subdivide the 5-year groups into single ages. Or multipliers for obtaining three single ages might be added together to obtain multipliers that would yield in one step an estimate for a desired 3-year age group. Or multipliers may be combined in a manner that enables one to derive estimates of average annual age-specific first marriage rates from data on the proportion of persons ever married by 5-year age groups, or to derive estimates of average annual age-specific birth rates from data on ratios of children under 5 years old to women by age. The possibilities for manipulation of the multipliers for demographic analysis are many and varied and not limited to the usual objective of subdividing grouped data on age into single years.

Application of Multipliers. — The general manner in which the multipliers (or coefficients) are used with given data to obtain an interpolated result is illustrated by the following example employing the Karup-King Third-Difference formula. Given these data,

Age group (years)	Population
15–19...................	35,700
20–24...................	30,500
25–29...................	32,600

let us estimate the population 20 years old. Age 20 is the "first fifth" of age group 20–24. Age group 20–24 is a middle group. The table of coefficients based on the Karup-King formula has the following values for interpolating a middle group to derive the first fifth (appendix table C–4):

	Coefficients to be applied to:		
	G_1	G_2	G_3
First fifth of G_2.................	+.064	+.152	−.016

The population aged 15–19 is taken as G_1, the population aged 20–24 as G_2, and the population aged 25–29 as G_3. The desired estimate (of the population 20 years old) is then computed as follows:

$$+.064(35,700) +.152(30,500) -.016(32,600)$$

$$= 2,285 + 4,636 - 522 = 6,399$$

Note that the Karup-King formula has four multipliers for point interpolation and three for subdivision of grouped data. Similarly some of the sets of multipliers for interpolation of groups are labeled as having come from six-term formulas but only five groups are employed in an interpolation.

Whenever possible, midpanel multipliers, i.e., the multipliers applicable to the middle group of three or five groups, should be used. End-panel multipliers make use of less information on one side of an interpolation range than on the other side and therefore are likely to give less reliable results than when the midpanel multipliers are used. For subdivision of the first group in a distribution (e.g., ages 0–4), the first-panel multipliers must be used and for subdivision of the last group (e.g., ages 70–74), the last-panel multipliers must be used. With the Sprague formula (table C–5) and the Beers formulas (table C–6), there are also special multipliers for the second panel from the beginning of the distribution (e.g., ages 5–9) and for the next-to-last panel from the end of the distribution (e.g., 65–69). Once the multipliers have been selected, they are applied in the same way as the midpanel multipliers. For example, let us estimate the population 8 years old on the basis of the Sprague formula and the following data:

Age group (years)	Population
0–4	74,300
5–9	68,700
10–14	60,400
15–19	63,900

Age 8 is the "fourth fifth" of the age group 5–9. Age group 5–9 is the next-to-first panel. The table of coefficients based on the Sprague formula has the following values for interpolating a next-to-first panel to derive the fourth fifth (appendix table C–5):

	Coefficients to be applied to:			
	G_1	G_2	G_3	G_4
Fourth fifth of G_2	−.0160	+.1840	+.0400	−.0080

The four population groups above are taken as G_1, G_2, G_3, and G_4, respectively. The desired estimate (of the population 8 years old) is then computed as follows:

$$-.0160(74,300) + .1840(68,700) + .0400(60,400)$$

$$-.0080(63,900) = -1,189 + 12,641 + 2,416 - 511 = 13,357$$

Subdivision of Unevenly Spaced Groups. — Interpolation coefficients may also be derived for subdividing unevenly spaced groups. Suppose we wish to divide in half a group which has the same width as the two following groups but is twice as wide as the two preceding groups. Thus, G_1 and G_2 might represent 5-year age groups while G_3, G_4, and G_5 represent 10-year age groups. The pattern of the available data to be subdivided into 5-year age groups is, therefore, 5–5–⑩–10–10, or 1–1–②–2–2:

Interpolated	Coefficients to be applied to:				
subgroup	G_1	G_2	G_3	G_4	G_5
First half of G_3	−.0677	+.2180	+.4888	−.0737	+.0097
Last half of G_3	+.0677	−.2180	+.5112	+.0737	−.0097

After a series of 5-year age groups is obtained by use of the above coefficients, the other sets of interpolation coefficients can be used further to subdivide the data into single years of age.

Interpolation multipliers can be derived for subdividing a group under many variations in the pattern of the available

data. The data may follow the pattern 1–1–②–2–5, 1–5–⑤–5–5, 1–1–⑤–5–5, 1–2–⑤–5–10, or other pattern. The midpanel (circled group) may be subdivided into fifths, tenths, halves, or other fraction.

Comparison and Selection of Osculatory-Interpolation Formulas. — As stated earlier, the choice of a method for interpolation is dependent on the nature of the data and on the purposes to be served. The several sets of interpolation coefficients that are presented in appendix C are based on formulas that differ in their underlying principles. There is no one "best" method for all purposes.

Use of ordinary formulas. — The Karup-King formula is the simplest one for which interpolation coefficients are presented in this book. It is "correct to second differences" and has an adjustment involving third differences. It uses four given points (or the four boundaries of three groups). It resembles the formula for an ordinary second-degree polynomial (expressed in differences) fitted to the first three points plus an adjustment involving the fourth point. If the third difference of the four given points is zero, then all four given points fall on the same second-degree curve and no adjustment results. The results will be exactly the same as if an ordinary second-degree polynomial had been fitted to the first three points and then used to interpolate.

The three formulas discussed reproduce the data. Specifically, the interpolated points fall on curves that pass through the given points, and the interpolated subdivisions of groups add up to the data for the given groups. Following are two examples of results from the use of the three methods described with certain kinds of regular well-behaved data. Results from rough data or data of erratic quality are considered later.

Example 1.

Given $y = x^2$ $y = x^4$

x	y	x	y
1	1	1	1
2	4	2	16
3	9	3	81
4	16	4	256
5	25	5	625
6	36	6	1,296

Find y for $x = 3.4$ by interpolation of the given values. Results:

	$y = x^2$	$y = x^4$
Karup-King formula	11.5600	133.7680
Sprague formula	11.5600	133.6336
Beers formula	11.5600	133.6336
True value	11.5600	133.6336

Example 1 demonstrates empirically that the two fifth-difference formulas (Sprague and Beers) will reproduce the results of polynomials of low degree (e.g., $y = x^2$) when the observed data are of that form, and also that the Karup-King third-difference formula can sometimes produce nearly correct results for a set of observed data in which fourth differences are not zero (as in the case of $y = x^4$).

Example 2.

Given the following $_5L_x$ values from the U.S. life table, 1959–61 for total males,

Age group (years)	Life table stationary population ($_5L_x$)
15–19	479,138
20–24	475,333
25–29	471,139
30–34	466,882
35–39	461,371

find the population aged 27 by interpolation. The results obtained by the three interpolation formulas under consideration, in comparison with the published value, are:

Karup-King formula	94,229
Sprague formula	94,219
Beers ordinary formula	94,221
Published value	94,227

In example 2, all methods shown give acceptable results. The original data happen to have larger fourth differences (not shown) than third differences; hence, in this particular example the Karup-King formula happens to yield a result closer to the published value than the other two formulas do. In some age ranges, another method may provide a closer result.

Use of modified formulas. — The modified formulas assume that the observed data are subject to error, and, in effect, they substitute weighted moving averages of the observed point or group data for these observed data. Thereby they obtain more smoothness in the interpolated results although at a cost of some modification of the original data.

The extent to which the modifications alter the original data can perhaps best be seen by adding together the five coefficients for subdividing a central group into five equal parts according to the various interpolation schemes. These sums are:

G_1	G_2	G_3	G_4	G_5

Beers' modified six-term minimized fourth-difference formula

−.0430	+.1721	+.7420	+.1720	−.0430

Grabill's modification of Sprague coefficients

+.0164	+.2641	+.4390	+.2641	+.0164

Formulas that reproduce group totals without modification

0	0	+1.0000	0	0

These figures represent consolidated coefficients which, if applied to the given G_1, G_2, G_3, G_4, and G_5 groups, would yield the sums of the five interpolated subdivisions of G_3. It is apparent from the difference in weights assigned to the middle panel that the Beers modified formula involves a less drastic modification of the original group values than Grabill's coefficients.

Figure C–3 illustrates the effect of interpolating 5-year age groups for Mexico, 1960, to single ages by use of the Karup-King formula and by use of Grabill's coefficients.[12] The figure shows that the enumerated data in single years fluctuate sharply as a result of the tendency of many persons to report ages that are multiples of five or two. The Karup-King formula smooths out only part of these undulations since the group totals are maintained. Use of Grabill's coefficients shows how completely the undulations can be removed by a drastic smoothing procedure.

An alternative procedure for subdividing the last few regular groups in the age distribution (e.g., 75–79, 80–84 for Mexico, 1960) involves, first, splitting up the open-end terminal group (e.g., 85 and over) into three groups (i.e., 85–89, 90–94, and 90 and over) and, then, applying midpanel coefficients for subdividing the groups just ahead of the terminal group. The precision of the results of subdividing the terminal group would have only a small effect on the interpolated single-year-of-age values for the preceding age groups. One device for subdividing the terminal group is to employ the distribution of L_x from an appropriate life table (e.g., life table for Mexico, 1959–61). Another is to fit a polynomial to the last several observed values and zero for 100 and over (e.g., third-degree polynomial to values for 75 and over, 80 and over, 85 and over, and 100 and over). The appropriate population by age at the preceding census may be "aged" to the current census year and the distribution of survivors may then be used to subdivide the current total of the terminal age group. At the beginning of the distribution, to the extent that the quality of the statistics permit, birth statistics, or birth statistics adjusted for deaths, may be employed to subdivide the 5-year totals into single ages.

ADJUSTMENT OF DISTRIBUTIONS TO MARGINAL TOTALS

There are many instances where available distributions of demographic data do not satisfy certain desired marginal totals. The distribution(s) in question may be a univariate distribution or multivariate cross-tabulations. The need to adjust to marginal totals, whether for a single distribution or for a two-dimensional table, may arise in connection with (1) the adjustment of sample data to agree with complete-count data or independent estimates, (2) the estimation of the frequencies in a distribution for a given year on the basis of data for prior years and a total or totals for the given year, (3) the adjustment of the detailed data for a given year(s) and area(s) to presumably more accurate marginal totals for the same year(s) and area(s) obtained from a different source, and (4) the adjustment of the frequencies in reported categories of the variables to absorb the categories designated as not reported.

Commonly, the detailed data and the marginal total or totals are all positive numbers, that is, neither zero nor negative. Zero cells may be encountered frequently in sample data and negative frequencies appear occasionally in demographic data, as for example in series on net migration. Although the procedures for adjusting distributions with negative cells are logical extensions of those with only positive or zero cells, somewhat different arithmetic steps are involved; so that we

---

[12] Wilson H. Grabill, U.S. Bureau of the Census. Grabill derived his coefficients by manipulating and recombining the Sprague coefficients. The Sprague coefficients needed to obtain single-year-of-age x from data for 5-year age groups were given a weight of 10. The Sprague coefficients for successively younger ages $x-1$, $x-2$, $x-3$, etc. were given successive weights of 9, 8, 7, 6, 5, 4, 3, 2, and 1. Those for successively older ages $x+1$, $x+2$, $x+3$, etc. were given successive weights of 9, 8, 7, 6, 5, 4, 3, 2, 1. The weighted coefficients for the 19 ages (age x and 9 ages on each side of age x) were added together and the sum divided by the total of the 1, 2, . . ., 9, 10, 9, . . ., 2, 1 weights.

Figure C–3. – **Population of Mexico, 1960, by Single Years of Age as Enumerated and as Interpolated From 5-Year Age Groups by the Use of the Karup-King Method and the Grabill Method**

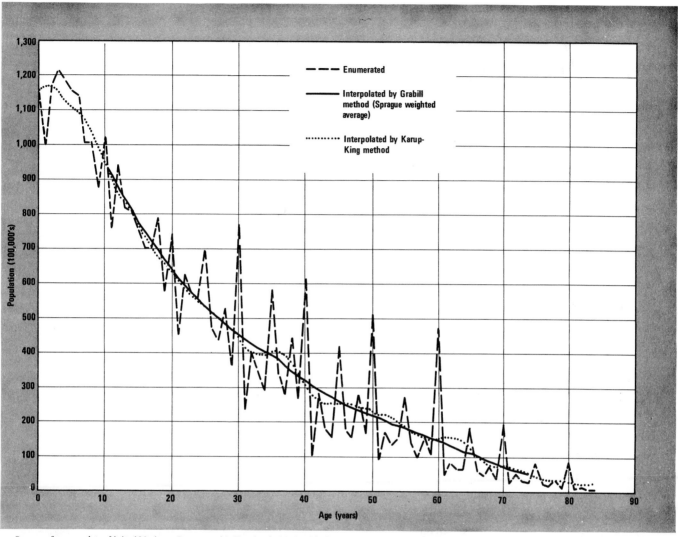

Source of census data: United Nations, *Demographic Yearbook, 1962*, table 6.

shall consider this case separately. We may then outline the types of situations considered here as follows:

I. Single distribution
    a. Frequencies all positive or zero
    b. Frequencies include negatives

II. Two-dimensional table
    a. Frequencies all positive or zero
    b. Frequencies include negatives

In the case of Ib and IIb the marginal totals may be positive, negative, or zero. The marginal totals are not likely to be negative or zero if the basic distribution has only positive frequencies or a combination of positive and zero frequencies.

Single Distribution

Distributions With All Positive or Zero Frequencies. – The simplest case involves a single distribution with only positive frequencies, or positive and zero frequencies, and a positive assigned total. It is illustrated by a distribution of preliminary

postcensal population estimates for States with an independent estimate of the national population or a distribution of births by order of birth including a category "order not stated." In the first case we wish to adjust the preliminary State figures to the independent national total; in the second case we wish to eliminate the "unknowns" by distributing them over the known categories in such a way that the adjusted frequencies will add to the required total. Assuming in the first case that we do not have any information regarding the errors in the preliminary estimates, we may assume that the discrepancy between the sum of the preliminary State estimates and the independent national estimate has a distribution proportionate to the preliminary State estimates. Similarly, assuming in the second case that we have no special information regarding the distribution of the unknowns, we may assume that the unknowns have the same relative distribution as the known categories.

The simplest way of applying the latter assumption to obtain adjusted figures is to multiply each known category in the distribution (n_i) by a factor representing the ratio of the required total (N) to the sum of the frequencies excluding the unknowns (n):

$$N_i = \frac{N}{n} \cdot n_i \qquad (25)$$

Such a proportionate adjustment of a distribution is sometimes called "raking." (This procedure gives exactly the same results as first distributing the unknowns in proportion to the distribution of the known categories and then adding the two distributions.) The proportionate adjustment of a distribution to meet a required total conforms to a least squares adjustment of the original frequencies; that is,

$$\sum_i \frac{(N_i - n_i)^2}{n_i} = \min. \qquad (26)$$

The adjustment of the reported distribution of births by order for Costa Rica in 1963 to include births of unknown order is shown by the following data:

Birth order	Reported births [a]	Adjusted births $=(1) \times 1.002514$
	(1)	(2)
Total, all orders......................	63,798	63,798
First.................................	10,263	10,289
Second...............................	9,114	9,137
Third................................	8,264	8,285
Fourth...............................	7,132	7,150
Fifth.................................	6,174	6,189
Sixth................................	5,121	5,134
Seventh..............................	4,278	4,289
Eighth and over.....................	13,292	13,325
Not stated...........................	160	
Total, excluding not stated........	63,638	

[a] Source: U.N. *Demographic Yearbook*, 1965, table 16.

The required total is 63,798 and the total of the distribution excluding the unknowns is 63,638. The adjustment factor is equal, therefore, to $63{,}798 \div 63{,}638$, or 1.002514. The results obtained (N_i) are unaffected by the number of categories in the distribution, e.g., the adjusted number of births of orders 6 and 7 combined would be the same whether the births of each order were adjusted separately or in combination.

Distributions With Some Negative Frequencies. — Occasionally a distribution which requires adjustment to a marginal total includes negative as well as positive values. This arises when a distribution of an element which "operates" negatively (e.g., deaths, out-migration) is superimposed on the distribution of an element which "operates" positively (e.g., births, in-migration). Distributions such as those of net migration or population change for the States of a country, or natural change for the counties of a State, may have negative cells.

The marginal total for a "plus-minus" distribution may be positive, negative, or zero. In this case the use of a single adjustment factor applied uniformly to all values would yield the required total but the original data would be subject to excessive modification.

A procedure that minimizes the adjustment requires the use of two factors, one for the positive items and one for the negative items. The formulas for the factors are as follows:

Factor for the positive values of n_i:

$$\frac{\Sigma|n_i| + (N - n)}{\Sigma|n_i|} \qquad (27)$$

Factor for the negative values of n_i:

$$\frac{\Sigma|n_i| - (N - n)}{\Sigma|n_i|} \qquad (28)$$

where $\Sigma|n_i|$ represents the sum of the absolute values (i.e., without regard to sign) of the original distribution, N the assigned total, and n the algebraic sum of the original observations. The factor for adjusting the positive items represents the ratio of (a) the sum of the absolute values in the distribution plus the net amount of adjustment required in the distribution to (b) the sum of the absolute values. The factor for adjusting the negative items in the distribution represents the ratio of (a) the excess of the sum of the absolute values over the net amount of adjustment required in the distribution to (b) the sum of the absolute values. The formulas are applied in the same way if the assigned total is zero. We may refer to this procedure as the plus-minus proportionate adjustment procedure.

The application of this procedure is illustrated in table C-1.

Table C-1. — **Application of Plus-Minus Proportionate Adjustment Procedure to a Single Distribution for Estimated Net Migration Between Geographic Divisions in the United States for 1965–66**

Geographic division	Preliminary estimates	Adjusted estimates	
		Divisions as units	States as units
	(1)	(2)	(3)
Total.........	+280,705	+449,463	+449,463
New England.......	+52,947	+71,750	+67,393
Middle Atlantic...	+17,357	+23,521	+33,825
East North Central	+19,589	+26,545	+35,855
West North Central	-57,323	-36,966	-39,010
South Atlantic....	+63,968	+86,684	+90,041
East South Central	-29,181	-18,818	-19,977
West South Central	-2,389	-1,541	+2,285
Mountain..........	-8,361	-5,392	+3,876
Pacific...........	+224,098	+303,680	+275,175

Calculation of adjustment factors	Divisions as units	States as units
n =	+280,705	+280,705
N =	+449,463	+449,463
$N-n$ =	+168,758	+168,758
$\Sigma\lvert n_i\rvert$ =	475,213	755,103
Σ pos. n_i =	377,959	461,930
Σ neg. n_i =	97,254	293,173
Factor for "+" items =	$\frac{475{,}213+168{,}758}{475{,}213}=1.355121$	$\frac{755{,}103+168{,}758}{755{,}103}=1.223490$
Factor for "-" items =	$\frac{475{,}213-168{,}758}{475{,}213}=.644879$	$\frac{755{,}103-168{,}758}{755{,}103}=.776510$

Source of basic data: U.S. Bureau of the Census, unpublished data.

Preliminary estimates of net migration for States in the United States in 1965–66, obtained by projecting the rate of migration for earlier years, were combined into estimates for nine geographic divisions to reduce the detail of our illustration. We want to adjust this distribution, which has a total of $+280{,}705$, to sum to $+449{,}463$, the estimated net immigration to the United States during 1965–66. The first factor, which is applied to the positive items, is $\frac{475{,}213 + 168{,}758}{475{,}213} = 1.355121$. The second factor which is applied to the negative items, is $\frac{475{,}213 - 168{,}758}{475{,}213} = .644879$. In effect, the procedure distributes the amount of adjustment among the items in proportion to their absolute values.

The plus-minus proportionate adjustment procedure described suffers from at least three weaknesses: First, the detail of the given distribution affects the results; that is, that a combination of cells adjusted separately would not show the same result as when the combined category is adjusted directly. If the adjustment procedure is applied to preliminary estimates for the 50 States and the District of Columbia instead of the estimates for the 9 geographic divisions, we obtain substantially different adjusted estimates of net migration for divisions when the adjusted State figures are added up. The figures are shown in column (3) of table C–1.

A second weakness of the procedure is that zero cells in the distribution cannot receive any of the adjustment. A third weakness is that, in the event that the net amount of adjustment required in the distribution $(N - n)$ exceeds the sum of the absolute values in the distribution $\Sigma |n_i|$, one of the adjustment factors will have a negative sign and, hence, will cause all the items in the distribution to which it is applied to reverse signs. The larger positives could become the larger negatives or the larger negatives could become the larger positives. This is untenable since the pattern of the original distribution would thereby be sharply altered. One procedure for avoiding this type of result is to "translate" the original data into a more acceptable form by adding or subtracting a fixed amount to or from each item in the distribution, apply the formulas, and then translate the numbers back by the same amount as used in the first translation.

Two-Dimensional Tables

Situations differ with regard to the availability of marginal totals for two-dimensional tables (bivariate distributions); hence, the adjustment procedures may vary somewhat. In some cases the marginal totals in two dimensions differ from the sums of the distributions. In other cases marginal totals in two dimensions are known but differ in only one dimension from the sums of the distributions. In still other cases marginal totals in neither dimension are known at first but emerge in the process of adjustment; and these totals differ from the sums of the original distributions. We also have the case of a bivariate distribution where only the grand total is fixed, none of the column or row totals being known, and the sum of all the cells differs from the required grand total. Any of these cases may be complicated by a mixture of positive and negative signs in the body of the table, with associated positive, negative, or zero marginal totals.

Tables With All Positive or Zero Frequencies.—The common situation where the sums of both the rows and columns differ from the required marginal totals is that where the cross-tabulations were obtained only for a sample of the population and marginal totals are available from a complete census count. Let us consider the specific case where we have sample statistics for the population by age and marital status and complete-count statistics only for age groups and marital status groups separately. We may employ the following symbols:

i = age group

j = marital status

n_{ij} = sample count for a particular marital status and age group

$n_{i.} = \sum_j n_{ij}$ = sample count for total number in an age group

$n_{.j} = \sum_i n_{ij}$ = sample count for total number in a marital status

$n = \sum_i \sum_j n_{ij}$ = sample count for all marital status and age groups

N_{ij} = estimated complete count for a particular marital status and age group

$N_{i.}$ = complete count for total number in an age group

$N_{.j}$ = complete count for total number in a marital status

$N = \sum_i \sum_j N_{ij}$ = complete count for all marital status and age groups

Then we can set up the following chart:

Age group	Sample figures on marital status				Complete count
	1	$.\quad j$	4	Total	
1		.		$n_{1.}$	$N_{1.}$
2		.		$n_{2.}$	$N_{2.}$
.		.		.	.
i		n_{ij}		$n_{i.}$	$N_{i.}$
5		.		$n_{5.}$	$N_{5.}$
Sample total..........	$n_{.1}$	$.\quad n_{.j}$	$n_{.4}$	n	
Complete count.....	$N_{.1}$	$.\quad N_{.j}$	$N_{.4}$		N

Method of iterative proportions.—The adjustment of the sample data to the complete-count marginal totals may be achieved by iterative sequential raking of the original matrix, alternating with rows and columns, i.e., first horizontally, then vertically, then horizontally, etc., or first vertically, then horizontally, then vertically, etc.[13] The procedure is illustrated in table C–2 with sample data on marital status cross-classified with age, and complete-count data on age and marital status, separately, for males, from the 1960 census of the United States. Each step in the calculations is like that for adjusting single distributions proportionately to an assigned total. We may call this procedure the method of iterative proportions or the Deming method.

We may begin with proportionate adjustment along each row to the marginal row totals $(n_{ij}^{(1)})$, then follow with proportionate adjustment of the results along each column to the marginal column totals $(n_{ij}^{(2)})$. The vertical adjustment throws the figures out of line with respect to the marginal row totals, and so we adjust the figures to the marginal row totals once again $(n_{ij}^{(3)})$. The readjustment of the rows to the marginal row totals then throws the figures out of line with respect to the marginal column totals, and so we adjust figures to the marginal column totals once again $(n_{ij}^{(4)})$. If this sequence of adjustments is continued, it produces complete convergence to a final unchanging matrix of numbers which add to the required row and column totals at once; at this point further raking merely replicates the numbers within a few digits.[14] The convergence to marginal row and column totals is usually rapid (though the rate of convergence has not been established mathematically), requiring only about two cycles (each consisting of a vertical adjustment and a horizontal adjustment) to

---

[13] W. Edwards Deming, *Statistical Adjustment of Data*, New York, John Wiley and Sons, September 1948, chapter 7, pp. 96–127. See also W. Edwards Deming and Frederick F. Stephan, "On a Least Squares Adjustment of a Sample Frequency Table When the Expected Marginal Totals are Known," *Annals of Mathematical Statistics*, 11(4):427–444, December 1940.

[14] C. T. Ireland and S. Kullback, "Contingency Tables with Given Marginals," *Biometreka*, 55(1):179–188, March 1968.

Table C–2. — Sample Data on the Male Population of the United States, by Age and Marital Status, 1960, Adjusted to Marginal Totals From the Complete Count by the Method of Iterative Proportions

[Adjustments were not made to eliminate minor discrepancies due to rounding]

Age (years)	Single	Married	Widowed	Divorced	$n_{i.}$	$N_{i.}$	$\dfrac{N_{i.}}{n_{i.}}$
	Basic data						
14–19............	7,831,612	262,133	1,947	5,869	8,101,561	8,022,608	.99025459
20–24............	2,807,784	2,417,552	4,780	53,112	5,283,228	5,272,340	.99793914
25–44............	2,755,476	19,527,926	103,230	526,128	22,912,760	22,934,692	1.00095720
45–64............	1,354,577	15,248,176	562,768	543,303	17,708,824	17,629,318	.99551037
65 and over......	564,373	5,174,635	1,399,185	170,792	7,308,985	7,503,097	1.02655800
$n_{.j}$	15,313,822	42,630,422	2,071,910	1,299,204	61,315,358	(X)	
$N_{.j}$	15,412,733	42,416,979	2,219,355	1,312,988	(X)	61,362,055	
$N_{.j} \div n_{.j}$[1]	(1.00645894	.99499318	1.07116381	1.01060957)			
	First adjustment (rows), $n_{ij}^{(1)}$				$n_{i.}^{(1)}$	$N_{i.}$	
14–19............	7,755,290	259,578	1,928	5,812	8,022,608	8,022,608	
20–24............	2,801,998	2,412,570	4,770	53,003	5,272,341	5,272,340	
25–44............	2,758,114	19,546,618	103,329	526,632	22,934,693	22,934,692	
45–64............	1,348,495	15,179,717	560,241	540,864	17,629,317	17,629,318	
65 and over......	579,362	5,312,063	1,436,345	175,328	7,503,098	7,503,097	
$n_{.j}^{(1)}$	15,243,259	42,710,546	2,106,613	1,301,639	61,362,057	(X)	
$N_{.j}$	15,412,733	42,416,979	2,219,355	1,312,988	(X)	61,362,055	
$N_{.j} \div n_{.j}^{(1)}$........	1.01111796	.99312659	1.05351814	1.00871901			
	Second adjustment (columns), $n_{ij}^{(2)}$				$n_{i.}^{(2)}$	$N_{i.}$	$\dfrac{N_{i.}}{n_{i.}^{(2)}}$
14–19............	7,841,513	257,794	2,031	5,863	8,107,201	8,022,608	.98956570
20–24............	2,833,151	2,395,987	5,025	53,465	5,287,628	5,272,340	.99710872
25–44............	2,788,779	19,412,266	108,859	531,224	22,841,128	22,934,692	1.00409630
45–64............	1,363,488	15,075,381	590,224	545,580	17,574,673	17,629,318	1.00310930
65 and over......	585,803	5,275,551	1,513,216	176,857	7,551,427	7,503,097	.99359989
$n_{.j}^{(2)}$	15,412,734	42,416,979	2,219,355	1,312,989	61,362,057	(X)	
$N_{.j}$	15,412,733	42,416,979	2,219,355	1,312,988	(X)	61,362,055	
	Third adjustment (rows), $n_{ij}^{(3)}$				$n_{i.}^{(3)}$	$N_{i.}$	
14–19............	7,759,692	255,104	2,010	5,802	8,022,608	8,022,608	
20–24............	2,824,960	2,389,060	5,010	53,310	5,272,340	5,272,340	
25–44............	2,800,203	19,491,784	109,305	533,400	22,934,692	22,934,692	
45–64............	1,367,727	15,122,255	592,059	547,276	17,629,317	17,629,318	
65 and over......	582,054	5,241,787	1,503,531	175,725	7,503,097	7,503,097	
$n_{.j}^{(3)}$	15,334,636	42,499,990	2,211,915	1,315,513	61,362,054	(X)	
$N_{.j}$	15,412,733	42,416,979	2,219,355	1,312,988	(X)	61,362,055	
$N_{.j} \div n_{.j}^{(3)}$........	1.00509285	.99804680	1.00336360	.99808060			
	Twentieth adjustment (columns), $n_{ij}^{(20)}$				$n_{i.}^{(20)}$	$N_{i.}$	$\dfrac{N_{i.}}{n_{i.}^{(20)}}$
14–19............	7,763,225	251,797	1,994	5,726	8,022,742	8,022,608	.99998330
20–24............	2,842,667	2,371,788	5,001	52,920	5,272,376	5,272,340	.99999317
25–44............	2,833,497	19,458,984	109,693	532,452	22,934,626	22,934,692	1.00000288
45–64............	1,384,617	15,103,647	594,432	546,552	17,629,247	17,629,318	1.00000405
65 and over......	588,727	5,230,763	1,508,235	175,339	7,503,063	7,503,097	1.00000453
$n_{.j}^{(20)}$	15,412,733	42,416,979	2,219,355	1,312,988	61,362,055	(X)	
$N_{.j}$	15,412,733	42,416,979	2,219,355	1,312,988	(X)	61,362,055	

X Not applicable.
[1] Not used in the calculations for adjusting the sample data.

Source of basic data: *U.S. Census of Population, 1960,* Vol. I, *Characteristics of the Population,* Part 1, *United States Summary,* tables 46, 49, and 176.

achieve close agreement in one dimension and precise agreement in the other.

In our illustration two cycles accounted for a substantial degree of convergence to the row totals, but even after 10 cycles perfect agreement with these totals was not achieved. The procedure is clearly laborious by hand calculation, but the operation is quite simple to carry out by electronic computer without concern for the number of cycles required. The calculation can be stopped at any required degree of agreement with marginal totals.

Deming has demonstrated that the procedure described is an approximation to a least squares solution to the problem (i.e., $\Sigma(N_{ij} - n_{ij})^2/n_{ij}$ is a minimum). Although the final results do not theoretically coincide with the least squares solution, in

Table C-3. — Procedure for Distributing Births of Order Not Stated and Age Not Stated, for Costa Rica, 1963

[Arbitrary adjustments were made to eliminate minor discrepancies due to rounding after the row adjustment and again after the column adjustment]

Birth order	All ages $(m_{i.})$	Age of mother (years)							Not stated	Sum excluding not stated $(n_{i.})$	Factor= $m_{i.}/n_{i.}$
		15-19[1]	20-24	25-29	30-34	35-39	40-44	45-49[2]			
					Recorded data, n_{ij}						
All orders..................	63,798	7,893	17,760	15,623	11,315	7,671	2,649	372	515	63,283	
First.......................	10,263	4,625	3,754	1,224	389	155	53	9	54	10,209	1.005289
Second......................	9,114	2,160	4,288	1,755	585	229	43	8	46	9,068	1.005073
Third.......................	8,264	838	3,961	2,173	854	299	61	17	61	8,203	1.007436
Fourth......................	7,132	199	2,856	2,426	1,065	428	81	13	64	7,068	1.009055
Fifth.......................	6,174	38	1,631	2,549	1,279	498	114	15	50	6,124	1.008165
Sixth.......................	5,121	10	713	2,144	1,407	636	147	14	50	5,071	1.009860
Seventh.....................	4,278	-	322	1,578	1,450	708	163	16	41	4,237	1.009677
Eighth and over.............	13,292	-	196	1,746	4,269	4,708	1,984	277	112	13,180	1.008498
Not stated..................	160	23	39	28	17	10	3	3	37	123	1.300813
					Row adjustment, $n_{ij}^{(1)}$						
All orders $(m_{.j}^1)$.............	63,798	7,944	17,896	15,756	11,416	7,739	2,672	375			
First.......................	10,263	4,650	3,774	1,230	391	156	53	9			
Second......................	9,114	2,171	4,310	1,764	588	230	43	8			
Third.......................	8,264	844	3,992	2,189	860	301	61	17			
Fourth......................	7,132	201	2,882	2,447	1,075	432	82	13			
Fifth.......................	6,174	38	1,644	2,570	1,290	502	115	15			
Sixth.......................	5,121	10	720	2,166	1,421	642	148	14			
Seventh.....................	4,278	-	325	1,593	1,464	715	165	16			
Eighth and over.............	13,292	-	198	1,761	4,305	4,748	2,001	279			
Not stated..................	160	30	51	36	22	13	4	4			
Sum excluding not stated $(n_{.j}^{(1)})$		7,914	17,845	15,720	11,394	7,726	2,668	371			
Factor = $m_{.j}^{(1)}/n_{.j}^{(1)}$..........		1.003791	1.002858	1.002290	1.001931	1.001683	1.001499	1.010782			
				Column adjustment, $n_{ij}^{(2)} = M_{ij}$							
All orders..................	63,798	7,944	17,896	15,756	11,416	7,739	2,672	375			
First.......................	10,296	4,668	3,785	1,233	392	156	53	9			
Second......................	9,139	2,179	4,322	1,768	589	230	43	8			
Third.......................	8,286	847	4,003	2,194	862	302	61	17			
Fourth......................	7,150	202	2,890	2,453	1,077	433	82	13			
Fifth.......................	6,188	38	1,649	2,576	1,292	503	115	15			
Sixth.......................	5,131	10	722	2,170	1,424	643	148	14			
Seventh.....................	4,287	-	326	1,597	1,467	716	165	16			
Eighth and over.............	13,321	-	199	1,765	4,313	4,756	2,005	283			

—Represents zero.   [1] Includes births to women under age 15.   [2] Includes births to women aged 50 and over.

Source of basic data: United Nations, *Demographic Yearbook, 1965*, table 16.

practice they usually conform to it rather closely. The mathematical proof that the same results are obtained whether the columns or the rows are "raked" first is not given by Deming but empirical testing indicates that the order of raking is immaterial.

In some cases the row and column totals are not fixed in advance but are developed in the process of adjusting the data. A general situation of this kind is that of a table in two dimensions with "unknown" categories in both dimensions which have to be distributed among the "known" categories. An abbreviated type of sequential raking is applied, covering only one cycle of adjustments and beginning either with the rows or the columns. This procedure is illustrated in table C-3, which shows data on births classified by age of mother and order of birth for Costa Rica in 1963. First, the figures were adjusted by rows so as to distribute births of unknown age over each order, including "order unstated." Second, the figures were adjusted by columns so as to distribute births of unknown order over each age group. No further adjustment is required, since the figures now fully incorporate the unknown frequencies and agree with the grand total. The order of adjustment in this problem affects the results; but, assuming that the frequencies of the unknown categories are not very large, the difference in results is inconsequential.

Tables With Some Negative Frequencies. — An earlier discussion indicated some of the complexities of the problem when single distributions with negatives are to be adjusted. Adjustment of bivariate distributions including negative values may be accomplished by an extension of procedures already described. For example, the method of iterative proportions may employ a two factor plus-minus proportionate procedure for each column and row.

COHORT ANALYSIS

Period Data Versus Cohort Data

We have referred several times to the distinction between period data and cohort data and have employed a form of analysis called cohort analysis. Because cohort analysis is useful as a general approach to many problems of demographic measurement and analysis, it seems desirable to consider it here in more detail. The discussion will emphasize the general features and the pervasive application and utility of the cohort approach.

A cohort consists of a group of individuals who experienced the same significant demographic event during a specified brief period of time, usually a year, and who may be identified as a group at successive later dates on the basis of this common demographic experience. Examples are a birth cohort, persons born during the same year or years; a marriage cohort, persons married during the same year or years; a first-parity cohort, women who had a first child in a given year; a school grade cohort, persons entering or leaving a significant grade of school (e.g., entering first grade, graduating from high school or college) during the same year or years; a migration cohort, persons moving into (or out of) an area (country or State) during the same year or years; a labor force cohort, persons who entered (or left) the labor force in a given year; and a cohort of deceased persons, persons who died in a given year The most common and the most important type of cohort considered in demographic studies is a birth cohort.

Since cohorts are observed in the successive years since the original event defining the cohort, cohort data include the observations relating to the cohort in these successive years. Cohorts may be identified in alternative terms at later dates in their history. For example, birth cohorts may be identified in terms of ages at later dates, marriage cohorts in terms of duration of married life (for the married population), migration cohorts in terms of length of residence or years since migration (for the migrant population), etc.

Cohort data may be absolute data or rates. Cohort data may be represented by (1a) a schedule of occurrences of events such as births, deaths, or marriages for successive years tabulated by age (birth cohorts) or (1b) the corresponding age-specific rates; (2a) age-specific data for various demographic characteristics for successive years, such as the numbers enrolled, in the labor force, or ever married (birth cohorts) or (2b) the corresponding age-specific rates; and (3a) data relating to some event (marriage, immigration, entry into labor force, etc.) occurring in successive years tabulated by years since the event occurred for successive years (marriage cohort, immigration cohort, worker cohort, etc.) or (3b) the corresponding rates. All data on demographic characteristics may represent cohort data in the sense that the elements in each set relate to the results of events, such as entry into or withdrawal from school, the labor force, or marriage, which have occurred to a birth or other type of cohort.

Period data, on the other hand, relate to a type of demographic event which occurred over a great number of years (i.e., involving many cohorts) but which is observed during a specified brief period of time (usually one year). Period data (or cross-sectional data) may be represented by the same types of data as cohort data, therefore; the basic difference is in how the data are grouped. Period data relate also to data on population characteristics observed in a single year involving many cohorts. Period data provide a current annual "picture" with respect to the event or characteristic under consideration

by age (period since birth) or duration (period since the event), whereas cohort data provide a "lifetime picture" for a cohort or a time series of cohorts with respect to the relevant variable.

Period Analysis Versus Cohort Analysis

Cohort analysis is defined by Ryder as ". . . quantitative description of data occurrences from the time a cohort is exposed to the risk of such occurrences."[15] Period analysis, on the other hand, involves a quantitative description and analysis of the data for the many cohorts observed during a specified time interval, such as a year, with respect to some variable.

Time series analysis of cohorts consists in examining the changing characteristics from cohort to cohort (e.g., birth cohorts) with respect to some variable (e.g., fertility, marriage, mortality). Summary measures describing each cohort, defined by the year in which the demographic event occurred, can be calculated and compared with similar measures for earlier and later cohorts, in order to observe and analyze the variations over time in the demographic behavior of the population.

Although cohort analysis has been applied in a number of fields of social science, the most extensive development and use of cohort analysis have occurred in the field of demography. Within the field of demography cohort analysis has been utilized primarily in the study of fertility. The difference between the "total fertility rate" and the "completed fertility rate" may be used here to illustrate the relationship between period analysis and cohort analysis. The period total fertility rate, which, as may be recalled, is obtained by summing the age-specific birth rates for a particular year, is a period measure since the age-specific rates refer to several different birth cohorts and to the same year. On the other hand, the cohort completed fertility rate, which is also computed by summing a set of age-specific birth rates, is a cohort measure since the age-specific rates refer to a single birth cohort and to many different years, i.e., successive ages in successive calendar years. In the following sketch the total fertility rate is represented by the sum of values on a vertical line and the completed fertility rate by the sum of values on a diagonal line:

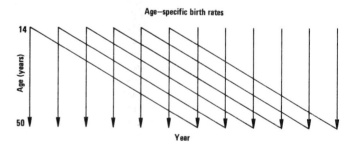

The issue of when to use cohort analysis and when to use period analysis is often dependent on the type of question to be answered. Owing to the problem of the availability of data, however, the dilemma is often not a real one. Adequate historical data are often not available for carrying out a cohort analysis of a particular type. On the other hand, the data are sometimes available for carrying out both types of analysis, and it is usually useful to do so. Because observations for several cohorts recorded in a single year may be combined as if they were the observations of a single cohort recorded over several

---

[15] Norman B. Ryder, "Cohort Analysis," in *The International Encyclopedia of the Social Sciences*, Vol. 2, 1968, p. 546.

years, period aggregations are often referred to as "synthetic" or hypothetical cohorts. We may distinguish, therefore, **real cohorts** from **synthetic cohorts;** the latter are represented by period data combined in such a way as to reflect hypothetical experience over a span of years or a lifetime. Accordingly, the total fertility rate may be described as a measure for a synthetic cohort and the completed fertility rate as a measure for a real cohort.

The utility of cohort analysis rests on the hypothesis that the rate of occurrence of certain demographic events (e.g., fertility, mortality, migration) in the lifetime of groups of individuals (e.g., a birth cohort, marriage cohort) is dependent on the previous experience or behavior of the group with respect to the occurrence of this type of event. It has permitted a better understanding and fuller explanation of annual changes in the event under study than period analysis alone, shedding light on some of the interacting factors affecting period data. It is also a valuable way of structuring past data for purposes of developing projections of the basic series. Cohort analysis is useful in evaluating census data; it can effectively be applied to data on characteristics which hardly change after a certain age (e.g., grade attainment) since the value for a particular birth cohort at a particular census can easily be derived from an earlier or later census for comparison with the figure in the census under study. Cohort analysis of census data may also be useful in historical reconstruction of the rate of occurrence of certain events in the past when records were not available.

Despite the considerations apparently favoring use of cohort analysis, analysis based upon period data has persisted for three reasons. First, the convenience of using period data often overshadows its weaknesses. Second, analysis based on period data supplies us with a "snapshot" of the current situation, in which there may be direct interest. (Measures for cohorts do not have a specific time reference and measures for completed cohorts tend to be marginally relevant to current experience.) Third, period data alone may be available for carrying out a "cohort" analysis, that is, for a synthetic cohort.

The basic data available for a given problem may be the frequencies or rates of occurrence of some event (e.g., birth, death, or marriage). Summary measures for the cohort (e.g., median age of childbearing, death, or marriage; median duration of marriage; median duration of work experience) may then be directly calculated. In other cases, the available data may refer to the current status of the population resulting from the event (e.g., number or percent enrolled, ever-married, or in the labor force) rather than the frequency or rate of occurrence of the event itself. In the latter situation we may describe the calculations for securing summary measures for the cohort as indirect.

The individual annual rates which are combined in order to describe a cohort's cumulative experience may not be in cohort form. A series of age-specific birth rates may be combined to represent a cohort's lifetime fertility experience, but the individual rates here are central birth rates, not "cohort" birth rates or birth probabilities. The two types of rates in this case are very nearly the same, however. Age-parity-specific birth probabilities, (life table) mortality rates and survival rates, and other true probabilities which relate events to the population actually exposed to risk are in cohort form. Effective cohort analysis can be often based on central rates, however.

Age, a key demographic variable, and time, a key dimension of cohort analysis, are so closely related that analysis in terms of cohorts defined by year of birth or changing age is important in the study of most demographic topics. If we want to note a given birth cohort's lifetime experience with respect to fertility, mortality, migration, etc., we simply examine the rates for this variable at each later age and date. We can also carry out an analysis for cohorts identified by some demographic experience other than birth (e.g., women marrying in a certain year) and note this cohort's subsequent demographic experience. It should be clear that cohort analysis of some demographic event or characteristic may be carried out for either birth cohorts or cohorts identified on the basis of other demographic events.

Sources of Cohort Data

Data useful for cohort analysis may be obtained from any of the standard sources of demographic data. Since most countries do not have satisfactory population registers or registration systems, surveys or censuses are a common source of cohort data. Depending on the survey design, surveys may provide panel data, retrospective data, or repetitive observations on the event or status for different "samples" of persons. The registration system provides principally the last type of data. All three kinds of data and the corresponding forms of analysis have been used extensively in demographic studies. Cohort analysis may be based on all three types of data, but the first two lend themselves particularly to this approach.

In **panel-study analysis**, information is collected from the same people at various time intervals in a number of surveys. This method is often referred to also as **longitudinal analysis** although the term "longitudinal" is sometimes used to encompass all three types of data listed above.[16] In **retrospective studies** data are obtained in one or more surveys from respondents who recall experiences at a series of earlier dates or ages pertinent to the topic under consideration.[17] Several important problems are associated with this type of study. First, the cohort experience is complete for some cohorts and incomplete for others. Hence, it becomes necessary to repeat such studies at periodic intervals if one is to secure information for a series of completed cohorts. Second, recall lapse is considerable; people simply forget many things that happened to them or misplace events in time. A third and quite important disadvantage of retrospective studies arises from the fact that data affected by deaths, dissolutions of marriages and households, and departures are frequently not available or are not satisfactorily reported. For example, the rates for cohorts for which completed experience is reported may be affected by differential mortality with respect to the variable under consideration. (The "true" rate of children ever born for women 60 to 64 years of age may be in error because of differential mortality in relation to family size.)

Fields of Cohort Analysis

Although cohort analysis has been most intensively applied in the study of fertility, it has also received limited application in the study of mortality, migration, and marriage and, to a lesser extent, in the study of educational and economic characteristics. If we also consider the applications involving

---

[16] See, for example, Nathan Goldfarb, *Introduction to Longitudinal Statistical Analysis: The Method of Repeated Observations From a Fixed Sample*, Glencoe, Ill., Free Press, 1960.

[17] See, for example, U.S. Bureau of the Census, *Current Population Reports, Population Characteristics*, Series P-20, No. 108, "Marriage, Fertility, and Childspacing: August 1959," July 12, 1961, and Series P-20, No. 186, "Marriage, Fertility, and Childspacing: June 1965," August 6, 1969.

Figure C–4. —Death Rates for Tuberculosis, All Forms, by Age, in the Years 1880 and 1930 and for the Birth Cohorts of 1870–80, for Massachusetts

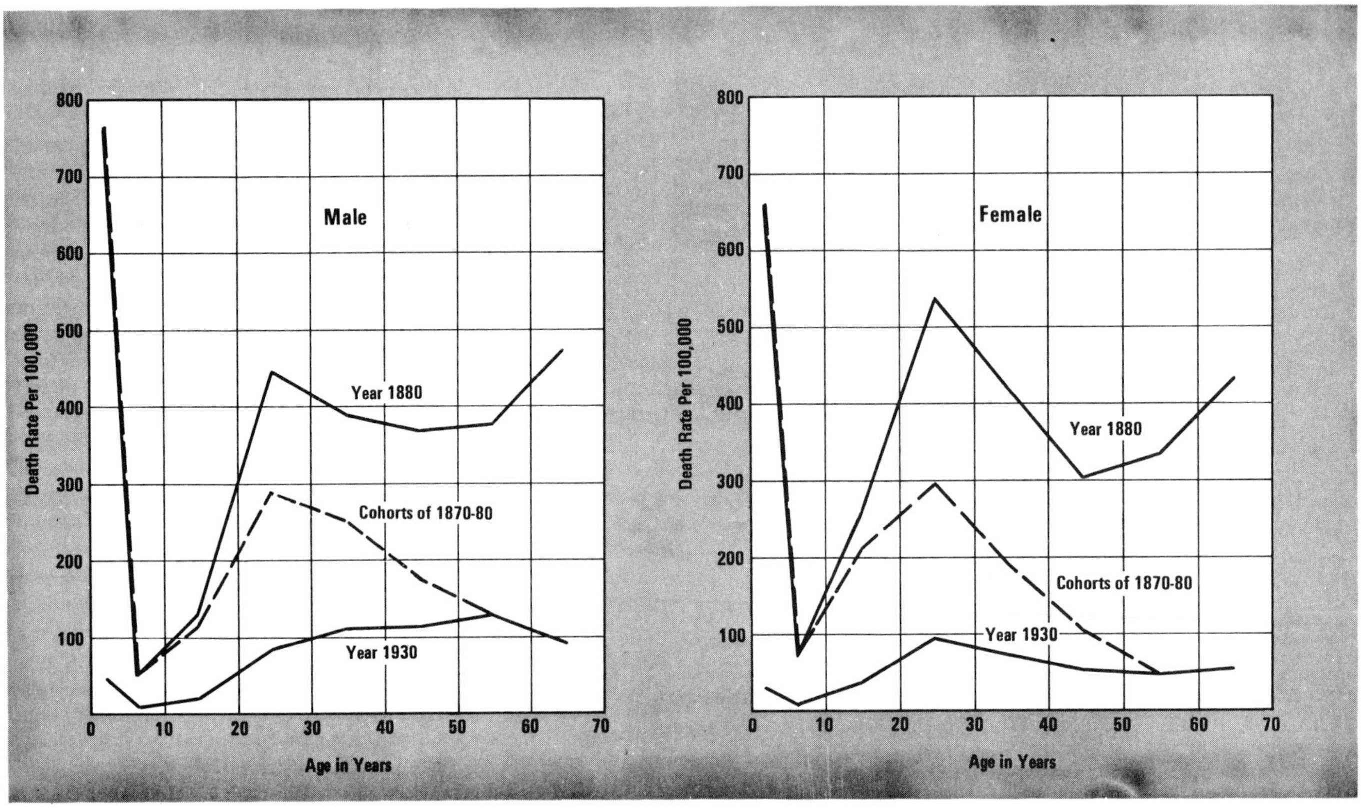

Source: Wade Hampton Frost, "The Age Selection of Mortality from Tuberculosis in Successive Decades," *Milbank Memorial Quarterly*, 18(1):64, January 1940.

analysis of synthetic cohorts, we should include all of the fields to which life table and multiple decrement table techniques have been applied. We might briefly consider some of these areas of application in turn.

Mortality.—The cohort method has been utilized in mortality studies in order to arrive at an understanding of changes in mortality with advancing age. The use of cohort analysis has revealed changes in age patterns of mortality which were not evident from analysis of period data.

There are basically three ways of compiling and analyzing cohort data on mortality — **complete cohort analysis, truncated cohort analysis,** and **analysis of segmented generations.** In complete cohort analysis a group of persons born in the same time period are followed until the death of the last survivor. This method is used in constructing so-called cohort or generation life tables.[18] In the cohort or generation life table, the mortality experiences of a particular cohort of births is followed over a period of many years from infancy to the oldest age lived by the cohort. A generation life table could be constructed, for example, which would show the mortality actually experienced throughout the lifetime of its individual members by the cohort of males born in 1850. Since the generation life table is based on the mortality rates by age for a group of persons born in the same time interval, it presents a historical record of mortality actually experienced by a single cohort.

The difficulty with this type of analysis is that a period of about 100 years must elapse before the death of the last survivor. This difficulty can be partially alleviated by truncating the cohort experience, i.e., covering the cohort experience only up to a certain age. For example, mortality up to ages 5, 50, and 70 for a number of cohorts may be analyzed. Another approach involves studying the experience of a cohort between two points in time. In such a study of a "segmented generation," data for a number of age cohorts existing at a particular point in time are analyzed between this point and a subsequent point in time.[19] This approach has the advantage of requiring data for a relatively short period of time without a concomitant loss of information.

Cohort analysis has been effectively applied in cause-of-death studies. This approach has made possible, for example, a clearer understanding of the changes in tuberculosis mortality in the United States.[20] Cross-sectional analysis using period data indicated a rise in the modal age of tuberculosis mortality. However, an examination of age-specific mortality rates for tuberculosis on a cohort basis revealed that the age pattern from cohort to cohort was quite similar and that the rise in the modal age of tuberculosis mortality was due to a decline in the level of mortality from tuberculosis from cohort to cohort (figure C–4).

[18] Paul H. Jacobson, "Cohort Survival for Generations Since 1840," *Milbank Memorial Fund Quarterly*, 42(3):36–53, July, 1964; L. I. Dublin and Mortimer Spiegelman, "Current Versus Generation Life Tables," *Human Biology*, 13(4):439–458, December 1941.

[19] Mortimer Spiegelman, "Segmented Generation Mortality," *Demography*, 6(2):117–124, 1969.
[20] Wade Hampton Frost, "The Age Selection of Mortality From Tuberculosis in Successive Decades," *The Milbank Memorial Fund Quarterly*, 18(1):61–66, January 1940.

Mortality data may also be subjected to analysis on a synthetic cohort basis. The conventional life table is the most common means of analyzing mortality data in this way. The conventional table employs a set of age-specific mortality rates all of which relate to a single year or group of years. Hence, if a life table is designed to relate to the 3-year period centering on 1960, the 3-year averages of age-sex-specific mortality rates for 1959–61 are the basis for this life table. The schedule of rates presents a picture of the level of mortality as of one brief period of time, e.g., one year, three years, etc. Although it may sum up the mortality level for this period, it presents a fictitious picture of mortality for any real cohort since no single cohort has experienced or will experience this particular schedule of mortality rates throughout its life span.

Fertility. — Fertility may be analyzed for birth cohorts of women, marriage cohorts of women, or both in combination.[21] The fertility of birth or marriage cohorts may be described by cumulative fertility rates, the completed fertility rate, and the median or mean value of the series (median or mean age of childbearing, or median or mean years of married life for mothers).

Divergence between the period total fertility rate and the cohort completed fertility rate based on age data may be due to the temporal change in the mean age at childbearing. Ryder notes, for example, that a declining cohort mean age at childbearing results in a period total fertility rate that is higher than the completed fertility rate for the cohort whose childbearing is centered in that period, and vice versa.[22] He explains this relationship as follows: "The fertility occurring in any year depends on the degree of overlap in the age spans of childbearing of the successive cohorts represented in that year. A progressively younger mean age of fertility implies an increase in the extent of this overlap, which is manifested in an apparent increase in the amount of fertility from period to period even in the absence of change in the cohort total fertility rate."

Since cohort analysis of fertility was discussed in some detail in chapter 16, reference should be made to that chapter for other illustrations pertinent to that field. Additional discussion may be found in the readings below.[23]

Migration. — In the analysis of migration, cohort experience may be defined either in terms of a birth cohort[24] or in terms of a migration cohort.[25]

In the former, migration is viewed in relation to the life cycle of the individual, and such migration characteristics of the cohort may be calculated as an array of its age-specific migration rates, cumulative migration rates over a part or all of the age range, and the median or mean age at migration. In this connection, migration may be viewed as a demographic "risk" like marriage, fertility, and mortality in the life cycle of an individual or cohort of individuals. Emphasis in cross-classification is on age of migrants.

When we are dealing with migration cohorts, on the other hand, the emphasis is on length of stay or duration of residence in a given area. Data for year-of-immigration cohorts or in-migration cohorts in a single census may be presented either in the form of duration of residence or year of migration. From residence histories secured in a census or survey, data may be obtained jointly for birth cohorts and migration cohorts, individuals may be classified both by duration of residence and age, and migration rates may be computed by age and duration of residence.

Other Fields. — Cohort analysis has been applied to the study of educational progression after entry into school, occupational careers after entry into the labor force, length of stay after entry into a hospital or institution, morbidity histories after first exposure to a condition, and many other demographic fields.[26]

[21] See, for example, E. G. Jacoby, "A Fertility Analysis of New Zealand Marriage Cohorts," *Population Studies*, 12(1):18–39, July 1958; and U.S. Bureau of the Census, *Current Population Reports*, Series P–20, No. 186.

[22] Ryder, op. cit., p. 547.

[23] Otis Dudley Duncan and Robert W. Hodge, "Cohort Analysis of Differential Natality," *International Population Conference*, International Union for the Scientific Study of Population, New York, Vol. 1, 1961, pp. 59–66; Wilson H. Grabill, Clyde V. Kiser, and Pascal K. Whelpton, *The Fertility of American Women*, chapter 9, New York, John Wiley & Sons, 1958; L. Osadnik, "Method of Cohort Analysis for Studying Fertility Factors," in Egon Szabady (ed.), *Studies on Fertility and Social Mobility*, Budapest, Akademiai Kiado, 1964, pp. 40–43; N. B. Ryder, "Problems of Trend Determination During a Transition in Fertility, *Milbank Memorial Fund Quarterly*, 34(1):5–21, January 1956; idem,

"La mesure des variations de la fécondité au cours du temps," *Population* (Paris), 11(1):29–46, January–March 1956; idem, "The Structure and Tempo of Current Fertility," in National Bureau of Economic Research, *Demographic and Economic Change in Developed Countries*, Princeton, N.J., Princeton University Press, 1960, pp. 117–136; idem, "The Process of Demographic Translation," *Demography*, 1:74–82, 1964; and Pascal K. Whelpton, *Cohort Fertility: Native White Women in the United States*, Princeton, N.J., Princeton University Press, 1954.

[24] See, for example, Hope T. Eldridge, "A Cohort Approach to the Analysis of Migration Differentials," *Demography*, 1(1):212–219, 1964; Henry S. Shryock, Jr., and Elizabeth A. Larmon, "Some Longitudinal Data on Internal Migration," *Demography*, 2:579–592, 1965; Karl E. Taeuber, "Cohort Migration," *Demography*, 3(2):416–422, 1966; and idem, "Cohort Population Redistribution and the Urban Hierarchy," *Milbank Memorial Fund Quarterly*, 43(2):450–462, October 1965.

[25] See, for example, Kenneth C. Land, "Duration of Residence and Prospective Migration: Further Evidence," *Demography*, 6(2):133–140, May 1969; Robert McGinnis, "A Stochastic Model of Migration," *American Sociological Review*, 33(5):712–722, October 1968; Peter A. Morrison, "Duration of Residence and Prospective Migration: The Evaluation of a Stochastic Model," *Demography*, 4(2):553–561, 1967; and idem, "Probabilities from Longitudinal Records," in Edgar F. Borgatta (ed.), *Sociological Methodology, 1969*, San Francisco, Jossey-Bass, Inc., 1969.

[26] See, for example, United Nations, *Methods of Analysing Census Data on Economic Activities of the Population*, Series A, Population Studies, No. 43, New York, 1968, pp. 33–39 and 93–100; and Morton Robins and Rose Sachs, "An Application of Biometry to Hospital Statistics: Hospital Stay Tables," *American Journal of Public Health*, 43(2):175–184, February 1953.

Table C-4. — Interpolation Coefficients Based on the Karup-King Formula

[The Karup-King formula is a four-term third-difference osculatory formula. It maintains the given values. Given points or groups must be equally spaced]

A. FOR INTERPOLATION BETWEEN GIVEN POINTS AT INTERVALS OF 0.2

Interpolated point	$N_{1.0}$	$N_{2.0}$	$N_{3.0}$	$N_{4.0}$
	Coefficients to be applied to--			
First interval				
$N_{1.0}$	+1.000	.000	.000	.000
$N_{1.2}$	+.656	+.552	-.272	+.064
$N_{1.4}$	+.408	+.856	-.336	+.072
$N_{1.6}$	+.232	+.984	-.264	+.048
$N_{1.8}$	+.104	+1.008	-.128	+.016
Middle interval				
$N_{2.0}$	.000	+1.000	.000	.000
$N_{2.2}$	-.064	+.912	+.168	-.016
$N_{2.4}$	-.072	+.696	+.424	-.048
$N_{2.6}$	-.048	+.424	+.696	-.072
$N_{2.8}$	-.016	+.168	+.912	-.064
Last interval				
$N_{3.0}$	.000	.000	+1.000	.000
$N_{3.2}$	+.016	-.128	+1.008	+.104
$N_{3.4}$	+.048	-.264	+.984	+.232
$N_{3.6}$	+.072	-.336	+.856	+.408
$N_{3.8}$	+.064	-.272	+.552	+.656
$N_{4.0}$	.000	.000	.000	+1.000

B. FOR SUBDIVISION OF GROUPS INTO FIFTHS

Interpolated subgroup	G_1	G_2	G_3
	Coefficients to be applied to--		
First panel			
First fifth of G_1	+.344	-.208	+.064
Second fifth of G_1	+.248	-.056	+.008
Third fifth of G_1	+.176	+.048	-.024
Fourth fifth of G_1	+.128	+.104	-.032
Last fifth of G_1	+.104	+.112	-.016
Middle panel			
First fifth of G_2	+.064	+.152	-.016
Second fifth of G_2	+.008	+.224	-.032
Third fifth of G_2	-.024	+.248	-.024
Fourth fifth of G_2	-.032	+.224	+.008
Last fifth of G_2	-.016	+.152	+.064

B. FOR SUBDIVISION OF GROUPS INTO FIFTHS--Continued

Interpolated subgroup	G_1	G_2	G_3
	Coefficients to be applied to--		
Last panel			
First fifth of G_3	-.016	+.112	+.104
Second fifth of G_3	-.032	+.104	+.128
Third fifth of G_3	-.024	+.048	+.176
Fourth fifth of G_3	+.008	-.056	+.248
Last fifth of G_3	+.064	-.208	+.344

C. FOR SUBDIVISION OF GROUPS INTO TENTHS OR HALVES

Interpolated subgroup	G_1	G_2	G_3
	Coefficients to be applied to--		
First tenth of G_2	+.0405	+.0640	-.0045
Second tenth of G_2	+.0235	+.0880	-.0115
Third tenth of G_2	+.0095	+.1060	-.0155
Fourth tenth of G_2	-.0015	+.1180	-.0165
Fifth tenth of G_2	-.0095	+.1240	-.0145
Sum of coefficients for first five-tenths = coefficients for first half of G_2	+.0625	+.5000	-.0625
Sixth tenth of G_2	-.0145	+.1240	-.0095
Seventh tenth of G_2	-.0165	+.1180	-.0015
Eighth tenth of G_2	-.0155	+.1060	+.0095
Ninth tenth of G_2	-.0115	+.0880	+.0235
Last tenth of G_2	-.0045	+.0640	+.0405
Sum of coefficients for last five-tenths = coefficients for last half of G_2	-.0625	+.5000	+.0625

NOTE. — Interpolation coefficients in tables C-4, C-5, and C-8 were computed by Wilson H. Grabill, U.S. Bureau of the Census, from the basic formulas. Interpolation coefficients in tables C-6 and C-7 are reproduced from the sources cited in these tables.

Table C–5.—Interpolation Coefficients Based on the Sprague Formula

[The Sprague formula is a six-term fifth-difference osculatory formula. It maintains the given values. Given points or groups must be equally spaced]

A. FOR INTERPOLATION BETWEEN GIVEN POINTS AT INTERVALS OF 0.2

Interpolated point	Coefficients to be applied to—					
	$N_{1.0}$	$N_{2.0}$	$N_{3.0}$	$N_{4.0}$	$N_{5.0}$	$N_{6.0}$
	First interval					
$N_{1.0}$	+1.0000	.0000	.0000	.0000	.0000	
$N_{1.2}$	+.6384	+.6384	−.4256	+.1824	−.0336	
$N_{1.4}$	+.3744	+.9984	−.5616	+.2304	−.0416	
$N_{1.6}$	+.1904	+1.1424	−.4896	+.1904	−.0336	
$N_{1.8}$	+.0704	+1.1264	−.2816	+.1024	−.0176	
	Next-to-first interval					
$N_{2.0}$	.0000	+1.0000	.0000	.0000	.0000	
$N_{2.2}$	−.0336	+.8064	+.3024	−.0896	+.0144	
$N_{2.4}$	−.0416	+.5824	+.5824	−.1456	+.0224	
$N_{2.6}$	−.0336	+.3584	+.8064	−.1536	+.0224	
$N_{2.8}$	−.0176	+.1584	+.9504	−.1056	+.0144	
	Middle interval					
$N_{3.0}$	.0000	.0000	+1.0000	.0000	.0000	.0000
$N_{3.2}$	+.0128	−.0976	+.9344	+.1744	−.0256	+.0016
$N_{3.4}$	+.0144	−.1136	+.7264	+.4384	−.0736	+.0080
$N_{3.6}$	+.0080	−.0736	+.4384	+.7264	−.1136	+.0144
$N_{3.8}$	+.0016	−.0256	+.1744	+.9344	−.0976	+.0128
	Next-to-last interval					
$N_{4.0}$		.0000	.0000	+1.0000	.0000	.0000
$N_{4.2}$		+.0144	−.1056	+.9504	+.1584	−.0176
$N_{4.4}$		+.0224	−.1536	+.8064	+.3584	−.0336
$N_{4.6}$		+.0224	−.1456	+.5824	+.5824	−.0416
$N_{4.8}$		+.0144	−.0896	+.3024	+.8064	−.0336
	Last interval					
$N_{5.0}$		.0000	.0000	.0000	+1.0000	.0000
$N_{5.2}$		−.0176	+.1024	−.2816	+1.1264	+.0704
$N_{5.4}$		−.0336	+.1904	−.4896	+1.1424	+.1904
$N_{5.6}$		−.0416	+.2304	−.5616	+.9984	+.3744
$N_{5.8}$		−.0336	+.1824	−.4256	+.6384	+.6384
$N_{6.0}$		.0000	.0000	.0000	.0000	+1.0000

B. FOR SUBDIVISION OF GROUPS INTO FIFTHS

Interpolated subgroup	Coefficients to be applied to—				
	G_1	G_2	G_3	G_4	G_5
	First panel				
First fifth of G_1	+.3616	−.2768	+.1488	−.0336	
Second fifth of G_1	+.2640	−.0960	+.0400	−.0080	
Third fifth of G_1	+.1840	+.0400	−.0320	+.0080	
Fourth fifth of G_1	+.1200	+.1360	−.0720	+.0160	
Last fifth of G_1	+.0704	+.1968	−.0848	+.0176	
	Next-to-first panel				
First fifth of G_2	+.0336	+.2272	−.0752	+.0144	
Second fifth of G_2	+.0080	+.2320	−.0480	+.0080	
Third fifth of G_2	−.0080	+.2160	−.0080	.0000	
Fourth fifth of G_2	−.0160	+.1840	+.0400	−.0080	
Last fifth of G_2	−.0176	+.1408	+.0912	−.0144	

B. FOR SUBDIVISION OF GROUPS INTO FIFTHS--Continued

Interpolated subgroup	Coefficients to be applied to—				
	G_1	G_2	G_3	G_4	G_5
	Middle panel				
First fifth of G_3	−.0128	+.0848	+.1504	−.0240	+.0016
Second fifth of G_3	−.0016	+.0144	+.2224	−.0416	+.0064
Third fifth of G_3	+.0064	−.0336	+.2544	−.0336	+.0064
Fourth fifth of G_3	+.0064	−.0416	+.2224	+.0144	−.0016
Last fifth of G_3	+.0016	−.0240	+.1504	+.0848	−.0128
	Next-to-last panel				
First fifth of G_4		−.0144	+.0912	+.1408	−.0176
Second fifth of G_4		−.0080	+.0400	+.1840	−.0160
Third fifth of G_4		.0000	−.0080	+.2160	−.0080
Fourth fifth of G_4		+.0080	−.0480	+.2320	+.0080
Last fifth of G_4		+.0144	−.0752	+.2272	+.0336
	Last panel				
First fifth of G_5		+.0176	−.0848	+.1968	+.0704
Second fifth of G_5		+.0160	−.0720	+.1360	+.1200
Third fifth of G_5		+.0080	−.0320	+.0400	+.1840
Fourth fifth of G_5		−.0080	+.0400	−.0960	+.2640
Last fifth of G_5		−.0336	+.1488	−.2768	+.3616

C. FOR SUBDIVISION OF GROUPS INTO TENTHS OR HALVES

Interpolated subgroup	Coefficients to be applied to—				
	G_1	G_2	G_3	G_4	G_5
First tenth of G_3	−.0076	+.0510	+.0660	−.0096	+.0002
Second tenth of G_3	−.0052	+.0338	+.0844	−.0144	+.0014
Third tenth of G_3	−.0022	+.0154	+.1036	−.0195	+.0027
Fourth tenth of G_3	+.0006	−.0010	+.1188	−.0221	+.0037
Fifth tenth of G_3	+.0027	−.0133	+.1272	−.0203	+.0037
Sum of coefficients for first five-tenths = coefficients for first half of G_3	−.0117	+.0859	+.5000	−.0859	+.0117
Sixth tenth of G_3	+.0037	−.0203	+.1272	−.0133	+.0027
Seventh tenth of G_3	+.0037	−.0221	+.1188	−.0010	+.0006
Eighth tenth of G_3	+.0027	−.0195	+.1036	+.0154	−.0022
Ninth tenth of G_3	+.0014	−.0144	+.0844	+.0338	−.0052
Last tenth of G_3	+.0002	−.0096	+.0660	+.0510	−.0076
Sum of coefficients for last five-tenths = coefficients for second half of G_3	+.0117	−.0859	+.5000	+.0859	−.0117

Table C–6.—Interpolation Coefficients Based on the Beers "Ordinary" Formula

[The Beers "ordinary" formula is a six-term formula which minimizes the fifth differences of the interpolated results. It maintains the given values. Given points or groups must be equally spaced]

A. FOR INTERPOLATION BETWEEN GIVEN POINTS AT INTERVALS OF 0.2

Interpolated point	Coefficients to be applied to—					
	$N_{1.0}$	$N_{2.0}$	$N_{3.0}$	$N_{4.0}$	$N_{5.0}$	$N_{6.0}$
First interval						
$N_{1.0}$	+1.0000	.0000	.0000	.0000	.0000	.0000
$N_{1.2}$	+.6667	+.4969	-.1426	-.1006	+.1079	-.0283
$N_{1.4}$	+.4072	+.8344	-.2336	-.0976	+.1224	-.0328
$N_{1.6}$	+.2148	+1.0204	-.2456	-.0536	+.0884	-.0244
$N_{1.8}$	+.0819	+1.0689	-.1666	-.0126	+.0399	-.0115
Next-to-first interval						
$N_{2.0}$	.0000	+1.0000	.0000	.0000	.0000	.0000
$N_{2.2}$	-.0404	+.8404	+.2344	-.0216	-.0196	+.0068
$N_{2.4}$	-.0497	+.6229	+.5014	-.0646	-.0181	+.0081
$N_{2.6}$	-.0389	+.3849	+.7534	-.1006	-.0041	+.0053
$N_{2.8}$	-.0191	+.1659	+.9354	-.0906	+.0069	+.0015
Middle interval						
$N_{3.0}$	.0000	.0000	+1.0000	.0000	.0000	.0000
$N_{3.2}$	+.0117	-.0921	+.9234	+.1854	-.0311	+.0027
$N_{3.4}$	+.0137	-.1101	+.7194	+.4454	-.0771	+.0087
$N_{3.6}$	+.0087	-.0771	+.4454	+.7194	-.1101	+.0137
$N_{3.8}$	+.0027	-.0311	+.1854	+.9234	-.0921	+.0117
Next-to-last interval						
$N_{4.0}$	.0000	.0000	.0000	+1.0000	.0000	.0000
$N_{4.2}$	+.0015	+.0069	-.0906	+.9354	+.1659	-.0191
$N_{4.4}$	+.0053	-.0041	-.1006	+.7534	+.3849	-.0389
$N_{4.6}$	+.0081	-.0181	-.0646	+.5014	+.6229	-.0497
$N_{4.8}$	+.0068	-.0196	-.0216	+.2344	+.8404	-.0404
Last interval						
$N_{5.0}$	.0000	.0000	.0000	.0000	+1.0000	.0000
$N_{5.2}$	-.0115	+.0399	-.0126	-.1666	+1.0689	+.0819
$N_{5.4}$	-.0244	+.0884	-.0536	-.2456	+1.0204	+.2148
$N_{5.6}$	-.0328	+.1224	-.0976	-.2336	+.8344	+.4072
$N_{5.8}$	-.0283	+.1079	-.1006	-.1426	+.4969	+.6667
$N_{6.0}$	.0000	.0000	.0000	.0000	.0000	+1.0000

B. FOR SUBDIVISION OF GROUPS INTO FIFTHS

Interpolated subgroup	Coefficients to be applied to—				
	G_1	G_2	G_3	G_4	G_5
First panel					
First fifth of G_1	+.3333	-.1636	-.0210	+.0796	-.0283
Second fifth of G_1	+.2595	-.0780	+.0130	+.0100	-.0045
Third fifth of G_1	+.1924	+.0064	+.0184	-.0256	+.0084
Fourth fifth of G_1	+.1329	+.0844	+.0054	-.0356	+.0129
Last fifth of G_1	+.0819	+.1508	-.0158	-.0284	+.0115
Next-to-first panel					
First fifth of G_2	+.0404	+.2000	-.0344	-.0128	+.0068
Second fifth of G_2	+.0093	+.2268	-.0402	+.0028	+.0013
Third fifth of G_2	-.0108	+.2272	-.0248	+.0112	-.0028
Fourth fifth of G_2	-.0198	+.1992	+.0172	+.0072	-.0038
Last fifth of G_2	-.0191	+.1468	+.0822	-.0084	-.0015
Middle panel					
First fifth of G_3	-.0117	+.0804	+.1570	-.0284	+.0027
Second fifth of G_3	-.0020	+.0160	+.2200	-.0400	+.0060
Third fifth of G_3	+.0050	-.0280	+.2460	-.0280	+.0050
Fourth fifth of G_3	+.0060	-.0400	+.2200	+.0160	-.0020
Last fifth of G_3	+.0027	-.0284	+.1570	+.0804	-.0117
Next-to-last panel					
First fifth of G_4	-.0015	-.0084	+.0822	+.1468	-.0191
Second fifth of G_4	-.0038	+.0072	+.0172	+.1992	-.0198
Third fifth of G_4	-.0028	+.0112	-.0248	+.2272	-.0108
Fourth fifth of G_4	+.0013	+.0028	-.0402	+.2268	+.0093
Last fifth of G_4	+.0068	-.0128	-.0344	+.2000	+.0404
Last panel					
First fifth of G_5	+.0115	-.0284	-.0158	+.1508	+.0819
Second fifth of G_5	+.0129	-.0356	+.0054	+.0844	+.1329
Third fifth of G_5	+.0084	-.0256	+.0184	+.0064	+.1924
Fourth fifth of G_5	-.0045	+.0100	+.0130	-.0780	+.2595
Last fifth of G_5	-.0283	+.0796	-.0210	-.1636	+.3333

Source: Henry S. Beers, "Discussion of Papers Presented in the Record, No. 68: 'Six-Term Formulas for Routine Actuarial Interpolation', by Henry S. Beers," *The Record of the American Institute of Actuaries*, 34, Part I (69):59–60, June 1945.

Table C–7. — Interpolation Coefficients Based on the Beers "Modified" Formula

[The Beers "modified" formula is a six-term formula which minimizes fourth differences of the interpolated results. This formula combines interpolation with some smoothing or graduation of given values; end panels maintain the given values, however. Given data must be equally spaced]

A. FOR INTERPOLATION BETWEEN GIVEN POINTS AT INTERVALS OF 0.2

Interpolated point	Coefficients to be applied to--					
	$N_{1.0}$	$N_{2.0}$	$N_{3.0}$	$N_{4.0}$	$N_{5.0}$	$N_{6.0}$
First interval						
$N_{1.0}$	+1.0000	.0000	.0000	.0000	.0000	.0000
$N_{1.2}$	+.6668	+.5270	−.2640	+.0820	−.0140	+.0022
$N_{1.4}$	+.4099	+.8592	−.3598	+.1052	−.0173	+.0028
$N_{1.6}$	+.2196	+1.0279	−.3236	+.0874	−.0136	+.0023
$N_{1.8}$	+.0862	+1.0644	−.1916	+.0464	−.0066	+.0012
Next-to-first interval						
$N_{2.0}$	.0000	+1.0000	.0000	.0000	.0000	.0000
$N_{2.2}$	−.0486	+.8655	+.2160	−.0350	+.0030	−.0009
$N_{2.4}$	−.0689	+.6903	+.4238	−.0442	+.0003	−.0013
$N_{2.6}$	−.0697	+.5018	+.5938	−.0152	−.0097	−.0010
$N_{2.8}$	−.0589	+.3233	+.7038	+.0578	−.0257	−.0003
Middle interval						
$N_{3.0}$	−.0430	+.1720	+.7420	+.1720	−.0430	.0000
$N_{3.2}$	−.0270	+.0587	+.7072	+.3162	−.0538	−.0013
$N_{3.4}$	−.0141	−.0132	+.6098	+.4708	−.0477	−.0056
$N_{3.6}$	−.0056	−.0477	+.4708	+.6098	−.0132	−.0141
$N_{3.8}$	−.0013	−.0538	+.3162	+.7072	+.0587	−.0270
Next-to-last interval						
$N_{4.0}$	.0000	−.0430	+.1720	+.7420	+.1720	−.0430
$N_{4.2}$	−.0003	−.0257	+.0578	+.7038	+.3233	−.0589
$N_{4.4}$	−.0010	−.0097	−.0152	+.5938	+.5018	−.0697
$N_{4.6}$	−.0013	+.0003	−.0442	+.4238	+.6903	−.0689
$N_{4.8}$	−.0009	+.0030	−.0350	+.2160	+.8655	−.0486
Last interval						
$N_{5.0}$	.0000	.0000	.0000	.0000	+1.0000	.0000
$N_{5.2}$	+.0012	−.0066	+.0464	−.1916	+1.0644	+.0862
$N_{5.4}$	+.0023	−.0136	+.0874	−.3236	+1.0279	+.2196
$N_{5.6}$	+.0028	−.0173	+.1052	−.3598	+.8592	+.4099
$N_{5.8}$	+.0022	−.0140	+.0820	−.2640	+.5270	+.6668
$N_{6.0}$	.0000	.0000	.0000	.0000	.0000	+1.0000

B. FOR SUBDIVISION OF GROUPS INTO FIFTHS

Interpolated subgroup	Coefficients to be applied to--				
	G_1	G_2	G_3	G_4	G_5
First panel					
First fifth of G_1	+.3332	−.1938	+.0702	−.0118	+.0022
Second fifth of G_1	+.2569	−.0753	+.0205	−.0027	+.0006
Third fifth of G_1	+.1903	+.0216	−.0146	+.0032	−.0005
Fourth fifth of G_1	+.1334	+.0969	−.0351	+.0059	−.0011
Last fifth of G_1	+.0862	+.1506	−.0410	+.0054	−.0012
Next-to-first panel					
First fifth of G_2	+.0486	+.1831	−.0329	+.0021	−.0009
Second fifth of G_2	+.0203	+.1955	−.0123	−.0031	−.0004
Third fifth of G_2	+.0008	+.1893	+.0193	−.0097	+.0003
Fourth fifth of G_2	−.0108	+.1677	+.0577	−.0153	+.0007
Last fifth of G_2	−.0159	+.1354	+.0972	−.0170	+.0003
Middle panel					
First fifth of G_3	−.0160	+.0973	+.1321	−.0121	−.0013
Second fifth of G_3	−.0129	+.0590	+.1564	+.0018	−.0043
Third fifth of G_3	−.0085	+.0260	+.1650	+.0260	−.0085
Fourth fifth of G_3	−.0043	+.0018	+.1564	+.0590	−.0129
Last fifth of G_3	−.0013	−.0121	+.1321	+.0973	−.0160
Next-to-last panel					
First fifth of G_4	+.0003	−.0170	+.0972	+.1354	−.0159
Second fifth of G_4	+.0007	−.0153	+.0577	+.1677	−.0108
Third fifth of G_4	+.0003	−.0097	+.0193	+.1893	+.0008
Fourth fifth of G_4	−.0004	−.0031	−.0123	+.1955	+.0203
Last fifth of G_4	−.0009	+.0021	−.0329	+.1831	+.0486
Last panel					
First fifth of G_5	−.0012	+.0054	−.0410	+.1506	+.0862
Second fifth of G_5	−.0011	+.0059	−.0351	+.0969	+.1334
Third fifth of G_5	−.0005	+.0032	−.0146	+.0216	+.1903
Fourth fifth of G_5	+.0006	−.0027	+.0205	−.0753	+.2569
Last fifth of G_5	+.0022	−.0118	+.0702	−.1938	+.3332

Source: Henry S. Beers, "Modified-Interpolation Formulas that Minimize Fourth Differences," *The Record of the American Institute of Actuaries*, **34**, Part I (69):19–20, June 1945.

Table C–8. — Interpolation Coefficients Based on Grabill's Weighted Moving Average of Sprague Coefficients

[See text for derivation. Used for drastic smoothing. Given groups must be equally spaced]

Interpolated subgroup	Coefficients to be applied to--				
	G_1	G_2	G_3	G_4	G_5
First fifth of G_3	+.0111	+.0816	+.0826	+.0256	−.0009
Second fifth of G_3	+.0049	+.0673	+.0903	+.0377	−.0002
Third fifth of G_3	+.0015	+.0519	+.0932	+.0519	++.0015
Fourth fifth of G_3	−.0002	+.0377	+.0903	+.0673	+.0049
Last fifth of G_3	−.0009	+.0256	+.0826	+.0816	+.0111

SUGGESTED READINGS

Part A. Interpolation

Aitken, A. C. "On Interpolation by Iteration of Proportional Parts, Without the Use of Differences." *Proceedings of the Edinburgh Mathematical Society,* Series 2, 3:56–76. 1932.

Beers, H. S. "Modified-Interpolation Formulas that Minimize Fourth Differences." *Record of the American Institute of Actuaries,* 34(69):14–20. Part I, June 1945.

———. "Six-Term Formulas for Routine Actuarial Interpolation." *Record of the American Institute of Actuaries,* 33(68):245–260. Part II, Nov. 1944.

Brass, W. "The Graduation of Fertility Distributions by Polynomial Functions." *Population Studies* (London), 14(2): 148–162. Nov. 1960.

Butler, R. and Kerr, E. *An Introduction to Numerical Methods,* pp. 131–212. New York, Pitman Publishing Corp., 1962.

Croxton, Frederick E., Cowden, Dudley J., and Klein, Sidney. *Applied General Statistics* (3rd ed.), pp. 263–274. Englewood Cliffs, N.J., Prentice-Hall, 1967.

Greville, T. N. E. "Recent Developments in Graduation and Interpolation." *Journal of the American Statistical Association,* 43(243):428–441. Sept. 1948.

———. "The General Theory of Osculatory Interpolation." *Transactions of the Actuarial Society of America,* 45(112): 202–265. Part II, Oct. 1944.

———. "Spline Functions, Interpolation, and Numerical Quadrature." In Anthony Ralston and Herbert S. Wilf (eds.), *Mathematical Methods for Digital Computers,* Vol. II, pp. 156–168. New York, John Wiley & Sons, 1967.

Jordan, C. W., Jr. *Life Contingencies.* Chicago, Ill., Society of Actuaries, 1952.

Keyfitz, Nathan. "Interpolation and Graduation." Chapter 10 in *Introduction to the Mathematics of Population.* Reading, Mass., Addison-Wesley, 1968.

———. "A Unified Approach to Interpolation and Graduation." *Demography,* 3(2):528–536. 1966.

Miller, Morton D. *Elements of Graduation.* Chicago, Ill., Actuarial Society of America and American Institute of Actuaries, 1946.

Milne, William E. *Numerical Calculus.* Princeton, N.J., Princeton University Press, 1949.

Wolfenden, H. H. *Population Statistics and Their Compilation* (rev. ed.), pp. 128–168. Chicago, University of Chicago Press, 1954.

Part B. Adjustment of Distributions to Marginal Totals

Deming, W. Edwards. *Statistical Adjustment of Data,* chapter VII. New York, John Wiley & Sons, 1948. Pp. 96–127.

Deming, W. Edwards, and Stephan, Frederick F. "On a Least Squares Adjustment of a Sampled Frequency Table When the Expected Marginal Totals Are Known." *The Annals of Mathematical Statistics,* 11(4):427–444. Dec. 1940.

Friedlander, D. "A Technique for Estimating a Contingency Table, Given the Marginal Totals and Some Supplementary Data." *Journal of the Royal Statistical Society* (London), Series A (General), 124(3):412–420. 1961.

Ireland, C. T., and Kullback, S. "Contingency Tables with Given Marginals." *Biometrika* (London), 55(1):179–188. March 1968.

Saw, Swee Hock. "A Problem of Estimating a Contingency Table Arising in Demographic Analysis." *Population Studies* (London), 19(3):311–315. March 1966.

Smith, J. H. "Estimation of Linear Functions of Cell Proportions." *Annals of Mathematical Statistics,* 18:231–254. 1947.

Stephan, F. F. "An Iterative Method of Adjusting Sample Frequency Tables when Expected Marginal Totals are Known." *Annals of Mathematical Statistics,* 13:166–178. 1942.

United Nations, Economic Commission for Latin America. "Application of the Registrar-General's Square Table to a Population Projection with Respect to Urban and Rural Population by Groups of Sex and Age." Santiago, Chile. Jan. 25, 1961.

Part C. Cohort Analysis

Douglas, J. W. B., and Blomfield, J. M. "The Reliability of Longitudinal Surveys." *Milbank Memorial Fund Quarterly,* 34(3):227–252. July 1956.

Dublin, Louis I., and Spiegelman, Mortimer. "Current Versus Generation Life Tables." *Human Biology,* 13(4):439–458. Dec. 1941.

Duncan, Otis Dudley, and Hodge, Robert W. "Cohort Analysis of Differential Natality." In *International Population Conference, New York, 1961.* Vol. 1, pp. 59–66. New York, International Union for the Scientific Study of Population, 1962.

Eldridge, Hope T. "A Cohort Approach to the Analysis of Migration Differentials." *Demography,* 1(1):212–219. 1964.

Frost, Wade Hampton. "The Age Selection of Mortality from Tuberculosis in Successive Decades." *Milbank Memorial Fund Quarterly,* 18(1):61–66. Jan. 1940.

Grabill, Wilson H., Kiser, Clyde V., and Whelpton, P. K. *The Fertility of American Women,* Chapter 9. New York, John Wiley & Sons, 1958.

Henry, Louis. "Analyse et mesure des phenomèmes démographiques par cohortes." *Population* (Paris), 21(3):465–482. May–June 1966.

Jacobson, Paul H. "Cohort Survival for Generations Since 1840." *Milbank Memorial Fund Quarterly,* 42(3):36–53. July 1964.

Jacoby, E. G. "A Fertility Analysis of New Zealand Marriage Cohorts." *Population Studies* (London), 12(1):18–39. July 1958.

SUGGESTED READINGS – Continued

Legaré, Jacques. "Quelques considérations sur les tables de mortalité de génération. Application à l'Angleterre et au Pays de Galles." *Population* (Paris), 21(5):915–938. Sept.–Oct. 1966.

Osadnik, L. "Method of Cohort Analysis for Studying Fertility Factors." In Egon Szabady (ed.), *Studies on Fertility and Social Mobility*. Budapest, Akadémiai Kiadó, 1964. Pp. 40–43.

Ryder, Norman B. "Cohort Analysis." In *International Encyclopaedia of the Social Sciences*. Vol. 2, pp. 546–550. 1968.

———. "La mesure des variations de la fécondité au cours du temps." *Population* (Paris), 11(1):29–46. January-March 1956.

Shryock, Henry S., and Larmon, Elizabeth A. "Some Longitudinal Data on Internal Migration." *Demography*, 2:579–592. 1965.

Spiegelman, Mortimer. "Segmented Generation Mortality." *Demography*, 6(2):117–124. 1969.

Taeuber, Karl E. "Cohort Migration." *Demography*, 3(2):416–422. 1966.

———. "Cohort Population Redistribution and the Urban Hierarchy." *Milbank Memorial Fund Quarterly*, 43(4):450–462. Part I, Oct. 1965.

———. "The Residential Redistribution of Farm-Born Cohorts." *Rural Sociology*, 32(1):20–36. March 1967

Uhlenberg, P. R. "A Study of Cohort Life Cycles: Cohorts of Native-Born Massachusetts Women, 1830–1920." *Population Studies* (London), 23(3):407–420. Nov. 1969.

United Nations. *Methods of Analysing Census Data on Economic Activities of the Population*, pp. 33–39 and 93–101. Series A, Population Studies, No. 43. 1968.

Whelpton, P. K. *Cohort Fertility: Native White Women in the United States*. Princeton, N.J., Princeton University Press, 1954.

———. "Cohort Analysis of Fertility." *American Sociological Review*, 14(6):735–749. Dec. 1949.

Author Index

Subject Index

sex ratios of, 109
spacing of. *See* Childspacing.
Bolivia, 57
Bombay, 380, 396
Boundary change, 64, 154, 211, 212, 379, 384, 386
Brazil, 40, 54, 57, 63, 85, 395, 396
survival rates, 265
British Commonwealth, 24
Bulgaria, 24, 63

C

Calcutta, 83
California, 18, 147
Cambodia, 63
Canada, 23, 50, 52, 53, 63, 67, 71, 159, 382
census statistics on economic characteristics, 203, 205, 207
evaluation of censuses, 37
farm population, 92
migration statistics, 392
population estimates, 426, 431, 432
projections of households, 468
projections of provincial distribution, 454-455
vital statistics, 40
Canal Zone, 155
Canvasser method. *See* Direct enumeration
Capital city, 63
Carrier-Farrag ratio method, 126
Carrier index, age preference, 118
Cause–of–death analysis, 229-232, 240-241, 552
Cause-specific death rates, 231-232
defined, 231
Cause-specific mortality rates, 240-241
Censal-ratio methods of population estimation, 426, 430
Census, 1, 3, 44-45
defined, 13
Census cohort-change method, 456
Census county division, 64, 70
Census coverage (*see also* Census errors, underenumeration), 53-60, 64, 148
Census-C.P.S. match. *See* Matching Studies.
Census errors:
age misreporting, 114-115, 119-120, 128
erroneous inclusions, 54, 56-57, 115
misreporting of characteristics other than age. *See* under each heading.
net census error. *See* Net undercount.
net undercount, 115, 127-128
net underenumeration, 54-60, 115, 217, 298
omissions, 54, 56, 57, 115
Census planning, 31-33
Census publicity, 31
Census survival rate method (*see also* residual method), 359, 360, 379, 381-382, 491-493
Census tracts, 67, 70
Censuses:
history of, 13, 14-18
periodicity of, 29
preparatory work, 31
topics for, 32-33
used in estimation methods -
for developing countries, 484, 491
for vital rates, 484, 501-502
Censuses of the United States by year. *See*

under U.S. population censuses, e.g., 1820, Census, 1910 Census.
Centenarians:
quality of census counts, 128-129
Center of population, 72-74
defined, 73
Central America, 50, 333, 335
Central city, 89
of SMSA, 68, 70
of urbanized area, 70
"Central place" theory, 102
Central rates, 228-235, 278, 288, 289, 293, 336, 338
Ceylon, 154, 156, 158, 378
Colombo district, 378
mortality, 227
Chandra Sekar-Deming method, 223, 503
Change, average, 213-215
Characteristics of husbands and wives, 173
Chicago, 70
Child–woman ratio, 151, 297-301
defined, 297
Childbearing ages, 278, 287, 297
Childlessness, 304, 305, 306
defined, 306
Children:
adopted, 304
age ratios, 124-125
mortality rates, 237-239, 360
number living, 302
sex ratios, 125
Children ever born, 302-308
defined, 302
measures based on, 305-308, 327
quality of statistics, 304-305, 496-497
uses of statistics, 311-312, 323-324
wording of census question, 303
Childspacing, 162, 289-291, 309
Chile, 57, 132, 133, 327, 395
mortality, 233, 235, 242, 243, 246
China (*see also* Taiwan), 13, 20, 63, 101, 110, 113
censuses, 13
Chinese age. *See* Age, Oriental reckoning
Cities:
functions of, 102
Citizen:
defined, 154
Citizens abroad, 51, 52
migration of, 362
Citizenship, 145, 151, 154-156
City block, 70-71
City directories, 25, 376-377, 421
Civilian population, 50, 53
Classification errors (*see also* under each heading), 114-115, 148
Closed population, 379, 389
Coale and Demeny model life tables (*see also* Life tables, model), 251, 300, 416, 445, 485
Coale's iterative method, 55, 58, 59
Coding, 35-36, 46
manual, 171
of number of own children, 301
Coefficient of population concentration, 99
Coefficient of urban concentration, 99
Cohort, 5, 8, 251, 550
birth, 175, 307-308, 309, 341, 397, 550
defined, 550
marriage, 5, 293, 307-308, 309, 329, 346
period-of-immigration, 550

real, 165, 239
school enrollment, 180, 550
synthetic, 166, 239, 324, 326, 338
Cohort analysis, 8, 550-553
age composition, 122-123, 136-140
defined, 550
fertility. *See* Cohort analysis, Natality
intercensal, 122-123, 187-188
migration, 402, 553
mortality, 257-258, 552-553
natality, 288-293, 303, 306, 307-308, 309, 553
Cohort-component method:
immigration estimation, 357
population estimation, 413-417, 432
population projection, 444-445, 456, 464, 466
projections of socioeconomic characteristics, 469-470, 474
households, 468
school enrollment, 471-472
Cohort fertility rate, 288-293
Cohort migration-survival method, 456-457
College towns:
population of, 51, 212-213
Colombia, 395
immigration rates, 366
mortality, 233
Colonial America, censuses in, 13
Color (*see also* Race):
defined, 146
Colorado, 147
Common-law marriages. *See* Consensual unions.
Commune, 83
Comparative mortality index, 243
Completed fertility rates, 289-292, 303, 308
cohort analysis, 550
in fertility projections, 448-450
terminal levels, 449
Completeness of enumeration (*See also* Evaluation of census coverage, Undercount, and Underenumeration):
in the United States, 57-59, 114-115
Completeness of vital registration, 45, 222-223, 274, 321
Component analysis:
age composition, 136-140
sex composition, 108-109
Component estimating equation (*see also* Balancing equation), 4, 56, 58
age composition, 136, 139
sex composition, 108
Component Method I of population estimation, 424, 433
Component Method II of population estimation, 424, 426, 433
Component method:
of population projection, 27, 442, 443
Components of change, 218
as related to age structure, 136-142, 329-330
Composite method of population estimation, 428-429
Compound interest law, 536
Computer usage, 36, 46, 73, 75, 129, 152, 171, 206, 258, 321, 331, 344, 379, 461, 475
Conference of European Statisticians, 172
Confidentiality of demographic record, 31, 45
Congressional district, 64, 65
Consensual unions, 40, 161, 162
defined, 333

Working-life tables, 203, 265-268
    abridged, 267-268
    complete, 265-267
    functions, 265-267
Workplace potential, 81
World Health Organization. *See* United Nations organizations.
World War I, 59, 212, 325
World War II, 52, 59, 147, 150, 212, 324, 325, 343, 398

Y

Year moved. *See* Duration of residence.
Years of school completed:
    defined, 186
"Young" population, 136
    defined, 132
"Younging" process, 132, 141, 358
Yugoslavia, 50, 145, 152, 182, 395

Z

Zip-code area, 65

STUDIES IN POPULATION

Under the Editorship of: H. H. WINSBOROUGH

Department of Sociology
University of Wisconsin
Madison, Wisconsin

Doreen S. Goyer. International Population Census Bibliography: *Revision and Update, 1945-1977.*

David L. Brown and John M. Wardwell (Eds.). New Directions in Urban–Rural Migration: *The Population Turnaround in Rural America.*

A. J. Jaffe, Ruth M. Cullen, and Thomas D. Boswell. The Changing Demography of Spanish Americans.

Robert Alan Johnson. Religious Assortative Marriage in the United States.

Hilary J. Page and Ron Lesthaeghe. Child-Spacing in Tropical Africa.

Dennis P. Hogan. Transitions and Social Change: *The Early Lives of American Men.*

F. Thomas Juster and Kenneth C. Land (Eds.). Social Accounting Systems: *Essays on the State of the Art.*

M. Sivamurthy. Growth and Structure of Human Population in the Presence of Migration.

Robert M. Hauser, David Mechanic, Archibald O. Haller, and Taissa O. Hauser (Eds.). Social Structure and Behavior: *Essays in Honor of William Hamilton Sewell.*

Valerie Kincade Oppenheimer. Work and the Family: *A Study in Social Demography.*

Kenneth C. Land and Andrei Rogers (Eds.). Multidimensional Mathematical Demography.

John Bongaarts and Robert G. Potter. Fertility, Biology, and Behavior: *An Analysis of the Proximate Determinants.*

Randy Hodson. Workers' Earnings and Corporate Economic Structure.

Ansley J. Coale and Paul Demeny. Regional Model Life Tables and Stable Populations, Second Edition.

Mary B. Breckenridge. Age, Time, and Fertility: *Applications of Exploratory Data Analysis.*

Neil G. Bennett (Ed.). Sex Selection of Children.

Rodolfo A. Bulatao and Ronald D. Lee (Eds.). Determinants of Fertility in Developing Countries. Volume 1: *Supply and Demand for Children;* Volume 2: *Fertility Regulation and Institutional Influences.*

Joseph A. McFalls, Jr., and Marguerite Harvey McFalls. Disease and Fertility.

Kenneth G. Manton and Eric Stallard. Recent Trends in Mortality Analysis.

H. ter Heide. Demographic Research and Spatial Policy.